HUTCHINS

PLACES OF THE WORLD

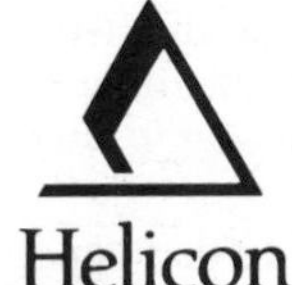

CONTENTS

INTRODUCTION

Hutchinson Places of the World is a unique work of reference providing detailed information on over 5,000 geographical places, including countries, states, regions, cities, towns, natural features, and landmarks around the world.

There are three sections in this book: the maps, the A–Z gazetteer (containing encyclopedic information about places), and the index. The book is designed so that any of the sections can be used as the start point.

Administrative maps and world atlas: This section consists of a set of administrative maps (maps of specific countries showing provinces or regions) followed by reference maps covering the whole world. On the maps, any name that has an entry in the gazetteer appears in red. Names are usually referenced to the largest-scale map on which they appear.

Local name forms are predominantly used on the maps. Where a popular English language conventional form exists, this may also be included in brackets immediately after the local form. Where space permits, alternate forms are included in brackets following the primary local form. All forms of a name will appear in both the gazetteer and the index. The names of international features such as seas and mountain ranges that are not part of any one country are given in English language form.

Gazetteer: The gazetteer is ordered in strict alphabetical sequence, as if there were no spaces or punctuation between words. Names containing the word 'St' are ordered as if spelt 'Saint'. Names containing generic terms are organized alphabetically by the main name e.g. Lagoda, Lake. Each entry has a concise description and all entries that can be referenced to a map are followed by a map reference.

The map reference is made up of a page number and a grid reference. These relate to the page and square on that page where the name occurs on the map. A name can be found by referring to the letters and numbers around the map. First find the letter which appears along the top and bottom of the map and then the number down the sides of the map. The name or feature will be found within the rectangle identified by that letter and number. A number in brackets preceding the grid reference indicates that the names can be found within an inset map. Those gazetteer entries that do not appear on a map have no map reference. Cross-references exist to guide you to alternative name forms as well as to take you to related place entries in the gazetteer.

Index: The index is a complete listing of all the names that appear in the maps. Entries that have an associated gazetteer entry appear in the index in bold. All places are referenced to the largest-scale map on which the name appears.

Helicon Publishing Ltd
3rd Floor, Clarendon House
Shoe Lane, Oxford OX1 2DP
United Kingdom
e-mail: admin@helicon.co.uk
Web site: www.helicon.co.uk

First published 2001
ISBN: 1-85986-383-3

British Library Cataloguing in Publication Data
A catalogue record for this book is available from the British Library

Typeset by Florence Production Ltd, Stoodleigh, Devon

Photo credits: Front cover left to right; Mount Fuji © Stone; Reykjavik cityscape © Stone; Aberdeen Harbour Hong Kong © The Image Bank. Spine; Boston Skyline © FPG

Printed and bound in Italy by Giunti Industrie Grafiche

Editorial Director
Hilary McGlynn

Head of Cartography
Chris Moore

Managing Editor
Elena Softley

Cartographic Production Manager
Caroline Beckley

Content Development Manager
Claire Lishman

Content Fulfilment Manager
Tracey Auden

Text Editor
Kath Pilgrem

Proofreader
Sue Donaldson

Cartographers
Ben Brown
Rachel Hopper
Adam Meara
Nikki Sargeant

Production Manager
John Normansell

Production Controller
Stacey Penny

Design Controller
Andrew Magee

Page and Cover Design
Robert Updegraff

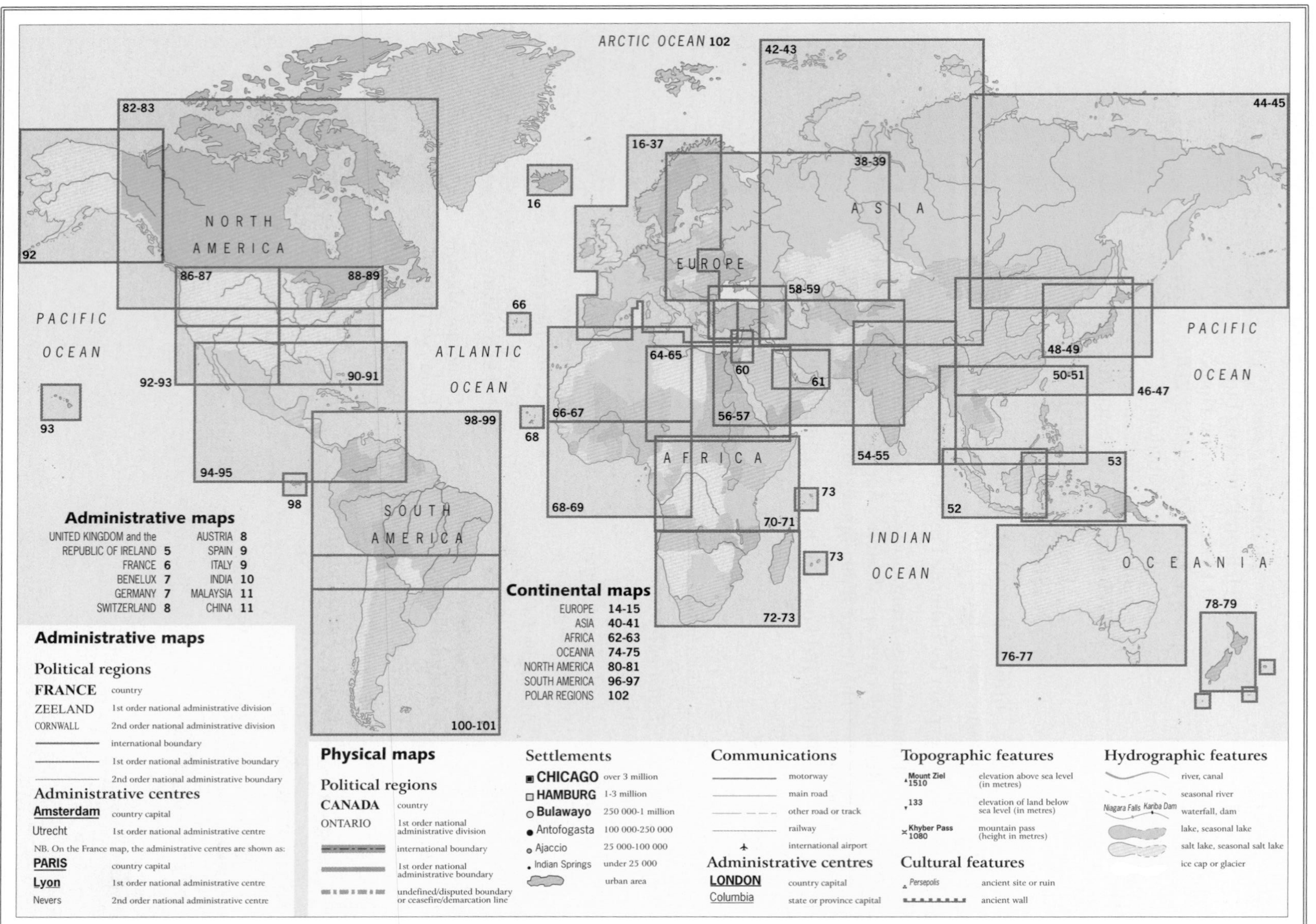
ARCTIC OCEAN 102
PACIFIC OCEAN
ATLANTIC OCEAN
INDIAN OCEAN
PACIFIC OCEAN
NORTH AMERICA
SOUTH AMERICA
EUROPE
ASIA
AFRICA
OCEANIA
82-83
92
86-87
88-89
92-93
90-91
93
94-95
98
98-99
100-101
66
68
16
16-37
64-65
66-67
68-69
60
56-57
61
70-71
72-73
73
73
42-43
38-39
58-59
54-55
44-45
48-49
50-51
46-47
52
53
76-77
78-79
Administrative maps
UNITED KINGDOM and the REPUBLIC OF IRELAND 5
FRANCE 6
BENELUX 7
GERMANY 7
SWITZERLAND 8
AUSTRIA 8
SPAIN 9
ITALY 9
INDIA 10
MALAYSIA 11
CHINA 11
Continental maps
EUROPE 14-15
ASIA 40-41
AFRICA 62-63
OCEANIA 74-75
NORTH AMERICA 80-81
SOUTH AMERICA 96-97
POLAR REGIONS 102
Administrative maps
Political regions
FRANCE country
ZEELAND 1st order national administrative division
CORNWALL 2nd order national administrative division
international boundary
1st order national administrative boundary
2nd order national administrative boundary
Administrative centres
Amsterdam country capital
Utrecht 1st order national administrative centre
NB. On the France map, the administrative centres are shown as:
PARIS country capital
Lyon 1st order national administrative centre
Nevers 2nd order national administrative centre
Physical maps
Political regions
CANADA country
ONTARIO 1st order national administrative division
international boundary
1st order national administrative boundary
undefined/disputed boundary or ceasefire/demarcation line
Settlements
CHICAGO over 3 million
HAMBURG 1-3 million
Bulawayo 250 000-1 million
Antofogasta 100 000-250 000
Ajaccio 25 000-100 000
Indian Springs under 25 000
urban area
Communications
motorway
main road
other road or track
railway
international airport
Administrative centres
LONDON country capital
Columbia state or province capital
Topographic features
Mount Ziel 1510 elevation above sea level (in metres)
133 elevation of land below sea level (in metres)
Khyber Pass 1080 mountain pass (height in metres)
Cultural features
Persepolis ancient site or ruin
ancient wall
Hydrographic features
river, canal
seasonal river
Niagara Falls Kariba Dam waterfall, dam
lake, seasonal lake
salt lake, seasonal salt lake
ice cap or glacier

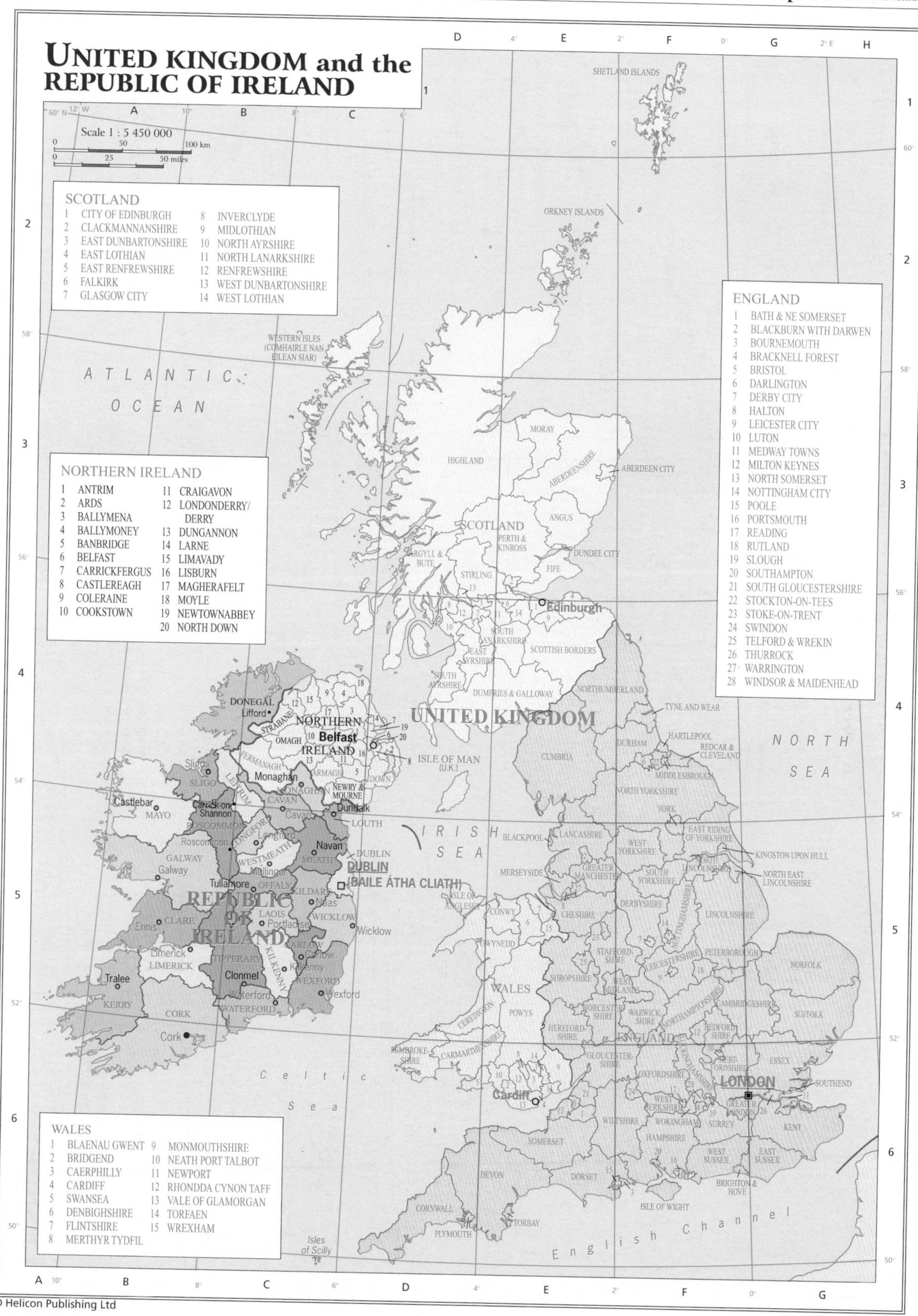

UNITED KINGDOM and the REPUBLIC OF IRELAND
Scale 1 : 5 450 000
SCOTLAND
1 CITY OF EDINBURGH
2 CLACKMANNANSHIRE
3 EAST DUNBARTONSHIRE
4 EAST LOTHIAN
5 EAST RENFREWSHIRE
6 FALKIRK
7 GLASGOW CITY
8 INVERCLYDE
9 MIDLOTHIAN
10 NORTH AYRSHIRE
11 NORTH LANARKSHIRE
12 RENFREWSHIRE
13 WEST DUNBARTONSHIRE
14 WEST LOTHIAN
NORTHERN IRELAND
1 ANTRIM
2 ARDS
3 BALLYMENA
4 BALLYMONEY
5 BANBRIDGE
6 BELFAST
7 CARRICKFERGUS
8 CASTLEREAGH
9 COLERAINE
10 COOKSTOWN
11 CRAIGAVON
12 LONDONDERRY/ DERRY
13 DUNGANNON
14 LARNE
15 LIMAVADY
16 LISBURN
17 MAGHERAFELT
18 MOYLE
19 NEWTOWNABBEY
20 NORTH DOWN
ENGLAND
1 BATH & NE SOMERSET
2 BLACKBURN WITH DARWEN
3 BOURNEMOUTH
4 BRACKNELL FOREST
5 BRISTOL
6 DARLINGTON
7 DERBY CITY
8 HALTON
9 LEICESTER CITY
10 LUTON
11 MEDWAY TOWNS
12 MILTON KEYNES
13 NORTH SOMERSET
14 NOTTINGHAM CITY
15 POOLE
16 PORTSMOUTH
17 READING
18 RUTLAND
19 SLOUGH
20 SOUTHAMPTON
21 SOUTH GLOUCESTERSHIRE
22 STOCKTON-ON-TEES
23 STOKE-ON-TRENT
24 SWINDON
25 TELFORD & WREKIN
26 THURROCK
27 WARRINGTON
28 WINDSOR & MAIDENHEAD
WALES
1 BLAENAU GWENT
2 BRIDGEND
3 CAERPHILLY
4 CARDIFF
5 SWANSEA
6 DENBIGHSHIRE
7 FLINTSHIRE
8 MERTHYR TYDFIL
9 MONMOUTHSHIRE
10 NEATH PORT TALBOT
11 NEWPORT
12 RHONDDA CYNON TAFF
13 VALE OF GLAMORGAN
14 TORFAEN
15 WREXHAM
ATLANTIC OCEAN
NORTH SEA
IRISH SEA
Celtic Sea
English Channel
SHETLAND ISLANDS
ORKNEY ISLANDS
WESTERN ISLES (COMHAIRLE NAN EILEAN SIAR)
HIGHLAND
MORAY
ABERDEENSHIRE
ABERDEEN CITY
ANGUS
SCOTLAND
PERTH & KINROSS
DUNDEE CITY
ARGYLL & BUTE
STIRLING
FIFE
Edinburgh
SOUTH LANARKSHIRE
SCOTTISH BORDERS
EAST AYRSHIRE
SOUTH AYRSHIRE
DUMFRIES & GALLOWAY
UNITED KINGDOM
NORTHUMBERLAND
TYNE AND WEAR
HARTLEPOOL
REDCAR & CLEVELAND
DURHAM
CUMBRIA
MIDDLESBROUGH
NORTH YORKSHIRE
YORK
EAST RIDING OF YORKSHIRE
KINGSTON UPON HULL
NORTH EAST LINCOLNSHIRE
NORTH LINCOLNSHIRE
ISLE OF MAN (U.K.)
BLACKPOOL
LANCASHIRE
WEST YORKSHIRE
MERSEYSIDE
GREATER MANCHESTER
SOUTH YORKSHIRE
DERBYSHIRE
NOTTINGHAMSHIRE
LINCOLNSHIRE
CHESHIRE
ISLE OF ANGLESEY
CONWY
GWYNEDD
STAFFORDSHIRE
LEICESTERSHIRE
PETERBOROUGH
NORFOLK
SHROPSHIRE
WEST MIDLANDS
WALES
POWYS
CEREDIGION
WORCESTERSHIRE
WARWICKSHIRE
NORTHAMPTONSHIRE
CAMBRIDGESHIRE
SUFFOLK
HEREFORDSHIRE
ENGLAND
BEDFORDSHIRE
BUCKINGHAMSHIRE
HERTFORDSHIRE
ESSEX
PEMBROKESHIRE
CARMARTHENSHIRE
GLOUCESTERSHIRE
OXFORDSHIRE
LONDON
SOUTHEND
Cardiff
GREATER LONDON
WEST BERKSHIRE
WILTSHIRE
WOKINGHAM
SURREY
KENT
HAMPSHIRE
SOMERSET
WEST SUSSEX
EAST SUSSEX
BRIGHTON & HOVE
DEVON
DORSET
ISLE OF WIGHT
CORNWALL
TORBAY
PLYMOUTH
Isles of Scilly
DONEGAL
Lifford
STRABANE
NORTHERN IRELAND
OMAGH
Belfast
FERMANAGH
ARMAGH
DOWN
NEWRY & MOURNE
Monaghan
MONAGHAN
Sligo
SLIGO
LEITRIM
CAVAN
Cavan
Castlebar
MAYO
Carrick-on-Shannon
Dundalk
LOUTH
ROSCOMMON
Roscommon
LONGFORD
Longford
Navan
MEATH
DUBLIN
DUBLIN (BAILE ÁTHA CLIATH)
GALWAY
Galway
WESTMEATH
Mullingar
Tullamore
OFFALY
KILDARE
Naas
REPUBLIC OF IRELAND
CLARE
Ennis
LAOIS
Portlaoise
WICKLOW
Wicklow
CARLOW
Carlow
Limerick
LIMERICK
TIPPERARY
KILKENNY
Kilkenny
WEXFORD
Wexford
Tralee
Clonmel
KERRY
Waterford
WATERFORD
CORK
Cork

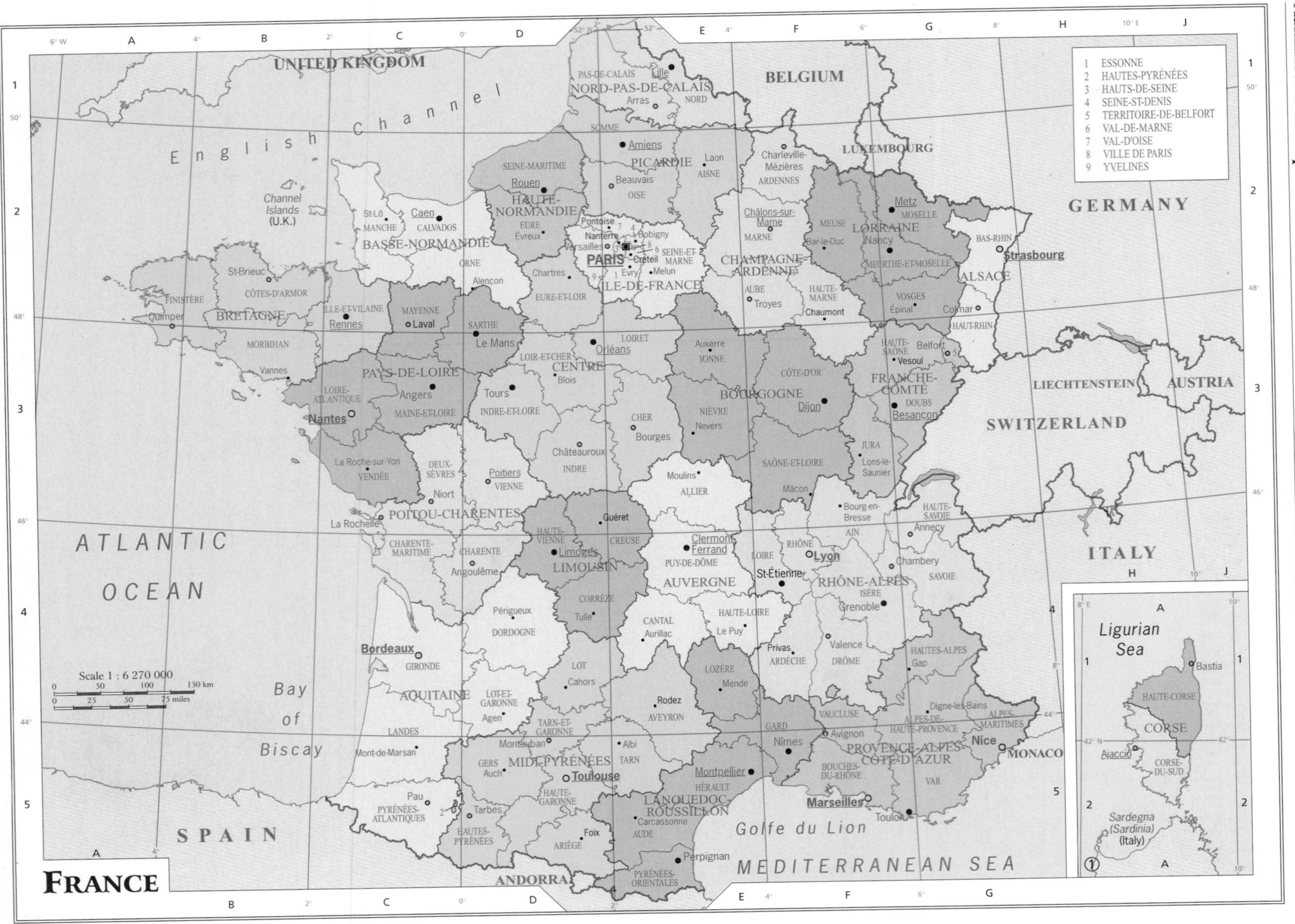
FRANCE
1 ESSONNE
2 HAUTES-PYRÉNÉES
3 HAUTS-DE-SEINE
4 SEINE-ST-DENIS
5 TERRITOIRE-DE-BELFORT
6 VAL-DE-MARNE
7 VAL-D'OISE
8 VILLE DE PARIS
9 YVELINES
Scale 1 : 6 270 000
150 km
75 miles
UNITED KINGDOM
BELGIUM
LUXEMBOURG
GERMANY
SWITZERLAND
LIECHTENSTEIN
AUSTRIA
ITALY
MONACO
ANDORRA
SPAIN
English Channel
Channel Islands (U.K.)
ATLANTIC OCEAN
Bay of Biscay
MEDITERRANEAN SEA
Golfe du Lion
Ligurian Sea
Sardegna (Sardinia) (Italy)
CORSE
HAUTE-CORSE
CORSE-DU-SUD
Bastia
Ajaccio
NORD-PAS-DE-CALAIS
PICARDIE
HAUTE-NORMANDIE
BASSE-NORMANDIE
BRETAGNE
PAYS-DE-LOIRE
CENTRE
ÎLE-DE-FRANCE
CHAMPAGNE-ARDENNE
LORRAINE
ALSACE
FRANCHE-COMTÉ
BOURGOGNE
POITOU-CHARENTES
LIMOUSIN
AUVERGNE
RHÔNE-ALPES
PROVENCE-ALPES-CÔTE-D'AZUR
LANGUEDOC-ROUSSILLON
MIDI-PYRÉNÉES
AQUITAINE
PARIS
Lille
Amiens
Rouen
Caen
Rennes
Nantes
Orléans
Châlons-sur-Marne
Metz
Strasbourg
Besançon
Dijon
Poitiers
Limoges
Clermont-Ferrand
Lyon
Marseilles
Montpellier
Toulouse
Bordeaux

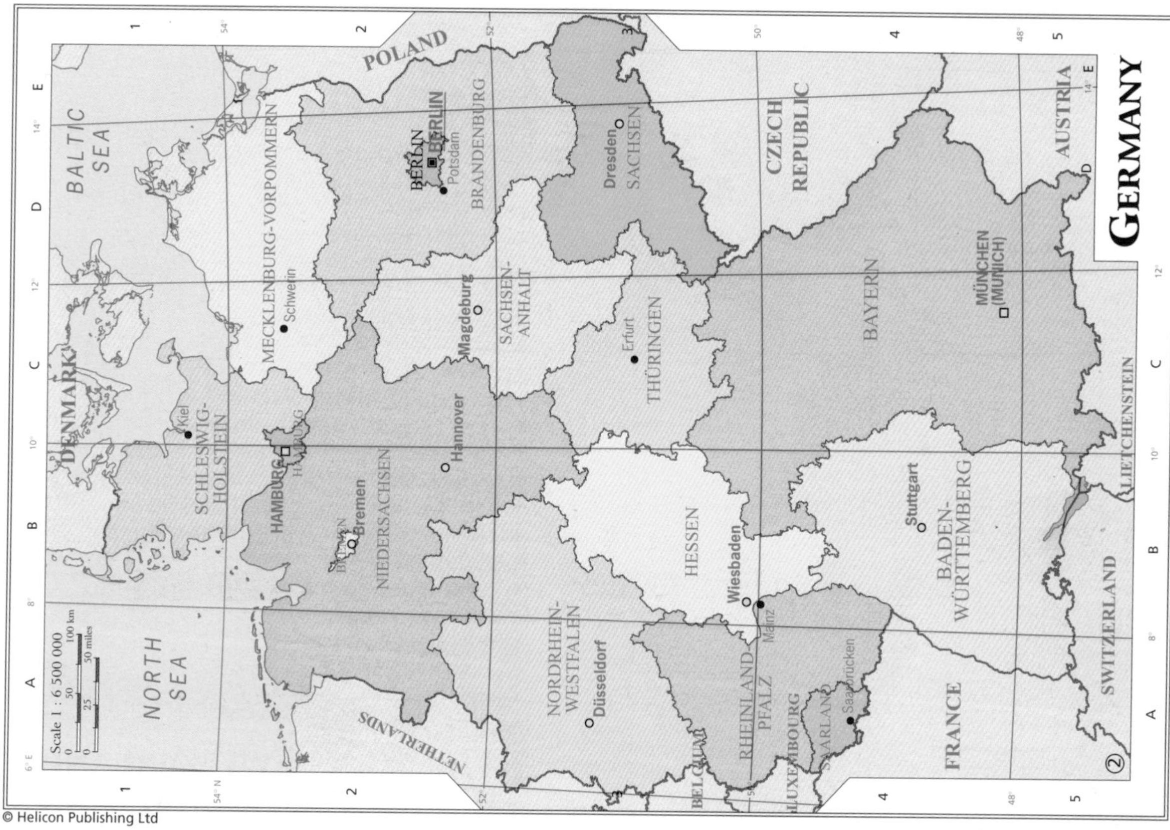

Germany

② Scale 1 : 6 500 000

0 25 50 miles

0 50 100 km

Benelux

① Scale 1 : 3 450 000

0 25 50 miles

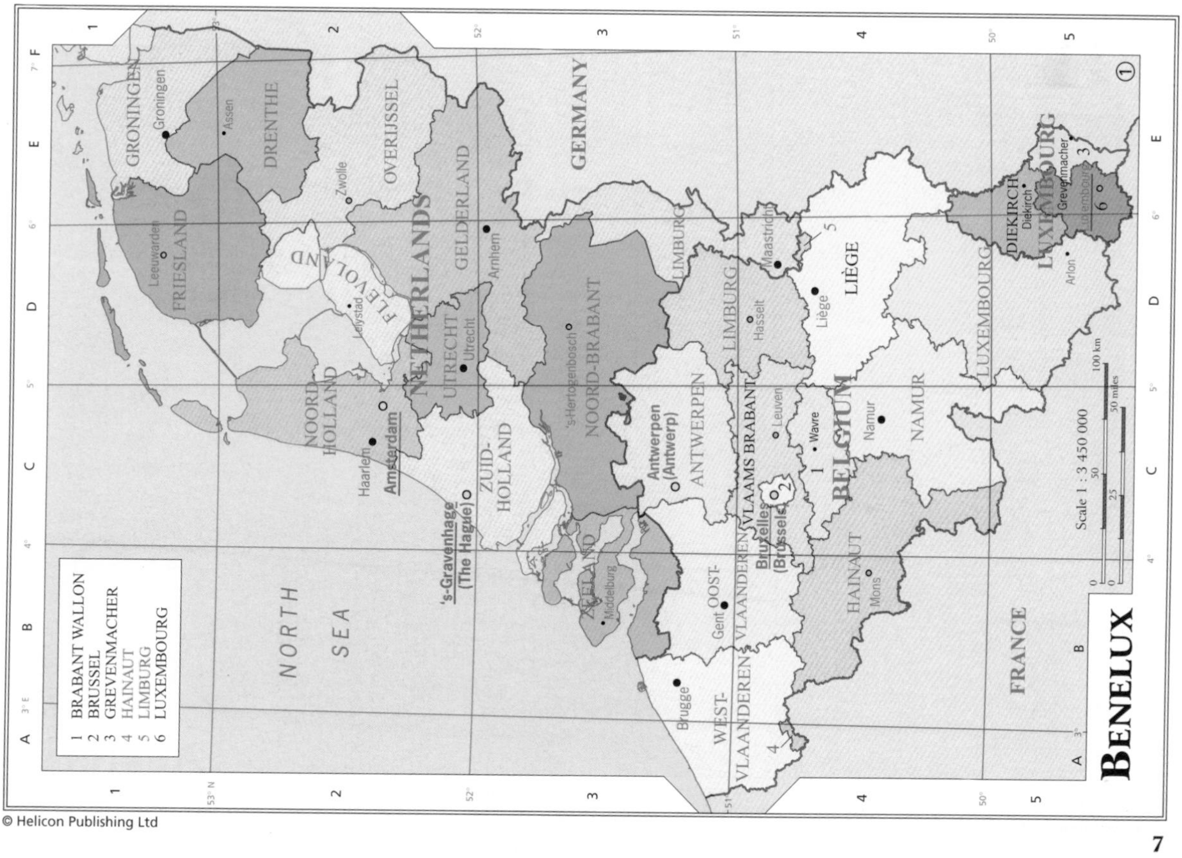

0 50 100 km

1 BRABANT WALLON
2 BRUSSEL
3 GREVENMACHER
4 HAINAUT
5 LIMBURG
6 LUXEMBOURG

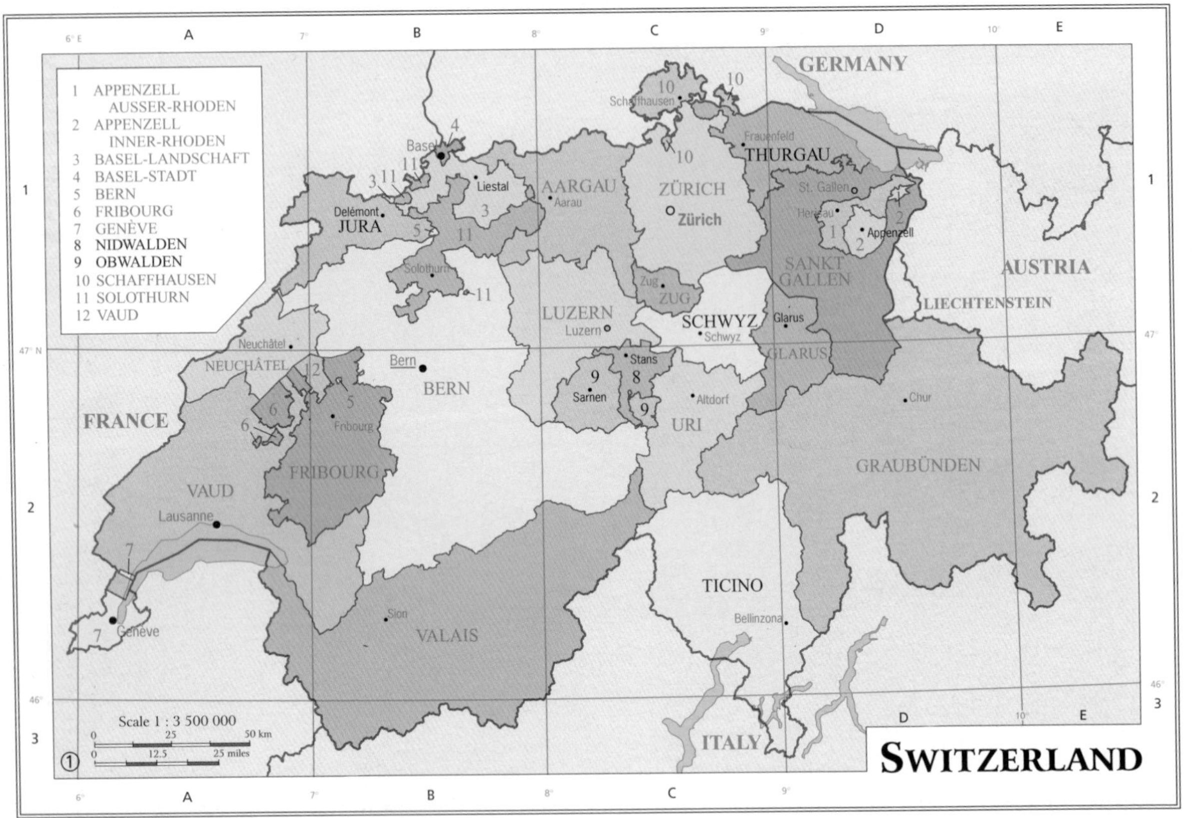

1 APPENZELL AUSSER-RHODEN
2 APPENZELL INNER-RHODEN
3 BASEL-LANDSCHAFT
4 BASEL-STADT
5 BERN
6 FRIBOURG
7 GENÈVE
8 NIDWALDEN
9 OBWALDEN
10 SCHAFFHAUSEN
11 SOLOTHURN
12 VAUD

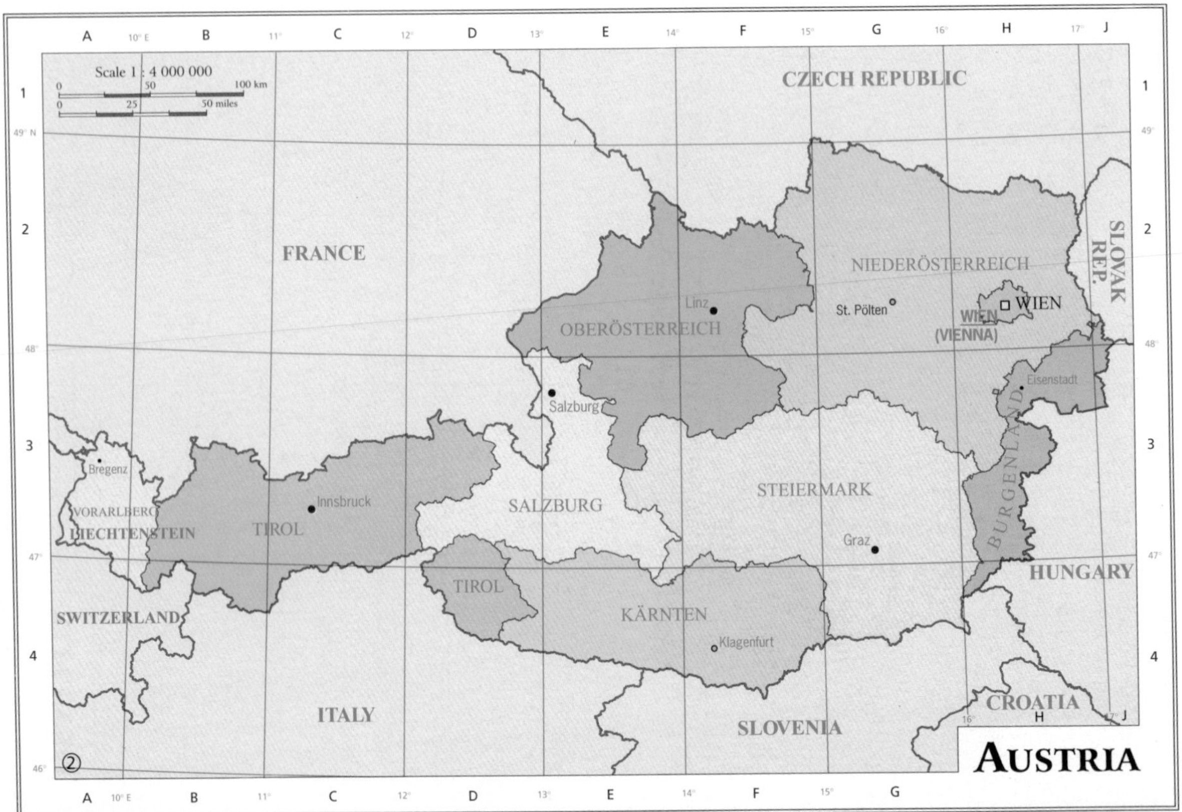

Scale 1 : 3 500 000
0 25 50 km
0 12.5 25 miles

① **SWITZERLAND**

Scale 1 : 4 000 000
0 50 100 km
0 25 50 miles

② **AUSTRIA**

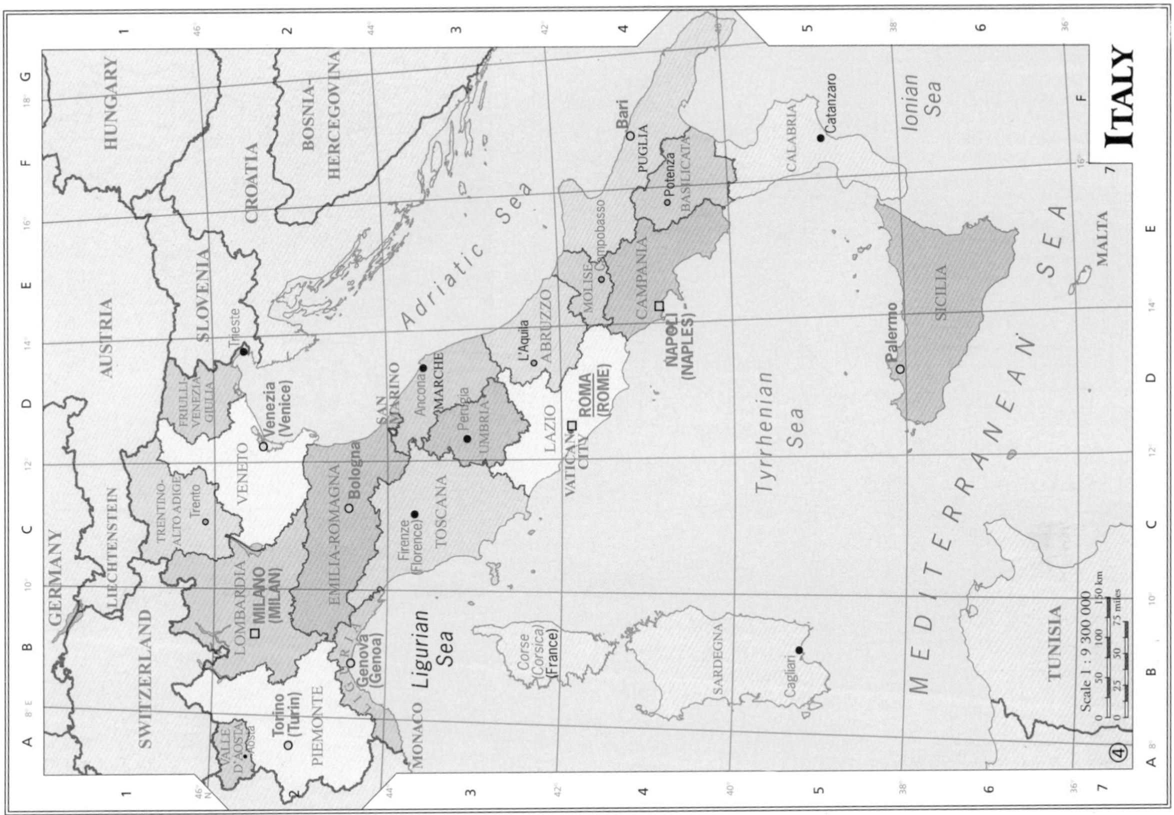

Italy

Scale 1 : 9 300 000

Spain

1 CANTABRIA
2 CASTILLA Y LEÓN
3 LA RIOJA
4 MADRID
5 VALENCIA

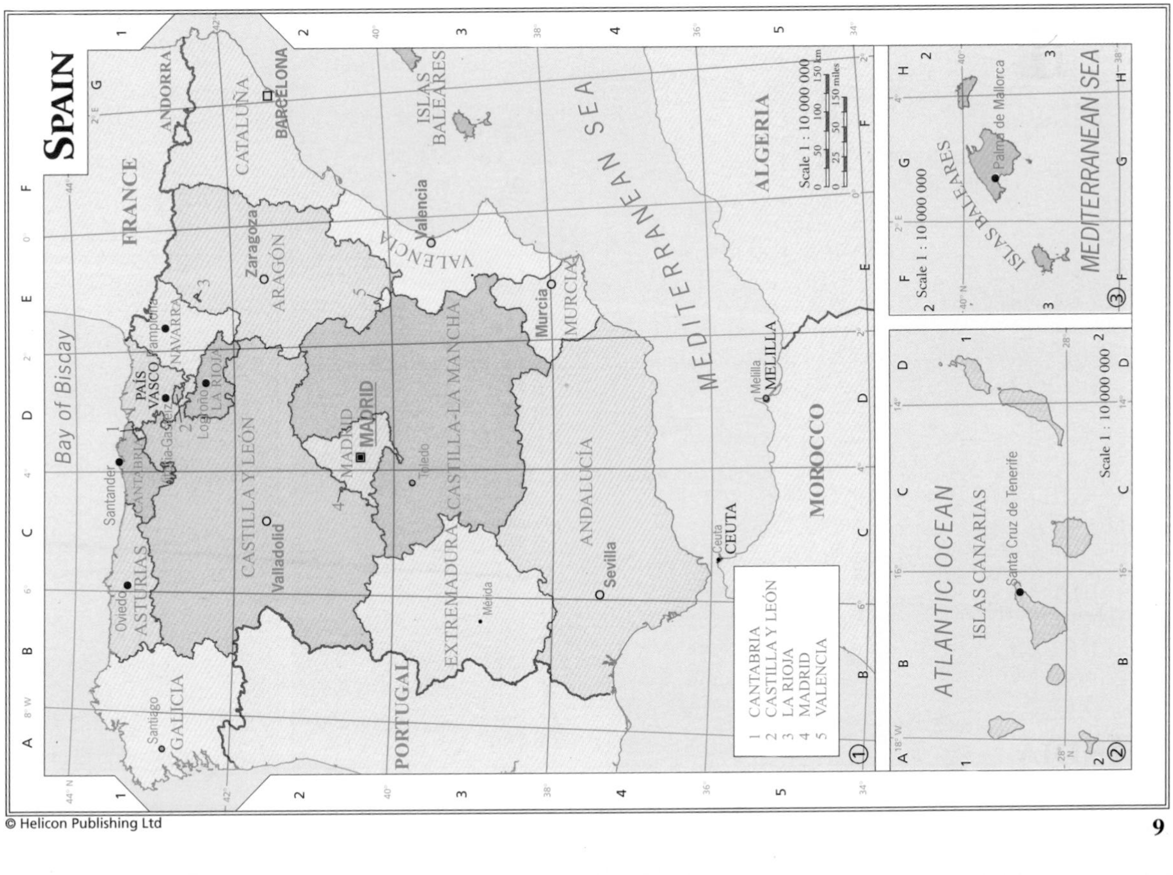

Scale 1 : 10 000 000

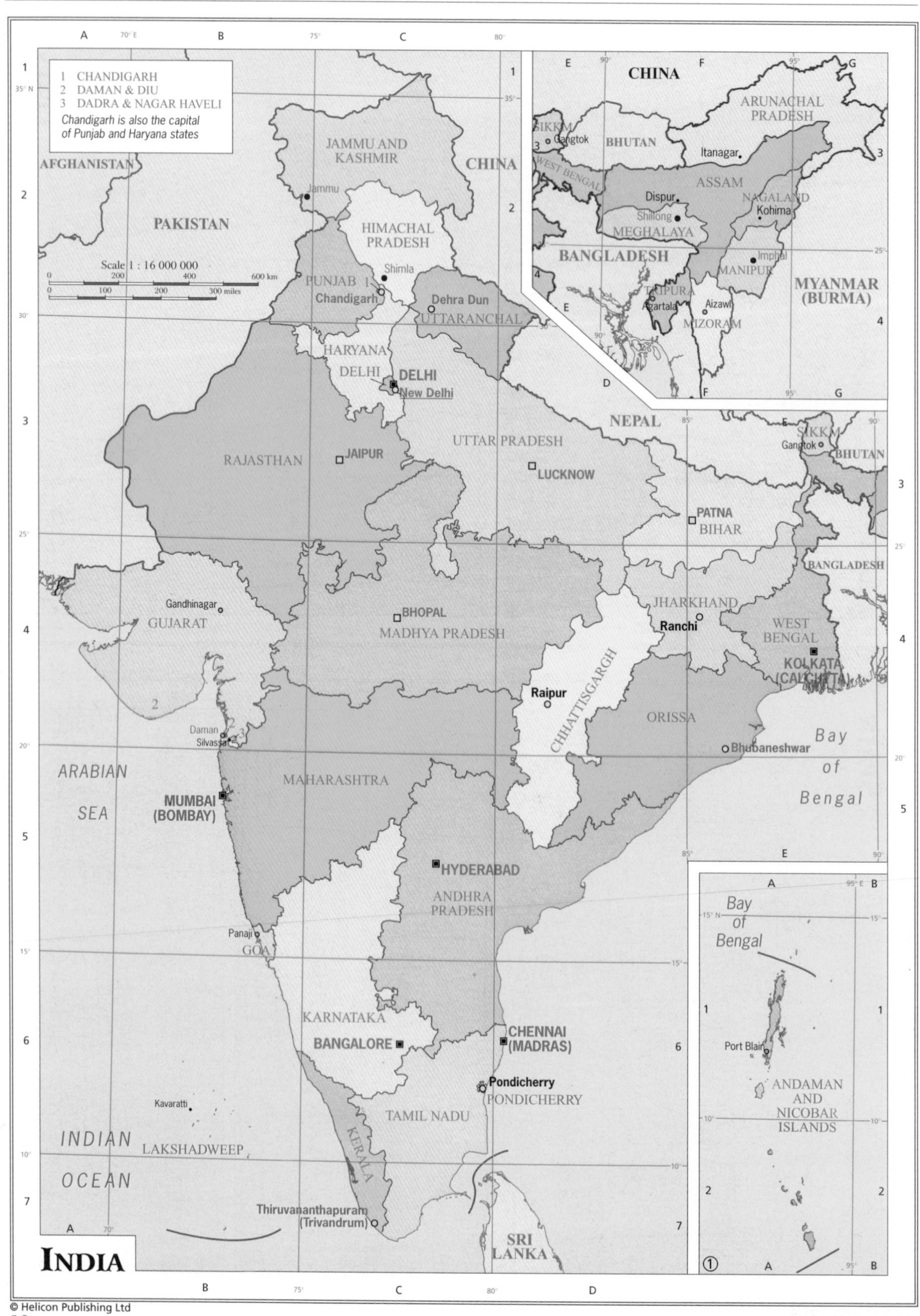
1 CHANDIGARH
2 DAMAN & DIU
3 DADRA & NAGAR HAVELI
Chandigarh is also the capital of Punjab and Haryana states
Scale 1 : 16 000 000
AFGHANISTAN
PAKISTAN
CHINA
JAMMU AND KASHMIR
Jammu
HIMACHAL PRADESH
Shimla
PUNJAB
Chandigarh
Dehra Dun
UTTARANCHAL
HARYANA
DELHI
New Delhi
RAJASTHAN
JAIPUR
UTTAR PRADESH
LUCKNOW
NEPAL
PATNA
BIHAR
SIKKIM
Gangtok
BHUTAN
BANGLADESH
Gandhinagar
GUJARAT
BHOPAL
MADHYA PRADESH
JHARKHAND
Ranchi
WEST BENGAL
KOLKATA (CALCUTTA)
CHHATTISGARH
Raipur
ORISSA
Bhubaneshwar
Daman
Silvassa
ARABIAN SEA
MUMBAI (BOMBAY)
MAHARASHTRA
Bay of Bengal
HYDERABAD
ANDHRA PRADESH
Panaji
GOA
KARNATAKA
BANGALORE
CHENNAI (MADRAS)
Pondicherry
PONDICHERRY
TAMIL NADU
KERALA
Kavaratti
LAKSHADWEEP
INDIAN OCEAN
Thiruvananthapuram (Trivandrum)
SRI LANKA
ARUNACHAL PRADESH
Itanagar
ASSAM
Dispur
NAGALAND
Kohima
Shillong
MEGHALAYA
Imphal
MANIPUR
MYANMAR (BURMA)
TRIPURA
Agartala
Aizawl
MIZORAM
Port Blair
ANDAMAN AND NICOBAR ISLANDS
INDIA

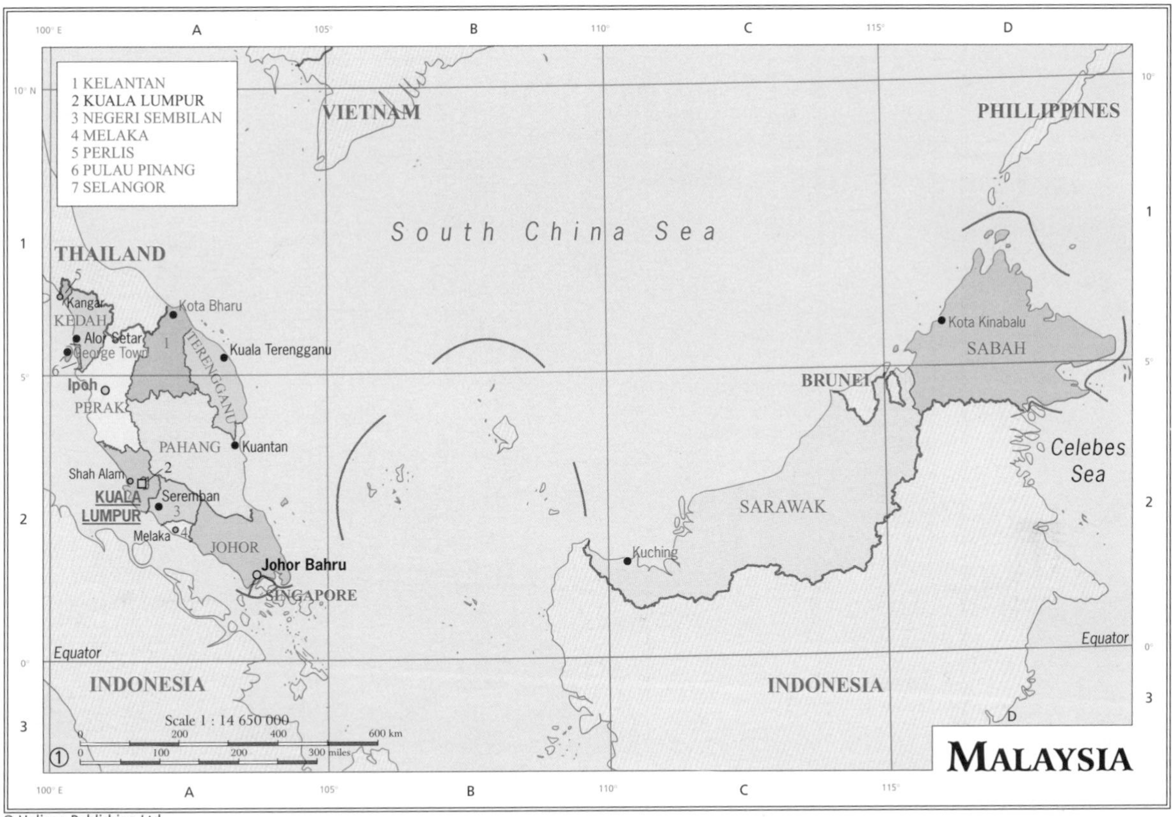

MALAYSIA

1 KELANTAN
2 KUALA LUMPUR
3 NEGERI SEMBILAN
4 MELAKA
5 PERLIS
6 PULAU PINANG
7 SELANGOR

VIETNAM
PHILLIPPINES
South China Sea
THAILAND
Kangar
KEDAH
Alor Setar
George Town
Kota Bharu
TERENGGANU
Kuala Terengganu
Ipoh
PERAK
PAHANG
Kuantan
Shah Alam
KUALA LUMPUR
Seremban
Melaka
JOHOR
Johor Bahru
SINGAPORE
Kota Kinabalu
SABAH
BRUNEI
Celebes Sea
SARAWAK
Kuching
Equator
INDONESIA
INDONESIA
Scale 1 : 14 650 000
0 200 400 600 km
0 100 200 300 miles

CHINA

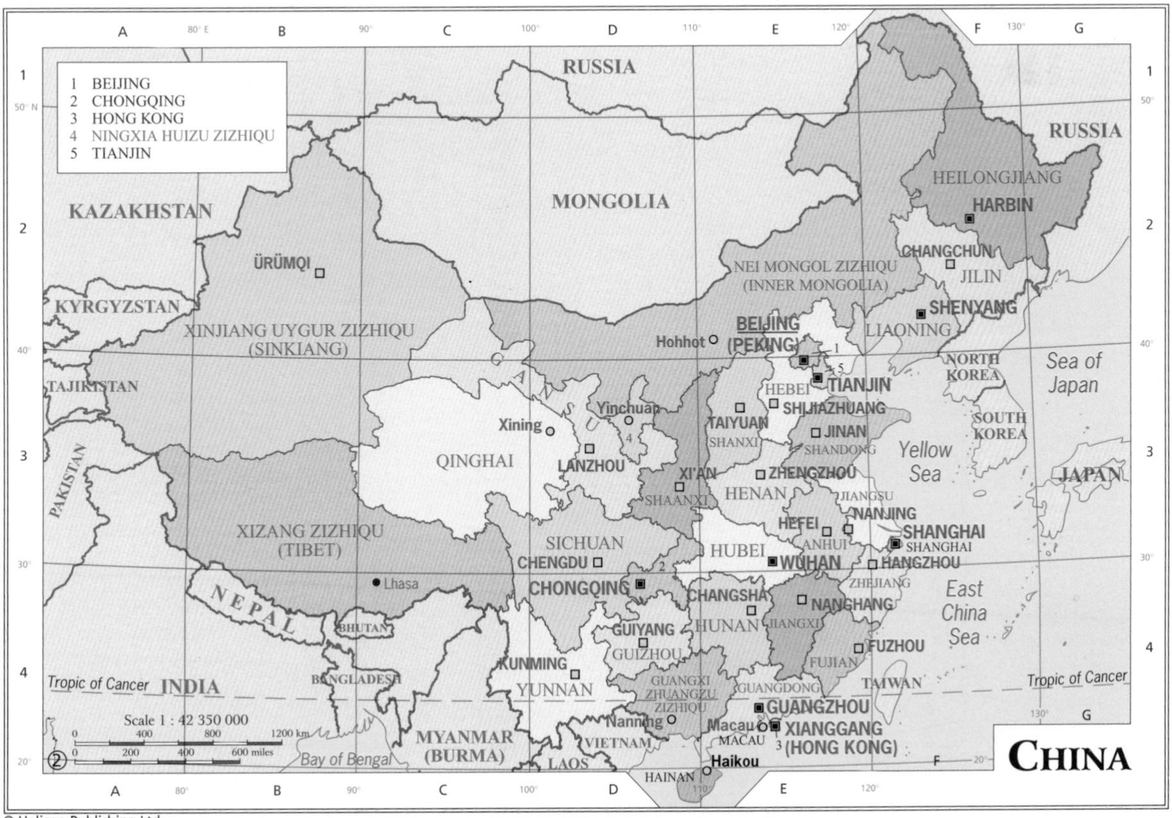

1 BEIJING
2 CHONGQING
3 HONG KONG
4 NINGXIA HUIZU ZIZHIQU
5 TIANJIN

RUSSIA
RUSSIA
KAZAKHSTAN
MONGOLIA
HEILONGJIANG
HARBIN
ÜRÜMQI
NEI MONGOL ZIZHIQU (INNER MONGOLIA)
CHANGCHUN
JILIN
KYRGYZSTAN
SHENYANG
LIAONING
XINJIANG UYGUR ZIZHIQU (SINKIANG)
BEIJING (PEKING)
Hohhot
NORTH KOREA
Sea of Japan
TAJIKISTAN
GANSU
TIANJIN
HEBEI
SHIJIAZHUANG
Yinchuan
Xining
TAIYUAN
SHANXI
JINAN
SHANDONG
SOUTH KOREA
PAKISTAN
QINGHAI
LANZHOU
Yellow Sea
XI'AN
SHAANXI
ZHENGZHOU
HENAN
JIANGSU
JAPAN
NANJING
HEFEI
SHANGHAI
SHANGHAI
ANHUI
XIZANG ZIZHIQU (TIBET)
SICHUAN
CHENGDU
HUBEI
WUHAN
HANGZHOU
ZHEJIANG
Lhasa
NEPAL
CHONGQING
CHANGSHA
NANCHANG
East China Sea
BHUTAN
GUIYANG
HUNAN
JIANGXI
FUZHOU
GUIZHOU
FUJIAN
KUNMING
Tropic of Cancer
INDIA
BANGLADESH
YUNNAN
GUANGXI ZHUANGZU ZIZHIQU
GUANGDONG
TAIWAN
Tropic of Cancer
GUANGZHOU
Nanning
Macau
MACAU
XIANGGANG (HONG KONG)
Scale 1 : 42 350 000
0 400 800 1200 km
0 200 400 600 miles
MYANMAR (BURMA)
VIETNAM
Bay of Bengal
LAOS
HAINAN
Haikou

Equatorial Scale 1 : 141 100 000

0 1000 2000 3000 4000 km

0 1000 2000 miles

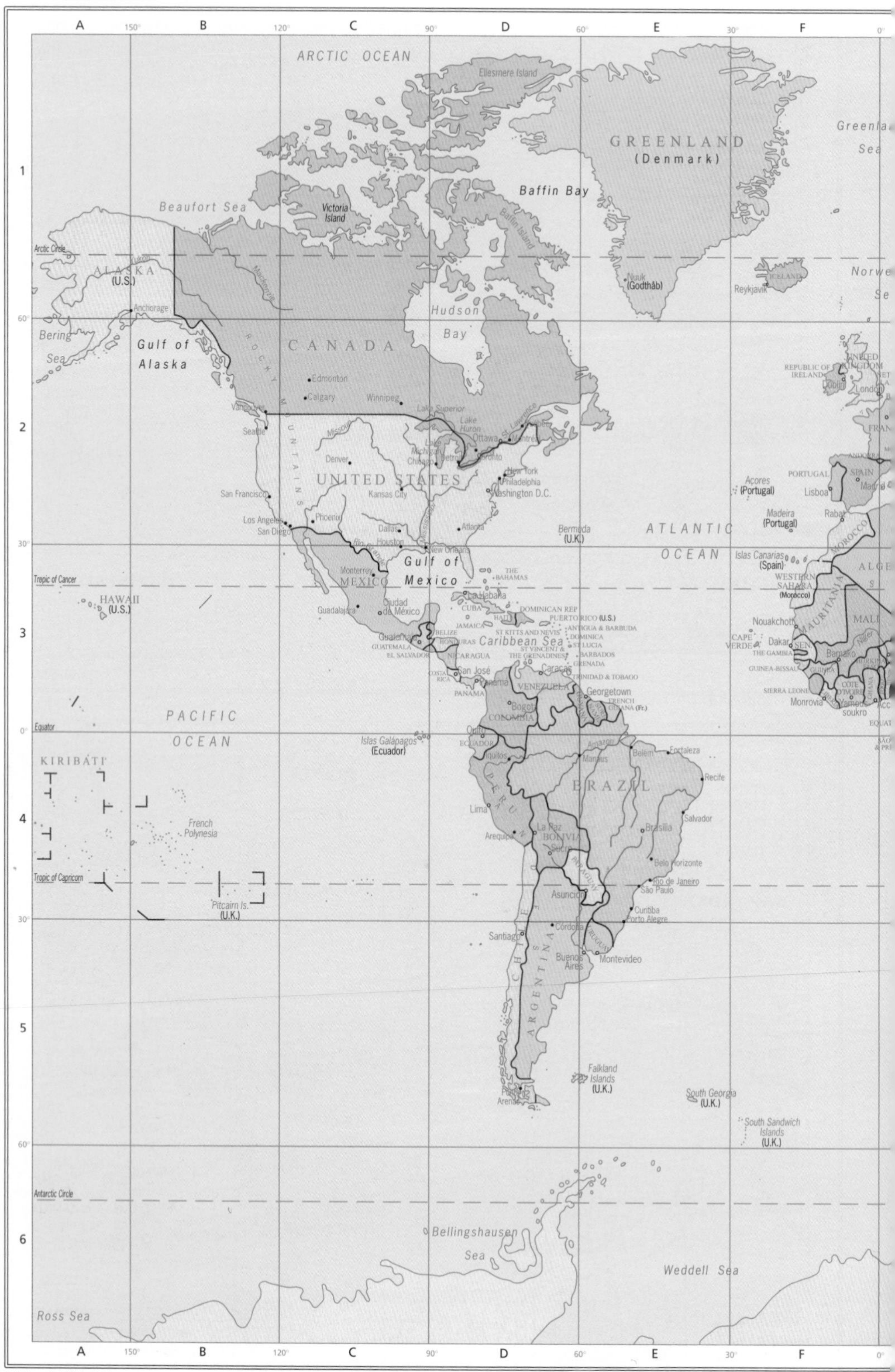

WORLD POLITICAL

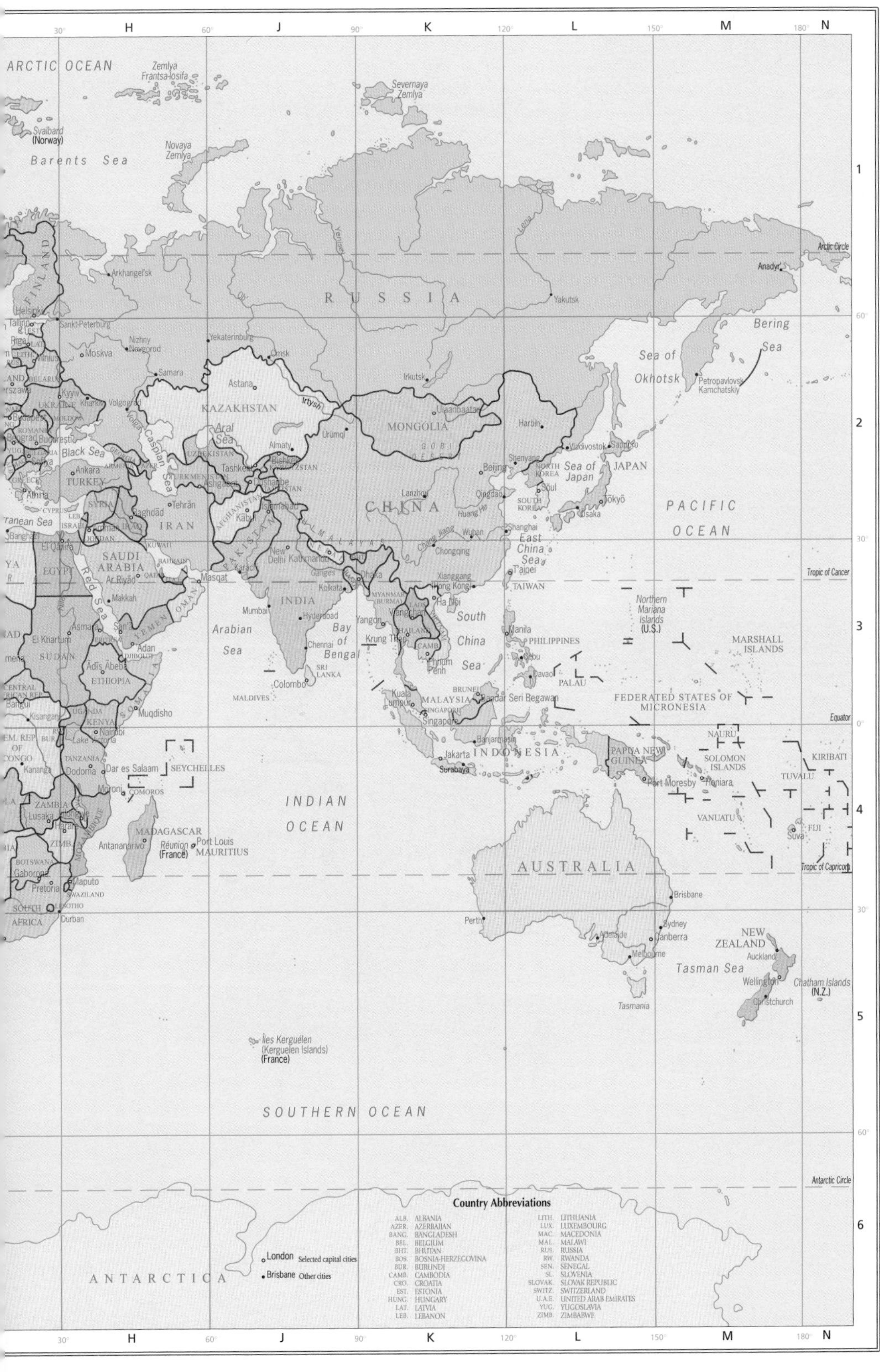

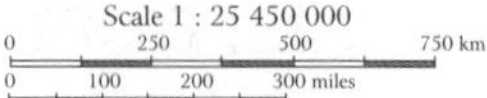

A 1 B C D E F
2
3
4
5
D E F

ICELAND
Reykjavík
Arctic Circle
Norwegian Sea
Faeroes (Denmark)
Rockall
Shetland Is. (U.K.)
Outer Hebrides
Orkney Is.
ATLANTIC OCEAN
SCOTLAND
Glasgow
Edinburgh
NORTHERN IRELAND
Belfast
REP. OF IRELAND
DUBLIN (BAILE ÁTHA CLIATH)
UNITED KINGDOM
WALES
BIRMINGHAM
ENGLAND
Cardiff
Plymouth
LONDON
English Channel
Channel Islands
North Sea
NORWAY
SWEDEN
Trondheim
Bergen
Stavanger
Oslo
Stockholm
Vänern
Göteborg
Sundsvall
Tromsø
Gulf of Bothnia
Gotland
Baltic Sea
DENMARK
Århus
København (Copenhagen)
Bornholm
Gdańsk
RUSSIA
Amsterdam
's-Gravenhage (The Hague)
NETHERLANDS
Bruxelles (Brussels)
BELGIUM
Luxembourg
LUXEMBOURG
HAMBURG
Hannover
BERLIN
GERMANY
Bonn
Frankfurt am Main
Elbe
Ems
Rhine
Oder
Odra
Wisła
WARSZAWA (WARSAW)
POLAND
PRAHA (PRAGUE)
CZECH REP.
MÜNCHEN (MUNICH)
Danube
WIEN (VIENNA)
AUSTRIA
SLOVAK REP.
Bratislava
BUDAPEST
HUNGARY
LIECHTENSTEIN
Vaduz
Bern
SWITZERLAND
4808 Mt. Blanc
PARIS
Seine
Loire
Strasbourg
FRANCE
Bay of Biscay
Bordeaux
Lyon
Massif Central
Rhône
Cabo Fisterra
PORTUGAL
Lisboa (Lisbon)
Cabo de São Vicente
Tagus
MADRID
SPAIN
Ebro
Pyrenees
Andorra la Vella
ANDORRA
Marseille
BARCELONA
Valencia
Islas Baleares (Balearic Islands)
Menorca
Mallorca
Eivissa
Strait of Gibraltar
Gibraltar (U.K.)
Ceuta (Spain)
Melilla (Spain)
RABAT
ALGER (ALGIERS)
Tunis
Genova (Genoa)
MILANO (MILAN)
MONACO
SAN MARINO
SLOVENIA
Ljubljana
Zagreb
CROATIA
BOSNIA-HERZEGOVINA
Sarajevo
BEOGRAD (BELGRADE)
YUGOSLAVIA
SOFIYA (SOFIA)
Skopje
MACEDONIA
Tirane (Tirana)
ALBANIA
Adriatic Sea
Apennines
ITALY
Corse (Corsica) (France)
Ajaccio
VATICAN CITY
ROMA (ROME)
Sardegna (Sardinia) (Italy)
NAPOLI (NAPLES)
Taranto
Cagliari
Tyrrhenian Sea
Mediterranean Sea
Palermo
Sicilia (Sicily)
Mte. Etna 3340
Valletta
MALTA
Kerkyra (Corfu)
GREECE
Ionian Sea
AFRICA
Tarābulus (Tripoli)
Banghāzī

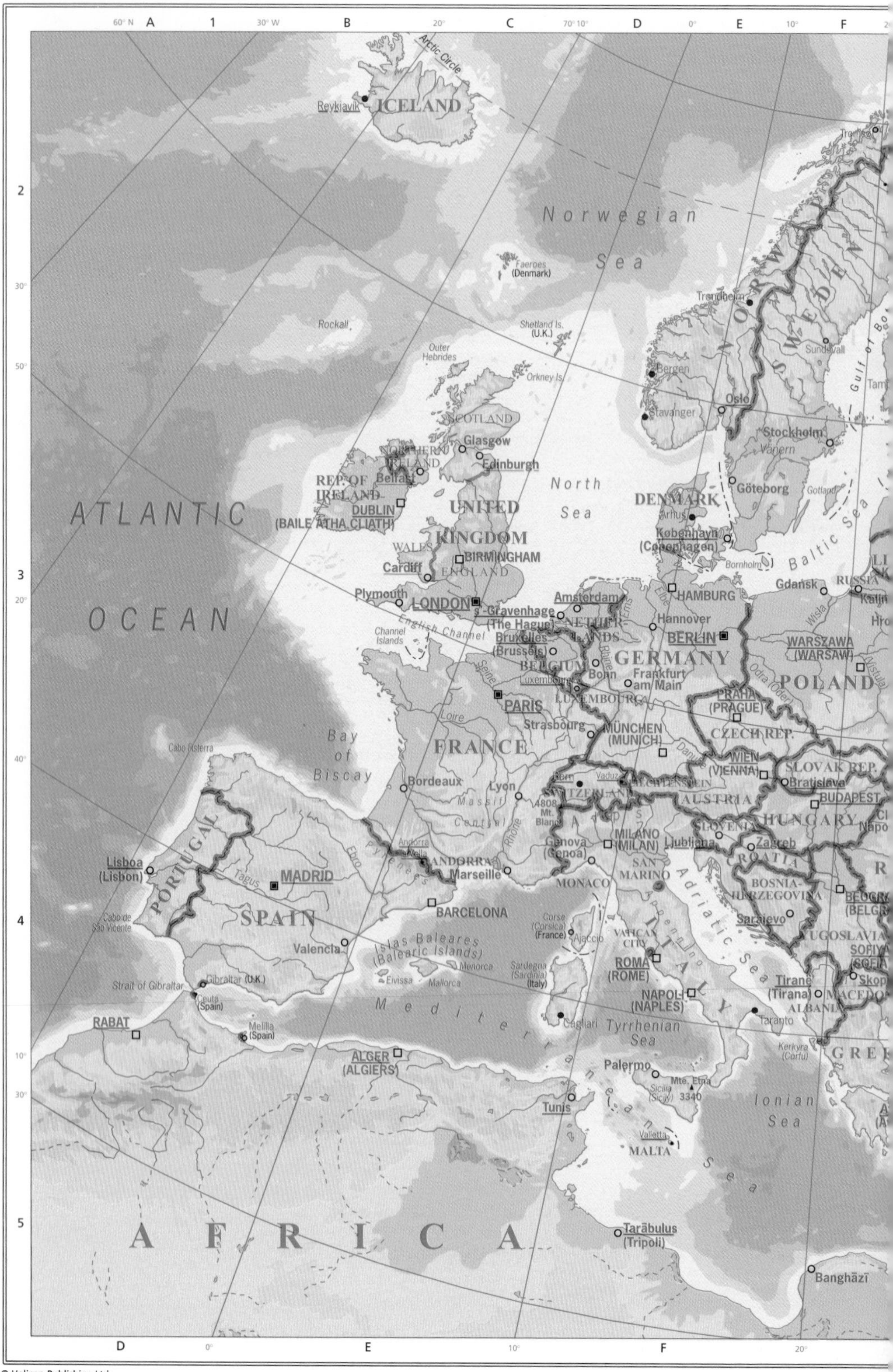

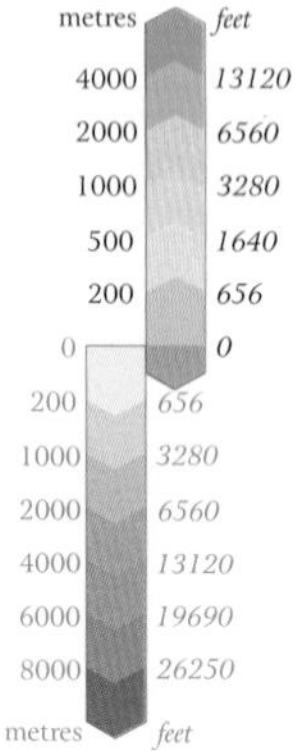

metres	feet
4000	13120
2000	6560
1000	3280
500	1640
200	656
0	0
200	656
1000	3280
2000	6560
4000	13120
6000	19690
8000	26250
metres	feet

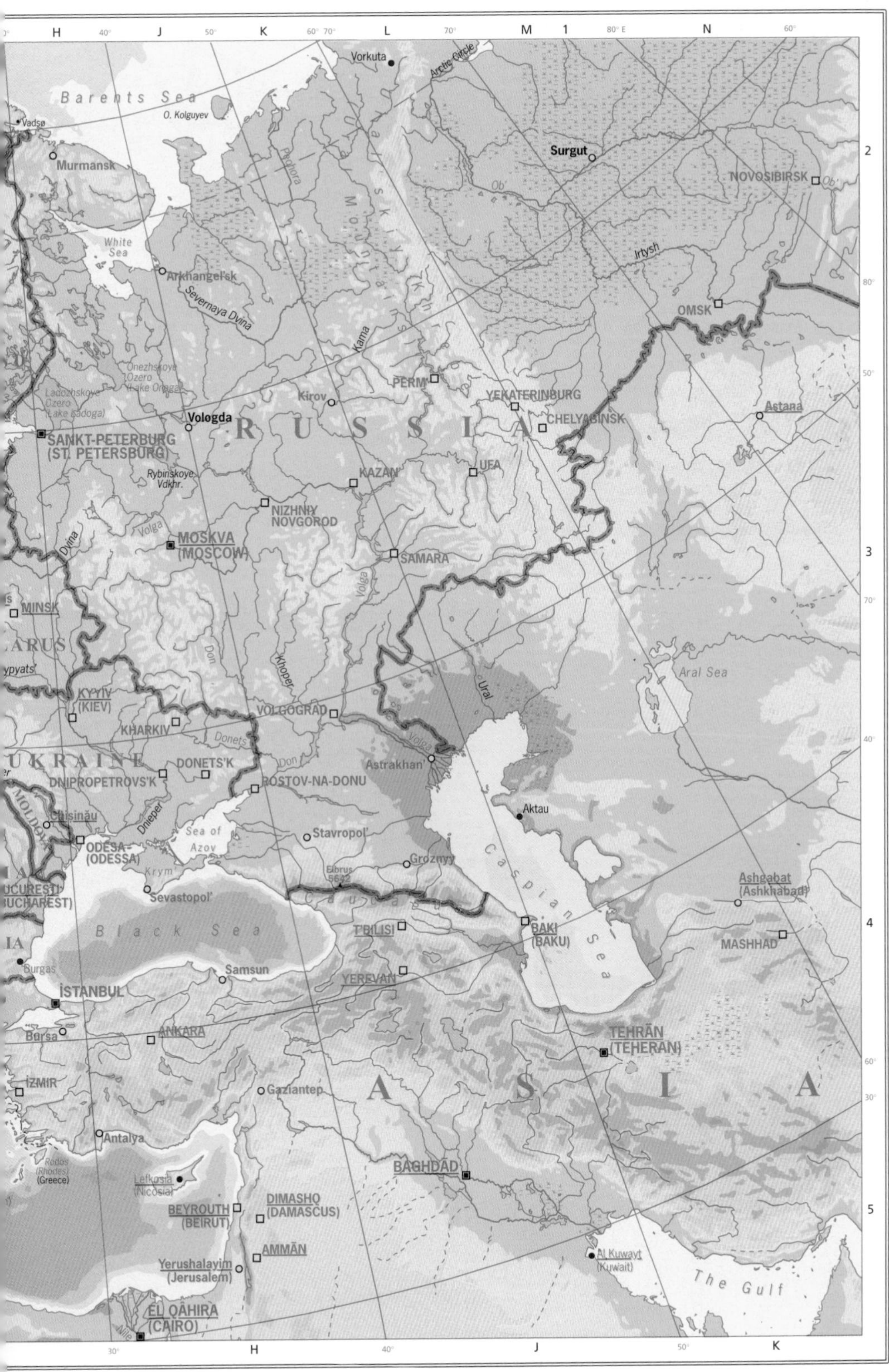
Barents Sea
O. Kolguyev
Vadsø
Murmansk
Vorkuta
Arctic Circle
Surgut
NOVOSIBIRSK
Ob'
White Sea
Arkhangel'sk
Severnaya Dvina
Pechora
Ural'skiy Khrebet
Mountains
Irtysh
OMSK
Karma
Onezhskoye Ozero (Lake Onega)
Ladozhskoye Ozero (Lake Ladoga)
PERM'
YEKATERINBURG
Kirov
CHELYABINSK
Astana
Vologda
RUSSIA
SANKT-PETERBURG (ST. PETERSBURG)
Rybinskoye Vdkhr.
KAZAN'
UFA
NIZHNIY NOVGOROD
Volga
MOSKVA (MOSCOW)
Dvina
SAMARA
MINSK
ARUS
Don
Khoper
Pripyats'
Aral Sea
Ural
KYYIV (KIEV)
VOLGOGRAD
KHARKIV
Donets
UKRAINE
DONETS'K
Astrakhan'
DNIPROPETROVS'K
ROSTOV-NA-DONU
MOLDOVA
Chişinău
Dnieper
Aktau
Sea of Azov
Stavropol'
ODESA (ODESSA)
Krym
Groznyy
Caspian Sea
Elbrus 5642
Caucasus
Ashgabat (Ashkhabad)
Sevastopol'
BUCUREŞTI (BUCHAREST)
Black Sea
T'BILISI
BAKI (BAKU)
MASHHAD
Burgas
Samsun
YEREVAN
İSTANBUL
Bursa
ANKARA
TEHRĀN (TEHERAN)
İZMIR
Gaziantep
ASIA
Antalya
Rodos (Rhodes) (Greece)
BAGHDĀD
Lefkosia (Nicosia)
DIMASHQ (DAMASCUS)
BEYROUTH (BEIRUT)
AMMĀN
Al Kuwayt (Kuwait)
Yerushalayim (Jerusalem)
The Gulf
EL QÂHIRA (CAIRO)
Nile

Scale 1 : 7 300 000

0 100 200 300 km

0 50 100 150 miles

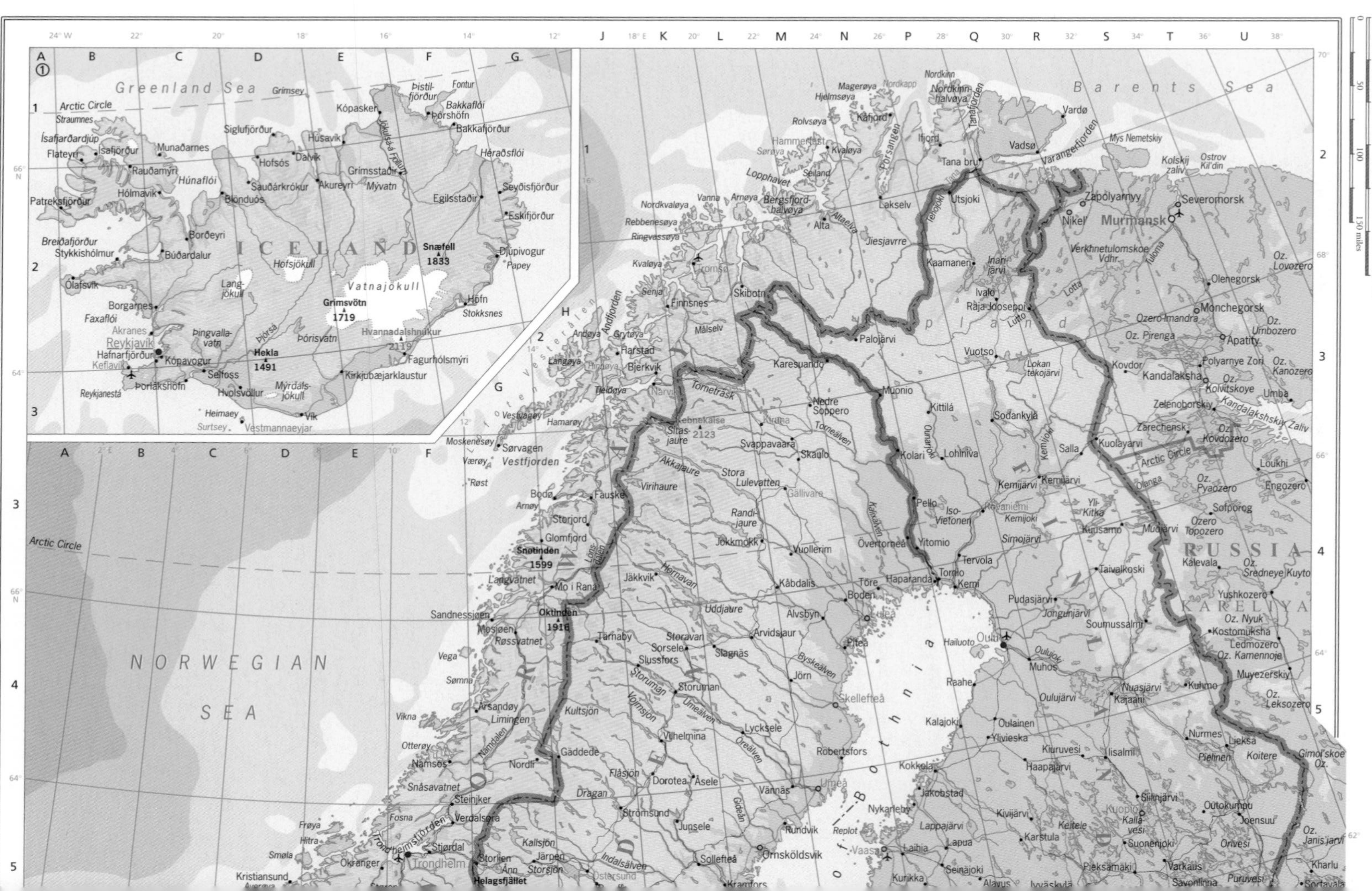

metres	feet
4000	13120
2000	6560
1000	3280
500	1640
200	656
0	0
200	656
1000	3280
2000	6560
4000	13120
6000	19690
8000	26250
metres	feet

SCANDINAVIA

- Denmark
- Estonia
- Finland
- Iceland
- Latvia
- Lithuania
- Norway
- Sweden

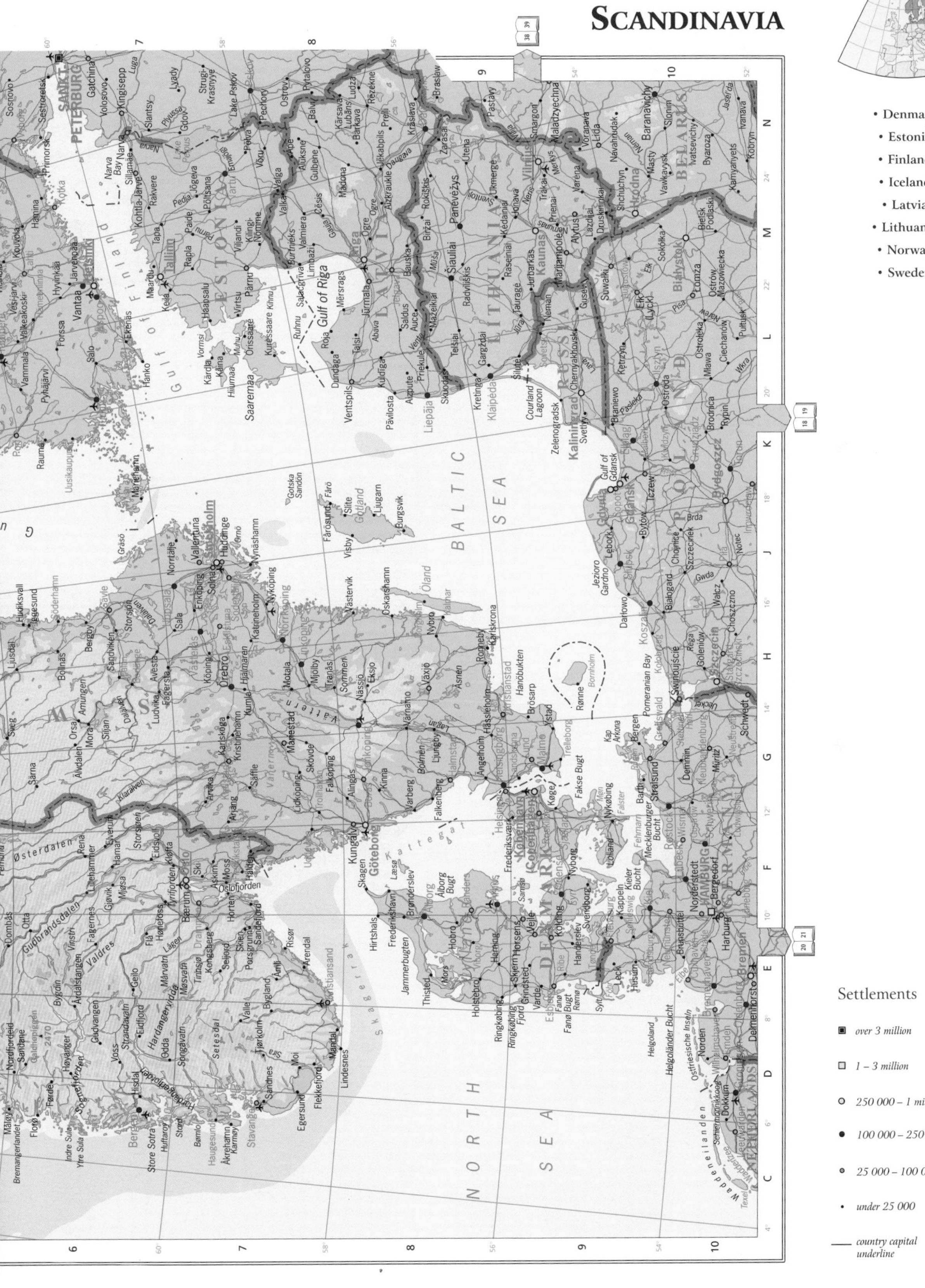

Settlements

- ▣ *over 3 million*
- □ *1 – 3 million*
- ○ *250 000 – 1 million*
- ● *100 000 – 250 000*
- ◦ *25 000 – 100 000*
- • *under 25 000*
- ___ *country capital underline*

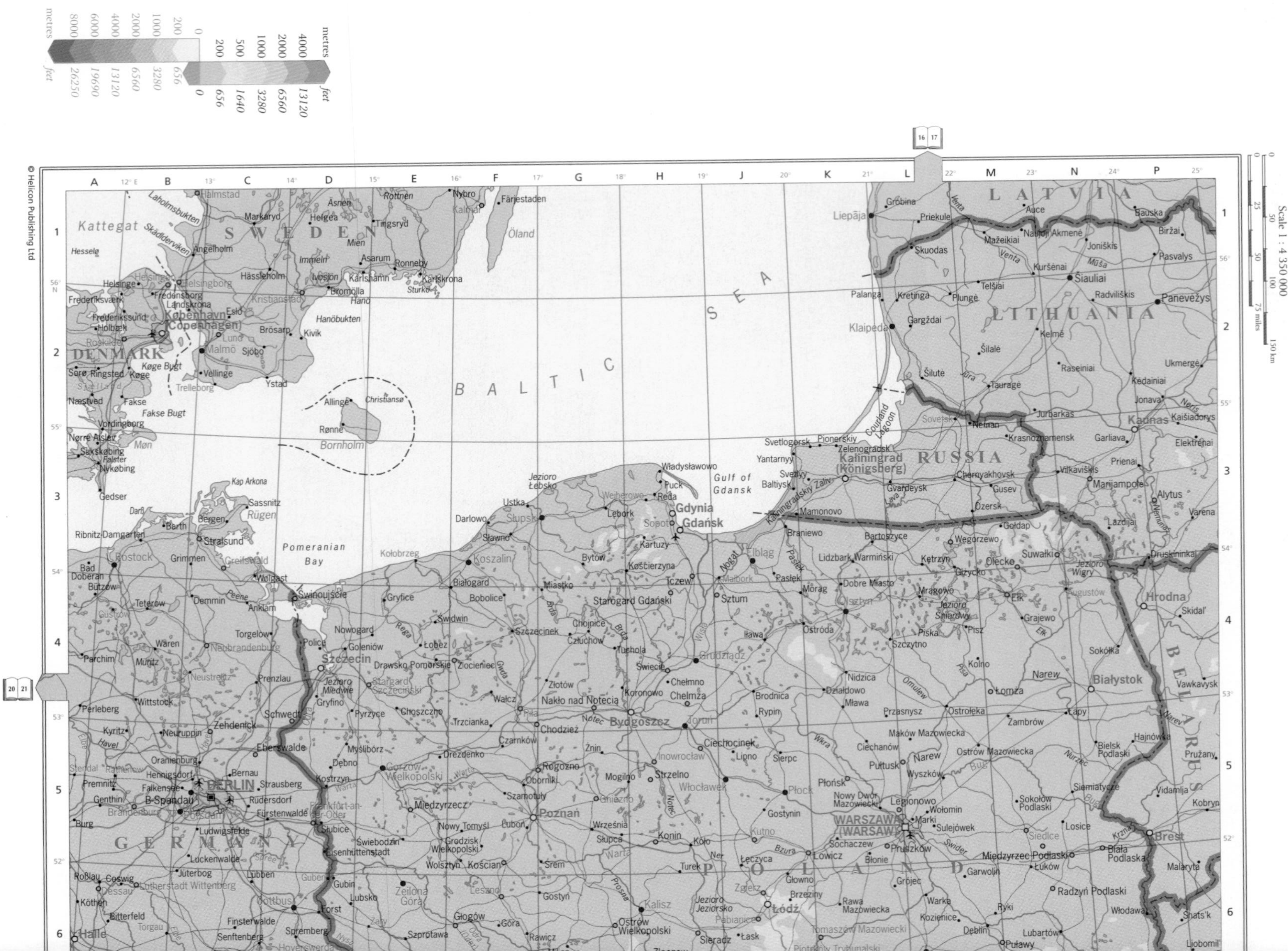

Scale 1 : 4 350 000
0 50 100 150 km
0 25 50 75 miles
metres feet
4000 13120
2000 6560
1000 3280
500 1640
200 656
0 0
200 656
1000 3280
2000 6560
4000 13120
6000 19690
8000 26250
metres feet
BALTIC SEA
LATVIA
LITHUANIA
RUSSIA
BELARUS
SWEDEN
DENMARK
POLAND
GERMANY
Gulf of Gdansk
Pomeranian Bay
Kattegat
Courland Lagoon
Bornholm
Öland
Rügen
Kaliningrad (Königsberg)
Gdańsk
Gdynia
Szczecin
Poznań
Bydgoszcz
Warszawa (Warsaw)
Łódź
Białystok
Kaunas
Klaipeda
Šiauliai
Panevėžys
Hrodna
Brest
BERLIN
København (Copenhagen)
Malmö
Rostock
Liepāja

CENTRAL EUROPE

- Czech Republic
- Hungary
- Poland
- Slovak Republic

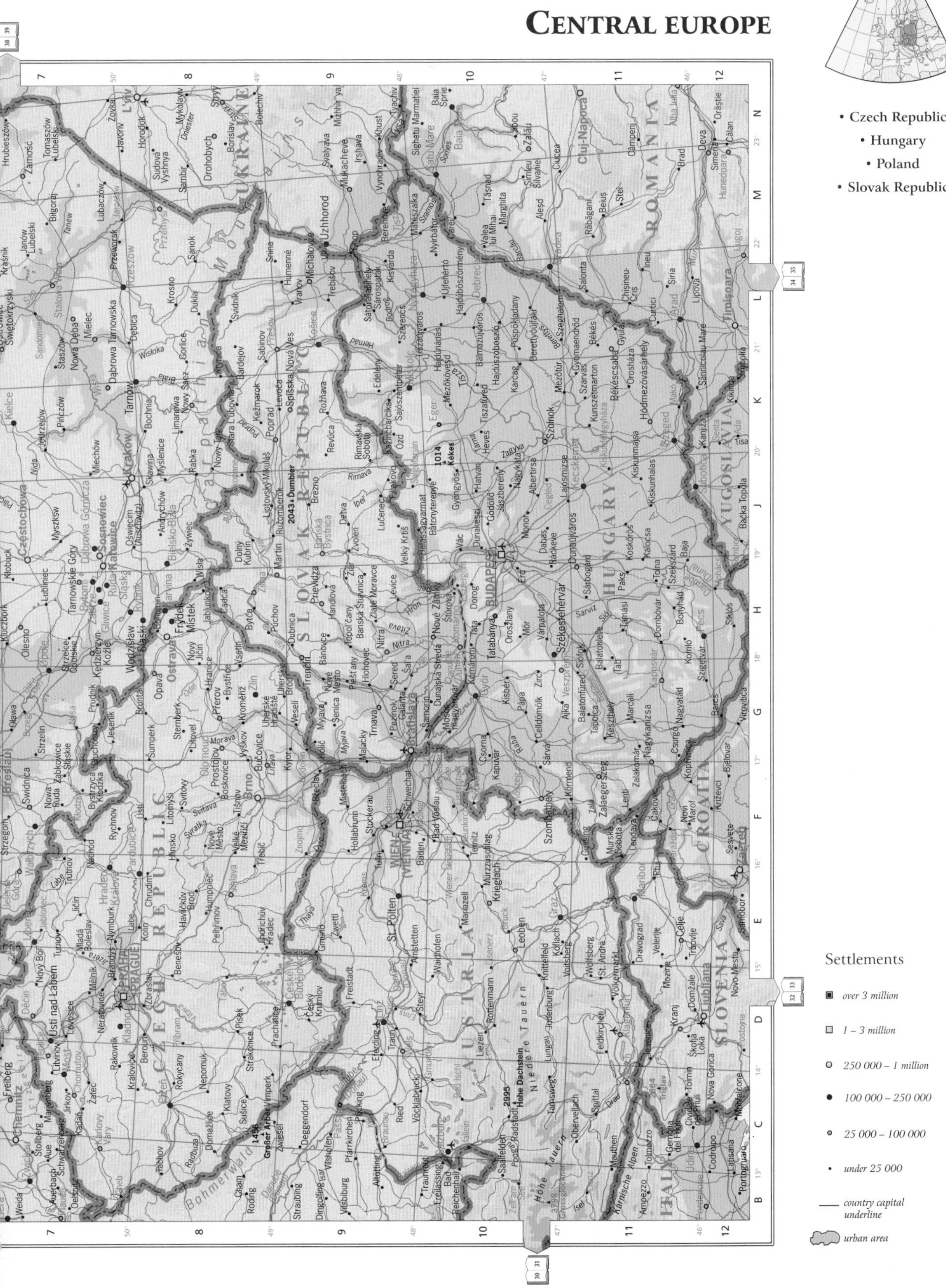

Settlements

- over 3 million
- 1 – 3 million
- 250 000 – 1 million
- 100 000 – 250 000
- 25 000 – 100 000
- under 25 000

country capital underline

urban area

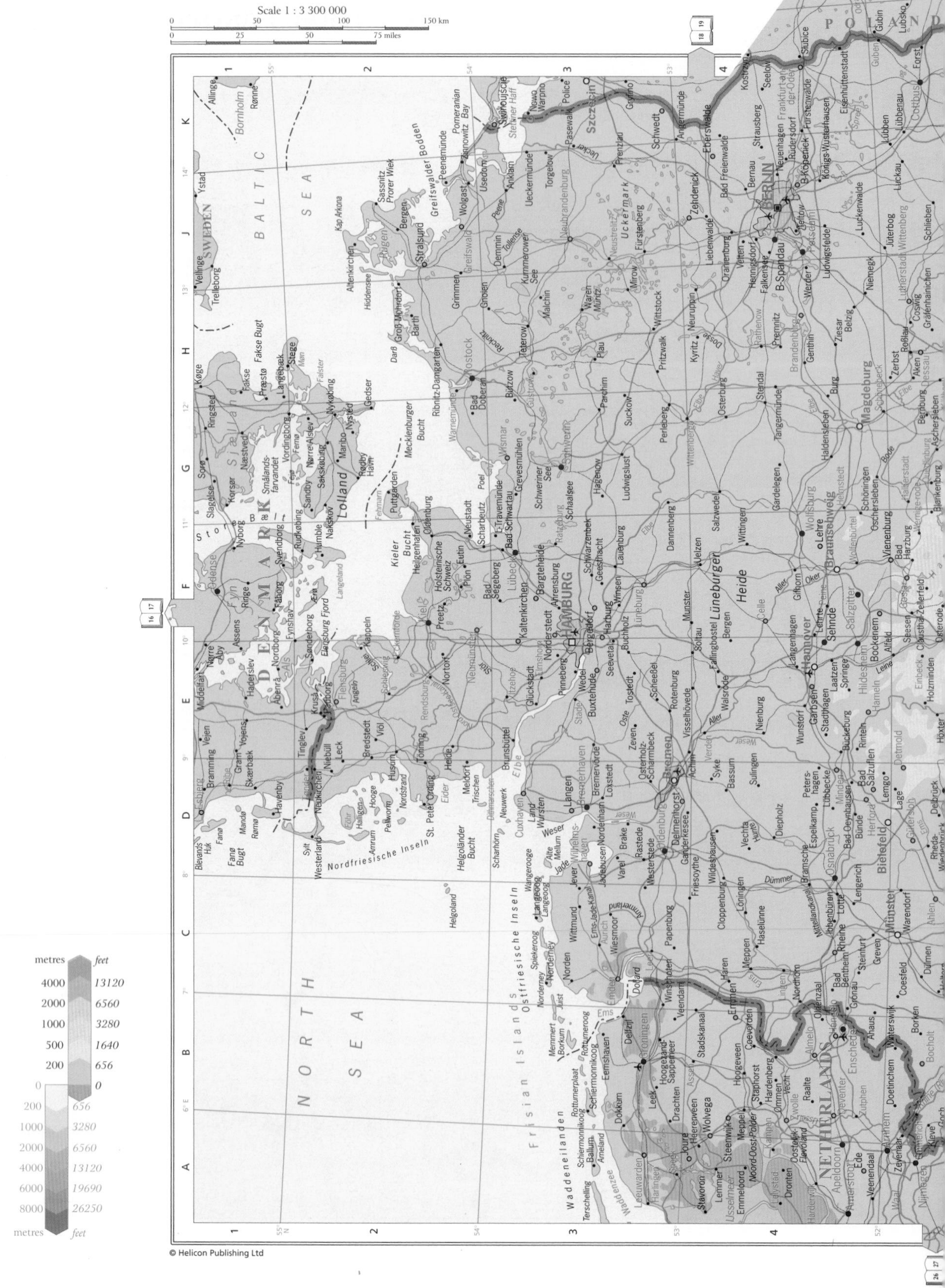

Scale 1 : 3 300 000
0 50 100 150 km
0 25 50 75 miles
metres
4000
2000
1000
500
200
0
200
1000
2000
4000
6000
8000
metres
feet
13120
6560
3280
1640
656
0
656
3280
6560
13120
19690
26250
feet
NORTH SEA
BALTIC SEA
DENMARK
SWEDEN
POLAND
NETHERLANDS
HAMBURG
BERLIN
Frisian Islands
Ostfriesische Inseln
Nordfriesische Inseln
Waddeneilanden
Lüneburger Heide
Kieler Bucht
Mecklenburger Bucht
Pomeranian Bay
Helgoländer Bucht
Bornholm
Lolland
Fyn
Sjælland

GERMANY

• Germany
• Luxembourg

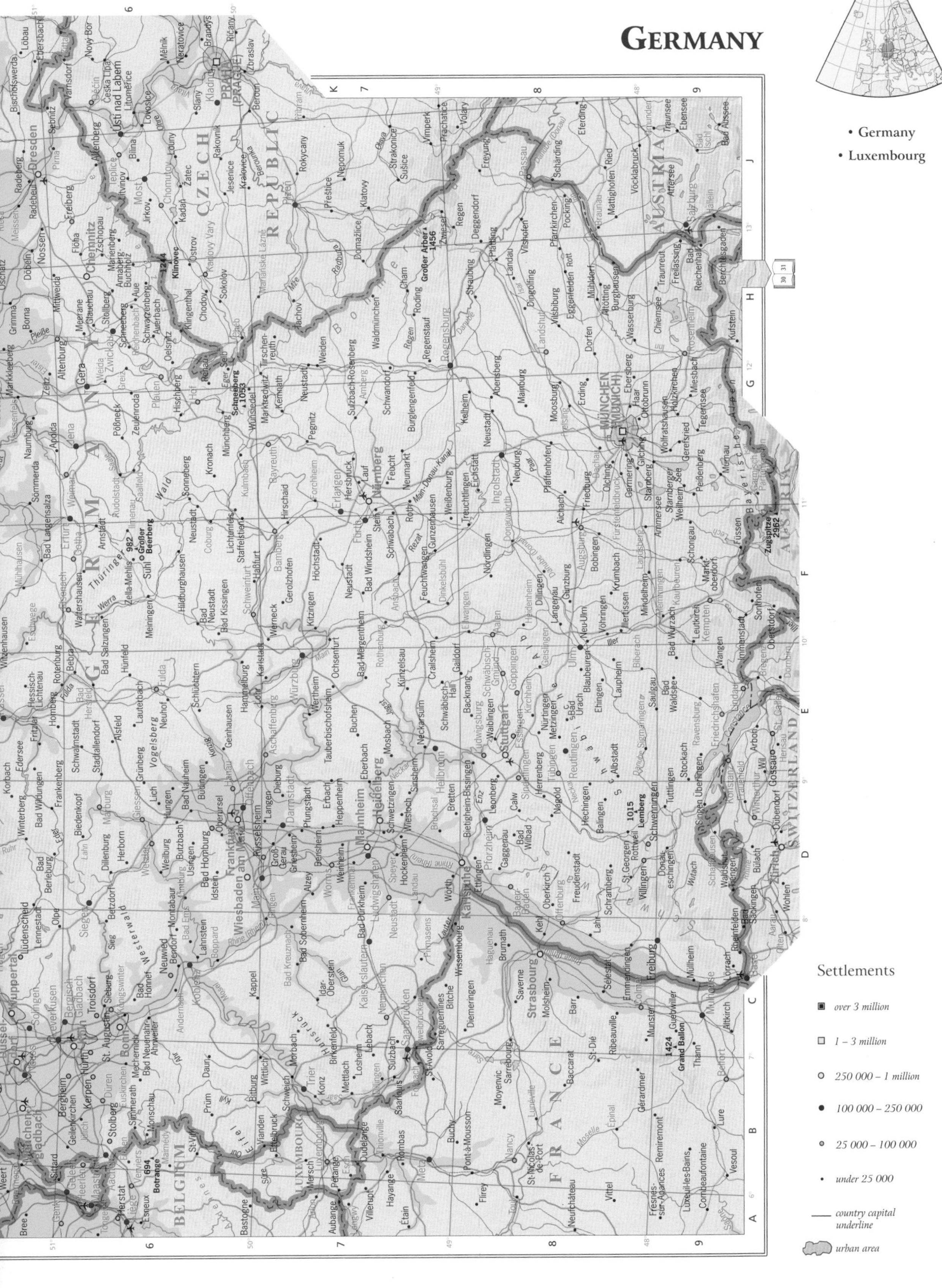

30 31

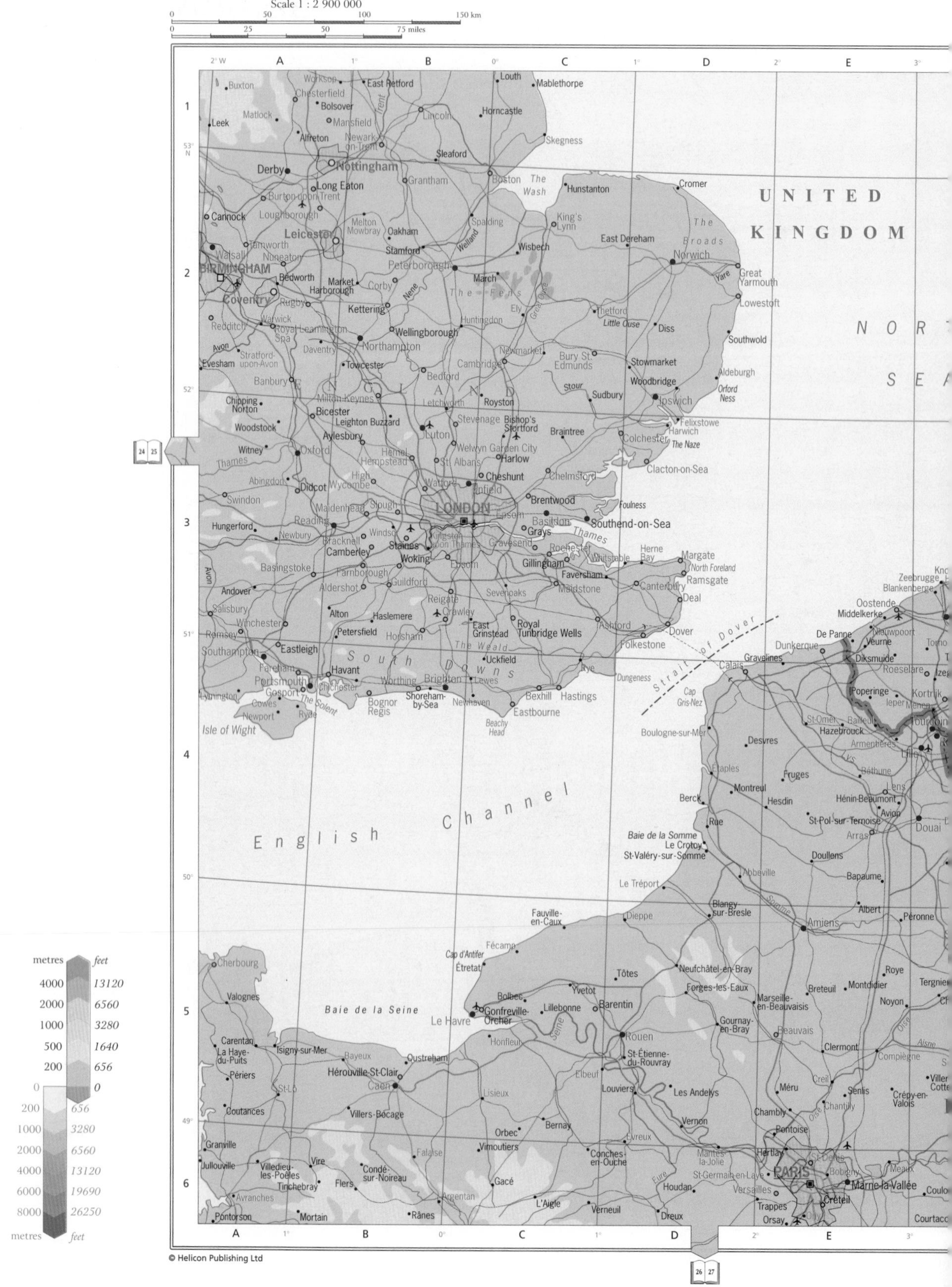

Scale 1 : 2 900 000
0 50 100 150 km
0 25 50 75 miles
A B C D E
1 2 3 4 5 6
UNITED KINGDOM
NORTH SEA
ENGLAND
English Channel
The Wash
The Broads
The Fens
South Downs
The Weald
Strait of Dover
Baie de la Seine
Baie de la Somme
Isle of Wight
The Solent
Thames
LONDON
PARIS
BIRMINGHAM
Nottingham
Leicester
Coventry
Derby
Buxton
Chesterfield
Worksop
East Retford
Louth
Mablethorpe
Lincoln
Horncastle
Skegness
Sleaford
Boston
Grantham
Hunstanton
Cromer
King's Lynn
East Dereham
Norwich
Great Yarmouth
Lowestoft
Peterborough
Wisbech
Spalding
Stamford
Oakham
Melton Mowbray
Loughborough
Long Eaton
Burton upon Trent
Cannock
Tamworth
Walsall
Nuneaton
Bedworth
Market Harborough
Corby
March
Ely
Thetford
Little Ouse
Diss
Southwold
Kettering
Wellingborough
Northampton
Huntingdon
Cambridge
Newmarket
Bury St. Edmunds
Stowmarket
Woodbridge
Aldeburgh
Orford Ness
Ipswich
Felixstowe
Harwich
The Naze
Sudbury
Stour
Rugby
Warwick
Royal Leamington Spa
Redditch
Stratford-upon-Avon
Evesham
Daventry
Towcester
Banbury
Bedford
Milton Keynes
Letchworth
Royston
Stevenage
Bishop's Stortford
Braintree
Colchester
Clacton-on-Sea
Chipping Norton
Bicester
Leighton Buzzard
Luton
Aylesbury
Woodstock
Witney
Oxford
Hemel Hempstead
St Albans
Welwyn Garden City
Harlow
Cheshunt
Watford
Enfield
Chelmsford
Abingdon
Didcot
High Wycombe
Swindon
Maidenhead
Slough
Brentwood
Basildon
Southend-on-Sea
Foulness
Reading
Hungerford
Newbury
Windsor
Staines
Bracknell
Camberley
Kingston upon Thames
Grays
Gravesend
Rochester
Gillingham
Whitstable
Herne Bay
Margate
North Foreland
Ramsgate
Deal
Canterbury
Faversham
Maidstone
Basingstoke
Farnborough
Woking
Epsom
Sevenoaks
Guildford
Aldershot
Reigate
Andover
Salisbury
Alton
Haslemere
Crawley
East Grinstead
Royal Tunbridge Wells
Ashford
Dover
Folkestone
Winchester
Romsey
Petersfield
Horsham
Southampton
Eastleigh
Uckfield
Rye
Dungeness
Fareham
Havant
Portsmouth
Gosport
Chichester
Worthing
Brighton
Lewes
Bexhill
Hastings
Lymington
Cowes
Newport
Ryde
Bognor Regis
Shoreham-by-Sea
Newhaven
Eastbourne
Beachy Head
Avon
Trent
Nene
Welland
Yare
Matlock
Leek
Bolsover
Mansfield
Alfreton
Newark-on-Trent
Zeebrugge
Blankenberge
Oostende
Middelkerke
Nieuwpoort
De Panne
Veurne
Diksmuide
Roeselare
Poperinge
Ieper
Kortrijk
Dunkerque
Gravelines
Calais
Cap Gris-Nez
Boulogne-sur-Mer
St-Omer
Hazebrouck
Armentières
Lille
Desvres
Étaples
Fruges
Béthune
Lens
Montreuil
Hesdin
Hénin-Beaumont
Berck
Rue
St-Pol-sur-Ternoise
Arras
Douai
Avion
Le Crotoy
St-Valéry-sur-Somme
Doullens
Abbeville
Bapaume
Le Tréport
Blangy-sur-Bresle
Somme
Albert
Péronne
Fauville-en-Caux
Dieppe
Amiens
Fécamp
Cap d'Antifer
Étretat
Tôtes
Neufchâtel-en-Bray
Roye
Bolbec
Yvetot
Forges-les-Eaux
Breteuil
Montdidier
Tergnier
Gonfreville-Orcher
Le Havre
Lillebonne
Barentin
Marseille-en-Beauvaisis
Noyon
Rouen
Gournay-en-Bray
Beauvais
Honfleur
St-Étienne-du-Rouvray
Clermont
Compiègne
Elbeuf
Oise
Aisne
Seine
Cherbourg
Valognes
Carentan
La Haye-du-Puits
Isigny-sur-Mer
Bayeux
Ouistreham
Périers
Hérouville-St-Clair
Caen
St-Lô
Lisieux
Louviers
Les Andelys
Méru
Creil
Senlis
Crépy-en-Valois
Chantilly
Coutances
Villers-Bocage
Vernon
Chambly
Bernay
Orbec
Évreux
Pontoise
Granville
Vimoutiers
Conches-en-Ouche
Mantes-la-Jolie
Herblay
St-Denis
Bobigny
Meaux
Jullouville
Villedieu-les-Poêles
Vire
Tinchebray
Flers
Condé-sur-Noireau
Falaise
Gacé
Eure
St-Germain-en-Laye
Versailles
Marne-la-Vallée
Houdan
Créteil
Trappes
Orsay
Avranches
Argentan
L'Aigle
Pontorson
Mortain
Rânes
Verneuil
Dreux
Courtacon
metres 4000 2000 1000 500 200 0 200 1000 2000 4000 6000 8000 metres
feet 13120 6560 3280 1640 656 0 656 3280 6560 13120 19690 26250 feet
24 25
26 27

Benelux

- Belgium
- Luxembourg
- Netherlands

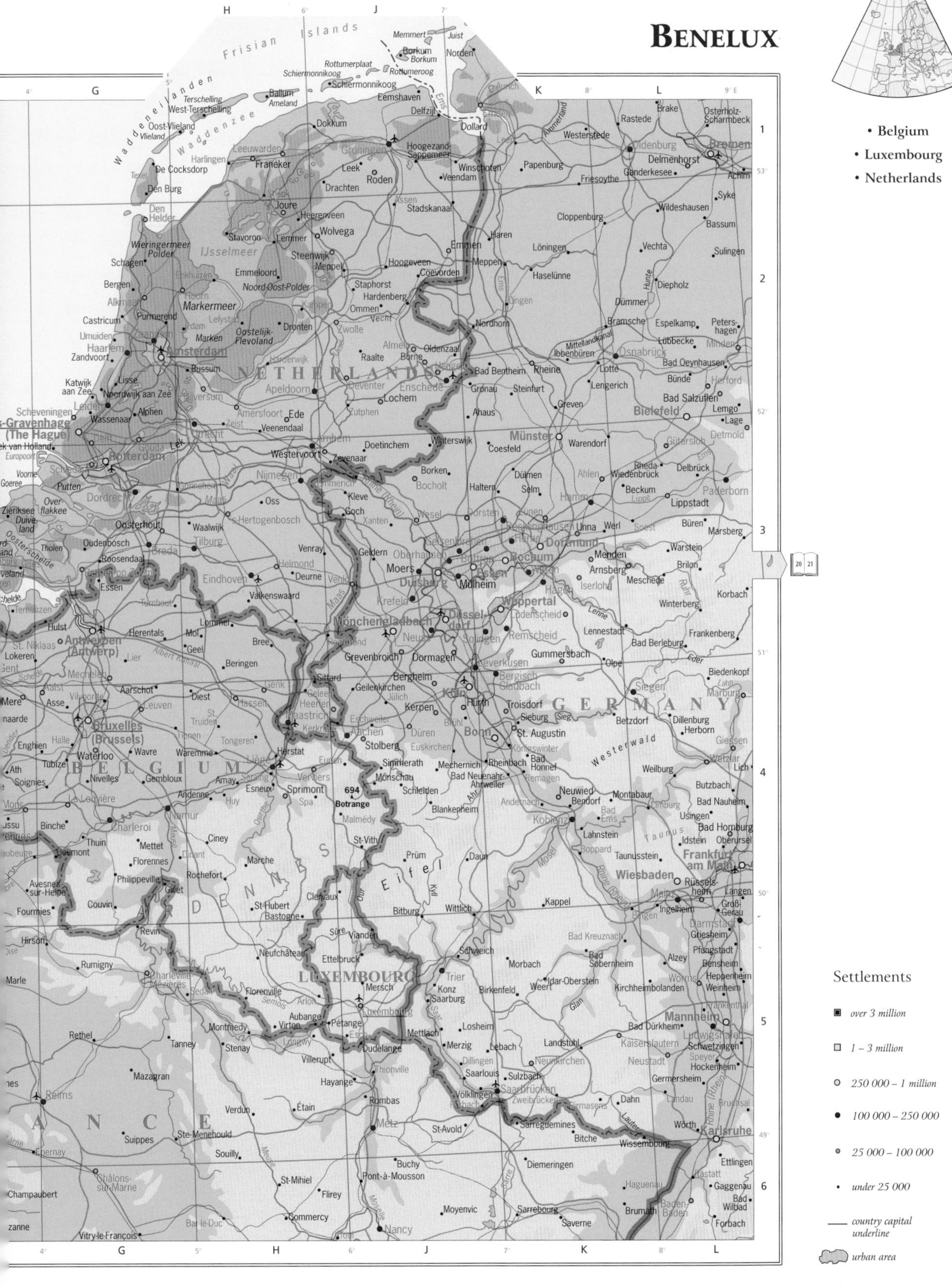

Scale 1 : 4 350 000

0 50 100 150 km

0 25 50 75 miles

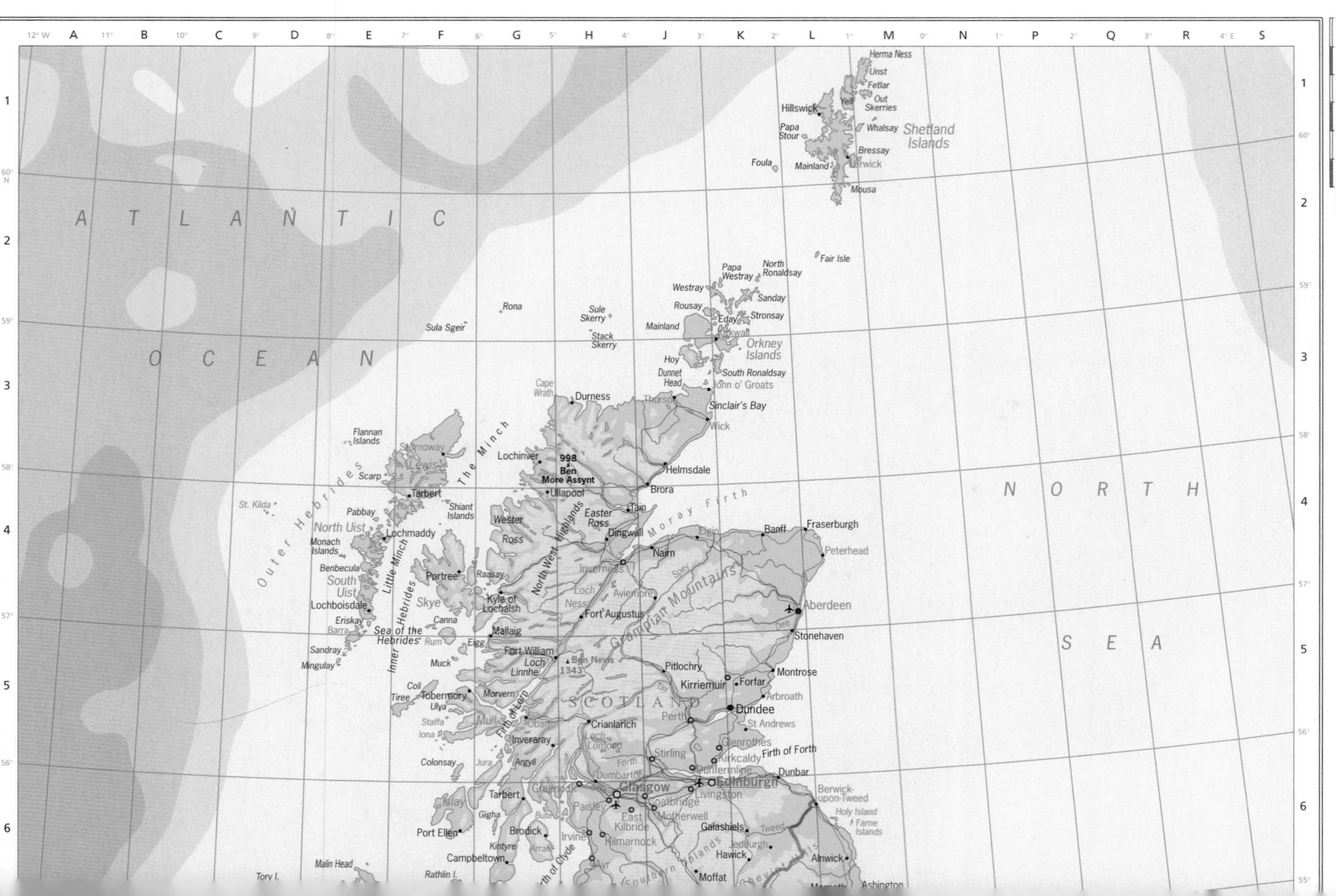

© Helicon Publishing Ltd

metres feet
4000 13120
2000 6560
1000 3280
500 1640
200 656
0 0
200 656
1000 3280
2000 6560
4000 13120
6000 19690
8000 26250
metres feet

British Isles

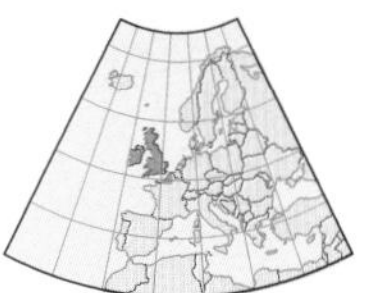

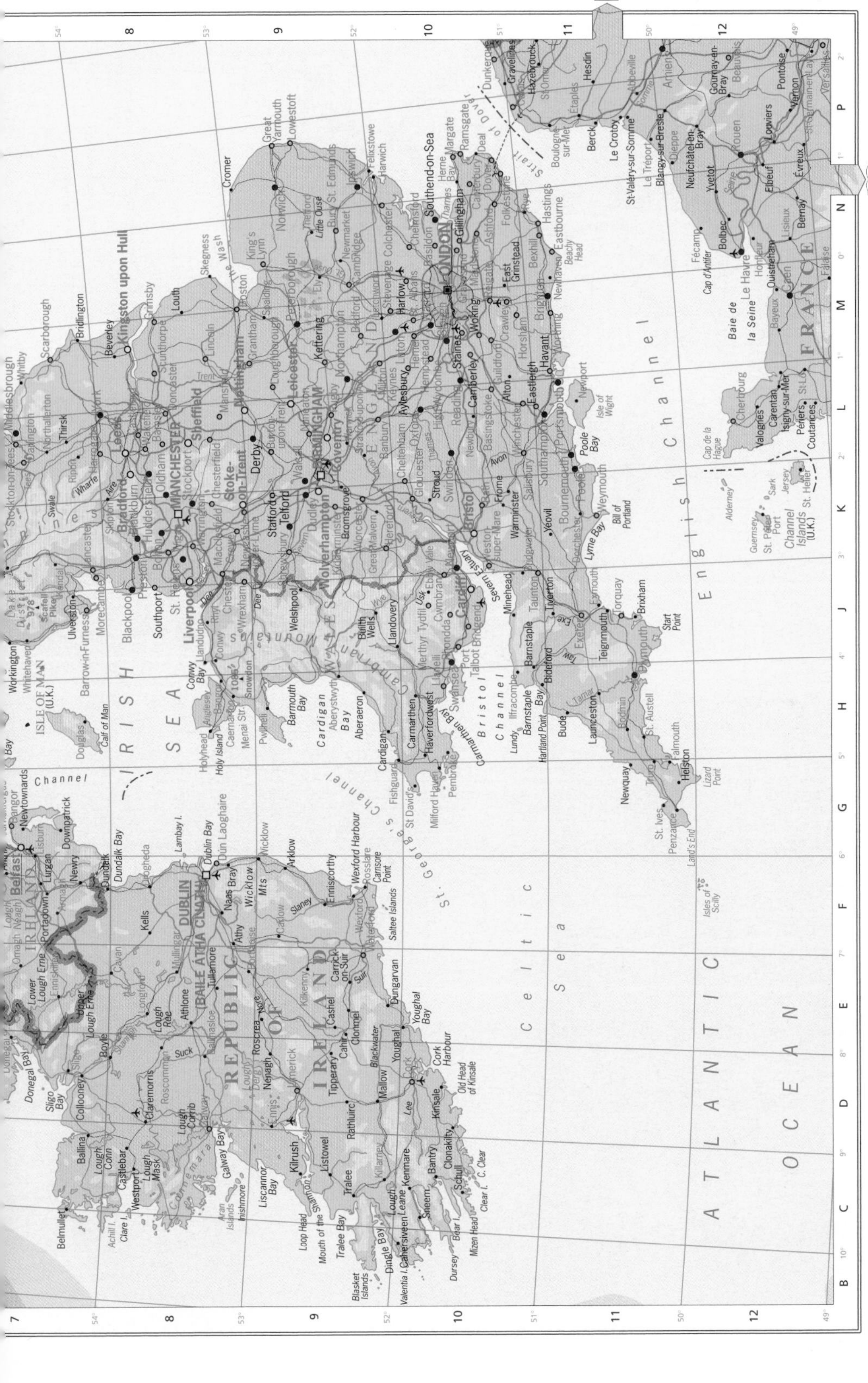

- Republic of Ireland
- United Kingdom

Settlements

- over 3 million
- 1 – 3 million
- 250 000 – 1 million
- 100 000 – 250 000
- 25 000 – 100 000
- under 25 000
- country capital underline
- state or province capital underline
- urban area

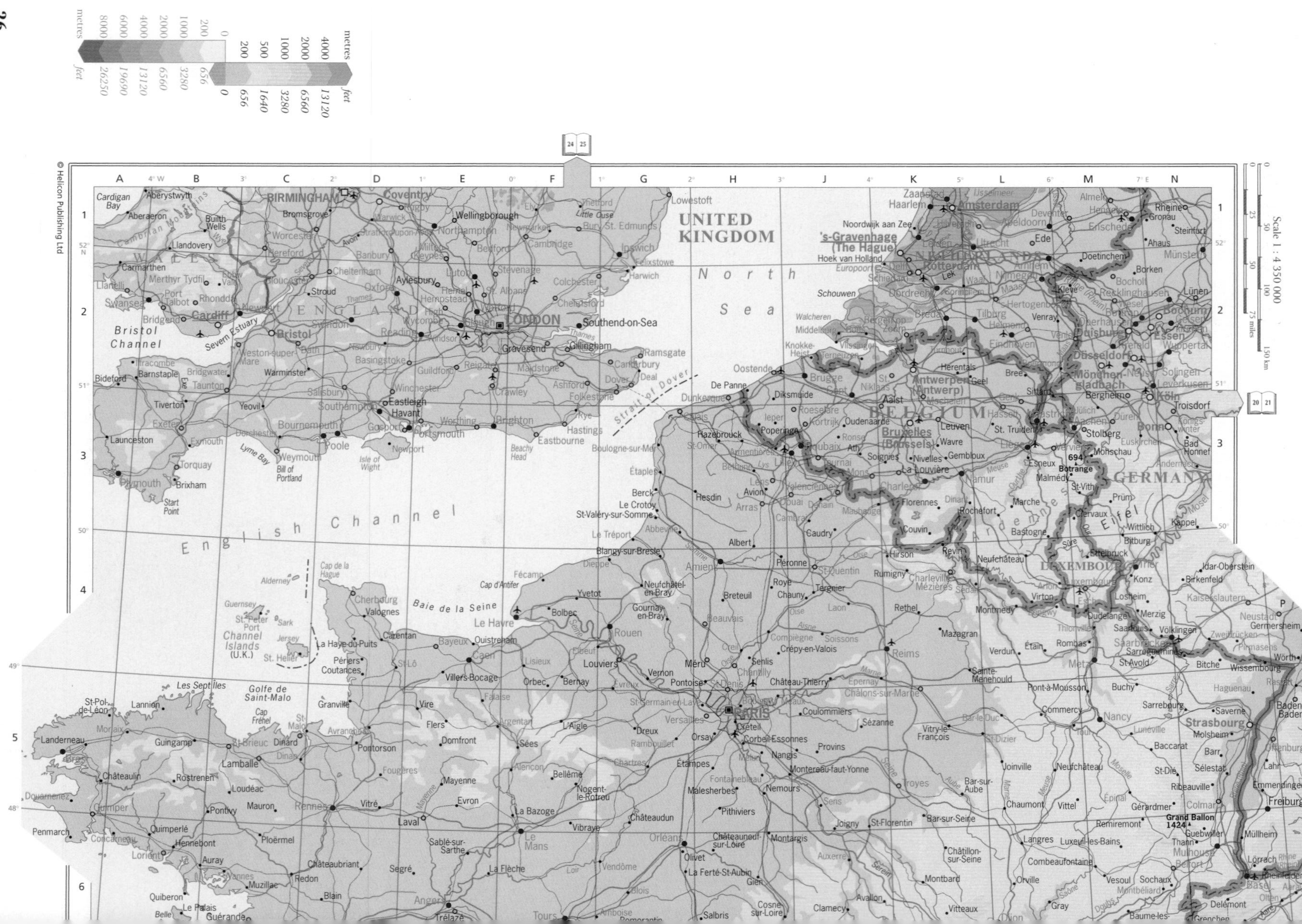
Scale 1 : 4 350 000
0 50 100 150 km
0 25 50 75 miles
metres feet
4000 13120
2000 6560
1000 3280
500 1640
200 656
0 0
200 656
1000 3280
2000 6560
4000 13120
6000 19690
8000 26250
metres feet
UNITED KINGDOM
North Sea
English Channel
Bristol Channel
Strait of Dover
Baie de la Seine
Golfe de Saint-Malo
Channel Islands (U.K.)
ENGLAND
WALES
BELGIUM
GERMANY
LUXEMBOURG
BIRMINGHAM
LONDON
PARIS
's-Gravenhage (The Hague)
Amsterdam
Rotterdam
Antwerpen (Antwerp)
Bruxelles (Brussels)
Köln
Düsseldorf
Strasbourg
Grand Ballon 1424
Botrange 694

FRANCE

- Andorra
- Channel Islands
- France
- Monaco

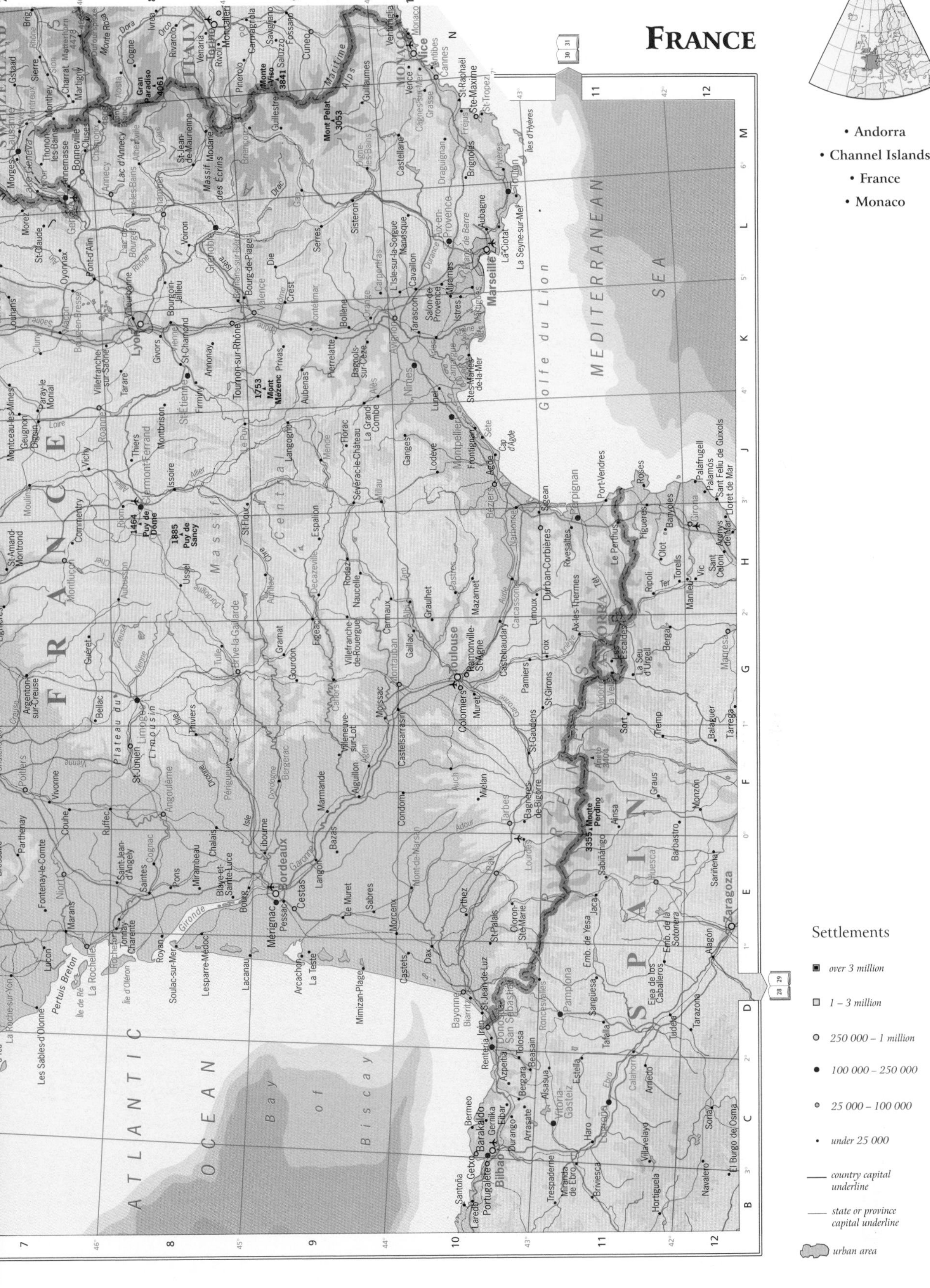

Settlements

- ▣ *over 3 million*
- □ *1 – 3 million*
- ○ *250 000 – 1 million*
- ● *100 000 – 250 000*
- ○ *25 000 – 100 000*
- • *under 25 000*
- ___ *country capital underline*
- ___ *state or province capital underline*
- *urban area*

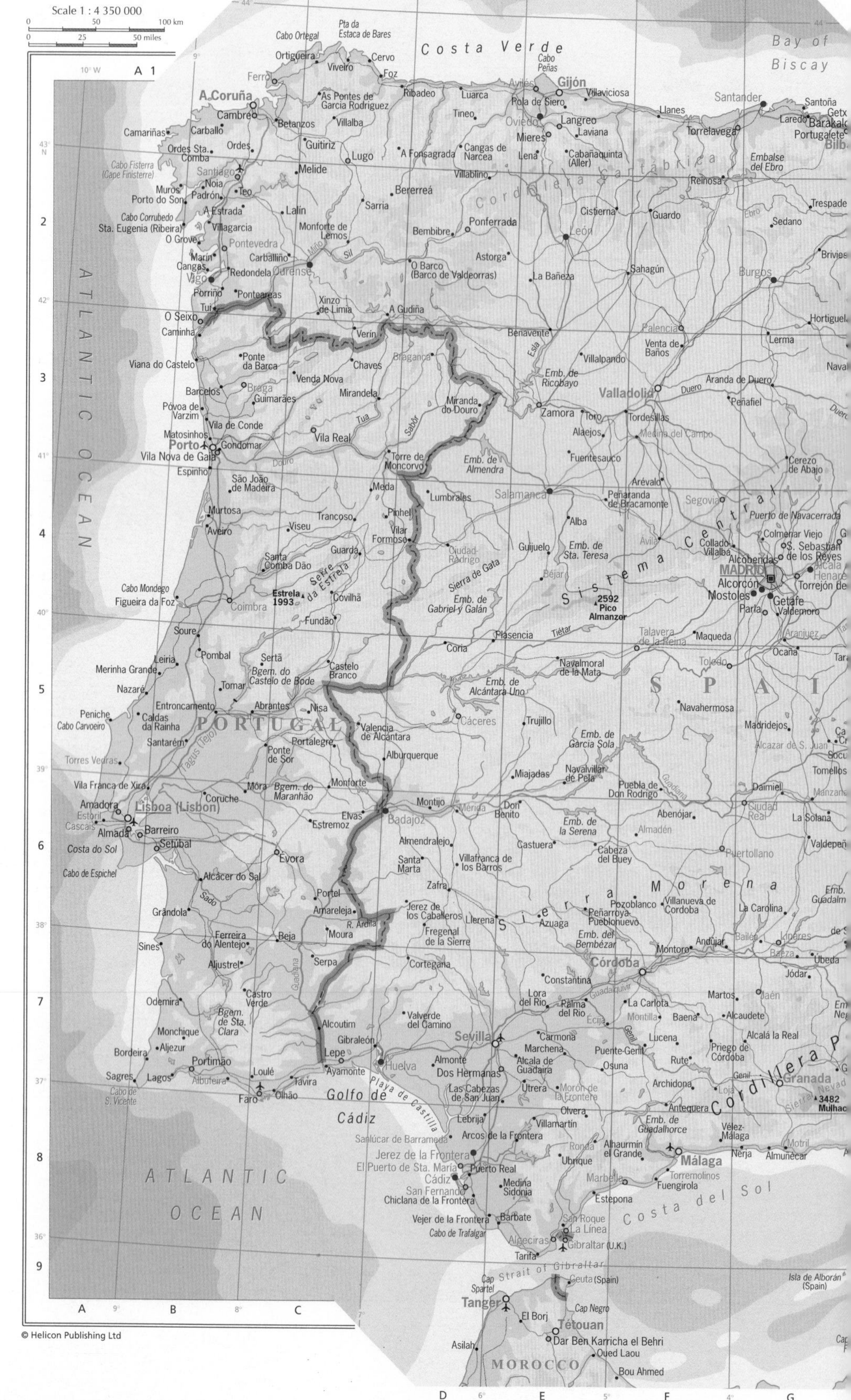

metres 4000 2000 1000 500 200 0 | feet 13120 6560 3280 1640 656 0

200 1000 2000 4000 6000 8000 metres | 656 3280 6560 13120 19690 26250 feet

SPAIN AND PORTUGAL

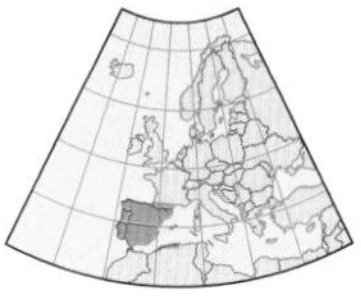

26 27

- Gibraltar
- Portugal
- Spain

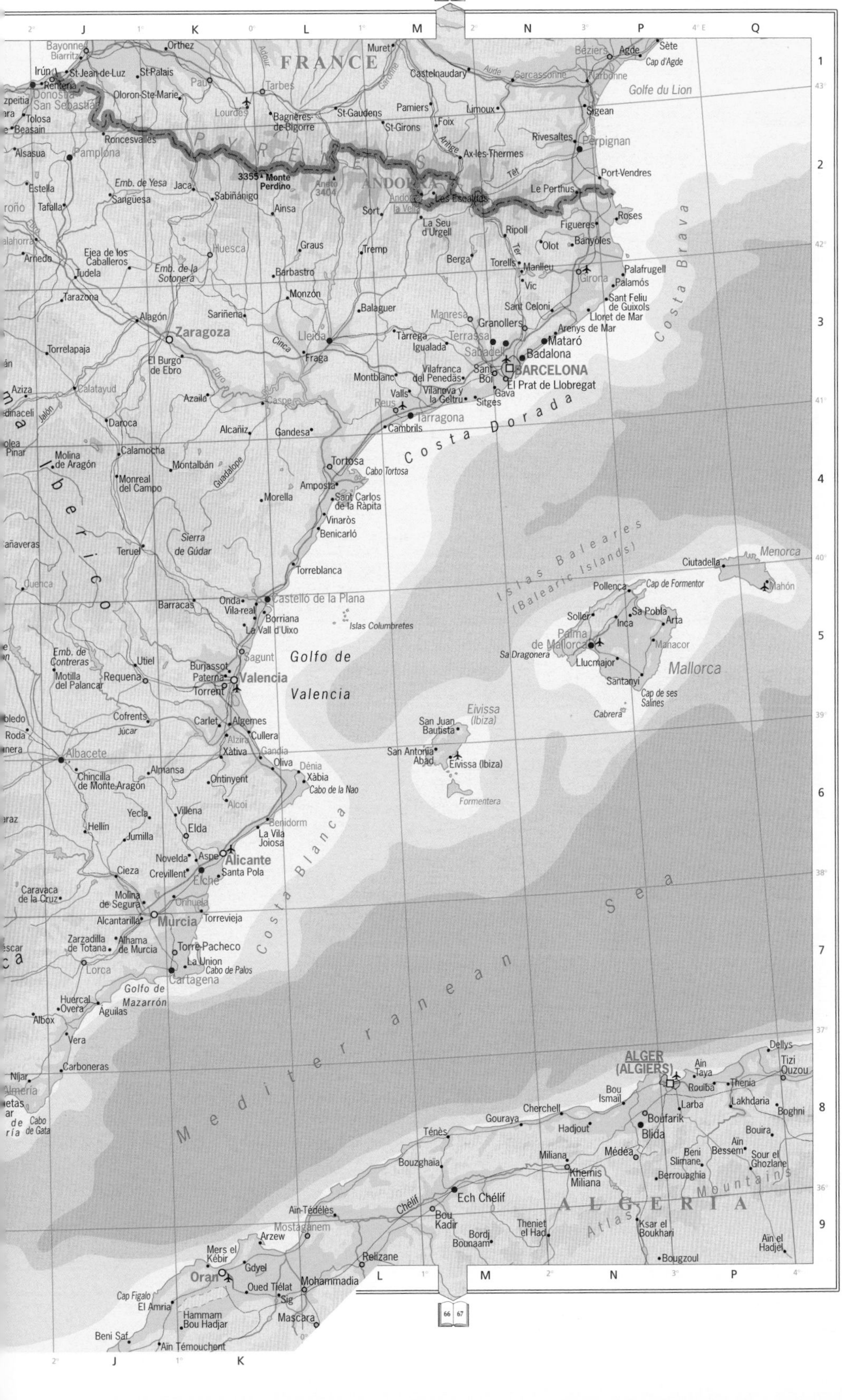

Settlements

- over 3 million
- 1 – 3 million
- 250 000 – 1 million
- 100 000 – 250 000
- 25 000 – 100 000
- under 25 000
- country capital underline

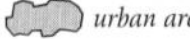 urban area

66 67

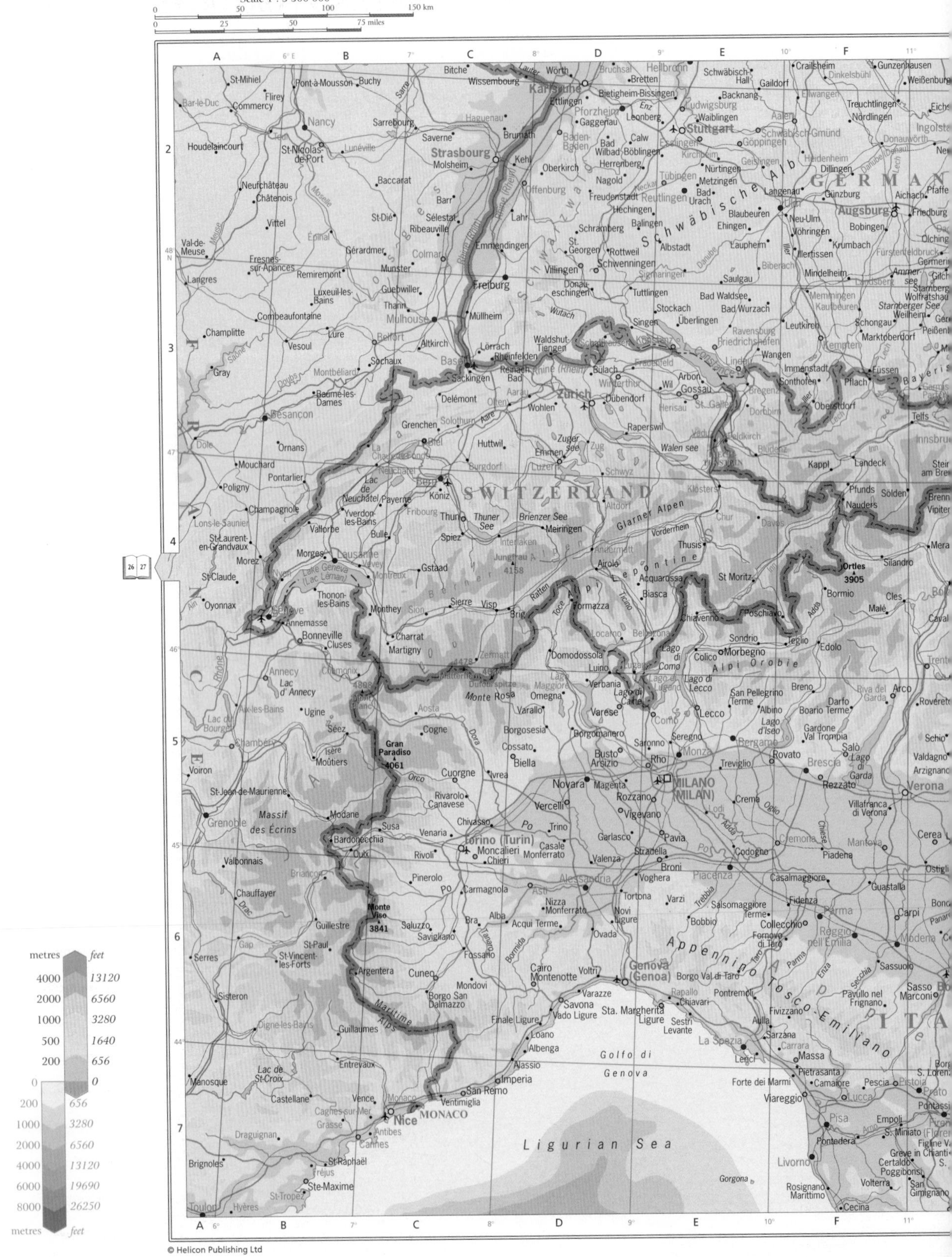

Scale 1 : 3 300 000
0 50 100 150 km
0 25 50 75 miles
A B C D E F
1 2 3 4 5 6 7
GERMANY
SWITZERLAND
FRANCE
ITALIA
Strasbourg
Karlsruhe
Stuttgart
Augsburg
Freiburg
Basel
Zürich
Bern
Lausanne
Genève
Lake Geneva (Lac Léman)
Lac de Neuchâtel
Thuner See
Brienzer See
Walen see
Jungfrau 4158
Monte Rosa
Gran Paradiso 4061
Monte Viso 3841
Ortles 3905
Berner Alpen
Glarner Alpen
Alpi Lepontine
Alpi Orobie
Schwäbische Alb
Schwarzwald
Massif des Écrins
Maritime Alps
MILANO (MILAN)
Torino (Turin)
Genova (Genoa)
Golfo di Genova
Ligurian Sea
Appennino Tosco-Emiliano
Nice
MONACO
Verona
Brescia
Bergamo
Novara
Grenoble
Besançon
Nancy
Lago Maggiore
Lago di Como
Lago d'Iseo
Lago di Garda
metres feet
4000 13120
2000 6560
1000 3280
500 1640
200 656
0 0
200 656
1000 3280
2000 6560
4000 13120
6000 19690
8000 26250
metres feet

The Alpine States

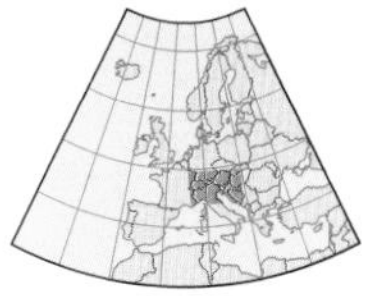

- Austria
- Liechtenstein
- Slovenia
- Switzerland

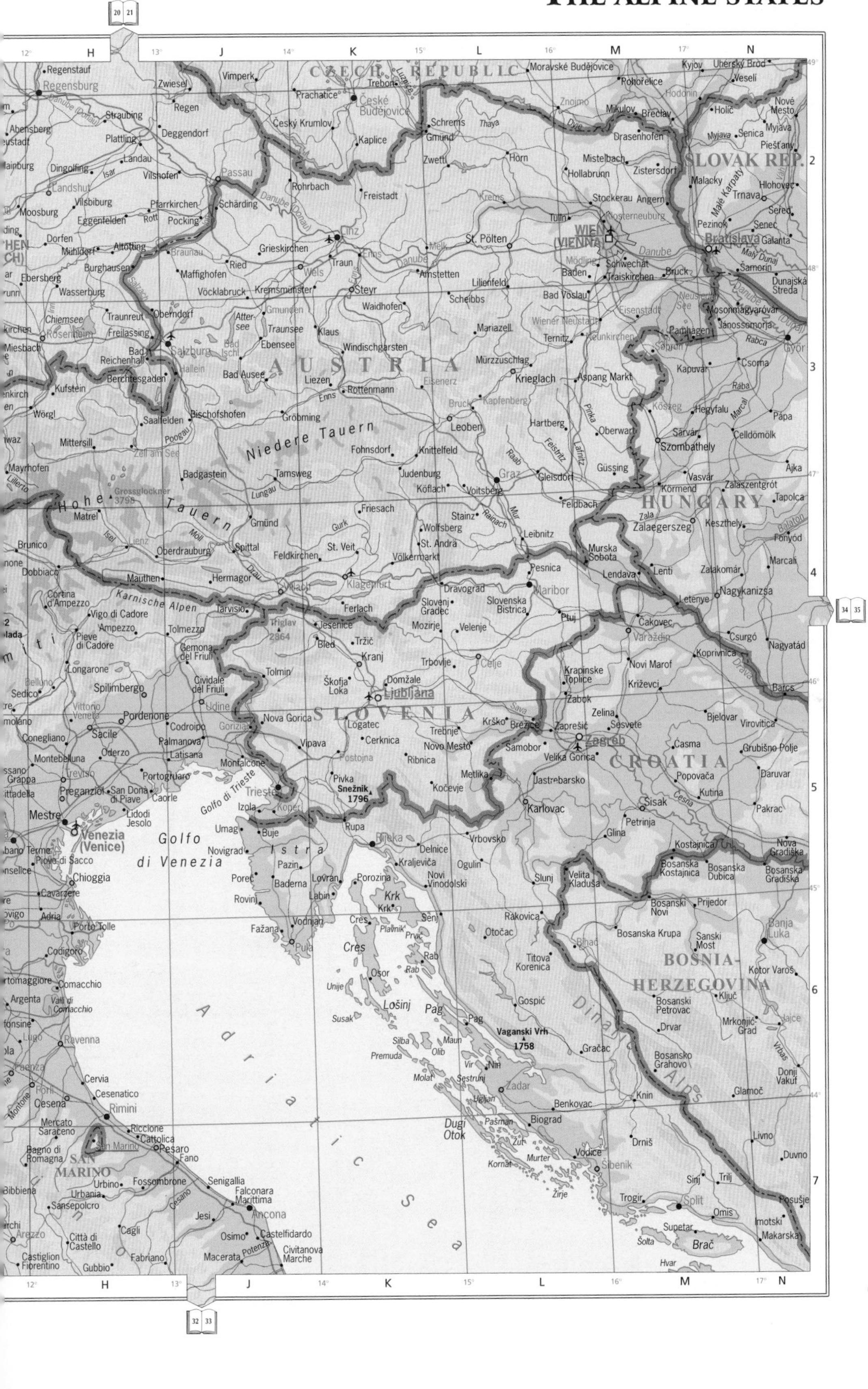

20 21

34 35

32 33

Settlements

- *over 3 million*
- *1 – 3 million*
- *250 000 – 1 million*
- *100 000 – 250 000*
- *25 000 – 100 000*
- *under 25 000*
- *country capital underline*

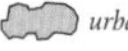
urban area

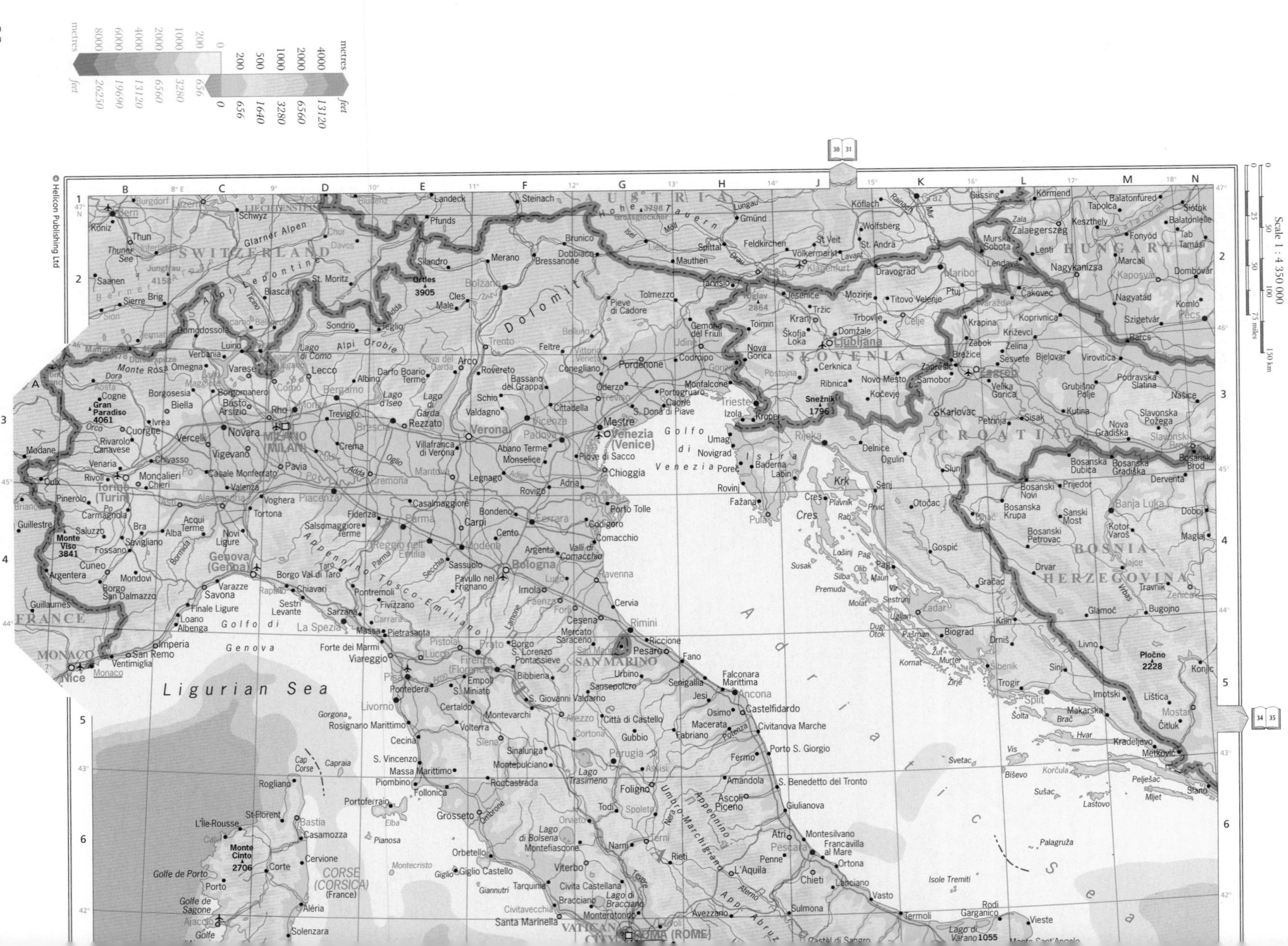
Scale 1 : 4 350 000
0 50 100 150 km
0 25 50 75 miles
metres feet
4000 13120
2000 6560
1000 3280
500 1640
200 656
0 0
200 656
1000 3280
2000 6560
4000 13120
6000 19690
8000 26250
metres feet
© Helicon Publishing Ltd
AUSTRIA
HUNGARY
SLOVENIA
CROATIA
BOSNIA-HERZEGOVINA
SWITZERLAND
LIECHTENSTEIN
SAN MARINO
MONACO
FRANCE
CORSE (CORSICA) (France)
Ligurian Sea
Adriatic Sea
Golfo di Venezia
Golfo di Genova
Dolomiti
Appennino Tosco-Emiliano
Appennino Umbro-Marchigiano
Alpi Orobie
Alpi Lepontine
Glarner Alpen
Hohe Tauern
ROMA (ROME)
Milano (Milan)
Torino (Turin)
Genova (Genoa)
Venezia (Venice)
Firenze (Florence)
Bologna
Ljubljana
Zagreb
Elba
Monte Cinto 2706
Monte Viso 3841
Gran Paradiso 4061
Ortles 3905
Triglav 2864
Snežnik 1796
Pločno 2228

ITALY

- Corsica
- Italy
- Malta
- San Marino
- Vatican City

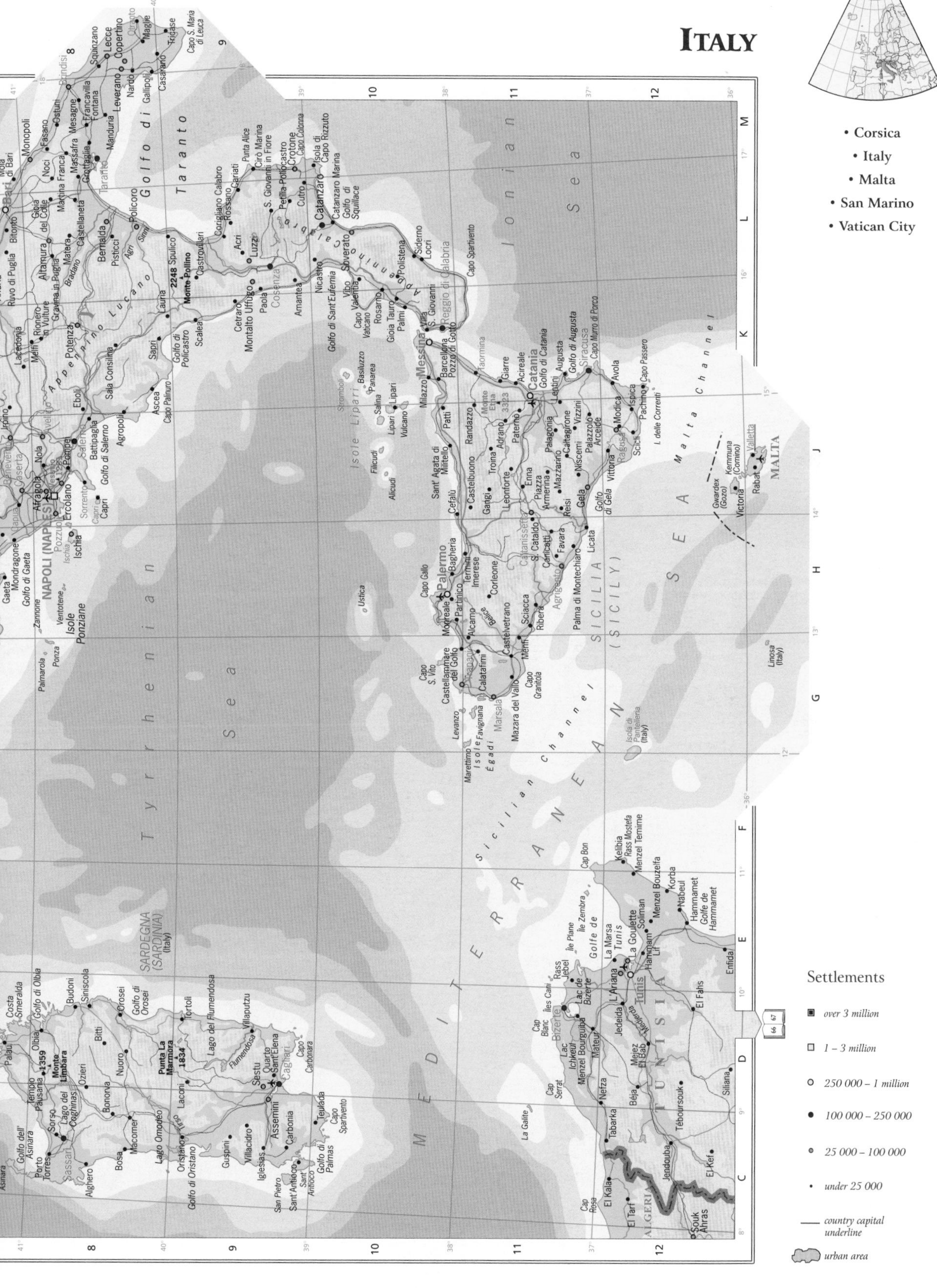

Settlements

- over 3 million
- 1 – 3 million
- 250 000 – 1 million
- 100 000 – 250 000
- 25 000 – 100 000
- under 25 000
- country capital underline
- urban area

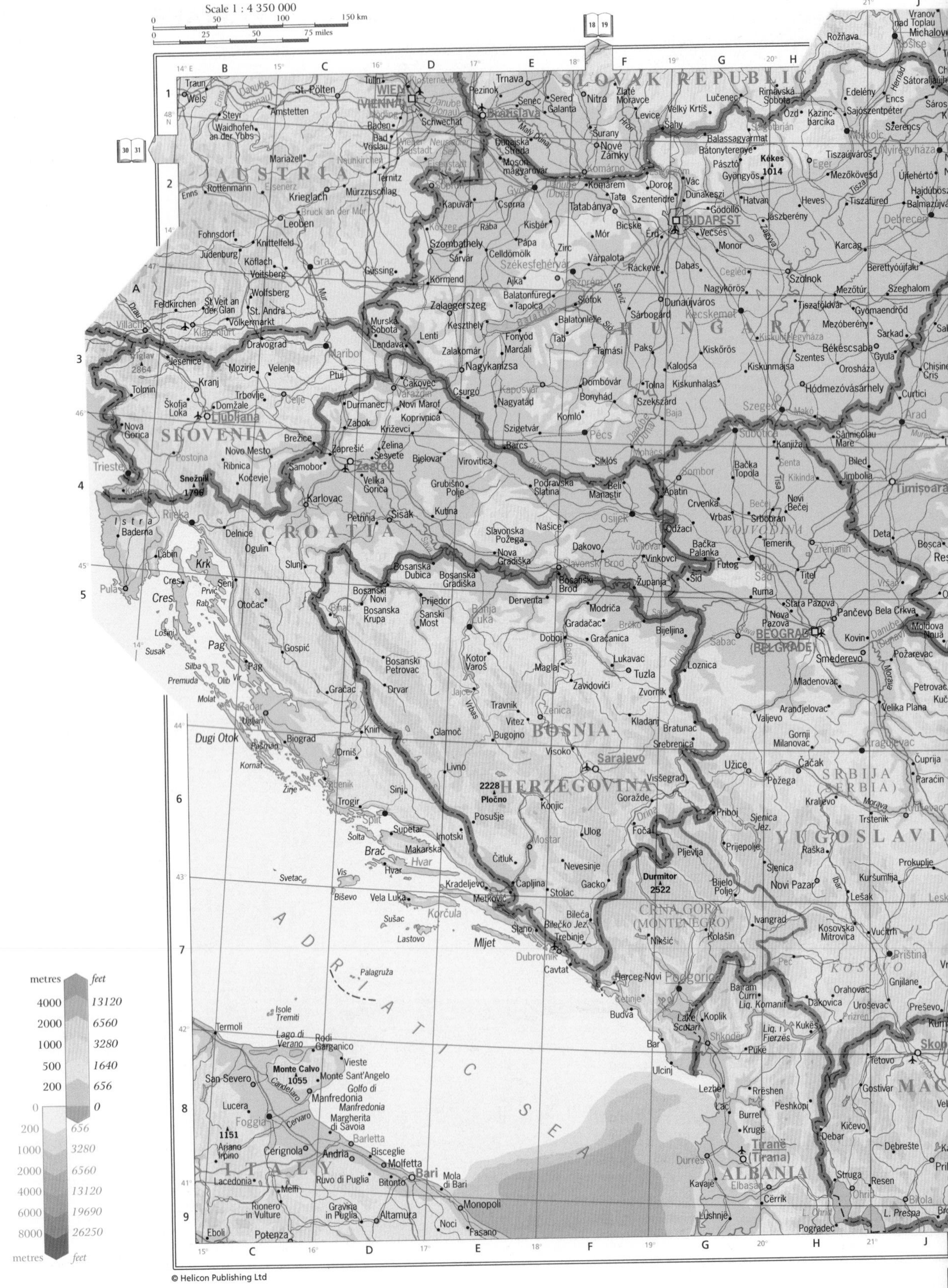
Scale 1 : 4 350 000
0 50 100 150 km
0 25 50 75 miles
18 19
30 31
SLOVAK REPUBLIC
AUSTRIA
HUNGARY
SLOVENIA
CROATIA
BOSNIA-HERZEGOVINA
VOJVODINA
SRBIJA (SERBIA)
YUGOSLAVIA
CRNA GORA (MONTENEGRO)
KOSOVO
ALBANIA
ITALY
ADRIATIC SEA
WIEN (VIENNA)
Bratislava
BUDAPEST
Ljubljana
Zagreb
Sarajevo
BEOGRAD (BELGRADE)
Podgorica
Tiranë (Tirana)
Skopje
Traun
Wels
Steyr
Amstetten
St. Pölten
Tulln
Waidhofen an der Ybbs
Baden
Bad Vöslau
Schwechat
Wiener Neustadt
Mariazell
Ternitz
Rottenmann
Kriegach
Mürzzuschlag
Leoben
Bruck an der Mur
Fohnsdorf
Judenburg
Knittelfeld
Köflach
Voitsberg
Graz
Güssing
Wolfsberg
Feldkirchen
St. Veit an der Glan
St. Andrä
Völkermarkt
Villach
Klagenfurt
Triglav 2864
Jesenice
Kranj
Tolmin
Škofja Loka
Domžale
Nova Gorica
Trieste
Postojna
Novo Mesto
Ribnica
Kočevje
Snežnik 1796
Rijeka
Istra
Baderna
Labin
Pula
Cres
Krk
Senj
Rab
Lošinj
Susak
Pag
Silba
Premuda
Olib
Molat
Ugljan
Dugi Otok
Pašman
Kornat
Žirje
Biograd
Zadar
Gospić
Gračac
Knin
Drniš
Šibenik
Trogir
Split
Sinj
Šolta
Brač
Hvar
Vis
Biševo
Svetac
Vela Luka
Korčula
Sušac
Lastovo
Mljet
Palagruža
Dubrovnik
Cavtat
Makarska
Imotski
Metković
Kardeljevo
Čapljina
Stolac
Mostar
Nevesinje
Gacko
Bileća
Bilečko Jez.
Trebinje
Slano
Herceg-Novi
Cetinje
Budva
Bar
Ulcinj
Nikšić
Kolašin
Durmitor 2522
Bijelo Polje
Pljevlja
Prijepolje
Priboj
Sjenica
Sjenica Jez.
Novi Pazar
Raška
Ivangrad
Kosovska Mitrovica
Vučitrn
Priština
Peć
Đakovica
Orahovac
Prizren
Uroševac
Gnjilane
Preševo
Kukës
Lake Scutari
Shkodër
Koplik
Bajram Curri
Liq. Komanit
Liq. i Fierzës
Puke
Lezhë
Laç
Rrëshen
Burrel
Krujë
Peshkopi
Durrës
Kavajë
Elbasan
Cërrik
Lushnjë
Pogradec
L. Ohrid
Ohrid
Struga
Resen
Bitola
L. Prespa
Debar
Kičevo
Gostivar
Tetovo
Debreste
Trnava
Pezinok
Senec
Sereď
Galanta
Nitra
Zlaté Moravce
Levice
Šurany
Hron
Malý Dunaj
Dunajská Streda
Mosonmagyaróvár
Nové Zámky
Komárno
Komárom
Veľký Krtíš
Šahy
Lučenec
Rimavská Sobota
Rožňava
Košice
Michalovce
Vranov nad Topľou
Hernád
Edelény
Encs
Ózd
Kazincbarcika
Sátoraljaújhely
Sárospatak
Sajószentpéter
Szerencs
Miskolc
Salgótarján
Balassagyarmat
Bátonyterenye
Pásztó
Gyöngyös
Kékes 1014
Eger
Tiszaújváros
Nyíregyháza
Mezőkövesd
Újfehértó
Hajdúböszörmény
Balmazújváros
Tiszafüred
Heves
Hatvan
Jászberény
Gödöllő
Vác
Dorog
Dunakeszi
Szentendre
Tata
Tatabánya
Győr
Sopron
Kapuvár
Csorna
Rába
Kisbér
Mór
Bicske
Érd
Vecsés
Monor
Zirc
Pápa
Kőszeg
Szombathely
Sárvár
Celldömölk
Várpalota
Székesfehérvár
Ráckeve
Dabas
Cegléd
Ajka
Veszprém
Körmend
Szolnok
Nagykőrös
Karcag
Debrecen
Berettyóújfalu
Mezőtúr
Szeghalom
Tiszaföldvár
Gyomaendrőd
Balatonfüred
Tapolca
Siófok
Dunaújváros
Kecskemét
Zalaegerszeg
Keszthely
Balatonlelle
Sárbogárd
Sió
Sárvíz
Mezőberény
Sarkad
Lenti
Murska Sobota
Lendava
Dravograd
Maribor
Ptuj
Mozirje
Velenje
Trbovlje
Celje
Brežice
Fonyód
Tab
Tamási
Paks
Kiskőrös
Kiskunfélegyháza
Békéscsaba
Gyula
Szentes
Orosháza
Zalakomár
Marcali
Nagykanizsa
Kalocsa
Kiskunmajsa
Kiskunhalas
Hódmezővásárhely
Csurgó
Kaposvár
Nagyatád
Dombóvár
Bonyhád
Tolna
Szekszárd
Baja
Szeged
Makó
Curtici
Arad
Chişineu Criş
Čakovec
Varaždin
Durmanec
Novi Marof
Zabok
Koprivnica
Križevci
Zaprešić
Zelina
Sesvete
Bjelovar
Samobor
Velika Gorica
Karlovac
Petrinja
Sisak
Kutina
Virovitica
Szigetvár
Komló
Pécs
Mohács
Barcs
Siklós
Subotica
Kanjiža
Sânnicolau Mare
Biled
Jimbolia
Timişoara
Grubišno Polje
Podravska Slatina
Beli Manastir
Osijek
Sombor
Bačka Topola
Senta
Kikinda
Apatin
Crvenka
Vrbas
Bečej
Novi Bečej
Srbobran
Odžaci
Temerin
Zrenjanin
Deta
Našice
Đakovo
Vukovar
Vinkovci
Bačka Palanka
Futog
Novi Sad
Titel
Šid
Slavonski Brod
Bosanski Brod
Županja
Ruma
Stara Pazova
Nova Pazova
Pančevo
Bela Crkva
Moldova Nouă
Kovin
Vršac
Delnice
Ogulin
Slunj
Otočac
Slavonska Požega
Nova Gradiška
Bosanska Gradiška
Bosanska Dubica
Bosanski Novi
Bosanska Krupa
Bihać
Prijedor
Sanski Most
Banja Luka
Derventa
Modriča
Gradačac
Brčko
Bijeljina
Doboj
Gračanica
Šabac
Loznica
Lukavac
Tuzla
Zvornik
Smederevo
Požarevac
Mladenovac
Bosanski Petrovac
Drvar
Kotor Varoš
Jajce
Maglaj
Zavidovići
Travnik
Zenica
Vitez
Kladanj
Bratunac
Srebrenica
Valjevo
Aranđelovac
Velika Plana
Gornji Milanovac
Kragujevac
Ćuprija
Paraćin
Glamoč
Bugojno
Livno
Visoko
2228 Pločno
Višegrad
Goražde
Užice
Čačak
Požega
Kraljevo
Trstenik
Kruševac
Konjic
Posušje
Foča
Ulog
Čitluk
Kuršumlija
Prokuplje
Lešak
Leskovac
Drina
Bosna
Vrbas
Sava
Drava
Mur
Tisza
Morava
Ibar
Danube (Donau)
Danube (Duna)
Termoli
Isole Tremiti
Lago di Varano
Rodi Garganico
Vieste
Monte Calvo 1055
Monte Sant'Angelo
San Severo
Candelaro
Manfredonia
Golfo di Manfredonia
Margherita di Savoia
Lucera
Foggia
Cervaro
1151
Ariano Irpino
Cerignola
Barletta
Andria
Bisceglie
Molfetta
Bari
Mola di Bari
Lacedonia
Melfi
Ruvo di Puglia
Bitonto
Rionero in Vulture
Gravina in Puglia
Altamura
Monopoli
Noci
Fasano
Potenza
Eboli
metres
feet
4000 13120
2000 6560
1000 3280
500 1640
200 656
0 0
200 656
1000 3280
2000 6560
4000 13120
6000 19690
8000 26250
© Helicon Publishing Ltd

The Balkans

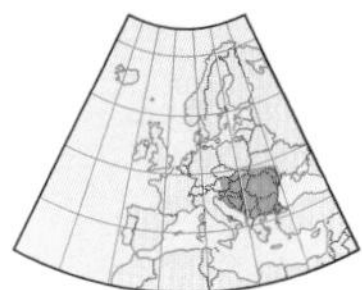

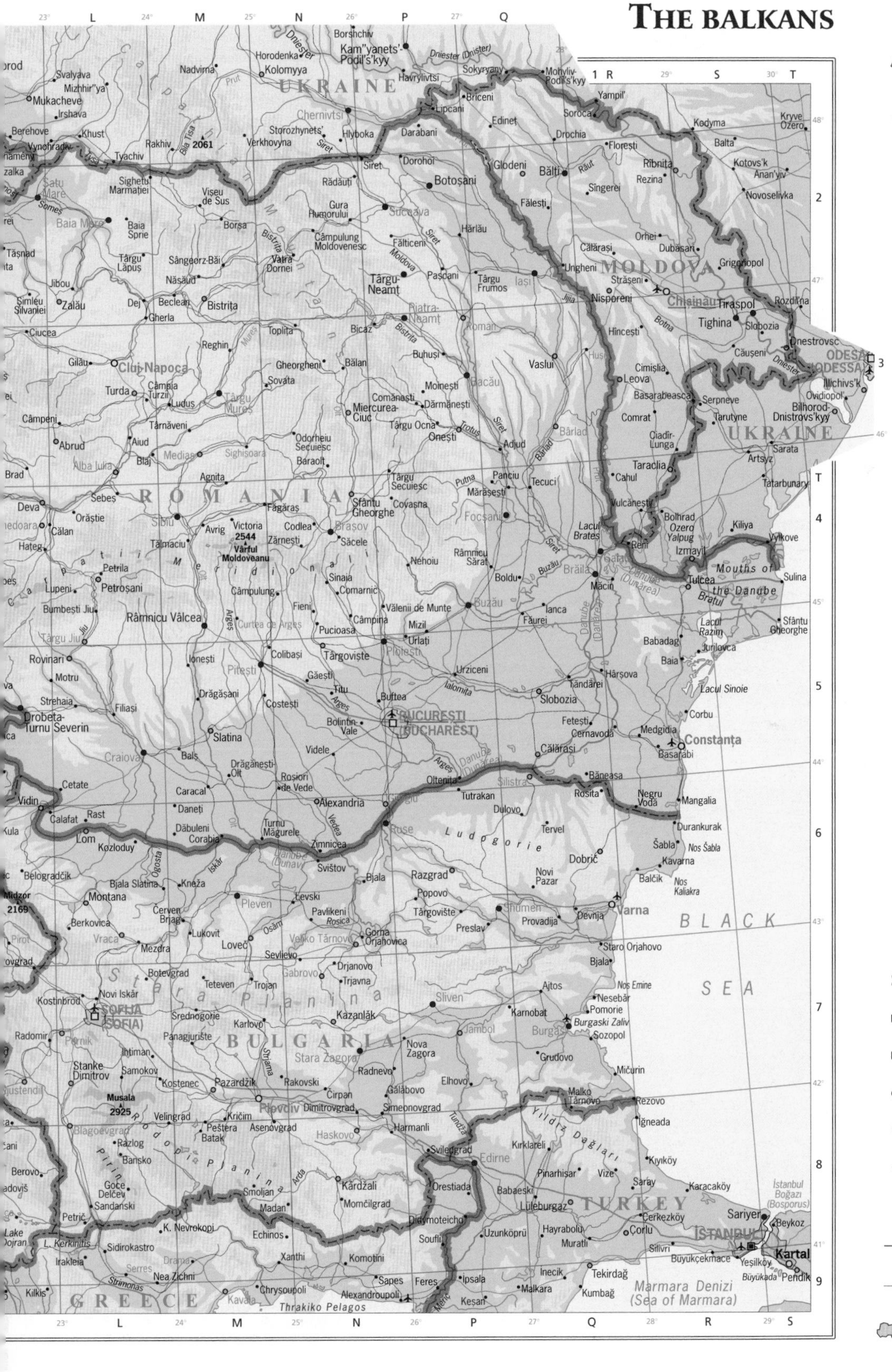

- Bosnia-Herzegovina
- Bulgaria
- Croatia
- Moldova
- Romania
- Yugoslavia

Settlements

- ▣ *over 3 million*
- □ *1 – 3 million*
- ○ *250 000 – 1 million*
- ● *100 000 – 250 000*
- ◦ *25 000 – 100 000*
- • *under 25 000*
- *country capital underline*
- *state or province capital underline*
- *urban area*

Settlements

- ■ *over 3 million*
- □ *1 – 3 million*
- ○ *250 000 – 1 million*
- ● *100 000 – 250 000*
- ◎ *25 000 – 100 000*
- • *under 25 000*
- *country capital underline*
- *state or province capital underline*
- *urban area*

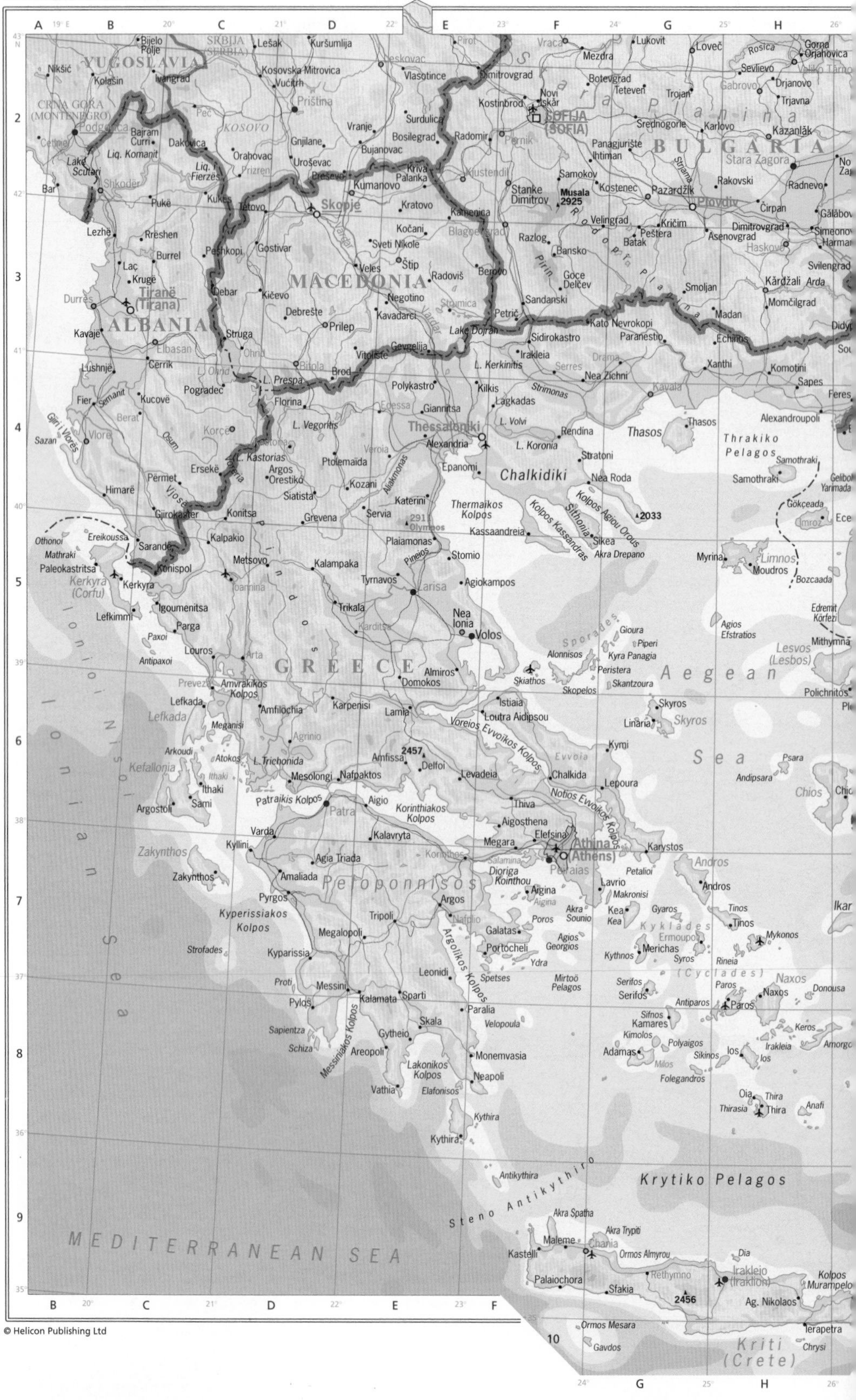

Scale 1 : 4 350 000

0 50 100 150 km

0 25 50 75 miles

34 | 35

YUGOSLAVIA

SRBIJA (SERBIA)

CRNA GORA (MONTENEGRO)

KOSOVO

MACEDONIA

ALBANIA

BULGARIA

GREECE

Peloponnisos

Chalkidiki

Thrakiko Pelagos

Aegean Sea

Ionian Sea

Krytiko Pelagos

Steno Antikythiro

MEDITERRANEAN SEA

Kriti (Crete)

Kyklades (Cyclades)

Sporades

Stara Planina

Rodopi Planina

Pindos

Tiranë (Tirana)

Skopje

SOFIJA (SOFIA)

Athina (Athens)

Thessaloniki

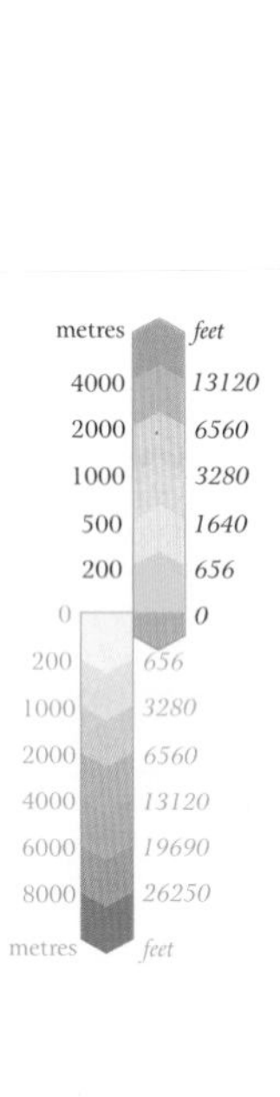

Greece and Western Turkey

- Albania
- Greece
- Macedonia
- Western Turkey

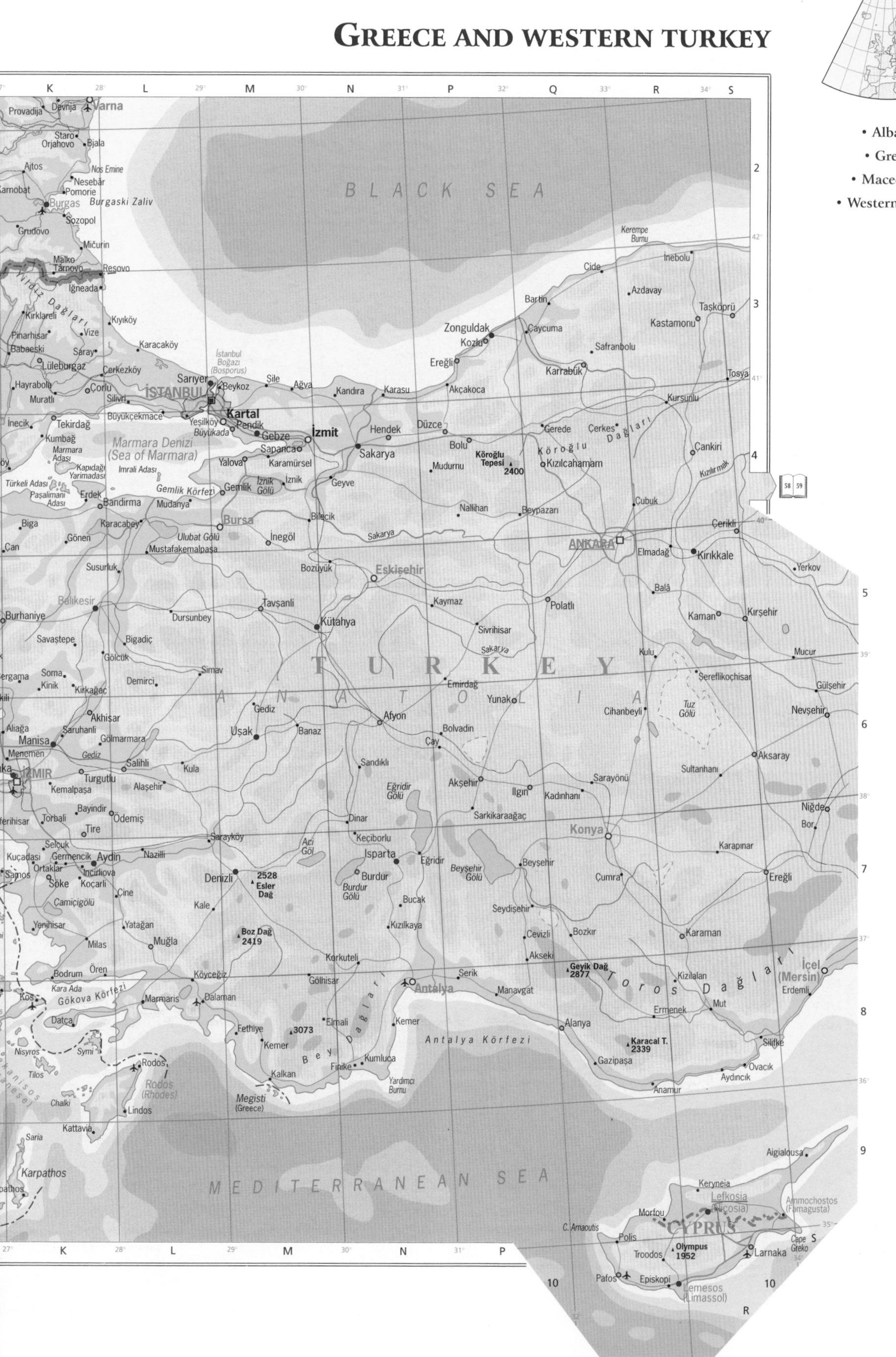

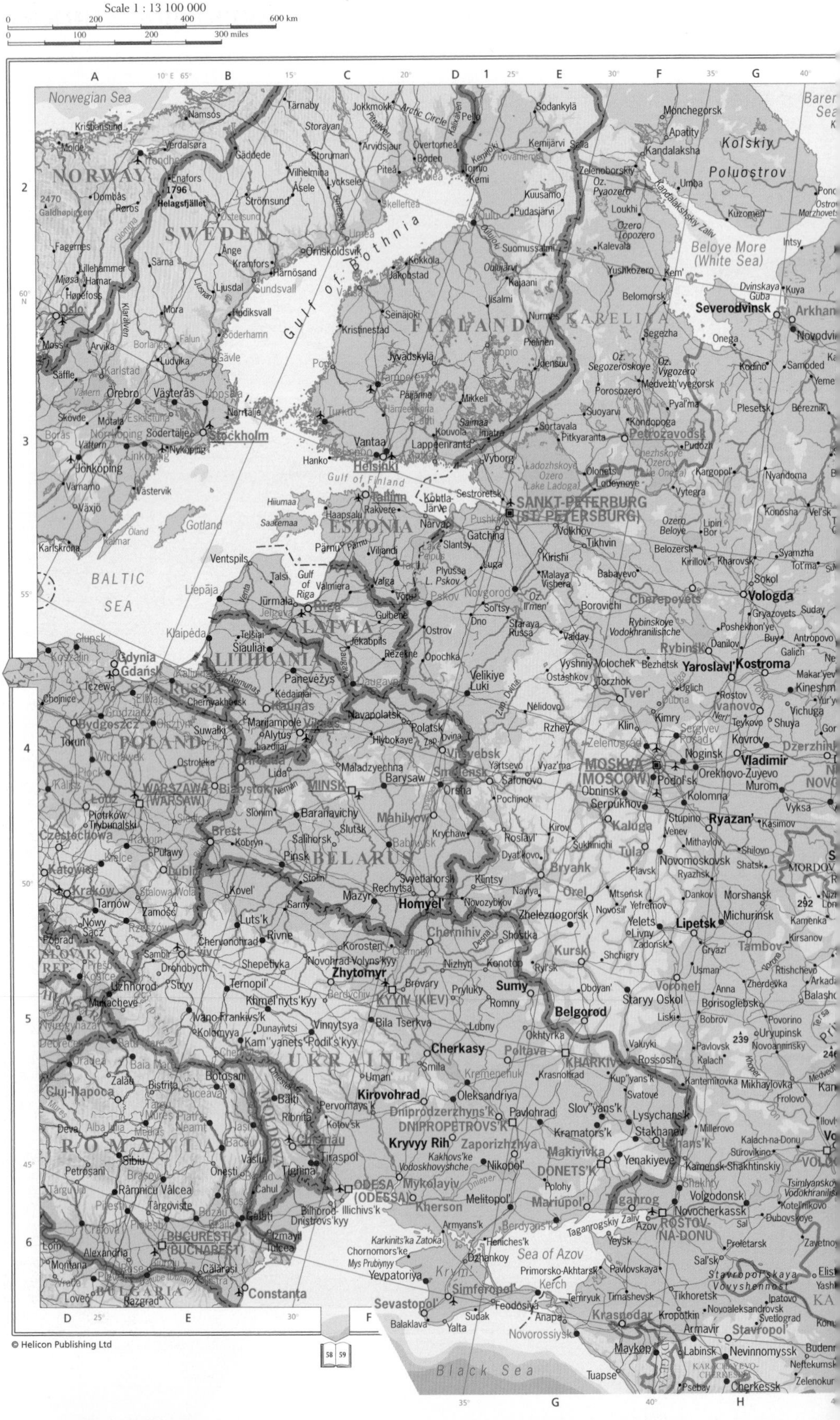

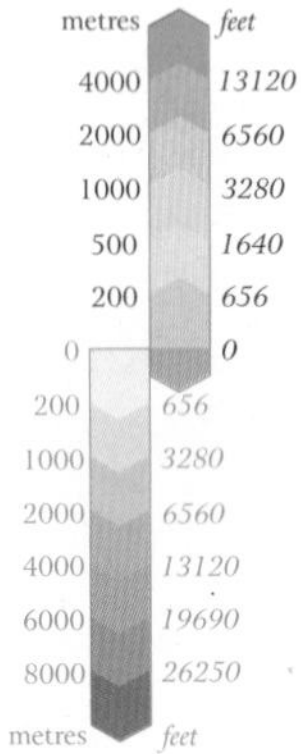

Scale 1 : 13 100 000

0 200 400 600 km

0 100 200 300 miles

16 17

58 59

metres	feet
4000	13120
2000	6560
1000	3280
500	1640
200	656
0	0
200	656
1000	3280
2000	6560
4000	13120
6000	19690
8000	26250
metres	feet

EUROPEAN RUSSIA

- Belarus
- European Russia
- Ukraine

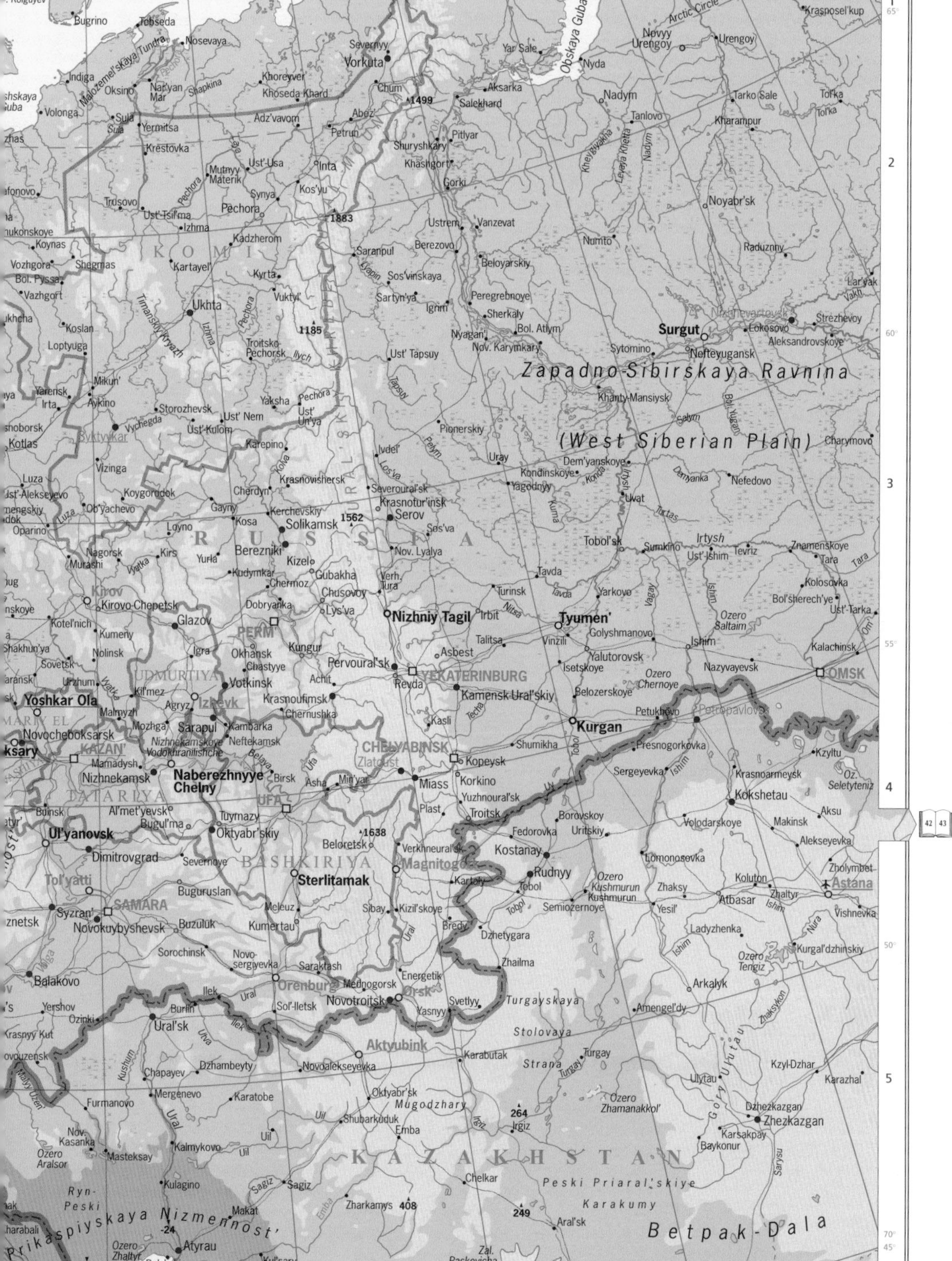

Settlements

- over 3 million
- 1 – 3 million
- 250 000 – 1 million
- 100 000 – 250 000
- 25 000 – 100 000
- under 25 000
- country capital underline
- state or province capital underline

Scale 1 : 41 450 000

0 500 1000 1500 2000 km

0 250 500 750 1000 miles

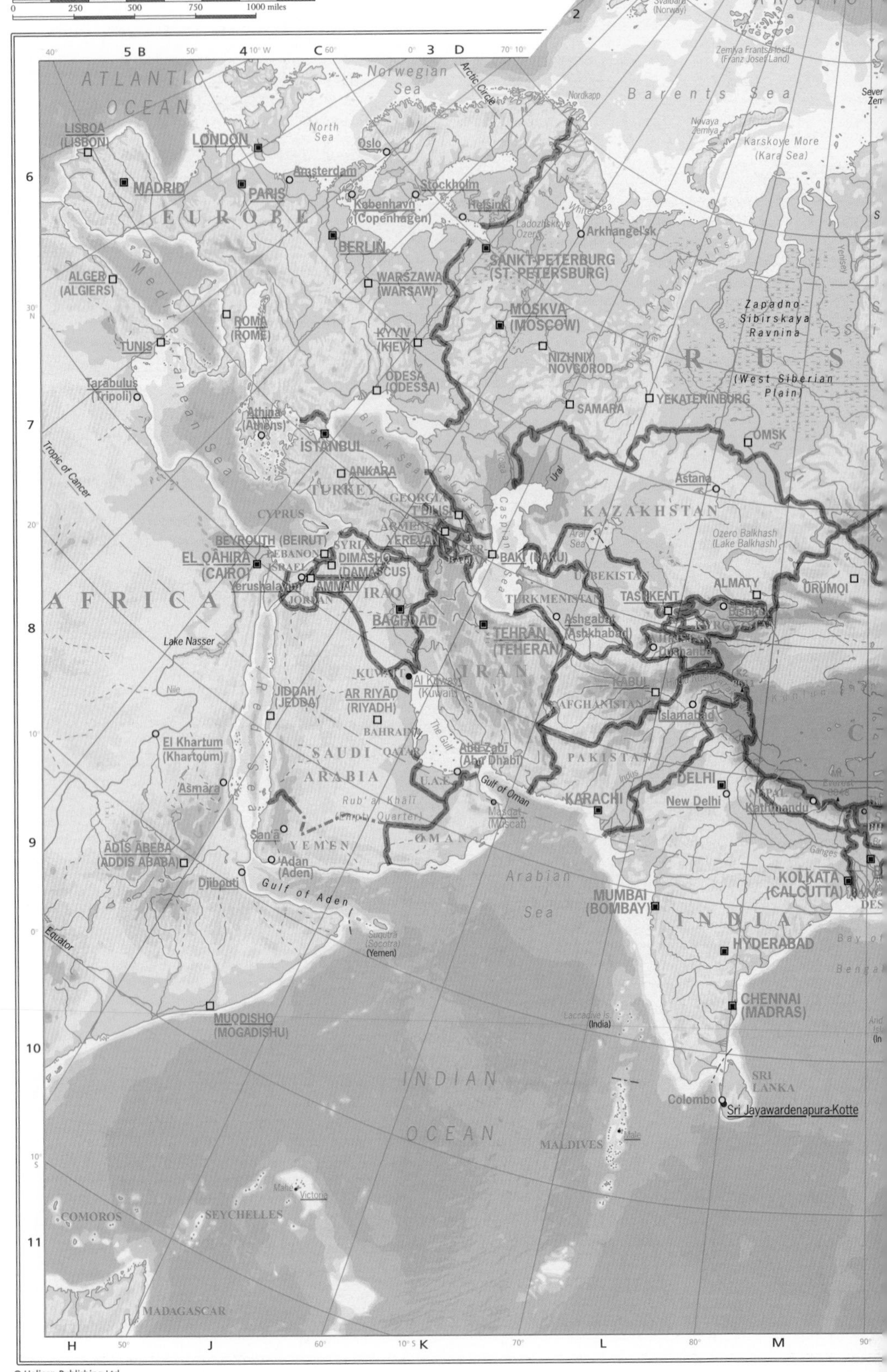

metres	feet
4000	13120
2000	6560
1000	3280
500	1640
200	656
0	0
200	656
1000	3280
2000	6560
4000	13120
6000	19690
8000	26250
metres	feet

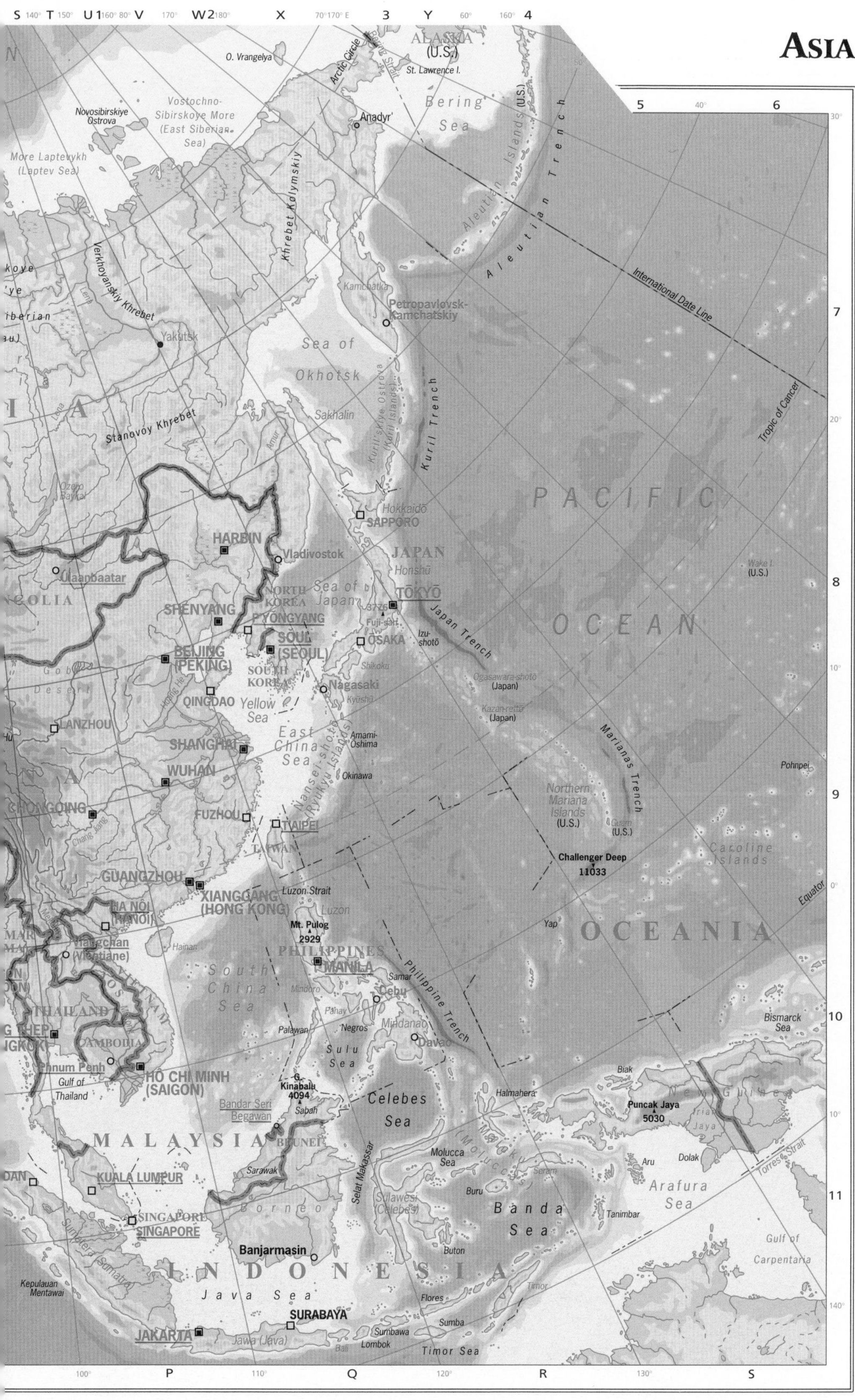

O. Vrangelya
ALASKA (U.S.)
St. Lawrence I.
Bering Sea
Anadyr'
Novosibirskiye Ostrova
Vostochno-Sibirskoye More (East Siberian Sea)
More Laptevykh (Laptev Sea)
Khrebet Kolymskiy
Verkhoyanskiy Khrebet
Yakutsk
Aleutian Islands (U.S.)
Aleutian Trench
International Date Line
Kamchatka
Petropavlovsk-Kamchatskiy
Sea of Okhotsk
Sakhalin
Kuril'skiye Ostrova (Kuril Islands)
Kuril Trench
Stanovoy Khrebet
Tropic of Cancer
PACIFIC OCEAN
Hokkaidō
SAPPORO
HARBIN
Vladivostok
JAPAN
Honshū
Ulaanbaatar
MONGOLIA
NORTH KOREA
Sea of Japan
TŌKYŌ
3776 Fuji-san
SHENYANG
P'YONGYANG
SOUL (SEOUL)
ŌSAKA
Izu-shotō
Japan Trench
BEIJING (PEKING)
Gobi Desert
SOUTH KOREA
Shikoku
Nagasaki
Kyūshū
QINGDAO
Yellow Sea
Ogasawara-shotō (Japan)
Kazan-rettō (Japan)
LANZHOU
East China Sea
Amami-Ōshima
Marianas Trench
SHANGHAI
WUHAN
Okinawa
Nansei-shotō (Ryukyu Islands)
Wake I. (U.S.)
Pohnpei
CHONGQING
FUZHOU
T'AIPEI
TAIWAN
Northern Mariana Islands (U.S.)
Guam (U.S.)
Caroline Islands
Challenger Deep 11033
GUANGZHOU
XIANGGANG (HONG KONG)
Luzon Strait
Luzon
Equator
HA NOI (HANOI)
Mt. Pulog 2929
Yap
OCEANIA
Viangchan (Vientiane)
Hainan
PHILIPPINES
MANILA
South China Sea
Mindoro
Samar
Cebu
Philippine Trench
THAILAND
VIETNAM
Panay
Negros
Mindanao
Davao
Bismarck Sea
CAMBODIA
Palawan
Sulu Sea
Phnum Penh
HO CHI MINH (SAIGON)
Gulf of Thailand
G. Kinabalu 4094
Sabah
Celebes Sea
Halmahera
Biak
New Guinea
Puncak Jaya 5030
Irian Jaya
Bandar Seri Begawan
BRUNEI
MALAYSIA
Maluku (Moluccas)
Molucca Sea
Aru
Dolak
Torres Strait
Sarawak
Selat Makassar
Seram
Buru
KUALA LUMPUR
Arafura Sea
SINGAPORE
Borneo
Sulawesi (Celebes)
Banda Sea
Tanimbar
Sumatera (Sumatra)
Banjarmasin
Gulf of Carpentaria
Buton
INDONESIA
Kepulauan Mentawai
Java Sea
Timor
Flores
SURABAYA
Sumba
JAKARTA
Jawa (Java)
Bali
Lombok
Sumbawa
Timor Sea

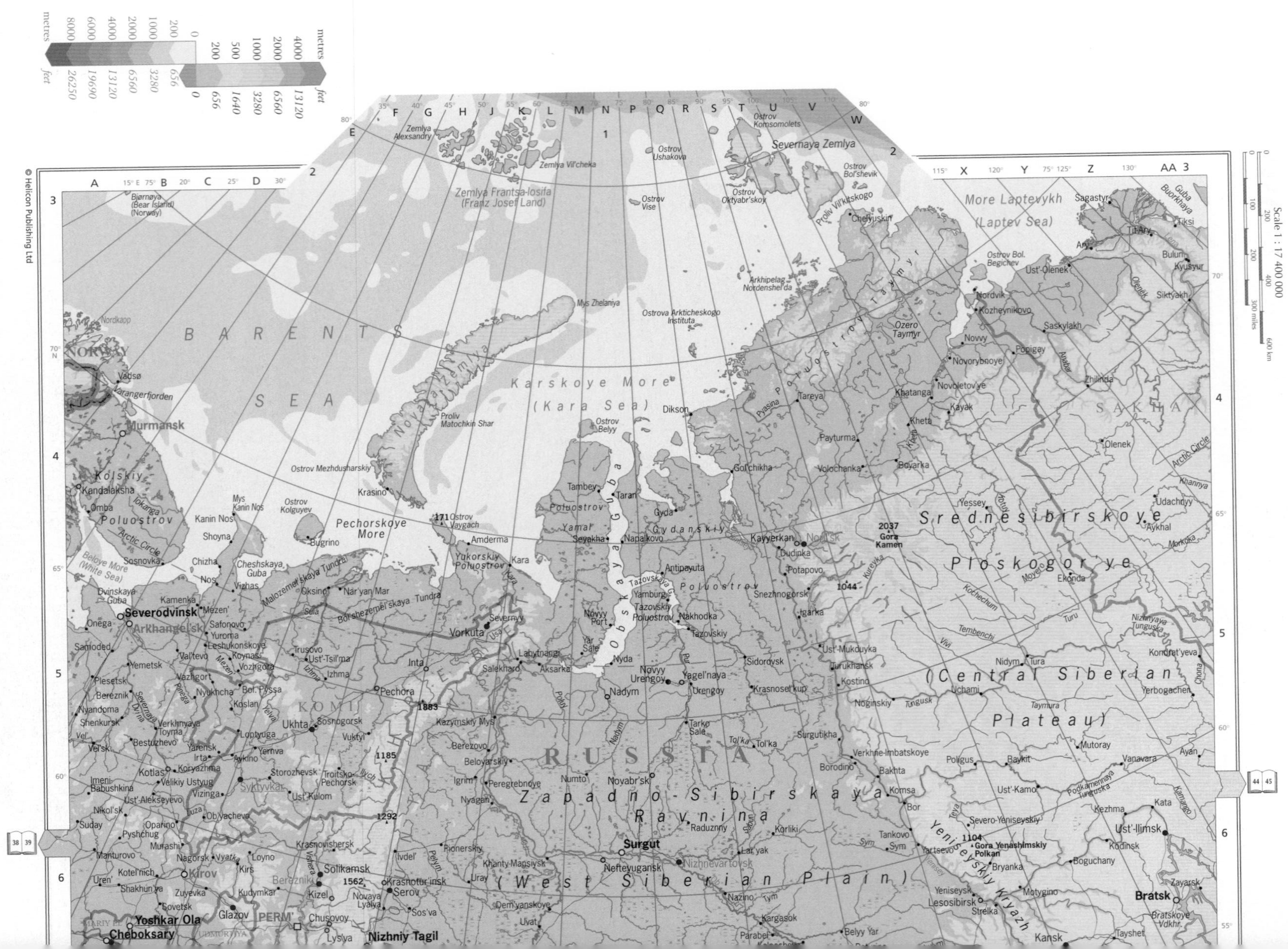

Scale 1 : 17 400 000
0 200 400 600 km
0 100 200 300 miles
BARENTS SEA
Karskoye More
(Kara Sea)
More Laptevykh
(Laptev Sea)
Pechorskoye More
Beloye More
(White Sea)
Zemlya Frantsa-Iosifa
(Franz Josef Land)
Severnaya Zemlya
Novaya Zemlya
RUSSIA
Zapadno-Sibirskaya Ravnina
(West Siberian Plain)
Srednesibirskoye Ploskogor'ye
(Central Siberian Plateau)
Poluostrov Taymyr
Poluostrov Yamal
Gydanskiy Poluostrov
Kol'skiy Poluostrov
Obskaya Guba
Murmansk
Arkhangel'sk
Severodvinsk
Vorkuta
Noril'sk
Dikson
Surgut
Nizhnevartovsk
Khanty-Mansiysk
Bratsk
Syktyvkar
Kirov
Perm'
Yoshkar Ola
Cheboksary
Nizhniy Tagil
Arctic Circle
metres feet
4000 13120
2000 6560
1000 3280
500 1640
200 656
0 0
200 656
1000 3280
2000 6560
4000 13120
6000 19690
8000 26250
metres feet

NORTHWEST ASIA

- Kazakhstan
- Kyrgyzstan
- Central Russia
- Tajikistan
- Turkmenistan
- Uzbekistan

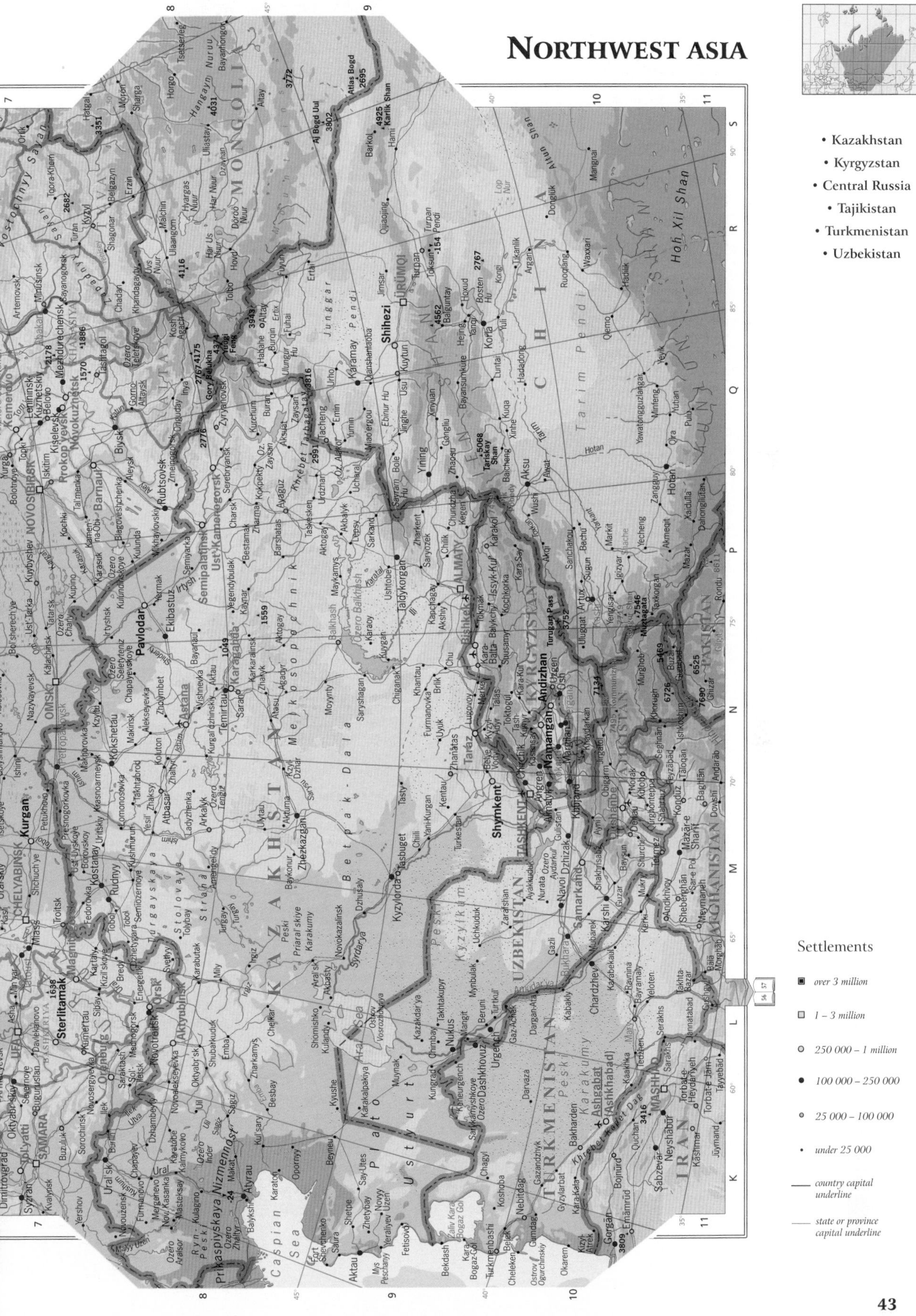

Settlements

- ■ *over 3 million*
- □ *1 – 3 million*
- ○ *250 000 – 1 million*
- ● *100 000 – 250 000*
- ◦ *25 000 – 100 000*
- • *under 25 000*
- *country capital underline*
- *state or province capital underline*

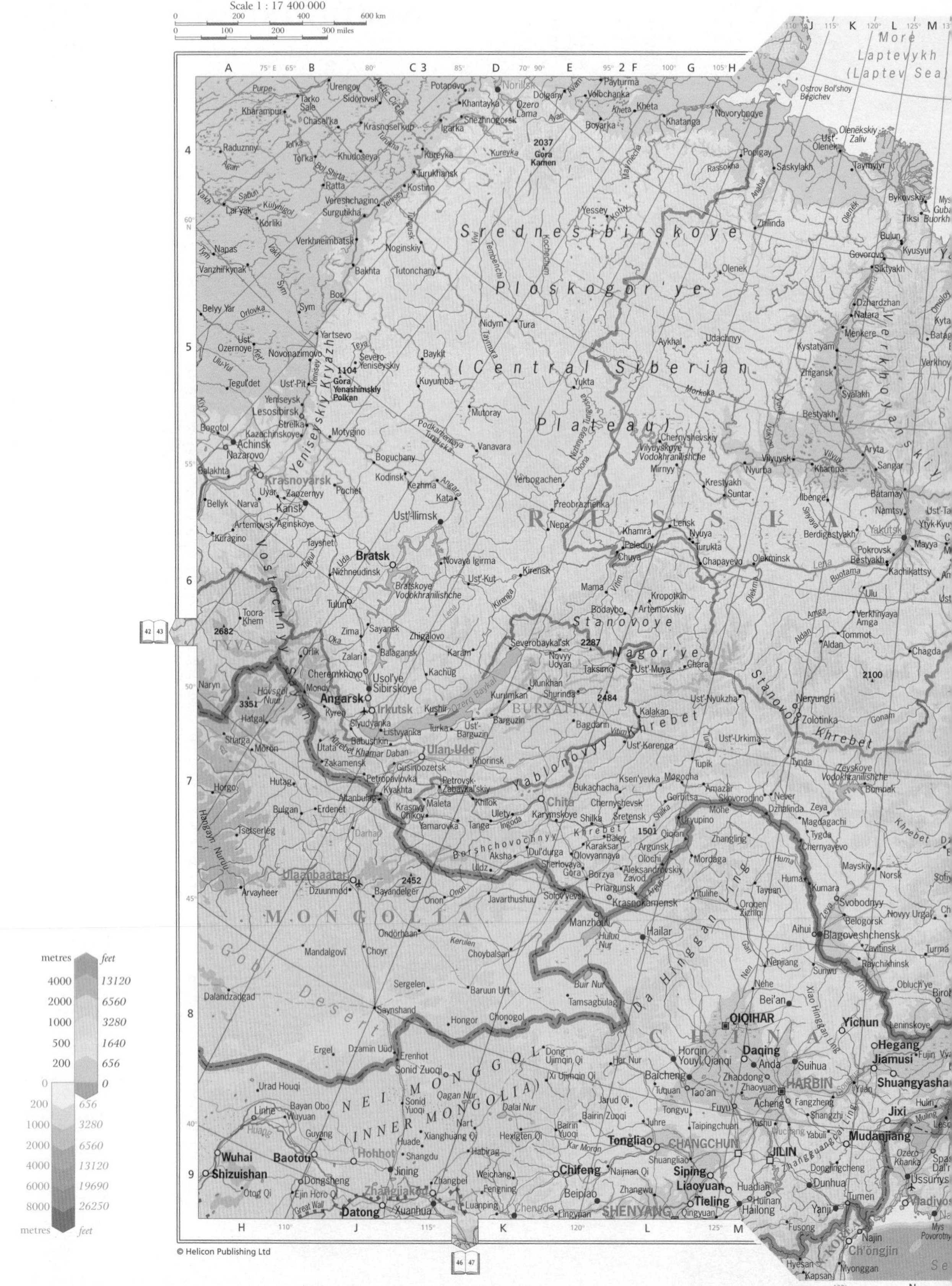

Scale 1 : 17 400 000
0 200 400 600 km
0 100 200 300 miles
More Laptevykh (Laptev Sea)
Srednesibirskoye Ploskogor'ye (Central Siberian Plateau)
RUSSIA
Verkhoyanskiy
Stanovoye Nagor'ye
Stanovoy Khrebet
Yablonovyy Khrebet
Borshchovochnyy Khrebet
Vostochnyy Sayan
Yeniseyskiy Kryazh
BURYATIYA
TYVA
MONGOLIA
Gobi Desert
CHINA
NEI MONGGOL (INNER MONGOLIA)
Da Hinggan Ling
Xiao Hinggan Ling
Zhangguangcai Ling
KOREA
Norilsk
Dudinka
Khatanga
Tiksi
Yakutsk
Mirnyy
Lensk
Olekminsk
Aldan
Neryungri
Krasnoyarsk
Achinsk
Kansk
Bratsk
Ust'-Ilimsk
Angarsk
Irkutsk
Ulan-Ude
Chita
Ozero Baykal
Bratskoye Vodokhranilishche
Vilyuyskoye Vodokhranilishche
Zeyskoye Vodokhranilishche
Blagoveshchensk
Ulaanbaatar
Hailar
Manzhouli
QIQIHAR
Daqing
HARBIN
Jiamusi
Yichun
Mudanjiang
CHANGCHUN
JILIN
SHENYANG
Chifeng
Hohhot
Baotou
Datong
Zhangjiakou
Wuhai
Shizuishan
Tongliao
Siping
Liaoyuan
Tieling
Great Wall
2037 Gora Kamen
1104 Gora Yenashimskiy Polkan
2682
3351
2287
2484
1501
2452
2100
metres feet
4000 13120
2000 6560
1000 3280
500 1640
200 656
0 0
200 656
1000 3280
2000 6560
4000 13120
6000 19690
8000 26250
© Helicon Publishing Ltd

NORTHEAST ASIA

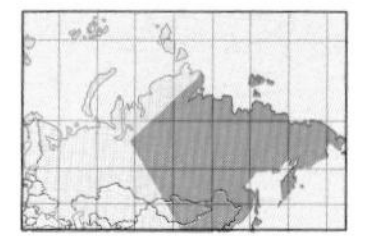

- Mongolia
- Eastern Russia

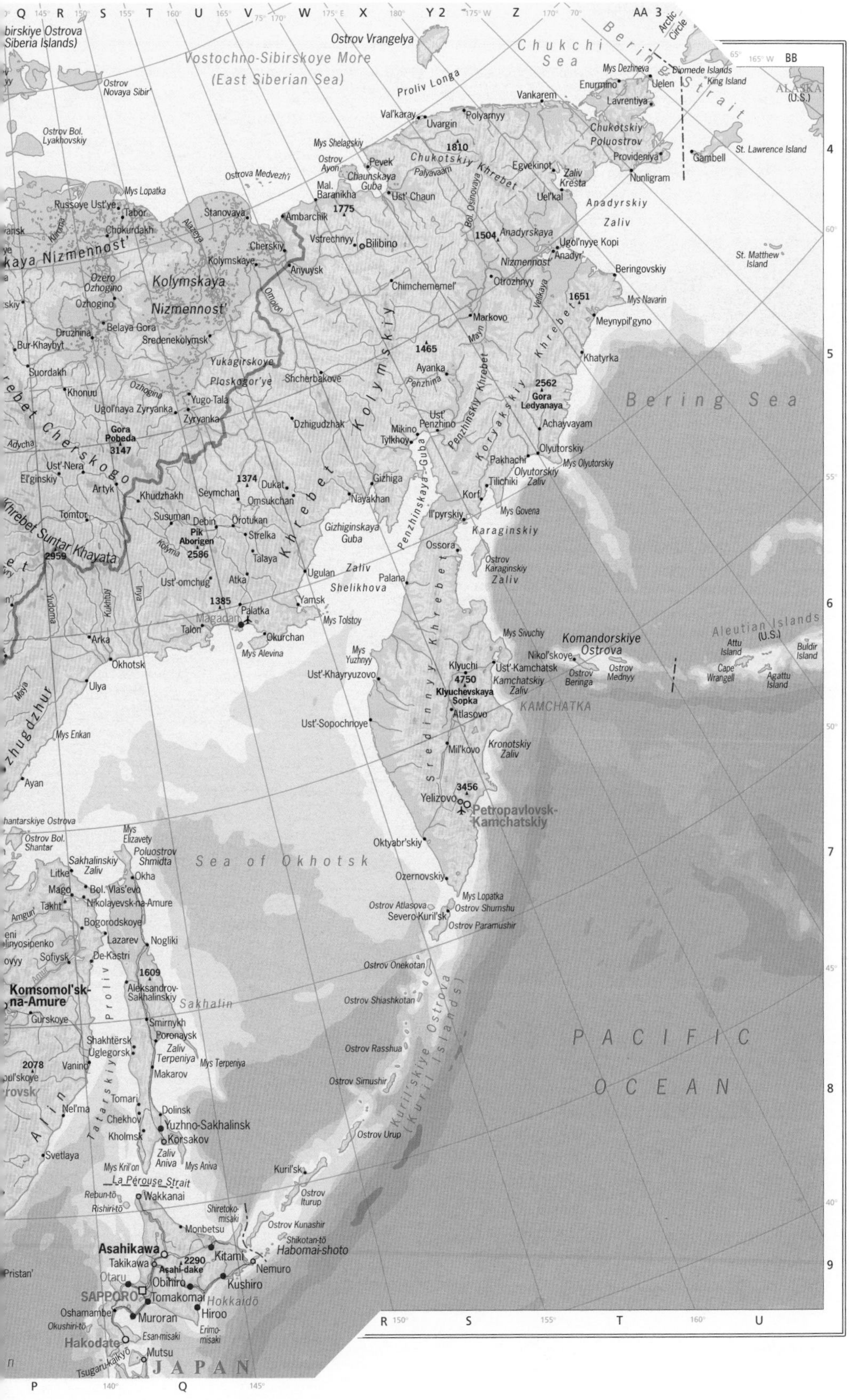

Settlements

- over 3 million
- 1 – 3 million
- 250 000 – 1 million
- 100 000 – 250 000
- 25 000 – 100 000
- under 25 000
- country capital underline
- state or province capital underline

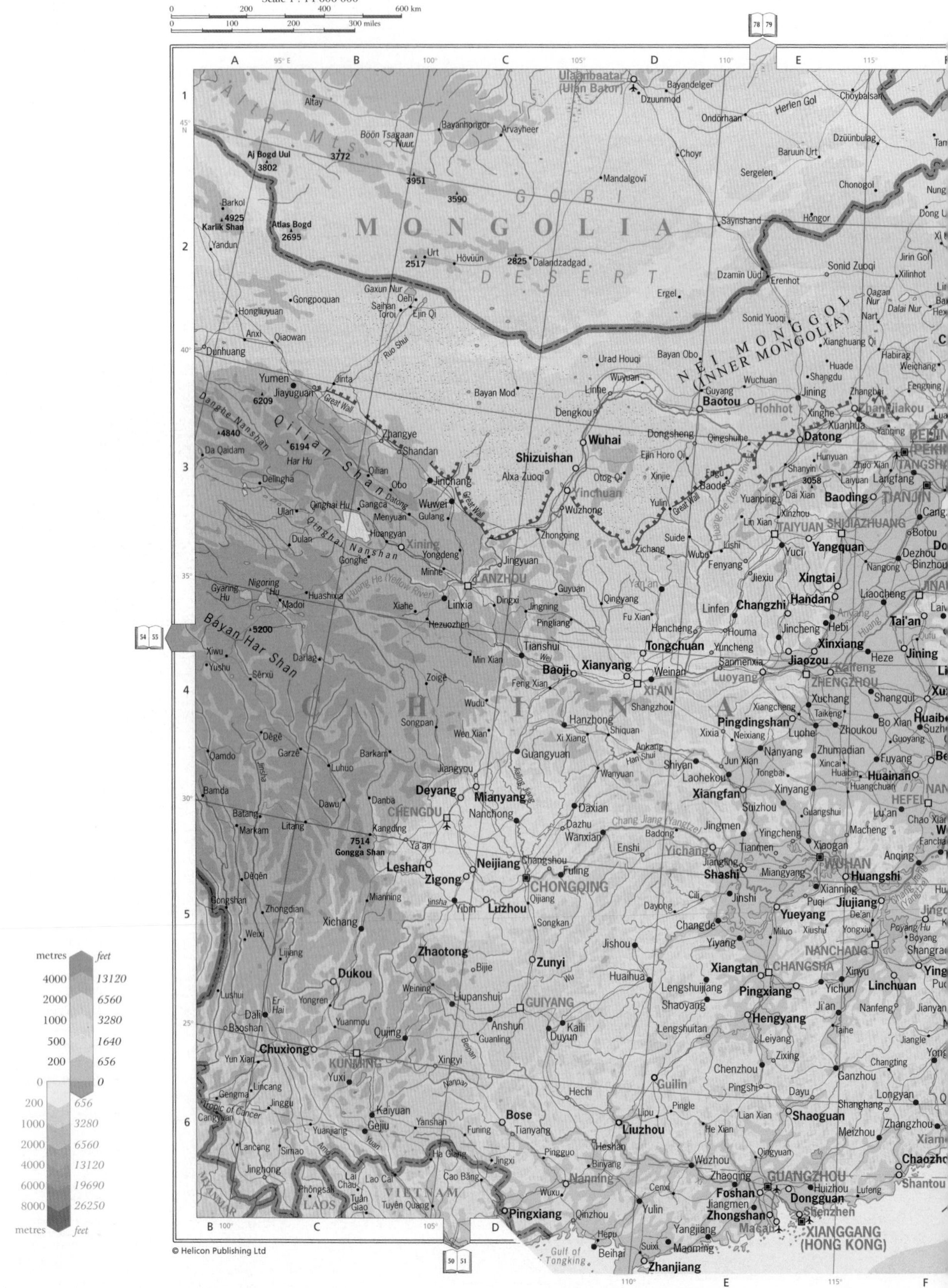
Scale 1 : 14 600 000
0 200 400 600 km
0 100 200 300 miles
78 79
54 55
50 51
A B C D E F
1 2 3 4 5 6
95° E 100° 105° 110° 115°
45° N 40° 35° 30° 25°
MONGOLIA
GOBI DESERT
CHINA
NEI MONGGOL (INNER MONGOLIA)
VIETNAM
LAOS
Altai Mts
Qilian Shan
Bayan Har Shan
Qinghai Nanshan
Danghe Nanshan
Great Wall
Huang He (Yellow River)
Chang Jiang (Yangtze)
Gulf of Tongking
Tropic of Cancer
Ulaanbaatar (Ulan Bator)
Aj Bogd Uul 3802
Atlas Bogd 2695
Karlik Shan 4925
Gongga Shan 7514
Qinghai Hu
Gyaring Hu
Ngoring Hu
Dunhuang
Yumen
Jiayuguan
Zhangye
Jinchang
Wuwei
Xining
LANZHOU
Linxia
Baotou
Hohhot
Datong
Zhangjiakou
Wuhai
Shizuishan
Yinchuan
Wuzhong
TAIYUAN
SHIJIAZHUANG
Baoding
TIANJIN
Yangquan
Xingtai
Handan
Changzhi
Tianshui
Baoji
Xianyang
XI'AN
Tongchuan
Luoyang
ZHENGZHOU
Jiaozuo
Xinxiang
Pingdingshan
Hanzhong
Guangyuan
Mianyang
Deyang
CHENGDU
Nanchong
Leshan
Zigong
Neijiang
CHONGQING
Luzhou
Yibin
Zhaotong
Dukou
Xichang
Zunyi
GUIYANG
KUNMING
Chuxiong
Dali
Gejiu
Kaiyuan
Bose
Nanning
Liuzhou
Guilin
Hengyang
CHANGSHA
Xiangtan
Zhuzhou
Yueyang
NANCHANG
Jiujiang
Yichang
Shashi
WUHAN
Huangshi
Xiangfan
HEFEI
Huainan
GUANGZHOU
Foshan
Dongguan
Zhongshan
Shenzhen
Macau
XIANGGANG (HONG KONG)
Shaoguan
Zhanjiang
Maoming
Beihai
Pingxiang
Shantou
Chaozhou
metres feet
4000 13120
2000 6560
1000 3280
500 1640
200 656
0 0
200 656
1000 3280
2000 6560
4000 13120
6000 19690
8000 26250
© Helicon Publishing Ltd

- Eastern China
- Japan
- North Korea
- South Korea
- Taiwan

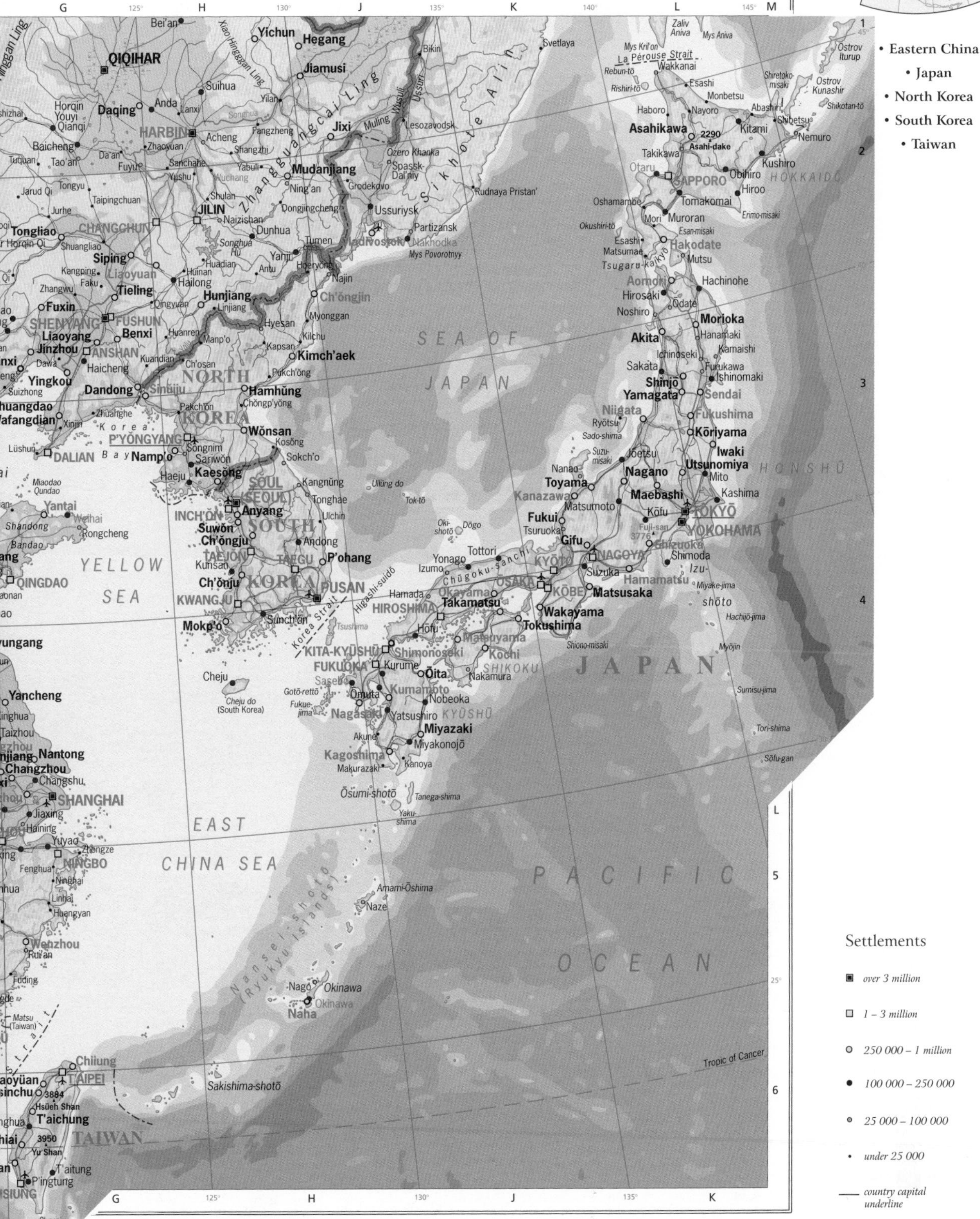

Settlements

- over 3 million
- 1 – 3 million
- 250 000 – 1 million
- 100 000 – 250 000
- 25 000 – 100 000
- under 25 000
- country capital underline

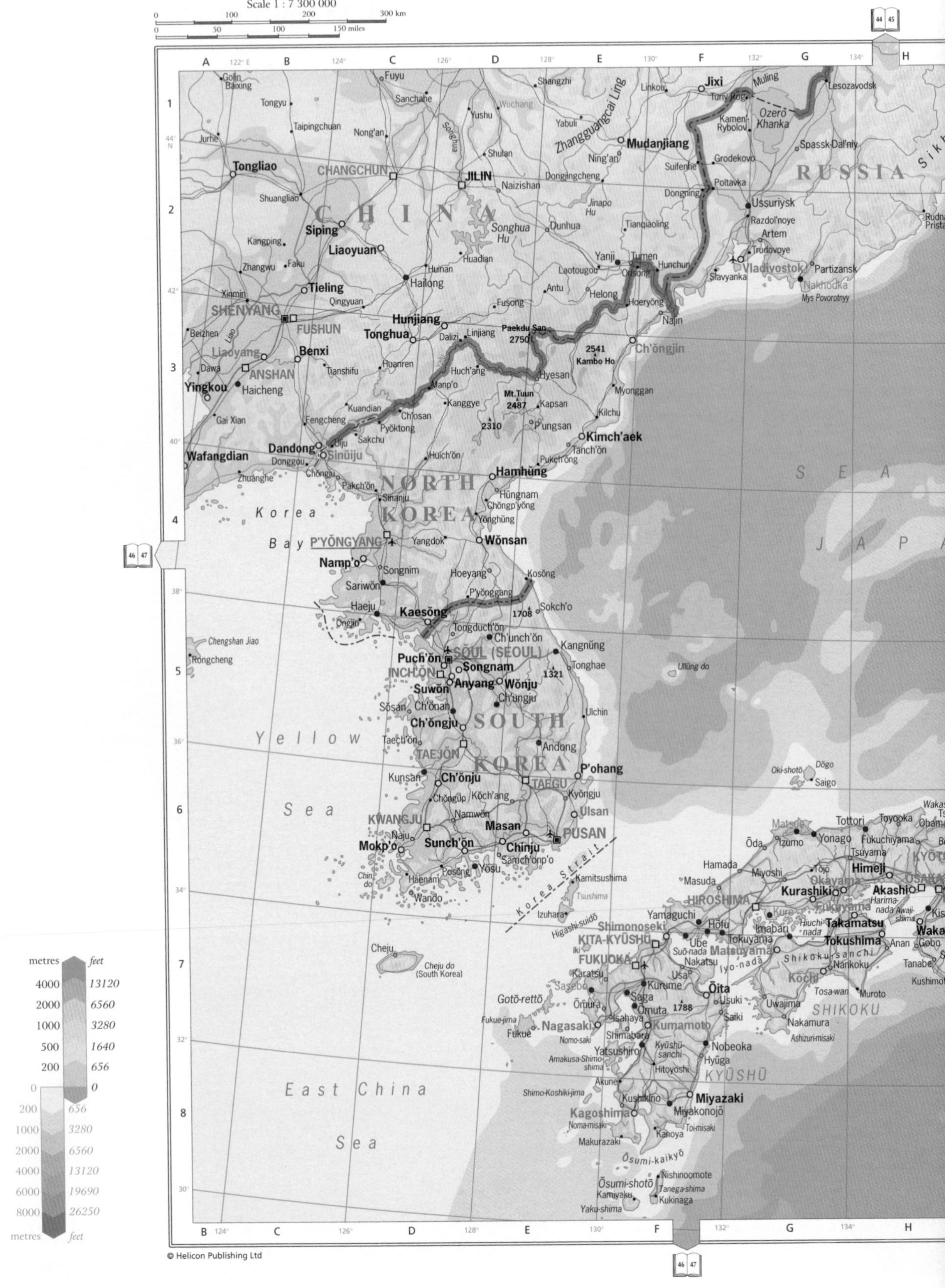

Scale 1 : 7 300 000
0 100 200 300 km
0 50 100 150 miles
C H I N A
NORTH KOREA
SOUTH KOREA
RUSSIA
SEA
JAPA
Korea Bay
Yellow Sea
East China Sea
Korea Strait
CHANGCHUN
JILIN
SHENYANG
FUSHUN
ANSHAN
Tongliao
Siping
Liaoyuan
Tieling
Hunjiang
Tonghua
Benxi
Yingkou
Haicheng
Wafangdian
Dandong
Sinŭiju
Mudanjiang
Jixi
Ussuriysk
Vladivostok
Nakhodka
Partizansk
Yanji
Tumen
Zhangguangcai Ling
Songhua Hu
Paekdu San 2750
Ch'ŏngjin
Kimch'aek
Hamhŭng
Wŏnsan
P'YŎNGYANG
Namp'o
Kaesŏng
SŎUL (SEOUL)
INCH'ŎN
Puch'ŏn
Songnam
Anyang
Suwŏn
Wŏnju
Ch'ŏngju
TAEJŎN
Ch'ŏnju
TAEGU
P'ohang
Ulsan
PUSAN
KWANGJU
Masan
Chinju
Mokp'o
Sunch'ŏn
Yŏsu
Cheju
Cheju do (South Korea)
Ullŭng do
Oki-shotō
Tottori
Himeji
Kurashiki
Akashi
Takamatsu
Tokushima
HIROSHIMA
Shimonoseki
KITA-KYŪSHŪ
FUKUOKA
Saga
Kurume
Ōita
Nagasaki
Kumamoto
Miyazaki
Kagoshima
Nobeoka
SHIKOKU
KYŪSHŪ
Gotō-rettō
Ōsumi-shotō
Tsushima
metres feet
4000 13120
2000 6560
1000 3280
500 1640
200 656
0 0
200 656
1000 3280
2000 6560
4000 13120
6000 19690
8000 26250

JAPAN AND KOREA

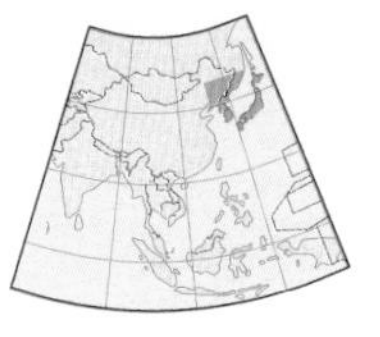

- Japan
- North Korea
- South Korea

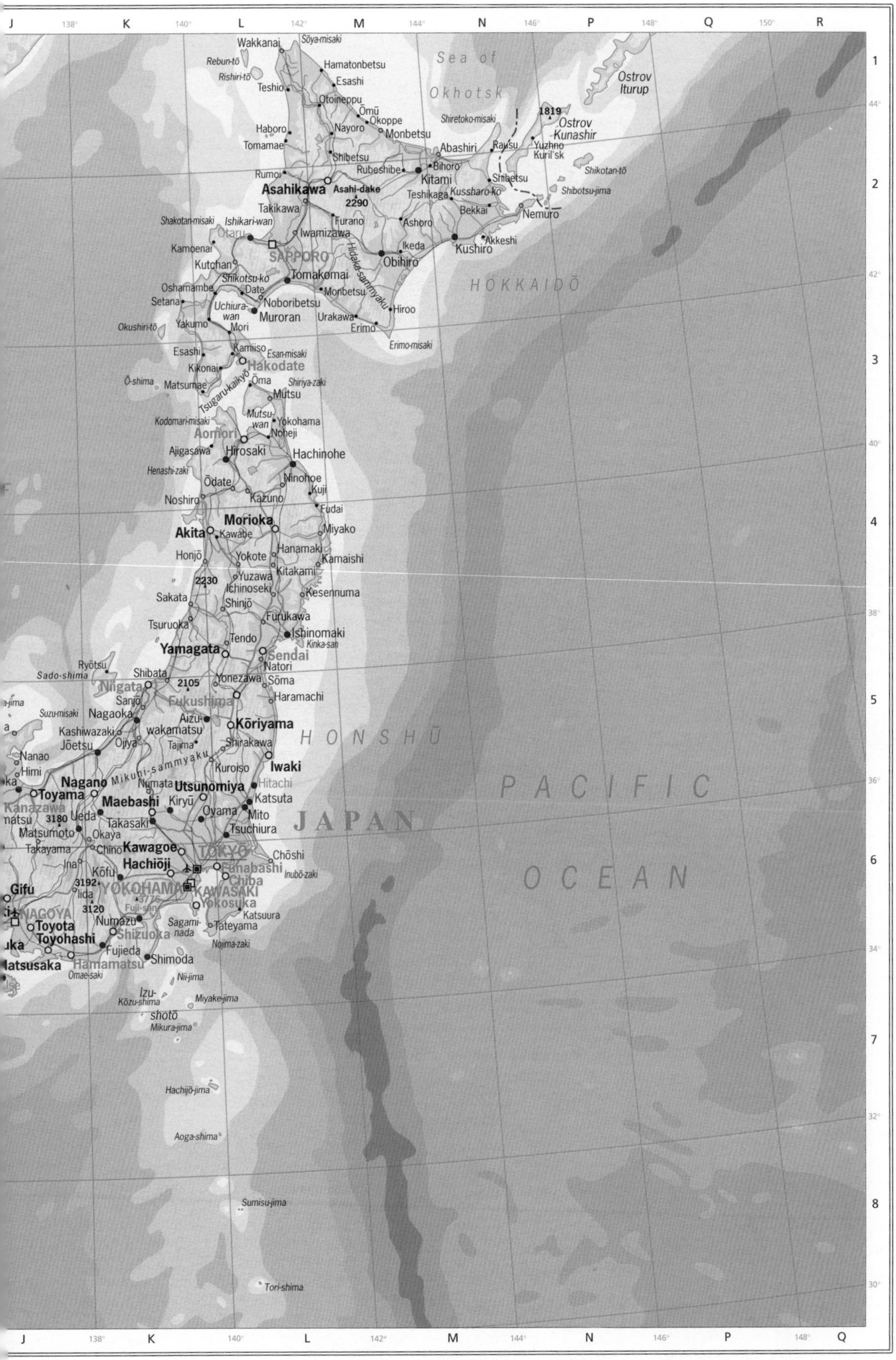

J 138° K 140° L 142° M 144° N 146° P 148° Q 150° R

1 2 3 4 5 6 7 8

44° 42° 40° 38° 36° 34° 32° 30°

Wakkanai
Sōya-misaki
Rebun-tō
Rishiri-tō
Hamatonbetsu
Esashi
Teshio
Otoineppu
Ōmū
Okoppe
Monbetsu
Nayoro
Haboro
Tomamae
Shibetsu
Rumoi
Rubeshibe
Asahikawa
Asahi-dake
2290
Takikawa
Furano
Shakotan-misaki
Ishikari-wan
Otaru
Iwamizawa
Kamoenai
SAPPORO
Kutchan
Tomakomai
Shikotsu-ko
Date
Oshamambe
Setana
Uchiura-wan
Noboribetsu
Muroran
Okushiri-tō
Yakumo
Mori
Urakawa
Erimo
Erimo-misaki
Hiroo
Hidaka-sammyaku
Monbetsu
Obihiro
Ikeda
Kushiro
Akkeshi
Ashoro
Bekkai
Nemuro
Teshikaga
Kussharo-ko
Kitami
Bihoro
Abashiri
Shiretoko-misaki
Rausu
Shibetsu
Yuzhno-Kuril'sk
Ostrov Kunashir
1819
Ostrov Iturup
Shikotan-tō
Shibotsu-jima
Sea of Okhotsk
HOKKAIDŌ
Esashi
Kamiiso
Esan-misaki
Kikonai
Hakodate
Ō-shima
Matsumae
Ōma
Shiriya-zaki
Tsugaru-kaikyō
Mutsu
Kodomari-misaki
Mutsu-wan
Yokohama
Aomori
Noheji
Ajigasawa
Hirosaki
Hachinohe
Henashi-zaki
Ōdate
Ninohoe
Noshiro
Kazuno
Kuji
Fudai
Morioka
Akita
Kawabe
Miyako
Honjō
Yokote
Hanamaki
Kamaishi
Kitakami
2230
Yuzawa
Ichinoseki
Sakata
Kesennuma
Shinjō
Furukawa
Tsuruoka
Tendo
Ishinomaki
Kinka-san
Yamagata
Sendai
Ryōtsu
Sado-shima
Shibata
Natori
2105
Yonezawa
Sōma
Niigata
Sanjō
Fukushima
Haramachi
Suzu-misaki
Nagaoka
Aizu-wakamatsu
Kōriyama
Kashiwazaki
Ojiya
Tajima
Shirakawa
Jōetsu
HONSHŪ
Nanao
Mikuni-sammyaku
Kuroiso
Iwaki
Himi
Nagano
Numata
Utsunomiya
Hitachi
Toyama
Maebashi
Kiryū
Katsuta
Kanazawa
3180
Ueda
Oyama
Mito
Takasaki
JAPAN
Matsumoto
Okaya
Tsuchiura
Takayama
Chino
Kawagoe
TŌKYŌ
Chōshi
Ina
Hachiōji
Funabashi
Inubō-zaki
Kōfu
Chiba
Gifu
3192
Iida
YOKOHAMA
KAWASAKI
3776
Fuji-san
Yokosuka
3120
NAGOYA
Numazu
Katsuura
Toyota
Sagami-nada
Tateyama
Toyohashi
Shizuoka
Nojima-zaki
Fujieda
Shimoda
Matsusaka
Hamamatsu
Omae-saki
Nii-jima
Izu-shotō
Kōzu-shima
Miyake-jima
Mikura-jima
Hachijō-jima
Aoga-shima
Sumisu-jima
Tori-shima
PACIFIC OCEAN

J 138° K 140° L 142° M 144° N 146° P 148° Q

Settlements

- over 3 million
- 1 – 3 million
- 250 000 – 1 million
- 100 000 – 250 000
- 25 000 – 100 000
- under 25 000
- country capital underline

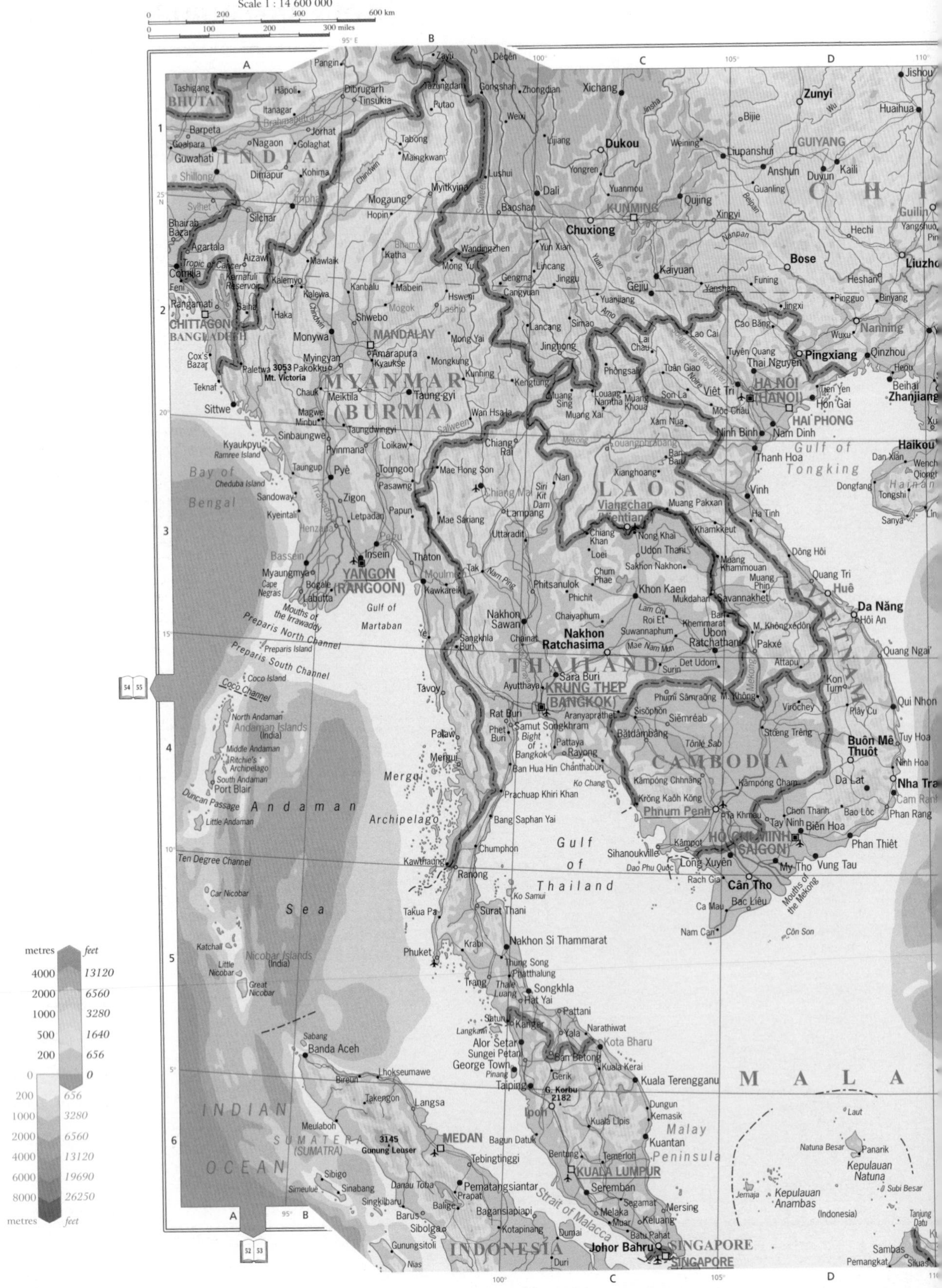

Scale 1 : 14 600 000
0 200 400 600 km
0 100 200 300 miles
metres feet
4000 13120
2000 6560
1000 3280
500 1640
200 656
0 0
200 656
1000 3280
2000 6560
4000 13120
6000 19690
8000 26250
metres feet
A B C D
1 2 3 4 5 6
95° E 100° 105° 110°
25° N 20° 15° 10° 5°
BHUTAN
INDIA
BANGLADESH
MYANMAR (BURMA)
LAOS
THAILAND
CAMBODIA
VIETNAM
CHINA
MALAYSIA
INDONESIA
SINGAPORE
Tashigang
Hapoli
Dibrugarh
Tinsukia
Itanagar
Brahmaputra
Barpeta
Goalpara
Guwahati
Nagaon
Jorhat
Golaghat
Shillong
Dimapur
Kohima
Imphal
Sylhet
Silchar
Bhairab Bazar
Agartala
Aizawl
Tropic of Cancer
Comilla
Feni
Karnafuli Reservoir
Rangamati
Saiha
CHITTAGONG
Cox's Bazar
Teknaf
Sittwe
Pangin
Zayü
Tazungdam
Putao
Tabong
Maingkwan
Chindwin
Myitkyina
Mogaung
Hopin
Bhamo
Katha
Mawlaik
Kalemyo
Kalewa
Haka
Kanbalu
Mabein
Mogok
Shwebo
Monywa
MANDALAY
Myingyan
Pakokku
3053 Mt. Victoria
Paletwa
Amarapura
Kyaukse
Chauk
Meiktila
Taung-gyi
Magwe
Minbu
Taungdwingyi
Sinbaungwe
Kyaukpyu
Ramree Island
Pyinmana
Loikaw
Taungup
Pyè
Toungoo
Cheduba Island
Sandoway
Zigon
Pasawng
Kyeintali
Letpadan
Papun
Henzada
Pegu
Bassein
Insein
Thaton
YANGON (RANGOON)
Myaungmya
Bogale
Cape Negrais
Labutta
Moulmein
Kawkareik
Gulf of Martaban
Mouths of the Irrawaddy
Bay of Bengal
Preparis North Channel
Preparis Island
Preparis South Channel
Coco Island
Coco Channel
North Andaman
Andaman Islands (India)
Middle Andaman
Ritchie's Archipelago
South Andaman
Port Blair
Duncan Passage
Little Andaman
Andaman Sea
Ten Degree Channel
Car Nicobar
Katchall
Nicobar Islands (India)
Little Nicobar
Great Nicobar
INDIAN OCEAN
Ye
Tavoy
Palaw
Mergui
Mergui Archipelago
Kawthaung
Ranong
Dêqên
Gongshan
Zhongdian
Weixi
Lijiang
Lushui
Dali
Baoshan
Wandingzhen
Yun Xian
Lincang
Gengma
Cangyuan
Jinggu
Salween
Mong Yu
Hsweni
Lashio
Mong Yai
Mongkung
Kunhing
Kengtung
Wan Hsa-la
Lancang
Simao
Jinghong
Xichang
Jinsha
Dukou
Yongren
Yuanmou
KUNMING
Chuxiong
Qujing
Xingyi
Weining
Liupanshui
Bijie
Zunyi
GUIYANG
Anshun
Duyun
Kaili
Guanling
Beipan
Nanpan
Yuan
Kaiyuan
Gejiu
Yuanjiang
Amo
Funing
Yanshan
Bose
Hechi
Guilin
Liuzhou
Heshan
Pingguo
Binyang
Jingxi
Nanning
Wuxu
Pingxiang
Qinzhou
Jishou
Huaihua
Beihai
Zhanjiang
Haikou
Hainan
Dongfang
Tongshi
Sanya
Lao Cai
Cao Bang
Lai Chau
Tuyên Quang
Thai Nguyen
Tuân Giao
Sông Hong (Red River)
Noire
Son La
Viêt Tri
HA NOI (HANOI)
Hon Gai
HAI PHONG
Nam Dinh
Ninh Binh
Thanh Hoa
Gulf of Tongking
Vinh
Ha Tinh
Dông Hoi
Quang Tri
Huê
Da Năng
Hôi An
Quang Ngai
Kon Tum
Qui Nhon
Plây Cu
Buôn Mê Thuôt
Tuy Hoa
Da Lat
Nha Trang
Cam Ranh
Phan Rang
Bao Lôc
Phan Thiêt
Biên Hoa
HO CHI MINH (SAIGON)
Vung Tau
My Tho
Mouths of the Mekong
Cân Tho
Long Xuyên
Rach Gia
Bac Liêu
Ca Mau
Nam Can
Côn Son
Phôngsali
Louang Namtha
Muang Sing
Muang Xai
Muang Khoua
Xam Nua
Louangphrabang
Mekong
Xianghoang
Viangchan (Vientiane)
Nong Khai
Muang Pakxan
Khamkkeut
Muang Khammouan
Savannakhet
Pakxé
Attapu
Chiang Rai
Mae Hong Son
Chiang Mai
Siri Kit Dam
Nan
Lampang
Mae Sariang
Uttaradit
Chiang Khan
Loei
Udon Thani
Sakhon Nakhon
Tak
Nam Ping
Phitsanulok
Chum Phae
Khon Kaen
Mukdahan
Phichit
Chaiyaphum
Roi Et
Nakhon Sawan
Nakhon Ratchasima
Suwannaphum
Ubon Ratchathani
Mae Nam Mun
Surin
Det Udom
Sangkhla Buri
Chainat
Sara Buri
Ayutthaya
KRUNG THEP (BANGKOK)
Aranyaprathet
Rat Buri
Samut Songkhram
Phet Buri
Bight of Bangkok
Pattaya
Rayong
Chanthaburi
Ban Hua Hin
Ko Chang
Prachuap Khiri Khan
Bang Saphan Yai
Chumphon
Gulf of Thailand
Ko Samui
Surat Thani
Takua Pa
Phuket
Krabi
Nakhon Si Thammarat
Thung Song
Phatthalung
Trang
Thale Luang
Songkhla
Hat Yai
Satun
Pattani
Yala
Narathiwat
Sisophon
Phumi Sâmraông
Siĕmréab
Bătdâmbâng
Tônlé Sab
Stœng Trêng
Kâmpóng Chhnăng
Kâmpóng Cham
Krŏng Kaôh Kŏng
Phnum Penh
Ta Khmau
Kâmpot
Sihanoukville
Dao Phu Quôc
Langkawi
Kangar
Alor Setar
Sungei Petani
George Town
Pinang
Kota Bharu
Kuala Kerai
Gerik
Taiping
G. Korbu 2182
Ipoh
Kuala Terengganu
Dungun
Kemasik
Kuala Lipis
Malay Peninsula
Kuantan
Bentong
Temerloh
KUALA LUMPUR
Seremban
Segamat
Mersing
Melaka
Muar
Keluang
Batu Pahat
Johor Bahru
Strait of Malacca
Sabang
Banda Aceh
Lhokseumawe
Bireun
Takengon
Langsa
Meulaboh
SUMATERA (SUMATRA)
3145 Gunung Leuser
MEDAN
Bagun Datuk
Tebingtinggi
Sibigo
Simeulue
Sinabang
Danau Toba
Pematangsiantar
Prapat
Singkilbaru
Balige
Barus
Sibolga
Bagansiapiapi
Kotapinang
Dumai
Gunungsitoli
Nias
Duri
Laut
Natuna Besar
Panarik
Kepulauan Natuna
Subi Besar
Jemaja
Kepulauan Anambas (Indonesia)
Tanjung Datu
Sambas
Pemangkat

Southeast Asia

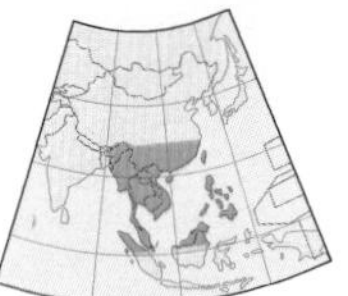

- Cambodia
- Laos
- Myanmar (Burma)
- Philippines
- Thailand
- Vietnam

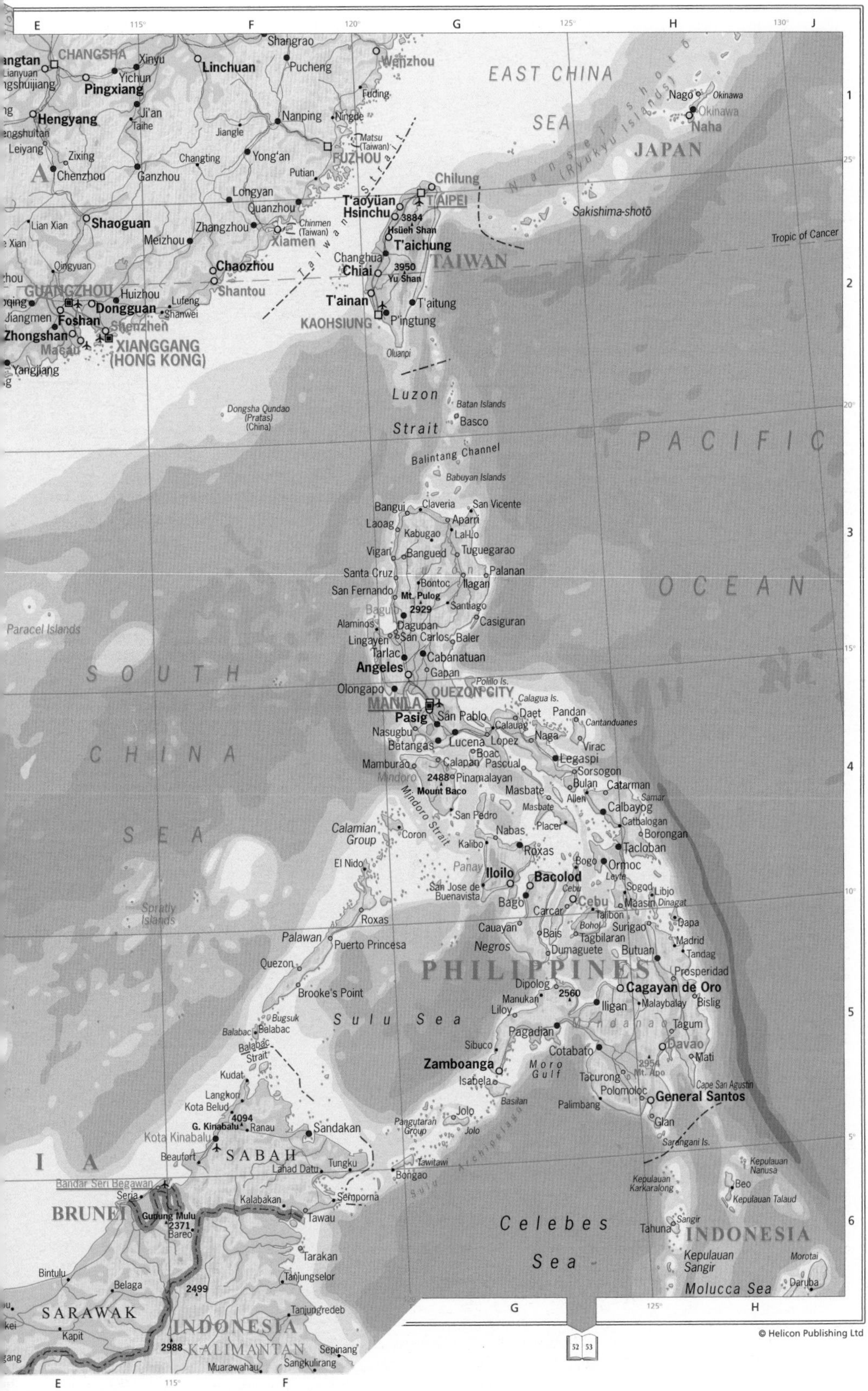

Settlements

- ▣ *over 3 million*
- ◻ *1 – 3 million*
- ○ *250 000 – 1 million*
- ● *100 000 – 250 000*
- ∘ *25 000 – 100 000*
- • *under 25 000*
- ___ *country capital underline*

Scale 1 : 14 600 000

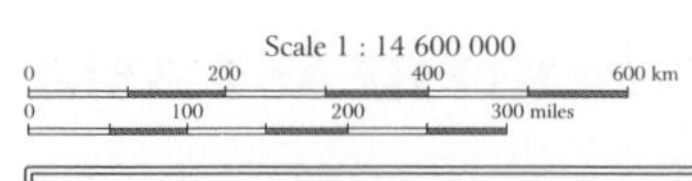

metres | feet
4000 | 13120
2000 | 6560
1000 | 3280
500 | 1640
200 | 656
0 | 0
200 | 656
1000 | 3280
2000 | 6560
4000 | 13120
6000 | 19690
8000 | 26250
metres | feet

South China Sea

Sulu Sea

Celebes Sea

Java Sea

Flores Sea

INDIAN OCEAN

THAILAND

MALAYSIA

BRUNEI

SINGAPORE

INDONESIA

PHILIPPINES

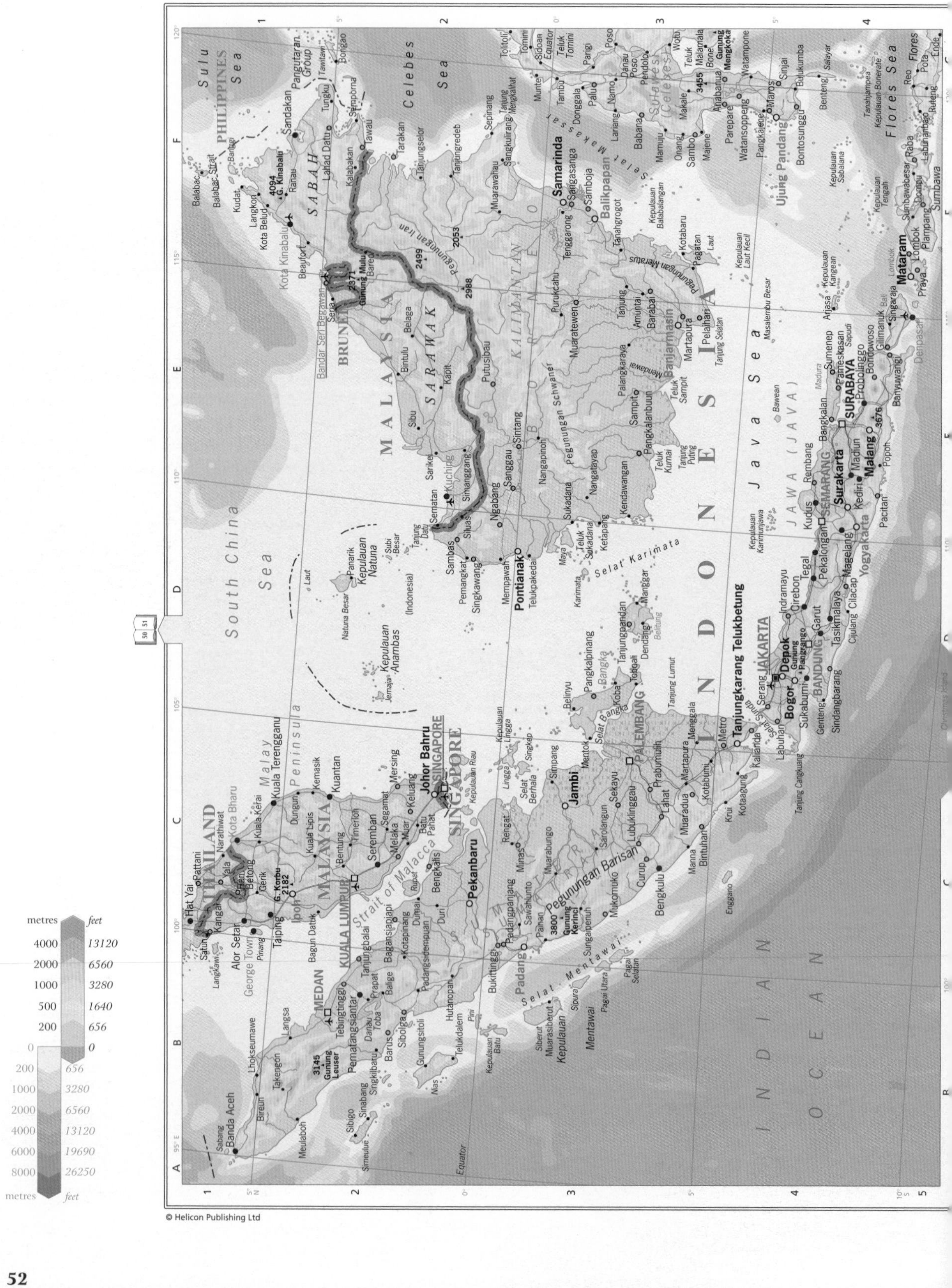

© Helicon Publishing Ltd

MALAYSIA AND INDONESIA

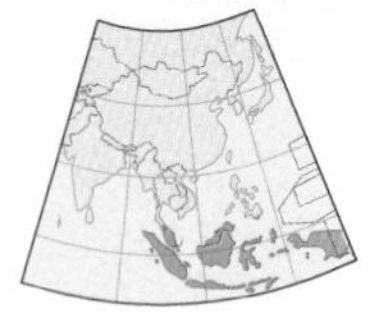

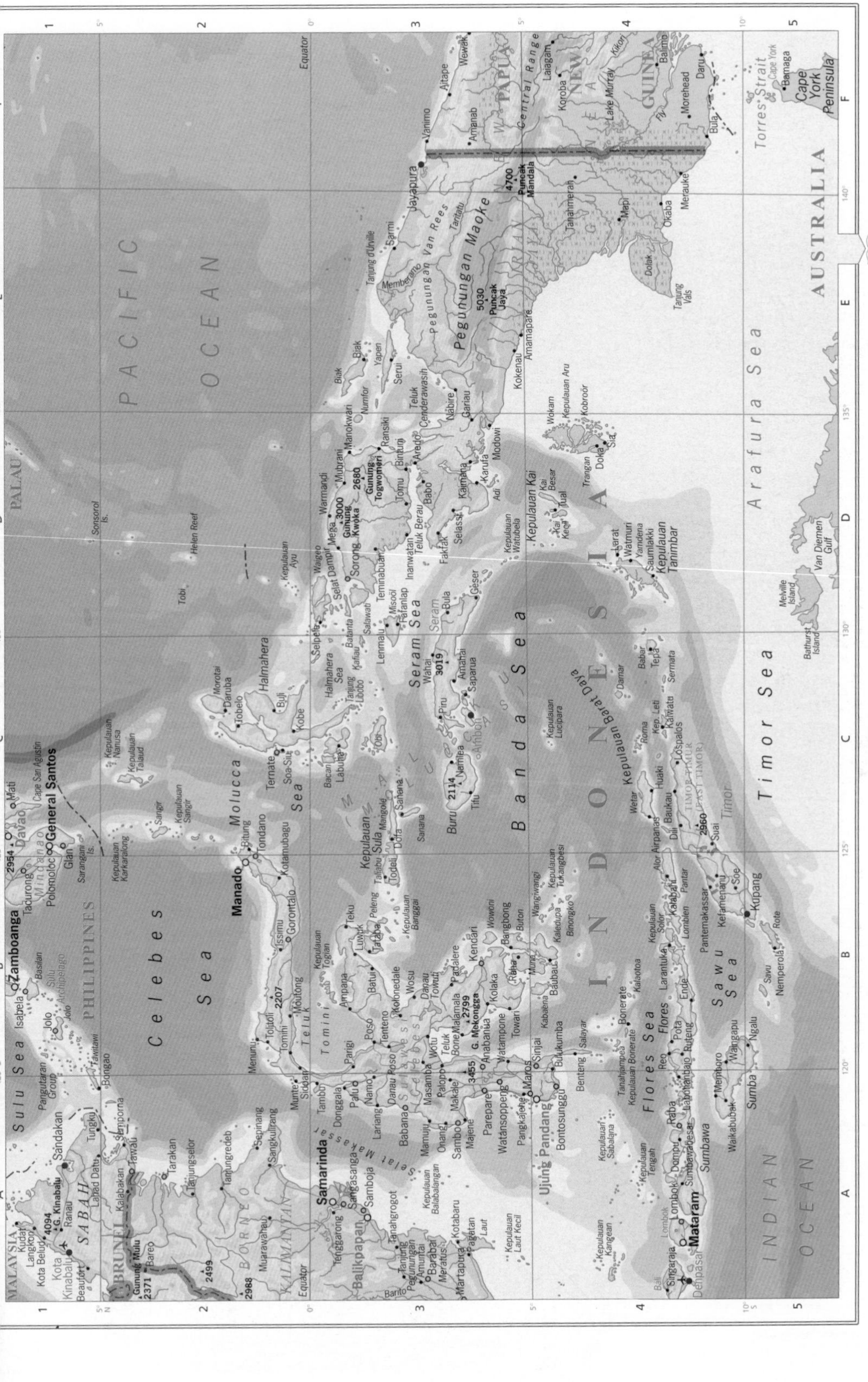

- Brunei
- Indonesia
- Malaysia
- Singapore

Settlements

- ▣ over 3 million
- ☐ 1 – 3 million
- ○ 250 000 – 1 million
- ● 100 000 – 250 000
- ◦ 25 000 – 100 000
- • under 25 000
- —— country capital underline

Scale 1 : 14 600 000

0 200 400 600 km

0 100 200 300 miles

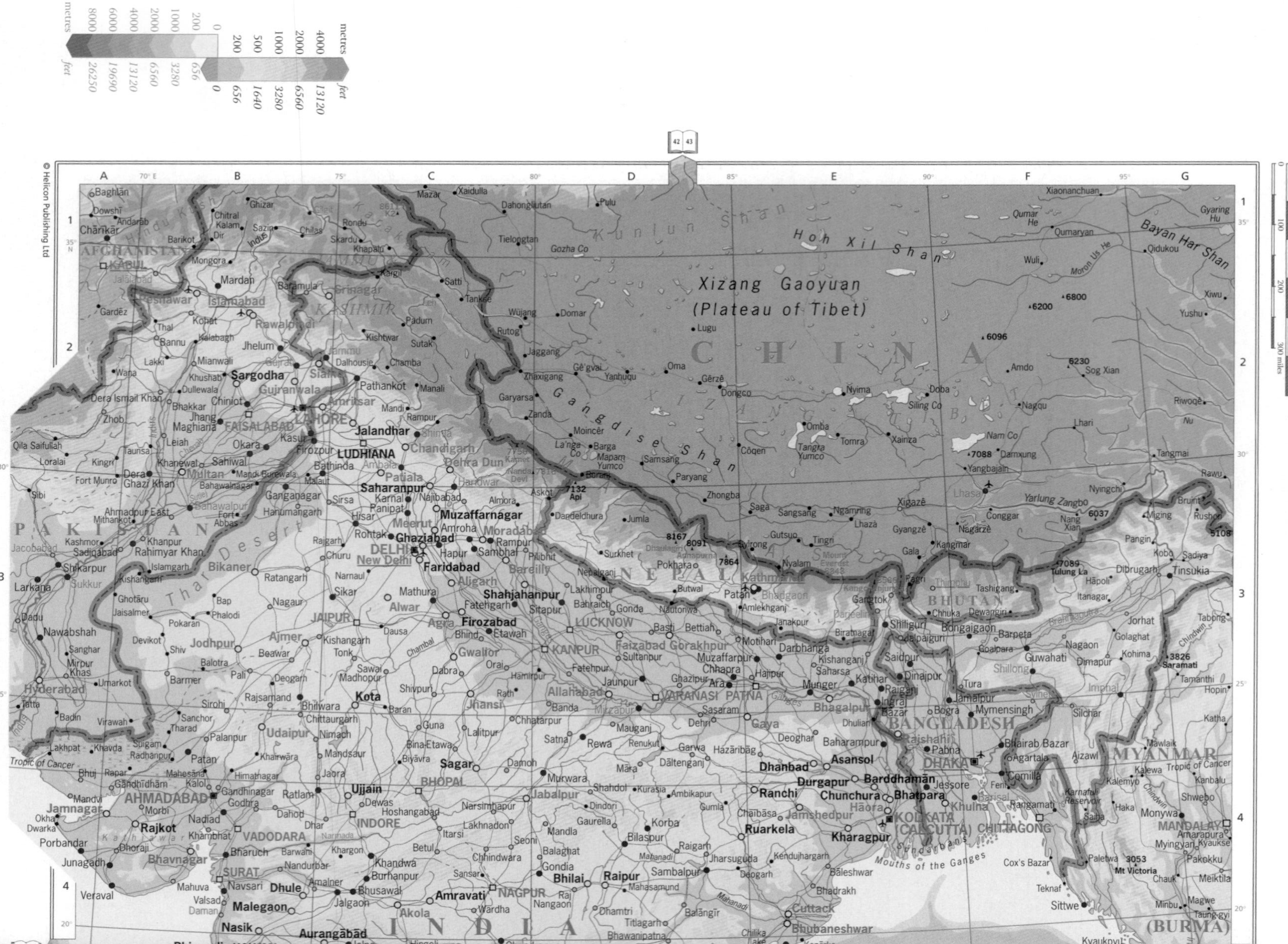

SOUTHERN ASIA

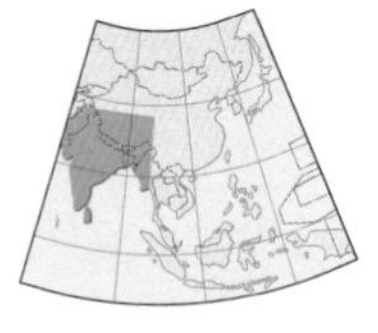

- Bangladesh
- Bhutan
- Southern China
- India
- Maldives
- Nepal
- Sri Lanka

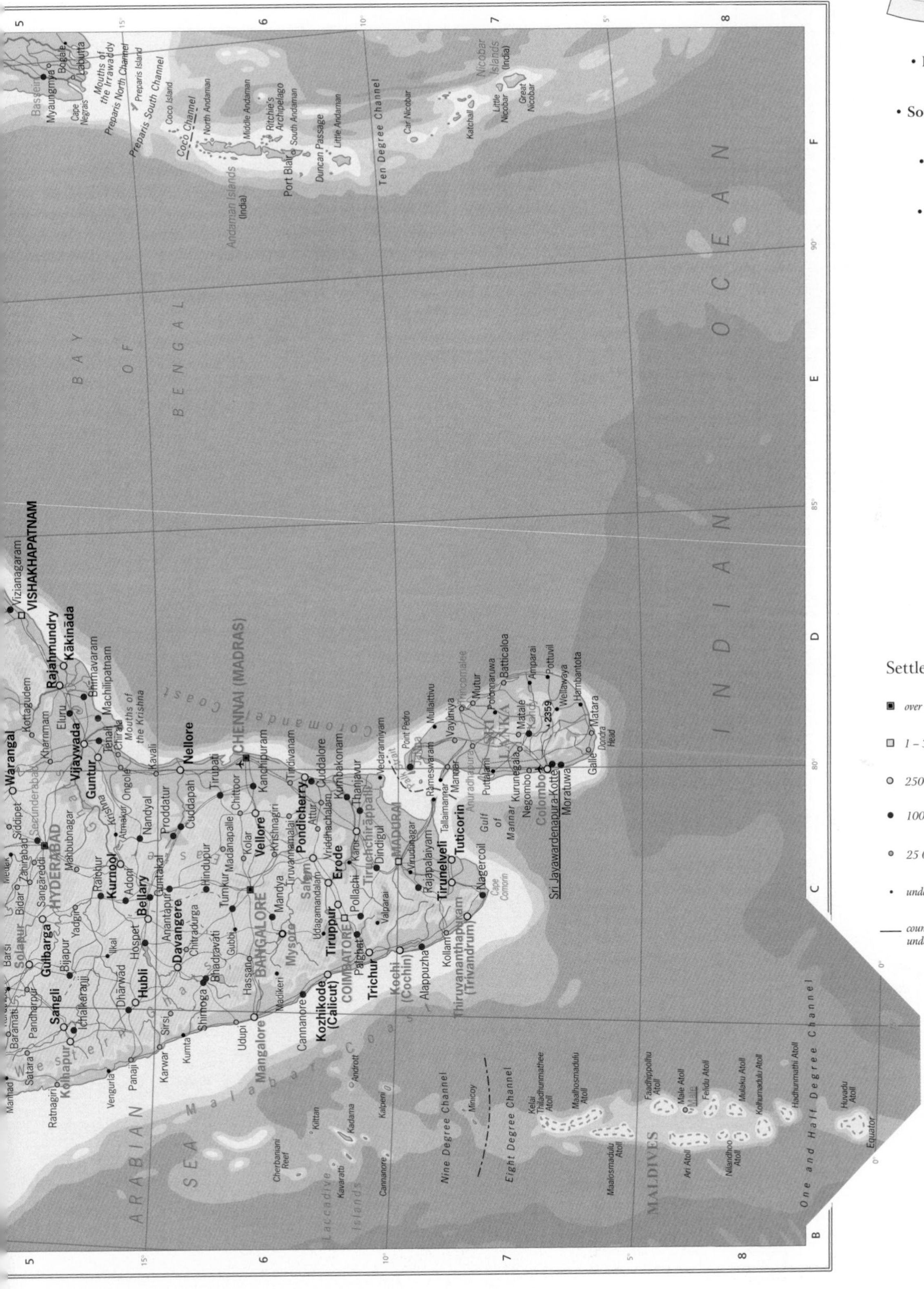

Settlements

- ▣ *over 3 million*
- □ *1 – 3 million*
- ○ *250 000 – 1 million*
- ● *100 000 – 250 000*
- ⊙ *25 000 – 100 000*
- • *under 25 000*
- ___ *country capital underline*

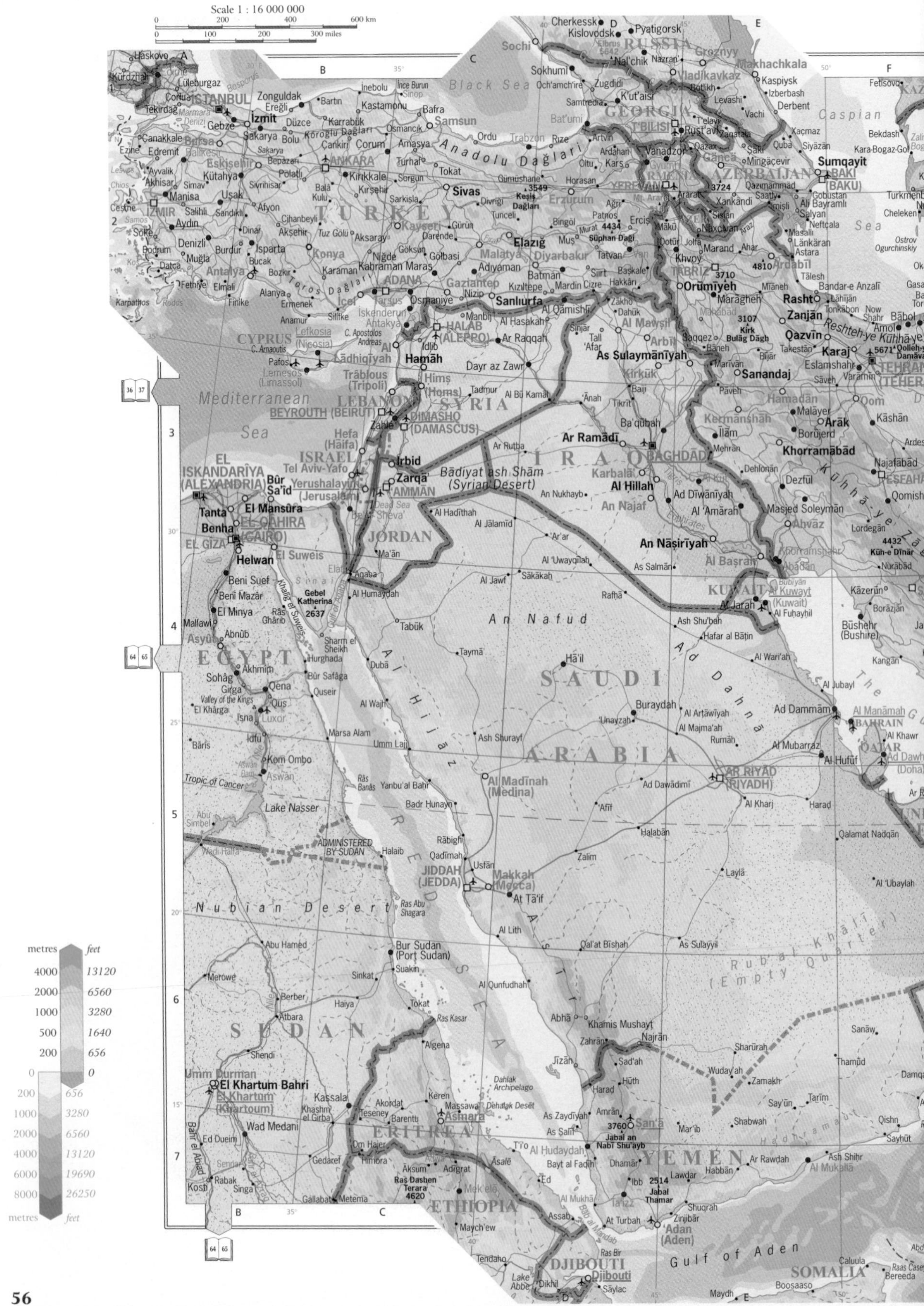

Scale 1 : 16 000 000
0 200 400 600 km
0 100 200 300 miles
Black Sea
Caspian Sea
Mediterranean Sea
Gulf of Aden
TURKEY
SYRIA
IRAQ
JORDAN
ISRAEL
LEBANON
CYPRUS
EGYPT
SUDAN
ERITREA
ETHIOPIA
DJIBOUTI
YEMEN
SAUDI ARABIA
KUWAIT
BAHRAIN
QATAR
GEORGIA
ARMENIA
AZERBAIJAN
RUSSIA
SOMALIA
Anadolu Dağlari
Toros Dağlari
An Nafud
Ad Dahnā
Al Hijāz
Bādiyat ash Shām (Syrian Desert)
Nubian Desert
Rub' al Khālī (Empty Quarter)
RED SEA
Lake Nasser
Tropic of Cancer
ADMINISTERED BY SUDAN
İSTANBUL
İzmit
ANKARA
İZMIR
ADANA
Samsun
Sivas
Kayseri
Konya
Elazığ
Şanlıurfa
Gaziantep
HALAB (ALEPPO)
Hamāh
Ar Raqqah
Dayr az Zawr
DIMASHQ (DAMASCUS)
BEYROUTH (BEIRUT)
Tel Aviv-Yafo
Yerushalayim (Jerusalem)
'AMMAN
Irbid
Zarqā'
BAGHDĀD
Ar Ramādī
Al Ḩillah
Karbalā'
An Najaf
An Nāşirīyah
Al Başrah
As Sulaymānīyah
Al Mawşil
Arbīl
Kirkūk
Al Kuwayt (Kuwait)
EL QÂHIRA (CAIRO)
EL ISKANDARÎYA (ALEXANDRIA)
Bûr Sa'îd
El Mansûra
Tanta
Benha
EL GÎZA
Helwan
El Suweis
Aswân
Luxor
Qena
Asyût
El Minya
Beni Suef
Sohâg
Gebel Katherina 2637
Tabūk
Ḩā'il
Buraydah
Al Madīnah (Medina)
Makkah (Mecca)
JIDDAH (JEDDA)
At Tā'if
AR RIYĀD (RIYADH)
Ad Dammām
Al Hufūf
Abhā
Ṣan'ā'
Adan (Aden)
Al Ḩudaydah
Al Mukallā
Bur Sudan (Port Sudan)
Kassala
El Khartum Bahri
El Khartum (Khartoum)
Umm Durman
Wad Medani
Asmara
Massawa
Ras Dashen Terara 4620
Jabal an Nabī Shu'ayb 3760
Djibouti
T'BILISI
YEREVAN
BAKI (BAKU)
Sumqayit
Tabrīz
Orūmīyeh
Rasht
Zanjān
Qazvīn
TEHRAN
Hamadān
Kermānshāh
Khorramābād
Ahvāz
Dezfūl
Arāk
Qom
Kāshān
Būshehr (Bushire)
metres
feet
4000 13120
2000 6560
1000 3280
500 1640
200 656
0 0
200 656
1000 3280
2000 6560
4000 13120
6000 19690
8000 26250

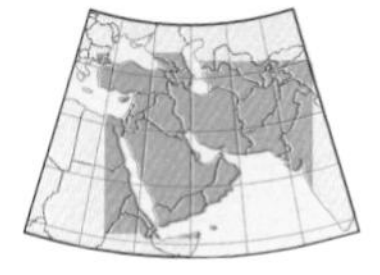

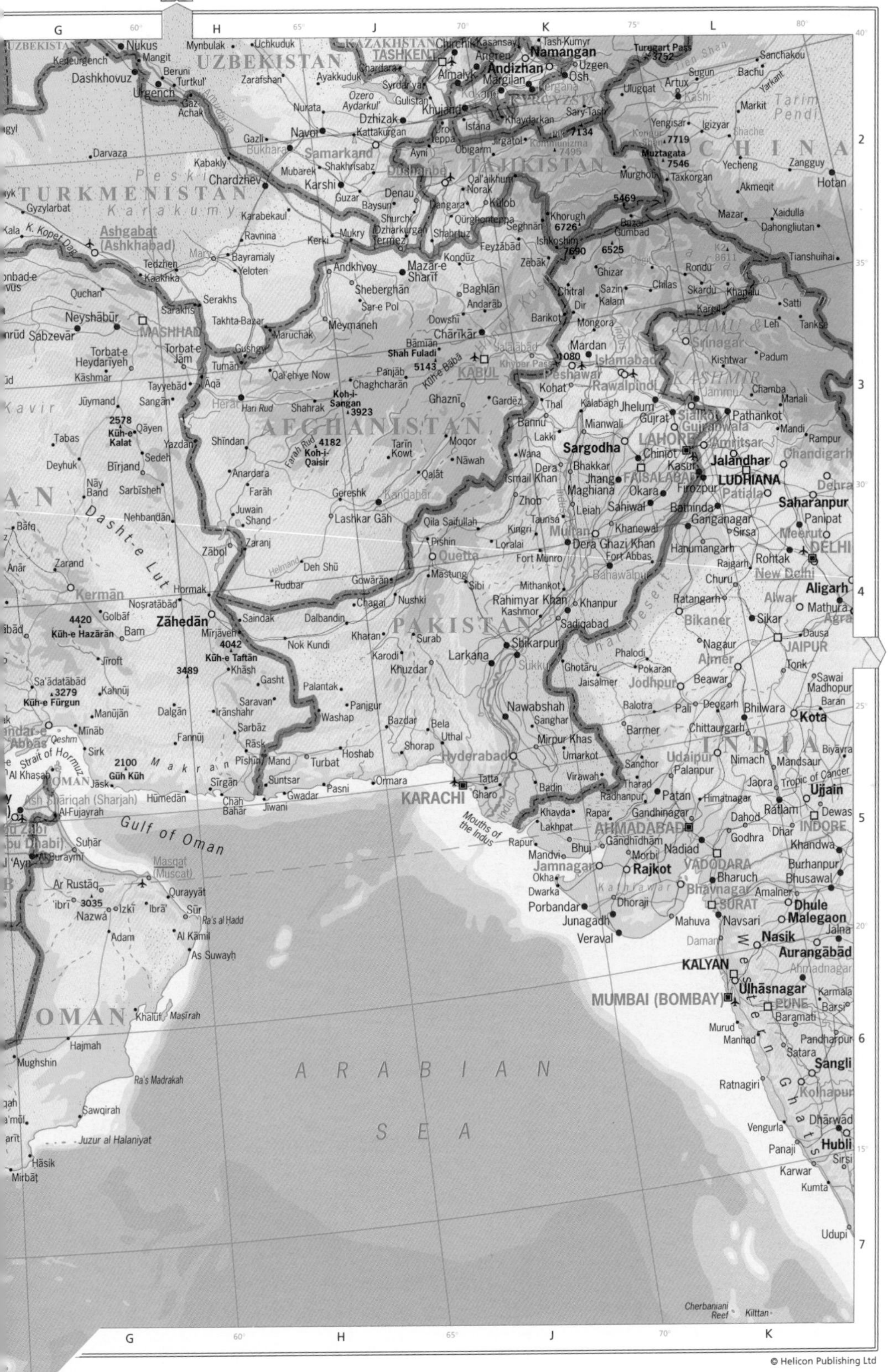

- Afghanistan
- Iran
- Iraq
- Oman
- Pakistan
- Saudi Arabia
- Syria
- Yemen

Settlements

- *over 3 million*
- *1 – 3 million*
- *250 000 – 1 million*
- *100 000 – 250 000*
- *25 000 – 100 000*
- *under 25 000*
- *country capital underline*

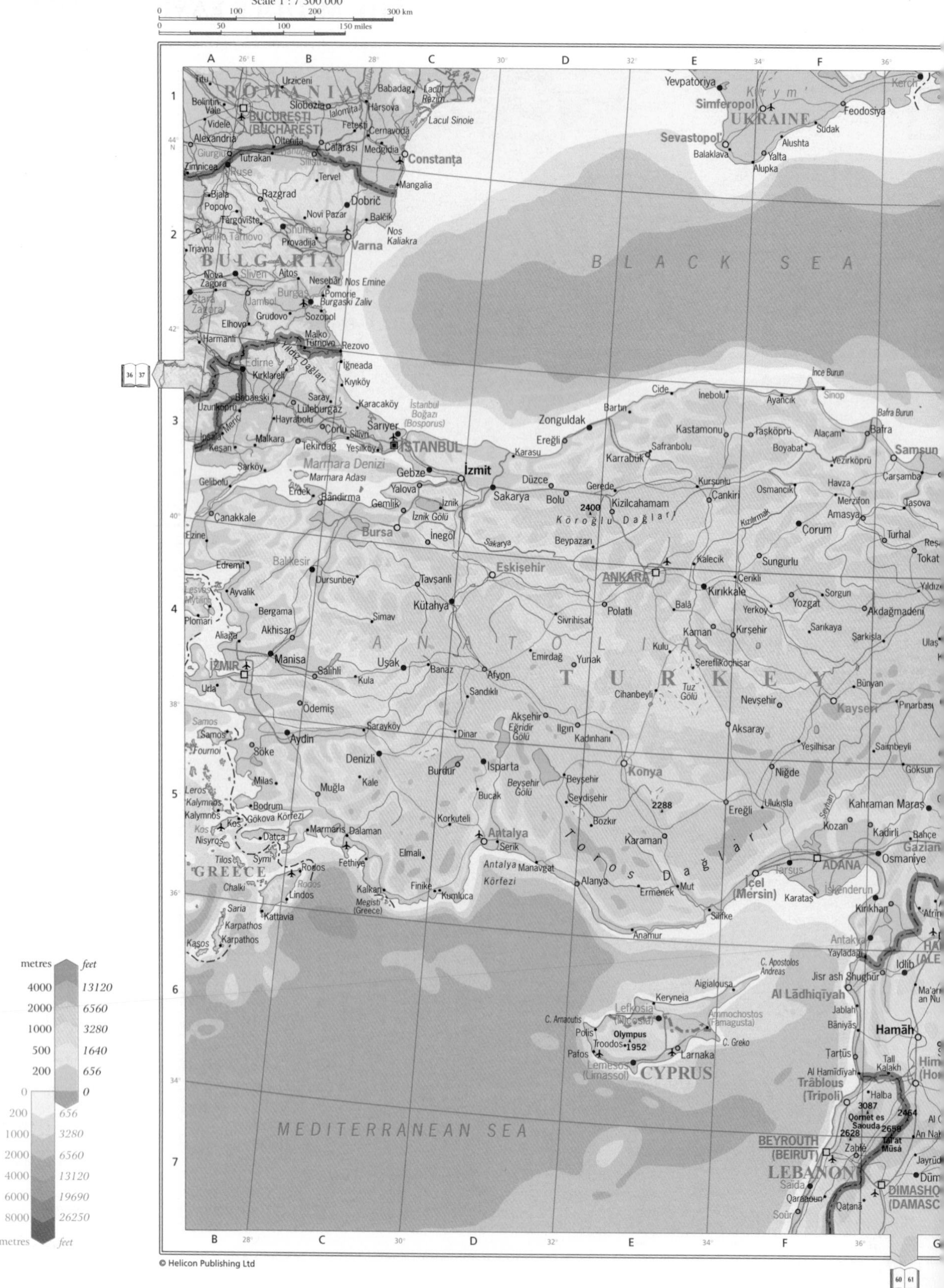

Scale 1 : 7 300 000
BLACK SEA
MEDITERRANEAN SEA
ROMANIA
BULGARIA
UKRAINE
TURKEY
ANATOLIA
GREECE
CYPRUS
LEBANON
BUCUREŞTI (BUCHAREST)
İSTANBUL
ANKARA
İZMIR
ADANA
BEYROUTH (BEIRUT)
metres
feet
© Helicon Publishing Ltd

TURKEY

- Armenia
- Azerbaijan
- Cyprus
- Georgia
- Turkey

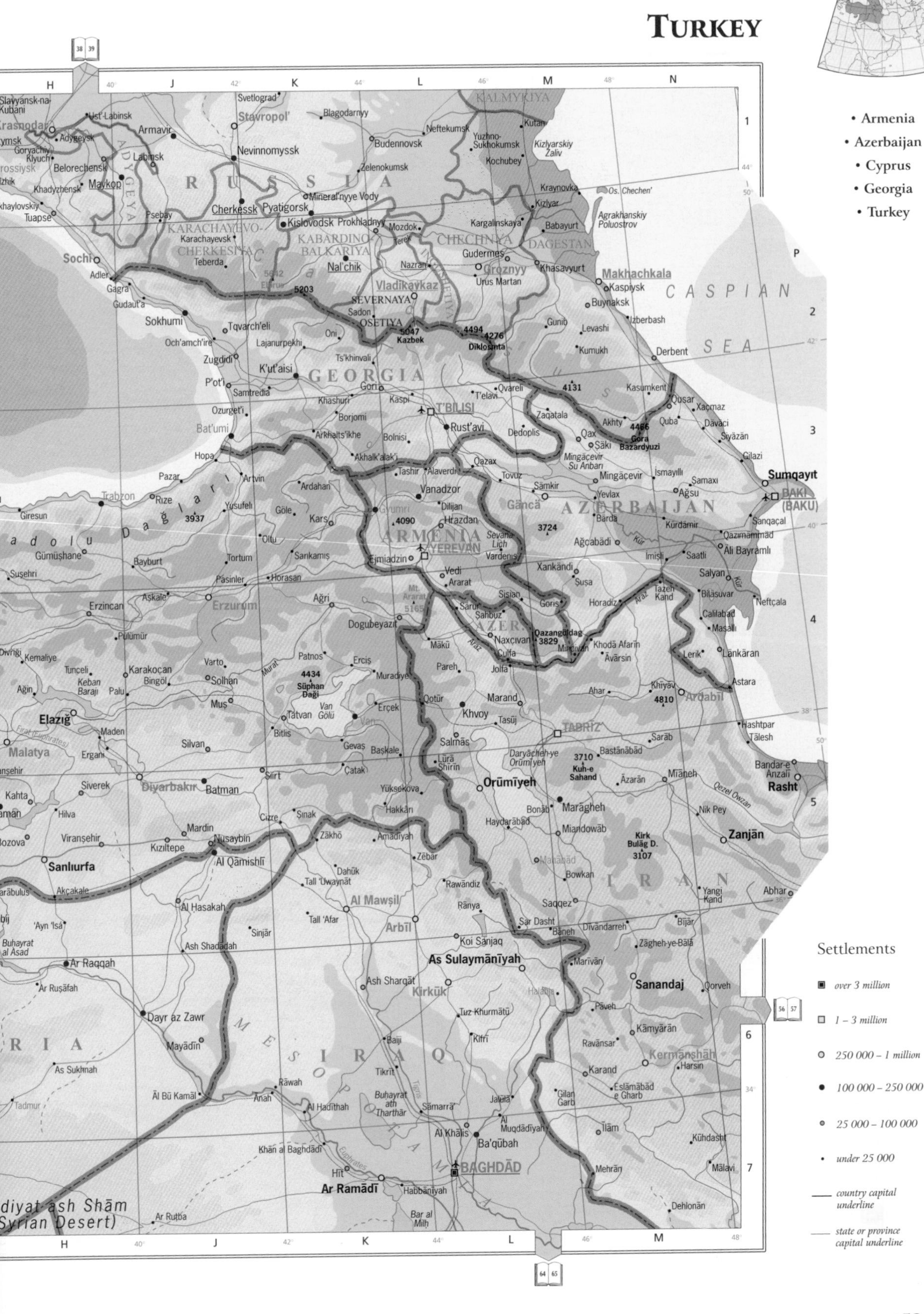

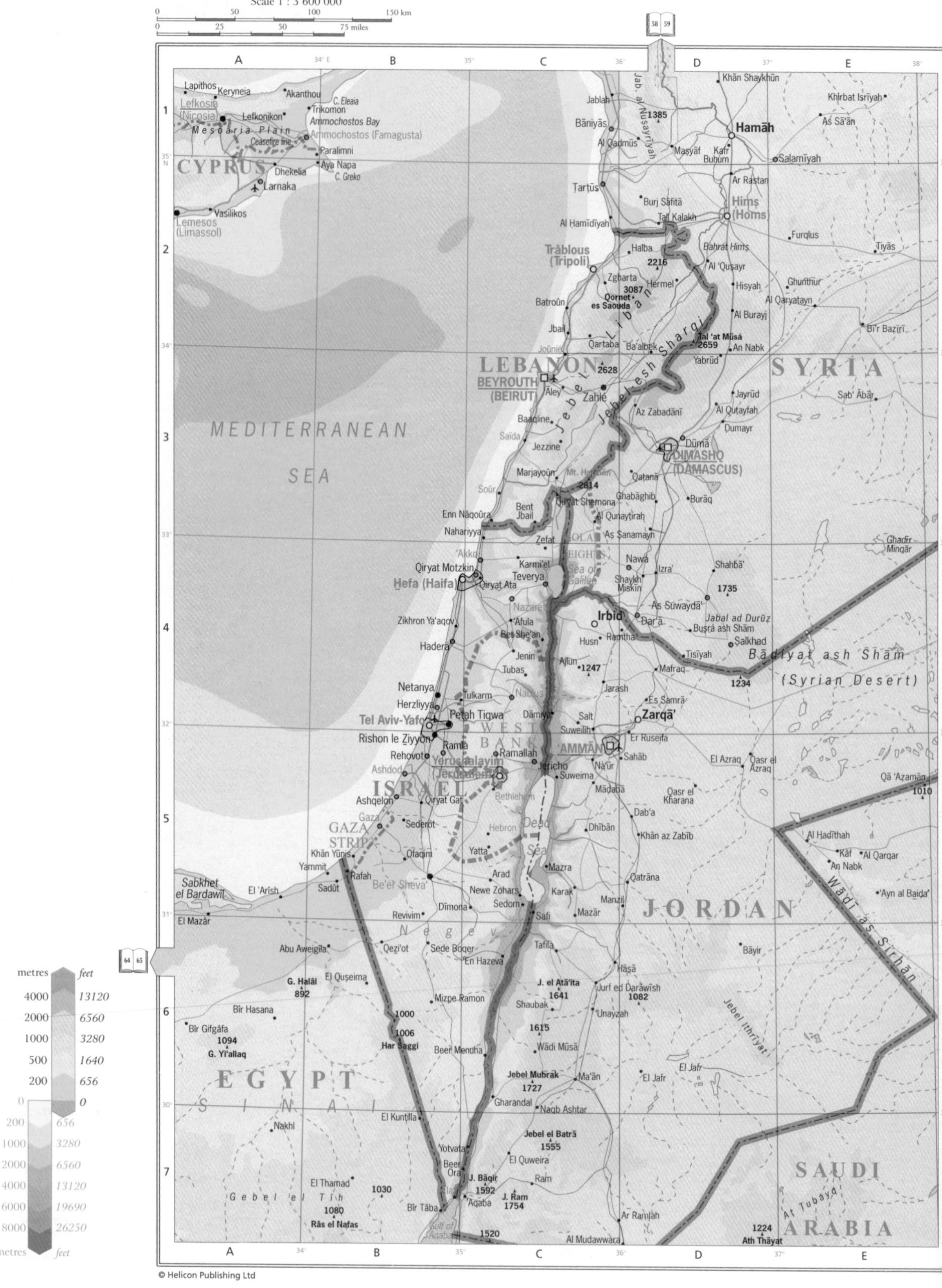

Scale 1 : 3 600 000
0 50 100 150 km
0 25 50 75 miles
58 59
A B C D E
1 2 3 4 5 6 7
34° E 35° 36° 37° 38°
CYPRUS
Lapithos
Keryneia
Akanthou
C. Eleaia
Lefkosia (Nicosia)
Lefkonikon
Trikomon
Ammochostos Bay
Mesaoria Plain
Ceasefire line
Ammochostos (Famagusta)
Paralimni
Dhekelia
Aya Napa
C. Greko
Larnaka
Vasilikos
Lemesos (Limassol)
MEDITERRANEAN SEA
LEBANON
SYRIA
ISRAEL
JORDAN
EGYPT
SINAI
SAUDI ARABIA
GAZA STRIP
WEST BANK
GOLAN HEIGHTS
Jablah
Jab. an Nuşayrīyah
1385
Bāniyās
Al Qadmūs
Maşyāf
Kafr Buhum
Hamāh
Khān Shaykhūn
Khirbat Isrīyah
As Sa'ān
Salamīyah
Ar Rastan
Ţarţūs
Burj Şāfitā
Hims (Homs)
Al Hamīdīyah
Tall Kalakh
Furqlus
Tiyās
Trâblous (Tripoli)
Halba
2216
Bahrat Hims
Al Quşayr
Zgharta
Hermel
Hisyah
Ghunthur
3087
Qornet es Saouda
Al Qaryatayn
Batroûn
Al Burayj
Bi'r Bazīrī
Jbail
Jebel Liban
Jebel esh Sharqi
Tal 'at Mūsá
2659
An Nabk
Joûnié
Qartaba
Ba'albek
Yabrūd
2628
BEYROUTH (BEIRUT)
Aley
Zahlé
Jayrūd
Sab' Ābār
Az Zabadānī
Al Quţayfah
Baaqline
Dumayr
Saida
Jezzine
Dūmā
DIMASHQ (DAMASCUS)
Marjayoûn
Mt. Hermon
2814
Qaţanā
Soûr
Ghabāghib
Bent Jbail
Qiryat Shemona
Buraq
Enn Nâqoûra
Al Qunayţirah
Nahariyya
Aş Şanamayn
Ghadir Minqār
Zefat
'Akko
Karmi'el
Nawá
Izra'
Shahbā'
Qiryat Motzkin
Teverya
Sea of Galilee
Shaykh Miskīn
Hefa (Haifa)
Qiryat Ata
1735
Nazareth
As Suwaydā'
Irbid
Jabal ad Durūz
Zikhron Ya'aqov
'Afula
Dar'ā
Buşrá ash Shām
Bet She'an
Husn
Ramtha
Şalkhad
Hadera
Jenin
Tisīyah
Bādiyat ash Shām
(Syrian Desert)
Ajlūn
1247
Mafraq
Tubas
1234
Netanya
Jarash
Tulkarm
Nablus
Es Samrā
Herzliyya
Tel Aviv-Yafo
Petah Tiqwa
Dāmiya
Salt
Zarqā'
Suweilih
Er Ruseifa
Rishon le Ziyyon
Ramla
'AMMĀN
Rehovot
Ramallah
Sahāb
El Azraq
Qasr el Azraq
Ashdod
Yerushalayim (Jerusalem)
Jericho
Nā'ūr
Suweima
Qā 'Azamān
Ashqelon
Qiryat Gat
Bethlehem
Mādabā
Qasr el Kharana
1010
Gaza
Sederot
Hebron
Dead Sea
Dab'a
Dhībān
Khān az Zabīb
Al Hadīthah
Khān Yūnis
Yammit
Ofaqim
Yatta
Kāf
Al Qarqar
An Nabk
Rafah
Be'er Sheva'
Arad
Mazra
Qaţrāna
Wādi as Sirhān
Sabkhet el Bardawīl
El 'Arīsh
Sadût
Newe Zohar
Karak
Manzil
'Ayn al Baida'
Dīmona
Sedom
Revivim
Safi
Mazār
El Mazâr
Negev
Abu Aweigila
Qezi'ot
Sede Boqer
Tafila
Bāyir
En Hazeva
Hāsā
El Quseima
G. Halāl
892
J. el Atā'ita
1641
Jurf ed Darāwīsh
1082
Mizpe Ramon
Shaubak
Bir Hasana
1000
'Unayzah
Jebel Ithrīyat
Bir Gifgâfa
1006
1615
1094
G. Yi'allaq
Har Saggi
Beer Menuha
Wādi Mūsā
El Jafr
Jebel Mubrak
1727
Ma'ān
El Jafr
Gharandal
Naqb Ashtar
El Kuntilla
Nakhl
Jebel el Batrā
1555
Yotvata
El Quweira
Beer Ora
J. Bāqir
1592
Ram
El Thamad
1030
J. Ram
1754
Gebel el Tih
1080
Râs el Nafas
Bîr Tâba
Elat
'Aqaba
Gulf of Aqaba
At Tubayq
1520
Ar Ramlah
Al Mudawwara
1224
Ath Thāyat
64 65
metres feet
4000 13120
2000 6560
1000 3280
500 1640
200 656
0 0
200 656
1000 3280
2000 6560
4000 13120
6000 19690
8000 26250
metres feet

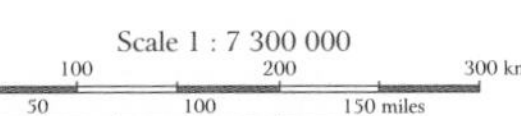

ISRAEL AND THE GULF STATES

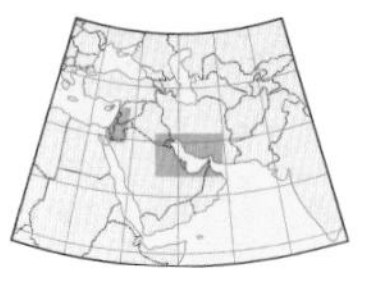

- Bahrain
- Israel
- Jordan
- Kuwait
- Lebanon
- Qatar
- United Arab Emirates

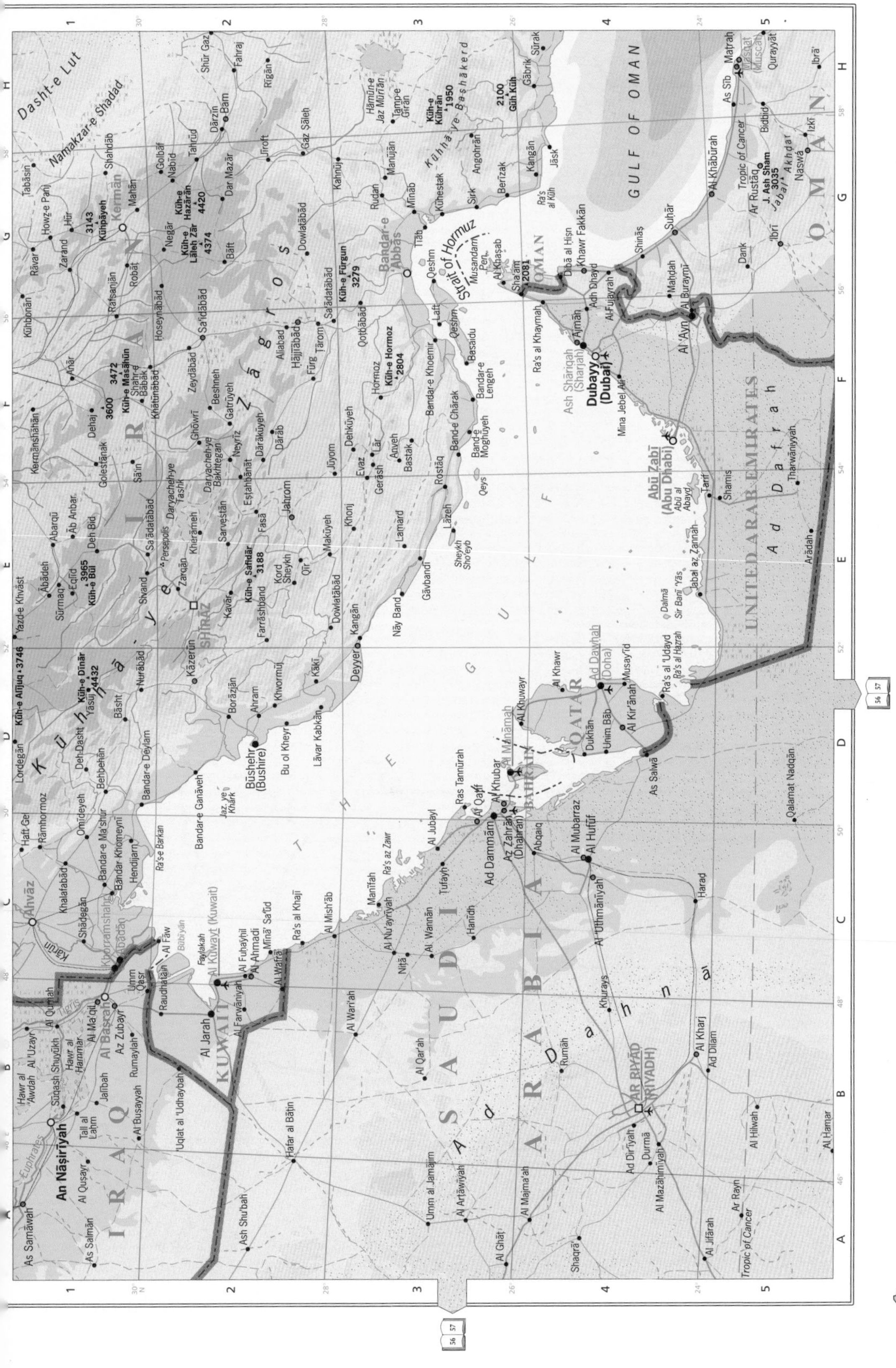

Settlements

- *over 3 million*
- *1 – 3 million*
- *250 000 – 1 million*
- *100 000 – 250 000*
- *25 000 – 100 000*
- *under 25 000*
- *country capital underline*
- *urban area*

metres	feet
4000	13120
2000	6560
1000	3280
500	1640
200	656
0	0
200	656
1000	3280
2000	6560
4000	13120
6000	19690
8000	26250
metres	feet

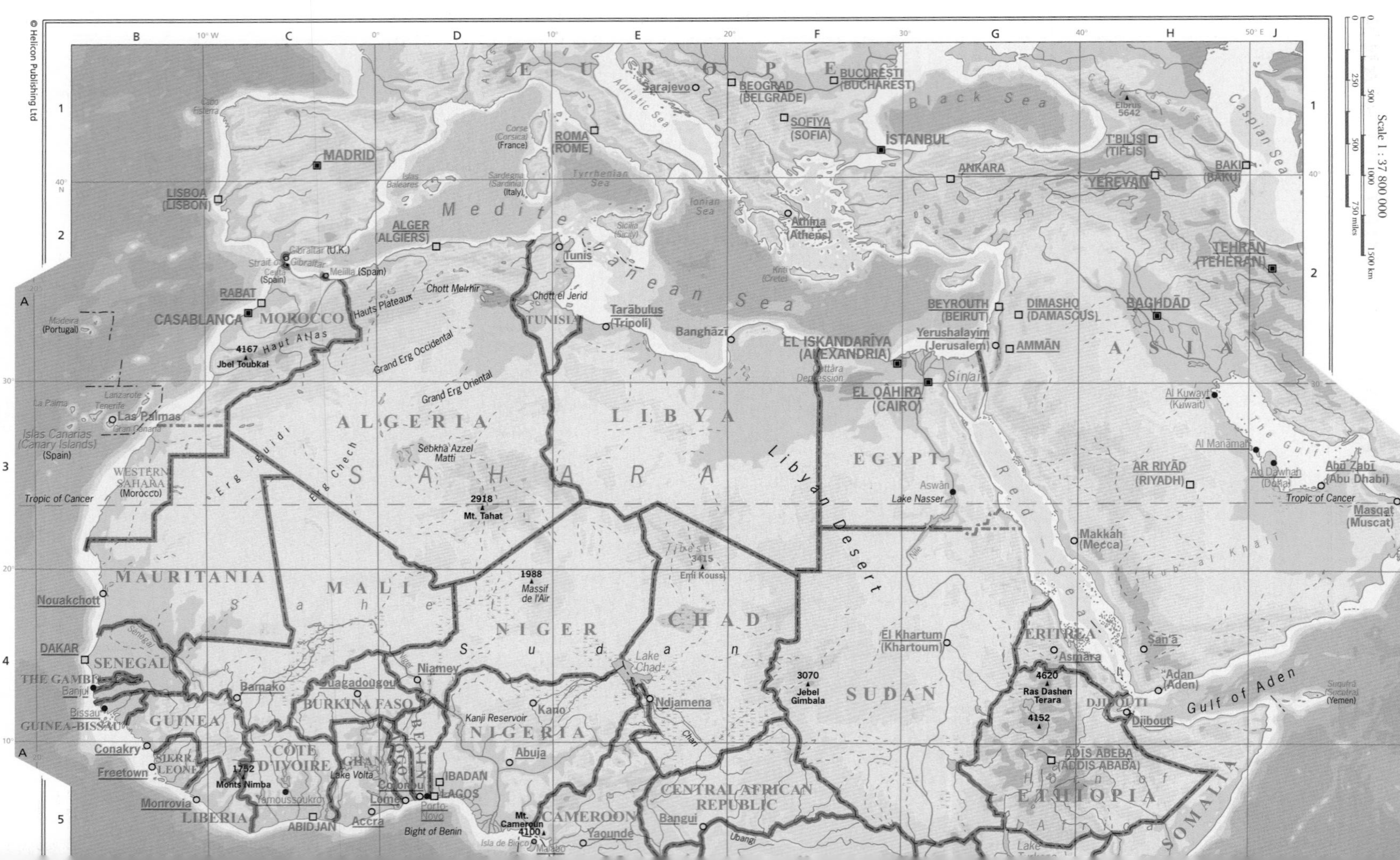

Africa

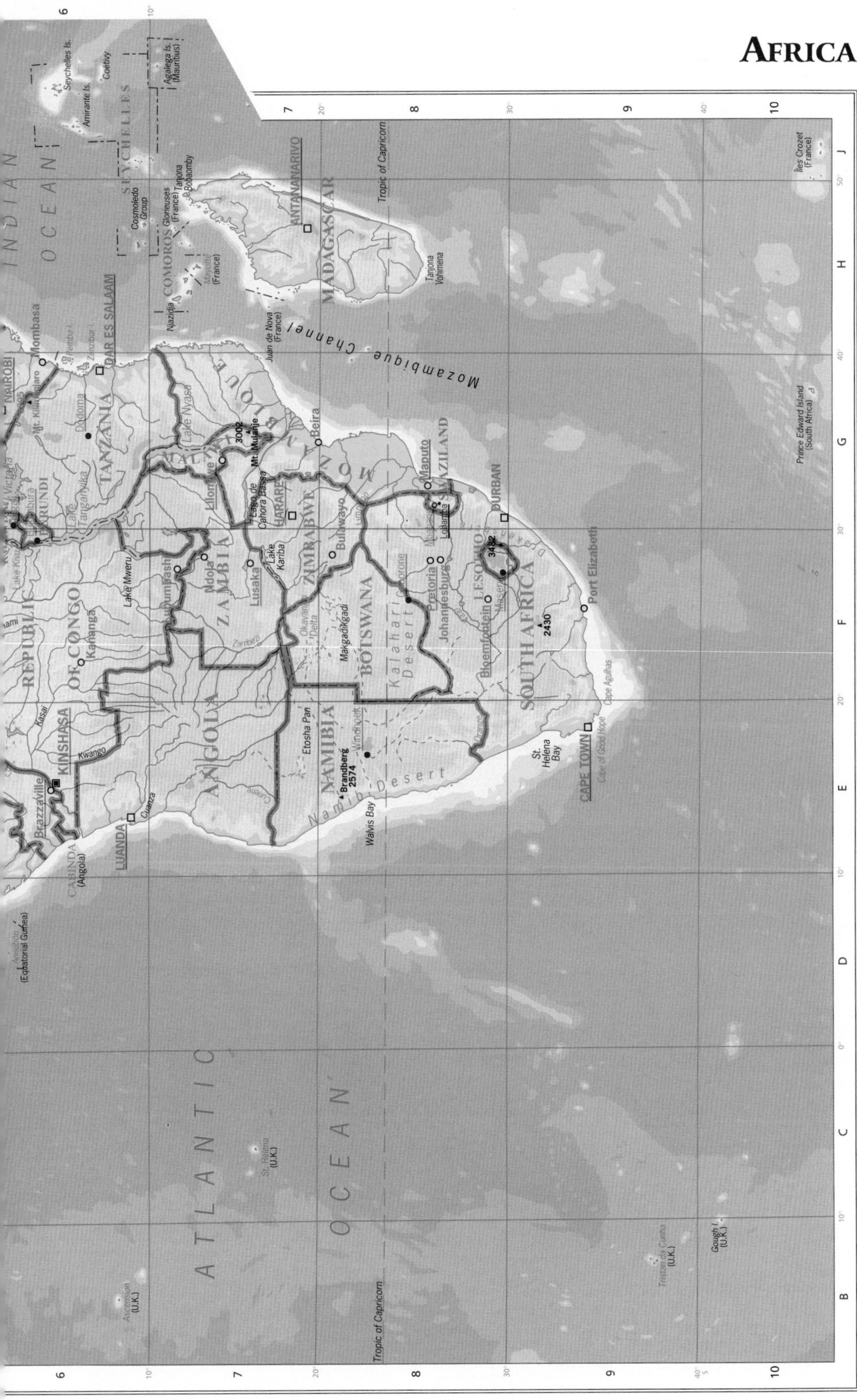

INDIAN OCEAN
ATLANTIC OCEAN
Mozambique Channel
Tropic of Capricorn
MADAGASCAR
ANTANANARIVO
SEYCHELLES
Seychelles Is.
Amirante Is.
Coëtivy
Agalega Is. (Mauritius)
Cosmoledo Group
Glorieuses (France)
Tanjona Bobaomby
COMOROS
Mayotte (France)
Njazidja
Juan de Nova (France)
Tanjona Vohimena
Îles Crozet (France)
Prince Edward Island (South Africa)
NAIROBI
Mombasa
Mt. Kilimanjaro
TANZANIA
Dodoma
DAR ES SALAAM
Zanzibar I.
Pemba I.
Lake Tanganyika
BURUNDI
Lake Nyasa
MALAWI
Lilongwe
Mt. Mulanje
3002
MOZAMBIQUE
Beira
Maputo
SWAZILAND
Mbabane
Lobamba
DURBAN
LESOTHO
Maseru
3482
SOUTH AFRICA
Port Elizabeth
2430
Bloemfontein
Johannesburg
Pretoria
CAPE TOWN
Cape of Good Hope
Cape Agulhas
St. Helena Bay
Orange
ZIMBABWE
HARARE
Bulawayo
Lago de Cahora Bassa
Lake Kariba
ZAMBIA
Lusaka
Ndola
Lake Mweru
Zambezi
BOTSWANA
Gaborone
Makgadikgadi
Kalahari Desert
Okavango Delta
NAMIBIA
Windhoek
Etosha Pan
Brandberg
2574
Namib Desert
Walvis Bay
ANGOLA
LUANDA
Cuanza
Cunene
DEMOCRATIC REPUBLIC OF CONGO
KINSHASA
Brazzaville
Kananga
Kasai
Kwango
CABINDA (Angola)
Annobón (Equatorial Guinea)
Ascension (U.K.)
St. Helena (U.K.)
Tristan da Cunha (U.K.)
Gough I. (U.K.)

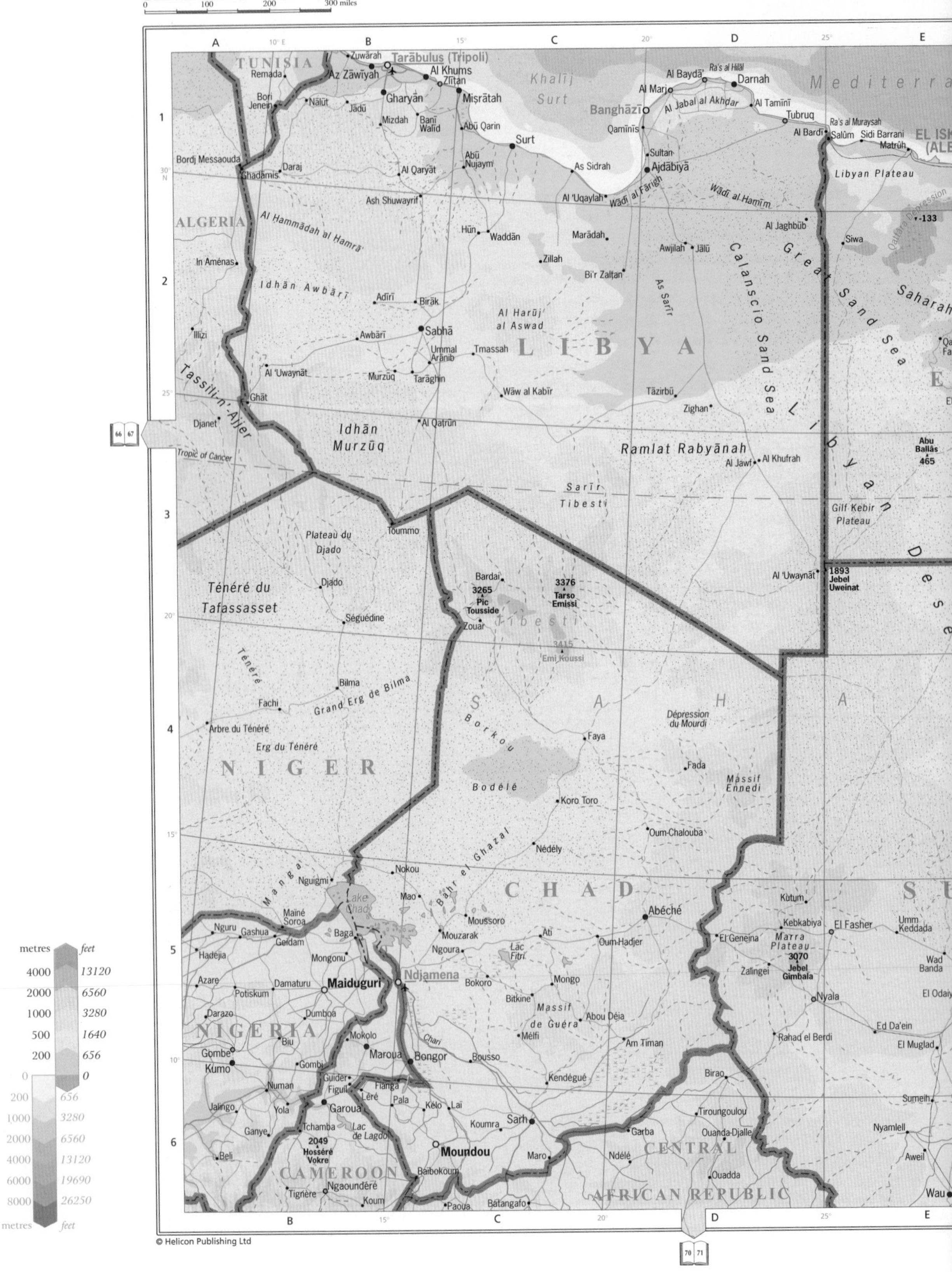

Scale 1 : 14 600 000
0 200 400 600 km
0 100 200 300 miles
A B C D E
1 2 3 4 5 6
TUNISIA
Zuwārah
Tarābulus (Tripoli)
Az Zāwīyah
Al Khums
Zlītan
Remada
Bori Jenein
Nālūt
Jādū
Gharyān
Mişrātah
Khalīj Surt
Al Bayḍā'
Ra's al Hilāl
Darnah
Mediterra
Al Marj
Banghāzī
Al Jabal al Akhḑar
Al Tamīmī
Tubruq
Ra's al Muraysah
Mizdah
Banī Walīd
Abū Qarin
Qamīnīs
Al Bardī
Salûm
Sidi Barrani
Matrûh
EL ISK
(ALE
Surt
Sultan
Bordj Messaouda
Daraj
Ghadāmis
Al Qaryāt
Abū Nujaym
As Sidrah
Ajdābiyā
Libyan Plateau
Ash Shuwayrif
Al 'Uqaylah
Wādī al Fārigh
Wādī al Hamīm
Qattara Depression
-133
ALGERIA
Al Hammādah al Hamrā'
Hūn
Waddān
Marādah
Awjilah
Jālū
Al Jaghbūb
Siwa
In Aménas
Zillah
Bi'r Zaltan
Great Sand Sea
Saharah
Idhān Awbārī
Adīrī
Birāk
As Sarīr
Calanscio Sand Sea
Al Harūj al Aswad
Sabhā
Illizi
Awbārī
LIBYA
Ummal Arānib
Tmassah
Al 'Uwaynāt
Murzūq
Tarāghin
Tassili-n'-Ajjer
Ghāt
Wāw al Kabīr
Tāzirbū
Zighan
Libyan
Djanet
Al Qaṭrūn
Idhān Murzūq
Ramlat Rabyānah
Abu Ballâs 465
Tropic of Cancer
Al Jawf
Al Khufrah
Sarīr Tibesti
Gilf Kebir Plateau
Plateau du Djado
Toummo
Desert
Al 'Uwaynāt
1893 Jebel Uweinat
Bardai
3265 Pic Tousside
3376 Tarso Emissi
Ténéré du Tafassasset
Djado
Séguédine
Zouar
Tibesti
3415 Emi Koussi
Ténéré
Bilma
Grand Erg de Bilma
Fachi
SAHARA
Dépression du Mourdi
Arbre du Ténéré
Erg du Ténéré
Borkou
Faya
NIGER
Fada
Massif Ennedi
Bodélé
Koro Toro
Oum-Chalouba
Bahr el Ghazal
Nédély
Manga
Nokou
Nguigmi
Lake Chad
Mao
CHAD
Kutum
SU
Mainé Soroa
Abéché
Moussoro
Kebkabiya
El Fasher
Umm Keddada
Nguru
Gashua
Geidam
Baga
Mouzarak
Ati
Oum-Hadjer
El Geneina
Marra Plateau
3070 Jebel Gimbala
Zalingei
Hadejia
Mongonu
Ngoura
Lac Fitri
Wad Banda
Azare
Potiskum
Damaturu
Maiduguri
Ndjamena
Bokoro
Mongo
Nyala
El Odaiya
Bitkine
Massif de Guéra
Abou Déia
Darazo
Dumboa
NIGERIA
Biu
Mokolo
Mélfi
Am Timan
Rahad el Berdi
Ed Da'ein
El Muglad
Gombe
Kumo
Gombi
Maroua
Bongor
Chari
Bousso
Numan
Guider
Figuil
Fianga
Léré
Pala
Kendégué
Birao
Jalingo
Yola
Garoua
Kélo
Laï
Tiroungoulou
Sumeih
Ganye
Tchamba
Lac de Lagdo
Koumra
Sarh
Garba
Ouanda-Djallé
Nyamlell
2049 Hosséré Vokre
Moundou
Maro
Ndélé
CENTRAL
Aweil
Beli
CAMEROON
Baïbokoum
Ouadda
Tignère
Ngaoundéré
Koum
Paoua
Batangafo
AFRICAN REPUBLIC
Wau
10° E 15° 20° 25°
30° N 25° 20° 15° 10°
66 67
70 71
metres feet
4000 13120
2000 6560
1000 3280
500 1640
200 656
0 0
200 656
1000 3280
2000 6560
4000 13120
6000 19690
8000 26250
metres feet

NORTHEAST AFRICA

- Chad
- Egypt
- Eritrea
- Libya
- Sudan

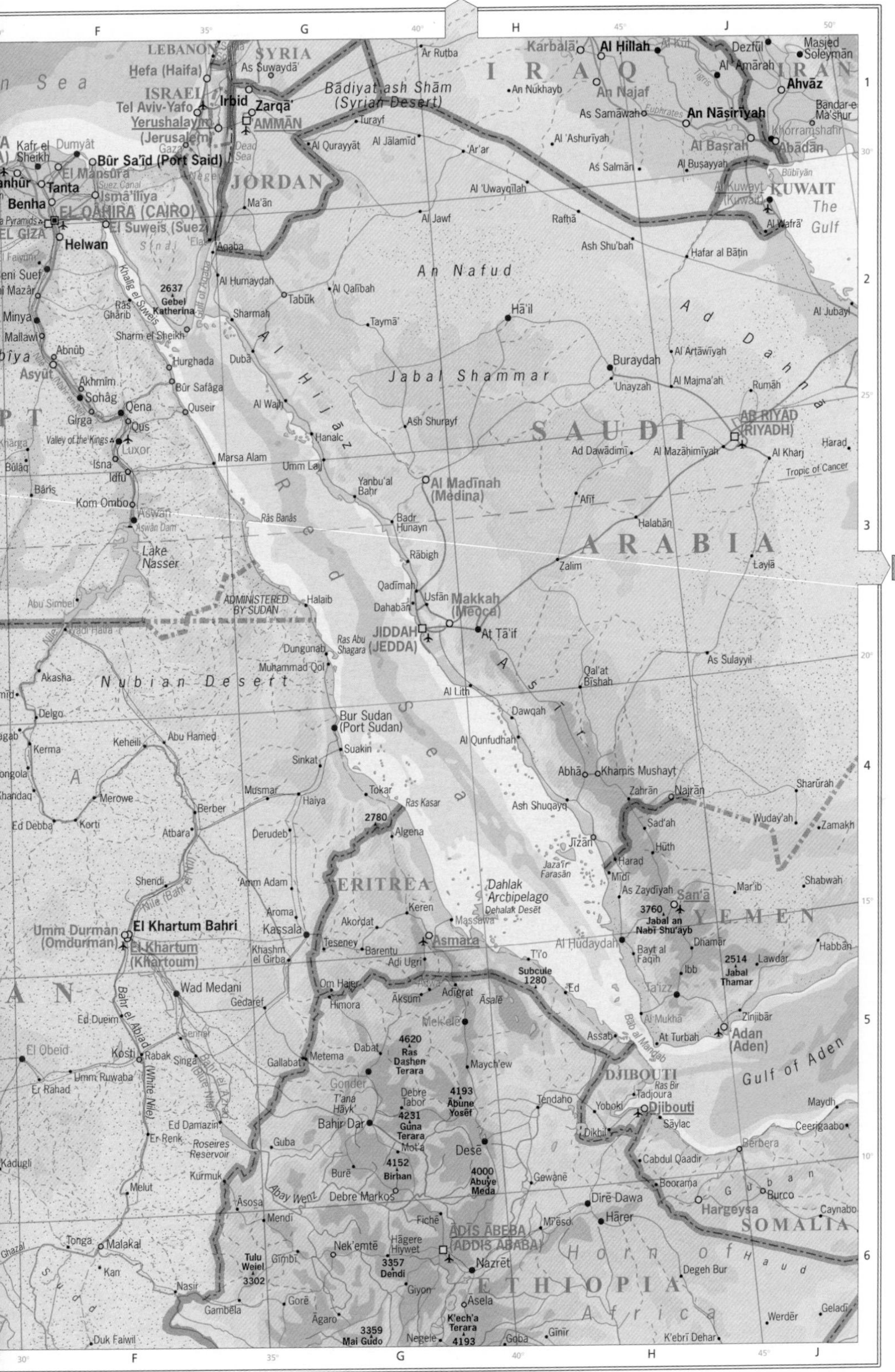

Settlements

- over 3 million
- 1 – 3 million
- 250 000 – 1 million
- 100 000 – 250 000
- 25 000 – 100 000
- under 25 000
- country capital underline

Scale 1 : 14 600 000

0 200 400 600 km

0 100 200 300 miles

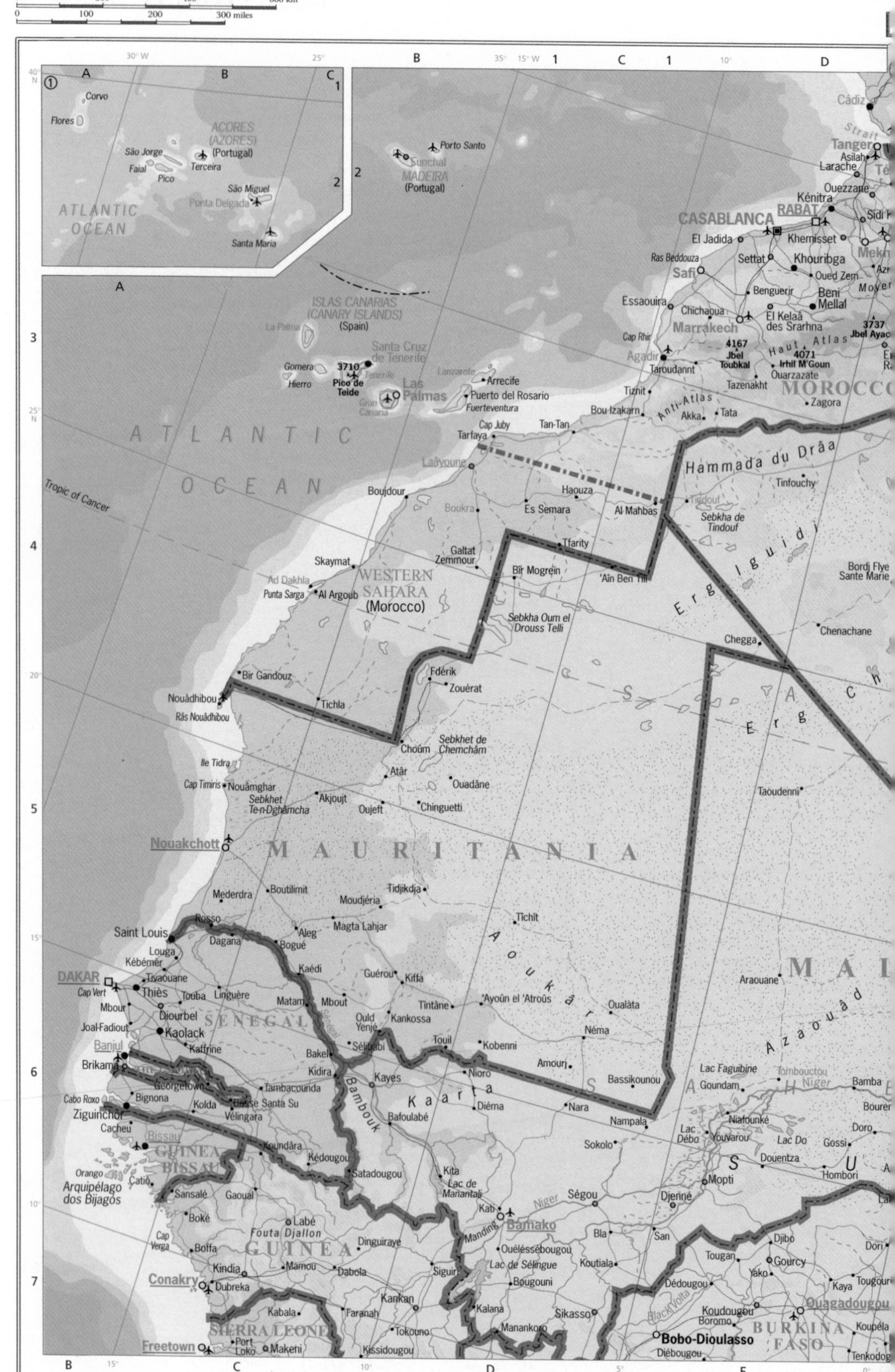

metres	feet
4000	13120
2000	6560
1000	3280
500	1640
200	656
0	0
200	656
1000	3280
2000	6560
4000	13120
6000	19690
8000	26250
metres	feet

NORTHWEST AFRICA

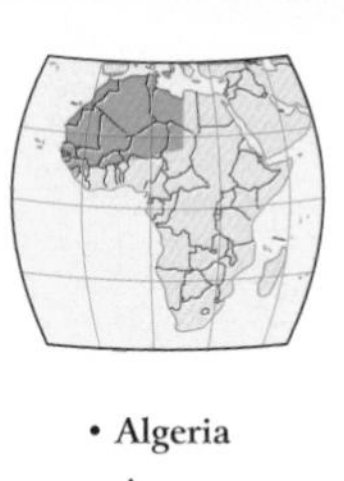

- Algeria
- Azores
- Canary Islands
- Madeira
- Mali
- Mauritania
- Morocco
- Niger
- Tunisia

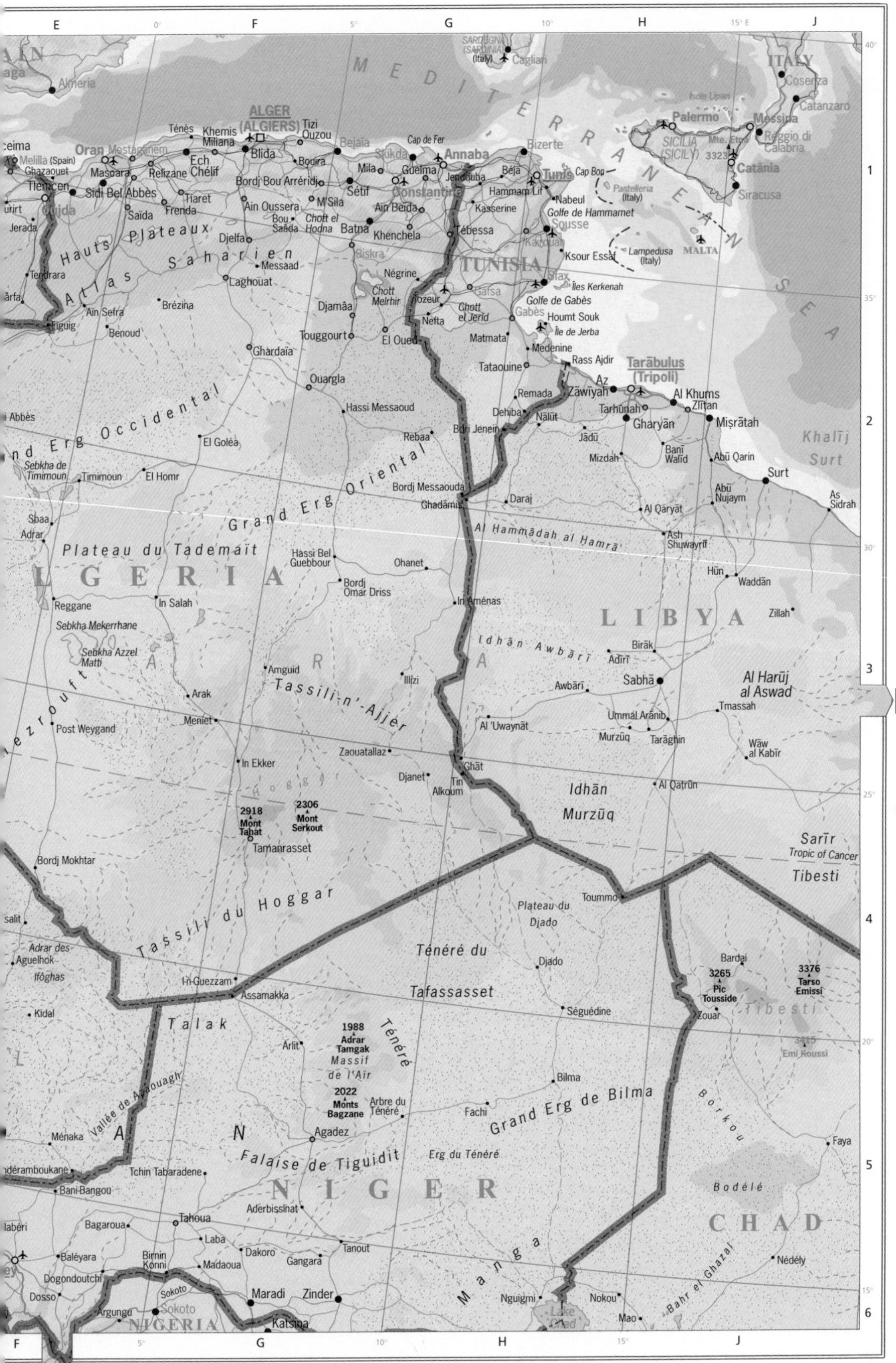

64 65

68 69

Settlements

- ▣ *over 3 million*
- □ *1 – 3 million*
- ○ *250 000 – 1 million*
- ● *100 000 – 250 000*
- • *25 000 – 100 000*
- · *under 25 000*
- —— *country capital underline*
- —— *state or province capital underline*

Scale 1 : 14 600 000

0 200 400 600 km

0 100 200 300 miles

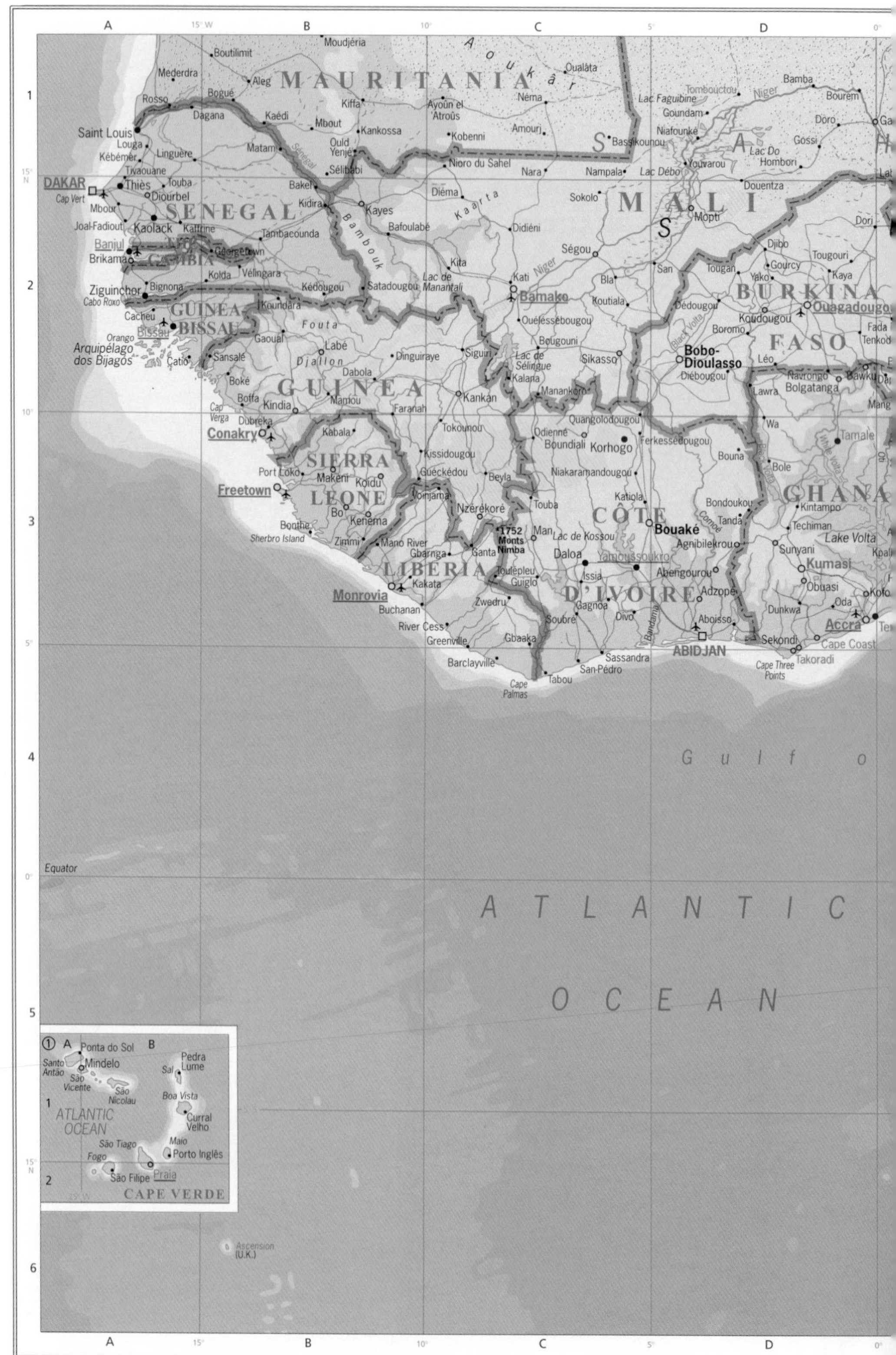

metres | feet
4000 | 13120
2000 | 6560
1000 | 3280
500 | 1640
200 | 656
0 | 0
200 | 656
1000 | 3280
2000 | 6560
4000 | 13120
6000 | 19690
8000 | 26250
metres | feet

66 67

WEST AFRICA

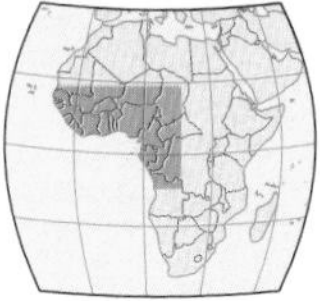

- Benin
- Burkina Faso
- Cameroon
- Cape Verde
- Congo
- Côte d'Ivoire
- Equatorial Guinea
- Gabon
- The Gambia
- Ghana
- Guinea
- Guinea-Bissau
- Liberia
- Nigeria
- São Tomé & Príncipe
- Senegal
- Sierra Leone
- Togo

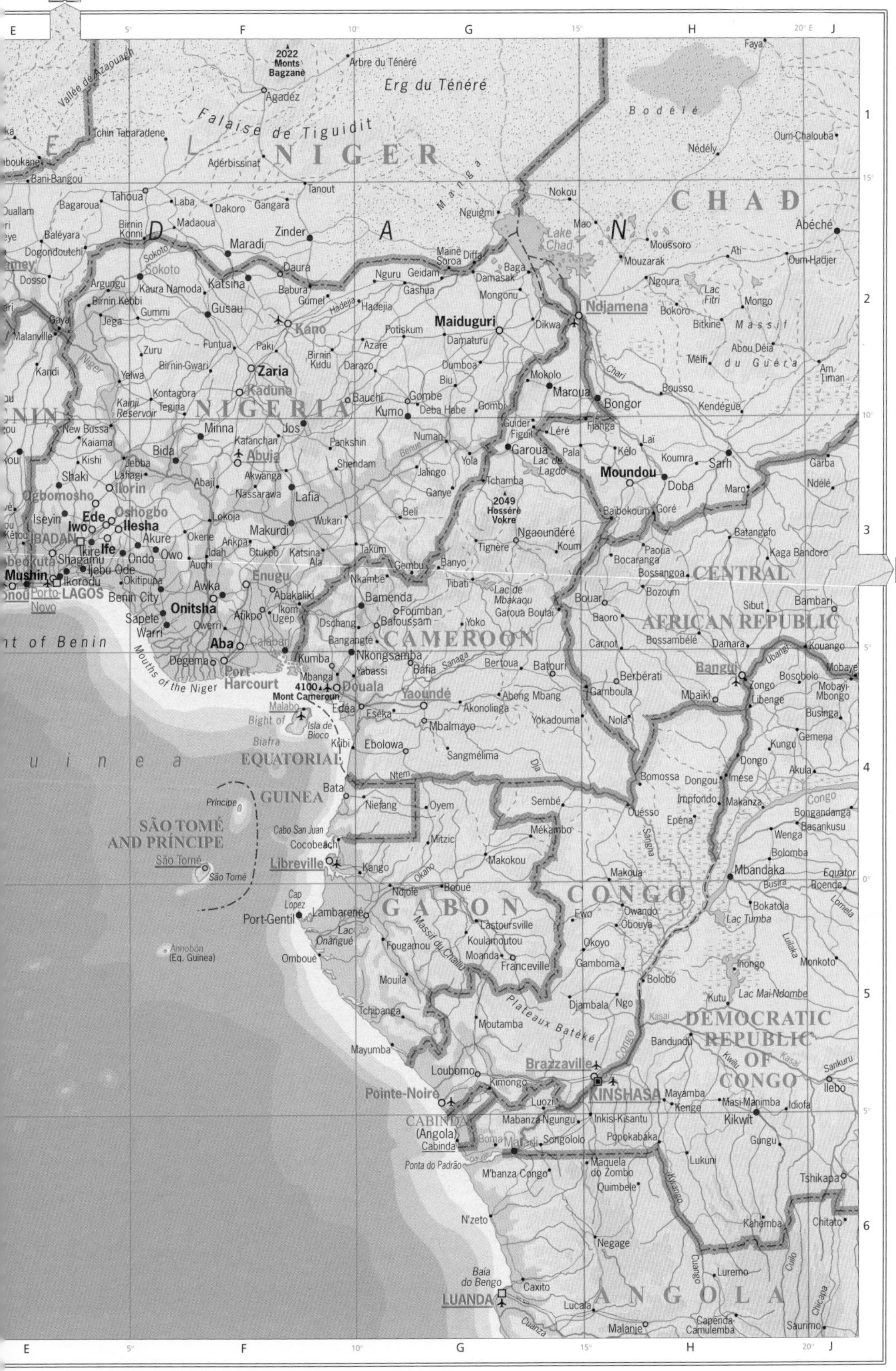

70 71

Settlements

- *over 3 million*
- *1 – 3 million*
- *250 000 – 1 million*
- *100 000 – 250 000*
- *25 000 – 100 000*
- *under 25 000*

country capital underline

state or province capital underline

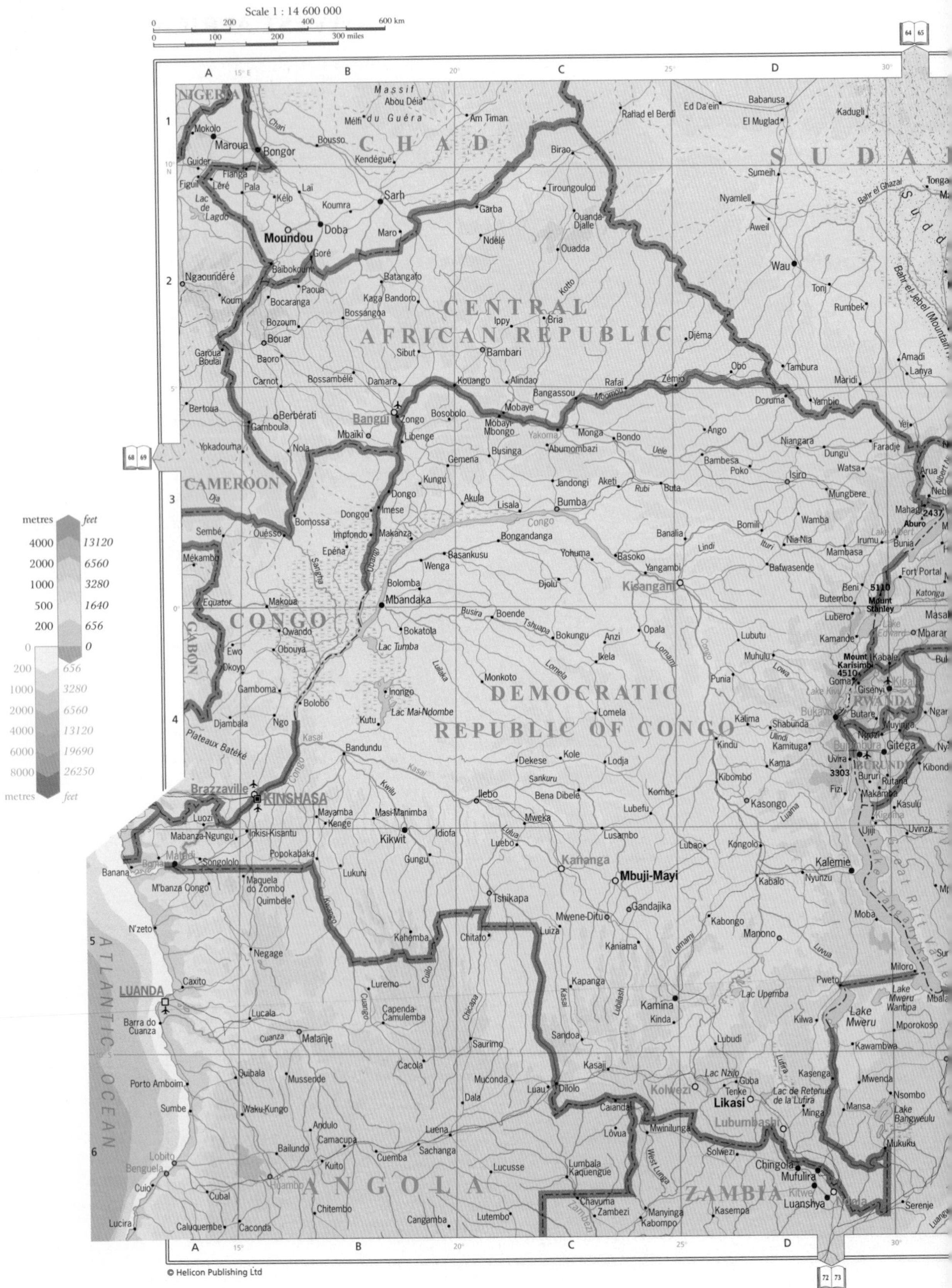

Scale 1 : 14 600 000
0
200
400
600 km
0
100
200
300 miles
64 65
68 69
72 73
A
B
C
D
1
2
3
4
5
6
15° E
20°
25°
30°
10° N
5°
0°
metres
feet
4000
13120
2000
6560
1000
3280
500
1640
200
656
0
0
200
656
1000
3280
2000
6560
4000
13120
6000
19690
8000
26250
metres
feet
NIGERIA
CHAD
SUDAN
CENTRAL AFRICAN REPUBLIC
CAMEROON
CONGO
GABON
DEMOCRATIC REPUBLIC OF CONGO
ANGOLA
ZAMBIA
RWANDA
BURUNDI
ATLANTIC OCEAN
Massif du Guéra
Abou Déia
Melfi
Am Timan
Mokolo
Maroua
Bongor
Bousso
Kendégué
Chari
Guider
Flanga
Léré
Figuil
Pala
Laï
Kélo
Lac de Lagdo
Koumra
Sarh
Moundou
Doba
Maro
Goré
Baïbokoum
Ngaoundéré
Paoua
Koum
Bocaranga
Bozoum
Bouar
Baoro
Carnot
Garoua Boulai
Bertoua
Berbérati
Gamboula
Yokadouma
Nola
Batangafo
Kaga Bandoro
Bossangoa
Bossambélé
Damara
Sibut
Bambari
Ippy
Bria
Kotto
Birao
Tiroungoulou
Garba
Ndélé
Ouanda Djallé
Ouadda
Djéma
Obo
Zémio
Rafaï
Mbomou
Bangassou
Alindao
Kouango
Mobaye
Bangui
Zongo
Mbaïki
Libenge
Bosobolo
Mobayi-Mbongo
Yakoma
Monga
Bondo
Abumombazi
Businga
Gemena
Kungu
Akula
Lisala
Dongo
Dongou
Imese
Bumba
Jandongi
Aketi
Rubi
Buta
Uele
Rahad el Berdi
Ed Da'ein
Babanusa
El Muglad
Kadugli
Sumeih
Nyamlell
Aweil
Wau
Tonj
Rumbek
Bahr el Ghazal
Sudd
Bahr el Jebel (Mountain Nile)
Tambura
Maridi
Amadi
Lanya
Doruma
Yambio
Yei
Ango
Niangara
Dungu
Faradje
Bambesa
Poko
Isiro
Watsa
Arua
Mungbere
Wamba
Nebbi
Mahagi
2437
Aburo
Lake Albert
Bunia
Irumu
Mambasa
Nia-Nia
Ituri
Bafwasende
Bomili
Banalia
Lindi
Yangambi
Basoko
Yohuma
Kisangani
Beni
5110
Mount Stanley
Butembo
Fort Portal
Katonga
Lubero
Lake Edward
Kamande
Mbarara
Kasese
Mount Karisimbi
4510
Goma
Gisenyi
Kigali
Lake Kivu
Bukavu
Butare
Muyinga
Ngozi
Gitega
Bujumbura
Uvira
3303
Bururi
Rutana
Fizi
Makamba
Kibondo
Kasulu
Kigoma
Ujiji
Uvinza
Sembé
Ouesso
Mékambo
Bomossa
Impfondo
Epéna
Makanza
Sangha
Ubangi
Congo
Bongandanga
Basankusu
Wenga
Bolomba
Djolu
Mbandaka
Equator
Makoua
Owando
Ewo
Obouya
Okoyo
Bokatola
Lac Tumba
Busira
Boende
Tshuapa
Bokungu
Anzi
Opala
Lomami
Lubutu
Ikela
Lomela
Muhulu
Lowa
Punia
Monkoto
Lulonga
Lukenie
Gamboma
Bolobo
Inongo
Lac Mai-Ndombe
Djambala
Ngo
Kutu
Plateaux Batéké
Kasai
Bandundu
Kalima
Shabunda
Ulindi
Kamituga
Kindu
Kama
Dekese
Kole
Lodja
Sankuru
Bena Dibele
Kibombo
Kombe
Kasongo
Luama
Lubefu
Brazzaville
KINSHASA
Luozi
Kwilu
Ilebo
Mweka
Mabanza-Ngungu
Inkisi-Kisantu
Mayamba
Kenge
Masi-Manimba
Kikwit
Idiofa
Luebo
Lulua
Lusambo
Lubao
Kongolo
Kalemie
Nyunzu
Banana
Matadi
Boma
Songololo
Popokabaka
Gungu
Lukuni
Kananga
Mbuji-Mayi
Kabalo
M'banza Congo
Maquela do Zombo
Quimbele
Tshikapa
Gandajika
Mwene-Ditu
Kabongo
Moba
N'zeto
Kwango
Kahemba
Chitato
Luiza
Kaniama
Manono
Lake Tanganyika
Great Rift Valley
Luvua
Negage
Cuilo
Caxito
LUANDA
Luremo
Cuango
Kapanga
Lubilash
Kamina
Kinda
Lac Upemba
Pweto
Moliro
Lake Mweru Wantipa
Lake Mweru
Mporokoso
Barra do Cuanza
Lucala
Capenda-Camulemba
Chicapa
Sandoa
Lubudi
Kilwa
Kawambwa
Cuanza
Malanje
Saurimo
Cacolo
Kasaji
Lac Nzilo
Guba
Tenke
Lufira
Kasenga
Mwenda
Quibala
Mussende
Porto Amboim
Muconda
Dala
Luau
Dilolo
Kolwezi
Likasi
Lac de Retenue de la Lufira
Minga
Mansa
Nsombo
Lake Bangweulu
Sumbe
Waku-Kungo
Caianda
Lubumbashi
Mwinilunga
Andulo
Camacupa
Luena
Lôvua
West Lunga
Solwezi
Chingola
Mufulira
Kitwe
Ndola
Luanshya
Mukuku
Lobito
Benguela
Bailundo
Cuemba
Sachanga
Kuito
Lucusse
Lumbala Kaquengue
Cuio
Huambo
Cubal
Chitembo
Chavuma
Zambezi
Manyinga
Kabompo
Kasempa
Serenje
Lucira
Caluquembe
Caconda
Cangamba
Lutembo

CENTRAL AFRICA

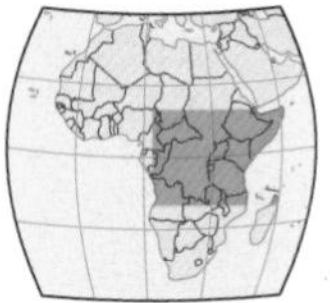

- Angola
- Burundi
- Central African Republic
- Democratic Republic of Congo
- Djibouti
- Ethiopia
- Kenya
- Rwanda
- Somalia
- Tanzania
- Uganda

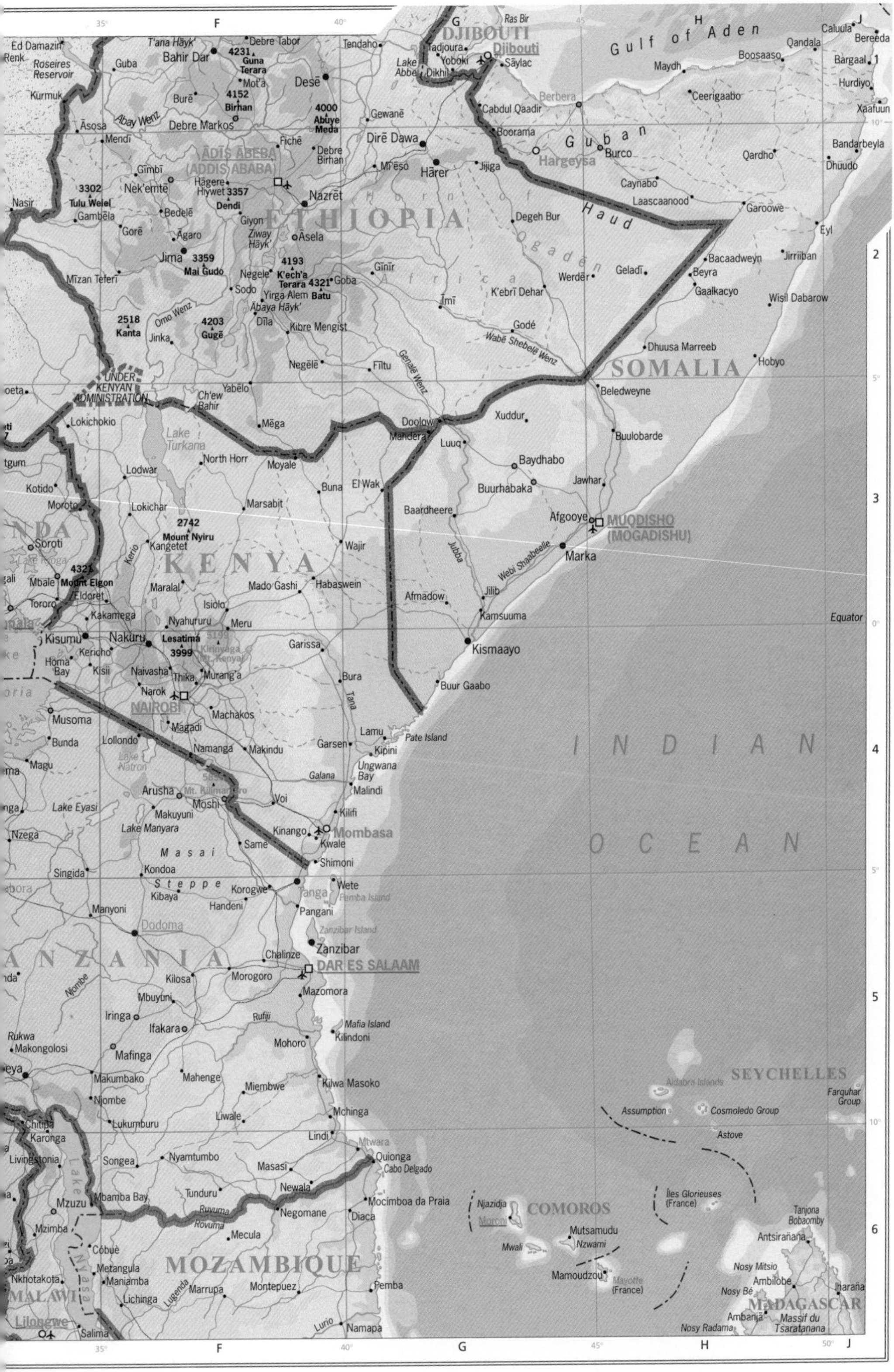

Settlements

- ▣ *over 3 million*
- □ *1 – 3 million*
- ○ *250 000 – 1 million*
- ● *100 000 – 250 000*
- ◉ *25 000 – 100 000*
- • *under 25 000*
- ___ *country capital underline*

Scale 1 : 14 600 000

0 200 400 600 km

0 100 200 300 miles

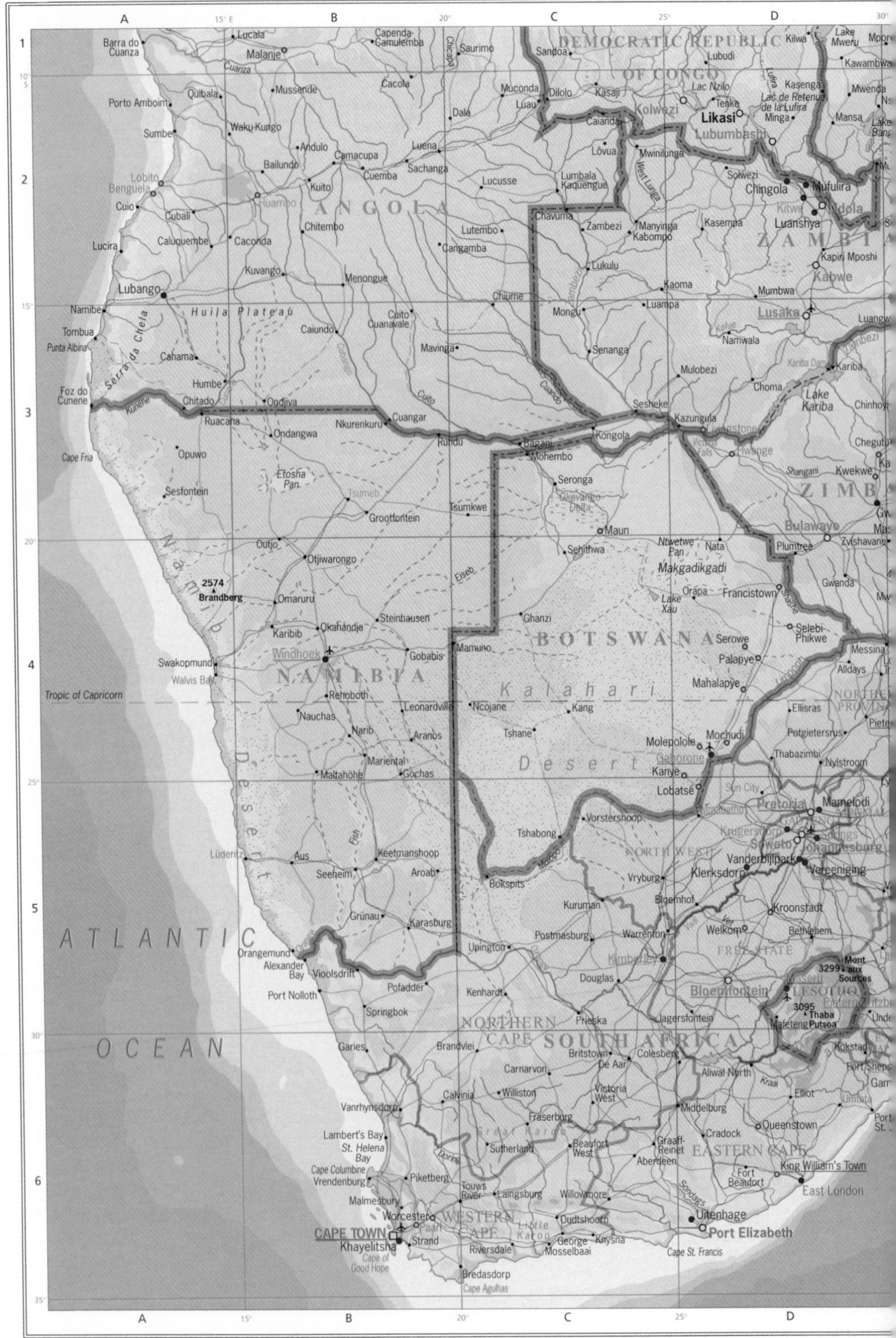

metres / feet

4000 13120

2000 6560

1000 3280

500 1640

200 656

0 0

200 656

1000 3280

2000 6560

4000 13120

6000 19690

8000 26250

metres / feet

SOUTHERN AFRICA

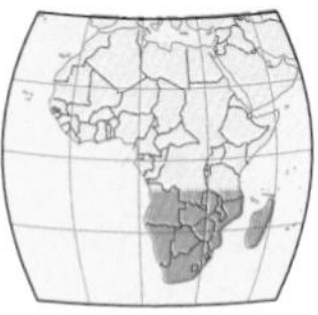

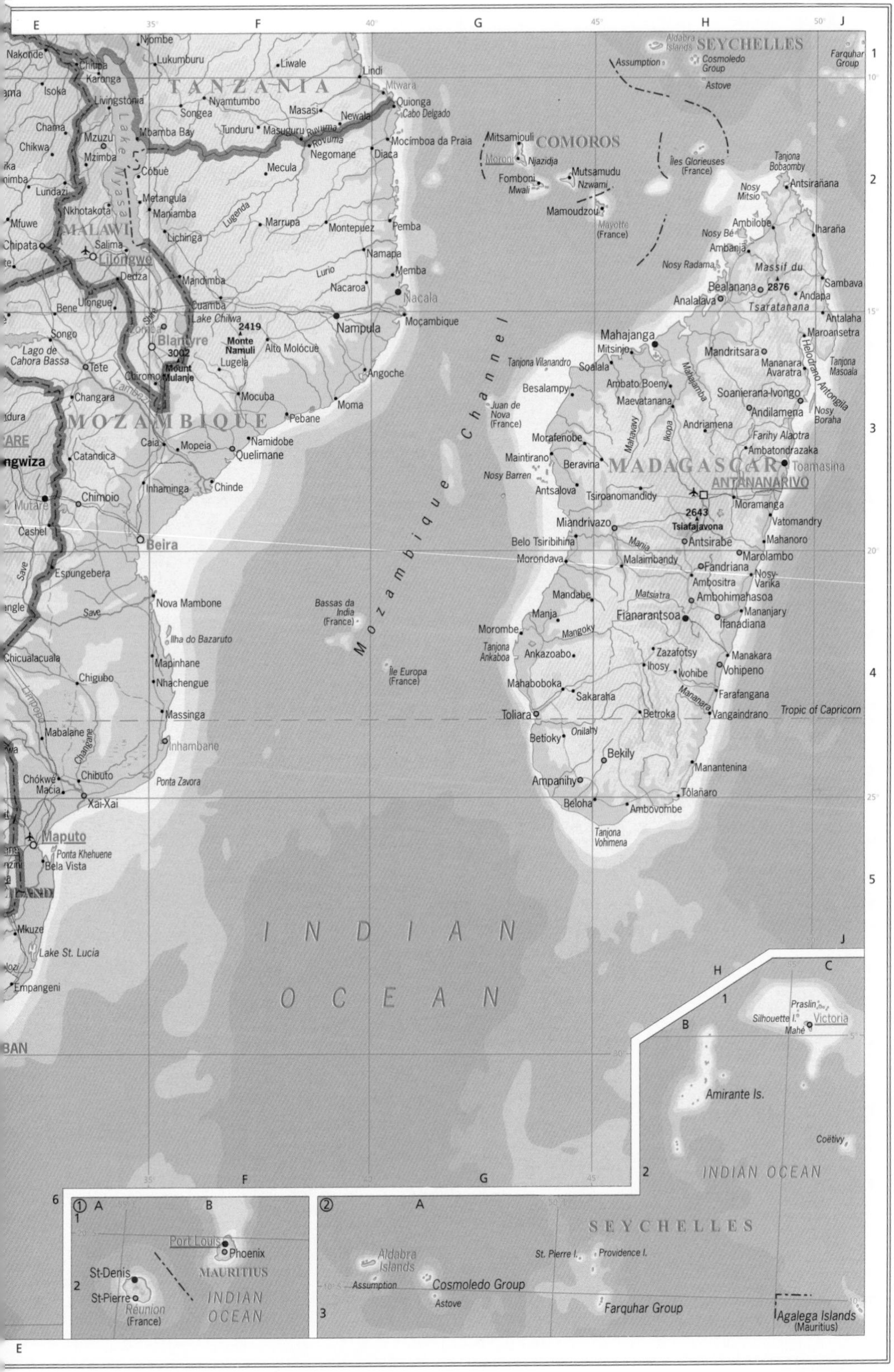

- Botswana
- Comoros
- Lesotho
- Madagascar
- Malawi
- Mauritius
- Mozambique
- Namibia
- Seychelles
- South Africa
- Swaziland
- Zambia
- Zimbabwe

E 35° F 40° G 45° H 50° J

1 2 3 4 5 6

TANZANIA
Nkhotakota
Nakonde
Chitipa
Karonga
Isoka
Livingstonia
Chama
Mzuzu
Mzimba
Chikwa
Lundazi
Mfuwe
Chipata
Lilongwe
Salima
Dedza
Mbamba Bay
Cóbuè
Metangula
Maniamba
Lichinga
Mandimba
Cuamba
Lake Chilwa
Ulongue
Bene
Songo
Lago de Cahora Bassa
Tete
Zomba
Blantyre
3002
Mount Mulanje
Chiromo
Changara
MALAWI
Lake Nyasa
Mbombe
Lukumburu
Liwale
Nyamtumbo
Songea
Tunduru
Masuguru
Masasi
Newala
Lindi
Mtwara
Quionga
Cabo Delgado
Mocímboa da Praia
Ruvuma
Negomane
Diaca
Mecula
Lugenda
Marrupa
Montepuez
Pemba
Namapa
Memba
Lurio
Nacaroa
Nacala
Nampula
Moçambique
2419
Monte Namuli
Alto Molócuè
Lugela
Angoche
Mocuba
Moma
Pebane
MOZAMBIQUE
Caia
Mopeia
Namidobe
Quelimane
Catandica
Chinde
Inhaminga
Chimoio
Mutare
Cashel
Beira
Save
Espungebera
Nova Mambone
Ilha do Bazaruto
Chicualacuala
Mapinhane
Chigubo
Nhachengue
Limpopo
Massinga
Mabalane
Changane
Inhambane
Chókwè
Chibuto
Macia
Xai-Xai
Ponta Zavora
Maputo
Ponta Khehuene
Bela Vista
Mkuze
Lake St. Lucia
Empangeni
Mozambique Channel
Bassas da India (France)
Île Europa (France)
Juan de Nova (France)
Aldabra Islands
SEYCHELLES
Assumption
Cosmoledo Group
Astove
Farquhar Group
Mitsamiouli
COMOROS
Moroni
Njazidja
Fomboni
Mwali
Mutsamudu
Nzwani
Mamoudzou
Mayotte (France)
Îles Glorieuses (France)
Tanjona Bobaomby
Nosy Mitsio
Antsirañana
Ambilobe
Iharaña
Nosy Be
Ambanja
Nosy Radama
Massif du Tsaratanana
2876
Sambava
Bealanana
Andapa
Analalava
Antalaha
Maroantsetra
Mahajanga
Mitsinjo
Mandritsara
Mananara Avaratra
Tanjona Masoala
Tanjona Vilanandro
Soalala
Ambato Boeny
Besalampy
Maevatanana
Mahajamba
Soanierana-Ivongo
Antongila
Nosy Boraha
Andilamena
Andriamena
Mahavavy
Ikopa
Fariny Alaotra
Ambatondrazaka
Morafenobe
Maintirano
Beravina
MADAGASCAR
Toamasina
Nosy Barren
ANTANANARIVO
Antsalova
Tsiroanomandidy
Moramanga
2643
Tsiafajavona
Vatomandry
Miandrivazo
Mahanoro
Belo Tsiribihina
Antsirabe
Mania
Marolambo
Morondava
Malaimbandy
Fandriana
Ambositra
Nosy-Varika
Matsiatra
Ambohimahasoa
Mandabe
Manja
Fianarantsoa
Mananjary
Ifanadiana
Morombe
Mangoky
Tanjona Ankaboa
Ankazoabo
Zazafotsy
Ihosy
Manakara
Ivohibe
Vohipeno
Mahaboboka
Sakaraha
Mananara
Farafangana
Toliara
Betroka
Vangaindrano
Tropic of Capricorn
Betioky
Onilahy
Bekily
Manantenina
Ampanihy
Tôlañaro
Beloha
Ambovombe
Tanjona Vohimena
INDIAN OCEAN

Inset 1 (A B C; H J):
Praslin
Silhouette I.
Victoria
Mahé
Amirante Is.
Coëtivy
INDIAN OCEAN

Inset ① (A B):
Port Louis
Phoenix
MAURITIUS
St-Denis
St-Pierre
Réunion (France)
INDIAN OCEAN

Inset ② (A):
SEYCHELLES
Aldabra Islands
Assumption
Cosmoledo Group
Astove
St. Pierre I.
Providence I.
Farquhar Group
Agalega Islands (Mauritius)

Settlements

- ▣ over 3 million
- □ 1 – 3 million
- ◎ 250 000 – 1 million
- ● 100 000 – 250 000
- ⊙ 25 000 – 100 000
- • under 25 000
- country capital underline
- state or province capital underline

Scale 1 : 51 030 000

0 500 1000 1500 km

0 250 500 750 miles

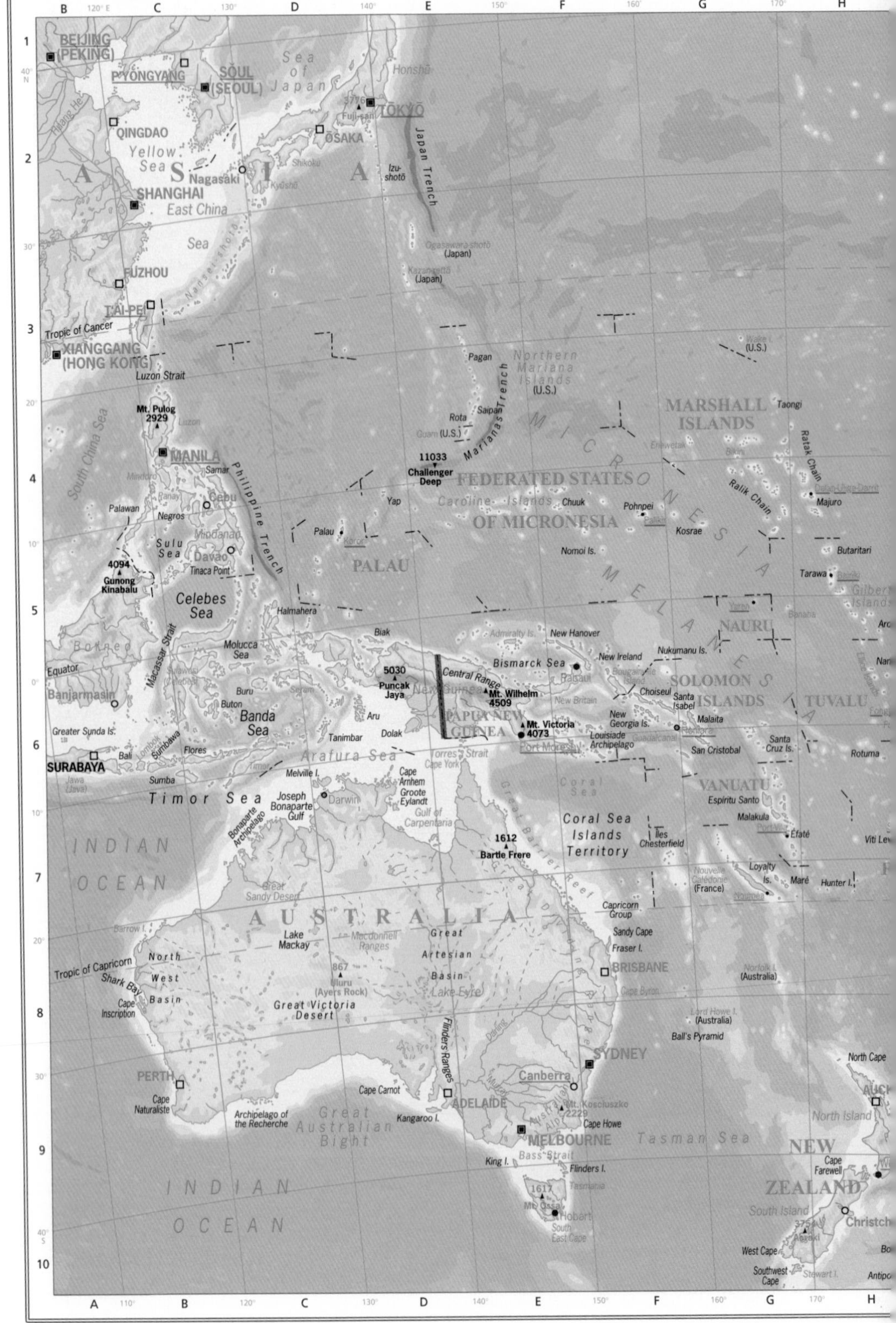

metres	feet
4000	13120
2000	6560
1000	3280
500	1640
200	656
0	0
200	656
1000	3280
2000	6560
4000	13120
6000	19690
8000	26250
metres	feet

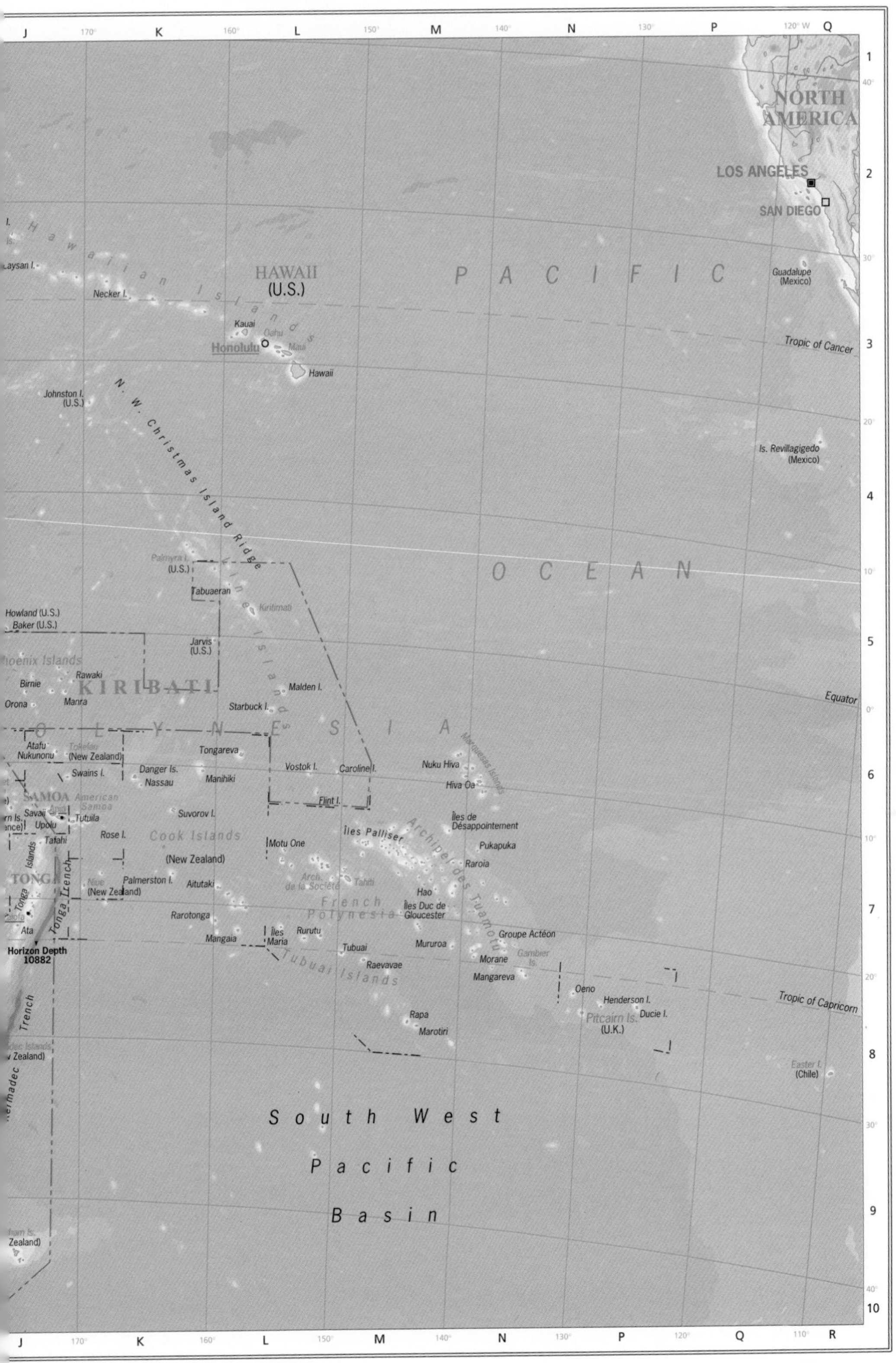
NORTH AMERICA
LOS ANGELES
SAN DIEGO
Guadalupe (Mexico)
Tropic of Cancer
Is. Revillagigedo (Mexico)
PACIFIC
OCEAN
HAWAII (U.S.)
Kauai
Oahu
Maui
Honolulu
Hawaii
Hawaiian Islands
Laysan I.
Necker I.
Johnston I. (U.S.)
N. W. Christmas Island Ridge
Palmyra I. (U.S.)
Tabuaeran
Kiritimati
Line Islands
Howland (U.S.)
Baker (U.S.)
Phoenix Islands
Jarvis (U.S.)
Birnie
Rawaki
KIRIBATI
Orona
Manra
Malden I.
Starbuck I.
Equator
POLYNESIA
Atafu
Nukunonu
Tokelau (New Zealand)
Swains I.
Danger Is.
Nassau
Tongareva
Manihiki
Vostok I.
Caroline I.
Flint I.
Nuku Hiva
Hiva Oa
Marquesas Islands
SAMOA
American Samoa
Savai'i
Upolu
Tutuila
Tafahi
Rose I.
Suvorov I.
Cook Islands (New Zealand)
Motu One
Îles Palliser
Îles de Désappointement
Pukapuka
Raroia
Archipel des Tuamotu
TONGA
Tonga Islands
Tonga Trench
Niue (New Zealand)
Palmerston I.
Aitutaki
Arch. de la Société
Tahiti
French Polynesia
Hao
Îles Duc de Gloucester
Rarotonga
Mangaia
Îles Maria
Rurutu
Tubuai
Tubuai Islands
Raevavae
Mururoa
Groupe Actéon
Morane
Gambier Is.
Mangareva
Ata
Horizon Depth 10882
Oeno
Henderson I.
Ducie I.
Pitcairn Is. (U.K.)
Tropic of Capricorn
Rapa
Marotiri
Easter I. (Chile)
Kermadec Trench
Kermadec Islands (New Zealand)
Chatham Is. (New Zealand)
South West Pacific Basin
J K L M N P Q R
170° 160° 150° 140° 130° 120° W 110°
40° 30° 20° 10° 0° 10° 20° 30° 40°
1 2 3 4 5 6 7 8 9 10

Scale 1 : 17 400 000

0 200 400 600 km

0 100 200 300 miles

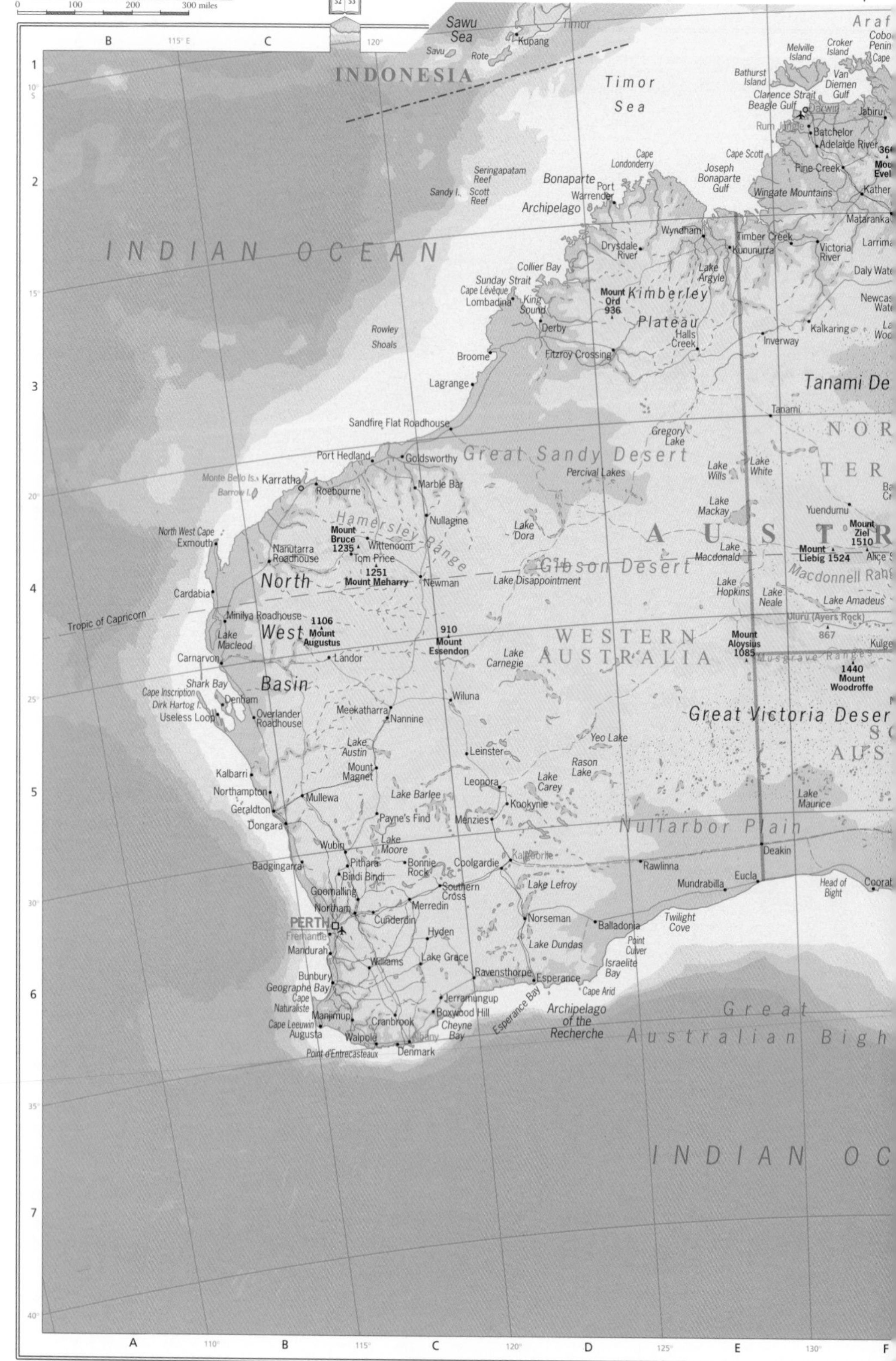

metres / feet

metres	feet
4000	13120
2000	6560
1000	3280
500	1640
200	656
0	0
200	656
1000	3280
2000	6560
4000	13120
6000	19690
8000	26250

metres / feet

Australia

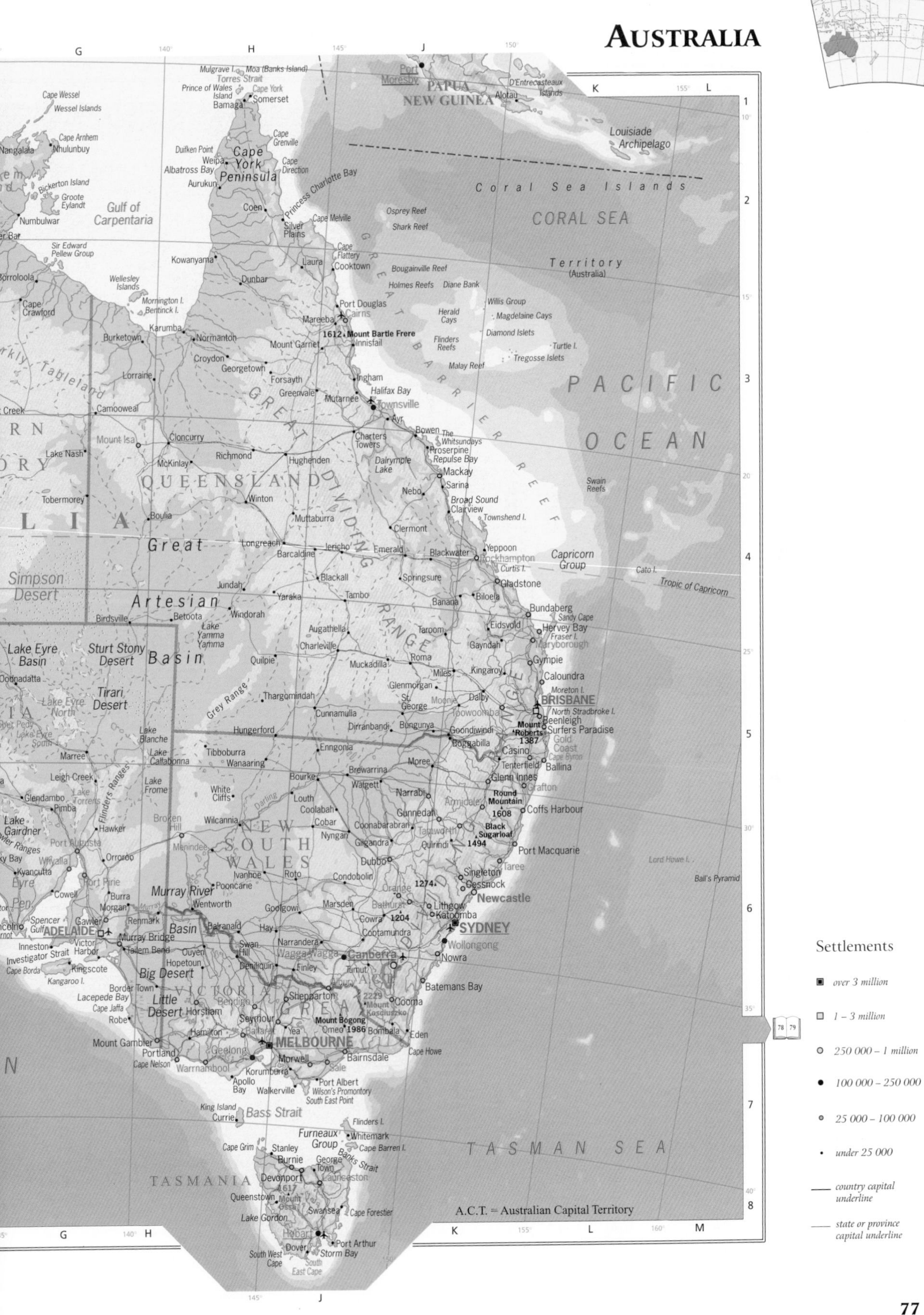

G
H
J
K
L
M
140°
145°
150°
155°
160°
10°
15°
20°
25°
30°
35°
40°
1
2
3
4
5
6
7
8
Torres Strait
Mulgrave I.
Moa (Banks Island)
Prince of Wales Island
Cape York
Bamaga
Somerset
Port Moresby
PAPUA NEW GUINEA
Alotau
D'Entrecasteaux Islands
Louisiade Archipelago
Cape Wessel
Wessel Islands
Cape Arnhem
Nhulunbuy
Nangalala
Bickerton Island
Groote Eylandt
Numbulwar
Gulf of Carpentaria
Duifken Point
Weipa
Albatross Bay
Aurukun
Cape York Peninsula
Cape Grenville
Cape Direction
Princess Charlotte Bay
Coen
Silver Plains
Cape Melville
Coral Sea Islands
CORAL SEA
Territory (Australia)
Osprey Reef
Shark Reef
Sir Edward Pellew Group
Borroloola
Kowanyama
Laura
Cape Flattery
Cooktown
Bougainville Reef
Holmes Reefs
Diane Bank
Wellesley Islands
Dunbar
Mornington I.
Bentinck I.
Cape Crawford
Port Douglas
Cairns
Mareeba
Willis Group
Magdelaine Cays
Herald Cays
Karumba
Burketown
Normanton
1612 Mount Bartle Frere
Innisfail
Mount Garnet
Diamond Islets
Flinders Reefs
Turtle I.
Tregosse Islets
Croydon
Georgetown
Malay Reef
Forsayth
Lorraine
Ingham
Halifax Bay
Greenvale
Mutarnee
Townsville
Ayr
GREAT BARRIER REEF
PACIFIC OCEAN
Camooweal
Mount Isa
Cloncurry
Charters Towers
Bowen
The Whitsundays
Proserpine
Repulse Bay
Lake Nash
Richmond
Hughenden
Dalrymple Lake
Mackay
Sarina
McKinlay
QUEENSLAND
Nebo
Swain Reefs
Tobermorey
Winton
Broad Sound
Clairview
Townshend I.
Boulia
Muttaburra
Clermont
Great Artesian Basin
Longreach
Barcaldine
Jericho
Emerald
Blackwater
Yeppoon
Rockhampton
Curtis I.
Capricorn Group
Cato I.
Tropic of Capricorn
Simpson Desert
Blackall
Springsure
Gladstone
Jundah
Yaraka
Tambo
Banana
Biloela
Birdsville
Betoota
Windorah
Lake Yamma Yamma
Bundaberg
Sandy Cape
Hervey Bay
Fraser I.
Maryborough
Eidsvold
Augathella
Taroom
Gayndah
Lake Eyre Basin
Sturt Stony Desert
Charleville
Quilpie
Roma
Gympie
Muckadilla
Miles
Kingaroy
Caloundra
Oodnadatta
Tirari Desert
Glenmorgan
Moreton I.
Lake Eyre North
Grey Range
Thargomindah
St George
Moonie
Dalby
Toowoomba
BRISBANE
North Stradbroke I.
Lake Eyre South
Cunnamulla
Dirranbandi
Bungunya
Goondiwindi
Mount Roberts 1387
Beenleigh
Surfers Paradise
Gold Coast
Hungerford
Lake Blanche
Boggabilla
Casino
Cape Byron
Marree
Tibooburra
Wanaaring
Lake Callabonna
Enngonia
Tenterfield
Ballina
Moree
Leigh Creek
Flinders Ranges
Lake Frome
Bourke
Brewarrina
Walgett
Glenn Innes
Grafton
Glendambo
Lake Torrens
Pimba
White Cliffs
Louth
Narrabri
Armidale
Round Mountain 1608
Coffs Harbour
Darling
Coolabah
Gunnedah
Lake Gairdner
Hawker
Broken Hill
Wilcannia
Cobar
Coonabarabran
Tamworth
Black Sugarloaf 1494
Gawler Ranges
Port Augusta
Menindee
NEW SOUTH WALES
Nyngan
Gilgandra
Quirindi
Port Macquarie
Whyalla
Orroroo
Dubbo
Taree
Lord Howe I.
Kyancutta
Eyre Pen.
Port Pirie
Ivanhoe
Roto
Condobolin
Singleton
Cessnock
Ball's Pyramid
Cowell
Burra
Murray River
Pooncarie
Orange
1274
Newcastle
Morgan
Wentworth
Bathurst
Lithgow
Katoomba
Spencer Gulf
Gawler
Renmark
Goolgowi
Marsden
Cowra
1204
SYDNEY
ADELAIDE
Murray Bridge
Balranald
Hay
Cootamundra
Wollongong
Inneston
Victor Harbor
Tailem Bend
Swan Hill
Narrandera
Wagga Wagga
Canberra
Nowra
Investigator Strait
Kingscote
Ouyen
Hopetoun
Deniliquin
Finley
Tumut
A.C.T.
Cape Borda
Kangaroo I.
Big Desert
Albury
Batemans Bay
Bordertown
VICTORIA
Shepparton
2229 Mount Kosciuszko
Cooma
Lacepede Bay
Little Desert
Bendigo
GREAT
Cape Jaffa
Horsham
Robe
Seymour
Mount Bogong
Omeo 1986
Bombala
Ballarat
Yea
Eden
Mount Gambier
Hamilton
Portland
Geelong
MELBOURNE
Cape Howe
Cape Nelson
Warrnambool
Morwell
Bairnsdale
Sale
Korumburra
Apollo Bay
Walkerville
Port Albert
Wilson's Promontory
South East Point
King Island
Currie
Bass Strait
Flinders I.
Furneaux Group
Whitemark
Cape Barren I.
Banks Strait
TASMAN SEA
Cape Grim
Stanley
Burnie
George Town
Devonport
Launceston
TASMANIA
1617 Mount Ossa
Queenstown
Swansea
Cape Forestier
Lake Gordon
Hobart
Port Arthur
Dover
Storm Bay
South West Cape
South East Cape
A.C.T. = Australian Capital Territory
78
79
Settlements
over 3 million
1 – 3 million
250 000 – 1 million
100 000 – 250 000
25 000 – 100 000
under 25 000
country capital underline
state or province capital underline

Scale 1 : 5 850 000

0 100 200 300 km

0 50 100 150 miles

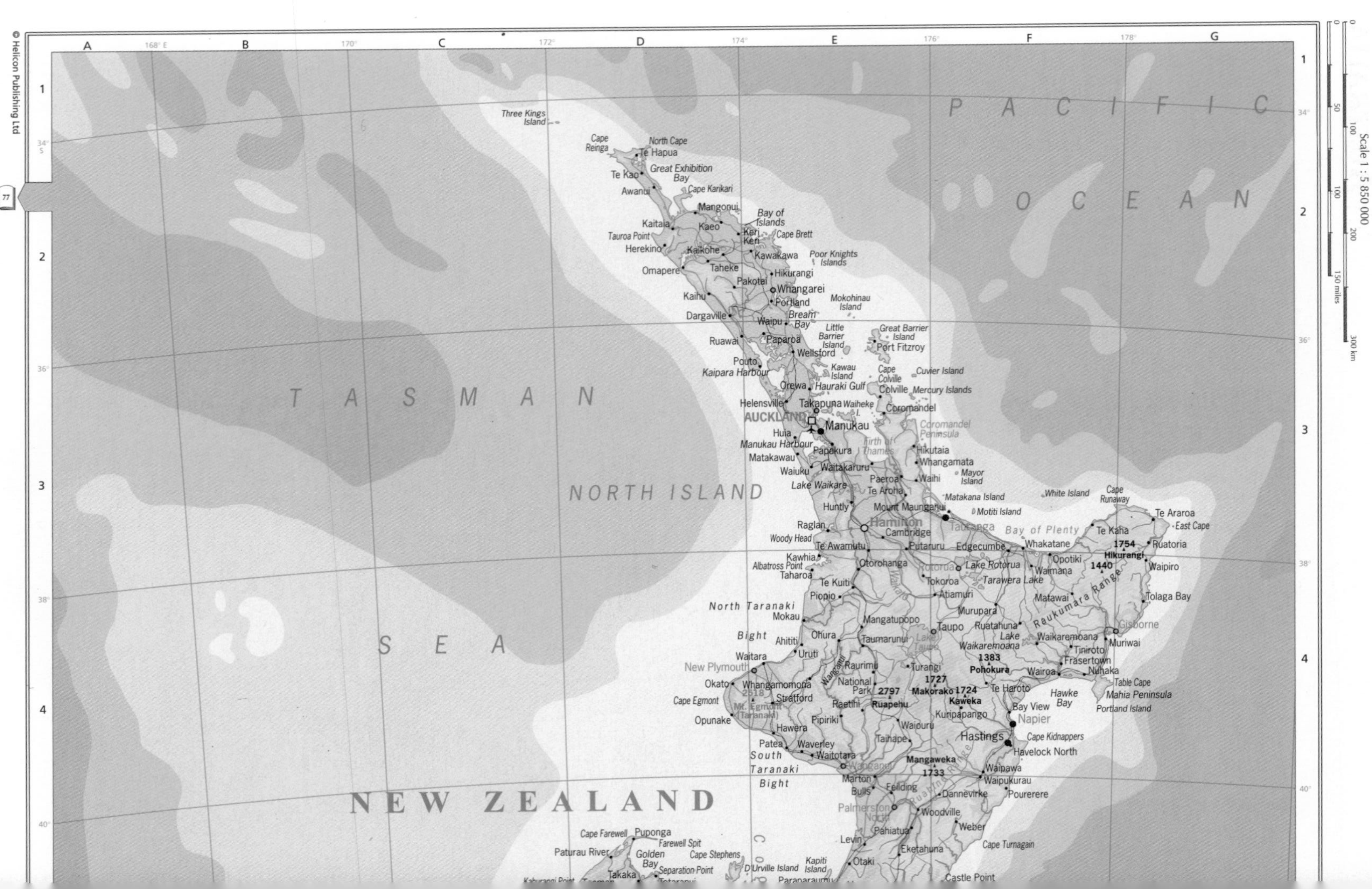

metres | feet
4000 | 13120
2000 | 6560
1000 | 3280
500 | 1640
200 | 656
0 | 0
200 | 656
1000 | 3280
2000 | 6560
4000 | 13120
6000 | 19690
8000 | 26250
metres | feet

76 77

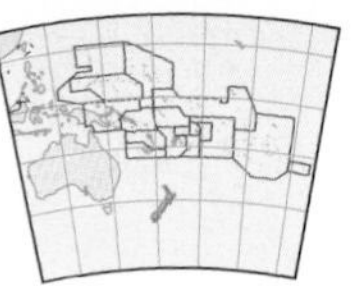

NEW ZEALAND

- Antipodes Islands
- Auckland Islands
- Campbell Island
- Chatham Islands

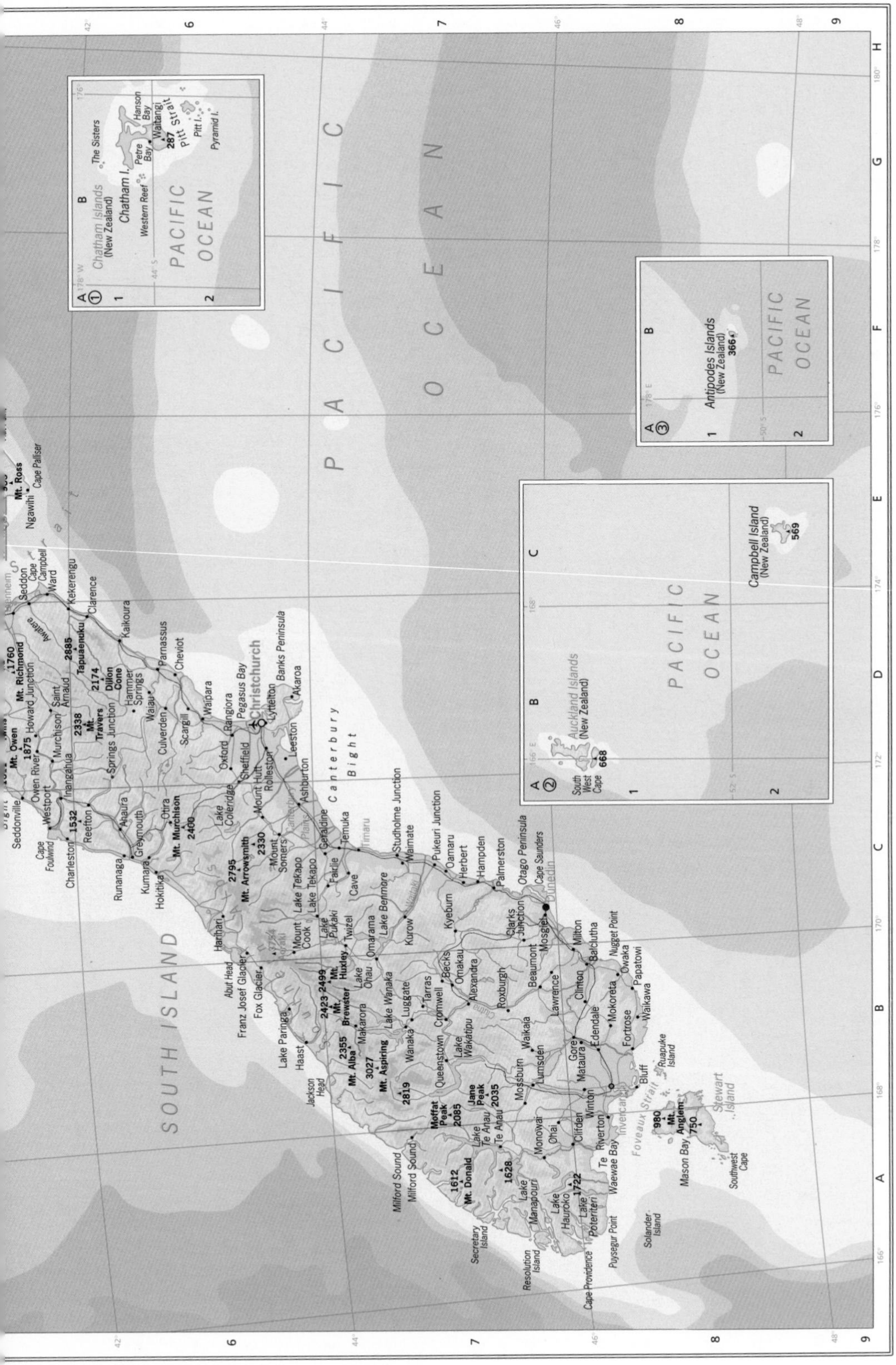

Settlements

- over 3 million
- 1 – 3 million
- 250 000 – 1 million
- 100 000 – 250 000
- 25 000 – 100 000
- under 25 000
- country capital underline
- country capital underline

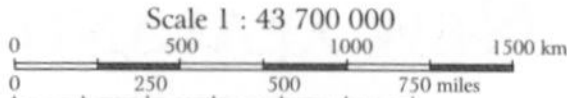

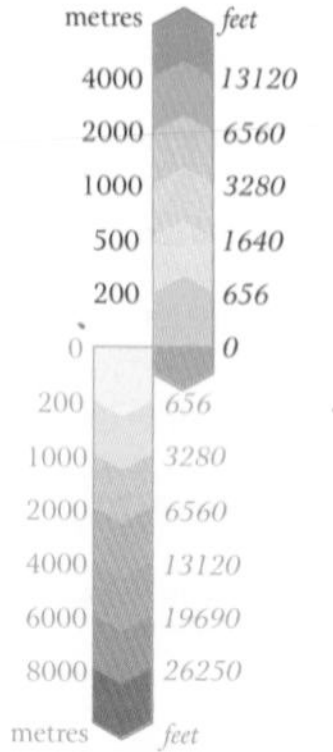

metres	feet
4000	13120
2000	6560
1000	3280
500	1640
200	656
0	0
200	656
1000	3280
2000	6560
4000	13120
6000	19690
8000	26250
metres	feet

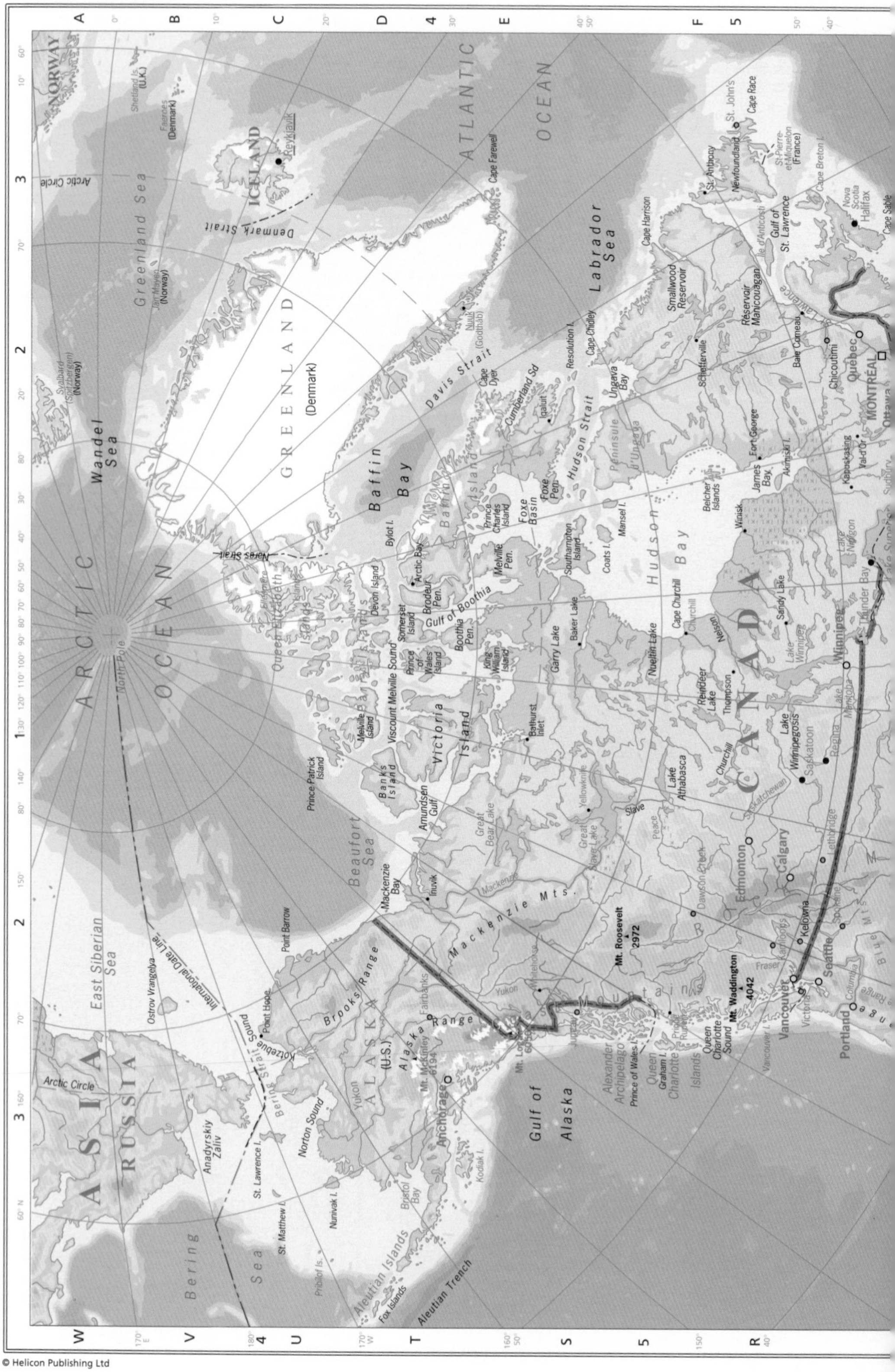

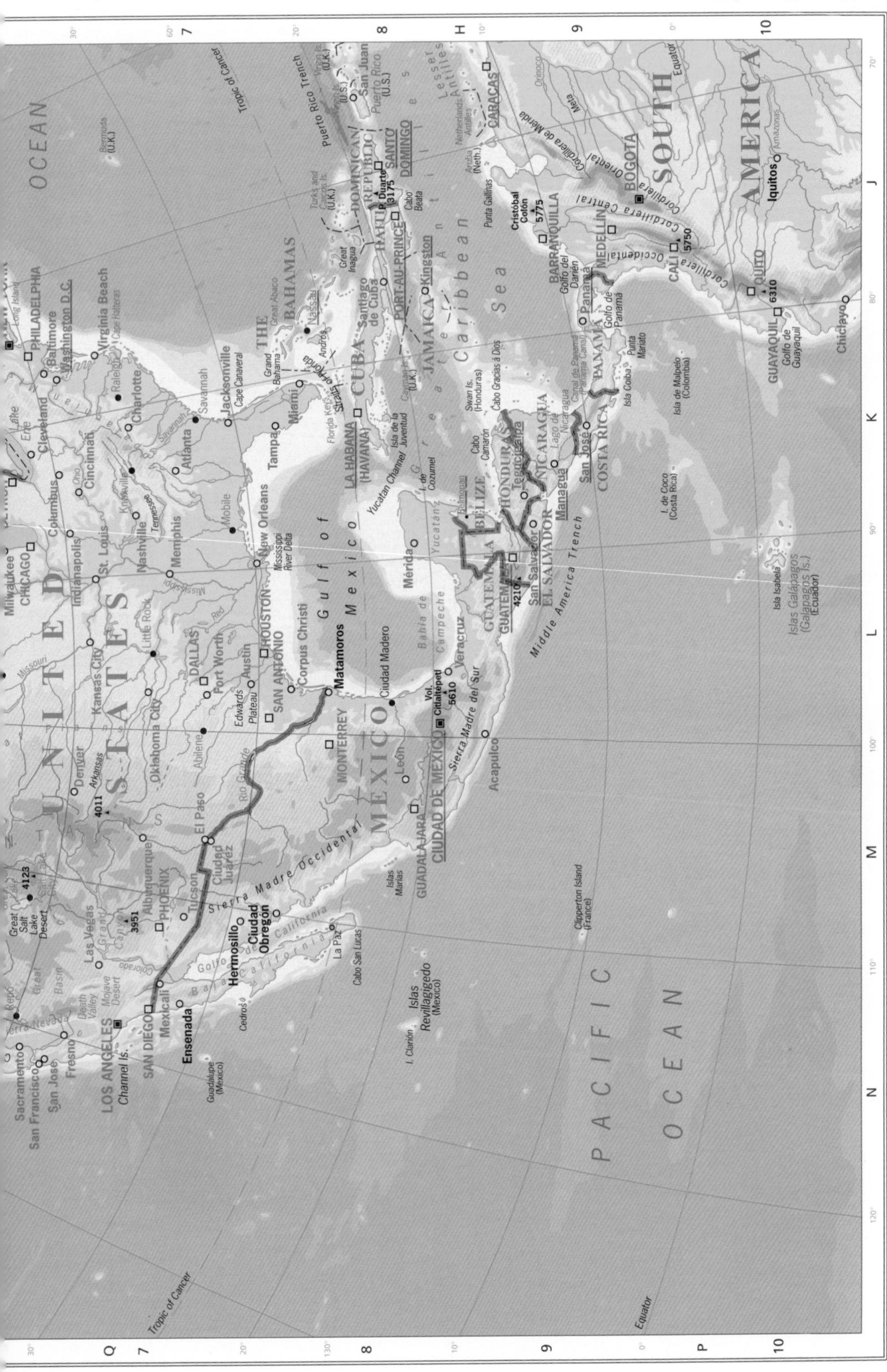

OCEAN
Bermuda (U.K.)
UNITED STATES
PHILADELPHIA
Baltimore
Washington D.C.
Virginia Beach
Cape Hatteras
Raleigh
Charlotte
Cleveland
Lake Erie
Cincinnati
Columbus
Atlanta
Savannah
Jacksonville
Cape Canaveral
Miami
Tampa
Knoxville
Nashville
Tennessee
Memphis
Mobile
New Orleans
Mississippi River Delta
Indianapolis
Milwaukee
CHICAGO
St. Louis
Kansas City
Missouri
Little Rock
Oklahoma City
DALLAS
Fort Worth
Austin
HOUSTON
SAN ANTONIO
Corpus Christi
Edwards Plateau
Abilene
Denver
Arkansas
4011
Albuquerque
El Paso
Rio Grande
Ciudad Juárez
PHOENIX
Tucson
Great Salt Lake Desert
4123
3951
Las Vegas
Grand Canyon
Colorado
Death Valley
Mojave Desert
Sacramento
San Francisco
San Jose
Fresno
LOS ANGELES
Channel Is.
SAN DIEGO
Mexicali
Ensenada
Guadalupe (Mexico)
Cedros
Hermosillo
Ciudad Obregón
Golfo de California
Baja California
La Paz
Cabo San Lucas
Sierra Madre Occidental
MEXICO
MONTERREY
Matamoros
Ciudad Madero
León
GUADALAJARA
CIUDAD DE MÉXICO
Vol. Citlaltépetl 5610
Veracruz
Sierra Madre del Sur
Acapulco
Islas Marías
Islas Revillagigedo (Mexico)
I. Clarión
Clipperton Island (France)
PACIFIC OCEAN
Tropic of Cancer
Equator
Gulf of Mexico
Bahía de Campeche
Mérida
Yucatán
I. de Cozumel
Yucatan Channel
BELIZE
Belmopan
GUATEMALA
4210
SAN SALVADOR
EL SALVADOR
HONDURAS
Tegucigalpa
NICARAGUA
Lago de Nicaragua
Managua
San José
COSTA RICA
Middle America Trench
I. de Coco (Costa Rica)
Isla Isabela
Islas Galápagos (Galapagos Is.) (Ecuador)
Isla de Malpelo (Colombia)
Isla Coiba
Punta Mariato
PANAMA
Golfo de Panamá
Panamá
Canal de Panamá (Panama Canal)
Golfo del Darién
Swan Is. (Honduras)
Cabo Camarón
Cabo Gracias á Dios
Caribbean Sea
Greater Antilles
Lesser Antilles
JAMAICA
Kingston
Cayman Is. (U.K.)
CUBA
LA HABANA (HAVANA)
Isla de la Juventud
Santiago de Cuba
Florida Keys
Straits of Florida
THE BAHAMAS
Nassau
Grand Bahama
Great Abaco
Andros
Great Inagua
Turks and Caicos Is. (U.K.)
HAITI
PORT-AU-PRINCE
DOMINICAN REPUBLIC
SANTO DOMINGO
P. Duarte 3175
Cabo Beata
Puerto Rico Trench
Puerto Rico (U.S.)
San Juan
Virgin Is. (U.S.) (U.K.)
Netherlands Antilles
Aruba (Neth.)
Punta Gallinas
CARACAS
Cristóbal Colón 5775
BARRANQUILLA
MEDELLÍN
BOGOTÁ
CALI
5750
Cordillera de Mérida
Cordillera Oriental
Cordillera Central
Cordillera Occidental
Meta
Orinoco
SOUTH AMERICA
Amazonas
Iquitos
QUITO
6310
GUAYAQUIL
Golfo de Guayaquil
Chiclayo

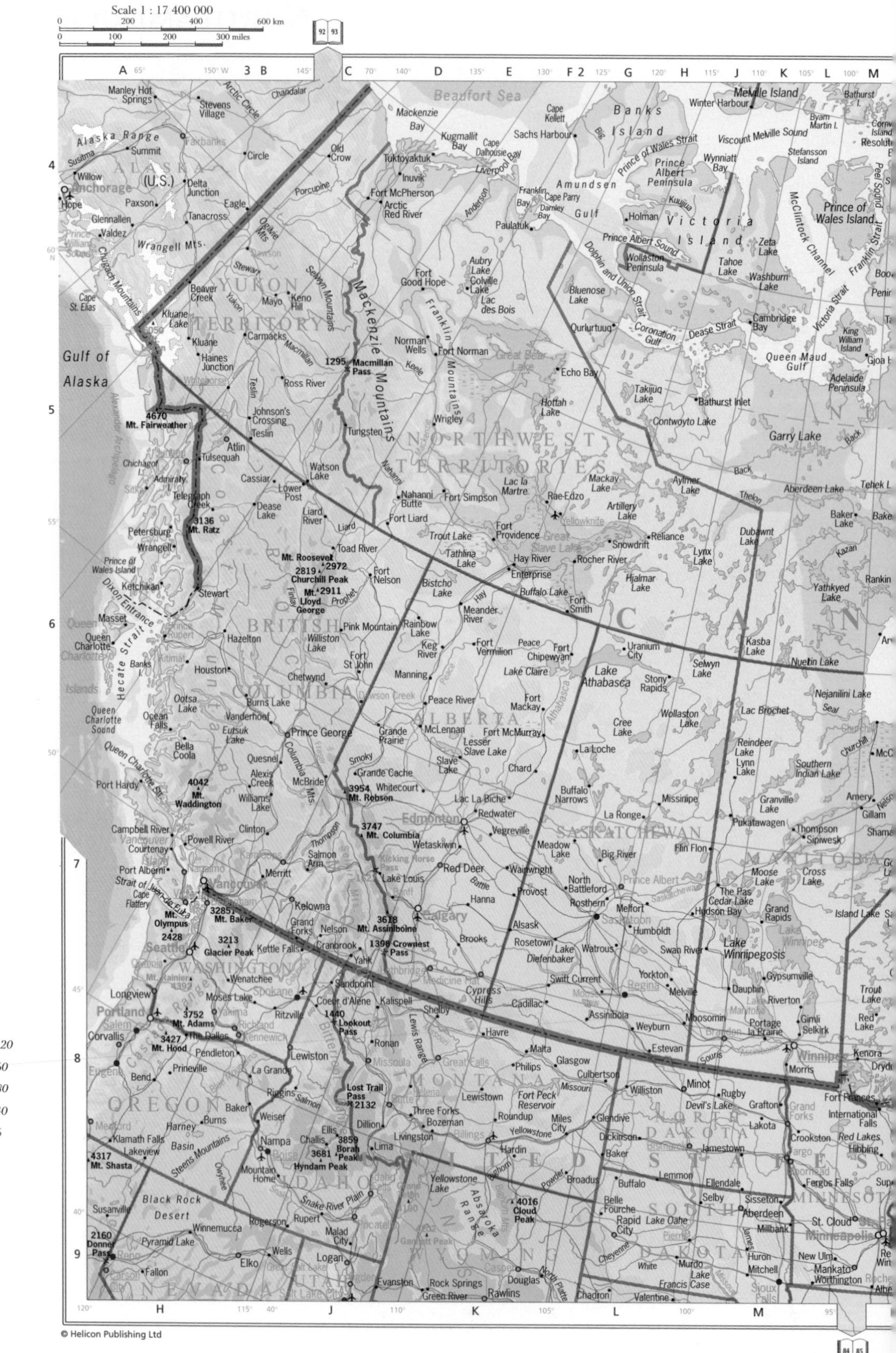

metres / feet
4000 13120
2000 6560
1000 3280
500 1640
200 656
0 0
200 656
1000 3280
2000 6560
4000 13120
6000 19690
8000 26250
metres / feet

CANADA

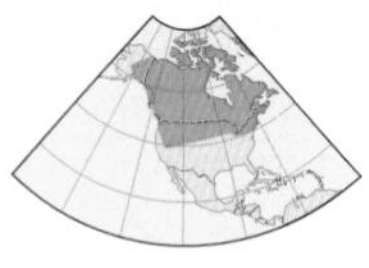

- Canada
- Greenland
- St. Pierre & Miquelon

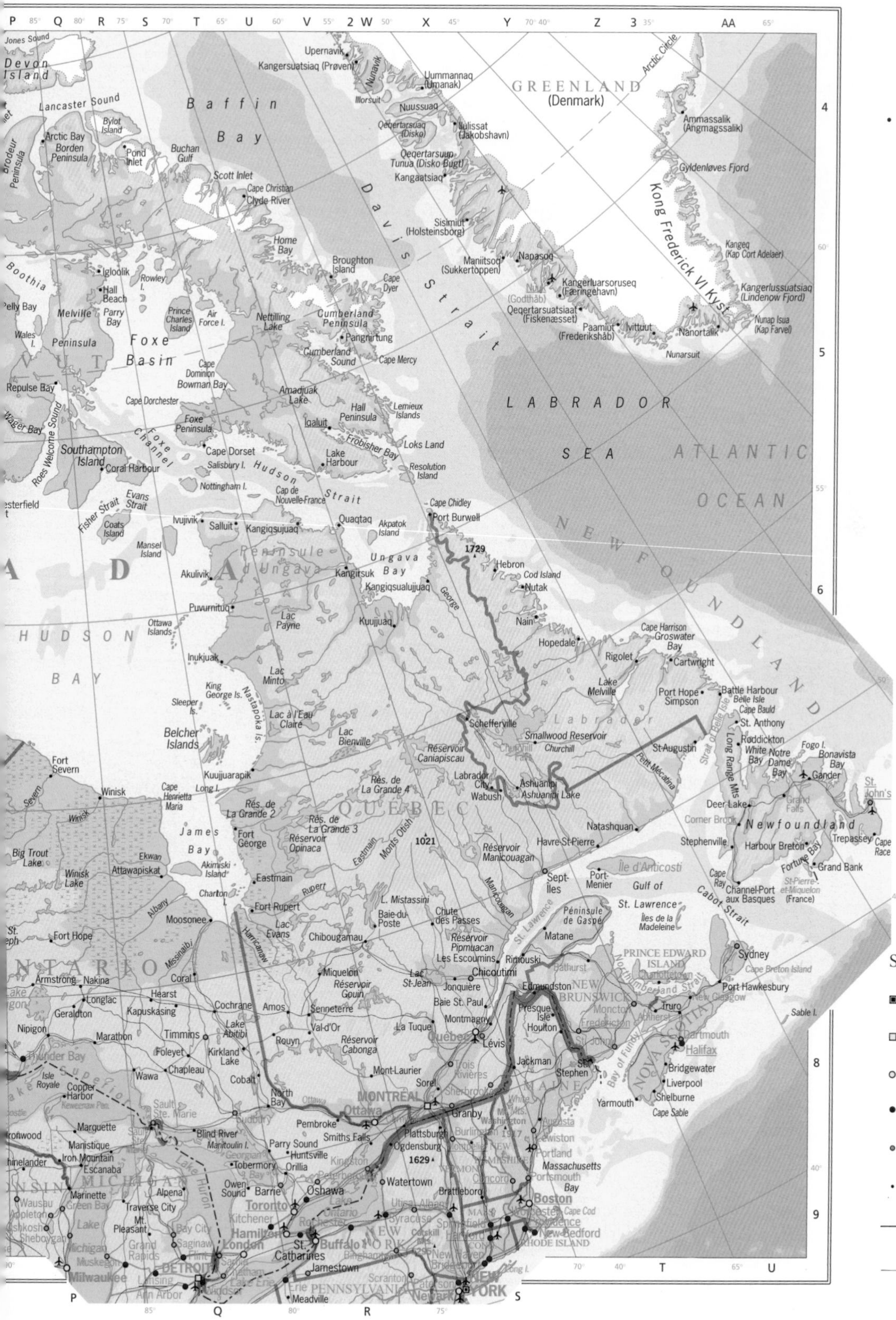

Settlements

- over 3 million
- 1 – 3 million
- 250 000 – 1 million
- 100 000 – 250 000
- 25 000 – 100 000
- under 25 000
- country capital underline
- state or province capital underline

Scale 1 : 19 550 000

0 200 400 600 km

0 100 200 300 miles

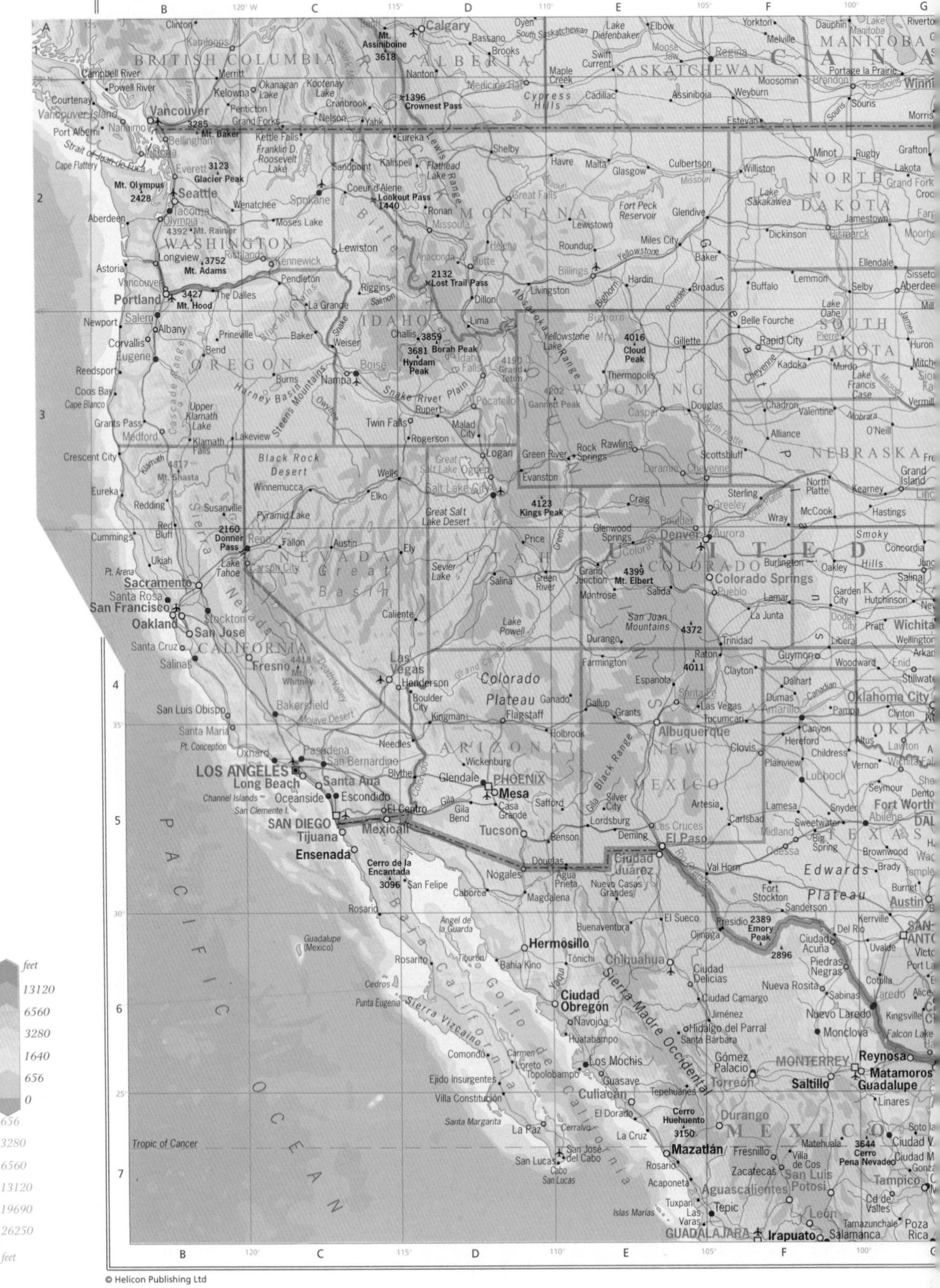

metres	feet
4000	13120
2000	6560
1000	3280
500	1640
200	656
0	0
200	656
1000	3280
2000	6560
4000	13120
6000	19690
8000	26250
metres	feet

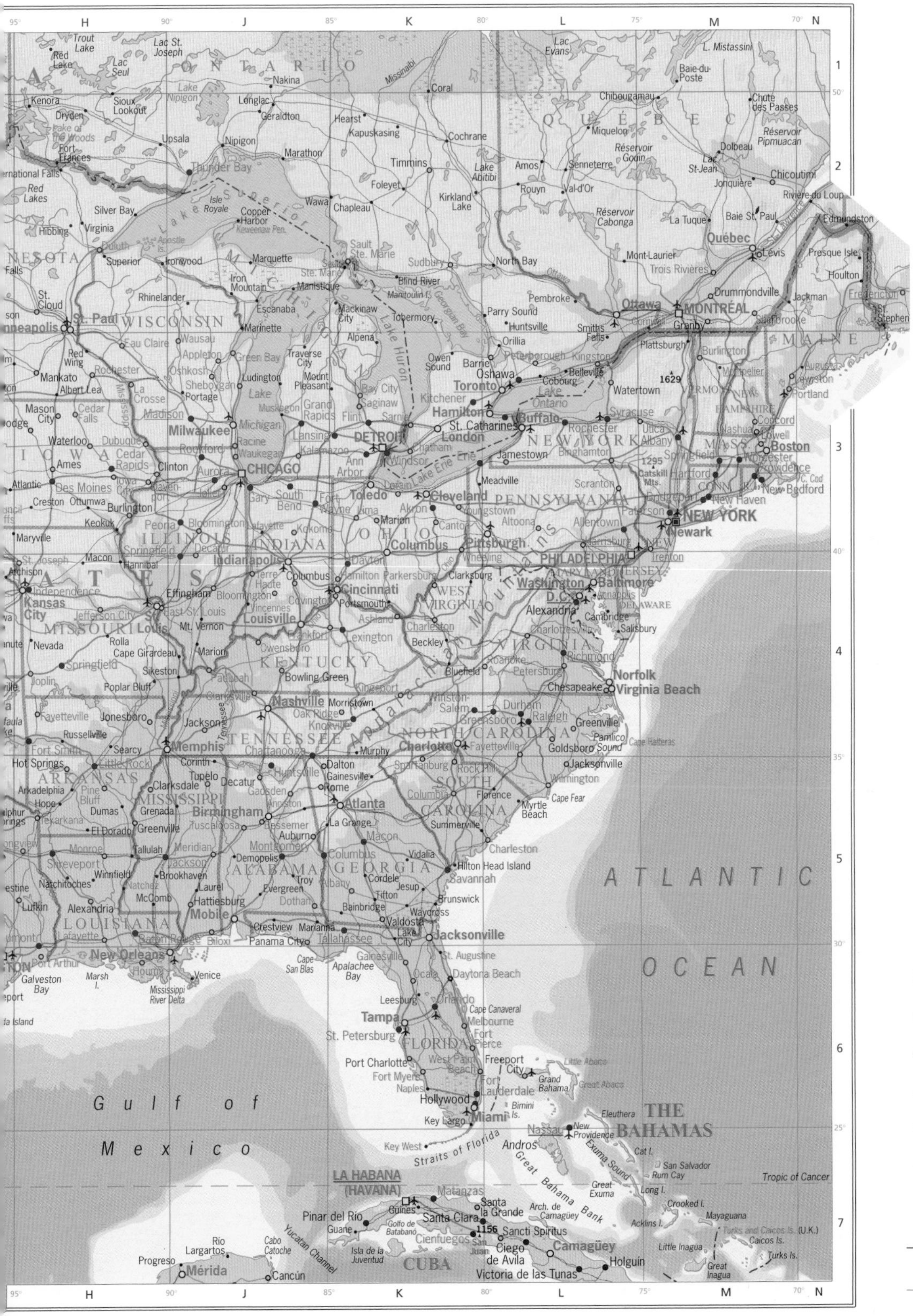

UNITED STATES

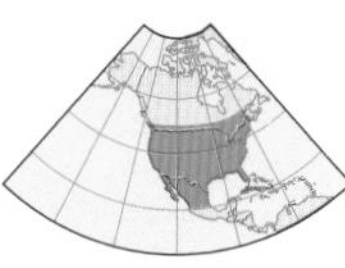

Settlements

- *over 3 million*
- *1 – 3 million*
- *250 000 – 1 million*
- *100 000 – 250 000*
- *25 000 – 100 000*
- *under 25 000*
- *country capital underline*
- *state or province capital underline*

Scale 1 : 9 100 000

0 100 200 300 km

0 50 100 150 miles

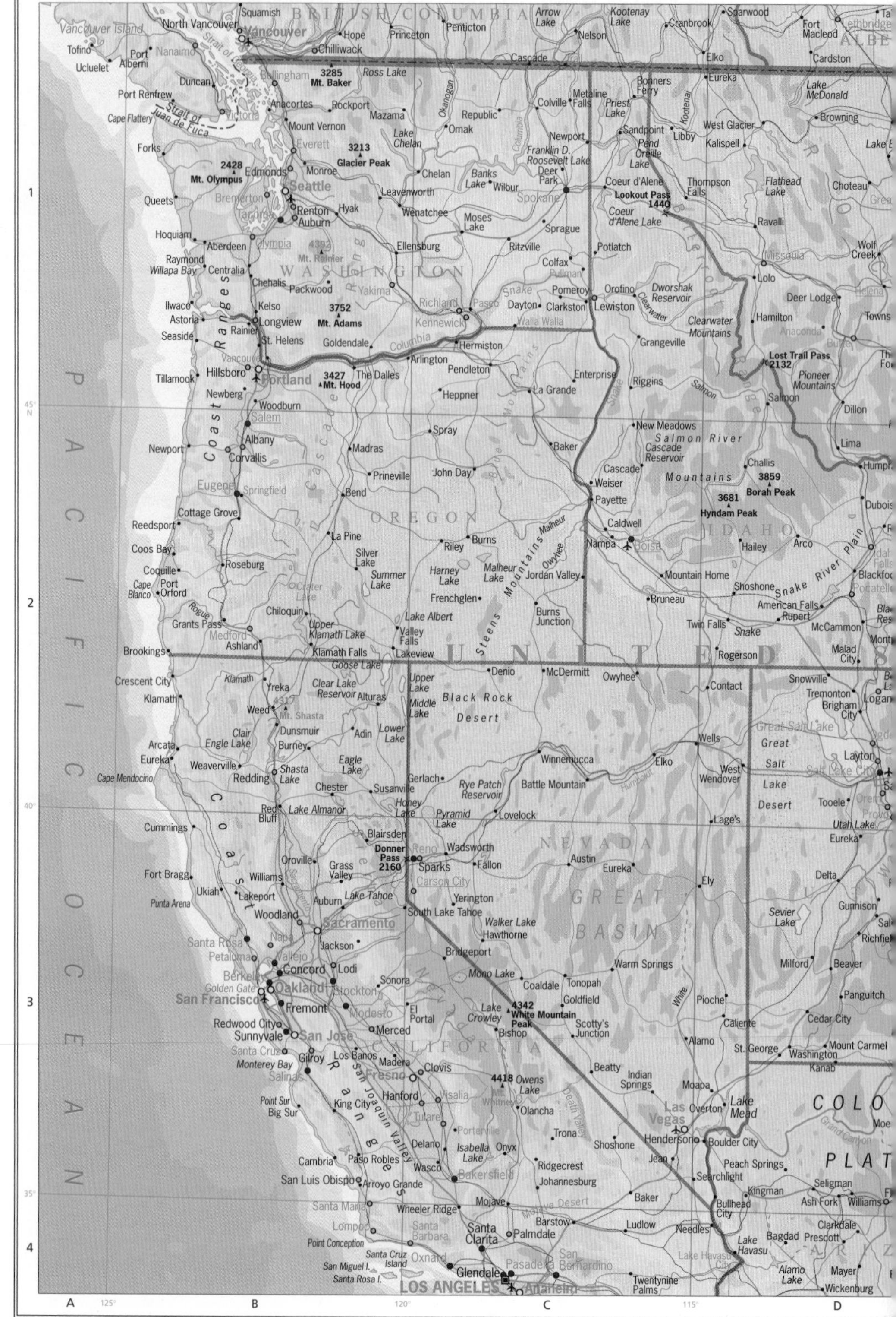

metres	feet
4000	13120
2000	6560
1000	3280
500	1640
200	656
0	0
200	656
1000	3280
2000	6560
4000	13120
6000	19690
8000	26250
metres	feet

Northwest United States

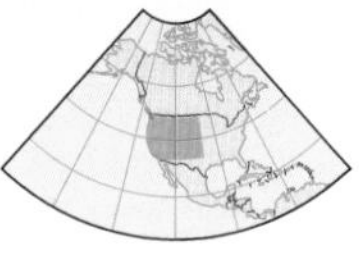

- Colorado
- Idaho
- Montana
- Nebraska
- Nevada
- North Dakota
- Oregon
- South Dakota
- Utah
- Washington
- Wyoming

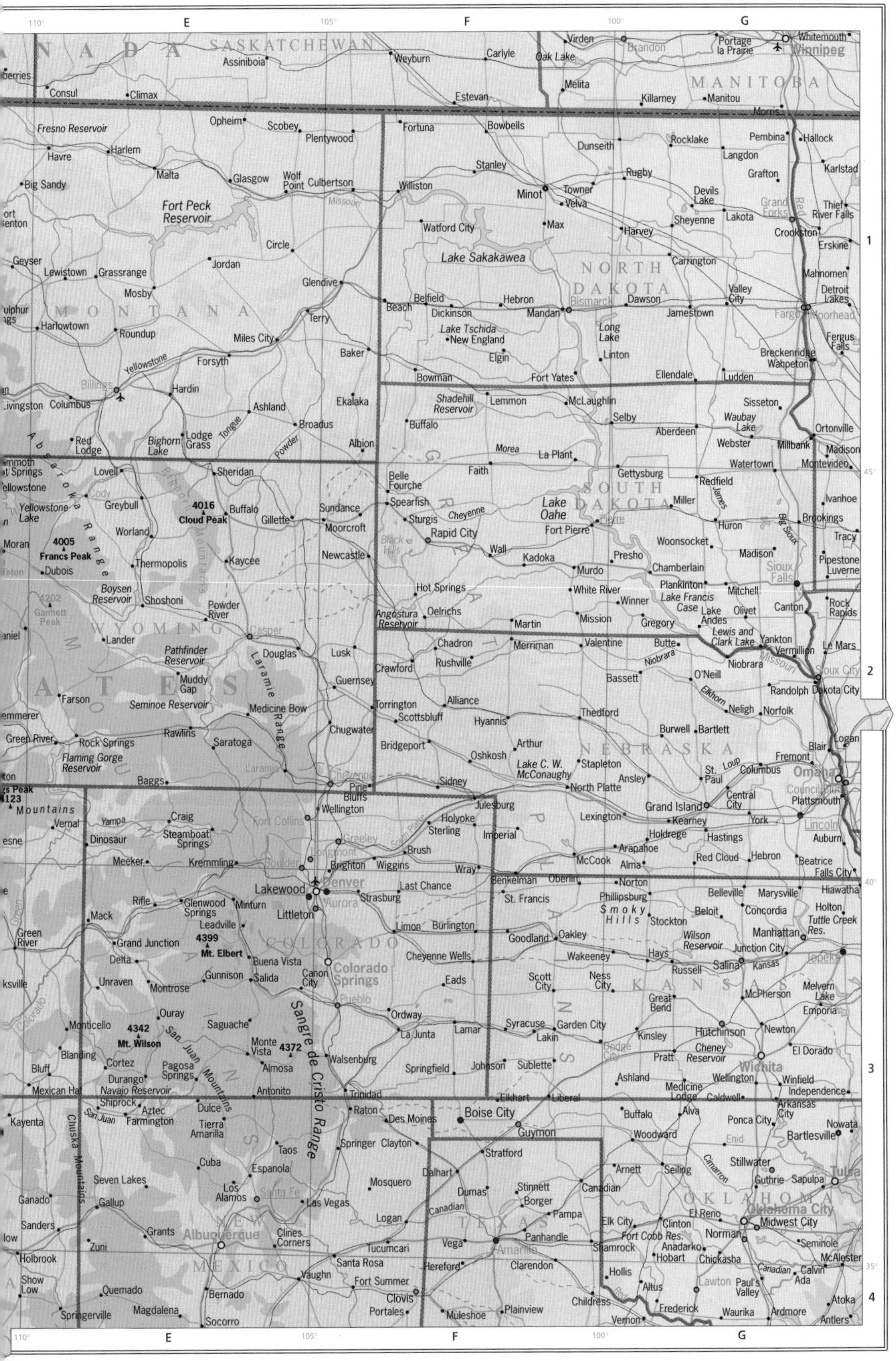

88 89

Settlements

- over 3 million
- 1 – 3 million
- 250 000 – 1 million
- 100 000 – 250 000
- 25 000 – 100 000
- under 25 000
- country capital underline
- state or province capital underline

Scale 1 : 9 100 000

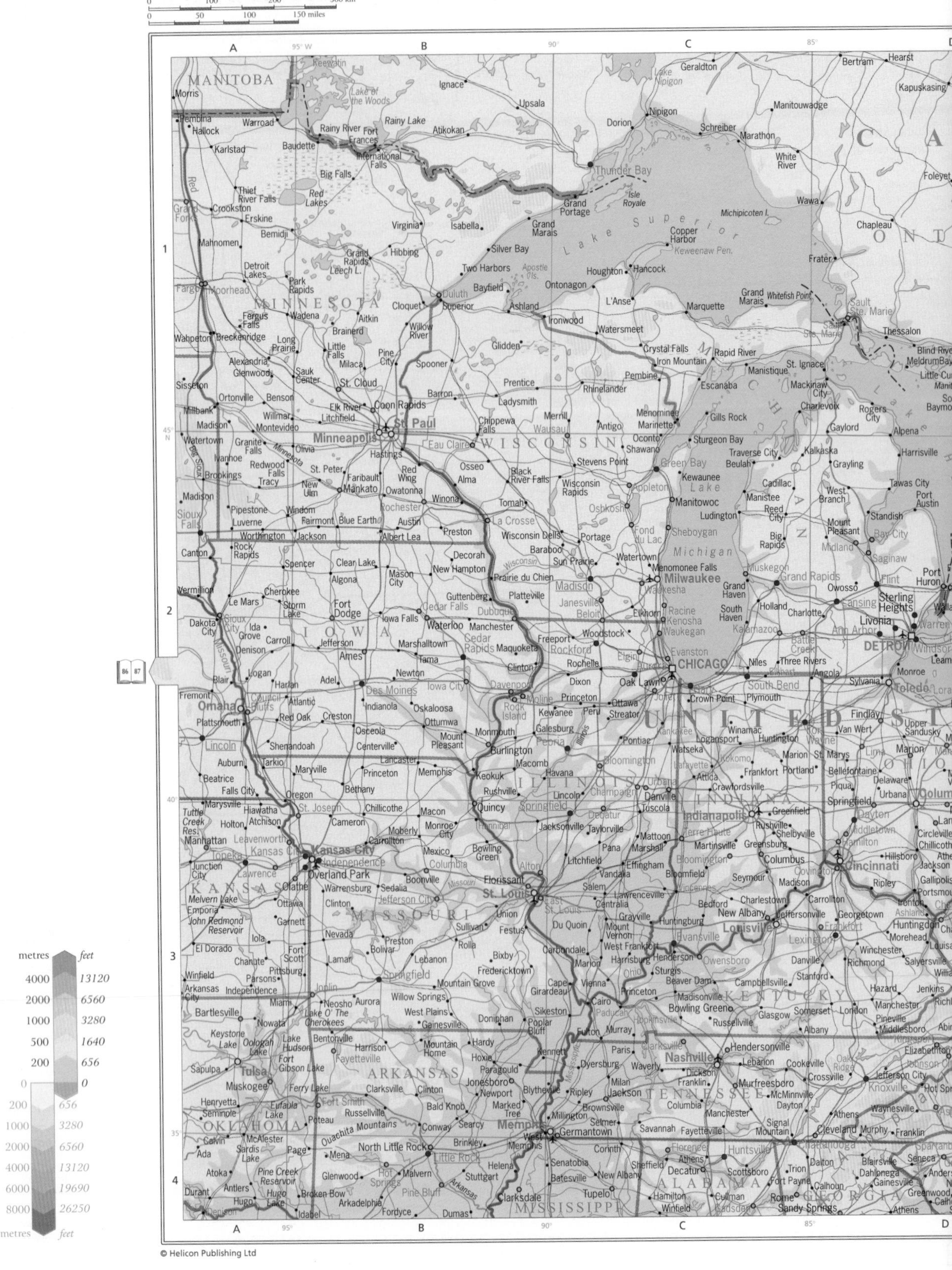

NORTHEAST UNITED STATES

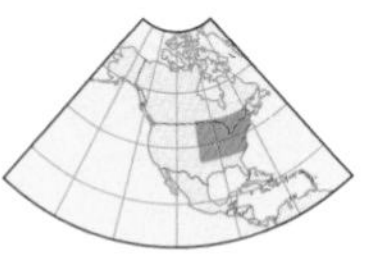

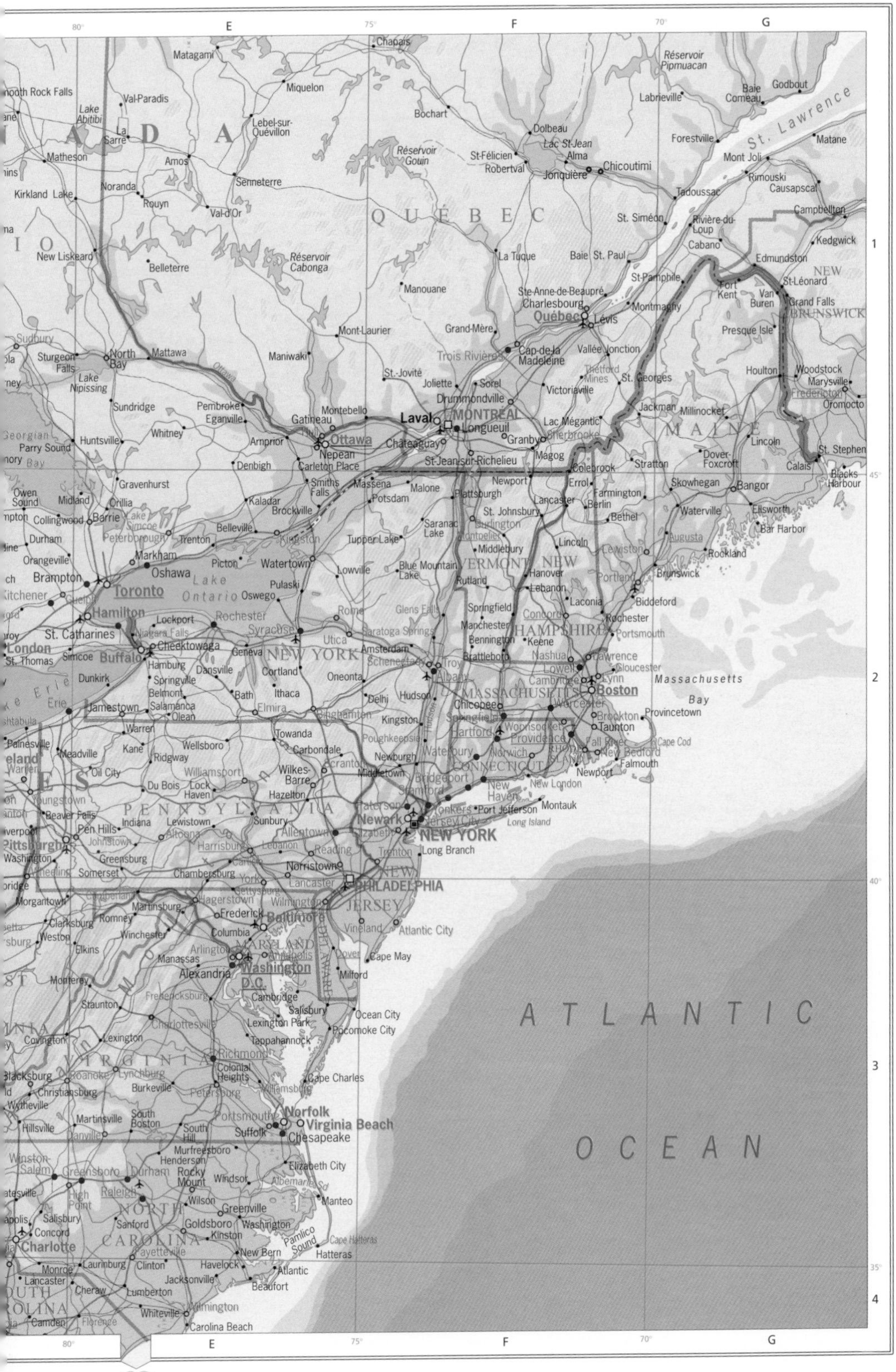

- Connecticut
- Delaware
- District of Columbia
- Illinois
- Indiana
- Iowa
- Maine
- Maryland
- Massachusetts
- Michigan
- Minnesota
- New Hampshire
- New Jersey
- New York
- Ohio
- Pennsylvania
- Rhode Island
- Vermont
- West Virginia
- Wisconsin

Settlements

- ▣ *over 3 million*
- □ *1 – 3 million*
- ○ *250 000 – 1 million*
- ● *100 000 – 250 000*
- ◦ *25 000 – 100 000*
- • *under 25 000*
- *country capital underline*
- *state or province capital underline*

90 91

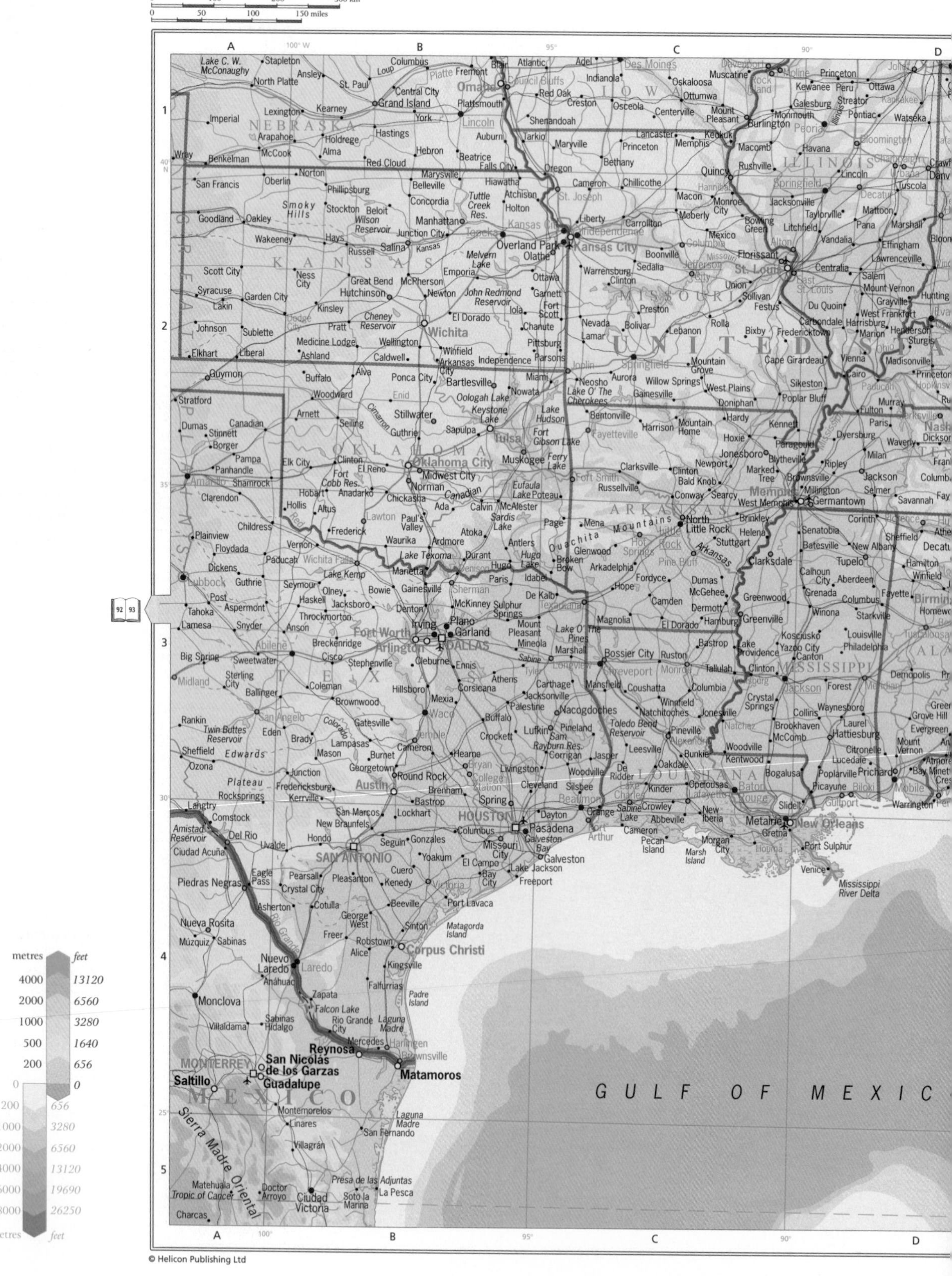
Scale 1 : 9 100 000
0 100 200 300 km
0 50 100 150 miles
A B C D
1 2 3 4 5
NEBRASKA
KANSAS
OKLAHOMA
TEXAS
IOWA
MISSOURI
ILLINOIS
ARKANSAS
LOUISIANA
MISSISSIPPI
UNITED STATES
MEXICO
GULF OF MEXICO
Omaha
Lincoln
Kansas City
Wichita
Tulsa
Oklahoma City
Fort Worth
DALLAS
HOUSTON
SAN ANTONIO
Austin
Corpus Christi
New Orleans
Baton Rouge
Little Rock
Memphis
St. Louis
Shreveport
Jackson
MONTERREY
Saltillo
Matamoros
Reynosa
Nuevo Laredo
Mississippi River Delta
Sierra Madre Oriental
Tropic of Cancer
metres feet
4000 13120
2000 6560
1000 3280
500 1640
200 656
0 0
200 656
1000 3280
2000 6560
4000 13120
6000 19690
8000 26250
metres feet

SOUTHEAST UNITED STATES

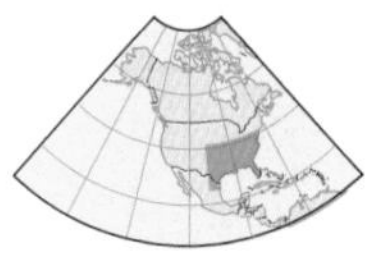

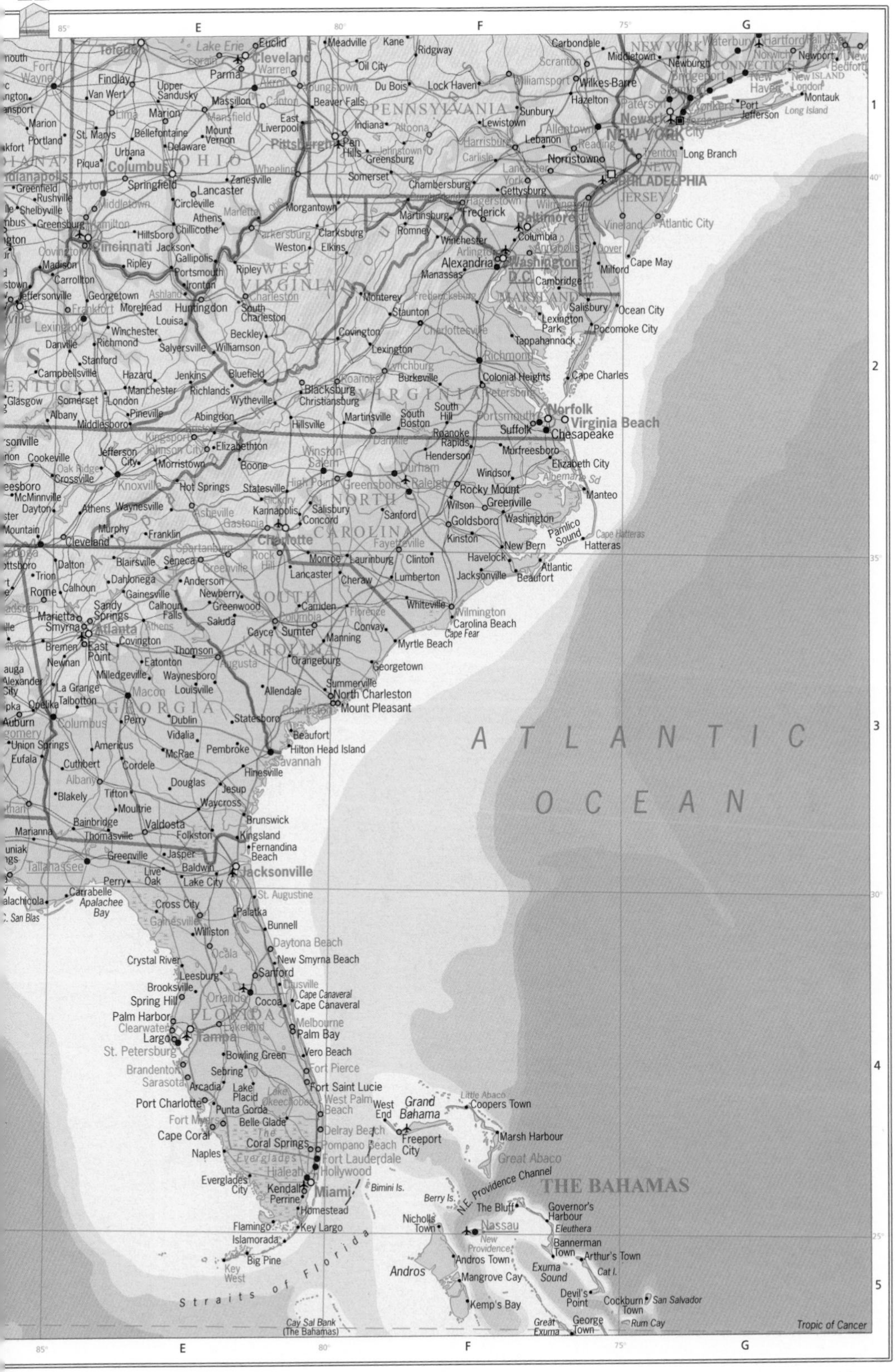

- Alabama
- Arkansas
- The Bahamas
- Florida
- Georgia
- Kentucky
- Louisiana
- Mississippi
- Missouri
- North Carolina
- South Carolina
- Tennessee
- Texas
- Virginia

ATLANTIC OCEAN

Settlements

- ▣ *over 3 million*
- □ *1 – 3 million*
- ○ *250 000 – 1 million*
- ● *100 000 – 250 000*
- ◉ *25 000 – 100 000*
- • *under 25 000*
- ___ *country capital underline*
- ___ *state or province capital underline*

Scale 1 : 9 100 000

0 100 200 300 km

0 50 100 150 miles

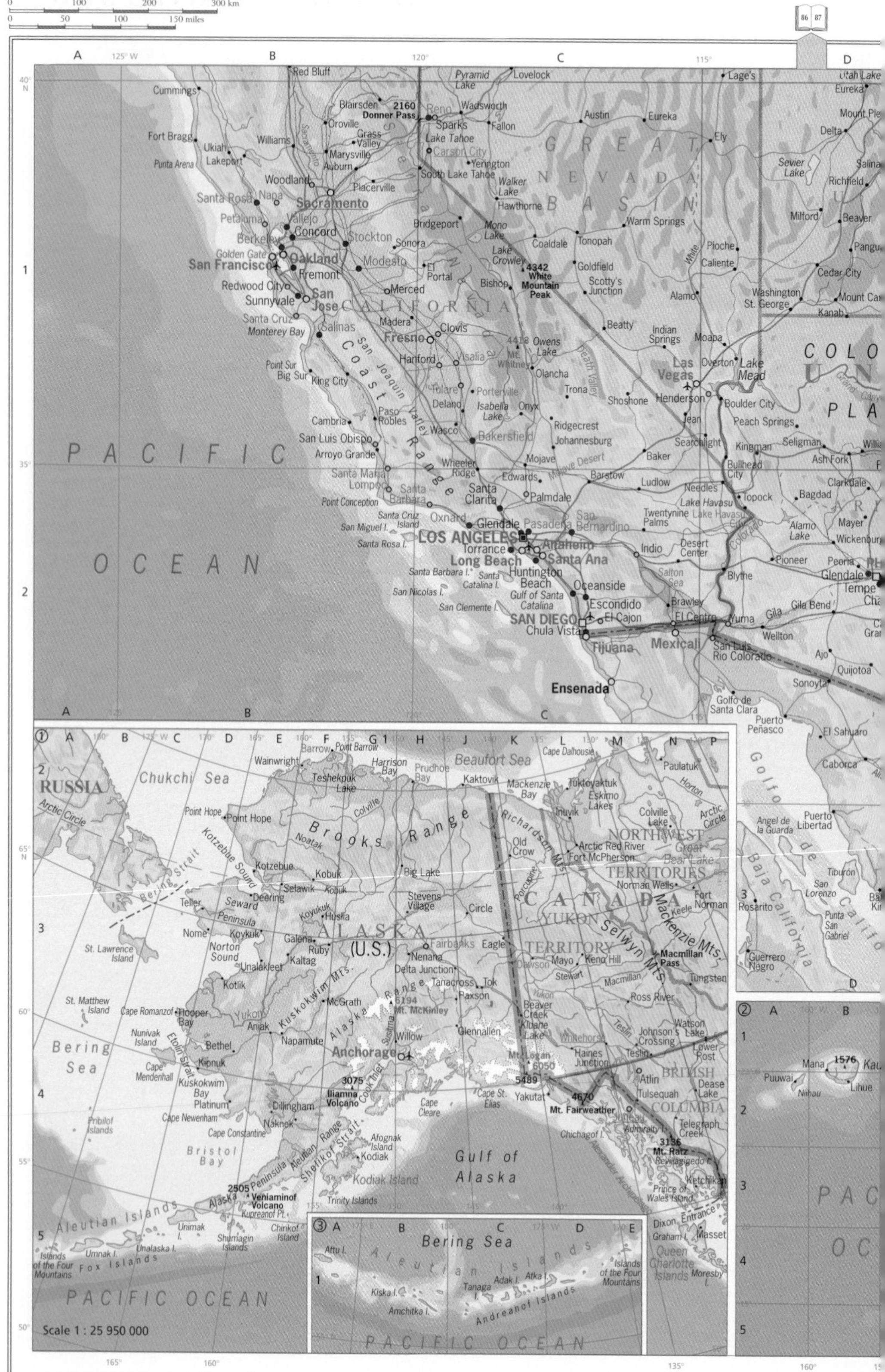

metres	feet
4000	13120
2000	6560
1000	3280
500	1640
200	656
0	0
200	656
1000	3280
2000	6560
4000	13120
6000	19690
8000	26250
metres	feet

SOUTHWEST UNITED STATES

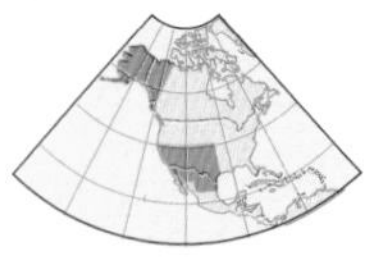

- Alaska
- Arizona
- California
- Hawaii
- Kansas
- New Mexico
- Oklahoma
- Texas

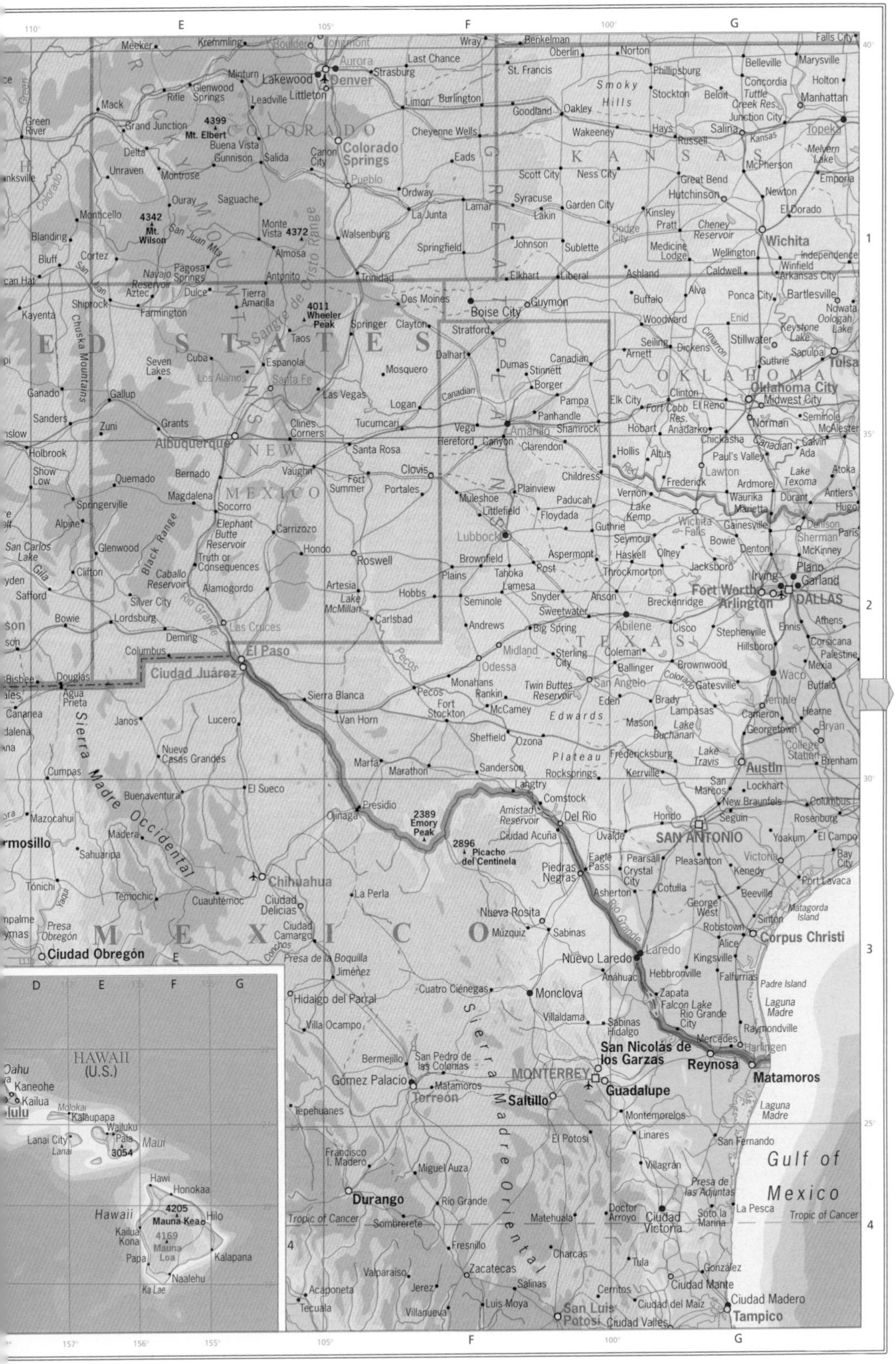

Settlements

- *over 3 million*
- *1 – 3 million*
- *250 000 – 1 million*
- *100 000 – 250 000*
- *25 000 – 100 000*
- *under 25 000*
- *country capital underline*
- *state or province capital underline*

90 91

Scale 1 : 20 300 000

0 200 400 600 km

0 100 200 300 miles

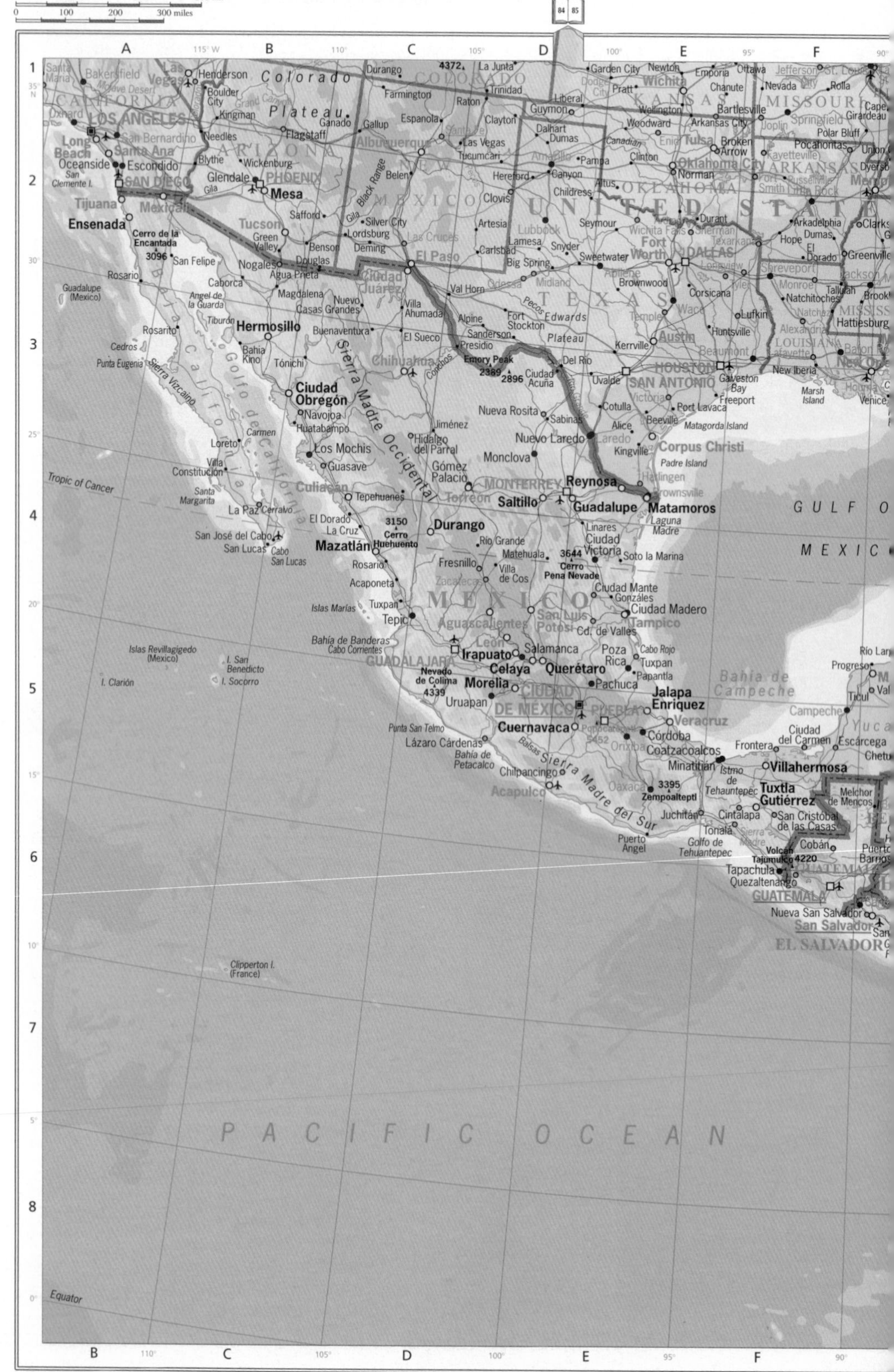

metres	feet
4000	13120
2000	6560
1000	3280
500	1640
200	656
0	0
200	656
1000	3280
2000	6560
4000	13120
6000	19690
8000	26250
metres	feet

CENTRAL AMERICA AND THE CARIBBEAN

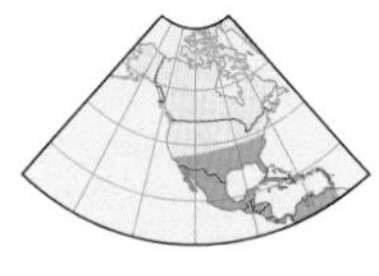

- Belize
- Caribbean Islands
- Costa Rica
- El Salvador
- Guatemala
- Honduras
- Mexico
- Nicaragua
- Panama

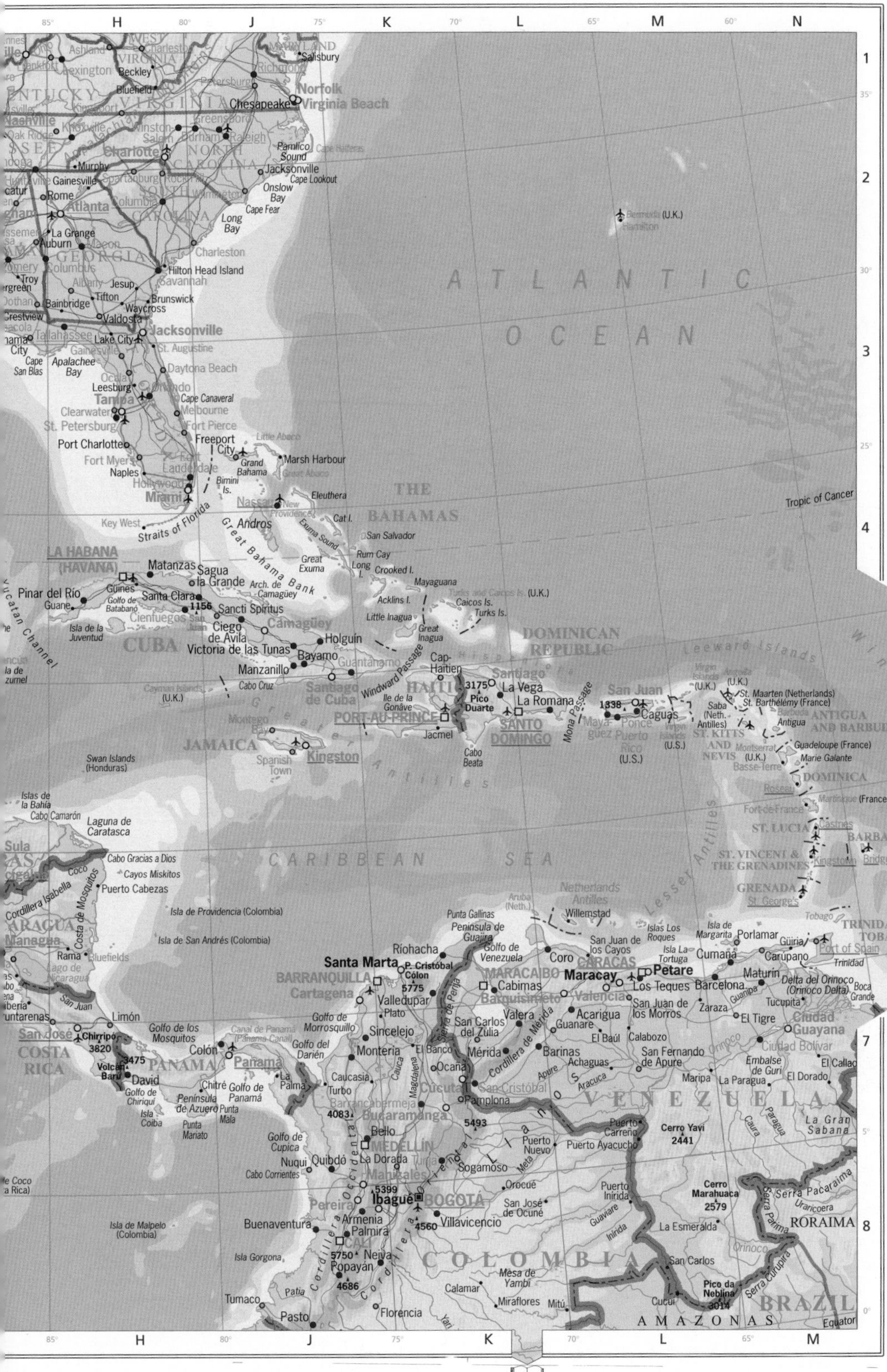

Settlements

- ▣ over 3 million
- □ 1 – 3 million
- ○ 250 000 – 1 million
- ● 100 000 – 250 000
- ◉ 25 000 – 100 000
- • under 25 000
- country capital underline
- state or province capital underline

98 99

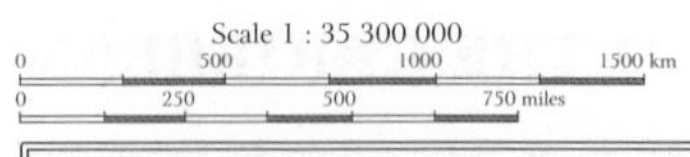

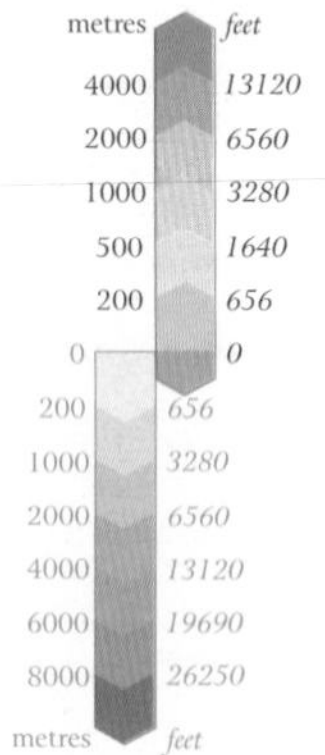

metres	feet
4000	13120
2000	6560
1000	3280
500	1640
200	656
0	0
200	656
1000	3280
2000	6560
4000	13120
6000	19690
8000	26250
metres	feet

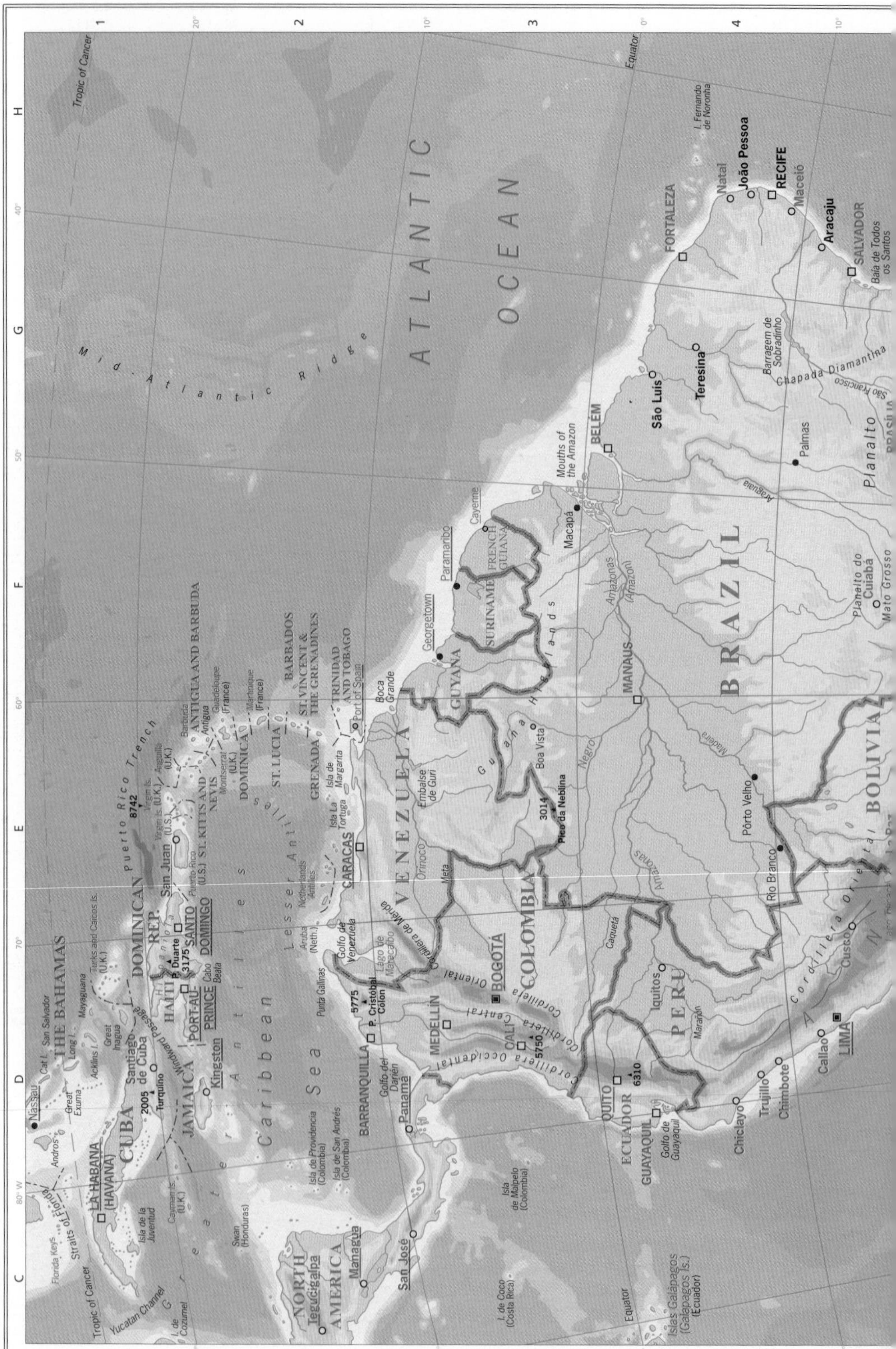

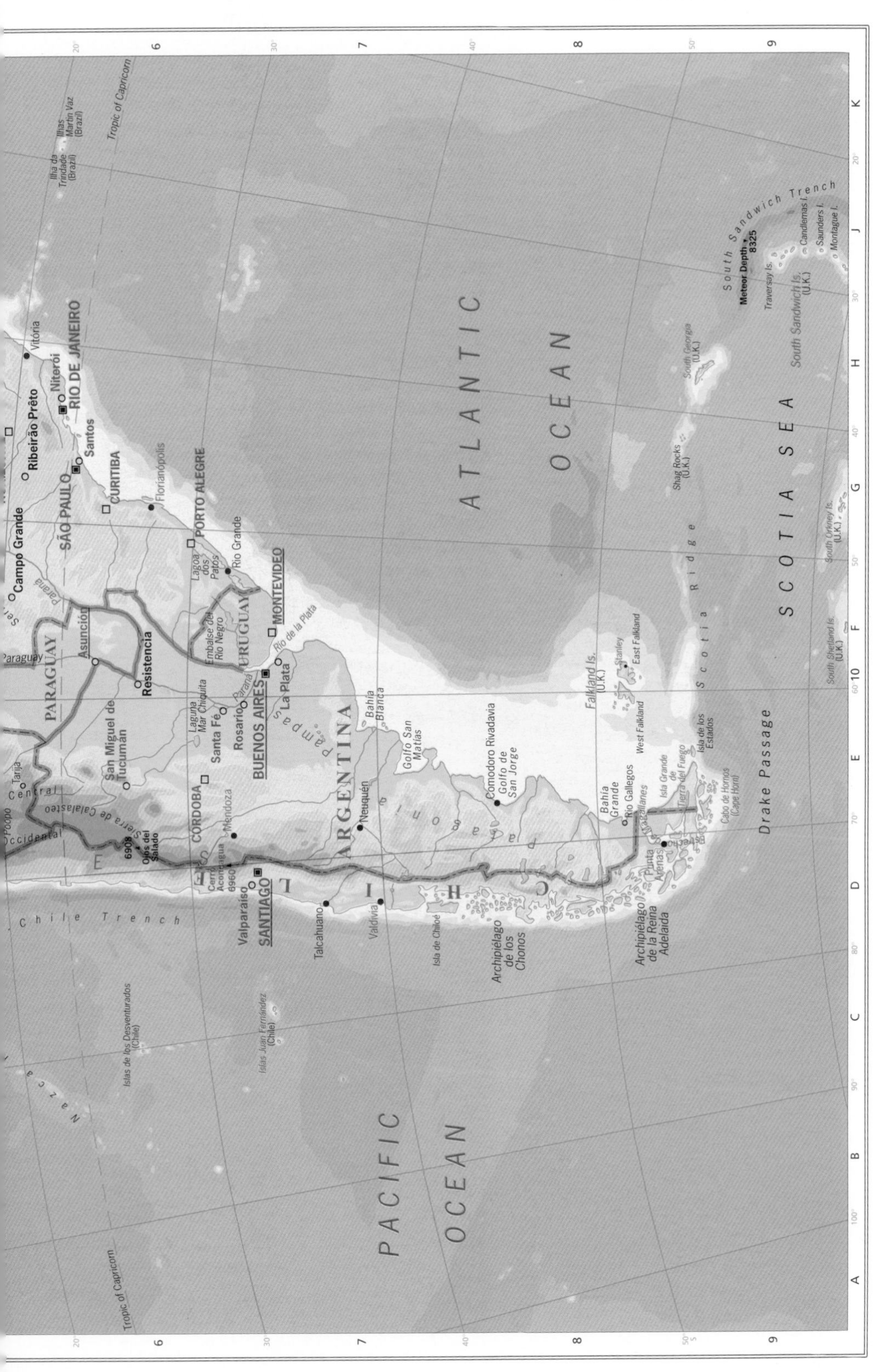
ATLANTIC
OCEAN
PACIFIC
OCEAN
SCOTIA SEA
Drake Passage
Scotia Ridge
South Sandwich Trench
Meteor Depth 8325
Chile Trench
Tropic of Capricorn
PARAGUAY
URUGUAY
ARGENTINA
CHILE
Pampas
Patagonia
Campo Grande
Ribeirão Prêto
Vitória
Niterói
RIO DE JANEIRO
SÃO PAULO
Santos
CURITIBA
Florianópolis
PORTO ALEGRE
Rio Grande
Lagoa dos Patos
MONTEVIDEO
Embalse del Río Negro
Asunción
Resistencia
Paraguay
Paraná
Río de la Plata
BUENOS AIRES
La Plata
Rosario
Santa Fé
Laguna Mar Chiquita
San Miguel de Tucumán
CÓRDOBA
Mendoza
Tarija
Sierra de Calalaste
Ojos del Salado
6908
Cerro Aconcagua 6960
Valparaíso
SANTIAGO
Talcahuano
Valdivia
Neuquén
Bahía Blanca
Golfo San Matías
Comodoro Rivadavia
Golfo de San Jorge
Isla de Chiloé
Archipiélago de los Chonos
Bahía Grande
Río Gallegos
Punta Arenas
Archipiélago de la Reina Adelaida
Isla Grande de Tierra del Fuego
Isla de los Estados
Cabo de Hornos (Cape Horn)
Falkland Is. (U.K.)
West Falkland
East Falkland
Stanley
Shag Rocks (U.K.)
South Georgia (U.K.)
South Orkney Is. (U.K.)
South Shetland Is. (U.K.)
South Sandwich Is. (U.K.)
Traversay Is.
Candlemas I.
Saunders I.
Montague I.
Ilha da Trindade (Brazil)
Ilhas Martin Vaz (Brazil)
Islas de los Desventurados (Chile)
Islas Juan Fernández (Chile)

Scale 1 : 20 300 000

0 200 400 600 km

0 100 200 300 miles

Settlements

- ▣ *over 3 million*
- ☐ *1 – 3 million*
- ⊙ *250 000 – 1 million*
- ● *100 000 – 250 000*
- ⊚ *25 000 – 100 000*
- • *under 25 000*
- ___ *country capital underline*
- ___ *state or province capital underline*

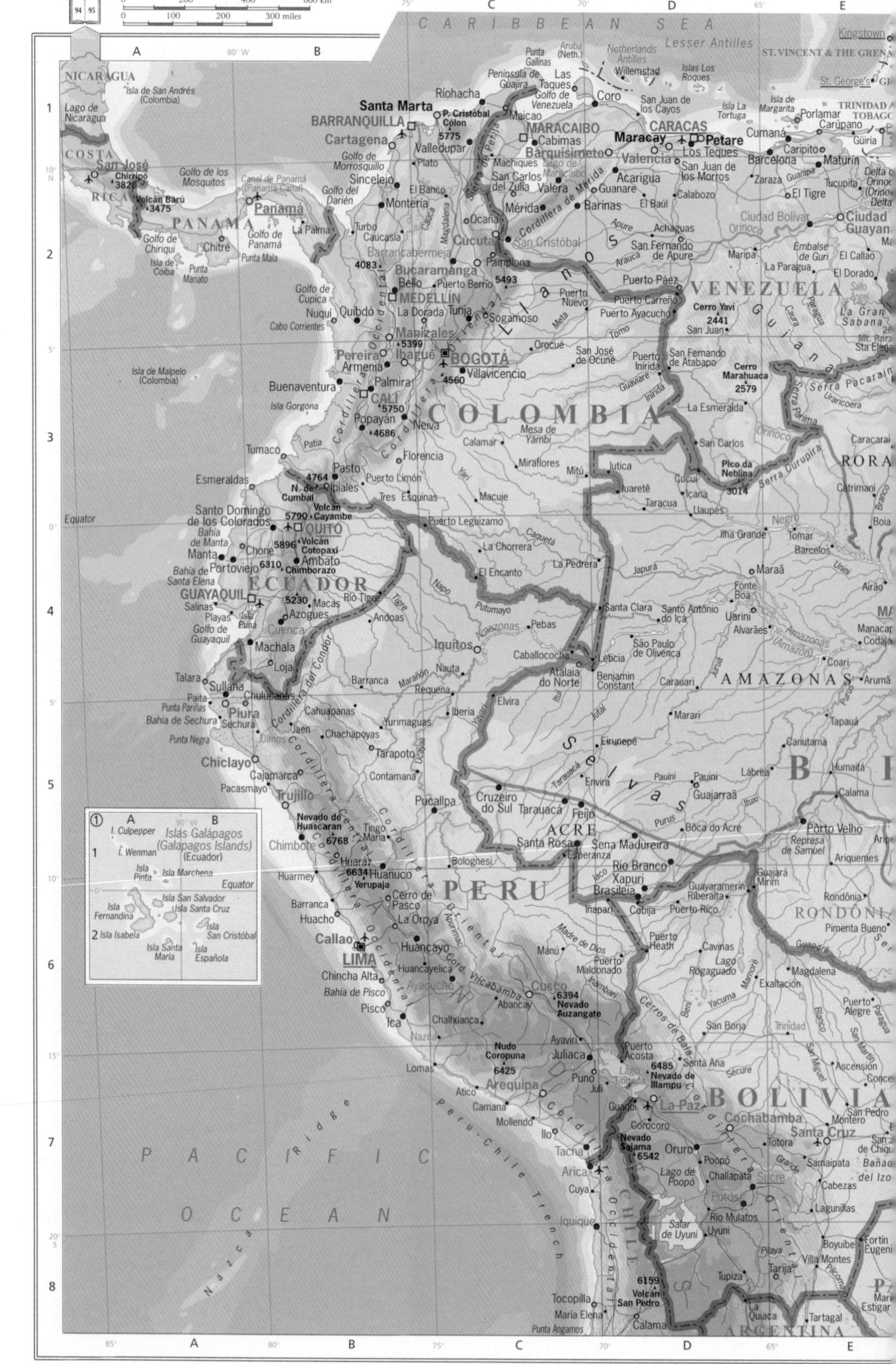

metres / feet

metres	feet
4000	13120
2000	6560
1000	3280
500	1640
200	656
0	0
200	656
1000	3280
2000	6560
4000	13120
6000	19690
8000	26250

metres / feet

NORTHERN SOUTH AMERICA

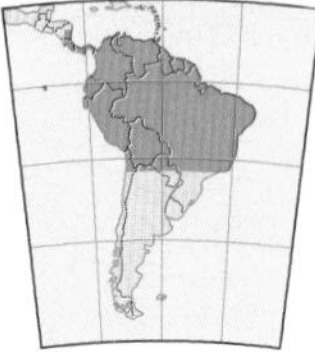

- Bolivia
- Brazil
- Colombia
- Ecuador
- French Guiana
- Galapagos Islands
- Guyana
- Peru
- Suriname
- Venezuela

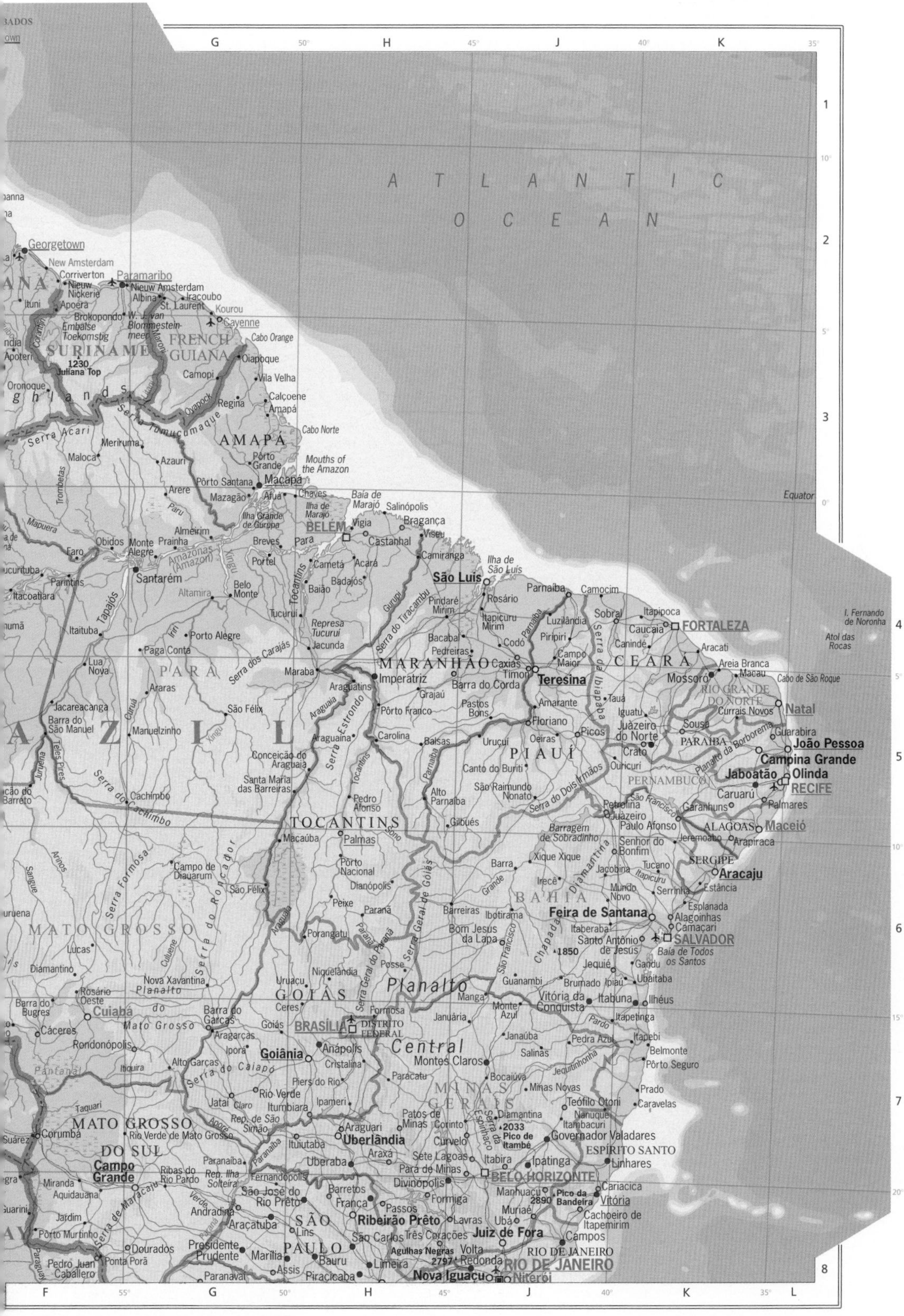

Scale 1 : 20 300 000

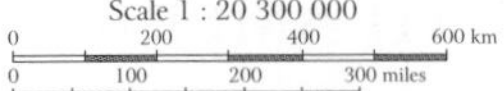

PACIFIC OCEAN

ARGENTINA

BOLIVIA

PERU

Drake Passage

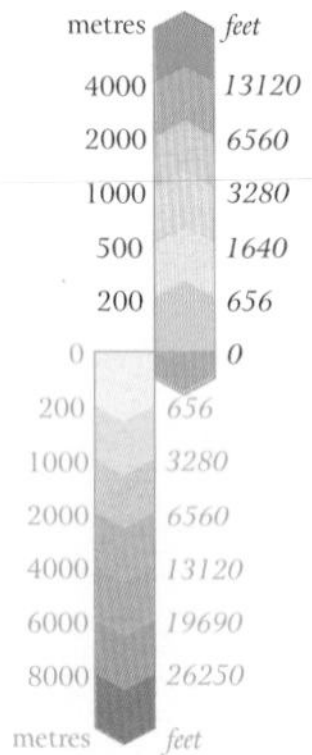

metres	feet
4000	13120
2000	6560
1000	3280
500	1640
200	656
0	0
200	656
1000	3280
2000	6560
4000	13120
6000	19690
8000	26250
metres	feet

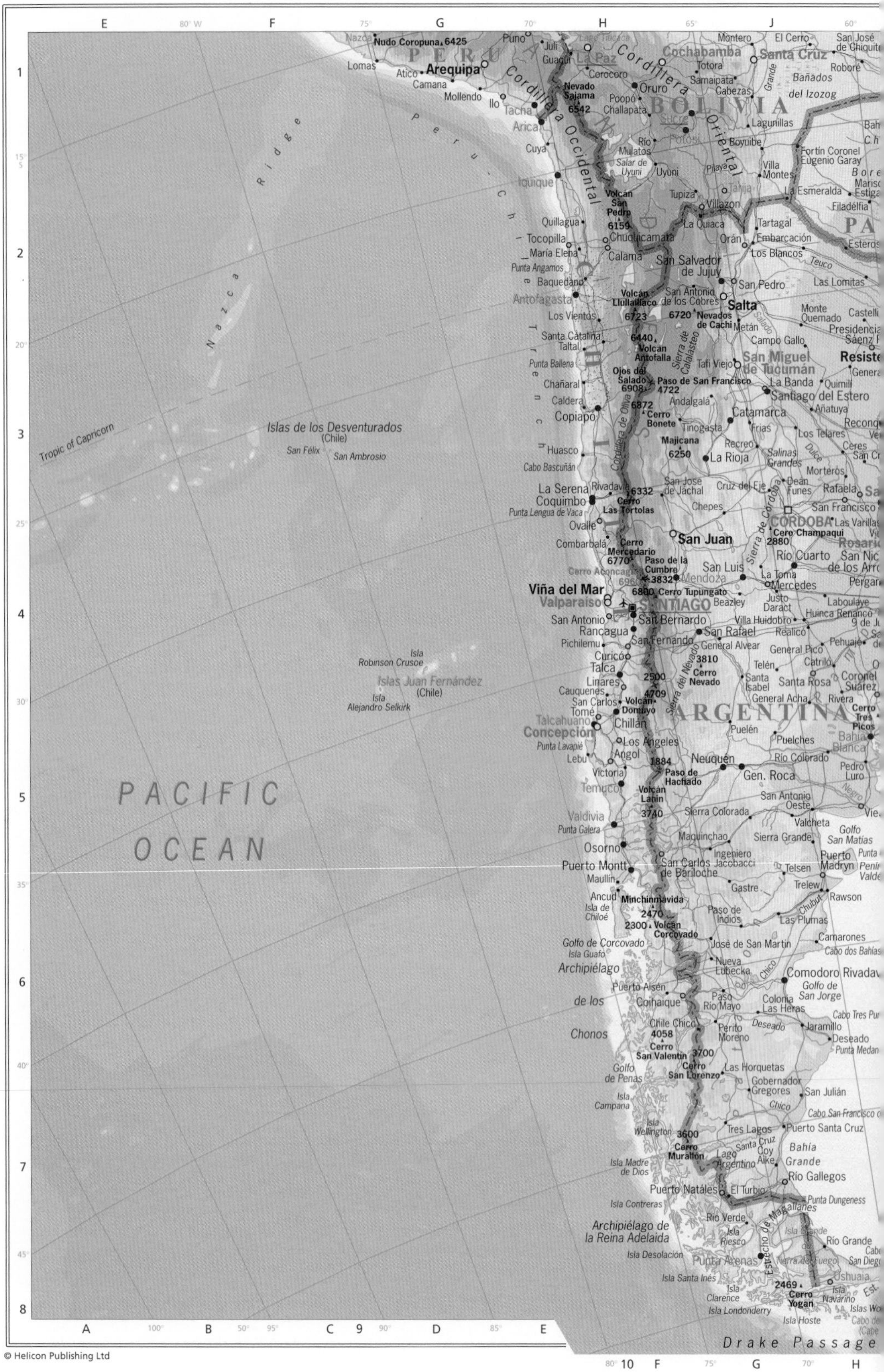

Southern South America

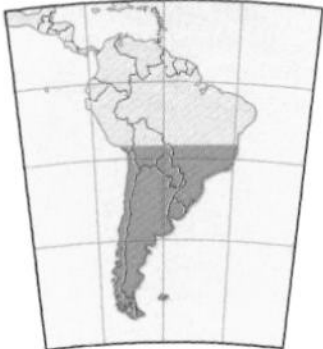

- Argentina
- Chile
- Falkland Islands
- Paraguay
- Uruguay

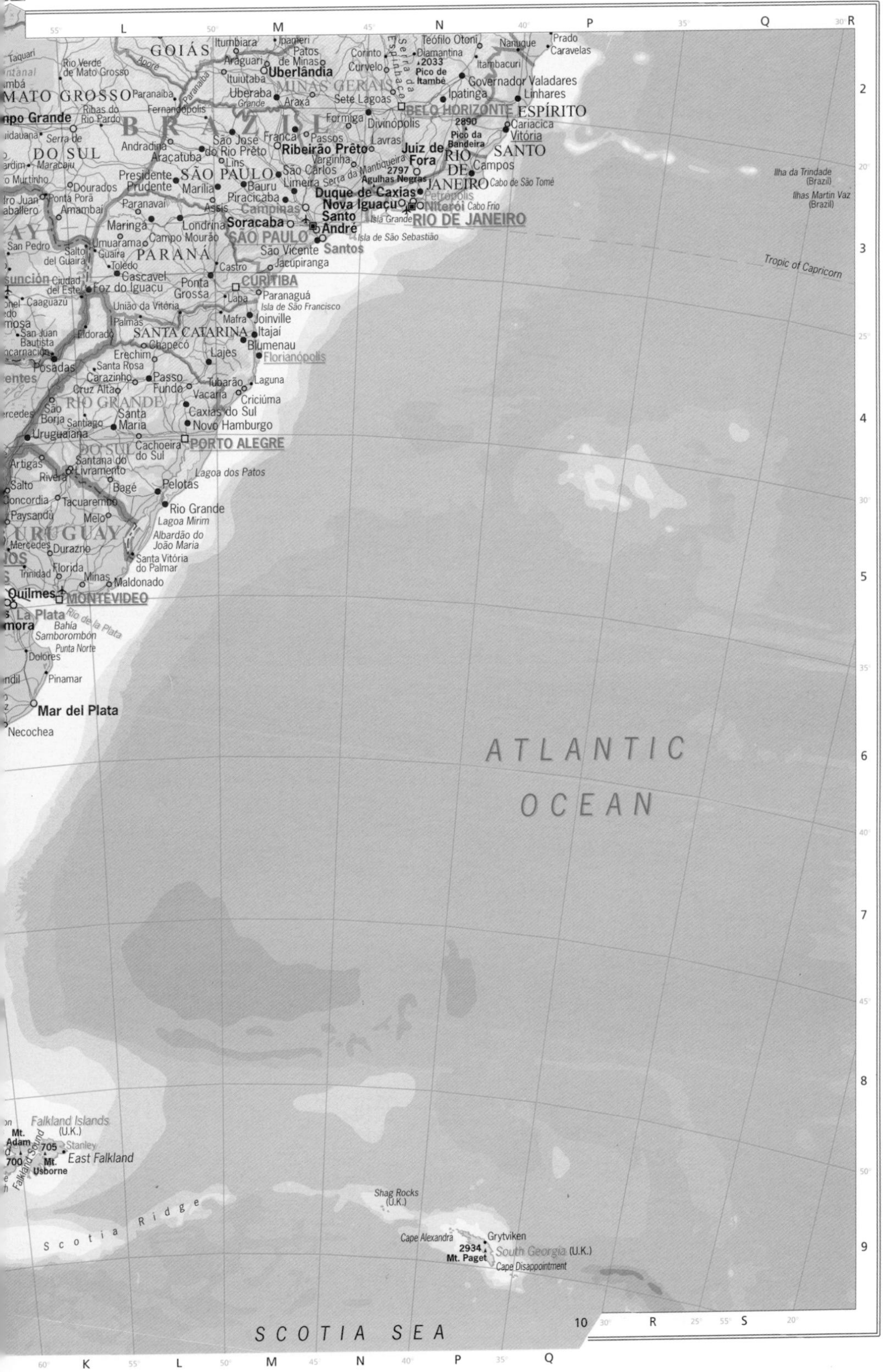

ATLANTIC OCEAN

SCOTIA SEA

Settlements

- ▣ over 3 million
- □ 1 – 3 million
- ○ 250 000 – 1 million
- ● 100 000 – 250 000
- ◉ 25 000 – 100 000
- • under 25 000
- country capital underline
- state or province capital underline

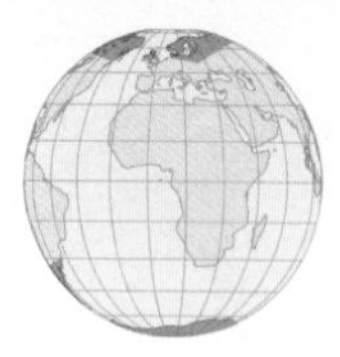

Polar regions

Scale 1 : 63 900 000
0 500 1000 1500 km
0 250 500 750 miles

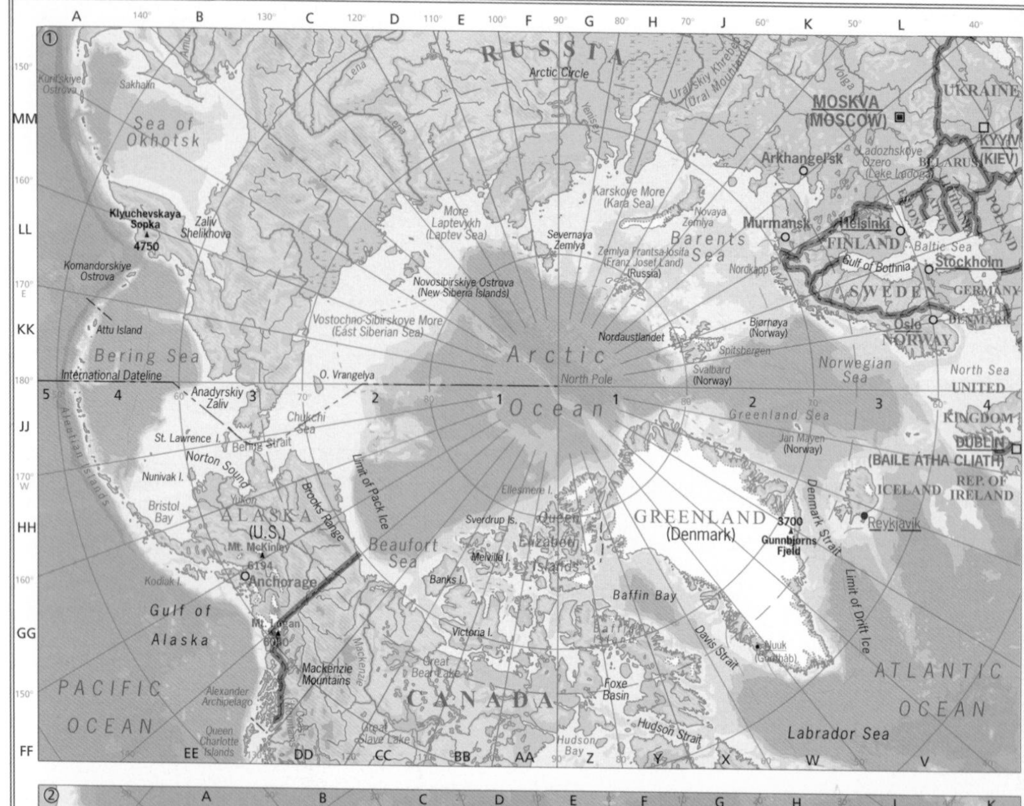

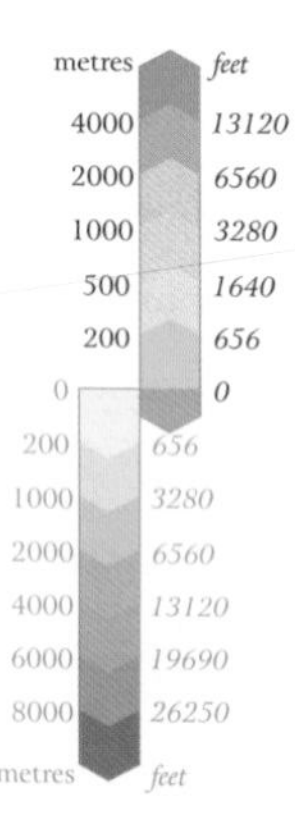

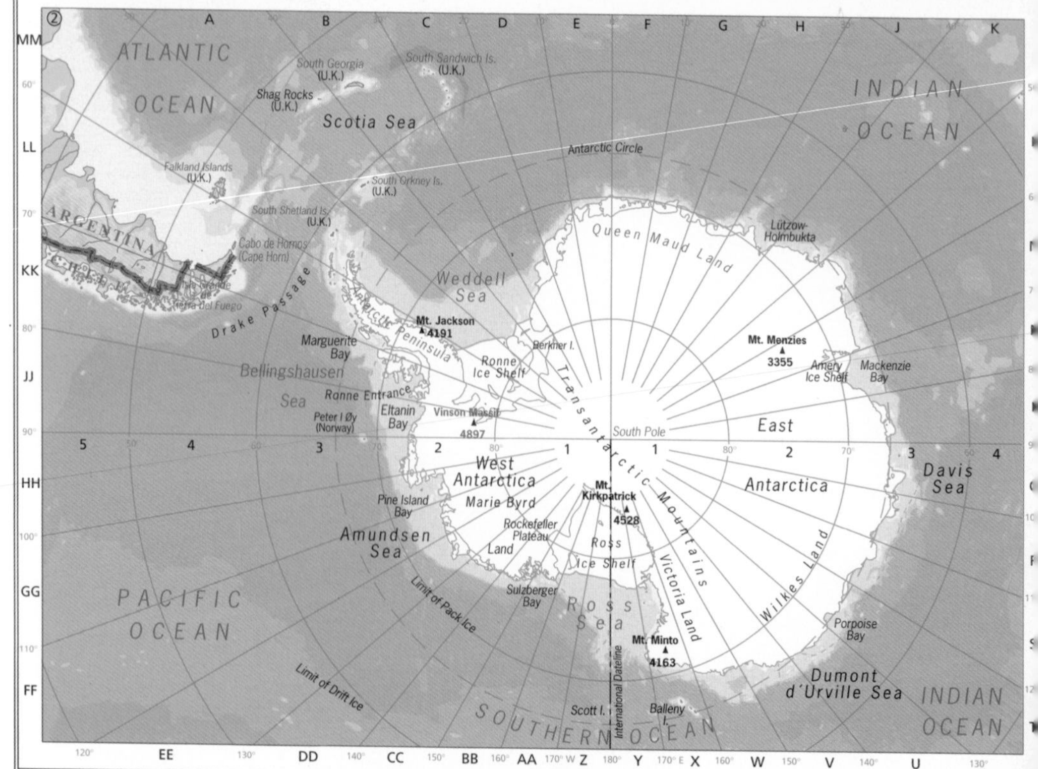

Aachen (French **Aix-la-Chapelle**), cathedral city and spa in the *Land* (state) of North Rhine-Westphalia, Germany, 64 km/40 mi southwest of Cologne, near the Dutch and Belgian borders; population (1995) 247,400. It has thriving electronic, glass, food, woollen textile, and rubber industries, and is one of Germany's principal railway junctions. It also lies at the centre of a coalmining district, although coal and lignite production is in decline. Aachen was the Roman **Aquisgranum** and was the site of baths in the 1st century. It has been visited for its thermal springs since Roman times, and the spa facilities continue today. Charlemagne, Holy Roman Emperor from 800, founded the cathedral in 796. 22 J4

Aalborg alternative form of ►Ålborg, a port in Denmark. 16 E8

Aalen industrial city in the *Land* (administrative region) of Baden-Württemberg, Germany, on the River Köcher, 65 km/40 mi east of Stuttgart; population (1995) 67,000. Textiles, iron, and optical goods are manufactured. 20 F8

Aalst (French **Alost**), industrial town in East Flanders, Belgium, on the River Dender 24 km/15 mi northwest of Brussels; population (1997) 76,200. There are brewing, textile, and clothing industries. 22 G4

Aarau commune and capital of the Swiss canton of Aargau, on the River Aar, 35 km/22 mi west of Zürich; population (1995) 15,900. Silk, leather, cotton goods, and mathematical instruments are made in the town. 30 D3

Aargau (French **Argovie**), canton of northern Switzerland; area 1,404 sq km/542 sq mi; population (2000 est) 545,300. Taking its name from the River Aar flowing through it, the canton is bounded on the north by the Rhine and on the south by the canton of Lucerne. The main towns are Aarau (the capital), Baden, and Brugg. The service sector is the largest employer, with 59% of the labour force. Industry and business are also important, with 37%. Only 3% work in the previously flourishing industry of agriculture and forestry. 8 (1)C1

Aarhus alternative form of ►Århus, a port in Denmark. 16 F8

Abaco alternative form of ►Abacos Islands, a group of islands in the northern Bahamas. 90 F4

Abacos (or **Abaco Islands**), group of islands in the northern Bahamas, comprising Great Abaco – population (1990) 10,100 – Little Abaco, and a number of keys (small, low-lying islands of sand and coral fragments); area 963 sq km/372 sq mi. The largest town in the group is Marsh Harbour on Great Abaco, and the largest town on Little Abaco is Fox Town. Tourism is an important industry. 90 F4

Abadan Iranian oil port in Khuzestan province, situated on an island on the east side of the Shatt-al-Arab waterway at the head of the Gulf, 675 km/420 mi southwest of Tehran; population (1997 est) 308,000. Abadan is the chief refinery and shipping centre for Iran's oil industry, nationalized in 1951. This measure was the beginning of the worldwide movement by oil-producing countries to assume control of profits from their own resources. Oil installations were badly damaged during the Iran–Iraq war from 1980–88. 61 C1

Abakan (formerly **Ust-Abakanskoye**), coal-mining city in the Russian Federation, capital of the Republic of Khakasia in the Krasnoyarsk territory of the Russian Federation, in southwestern Siberia, and cultural centre of the Minusinsk Basin; population (1996 est) 163,000. Lying at the confluence of the Yenisey and Abakan rivers 275 km/171 mi southwest of Krasnoyarsk, and the starting point of the Abakan-Taishet railway, a branch of the Trans-Siberian railway line, Abakan is a transportation centre. It also has textile, metalworking, timber, and food-production industries. Marble is quarried in the region. 42 S7

Abbeville (derived from Latin *villa* 'country estate', *abbatis* 'of the abbot'), town in the Somme *département*, Picardy region, in northern France, 19 km/12 mi inland from the mouth of the River Somme and 40 km/25 mi northwest of Amiens; population (1999 est) 24,500. Abbeville has a large agricultural market; manufacturing industries include textiles, metal goods, and brewing. The town's 15th-century church of St-Vulfran is notable for its flamboyant facade and twin towers. 22 D4

Abeokuta agricultural trade centre and capital of Ogun State, Nigeria, on the Ogun River, 103 km/64 mi north of Lagos; population (1992 est) 386,800. Products include rice, yams, cassava, maize, palm oil, cotton, fruits, and vegetables. The main industries are zinc oxide manufacturing, light engineering, textiles, fruit canning, plastics, brewing, and timber processing. 68 E3

Aberdeen City city and unitary authority in northeast Scotland. The unitary authority was created in 1996 from the district of the same name that was part of Grampian region from 1975; before that it was part of Aberdeenshire. The city of Aberdeen, as well as being the administrative headquarters of the Aberdeen City unitary authority, is the administrative headquarters of ►Aberdeenshire unitary authority. It is now Scotland's third largest city.
area 185 sq km/71 sq mi
physical low-lying coastal area on the banks of the rivers Dee and Don; it has 3 km/2 mi of sandy beaches
features St Andrew's Episcopal Cathedral (consecrated in 1816), King's College (from 1500) and Marischal College (founded in 1593) which together form Aberdeen University, Brig O'Balgownie (1314–18), Municipal Buildings (1867), St Machar Cathedral (from 1370). Aberdeen's granite buildings have given it the name of 'Silver City'
agriculture white and salmon fishing
industries North Sea oil (it is the main centre in Scotland and Europe for offshore oil exploration and there are shore-based maintenance and service depots for the North Sea oil rigs, oil and gas service industries, paper manufacturing, textiles, engineering, food processing, chemicals, fish processing
population (1998 est) 213,100
24 K4/5 F3

Aberdeenshire unitary authority in northeast Scotland, created in 1996 from three districts within the former Grampian region; its administrative headquarters, Aberdeen, lies outside the authority.
area 6,308 sq km/2,436 sq mi
towns Banff, Fraserburgh, Huntly, Peterhead, Stonehaven, Inverurie
physical area of contrast with mountainous western interior, intensively farmed core, and coastal

plain; Cairngorm Mountains; rivers Deveron, Ythan, Don, and Dee
features Balmoral Castle; Braemar Games
industries oil and gas, papermaking, whisky distilling, seafood, tourism
agriculture fishing, beef cattle, cereal crops
population (1998 est) 226,300; the population has risen by 20% since 1981, largely as a result of employment-driven immigration, mainly because of the oil and gas industry 5 E3

Aberystwyth commercial, tourist, and educational centre in Ceredigion, mid-Wales, situated at the mouths of the rivers Ystwyth and Rheidol in Cardigan Bay; population (2000 est) 13,000, of which 45% is Welsh-speaking. It is the site of the University College of Wales (1872), which dominates the town, and during term time the population swells to around 20,000 (2000 est). Aberystwyth is the largest town in mid-Wales and houses many of Wales's national organizations and institutions. 24 H9

Abidjan port and former capital (until 1983) of the Republic of Côte d'Ivoire; population (1995 est) 2,722,000. There is an airport, communication by rail, as well as by sea, and the city has become increasingly important for its industries which include metallurgy, farm machinery, car and electrical assembly. Products include coffee, palm oil, cocoa, and timber (mahogany). There are tourist markets trading in handicrafts and traditional medicines. 68 D3

Abilene city in Jones County and Taylor County (of which it is the administrative headquarters), north-central Texas, USA; population (1994 est) 110,100. It is situated on the Osage Plains, near the Clear Fork of the Brazos River, 220 km/137 mi west-southwest of Fort Worth. It is the centre for a natural gas and oil-producing area. The manufacture of aero-space equipment and cotton and livestock farming are also important to the regional economy. Clothing, concrete, structural steel, electrical appliances, processed food, and musical instruments are among other local products. 92 G2

Abingdon market town in Oxfordshire, south-central England, at the confluence of the River Ock with the Thames, 10 km/6 mi south of Oxford; population (1991) 35,200. Light industries include brewing, printing, and the manufacture of electronic and scientific instruments. The remains of a Benedictine abbey, founded in 675, include the 14th-century Checker Hall, now restored as an Elizabethan-type theatre, the 16th-century Long Gallery, and the 15th-century gateway. 22 A3

Åbo Swedish name for ►Turku, a port in southwest Finland. 16 M6

Abomey town and centre of the palm oil industry in Benin; population (1992) 65,700. It is connected by rail to Cotonou 100 km/62 mi to the south. Abomey was founded in the 17th century and was the capital of the Yoruba state of Dahomey, which flourished in the 17th–19th centuries, until its capture by the French in 1893. The town is a tourist centre. 68 E3

Abruzzi (or **Abruzzo**), mountainous region of southern central Italy, comprising the provinces of L'Aquila, Chieti, Pescara, and Teramo; area 10,800 sq km/4,169 sq mi; population (1999 est) 1,277,300. L'Aquila, the capital, and Pescara are the principal towns. Gran Sasso d'Italia, 2,914 m/9,564 ft, is the highest point of the ►Apennines. 9 (4)D3

Abruzzo alternative name for ►Abruzzi, a mountainous region of southern central Italy. 9 (4)D3

Abu Dhabi (Arabic **Abū Zabī**), sheikhdom in southwest Asia, on the Gulf, capital of the ►United Arab Emirates; area 67,350 sq km/26,000 sq mi; population (1995) 928,400. Formerly under British protection, it has been ruled since 1971 by Sheikh Sultan Zayed bin al-Nahayan, who is also president of the Supreme Council of Rulers of the United Arab Emirates. 61 F4

Abuja capital of Nigeria (formally designated as such 1982, although not officially recognized until 1992); population of Federal Capital District (1991) 378,700; population of city alone (1992 est) 305,900. Shaped like a crescent, the city was designed by Japanese architect Kenzo Tange; building of the city began in 1976 as a replacement for Lagos, and is still largely under construction. The main functions of the city are administrative, with only light industry. 68 F3

Abu Simbel site of two ancient temples cut into the rock on the banks of the Nile in southern Egypt during the reign of Rameses II, commemorating him and his wife Nefertari. The temples were moved in sections in 1966–67 and rebuilt 60 m/200 ft above their original location before the site was flooded by the waters of the ►Aswan High Dam. 64 F3

Abū Zabī alternative form of ►Abu Dhabi, capital of the United Arab Emirates. 61 F4

Abyssinia former name of ►Ethiopia. 62 G5

Acapulco (or **Acapulco de Juarez**), port and holiday resort in southern Mexico; population (1995 est) 657,000. The city lies in Guerrero state 310 km/193 mi southwest of Mexico City. It is one of the leading Mexican tourist resorts on what is known as the 'Mexican Riviera'. There is deep-sea fishing, and tropical products are exported. 94 E5

Accra capital and port of Ghana; population (1998) 1,446,000. It is an important political, commercial, and administrative centre. The port trades in cacao, gold, diamonds, and timber. Scrap metal is a major import and is the basis of local engineering businesses. Other industries include light engineering, brewing, and tobacco and food processing. Osu (Christiansborg) Castle is the presidential residence. The National Museum of Ghana is here, and the University of Ghana is at nearby Legon. 68 D3

Achill Island (or **Eagle Island**), largest of the Irish islands, off County Mayo; area 148 sq km/57 sq mi; population (1996) 976. It has mountain peaks rising to over 610 m/2,000 ft, moorland, and fine cliffs and beaches. It has been connected to the mainland by a bridge since 1888. 24 B8

Aconcagua extinct volcano in the Argentine Andes; the highest peak in the Americas, 6,960 m/22,834 ft. It was first climbed by Edward Fitzgerald's expedition in 1897. 100 G5

Açores alternative name for the ►Azores, a group of nine islands in the North Atlantic. 66 (1)B2

A Coruña (or **Corunna**), seaport and capital of A Coruña province in Galicia, northwest Spain, lying on a narrow peninsula between two bays; population (1998 est) 243,100. It is one of the most important European fishing ports, and is well known for its high quality fish and seafood. Other industries include tobacco, sugar refining, textiles, and glass. A Coruña has a large and busy harbour, and a 2 km/1.5 mi stretch of sandy beach in the heart of the city. The Spanish Armada sailed from A Coruña in 1588, and the town was sacked by English explorer Francis Drake in 1589. **28** B1

Acre former name of the Israeli seaport of ►Akko. **60** C4

ACT abbreviation for ►Australian Capital Territory. **76** J7

'Adan alternative form of ►Aden, the main port of Yemen. **56** E7

Adana capital of Adana (Seyhan) province, southern Turkey; population (1990) 916,150. It is a major cotton-growing centre and Turkey's fourth-largest city. It is located in the middle of the Curukova plain, in the most fertile area of the country. **58** F5

Ad Dakhla port and southern region in Western Sahara; population (1994) 29,800. The town was first established as a Spanish trading port in 1476, when it was known as **Villa Cisneros**. **66** B4

Ad Dawhah Arabic for ►Doha, the capital and chief port of Qatar. **61** D4

Addis Ababa (or **Ādīs Ābeba**; Amharic 'new flower'), capital of Ethiopia; population (1992) 2,213,000. The city is at an altitude of 2,500 m/8,200 ft. It was founded in 1887 by Menelik II, chief of Shoa, who ascended the throne of Ethiopia in 1889. His former residence, Menelik Palace, is now occupied by the government. Industries include light engineering, food processing, brewing, livestock processing, chemicals, cement, textiles, footwear, clothing, and handicrafts. **70** F2

Adelaide capital and chief port of ►South Australia; population (1996) 978,100. Adelaide is situated on the River Torrens, 11 km/7 mi from the Gulf of St Vincent. Industries include oil refining, shipbuilding, electronics, and the manufacture of electrical goods and cars. Grain, wool, fruit, and wine are exported from Port Adelaide, 11 km/7 mi northwest of the city. Adelaide was founded in 1836 and named after the queen of William IV. The city's fine buildings include Parliament House, Government House, the Anglican cathedral of St Peter, and the Roman Catholic cathedral of St Francis Xavier (built 1856–1926). **76** G6

Adelsberg German name for ►Postojna, a town in Slovenia. **30** K5

Aden (Arabic **'Adan**), main port and commercial centre of Yemen, on a rocky peninsula at the southwest corner of Arabia, commanding the entrance to the Red Sea; population (1995) 562,000. The city's economy is based on oil refining, fishing, and shipping. A British territory from 1839, Aden became part of independent South Yemen in 1967; it was the capital of South Yemen until 1990. **56** E7

Adige (German **Etsch**; ancient **Athesis**), second longest river (after the Po) in Italy, 410 km/255 mi in length. It rises in the Rhaetian Alps, crosses the Lombardy Plain and enters the Adriatic just north of the Po delta. **30** G5

Adirondacks mountainous area in northeast New York State, USA, rising to 1,629 m/5,344ft at Mount Marcy; the source of the Hudson and Ausable rivers. The Adirondacks region is named after an American Indian people; it is now a summer resort area with good sports facilities, and is noted for its beautiful scenery.

Ādīs Ābeba alternative spelling of ►Addis Ababa, the capital of Ethiopia. **70** F2

Admiralty Islands group of small islands in the southwest Pacific, part of Papua New Guinea; area 2,071 sq km/800 sq mi; population (1995 est) 35,200. The islands form part of the ►Bismarck Archipelago and with the North Western Islands constitute the Manus district of Papua New Guinea. The largest island (about 80 km/50 mi long) is Manus of which Lorengau is the chief town. Exports are copra and pearls. The islands became a German protectorate in 1884 and an Australian mandate in 1920. **74** E6

Adour river in southwest France; length 335 km/208 mi. The Adour rises in the Hautes-Pyrénées and flows through the *départements* of Gers and Landes to the Bay of Biscay, near Bayonne. **26** F10

Adowa alternative form of ►Ādwa, the former capital of Ethiopia. **64** G5

Adriatic Sea large arm of the Mediterranean Sea, lying northwest to southeast between the Italian and the Balkan peninsulas. The western shore is Italian; the eastern includes Croatia, Montenegro, and Albania, with two small strips of coastline owned by Slovenia and Bosnia Herzogovina. The Strait of Otranto, between Italy and Albania, links the Adriatic with the Ionian Sea to the south. The chief ports are Venice, Brindisi, Trieste, Ancona, and Bari in Italy, and Rijeka in Croatia. The sea is about 805 km/500 mi long; area 135,250 sq km/52,220 sq mi. **32** H4

Ādwa (or **Adowa** or **Aduwa**), town of Tigré province, Ethiopia, and former capital until 1889 when Addis Ababa became the new capital; population (1994) 24,500. Situated at an altitude of 1,910 m/6,270 ft, it lies about 130 km/80 mi south of Asmara, Eritrea, and 180 km/110 mi southwest of the port of Massawa. The town is an important market centre for grain, animal hides, coffee, and honey. **64** G5

Adygeya (or **Adigey**, or **Adygei**), republic in northwestern ►Caucasia, entirely surrounded by the Krasnodar krai (territory), in the southwest of the Russian Federation; area 7,600 sq km/2,934 sq mi; population (1996) 450,000 (68% Russians, 22% Adygeans). The capital is ►Maikop. It occupied its present borders from 1936, and attained the status of a republic when the Soviet Union collapsed in 1991. It is situated on the left bank of the River Kuban, with plains in the north (providing plentiful fertile black soil called *chernozem*) and the foothills of the Caucasus Mountains in the south. **58** J1

Aegean Islands region of Greece comprising the Dodecanese islands, the Cyclades islands, Lesvos, Samos, and Chios; area 9,122 sq km/3,523 sq mi; population (1991) 460,800.

Aegean Sea branch of the Mediterranean between Greece and Turkey, extending as far south as Crete; the Dardanelles connect it with the Sea of Marmara, in turn linked with the Black Sea via the Bosporus. It is about 600 km/372 mi long and 290 km/180 mi wide, and covers some 214,000 sq km/82,625 sq mi, with a maximum depth of 3,540 m/11,600 ft. Tides are minimal, with a range of only about 40 cm/15 in. The numerous islands in the Aegean Sea include Crete, the Cyclades, the Sporades, and the Dodecanese. There is political tension between Greece and Turkey over sea limits claimed by Greece around such islands as Lesvos, Chios, Samos, and Kos. **36** H5

Aegina (or **Aigina**), Greek island in the Gulf of Aegina about 32 km/20 mi southwest of Peiraias; area 83 sq km/ 32 sq mi; population (1991) 12,400. In 1811 remarkable sculptures were recovered from a Doric temple in the northeast and taken to Munich, Germany. **36** F7

Aeolian Islands another name for the ►Lipari Islands. **32** J10

Afars and the Issas, French Territory of the former French territory that became the Republic of ►Djibouti in 1977.

Afghanistan mountainous, landlocked country in south-central Asia, bounded north by Tajikistan, Turkmenistan, and Uzbekistan, west by Iran, and south and east by Pakistan, India, and China.

Afghanistan covers an area of 652,225 sq km/251,825 sq mi, and has a population of 22,720,000 (2000 est). The capital is ►Kabul; other major towns and cities include Kandahar, Herat, Mazar-i-Sharif, Jalalabad, Kondūz, and Qal'eh-ye Now. The head of state and government is Muhammad Rabbani (from 1996). The currency is the afgháni, and the languages spoken include Pashto, Dari (both official), Uzbek, Turkmen, Balochi, and Pashai. **56** H3

Africa second largest of the seven continents. Africa is connected with Asia by the isthmus of Suez, and separated from Europe by the Mediterranean Sea. The name Africa was first given by the Romans to their African provinces with the city of Carthage, and it has since been extended to the whole continent.
area 30,097,000 sq km/11,620,451 sq mi (three times the area of Europe)
largest cities (population over 2 million; population given in millions) Abidjan (2.9), Addis Ababa (2.6), Alexandria (3.7), Algiers (3.8), Cairo (9.9), Casablanca (3.2), Johannesburg (2.2), Khartoum (2.3), Kinshasa (4.4), Lagos (10.9), Luanda (2.2), Maputo (2.4)
features Great Rift Valley, containing most of the great lakes of East Africa (except Lake Victoria); Atlas Mountains in the northwest; Drakensberg mountain range in the southeast; Sahara Desert (world's largest desert) in the north; Namib, Kalahari, and Great Karoo deserts in the south; Nile, Congo, Niger, Zambezi, Limpopo, Volta, and Orange rivers
physical dominated by a uniform central plateau comprising a southern tableland with a mean altitude of 1,070 m/3,000 ft that falls northwards to a lower elevated plain with a mean altitude of 400 m/1,300 ft. Although there are no great alpine regions or extensive coastal plains, Africa has a mean altitude of 610 m/2,000 ft, two times greater than Europe. The highest points are Mount Kilimanjaro 5,900 m/19,364 ft, and Mount Kenya 5,200 m/17,058 ft; the lowest point is Lac Assal in Djibouti −144 m/ −471 ft. Compared with other continents, Africa has few broad estuaries or inlets and therefore has proportionately the shortest coastline (24,000 km/15,000 mi). The geographical extremities of the continental mainland are Cape Hafun in the east, Cape Almadies in the west, Ras Ben Sekka in the north, and Cape Agulhas in the south. The Sahel is a narrow belt of savannah and scrub forest which covers 700 million hectares/1.7 billion acres of west and central Africa; 75% of the continent lies within the tropics
population (1998 est) 748.9 million; more than double the 1970 population of 364 million, and rising to an estimated 1 billion by 2010; annual growth rate 3% (10 times greater than Europe); 27% of the world's undernourished people live in sub-Saharan Africa, where an estimated 25 million are facing famine **62**

Africa, Horn of projection on the east coast of Africa constituted by Somalia and adjacent territories. **70** G2

Agadir tourist resort and seaport in south Morocco, near the mouth of the River Sus; population (1993) 137,000. It was rebuilt after being destroyed by an earthquake in 1960. Agadir is one of the main fishing ports of Morocco; other industries include food processing, tourism, and crafts. The mild winter climate is suitable for the growth of citrus fruits, and cobalt, manganese, lead, and zinc are mined nearby. **66** D2

Agen French town, capital of the *département* of Lot-et-Garonne, 120 km/75 mi southeast of Bordeaux, on the River Garonne; population (1990) 32,200. It has chemical industries, and a long tradition of making preserved fruit, especially prunes. **26** F9

Agra city in Uttar Pradesh, northern India, on the River Jumna (or Yamuna), 160 km/100 mi southeast of Delhi; population (1991) 892,000. It is a centre for commerce, tourism, and industry. There are many small-scale engineering plants, and carpets, leather goods, gold and silver embroidery, and engraved marble are produced. The capital of the Mogul empire from 1566–69 and 1601–1658, it is the site of the ►Taj Mahal, built during the latter period. Other notable buildings include the Moti Masjid (Pearl Mosque), the Jama Masjid (Great Mosque), and the Red Fort, with red sandstone walls over 20 m/65 ft high and 2.5 km/1.5 mi long. It has a university (1927). **54** C3

Agrigento town in Sicily, known for Greek temples; population (1990) 56,660. **32** H11

Agrinio town in Greece in the department of Aetolia and Acarnania, Central Greece and Euboea region, 29 km/18 mi north of Missolonghi; population (1993) 40,000. It is at the centre of a tobacco- and currant-producing region. **36** D6

Aguascalientes city and capital of the state of Aguascalientes, in central Mexico; population (1990 est) 479,700. Industries include pottery, metal smelting, tanning, and cotton textiles. The mild climate and local hot mineral springs have helped the development of Aguascalientes as a health resort. **94** D4

Agulhas, Cape (Portuguese 'needles'), southernmost cape in Africa, in Western Cape Province, South Africa. It is dangerous to shipping, on account of fogs, rocks, and uncertain currents. The meridian of the Cape marks the boundary between the Atlantic and the Indian oceans. **72** C6

Ahaggar (or **Hoggar**), large plateau region in the central Sahara, about 1,500 km/932 mi south of Algiers. It averages about 900 m/2,953 ft above sea level, its highest point being Mount Tahat (2,918 m/9,573 ft). The plateau is the home of the formerly nomadic Tuareg. 66 G4

Ahlen city in the *Land* (administrative region) of North Rhine-Westphalia, Germany, 96 km/60 mi northeast of Düsseldorf; population (1995) 55,700. The principal industries are cotton textiles, coal mining, and metal-working. 22 K3

Ahmadabad (or **Ahmedabad**), city in Gujarat, India, situated on the Sabarmati River, 430 km/260 mi north of Mumbai (formerly Bombay); population (1991) 3,298,000. The former state capital and Gujarat's largest city, it is a major industrial centre specializing in cotton manu-facturing. It has many sacred buildings of the Hindu, Muslim, and Jain faiths, as well as buildings designed by 20th-century architects, such as Le Corbusier, reflecting commercial success. 54 B4

Ahmadnagar city in Maharashtra, India, 195 km/120 mi east of Mumbai (formerly Bombay), on the left bank of the River Sina; population (1991) 222,000. It is a centre of a cotton trade, and copper and brassware are manufactured. 54 B5

Ahvaz (or **Ahwaz**), capital of Khuzestan province, southwest Iran; population (1991) 724,700. Situated on the River Karun, the city is an important administrative and supply centre for the southwestern Iranian oilfields, and is connected by rail to Tehran and the Gulf. There are also textile and petrochemical industries. Ahvaz was badly damaged during the Iran–Iraq war (1980–88). 56 E3

Ahvenanmaa Finnish name for the ►Åland Islands, a group of some 6,000 islands in the Baltic Sea. 16 K6

Aigina alternative spelling of ►Aegina, a Greek island southwest of Peiraias. 36 F7

Ain *département* in the ►Rhône-Alpes region of France; area 5,756 sq km/ 2,222 sq mi; population (1999 est) 515,300. It takes its name from the River ►Ain, which rises in the *départe-ment* in the Jura Mountains and flows south through steep gorges into the River Rhône. On the northern and western plains cereals, fruit, and vegetables are grown. Livestock is bred, including poultry. The southern region is swampy, but vineyards exist in the valleys of the Jura Mountains in the east. Most employment is now in the service sector, but manufacturing is also important, and the region has hydroelectric schemes and a plastics industry. The administrative centre is ►Bourg-en-Bresse. 6 F4

Ain river of east-central France; length 190 km/118 mi. It rises in the Jura Mountains and flows southwest through the *départements* of Jura and ►Ain, passing through the towns of Pont-du-Navoy, Pont-de-Poitte, Poncin, and Pont-d'Ain. The river joins the ►Rhône 29 km/18 mi east of Lyon. There are several hydroelectric power stations along its route, creating attractive lakes in its gorges. 26 L7

Aintab Syrian name of ►Gaziantep, a city in Turkey. 58 G5

Aisne *département* in the ►Picardy region of France, area 7,378 sq km/ 2,849 sq mi; population (1999 est) 535,300. Aisne is an agricultural *département*, producing cereals, potatoes, sugar beet, vines, and livestock. The terrain is generally flat; wooded in the north and undulating in the south. About half the workforce is employed in the service sector and one-third in manufacturing industries, producing precision tools, electrical appliances, textiles, chemicals, and cosmetics. Three rivers flow through the region: the Marne, Oise, and Aisne. The administrative centre is ►Laon. 6 E2

Aisne river of northern France which rises in the ►Meuse *département* and flows northwest and then west until it joins the River Oise near Compiègne; length 282 km/175 mi. The Aisne passes through the *départements* of Marne, Ardennes, ►Aisne, and Oise. The Aire and Vesle rivers are its main tributaries. 22 F5

Aix-en-Provence city and spa in the *département* of ►Bouches-du-Rhône, southeast France, 29 km/18 mi north of Marseille; population (1990) 126,800. The town dates from Roman times and was the capital of the former province of Provence. The city still maintains some of its important traditional functions, namely the courts, university, and spas. These activities are now undergoing regeneration. 26 L10

Aix-la-Chapelle French name of ►Aachen, an ancient city in Germany. 22 J4

Aix-les-Bains French spa town in the *département* of Savoie, near Lake Bourget, 13 km/8 mi north of Chambéry; population (1990) 24,800, conurbation 35,400. The warm sulphur and alkaline springs have attracted visitors since Roman times, and are still popular. The Romans called the town **Aquae Gratianae**, and parts of some Roman buildings remain. 26 L8

Ajaccio capital and second-largest port of the French island of Corsica; population (1990) 59,300. Founded by the Genoese in 1492, it was the birthplace of Napoleon; it has been French since 1768. 32 C7

Ajman smallest of the seven states making up the ►United Arab Emirates; area 250 sq km/96 sq mi; population (1995) 118,800.

Ajmer city in Rajasthan, India, 350 km/220 mi southwest of Delhi; population (1991) 402,000. It is a commercial and industrial centre, with textile and leather manufacturing, railway workshops, and grain process-ing. Situated in a deep valley in the Aravalli Mountains at the foot of Mount Taragarh (870 m/2,800 ft), it is surrounded by a stone wall and dominated by the Taragarh fort built by the emperor Akbar. 54 B3

Akaba, Gulf of alternative translit-eration of ►Aqaba, a gulf of the Red Sea and a city of the same name at its head. 60 C7

Akko (formerly **Acre**; New Testament **Ptolemais**), seaport in northwest Israel, situated on the Mediterranean Sea; population (1995 est) 48,300. The city was built on a small promontory which, with Mount Carmel to the south, forms a semicircular bay. From being part of British-mandated Palestine, it became part of Israel in 1948. Industries include tourism, fishing, and light manufactures. 60 C4

Akola city in Maharashtra, India, 480 km/298 mi northeast of Mumbai (formerly Bombay); population (1991) 328,000. It is a major cotton and grain centre. 54 C4

Akranes town on Faxa Bay, Iceland, in the county of Vesturland, north of Reykjavik; population (1993) 5,200. It is a trading and fishing centre with a cement factory using offshore shell deposits for lime. 16 (1)B2

Akron (Greek 'summit'), city in northeastern Ohio, USA, on the Cuyahoga River, 56 km/35 mi southeast of Cleveland; population (1994 est) 221,900. Industries include chemical, plastic, and aerospace products. Known as the 'Rubber Capital of the World', it is home to the headquarters of several tyre and rubber companies, although production had ended here by 1982. **88** D2

Aktyubinsk former name (1869–1991) of ▶Aqtöbe, a city in northwestern Kazakhstan. **38** L4

Alabama state in southern USA. It is nicknamed Heart of Dixie or the Camellia State. Alabama was admitted to the Union in 1819 as the 22nd US state. Historically it was a plantation state associated with slavery and, in the 20th century, the civil-rights movement. It is bordered to the east by Georgia, with the Chattahoochee River forming the lower half of the boundary, to the north by Tennessee, and to the west by Mississippi, with the Tennessee River forming a small part of the boundary in the northwest. To the south is the Florida panhandle and a 100km-/60 mi-long stretch of coast on the Gulf of Mexico, bisected by Mobile Bay.
population (2000 est) 4,447,100
area 134,700 sq km/52,000 sq mi
capital ▶Montgomery
towns and cities Birmingham, Mobile, Huntsville, Tuscaloosa
industries and products cotton (still important though no longer prime crop), soybeans, peanuts, wood products, marble, coal, oil, natural gas, livestock, poultry, fishing, iron, steel, aluminium, chemicals, textiles, paper, power generation, aerospace industry. **90** D3

Åland island in the Gulf of Bothnia, Finland; largest of the ▶Åland Islands. **16** K6

Åland Islands (Finnish **Ahvenanmaa** 'land of waters'), group of some 6,000 islands in the Baltic Sea, at the southern extremity of the Gulf of Bothnia; area 1,481 sq km/572 sq mi; population (1992) 25,000. Only 80 are inhabited; the largest island, Åland, has a small town, Mariehamn. The main sectors of the island economy are tourism, agriculture, and shipping. **16** K6

Alaska largest state of the USA, located on the northwest extremity of North America, and separated from the lower 48 states by British Columbia. It is nicknamed Last Frontier. Alaska was admitted to the Union in 1959 as the 49th US state. Historically and commercially the state has been associated with mineral exploitation. It is bordered to the east by the Yukon Territory, Canada, and to the southeast, along its panhandle, by the Yukon Territory and British Columbia, Canada. Northern Alaska lies on the Beaufort Sea, part of the Arctic Ocean. To the northwest is the Chukchi Sea, narrowing to about 80 km/50 mi at the Bering Strait, which separates the Alaskan Seward Peninsula from Russian East Asia. The Bering Sea is bounded to the south by Alaska's long ▶Aleutian Island chain, extending in an east–west arc across the North Pacific Ocean from the Alaska Peninsula. To the peninsula's east is the Gulf of Alaska.
population (2000 est) 626,900; including 15% American Indians, Aleuts, and Inuit
total area 1,530,700 sq km/591,004 sq mi
land area 1,478,457 sq km/570,833 sq mi
capital ▶Juneau
towns and cities Anchorage, Fairbanks, Fort Yukon, Holy Cross, Nome, College, Sitka
industries and products oil, natural gas, coal, copper, iron, gold, tin, fur, salmon fisheries and canneries, lumber; tourism is a large and growing industry (tourists outnumber residents each year). **92** (1)F2

Albacete market town and capital of Albacete province, southeast Spain; population (1994) 141,000. It produces cutlery, clothes, and footwear. Albacete is the largest city in the Castilla-La Mancha region. **28** J5

Alba Iulia (German **Karlsburg**), city on the River Mures, western central Romania; population (1992) 71,300. It was founded by the Romans in the 2nd century. The Romanian kings were crowned here, Michael the Brave (1601), and the kings of modern Romania until World War II. **34** L3

Albania country in southeastern Europe, bounded north by Yugoslavia, east by the Former Yugoslav Republic of Macedonia, south by Greece, and west and southwest by the Adriatic Sea.

Albania covers an area of 28,748 sq km/11,099 sq mi, and has a population of 3,113,000 (2000 est). The capital is ▶Tirana; other major towns and cities include Durrës, Shkodër, Elbasan, Vlorë, and Korçë. The head of state is Rexhep Mejdani (from 1997); the head of government is Ilir Meta (from 1999). The currency is the lek, and the languages spoken include Albanian (official) and Greek. **36** B3

Albany town and port in Western Australia, on King George Sound, 409 km/254 mi southeast of Perth; population (1996) 20,493. It suffered from the initial development of the port of ▶Fremantle, but has grown with the greater exploitation of the surrounding area. It is the centre for an agricultural region and the main port for southwestern Western Australia, exporting cattle, wool, and grain; industries include woollen milling, meat processing, and fish canning. Albany is also a popular tourist resort, with a temperate climate and fine beaches. **76** C6

Albany city and administrative headquarters of Dougherty County, southwest Georgia, USA; population (1992) 79,600. It is located 125 km/78 mi southeast of Columbus, on the Flint River. Albany is a commercial centre for the production of pecans and peanuts, lumber, chemicals, textiles, and other industrial products, including aircraft. The city is home to Albany State College (1903), Darton College (1963), and the Albany Naval Air Station. Founded in 1836 and incorporated in 1838, Albany was an important cotton market in the 19th century. Tourists are drawn to Radium Springs, a resort 4 mi/6 km to the south, featuring Georgia's largest natural spring. **90** E3

Albany capital of ▶New York State, USA, situated on the west bank of the Hudson River, about 225 km/140 mi north of New York City; population (1994 est) 105,000. Albany is an important deep-water port and state administrative centre; its chief industries include clothing, brewing, and engineering. Albany was originally a Dutch trading post called Fort Nassau (1614). It was renamed Albany in 1664 when the English took control. The completion of the Champlain Canal (1822) and the Erie Canal (1825), as well as the Hudson Railroad (1831), helped to foster its subsequent economic development. 88 F2

Al Başrah Arabic name for ▶Basra, the principal city in southeast Iraq; and Basra, a governate of Iraq. 56 E3

Albemarle Sound large inlet of the Atlantic Ocean on the coast of North Carolina; length 88 km/55 mi; width 8–24 km/5–15 mi. It is crossed by the Intracoastal Waterway and is joined by the estuaries of the Chowan, Perquimans, Little, Pasquotank, and North rivers to the north and the Roanoke, Scuppernong, and Alligator rivers to the south. The sound is separated from the ocean by Bodie Island. In an area of pine and hardwood forests, marshes, and swampland, the sound is bordered by several wildlife refuges. Elizabeth City, on the Pasquotank, is the sound's chief port. 88 E3

Alberta province of western Canada; area 661,200 sq km/255,223 sq mi; population (1996 est) 2,697,000. Its capital is ▶Edmonton, and the main towns and cities include Calgary, Lethbridge, Medicine Hat, and Red Deer. Oil extraction is the most important economic activity in Alberta, with the province accounting for most of the country's oil production. It became a province in 1905. 82 H6

Albert Canal (or **Albert Kanaal**), canal, 128 km/80 mi long, designed as part of Belgium's frontier defences. It links the River Meuse and the industrial basin of Liège with the port of Antwerp and the River Schelde. It was built 1930–39 and named after the Belgian monarch, Albert I. 22 G3

Albert Kanaal alternative name of ▶Albert Canal, a canal in Belgium. 22 G3

Albert, Lake lake on the border of Uganda and the Democratic Republic of Congo in the Great ▶Rift Valley; area 5,600 sq km/2,162 sq mi. The first European to see it was the British explorer Samuel Baker, who named it Lake Albert after the Prince Consort. From 1973 to 1997 it was called Lake Mobutu after President Mobutu of Zaire (now Democratic Republic of Congo). 70 D3

Albertville resort town at the entrance to the Val d'Arly in the *département* of Savoie, southeast France, situated on the right bank of the River Arly not far from its confluence with the Isère; population (1999) 17,300. It was the scene of the 1992 Winter Olympics. 26 M8

Albi administrative centre of the ▶Tarn *département* in the Midi-Pyrénées region of southwest France, situated on the River ▶Tarn, 72 km/45 mi northeast of Toulouse; population (1990) 48,700. Industries include the manufacture of textiles, glass, cement, dyes, and flour. Albi was the centre of the Albigensian heresy in the 12th and 13th centuries, and was the birthplace of the artist Henri Toulouse-Lautrec. 26 H10

Ålborg (or **Aalborg**), port in Denmark 32 km/20 mi inland from the Kattegat, on the south shore of the Limfjord; population (1995) 159,000. One of Denmark's oldest cities, it has a castle and the Budolfi cathedral (named after the English St Botolph), dating mainly from about 1400. It is the capital of Nordjylland county in Jylland (Jutland); the port is linked to Nørresundby on the north side of the fjord by a tunnel built in 1969. Major industries include shipbuilding, cement, and textiles. 16 E8

Albufeira fishing port and resort on the Algarve coast of southern Portugal, 43 km/27 mi west of Faro; it is known as the 'St Tropez of the Algarve'; population (1991) 6,300. There are Moorish remains here. 28 B7

Albuquerque largest city of New Mexico, USA, situated east of the Rio Grande, in the Pueblo district; seat of Bernalillo County; population (1994 est) 412,000. Albuquerque is a resort and industrial centre specializing in electronic products and aerospace equipment, and is a centre for livestock rearing. Founded in 1706, it was named after the viceroy of New Spain, Afonso de Albuquerque, and was incorporated as a city in 1891. 92 E1

Albury town in New South Wales, Australia, situated on the north bank of the River Murray, on the border with Victoria, 573 km/356 mi southwest of Sydney; population (1996) 41,491. Its twin town, on the south side of the River Murray in the state of Victoria, is Wodonga, and the two towns are commonly referred to as Albury-Wodonga. The towns were designated a growth centre in the 1970s and serve the surrounding agricultural region. Industries include food processing, electrical engineering, meatworks, and the production of pet food, beverages, clothing, stationery, timber, wine, and wool. 76 J7

Alcalá de Henares (ancient **Complutum**), town in the autonomous community of Madrid, Spain, on the River Henares; population (1995) 166,900. Situated to the east of the capital, near Madrid-Barajas airport, the town has attracted many international companies and is a centre for research and development. Miguel de Cervantes, author of *Don Quixote*, was born here. 28 G4

Alcatraz small island in San Francisco Bay, California, USA. Once used as a Spanish military post, the island was later the site of a US military prison (1868–1934), and then served as a Federal maximum security prison (1934–63). The dangerous tides allowed few successful escapes. In the 1970, the island was occupied for a time by Native American activists. The island is now part of the Golden Gate National Recreation Area and is a popular tourist attraction.

Alcázar de San Juan town in the province of Ciudad Real, central Spain, on the plain of La Mancha; population (1995) 26,700. It is an important rail junction and has a substantial wine trade. 28 G5

Alcoi alternative spelling of ▶Alcoy, a town in eastern Spain. 28 K6

Alcoy (or **Alcoi**), town in the province of Alicante, eastern Spain, on the River Serpis; population (1991) 65,500. Products include textiles, paper, and machinery. During an annual festival held here, a mock battle is staged between Moors and Christians. 28 K6

Aldabra Islands high limestone island group in the ▶Seychelles, some 420 km/260 mi northwest of Madagascar; area 154 sq km/59 sq mi. The islands are uninhabited. A nature reserve since 1976, it has rare plants and animals, including the giant tortoise. 72 (2)A2

Aldeburgh small town and coastal resort in Suffolk, eastern England, 33 km/20 mi from Ipswich; population (1991) 2,700. It maintains a small fishing fleet, serving the local market. The Aldeburgh Festival, founded in 1948 by the English composer Benjamin Britten, is held annually at the Snape Maltings, 8 km/5 mi west of the town. It is the home of the Britten–Pears School for Advanced Musical Studies. 22 D2

Alderney third largest of the ►Channel Islands, with its capital at St Anne's; area 8 sq km/3 sq mi; population (1991) 2,300. Tourism flourished on Alderney from the early 20th century. The main employers are now tourist-related businesses and services providing building and maintenance work for locals and immigrants. There is also a small finance industry. 26 C4

Aldershot town in Hampshire, southern England, 56 km/35 mi southwest of London; population (1991) 51,400. Industrial products include electronics, and vehicle components for cars and tankers. It contains the largest permanent military training camp in the UK, dating from 1854. The Ministry of Defence owns over 850 ha/2,100 acres here, mainly north of the town centre. 22 B3

Aleksandrovsk former name (to 1921) of ►Zaporizhzhya, a major port on the ►Dnieper River in Ukraine. 38 G5

Alençon administrative centre of ►Orne *département* in the Basse-Normandie region of France, situated in a rich agricultural plain southeast of Caen; population (1990) 31,100. The town lies at the confluence of the Sarthe and Briante rivers. Lace-making was once an important industry, particularly the production of a fine lace called *point d'Alençon* which began in 1665. Some is still produced but its economic importance has now been superseded by printing works and the manufacture of electrical goods and vehicle parts. 26 F5

Aleppo (Arabic **Halab**), ancient city in northwest Syria, situated on the River Kuweik on the edge of the Syrian Desert; population (1993) 1,494,000. It is the administrative centre of the governorate of Aleppo; population (1996 est) 3,694,000. Silk and cotton goods, leather, grain, carpets, tobacco, and metalwork are produced. Chief industries are cotton and wool textile manufacturing. There has been a settlement on the site for at least 4,000 years. 58 G5

Alès French town in the *département* of Gard, 45 km/28 mi southwest of Nîmes, on the River Gardon in the foothills of the Cévennes; population (1999) 39,300. Its high-technology industries make it an important industrial centre; silk, glass, and metal goods are also manufactured. 26 K9

Alessandria town in Piedmont, northern Italy, on the River Tanaro, 55 km/34 mi north of Turin; population (1999 est) 99,700. Alessandria is the market centre of a corn- and wine-producing district. Industries include engineering, chemical industries, and the manufacture of confectionery, footwear, hats, and cottons. It was founded in 1168 by Pope Alexander III as a defence against Frederick (I) Barbarossa. 30 D6

Ålesund (or **Aalesund**), coastal town in Møre og Romsdal county, western Norway; population (1995) 37,100. It is the main centre of the Norwegian herring-fishing industry. The town is built on five islands, linked by bridges. 16 D5

Aleutian Islands volcanic island chain in the North Pacific, stretching 1,900 km/1,200 mi southwest of Alaska, of which it forms part, towards Kamchatka; population (1990) 12,000. There are 14 large and more than 100 small islands running along the Aleutian Trench; the largest island is Unimak (with an area of 3,500 sq km/1,360 sq mi), which contains two active volcanoes. The islands are mountainous, barren, and treeless; they are ice-free all year but are often foggy, with only about 25 days of sunshine recorded annually. The only industries are fishing, seal hunting, and sheep farming; the main exports are fish and furs. Unalaska is the chief island for trade as it has a good harbour. 92 (3)B1

Alexander Archipelago group of over a thousand islands in the Gulf of Alaska, off the coast south of Juneau in southeast Alaska. Fishing, fish processing, lumbering, and trapping are the main industries. The largest islands in the group are Baranof Island, Prince of Wales, Chichagof, Admiralty, Kuin, Kupreanof, and Revillagigedo, which are densely forested. The principal towns are Ketchikan (on Revillagigedo Island), ►Sitka (on Baranof Island), Petersburg (on Mitkof Island), and Wrangell (on Wrangell Island). Long inhabited by the Tlingit, the islands were first explored by Russians in 1741. Most are now part of the Tongass National Forest. 92 (1)K4

Alexandretta former name (to 1939) of ►Iskenderun, a port in southern Turkey. 58 G5

Alexandria (or **El Iskandarîya**), city, chief port, and second-largest city of Egypt, situated between the Mediterranean and Lake Maryut; population (1992) 3,380,000. It is linked by canal with the Nile. There is oil refining, gas processing, and trade in cotton and grain. Founded in 331 BC by Alexander the Great, Alexandria was the capital of Egypt for over 1,000 years. 64 E1

Alexandria town and administrative headquarters of Rapides Parish, central Louisiana; population (1990) 49,200. It is situated on the Red River, opposite Pineville, and 150 km/93 mi northwest of Baton Rouge. It is a centre for livestock, meatpacking, and timber. Cotton, sugar cane, and livestock are processed in and shipped from Alexandria. The town's other industries include oil refining and the manufacture of concrete, brick and tile roofing, naval stores, valves, pipe fittings, and road machinery. Alexandria is home to an undergraduate branch of Louisiana State University (1960). 90 C3

Al Fāw (or **Fao**), oil port on a peninsula at the mouth of the Shatt-al-Arab in Iraq. Iran launched a major offensive against Iraq in 1986, capturing Al Fāw for two years. 61 C2

Al Fayyum (or **El Faiyûm**), city and governorate in northern Egypt, 90 km/56 mi southwest of Cairo; population of city (1992) 250,000, and governorate (1997) 1,989,881. Noted for its waterwheels which distribute the waters of the Bahr Yusef Canal, Al Fayyum is a fertile agricultural region producing fruit, cotton, sugar, and attar of roses. Fishing is an important industry together with the raising of livestock. Originally named Crocodilopolis, the oasis of Al Fayyum was the main place of worship of the crocodile god, Sobek, and realistic mummy portraits dating from the 1st–4th centuries have been found in the area. 64 F2

Alföld (or **Great Plain of Hungary**), large area of flat open land in Hungary, between the River Danube and the Carpathian Mountains. Cereals, vines, tobacco, hemp, and flax are produced, and livestock is reared. The plain occupies about half the area of Hungary, with small areas in Romania, Croatia, and Yugoslavia. The Great Alföld (Hungarian **Nagyalföld**) is the central plain; the Little Alföld (Hungarian **Kisalföld**) is in northwest Hungary, northwest of the Transcaucasian highlands. 34 H2

Algarve (Arabic *al-gharb* 'the west'), ancient kingdom in southern Portugal, bordered on the east by Spain, and on the west and south by the Atlantic Ocean; it is co-extensive with the modern district of Faro, the provincial capital of the Algarve; area 5,071 sq km/ 1,958 sq mi; population (1995 est) 346,000. The population increased during the 1980s and 1990s as a result of inward migration. Tourism is the largest employer in the area. It is based mainly on beach resorts although there has been a shift towards cultural and golfing tourism. Regional agriculture is mainly citrus fruits, due to the use of modern irrigation methods, although traditional crops of grapes, olives, figs, almonds, and carobs are still grown. The hilly areas are heavily forested with pine and eucalyptus trees.

Algeciras port in Cádiz province, southern Spain, situated to the west of Gibraltar across the Bay of Algeciras; population (1994) 104,000. An industrial complex, built around a petrol refinery, was established in the Bay of Algeciras in the 1970s. Algeciras was originally founded by the Moors in 713; it was taken from them by Alfonso XI of Castile in 1344. 28 E8

Alger alternative name of ▶Algiers, the capital of Algeria. 66 F1

Algeria country in North Africa, bounded east by Tunisia and Libya, southeast by Niger, southwest by Mali and Mauritania, northwest by Morocco, and north by the Mediterranean Sea.

Algeria covers an area of 2,381,741 sq km/919,590 sq mi, and has a population of 31,471,000 (2000 est). The capital is ▶Algiers (Arabic al-Jaza'ir); other major towns and cities include Oran, Annaba, Blida, Sétif, and Constantine. The head of state is Abdel Aziz Bouteflika (from 1999); the head of government is Ali Benflis (from 2000). The currency is the Algerian dinar, and the languages spoken include Arabic (official), Berber, and French. 66 E3

Algiers (Arabic **al-Jaza'ir**; French **Alger**), capital of Algeria, situated on the narrow coastal plain between the Atlas Mountains and the Mediterranean; population (1995) 2,168,000. It distributes grain, iron, phosphates, wines, and oil from central Algeria. The main industries are oil refining, petrochemicals, and metal working. The city is a popular winter resort. 66 F1

Al Hudaydah alternative name for ▶Hodeida, a port in Yemen. 64 H5

Alicante seaport and administrative capital of Alicante province, southeast Spain, situated on the Mediterranean coast, 123 km/77 mi south of Valencia; population (1991) 261,300. It exports wine, olive oil, and fruit, and has ferry services to the Balearic Islands. There is food-processing and other light industry; products include leather, textiles, and pottery. Almond paste sweetmeats (*turrones*) are a local speciality. 28 K6

Aligarh city in Uttar Pradesh, India, 120 km/75 mi southeast of Delhi; population (1991) 480,000. Industries include agricultural manufacturing and processing, engineering, and textiles. Aligarh Muslim University was founded here in 1875 as the Anglo-Oriental College. 54 C3

Al Kuwayt Arabic name for ▶Kuwait, capital of Kuwait. 61 C2

Alkmaar town in the province of North Holland, Netherlands, 32 km/20 mi northwest of Amsterdam; population (1996) 93,100. Alkmaar is situated on the North Holland Canal and is intersected by a network of canals. The town has a Gothic church, built 1470–1512, and a Gothic town hall, and holds a large weekly cheese market. 22 G2

Al Kut (or **Kut-al-Imara**), city and river port in eastern Iraq, 164 km/ 103 mi southeast of Baghdad, on the River Tigris; population (2001 est) 352,800. It is a grain market for the surrounding agricultural area and a carpet-manufacturing centre. 56 E3

Al Lādhiqīyah Arabic name for ▶Latakia, a port in northwest Syria. 58 F6

Allahabad ('city of god'), historic city in Uttar Pradesh state, India, 580 km/360 mi southeast of Delhi, on the Yamuna River where it meets the Ganges and the mythical underground Seraswati River; population (1991) 806,000. A growing commercial centre, its main industries are textiles and food processing. A Hindu religious event, the festival of the jar of nectar of immortality (Kumbh Mela), is held here every 12 years with the participants washing away sin and sickness by bathing in the rivers; in 1989 15 million pilgrims attended. It is also the site of the Asoka Pillar, dating from 232 BC, on which are carved edicts of the Emperor Asoka. 54 D3

Allegheny Mountains (or the **Alleghenies**), mountain range over 800 km/500 mi long extending from Pennsylvania to Virginia, USA, rising to more than 1,500 m/4,900 ft and averaging 750 m/2,500 ft. The Alleghenies are rich in hardwood timber and bituminous coal, and also contain iron ore, natural gas, clay, and petroleum. The mountains initially hindered western migration, with the first settlement to the west being Marietta in 1788.

Allentown city and administrative headquarters of Lehigh County, east Pennsylvania, USA; population (1994 est) 105,000. It is situated on the Lehigh River, 64 km/40 mi to the northwest of Philadelphia. It is the leading commercial and industrial centre of the Lehigh Valley. Located in the fertile Pennsylvania Dutch farming region, it serves as a retail and wholesale trading centre for local potatoes, orchard fruits, corn, and wheat. A major producer of cement, it also manufactures Mack trucks, electronic equipment, appliances, machinery, tools, textiles, and other products. It has, however, suffered substantial industrial decline in recent decades. 88 E2

Allier *département* in the ▶Auvergne region of central France; area 7,340 sq km/2,835 sq mi; population (1990) 358,326. The principal rivers are the ▶Allier, the Cher, and the Loire (which forms the eastern boundary of the *département*). Allier is primarily agricultural, though manufactured goods produced here include cutlery, glass, and textiles. The departmental capital is ▶Moulins; other important urban centres are the spa town of Vichy, Montluçon, and Lapalisse. 6 E3

Allier river in central France, rising in the Margeride Mountains in the *département* of Lozère and flowing northwest to join the Loire 6 km/

3.7 mi west of Nevers; length 410 km/254 mi. The river gives its name to the *département* of ►Allier. Vichy and ►Moulins are the main towns on the river, which forms the boundary between the *départements* of Cher and Nièvre. 26 J8

Almadén mining town in Ciudad Real province, Castilla–La Mancha, central Spain; population (1991) 7,700. It has the world's largest supply of mercury, which has been worked since the 4th century BC. 28 F6

Al Madīnah Arabic name for ►Medina, the second holiest city of Islam, situated in Saudi Arabia. 64 G3

Al Manamah capital and free trade port of Bahrain, on Bahrain Island; population (1991) 137,000. It handles oil and entrepôt trade. 61 D3

Almaty (formerly **Vernyi** 1854–1921, **Alma-Ata** 1921–94), former capital of ►Kazakhstan to 1998, in the southeast of the country on the Almaatinka River, and capital of Almaty oblast; population (1996) 1,500,000. Its industries include engineering, printing, tobacco processing, textile manufacturing, and the production of leather goods. The city is at the centre of a large fruit-growing region, and food processing (meat packing, flour milling, wine bottling) is also a major source of employment in the city. 42 P9

Al Mawşil Arabic name for ►Mosul, an industrial city in Iraq. 58 K5

Almelo town in the province of Overijssel, Netherlands, 45 km/28 mi southeast of Zwolle; population (1996) 65,300. Almelo is situated on the Overijssel Canal, and has textile mills. 22 J2

Almería Spanish city, chief town of a province of the same name on the Mediterranean; population (1994) 167,000. The province is famous for its white grapes, and in the Sierra Nevada are rich mineral deposits. 28 H8

Al Mukalla seaport capital of the Hadhramaut coastal region of Yemen, on the Gulf of Aden 480 km/300 mi east of Aden; population (1995 est) 154,400. 56 E7

Al Mukha (or **Mocha**), seaport of the Republic of Yemen near the mouth of the Red Sea, formerly handling coffee exports; population about 8,000. 64 H5

Alost French name for the Belgian town of ►Aalst. 22 G4

Alpes-de-Haute-Provence *département* in the ►Provence-Alpes-Côte d'Azur region of France; area 6,944 sq km/2,681 sq mi; population (1990) 64,300. It is situated in the ►Alps, and winter sports are very important to the economy. The main rivers are the Durance and the Verdon. Forestry and livestock farming are the main occupations in the uplands, and the growing of vines, fruit, potatoes, and olives in the valleys. The principal towns are ►Digne (the administrative centre), Barcelonnette, and Castellane. 6 G4

Alpes-Maritimes *département* in the ►Provence-Alpes-Côte d'Azur region of France; area 4,292 sq km/1,657 sq mi; population (2000) 1,048,100. The main river is the Var. There are many mineral springs in the region. Tourism is the main activity, with resorts on the coast, the ►Côte d'Azur, including ►Nice (the administrative centre), Antibes, Menton, and Cannes. Flowers, fruit, and early vegetables are grown and there are olive oil, perfume, and metal industries. On the coast there is fishing for tuna and sardines. 6 G4

Alps the highest and most extensive mountain range in Europe. The Alps run in an arc from the Mediterranean coast of France in the west through northern Italy, Switzerland, southern Germany, and Austria to the outskirts of Vienna and the River Danube in the east – a total distance of some 960 km/597 mi. Alpine ranges also extend down the Adriatic coast into Slovenia and Croatia. The Alps form a natural frontier between several countries in south-central Europe. The highest peak, at 4,808 m/15,774 ft, is ►Mont Blanc, on the Franco-Italian border. The Alps are the source of many of Europe's major rivers – or their tributaries – including the Rhine, the Rhône, the Po, and the Danube. As well as agriculture, an important economic activity in the Alps is tourism: winter visitors come for the skiing offered at numerous resorts; summer tourism centres on sightseeing and walking in this area of outstanding natural beauty. The Alps are also a widely exploited source of hydroelectric power. Much Alpine woodland has been severely damaged by acid rain. 30 B5

Alps, Australian see ►Australian Alps. 74 E9

Alps, Southern see ►Southern Alps. 78 B6

Al-Quds Arabic name for ►Jerusalem, Israel. 60 C5

Als (German **Alsen**), Danish island in Little Belt, Baltic Sea, off the coast of Slesvig; area 310 sq km/120 sq mi. The island, about 30 km/19 mi long and 5–20 km/3–12 mi wide, is part of the county of Sønderborg. Since 1930 it has been connected with mainland Denmark by a bridge across Als Sound. A number of engineering and systems-engineering companies are located here. Sønderborg is the principal town. 20 E1

Alsace region of France; area 8,300 sq km/3,204 sq mi; population (1990) 1,624,400. It consists of the *départements* of ►Bas-Rhin and ►Haut-Rhin; its administrative centre is ►Strasbourg. Alsace has much rich agricultural land, and is noted for its white wines. 6 G2

Altai (or **Altay**; formerly **Oirot Autonomous Oblast** 1922–48), republic of the Russian Federation, within the Altai krai (territory) of southern Siberia; area 92,600 sq km/35,752 sq mi; population (1996) 202,000 (24% urban). The oblast was formed in 1922, and Altai was formed after the break-up of the Soviet Union in 1992. Its capital is Gorno-Altaisk. 42 R7

Altai Mountains (or **Altay Mountains**), mountain system running through ►Kazakhstan, the ►Altai Republic of the Russian Federation, western ►Mongolia, and northern ►China. It is divided into two parts: the Russian Altai, which includes the highest peak, Mount Belukha, 4,506 m/14,783 ft, on the border with Kazakhstan; and the Mongolian or Great Altai. Alpine pastureland is found on the upper slopes below the snowline, while the lower slopes are heavily wooded with a variety of conifer and birch trees. 46 A1

Altamira Amazonian city in the state of Pará, northeastern Brazil, situated at the junction of the Trans-Amazonian Highway with the Xingu River, 700 km/400 mi southwest of

Belém; population (1991) 157,900. 98 G4

Altay alternative spelling for ►Altai in Russia. 42 R7

Altdorf capital of the Swiss canton ►Uri at the head of Lake Lucerne, Switzerland, 30 km/19 mi southwest of Lucerne on the St Gotthard railway line; population (1994) 10,100. It was the scene of the exploits of the legendary Swiss archer William Tell. 30 D4

Alton city in Madison County, west Illinois, USA; population (1990) 32,900. It is situated on the Mississippi River, 32 km/20 mi northeast of St Louis, Missouri. The site of oil refineries and shipping facilities, this highly industrialized port city also has factories producing glass, paper, plastic, ammunition, hardware, steel products, boats, and clothing. Limestone and building stone are quarried in the vicinity. 88 B3

Altoona city in Blair County, south-central Pennsylvania, USA; population (1992) 52,500. It is situated in the Allegheny Mountains, 135 km/84 mi northwest of Harrisburg and 129 km/80 mi east of Pittsburgh. It is a railroad manufacturing and repair centre, and there is also coalmining here. The city is a business centre for the area's agricultural products and produces silk, apparel, food, among other manufactures. Altoona is the site of a campus of Pennsylvania State University. 88 E2

Alwar city in Rajasthan, India, 135 km/90 mi southwest of Delhi, chief town of the district (formerly princedom) of the same name; population (1991) 211,000. It has palaces, temples, and tombs. Flour milling and trade in cotton goods and millet are major occupations. 54 C3

Alzira town in the province of Valencia, eastern Spain; population (1995) 41,100. Built on the site of the Roman town of **Saetabicula**, Alcira has a 14th-century bridge, a Gothic church, and several old mansions. It is the centre of a rich orange-growing district. 28 K5

Amarillo city in the panhandle of northern Texas, USA, situated on the high plains, west of the Canadian and Red rivers; population (1994 est) 165,000. The centre of the world's largest cattle-producing area, it processes the live animal into frozen supermarket packets in a single continuous operation on an assembly line. Other industries include oil, farming, copper and zinc refining, and the manufacture of helicopters. There is a large helium plant here, and the city is also a centre for the assembly of nuclear warheads. 92 F1

Amazon (Portuguese and Spanish **Rio Amazonas**, Indian **Amossona** 'destroyer of boats'), river in South America, the second longest in the world; length 6,516 km/4,050 mi. The Amazon ranks as the largest river in the world in terms of the volume of water it discharges (around 95,000 cu m/3.3 million cu ft every second), its number of tributaries (over 500), and the total basin area that it drains (7 million sq km/2.7 million sq mi – almost half the landmass of South America). It has 48,280 km/30,000 mi of navigable waterways. The river empties into the Atlantic Ocean on the Equator, through an estuary 80 km/50 mi wide. Over 5 million sq km/2 million sq mi of the Amazon basin is virgin rainforest, containing 30% of all known plant and animal species. This is the wettest region on Earth, with an average annual rainfall of 2.54 m/8.3 ft. 96 F4

Amazonia those regions of Brazil, Colombia, Ecuador, Peru, and Bolivia lying within the basin of the Amazon River.

Ambala (or **Umballa**), city and railway junction in northern India, situated 176 km/110 mi northwest of Delhi; population (1991) 140,050. Food processing, flour milling, and cotton ginning are the foremost industries. It is an archaeological site with prehistoric artefacts. 54 C2

Amberg city in Bavaria, Germany, 153 km/95 mi north of Munich; population (1993) 43,700. Iron has been worked in the city since the 14th century. Other industries include extraction of brown coal (lignite) and brewing. Amberg has medieval walls, a castle, and several old churches, including the 15th-century Gothic St Martinskirche. 20 G7

Amboise French town in the *département* of Indre-et-Loire, on the River Loire, 25 km/16 mi east of Tours; population (1990) 11,500. Agricultural machinery and textiles are manufactured; tourism is important. Its château, one of the finest of the Loire châteaux, rises on a terrace above the river, and was a residence of the Valois dynasty. Other buildings in the town include a 12th-century church. 26 G6

Ambon (or **Amboina**), small island in Maluku, Republic of Indonesia; population (1990) 206,300. The island has an area of 813 sq km/314 sq mi. It lies in the Banda Sea and is mountainous, but with a fertile coastal plain which produces cloves, nutmeg, rice, sugar, and copra. The town of Ambon, formerly a historic centre of Dutch influence, has shipyards and is a port for trade in spices. Before social and political unrest erupted in 1999, it was a popular diving centre for tourists. 53 C3

America landmass in the Western hemisphere of the Earth, comprising the continents of ►North America and ►South America, with ►Central America in between. This great land mass extends from the Arctic to the Antarctic, from beyond 75° N to past 55° S. The area is about 42,000,000 sq km/16,000,000 sq mi, and the estimated population is over 808 million (1998). Politically, it consists of 36 nations and US, British, French, and Dutch dependencies.

American Samoa see ►Samoa, American. 74 J7

Amersfoort city in the province of Utrecht, Netherlands, 19 km/12 mi northeast of Utrecht; population (1997) 117,900. Industries include electrical engineering, car assembly, chemicals, and food processing. 22 H2

Amherst town in Hampshire County, west-central Massachusetts, USA; population (1998 est) 35,250. It is situated in the Pioneer Valley, 35 km/22 mi northeast of Springfield. Settled in 1703, the town was agricultural before the Revolution, and then developed various industries including hat, textile, and carriage making in the 19th century. The local cattle industry gave rise to the foundation of an agricultural college, which became the University of Massachusetts (1863). Amherst is now primarily an educational centre, with around 23,000 students at the University of Massachusetts, and the presence of four other institutions, which together operate as the Five Colleges Consortium. 82 U7

Amiens administrative centre of ►Somme *département* and the major town in the ►Picardy region of northeast France, 130 km/81 mi north of Paris at the confluence of the rivers ►Somme and Avre; population (1999 est) 135,500. Situated in an area irrigated by canals, it has been a market-gardening region and textile centre since the Middle Ages, and has produced velvet since the 16th century. Other industries include clothing, tyres, chemicals, and machinery. Amiens is the seat of the University of Picardy. Amiens gave its name to the battles of August 1918, when British field marshal Douglas Haig launched his victorious offensive in World War I. **22** E5

Amman capital and chief industrial centre of Jordan, 80 km/50 mi northeast of Jerusalem; population (1994 est) 1,300,000. It is a major communications centre, linking historic trade routes across the Middle East. **60** C5

Ammochostos alternative name for ►Famagusta, a seaport on the east coast of Cyprus. **58** E6

Amoy port in Fujian province, southeast China, known by its Mandarin name of ►Xiamen. **50** F2

Amritsar industrial city in the Punjab, India; population (1991) 709,000. It is the holy city of Sikhism, with the Guru Nanak University (named after the first Sikh guru), and the Golden Temple, surrounded by the sacred pool Amrita Saras. The Jallianwalah Bagh area of the city was the scene of the Amritsar Massacre in 1919, when the British Gen Dyer ordered troops to fire on a crowd agitating for self-government; 379 were killed and 1,200 wounded. In 1984, armed Sikh demonstrators were evicted from the Golden Temple by the Indian army, in Operation Bluestar, led by Gen Dayal. Over 300 were killed. Later in 1984, Indian prime minister Indira Gandhi was assassinated in reprisal by Sikh extremists wanting an independent Sikh state in Punjab. The whole of Punjab was put under presidential control in 1987 following riots. Rajiv Gandhi ordered further attacks on the Golden Temple in 1988. **54** B2

Amsterdam official capital of the Netherlands; population (1999 est) 727,000. The Netherlands' second most important port after Rotterdam, Amsterdam is connected to the North Sea by the North Sea Canal, completed in 1876. A new canal leading to the River Waal, south of Utrecht, was completed in 1952 to improve the connection between Amsterdam and the River Rhine. Industries include diamond cutting and polishing, sugar refining, clothes manufacture, printing, chemicals, shipbuilding, brewing, and distilling. Amsterdam is also an international centre of banking and insurance. **22** G2

Amu Darya (or **Amudar'ya**; formerly **Oxus**), river in central Asia, flowing 2,530 km/1,578 mi from the Pamirs to the ►Aral Sea. **42** L9

Amur river in east Asia, which with its tributary, the Ussuri, forms the boundary between Russia and China for much of its course. Formed by the Argun and Shilka rivers, the Amur flows for over 4,400 km/2,730 mi and enters the Sea of Okhotsk. At its mouth at Nikolevsk it is 16 km/10 mi wide. **44** P6

Anaconda town and administrative headquarters of Deer Lodge County, southwest Montana, USA; population (1990) 10,300. It lies 42 km/26 mi northwest of Butte, on Warm Springs Creek, in a mineral-rich mountainous region, and stands 1,615 m/5,300 ft above sea level. The city was founded as Copperopolis in 1883 by the Anaconda Copper Mining Company and was incorporated as Anaconda in 1888. At one time it had the world's largest copper smelter; smelting and phosphate production were the town's economic mainstays until Atlantic Richfield closed the smelter in 1980, causing severe economic hardship. **86** D1

Anaheim city in southwestern California, USA, 50 km/31 mi southeast of Los Angeles; part of Orange County; population (1994 est) 282,100. Industries include electronic and farm equipment, batteries, and processed foods. Disneyland amusement park (opened 1955) is located here. **92** C2

Anatolia (Turkish **Anadolu**), Asian part of Turkey, consisting of a mountainous peninsula with the Black Sea to the north, the Aegean Sea to the west, and the Mediterranean Sea to the south. **36** M6

Anchorage port and largest city in Alaska, USA, at the head of Cook Inlet; population (1994 est) 253,600. It is an important centre of administration, communication, and commerce for much of central and western Alaska. Local industries include oil and gas extraction, tourism, and fish canning. **92** (1)H3

Ancona (Greek *angkon* 'elbow', referring to the town's position on a promontory), Italian city, naval base, and ferry port; capital of Marche region, situated on the Adriatic Sea 140 km/87 mi northeast of Rome; population (1992) 100,700. There are shipbuilding, chemical, fishing, and engineering industries here. Ancona has a Romanesque cathedral and a former palace of the popes. **32** H5

Andalusia (Spanish **Andalucía**), autonomous community of southern Spain, including the provinces of Almería, Cádiz, Córdoba, Granada, Huelva, Jaén, Málaga, and Seville; area 87,300 sq km/33,698 sq mi; population (1991) 6,940,500. The Guadalquivir River flows through Andalusia, which is bounded on the north by the Sierra Morena mountain range. The region is fertile, and produces oranges and wine (especially sherry); horses are bred here also, and copper is mined at Rio Tinto. ►Seville, an inland port, is the administrative capital and the largest industrial centre; Málaga, Cádiz, and Algeciras are the chief ports and also important industrial centres. The Costa del Sol on the south coast has many tourist resorts, including Marbella and Torremolinos; the Sierra Nevada mountain range in the southeast is a winter ski destination. **9** (1)C4

Andaman and Nicobar Islands two groups of islands in the Bay of Bengal, 1,200 km/745 mi off the east coast of India, forming a Union Territory of the Republic of India; capital Port Blair; area 8,300 sq km/3,204 sq mi; population (1994 est) 322,000. Much of the islands is densely forested and the economy is based on fishing, timber, rubber, fruit, nuts, coffee, and rice. **10** (1)A1

Andaman Islands group of Indian islands, part of the Union Territory of ►Andaman and Nicobar Islands. **50** A4

Andermatt village in the Swiss canton of Uri, some 25 km/16 mi south of Altdorf; population (1995) 1,400. It is a tourist and winter-sports centre in the upper Reuss valley (altitude 1,432m/4,690 ft), and is situated at the junction of the roads leading to the Furka Pass and the St Gotthard Pass. **30** D4

Andernach city in the *Land* (administrative region) of Rhineland-Palatinate, Germany, on the River Rhine, 17 km/11 mi northwest of Koblenz; population (1994) 30,000. Its manufacturing industries include chemicals and cement. Pumice stone is mined in the surrounding region. Andernach is a tourist centre. **22** K4

Andes great mountain system or cordillera that forms the western fringe of South America, extending through some 67° of latitude and the republics of Colombia, Venezuela, Ecuador, Peru, Bolivia, Chile, and Argentina. It is the longest mountain range in the world, 8,000 km/5,000 mi, and its peaks exceed 3,600 m/12,000 ft in height for half that length. **96** D5

Andhra Pradesh state in east central India; area 275,100 sq km/ 106,216 sq mi; population (1994 est) 71,800,000. The main cities and towns are ►Hyderabad (capital), Secunderabad, Visakhapatnam, Vijayawada, Kakinda, Guntur, and Nellore. The state is situated on coastal plains with extensive river valleys (Krishna and Godavari) reaching into the Eastern Ghats; smaller rivers are the Pennar and Cheyyar; the Deccan plateau lies inland. The main industries are mica, coal, iron ore, oil refining, shipbuilding, and fertilizers; the main agricultural products are rice, millet, sugar cane, tobacco, groundnuts, sorghum, and cotton. The languages spoken are Telugu, Urdu, and Tamil. **10** C5

Andorra landlocked country in the east Pyrenees, bounded north by France and south by Spain.

Andorra covers an area of 468 sq km/181 sq mi, and has a population of 78,000 (2000 est). The capital is ►Andorra la Vella; other major towns and cities include Les Escaldes and Escaldes-Engordany (a suburb of the capital). The heads of state are Joan Marti i Alanis (bishop of Urgel, Spain; from 1971) and Jacques Chirac (president of France; from 1995); the head of government is Marc Forne Molne (from 1994). The currencies are the French franc and the Spanish peseta, and the languages spoken include Catalan (official), Spanish, and French. **28** L2

Andorra la Vella (French **Andorre la Vielle**, Spanish **Andorra la Vieja**), capital of the principality of Andorra; population (1996) 22,400. It is in the Gran Valira valley, on the eastern slopes of the Pic d'Enclar, in the Pyrenees, at 1,029 m/3,376 ft. **28** M2

Andros Greek island, most northerly of the ►Cyclades; area 383 sq km/ 147 sq mi; population (1981) 9,000. It is 40 km/25 mi long, 12 km/7.5 mi wide, mountainous, and produces wine. The principal town, Andros, is situated on the east coast. **36** G7

Aneto, Pico highest peak of the Pyrenees mountains, rising to 3,404 m/11,168 ft in the Spanish province of Huesca. **28** L2

Angel Falls (or **Salto Angel**), waterfall on the River Caroní in the tropical rainforest of Bolívar Region, southeast Venezuela. It is the highest cataract in the world with a total height of 978 m/3,210 ft. The falls plunge from the lip of the Auyán–Tepúplateau (Guinana Highlands). They were named after the aviator and prospector James Angel who flew over the falls and crash-landed nearby in 1935. **98** E2

Angers ancient French city and administrative centre of ►Maine-et-Loire *département*, situated on the River Maine just north of its conjunction with the Loire and just south of its conjunction with the Mayenne; population (1990) 146,100. Its main products include electrical machinery, wine, and Cointreau liqueur. It is a bishopric with a 12th–13th-century Gothic cathedral. The 12th–13th-century castle is a tourist attraction. **26** E6

Anglesey (Welsh **Ynys Môn** (island); **Sir Ynys Môn** (authority)), island and unitary authority off the northwest coast of Wales.
area 720 sq km/278 sq mi (34 km/ 21 mi long and 31 km/19 mi broad)
towns Llangefni (administrative headquarters), Holyhead, Beaumaris, Amlwch
features separated from the mainland by the Menai Strait, which is crossed by the Britannia tubular railway bridge and Telford's suspension bridge, originally built between 1819 and 1826 but rebuilt since; rich fauna, notably bird life, and flora; many buildings and relics of historic interest
industries manufacture of toys and electrical goods; bromine extraction from the sea
agriculture sheep farming, varied agriculture
population (1996) 71,100 **24** H8/**5** D5

Angola country in southwest Africa, bounded west by the Atlantic Ocean, north and northeast by Congo (formerly Zaire), east by Zambia, and south by Namibia. The Cabinda enclave, a district of Angola, is bounded west by the Atlantic Ocean, north by the Congo River, and east and south by Congo.

Angola covers an area of 1,246,700 sq km/481,350 sq mi, and has a population of 12,878,000 (2000 est). The capital (and chief port) is ►Luanda; other major towns and cities include Lobito, Benguela, Huambo, Lubango, Malange, Namibe, and Kuito. The head of state is José Eduardo dos Santos (from 1979), who is also head of government (from 1999). The currency is the kwanza, and the languages spoken include Portuguese (official), Bantu, and other native dialects. **62** E7

Angostura former name of ►Ciudad Bolívar, a port in Venezuela. **98** E2

Angoulême French town, capital of the *département* of Charente, on the Charente River; population (1990) 46,100. It has a cathedral, and a castle and paper mills dating from the 16th century. **94** M5

Anguilla island in the eastern Caribbean; area 160 sq km/62 sq mi; population (2000) 11,800. Anguilla is a popular tourist destination because of its white coral-sand beaches. However, 80% of its coral reef has been lost through tourism (pollution and souvenir sales). Its main exports include lobster and salt. The currency used is the East Caribbean dollar. It has been a separate dependency of the UK since 1980, with its capital being The Valley. Both English and Creole languages are spoken. **94** M5

Angus unitary authority on the east coast of Scotland. A former county, it was part of Tayside region 1975–96.
area 2,187 sq km/844 sq mi
towns Arbroath, Brechin, Carnoustie, Forfar (administrative headquarters), Kirriemuir, and Montrose

physical the ►Grampian Mountains in the north are dissected by the fertile valleys of the rivers Isla, Clova, Prosen, Water of Saughs, and North Esk; the wide Vale of Strathmore separates the Grampian Mountains from the low-lying Sidlaw Hills in the south
features Pictish and Iron Age remains
industries textiles, light engineering (declining), fish processing
agriculture some fishing (mainly in Arbroath), cereal production
population (1996) 111,300 **5** E3

Anhui (or **Anhwei**), province of eastern China, bounded to the north by Shandong, to the east by Jiangsu, to the southeast by Zhejiang, to the south by Jiangxi, to the southwest by Hubei, and to the northwest by Henan provinces; area 139,900 sq km/54,000 sq mi; population (1996) 60,700,000. Anhui is an intensively cultivated province, with grain the most important crop. Its capital is ►Hefei, while important towns and cities include Anqing, Bengbu, Huainan, and Wuhu. **11** (2)E3

Anhwei alternative spelling of ►Anhui. **11** (2)E3

Ankara (formerly **Angora**), capital of Turkey; population (1990) 2,559,500. Industries include cement, textiles, and leather products. It replaced Istanbul (then in Allied occupation) as capital in 1923. **58** E4

Annaba (formerly **Bône**, 'city of jujube trees'), seaport in Algeria; population (1991 est) 343,000. It is situated in Annaba department close to the Tunisian border. Major industries include an iron and steel complex, railway workshops, and aluminium and chemical works. Iron ore, phosphates, and other minerals, extracted in Tébessa to the southeast, are exported, as are wine and cork. **66** G1

An Najaf walled city near the Euphrates in Iraq, 144 km/90 mi south of Baghdad; population (2001 est) 529,400. The tomb of Ali, son-in-law of Muhammad, is a Shiite Muslim shrine and a starting point of the route to Mecca. **56** D3

Annapolis seaport and capital of ►Maryland, USA, near the mouth of the River Severn on Chesapeake Bay; seat of Anne Arundel County; population (1996) 33,200. Annapolis, whose industries include radar equipment, is a commercial centre for the surrounding agricultural area, and its waterside setting and historic buildings make it a popular recreational and tourist area, lying as it does only 40 km/25 mi from both Washington and Baltimore. Founded in 1694 as the capital of colonial Maryland, Annapolis was named after Princess (later Queen) Anne in 1697, and incorporated as a city in 1796. **88** E3

Annapurna mountain 8,075 m/ 26,502 ft in the Himalayas, Nepal. The north face was first climbed by a French expedition (Maurice Herzog) in 1950 and the south by a British team in 1970. **54** D3

Ann Arbor city in Washtenaw County, southeast Michigan, USA; population (1994 est) 109,000. It is located 61 km/38 mi west of Detroit on the River Huron. Ann Arbor is situated in an agricultural and fruit-growing area, and is a centre for medical, aeronautical, nuclear, and chemical research, and for the manufacture of computer software and electronic equipment. Other local manufactures include ball bearings, coil springs, machinery, tools, cameras, and scientific instruments. **88** D2

Annecy capital of the *département* of ►Haute-Savoie in the Rhône-Alpes region, southeast France, at the northern end of Lake Annecy; population (1990) 51,100, conurbation 123,000. Its industries include the production of textiles, paper, and precision instruments; there is a nuclear research centre. Annecy is the seat of a bishopric and has a 16th-century cathedral. It is also a popular tourist resort. The old part of the city has a network of canals. **30** B5

Anniston city and administrative headquarters of Calhoun County, northeast Alabama, USA; population (1990) 26,600. It is located 85 km/ 53 mi northeast of Birmingham, in the foothills of the Appalachian Mountains. Anniston is in an iron mining and cotton growing area, and produces appliances, pipes, chemicals, machine parts, industrial castings, yarn, and clothing. Anniston is also a military service centre; the Anniston Army Depot, 16 km/10 mi to the west, is a major ordnance facility, and Fort McClellan (1917), with units to the northeast and northwest, is the headquarters of the Army's military police and its Chemical Corps. **90** D3

Annobón (formerly **Pagalu** 1973–79), mountainous, volcanic island in Equatorial Guinea, 600 km/375 mi southwest of Bata; area 17 sq km/7 sq mi; population (1990 est) 2,600. The island rises to a height 671 m/2,200 ft. The chief town is San Antonio. Annobón's inhabitants are descended from slaves of the Portuguese and still speak a Portuguese patois. The island was Portuguese from the end of the 5th century until it was ceded to Spain in 1778. It remained a Spanish possession until 1968 when independence was declared by the central government on Fernando Póo, present-day Bioko. Fishing and lumbering are the chief economic activities. **68** F5

Ansbach (formerly **Anspach**), city in Bavaria, Germany, 150 km/93 mi northwest of Munich, on the River Rezat; population (1995) 39,600. Important industries include plastics processing, food, and electrotechnology, although the majority of inhabitants work in the service sector. **20** F7

Anshan (or **An-shan**), city in Liaoning province, 89 km/55 mi southeast of Shenyang; population (1994) 1,430,000. The iron and steel centre started here in 1918 was expanded by the Japanese, dismantled by the Russians, and restored by the communist government of China. It produces some 8 million tonnes of steel annually. Other products include chemicals, tractors, and machinery. **48** B3

Antakya (or **Hatay**), city in southeastern Turkey, site of the ancient Antioch; population (1994) 137,200. **58** G5

Antalya Mediterranean port on the west coast of Turkey and capital of a province of the same name; population (1990) 378,200. The port trades in grain and timber. Industries include canning and flour milling. It is a popular coastal resort. **36** N8

Antananarivo (formerly **Tananarive**), capital and administrative centre of Madagascar, on the interior plateau, with a rail link to Tamatave; population (1993) 1,052,800. Industries include food processing, leather goods, clothing, wood pulp and paper manufacturing, and brewing. **72** H3

Antarctica continent surrounding the South Pole, arbitrarily defined as the region lying south of the Antarctic Circle. Occupying 10% of the world's surface, it is almost one-and-a-half times the size of the USA. Antarctica contains 90% of the world's ice, representing nearly three-quarters of its fresh water. It is thought that if all the ice suddenly melted, the world sea level would rise by 60 m/197 ft.
area 13,000,000 sq km/5,019,300 sq mi; ice shelves which fill the surrounding seas add a further 1,300,000 sq km/501,930 sq mi to this figure
features Mount Erebus on Ross Island is the world's southernmost active volcano; the Ross Ice Shelf is formed by several glaciers coalescing in the Ross Sea
physical Antarctica can be divided into two regions, separated by the Transantarctic Mountains, which extend for 3,500 km/2,175 mi and whose peaks, many of them exceeding 3,000 m/9,850 ft in height, protrude through the ice. The larger region, known as Greater or East Antarctica, is comprised of ancient rocks lying mostly at sea level, which are approximately 3,800 million years old. In contrast, Lesser or West Antarctica is 150–200 million years old and has mountain ranges buried under the ice. These include the Antarctic Peninsular and the Ellsworth Mountains, in which the highest peak in Antarctica, the Vinson Massif, is located; height 5,140 m/16,863 ft. The few peaks that are visible above the ice are known as nunataks. Two vast seas, the Ross Sea and the Weddell Sea, cut into the continent. Between them lies the mountainous Antarctic Peninsula, which was originally connected to South America before continental drift
population no permanent residents and no indigenous inhabitants; settlement is limited to scientific research stations with maximum population of 10,000 (including 3,000 tourists) during the summer months. Sectors of Antarctica are claimed by Argentina, Australia, Chile, France, the UK, Norway, and New Zealand

Antarctic Ocean popular name for the reaches of the Atlantic, Indian, and Pacific oceans extending south of the Antarctic Circle (66° 32′S). The term is not used by the International Hydrographic Bureau.

Antarctic Peninsula mountainous peninsula of West Antarctica extending 1,930 km/1,200 mi north toward South America. **102** (2)LL3

Antibes resort, which includes Juan les Pins, on the French Riviera, in the *département* of Alpes Maritimes; population (1990) 70,600. There is a Picasso collection in the 17th-century castle museum. **30** C7

Anticosti Island English name for L'Île d'▶Anticosti, an island in the Gulf of St Lawrence, Québec, Canada. 82 U7

Anticosti, L'Île d' (English **Anticosti Island**), island in the Gulf of St Lawrence at the mouth of the St Lawrence River, 80 km/50 mi northeast of the Gaspé Peninsula and 600 km/375 mi northeast of Québec City, in east Québec, Canada; area 7,923 sq km/3059 sq mi; population (1991 est) 300. Heavily forested, with many small rivers, canyons, and a bog, it extends 220 km/135 mi from west-northwest–east-southeast, and rises to 191 m/625 ft in the north. Wildlife is abundant, and there are several hunting reserves. Port-Menier, near the western tip, is its only settlement, although there are also scattered campsites. 82 U7

Antigua and Barbuda country comprising three islands in the eastern Caribbean (Antigua, Barbuda, and uninhabited Redonda).

Antigua and Barbuda covers an area of 440 sq km/169 sq mi (Antigua 280 sq km/108 sq mi, Barbuda 161 sq km/62 sq mi, plus Redonda 1 sq km/0.4 sq mi), and has a population of 68,000 (2000 est). The capital (and chief port) is ▶St John's (on Antigua); other major towns and cities include Codrington (on Barbuda). The head of state is Queen Elizabeth II (from 1981), represented by Governor General James B Carlisle (from 1993); the head of government is Lester Bird (from 1994). The currency is the East Caribbean dollar, and the languages spoken include English (official) and local dialects. **94** M5

Antilles group of West Indian islands, divided N–S into the **Greater Antilles** (Cuba, Jamaica, Haiti–Dominican Republic, Puerto Rico) and **Lesser Antilles**, subdivided into the Leeward Islands (Virgin Islands, St Kitts and Nevis, Antigua and Barbuda, Anguilla, Montserrat, and Guadeloupe) and the Windward Islands (Dominica, Martinique, St Lucia, St Vincent and the Grenadines, Barbados, and Grenada). **94** J5/**94** L6

Antofagasta port of northern Chile, and capital of Antofagasta region on the Pacific coast; population (1992) 226,750. The area of the region is 125,300 sq km/48,366 sq mi. Nitrates and copper from the Atacama Desert are exported, and there are ore refineries and foundries. Antofagasta was founded in 1870. **100** G3

Antrim historic county of Northern Ireland.
area 2,830 sq km/1,092 sq mi
towns and cities ▶Belfast (county town), Larne (port), Antrim, Ballymena, Lisburn, Carrickfergus
physical peat bogs; Antrim borders Lough Neagh, and is separated from Scotland by the North Channel, which is only 21 km/13 mi wide at Torr Head, the narrowest point; the main rivers are the Bann and the Lagan
features ▶Giant's Causeway, a World Heritage Site, on the north coast; Antrim Mountains (highest point Trostan 554 m/1,817 ft) and the Glens of Antrim (an area of outstanding natural beauty); Kebble National Nature Reserve on Rathlin Island, off the coast near Ballycastle; Bushmills Distillery, in the village of Bushmills, has the oldest known licence for distilling whisky; a number of early fortifications, castles, and medieval ecclesiastical remains in the county; the village of Cushendun was built by Clough Williams-Ellis; Carnfunnock Country Park
industries shipbuilding; whisky
agriculture potatoes, oats, livestock
population (1991) 655,000

Antwerp (Flemish **Antwerpen**), province of Belgium, bounded to the north and northeast by the Netherlands, to the south by Brabant, to the southeast by Limbourg, and to the southwest by East Flanders; area 2,859 sq km/1,104 sq mi; population (1997) 1,635,600. **7** (1)C3

Antwerp (Flemish **Antwerpen**; French **Anvers**), port in Belgium on the River Schelde, capital of the province of Antwerp, 43 km/27 mi north of Brussels; population (1997) 453,000. It is Belgium's second city and is also the largest town in Flanders, the Flemish-speaking part of Belgium. One of the world's busiest ports, it claims to be the world's leading centre for diamond cutting and dealing. Other industries include shipbuilding, oil refining, petrochemicals, and textiles. The home of the 17th-century Flemish artist Rubens is preserved, and several of his works are in the Gothic cathedral. 22 G3

Antwerpen alternative name for ►Antwerp, a port and province in Belgium. 22 G3/7 (1)C3

Anuradhapura ancient holy city in Sri Lanka; population (district, 1993 est) 741,000. It was the capital of the Sinhalese kings of Sri Lanka 5th century BC–8th century AD; rediscovered in the mid-19th century. Sacred in Buddhism, it claims a Bo tree descended from the one under which the Buddha became enlightened. 54 D7

Anvers French form of ►Antwerp, a province in northern Belgium. 22 G3

Anyang (ancient **Yin**; later **Changteh** or **Zhangde**), city in Henan province, east China; population (1994) 1,038,000. It lies on the Beijing–Guangzhou railway. Iron- and steel-smelting are the principal industries, using local coal and iron ore from the nearby Hanxing mining area, and the Hebi coalmining complex. Engineering and the manufacture of textiles are also important. The city was a capital of the Shang dynasty (16th–11th centuries BC). Rich archaeological remains have been uncovered since the 1920s. 46 E3

Anzhero-Sudzhensk town in the Kemerovo oblast (region) of the Russian Federation, in southern Siberia, 80 km/50 mi east of Tomsk, situated on the Trans-Siberian Railway; population (2001 est) 94,200. It is a major centre of coal mining in the ►Kuznetsk Basin; mining equipment, chemicals, and pharmaceuticals are manufactured here. 42 R6

Anzio (Roman **Antium**), port and resort on the west coast of Italy in Lazio region, 57 km/36 mi southeast of Rome; population (1990) 34,700. Fishing is an important industry, and it is the ferry point for the Pontine Islands. The resort is a popular watersports centre; the Festa del Moree sailing races take place here in June. The town was the birthplace of the Roman emperor Nero. 32 G7

Aomori (or **Awomori**), port at the head of Mutsu Bay, on the north coast of Honshu island, Japan, 40 km/25 mi northeast of Hirosaki; population (1994) 293,000. Aomori is the chief town of the prefecture of the same name and has been a port since the establishment of harbour facilities in 1624. The port handles a large local trade in fish, rice, and timber. 48 L3

Aoraki (formerly **Mount Cook**), highest point, 3,764 m/12,349 ft, of the ►Southern Alps, a range of mountains running through New Zealand. 78 C6

Aosta industrial town and tourist centre on the River Dora Baltea, northwest Italy, 79 km/49 mi northwest of Turin; population (1990 est) 36,100. It is the capital of ►Valle d'Aosta (French-speaking) autonomous region. The town's economy is based on tourism, commerce, and administration, as well as industry, which includes iron, food, chemical, textiles, publishing, and wood-working businesses. Aosta is a commercial and financial centre for the region. There are extensive Roman remains, and the town stands at the junction of the two St Bernard passes. 30 C5

Aotearoa (Maori 'Land of the Long White Cloud'), Maori name for ►New Zealand, reflecting the fact that clouds tend to build up in alignment with the main mountain ranges of the islands. 78 B5

Aozou Strip disputed territory 100 km/60 mi wide on the Chad–Libya frontier, occupied by Libya 1973, part of Chad according to an International Court of Justice ruling 1994. Lying to the north of the Tibesti massif, the area is rich in uranium and other minerals.

Apeldoorn commercial city in Gelderland province, east central Netherlands, 27 km/17 mi northeast of Arnhem; population (1997) 151,700. Industries include metallurgy, papermaking, textiles, computer manufacture, and chemicals. Het Loo, which is situated nearby, has been the summer residence of the Dutch royal family since the time of William of Orange. 22 H2

Apennines (or **Appennino**), chain of mountains stretching the length of the Italian peninsula. An older and more weathered continuation of the Maritime Alps, from Genoa the Apennines swing across the peninsula to Ancona on the east coast, and then back to the west coast and into the 'toe' of Italy. The system is continued over the Strait of Messina along the north Sicilian coast, then across the Mediterranean Sea in a series of islands to the Atlas Mountains of North Africa. The highest peak is Monte Corno in Gran Sasso d'Italia at 2,914 m/9,560 ft. 32 G5

Apia capital and chief port of ►Samoa, on the north coast of Upolu; population (1994) 34,000. Copra, bananas, and cocoa beans are exported. There is a high-power radio communication station and an observatory. The town is built on a curving bay, backed by a conical hill, Vaea, the burial place of Scottish writer Robert Louis Stevenson, beyond which is a mountain range. 74 J7

Apo, Mount active volcano and highest peak in the Philippines, rising to 2,954 m/9,692 ft on the island of Mindanao. 50 H5

Apostle Islands (or **Twelve Apostles**), group of 22 islands in southwest Lake Superior, Ashland County, north Wisconsin; area approximately 518 sq km/200 sq mi. The largest and only commercially developed island is Madeline Island. La Pointe, the only island settlement, has a year-round population of (1992) 200, that swells in the summer. The picturesque islands and nearby mainland are designated a national lakeshore area, popular with campers, tourists, and outdoors enthusiasts. 88 B1

Appalachian Mountains mountain system in eastern North America, stretching about 2,400 km/1,500 mi from Alabama to Québec. The chain, composed of ancient eroded rocks and rounded peaks, includes the Allegheny, Catskill, and Blue Ridge Mountains. Its width in some parts reaches 500 km/311 mi. Mount Mitchell, in the Blue Ridge Mountains, is the highest peak at 2,045 m/6,712 ft. The eastern edge has a fall line to the coastal plain where Philadelphia, Baltimore, and Washington stand. The Appalachians are heavily forested and have deposits of coal and other minerals. 90 E3

Appennino alternative name for ►Apennines, a chain of mountains stretching the length of the Italian peninsula. 32 G5

Appenzell canton in northeast Switzerland, enclosed within the canton of St Gallen; area 415 sq km/ 160 sq mi. Appenzell is subdivided into the regions of Outer Rhodes, or Ausser-Rhoden (capital Herisau) and Inner Rhodes, Inner-Rhoden, (population (1996) 14,800; (capital Appenzell). Textiles are manufactured in Outer Rhodes, whereas Inner Rhodes is mainly agricultural. The highest point in the canton is Mount Santis (2,500 m/8,202 ft), which is surmounted by an observatory. 8 (1)D1

Appleton city in Calumet, Winnebago, and Outagamie (of which it is administrative headquarters) counties, east-central Wisconsin, USA; population (1992) 68,500. It is located 30 km/19 mi northwest of Oshkosh and 145 km/90mi north of Milwaukee, on the River Fox, near Lake Winnebago, in a rich dairying and farming region. Paper, pulp, paper and wood products, canned goods, beer, and building materials are among the city's manufactures. 88 C2

Aqaba, Gulf of gulf extending northwards from the Red Sea for 160 km/100 mi to the Negev; its coastline is uninhabited except at its head, where the frontiers of Israel, Egypt, Jordan, and Saudi Arabia converge. The two ports of Elat (Israeli **Elath**) and Aqaba, Jordan's only port, are situated here. A border crossing near the two ports was opened in 1994, for non-Israelis and non-Jordanians, to encourage the eastern Mediterranean tourist industry. 60 C7

Aqtöbe (formerly Russian **Aktyubinsk** 1869–1993), industrial city on the River Ilek, in northwestern Kazakhstan, 100 km/62 mi south of the border with the Russian Federation, capital of Aqtöbe oblast (region); population (1995) 260,000. Chemicals, metals, electrical equipment are among the goods manufactured here. Aqtöbe stands on the main Moscow–Tashkent railway line; it is also located near to a large natural-gas field at Karachaganak. 38 L4

Aquitaine region of southwest France; administrative capital ▶Bordeaux; area 41,300 sq km/ 15,942 sq mi; population (1990) 2,795,800. It comprises the *départements* of ▶Dordogne, ▶Gironde, ▶Landes, ▶Lot-et-Garonne, and ▶Pyrénées-Atlantiques. Red wines (Margaux, St Julien) are produced in the Médoc district, bordering the Gironde. Aquitaine was an English possession 1152–1453. 6 C4

Arab Emirates see ▶United Arab Emirates. 56 F5

Arabia (or **Arabian Peninsula**; Arabic **Jazirat al-Arab**, the 'peninsula of the Arabs'), peninsula between the Gulf and the Red Sea, in southwest Asia; area 2,600,000 sq km/1,000,000 sq mi. The length from north to south is about 2,400 km/1,490 mi and the greatest width about 1,600 km/ 994 mi. The peninsula contains the world's richest gas reserves and half the world's oil reserves. It comprises the states of Bahrain, Kuwait, Oman, Qatar, Saudi Arabia, the United Arab Emirates, and Yemen.

Arabian Gulf alternative name for the ▶Gulf. 61 C2

Arabian Sea northwestern branch of the ▶Indian Ocean, covering 3,859,000 sq km/1,489,970 sq mi, with India to the east, Pakistan and Iran to the north, and the Arabian Peninsula and Somalia to the west. It is linked with the Red Sea via the Gulf of Aden, and with the Gulf via the Gulf of Oman. Its mean depth is 2,730 m/8,956 ft. The chief river flowing into the Arabian Sea is the Indus, which is linked with a large submarine canyon in the continental shelf. The sea is rich in fish. 56 H6

Arad Romanian city on the River Mures, 160 km/100 mi northeast of Belgrade; population (1993) 189,000. It is a major trading centre with many industries. 34 J3

Arafat (or **Jabal al-Rahma** 'the mount of mercy'; Arabic *arifa* 'to know', 'to recognize'), hill and plain about 24 km/15 mi east and about 19 km/12 mi southwest of Mecca. The hill is granite and rises to 61 m/200 ft. The most important ceremony of the Muslim pilgrimage, or hajj, occurs on this hill.

Arafura Sea area of the Pacific Ocean between northern Australia and Indonesia, bounded by the Timor Sea in the west and the Coral Sea, via the Torres Strait, in the east; 1,290 km/ 800 mi long and 560 km/350 mi wide. It lies on the Arafura Shelf, and is 50–80 m/165–265 ft deep. The Indonesian Aru islands lie to the north. To the northwest, the Aru Trough (3,650 m/12,000 ft deep) separates the Arafura Sea from the Banda Sea. 53 D5

Aragón autonomous community and former kingdom of northeast Spain, including the provinces of Huesca, Teruel, and Zaragoza; area 47,700 sq km/18,420 sq mi; population (1994) 1,183,600. Products include cereals, rice, olive oil, almonds, figs, grapes, and olives; merino wool is a major export. The principal river of Aragón is the ▶Ebro, which receives numerous tributaries both from the mountains of the south and from the Pyrenees in the north. Aragón was an independent kingdom from 1035 to 1479. The capital of modern Aragón is ▶Zaragoza. 9 (1) E2

Aral Sea (Russian **Aralskoye More**), inland sea divided between Kazakhstan and Uzbekistan, the world's fourth-largest lake; former area 62,000 sq km/24,000 sq mi, but decreasing. Water from its tributaries, the Amu Darya and Syr Darya, has been diverted for irrigation and city use, and the sea is disappearing, with long-term consequences for the climate. 42 K8

Aran Islands group of three limestone islands in the mouth of Galway Bay, which is about 32 km/ 20 mi wide. They lie 48 km/30 mi from Galway, on the west coast of the Republic of Ireland; the principal town is Kilronan on Inishmore. The islands form a natural breakwater, and comprise Inishmore (Irish **Inis Mór**), area 3,092 ha/7,637 acres, population (1996) 838; Inishmaan (Irish **Inis Meáin**), area 912 ha/2,253 acres, population (1996) 191; and Inisheer (Irish **Inis Óirr**), area 567 ha/ 1,400 acres, population (1996) 274. The chief industries are tourism, fishing, and agriculture. 24 B8

Aranjuez Spanish town on the River Tagus, 40 km/25 mi southeast of Madrid; population (1991) 35,900. The palace was a royal residence for centuries. 28 G4

Ararat, Mount double-peaked mountain in Turkey near the Iranian border; Great Ararat, at 5,137 m/ 16,854 ft, is the highest mountain in Turkey. According to the Old testament, it was the resting place of

Noah's Ark after the Flood. **58** L4

Arbīl alternative name for Irbil, a governate in northern Iraq; and ►Irbil, its capital. **58** K5

Arbroath fishing town in Angus, on the east coast of Scotland, 26 km/16 mi northeast of Dundee, at the mouth of Brothock Water; population (1991) 23,500. In 1320 the Declaration of Arbroath was signed by the Scottish Parliament in Arbroath Abbey, proclaiming Scotland's independence to the pope. The town has a number of oil-related firms, a fishing industry and produces smoked haddock (Arbroath smokies). The harbour at Arbroath has been working for many years. Although not as busy as in past times, many fishing boats still sail from Arbroath. **24** K5

Archangel English name for ►Arkhangelsk, a port in the north of the Russian Federation. **38** H2

Archipel de la Société French name for the ►Society Islands, an archipelago in French Polynesia. **74** L7

Archipel des Tuamotu French name for ►Tuamotu Archipelago, part of French Polynesia. **74** M7

Arctic, the that part of the northern hemisphere surrounding the North Pole; arbitrarily defined as the region lying north of the Arctic Circle (66° 30' north) or north of the treeline; area 36,000,000 sq km/14,000,000 sq mi; population around 1 million.

Arctic Ocean ocean surrounding the North Pole; area 14,000,000 sq km/5,405,400 sq mi. Because of the Siberian and North American rivers flowing into it, it has comparatively low salinity and freezes readily. **102** (1)A1

Ardabil (or **Ardebil**), city in northwest Iran, near the frontier with Azerbaijan; population (1991) 311,000. Ardabil exports dried fruits, carpets, and rugs. **58** N4

Ardèche *département* in the ►Rhône-Alpes region of France; area 5,523 sq km/2,132 sq mi; population (1990) 135,900. The area is mountainous, containing part of the Vivarais mountain range in the west. The River ►Ardèche flows through the southeast of the *département*. The Ardèche gorges, cut through limestone, attract many tourists. Cereals, vines, chestnuts, mushrooms, and olives are produced. Coal and iron are mined. The administrative centre is Privas; population (1990) 14,000. **6** F4

Ardèche river in southeast France, rising in the ►Cévennes and flowing southeast to the Rhône; length 120 km/75 mi. Near Vallon it flows under the Pont-d'Arc, a natural bridge. It gives its name to the *département* of ►Ardèche. The spectacular gorges of the Ardèche, cut into limestone, can be explored by canoe or from a twisting tourist road.

Ardennes *département* in the Champagne-Ardennes region of France, bordering on Belgium; area 5,218 sq km/2,105 sq mi; population (1990) 146,300. It is wooded and hilly: the Forest of ►Ardennes covers about one-third of the *département*. The west and the valleys of the Rivers Aisne and Meuse are fertile. Livestock farming and cereal production are important. There is a nuclear power station, located 6 km/3.7 mi south of Givet, car manufacturing, and a growing service sector. The administrative centre is ►Charleville-Mézières; population (1990) 90,000. **6** F2

Ardennes hilly, wooded plateau in northeast France, southeast Belgium, and northern Luxembourg, cut through by the River ►Meuse. The area gives its name to the region of ►Champagne-Ardenne and the *département* of the ►Ardennes in France. The highest hills are about 590 m/1,936 ft. Cattle and sheep are raised and the area is rich in timber and minerals. There was heavy fighting here in both world wars, notably in the Battle of the Bulge (1944–1945, also known as the Ardennes offensive). In World War I it was the route of the main German advance in 1914. **22** G4

Arecibo town in northwest Puerto Rico; population (2001 est) 110,200. It is located on the Atlantic Ocean at the mouth of the Arecibo River, 77 km/48 mi to the west of San Juan. This colonial centre was originally settled in the mid-16th century. A port with some light manufacturing facilities and the site of a college of the University of Puerto Rico, it is also a trade, processing, and distribution hub for an agricultural region producing sugar cane, coffee, fruit, and tobacco. The rum and pharmaceutical industries are also important. The Cueva del Indio caves and the Río Camuy Cave Park are nearby, and 19 km/12 mi to the south is Cornell University's Arecibo Observatory, the site of the world's largest single-dish radio telescope.

Arequipa capital of Arequipa department in the western Andes of southern Peru; it stands at a height of 2,363 m/7,753 ft in a fertile valley at the base of the dormant volcano El Misti (5,822 m/19,100 ft); industries include textiles, soap, and leather goods; population (1993) 619,200. It is the second-largest city of Peru and the cultural focus of southern Peru. Arequipa was founded by Pizarro in 1540 on the site of an ancient Inca city, and has a cathedral, founded in 1621, and a university. **98** C7

Arezzo (ancient **Arretium**), town in Tuscany, Italy, 80 km/50 mi southeast of Florence; population (1999) 91,300. Originally an Etruscan settlement, Arezzo is an important market for agricultural and animal products from the fertile surrounding districts, and trades in textiles and clothing, shoes, olive oil, antiques, and gold and jewellery crafts. There is a fresco series by the Renaissance painter Piero della Francesca. **32** F5

Argentan French town in the *département* of Orne, Normandy, on the River Orne; population (1990) 17,000. The town is a centre for horse-breeding and leather production. **22** B6

Argentina country in South America, bounded west and south by Chile, north by Bolivia, and east by Paraguay, Brazil, Uruguay, and the Atlantic Ocean.

Argentina covers an area of 2,780,400 sq km/1,073,518 sq mi, and has a population of 37,032,000 (2000 est). The capital is ►Buenos Aires; other major towns and cities include Rosario, Córdoba, San Miguel de Tucumán, Mendoza, Santa Fé, and La Plata. The head of state and government is Fernando de la Rua (from 1999). The currency is the peso, and the languages spoken include Spanish (official), Italian, English, German, and French. **100** H6

Argyll and Bute unitary authority in western Scotland, created in 1996 from the district of the same name and part of Dumbarton district, which were both parts of Strathclyde region; it includes the islands of Gigha, Bute,

Mull, Islay, Jura, Tiree, Coll, Colonsay, Iona, and Staffa.
area 7,016 sq km/2,709 sq mi
towns Campbeltown, Dunoon, Helensburgh, Inveraray, Lochgilphead (administrative headquarters), Oban, Rothesay
physical rural area consisting of mainland and islands; the coast is heavily indented. Inland the area is mountainous; highest peak, Ben Cruachan (1,126 m/3,693 ft). Lochs Fyne and Long are the largest sea lochs; freshwater lochs include Loch Awe and Loch Lomond; Fingal's Cave (Staffa); Corryvrekan Whirlpool (Jura-Scarba); Ben Arthur (The Cobbler), 884 m/2,900 ft
features Bronze, Stone, and Iron Age remains
industries limited manufacture, seaweed processing, fish, timber harvesting
agriculture sheep, forestry
population (1996) 89,300 **5** D3

Århus (or **Aarhus**), second-largest city of Denmark, on the east coast overlooking the Kattegat; population (1995) 277,500. It is the capital of Århus county in Jylland (Jutland) and a shipping and commercial centre. **16** F8

Arica port and northernmost city of Chile, in Tarapacá region 20 km/12 mi from the Peruvian border; it is situated in a rainless district north of the Atacama Desert; population (1992) 169,200. The city is linked by railway to La Paz, Bolivia, and much of Bolivia's foreign trade is shipped via Arica. Fishmeal is an important industry. Tourism and the creation of a duty-free zone have boosted the local population. There is an increasing international trade in minerals. The world's largest open-pit copper mine at nearby Chuquicamata dominates the country's mining sector. **98** C7

Ariège *département* in the Midi-Pyrénées region of France: area 4,890 sq km/1,888 sq mi; population (1990) 136,600. Spurs of the Pyrenees Mountains cover a large part of the *département* and there are many fertile valleys. In the east flows the River ►Ariège, and in the west, the Salat, both tributaries of the Garonne. Cereals, vines, vegetables, potatoes, and hemp are produced, and livestock is farmed. Iron, zinc, manganese, and bauxite are mined, and there are chemical, metallurgical, textile, and paper industries. The administrative centre is Foix. **6** D5

Ariège river in southern France, which rises in the Pyrenees and flows north to join the Garonne 10 km/6 mi south of Toulouse; length 170 km/106 mi. It gives its name to the *département* of ►Ariège. **26** G11

Arizona state in southwestern USA. It is nicknamed Grand Canyon State. Arizona was admitted to the Union in 1912 as the 48th US state. The state is renowned for its natural wonders, including Monument Valley and the Grand Canyon, and is strongly associated with such indigenous peoples as the Navajo and Hopi. It is bordered to the east by New Mexico, to the south by the Mexican state of Sonora, to the west by the Mexican state of Baja California and the US states of California and Nevada, and to the north by Utah and, at the 'Four Corners' to the northeast, Colorado.
population (2000 est) 5,130,600; including 5.6% American Indians (Navajo, Hopi, Apache), who by treaty own 25% of the state
area 294,100 sq km/113,500 sq mi
capital ►Phoenix
towns and cities Tucson, Scottsdale, Tempe, Mesa, Glendale, Flagstaff
industries and products cotton under irrigation, livestock ranching, copper (more than half of US annual output), silver, uranium mining, molybdenum, electronics, aircraft **92** D2

Arkansas state in southern central USA. It is nicknamed Land of Opportunity. Arkansas was admitted to the Union in 1836 as the 25th US state. Historically it was a cotton plantation state, dependent on slavery. It was the site of civil-rights struggles in the 1950s and 1960s and became closely associated with the political intrigues surrounding Bill Clinton during his presidential tenure in the 1990s. Arkansas is bordered to the south by Louisiana, to the southwest by Texas, to the west by Oklahoma, to the north by Missouri, and to the east by Tennessee and Mississippi; the Red, St Francis, and Mississippi rivers form part of its natural borders.
population (2000 est) 2,673,400
area 137,800 sq km/53,191 sq mi
capital ►Little Rock
towns and cities Fort Smith, Pine Bluff, Fayetteville, North Little Rock
industries and products cotton, soybeans, rice, oil, natural gas, timber, processed foods, electronics, financial sector, military bases **90** C3

Arkhangelsk (English **Archangel**), port in the north of the Russian Federation, capital of Archangelsk oblast; population (1990) 419,000. Situated on the Northern Dvina River, it is Russia's chief timber-exporting port, and the centre of a rapidly growing mineral extraction industry (oil drilling and diamond mining). Formerly blocked by ice for half the year, it has been kept open constantly since 1979 by icebreakers. Archangel was made an open port by Boris Godunov and was of vital significance for trade until Peter the Great built St Petersburg. **38** H2

Arles town in Bouches-du-Rhône *département*, southwest France, on the Arles canal and the left bank of the Rhône, at the head of the Camargue delta; population (1990) 52,600. Its main economic activities are tourism and agriculture, and it is in an important fruit- and vine-growing district. Roman relics include an aqueduct, baths, and a 21,000-spectator amphitheatre now used for bullfighting and plays. The Romanesque-Provençal church of St Trophime, formerly an archiepiscopal cathedral, has fine cloisters and a notable 12th-century portal. The Dutch painter Vincent van Gogh lived here 1888–89, during which time he painted some of his major works. **26** K10

Arlington city in Tarrant County, northeast Texas, USA; population (1994 est) 287,000. It is located between Dallas and Fort Worth on the Trinity River and Lake Arlington. An agricultural centre, it developed quickly after 1954, when General Motors opened a large auto plant. The city now makes cars, trailers, steel and iron, aerospace equipment, chemicals, rubber and paper products, cans, and machinery. The University of Texas at Arlington began as a state college in 1895, and Arlington Baptist College was established in 1939. The city's population almost tripled between 1970 and 1990. **90** B3

Arlington county in Virginia, and a suburb of Washington, DC, USA, which it faces across the Potomac River; area 67 sq km/26 sq mi; population (1994 est) 175,000. Arlington is a busy residential community with clusters of modern high-rise buildings and some light manufacturing (electrical components, scientific instruments, machinery). It is home to Strayer College (1892) and Marymount University (1950), as well as numerous federal buildings. **88** E3

Arlon capital of the province of Luxembourg in southeast Belgium; population (2001 est) 22,600. Arlon is one of the oldest settlements in Belgium, and was a trading centre in Roman times. Trade is now conducted from the town in corn, ironware, tobacco, and crockery. **22** H4

Armagh (Irish **Ard Macha**, 'the height of Macha' a legendary queen), historic county of Northern Ireland.
area 1,250 sq km/483 sq mi
towns and cities ►Armagh (county town), Lurgan and Portadown (merged to form Craigavon), Keady
physical smallest county of Northern Ireland; flat in the north, with many bogs and mounds formed from glacial deposits; low hills in the south, the highest of which is Slieve Gullion (577 m/1,893 ft); principal rivers are the Bann, the Blackwater and its tributary, the Callan
features 17th-century manor Ardress House; Oxford Island Nature Reserve; Gosford Forest Park
agriculture mostly good farmland with apple orchards; potatoes; flax; livestock in the south; fruit and market gardening in the north
industries linen manufacture; milling; light engineering; concrete; potato crisps
population (1991) 51,800

Armagh city and county town of County ►Armagh, Northern Ireland; population (1991) 12,700. Industries include textiles, including linen; the manufacture of shoes, optical instruments, and chemicals; and engineering and food processing. The city became the religious centre of Ireland in the 5th century when St Patrick was made archbishop. **24** F7

Armenia country in western Asia, bounded east by Azerbaijan, north by Georgia, west by Turkey, and south by Iran.

Armenia covers an area of 29,800 sq km/11,505 sq mi, and has a population of 3,520,000 (2000 est). The capital is ►Yerevan; other major towns and cities include Gyumri (formerly Leninakan), Vanadzor (formerly Kirovakan), Hradzan, and Aboyvan. The head of state is Robert Kocharian (from 1998); the head of government is Andranik Markaryan (from 2000). The currency is the dram, and the main language spoken is Armenian (official). **58** K3

Armentières town in northern France on the River Lys in the *département* of Nord; population (1990) 26,200. Major industries include the manufacture of textiles, especially linen, printing, brewing, dyes, and clothing. The song 'Mademoiselle from Armentières' originated during World War I, when the town was held by the British and was a popular rest centre for British troops. The town was flattened by bombardment during the German Spring Offensive of 1918, but has been rebuilt. **22** E4

Armidale city in New South Wales, southeastern Australia, 567 km/352 mi north of Sydney; population (1996) 21,330. Armidale is the site of the University of New England (1954), and it is the service centre of a sheep-rearing district; industries include sawmills, joinery, and plywood milling. Among the 19th-century buildings of the city are mansions of the squatters (early settlers). **76** K6

Arnhem capital of Gelderland province, Netherlands, situated on the Rhine near where it joins the IJssel, 56 km/35 mi southeast of Utrecht; population (1997) 135,000. It produces salt, chemicals, and pharmaceuticals. During World War II, the town was the scene of fierce fighting in 1944, when much of the old town was destroyed. **22** H3

Arnhem Land plateau of the central peninsula in northeast Northern Territory, Australia, west of the Gulf of Carpentaria; approximate area 80,776 sq km/31,188 sq mi. Arnhem Land was named after a Dutch ship which dropped anchor here in 1618. The chief town is Nhulunbuy; population (1996) 3,695. It is the largest of the Aboriginal reserves, and was declared Aboriginal land in 1976. Many of the inhabitants live in small settlements and maintain a traditional way of life. Bauxite and uranium mining and the supporting industries provide the main economic base of the area. **76** F2

Arno Italian river 240 km/150 mi long, rising in the Apennines, and flowing westwards to the Mediterranean Sea. Florence and Pisa stand on its banks. A flood in 1966 damaged virtually every Renaissance landmark in Florence. **30** F7

Arran large island in the Firth of Clyde, lying between the Kintyre peninsula and the mainland of North Ayrshire, Scotland; area 427 sq km/165 sq mi; population (1991) 4,500. The economy is largely based on tourism and craft industries, though other industries include whisky distilling and food processing. The island, which is mountainous to the north and undulating to the south, is a popular holiday resort. The chief town is Brodick. **24** G6

Arras French town on the Scarpe River northeast of Paris; population (1990) 42,700. It is the capital of Pas-de-Calais *département*, and was formerly known for tapestry. It was the birthplace of the French revolutionary leader Robespierre. **22** E4

Ar Riyād Arabic name for ►Riyadh, the capital of Saudi Arabia. **56** E5

Arta town and department in the Epirus region of Greece; area (department) 1,612 sq km/622 sq mi; population (town) (1993) 80,500. It derives its name from the River Arakhthos, on which it stands and which formed the boundary between Greece and Turkey until 1912. Its chief industries are woollen and cotton textiles, embroidery, and leatherworking. The town has a Byzantine castle and a medieval bridge. **36** C5

Aruba island in the Caribbean, the westernmost of the Lesser Antilles, 30 km/19 mi north of the Paraguana Peninsula in Venezuela; area 193 sq km/75 sq mi; population (1996) 66,687 (half of Indian descent). The chief town is Oranjestad. Aruba is an overseas territory of the Netherlands. Languages spoken are Dutch (official) and Papiamento (a Creole language). Tourism is a mainstay of the economy. **94** K6

Arunachal Pradesh state of India, in the Himalayas on the borders of Tibet and Myanmar; area 83,700 sq km/32,316 sq mi; population (1994 est) 965,000 (over 80 ethnic groups). Formerly part of the state of Assam, Arunachal Pradesh became a state of India in 1987. The main towns include Bomdila and Ziro, while its capital is Itanagar. **10** F3

Aryana ancient name of ►Afghanistan. 56 H3

Ascension British island of volcanic origin in the South Atlantic, a dependency of ►St Helena since 1922; area 88 sq km/34 sq mi; population (2001 est) 1,100 (excluding military personnel). The chief settlement is Georgetown. 62 B6

Aschaffenburg city in Bavaria, Germany, on the River Main, 40 km/25 mi southeast of Frankfurt am Main; population (1993) 65,200. Aschaffenburg is a river port and holiday resort. Its main industries are metalworking and the manufacture of clothing and paper. 20 E7

Ashdod (Hebrew **Isdud** or **Esdud**), deep-water port of Israel, on the Mediterranean 32 km/20 mi south of Tel Aviv; population (1994) 120,100. It is 3 km/1.9 mi from the ancient city of the same name. 60 B5

Asheville town and administrative headquarters of Buncombe County, North Carolina, USA; population (1992) 62,800. It is situated in the Blue Ridge Mountains, at the junction of the French Broad and Swannanoa rivers. Asheville manufactures furniture, paper products, textiles and electronics, and is a popular tourist resort. A branch of the University of North Carolina is one of several institutions in Asheville. The city is known as a centre for mountain music and culture. 88 D3

Ashford market town in Kent, southeast England, on the Great Stour River, 22 km/14 mi southwest of Canterbury; population (1991) 52,000. It expanded in the 1980s as a new commercial and industrial centre for the southeast. Industries include a railway works, light engineering, brewing, and the manufacture of agricultural goods. A terminus of the Channel Tunnel is sited here. 22 C3

Ashgabat (formerly **Poltoratsk** 1919–27, **Ashkhabad** 1927–92), capital of ►Turkmenistan; population (1996) 450,000. Industries include the manufacture of glass, carpets (hand-woven 'Bukhara' carpets and rugs are made here), and cotton. The city is in a spectacular natural setting, between the ►Kara-Kum Desert and the Kopet-Dag mountain range. 56 G2

Ashkhabad former name (to 1992) of ►Ashgabat. 56 G2

Ashland town in Boyd County, northeast Kentucky; population (1990) 23,600. It is situated on the Ohio River, to the east of Louisville, and 19 km/12 mi west-northwest of Huntington, West Virginia, in a region rich in coal, clay, oil, limestone, and natural gas. Iron and steel, firebrick, chemicals, oil products, coke, mining equipment, and leather are manufactured in the town. 88 D3

Ashmore and Cartier Islands group of uninhabited Australian islands comprising Middle, East, and West Islands (the Ashmores), and Cartier Island, in the Indian Ocean, about 190 km/120 mi off the northwest coast of Australia; area 5 sq km/2 sq mi. They were transferred to the authority of Australia by Britain in 1931. Formerly administered as part of the Northern Territory, they became a separate territory in 1978. West Ashmore has an automated weather station. Ashmore reef was declared a national nature reserve in 1983.

Ash Shāriqah Arabic name for ►Sharjah, an emirate situated on the Gulf northeast of Dubai. 61 F4

Ashtabula city and port in Ashtabula County, northeast Ohio, USA; population (1990) 21,800. It is located 89 km/55 mi northeast of Cleveland. The town was founded *c.* 1801 on Lake Erie at the mouth of the Ashtabula River. It was a vital stop on the Underground Railroad (a network for escaped slaves) in the 1850s. Ashtabula is a trading centre for the Lake Erie resort area, and has long been a major port. Particularly important in the interstate transport of iron and coal, it also handles international shipping. Local manufactures include automobile, truck, and tractor parts, chemicals, farm and garden tools, fibreglass, and reinforced plastics. 88 D2

Ashton-under-Lyne town and administrative headquarters of Tameside metropolitan borough, Greater Manchester, northwest England, on the River Tame; population (1994 est) 177,000. Industries include light engineering and cotton milling, and the manufacture of rubber, textiles, tobacco, plastics, leather goods, and footwear.

Asia largest of the continents, occupying one-third of the total land surface of the world. The origin of the name is unknown, though it seems probable that it was at first used with a restricted local application, gradually extended to the whole continent.

area 44,000,000 sq km/17,000,000 sq mi

largest cities (population over 5 million) Bangkok, Beijing, Chennai (formerly Madras), Delhi, Dhaka, Hong Kong, Hyderabad, Istanbul, Jakarta, Kolkata (formerly Calcutta), Karachi, Lahore, Manila, Mumbai (formerly Bombay), Osaka, Seoul, Shanghai, Shenyang, Tehran, Tianjin, Tokyo

features Mount Everest, at 8,872 m/29,118 ft is the world's highest mountain; the Dead Sea at −394 m/−1,293 ft is the world's lowest point below sea level; rivers (over 3,200 km/2,000 mi) include Chang Jiang (Yangtze), Huang He (Yellow River), Ob-Irtysh, Amur, Lena, Mekong, Yenisey; lakes (over 18,000 sq km/7,000 sq mi) include the Caspian Sea (the largest lake in the world), the Aral Sea, Lake Baikal (largest freshwater lake in Eurasia), Balkhash; deserts include the Gobi, Takla Makan, Syrian Desert, Arabian Desert, Negev

physical lying in the eastern hemisphere, Asia extends from the Arctic Circle to just over 10° south of the Equator. The Asian mainland, which forms the greater part of the Eurasian continent, lies entirely in the northern hemisphere and stretches from Cape Chelyubinsk at its northern extremity to Cape Piai at the southern tip of the Malay Peninsula. From Dezhneva Cape in the east, the mainland extends west over more than 165° longitude to Cape Baba in Turkey

climate showing great extremes and contrasts, the heart of the continent becomes bitterly cold in winter and extremely hot in summer. When the heated air over land rises, moisture-laden air from the surrounding seas flows in, bringing heavy monsoon rains to all Southeast Asia, China, and Japan between May and October

industries 62% of the population are employed in agriculture; Asia produces 46% of the world's cereal crops (91% of the world's rice); other crops include mangoes (India), groundnuts (India, China), 84% of the world's copra (Philippines, Indonesia), 93% of the world's rubber (Indonesia,

Malaysia, Thailand), tobacco (China), flax (China, Russia), 95% of the world's jute (India, Bangladesh, China), cotton (China, India, Pakistan), silk (China, India), fish (Japan, China, Korea, Thailand); China produces 55% of the world's tungsten; 45% of the world's tin is produced by Malaysia, China, and Indonesia; Saudi Arabia is the world's largest producer of oil

population (1998 est) 3,585 million; the world's largest population, amounting to more than half the total number of people in the world; between 1950 and 1990 the death rate and infant mortality were reduced by more than 60%; annual growth rate 1.6%; (exceeded only by Africa)

language predominantly tonal languages (Chinese) and Japanese in the east, Indo-Iranian languages (Hindi, Urdu, Persian) in South Asia, Altaic languages (Mongolian, Turkish) in West and Central Asia, Semitic languages (Arabic, Hebrew) in the southwest

religion the major religions of the world had their origins in Asia – Judaism and Christianity in the Middle East; Islam in Arabia; Buddhism, Hinduism, and Sikhism in India; Confucianism in China; and Shintoism in Japan **40**

Asia Minor historical name for ▶Anatolia, the Asian part of Turkey.

Asmara (or **Asmera**), capital of Eritrea, 64 km/40 mi southwest of Massawa on the Red Sea and 2,300 m/7,546 ft above sea level; population (1991) 367,300. Products include beer, clothes, and textiles. The University of Asmara is here, together with a naval school, a cathedral and many modern buildings. The population is half Christian and half Muslim. **64** G4

Aspen resort town in central Colorado, on the Roaring Fork River, lying at an altitude of 2,417 m/7,930 ft; seat of Pitkin County; population (1991) 3,700. Aspen offers winter skiing and summer river rafting, and is a cultural centre with an annual music festival.

Assam state of northeast India; area 78,400 sq km/30,270 sq mi; population (1994 est) 24,200,000. The state includes 12 million Assamese (Hindus), 5 million Bengalis (chiefly Muslim immigrants from Bangladesh), Nepalis, and 2 million indigenous people (Christian and traditional religions). Assamese is the official language. Half of India's oil is produced here, while coal, petrochemicals, paper, and cement are the other main industries. Half of India's tea is grown in Assam, with rice, jute, sugar, and cotton also being popular crops. Its main towns and cities are Guwahati, Dibrugarh, Silchar, while the capital is Dispur, a suburb of Guwahati. **10** F3

Assen capital of the province of Drenthe, Netherlands, 25 km/16 mi south of Groningen; population (1996) 53,300. **22** J2

Assiniboine river in Saskatchewan and Manitoba, southern Canada; length about 1,070 km/665 mi. It rises in southeast Saskatchewan and joins the Red River of the North at Winnipeg. Its tributaries are the Little Souris, Qu'Appelle, Rapid, White Sand, and Beaver Creek. An American Indian people gave their name to the river. **82** M7

Assisi (ancient **Asisium**), medieval town in Umbria, Italy, 19 km/12 mi southeast of Perugia, in the foothills of the Apennines overlooking the Tiber Valley; population (1990) 24,800. It is a long-established place of pilgrimage. St Francis was born here and is buried in the Franciscan monastery, completed in 1253. Its basilica was adorned with frescoes by Giotto, Cimabue, Cavallini, and others. Assisi was severely damaged by two earthquakes which hit central Italy in late September 1997, crippling the town's economy which had been dependent on tourism. **32** G5

Assiut alternative transliteration of ▶Asyut, a city in Egypt. **64** F2

Assuan alternative transliteration of ▶Aswan, a city in Egypt. **64** F3

Astana (formerly Russian **Akmolinsk** 1824–1961, **Tselinograd** 1961–93, **Akmola** 1993–98), commercial and industrial city in northern ▶Kazakhstan, capital of Akmola oblast (region), and from 1998 the capital of Kazakhstan; population (1995) 300,000. Astana is situated on the River Ishim, 190 km/118 mi northwest of Qaraghandy, and stands at the centre of a major wheat-farming area of steppe (developed in the 'Virgin Land Campaign' of the 1950s). Agricultural machinery, textiles, and chemicals are produced here. Astana is also located at an important railway junction, and is the site of rolling-stock repair depots. The city replaced ▶Almaty as the republic's capital in 1998. **42** N7

Asti town in Piedmont, southeast of Turin, Italy; population (1990) 74,500. Asti province is famed for its sparkling wine. Other products include chemicals, textiles, and glass. The town has annual jousting and gastronomic festivals. **30** D6

Astrakhan capital city, economic and cultural centre of Astrakhan oblast (region), southwestern Russian Federation; population (1996 est) 488,000. Astrakhan is sited in the Volga delta on the northeastern shore of the Caspian Sea. It is one of the Russian Federation's principal ports, which developed rapidly in the 1870s with the growth of the oil industry at Baku (now in Azerbaijan). There is a major fishing and canning industry here, together with shipbuilding and cotton manufacturing. **38** J5

Asturias autonomous community of northern Spain; area 10,600 sq km/4,092 sq mi; population (1991) 1,091,100. Agricultural products include maize, fruit, cider, and dairy products, and sheep and other livestock are reared. In the past Asturias produced half of Spain's coal; most of the coal mines have since closed down. ▶Oviedo (the capital) and Gijón are the main industrial towns. **9** (1)B1

Asunción capital and chief port of Paraguay, situated on the east bank of the Paraguay River, near its confluence with the River Pilcomayo; population (1992) 502,400 (metropolitan area 637,700); there are textile, footwear, and food processing industries. The climate is subtropical, and cattle are raised in the surrounding area, and maize, cotton, sugar, fruit, and tobacco are grown. **100** K4

Aswan (ancient Pharaonic name **Syene**), winter resort in southern Egypt, and capital of the governorate of the same name; population (1992) 220,000. It is near the ▶Aswan High Dam. The main industries are tourism and textiles. **64** F3

Aswan High Dam major rock-fill dam across the Nile near Aswan, Egypt. It is 111 m/364 ft high and 3,600 m/11,815 ft long, with a volume of 42,600,000 cubic metres/

55,809,000 cubic yards. Financed by the USSR, construction of the dam began in 1960 and was completed in 1970. The dam keeps the level of the Nile constant throughout the year without flooding. It has the capacity to generate 2,100 megawatts of electricity. 64 F3

Asyut commercial centre in southern Egypt, near the west bank of the Nile, 322 km/200 mi south of Cairo; population (1992) 321,000. It is the capital of the governorate (area 1,600 sq km/618 sq mi; population (1996) 2,802,185) of the same name. One of the Nile dams is located here. An ancient Graeco-Egyptian city, it has many rock tombs of 11th and 12th-dynasty nobles. 64 F2

Atacama Desert arid coastal region of northern Chile, with an area of about 80,000 sq km/31,000 sq mi, and extending south from the Peruvian border for 965 km/600 mi. It consists of a series of salt pans within a plateau region. Its rainless condition is caused by the Peru Current offshore; any moist airstreams from the Amazon basin are blocked by the Andean Mountains. The desert has silver and copper mines, and extensive nitrate and iodine deposits. The main population centres are the ports of Antofagasta and Iquique.

Atatürk Dam dam on the River Euphrates, in the province of Gaziantep, southern Turkey, completed in 1989. The lake, 550 km/340 mi southeast of Ankara, covers 815 sq km/ 315 sq mi (when full, it holds four times the annual flow of the Euphrates). In 1990 it was filled for the first time, submerging 25 villages, all of whose 55,000 inhabitants were relocated.

Athabasca river and lake in Alberta and Saskatchewan, Canada. The river, a tributary of the Mackenzie, is 1,230 km/764 mi long. Its source is a small lake at the base of Mount Brown in the Rocky Mountains, called the Committee's Punch-Bowl. The river flows northeast across Alberta where it enters Lake Athabasca; length 320 km/ 199 mi; area 7,936 sq km/3,064 sq mi. Large tar-sand deposits (source of the hydrocarbon mixture 'heavy oil') have been found to the southwest of the lake, near Fort McMurray. 82 J5

Athens city and administrative headquarters of Clarke County, northeast Georgia; population (1990) 45,700. It is situated on the Oconee River, 100 km/62 mi east-northeast of Atlanta. A trading centre for the region's poultry and cattle, Athens also manufactures textiles, clothing, cottonseed products, clocks, and electrical equipment. 90 E3

Athens (Greek **Athinai**), capital city of Greece and of ancient Attica; population (1991) 784,100, metropolitan area (1991) 3,096,800. Situated 8 km/5 mi northeast of its port of Peiraias on the Gulf of Aegina, it is built around the rocky hills of the Acropolis 169 m/555 ft and the Areopagus 112 m/368 ft, and is overlooked from the northeast by the hill of Lycabettus, 277 m/909 ft high. It lies in the south of the central plain of Attica, watered by the mountain streams of Cephissus and Ilissus. It has less green space than any other European capital (4%) and severe air and noise pollution. 36 F7

Athos mountainous peninsula on the Macedonian coast of Greece. Its peak is 2,033 m/6,672 ft high. The promontory is occupied by a group of 20 Orthodox monasteries, inhabited by some 3,000 monks and lay brothers. A council of representatives from the monasteries runs the affairs of the peninsula as a self-governing republic under the protection of the Greek government.

Atlanta capital and largest city of ►Georgia, USA, situated 300 m/984 ft above sea level in the foothills of the Blue Ridge Mountains; seat of Fulton County; population (1994 est) 396,100, metropolitan area (1992) 3,143,000. It is the headquarters of Coca-Cola, and there are Ford and Lockheed motor-vehicle and aircraft assembly plants. The CNN Center, headquarters of Cable News Network, is located here. Atlanta hosted the 1996 Olympic Games. 90 E3

Atlantic City seaside resort on Absecon Island, southeastern New Jersey, USA; population (1990) 38,000. Formerly a family resort, Atlantic City is now a centre for casino gambling, which was legalized here in 1978. Other industries include shell fishing. 88 F3

Atlantic Ocean ocean lying between Europe and Africa to the east and the Americas to the west; area of basin 81,500,000 sq km/31,500,000 sq mi; including the Arctic Ocean and Antarctic seas, 106,200,000 sq km/ 41,000,000 sq mi. It is generally divided by the Equator into the North Atlantic and South Atlantic. It was probably named after the legendary island continent of Atlantis. The average depth is 3 km/2 mi; greatest depth is at the Milwaukee Depth in the Puerto Rico Trench 8,648 m/ 28,374 ft. The Mid-Atlantic Ridge, of which the Azores, Ascension, St Helena, and Tristan da Cunha form part, divides it from north to south. Lava welling up from this central area annually increases the distance between South America and Africa. The North Atlantic is the saltiest of the main oceans and has the largest tidal range. 14 C3

Atlas Mountains mountain system of northwest Africa, stretching 2,400 km/1,500 mi from the Atlantic coast of Morocco to the Gulf of Gabes, Tunisia, and lying between the Mediterranean on the north and the Sahara on the south. The highest peak is Mount Toubkal 4,165 m/13,665 ft. 28 N9

Aube *département* in the ►Champagne-Ardenne region of northeast France; area 6,002 sq km/2,318 sq mi; population (1990) 289,466. The eastern part of Aube is drained by the River ►Aube, and its western part by the Seine. It is at the centre of the Champagne wine-producing region. Other important agricultural activities include livestock farming in the northeast, and the cultivation of wheat, oats, and vegetables in the southwest. The principal towns are the capital, ►Troyes, Bar-sur-Aube, and Nogent-sur-Seine. Industry is based on the quarrying of chalk, potter's clay, building stone, and limestone, and on cotton-spinning and weaving. 6 F2

Aube river of northeast France, a tributary of the Seine, length 248 km/ 155 mi; it gives its name to the *département* of ►Aube. 26 K5

Aubusson town in the *département* of Creuse, on the River Creuse, central France; 20 km/12 mi southeast of Gueret; population (1990) 5,320. Its carpet and tapestry industry dates from the 15th century. 26 H8

Auch capital of the French *département* of Gers, on the River Gers: population (1990) 25,000. It has footwear industries and trade in brandy, horses, and poultry. Auch is the seat of an archbishop, and has a Gothic cathedral. **26** F10

Auckland largest city of North Island, New Zealand, in the north of the island, in an area of impressive volcanic scenery; population (1996) 997,900. It fills the isthmus that separates its two harbours (Waitemata and Manukau), and its suburbs spread north across the Harbour Bridge. It is the country's chief port and leading industrial centre, having iron and steel plants, engineering, car assembly, textiles, food processing, sugar refining, and brewing. Auckland was officially founded as New Zealand's capital in 1840, remaining so until 1865. **78** E3

Auckland Islands six uninhabited volcanic islands about 300 km/186 mi south of South Island, New Zealand; area about 600 sq km/230 sq mi. The biggest islands are Enderby, Adams, and Auckland. **78** (2)B1

Aude *département* in the Languedoc-Roussillon region of southern France, bounded by the Gulf of Lion to the east, the sparse uplands of the Corbières to the south, and the Massif Central to the north; area 6,139 sq km/ 2,271 sq mi; population (1990) 298,710. The River ►Aude flows through it. The capital and chief tourist centre is ►Carcassonne. Other important towns include Narbonne, Limoux, and Couiza. Aude is predominantly a wine-growing region, although there is some livestock rearing, as well as the growing of fruit, wheat, and oats, mainly to the west of the capital. **6** E5

Aude river in southeast France, which rises in the *département* of Pyrénées-Orientales, then flows north and east through the *département* of ►Aude and into the Mediterranean Sea near Narbonne; length 210 km/130 mi. ►Carcassonne is the main town through which it passes. **26** H10

Augsburg industrial city in Bavaria, Germany, at the confluence of the Wertach and Lech rivers, 52 km/32 mi northwest of Munich; population (1995) 261,000. Products include textiles, cash registers, diesel engines, motor vehicles, electrical goods, and aircraft. It is named after the Roman emperor Augustus, who founded it in 15 BC. **30** F2

Augusta city in east-central Georgia, USA, 260 km/161 mi east of Atlanta on the Savannah River, near the South Carolina border; seat of Richmond County; population (1990) 44,600. It is a manufacturing city and a terminal for river barges; industries include textiles and building materials. Augusta is also a golfing resort and the site of the annual Masters golf tournament. Established as a fur-trading post in 1736, Augusta was the scene of several battles during the American Revolution, and served as Georgia's capital 1786–95. **90** E3

Augusta capital of ►Maine, USA, and administrative headquarters of Kennebec County, southwest Maine; population (1992) 21,800. It is located on the Kennebec River, 40 km/ 25 mi northeast of Lewiston and 80 km/50 mi northeast of Portland. It is a trade centre for surrounding farms and resorts, and has some light industry. Tourism, computer equipment, and paper products are important to the local economy. **88** G2

Augustów town in Białystok province, Poland, on the **Augustów canal** between the rivers Vistula and Neman, 80 km/50 mi north of Białystok; population (1993 est) 29,000. Industries include wood, leather, and flour. **18** M4

Aurich town in the *Land* (administrative region) of Lower Saxony, Germany, on the Ems–Jade canal, 188 km/117 mi northwest of Hannover; population (1994) 5,000. Products include electrical goods, clothing, and food. **22** K1

Aurillac capital of the French *département* of Cantal, on the River Jordanne; population (1990) 33,000. It has a livestock market. Historically, Aurillac was important for its monastic school. There are many old buildings and monuments in the town. **26** H9

Aurora city in central Colorado, USA, 10km/6mi east of Denver, of which it is mainly a residential suburb; population (1994 est) 251,000. Previously known as Fletcher, it was renamed, after the Roman goddess of the dawn, when it was incorporated as a city in 1907. It is the third city of Colorado, after Denver and Colorado Springs, and has industries that include telecommunications and aerospace equipment, fishing tackle and travel goods. Aurora is a main gateway to winter-sports centres and other tourist attractions in the Rocky Mountains. **86** F3

Aurora city in Kane and Du Page counties, northeast Illinois, USA; population (1994 est) 112,000. It is located on the Fox River, 60 km/37 mi west-southwest of Chicago. An important regional economic centre, it manufactures and distributes such products as auto parts, office furniture and equipment, electrical and pneumatic tools, clothing, pumps, toys, tractors, and road paving machinery. Local institutions include Aurora University (1893) and the Bellarmine School of Theology of Loyola University. **88** C2

Auschwitz (Polish **Oświęcim**), town near Kraków in Poland, the site of a notorious concentration camp used by the Nazis in World War II to exterminate Jews and other political and social minorities, as part of the 'final solution'; population (1992 est) 45,100. Each of the four gas chambers could hold 6,000 people. **18** J7

Austin capital of ►Texas, USA, on the Colorado River; seat of Travis County; population (1994 est) 514,000. It is the market centre for the surrounding agricultural region, and an important centre for electronic and scientific research and the manufacture of high technology goods, including semiconductors and computers. **90** B3

Australasia and Oceania two geographical terms; **Australasia** is applied somewhat loosely to the islands of the South Pacific, including Australia, New Zealand, and their adjacent islands, while **Oceania** is a general or collective name for the groups of islands in the southern and central Pacific Ocean, comprising all those intervening between the southeastern shores of Asia and the western shores of America. The 10,000 or more Pacific Islands offer a great diversity of environments, from almost barren, waterless coral atolls to vast, continental islands.

area (land) 8,500,000 sq km/ 3,300,000 sq mi

largest cities (population over 500,000) Sydney, Melbourne, Brisbane, Perth, Adelaide, Auckland *features* the Challenger Deep in the Mariana Trench −11,034 m/ −36,201 ft is the greatest known depth of sea in the world; Ayers Rock in Northern Territory, Australia, is the world's largest monolith; the Great Barrier Reef is the longest coral reef in the world; Mount Kosciusko 2,229 m/ 7,316 ft in New South Wales is the highest peak in Australia; Aoraki 3,764 m/12,349 ft is the highest peak in New Zealand

physical Oceania can be broadly divided into groups of volcanic and coral islands on the basis of the ethnic origins of their inhabitants: Micronesia (Guam, Kiribati, Mariana, Marshall, Caroline Islands), Melanesia (Papua New Guinea, Vanuatu, New Caledonia, Fiji Islands, Solomon Islands), and Polynesia (Tonga, Samoa, Line Islands, Tuvalu, French Polynesia, Pitcairn); the highest point is Mount Wilhelm, Papua New Guinea 4,509 m/14,793 ft; the lowest point is Lake Eyre, South Australia −16 m/−52 ft; the longest river is the Murray in southeast Australia 2,590 km/1,609 mi; Australia is the largest island in the world. Most of the small islands are coral atolls, though some are of volcanic origin

population (1998 est) 29.6 million, rising to over 39 million by 2025; annual growth rate from 1980 to 1985 1.5%; Australia accounts for 65% of the population

language English, French (French Polynesia, New Caledonia, Wallis and Futuna, Vanuatu); a wide range of indigenous Aboriginal, Maori, Melanesian, Micronesian, and Polynesian languages and dialects (over 700 in Papua New Guinea) are spoken

religion predominantly Christian; 30% of the people of Tonga adhere to the Free Wesleyan Church; 70% of the people of Tokelau adhere to the Congregational Church; French overseas territories are largely Roman Catholic

Australia country occupying all of the Earth's smallest continent, situated south of Indonesia, between the Pacific and Indian oceans.

Australia covers an area of 7,682,850 sq km/2,966,136 sq mi, and has a population of 18,886,000 (2000 est). The capital is ►Canberra; other major towns and cities include Adelaide, Alice Springs, Brisbane, Darwin, Melbourne, Perth, Sydney, Hobart, Newcastle, and Wollongong. The head of state is Queen Elizabeth II (from 1952), represented by Governor General Sir William Deane (from 1996); the head of government is John Howard (from 1996). The currency is the Australian dollar, and the languages spoken include English (official) and Aboriginal languages. 76 E4

Australian Alps southeastern and highest area of the Eastern Highlands of Australia, extending for about 433 km/269 mi through Victoria and New South Wales in a northeasterly direction, and forming a continuation of the ►Great Dividing Range. They include the Snowy Mountains and Mount ►Kosciusko, Australia's highest mountain, 2,229 m/7,316 ft. The Alps are popular for winter sports. 74 E9

Australian Capital Territory federal territory of southeastern Australia, an enclave in the state of New South Wales; it includes ►Jervis Bay Territory (the site of Canberra's port on the coast) for administrative purposes; area 2,400 sq km/926 sq mi (Jervis Bay 73 sq km/28 sq mi); population (1996) 297,000. Government administration and defence employs almost half of the population of Australian Capital Territory; retail, property, and business services are also important. ►Canberra is the main city in the territory. 76 J7

Austral Islands alternative name for ►Tubuai Islands, part of ►French Polynesia. 74 L8

Austria landlocked country in central Europe, bounded east by Hungary, south by Slovenia and Italy, west by Switzerland and Liechtenstein, northwest by Germany, north by the Czech Republic, and northeast by the Slovak Republic.

Austria covers an area of 83,859 sq km/32,367 sq mi, and has a population of 8,211,000 (2000 est). The capital is ►Vienna; other major towns and cities include Graz, Linz, Salzburg, Innsbruck, and Klagenfurt. The head of state is Thomas Klestil (from 1992); the head of government is Wolfgang Schüssel (from 2000). The currency is the schilling, and the main language spoken is German (official). 30 J3

Autun French town in the *département* of Saône-et-Loire, 80 km/50 mi northwest of Mâcon, on the River Arroux; population (1990) 23,000. It has metallurgical, furniture, and oil-shale industries; tourism is important. Roman remains include a pyramid, a theatre, and two gates. There is also a cathedral, mainly of the 12th century. 26 K7

Auvergne ancient province of central France and modern region comprising the *départements* of ►Allier, ►Cantal, ►Haute-Loire, and ►Puy-de-Dôme; administrative centre ►Clermont-Ferrand; area 26,000 sq km/10,000 sq mi; population (1990) 1,321,200. It is a mountainous area, composed chiefly of volcanic rocks in several masses. Products include cattle, sheep, tyres, and metal goods. 6 E4

Auxerre administrative centre of ►Yonne *département*, in France, 170 km/106 mi southeast of Paris, on the River Yonne; population (1990) 40,600. Auxerre is a market town and produces wines and metal goods. 26 J6

Avellino province of southern Italy in central ►Campania region; capital Avellino (population (1990) 55,800); area 2,802 sq km/1,082 sq mi; population (1992) 440,100. 32 J8

Aveyron *département* in the ►Midi-Pyrénées region of France; area 8,734 sq km/3,372 sq mi; population (1990) 138,400. The region is mountainous with the Rivers Lot, Tarn, and Aveyron flowing through it. The land is used primarily for stock rearing, but some cereals, vines, tobacco, and fruit trees are cultivated. Textiles, iron, and cheese are manufactured. Hydroelectric plants provide some power for the region. The administrative centre is Rodez. 6 E4

Aviemore all-year sports and tourist centre, in the Highland unitary authority, Scotland, 45 km/28 mi southeast of Inverness and adjacent to the Cairngorm Mountains; population (1991) 2,200. The centre specializes in winter sporting activities. 24 J4

Avignon city in Provence, France, administrative centre of ►Vaucluse *département*, on the River Rhône, 80 km/50 mi northwest of Marseille; population (1990) 89,400, conurbation 180,000. Tourism and food processing are important; other industries include the manufacture of leather, textiles, soaps, machinery, and chemicals. Avignon has a significant trade in wine. There is an atomic plant at Marcoule nearby. An important Gallic and Roman city, it has a 12th-century bridge (only half of which still stands), a 13th-century cathedral, 14th-century walls, and the Palais des Papes, the enormous fortress-palace of the popes, one of the most magnificent Gothic buildings of the 14th century. **26** K10

Avila capital of Avila province, central Spain, situated 1,100 m/3,609 ft above sea level 90 km/56 mi northwest of Madrid, and enclosed on three sides by mountains; population (1991) 46,000. It has a Gothic cathedral and the convent and church of St Teresa, who was born here. The medieval town walls are among the best preserved in Europe. **28** F4

Avilés town in the autonomous community of ▶Asturias, northern Spain; population (1995) 88,500. Avilés forms part of the most important iron- and steel-producing region in Spain (although its main factory, Ensidesa, built in the 1950s at Gen Franco's instruction, is now out-of-date and unprofitable); there are coal mines nearby, and zinc, aluminium, and glass are also produced. **28** E1

Avon (or **Upper Avon**, or **Warwickshire Avon**; Celtic *afon* 'river'), river in southern England; length 154 km/96 mi. It rises in the Northamptonshire uplands near Naseby and flows southwest through Warwick, Stratford-upon-Avon, and Evesham, before joining the River Severn near Tewkesbury, Gloucestershire. **22** A2

Avranches French town in the *département* of Manche, Normandy; population (1990) 9,500. Situated on a hill, the town overlooks the estuary of the River Sée. Horse breeding, leatherwork, and fishing are the main industries. **26** D5

Ayacucho capital of Ayacucho department in the Andean Mountains of central Peru, 578 km/359 mi southeast of Lima, at an altitude of 2,746 m/3,009 ft; population (1993) 105,900. The city is in the centre of a mining region, and silver, copper, and nickel are mined nearby; it is also a tourist centre. Ayacucho was founded in 1539 by the Spanish conquistador Pizarro. The last great battle against Spanish troops in the war of independence was fought in 1824 at Quina 40 km/25 mi northeast of Ayacucho. The Maoist guerrilla group the Sendero Luminoso originated in Ayacucho in the 1960s. **98** C6

Ayers Rock (Aboriginal **Uluru**), vast ovate mass of pinkish rock in Northern Territory, Australia; 335 m/1,110 ft high and 9 km/6 mi around. For the Aboriginals, whose paintings decorate its caves, it has magical significance. **76** F5

Ayr administrative headquarters of ▶South Ayrshire, southwest Scotland, at the mouth of the River Ayr; population (1991) 48,000. Ayr has strong associations with the Scottish poet Robert Burns. Industries include fishing, electronics, shipbuilding in Troon, and aircraft parts. **24** H6

Azerbaijan country in western Asia, bounded south by Iran, east by the Caspian Sea, west by Armenia and Georgia, and north by Russia.

Azerbaijan covers an area of 86,600 sq km/33,436 sq mi, and has a population of 7,734,000 (2000 est). The capital is ▶Baku; other major towns and cities include Gäncä, Sumqayit, Nakhichevan, Xankändi, and Mingechaur. The head of state is Geidar Aliyev (from 1993); the head of government is Artur Rasizade (from 1996). The currency is the manat, and the languages spoken include Azeri (official) and Russian. **58** M3

Azores (or **Açores**), group of nine islands in the North Atlantic, forming an autonomous region belonging to Portugal; area 2,247 sq km/867 sq mi; population (1991) 237,800. The islands are outlying peaks of the Mid-Atlantic Ridge and are volcanic in origin. Products include sugar cane, coffee, tobacco, fruit, and wine. There are many hot springs, and the countryside is mountainous and rugged. The climate is moist but mild, and some of the islands are used as winter resorts. The administrative capital is ▶Ponta Delgada on the main island, São Miguel; the other islands are Santa Maria, Terceira, Graciosa, São Jorge, Pico, Faial, Flores, and Corvo. **66** (1)B2

Azov, Sea of (Russian **Azovskoye More**, Latin **Palus Maeotis**), inland sea between Ukraine and Russia, forming a gulf in the northeast of the Black Sea, to which it is connected by the narrow Kerch Strait. It has an area of 37,555 sq km/14,500 sq mi, and is extremely shallow, with an average depth of only 8 m/26 ft, and nowhere exceeding 16 m/52 ft. The sea is frozen for four to six months every year. Principal ports include Rostov-na-Donu, Mariupol, Kerch', and Taganrog. The main rivers flowing into the Sea of Azov are the ▶Don, and the Kuban. **38** G5

Az Zahrān Arabic for ▶Dhahran, a city in eastern Saudi Arabia. **61** D3

Bab al Mandab strait joining the Red Sea and the Gulf of Aden, and separating Arabia and Africa. The name, meaning 'gate of lamentation', refers to its currents. **56** D7

Babruysk alternative spelling of ▶Bobruisk, a city in Belarus. **38** E4

Bacău industrial city in Romania, 250 km/155 mi north northeast of Bucharest, on the River Bistrita; population (1993) 207,000. It is the capital of Bacău county, a leading oil-producing region. **34** P3

Badajoz capital of Badajoz province in Extremadura, southwest Spain, situated on the River Guadiana at the Portuguese frontier; population (1991) 121,900. Textiles, pottery, leather, and soap are manufactured. Badajoz has a 16th-century bridge and a 13th-century cathedral. **28** D6

Bad Ems (or **Ems**), town and spa in the Rhineland-Palatinate, Germany, on the River Lahn, 17 km/11 mi south of Koblenz; population (1994) 10,200. The peak years of the spa were 1875–1914. **22** K4

Baden-Baden Black Forest spa town in Baden-Württemberg, southwestern Germany, in the Oos valley, 68 km/42 mi west of Stuttgart; population (1995) 52,600. Its mineral springs have been known since Roman times, and it became a fashionable spa in the 19th century. The town has a conference centre, the Kongresshaus (1968). **30** D2

Baden-bei-Wien Austrian spa town in the province of Lower Austria, 25 km/16 mi south of Vienna; population (1995) 23,500. It has sulphur springs, and was known to the Romans as **Aquae Panoniae**. **18** F9

Baden-Württemberg administrative region (German *Land*) of Germany, bounded to the west by France, to the south by Switzerland, to the east by Bavaria, and to the west by the Rhine valley. area 35,800 sq km/13,820 sq mi; population (1995) 10,350,000. The capital is ▶Stuttgart; main cities and towns are Mannheim, Karlsruhe, Freiburg im Breisgau, Heidelberg, Heilbronn, Pforzheim, and Ulm. The area has the Rhine as boundary to the south and west and it contains the ▶Black Forest; the source of the River Danube is at Donaueschingen in the Black Forest mountains. Principal industries are luxury motor vehicles, jewellery, watches, clocks, musical instruments, textiles, chemicals, iron, steel, electrical equipment, surgical instruments, and precision engineering. Agricultural activities centre on wine production, animal husbandry, and fruit growing. **7** (2)B4

Bad Hersfeld city and spa for liver and diabetic cures in Hesse, Germany, on the River Fulda, 115 km/71 mi northeast of Wiesbaden; population (1994) 31,200. Textiles, cables, and carpets are manufactured here. Festival plays are held annually in Bad Hersfeld's 11th-century abbey church. **20** E6

Bad Ischl alternative name for ▶Ischl, a town in Upper Austria. **30** J3

Bad Kreuznach city and spa in Germany; see ▶Kreuznach. **22** K5

Baeza town in the province of Jaén, southern Spain; population (1991) 15,100. It was a flourishing city under the Moors, but was sacked in 1228 by the Castilians. There are several fine churches and a Gothic cathedral. The town was once the seat of a university (1533–1807). **28** G6

Baffin Island island in Northwest Territories, Canada, situated across the entrance to ▶Hudson Bay; area 507,450 sq km/195,930 sq mi.

features Baffin Island is the largest island in the Canadian Arctic; the mountains here rise above 2,000 m/ 6,000 ft, and there are several large lakes. The northernmost part of the strait separating Baffin Island from Greenland forms Baffin Bay; the southern end is Davis Strait. The predominantly Inuit population is settled mainly around Lake Harbour and Frobisher Bay in the south.

It is named after William Baffin, who carried out research here in 1616 during his search for the ▶Northwest Passage. Sir Martin Frobisher, who landed here in 1576, discovered gold, and there are coal and iron-ore deposits. **82** R2

Baghdad historic city and capital of Iraq, and capital of the governorate of Baghdad, on the River Tigris; population (1995 est) 5,385,000. Industries include oil refining, distilling, tanning, tobacco processing, and the manufacture of textiles and cement. Founded in 762, it became Iraq's capital in 1921. During the Gulf War of 1991, the UN coalition forces bombed it in repeated air raids. **56** D3

Bago (or **Pegu**), city in southern Myanmar on the River Pegu, 76 km/ 47 mi northeast of Yangon; population (1998 est) 198,900. It was founded in 573 and was once an important seaport, but was destroyed in 1757 by the Burmese. It now lies on the Yangon–Mandalay railway. Bago has many rice and saw mills, as it is surrounded by an area of rice production and the forested slopes of the Pegu Yoma. It has long been a centre of Buddhist culture. **50** B3

Baguio summer resort on Luzon island in the Philippines, 200 km/ 125 mi north of Manila, 1,370 m/ 4,500 ft above sea level; population (1995 est) 174,000. It is the official summer residence of the Philippine president. **50** G3

Bahamas country comprising a group of about 700 islands and about 2,400 uninhabited islets in the Caribbean, 80 km/50 mi from the southeast coast of Florida. They extend for about 1,223 km/760 mi from northwest to southeast, but only 22 of the islands are inhabited.

The Bahamas covers an area of 13,880 sq km/5,383 sq mi, and has a population of 307,000 (2000 est). The capital is ▶Nassau (on New Providence island); other major towns and cities include Freeport (on Grand Bahama). The head of state is Queen Elizabeth II (from 1973), represented by Governor General Orville Turnquest (from 1995); the head of government is Hubert Ingraham (from 1992). The currency is the Bahamian dollar, and the languages spoken include English (official) and Creole. **90** F4

Bahawalpur (or **Bhawalpur**), city in Punjab, Pakistan, situated on the Sutlej River 350 km/220 mi southwest of Lahore; population (1981) 178,000. Once the capital of the former Indian princely state of Bahawalpur, it is now an industrial city relying on cotton ginning, rice and flour milling, and the production of hand-woven textiles. A rich agricultural area surrounds the city, irrigated by canals. The Islamia University was established in 1975. **54** B3

Bahía federal unit (state) of eastern Brazil, with an eastern coastline on the Atlantic; area 566,978 sq km/ 218,911 sq mi; population (1991)11,868,000; capital ▶Salvador. Bahía has a low coastal plain rising to a central plateau in the west, which is crossed southwest to northeast by the São Francisco River. This plateau region is dominated by an arid climate – the Sertão region. The Paulo Afonso Falls, in the lower reaches of the river, provide a major source of hydroelectric power. There are oil, cotton, and chemical industries, and industrial diamonds are mined. The coastal belt is hot, humid, and very fertile, and produces sugar cane, cacao, and tobacco. Cattle-raising is the main economic activity in the higher parts of the state. Brazil's chief oilwells are on the coast near Salvador. Other main towns include Ilhéus, Porto Seguro, Itabuna, and Jequié. **98** J6

Bahía Blanca port in Buenos Aires department, southeast Argentina, situated on the River Naposta, 5 km/3 mi from its mouth on the South Atlantic Ocean; population (1991) 271,500. It is a rail terminus and major distribution centre, serving a considerable area of central Argentina. Main exports include grain from Buenos Aires province, and produce from the Rio Negro valley including meat, and wool. The naval base of Puerto Belgrano, South America's largest, is here. **100** J6

Bahrain country comprising a group of islands in the Gulf, between Saudi Arabia and Iran.

Bahrain covers an area of 688 sq km/ 266 sq mi, and has a population of 617,000 (2000 est). The capital is ▶Al Manamah (on Bahrain island); other major towns and cities include Muharraq, Jiddhafs, Isa Town, Rifa'a, and Sitra. The head of state is Sheikh Hamad bin Isa al-Khalifa (from 1999); the head of government is Sheikh Khalifa bin Salman al-Khalifa (from 1970). The currency is the Bahraini dinar, and the languages spoken include Arabic (official), Farsi, English, and Urdu. 61 D4

Bahr el-Azraq Arabic name for the ►Blue Nile. 64 F5

Bahr el Nîl Arabic name in Sudan for the River ►Nile, a river in Africa. 64 F4

Baia Mare industrial town and capital of Maramureş county, northern Romania, 100 km/63 mi north of Cluj Napoca; population (1993) 150,000. Chemicals, metals, machinery, food, and fertilizers are produced. The nearby area is rich in ores containing gold, silver, copper, lead, zinc, and uranium. 34 L2

Baikal, Lake (Russian **Ozero Baykal**), freshwater lake in southern Siberia, Russia, the largest in Asia, and the eighth largest in the world (area 31,500 sq km/12,150 sq mi). Lake Baikal is also the world's deepest lake (up to 1,640 m/5,700 ft) and its oldest, having existed for over 25 million years. It extends for some 636 km/395 mi, and has an average width of 48 km/30 mi. Fed by more than 300 rivers, the main one of which is the Selenga, it is drained only by the Lower Angara. Lake Baikal is famous for its great clarity and the diversity of its fauna. 44 H6

Baile Átha Cliath (Irish 'the town of the ford of the hurdles'), official Irish name of ►Dublin, capital of the Republic of Ireland, from 1922. 24 F8

Bailén market town in the province of Jaén, southern Spain; population (1995) 17,800. There is some food processing. Here, during the Peninsular War, the French suffered their first serious reverse against the Spaniards in 1808. 28 G6

Bailleul French town in the *département* of Nord, 25 km/16 mi northwest of Lille; population (1990) 13,900. The chief industries are the manufacture of linen and lace. The town hall dates from the 15th century. The town was largely destroyed during World War I and was rebuilt. 22 E4

Bairiki port and capital of Kiribati on Tarawa atoll; population (1998 est) 2,565. Mother-of-pearl and copra are exported. 74 H5

Baja market town in southern Hungary, near the River Danube, 96 km/60 mi southwest of Kecskemét; population (1991) 38,900. Industries include engineering, chemicals, and textiles. 34 F3

Baja California mountainous peninsula that forms the twin northwestern states of Lower (Spanish *baja*) California, Mexico; Baja California Norte in the north, and Baja California Sur in the south. 84 C5

Bakersfield city in south-central California, USA, 115 km/71 mi northeast of Santa Barbara and 175 km/109 mi north of Los Angeles, situated on the Kern River at the southern end of the San Joaquin Valley; the seat of Kern County; population (1994 est) 191,000. Local industries include petroleum products, textiles and food processing. Oil was first discovered here in 1899. 92 C1

Baki alternative spelling of ►Baku, the capital of Azerbaijan. 56 E1

Baku (or **Baki**), capital city of the republic of ►Azerbaijan, located on the Apsheron Peninsula on the western shore of the Caspian Sea; population (1997) 1,700,000. Baku is an important industrial city and port. It has been a major centre of oil extraction and refining since the 1870s; the oilfields here are linked by pipelines with the Georgian Black Sea port of Batumi, while petroleum exports to Russia are shipped across the Caspian to Astrakhan. Heavy engineering enterprises in the city produce equipment for the oil industry and ships; light industries include leather tanning and food processing. Baku has a hot climate and is subject to strong northwest winds. 56 E1

Balaton, Lake (German **Plattensee**), lake in western Hungary, south of the Bakony Mountains. It is the largest freshwater lake in central Europe, 77 km/48 mi long and covering 635 sq km/245 sq mi. Its waters drain into the River Danube. Vineyards cover the hilly northern shores, but the southern shore is generally flat. Fruit and wine are produced in the surrounding region, and there are numerous lakeside holiday resorts. 34 E3

Bâle French form of Basle or ►Basel, a city in Switzerland. 30 C3

Balearic Islands (Spanish **Islas Baleares**), group of Mediterranean islands forming an autonomous region of Spain, comprising ►Mallorca, ►Menorca, ►Ibiza, Cabrera, and ►Formentera; area 5,014 sq km/1,936 sq mi; population (1991) 709,100. The capital is ►Palma de Mallorca. Tourism is a mainstay of the economy; other industries include figs, olives, oranges, wine, brandy, coal, iron, and slate. 28 N5/9 (3)A2

Bali island of Indonesia, east of Java, one of the Sunda Islands; area 5,800 sq km/2,240 sq mi; population (1995 est) 3,037,000. The capital is Denpasar. The island features volcanic mountains, the highest peak is Gunung Agaung (3,142 m/10,308 ft). Industries include gold and silver work, woodcarving, weaving, copra, salt, coffee, and tourism; arts include Balinese dancing, music (the Gamelan), and drama. Bali's Hindu culture goes back to the 7th century; the Dutch gained control of the island by 1908. 52 F4

Balikesir city in northwestern Turkey, capital of Aydin province; population (1990) 170,600. There are silver mines nearby. 36 K5

Balikpapan port in Indonesia, on the east coast of southern Kalimantan, Borneo; population (urban area, 1990 est) 1,200,000. It is an oil-refining centre. 52 F3

Balkan Mountains (Bulgarian **Stara Planina**; ancient **Haemus**), range of mountains in central Europe, extending 603 km/377 mi east–west through central Bulgaria from the Black Sea to Yugoslavia, about 25–40 km/16–25 mi in breadth. The highest peak is Mount Botev (2,376 m/7,920 ft). Its chief crossing, the Šipka Pass, runs from Veliko Turnovo south to Kazanlâk. 36 F1

Balkans (Turkish 'mountains'), peninsula of southeastern Europe, stretching into Slovenia between the Adriatic and Aegean seas, comprising Albania, Bosnia-Herzegovina, Bulgaria, Croatia, Greece, Romania, the part of Turkey in Europe, and Yugoslavia. It is joined to the rest of Europe by an isthmus 1,200 km/750 mi wide between Rijeka, Croatia, on the west and the mouth of the Danube on the Black Sea to the east. The great ethnic diversity resulting from successive waves of invasion has made the Balkans a byword for political dissension, and the 1990s saw the break-up of the Federation of Yugoslavia along ethnic lines. To '**Balkanize**' is to divide into small warring states.

Balkhash city on the northern shore of Lake Balkhash in Kazhakstan; population (1990) 86,600. It was founded in 1928. Chief industries include copper mining and salt extraction. 42 N8

Balkhash, Lake (or **Ozero Balkhash**), lake in eastern ▶Kazakhstan, the eastern half of which is salty, and the western half fresh; area 17,400 sq km/6,715 sq mi. Lake Balkhash is 600 km/375 mi long and is fed by several rivers, including the Karatal, Lepsy and Ili, but has no outlet. It is very shallow, especially in the east, and is frozen throughout the winter (November–mid-April). 40 L5

Ballarat city in Victoria, Australia, 112 km/70 mi northwest of Melbourne; population (1996) 64,831. It is Victoria's largest inland city, the third-largest city of the state, and is an important railway junction. Industries include the manufacture of agricultural machinery, fibreglass, and paper; champagne, cheese, and lavender are produced in the area and tourism is important. Ballarat was founded in the 1851 gold rush, after the discovery of the largest gold reserves in Australia. 76 H7

Ballinasloe market town in Galway, Republic of Ireland, on the River Suck, a tributary of the Shannon; population (1996) 5,600. Industries include limestone quarrying, footwear, bonemeal, and electrical equipment. The annual horse, cattle, and sheep fair, founded in the 17th century, is held in the week of the second Tuesday of October, and is the largest in Ireland. 24 D8

Baltic Sea shallow sea, extending northeast from the narrow Skagerrak and Kattegat, between Sweden and Denmark, to the Gulf of Bothnia between Sweden and Finland. Its coastline is 8,000 km/5,000 mi long; the sea is 1,500 km/930 mi long and 650 km/404 mi wide, and its area, including the gulfs of Riga, Finland, and Bothnia, is 422,300 sq km/163,000 sq mi. Its average depth is 65 m/213 ft, but it is 460 m/1,500 ft at its deepest.

Its shoreline is shared by Denmark, Germany, Poland, the Baltic States, Russia, Finland, and Sweden. 16 J8

Baltic States collective name for the states of ▶Estonia, ▶Latvia, and ▶Lithuania. They were formed as independent states after World War I out of former territories of the Russian Empire. The government of the USSR recognized their independence in peace treaties signed in 1920, but in 1939 forced them to allow occupation of important military bases by Soviet troops. In the following year, the Baltic states were absorbed into the Soviet Union as constituent republics. They regained their independence in September 1991 after the collapse of the Soviet Union.

Baltimore industrial port and largest city in Maryland, on the western shore of Chesapeake Bay, 50 km/31 mi northeast of Washington, DC, USA; population (1994 est) 703,000; metropolitan area (1992 est) 2,434,000. Industries include shipbuilding, oil refining, food processing, and the manufacture of steel, chemicals, and aerospace equipment. The city was named after the founder of Maryland, Lord Baltimore (1579–1632). Baltimore dates from 1729 and was incorporated as a city in 1797. 88 E3

Bamako (or **Bamaku**, or **Bammaco**), capital and port of ▶Mali, lying on the upper Niger River, in the southwest of the country; population (2001 est) 947,100. As a major river port, it is vital to the economy of a landlocked country. It also has an international airport and is linked by rail to the port of Dakar in Senegal on the Atlantic coast. Industries include ceramics, pharmaceuticals, chemicals, textiles, farm machinery, batteries, river fishing, and food and tobacco processing. 68 C2

Bamberg city and industrial port in Bavaria, Germany, on the River Regnitz, near its junction with the Main, 45 km/28 mi west of Bayreuth; population (1995) 70,000. The economy is based on engineering, brewing, and the production of cotton textiles, leather, carpets, and electrical goods. It has an early 13th-century Romanesque cathedral. In 1460 one of the first books in the German language was printed here. 20 F7

Banaba (formerly **Ocean Island**), island in the Republic of ▶Kiribati; population (1990) 300. 74 G6

Banaras alternative transliteration of ▶Varanasi, a holy Hindu city in Uttar Pradesh, India. 54 D3

Banbury market town in Oxfordshire, central England, on the River Cherwell, 40 km/25 mi north of Oxford, and administrative centre for Cherwell District Council; population (1991) 39,900. Industries include food processing (Kraft Jacobs Suchard), traditional brewing (Hook Norton, Merivales), printing, and the manufacture of car components, electrical goods, and aluminium. The **Banbury Cross** of the nursery rhyme 'Ride a Cock Horse to Banbury Cross' was destroyed by the Puritans in 1602, but replaced in 1859. 24 L9

Banca alternative form of the Indonesian island ▶Bangka. 52 D3

Bandar-e 'Abbas port and capital of Hormozgan province, Iran, on the Strait of Hormuz in the Gulf; population (1991) 249,500. Formerly called Gombroon, it was renamed and made prosperous by Shah Abbas I (1571–1629). It is a naval base and winter resort. 61 G3

Bandar Seri Begawan (formerly **Brunei Town** until 1970), capital and largest town of Brunei, 14 km/9 mi from the mouth of the Brunei River; population (1992) 55,000. Industries include oil refining and construction. 52 E2

Bandung commercial city and capital of Jawa Barat (West Java) province on the island of Java, Indonesia; population (1995 est) 1,921,000. Bandung is the third-largest city in Indonesia and was the administrative centre when the country was the Netherlands East Indies. It lies 180 km/112 mi southeast of Jakarta, the capital, and has an airport, two universities (established in 1955 and 1957), and a centre for nuclear research (established in 1964). Industries include textiles, chemicals, and plastics. 52 D4

Banff town and resort in southwest Alberta, Canada; population (1991) 5,700. It is located on the Bow River, 100 km/62 mi northwest of Calgary and 90 km/56 mi east of Vancouver, near Lake Louise. Local industries include brewing and iron founding. The town is a centre for Banff National Park (Canada's first, founded 1885) in the Rocky Mountains; area 6,641 sq km/2,565 sq mi. 82 H6

Bangalore capital of ▶Karnataka state, southern India, lying 950 m/3,000 ft above sea level; population (1991) 4,087,000. Industries include electronics, aircraft and machine-tools construction, and coffee. Bangalore University and the University of Agriculture Sciences were founded in 1964, and the National Aeronautical Institute in 1960. 54 C6

Banghāzī alternative spelling of ▶Benghazi, an industrial port in northern Libya. 64 D1

Bangka (or **Banka** or **Banca**), Indonesian island off the east coast of Sumatra; area 12,000 sq km/4,600 sq mi; population (1990) 627,000. The capital is Pangkalpinang. It is one of the world's largest producers of tin. 52 D3

Bangkok (Thai **Krung Thep**; Thai 'City of Angels'), capital and chief port of Thailand, on the east bank of the River Chao Phraya, 40 km/24 mi from the Gulf of Thailand; population (1996 est) 5,585,000. The population tripled 1966–96. The population of the whole metropolitan area, including the industrial centre of Thon Buri across the river, was estimated at 8,896,500 in 1995. The city is the headquarters of the Southeast Asia Treaty Organization (SEATO) and hosted the 13th Asian Games in 1998. 50 C4

Bangladesh country in southern Asia, bounded north, west, and east by India, southeast by Myanmar, and south by the Bay of Bengal.

Bangladesh covers an area of 144,000 sq km/55,598 sq mi, and has a population of 129,155,000 (2000 est). The capital is ▶Dhaka; other major towns and cities include Rajshahi, Khulna, Chittagong, Sylhet, Rangpur, and Narayanganj. The head of state is Shahabuddin Ahmed (from 1996); the head of government is Sheikh Hasina Wazed (from 1996). The currency is the taka, and the languages spoken include Bengali (official) and English. 54 E4

Bangor (Irish *beanna* 'peaks', *chor* 'curve'), Belfast commuter town in County Down, Northern Ireland, on the shore of Belfast Lough, 20 km/12 mi northeast of Belfast; population (1991) 52,400. It is the site of a famous missionary abbey of the Celtic church, founded by St Comgall in AD 555 and sacked by the Danes in the 9th century. The abbey was the home of St Columbanus and St Gall. A Protestant church, the Abbey Church, was built on the site in 1617 by Thomas Hamilton. Bangor Castle, built by Robert Ward in 1852, is now the town hall and a heritage centre. Bangor has the largest marina in Ireland, and the Royal Ulster Yacht Club is based here. 24 G7

Bangor cathedral, university, and market town in Gwynedd, north Wales, on the Menai Strait 15 km/9 mi northeast of Caernarfon; population (1991 est) 12,338. Industries include chemicals, electrical goods, and engineering. Trade in slate from Penrhyn quarries was at one time very important, but has declined in recent decades. Bangor is set in a region of outstanding natural beauty, with Snowdonia National Park and the Gwynedd coastline on its doorstep. 24 H8

Bangui capital and main river port of the ▶Central African Republic, on the River Ubangi; population (1995 est) 698,000. The city is the centre for the country's light industries, including beer, cigarettes, office machinery, and timber and metal products. Bangui has an airport and contains the main depot for the storage and transportation of imported petroleum products. The city, which also serves as an outlet for the Republic of Chad, has a considerable trade in cotton and coffee. 70 B3

Banja Luka ('Baths of St Luke'), town in Bosnia-Herzegovina, situated on the River Vrbas; population (1991) 143,000. It is a commercial and manufacturing centre. The town dates from Roman times. It has a spa, is the seat of Roman Catholic and Orthodox bishops, and has several mosques. It lies at the centre of Bosnian Serb territory. 34 E5

Banjarmasin river port in Indonesia, on the island of Borneo; capital of Kalimantan Selatan province; population (1990) 443,700. It exports rubber, timber, and precious stones. 52 E3

Banjul capital and chief port of Gambia, on an island at the mouth of the River Gambia; population of urban area (1995 est) 186,000; city (1995 est) 58,700 . It is located 195 km/121 mi southeast of Dakar (capital of Senegal). Established in 1816 as a settlement for freed slaves, it was known as Bathurst until 1973. The city has an airport and industries include peanut processing and exporting, brewing, and tourism (centred at the nearby resorts of Bakau, Fajara, Kotu, and Kololi). 68 A2

Banka alternative form of ▶Bangka, an Indonesian island. 52 D3

Banská Bystrica (German **Neusohl**; Hungarian **Beszterczebanya**), industrial town (textiles) in the Slovak Republic, on the River Hron, 160 km/100 mi northeast of Bratislava; population (2001 est) 84,400. It is the capital of the Central Slovakia region, and has Gothic churches. 18 J9

Barbados island country in the Caribbean, one of the Lesser Antilles. It is about 483 km/300 mi north of Venezuela.

Barbados covers an area of 430 sq km/166 sq mi, and has a population of 270,000 (2000 est). The capital is ▶Bridgetown; other major towns and cities include Speightstown, Holetown, and Oistins. The head of state is Queen Elizabeth II (from 1966), represented by Governor General Sir Clifford Straughn Husbands (from 1996); the head of government is Owen Arthur (from 1994). The currency is the Barbados dollar, and the languages spoken include English (official) and Bajan (a Barbadian English dialect). 98 F1

Barbuda one of the islands that form the state of ▶Antigua and Barbuda. 94 M5

Barcelona port and capital of Barcelona province and of the autonomous community of ▶Cataluña, northeast Spain; population (1994) 1,631,000. Industries include textiles, engineering, and chemicals. As the chief centre of Catalan nationalism, Barcelona was prominent in the overthrow of the monarchy in 1931 and was the last city of the republic to surrender to Gen Franco in 1939. The city hosted the Summer Olympics in 1992. 28 N3

Bareilly industrial city in Uttar Pradesh, northern India; population (1991) 591,000. Its industries include sugar production, and furniture and rope manufacture. Founded in 1537, it was a Mogul capital in 1657 and ceded to the British in 1801. In 1857 it was a centre of the Indian Mutiny. Bareilly College was founded in 1837. The Indian Veterinary Research Institute is at Izatnagar, part of the Bareilly urban area. 54 C3

Barents Sea section of the Eastern ▶Arctic Ocean which has oil and gas reserves. In World War II, the Battle of the Barents Sea on 30 December 1942 was a humiliating defeat for the German Navy. 42 E3

Bari (ancient **Barium**), capital of Apulia region, southern Italy, and industrial and ferry port on the Adriatic Sea; population (1992) 342,100. It is the site of Italy's first nuclear power station. Part of the town is known as Tecnopolis, the Italian equivalent of ▶Silicon Valley. 32 L7

Barisal river port and capital city of Barisal region, southern Bangladesh; population (1991) 173,000. It trades in jute, rice, fish, and oilseed. 54 F4

Barkly Tableland large-scale, open-range, cattle-raising area in Northern Territory and Queensland, Australia. 76 F3

Bârlad alternative spelling of ▶Bîrlad, a town in eastern Romania. 34 Q3

Bar-le-Duc French town, capital of the *département* of Meuse, on the River Ornain, and on the Marne–Rhine canal; population (1990) 18,600. The chief industries are the manufacture of hosiery and metal goods. Once capital of the duchy of Bar, the town has many old churches and mansions and the remains of the ducal palace. 22 H6

Barletta industrial port on the Adriatic Sea, Italy, 50 km/31 mi northwest of Bari; population (1990) 88,200. It is an agriculture centre trading in wine and fruit, and manufactures automobiles, chemicals, and soap. There is a Romanesque cathedral built in 1150 and a 13th-century castle. 32 L7

Barnaul city in the Russian Federation, on the River Ob in southwestern Siberia, 210 km/130 mi south of Novosibirsk; population (1996 est) 594,000. Barnaul is the capital of Altai krai (territory) and its economic and cultural centre. There are large textile, engineering, chemical, and food industries in the city. 42 Q7

Barnsley town and administrative headquarters of Barnsley metropolitan borough, South Yorkshire, England, on the River Dearne, 26 km/16 mi north of Sheffield; population (1991) Barnsley 75,100; Dearne Valley urban area 211,500. It lies on one of Britain's richest coal fields, although the industry is in decline. Manufactured products include steel, glass, paper, carpets, cakes (Lyons), sports equipment (Dunlop-Slazenger), and clothing. The headquarters of the National Union of Mineworkers are sited here. 24 L8

Baroda former name of ▶Vadodara, a city in Gujarat, India. 54 B4

Barquisimeto capital of Lara state, northwestern Venezuela; population (1990) 625,450; metropolitan area (1990) 745,400. It is the centre of a coffee-growing area and trades in coffee, cacao, sugar, and rum. Industries include flour milling, tanning, textiles, leather goods, footwear, biscuits, cigarettes, and cement. The city lies on both the Trans-Andean and Pan-American highways. It was first established in the 16th century, but needed to be rebuilt after a disastrous earthquake in 1812. 98 D1

Barra southern island of the larger Outer ▶Hebrides, Scotland, part of the Western Isles unitary council area; area 90 sq km/35 sq mi; population (1991) 1,280. It is separated from South Uist by the Sound of Barra. The principal town is Castlebay. The main industries are fishing and tourism. 24 E4

Barrancabermeja port and oil-refining centre on the Magdalena River in the department of Santander, northeastern Colombia; population (1995 est) 180,700. The city was founded by the Spaniards in 1536 on the site of an Indian settlement and renamed after the red cliffs of the Magdalena River. Cattle ranching and the marketing of beef continue to be important activities of the area, though economic growth was boosted by the discovery of oil in 1921. The city has a major oil refinery in the De Mares fields, which is now linked by pipeline to Cartagena on the Caribbean coast. There is also an international airport. 98 C2

Barranquilla major port and capital of Atlántico department on the Caribbean coast of northern Colombia, founded in 1629, on the western bank of the Magdalena River; population (1994) 1,049,000. Industries include chemicals, tobacco, food-processing, textiles, furniture, and footwear. Coffee, coal, oil and nickel are exported. It is a commercial centre and a terminal for river traffic. 94 K6

Barrow town and administrative headquarters of North Slope Borough, north-central Alaska, USA; population (1990) 3,500. It is situated on the Arctic Ocean, 18 km/11 mi southwest of the northernmost point in the USA, Point Barrow (71° 23' N). It lies 820 km/510 mi north-northwest of Fairbanks. Barrow is the northernmost town in the USA, and is home to the world's largest Inuit settlement. The town serves as a regional trading centre. Whaling, trapping, crafts, government work, and the oil industry have all been important to the local economy. 92 (1)F1

Barrow-in-Furness industrial port in Cumbria, northwest England, at the south end of the Furness peninsula; population (1991) 48,900. Industries include shipbuilding, engineering, the manufacture of chemicals and nuclear submarines, and those industries associated with offshore gasfields; the British Gas pipeline from Morecambe Field in the Irish Sea comes ashore here. Other industries include footwear and paper manufacture. 24 J7

Barrow Island arid island and wildlife sanctuary in the Indian Ocean, between the Monte Bello Islands and the northwest coast of Western Australia; area 202 sq km/78 sq mi. The island is uninhabited. 76 B4

Basel (French **Bâle**; or **Baselland**), canton of Switzerland, divided into two half-cantons: Basel-Stadt (area 37 sq km/14 sq mi; population (1995) 195,800) and Basel-Landschaft (area 428 sq km/165 sq mi; population (1995) 252,300; capital Liestal). Outside the city, the canton is primarily agricultural land, with fruit growing, dairy farming, and cattle breeding the main economic activities. 8 (1)B1

Basel (or **Basle**; French **Bâle**), commercial and industrial city, capital of Basel canton, Switzerland, situated on the Rhine at the point where the French, German, and Swiss borders meet; population (1994) 179,600. Manufactured goods include dyes, textiles, vitamins, agrochemicals, dietary products, and genetic products. Basel was a strong military station under the Romans. In 1501 it joined the Swiss confederation and later developed as a centre for the Reformation. 30 C3

Bashkortostan (formerly **Bashkiria** or **Bashkir Autonomous SSR**), autonomous republic of the Russian Federation; area 143,600 sq km/ 55,444 sq mi; population (1990) 3,964,000 (40% Russian, 30% Tatar, 25% Bashkir). The capital is ►Ufa. The Ural Mountains are in the east and River Kama in the northwest; other rivers are Belaya, Ufa, Dema, and Zilim. Chief industries are oil, natural gas, minerals (gold and iron ore), chemicals, engineering, timber, and paper. The languages Russian and Bashkir are spoken. 38 K4

Basildon industrial town in Essex, eastern England, 19 km/12 mi southwest of Chelmsford; population (1994 est) 101,000. It was designated a new town in 1949 to accommodate overspill population from London. Industries include printing, engineering, and the manufacture of chemicals and clothing. 22 C3

Basilicata (Roman **Lucania**), mountainous region of southern Italy, comprising the provinces of Potenza and Matera; area 10,000 sq km/3,860 sq mi; population (1992 est) 610,800. Its capital is Potenza. Agriculture is important; durum wheat, olives, and grapes are cultivated, and sheep and goats raised. 9 (4)E4

Basingstoke town in Hampshire, England, 72 km/45 mi west-southwest of London; population (1998) 90,000. It is a financial centre, containing the headquarters of the Automobile Association and Sun Life Insurance. Industries include light engineering, food processing, printing, publishing, and the manufacture of cosmetics (Wella, Alberto-Culver), scientific instruments, medical equipment, agricultural machinery, and electronics. 24 L10

Basle alternative form of ►Basel, a city in Switzerland. 30 C3

Basque Country (Basque **Euskal Herria**), homeland of the Basque people in the western Pyrenees, divided by the Franco-Spanish border. The Spanish Basque Country (Spanish *País Vasco*) is an autonomous region (created in 1979) of central northern Spain, comprising the provinces of Vizcaya, Alava, and Guipúzcoa (Basque *Bizkaia*, *Araba*, and *Gipuzkoa*); area 7,300 sq km/2,818 sq mi; population (1991) 2,104,000. The French Basque Country (French *Pays Basque*) is the area occupied by Basques in the *département* of Pyrénées-Atlantiques. It is estimated that there are about 170,000 Basques in France.

Basra (or **Basrah**, or **Busra**; Arabic **Al Başrah**), principal city in southeast Iraq, 97 km/60 mi from the Gulf; population (1995 est) 822,000. Basra lies at the head of the tidal Shatt al-Arab waterway (formed by the confluence of the Tigris and Euphrates rivers). Founded in the 7th century and now Iraq's main port on the Shatt-al-Arab River, exports include wool, oil, cereal, and dates. It can be reached by large ocean-going vessels, though it lies 120 km/75 mi from the Gulf. Aerial bombing during the 1991 Gulf War damaged bridges, factories, power stations, water-treatment plants, sewage-treatment plants, and the port. 56 E3

Bas-Rhin *département* in the Alsace region of France, situated on the Rhine and the German frontier; area 4,787 sq km/1,848 sq mi; population (1990) 489,600. In the west are the Vosges Mountains, covered with forests of beech and fir trees. The central plains are very fertile, producing cereals, tobacco, hops, beets, and fruit. Livestock is also farmed. Vines are grown on the hill slopes. The *département* has oil and manufacturing industries such as machinery, textiles, chemicals, and pottery. The administrative centre is ►Strasbourg, which houses many buildings of the European Union. 6 G2

Basse-Normandie (English Lower Normandy), coastal region of northwest France lying between Haute-Normandie and Brittany (Bretagne). It includes the *départements* of calvados, Manche, and ►Orne; area 17,600 sq.km/6,794 sq mi; population (1990) 1,391,300. Its administrative cnre is ►Caen. Apart from stock farming, dairy farming, and textiles, the area produces apples, cider, and Calvados apple brandy. Tourism is important. 6 C2

Basse-Terre port on Basse-Terre, one of the Leeward Islands; population (1990 est) 14,100; capital of the French overseas *département* of ►Guadeloupe. It has a significant tourist industry. 94 M5

Basse-Terre one of two main islands of the French overseas *département* of Guadeloupe in the Leeward Islands, West Indies; area 848 sq km/327 sq mi; population (1995 est) 169,000. A narrow stretch of water, the Rivière Salée, divides it from Grande-Terre, the other main island. It has an active volcano, Grande Soufrière, rising to 1,484 m/4,870 ft.

Basseterre capital and port of St Kitts and Nevis, in the Leeward Islands; population (1990 est) 15,000. Industries include data processing, rum, clothes, and electrical components.

Bass Strait sea channel separating the mainland of Australia from Tasmania. The strait is 322 km/ 200 mi long, with an average width of 255 km/158 mi. Oil was discovered here in 1965 and first extracted in 1969. The region now has 18 oil and gas fields. 76 H7

Bastia (Italian *bastiglia* 'fortress'), port and commercial centre on the northeast coast of Corsica, France, administrative centre of the *département* of ►Haute-Corse, 96 km/60 mi northeast of Ajaccio; population (1990) 52,390. Industries include the manufacture of processed foods. Founded in 1380 by the Genoese, who built a fortress to protect it, Bastia was the capital of Corsica until 1791. The town has several fine churches, notably San Giovanni Battiste with its classical facades. 32 D6

Basutoland former name (to 1966) for ►Lesotho, a kingdom in southern Africa. 72 D5

Batavia former name (to 1949) for ►Jakarta, capital of Indonesia on the island of Java. 52 D4

Bath historic city and administrative headquarters of ►Bath and North East Somerset unitary authority, southwest England, 171 km/106 mi west of London; population (1991 est) 78,700. Industries include printing, plastics, engineering, and tourism. Bath was the site of the Roman town of Aquae Sulis, and in the 18th century flourished as a fashionable spa, with the only naturally occurring hot mineral springs in Britain. Although the baths were closed to the public in 1977, a Millennium Spa Project, due to open in summer 2002, is intended to bring back public bathing to Bath's hot springs. 24 K10

Bath and North East Somerset unitary authority in southwest England created in 1996 from part of the former county of Avon.

area 351 sq km/136 sq mi
towns and cities ►Bath (administrative headquarters), Keynsham, Chew Magna, Paulton, Radstock, Peasedown St John, Midsomer Norton
features River Avon and tributaries; Chew Valley Lake; Beckford's Tower (Bath) built in 1827 for William Beckford; Roman baths; Regency architecture; Stanton Drew Bronze Age stone circles, including second largest in Great Britain.
industries tourism, central government administration, clothing manufacture
population (1996) 158,700 5 E6

Bathurst city in the central tableland district of New South Wales, Australia, on the south bank of the River Macquarie, 204 km/127 mi west of Sydney; population (1996) 26,029. Bathurst is the service centre for the surrounding agricultural region which produces cattle, sheep, orchard fruits, and vegetables. The town dates from the 1851 gold rush. **76** J6

Bathurst city, port, and administrative headquarters of Gloucester County, north New Brunswick, Canada; population (1991) 15,900. It is situated at the mouth of the Nepisiguit River. Industries include copper and zinc mining; products include paper and timber. **82** T7

Bathurst former name (to 1973) of ►Banjul, capital of Gambia. **68** A2

Baton Rouge deep-water port on the Mississippi River, USA, the capital of ►Louisiana; population (1994 est) 227,000. Industries include oil refining, petrochemicals, and iron; the port has become one of the largest in the country. **90** C3

Battle Creek city in Calhoun County, south Michigan, USA; population (1992) 54,400. It is located 36 km/22 mi east of Kalamazoo. Seventh-Day Adventists built a sanitorium here in 1866. One of its directors, John Kellogg (1876–1943), experimented with foods, leading to the production of ready-to-eat cereals, which became the city's biggest industry. This enterprise turned international after W K Kellogg (his younger brother) and C W Post marketed the ideas and products on a wider scale. Auto parts, farm equipment, paper products, and trucks are also manufactured in the city. **88** C2

Batumi (or **Batum**), Black Sea port and capital of the autonomous republic of Adjaria, in southwestern Georgia; population (2001 est) 143,800. Key industries include oil refining and transhipment, shipbuilding, food canning, engineering, and the manufacture of clothing and pharmaceuticals. Batumi receives its crude oil via a pipeline running from the oilfields at Baku in Azerbaijan. Tea and citrus fruits are grown in the region around the city. **58** J3

Bavaria (German **Bayern**), administrative region (German *Land*) in southeast Germany; bordered on the west by Hesse and Baden Württemberg, on the north by Thuringia and Saxony, on the northeast by the Czech Republic, and on the south and southeast by Austria; area 70,600 sq km/27,252 sq mi; population (1995) 12,100,000. Bavaria is the largest of the German *Länder*.The capital is ►Munich; main cities and towns are Nuremberg, Augsburg, Würzburg, Regensburg, Passau, Fürth, and Ingolstadt. Bavaria's main industries are electronics, electrical engineering, optics, automobile assembly, aerospace, brewing, chemicals, plastics, oil refining, textiles, and glass. In the agricultural sector the main products are wheat, rye, barley, oats, potatoes, and sugar beet; there is also livestock farming and forestry. **7** (2)C4

Bay City city, port, and administrative headquarters of Bay County, east Michigan; population (1990) 38,900. It is situated near the mouth of the Saginaw River, 21 km/13 mi north of Saginaw and 6 km/4 mi southwest of Lake Huron. Bay City is a major deep-water port for the Great Lakes and oceangoing ships, and is the centre of an agricultural region; its industries include shipbuilding and the manufacture of iron, magnesium castings, petrochemicals, auto and aircraft equipment, and processed foods. **88** D2

Bayern German name for ►Bavaria, a region of Germany. **7** (2)C4

Bayeux town in the *département* of Calvados, northern France, on the River Aure, 27 km/17 mi northwest of Caen; population (1990) 14,700. The town has an agricultural market, and industries include the production of pottery, lace, and processed foods. Its museum houses the 11th-century Bayeux Tapestry. There is a 13th-century Gothic cathedral. **22** B5

Bayonne river port in the French *département* of Pyrénées-Atlantique in southwest France, situated at the confluence of the Adour and Nive rivers, 5 km/3 mi from the sea; population (1990) 41,800. Historically a centre for the making of swords and knives, the town claims the invention of the bayonet. Bayonne is a centre of Basque culture, and lies in an important tourist area. Industries include the manufacture of leather, fertilizers, steel, and aircraft. It also has distilleries and ham-curing factories, and trades in timber. **26** D10

Bayreuth town in Bavaria, south Germany, on the Red Main River, 65 km/40 mi northeast of Nuremberg; population (1995) 72,700. There are cotton textile, porcelain, cigarette, and optical industries. Bayreuth was the home of the composer Richard Wagner, and the Wagner theatre was established in 1876 as a performing centre for his operas. Opera festivals are held here every summer. **20** G7

Beachy Head (French *beau chef*, 'beautiful head'), chalk headland on the south coast of England, between Seaford and Eastbourne in East Sussex. Rising to 163 m/535 ft, it is the eastern end of the South Downs. The lighthouse at the foot of the cliff is 38 m/125 ft high. **22** C4

Beaufort Sea section of the Arctic Ocean off Alaska and Canada, named after the British admiral Francis Beaufort. Oil drilling is allowed only in the winter months because the sea is the breeding and migration route of bowhead whales, the staple diet of the local Inuit people. **80** Q2

Beaumont city and port in southeastern Texas, USA, on the Neches River, 124 km/77 mi northeast of Houston; seat of Jefferson County; population (1994 est) 115,000. It is an oil-processing centre for the surrounding oilfields and a shipping point via the Sabine–Neches canal to the Gulf of Mexico; other industries include steel mills, shipbuilding, and paper production. Its port handles 15–20 million tonnes of cargo annually, mostly the shipment of petroleum products along the coast to the eastern seaboard of the USA. **90** C3

Beaune city in the *département* of Côte-d'Or, France, 35 km/22 mi southwest of Dijon; population (1990) 22,100. It is the centre of the Burgundian wine trade, and has a wine museum, the Musée du Vin de Bourgogne. Industries include the production of agricultural equipment, casks, oil, white metal, and mustard. Notable buildings in the town include two 12th-century churches and a hospital, St Etienne, founded in 1443. 26 K6

Beauvais (ancient **Bellovacum** or **Caesaromagus**), administrative centre of the ►Oise *département* in the ►Picardy region of north-central France, situated 76 km/47 mi northwest of Paris at the confluence of the Thérain and Avelon rivers; population (1999 est) 57,400. It is a trading centre for fruit, dairy produce, and agricultural machinery. Manufacturing industries include carpets, woollens, brushes, tractors, and furniture covers. Beauvais is the seat of a bishopric, has the tallest Gothic cathedral in France (St Pierre, 68 m/223 ft), and was renowned for tapestries. The town centre suffered serious bomb damage in 1940, and was rebuilt in a modern style. 22 E5

Becej town in Vojvodina province, northern Serbia, Yugoslavia, on the Marshal Tito canal; population (1991) 26,600. It has mineral springs and is a commercial and agricultural centre. 34 H4

Bechuanaland former name (to 1966) of ►Botswana. 72 C4

Bedford town and administrative headquarters of ►Bedfordshire, southern England, on the River Ouse, about 80 km/50 mi north of London; population (1991 est) 73,900. Industries include light engineering, food-processing, aircraft services, and the manufacture of agricultural machinery, diesel engines, pumps, bricks, communications systems, and electronic components. The writer John Bunyan is said to have written part of *The Pilgrim's Progress* (1678) while imprisoned in the town. 24 M9

Bedfordshire county of south central England (since April 1997 Luton has been a separate unitary authority).
area 1,192 sq km/460 sq mi
towns and cities ►Bedford (administrative headquarters), Dunstable
physical the Great Ouse River and its tributary, the Ivel; the county is low lying with the Chiltern Hills in the southwest
features Whipsnade Wild Animal Park, near Dunstable (200 ha/494 acres)
agriculture cereals (especially wheat and barley); vegetables
industries agricultural machinery; cement manufacture (using local chalk); clay; electrical goods; motor vehicles; packaging; brickworks at Stewartby
population (1996) 548,800 5 F5

Beds abbreviation for ►Bedfordshire. 5 F5

Beersheba (or **Be'ér Sheva'**; Arabic *Bir-es-Saba*, 'seven wells'), industrial city in the south of Israel, 80 km/50 mi from Jerusalem; population (1995) 152,600. It is the chief centre of the Negev Desert and has been a settlement from the Stone Age. 60 B5

Be'ér Sheva' alternative name for ►Beersheba, an industrial city in southern Israel. 60 B5

Beijing (formerly **Peking**; 'northern capital'), capital of ►China; parts of the northeast municipal boundary coincide with sections of the Great Wall of China; population (1994) 7,084,000. The municipality of Beijing has an area of 17,800 sq km/6,871 sq mi and a population (1996) of 12,590,000. Industries include engineering and the production of steel, vehicles, textiles, and petrochemicals. 46 F3/11 (2)E2

Beira port at the mouth of the Pungwe and Busi rivers, Mozambique; population (1991) 294,200. It exports minerals, cotton, and food products, and is an important site for the domestic import of petroleum products. A railway through the **Beira Corridor** links the port with Zimbabwe, and there is an oil pipeline to Harare. 72 E3

Beirut (or **Beyrouth**), capital and port of ►Lebanon, 90 km/60 mi northwest of Damascus, situated on a promontory into the eastern Mediterranean with the Lebanon Mountains behind it; population (1993) 1,200,000. It was devastated by civil war in the 1970s and 1980s. The city dates back to at least 1400 BC. 60 C3

Bejaia (French **Bougie**), port in Algeria, 193 km/120 mi east of Algiers; population (1998 est) 165,800. The town lies on the edge of a fertile plain, with an annual rainfall of 1,000 mm/40 in at the mouth of the Wadi Soummam and near Mount Gouraya (660 m/2,165 ft). Trade includes exports in wood, hides, iron ore, phosphates, tobacco, hydrocarbons, and liqueurs. Bejaia dates from Roman times, and has also been controlled by pirates, by the Spanish and by the Turks. It grew to greater importance from 1833 after French occupation, and especially after the improvement of the harbour between 1905 and 1909. The installation of the oil pipeline from Hassi Messaoud in 1959 brought a further boost to its prosperity as it became a major exporting port for Saharan oil, and exports now also include olive oil, wine and cork. 66 G1

Béjar town in the province of Salamanca, western Spain; population (1995) 17,200. It is situated on a high ridge and has a 13th-century church, and a 16th-century palace belonging to the dukes of Béjar. Béjar is an important textile centre. 28 E4

Belarus (or **Byelorussia**, or **Belorussia**), country in east-central Europe, bounded south by Ukraine, east by Russia, west by Poland, and north by Latvia and Lithuania.

Belarus covers an area of 207,600 sq km/80,154 sq mi, and has a population of 10,236,000 (2000 est). The capital is ►Minsk (Belorussian Mensk); other major towns and cities include Gomel, Vitsyebsk, Mahilyow, Bobruisk, Hrodna, and Brest. The head of state is Alexandr Lukashenko (from 1994); the head of government is Uladzimir Yarmoshyn (from 2000). The currency is the Belarus rouble, or zaichik, and the languages spoken include Belorussian (official), Russian, and Polish. 14 G2

Belau former name for the Republic of ►Palau. 74 D5

Belém (or **Belém do Pará**), port and capital of Pará federal unit (state) in northern Brazil, on the River Pará near the mouth of the River Tocantins and River Amazon, 144 km/89 mi from the Atlantic; population (1991) 1,235,600 (metropolitan area 1,620,600). Belém lies just 161 km/100 mi south of the Equator, and is hot (mean temperature 27°C/80°F) and humid with a high rainfall. It is the chief trade and distribution centre of northern Brazil, and main exports include rubber, nuts, tropical hardwoods, and jute. **98** H4

Belfast (Irish *Beal Feirste* 'the mouth of the Farset'), capital city and industrial port of ►Northern Ireland, situated in County Antrim and County Down, at the mouth of the River Lagan on Belfast Lough; population (2001 est) 257,400. It is the county town of County ►Antrim, and has been the capital of Northern Ireland since 1920. Industries include shipbuilding, textiles, and engineering. **24** G7

Belfort administrative centre of the *département* of the Territoire de Belfort in the Franche-Comté region of northeast France (see ►Belfort, Territoire de), situated on the River Savoureuse, in the Trouée de Belfort (**Belfort Gap**), a 24 km/15 mi pass between the Vosges and Jura mountains; population (1990) 51,900. The town is the centre of a trade in wine and grain, and its industries include chemicals, engineering, plastics, and textiles. **30** B3

Belfort, Territoire de *département* in the ►Franche-Comté region of eastern France; area 610 sq km/236 sq mi; population (1990) 134,313. Situated in the narrow area of lowland between the ►Vosges and ►Jura mountains, it is the smallest such administrative area in France, outside the suburban *départements* around Paris. The capital, and only major town, is ►Belfort. **6** G3

Belgian Congo former name (1908–60) of the Democratic Republic of ►Congo; known 1960–97 as Zaire. **68** G5

Belgium country in Western Europe, bounded to the north by the Netherlands, to the northwest by the North Sea, to the south and west by France, and to the east by Luxembourg and Germany.

Belgium covers an area of 30,510 sq km/11,779 sq mi, and has a population of 10,161,000 (2000 est). The capital is ►Brussels; other major towns and cities include Antwerp, Ghent, Liège, Charleroi, Bruges, Mons, Namur, and Louvain. The head of state is King Albert II (from 1993); the head of government is Guy Verhofstadt (from 1999). The currency is the Belgian franc, and the languages spoken include Flemish (a Dutch dialect, known as *Vlaams*; official), French (especially the dialect Walloon; official), and German. **22** G4

Belgrade (Serbo-Croat **Beograd**), port and capital of the Federal Republic of Yugoslavia, and of its constituent republic of Serbia, at the confluence of the Danube and Sava rivers: population (1991) 1,168,500. It is linked to the port of Bar on the Adriatic Sea. Industries include light engineering, food processing, textiles, pharmaceuticals, and electrical goods. **34** H5

Belitung Indonesian island in the Java Sea, between Borneo and Sumatra, one of the Sunda Islands; area 4,830 sq km/1,860 sq mi; population (1990) 193,000. The chief port is Tanjungpandan. Tin mining is the chief industry. **52** D3

Belize country in Central America, bounded north by Mexico, west and south by Guatemala, and east by the Caribbean Sea.

Belize covers an area of 22,963 sq km/8,866 sq mi, and has a population of 241,000 (2000 est). The capital is ►Belmopan; other major towns and cities include Belize City, Dangriga, Orange Walk, Corozal, and San Ignacio. The head of state is Queen Elizabeth II (from 1981), represented by Governor General Dr Norbert Colville Young (from 1993); the head of government is Said Musa (from 1998). The currency is the Belize dollar, and the languages spoken include English (official), Spanish (widely spoken), and Creole dialects. **94** G5

Belize City former capital (until 1970) and chief port of Belize, situated at the mouth of the Belize River on the Caribbean coast; population (1991) 44,000. It is Belize's largest city and capital of Belize district. Exports include sugar, timber, citrus fruits, coconuts, and maize. The port also serves parts of Mexico. The city was severely damaged by hurricanes in September 1931 and in October 1961, after which it was decided to move the capital inland, to Belmopan. **94** G5

Belle Isle, Strait of channel between Newfoundland and Labrador in northeast Canada. It forms an entrance to the Gulf of St Lawrence from the Atlantic Ocean. The channel is about 130 km/81 mi long and 16–30 km/10–19 mi wide. During the winter months it is often blocked with ice. Its strong currents include part of the southward Labrador Current, which brings icebergs south from the Labrador Sea. It takes its name from Belle Isle, to the northeast. **82** V6

Bellingham city, port, and administrative headquarters of Whatcom County, northwest Washington, USA; population (1992) 55,300. It is situated on Bellingham Bay, in the Strait of Georgia, 53 km/33 mi southeast of Vancouver, British Columbia. It is a port of entry for the logging and paper industry; there are also shipbuilding and food processing industries here, as well as various agricultural activities. The city is a major supply point for salmon canneries. Tourism is also important to the local economy. It is home to Western Washington University (1899). Bellingham was settled in 1852, and became a staging area for the Frazer River gold rush of 1857–58; it was incorporated in 1903. **86** B1

Bellingshausen Sea section of the South Pacific off the Antarctic coast. It is named after the Russian explorer Fabian Gottlieb von Bellingshausen. **102** (2)JJ4

Bellinzona town in Switzerland on the River Ticino, 16 km/10 mi from Lake Maggiore; capital of Ticino canton; population (1995) 17,150. It is a traffic centre for the St Gotthard Pass, and also a tourist centre. **30** E4

Belmopan capital of ►Belize from 1970; situated in central Belize, 80 km/50 mi inland in Central America, between the Belize and Sibun Rivers. It is 80 km/50 mi southwest of Belize City, near the junction of the Western Highway and the Hummingbird Highway to Dangriga; population (1995 est) 5,900. Principal exports from the region are sugar cane, citrus fruits, bananas, and coconuts. Belmopan was established in 1970 in the mountainous interior to replace Belize City as the administrative centre of the country following hurricane damage to the latter in 1961. The traditional Maya-style architecture prevails. **94** G5

Belo Horizonte industrial city and capital of ►Minas Gerais federal unit (state) in southeast Brazil; population (1991) 2,103,300 (metropolitan area 4,620,600). It stands on a plateau

823 m/2,700 ft high, surrounded by the Sierra do Espinhaço mountain range, and is the centre of a mining and agricultural district. Industries include steel, engineering, textiles, and cement. Gold, diamonds, and manganese are mined nearby and are processed in Belo Horizonte. 98 J7

Beloit city in Rock County, southeast Wisconsin; population (1994 est) 36,600. It is situated on the Rock River, 65 km/40 mi southeast of Madison105 km/65 mi southwest of Milwaukee. Beloit's industries include the manufacture of papermaking and woodworking machinery, diesel engines, electric motors and generators, pumps, and tools. The city is also a commercial centre for the fertile farming and dairying area that surrounds it. The city is the home of Beloit College (1846). 88 C2

Belorussia see ►Belarus. 14 G2

Beloye More Russian name for the ►White Sea, a gulf of the Arctic Ocean on the northwest coast of Russia. 38 G1

Benares alternative transliteration of ►Varanasi, a holy Hindu city in Uttar Pradesh, India. 54 D3

Bendigo city in Victoria, Australia, about 150 km/93 mi north of Melbourne; population (1996) 59,936. It is the centre of a pastoral wine-producing and wheat-growing area, and has manufacturing industries. It has two cathedrals, the Anglican St Paul's and the Roman Catholic Sacred Heart Cathedral. 76 H7

Benevento (ancient **Beneventum**), historic town in Campania, southern Italy, situated at the principal crossing place over the Apennines, 55 km/34 mi northeast of Naples; population (1990) 64,700. The town produces Strega liqueur, machinery, leather goods, and confectionery. There is a Roman theatre, and a triumphal arch of the Emperor Trajan, built in AD 114. 32 J7

Bengal, Bay of part of the Indian Ocean lying between the east coast of India and the west coast of Myanmar (Burma) and the Malay Peninsula.

The Irrawaddy, Ganges, and Brahmaputra rivers flow into the bay. The principal islands are to be found in the Andaman and Nicobar groups. 54 E5

Benghazi (or **Banghāzī**), historic city and industrial port in northern Libya on the Gulf of Sirte; population (1995 est) 545,000. It is the second largest Libyan city and lies 645 km/400 mi east of the largest, Tripoli. There are oil refining and engineering industries. It was controlled by Turkey between the 16th century and 1911, and by Italy from 1911 to 1942; it was a major naval supply base during World War II. The university was founded in 1955. 64 D1

Benguela (formerly **São Filipe de Benguela**), port in Angola, south of Luanda; population (1991) 155,000. It was founded in 1617 by the Portuguese, and was a major base for Portuguese colonisation in Africa. There is a railway running inland to the copper mines of the Democratic Republic of Congo and Zambia. Local industries include sugar milling, fish processing, soap, tools, and pottery. The port exports mainly agricultural goods, including coffee, tobacco, sugar, and sisal, and some manganese from mines to the south of the city is also exported. 72 A2

Benidorm fishing port and tourist resort on the Mediterranean Costa Blanca, in Alicante province, eastern Spain; population (1991) 74,900. Large numbers of elderly people from northern Europe spend time here in the winter. 28 K6

Benin country in west Africa, bounded east by Nigeria, north by Niger and Burkina Faso, west by Togo, and south by the Gulf of Guinea.

Benin covers an area of 112,622 sq km/43,483 sq mi, and has a population of 6,097,000 (2000 est). The capitals are ►Porto-Novo (official) and ►Cotonou (de facto); other major towns and cities include Abomey, Natitingou, Parakou, Kandi, Ouidah, Djougou, Bohicou, and Cotonou. The head of state is Mathieu Kerekou (from 1996); the head of government is vacant (from 1998). The currency is the franc CFA, and the languages spoken include French (official), Fon, Yoruba (both in the south), and six major tribal languages in the north. 68 E2

Ben Nevis highest mountain in the British Isles (1,344 m/4,409 ft), 7 km/4 mi southeast of Fort William, Scotland. 24 H5

Benue river in Nigeria, largest tributary of the River Niger; it is navigable for most of its length of 1,400 km/870 mi. 68 G3

Beograd Serbo-Croat form of ►Belgrade, the capital of Yugoslavia. 34 H5

Berat town in Albania, capital of Berat district, on the River Osum; population (1990) 43,800. It is thought to be the site of the ancient Greek Antipatria, and was named Berat in the 14th century. The Stalin oil field is 11 km/7 mi to the north. 36 B4

Berbera seaport in Somalia, with the only sheltered harbour on the south side of the Gulf of Aden; population (1990) 70,000. It is in a strategic position on the oil route and has a deep-sea port, completed in 1969. The port, which was the capital of British Somaliland until 1941, is the terminus of roads from Hargeysa and Burko, and an airport now adds to its accessibility. Berbera exports sheep, gum arabic, frankincense, and myrrh. Its seaborne trade is chiefly with Aden which lies in Yemen 240 km/150 mi to the north. 64 H5

Berdichev (or **Berdychiv**), city in western Ukraine, 48 km/30 mi south of Zhitomir; population (2001 est) 90,100. Industries include engineering and food processing. 38 E5

Berdyansk city and port in southeastern Ukraine; population (1990) 134,000. It is situated on the **Berdyansk Gulf** on the north coast of the Sea of Azov. 38 G5

Berdychiv alternative spelling of ►Berdichev, a city in western Ukraine. 38 E5

Berezniki city in the Perm oblast (region) of the Russian Federation; population (1996 est) 183,000. Berezniki is situated on the left bank of the River Kama in the foothills of the Ural Mountains. It is one of the chief industrial centres of the Urals, with a vast chemical combine that processes local sodium, potassium, and magnesium salts. Paper is also manufactured here. 38 L3

Bergamo town in Lombardy, Italy, 48 km/30 mi northeast of Milan, at the foot of the Alps between the rivers Brembo and Serio; population (1992) 115,100. Industries include printing, engineering, and the manufacture of cement, textiles (silk), and electrical goods. The Academia Carrara holds a collection of paintings. 30 E5

Bergen industrial port and capital of Hordaland county on the southwest coast of Norway; population (1994) 195,000. Industries include shipbuilding, engineering, and fishing. Often called the 'gateway to the fjords', Bergen is a major centre for tours of the fjords of Norway's west coast. Founded in 1070, Bergen was a member of the Hanseatic League. 16 C6

Bergen op Zoom fishing port in the province of North Brabant, southwest Netherlands, at the junction of the East Schelde and Zoom rivers; population (1997) 47,000. The local economy is dependent on the sugar industry, asparagus trade, oyster fisheries, and chemicals, cigarette, and precision goods production. 22 G3

Bergerac French market town in the *département* of Dordogne, on the River Dordogne; population (1990) 28,000. It is the centre of the French tobacco-growing area, and trades also in truffles, wine, and maize. It has distilleries and spinning mills. Bergerac was a Huguenot religious stronghold. 26 F9

Bergisch Gladbach industrial city 14 km/9 mi northwest of Cologne in North Rhine–Westphalia, Germany; population (1995) 105,200. It produces paper, metal products, and chemicals. 20 C4

Bering Sea section of the Pacific Ocean north of the Aleutian Islands, between Siberia and Alaska; area 2.28 million sq km/880,000 sq mi. It connects with the Chukchi Sea, to the north, via the Bering Strait, extending for 87 km/54 mi from east–west, between the Chukchi Peninsula of Siberia and the Seward Peninsula of Alaska. It is named after the Danish explorer Vitus Bering, who explored the Bering Strait. 92 (1)C4

Bering Strait strait between Alaska and Siberia, linking the North Pacific and Arctic oceans. 92 (1)C2

Berkeley city in central California, USA, 10 km/6 mi north of Oakland, on the northeastern side of San Francisco Bay; population (1992) 101,100. It has an industrial waterfront, but is best known as the main campus of the University of California and for nuclear research at the Lawrence Berkeley Laboratory. During the 1960s, the campus was the site of major political demonstrations against US military involvement in Vietnam. 92 B1

Berks abbreviation for ▶Berkshire.

Berkshire (or **Royal Berkshire**), former county of south-central England; from April 1998 split into six unitary authorities: ▶West Berkshire, ▶Reading, ▶Slough, ▶Windsor and Maidenhead, ▶Wokingham and ▶Bracknell Forest.

Berlin industrial city and capital of the Federal Republic of ▶Germany, lying on the River Spree; population (1995) 3,470,200. Products include machine tools, engineering goods (including cars), electrical goods, paper, food and drink, and printed works. After the division of Germany in 1949, East Berlin became the capital of East Germany and Bonn was made the provisional capital of West Germany. The Berlin Wall divided the city from 1961 until it was dismantled in 1989. Following the reunification of Germany on 3 October 1990, East and West Berlin were once more reunited as the 16th *Land* (state) of the Federal Republic. 20 J4

Bermuda British colony in the Northwest Atlantic Ocean; area 54 sq km/21 sq mi; population (1994) 60,500. The colony consists of about 150 small islands, of which 20 are inhabited, linked by bridges and causeways. The capital and chief port is ▶Hamilton. Bermuda is Britain's oldest colony, officially taken by the crown in 1684. Under the constitution of 1968, it is fully self-governing, with a governor (Thorold Masefield from 1997), senate, and elected House of Assembly (premier from 1998 Jennifer Smith, Progressive Labour Party). Industries include growing Easter lilies and pharmaceuticals; tourism, banking, and insurance are also important. The currency used in the colony is the Bermuda dollar, the main language spoken is English, and the main religion is Christianity. 80 H6

Bermuda Triangle sea area bounded by Bermuda, Florida, and Puerto Rico, which gained the nickname 'Deadly Bermuda Triangle' in 1964 when it was suggested that unexplained disappearances of ships and aircraft were exceptionally frequent there. Analysis of the data has not confirmed the idea.

Bern (French **Berne**), canton of western Switzerland; area 6,887 sq km/ 2,659 sq mi; population (1995) 942,000. Bern is the country's second-largest and second-most populous canton. Tourism is an important industry in this picturesque region; among the peaks of the Bernese ▶Alps to the southeast are the Finsteraarhorn, ▶Jungfrau, Eiger, and Schreckhorn; lakes in the canton include Thun, Brienz, and Bienne (see ▶Biel). 8 (1)B2

Bern (French **Berne**), capital of ▶Switzerland and of ▶Bern canton, in the west of the country on the River Aare; population (1994) 134,100. Industries include the manufacture of textiles, chocolate, pharmaceuticals, and light metal and electrical goods. There is a magnificent Gothic cathedral, dating from the 15th century. Bern joined the Swiss confederation in 1353 as its eighth member, and became the capital in 1848. 30 C4

Berner Alpen alternative name for the ▶Bernese Alps, a mountainous area in the south of Bern canton, Switzerland. 30 C4

Bernese Alps (or **Berner Alpen**), mountainous area in the south of Bern canton, Switzerland. It includes the Jungfrau, Eiger, and Finsteraarhorn peaks. Interlaken is the chief town. 30 C4

Berre, Etang de saltwater lagoon in the French *département* of Bouches-du-Rhône, west of ▶Marseille; area 155 sq km/60 sq mi. It is joined to the Mediterranean by the Martigues canal and the Rove tunnel. The port and industrial development at Fos is located to the west of the lagoon. 26 L10

Berwick-upon-Tweed town and port in Northumberland, northeast England, at the mouth of the River Tweed, and administrative centre of Berwick-upon-Tweed Borough Council; population (1991 est) 13,500. Lying 5 km/3 mi southeast of the Scottish border, it is England's most northerly town. Salmon-netting and sea fishing are important, salmon and grain being the port's principal exports. Other industries include agricultural engineering and the manufacture of foodstuffs and fertilizers. 24 L6

Besançon administrative centre of the *département* of ►Doubs and of the ►Franche-Comté region of France, situated on the River Doubs 75 km/ 47 mi east of Dijon; population (1990) 119,200. It is an agricultural trading centre and industries include the manufacture of textiles, processed foods, cars, and paper. It is the principal French centre for the manufacture of watches and clocks, and has a school of watch-making. It also has a university. The first factory to produce artificial fibres was established here in 1890. The French writer Victor Hugo (1802–1885) and the Lumière brothers Auguste Marie (1862–1954), and Louis Jean (1864–1948), inventors of cinematography, were born here. 26 M6

Bessemer industrial city in Jefferson County, Alabama, southeast USA, 25 km/16 mi southwest of Birmingham; population (1990) 33,500. Situated close to rich coal and iron deposits, Bessemer was planned for steel production. Today, chemical and railway-car manufacturing are important. The town is named after Henry Bessemer, the inventor of the Bessemer process of steelmaking. 90 D3

Beszterczebanya Hungarian name for ►Banská Bystrica, a town in the Slovak Republic. 18 J9

Bethlehem (Arabic **Beit-Lahm**), city on the west bank of the River Jordan, 8 km/5 mi south of Jerusalem; population (1997 est) 135,000. It was occupied by Israel in 1967 and came under control of the Palestinian Authority in December 1995. In the Bible it is mentioned as the birthplace of King David and Jesus, and in 326 the Church of the Nativity was built over the grotto said to be the birthplace of Jesus. 60 C5

Béthune town in the Pas-de-Calais *département* of northern France, 30 km/ 19 mi southwest of Lille; population (1990) 25,200, conurbation 120,000. Industries include the manufacture of textiles, machinery, beet sugar, and tyres. Béthune was formerly a centre for coal mining, but this has now been supplanted by the automobile industry. The church of St Vaast is of historical interest, part of it dating from the 13th century; it also has a 14th–16th-century belfry. 22 E4

Beuthen German name for ►Bytom, a town in Poland. 18 H7

Beverly Hills residential city and a part of greater Los Angeles, southwest California, USA; population (1990) 31,900. It is located 16 km/10 mi northwest of Los Angeles city centre, and extends up the Santa Monica Mountains to the north. It is entirely surrounded by the city of Los Angeles. There is some light manufacturing, chiefly of electronic equipment, and a number of corporate and financial offices. Rodeo Drive is a popular shopping area for prestige goods.

Bexhill-on-Sea seaside resort in East Sussex, England, 10 km/6 mi southwest of Hastings, and administrative centre of Rother District Council; population (2000 est) 42,500. Manufactured products include industrial machinery and scientific instruments. 22 C4

Beyrouth alternative name for ►Beirut, the capital of Lebanon. 60 C3

Béziers town in the Hérault *département* in southern France, situated at the crossing of the River Orb and the Canal du Midi, 60 km/37 mi southwest of Montpellier; population (1990) 72,300. It is a centre of the wine and spirit trade. Other industries include the manufacture of textiles, chemicals, and confectionery. Features include pre-Roman and Roman remains, the Gothic church of St-Nazaire, formerly a cathedral, and ancient city walls. 26 J10

Bhadgaon (or **Bhatgaon**), alternative name for ►Bhaktapur, a town in Nepal. 54 E3

Bhagalpur city in the state of Bihar, northern India, on the River Ganges; population (1991) 253,000. Silk and textiles are manufactured. There are several Jain temples. 54 E3

Bhaktapur (or **Bhadgaon**), town in Nepal, 11 km/7 mi southeast of Kathmandu; population (2001 est) 75,400. It has been a religious centre since the 9th century; there is a palace. 54 E3

Bhamo town in Myanmar (Burma), near the Chinese frontier, on the Irrawaddy River. It is the inland limit of steam navigation and is mainly a trading centre. 50 B2

Bharat Hindi name for ►India. 54 C4

Bhavnagar port in Gujarat, northwest India, in the Kathiawar peninsula on the Gulf of Khambat, an inlet of the Arabian Sea; population (1991) 404,000. It is a centre for textile industry. It was capital of the former Rajput princely state of Bhavnagar. 54 B4

Bhopal industrial city and capital of ►Madhya Pradesh, central India, 525 km/326 mi southwest of Allahabad; population (1991) 1,064,000. Textiles, chemicals, electrical goods, and jewellery are manufactured. Nearby Bhimbetka Caves, discovered in 1973, have the world's largest collection of prehistoric paintings, about 10,000 years old. In 1984 some 2,600 people died from an escape of the poisonous gas methyl isocyanate from a factory owned by US company Union Carbide; another 300,000 suffer from long-term health problems. 54 C4

Bhubaneshwar city in northeast India, 360 km/225 mi southwest of Kolkata (formerly Calcutta); capital of ►Orissa state; population (1991) 412,000. A place of pilgrimage and centre of Shiva worship, it has temples of the 6th–12th centuries. It is a popular tourist destination. Utkal University was founded here in 1843. 54 E4

Bhutan mountainous, landlocked country in the eastern Himalayas (southeast Asia), bounded north and west by Tibet (China) and to the south and east by India.

Bhutan covers an area of 47,500 sq km/18,147 sq mi, and has a population of 2,124,000 (2000 est). The capital is ►Thimphu; other major towns and cities include Paro, Punakha, Mongar, Phuntsholing, Wangdiphodrang, and Tashigang. The head of state is Jigme Singye Wangchuk (from 1972); the head of government is Lyonpo Jigme Thimley (from 1998). The currency is the ngultrum, although the Indian rupee is also accepted, and the languages spoken include Dzongkha (a Tibetan dialect; official), Tibetan, Sharchop, Bumthap, Nepali, and English. 54 E3

Biafra, Bight of (or **Bonny, Bight of**), area of sea off the coasts of Nigeria and Cameroon. 68 F4

Białystok city in eastern Poland; capital of Białystok region; population (1993) 273,300. Industries include textiles, chemicals, and tools. Founded in 1310, the city belonged to Prussia 1795–1807 and to Russia 1807–1919. 18 N4

Biarritz seaside and spa town in Pyrénées-Atlantique *département*, Languedoc-Roussillon region, southwest France, on the Bay of Biscay close to the Spanish border; population (1990) 28,900. Biarritz has a mild climate, attractive beaches, casinos, and many luxurious villas and hotels. Its continued popularity as an international tourist centre has sustained its prosperity. 26 D10

Biberach an der Riss city in the *Land* (administrative region) of Baden-Württemberg, Germany, on the River Riss, 85 km/53 mi southeast of Stuttgart; population (1993) 170,900. The city has machinery, chemical, and pharmaceutical industries.

Biberach was one of the city-states of the Holy Roman Empire. It has medieval towers and gates, old mansions, and a 13th–18th-century church. 30 E2

Biel (French **Bienne**), town in the canton of Bern, northwest Switzerland, 27 km/17 mi northwest of the city of Bern; population (1990) 52,700. Its main industries include watch-making, engineering, and the manufacture of scientific instruments. 30 C3

Bielefeld city in North Rhine-Westphalia, Germany, 55 km/34 mi east of Münster, at the foot of the Teutoburger Wald; population (1995) 324,000. Manufactured products include linen textiles, drinks, chemicals, glass, machinery, and motor vehicles, and other industries include publishing. 20 D4

Bielitz German name for ▶Bielsko-Biała, a town in Poland. 18 J8

Bielsko-Biała (German **Bielitz**), industrial town (woollen textiles, engineering, chemicals) in Katowice province, Poland, on the River Biała, 47 km/29 mi south of Katowice; population (1997) 179,800. 18 J8

Bienne French form of ▶Biel, a town in Switzerland. 30 C3

Bighorn Mountains range of the ▶Rocky Mountains lying principally in north Wyoming, USA, to the east of the Bighorn River. The range runs northwest to southeast for nearly 200 km/125 mi, and has a number of summits over 3,000 m/9,800 ft, culminating in Cloud Peak, 4,019 m/13,186 ft. At the north end of the range, in the Crow Reservation in southeast Montana, is the site of the Battle of Little Bighorn 1876. The foothills are used for livestock grazing and irrigated agriculture. The mountains are a popular visitor destination. 86 E2

Bihac town in Bosnia-Herzegovina, on the River Una; population (1991) 71,000. It was under Turkish rule 1592–1878, and has many old buildings, including a mosque that was formerly a Gothic church. Bihac was designated a 'safe haven' by the United Nations 1993. 30 L6

Bihar (or **Behar**), state of northeast India; area 99,199 sq km/38,301 sq mi; population (2001 est) 77,633,000. The capital is ▶Patna. The River Ganges runs west–east in the north of the state, through intensely cultivated alluvial plains which are prone to drought and floods. In the largely forested south are the Rajmahal Hills and Chota Nagpur plateau. The chief industries are copper, iron, and coal; Bihar accounts for 40% of India's mineral production. Important agricultural products are rice, jute, sugar cane, cereals, oilseed, tobacco, and potatoes. The languages spoken are Hindi and Bihari. As part of the Bihar Reorganization Act, the region was split in 2000 to form the new state ▶Jharkland 10 E3

Bikaner city in Rajasthan, northern India, 200 km/125 mi north of Jodhpur on the edge of the Thar Desert; population (1991) 415,000. Once capital of the Rajput state of Bikaner, it is now a centre for carpet-weaving. It developed originally as a centre for cross-desert trade. From 1571 to 1611 the Junagarh fort was built on a rocky outcrop as a defensive measure by the ruling princes; it contains a fine collection of Sanskrit and Persian books. 54 B3

Bikini Atoll atoll in the ▶Marshall Islands, western Pacific, where the USA carried out 23 atomic- and hydrogen-bomb tests (some underwater) from 1946 to 1958. 74 G4

Bilbao industrial port and capital of Vizcaya province in the Basque Country, northern Spain; it is surrounded by mountains, and situated on an inlet of the Bay of Biscay, and on the Nervion River; population (1994) 372,000. Bilbao is a commercial centre and one of the chief ports in Spain; industries include iron and steel production, shipbuilding, chemicals, cement, and food-processing. 28 H1

Billings city in south-central Montana, USA, on the northern bank of the Yellowstone River; seat of Yellowstone County; population (1992) 84,000. It is the largest city in the state and a centre for the transportation of livestock and animal products; its industries include vegetable and grain processing, and oil-refining. Nearby, to the southeast, is Bighorn American Indian reservation where the Battle of the Little Bighorn took place in 1876. Billings is a gateway to ▶Yellowstone National Park, the largest nature reserve in the USA (established 1872). 86 D1

Billiton alternative name for the Indonesian island of ▶Belitung. 52 D3

Biloxi city and port in Harrison County, southeast Mississippi, USA, on the Gulf of Mexico coast; population (1990) 46,300. Its chief industries include tourism, boat building, fishing gear, and seafood canning. 90 D3

Bingen city in the Rhineland-Palatinate, Germany, on the River Rhine, 32 km/20 mi west of Mainz; population (1994) 23,900. Bingen is a centre for tourism and has a long history of wine trading. 20 C7

Binghamton city and administrative headquarters of Broome County, south-central New York State, USA; population (1992) 52,050. It is situated at the confluence of the Chenango and Susquehanna rivers, near the Pennsylvania border. Industries manufacture shoes, textiles, aviation training equipment, machinery, photo supplies, and computers. Agriculture, primarily the raising of poultry, livestock, and dairy cattle, and the production of poultry and dairy products, lumber, and pulpwood, is important in the surrounding region. With Johnson City and Endicott, directly to the west on the Susquehanna River, Binghamton forms the manufacturing region known as the Triple Cities. 88 E2

Bío-Bío longest river in Chile; length 370 km/230 mi. It rises in the southern Andes, on the border with Argentina, and flows northwest to the Pacific Ocean near Concepción. The name is an Araucanian term meaning 'much water'. Orchards, cereal crops and Chile's main vineyards are located in the Bío-Bío river valley.

Bipontium Roman name for ►Zweibrücken, a town in Germany. 22 K5

Birkenhead seaport and industrial town in the Wirral, Merseyside, England, opposite Liverpool on the Wirral peninsula, on the west bank of the Mersey estuary; population (1994 est) 93,100, urban area 270,200. It developed as a shipbuilding town with important dock facilities, but other principal industries now include engineering and flour-milling. The Mersey rail tunnel (1886), the Queensway road tunnel (1934), and a passenger ferry service link Birkenhead with Liverpool.

Bîrlad (or **Bârlad**), town in Vaslui county, eastern Romania, on the River Bîlad, 100 km/63 mi northwest of Galaţi (Galatź); population (2001 est) 78,000. Industries include soap and candles. An annual horse fair is held in the town. 34 Q3

Birmingham industrial city and administrative headquarters of ►West Midlands metropolitan county, central England, second-largest city in the UK, 177 km/110 mi northwest of London; population (1994 est) 1,220,000, metropolitan area 2,632,000. It is a major manufacturing, engineering, commercial, and service centre. The city's concert halls, theatres, and three universities also make it an important cultural and educational centre. Its chief products are motor vehicles, vehicle components and accessories, machine tools, aerospace control systems, weapons, electrical equipment, plastics, chemicals, food, chocolate (Cadbury), jewellery, and glass. 24 L9

Birmingham city in north-central Alabama, USA, situated in the Jones Valley, at the southern end of the Appalachian Mountains; seat of Jefferson County; population (1994 est) 265,000. It is the largest city in ►Alabama and is an industrial centre for iron, steel (Birmingham was once the steelmaking centre of the South), chemicals, building materials, railroad and aircraft equipment, computers, and cotton textiles; its port is connected to the Gulf of Mexico via the Black Warrior River. 90 D3

Biscay, Bay of bay of the Atlantic Ocean between northern Spain and western France, known for rough seas and high tides. It is traditionally a rich fishing area. 26 C9

Bishkek (formerly **Pishpek** 1878–1926; **Frunze** 1926–92), capital of ►Kyrgyzstan; population (1996) 670,000. Bishkek is situated in the valley of the River Chu north of the Kyrgyz Alatau mountain range, 180 km/112 mi west of Almaty in Kazakhstan. Among the goods produced here are textiles, agricultural machinery, and electrical goods. Food industries include meat-packing and tobacco production. 42 N9

Biskra oasis town in Batna department, Algeria, on the edge of the Sahara Desert, about 200 km/124 mi southwest of Constantine; population (1998 est) 181,600. Biskra is the centre of the Zab (Ziban) group of oases and lies to the south of a large depression between the Aurés Massif and the Tell Atlas, en route from Constantine to Touggourt. Dates, olives, figs, apricots, wheat, and barley are produced. The mild winters, and the airport and hot sulphur springs in the modern spa of Hammam Salahine 5 km/3 mi to the northwest, have helped the rise of the tourist industry. 66 G2

Bismarck state capital of ►North Dakota, USA, and seat of Burleigh County, situated on the east bank of the Missouri River in the south-central part of the state; population (2000 est) 55,500. It is a shipping point for the region's agricultural and livestock products, and for oil products from nearby oil wells. Originally an American Indian site, Bismarck was first settled by Europeans in 1872 and was incorporated in 1875. 84 F2

Bismarck Archipelago group of over 200 islands in the southwest Pacific Ocean, part of ►Papua New Guinea; area 49,660 sq km/19,200 sq mi. The largest island is New Britain. Coconut fibre, copra, cotton, rubber, coffee, tortoiseshell, trepang (sea cucumbers), mother-of-pearl, and fruit are the chief products. The population is mostly Papuan.

Bissau capital and chief port of ►Guinea-Bissau, on an island at the mouth of the Geba River; population (1992) 145,000. Originally a Portuguese fortified slave-trading centre (1687), Bissau became a free port in 1869. Industries include agricultural processing, fishing, textiles, and crafts. There are refrigeration units at the port, and there is an international airport and a university. Bissau replaced Bolama as the capital in 1941. 66 B6

Bitola (Serbo-Croat **Bitolj**; Turkish **Monastir**), city in the Former Yugoslav Republic of Macedonia, 32 km/20 mi north of the Greek border; population (1991) 78,000. Agricultural produce, tobacco, and wool are traded, and textiles and filigree work manufactured. 36 D3

Bitterroot Range range of mountains forming part of the boundary between Idaho and Montana, northwest USA; maximum altitude 3,000 m/9,840 ft. The range is an outlying part of the ►Rocky Mountains, branching off south where the main range turns east through Montana. The mountains are part of the Bitterroot national forest. 86 C1

Bizerte (or **Bizerta**), port in Tunisia, 90 km/56 mi northwest of Tunis; population (1994) 98,900. Chief industries include fishing, oil refining, and metal works. It was formerly a French protectorate. 66 G1

Björneborg Swedish name of ►Pori, a town in Finland. 16 L6

Blackburn industrial city and administrative headquarters of ►Blackburn with Darwen unitary authority, northwest England, on the Leeds–Liverpool canal, 32 km/20 mi northwest of Manchester; population (1991 est) 106,000. Blackburn lies in the Calder valley, between the Rossendale upland to the south and hilly country to the north, and until 1998 it was part of the county of Lancashire. Historically the city was a centre of the cotton-weaving industry, but other important trades now include engineering, brewing, and high-tech industries. Paint, office furniture, motor manufacturer supplies, rubber, plastics, electronics, radio and television components, leather, chemicals, paper, carpets, and compact discs (Polygram) are produced. 24 K8

Blackburn with Darwen unitary authority (borough status) in

northwest England created in 1998, formerly part of Lancashire.
area 136 sq km/53 sq mi
towns and cities ►Blackburn (administrative headquarters), Darwen
features Leeds–Liverpool canal; River Darwen; Darwen Hill and Tower (372 m/1,220 ft); western foothills of Rossendale uplands; Lewis Textile Museum (Blackburn) includes working model of spinning jenny; Blackburn Museum and Art Gallery has largest display of European icons in Britain
industries engineering, brewing, chemicals, high technology industries, textiles, leather, electronics, paint, paper, carpets, compact discs
population (1996) 139,400 **5** E5

Black Country central area of England, to the west and north of Birmingham, incorporating the towns of ►Dudley, ►Walsall, ►Wolverhampton, and Sandwell. Heavily industrialized, it gained its name in the 19th century from its belching chimneys and mining spoil. Anti-pollution laws and the decline of heavy industry have changed the region's landscape. Coalmining in the area ceased in 1968.

Black Forest (German **Schwarzwald**), mountainous region of coniferous forest in Baden-Württemberg, western Germany; length 160 km/100 mi, greatest breadth 57 km/35 mi. Bounded to the west and south by the Rhine, which separates it from the Vosges, it rises to 1,493 m/4,898 ft in the Feldberg. It extends to the Swiss border in the south and to the Neckar valley in the north. Parts of the forest have recently been affected by acid rain. The region is a popular tourist destination and lumbering is an important industry. **30** D3

Black Hills mountains in western South Dakota and northeastern Wyoming, USA. They rise out of the Great Plains, 300–400 km/186–248 mi east of the Rocky Mountains front. The Black Hills occupy about 15,500 sq km/6,000 sq mi and rise to 2,207 m/7,242 ft at Harney Peak, South Dakota. The region includes a national forest (area 4,921 sq km/1,900 sq mi) and Mount ►Rushmore, which has the face of four former presidents carved into a granite cliff (height 1,745 m/5,725 ft). **86** F2

Blackpool seaside resort and unitary authority in northwest England, 45 km/28 mi north of Liverpool; part of the county of Lancashire until April 1998.
area 35 sq km/14 sq mi
physical with its neighbours Lytham St Anne's to the south and Fleetwood to the north, Blackpool is part of a urban ribbon along the northwest coast, between the Ribble estuary and Morecambe Bay
features 11 km/7 mi of promenades, known for their autumn 'illuminations' of coloured lights; Blackpool Tower (built in 1894 and modelled on the Eiffel Tower in Paris, France), 157 m/518 ft high; the Pleasure Beach, an amusement park that includes one of Europe's largest and fastest roller-coasters, 75 m/235 ft high and 1.5 km/1 mi long (opened in 1994); three 19th-century piers
industries The service sector provides over 86% of jobs, mainly in tourism and tourism-related industries, although public administration, banking, finance, and insurance are also important. Manufacturing includes light engineering, food and drink, plastics, and aircraft components
population (1997 est) 151,200; urban area (1991 est) 261,400 **24** J8/**5** E5

Black Sea (Russian **Chernoye More**), inland sea in southeast Europe, linked with the seas of Azov and Marmara, and via the Dardanelles strait with the Mediterranean; area 423,000 sq km/163,320 sq mi; maximum depth 2,245 m/7,365 ft, decreasing in the Sea of Azov to only 13.5 m/44 ft. It is bounded by Ukraine, Russia, Georgia, Turkey, Bulgaria, and Romania, and the rivers Danube, Volga, Bug, Dniester and Dnieper flow into it, keeping salinity levels low. Uranium deposits beneath it are among the world's largest. About 90% of the water is polluted, mainly by agricultural fertilizers. **58** D2

Black Volta one of the two main upper branches of the River ►Volta, running through Burkina Faso and Ghana. **68** D3

Blaenau Gwent unitary authority in south Wales, created in 1996 from part of the former county of Gwent.
area 109 sq km/42 sq mi
towns ►Ebbw Vale (administrative headquarters), Tredegar, Abertillery
features Mynydd Carn-y-Cefn (550 m/1,800 ft); rivers Sirhowy and Ebbw; part of the Brecon Beacons National Park is here
population (1996) 73,000 **5** E6

Blagoevgrad town in Sofiya region, southwest Bulgaria, near the River Struma, 77 km/48 mi southwest of Sofia; population (1990) 86,200. It is a commercial centre and health resort. The town was built on the ruins of the ancient Thracian town of Scaptopara (founded 3rd century BC). **36** F3

Blankenberge seaside resort on the coast of the province of West Flanders, Belgium, 21 km/13 mi northeast of Ostend; population (1991) 16,600. The town has fishing and shipbuilding industries. **22** F3

Blantyre chief industrial and commercial centre of Malawi, in the Shire highlands at the foot of Mchiru Mountain; population (1993) 399,000. The largest city in Malawi, it produces tea, coffee, rubber, tobacco, textiles, and wood products. **72** F3

Blenheim centre of a wine-making and sheep-farming region in the northeast of South Island, New Zealand; population (1996) 25,900. It lies on the Wairau River, 29 km/18 mi from Picton, its port. **78** D5

Bloemfontein (Afrikaans 'fountain of flowers'), capital of the ►Free State (formerly Orange Free State) and judicial capital of the Republic of South Africa; population (1991) 300,150. Founded in 1846 and declared a municipality in 1880, the city produces canned fruit, glassware, furniture, plastics, and railway engineering. The city's climate makes it a popular health resort. **72** D5

Blois administrative centre of ►Loir-et-Cher *département* in central France, situated on the River ►Loire 55 km/34 mi southwest of Orléans; population (1990) 51,500, conurbation 65,000. It is the seat of a bishopric and has a historic château. Industries include the manufacture of porcelain, jewellery, footwear, and chocolates. There is a strong trade in wine, timber, and grain, and the town is a major tourist centre of the Loire valley. **26** G6

Bloomington city and administrative headquarters of McLean County, central Illinois, USA; population (1992) 54,100. It lies in the middle of a rich farming and livestock area,

56 km/35 mi southeast of Peoria. It is a manufacturing centre for electrical appliances and food-processing. Local products include agricultural items such as hybrid seed corn and farm machinery, along with builders' supplies, electronic tubes, heating and ventilating equipment, and vacuum cleaners. 88 C2

Bloomington city and administrative headquarters of Monroe County, south-central Indiana, USA; population (1992) 61,500. It is located 75 km/47 mi southwest of Indianapolis. It is an exporter of limestone from nearby quarries, and is a centre for the manufacture of electrical products and lifts. Bloomington was laid out in 1818 and incorporated in 1877. It is home to Indiana University (1820). 88 C3

Bludenz Austrian town in the province of Vorarlberg, on the River Ill 40 km/25 mi south of Bregenz; population (1995) 13,600. It has a 16th-century church and a Renaissance castle, and is a tourist resort. 30 E3

Bluefields one of three major port facilities on the east coast of Nicaragua, situated on an inlet of the Caribbean Sea; population (1990 est) 18,000. 94 H6

Blue Mountains mountain range in northeast Oregon and southeast Washington, USA. The average elevation of the range is 2,000 m/ 6,500 ft; the highest peak is Rock Creek Butte (2,775 m/9,105 ft). The mountains are the source of tributaries to the Columbia River. Densely forested, the mountains are used for stock grazing and some lumbering. They are also popular with outdoors enthusiasts. 86 C2

Blue Nile (Arabic **Bahr el-Azraq**), river rising at a spring site upstream of Lake Tana in Ethiopia, 2,150 m/ 7,054 ft above sea level. Flowing west then north for 1,460 km/907 mi, it eventually meets the White Nile at Khartoum. A length of 800 km/500 mi is navigable at high water. Some 80% of Sudan's electricity is provided by hydroelectric schemes at Roseires and Sennar, and these dams provide irrigation water for over 10,000 sq km/ 3,860 sq mi of the Gezira Plain. 64 F5

Blue Ridge Mountains mountain range in southeastern USA, and part of the ►Appalachian Mountains system. The Blue Ridge Mountains run from northwest Georgia to West Virginia. The highest summit (and also the highest point in eastern USA) is Mount Mitchell; height 2,037 m/ 6,684 ft.

Bobigny administrative centre of ►Seine-St-Denis *département*, situated in the northeastern suburbs of Paris, France; population (1990) 44,700. It is largely residential, with a range of commercial, office, and light industrial functions. 22 E6

Bóbr (German **Bober**), river in Poland, a tributary of the River ►Oder; length 267 km/167 mi. It rises at the southeastern edge of the Karkonosze mountains and flows north to join the Oder near Krosno. 18 E6

Bobruisk (or **Babruysk**), city in Belarus, on the Beresina River, 130 km/81 mi southeast of Minsk; population (2001 est) 227,700. Industries include timber, machinery, tyres, and chemicals. 38 E4

Bocholt city in the *Land* (administrative region) of North Rhine-Westphalia, Germany, 69 km/ 43 mi northwest of Düsseldorf, near the Dutch border; population (1994) 70,200. The town's industries include textiles and engineering. Bocholt has a Renaissance Rathaus (town hall) and a 15th-century church. 20 B5

Bochum city in the Ruhr district in North Rhine-Westphalia, Germany, 40 km/25 mi northeast of Düsseldorf; population (1995) 400,500. Industries include metallurgy, brewing, food-processing, and the manufacture of vehicles, chemicals, and electrical goods. 20 C5

Bodensee German name for Lake ►Constance, north of the Alps. 30 E3

Bodmin market town in Cornwall, southwest England, 48 km/30 mi northwest of Plymouth; population (1991 est) 13,100. It is the commercial centre of a farming area. Other activites include some light industry and the provision of tourist services. Bodmin Moor, to the northeast, is a granite upland culminating in Brown Willy, the highest point in Cornwall at 419 m/1,375 ft. 24 H11

Bognor Regis (Saxon **Bucgan-Ora**), seaside resort in West Sussex, southern England, 105 km/66 mi southwest of London; population (1991 est) 56,700. Hi-tech industries such as refrigeration and light engineering have developed alongside Bognor's traditional tourist economy. The word 'Regis' was added to its name after George V visited nearby Aldwick to convalesce in 1929. 22 B4

Bogotá (officially **Santa Fé de Bogotá**), capital of ►Colombia, and of Cundinamarca department, situated at 2,640 m/8,660 ft above sea level, on the edge of the Eastern Cordillera plateau of the Andes; population (1994) 5,132,000. Main industries include textiles, chemicals, food processing, and tobacco. Bogotá is Colombia's largest city, and the financial, commercial, and cultural centre of the country. It has several universities and museums. 98 C3

Bohemia area of the Czech Republic, a fertile plateau drained by the Elbe and Vltava rivers. It is rich in mineral resources, including uranium, coal, lignite, iron ore, silver, and graphite. The main cities are Prague and Plzeň. The name Bohemia derives from the Celtic Boii, its earliest known inhabitants.

Bohemian Forest (German **Böhmer Wald**, Czech **Český Les**), wooded mountain range, a southern extension of the Bohemian massif, stretching over 200 km/125 mi from Bavaria, Germany, to the southwestern Czech Republic. The River Vltava, rising in the Czech Republic, cuts deeply through the rugged and sparsely populated terrain. The highest point in Bavaria is the Gross Arber (1,457 m/4,856 ft) and in the Czech Republic, the Plechý or Plöchenstein (1,378 m/4,593 ft). 20 H7

Böhmer Wald German name of the ►Bohemian Forest, a wooded mountain range in central Europe. 20 H7

Boise (French *boisé*, 'wooded'), state capital of ►Idaho, USA, and administrative headquarters of Ada County, southwest Idaho, situated on the Boise River in the western foothills of the Rocky Mountains; population (2000 est) 185,800 (urban agglomeration 400,000). It serves as a processing centre for the farm and livestock products of the region. Other industries include lumber milling, food processing, meatpacking, and the manufacture of electronic equipment, mobile homes, wood and steel products, and farm machinery. It was incorporated in 1864. 86 C2

Bois-le-Duc French form of ►'s-Hertogenbosch, a town in North Brabant, the Netherlands. 22 H3

Bokhara variant spelling of ►Bukhara, a city in Uzbekistan. 56 H2

Bolesławiec (German **Bunzlau**), town in Wrocław province, Poland, on the River Bóbr, 103 km/64 mi west of Wrocław; population (1994 est) 44,500. Pottery has been produced since the 16th century, and copper is mined locally. 18 E6

Bolivia landlocked country in central Andes mountains in South America, bounded north and east by Brazil, southeast by Paraguay, south by Argentina, and west by Chile and Peru.

Bolivia covers an area of 1,098,581 sq km/424,162 sq mi, and has a population of 8,329,000 (2000 est). The capitals are ►La Paz (seat of government) and ►Sucre (legal capital and seat of the judiciary); other major towns and cities include Santa Cruz, Cochabamba, Oruro, El Alto, Potosí, and Tarija. The head of state and government is Hugo Banzer Suarez (from 1997). The currency is the boliviano, and the languages spoken include Spanish (official), Aymara, and Quechua. 98 D7

Bologna (Etruscan **Felsina**; Roman **Bononia**), industrial town and capital of Emilia-Romagna, Italy, at the foot of the Apennines, 80 km/50 mi north of Florence; population (1992) 401,300. It is a major rail hub. Industries include engineering, food-processing, and the manufacture of electrical components and chemicals. An important venue for specialist trade fairs (perfume, camping), it also hosts an annual international children's book fair. 30 G6

Bolton (formerly **Bolton-le-Moors**), town and seat of Bolton metropolitan borough, Greater Manchester, northwest England, on the River Croal, 18 km/11 mi northwest of Manchester; population (2001 est) 138,600. Industries include engineering and the manufacture of chemicals, paper, and textiles. Bolton developed rapidly in the 18th century as a cotton-spinning town. 24 K8

Bolzano (German **Bozen**), town in Italy, in Trentino-Alto Adige region, near the confluence of the rivers Isarco and Adige in the Alps; population (1990) 100,000. Bolzano is known as the gateway to the Tyrol; it belonged to Austria until 1919 and its inhabitants are mostly German-speaking. The town is an important agricultural market, and a focus for winter sports. Industries include tourism, distilling, food-processing, and the manufacture of aluminium, pianos, textiles, and car components. 30 G4

Boma port in the Democratic Republic of Congo, on the estuary of the Congo River, 88 km/55 mi from the Atlantic; population (1991) 246,000. The oldest European settlement in the Democratic Republic of Congo, it was a centre of the slave trade, and capital of the Belgian Congo until 1927. Boma is now an important commercial centre. 68 G6

Bombay former name (until 1995) of ►Mumbai, capital of Maharashtra state in west central India. The city's name was officially changed in July 1995, and was renamed Mumbai after the goddess Mumba, the name in the local Marathi language for Parvati, the wife of the Hindu god Shiva. 54 B5

Bône (or **Bohn**), former name of ►Annaba, a port in Algeria. 66 G1

Bonin and Volcano islands Japanese islands in the Pacific, north of the Marianas and 1,300 km/800 mi east of the Ryukyu Islands. They were under US control from 1945 to 1968. The **Bonin Islands** (Japanese *Ogasawara-shotō*) number 27 (in three groups), the largest being Chichijima: area 104 sq km/40 sq mi, population (1991) 2,430. The **Volcano Islands** (Japanese *Kazan-rettō*) consist of three, including ►Iwo Jima, scene of some of the fiercest fighting of World War II; total area 28 sq km/11 sq mi. They have no civilian population, but a 200-strong maritime self-defence force and a 100-strong air self-defence force are stationed there. 40 T7

Bonn industrial city in North Rhine-Westphalia, Germany, 18 km/15 mi southeast of Cologne, on the left bank of the Rhine; population (1995) 291,700. Industries include the manufacture of chemicals, textiles, plastics, and aluminium. Bonn was the seat of government of West Germany 1949–90 and of the Federal Republic of Germany from 1990. In 1991 the Bundestag voted to move the capital to Berlin. This has taken place in phases, with the major phase in 1998. 20 C6

Bonny, Bight of alternative name for the Bight of ►Biafra. 68 F4

Boppard town in the Rhineland-Palatinate, Germany, on the left bank of the Rhine, 20 km/12 mi south of Koblenz; population (1994) 16,100. It is the centre of a wine-growing area. Built on the site of the Roman Bodobriga, Boppard was one of the city-states of the Holy Roman Empire and has a 12th–13th-century church. 20 C6

Borås city in southwestern Sweden; population (1992) 102,800. Chief industries include textiles and engineering. 16 G8

Bordeaux administrative centre of the *département* of ►Gironde and of the ►Aquitaine region, southwest France, situated on the River Garonne, 100 km/62 mi from the Atlantic; population (1990) 213,300, conurbation 685,000. Bordeaux is accessible to seagoing ships and is a major port; it is a centre for the wine trade, oil refining, chemicals, and the aircraft and aeronautics industries. Other industries include shipbuilding, sugar refining, and the manufacture of electrical goods, motor vehicles, and processed foods. Bordeaux was under the English crown for three centuries until 1453. In 1870, 1914, and 1940 the French government was moved here because of German invasions. 26 E9

Borders former region of Scotland (1975–96) which consisted of four districts and was replaced in 1996 by the ►Scottish Borders unitary authority.

Borlänge town in southeastern Kopparberg county, central Sweden, on the western River Dal; population (1995) 48,500. It is an important steel and paper-mill centre. 16 H6

Borneo third-largest island in the world, one of the Sunda Islands in the West Pacific; area 754,000 sq km/290,000 sq mi; population (1990) 9,102,900. It comprises the Malaysian territories of ►Sabah and ►Sarawak; ►Brunei; and, occupying by far the largest part, the Indonesian territory of ►Kalimantan. It is mountainous and densely forested. A forest fire in early 1998 destroyed 30,000 sq km/11,583 sq mi of forest.

In coastal areas the people of Borneo are mainly of Malaysian origin, with a few Chinese, and the interior is inhabited by the indigenous Dyaks. It was formerly under both Dutch and British colonial influence until Sarawak was formed in 1841. 52 E3

Bornholm Danish island in the Baltic Sea, 35 km/22 mi southeast of the nearest point of the Swedish coast; area 587 sq km/227 sq mi; population (1993) 45,200. The capital is Rönne. It constitutes a county of the same name. 16 H9

Borobudur site of a Buddhist shrine near ▶Yogyakarta, Indonesia.

Bosna river in Bosnia-Herzegovina, 257 km/160 mi long. It rises south of Sarajevo, and flows north to join the Sava. 34 F5

Bosnia-Herzegovina (Serbo-Croat **Bosna-Hercegovina**), country in central Europe, bounded north and west by Croatia, east by the Yugoslavian republic of Serbia, and east and south by the Yugoslavian republic of Montenegro.

Bosnia-Herzegovina covers an area of 51,129 sq km/19,740 sq mi, and has a population of 3,972,000 (2000 est). The capital is ▶Sarajevo; other major towns and cities include Banja Luka, Mostar, Prijedor, Tuzla, Zenica, Bihać, and Goražde. The head of state is Zivko Radisic (from 2000), chairman of the rotating collective presidency; the head of government is Martin Raguz (from 2000). The currency is the dinar, and the languages spoken include Serbian, Croat, and Bosnian. 34 E5

Bosporus (Turkish **Istanbul Boğazi** or **Karadeniz Boğazi**), strait 27 km/17 mi long, joining the Black Sea with the Sea of Marmara and forming part of the water division between Europe and Asia; its name may be derived from the Greek legend of Io. Istanbul stands on its west side. The **Bosporus Bridge** (1973), 1,621 m/5,320 ft, links Istanbul and Anatolia (the Asian part of Turkey). In 1988 a second bridge across the straits was opened, linking Asia and Europe. 36 M3

Boston port and market town in Lincolnshire, eastern England, on the River Witham, 50 km/31 mi southeast of Lincoln; population (1991) 34,600. Industries include food-processing and shell-fishing, and the manufacture of labels and tags. Trade is conducted particularly with Scandinavia and the Baltic countries, agricultural products forming the bulk of the traffic. 24 M9

Boston industrial port and commercial centre, capital of Massachusetts, USA, on Massachusetts Bay; population (1998 est) 555,400; metropolitan area (1992) 5,439,000. Its economy is dominated by financial and health services and government. It is also a publishing and academic centre. The subway system, begun in 1897, was the first in the USA. Boston's baseball team, the Red Sox, is based at Fenway Park. Boston was founded by Puritans in 1630 and has played an important role in American history. 88 F2

Botany Bay inlet on the east coast of New South Wales, Australia, 8 km/5 mi south of Sydney. It is the outlet of the River Georges. The English explorer Captain James Cook landed here in 1770. In 1787 the bay was chosen as the site for a British penal colony, but proved unsuitable, and the colony was located at Port Jackson.

Botswana landlocked country in central southern Africa, bounded south and southeast by South Africa, west and north by Namibia, and northeast by Zimbabwe.

Botswana covers an area of 582,000 sq km/224,710 sq mi, and has a population of 1,622,000 (2000 est). The capital is ▶Gaborone; other major towns and cities include Mahalapye, Serowe, Francistown, Selebi-Phikwe, Molepolol, Kange, and Maun. The head of state and government is Festus Mogae (from 1998). The currency is the franc CFA, and the languages spoken include English (official) and Setswana (national). 72 C4

Bottrop industrial city in North Rhine-Westphalia, Germany, in the Ruhr 42 km/26 mi northeast of Düsseldorf; population (1995) 119,900. It produces chemicals, textiles, and electrical equipment. 22 J3

Bouches-du-Rhône *département* in the ▶Provence-Alpes-Côte d'Azur region in France, containing the delta of the River ▶Rhône; area 5,112 sq km/ 1,974 sq mi; population (1990) 851,200. The western portion, known as the ▶Camargue, is a low-lying plain, and in the north and east the Alpes Maritimes slope down to the Rhône basin. There are heavy industries, especially petrochemicals, in ▶Marseille, the administrative centre. Other towns include Aix-en-Provence and Arles. 6 F5

Bougainville autonomous island, formerly of Papua New Guinea, which with Buka Island and other smaller islands forms the province of North Solomon; area 10,620 sq km/4,100 sq mi; population (1994) 160,000. It is the largest of the Solomon Islands archipelago. The capital is Arawa. The land is volcanic and mountainous, with the Emperor Range in the north and Crown Prince Range in the south; the highest peak is the active volcano Mount Balbi, 3,110 m/10,205 ft. The chief industries are copper, gold, and silver; copra, ivory nuts, and tortoiseshell are exported. 74 F6

Bougie former name (to 1962) of ▶Bejaia, a port in Algeria. 66 G1

Boukra (or **Bou Kraa**), principal phosphate-mining centre of Western Sahara, linked by conveyor belt to the Atlantic coast near Laâyoune; population (1994) 2,000. 66 C3

Boulder city and administrative headquarters of Boulder County, north-central Colorado, USA, 40 km/25 mi northwest of Denver, in the eastern foothills of the Rocky Mountains; population (1992) 85,600. The site of the University of Colorado (1876) and a centre of scientific research, especially space research. Local manufactures include aircraft, business machines, pharmaceuticals, and chemicals. Agriculture and mining are also important to the local economy. The city is a tourist and New Age centre and a mecca for such sports as skiing; mountain, ice, and rock climbing; skydiving; hot-air ballooning; and kayaking. Visitors are also drawn to the University's annual Shakespeare Festival. 86 E2

Boulogne-sur-Mer port on the English Channel in the *département* of Pas-de-Calais, northern France, situated at the mouth of the River Liane; population (1990) 44,200, conurbation 95,000. The city is a ferry port (connecting with the English ports of Dover and Folkestone) and seaside resort. Industries include oil refining, food processing, boatbuilding and the manufacture of textiles, It is the chief fishing port of France and carries an important import/export trade. Boulogne was a medieval countship, but was united with the French crown by Louis XI in 1477. 22 D4

Bourbon former name (1649–1815) of the French island of ▶Réunion, in the Indian Ocean. 72 (1)B2

Bourg (-en-Bresse), capital of the French *département* of Ain, on the River Reyssouze, 75 km/47 mi west of Geneva; population (1990) 43,000. There is trade in horses, cattle, poultry, grains, and copper goods. The suburb of Brou has a 16th-century church, containing royal tombs and stained glass. 26 L7

Bourges administrative centre of the *département* of ▶Cher in central France, 200 km/125 mi south of Paris and situated on the Canal du Berry and the River Yèvre; population (1990) 78,800, conurbation 95,000. The surrounding district, once marshy, now has rich market gardens. Industries include the manufacture of aircraft, armaments, textiles, processed foods, and tyres. It is a trading town, particularly in wool and wine. The 13th-century archiepiscopal Gothic cathedral, St-Etienne, contains beautiful medieval glass. 26 H6

Bourget, Lac du lake in the *département* of Savoie, France, near Aix-les-Bains. It is 17 km/11 mi long and about 3 km/2 mi wide. 30 A5

Bourgogne French name of ▶Burgundy, a region of eastern France. 6 E3

Bournemouth seaside resort and unitary authority in southern England. The town lies on Poole Bay, 40 km/25 mi southwest of Southampton, and was part of the county of Dorset until 1997.
area 46 sq km/18 sq mi
features a 10 km/6 mi stretch of sands, as well as parks, winter gardens, and two piers, one of which is 305 m/1,000 ft long. The Bourne stream, bordered by gardens, runs through the town centre. The Pavilion, opened in 1929, includes a theatre and dance hall. Bournemouth University was founded in 1992 (formerly Bournemouth Polytechnic). Bournemouth Airport is at Hurn to the north of the town.
industries tourism, financial services, and the manufacture of communications systems (Siemens); an International Conference Centre is situated here
population (1997) 161,500 24 L11/5 F6

Bozen German form of ▶Bolzano, a town in Italy. 30 G4

Brabant (Flemish **Braband**), former duchy of Western Europe, comprising the Dutch province of ▶North Brabant and the Belgian provinces of Brabant and Antwerp. They were divided when Belgium became independent in 1830. The present-day Belgian Brabant comprises two provinces: Flemish Brabant (area 2,106 sq km/813 sq mi; population (1997) 1,004,700) and Walloon Brabant (area 1,091 sq km/421 sq mi; population (1997) 341,600). Belgian Brabant is very densely populated, and rich both in agriculture and industry. The principal towns are Louvain in Flemish Brabant, Nivelles in Walloon Brabant, and the Belgian capital, Brussels. 7 (1)C4

Bracknell town and, since April 1998, administrative headquarters of ▶Bracknell Forest unitary authority in southern England, 16 km/10 mi southeast of Reading; population (1991 est) 93,800. It was designated a new town in 1949; before 1998 it was part of the county of Berkshire. The headquarters of the Meteorological Office are here. Bracknell is one of the world's two global area forecasting centres which monitor upper-level winds and temperatures for all air traffic; the other is in Washington, DC, USA. Industries include engineering, electronics, and the manufacture of biscuits (Burtons). 22 B3

Bracknell Forest unitary authority (borough status) in central south England, created in 1998 from part of the former county of Berkshire.
area 109 sq km/42 sq mi
towns ▶Bracknell (administrative headquarters), Sandhurst, Crowthorne
features Royal Military Academy at Sandhurst (established in 1799 for officer training); the Meteorological Office at Bracknell (one of two global forecasting centres for the world's airlines); Transport Research Laboratory
industries high technology industries, engineering, electronics, manufacture of clothing and furniture, bakery products
population (1997) 109,600 5 F6

Bradford industrial city and metropolitan borough in West Yorkshire, England, 14 km/9 mi west of Leeds; population (1994 est) 357,000. The manufacture of textiles, traditionally the base of Bradford's prosperity, declined in the 1970s but remains important. Other principal industries include printing, precision and construction engineering, and the manufacture of chemicals and electronics. Stone quarrying, brewing, photo-engraving, financial services, and publishing also contribute to the city's economy. 24 L8

Braga city and capital of Braga district, northern Portugal, 48 km/30 mi northeast of Porto; population (1993) 144,100. Industries include textiles, electrical goods, and vehicle manufacture. There are Roman remains here, and parts of the medieval walls still exist. Braga has a Gothic cathedral (12th–17th centuries), and the archbishop is primate of Portugal. 28 B3

Bragança capital of Bragança district, northeast Portugal, 176 km/110 mi northeast of Porto; population (1991) 16,500. It was the original family seat of the **House of Braganza**, which ruled Portugal 1640–1910. 28 D3

Brahmaputra river in Asia 2,900 km/1,800 mi long, a tributary of the ▶Ganges, rising in the ▶Himalayas range, and flowing through Tibet, India, and Bangladesh. 54 F3

Brăila port in Romania on the River Danube; 170 km/106 mi from its mouth; population (1993) 236,000. It is a naval base. Industries include the manufacture of artificial fibres, iron and steel, machinery, and paper. It was controlled by the Ottoman Empire from 1544 to 1828. 34 Q4

Brandenburg administrative *Land* (administrative region) of Germany; area 25,000 sq km/10,000 sq mi; population (1995) 2,500,000. The capital is ▶Potsdam; other main cities and towns are Cottbus, Brandenburg, and Frankfurt-an-der-Oder. The main industries are iron and steel, paper, pulp, metal products, and semiconductors. 7 (2)D2

Brandenburg town in Germany, on the River Havel; 60 km/36 mi west of Berlin; population (1995) 87,000. Industries include textiles, engineering, and telecommunications. It has a 12th-century Romanesque cathedral, with Gothic (15th-century) modifications. There is also a Gothic town hall with a 5-m/16-ft high statue of Roland (1474), nephew of Charlemagne, as a symbol of municipal authority. The old town contains interesting 18th-century buildings and parts of the town walls survive. **20** H4

Brandon town in Manitoba, Canada, situated on the Assiniboine River, approximately 211 km/131 mi west of Winnipeg; population (1991) 38,600. The town is in one of the most richly cultivated parts of the province, wheat being the chief crop. It is an important centre for research and development in agricultural techniques and agricultural genetics. The city's industries include oil refining and meat, poultry, and dairy processing. Chemicals, pharmaceuticals, fertilizers, agricultural machinery, textiles, boats, and furniture are manufactured here. It is Manitoba's second-largest town and a booming tourist centre. The city is home to Brandon University (1899) and Assiniboine Community College. **82** M7

Brasília capital of Brazil from 1960, situated on the central plateau 1,150 m/3,773 ft above sea level; population (1991) 1,601,100. The main area of employment is in government service; only light industry is allowed within the city. Brasília is also capital of the Federal District, which has an area of 5,794 sq km/2,317 sq mi. The city was originally designed to accommodate up to 500,000 people, which has now been exceeded, and many people live in shanty towns located outside the main city. **98** H7

Braşov (Hungarian **Brassó**; German **Krondstadt**), industrial city (machine tools, industrial equipment, chemicals, cement, woollens) in central Romania at the foot of the Transylvanian Alps; population (1993) 324,000. It belonged to Hungary until 1920. **34** N4

Bratislava (German **Pressburg**), industrial port (engineering, chemicals, oil refining) and capital of the Slovak Republic, on the River Danube; population (1991) 441,500. It was the capital of Hungary from 1526 to 1784 and capital of Slovakia (within Czechoslovakia) until 1993. **18** G9

Braunau am Inn Austrian town in the province of Upper Austria, situated on the border with the German *Land* (administrative region) of Bavaria, 92 km/57 mi west of Linz; population (1995) 17,600. Braunau was the birthplace of the German Nazi leader Adolf Hitler. The town contains several Gothic churches, and in its industrial suburb has the largest aluminium plant in Austria. **30** J2

Braunschweig German form of ►Brunswick, a city in Lower Saxony, Germany. **20** F4

Brazil largest country in South America (almost half the continent), bounded southwest by Uruguay, Argentina, Paraguay and Bolivia; west by Peru and Colombia; north by Venezuela, Guyana, Suriname, and French Guiana; and northeast and southeast by the Atlantic Ocean.

Brazil covers an area of 8,511,965 sq km/3,286,469 sq mi, and has a population of 170,115,000 (2000 est). The capital is ►Brasília; other major towns and cities include São Paulo, Belo Horizonte, Nova Iguaçu, Rio de Janeiro, Belém, Recife, Porto Alegre, Salvador, Curitiba, Manaus, and Fortaleza. The head of state and government is Fernando Henrique Cardoso (from 1995). The currency is the real, and the languages spoken include Portuguese (official), Spanish, English, French, and 120 Indian languages. **96** F4

Brazzaville river port and capital of the Republic of the Congo (Congo-Brazzaville), on the west bank of the Congo River, opposite Kinshasa; population (1995 est) 937,600. Industries include foundries, railway repairs, shipbuilding, beverages, textiles, food processing, shoes, soap, furniture, and bricks. Tourism is important, with arts and crafts markets in the Plateau district of the city. There is a cathedral built in 1892 and the Pasteur Institute founded in 1908. The city stands on Malebo Pool (Stanley Pool). **70** B4

Brcko town in Serbian Bosnia-Herzegovina, on the River Sava; population (1991) 87,300. It grew up around a Turkish fort, and is the centre of a rich agricultural region. **34** F5

Breda city in North Brabant, the Netherlands, situated at the confluence of the rivers Mark and Aa, 40 km/25 mi southwest of Hertogenbosch; population (1997) 156,400. Engineering, brewing, millinery, dyeing, woollen and linen textiles manufacture, and carpet production are the main industries. It was here that Charles II of England made the Declaration of Breda that paved the way for his restoration in 1660. **22** G3

Bregenz capital of the Austrian province of ►Vorarlberg, at the eastern end of Lake Constance; population (1995) 27,500. It is a tourist resort, trades in agricultural produce, and has textile, chemicals, and woodworking industries. The Romans called it Brigantium. Part of the 13th-century walls remain, and the church is partly 11th century. A music festival is held annually in the city. **30** E3

Bremen industrial port and capital of the *Land* (administrative region) of ►Bremen, Germany, on the River Weser 69 km/43 mi from the open sea; population (1995) 549,000. Industries include iron, steel, oil refining, the manufacture of chemicals, aircraft, and cars, ship repairing, marine engineering, and electronics. The Bremer Vulkan Shipyards closed in 1996. Nearby ►Bremerhaven serves as an outport. **20** D3

Bremen German *Land* (administrative region) of Germany, consisting of the cities of Bremen and Bremerhaven; area 400 sq km/154 sq mi; population (1994 est) 683,000. **7** (1)B2

Bremerhaven (formerly **Wesermünde** until 1947), port at the mouth of the River Weser on the North Sea 57 km/35 mi north of Bremen, in the *Land* (administrative region) of Bremen, Germany; population (1995) 130,800. Bremerhaven, the outport for Bremen, has extensive docks which handle ocean-going vessels, ship repair yards, and one of the largest fishing harbours in mainland Europe. **20** D3

Bremerton city and port in Kitsap County, west Washington, USA, 30 km/19 mi west of Seattle and 50 km/31 mi northwest of Tacoma, on an inlet of Puget Sound near the Olympic Mountains; population (1990) 38,100. The town serves as a port for the area's fish, dairy, and lumber products; the Puget Sound Naval Shipyard (1891) is also important to the economy, as is tourism. The city is home to the

Bremerton Naval Museum and Olympic College (1946). 86 B1

Brescia (ancient **Brixia**), town in Lombardy, northern Italy, 84 km/52 mi east of Milan; population (1992) 192,900. Industries include precision engineering, brewing, and the manufacture of iron, steel, machine tools, transport equipment, firearms, metal products, and textiles. It has medieval walls and two cathedrals (12th and 17th century). 30 F5

Breslau German name of ▶Wrocław, a city in Poland. 18 G6

Brest city and river port in Belarus, capital and economic and cultural centre of Brest oblast, at the junction of the Western Bug and Mukhavets rivers and near the Polish frontier; population (1990) 268,800. Situated 346 km/215 mi southwest of Minsk, it is a busy centre of rail and road communications, and its industries include engineering and textile manufacture. Brest was a Polish town until 1795 when it was acquired by Russia, and reverted to Poland (1919–39). In World War I, the Russian truce with the Central Powers, the Treaty of Brest-Litovsk (an older Russian name for the city), was signed here in March 1918. 38 D4

Brest naval base and industrial port in the *département* of ▶Finistère, situated on two hills separated by the River Penfeld at **Rade de Brest** (Brest Roads), a great bay whose only entrance is a narrow channel, at the western extremity of Brittany in northwest France; population (1990) 201,500. The town has a naval academy, several schools of nautical science, a university and an oceanographic research centre. Industries include electronics, shipbuilding, and the manufacture of chemicals and paper. Occupied as a U-boat base by the Germans from 1940–44, part of the old city was destroyed by Allied bombing and the retreating Germans. 26 A5

Bretagne French form of ▶Brittany, a region of northwest France. 6 B2

Briançon French town on the River Durance, in the *département* of Hautes-Alpes; population (1990) 35,600. It has a trade in *craie de Briançon* (French chalk) and cheese. The highest town in Europe, it is a tourist resort – the older part of the town, strongly fortified by the 17th-century military engineer Vauban, is over 1,200 m/3,900 ft above sea level. 30 B6

Bridgend unitary authority in south Wales created in 1996 from part of the former county of Mid Glamorgan.
area 40 sq km/15 sq mi
towns Bridgend (administrative headquarters), Porthcawl (resort and residential area), Maesteg
physical most of the authority consists of the western end of a lowland plateau, Bro Morgannwg, a rich agricultural area of mixed farming and large villages; in the north is the Cymer Forest and Mynydd Caerau (556 m/1,824 ft)
industries civil engineering; chocolate manufacture
population (1996) 128,300 24 J10/5 E6

Bridgeport city and port in southwestern Connecticut, USA, on the shore of Long Island Sound at the mouth of the Pequonnock River, 100 km/62 mi northeast of New York City; population (1994 est) 133,000. Industries include metal goods, electrical appliances, and helicopters, but many factories closed in the 1970s. Bridgeport was settled in 1639 and was incorporated as a city in 1836. 88 F2

Bridgetown port and capital of Barbados; population (1995 est) 6,100. It lies in the southwest of the island on Carlisle Bay and to the northwest includes a recently built deep water harbour, through which the products of traditional sugar manufacturing are exported. Tourism is also an important industry, and to the north of Bridgetown is the resort of Paradise Beach. Bridgetown was founded in 1628. 98 F1

Bridgwater market town and port in Somerset, southwest England, on the River Parrett northeast of Taunton; population (1991) 34,600. Industries include engineering and the manufacture of furniture, wickerwork, industrial chemicals, fibre fabrics, cellophane, footwear, plastics, electrical goods, and preserves. A nuclear power-station stands to the northwest at Hinkley Point on **Bridgwater Bay**. The site of the Battle of Sedgemoor is 5 km/3 mi to the east. 24 J10

Brieg German name for ▶Brzeg, a town in Poland. 18 G7

Brighton seaside resort in ▶Brighton and Hove unitary authority, on the south coast of England; population (1994 est) 155,000. The city was part of the county of East Sussex until 1997. It is an education and service centre with two universities, language schools, and tourist and conference business facilities. 22 B4

Brighton and Hove unitary authority in southern England, created in 1997.
area 84 sq km/32 sq mi
towns Brighton, Hove (administrative headquarters), Woodingdean, Rottingdean, Portslade-by-Sea
features English Channel; South Downs; Royal Pavilion (Brighton) redesigned and enlarged by English architect John Nash in the 19th century; Palace Pier and West Pier (Brighton); Booth Museum of Natural History (Brighton); British Engineerium (Hove)
industries financial services (including American Express), tourism, conference facilities, language schools
population (1996) 248,000 5 F6

Brindisi (Latin **Brundisium**), port and naval base on the Adriatic Sea, in Apulia, on the 'heel' of Italy, 100 km/62 mi southeast of Bari; population (district, 1996 est) 413,300. It is a ferry point for Greece. Industries include food-processing and the manufacture of petrochemicals. It is one of the oldest Mediterranean ports, at the end of the Appian Way from Rome. 32 M8

Brisbane capital and chief port of the state of ▶Queensland, Australia; population (1996) 1,291,157. Brisbane is situated on the east coast of Australia, 14 km/9 mi inland of the mouth of the River Brisbane, about 29 km/18 mi south of Moreton Bay. It is the third-largest city in Australia, and the financial and commercial centre for Queensland; it has diverse industries including shipbuilding, engineering, brewing, food processing, tobacco production, tanning, the manufacture of agricultural machinery, shoes, and clothing. Tourism is also important. A pipeline from ▶Moonie carries oil for refining. Brisbane has three universities, Queensland University (1909), Griffith University (1975), and Queensland University of Technology (1989). 76 K5

Bristol industrial port and unitary authority in southwest England, at the

junction of the rivers Avon and Frome; it was part of the former county of Avon to 1996.
area 109 sq km/42 sq mi
features British engineer and inventor Isambard Kingdom Brunel's Temple Meads railway station; a 12th-century cathedral; National Lifeboat Museum; Clifton Suspension Bridge (completed in 1864), designed by Brunel; University of Bristol (founded in 1909) and University of the West of England (established in 1992), formerly the Bristol Polytechnic; Ashton Court mansion
industries engineering, microelectronics, tobacco, printing, metal refining, banking, insurance, sugar refining, and the manufacture of aircraft engines, chemicals, paper, soap, Bristol 'blue' glass, and chocolate
population (1996) 374,300, urban area (1991) 516,500 **24** K10/5 E6

Bristol city located in Virginia and Tennessee (the state border runs through the middle of State Street in the centre of the city), USA, 130 km/81 mi northeast of Knoxville; population (1990) 23,400 (Tennessee), 18,400 (Virginia). This dual city is divided politically, but is one unit economically. Bristol is a centre for the surrounding tobacco- and grain-farming region. Industries include lumber, steel, office machines, pharmaceuticals, missile parts, and textiles. The nearby Tennessee Valley Authority dams and lakes are a source of power, jobs, recreation, and tourist trade. **90** E2

Bristol Bay large inlet of the Bering Sea in southwest Alaska. It is bordered by the Ahklun Mountains and the Nushagak Peninsula, to the north, and the Alaska Peninsula, to the south. The name is also applied to the empty, lake-strewn area of Alaska on the north shore of the bay. Approximately 320 km/200 mi wide at its mouth, the shallow bay is a major centre for salmon fishing and canning. The Kvichak, Nushagak, and Togiak rivers empty into Bristol Bay. The bay was explored and named by the English navigator James Cook in 1778. **92** (1)E4

Britain island off the northwest coast of Europe, one of the ▶British Isles. It comprises England, Scotland, and Wales (together officially known as ▶Great Britain), and is part of the ▶United Kingdom. The name is also sometimes used loosely to denote the United Kingdom. It is derived from the Roman name for the island **Britannia**, which in turn is derived from the ancient Celtic name for the inhabitants, *Bryttas*.

British Columbia most westerly, and only Pacific, province of Canada. It is bordered on the east by Alberta, with the Continental Divide in the Rocky Mountains forming its southeastern boundary. To the south, it has a frontier along the 49th Parallel with the US states of Montana, Idaho, and Washington. To the north, along the 60th Parallel, lie the Northwest Territories and Yukon Territory. British Columbia borders in the northwest on the panhandle of Alaska for about half its length (the other half forming the frontier with Yukon Territory); area 947,800 sq km/365,851 sq mi; population (1996) 3,724,500. The capital is ▶Victoria; other main cities and towns are Vancouver, Prince George, Kamloops, Kelowna, Surrey, Richmond, and Nanaimo. British Columbia is mostly mountainous and over half the land is forested; it has a deeply indented coastline, over 80 major lakes, and numerous rivers, including the Fraser and Columbia. Chief industries are lumbering and the manufacture of finished wood products, fishing, mining (coal, copper, iron, lead), extraction of oil and natural gas, and hydroelectric power generation; there is also fruit and vegetable growing. **82** F5

British Honduras former name (to 1973) of ▶Belize. **94** G5

British Indian Ocean Territory British colony in the Indian Ocean directly administered by the Foreign and Commonwealth Office, consisting of the Chagos Archipelago some 1,900 km/1,200 mi northeast of Mauritius; area 60 sq km/23 sq mi. Copra, salted fish, and tortoiseshell are produced. There is a US naval and air base on Diego Garcia. In 2000 a number of Ilois, British subjects who had lived on the Chagos islands, claimed that the British government had unlawfully removed them from the islands during the period 1967–73 to allow the US military base to be built. The High Court ruled that they had been unlawfully ejected, and the remaining Ilois, many in exile in Mauritius, began planning to return to the islands.

British Isles group of islands off the northwest coast of Europe, consisting of Great Britain (England, Wales, and Scotland), Ireland, the Channel Islands, the Orkney and Shetland islands, the Isle of Man, and many other islands that are included in various counties, such as the Isle of Wight, Scilly Isles, Lundy Island, and the Inner and Outer Hebrides. The islands are divided from Europe by the North Sea, Strait of Dover, and the English Channel, and face the Atlantic to the west.

British Virgin Islands part of the ▶Virgin Islands group in the West Indies. **96** E2

Brittany (French **Bretagne**; Breton **Breiz**), modern region of northwest France and former province, on the Breton peninsula between the Bay of Biscay and the English Channel; area 27,200 sq km/10,499 sq mi; population (1990) 2,795,600. A farming region, it includes the *départements* of Côtes-d'Armor, ▶Finistère, Ille-et-Vilaine, and Morbihan. The administrative centre is ▶Rennes, and other towns include Brest, Lorient, Nantes, St-Brieuc, Vannes, and Quimper. **6** B2

Brive (**-la-Gaillarde**), French town in the *département* of Corrèze, on the River Corrèze; population (1990) 53,000. It manufactures shoes, textiles, and paper, and has a metal-processing industry. There is trade in chestnuts, wine, and truffles. It was known to the Romans as Brive Curretiae and has a 12th-century church. **26** G8

Brno industrial city (chemicals, arms, textiles, machinery) in the Czech Republic; population (1993) 390,000. Now the second-largest city in the Czech Republic, Brno was formerly the capital of the Austrian crown land of Moravia. **18** F8

Broads, Norfolk area of navigable lakes and rivers in England; see ▶Norfolk Broads. **22** D2

Brockton city in Plymouth County, southeast Massachusetts, 32 km/20 mi south of Boston; population (1998 est) 98,200. Industries include footwear, tools, printing, electronic equipment, and clothing. Brockton was built on land sold by American Indians in 1649 to the English settlers Myles Standish and John Alden, which became part of the Plymouth colony.

It was part of Bridgewater until 1821. It is home to Massasoit Community College (1966). **88** F2

Broken Hill mining town in New South Wales, Australia, near the border with South Australia, 1,158 km/720 mi west of Sydney and 517 km/321 mi northeast of Adelaide; population (1996) 20,963. Broken Hill was founded in 1894 as a mining centre, and the town is the second-largest producer of silver, and the largest producer of lead and zinc, in Australia. It is the base of the Royal Flying Doctor Service, and the School of the Air. **76** H6

Broken Hill former name (to 1967) of ►Kabwe, a city in Zambia. **72** D2

Bromberg German name of ►Bydgoszcz, a port in Poland. **18** H4

Bronx, the northernmost borough of ►New York City, USA, northeast of the Harlem River; area 109 sq km/42 sq mi; population (1990) 1,169,000. The Bronx is bounded on the north by Westchester County, on the east by Long Island Sound, on the south by the East River, and on the west by the Harlem and Hudson rivers; it is largely a residential area. The New York Zoological Society and Gardens are here, popularly called the Bronx Zoo and the Bronx Botanical Gardens. Yankee Stadium, home of the New York Yankees baseball team, is also located here.

Brooklyn borough of New York City, USA, occupying the southwestern end of Long Island; area 184 sq km/71 sq mi; population (1990) 2,300,700. It is linked to Manhattan Island by the Brooklyn Bridge (1883), the Williamsburg Bridge (1903), the Manhattan Bridge (1909), and the Brooklyn–Battery Tunnel (1950); and to Staten Island by the Verrazano-Narrows Bridge (1964). Brooklyn is a densely populated residential area, and its waterfront is highly industrialized. There are more than 60 parks here, of which Prospect Park is the largest. There is also a museum, a botanical garden, and a beach and amusement area at Coney Island.

Brownsville city in southern Texas, USA, on the Rio Grande just before it flows into the Gulf of Mexico, 212 km/132 mi south of Corpus Christi and 8 km/5 mi north of Matamoros, Mexico; seat of Cameron County; population (1994 est) 113,000. It is a port of entry to the USA, and serves as a commercial and processing centre for the irrigated farmlands of the lower Rio Grande Valley; industries include electronic and aerospace equipment, petrochemicals, and food products; tourism is also important. **90** B4

Bruchsal city in Baden-Württemberg, Germany, 20 km/12 mi northwest of Karlsruhe; population (1994) 40,100. Bruchsal is an important railway junction and has metallurgical, chemical, paper, and textile industries. The city has a baroque castle and church. **20** D7

Bruck an der Mur town in the Austrian province of Styria, 40 km/25 mi northwest of Graz; population (1991) 14,100. It has a number of old buildings, and manufactures iron and steel, paper, and cables. **34** C2

Bruges (Flemish **Brugge**), historic city in northwest Belgium; capital of West Flanders province, about 96 km/60 mi northwest of Brussels and 16 km/10 mi from the North Sea, to which it is connected by canal; population (1997) 115,500. The port handles coal, iron ore, oil, and fish; local industries include lace, textiles, paint, steel, beer, furniture, motors, and tourism. Bruges was the capital of medieval ►Flanders and was mainland Europe's major wool producing town as well as its chief market town. **22** F3

Brugge Flemish form of ►Bruges, a city in Belgium. **22** F3

Brühl city in North Rhine-Westphalia, Germany, 12 km/7 mi southwest of Cologne, and administratively a suburb of ►Bonn; population (1994) 43,400. Manufactures include food, iron, chemicals, and paper. Brühl has an 18th-century rococo palace, now used by the federal chancellor to host receptions. **22** J4

Brunei country comprising two enclaves on the northwest coast of the island of Borneo, bounded to the landward side by Sarawak and to the northwest by the South China Sea.

Brunei covers an area of 5,765 sq km/2,225 sq mi, and has a population of 328,000 (2000 est). The capital (and chief port) is ►Bandar Seri Begawan; other major towns and cities include Seria and Kuala Belait. The head of state and government is Sultan Hassanal Bolkiah Mu'izzaddin Waddaulah (from 1967). The currency is the Bruneian dollar, although the Singapore dollar is also accepted, and the languages spoken include Malay (official), Chinese (Hokkien), and English. **52** E2

Brunei Town former name (to 1970) of ►Bandar Seri Begawan, the capital of Brunei. **52** E2

Brünn German form of ►Brno, a city in the Czech Republic. **18** F8

Brunswick (German **Braunschweig**), industrial city in Lower Saxony, Germany, on the River Oker, 56 km/35 mi southeast of Hannover; population (1995) 253,600. Industries include chemical engineering, precision engineering, food processing, and the manufacture of motor vehicles, photographic equipment, and musical instruments. It was one of the chief cities of north Germany in the Middle Ages and an important trading city in the Hanseatic League trade federation. It was capital of the duchy of Brunswick from 1671. The city's rich architectural heritage includes a Romanesque cathedral (1173–1195). **20** F4

Brusa alternative form of ►Bursa, a city in Turkey. **36** M4

Brussels (Flemish **Brussel**; French **Bruxelles**), city and capital of Belgium and the province of Brabant, situated almost in the centre of the country in the Senne river valley; city population (1997) 133,800; metropolitan/capital region population (1997) 950,600. Industries include lace, textiles, machinery, and chemicals. It is the headquarters of the European Union (EU) and, since 1967, of the international secretariat of NATO. It contains the Belgian royal seat, the chief courts, the chamber of commerce, and is the centre of the principal banks of the country. Founded on an island in the River Senne *c.* 580, Brussels became a city in 1312, and was declared capital of the Spanish Netherlands in 1530 and of Belgium in 1830. **22** G4

Brüx German name for ►Most, a town in the Czech Republic. **20** J6

Bruxelles French form of ►Brussels, the capital of Belgium. **22** G4

Bryan city and administrative headquarters of Brazos County, east-central Texas, USA; population (1992) 56,600. It is located 140 km/87 mi northwest of Houston, in a region that produces cattle, sorghums, poultry, cotton, and other crops.

Bryan processes agricultural goods and its other industries include the manufacture of aluminium products, chemicals, concrete, electrical equipment, furniture, rubber goods, and dairy products. **90** B3

Bryansk (or **Briansk**; formerly **Bryn**), capital city of Bryansk oblast (region), in the western Russian Federation, 350 km/217 mi southwest of Moscow on the Desna River; population (1996 est) 463,000. Bryansk is a major industrial city, producing steel, locomotives, rolling stock and other machinery, cement, clothing, timber and foodstuffs. Electricity is supplied locally by a large peat-fired power station, and natural gas is piped from Dashava and Shebelinka. Bryansk is the site of a major Russian military base. **38** F4

Brzeg (German **Brieg**), industrial town (chemicals, electrical goods) in Opole province, Poland, on the River Oder, 43 km/27 mi northwest of Opole; population (1990) 5,400. In 1311 it became a duchy of the Piast dynasty (first historical rulers of Poland), passing to the Habsburgs 1675. Before 1945 it was part of Silesia with a large German population. **18** G7

Brześć nad Bugiem Polish name of ►Brest, a city in ►Belarus. **38** D4

Bubiyan uninhabited island off Kuwait, occupied by Iraq in 1990; area about 1,000 sq km/380 sq mi. On 28 February 1991, following Allied success in the Gulf War, Iraqi troops were withdrawn. **61** C2

Bucaramanga industrial and commercial city, and capital of Santander department in north-central Colombia; population (1992) 350,000. Coffee, tobacco, cacao, and cotton are produced in the surrounding region. The city is noted for its cigars. Bucaramanga was founded by the Spanish in 1622. **98** C2

Bucharest (Romanian **Bucuresti**), capital and largest city of Romania; population (1993) 2,343,800. The conurbation of Bucharest district has an area of 1,520 sq km/587 sq mi. It was originally a citadel built by Vlad the Impaler to stop the advance of the Ottoman invasion in the 14th century. Bucharest became the capital of the princes of Wallachia in 1698 and of Romania in 1861. Savage fighting took place in the city during Romania's 1989 revolution. **34** P5

Buckinghamshire county of south-east central England.
area 1,565 sq km/604 sq mi
towns Aylesbury (administrative headquarters), Beaconsfield, Buckingham, High Wycombe, Olney
physical Chiltern Hills; Vale of Aylesbury
features Chequers (country seat of the prime minister); the church of the poet Thomas Gray's 'Elegy' at Stoke Poges; Cliveden, a country house (now a hotel) designed by architect Charles Barry; Bletchley Park, home of World War II code-breaking activities; homes of the poets William Cowper at Olney and John Milton at Chalfont St Giles, and of the Tory prime minister Disraeli at Hughenden Valley; Stowe landscape gardens
industries engineering; furniture (chiefly beech); paper; printing; railway workshops; motor cars
agriculture about 75% of the land is under cultivation; cereals (barley, wheat, oats); cattle, pigs, poultry, sheep
population (1996) 671,700 **5** F6

Bucks abbreviation for ►Buckinghamshire. **5** F6

Bucuresti alternative form of ►Bucharest, the capital of Romania. **34** P5

Budapest capital of Hungary, industrial city (chemicals, textiles) on the River Danube; population (1993 est) 2,009,000. Buda, on the right bank of the Danube, became the Hungarian capital in 1867 and was joined with Pest, on the left bank, in 1872. **34** G2

Budějovice a town in the Czech Republic; see ►České Budějovice. **30** K2

Budweis German form of ►České Budějovice, a town in the Czech Republic. **30** K2

Buenos Aires industrial city, chief port, and capital of ►Argentina, situated in the 'Capital Federal' – a separate federal district, on the south bank of the Río de la Plata, at its estuary; population (1992 est) 11,662,050. Industries include motor vehicles, engineering, oil, chemicals, textiles, paper, and food processing. Main exports are grain, beef, and wool, which are produced in the surrounding pampas. The administrative Federal District of Buenos Aires has an area of 200 sq km/77 sq mi, with a population of (1991) 2,960,976. Buenos Aires is the financial and cultural centre of Argentina, and has many museums and libraries. It is a major railway terminus, and has an international airport 35 km/22 mi southwest of the city centre. **100** K5

Buffalo city and port in New York State, USA, at the eastern end of Lake Erie at the head of the Niagara River; seat of Erie County; population (1994 est) 313,000; metropolitan area (1994 est) 1,189,000. It is linked with New York City via the New York State Barge Canal (formerly the Erie Canal), and is an important inland port, especially as a centre for trade between the USA and Canada. An industrial city, Buffalo was hard hit by the closure of steel mills and car factories in the late 1970s and early 1980s. **88** E2

Bug name of two rivers in eastern Europe: the **Western Bug** rises in southwestern Ukraine and flows north to the Vistula, with a length of 768 km/480 mi; the **Southern Bug** rises in western Ukraine and flows southeast into the Dnieper estuary on the Black Sea, with a total length of 806 km/501 mi. **18** L5

Bujumbura (formerly **Usumbura** until 1962), capital of Burundi, located at the northeastern end of Lake Tanganyika; population (1996 est) 300,000. Bujumbura is the main banking and financial centre of Burundi; industries include food processing and paint manufacture. It was founded in 1899 by German colonists, and a university was established in 1960. **70** D4

Bukavu (formerly **Costermansville** until 1966), port in eastern Democratic Republic of Congo, on Lake Kivu; population (1994 est) 201,600. Mining is the chief industry. It is the capital of Kivu region and is in one of the most densely populated parts of the country with a diverse agricultural hinterland. **70** D4

Bukhara (or **Bokhara** or **Bukhoro**), city in south-central ►Uzbekistan, on the Zerevshan River 220 km/137 mi east of Samarkand; population (1995) 250,000. A historic city with over 140 protected buildings, it was once the heart of Muslim Central Asia, and second only to Mecca as an Islamic holy site. It is the capital of the Bukhara region of Turkestan, which has given its name to a type of handwoven carpet. Textiles, including rugs and carpets, are manufactured here (though 'Bukhara' carpets are now principally made in ►Ashgabat, in Turkmenistan). Natural gas is extracted in the surrounding region, and cotton is grown extensively. 56 H2

Bulawayo (Ndebele 'place of slaughter'), industrial city and railway junction in Zimbabwe; population (1992) 620,900. The city lies at an altitude of 1,355 m/4,450 ft on the River Matsheumlope, a tributary of the Zambezi, and was founded on the site of the kraal (enclosed village), burned down in 1893, of the Matabele chief Lobengula. It produces cement and agricultural and electrical equipment. The former capital of Matabeleland, Bulawayo developed with the exploitation of gold mines in the neighbourhood. It is the second-largest city in Zimbabwe. 72 D4

Bulgaria country in southeast Europe, bounded north by Romania, west by Yugoslavia and the Former Yugoslav Republic of Macedonia, south by Greece, southeast by Turkey, and east by the Black Sea.

Bulgaria covers an area of 110,912 sq km/42,823 sq mi, and has a population of 8,225,000 (2000 est). The capital is ►Sofia; other major towns and cities include Plovdiv, Varna, Ruse, Burgas, Stara Zagora, and Pleven. The head of state is Petar Stoyanov (from 1997); the head of government is Ivan Kostov (from 1997). The currency is the lev, and the languages spoken include Bulgarian (official) and Turkish. 34 M7

Bunzlau German name for ►Bolesławiec, a town in Poland. 18 E6

Burgas Black Sea port and resort in Bulgaria; population (1990) 204,900. 34 Q7

Burgdorf (French **Berthoud**), commune in the canton of Bern, Switzerland, on the River Emme, 18 km/11 mi northeast of Bern; population (1995) 14,500. It has silk and clothmaking industries and a trade in cheese. In the castle at Burgdorf, the educational reformer Johann Heinrich Pestalozzi established his first school 1798–1804. 30 C3

Burgenland federal state of southeast Austria, extending south from the Danube along the western border of the Hungarian plain; area 4,000 sq km/1,544 sq mi; population (1994) 273,600. It is a largely agricultural region adjoining the ►Neusiedler See, and produces timber, fruit, sugar, wine, lignite, antimony, and limestone. Its capital is Eisenstadt. In the north it is generally flat, but in the south are spurs of the Alps, with the valleys of the River Raab and its tributaries. 8 (2)H3

Burgos capital of Burgos province in Castilla–León, Spain, 217 km/135 mi north of Madrid; population (1991) 160,400. Products include textiles, motor parts, and chemicals, and the town has a large trade in agricultural products. Founded in the 9th century, Burgos was capital of the old kingdom of Castile. The national hero El Cid is buried in the cathedral, which was built between 1221and 1567, and is regarded as one of the most important examples of Gothic architecture in Europe. 28 G2

Burgundy (French **Bourgogne**), modern region and former duchy of east-central France that includes the *départements* of ►Ain, ►Côte-d'Or, ►Nièvre, ►Saône-et-Loire, and ►Yonne; area 31,600 sq km/12,198 sq mi; population (1990) 1,609,700. Its administrative centre is ►Dijon. 6 E3

Burkina Faso (formerly **Upper Volta**), landlocked country in west Africa, bounded east by Niger, northwest and west by Mali, and south by Côte d'Ivoire, Ghana, Togo, and Benin.

Burkina Faso covers an area of 274,122 sq km/105,838 sq mi, and has a population of 11,937,000 (2000 est). The capital is ►Ouagadougou; other major towns and cities include Bobo-Dioulasso, Koudougou, Banfora, Ouahigouya, and Tenkodogo. The head of state is Blaise Compaoré (from 1987); the head of government is Paramanga Ernest Yonli (from 2000). The currency is the franc CFA, and the languages spoken include French (official) and 50 Sudanic languages. 68 D2

Burlington city in northwestern Vermont, USA, on the eastern shore of Lake Champlain; seat of Chittenden County; population (1990) 39,100. It is a port of entry to the USA; industries include computer parts, steel, marble, lumber, dairy products, and tourism. It is the largest city in Vermont. Burlington was settled in 1773, and incorporated in 1865. 88 F2

Burma former name (to 1989) of ►Myanmar. 50 B2

Bursa city in northwestern Turkey, southeast of the Sea of Marmara, situated on the lower slopes of Mt Uladag (2,443 m/8,015 ft), with a port at nearby Mudanya; population (1990 est) 834,600. It was the capital of the Ottoman Empire from 1326 until 1423. Industries include the silk trade and towel manufacture. 36 M4

Bûr Sa'îd Arabic name for ►Port Said, a port in Egypt. 64 F1

Burton upon Trent town in Staffordshire, central England, on the River Trent, northeast of Birmingham; population (1991) 60,500. It is a former cotton-spinning town. Brewing is the principal industry, with five major breweries in operation; Marmite savoury spread is produced from the yeast by-products. Engineering, food-processing, and the manufacture of tyres and rubber goods are also important. The Benedictine monks of Burton Abbey (founded in 1002) began the town's tradition of brewing in the 11th century. 24 L9

Burundi country in east central Africa, bounded north by Rwanda, west by the Democratic Republic of Congo, southwest by Lake Tanganyika, and southeast and east by Tanzania.

Burundi covers an area of 27,834 sq km/10,746 sq mi, and has a population of 6,695,000 (2000 est). The capital is ►Bujumbura; other major towns and cities include Kitega, Bururi, Ngozi, Muhinga, Ruyigi, and Kayanza. The head of state and government is Pierre Buyoya (from 1996). The currency is the Burundi franc, and the languages spoken include Kirundi, French (both official), and Kiswahili. 70 D4

Buryat (or **Buryatiya**; formerly **Buryat–Mongol Autonomous Soviet Socialist Republic** 1923–58), republic in the eastern Siberian region of the Russian Federation; area 351,300 sq km/135,637 sq mi; population (1996) 1,053,000 (70% Russian, 24% Buryat). The main cities are ►Ulan-Ude (capital),

Kyakhta, and Gusinoozersk. Buryat is bordered on the south by Mongolia, and occupies the eastern and northern shores of Lake ►Baikal. The land is largely mountainous and covered by coniferous forests; the Sayan Mountains are in the far west. Mineral deposits include rare metals (tungsten, molybdenum, gold), together with lignite (brown coal), iron ore, and graphite. The chief industries are aerospace engineering, mining, food processing, fishing, lumbering, and the breeding of sheep and cattle. The republic is crossed by the Trans-Siberian Railway. 44 J6

Bury St Edmunds market town in Suffolk, eastern England, on the River Lark, 43 km/27 mi east of Cambridge; population (1991) 31,200. Industries include engineering, brewing, sugar-beet refining, printing, and the manufacture of agricultural machinery, electronic equipment, cameras, lamps, and confectionery. It was named after St Edmund, last Saxon king of East Anglia, and there are remains of a large Benedictine abbey. 22 C2

Bute island in the Firth of Clyde, Scotland; area 120 sq km/46 sq mi. The chief town and resort is Rothesay; population (2000 est) 7,400. Farming and tourism are the principal industries. It is separated from the mainland in the north by a winding channel, the **Kyles of Bute**. With Arran and the Cumbraes it comprised the former county of Bute. In 1975 Bute merged into the district of Argyll and Bute in Strathclyde region, and is now part of the Argyll and Bute unitary authority. 24 G6

Butte (or **Butte-Silver Bow**), city in southwestern Montana, USA, on the western slope of the Rocky Mountains; population (1990) 33,900. Butte was founded in 1864 during a rush for gold (which was soon exhausted); copper was found some 20 years later on what was then called 'the richest hill on earth'. Declining copper prices forced many mines to close in the 1970s and 1980s (the last mine in Butte closed in 1983), but the city has developed industries of its own, including energy research and tourism, and now acts as a service centre for the region (where copper and molydenum mining are still important). 86 D1

Buxton town and former spa in Derbyshire, central England, 54 km/34 mi northwest of Derby, on the River Wye, surrounded by the Peak District National Park; population (1991) 19,900. Known from Roman times for its hot springs, it is now a tourist centre and a source for bottled mineral water. 22 A1

Buzău capital of Buzău county, Romania, on the River Buzău, 68 km/43 mi northeast of Bucharest; population (1993) 149,000. It is a market town at the centre of a cereal and livestock region, trading in grain, timber, and petroleum, and produces metalwork and wire. Buzău has a 12th-century bishop's palace and a 15th-century cathedral. 34 P4

Bydgoszcz (formerly **Bromberg** until 1919), industrial river port in northern Poland, 105 km/65 mi northeast of Poznań on the River Warta; population (1993) 383,600. As Bromberg, it was under Prussian control from 1772 until 1919. 18 H4

Byelorussia alternative name for ►Belarus. 14 G2

Bytom (German **Beuthen**), mining town in Katowice province, Poland, 13 km/8 mi northwest of Katowice; population (1990) 231,200. The district is a large source of coal, lead, and zinc, mined since medieval times, and has engineering industries. Formerly in Upper Silesia, Bytom remained German after the plebiscite of 1921 but was incorporated with Poland 1945. 18 H7

Cabinda (or **Kabinda**), coastal exclave, a province of ►Angola, bounded on the east and south by the Democratic Republic of Congo, on the north by the Republic of the Congo, and on the west by the Atlantic Ocean; area 7,770 sq km/3,000 sq mi; population (1992) 152,100. The capital is Cabinda. There are oil reserves. Products include timber and phosphates. Attached to Angola in 1886, the exclave has made claims to independence. 68 G6

Cabo de Hornos Spanish name for Cape ►Horn, the southernmost point of South America, in Chile. 100 H10

Cabo Fisterra Spanish for ►Cape Finisterre, a cape on the northwest coast of Spain. 28 A2

Cáceres capital of the province of Cáceres, western Spain; population (1991) 71,500. Built on a hill, the old part of the town, surrounded by Roman and Moorish walls, is preserved as a national monument. It has many churches and towers, and medieval and Renaissance mansions. Products include textiles, soap, olive oil, and leather goods, and agriculture is important. 28 D5

Cádiz Spanish city and naval base, capital and seaport of the province of Cádiz, sited on a peninsula on the south side of Cádiz Bay, an inlet of the Atlantic Ocean, 103 km/64 mi south of Seville; population (1991) 153,600. There are ferries to the Canary Islands and Casablanca, Morocco, and shipbuilding and repairs are important, as are fishing and tourism. After the discovery of the Americas in 1492, Cádiz became one of Europe's most vital trade ports.
The English adventurer Francis Drake burned a Spanish fleet here in 1587 to prevent the sailing of the Armada. The city has an 18th-century cathedral. 28 D8

Caen administrative centre of ►Calvados *département* and of the ►Basse-Normandie region, France, on the River Orne, 200 km/124 mi from Paris; population (1990) 115,600, conurbation 189,000. It is a busy port, connected by 11 km/7 mi of canal to the English Channel. The town is also a business centre, with ironworks, manufacturing, electrical, and electronic industries, and produces a building stone that has been used widely since the 11th century. In World War II Caen was one of the main objectives of the D-Day landings and was finally captured by British forces on 9 July 1944 after five weeks' fighting, during which it was badly damaged. Despite this, the town retains many historic buildings, especially churches. The central part of the town was rebuilt in the 1950s. 22 B5

Caernarfon (or **Carnarvon**), administrative centre of ►Gwynedd, north Wales, situated on the southwest shore of the ►Menai Strait; population (1991) 9,700. Formerly the Roman station of **Segontium** (Caer Seint), it is now a market town, port, and tourist centre. Industries include the manufacture of plastics and metal-working. 24 H8

Caerphilly unitary authority in south Wales, created in 1996 from parts of the former counties of Mid Glamorgan and Gwent.
area 270 sq km/104 sq mi

towns Hengoed (administrative headquarters), Caerphilly, Bargoed, Newbridge, Rhymney
physical rivers Rhymney and Sirhowy
industries iron and steel production and coal mining have been replaced by a wide range of light industries
population (1996) 172,000 5 E6

Cagliari (ancient **Carales**), capital and port of Sardinia, Italy, on the south coast of the island on the Gulf of Cagliari; population (1992) 180,300. Industries include oil-refining and the manufacture of petrochemicals. 32 D9

Cagnes-sur-Mer residential and commercial town in the *département* of Alpes-Maritimes, southeast France, situated 10 km/6 mi southwest of Nice; population (1990) 41,300. Its 13th–17th-century château contains mementoes of the Impressionist painter Pierre-Auguste Renoir, who lived here from 1900 to 1919. 30 C7

Cahors French town, capital of the *département* of Lot, 100 km/62 mi north of Toulouse, on the River Lot; population (1990) 20,800. The town's main industries are the manufacture of leather and pottery. It is the centre of a wine-producing area. With many medieval buildings, Cahors is a popular tourist destination. 26 G9

Cairns seaport and resort in northeast Queensland, Australia, 1,733 km/1,077 mi north of Brisbane; population (1996) 92,273. Cairns is the administrative centre for North Queensland, and grazing is an important industry. Sugar, timber, fish, and agricultural produce are exported. Cairns is a centre for tours to the ▶Great Barrier Reef. 76 J3

Cairo (Arabic **El Qahira**, 'the victorious'), capital of Egypt, and the largest city in Africa and in the Middle East, situated on the east bank of the River Nile 13 km/8 mi above the apex of the delta and 160 km/100 mi from the Mediterranean; population (1995 est) 6,955,000. Industries include the manufacture of textiles, cement, vegetable oils, tourism and steel. At Helwan, 24 km/15 mi to the south, an industrial centre is powered by electricity from the ▶Aswan High Dam. 64 F1

Calabar port and capital of Cross River State, southeast Nigeria, on the Calabar River (a tributary of the Cross River), 64 km/40 mi from the Atlantic Ocean; population (1992) 157,800. Rubber, timber, vegetable oils, and cement are exported. It was a centre of the slave trade in the 18th and 19th centuries. 68 F3

Calabria mountainous region occupying the 'toe' of Italy, comprising the provinces of Catanzaro, Cosenza, and Reggio di Calabria; area 15,100 sq km/5,800 sq mi; population (1992) 2,074,800. Its capital is Catanzaro and the principal towns are Crotone and Reggio di Calabria. 9 (4)F5

Calahorra town in the autonomous community of ▶La Rioja, northern Spain; population (1991) 18,600. Products include wine, olive oil, and cattle. Its ancient Roman name was **Calagurris**. From 76–72 BC the town was successfully defended against Pompey by Quintus Sertorius at the head of a rising against Rome by the indigenous Celto-Iberians and Roman exiles. 28 J2

Calais port in ▶Pas-de-Calais *département*, in northern France, 238 km/148 mi north of Paris; population (1990) 75,800, conurbation 100,000. Situated on the Strait of Dover, Calais lies on the shortest crossing of the English Channel, being just 34 km/21 mi southeast of Dover, and is a major ferry port. Its strategic position has made it the victim of several conquests from 1347, when it was conquered by the English, until 1940 when it was occupied by German forces. The entry to the Channel Tunnel, completed in 1994, is at Sangatte, 6 km/3.7 mi from Calais, and the station on the new high-speed railway is at Fréthun, a similar distance from the town centre. 22 D4

Calatayud (Arabic *Kalut Ayub* 'Ayub's castle'), market town in the province of Zaragoza, northeast Spain; population (1995) 17,300. Founded by the Moors, Calatayud has a collegiate church, originally a mosque, and a 12th-century church founded by the Templars. There are rock dwellings here, and the ruins of the Roman town of **Bibilis** lie 3 km/2 mi to the east. 28 J3

Calcutta former name (until 2000) of ▶Kolkata, city in India. 54 E4

Caldey Island (or **Caldy Island**; Welsh **Ynys Bŷr**), island in Carmarthen Bay off the coast of Pembrokeshire, southwest Wales, 4 km/2.5 mi south of Tenby; area 1.82 sq km/0.7 sq mi. It is separated from the mainland by **Caldey Sound**. It was inhabited by Celtic monks as early as the 6th century. A monastery, built by Anglican Benedictines in 1906, is now occupied by Trappist monks from a sect which came here from Belgium in 1929. There is a small village, and a lighthouse on the southern cliffs.

Caledonian Canal waterway in northwest Scotland, 98 km/61 mi long, linking the Atlantic and the North Sea. Situated between the Moray Firth and Loch Linnhe, the canal was constructed as a transport route to save the long sail around Scotland. It is one of Scotland's largest marina facilities. Of its total length, only a 37 km/22 mi stretch is artificial, the rest being composed of lochs Lochy, Oich, and Ness.

Calgary city in southern ▶Alberta, southwest Canada, on the east of the Rocky Mountains, at the junction of the Bow and Elbow rivers; population (1991) 710,700. It is the largest city in Alberta and the regional capital of the southern part of the province. Calgary was founded in 1875 (incorporated as a city in 1893) and grew rapidly with the growth of the petroleum industry. Other industries include tourism, agriculture, petrochemicals, and service industries. 82 J6

Cali industrial city and capital of Valle del Cauca department, western Colombia, 300 km/185 mi southwest of Bogotá, situated in the fertile valley of the River Cauca, 975 m/3,200 ft above sea level; population (1994) 1,687,000. The River Cali, a tributary of the Cauca, flows through the city. Cali is the western region's largest urban centre and third-largest city in Colombia. It is the commercial centre of a rich farming area which produces sugar, coffee, cotton, and rice, and there is a trade in livestock, tobacco, and minerals. Industries include textiles, sugar, and engineering. The colonial town of Cali was founded in 1536 by the Spanish conquistadors, but began to expand as an industrial centre only from 1900, following the arrival of the railway. 98 B3

California western state of the USA. It is nicknamed the Golden State, originally because of its gold mines, and more recently because of its orange groves and sunshine. California was admitted to the Union in 1850 as

the 31st US state. It is bordered to the south by the Mexican state of Baja California, to the east by Arizona and Nevada, to the north by Oregon, and to the west by the Pacific Ocean.
population (2000 est) 33,871,600, the most populous state of the USA (69.9% white; 25.8% Hispanic; 9.6% Asian and Pacific islander, including many Vietnamese, 7.4% African-American; 0.8% American Indian)
area 411,100 sq km/158,685 sq mi
capital ▶Sacramento
towns and cities Los Angeles, San Diego, San Francisco, San Jose, Fresno
industries and products leading agricultural state with fruit (peaches, citrus, grapes in the valley of the San Joaquin and Sacramento rivers), nuts, wheat, vegetables, cotton, and rice, all mostly grown by irrigation, the water being carried by concrete-lined canals to the Central and Imperial valleys; . beef cattle; timber; fish; oil; natural gas; aerospace technology; electronics (Silicon Valley); financial sector; food processing; films and television programmes; tourism; leisure industry; great reserves of energy (geothermal) in the hot water that lies beneath much of the state **84** B4

California, Gulf of (or **Golfo de California**), gulf on the Pacific coast of Mexico, separating the mainland of Mexico from the peninsula of ▶Baja California. It is 1,125 km/700 mi long and 80–210 km/50–130 mi wide, reaching a depth of about 2,595 m/ 8,500 ft. The chief ports on the gulf are Guaymas on the mainland and La Paz on the peninsula. The Colorado River reaches the sea at the gulf. **94** B3

California, Lower British name for ▶Baja California, Mexico. **84** C5

Calisia ancient name of ▶Kalisz, a town in Poland. **18** H6

Callao chief commercial port of Peru, situated 12 km/7 mi northwest of Lima on Callas Bay; sheltered by the island of San Lorenzo; population (1993) 369,800. The port handles 75% of Peru's imports and 25% of its exports. Exports include petroleum, minerals (copper, iron, silver, zinc, lead) cotton, sugar, and coffee. Manufactures include petrochemical products, fertilizers, textiles, and leather goods. It has a dockyard and an oil refinery. Callao is now contiguous with Lima and the combined metropolitan area produces 75% of Peru's manufactures. **98** B6

Calpe name of ▶Gibraltar in ancient Phoenician and Carthaginian times. **28** E8

Caltanissetta town in Sicily, Italy, 96 km/60 mi southeast of Palermo; population (1990) 62,900. It is situated on a high plateau in the middle of the island. Sulphur, potash, and magnesium are mined nearby, and there are mining schools in the town. It has a baroque cathedral (16th–17th centuries), and a ruined castle. The annual Agricultural and Industrial Fair of Central Sicily is held here in May. **32** J11

Calvados maritime *département* in the northwestern region of ▶Basse-Normandie, France; area 5,536 sq km/2,137 sq mi; population (1997) 642,200. Calvados is best known for the local brandy of the same name, which is distilled from cider apples grown in the extensive orchards of the Pays d'Auge, east of the departmental capital of ▶Caen. Other major towns are ▶Bayeux, ▶Lisieux, and the holiday resort of Deauville. **6** C2

Calvi seaport of northwest Corsica, France, 72 km/45 mi north of Ajaccio, in the *département* of Haute Corse; population (1993 est) 4,900. The town's main industries are tourism and wine. It was founded in 1268, and was an important Genoese stronghold. **32** C6

Camagüey city in Cuba; population (1993 est) 294,000. It is the capital of Camagüey province in the centre of the island, linked by rail and road to Havana in the west and Santiago de Cuba in the east. Founded about 1514 as Santa María de Puerto Príncipe, it was the capital of the Spanish West Indies during the 19th century. **94** J4

Camargue area of the ▶Rhône delta enclosed by the two principal arms of the river, south of Arles, France; area about 780 sq km/300 sq mi. One-third of the area is lake or marshland and dykes have been constructed to prevent widespread flooding. Black bulls and white horses are bred, and rice and vines are grown. A nature reserve, known for its bird life, forms the southern part. **26** K10

Cambodia (formerly **Khmer Republic** 1970–76, **Democratic Kampuchea** 1976–79, and **People's Republic of Kampuchea** 1979–89), country in southeast Asia, bounded north and northwest by Thailand, north by Laos, east and southeast by Vietnam, and southwest by the Gulf of Thailand.

Cambodia covers an area of 181,035 sq km/69,897 sq mi, and has a population of 11,168,000 (2000 est). The capital is ▶Phnom Penh; other major towns and cities include Battambang, Kompong Cham, Siemreap, Prey Vêng, and Preah Seihânu. The head of state is King Norodom Sihanouk (from 1991); the head of government is Hun Sen (from 1998). The currency is the Cambodian riel, and the languages spoken include Khmer (official) and French. **50** C4

Cambrai (Flemish **Kambryk**), city in Nord *département*, northern France, situated at the junction of the canalized River Escaut (Schelde) and the St-Quentin canal, 59 km/37 mi southeast of Lille; population (1990) 34,200. Industries include the manufacture of textiles, metal goods, soap, processed foods, confectionery, and beer. Flour and chicory are also produced. Cambric, a fine linen, was first made here, and lace-making is a traditional industry. Under the **Peace of Cambrai** or Ladies' Peace (1529), France regained Burgundy but lost Flanders, Artois, and Milan. **22** F4

Cambrian Mountains region of hills, plateaux, and deep valleys in Wales, 175 km/ 110 mi long, linking Snowdonia in the northwest and the Brecon Beacons and Black Mountains in the south. **24** H10

Cambridge city and administrative headquarters of ▶Cambridgeshire, eastern England, on the River Cam, 80 km/50 mi north of London; population (1997 est) 118,200. It is the seat of Cambridge University (founded in the 13th century). Industries include the manufacture of computers and electronic products, scientific instruments, and paper, printing, publishing, financial services, and insurance, as well as technological, medical, and telecommunications research. Tourism is also an important industry; there are about 3.5 million visitors each year. **24** N9

Cambridge city in northeastern Massachusetts, USA, on the north bank of the Charles River, just above the river's entry into Boston Bay; seat (with Lowell) of Middlesex County; population (1998 est) 93,500. Its seaward side is industrial and includes a part of the Boston port area; industries include paper and publishing, electronic equipment and scientific instruments. Cambridge is the seat of several important colleges: Harvard University (1636), Massachusetts Institute of Technology (1861), and Radcliffe College (1879). The John F Kennedy School of Government is part of Harvard University. One quarter of the residents are students and one sixth of the workforce is employed in higher education. Although noted for its educational institutions, Cambridge used also to be an industrial town. Some of the old factories have been taken over by high tech facilities. 88 F2

Cambridgeshire county of eastern England, which has contained the unitary authority Peterborough since April 1998.
area 3,410 sq km/1,316 sq mi
towns and cities ▶Cambridge (administrative headquarters), Ely, Huntingdon, March, Wisbech, St Neots, Whittlesey
physical county is flat with fens, whose soil is very fertile; Bedford Level (a peaty area of the fens); rivers: Nene, Ouse (with tributaries Cam, Lark, and Little Ouse), Welland
features Cambridge University
agriculture one of the chief cereal and sugar-beet producing districts of England; fruit and vegetables; there is also dairy farming and sheep-rearing
industries brewing, paper, electronics, food processing, mechanical engineering; scientific and pharmaceutical research
population (1996) 703,100 5 F5

Cambs abbreviation for ▶Cambridgeshire, an English county. 5 F5

Cameroon country in west Africa, bounded northwest by Nigeria, northeast by Chad, east by the Central African Republic, south by the Republic of the Congo, Gabon, and Equatorial Guinea, and west by the Atlantic.

Cameroon covers an area of 475,440 sq km/183,567 sq mi, and has a population of 15,085,000 (2000 est). The capital is ▶Yaoundé; other major towns and cities include Garoua, Douala, Nkongsamba, Maroua, Bamenda, Bafoussam, and Ngaoundéré. The head of state is Paul Biya (from 1982); the head of government is Peter Musonge Mafani (from 1996). The currency is the franc CFA, and the languages spoken include French, English (both official; often spoken in pidgin), Sudanic languages (in the north), and Bantu languages (elsewhere). 68 G3

Campagna Romana lowland stretch of the Italian peninsula, including and surrounding the city of Rome. Lying between the Tyrrhenian Sea and the Sabine Hills to the northeast, and the Alban Hills to the southeast, it is drained by the lower course of the River Tiber and a number of small streams, most of which dry up in the summer. Prosperous in Roman times, it later became virtually derelict through overgrazing, lack of water, and the arrival in the area of the malaria-carrying *Anopheles* mosquito. Extensive land reclamation and drainage in the 19th and 20th centuries restored its usefulness.

Campania region of southern Italy, comprising the provinces of Avellino, Benevento, Caserta, Naples, and Salerno; area 13,600 sq km/5,250 sq mi; population (1992) 5,668,900. The administrative capital is ▶Naples; industrial centres include Benevento, Caserta, and Salerno. Agriculture is important; wheat, citrus fruits, wine, vegetables, tobacco, and hemp are produced. The volcano ▶Vesuvius is near Naples, and there are ancient sites at Pompeii, Herculaneum, and Paestum. 9 (4)E4

Campeche, Bay of (or **Bahía de Campeche**), southwestern area of the Gulf of Mexico, site of a major oil-pollution disaster from the field off Yucatán peninsula in 1979. 94 F5

Campinas city of São Paulo, Brazil, situated on the central plateau; population (1991) 960,000. It is a coffee-trading centre. There are also metallurgical and food industries, sugar refining, and textiles. 100 M3

Campobasso capital of Molise region, Italy, about 190 km/120 mi southeast of Rome; population (1990) 51,300. An agricultural market town, its principal industry is the manufacture of cutlery. The old town is built on a hill crowned with a 16th-century castle, and there are several fine churches. The Sagra di Misteri festival is held annually at Corpus Christi in June. 32 J7

Cam Ranh port in South Vietnam; population (2001 est) 161,000. The port has one of the best deepwater harbours in southeast Asia and lies on a peninsula sheltering Cam Ranh Bay, which is part of the South China Sea. 50 D4

Canada country occupying the northern part of the North American continent, bounded to the south by the USA, north by the Arctic Ocean, northwest by Alaska, east by the Atlantic Ocean, and west by the Pacific Ocean.

Canada covers an area of 9,970,610 sq km/3,849,652 sq mi, and has a population of 31,147,000 (2000 est). The capital is ▶Ottawa; other major towns and cities include Toronto, Montréal, Vancouver, Edmonton, Calgary, Winnipeg, Québec, Hamilton, Saskatoon, Halifax, London, Kitchener, Mississauga, Laval, and Surrey. The head of state is Queen Elizabeth II (from 1952), represented by Governor General Adrienne Clarkson (from 1999); the head of government is Jean Chrétien (from 1993). The currency is the Canadian dollar, and the languages spoken include English, French (both official), American Indian languages, and Inuktitut (Inuit). 80 M4

Çanakkale Boğazi Turkish for the ▶Dardanelles, a strait connecting the Sea of Marmara with the Aegean Sea. 36 J4

Canary Islands (Spanish **Islas Canarias**), group of volcanic islands and autonomous Spanish community 100 km/60 mi off the northwest coast of Africa, comprising the provinces of Las Palmas and Santa Cruz de Tenerife; area 7,300 sq km/2,818 sq mi; population (1991) 1,456,500. Products include bananas and tomatoes, both grown for export. Tourism is the major industry.
features The chief centres are Santa Cruz on Tenerife (which also has the highest Spanish peak, Pico de Teide; height 3,718 m/12,198 ft), and Las Palmas on Gran Canaria. The province of Santa Cruz comprises Tenerife, La Palma, Gomera, and Hierro; the province of Las Palmas comprises Gran Canaria, Lanzarote, and Fuerteventura. There are also six uninhabited islets (Graciosa, Alegranza, Montaña Clara, Roque del Oeste, Roque del Este, and Lobos). The Northern Hemisphere Astronomical Observatory

(1981) is on the island of La Palma. Observation conditions are exceptionally good here because there is little moisture, no artificial-light pollution, and little natural airglow.

The aboriginal inhabitants of the Canary Islands were called Guanches, and the Organization of African Unity (OAU) supports the creation of an independent state, the Guanch Republic, and the revival of the Guanch language. 62 A3/9 (2)B1

Canberra capital of ▶Australia and seat of the federal government, situated in the ▶Australian Capital Territory in southeast Australia; population (1996) 299,243. Canberra is enclosed within the state of New South Wales, 289 km/180 mi southwest of Sydney and 655 km/407 mi northeast of Melbourne, on the River Molonglo, a tributary of the Murrumbidgee. It succeeded Melbourne as capital of Australia in 1927. It is an administrative, cultural, and tourist centre. The new Parliament House (1988) is located here, as well as government offices, foreign embassies, and many buildings of national importance. 76 J7

Cancún resort in Yucatan, Mexico, created on a barrier island 1974 by the Mexican government to boost tourism; population around 30,000 (almost all involved in the tourist industry). It is Mexico's most popular tourist destination, with beaches, a coral reef, and a lagoon. The hotel zone is on the sandbar, and the workers live in Cancún city on the mainland. 94 G4

Cannes resort in Alpes-Maritimes *département* of southern France, on the Mediterranean coast and 21 km/13 mi southwest of Nice; population (1990) 69,400, Grasse-Cannes-Antibes conurbation 335,000. Formerly a small fishing village and seaport, in 1834 Cannes attracted the patronage of the English Lord Brougham and other distinguished visitors and soon grew into a popular winter and summer holiday resort on the French ▶Riviera. The prestigious Cannes Film Festival is held here annually. The city has textile and aircraft industries, and a strong trade in olive oil, soap, fish, fruit, and flowers. 30 C7

Cantabria autonomous community of northern Spain; area 5,300 sq km/ 2,046 sq mi; population (1991) 527,300. From the coastline on the Bay of Biscay it rises to the Cantabrian Mountains. There is some mining here, as well as engineering and food industries, particularly dairy products. The capital is ▶Santander. 9 (1)C1

Cantabrian Mountains (Spanish **Cordillera Cantábrica**), mountain range running along the north coast of Spain, continuing the line of the Pyrenees westwards for about 480 km/300 mi, parallel to the coast of the Bay of Biscay. They rise to 2,648 m/8,688 ft in the Picos de Europa massif. The mountains contain coal and iron deposits, but they are of poor quality; little is mined today. 28 D2

Cantal *département* in the ▶Auvergne region of central France; area 5,741 sq km/ 2,217 sq mi; population (1990) 158,761. Cantal takes its name from the mountain range in the region. The principal towns are Aurillac (the capital), Mauriac, and St-Flour. Agriculture is the main economic activity; cattle are reared, and rye, potatoes, and chestnuts are grown. A hard cheese named after the *département* is produced. 6 E4

Canterbury (Old English *Cantwarabyrig* 'fortress of the men of Kent'), historic cathedral city in Kent, southeast England, on the River Stour, 100 km/62 mi southeast of London; population (1991) 36,500. The city is the metropolis of the Anglican Communion and seat of the archbishop of Canterbury. The University of Kent is here. It is a popular tourist destination. Paper, paper products, and electrical goods are manufactured. 22 D3

Canterbury Plains area of rich grassland between the mountains and the sea on the east coast of South Island, New Zealand; area 10,000 sq km/4,000 sq mi. Canterbury lamb is produced here. 78 C6

Canton alternative spelling of ▶Guangzhou or Kwangchow, capital of Guangdong province in China. 50 E2

Canton city and administrative headquarters of Stark County, northeast Ohio, USA, on Nimishillen Creek, 84 km/52 mi southeast of Cleveland; population (1992) 84,800. The city is a major producer of metal alloys and steel. Other manufactures include metalworking presses, roller bearings, brick, and rubber gloves. It is also a trade centre for the surrounding dairy and poultry farming region. 90 E1

Cape Breton Island island forming the northeastern part of the province of Nova Scotia, Canada, at the eastern extremity of Nova Scotia Island; area 10,311 sq km/3,981 sq mi; population (1991) 161,800. Divided by a waterway, it has road and rail links with the mainland across the Strait of Canso. Cape Breton Island has coal resources and steelworks, and there has been substantial development in the strait area, with docks, oil refineries, and newsprint production from local timber, as well as tourism and fishing. The first European to discover the island was the English navigator, John Cabot, who arrived here in 1497. 82 U7

Cape Byron eastern extremity of Australia, in New South Wales, just south of the border with Queensland. 76 K5

Cape Coast port and capital of Central Region, Ghana, 130 km/80 mi west of Accra; population (1998 est) 89,800. In 1962 it was superseded as the main port by Tema. The town is built on a natural breakwater on the Gulf of Guinea, adjoining Cape Coast Castle. It exports cocoa, palm oil and kernels, coconuts, citrus fruits, corn, and cassava, and is a fishing port, exporting frozen fish. It is the site of Ghana's earliest university, the University of Cape Coast. 68 D3

Cape Cod hook-shaped peninsula in southeastern Massachusetts, USA, separated from the rest of the state by the Cape Cod Canal; length 100 km/62 mi; width 1.6–32 km/ 1–20 mi. Its beaches and woods make it a popular tourist area. The islands of Martha's Vineyard and ▶Nantucket are just south of the cape. 88 G2

Cape Finisterre (Spanish **Cabo Fisterra**; Latin *finis terrae* 'land's end'), Atlantic cape on the northwest coast of Spain, in the province of La Coruña. It is the most westerly point in Spain, and its lighthouse is visible to vessels from a distance of 80 km/50 mi. 28 A2

Cape Hatteras cape on the coast of eastern North Carolina, USA, which projects into the Atlantic from **Hatteras Island**. This is where the cold waters of the North Atlantic meet the warm Gulf Stream, causing great turbulence; more than 700 shipwrecks are said to have occurred here, and the area is nicknamed 'the Graveyard of the Atlantic'. Cape Hatteras lighthouse was built in 1870, and in 1999 had to be relocated 488 m/1,600 feet from the shoreline as a result of beach erosion. A new lighthouse offshore and global positioning satellite navigation have rendered the old lighthouse obsolete, but it remains the nation's tallest brick lighthouse and a tourist attraction. **90** F2

Cape St Vincent (Portuguese **Cabo de São Vicente**), cape of the Algarve region, southwest Portugal. It is the extreme southwesterly point of continental Europe. England defeated the French and Spanish fleets off the cape in 1797, during the Napoleonic Wars. **28** A7

Cape Town (Afrikaans **Kaapstad**), port and oldest city (founded in 1652) in South Africa, situated at the northern end of the Cape Peninsula, on Table Bay; population (1991) 854, 616 (urban area); (1991) 2, 350, 200 (peninsula). Industries include horticulture and trade in wool, wine, fruit, grain, and oil. Tourism is important. It is the legislative capital of the Republic of South Africa and capital of ▶Western Cape province. **72** B6

Cape Verde country formed by a group of islands in the Atlantic, west of Senegal (West Africa).

Cape Verde covers an area of 4,033 sq km/1,557 sq mi, and has a population of 428,000 (2000 est). The capital is ▶Praia; other major towns and cities include Mindelo and Santa Maria. The head of state is Antonio Mascarenhas Monteiro (from 1991); the head of government is Gualberto do Rosário (from 2000). The currency is the Cape Verde escudo, and the languages spoken include Portuguese (official) and Creole. **68** (1)B2

Cape Wrath (Norse **huaf** 'point of turning'), headland at the northwest extremity of Scotland, extending 159 m/523 ft into the Atlantic Ocean. It is one of a series of rugged cliffs formed of gneiss. Its lighthouse dates from 1827. **24** G3

Cape York peninsula, the northernmost point (10° 41' S) of the Australian mainland, named by English explorer Capt James Cook in 1770. The peninsula is about 800 km/500 mi long and 640 km/400 mi wide at its junction with the mainland. Its barrenness deterred early Dutch explorers, although the south is being developed for cattle (Brahmin type). In the north there are large bauxite deposits. **76** H2

Capri Italian island at the southern entrance of the Bay of Naples; 32 km/20 mi south of Naples; area 13 sq km/5 sq mi; population (1987 est) 7,800. It has two towns, Capri and Anacapri, a profusion of flowers, beautiful scenery, and an ideal climate. The Blue Grotto on the north coast is an important tourist attraction. **32** J8

Capua Italian town in Caserta province on the Volturno River, in a fertile plain north of Naples; population (1990) 19,500. There was heavy fighting here 1943 during World War II, which almost destroyed the Romanesque cathedral. **32** J7

Caracas chief city and capital of ▶Venezuela, situated in the Central Highlands of the Andes Mountains 900 m/2,950 ft above sea level, 13 km/8 mi south of its port La Guaira on the Caribbean coast; population of metropolitan area (1996) 3,007,000 (Federal District 2,265,900). Main industries include oil refining, textiles, chemicals, and food processing. During the oil boom of the 1950s Caracas developed rapidly, its rate of growth greater than that of any other South American capital; much of the old colonial town was largely effaced during this period. It is now a large modern industrial and commercial centre, notably for oil companies, developed since the 1950s. **98** D1

Carcassonne city in southwest France, administrative centre of ▶Aude *département*, 100 km/62 mi southeast of Toulouse on the Canal du Midi and the River Aude, which divides it into the ancient and modern town; population (1990) 45,000. The medieval fortification, restored by Viollet-le-Duc in the 19th century, crowns a hill on the right bank of the river. Surrounded by two ramparts and 53 towers, it is the finest in France and one of the most impressive medieval fortified towns surviving. **26** H10

Cardiff unitary authority in south Wales, created in 1996 from part of the former county of South Glamorgan; administrative headquarters is ▶Cardiff.

area 139 sq km/54 sq mi

population (1996 est) 315,000 **5** E6

Cardiff (Welsh **Caerdydd**), seaport, capital of Wales (from 1955), and administrative centre of ▶Cardiff unitary authority, situated at the mouth of the Taff, Rhymney, and Ely rivers; population (1996 est) 315,000. It is the seat of government for the National Assembly for Wales, which was established in 1998 following the Government of Wales Act passed by the British parliament. Industries are predominantly in the service sector, with 82% of the workforce employed in services, the largest of which are education, business services, health, and consumer services. Manufacturing has declined in importance, although electronics and motor components remain important industries. **24** J10

Caribbean Sea western part of the Atlantic Ocean between Cuba to the north and the northern coasts of South America to the south. Central America is to the west and to the east are the West Indies. The sea is about 2,740 km/1,700 mi long and 650–1,500 km/400–900 mi wide; area 2,640,000 sq km/1,019,304 sq mi. It is linked with the Gulf of Mexico via the Yucatan Strait. It is from here that the Gulf Stream turns towards Europe. **94** J6

Carinthia (German **Kärnten**), federal state of alpine southeast Austria, bordering Italy and Slovenia in the south; area 9,500 sq km/3,667 sq mi; population (1994) 559,700. The capital is Klagenfurt. Mining, stock-raising and forestry are important. It was an independent duchy from 976 and a possession of the Habsburg dynasty from 1276 until 1918. **8** (2)E4

Carlisle city and administrative headquarters of Cumberland County, south Pennsylvania, in the Cumberland Valley, 34 km/21 mi southwest of Harrisburg; population (1990) 18,400. Industries include carpets, shoes, tyres, electronics, and steel products. Carlisle was the site of a noted Indian school (1879–1918) attended by athlete Jim Thorpe, and has long been the home of Dickinson College (1773). The US Army War College moved to the city in 1951. **88** E2

Carlisle city and administrative headquarters of ▶Cumbria, northwest England, on the River Eden at the western end of Hadrian's Wall, 14 km/9 mi south of the Scottish border; population (1991) 72,400. It is a leading railway and service centre. Industries include engineering and brewing (Scottish and Newcastle), and the manufacture of textiles, agricultural machinery, metal goods, confectionery, and processed foods. Carlisle was the Roman settlement of **Luguvalium**. There is a Norman cathedral and a restored castle dating from 1092. **24** K7

Carlow second-smallest county in the Republic of Ireland, in the province of Leinster; county town Carlow; area 900 sq km/347 sq mi; population (1996) 41,600. The land is mostly flat except for the Blackstairs mountains in the south (rising to 796 m/2,612 ft in Mount Leinster). The land in the west is fertile and well suited to dairy farming. Products include barley, wheat, and sugar beet. **5** C5

Carmarthenshire (Welsh **Sir Gaerfyrddin**), unitary authority in south Wales; a former county, it was part of Dyfed between 1975 and 1996.
area 2,390 sq km/923 sq mi
towns Carmarthen (administrative headquarters), ►Llanelli
physical rivers Tywi, Taf, Teifi; Black Mountain range in the east, southern spur of the Cambrian Mountains in the north, including Mynydd Mallaen (459 m/1,1,506 ft); along the coast are extensive sands and marshes. Carmarthenshire is dominated by the Vale of Tywi, but there are numerous grassy hills, mostly under 300 m/1,000 ft; the valleys are fertile and the hillsides afford good pasturage
features Brecon Beacons National Park on the eastern border; Museum of the Woollen Industry at DreFach-Felindre; home of Welsh poet Dylan Thomas in the village of Laugharne, 6 km/3.7 mi southeast of St Clears
agriculture dairy farming, stock-raising
population (1996) 68,900 **5** D6

Carnarvon alternative spelling of ►Caernarfon, a town in Wales. **24** H8

Carolina either of two separate states of the USA; see ►North Carolina and ►South Carolina. **90** F2/**90** E3

Caroline Islands scattered archipelago in Micronesia, Pacific Ocean, consisting of over 500 coral islets; area 1,200 sq km/463 sq mi. The chief islands are Ponape, Kusai, and Truk in the eastern group, and Yap and Palau in the western group. **74** E5

Carpathian Mountains central European mountain system, forming a semicircle through Slovakia–Poland–Ukraine–Moldova–Romania, 1,450 km/900 mi long. The central **Tatra Mountains** on the Slovak–Polish frontier include the highest peak, Gerlachovka, 2,663 m/8,737 ft. **18** J8

Carpentaria, Gulf of shallow gulf opening out of the Arafura Sea, between the capes of Arnhem and York, north of Australia; 600 km/373 mi long, 490 km/304 mi wide. The first European to reach it was the Dutch navigator Abel Tasman in 1606 and it was named in 1623 in honour of Pieter Carpentier, governor general of the Dutch East Indies. **76** G2

Carpentras French town in the *département* of Vaucluse, on the River Auzun, and on the **Carpentras Canal**, 21 km/13 mi northeast of Avignon; population (1990) 24,200. It has a textile industry, tourism, and an important market for fruit and vegetables. Carpentras has a late-Gothic church (1405–1519), an 18th-century synagogue, and an 18th-century aqueduct. There is also a Roman triumphal arch. **26** L9

Carrara town in Tuscany, Italy, in the foothills of the Apennines, 60 km/37 mi northwest of Livorno; population (1990) 68,500. It is known for its quarries of fine white marble, which were worked by the Romans, abandoned in the 5th century, and came into use again with the revival of sculpture and architecture in the 12th century. **30** F6

Carrickfergus seaport on Belfast Lough, County Antrim, Northern Ireland; population (1991) 32,800. There is some light industry, and the town is a major centre for the man-made fibre industry. The well-preserved **Carrickfergus Castle** was begun in 1180, and now houses a museum. The church of St Nicholas dates partly from the 12th century. The port has a large marina and a sailing school. Carrickfergus was the main port of medieval Ulster but declined from the 17th century onwards, with the development of Belfast. **24** G7

Carson City capital of ►Nevada, situated at an altitude of 1,425 m/4,674 ft near the eastern foothills of the Sierra Nevada, 48 km/30 mi south of Reno, and to the east of Lake Tahoe; population (1992) 42,800. The mainstays of the economy are tourism and legalized gambling, although mining and livestock raising are still important. **86** C3

Cartagena (or **Cartagena de los Indes**), historic port, industrial centre, and capital of Bolívar department, on the Caribbean coast of northwest Colombia; population (1994) 726,000. There are petrochemical, textile, and pharmaceutical industries; oil and coffee are exported. The city is also a fashionable tourist resort, with beaches, lakes, and inland lagoons. There is a 16th-century cathedral, and several 16th- and 17th-century churches, and a state university, founded in 1827. The fortress 'Castillo de San Felipe de Barajas' was constructed over a period of 150 years, commencing in 1639. **98** B1

Cartagena city in the autonomous community of ►Murcia, southeast Spain, on the Mediterranean coast; population (1994) 180,000. It is a seaport and naval base, and has metallurgical, petrochemical, and glass industries. As the naval base of the Republicans it was bombed during the Spanish Civil War (1936–39). It has a 13th-century cathedral and Roman remains. **28** K7

Casablanca (Arabic **Dar el-Beida**), port, commercial, and industrial centre on the Atlantic coast of Morocco; population (1993) 2,943,000. Casablanca is one of the major ports of Africa, and the industrial centre of Morocco. It trades in fish, phosphates, and manganese. The Great Hassan II Mosque, completed in 1989, is the world's largest; it is built on a platform (40,000 sq m/430,000 sq ft) jutting out over the Atlantic, with walls 60 m/200 ft high, topped by a hydraulic sliding roof, and a minaret 175 m/574 ft high. **66** D2

Cascade Range volcanic mountains in northwestern USA and southwestern Canada, extending 1,120 km/700 mi from northern California through Oregon and Washington as far as the Fraser River in British Columbia, Canada. The highest peaks are Mount Rainier (4,392 m/14,408 ft), Mount Shasta (4,317 m/

14,163 ft), and Mount Adams (3,751 m/12,306 ft). They are the most active volcanoes in the USA, excluding Alaska and Hawaii; Mount St Helens in Washington (2,549 m/ 8,364 ft), which had been dormant since 1857, erupted violently in 1980 and 1982. **86** B2

Cascais fishing port and resort town on the Costa do Sol, 25 km/16 mi west of Lisbon, Portugal; population (1991) 14,500. The 17th-century citadel is the summer palace of the president. There are local fishing and canning industries here; tourism is important. **28** A6

Caserta town in Campania, southern Italy, 33 km/21 mi northeast of Naples; population (1990) 69,300. It trades in textiles, chemicals, olive oil, wine, and grain. The base for Garibaldi's campaigns in the 19th century, it was the Allied headquarters in Italy from 1943 to 1945, when the German forces surrendered here to Field Marshal Alexander. **32** J7

Caspe town in the province of Zaragoza, northwest Spain, near the confluence of the Guadalope and Ebro rivers; population (1991) 8,000. Its castle was the scene of the **Compromise of Caspe** in 1410, when the delegates of Aragón, Catalonia, and Valencia chose Ferdinand of Antequera as king (Ferdinand I of Aragón). **28** K3

Casper city and administrative headquarters of Natrona County, east-central Wyoming, on the North Platte River, 212 km/132 mi northwest of Cheyenne; population (1990) 46,700. Oil and natural gas refining are Casper's main industries. Others include shipping and wholesaling livestock and agricultural products, and servicing the local mining of uranium, coal, and bentonite. The city is also a financial and medical centre. It is home to Casper College (1945) and the Tate Earth Science and Mineralogical Museum. **86** E2

Caspian Sea world's largest inland sea, on the border between Europe and Asia east of the Black Sea, divided between Iran, Azerbaijan, Russia, Kazakhstan, and Turkmenistan. It extends north–south for 1,200 km/ 745 mi, and its average width is 300 km/186 mi; area about 400,000 sq km/155,000 sq mi, with a maximum depth of 1,000 m/3,250 ft. An underwater ridge divides it into two halves, of which the shallow northern half is almost salt-free. There are no tides, but violent storms make navigation hazardous. The chief ports are Astrakhan (Russia), Baku (Azerbaijan), and Bandar Shah (Iran). The River Volga supplies 80% of freshwater inflow; the Ural, Emba, Terek, Kura, and Atrek rivers also flow into the Caspian Sea. Prolonged drought, drainage in the north, and regulation of the Volga and Kura rivers reduced the area from 430,000 sq km/ 166,000 sq mi in 1930 to 382,000 sq km/147,000 sq mi in 1957, and left the sea approximately 28 m/90 ft below sea level. In June 1991 opening of sluices in the dams caused the water level to rise dramatically, threatening towns and industrial areas. **14** J3

Cassel alternative spelling of ►Kassel, an industrial town in Germany. **20** E5

Cassino town in Lazio, southern Italy, 80 km/50 mi northwest of Naples, at the foot of Monte Cassino; population (1990) 34,600. Industries include engineering, printing, food-processing, and papermaking. Cassino was the scene of heavy fighting in 1944 during World War II, as it was a key position blocking the Allied advance to Rome, and most of the town was destroyed. It was rebuilt slightly south of its former site. The abbey on the summit of Monte Cassino, founded by St Benedict in 529, was rebuilt in 1956 to repair the damage caused by Allied bombardment. **32** H7

Castellón de la Plana (or **Castelló de la Plana**), port and capital of Castellón province, in the Valencian Community, eastern Spain; population (1991) 133,200. It is in the centre of a fertile, irrigated district; oranges, onions, carobs, and brandy are produced. Manufactures include textiles, paper, porcelain, and chemicals. There are several churches, a museum, a library, and gardens. **28** K5

Castilla–La Mancha autonomous community of central Spain; area 79,200 sq km/30,600 sq mi; population (1991) 1,658,400. It includes the provinces of Albacete, Ciudad Real, Cuenca, Guadalajara, and Toledo. Irrigated land produces mainly cereals and vines, especially in the Valdepeñas region, and merino sheep are raised. The capital is ►Toledo. **9** (1)D3

Castilla–León (or **Castilla y Léon**), autonomous community of central Spain; area 94,100 sq km/ 36,300 sq mi; population (1991) 2,545,900. It includes the provinces of Avila, Burgos, León, Palencia, Salamanca, Segovia, Soria, Valladolid, and Zamora. Irrigated land produces wheat and rye; cattle, sheep, and fighting bulls are bred in the uplands. There are important food industries in Burgos, Palencia, and Segovia provinces. The capital is ►Valladolid, which is the main manufacturing centre, particularly for engineering and the production of motor vehicles. **9** (1)C2

Castleford town in Wakefield, West Yorkshire, northern England, at the confluence of the rivers Aire and Calder, 15 km/10 mi southeast of Leeds; population (1991) 36,000. A former coalmining town, Castleford now manufactures chemicals, glass, clothing, flour, confectionary, and bakery products. Under the Romans it was **Lagentium** on the Roman road Watling Street. **24** L8

Castres French town in the *département* of Tarn, on the River Agout; population (1990) 47,000. Its main industries include textiles, metals, pottery, and mechanical and electrical goods. It grew up around a 7th-century Benedictine abbey. In the 16th century it was a Huguenot religious stronghold. **26** H10

Castries (or **Port Castries**), capital and port of ►St Lucia, on the northwest coast of the island; population (2001 est) 59,600. From its almost enclosed harbour, it exports sugar cane, bananas, limes, coconuts, cacao, and rum. The town processes foodstuffs and drinks, and manufacturing industries include tobacco, textiles, wood, rubber and metal products, chemicals, and printing. The nearby Vigie airport has helped the rise of a significant tourist industry. **94** M6

Catalonia alternative spelling for ►Cataluña. **9** (1)F2

Cataluña (or **Catalonia**; **Catalan Catalunya**), autonomous community of northeast Spain; area 31,900 sq km/ 12,300 sq mi; population (1991) 6,059,500. It includes the provinces of Barcelona, Girona (formerly Gerona), Lleida (formerly Lérida), and

Tarragona. Olives, vines, cereals, and nuts are grown, and some livestock is raised. Cataluña is the main industrial region of Spain. Originally based on the textile industry, the region has diversified into engineering, chemicals, paper, publishing, and many service industries; hydroelectric power is also produced. The capital is ▶Barcelona. 9 (1)F2

Catalunya Catalan name for ▶Cataluña. 9 (1)F2

Catania industrial port in Sicily on the Gulf of Catania, just south of Mount Etna; population (1992) 329,900. It is the capital of the province of Catania. It exports local sulphur; there are also shipbuilding, textile, and light industries. 32 K11

Catskill Mountains mountain range in southeastern New York State, USA, forming part of the ▶Appalachian Mountains west of the Hudson River; the highest peaks are Slide Mountain (1,281 m/4,204 ft) and Hunter Mountain (1,227 m/4,026 ft). The Catskill Mountains have long been a vacation and resort centre for residents of New York City, offering hiking, skiing, hunting and fishing. The valleys, which have been carved out by glaciation, are the chief scenic attractions; much of the mountain area is included in the Catskill Forest Preserve. 84 M3

Caucasia (or **Caucasus**; Russian **Kavkaz**), geographical area between the Black, Azov, and Caspian Seas, covering some 440,000 sq km/ 169,884 sq mi. Northern Caucasia lies within the Russian Federation, while ▶Transcaucasia encompasses the territory of Georgia, Azerbaijan, and Armenia. The northern Caucasus region is home to some 40 different ethnic groups.

Caucasus mountain range extending from the Taman Peninsula on the Black Sea to the Apsheron Peninsula on the Caspian Sea, a total length of 1,200 km/750 mi. The Caucasus, which form the boundary between Europe and Asia, is divided into the **Greater Caucasus** (northern) and **Little Caucasus** (southern) chains. The range crosses the territory of the Russian Federation, Georgia, Armenia, and Azerbaijan. At 5,642 m/18,510 ft, Elbrus (in the Greater Caucasus) is the highest peak in Europe. 58 K2

Cavan county of the Republic of Ireland, in the province of Ulster; county town Cavan; area 1,890 sq km/730 sq mi; population (1996) 52,900. The chief rivers are the Woodford, the Shannon (rising on the south slopes of Cuilcagh mountain; 667 m/2,188 ft), and the Erne, which divides Cavan into two parts: a narrow, mostly low-lying peninsula, 30 km/19 mi long, between Leitrim and Fermanagh; and an eastern section of wild and bare hill country. The chief towns are Cavan, population (1996) 3,500, and Kilmore, seat of Roman Catholic and Protestant bishoprics. The soil is generally poor and the climate moist and cold. 5 C5

Cawnpore former spelling of ▶Kanpur, a city in India. 54 D3

Cayenne capital and chief port of the overseas *département* of ▶French Guiana in South America; situated on Cayenne Island on the Atlantic coast at the mouth of the River Cayenne; population (1995 est) 51,300. The main occupation is fishing, of which fresh and processed shrimp constitute nearly 75% of total exports by value. Rum, pineapples, hardwoods, and cayenne pepper – a main constituent of hot curries – are also exported. Many imports pass through the port as the country is very much import-dependent. 98 G3

Cayman Islands British island group in the West Indies; area 260 sq km/ 100 sq mi; population (1993 est) 31,150 (mostly on Grand Cayman). The Caymans comprise three low-lying islands: Grand Cayman, Cayman Brac, and Little Cayman. The capital is ▶George Town (on Grand Cayman). The islands export seawhip coral (a source of prostaglandins), shrimps, honey, and jewellery. 94 H5

Cebu chief city and port of the island of Cebu in the Philippines; population (1995 est) 681,000; area of the island 5,086 sq km/1,964 sq mi. The first Spanish settlement in the Philippines, Cebu was founded as San Miguel in 1565 and became the capital of the Spanish Philippines. Industries include food processing and textiles, and there is a considerable trade in copra and tobacco. 50 G4

Cedar Falls city in Black Hawk County, northeast Iowa, on the Cedar River, 15 km/9 mi northwest of Waterloo; population (1998 est) 34,730. The city is mainly residential, with some industry, including the manufacture of pumps, farm equipment, golf supplies, humidifiers, and air conditioning and heating grilles. It is the home of the University of Northern Iowa (1876). 88 B2

Cedar Rapids city and administrative headquarters of Linn County, east-central Iowa, on the Cedar River, 170 km/105 mi northeast of Des Moines; population (1994 est) 113,000. It produces communications equipment, construction machinery, and processed foods, and is the hub of the surrounding agricultural region. In 1998 there were 22,000 manufacturing workers in the region. Quaker Oats operates a huge cereal mill here. Noted as a centre of Czech and Slovak migration, the Czech village in the city centre dates from 1906 and is now a tourist attraction. 88 B2

Cegléd town in Pest county, Hungary, 72 km/45 mi southeast of Budapest;population (2001 est) 36,100. It is a trading centre for agricultural and market-garden produce, and has distillery, flour, and brick industries. 34 G2

Celebes English name for ▶Sulawesi, an island of Indonesia. 53 A3

Celje town in Slovenia, on the River Savinja; population (1991) 40,200. It is a zinc-smelting and metallurgical centre. It was founded by the Romans as Claudia Celea, and retains its medieval walls. 34 C3

Celle (or **Zelle**), city in Lower Saxony, Germany, on the River Aller, 35 km/22 mi northeast of Hannover; population (1994) 177,500. The city has a stud farm founded in 1735, and manufactures dyes, foodstuffs, furniture, leather, and electrical goods. The city dates from 1292. 20 F4

Celtic Sea sea area bounded by Wales, Ireland, and southwest England; the name is commonly used by workers in the oil industry to avoid nationalist significance. The Celtic Sea is separated from the Irish Sea by St George's Channel. 24 E10

Central former region of Scotland (1975–96); replaced in 1996 by Clackmannanshire, Falkirk, and Stirling unitary authorities.

Central African Republic landlocked country in Central Africa, bordered northeast and east by Sudan, south by the Democratic Republic of

Congo and the Republic of the Congo, west by Cameroon, and northwest by Chad.

The Central African Republic covers an area of 622,436 sq km/ 240,322 sq mi, and has a population of 3,615,000 (2000 est). The capital is ►Bangui; other major towns and cities include Berbérati, Bouar, Bambari, Bossangoa, Carnot, and Kaga-Bandoro. The head of state is Ange-Felix Patasse (from 1993); the head of government is Anicet Georges Dologuele (from 1999). The currency is the franc CFA, and the languages spoken include French (official), Sangho (national), Arabic, Hunsa, and Swahili. **70** C2

Central America the part of the Americas that links Mexico with the Isthmus of Panama, comprising ►Belize and the republics of ►Costa Rica, ►El Salvador, ►Guatemala, ►Honduras, ►Nicaragua, and ►Panama.

It is also an isthmus, crossed by mountains that form part of the Cordilleras, rising to a maximum height of 4,220 m/13,845 ft. There are numerous active volcanoes. The principal river is the Usumacinta, which rises in Guatemala and flows north for 965 km/600 mi, crossing Mexico, and empties into the Bay of Campeche in the Gulf of Mexico. Central America has an area of about 523,000 sq km/200,000 sq mi, and a population estimated (1995) at 33,132,000, comprising mostly Indians or mestizos (of mixed white–Indian ancestry), with the exception of Costa Rica, which has a predominantly white population. Tropical agricultural products, raw materials, and other basic commodities are exported.

Central Asian Republics geographical region covering the territory of five nation-states: ►Kazakhstan, ►Kyrgyzstan, ►Tajikistan, ►Turkmenistan, and ►Uzbekistan. These republics were part of the Soviet Union before gaining their independence in 1991. Central Asia is bordered on the north by the Russian Federation, on the south by Iran and Afghanistan, and on the east by the Chinese region of Xinjiang Uygur. The western boundary of Central Asia is marked by the Caspian Sea. The topography of the region is characterized by several major mountain ranges, including the ►Tien Shan range and the Pamirs, and extensive deserts, principally the ►Kara-Kum and ►Kyzyl-Kum. The people of Central Asia are predominantly Muslim.

Central Lowlands one of the three geographical divisions of Scotland, being the fertile and densely populated plain that lies between two geological fault lines, which run nearly parallel northeast–southwest across Scotland from Stonehaven to Dumbarton and from Dunbar to Girvan.

Centre region of north-central France; area 39,150 sq km/15,116 sq mi; population (1990) 2,371,000. Centre includes the *départements* of ►Cher, ►Eure-et-Loir, ►Indre, ►Indre-et-Loire, ►Loire-et-Cher, and ►Loiret. The administrative centre is ►Orléans. **6** D3

Cephalonia English form of ►Kefallonia, the largest of the Ionian Islands, off the west coast of Greece. **36** C6

Ceram alternative form of ►Seram, an Indonesian island. **53** D3

Ceredigion unitary authority in southwest Wales, created in 1996 from part of the former county of Dyfed, of which it was a district.
area 1,793 sq km/ 692 sq mi
towns Aberaeron (administrative headquarters), ►Aberystwyth, Cardigan, Lampeter, Llandyssul, Tregaron
physical part of the Cambrian Mountains, including Plynlimon Fawr (752 m/2,468 ft); rivers Teifi, Rheidol, Ystwyth, Aeron, and Tywi
features remains of Roman roads and military stations, and inscribed stones; ruins of 12th-century Strata Florida Abbey southeast of Aberystwyth; Devil's Bridge (spanning the Rheidol Falls)
industries tourism, woollens production
agriculture sheep-rearing, dairy production
population (1996) 68,900 **5** D5

České Budějovice (German **Budweis**), town in the Czech Republic, on the River Vltava; population (1991) 123,400. It is a commercial and industrial centre for southern Bohemia, producing beer, timber, and metal products. **30** K2

Cetinje town in Montenegro, Yugoslavia, 19 km/12 mi southeast of Kotor; population (1991) 15,950. Founded in 1484 by Ivan the Black, it was the capital of Montenegro until 1918. It has a palace built by Nicholas, the last king of Montenegro. **34** F7

Ceuta Spanish seaport and military base in Morocco, 27 km/17 mi south of Gibraltar and overlooking the Mediterranean approaches to the Straits of Gibraltar; area 18 sq km/ 7 sq mi; population (1991) 67,600. It trades in petrol products. **66** D1

Cévennes series of mountain ranges on the southern, southeastern, and eastern borders of the Massif Central in France, forming the watershed between the Rhône and Garonne rivers. The total length of the ranges is 530 km/330 mi. The highest peak is Mount Mézenc (1,754 m/5,755 ft).

Ceylon former name (to 1972) of ►Sri Lanka. **54** D7

Chad landlocked country in central North Africa, bounded north by Libya, east by Sudan, south by the Central African Republic, and west by Cameroon, Nigeria, and Niger.

Chad covers an area of 1,284,000 sq km/495,752 sq mi, and has a population of 7,651,000 (2000 est). The capital is ►Ndjamena (formerly Fort Lamy); other major towns and cities include Sarh, Moundou, Abéché, Bongor, Doba, Kélo, and Koumra. The head of state is Idriss Deby (from 1990); the head of government is Nagoum Yamassoum (from 1999). The currency is the franc CFA, and the languages spoken include French, Arabic (both official), and over 100 African languages. **64** C5

Chad, Lake lake on the northeastern boundary of Nigeria and the eastern boundary of Chad. It once varied in extent between rainy and dry seasons from 50,000 sq km/19,000 sq mi to 20,000 sq km/7,000 sq mi, but a series of droughts between 1979 and 1989 reduced its area to 2,500 sq km/ 965 sq mi in 1993. It is a shallow lake (depth does not exceed 5–8 m/ 16–26 ft), with the northern part being completely dry and the southern area being densely vegetated, with swamps and open pools. The lake was first seen by European explorers in 1823. **64** B5

Chagos Archipelago island group in the Indian Ocean; area 60 sq km/ 23 sq mi. Formerly a dependency of Mauritius, it now forms the ►British

Indian Ocean Territory. The chief island is Diego Garcia, from which several thousand inhabitants were forcibly removed, primarily to Mauritius, between 1967 and 1973 to make way for a US-British strategic base. In November 2000, the British High Court ruled that the British government should return the islanders to their homes.

Châlons-sur-Marne (Latin **Catalaunum**), administrative centre of the *département* of ►Marne and of the ►Champagne-Ardenne region in northeast France, 150 km/93 mi east of Paris on the right bank of the River Marne; population (1990) 51,500. Industries include champagne production, brewing, electrical engineering, and textiles manufacture. Châlons was formerly known for its worsted cloth. Tradition has it that Attila the Hun was defeated near here in his attempt to invade France at the **Battle of the Catalaunian Plains** in 451 AD by the Roman general Aëtius and the Visigoth Theodoric. **22** G6

Chambéry administrative centre of ►Savoie *département* in eastern France; population (1990) 55,600. It lies in a gorge linking the River Isère to Lake Bourget, 85 km/53 mi east of Lyons, and is the seat of an archbishopric, a market centre, and also a holiday and health resort. The town produces metallurgical goods and construction materials as well as food and wine. It gives its name to a French vermouth. Chambéry was the capital of the former Duchy of Savoy. **30** A5

Chamonix holiday resort at the foot of Mont Blanc, in the French Alps; population (1990) 10,100. It was the site of the first Winter Olympics in 1924. **30** B5

Chamonix-Mont-Blanc French village in the *département* of Haute-Savoie, on the River Arve, 65 km/40 mi southeast of Geneva; population (1990) 10,000; altitude about 1,000 m/3,400 ft. It is in a scenic, narrow valley, north of the Mont Blanc mountain range. The most commonly used approach to ►Mont Blanc, it is a mountaineering and winter-sports centre. **30** B5

Champagne-Ardenne region of northeast France; area 25,600 sq km/9,900 sq mi; population (1990) 1,347,800. Its largest town is ►Reims, but its administrative centre is ►Châlons-sur-Marne. It comprises the *départements* of ►Ardennes, ►Aube, ►Marne, and ►Haute-Marne. The land is fertile in the west and supports sheep and dairy farming; its vineyards produce the famous champagne wines. The region also includes part of the ►Ardennes forest. **6** F2

Champaign city in Champaign County, east-central Illinois, USA, 8 km/5 mi west of Urbana; population (1992) 64,350. Originally a trading centre for the area's corn, soybeans, and livestock, Champaign today has industries producing drop forgings, alloy castings, air conditioners, bleachers, cement products, athletic equipment, and soybean oil. Together with Urbana, Champaign is the site of the University of Illinois. The city was incorporated in 1860. **88** C2/**10** C2

Chandigarh city of north India, in the foothills of the Himalayas; population (1991) 511,000. It is also a Union Territory; area 114 sq km/44 sq mi; population (1991) 640,725. Planned by the architect Le Corbusier, the city was inaugurated in 1953 to replace Lahore (capital of British Punjab), which went to Pakistan at partition in 1947. Since 1966, when Chandigarh became a Union Territory, it has been the capital city of the states of both Haryana and Punjab, pending the construction of a new capital for the former. **54** C2/**10** C2

Changchiakow alternative transcription of ►Zhangjiakou, a trading centre in Hebei province, China. **46** E2

Changchun industrial city and capital of ►Jilin province, northeast China; population (1994) 2,237,000. It is the centre of an agricultural district, and manufactures machinery, vehicles, and railway equipment. **48** C2

Chang Jiang (or **Yangtze Kiang**, 'long river'), longest river of China and third longest in the world, flowing about 6,300 km/3,900 mi from Qinghai to the Yellow Sea. It is a major commercial waterway. Work began on the Three Gorges Dam on the river in December 1994. **46** D4

Changsha port and capital of ►Hunan province, China, on the River Xiang (or Hsiang); population (1994) 1,422,600. It trades in rice, tea, timber, coal, and non-ferrous metals; works antimony, lead, silver, and aluminium; and produces machinery, chemicals, electronics, vehicles, textiles, plastics, porcelain, and embroidered goods. **46** E5

Chaniá (Greek **Khaniá**), city on the northwest coast of the Greek island of Crete, capital until 1971, when it was replaced by Iraklion; population (1991) 65,500. It was founded in 1252 by the Venetians and is surrounded by a wall. Vegetable oils, soap, and leather are exported. Heavy fighting took place here during World War II, after the landing of German parachutists in May 1941. **36** G9

Channel, English see ►English Channel. **24** J12

Channel Islands group of islands in the English Channel, off the northwest coast of France; they are a possession of the British crown. They comprise the islands of Jersey, Guernsey, Alderney, Great and Little Sark, with the lesser Herm, Brechou, Jethou, and Lihou; area 194 sq km/75 sq mi; population (1991) 145,600. Chief industries are farming, fishing, and tourism; flowers, early potatoes, tomatoes, butterflies, and dairy cattle are exported. The official language is French (Norman French) but English is more widely used. The islands are a tax haven. The currency is English sterling, as well as local coinage. **24** K12

Chantilly town in the Oise *département*, northern France, situated beside the forest of Chantilly on the River Nonette, 40 km/25 mi north of Paris; population (1999 est) 35,100. It is the centre of French horse racing. The racing industry is a large source of local employment, and helps to attract over 350,000 tourists a year to the town. Chantilly formerly produced lace and a type of whipped cream, giving its name to both. It also manufactured porcelain. Many of its residents commute to work in Paris. **22** E5

Chao Phraya chief river (formerly Menam) of Thailand, flowing 1,200 km/750 mi into the Bight of Bangkok, an inlet of the Gulf of Thailand. **50** C4

Charente *département* in the ►Poitou-Charentes region of France; area 5,952 sq km/2,298 sq mi; population (1990) 176,300. The Rivers Vienne and Dronne flow through it, the latter forming the southern boundary. The countryside is undulating, crossed by the low spurs of the Limousin, Guienne, and Poitou hills. There are extensive areas of forest and heath land. The *département* is predominantly agricultural, producing cereals, potatoes, beet, hemp, flax, and truffles, but the most important product is the brandy made from grapes grown in the Charente basin. Industries include the manufacture of paper, textiles, chemicals, and metals. The administrative centre is ►Angoulême. **6** D4

Charente-Maritime *département* in the ►Poitou-Charentes region of France, including the islands of Ré and Oléron; area 6,848 sq km/2,644 sq mi; population (1990) 271,400. The rivers Charente, Boutonne, Sèvre Niortaise, and Touvre, flow through the *département*, which borders on the Gironde estuary. Cereals, vines, sugar beet, potatoes, hemp, and fruit are grown, and livestock are reared. Industries include the production of brandy, iron, chemicals, and salt. The administrative centre is ►La Rochelle. **6** C4

Charleroi city and railway junction in Hainaut province, Belgium, 48 km/30 mi south of Brussels, on the River Sambre; population (1997) 204,900. Set in a former coal-mining area, its industries include electronics and electrical engineering, as well as the production of steel, pharmaceuticals, glass, and cement. The Charleroi canal connects the town with Brussels. Founded by the Spanish in 1666, the city was named after Charles II of Spain. **22** G4

Charleston city and main port in southeastern South Carolina, USA, situated on a peninsula between the Ashley and Cooper rivers, at the head of a bay 11 km/7 mi from the Atlantic Ocean; seat of Charleston County; population (1992) 81,300. Industries include tourism, paper, chemicals, and petrochemicals. The original settlement dates from 1670, and Charleston was incorporated as a city in 1783; there are many historic houses and fine gardens here. **90** F3

Charleston capital of ►West Virginia and administrative headquarters of Kanawha County, situated on the Kanawha River, 85 km/53 mi upstream from its junction with the Ohio River; population (1992) 57,100. It is the commercial centre of a region that produces coal, natural gas, salt, clay, timber, and oil; it also has major chemical and glass industries. **90** E2

Charleville-Mézières French town, capital of the *département* of Ardennes, situated on a bend in the River Meuse; population (1990) 58,000. Industries include machine tools and hardware. **22** G5

Charlotte city in southern North Carolina, USA, on the border with South Carolina; seat of Mecklenburg County; population (1994 est) 438,000; metropolitan area (1994 est) 1,260,000. Industries include printing, microelectronics, data processing, textiles, chemicals, machinery, and food products. Settled around 1750, it was the gold-mining centre of the USA until gold was discovered in California (1849). Charlotte enjoyed rapid growth in the 1970s, and is the largest city in the state. **90** E2

Charlottesville city and administrative headquarters of (but administratively separate from) Albemarle County, central Virginia, USA; population (1990) 40,300. It is situated in the foothills of the Blue Ridge Mountains, 105 km/65 mi northwest of Richmond on the Rivanna River. It is the centre of an agricultural district; industries include tourism, processed food, and electronic equipment. It is the home of the University of Virginia, established in 1819 by Thomas Jefferson. Piedmont Virginia Community College (1969) is also located in the city. **90** F2

Charlottetown capital and main port of ►Prince Edward Island, Canada, situated on the south side of the island on Hillsborough Bay; population (1996) 32,500. Charlottetown was incorporated as a city in 1995 and serves as the administrative headquarters of Queens County. It trades in textiles, fish, timber, vegetables, and dairy produce; the main industries are food-processing, shipbuilding, and tourism. **82** U7

Chartres (Latin **Autricum**), administrative centre of the *département* of ►Eure-et-Loir in northwest France, 96 km/59 mi southwest of Paris on the River Eure; population (1990) 41,850; conurbation 84,000. The city is an agricultural centre for the fertile Plaine de la Beauce. The twin-spired cathedral of Notre Dame, completed about 1240, is a masterpiece of Gothic architecture and a world heritage monument; the city also has other medieval churches and some fine old houses, and attracts large numbers of tourists. **26** G5

Châteauroux French town, capital of the *département* of Indre, on the River Indre; population (1990) 52,900. The town manufactures textiles, agricultural machinery, foodstuffs, and tobacco. Châteauroux is named after a 10th-century château built by Raoul of Deols. The present château was built in the 15th century. Situated close to the Loire valley, the town attracts tourists. **26** G7

Chatellerault French town in the *département* of Vienne, on the River Vienne, 35 km/22 mi northeast of Poitiers; population (1990) 35,000. It manufactures small arms, cutlery, and preserves, and has a distillery. The town has a 15th-century château, two 12th-century churches, and a number of historic houses. **26** F7

Chatham city and administrative headquarters of Kent County, Ontario, eastern Canada, situated on the Thames River, 19 km/12 mi from its mouth, 103 km/64 mi southwest of London and 290 km/180 mi west of Toronto; population (1991) 43,600. The city is the commercial and industrial centre of an agricultural (fruit, grain, tobacco, and dairy goods) and natural gas producing region. Textiles, lumber, auto parts, plastics, and fabricated metals are manufactured. Tourism is also important to the local economy. **88** D2

Chatham Islands two Pacific islands (Chatham and Pitt), forming a county of South Island, New Zealand, lying 580 km/360 mi to the east; area 960 sq km/371 sq mi; population (1995 est) 790. The chief settlement is Waitangi. There is seal fishing, and cattle and sheep are raised. The chief export is wool. **78** (1)B1

Chattanooga city in southeastern Tennessee, USA, on the Tennessee River near the Georgia border; seat of Hamilton County; population (1994 est) 152,00. Chattanooga is the centre of the Tennessee Valley Authority area. White settlement began here in 1835, and the city developed as a river port and a centre for salt-trading; it now produces chemicals, textiles, and metal products. Chattanooga was laid out in 1851 and incorporated as a city in 1893. 84 J4

Cheb (German **Eger**), industrial town (textiles) in western Bohemia, in the Czech Republic, on the River Ohře near the German border, 145 km/91 mi west of Prague; population (1991) 31,800. The town has a 13th-century church and was formerly a strategic fortress. 20 H6

Chechnya (or **Chechenia** or **Chechen Republic**), breakaway part of the former Russian autonomous republic of Checheno-Ingush, on the northern slopes of the ►Caucasus Mountains; official name **Noxcijn Republika Ickeriy** from 1994; area 17,300 sq km/6,680 sq mi; population (1990) 1,290,000 (Chechen 90%). The capital is ►Groznyy. Chief industries are oil extraction (at one of the largest Russian oilfields), engineering, chemicals, building materials, and timber. Most of the inhabitants are Sunni Muslim. 58 L2

Chefoo former name of ►Yantai, a port in Shandong province, China. 46 G3

Chełm (Russian **Kholm**), industrial town (metals, bricks, cement, food processing) in Lublin province, Poland, 64 km/40 mi east of Lublin; population (1990) 66,400. 18 N6

Chelmsford market town and administrative headquarters of ►Essex, southeast England, at the confluence of the Chelmer and Can rivers, 48 km/30 mi northeast of London; population (1991) 97,500. Corn and cattle are traded. Industries include agricultural services, radio communications, engineering, flour-milling, brewing, and the manufacture of electrical equipment, plastics, and soft drinks (Britvic). 22 C3

Cheltenham spa town at the foot of the Cotswold Hills, Gloucestershire, England, 12 km/7 mi northeast of Gloucester; population (1991) 91,300. The town has light industries including aerospace electronics and food-processing (Kraft). Tourism and the conference business are also important. Annual events include the Cheltenham Festival of Literature in October, the International Festival of Music, the National Hunt Festival in March, and the Cheltenham Cricket Festival. There is a steeplechase course in Prestbury Park, known for the annual Cheltenham Gold Cup. 24 L10

Chelyabinsk capital city, economic and cultural centre of Chelyabinsk oblast (region), Russian Federation, 240 km/150 mi south of Yekaterinburg on the Miass River; population (1996 est) 1,083,000. Chelyabinsk is a major industrial centre in the Urals and an important rail centre. The main branches of industry are engineering (tractors, aircraft, machine tools), and metallurgy (steel, ferro-alloys, zinc). There is a large lignite-fired power station nearby. The important Chelyabinsk coal basin (first exploited in 1906) lies 15 km/9 mi to the east of the city. Waste from the city's plutonium plant makes it possibly the most radioactive place in the world. 38 M3

Chemnitz industrial city in Saxony, Federal Republic of Germany, on the River Chemnitz, 65 km/40 mi south-southeast of Leipzig; population (1995) 271,400. Industries include engineering and the manufacture of textiles and chemicals. As a former district capital of East Germany it was named Karl-Marx-Stadt from 1953 to 1990. The city has a university (Chemnitz-Zurickan). 20 H6

Chemulpo former name for ►Inchon, port and summer resort on the west coast of South Korea. 48 D5

Chenab tributary of the River ►Indus, rising in the Himalayas and flowing through Kashmir, India and Pakistan. It is one of the five rivers which give Punjab its name. 54 B2

Chengchow alternative transcription of ►Zhengzhou, the capital of Henan province, China. 46 E4

Chengde (or **Chengteh**), city in Hebei province, north China, 173 km/109 mi northeast of Beijing; population (1994) 250,000. It is a market town for agricultural and forestry products and an important tourist centre. Its palaces and temples were built in 1703 by Emperor Kangxi as a summer resort for the Manchu rulers. As **Jehol**, it was capital of a former province of the same name from 1913 to 1956. 46 F2

Chengdu (or **Chengtu**), ancient city and capital of ►Sichuan province, China; population (1994) 3,016,000. It is a busy rail junction and has railway workshops. Industries include food-processing, engineering, electronics, and the manufacture of textiles and petrochemicals. There are well-preserved temples of the 8th-century poet Tu Fu and other historical figures. 46 C4

Chengteh alternative transliteration of ►Chengde, a city in Hebei province, China. 46 F2

Chengtu alternative transliteration of ►Chengdu, capital of Sichuan province, China. 46 C4

Chennai (formerly **Madras**, to 1996), industrial port and capital of Tamil Nadu, India, on the Bay of Bengal; population (1991) 5,361,000. An all-weather artificial harbour handles cotton goods, oilseeds, hides and skins, and industrial raw materials. Main industries include cotton, cement, chemicals, railway, car and bicycle manufacture, oil refining, iron, and steel. Fort St George (1639) remains from the East India Company when Chennai was the chief port on the east coast; the fort now contains government offices and St Mary's Church, the first English church built in India (1680). Chennai was occupied by the French from 1746 to 1748 and shelled by the German ship *Emden* in 1914, the only place in India attacked in World War I. The University of Madras was founded in 1857, and there is a technical institute (1959). 54 D6

Cher *département* in the ►Centre region of France; area 7,227 sq km/2,790 sq mi; population (1990) 157,100. The landscape is generally level and wooded. It is a fertile agricultural region with a temperate climate, producing corn, vines, fruit trees, sugar beet, hemp, and flax, and there is livestock farming. About a quarter of the population is employed in manufacturing – wool, electrical, and metal goods, porcelain, glass, and confectionery – and three-fifths in tertiary occupations. The principal towns are ►Bourges (the administrative centre) and Vierzon. 6 E3

Cher French river, rising in the Massif Central on the eastern edge of Creuse *département*, and flowing northwest to join the River ►Loire 19 km/12 mi below Tours; length 350 km/217 mi. It is navigable from Vierzon. It gives its name to the *département* of ►Cher. 26 G6

Cherbourg French port and naval station at the northern end of the

Cotentin peninsula, in Manche *département*; population (1990) 28,800, conurbation 92,000. There is an institute for studies in nuclear warfare, and the town's dry-docks house some of the largest shipbuilding yards in France. Other industries include the manufacture of hosiery and engineering goods. Transatlantic maritime trade has declined greatly. 26 D4

Cherepovets city in the Vologda oblast (region) of the northwestern Russian Federation, 375 km/233 mi north of Moscow; population (1996 est) 321,000. The city is located near the northern shore of the Rybinsk Reservoir on the Volga, and stands at the southern end of the Volga–Baltic Waterway. Iron and steel foundries, chemical plants, and shipyards are the city's principal industries. Cherepovets was founded in 1777 on the site of a monastery. 38 G3

Chernihiv (or **Chernigov**), city and port on the River Desna in northern Ukraine, 125 km/78 mi north of Kiev; population (1998) 311,000. The city is capital of the province of the same name. Lumbering, textiles, chemicals, distilling, and food-canning are among its industries. It has two cathedrals, dating from the 11th and 12th centuries. 38 F4

Chernivtsi (or **Chernovtsy**), city in southwestern Ukraine, capital of the oblast (region) of the same name; population (1990) 257,000. The city is located in the foothills of the Carpathian Mountains, which extend south from here into Romania. Industries include the manufacture of textiles, clothing, machinery, and rubber, together with sawmilling and engineering. 38 E5

Chernobyl town in northern Ukraine, 100 km/62 mi north of Kiev; site of a former nuclear power station. The town is now abandoned. On 26 April 1986, two huge explosions occurred at the plant, destroying a central reactor and breaching its 1,000-tonne roof. In the immediate vicinity of Chernobyl, 31 people died (all firemen or workers at the plant) and 135,000 were permanently evacuated. It has been estimated that there will be an additional 20,000–40,000 deaths from cancer over 60 years; 600,000 people are officially classified as at risk. According to World Health Organization (WHO) figures from 1995, the incidence of thyroid cancer in children increased 200-fold in Belarus as a result of fallout from the disaster. The last remaining nuclear reactor at Chernobyl was shut down in December 2000. 38 F4

Ches. abbreviation for ▶Cheshire, an English county. 5 E5

Chesapeake Bay largest of the inlets on the Atlantic coast of the USA, bordered by eastern Maryland and eastern Virginia. Chesapeake Bay extends southwards from Havre de Grace in northeast Maryland, and enters the Atlantic between Cape Charles and Cape Henry in Virginia; it is about 320 km/200 mi in length and 6–64 km/4–40 mi in width. There are several deep-water ports located on the bay: Newport News, Norfolk, Portsmouth, and Baltimore.

Cheshire county of northwest England, which has contained the unitary authorities Halton and Warrington since April 1998.
area 2,320 sq km/896 sq mi
towns and cities ▶Chester (administrative headquarters), Crewe, Congleton, Macclesfield
physical chiefly a fertile plain, with the Pennines in the east; rivers: Mersey, Dee, Weaver; a sandstone ridge extending south through central Cheshire together with Delamere Forest constitute a woodland and heath landscape
features salt mines and geologically rich former copper workings at Alderley Edge; discovery of Lindow Man, the first 'bogman' to be found in mainland Britain, dating from around 500 BC; Quarry Bank Mill at Styal is a cotton-industry museum
agriculture arable farming in the north; cheese and dairy products in the centre and south of the county
industries aerospace industry, chemicals, pharmaceuticals, salt, silk and textiles, vehicles
population (1996) 980,000 5 E5

Chester city and administrative headquarters of ▶Cheshire, England, on the River Dee 26 km/16 mi south of Liverpool, two miles from the border with Wales; population (1991) 80,100; Chester urban area 89,629. There are engineering, aerospace (Airbus), metallurgical, and clothing industries, and car components are manufactured. It is a centre of trade and commerce, and tourism is also important. 24 K8

Chesterfield market town in Derbyshire, central England, on the River Rother, to the east of the Peak District National Park, 40 km/25 mi north of Derby; population (1991) 71,900. Industries include iron-founding, engineering, and the manufacture of chemicals, pottery, and glass. Extensive coalfields are mined to the southeast. 24 L8

Cheviot Hills range of hills, 56 km/35 mi long, mainly in Northumberland but also extending into the Scottish Borders; they form the border between England and Scotland for some 48 km/30 mi. The Cheviots lie between the Pennines and the Southern Uplands ranges. The highest point is the **Cheviot** at 816 m/2,676 ft. For centuries the region was a battleground between the English and the Scots. The area gives its name to a breed of white-faced sheep. 24 K6

Cheyenne capital of ▶Wyoming, USA, located in the southeastern part of the state, just north of the Colorado border on Crow Creek, in the foothills of the Laramie range of the Rocky Mountains; seat of Laramie County; population (1992) 51,900. An agricultural and transportation centre, its industries include oil refining, chemicals, electrical goods, machinery, meat packing, and food processing; tourism is also important to the economy. 86 F2

Chhattisgarh state of central India, bordered by Maharashtra, Andhra Pradesh, Orissa, Jharkhand, Uttar Pradesh, and Madhya Pradesh; area 146,361 sq km/56,510 sq mi; population (2001 est) 17,600,000. It was carved from Madhya Pradesh and was incorporated in November 2000. The capital is Raipur. The state touches the northern-most part of the Deccan Plateau, and is crossed by the Mahandi River. Nicknamed the "Rice Bowl of India", Chhattisgarh provides rice to 600 rice mills within the state and Madhya Pradesh. 80% of the population is involved in agricultural activities. Industries also include forest production, bauxite, limestone, iron ore, diamonds, gold, and china clay. One-third of the population is tribal, and the principal languages are Hindu and Chhattisgarhi. 10 D4

Chiang Mai (or **Chiengmai**), city in northern Thailand; population (1993) 161,500. There is a trade in teak and lac (as shellac, a resin used in varnishes and polishes) and many handicraft industries. It is the former capital of the Lan Na Thai kingdom. **50** B3

Chiba city and port in Kanto region, on Tokyo Bay, east Honshu island, Japan, 40 km/25 mi west of Tokyo; population (1995 est) 856,900. Industries include shipbuilding, petrochemicals, paper, steel, and textiles. The centre of a coastal industrial zone, with one of the busiest ports in Japan, it imports crude oil, iron ore, and coking coal, and exports steel. Much of the building in the city has taken place on reclaimed land. **48** L6

Chicago (Ojibway 'wild onion place') financial and industrial city in Illinois, USA, on Lake Michigan. It is the third-largest US city; population (1992) 2,768,500; metropolitan area (1992) 8,410,000. Industries include iron, steel, chemicals, electrical goods, machinery, meatpacking and food processing, publishing, and fabricated metals. The once famous stockyards are now closed. Chicago grew from a village in the mid-19th century. The world's first skyscraper was built here in 1885 and some of the world's tallest skyscrapers, including the Sears Tower (443 m/1,454 ft high), are in Chicago. **88** C2

Chichester city and market town and administrative headquarters of ▶West Sussex, southern England, 111 km/69 mi southwest of London; population (1991) 27,100. It lies in an agricultural area, and has a harbour. It was a Roman town, **Noviomagus Regnensium**, and the nearby remains of Fishbourne Palace (about AD 80) is one of the finest Roman archeological sites outside Italy. **22** B4

Chiclayo commercial centre and capital of Lambayeque department, northwest Peru, 640 km/400 mi northwest of Lima; population (1993) 411,500. One of Peru's fastest growing cities, it lies at the centre of an agricultural area producing rice, cotton, and sugar cane, which are all grown and processed in the region. There are tanning, distilling, and dairy industries. The city comprises a mixture of Creole, Spanish, and Indian architecture, and there is a 19th-century cathedral, designed by the English architect Andrew Townsend. **98** B5

Chiengmai alternative name for ▶Chiang Mai, a town in Thailand. **50** B3

Chihuahua capital of Chihuahua state, Mexico, 1,285 km/800 mi northwest of Mexico City; population (1995 est) 589,000. It lies in the north of Mexico at an altitude of 1428 m/4685 ft, and was founded in 1707. It is the centre of both a mining district and a cattle raising area and has textile mills and smelting industries. The area is noted for breeding the small, short-haired dogs named after the town and state. **92** E3

Chile South American country, bounded north by Peru and Bolivia, east by Argentina, and south and west by the Pacific Ocean.

Chile covers an area of 756,950 sq km/292,258 sq mi, and has a population of 15,211,000 (2000 est). The capital is ▶Santiago; other major towns and cities include Concepción, Viña del Mar, Valparaíso, Talcahuano, Puente Alto, Temuco, and Antofagasta. The head of state and government is Ricardo Lagos Escobar (from 2000). The currency is the Chilean peso, and the main language spoken is Spanish (official). **96** D8

Chiltern Hills range of chalk hills extending for some 72 km/45 mi in a curve from a point north of Reading to the Suffolk border. Coombe Hill, near Wendover, 260 m/852 ft high, is the highest point.

Chi-lung alternative name for ▶Keelung, an industrial port in Taiwan. **46** G5

Chimbote largest fishing port in Peru, situated at the mouth of the Rio Santa, 450 km/280 mi north of Lima, in the department of Ancash; population (1993) 297,000. The main industries are fishmeal and steel; fish products are exported. The port was built in the natural harbour on Ferrol Bay in the 1950s, primarily to serve the developing national steel industry. **98** B5

China the largest country in East Asia, bounded to the north by Mongolia; to the northwest by Tajikistan, Kyrgyzstan, Kazakhstan, and Afghanistan; to the southwest by India, Nepal, and Bhutan; to the south by Myanmar, Laos, and Vietnam; to the southeast by the South China Sea; to the east by the East China Sea, North Korea, and Yellow Sea; and to the northeast by Russia.

China covers an area of 9,572,900 sq km/3,696,000 sq mi, and has a population of 1,277,558,000 (2000 est). The capital is ▶Beijing (or Peking); other major towns and cities include Shanghai, Hong Kong, Chongqing, Tianjin, Guangzhou (English Canton), Shenyang (formerly Mukden), Wuhan, Nanjing, Harbin, Chengdu, and Xi'an. The head of state is Jiang Zemin (from 1993); the head of government is Zhu Rongji (from 1998). The currency is the yuan, and the main language spoken is Chinese (dialects include Mandarin (official), Yue (Cantonese), Wu (Shanghaiese), Minbai, Minnah, Xiang, Gan, and Hakka). **40** N6

China Sea area of the Pacific Ocean bordered by China, Vietnam, Borneo, the Philippines, and Japan. Various groups of small islands and shoals, including the Paracels, 500 km/300 mi east of Vietnam, have been disputed by China and other powers because they lie in oil-rich areas. The chief rivers which flow into the South China Sea are the Red River and Mekong; the main ports include Canton, Hong Kong, Manila, Bangkok, Singapore, and Ho Chi Minh City. **50** E4/**46** H4

Chios (Greek **Khios**), one of the ▶Aegean Islands in Greece, 13 km/8 mi off the west coast of Turkey at the entrance to the Gulf of Smyrna; area 830 sq km/320 sq mi; population (1993) 50,200. The island is 50 km/31 mi long and 12–25 km/8–16 mi wide. Grapes, oranges, lemons, figs, olives, aniseed, and tobacco are grown. The capital of the island is also called Chios. **36** H6

Chişinău (Russian **Kishinev**), capital of Moldova, situated in a rich agricultural area; population (1990) 676,000. It is a commercial and cultural centre; industries include cement, food processing, tobacco, and textiles. **34** R2

Chita capital city, economic and cultural centre of Chita oblast (region) in southeastern Siberia, Russian Federation; population (1996 est) 321,000. It has light engineering works, coal mines, sawmills, and leather and tanning industries. Chita is also an important transportation centre, being situated on the Trans-Siberian Railway near its junction with the Trans-Manchurian Railway; major locomotive repair works are located here. 44 J6

Chittagong city and port in Bangladesh, 16 km/10 mi from the mouth of the Karnaphuli River, on the Bay of Bengal; population (1991) 1,364,000. Industries include steel, engineering, chemicals, and textiles. 54 F4

Chkalov former name (1938–57) of ►Orenburg, a town in the Russian Federation. 38 L4

Chomutov (German **Komotau**), industrial town (engineering, coal) in northern Bohemia, the Czech Republic, at the foot of the Erzgebirge, 80 km/50 mi northwest of Prague; population (1991) 53,200. 18 C7

Chongjin port and capital of North Hamgyong province on the northeast coast of North Korea; population (1998 est) 628,800. Timber, iron, textiles, ships, and chemicals are exported; there is steel, pig-iron, and fish processing. 48 E3

Chongqing (or **Chungking**, or **Pahsien**), city in Sichuan province, China, at the confluence of the Chang Jiang and Jialing Jiang rivers; population (1994) 5,408,000. It is the largest of China's four municipalities directly under the central government, with an area of 82,000 sq km/31,700 sq mi and population (1996) of 30,000,000. Industries include coalmining, food-processing, and the manufacture of iron, steel, chemicals, synthetic rubber, automobiles, electrical equipment, and textiles. 40 P7

Chorzów industrial town in Katowice province, Upper Silesia, southern Poland, 6 km/4 mi northwest of Katowice; population (1990) 131,900. It is a centre for coal, iron, steel, chemicals (nitrates), and engineering.

Christchurch city on South Island, New Zealand, 11 km/7 mi from the mouth of the Avon River; population (1996) 331,400. It is the principal commercial centre of the Canterbury Plains and the seat of the University of Canterbury. Industries include fertilizers, chemicals, canning, meat processing, rail workshops, and shoes. 78 D6

Christiania former name of the Norwegian capital of ►Oslo (1624–1924), after King Christian IV who replanned it after a fire in 1624. 16 F7

Christmas Island island in the Indian Ocean, 360 km/224 mi south of Java; area 140 sq km/54 sq mi; population (1994 est) 2,500. Found to be uninhabited when reached by English explorer Capt W Mynars on Christmas Day 1643, it was annexed by Britain in 1888, occupied by Japan between 1942 and 1945, and transferred to Australia in 1958. After a referendum in 1984, it was included in Northern Territory. Its phosphate mine was closed in 1987. Tourism and casinos are being developed. 52 D5

Christmas Island former name of ►Kiritimati, an island in Kiribati, in the central Pacific. 74 L5

Chubu mountainous coastal region of central Honshu island, Japan, area 66,774 sq km/25,781 sq mi; population (1992) 21,162,000. The chief city is ►Nagoya. The region, which produces tea, fruits, and fish, contains the Niigata plain, one of the country's largest rice-producing areas. Mount Fuji and several of Japan's longest rivers can also be found in the region.

Chufu alternative transcription of ►Qufu, a town in Shandong province, China. 46 F3

Chugoku southwest region of Honshu island, Japan, area 31,881 sq km/12,309 sq mi; population (1992) 7,754,000. The chief city is ►Hiroshima. Citrus and grapes are grown in the region. Formerly rich fishing grounds have been damaged by industrial pollution.

Chukchi Sea (or **Chuckchi**; Russian **Cukotskoje More**), part of the Arctic Ocean, situated north of the Bering Strait between Asia and North America. 92 (1)C2

Chungking alternative transcription of ►Chongqing, a city in Sichuan province, China. 40 P7

Chur (French **Coire**; Italian **Coira**; Romansch **Cuera**), capital of the canton of ►Graubünden, Switzerland, in the valley of the upper Rhine; population (1995) 30,100. The Roman Catholic cathedral of St Lucius was begun in 1178. 30 E4

Churchill town in the province of Manitoba, Canada, situated on Hudson Bay; population (1991) 1,143. Although the port is ice-free only three months a year, Churchill handles about 500,000 tonnes of grain annually, as well as fuel oil and bulk cargo. It is a major centre for Arctic research, health programmes, and education. The Hudson's Bay Company established a post here and named it for Lord Churchill (later 1st duke of Marlborough). 82 N5

Churchill Falls waterfalls on the Churchill River, the longest river in Labrador, eastern Canada, situated about 400 km/249 mi from the mouth of the river. They fall 75 m/245 ft. A power plant at the falls supplies hydroelectric power, much of which is transmitted from the province to other parts of Canada and the USA. 82 U6

Chuvash (or **Chuvashiya**), autonomous republic in the western Russian Federation; area 18,300 sq km/7,066 sq mi; population (1990) 1,340,000 (68% Chuvash, 25% Russian). The main cities are Cheboksary (capital), Alatyr, and Shumerla. Chuvash lies south of the Volga River, 560 km/350 mi east of Moscow. The main industries are textiles, lumbering, electrical and engineering industries, phosphates, and limestone; there is grain and fruit farming. 38 J3

Cienfuegos port and naval base in Cuba; population (1990 est) 123,600. It trades in sugar, fruit, and tobacco. It was founded in 1819, destroyed by a storm in 1825, and rebuilt. 94 H4

Cincinnati city and port in southwestern Ohio, USA, on the northern bank of the Ohio River; seat of Hamilton County; population (1994 est) 358,000; metropolitan area (1994 est) 1,894,000. The city is an important inland port on the Ohio–Mississippi system, and a major manufacturing centre; its chief industries include aircraft and car machinery, clothing, furniture making, wine, chemicals, and meatpacking. Founded in 1788, Cincinnati was incorporated as a city in 1819. It attracted large numbers of European immigrants, particularly Germans, during the 19th century. 88 D3

Ciudad Bolívar (formerly **Angostura**), inland port and capital of Bolívar state in southeast Venezuela, situated on the south bank of the Orinoco River, 400 km/250 mi from its mouth; population (1992 est) 253,100. The city is linked with Soledad across the river by the Angostura Suspension Bridge (1967), which has as a span of 712 m/2,336 ft and is the only bridge to span the Orinoco. Exports include gold, diamonds, timber, cattle, and horses. It is popular with tourists as a base for tours to the Angel Falls and the National Parks. **98** E2

Ciudad de México Spanish name for ►Mexico City, the capital of Mexico. **94** E5

Ciudad Guayana city in Venezuela, on the south bank of the River Orinoco, population (1995 est) 605,000. This new city in Bolívar state was founded in 1961 by linking the towns of San Félix (the site of the commercial port), Puerto Ordaz, Palúa (the site of a major iron-ore terminal), and Matanzas. Main industries include iron and steel. It is planned that eventually the city will hold a population of one million. **98** E2

Ciudad Juárez city on the Rio Grande, in Chihuahua, northern Mexico, on the US border; population (1995 est) 885,000. It lies at an altitude of 1133 m/3717 ft opposite to El Paso, which is on the US side of the river in Texas. In 1865 the city was the headquarters of Benito Juárez who founded Mexico as a modern state. It is a centre for cotton. **92** E2

Ciudad Real capital of Ciudad Real province, central Spain, 170 km/105 mi south of Madrid; population (1991) 56,300. There is a trade in livestock and products include textiles, pharmaceuticals, and brandy. The town has a 13th-century Gothic cathedral, and a 16th-century synagogue. Ciudad Real is the seat of the bishop-prior of the military orders. **28** G6

Ciudad Rodrigo town in the province of Salamanca, western Spain, on a hill above the River Agueda; population (1991) 14,900. Products include leather, soap, and pottery. It was captured by the English in 1706, during the War of Spanish Succession, and again in 1812, during the Peninsular War, by the Duke of Wellington; he was subsequently created duke of Ciudad Rodrigo. **28** D4

Ciudad Trujillo former name (1936–61) of ►Santo Domingo, capital city and seaport of the Dominican Republic. **94** L5

Civitavecchia (Roman **Trajani Portus**), ancient port in Lazio region, on the west coast of Italy, 64 km/40 mi northwest of Rome; population (1990) 51,200. Built by the Roman emperor Trajan, it is the main port of the region, and the nearest to Rome. Industries include fishing and the manufacture of cement, calcium carbide, aluminium, and foodstuffs. **32** F6

Clackmannanshire unitary authority in central Scotland, bordering the north side of the Firth of Forth. A former county (until 1974), it was a district of Central region (1975–96).
area 161 sq km/62 sq mi
towns Alloa (administrative headquarters), Tillicoultry
physical compact geographical area comprising the extensive flat flood plain of the River Devon, which rises dramatically at the Ochil Hills to Ben Cleuch (721 m/2,365 ft)
industries brewing, distilling, manufacture of bottles and knitwear
agriculture intensive on flood plain of Forth; less intensive on Ochil Hills
population (1996) 47,700 **5** E3

Clacton-on-Sea seaside resort in Essex, eastern England, 26 km/16 mi southeast of Colchester; population (1991) 45,100. It possesses nearly 11 km/7 mi of sandy beaches, a pier, and a promenade and marine parade over 6 km/4 mi long. **22** D3

Clare county on the west coast of the Republic of Ireland, in the province of Munster, situated between Galway Bay in the north and the Shannon estuary in the south; county town ►Ennis; area 3,190 sq km/1,231 sq mi; population (1996) 94,000. Other towns include Kilrush, Kilkee, and Shannon, an important 'new' town noted for its light industry, and electronics and aerospace industries. Dairying and cattle rearing are the principal farming activities; there are also important salmon fisheries and extensive oyster beds. Slate and black marble are quarried and worked; lead is also found. The Shannon is a source of hydroelectricity: there is a power station at Ardnacrusha, 5 km/3 mi north of Limerick. **5** B5

Clarksville city and administrative headquarters of Montgomery County, north Tennessee, USA, at the junction of the Cumberland and Red rivers, 70 km/44 mi northwest of Nashville; population (1992) 84,400. It is the processing centre for a beef and dairy cattle region. A leading market centre for dark-fired tobacco, it produces rubber, flour, clothing, machine parts, and heating and cooling equipment. Austin Peay University (1927) is in the city. **90** D2

Clearwater city and administrative headquarters of Pinellas County, west-central Florida; population (1992) 98,100. It is situated on the Pinellas Peninsula, Clearwater Harbor (a lagoon of the Gulf of Mexico), and Old Tampa Bay, and lies 32 km/20 mi west of Tampa and 25 km/16 mi northwest of St Petersburg. Industries include tourism, computer components, citrus fruits, fishing, electronics, and flowers. **90** E4

Clermont-Ferrand city and administrative centre of ►Puy-de-Dôme *département* in the Auvergne region of France, situated 140 km/87 mi east of Limoges; population (1990) 140,200. It is a centre for agriculture; the rubber industry here, including the manufacture of car tyres, is the largest in France. The town has over 20 mineral springs and supplies a large amount of the country's bottled water. Other products include chemicals, preserves, foodstuffs, and clothing. **26** J8

Cleveland former county of northeast England, formed in 1974 from parts of Durham and northeast Yorkshire. It was abolished in 1996 when the unitary authorities of Hartlepool, Middlesbrough, Redcar and Cleveland, and Stockton-on-Tees were created.

Cleveland port and city in north-eastern Ohio, USA, on Lake Erie, at the mouth of the Cuyahoga River; seat of Cuyahoga County; population (1990) 505,600. Cleveland is the centre of a seven-county (including Cuyahoga, Portage, Summit, Lake, Medina, and Lorain) metropolitan area (CMSA) with a population (1990) of 2,759,800. An industrial, commercial, and transportation centre, and formerly one of the leading iron and steel producers in the USA, the city has also figured prominently in petroleum, chemicals, automobile manufacturing, and electric power. Cleveland's current industries include chemical and food processing, steel (although not on the scale seen previously), and the manufacture of electrical products and auto parts. Printing and publishing and international and Great Lakes ore shipping also figure prominently. It is also the location of dozens of corporate headquarters. 88 D2

Cluj-Napoca (German **Klausenberg**), city in Transylvania, Romania, located on the River Somes; population (1993) 322,000. It is a communications centre for Romania and the Hungarian plain. Industries include machine tools, furniture, and knitwear. 34 L3

Cluny town in Saône-et-Loire *département* of Burgundy, eastern France, on the River Grosne, 23 km/14 mi northwest of Mâcon; population (1990) 4,700. Its abbey, now in ruins, was from 910 to 1790 the foundation house of the Cluniac order, originally a reformed branch of the Benedictines. Cluny was once a lace-making centre; it has a large cattle market, and attracts many tourists. 26 K7

Clutha longest river in South Island, New Zealand, 322 km/201 mi long. It rises in Lake Wanaka in the Southern Alps, crosses the province of Otago, and flows to meet the sea near Kaitangata. It has hydroelectric installations. 78 B7

Clwyd former county of north Wales, created in 1974 and, in 1996, divided between ►Conwy, ►Denbighshire, ►Flintshire, ►Powys, and ►Wrexham unitary authorities.

Clyde third-longest river and firth (estuary) in Scotland, and longest in southern Scotland; 171 km/106 mi long. Formerly one of the world's great industrial waterways, and famed for its shipbuilding, its industrial base has declined in recent years and the capacity of the ports on the Clyde has reduced. 24 H6

Coast Ranges mountain system in North America, extending from the Coast Mountains of Alaska and British Columbia, through Washington, Oregon, and California, to Baja California in northwest Mexico, and running almost parallel to the Pacific coast. The mountains of the Coast Ranges are very irregular in height and geological structure, with a number of peaks rising to 2,000 m/6,600 ft; while in the St Elias Mountains of the Yukon is the second highest peak in North America, Mt Logan (6,050 m/19,850 ft). In some places, especially the south, the mountains are almost bare, while in other places they are densely covered with forests. To the south, the ranges are subject to occasional earthquakes. 82 E5

Coatbridge town in North Lanarkshire unitary authority, Scotland, 13 km/8 mi east of Glasgow; population (1991) 43,600. Coatbridge lies in the centre of a mineral district with a mining and manufacturing tradition. 24 J6

Coblenz alternative spelling of the German city ►Koblenz. 20 C6

Coburg town in Bavaria, Germany, on the River Itz in the foothills of the Thuringian Forest, 80 km/50 mi southeast of Gotha; population (1995) 43,900. Industries include machinery, toys, porcelain, and tourism. Formerly the capital of the duchy of Coburg, it was part of Saxe-Coburg-Gotha from 1826 to 1918, and a residence of its dukes. 20 F6

Cochabamba capital of Cochabamba department, central Bolivia, situated on the River Rocha, at an altitude of 2,550 m/8,370 ft on the slopes of the eastern Cordillera, 230 km/144 mi southeast of La Paz; population (1992) 404,100. It is the third-largest city in Bolivia. The Cochabamba valley is a rich agricultural region producing grain, beef, fruit, and timber for the towns of the Altiplano Plateau. The city is a commercial and communications centre with a railway terminus and an international airport, and distributes much of the agricultural produce of eastern Bolivia. It has food-processing and vehicle assembly plants, and there is an oil refinery, linked by pipeline with the Camiri oilfields. 98 D7

Cocos Islands (or **Keeling Islands**), group of 27 small coral islands in the Indian Ocean, about 2,700 km/1,678 mi northwest of Perth, Australia; area 14 sq km/5.5 sq mi; population (1996) 655. An Australian external territory since 1955, the islanders voted to become part of Australia in 1984, and in 1992 they became subject to the laws of Western Australia. The main product is copra (dried kernels of coconut, used to make coconut oil), and the islands are a site for ecotourism.

Cody resort town and administrative headquarters of Park County, northwest Wyoming, USA, on the Shoshone River, situated at the entrance to Yellowstone National Park; population (1990) 7,897. It is in a region of dude and working ranches and oilfields, irrigated by the Buffalo Bill Dam and Reservoir, 10 km/6 mi to the southwest. Tourism is the mainstay of Cody's economy. Other industries include oil refining and gypsum board manufacture. The town was founded by US showman William F Cody ('Buffalo Bill') in 1901. 86 E2

Cognac town in the Charente *département* of western France, on the River Charente 40 km/25 mi west of Angoulême; population (1990) 30,000. Situated in a vine-growing district, Cognac has given its name to a brandy. Under French law the name may be applied only to brandy produced within a strictly limited area around the town. Bottles, corks, barrels, and crates are manufactured here. Francis I of France was born in its castle in 1494. 26 E8

Coimbatore city in Tamil Nadu, southern India, on the Noyil River in the Nilgiri Hills, 480 km/300 mi southwest of Chennai (formerly Madras); population (1991) 816,000. It has important cotton and agricultural processing industries, along with the manufacture of agricultural machinery. These industries have expanded since the development of hydroelectric power from the Pykara Falls in the 1930s. There is an agricultural research university and the Indian Air Force Administrative College. 54 C6

Coimbra capital of Coimbra district, west-central Portugal, situated on the right bank of the Mondego River, 180 km/112 mi north of Lisbon, and 32 km/19 mi from the Atlantic coast; population (1991) 96,200. There are textile and pottery industries, and paper and biscuits are produced. Coimbra is a bishopric, and was capital of Portugal from 1139 to 1260. There is a 12th-century Romanesque church (formerly a cathedral) incorporating part of an older mosque, and a 16th-century church, which has been the cathedral since 1772. The university, founded in Lisbon in 1290, was transferred to Coimbra in 1537. 28 B4

Colchester city and river port in Essex, eastern England, on the River Colne, 80 km/50 mi northeast of London; population (1991) 96,100. It is the market centre of an agricultural and shell-fishing area, roses and oysters being notable products.

Industries include engineering, printing, and the manufacture of clothing. The oldest recorded town in England, Colchester was the capital of the kingdom of the chieftain Cymbeline, until his death in about AD 40. As **Camulodunum**, it was the first capital of Roman Britain, and it was burned by Boudicca queen of the Britons in AD 60. 22 C3

College Station city in Brazos County, east-central Texas, 132 km/ 82 mi northwest of Houston, and adjoining the city of Bryan; population (1992) 55,700. Texas A & M (Agricultural and Mechanical) University was founded here in 1876; the community grew up around the college, and was incorporated as a city in 1939. The university is known for research in engineering, fisheries, agriculture, nuclear technology, and animal pathology. 90 B3

Colmar administrative centre of the ►Haut-Rhin *département* of eastern France, on the River Lauch between the River Rhine and the Vosges mountains; population (1990) 64,900, conurbation 84,000. It is the centre of a wine-growing and market-gardening area. Industries include engineering, food processing, and textiles. The church of St Martin is 13th–14th century, and the former Dominican monastery, now the Unterlinden Museum, contains a Grünewald altarpiece. Many tourists visit the town. 30 C2

Cologne (German **Köln**), industrial and commercial port in North Rhine-Westphalia, Germany, on the left bank of the Rhine, 35 km/22 mi southeast of Düsseldorf; population (1995) 964,200. Cologne is an important transhipment centre, and a major industrial centre for the manufacture of cars (Ford), machinery, electrical goods, chemicals, clothing, and food; other industries include environmental and chemical engineering, and waste management. 20 B6

Colombia country in South America, bounded north by the Caribbean Sea, west by the Pacific Ocean, northwestern corner by Panama, east and northeast by Venezuela, southeast by Brazil, and southwest by Peru and Ecuador.

Colombia covers an area of 1,141,748 sq km/440,828 sq mi, and has a population of 42,321,000 (2000 est). The capital is ►Bogotá; other major towns and cities include Medellín, Cali, Barranquilla, Cartagena, Bucaramanga, Cúcuta, and Ibagué. The head of state and government is Andres Pastrana (from 1998). The currency is the Colombian peso, and the main language spoken is Spanish (official). 98 C3

Colombo chief commercial city and seaport, and former capital, of Sri Lanka, on the west coast near the mouth of the Kelani River; population (2001 est) 777,000. It trades in tea, rubber, and cacao. It has iron- and steelworks and an oil refinery. It was succeeded as capital by nearby Sri Jayewardenepura Kotte in 1982. 54 C7

Colón chief port and second-largest city of Panama, built on Manzanillo Island at the Caribbean entrance of the Panama Canal, 80 km/50 mi northwest of Panamá, capital city of Colón province; population (1995 est) 146,000. It is a major commercial centre and has an oil refining industry. It is second only to Hong Kong among the major free ports of the world. The free trade zone, created in 1948, has a pivotal role in bolstering the Panamanian economy. 94 J7

Colón, Archipiélago de official name of the ►Galapagos Islands. 98 (1)B1

Colorado river in southwestern USA and northwestern Mexico, rising in the Rocky Mountains and flowing 2,333 km/1,447 mi to the Gulf of California through Colorado, Utah, Arizona (including the Grand Canyon), and extending into northern Mexico. The many dams along its course, including Hoover Dam and Glen Canyon Dam, provide hydroelectric power and irrigation water, but have destroyed wildlife and scenery; they have also created a series of lakes including Lake Powell, Lake Mead, and Lake Havasu.

To the west of the river in southeastern California is the **Colorado Desert**, an arid area of some 5,000 sq km/1,931 sq mi. 92 E1

Colorado state of the western central USA. It is nicknamed the Centennial State. Colorado was admitted to the Union in 1876 as the 38th US state. Its expansion from World War II onwards has been closely associated with the US's military-industrial surge, with the state serving as the home of numerous military facilities and weapons plants. Colorado is bordered to the east by Kansas and Nebraska, to the north by Nebraska and Wyoming, and to the west by Utah. To the southwest, at the 'Four Corners', it meets Utah, Arizona, and New Mexico. New Mexico extends east along its southern border, which in the southeast also adjoins part of the Oklahoma panhandle.

population (2000 est) 4,301,300
area 269,700 sq km/104,104 sq mi
capital ►Denver
towns and cities Colorado Springs, Aurora, Lakewood, Fort Collins, Greeley, Pueblo, Boulder, Arvada
industries and products cereals, meat and dairy products, oil, coal, molybdenum, uranium, iron, steel, scientific instruments, machinery 86 E3

Colorado Springs city and resort in central Colorado, USA, 120 km/75 mi southeast of Denver, at the foot of the Front Range of the Rocky Mountains; seat of El Paso County; population (1994 est) 316,000. The main industries here are defence- and computer-related. It is situated at an altitude of about 1,800 m/6,000 ft and surrounded by magnificent scenery. 86 F3

Columbia capital of ►South Carolina, USA, on the Congaree River; seat of Richland County; population (1994 est) 104,000. Columbia is the centre of an agricultural region; industries include the manufacture of textiles, plastics, electrical goods, fertilizers, and hosiery, but the chief product is fuel assemblies for nuclear reactors. Columbia was laid out as the state capital in 1786, and was incorporated as a city in 1854. 90 E3

Columbia city and administrative headquarters of Boone County, central Missouri, USA, 38 km/24 mi northeast of Jefferson City; population (1992) 73,100. It is mainly an educational and medical research centre; also vital to the city's economy are the insurance industry, medical institutions, agricultural products, and coal. It is the home of Stephens College for women (1833), the University of Missouri (1839), the first university west of the Mississippi River, and Columbia College (1851). 90 C2

Columbia river in western North America; length over 2,005 km/ 1,245 mi. It rises in Columbia Lake on the western slope of the Rocky Mountains in British Columbia, Canada, 130 km/81 mi north of the USA border. It flows through Washington State along the northern border of Oregon, until it reaches the Pacific below Astoria; its estuary is about 55 km/34 mi long and from 5–11 km/3–7 mi wide, and its mouth is the only deep-water harbour between San Francisco and Cape Flattery. **86** C1

Columbia, District of federal district of the USA, see ►District of Columbia and ►Washington, DC.

Columbus city and port in west-central Georgia, USA, 175 km/109 mi south of Atlanta, across the Chattahoochee River from Phenix City, Alabama; seat of Muscogee County; population (1994 est) 186,000. Columbus is one of the largest textile centres in the South, and its industries are powered by hydroelectric plants situated on the Chattahoochee River; other industries include processed food, machinery, iron and steel, and peanuts. It is a distribution centre for surrounding farmlands, and lies just north of Fort Benning, the US Army infantry base. **90** E3

Columbus capital of ►Ohio, USA, on the Scioto and Olentangy rivers; seat of Franklin County; population (1994 est) 636,000; metropolitan area (1992) 1,394,000. There are coalfields and natural gas resources nearby, and local industries include the manufacture of cars, aircraft, space equipment, missiles, and electrical goods; it is also a centre for government, banking and insurance. Columbus was founded in 1812, became the state capital in 1816, and was incorporated as a city in 1834. **90** E1

Como (ancient **Comum**), town and resort in Lombardy, Italy, on the southwest shore of Lake Como at the foot of the Alps, 40 km/25 mi north of Milan; population (1990) 88,800. Products include motorcycles, glass, silk, furniture, optical instruments, and metal goods. Tourism is important. Como has a marble cathedral, built between 1396 and 1732. **30** E5

Comorin, Cape southernmost cape of the Indian subcontinent, in Tamil Nadu, where the Indian Ocean, Bay of Bengal, and Arabian Sea meet. **54** C7

Comoros country in the Indian Ocean between Madagascar and the east coast of Africa, comprising three islands – Njazidja (Grande Comore), Nzwani (Anjouan), and Mwali (Moheli). A fourth island in the group, Mayotte, is a French dependency. Together the islands are known as the Comoros Islands.

The Comoros covers an area of 1,862 sq km/718 sq mi, and has a population of 694,000 (2000 est). The capital is ►Moroni; other major towns and cities include Mutsamudu, Domoni, Fomboni, and Mitsamiouli. The head of state is Azali Assoumani (from 1999); the head of government is Hamada Madi (from 2000). The currency is the Comorian franc, and the languages spoken include Arabic, French (both official), Comorian (a Swahili and Arabic dialect), and Makua. **72** G2

Compiègne town in the Oise *département*, France, on the River Oise near its confluence with the River Aisne; population (1990 est) 44,700. Compiègne is a fashionable holiday town for the Paris region; it also has machinery, printing, rubber, and timber industries. The enormous château, founded in Merovingian times, was rebuilt by Louis XV. The same railway coach in the forest of Compiègne was the setting for both the armistice of 11 November 1918, which ended World War I, and the truce agreed by the Pétain government with Hitler in 1940. **22** E5

Conakry capital and chief port of the Republic of Guinea; population (1992) 950,000. It is on the island of Tumbo, and is linked with the Kaloum Peninsula by a causeway and (from 1914) by rail with Kankan, 480 km/300 mi to the northeast. One of the major exports is alumina (treated bauxite), which is mined at Fria; iron ore is mined on the nearby Kaloum Peninsula. Other industries include plastics, fisheries, motor vehicle assembly, and tourism. Agricultural products include bananas, oranges, pineapples, palm products, and coffee. **68** B3

Concarneau seaport in the *département* of Finistère, in Britanny, France, 20 km/12 mi southeast of Quimper; population (1990) 20,000. It is a centre of the sardine, mackerel, and tuna fisheries, and there are boatyards. The old town, which is surrounded by ramparts, dates from the 14th century; it lies on an island near the bay of La Forêt, while the newer portion, Ste-Croix, is on the opposite shore. **26** B6

Concepción industrial city in south-central Chile, 350 km/217 mi southwest of Santiago; situated on the north bank of the Bió-Bió River, near its mouth; population (1992) 330,400. It is Chile's third-largest city and capital of Concepción province and of Bió-Bió region. The city is a tourist resort and lies in a rich agricultural district. Industries include coal (from the pits on the Péninsula de Lebú, steel, glass, cement, paper, and textiles. Most of Chile's coal is mined in the vicinity. Its port, Talcahuano, 15 km/9 mi to the north, is Chile's most important naval base. **100** G6

Concord capital of ►New Hampshire, USA, in the south-central part of the state, on the Merrimack River, 30 km/19 mi north of Manchester; population (1992) 35,600. It also serves as the administrative headquarters of Merrimack County. Concord is the centre of an agricultural region; industries include granite, leather goods, electrical equipment, printed products, and wood products. **88** F2

Coney Island seaside resort in Brooklyn, in the southwest of Long Island, New York, USA; it is 10 km/6 mi long and about 1 km/0.6 mi at its widest point. Coney Island (now a peninsula) has been popular for its amusement parks and ocean bathing since the 1840s, but this has since declined; it is now mainly a residential area.

Congo former name (1960–71) of the Democratic Republic of Congo, (see Congo, Democratic Republic of) known as Zaire from 1971 to 1997. It is also the shortened form of the Republic of the Congo (see Congo, Republic of). **68** G5

Congo, Democratic Republic of country in central Africa, formerly Zaire (1971–97), bounded west by the Republic of the Congo, north by the Central African Republic and Sudan, east by Uganda, Rwanda, Burundi, and Tanzania, southeast by Zambia, and southwest by Angola. There is a short coastline on the Atlantic Ocean.

The Democratic Republic of Congo covers an area of 2,344,900 sq km/905,366 sq mi, and has a population of 51,654,000 (2000 est). The capital is ►Kinshasa; other major towns and cities include Lubumbashi, Kananga, Mbuji-Mayi, Kisangani, Kolwezi, Likasi, and Boma. The head of state and government is Joseph Kabila (from 2001). The currency is the congolese franc, and the languages spoken include French (official), Swahili, Lingala, Kikongo, Tshiluba (all national languages), and over 200 other languages. **70** C4

Congo, Republic of country in west-central Africa, bounded north by Cameroon and the Central African Republic, east and south by the Democratic Republic of Congo, west by the Atlantic Ocean, and northwest by Gabon.

The Republic of Congo covers an area of 342,000 sq km/132,046 sq mi, and has a population of 2,943,000 (2000 est). The capital is ►Brazzaville; other major towns and cities include Pointe-Noire, Nkayi, Loubomo, Bouenza, Mossendjo, Ouesso, and Owando. The head of state and government is Denis Sassou-Nguessou (from 1997). The currency is the franc CFA, and the languages spoken include French (official), Kongo, Monokutuba and Lingala (both patois), and other dialects. **68** G5

Congo River (or **Zaire River**), second-longest river in Africa, rising near the Zambia–Democratic Republic of Congo border (and known as the **Lualaba River** in the upper reaches) and flowing 4,500 km/2,800 mi to the Atlantic Ocean, running in a great curve that crosses the Equator twice, and discharging a volume of water second only to the River Amazon. The chief tributaries are the Ubangi, Sangha, and Kasai. **62** E6

Connacht (or **Connaught**), historic province of the Republic of Ireland, comprising the counties of Galway, Leitrim, Mayo, Roscommon, and Sligo; area 17,130 sq km/6,612 sq mi; population (1996) 433,200. The chief towns are Galway, Roscommon, Castlebar, Sligo, and Carrick-on-Shannon. Mainly lowland, it is agricultural and stock-raising country, with poor land in the west.

Connecticut state in New England, USA. It is nicknamed Constitution State or the Nutmeg State. Connecticut ratified the US Constitution in 1788, becoming the 5th state in the Union. It is bordered to the north by Massachusetts, to the east by Rhode Island, with the Pawcatuck River forming part of the boundary, to the west and southwest by New York, and to the south by the Long Island Sound, across which are New York's Long and Fishers islands.
population (2000 est) 3,405,600
area 13,000 sq km/5,018 sq mi
capital ►Hartford
towns and cities Bridgeport, New Haven, Waterbury, Stamford
industries and products dairy, poultry, and market-garden products; tobacco, watches, clocks, silverware, helicopters, jet engines, nuclear submarines, hardware and locks, electrical and electronic equipment, guns and ammunition, optical instruments. Hartford is the centre of the nation's insurance industry **88** F2

Connemara (Irish **Comnhaicne Mara** 'the tribe of the sea'), northwestern part of County Galway, Republic of Ireland, an area of rocky coastline and glaciated, quartzite, mountainous scenery, dominated by the Twelve Bens, the highest of which is Benbaun (730 m/2,395 ft). The principal town is Clifden. Connemara is noted for its quarries of green marble. There is fishing and tourism, and Connemara ponies are bred; kelp gathering and weaving were also once important, and continue as minor cottage industries. **24** C8

Constance (German **Konstanz**), town in Baden-Württemberg, south Germany, on the section of the River Rhine joining Lake ►Constance and Lake Untersee, 160 km/100 mi south of Stuttgart; population (2001 est) 77,900. Suburbs stretch across the frontier into Switzerland. Constance has traditional industries such as mechanical engineering, clothing manufacture, and printing, and it has also attracted modern pharmaceutical and computer industries. There is a modern university, and the town is a tourist centre and health spa. It is also a port on Lake Constance. **30** E3

Constance, Lake (German **Bodensee**), lake bounded by Germany, Austria, and Switzerland, through which the River Rhine flows; area 539 sq km/208 sq mi. It is about 72 km/45 mi long, 13 km/8 mi wide, and lies 396 m/1,300 ft above sea level. At its northwestern end it divides into two lakes, the northern being the Überlingen, and the southern – joined to Lake Constance by a short stretch of the River Rhine – the Untersee. The main part of the lake is known as the Obersee. **30** E3

Constanţa chief Romanian port on the Black Sea, capital of Constanţa region, and second-largest city of Romania; population (1993) 349,000. It has refineries, shipbuilding yards, and food factories. **58** C1

Constantine (or **Qacentina**, from 1981), city and capital of the department of Constantine in northeast Algeria; population (1998 est) 596,100. It lies 320 km/200 mi south of the capital, Algiers. Products include carpets and leather and woollen goods. The oldest city in Algeria, it was the capital of the Roman province of Numidia, but declined and was ruined, then restored in 313 by Constantine the Great, whose name it bears. It was subsequently ruled by Vandals, Arabs, and Turks and was captured by the French in 1837. **66** G1

Conwy unitary authority in north Wales, created in 1996 from parts of the former counties of Clwyd and Gwynedd.
area 1,107 sq km/427 sq mi
towns ►Conwy (administrative headquarters), Abergele, ►Llandudno, Llanrwst
physical rivers Conwy and Elwy
features Snowdonia National Park; coastline of sandy beaches, including the seaside resort of Colwyn Bay
industries tourism
population (1996) 113,000 **5** E5

Conwy port, market town and administrative centre of the ►Conwy unitary authority, north Wales, situated on a steep slope at the estuary of the River Conwy; population (1991) 3,600. It was known until 1972 by the anglicized form **Conway**. Still surrounded by walls, it has the ruins of a castle rebuilt by Edward I in 1284 to subjugate the Welsh. Bodnant Gardens are situated on the east bank of the river. **24** J8

Coober Pedy (Aboriginal 'white man in a hole'), town in the Great Central Desert, Australia; 700 km/437 mi northwest of Adelaide, southern Australia; population (1991) 2,500. The town lies on the Sturt Highway, at the edge of the desert, on the world's leading opal field. Water has to be imported, and, because of the extreme heat (temperatures reach

60°C/140°F), most of the town has been built below ground. **76** F5

Cook Islands group of six large and a number of smaller Polynesian islands 2,600 km/1,600 mi northeast of Auckland, New Zealand; area 290 sq km/112 sq mi; population (2001 est) 14,300. Their main products include citrus fruit, copra, bananas, pearl-shell, cultivated (black) pearls, and crafts. The islands became a self-governing overseas territory of New Zealand in 1965. **74** K7

Cook Strait strait dividing North Island and South Island, New Zealand, about 30 km/19 mi in width at its narrowest point. A submarine cable carries electricity from South to North Island. **78** E5

Copenhagen (Danish **København**), capital of Denmark, on the islands of Zealand and Amager; population (1995) 1,353,300 (including suburbs). **16** G9

Coral Sea (or **Solomon Sea**), part of the ►Pacific Ocean bounded by northeastern Australia, New Guinea, the Solomon Islands, Vanuatu, and New Caledonia; area 4,790,000 sq km/ 1,849,000 sq mi, with an average depth of 2,400 m/7,870 ft, with three deep trenches on its eastern edge. It contains numerous coral islands and reefs. The Coral Sea Islands are a territory of Australia; they comprise scattered reefs and islands over an area of about 1,000,000 sq km/ 386,000 sq mi. They are uninhabited except for a meteorological station on Willis Island. The ►Great Barrier Reef lies along its western edge, just off the east coast of Australia. **76** K2

Corby town in Northamptonshire, England, 32 km/19 mi north of Northampton; population (1991) 49,100. A small village until the 1930s, Corby expanded rapidly as a centre of the steel-making industry. The steelworks closed in 1979 and the town is now an enterprise zone producing plastics. Other industries include engineering, electronics, and the manufacture of steel tubing, crisps, and cosmetics (Avon). **22** B2

Cordillera Cantábrica Spanish for the ►Cantabrian Mountains, a mountain range running along the north coast of Spain. **28** D2

Cordilleras, The mountainous western section of North America, with the Rocky Mountains and the coastal ranges parallel to the contact between the North American and the Pacific plates.

Córdoba industrial city and capital of Córdoba province, central Argentina, situated on the Primero (or Suquiá) River, 400 m/1,310 ft above sea level at the foot of the Sierra Chica, between the Pampas on the east and the Andes on the west; population (1992 est) 1,179,400. Main industries include cement, glass, textiles, and motor vehicles. The city's water and electric power come from the lake created by the San Roque Dam on the River Primero. **100** J5

Córdoba capital of Córdoba province, southern Spain, on the River Guadalquivir; population (1991) 300,200. Paper, textiles, silverware, and copper products are manufactured, and there is a large trade in agricultural produce, wine, olive oil, and lead. It has many Moorish remains, including the mosque, now a cathedral, founded by 'Abd-ar-Rahman I in 785, which is one of the largest Christian churches in the world. Córdoba was probably founded by the Carthaginians; it was held by the Moors from 711 to 1236. **28** F7

Corfu (Greek **Kérkyra**), northernmost and second largest of the Ionian islands of Greece, off the coast of Epirus in the Ionian Sea; area 1,072 sq km/414 sq mi; population (1991) 105,000. Its businesses include tourism, fruit, olive oil, and textiles. Its largest town is the port of Corfu (Kérkyra), population (1991) 36,900. Corfu was colonized by the Corinthians about 700 BC. Venice held it 1386–1797, Britain 1815–64. **36** B5

Corinth (Greek **Korinthos**), port in Greece, on the Isthmus of Corinth connecting the Peloponnese with the mainland; population (2001 est) 32,800. The rocky isthmus is bisected by the 7 km/4 mi Corinth canal, opened in 1893. It is a major export centre for tobacco and local fruit. The site of the ancient city-state of Corinth lies 7 km/4 mi southwest of the port. **36** E7

Cork largest county of the Republic of Ireland, in the province of Munster; county town ►Cork; area 7,460 sq km/2,880 sq mi; population (1996) 420,500. Cork is mainly agricultural, but there is some copper and manganese mining, marble quarrying, salmon farming, and river and sea fishing; industries include chemical, and computer hardware and software. There are natural gas and oil fields off the south coast at Kinsale. Angling is a popular sport, and tourism is concentrated in Kinsale, Bantry, Glengarriff, and Youghal; one of the most popular visitor attractions is Charles Fort, Kinsale. Cork is rich in Christian and pre-Christian antiquities. **5** B5

Cork third-largest city in Ireland; port and county town of County ►Cork, important industrial and trading centre on the River Lee, at the head of the long inlet of Cork harbour, 21 km/13 mi from the sea; population (1996) 180,000. The lower harbour, at Cobh, can berth liners. The city has breweries, distilleries, container ports, and iron foundries. Other industries include cars, chemicals, food processing, oil refining, pharmaceuticals, pottery, steel, and tanning; manufacturing includes rubber and metal products, and computer hardware and software. St Fin Barre founded a school and an abbey here in the 7th century. The area was subsequently settled by Danes, who were in turn dispossessed by the Normans in 1172. **24** D10

Corner Brook second-largest city in western Newfoundland, Canada; population (1991) 22,400. It is located at the head of the Humber River estuary, near the Bay of Islands, in a lumbering region close to the Long Range Mountains. The city includes large newsprint plants and also has cement, gypsum, iron, furniture, and plasterboard industries. Huge pulp and paper mills have been operating here since the 1920s, and commercial fishing is also important to the local economy. Corner Brook is a service centre for the western region of Newfoundland. A plant at Deer Lake, to the northeast, supplies hydroelectric power. Sir Wilfred Grenfell College (1975), a branch of Memorial University, is in the city. **82** V7

Cornwall city in eastern Ontario, Canada, 88 km/55 mi southeast of Ottawa, at the foot of the long Sault Rapids of the St Lawrence River; seat of the united counties of Stormont, Dundas, and Glengarry; population (1991) 47,100. Industries include textile and chemical plants, furniture, pulp and paper, and there is a large power plant in the city. **84** M2

Cornwall county in southwest England including the Isles of ►Scilly (Scillies).
area (excluding Scillies) 3,550 sq km/ 1,370 sq mi
towns and cities ►Truro (administrative headquarters), Camborne, Launceston; Bude, Falmouth, Newquay, Penzance, St Ives (resorts)
physical Bodmin Moor (including Brown Willy 419 m/1,375 ft); Land's End peninsula; rivers Camel, Fal, Fowey, Tamar
features St Michael's Mount; Poldhu, site of first transatlantic radio signal (1901); the Stannary or Tinners' Parliament; Tate Gallery, St Ives; the Mineral Tramways Project to preserve the mining landscape, once the centre of the world's hard-rock mining industry; Eden Project, two 'biomes' (tropical rainforest and Mediterranean)
agriculture crops are early in some places: fruit, oats, and vegetables, including swedes, turnips, and mangolds; spring flowers; cattle and sheep rearing; dairy farming; fishing
industries tourism; electronics; kaolin
population (1996) 483,300 **5** D6

Coromandel Coast east coast of Tamil Nadu and Andhra Pradesh states, India, between the Krishna river delta in the north and Point Calimere in the south. **54** D6

Coromandel Peninsula peninsula on North Island, New Zealand, east of Auckland. **78** E3

Corpus Christi city and port in southeastern Texas, USA, at the mouth of the Nueces River on Corpus Christi Bay, 210 km/130 mi southeast of San Antonio; seat of Nueces County; population (1994 est) 275,000. Its port is linked to the Gulf of Mexico via the Aransas Pass. The main industries here are oil refining, fabricated metals, electronic equipment, commercial fishing, and the processing and shipping of agricultural products. **90** B4

Corrèze *département* in the Limousin region of France, area 5,860 sq km/ 2,263 sq mi; population (1990) 115,800. The region is tableland, with hills in the northeast, and is broken up by fertile river valleys. The chief rivers are the Dordogne, Vézère, and Corrèze. The main agricultural activity is the raising of cattle, pigs, and poultry. Industries include the manufacture of metals, textiles, leather, and food. There is an important hydroelectric power station at Bort-les-Orgues. The administrative centre is ►Tulle. **5** D4

Corrientes city, river port, and capital of Corrientes province, northeast Argentina, 1,025 km/640 mi northeast of Buenos Aires; situated on the Paraná River 40 km/25 mi below its confluence with the River Paraguay; population (1991) 267,700. It is in a stock-raising district, and serves as a distribution centre and export outlet for the region's agricultural produce. There are sawmills, and tanning and textile industries. **100** K4

Corse French name for the French island of ►Corsica. **32** D6/**6** (1)A1

Corse-du-Sud *département* covering the southern half of the French island of ►Corsica; area 4,014 sq km/1,550 sq mi; population (1990) 118,515. Together with Haute-Corse, it comprises the administrative region of Corsica. The border with the northern *département* runs for the most part along the central mountain spine of the island; peaks over 2,000 m/ 6,500 ft in Corse-du-Sud include Monte d'Oro (2,391m/7,844 ft) and L'Incudine (2,136 m/7,008 ft). Sheep and goats are grazed on this mainly rocky, infertile terrain, while olives, olive oil, and wine (from the lower slopes) are the principal agricultural produce. The island's capital, ►Ajaccio, on the west coast, is also the departmental capital; other towns include Sartène, Sari-d'Orcino, Vico, and Piana. **6** (1)A2

Corsica (French **Corse**), island region of France, in the Mediterranean off the west coast of Italy, north of Sardinia; it comprises the *départements* of ►Haute Corse and ►Corse du Sud; area 8,700 sq km/3,358 sq mi; population (1990) 250,400 (including just under 50% native Corsicans). The capital is ►Ajaccio. The island is largely mountainous and characterized by maquis vegetation (drought-tolerant shrubs such as cork oak and myrtle). The main products are wine and olive oil; tourism is the island's economic mainstay. The languages spoken are French (official) and Corsican, an Italian dialect. The French Emperor Napoleon was born in Ajaccio in 1769, the same year that Corsica became a province of France. The island is the main base of the Foreign Legion. **32** D6/**6** (1)A1

Cortona town in Tuscany, northern Italy, 22 km/13 mi southeast of Arezzo; population (1990) 22,600. It has a thriving business in antiques and crafts. One of Europe's oldest cities, it is encircled by walls built by the Etruscans, and has a medieval castle and an 11th-century cathedral. **32** F5

Cos alternative spelling of ►Kos, a Greek island. **36** K8

Cosenza (ancient **Consentia**), town in Calabria, southern Italy, at the junction of the rivers Crati and Busento, 140 km/87 mi northeast of Reggio di Calabria; population (1990) 104,500. It is an agricultural market centre trading in olive oil, fruit, and cereals. Silk, leather goods, furniture, and pottery are manufactured. There is a cathedral (12th–13th centuries). The town is the capital of Cosenza province. **32** L9

Cossyra ancient name for ►Pantelleria, Italian island in the Mediterranean. **66** H1

Costa Blanca (Spanish 'White Coast'), Mediterranean coastline of Alicante and Murcia provinces, eastern Spain. The chief ports are Alicante and Cartagena; popular resorts include Benidorm and Torrevieja. **28** K7

Costa Brava (Spanish 'Wild Coast'), Mediterranean coastline of northeast Spain, stretching from Port-Bou on the French border southwards to Blanes, northeast of Barcelona. It is noted for its irregular rocky coastline, small fishing villages, and resorts such as Puerto de la Selva, Palafrugell, Playa de Aro, and Lloret del Mar. **28** P3

Costa del Sol (Spanish 'Coast of the Sun'), Mediterranean coastline of Andalusia, southern Spain, stretching for nearly 300 km/190 mi from Gibraltar to Almería. Málaga is the principal port and Marbella, Torremolinos, and Nerja are the chief tourist resorts. **28** F8

Costa de Mosquitos Spanish name for ►Mosquito Coast, an area of coast in Honduras and Nicaragua. **94** H6

Costa Rica country in Central America, bounded north by Nicaragua, southeast by Panama, east by the Caribbean Sea, and west by the Pacific Ocean.

Costa Rica covers an area of 51,100 sq km/19,729 sq mi, and has a population of 4,023,000 (2000 est). The capital is ►San José; other major towns and cities include Alajuela, Cartago, Limón, Puntarenas, San Isidro, and Desamparados. The head of state and government is Miguel Angel Rodriguez Echeverria (from 1998). The currency is the colón, and the main language spoken is Spanish (official). 94 G7

Côte d'Azur Mediterranean coast from Menton to St-Tropez in the *départements* of Alpes-Maritimes and Var, France, renowned for its beaches; it is part of the region ►Provence-Alpes-Côte d'Azur. The chief resorts are Antibes, Cannes, Nice, Juan-Les-Pins, and Monte Carlo in Monaco.

Côte d'Ivoire country in West Africa, bounded north by Mali and Burkina Faso, east by Ghana, south by the Gulf of Guinea, and west by Liberia and Guinea.

Côte d'Ivoire covers an area of 322,463 sq km/124,502 sq mi, and has a population of 14,786,000 (2000 est). The capital is ►Yamoussoukro; other major towns and cities include Abidjan, Bouaké, Daloa, Man, Korhogo, and Gagnoa. The head of state is Laurent Gbagbo (from 2000); the head of government is Affi N'Guessan (from 2000). The currency is the franc CFA, and the languages spoken include French (official) and over 60 ethnic languages. 68 C3

Côte d'Or *département* in the ►Bourgogne region of France; area 8,765 sq km/3,384 sq mi; population (1990) 240,500. The River Seine rises in Châtillonnais, a densely wooded northwestern district. The plains and valleys are fertile, and there is rich pasture land. The main wealth of the region lies in its vineyards where Burgundy wines are produced. Other products are wheat, barley, potatoes, hops, and some tobacco. Sheep and cattle rearing takes place in the western districts. Iron is mined, and there are chemical, metallurgical, food, and service industries. The administrative centre is ►Dijon. Other towns include ►Beaune and Montbard. 6 F3

Côtes d'Armor maritime *département* in the ►Brittany region of France, formerly known as Côtes-du-Nord; area 6,877 sq km/2,655 sq mi; population (1990) 540,000. Wheat, oats, flax, potatoes, apples, and plums are grown; the *département* is the nation's leading producer of pork and poultry; and there is horse breeding. Slate, lime, and china clay are found, and there are flour mills, tanneries, ironworks, and shipyards. The fishing industry is of great importance; scallions are caught in St-Brieuc bay. The administrative centre is ►St-Brieuc. 6 B2

Cotonou chief port and largest city of Benin, on the Bight of Benin; population (1994) 537,000. Palm products and timber are exported, and textiles are manufactured. Although not the official capital, it is the seat of the president, and the main centre of commerce and politics. 68 E3

Cotopaxi active Andean volcano in north-central Ecuador on the border of the Cotopaxi, Napo, and Pichincha provinces. It is located 48 km/30 mi south of ►Quito at an altitude of 5,897 m/19,347 ft above sea level. It one of the highest volcanoes in the world. It is now contained within a 340 sq km/131 sq mi national park, established in 1975.

Cotswold Hills (or **Cotswolds**), range of limestone hills in Gloucestershire, South Gloucestershire, and Bath and North East Somerset, England, 80 km/50 mi long, between Bath and Chipping Camden. They rise to 333 m/1,086 ft at Cleeve Cloud, near Cheltenham, but average about 200 m/600 ft. The area is known for its picturesque villages, built with the local honey-coloured stone. Tourism is important.

Cottbus industrial city in the *Land* (administrative region) of Brandenburg, Germany; population (1995) 124,600. Industries include engineering and the manufacture of textiles, carpets, and glassware. Formerly the capital of the East German district of Cottbus 1952–90, the city is the economic and cultural centre of Lower Lusatia and the southeastern gateway to the Spreewald. 18 D6

Council Bluffs city and administrative headquarters of Pottawattamie County, southwest Iowa, on the Missouri River opposite Omaha, Nebraska; population (1998 est) 56,400. It became the eastern terminus of the Union Pacific Railroad in 1859. It has an agricultural economy, though the railway has stimulated some industrial development, including food processing and metal manufacturing. The largest employers are the three casinos which each provide over 1,000 jobs. 86 F2

Courtrai (Flemish **Kortrijk**), town in Belgium on both sides of the River Lys, in West Flanders, 42 km/26 mi southwest of Ghent; population (1997) 75,600. It is connected by canal with the coast, and by river and canal with Antwerp and Brussels. Locally grown flax has historically been the basis of its textile industry, ranging from carpet to linen production. Light industries have developed, including engineering and electronics, jewellery manufacture and printing. 22 F4

Coventry industrial city in the West Midlands, England, on the River Sherbourne, 29 km/18 mi southeast of Birmingham; population (1994 est) 303,000. Principal industries are engineering and the manufacture of electronic equipment, machine tools, agricultural machinery, manmade fibres, aerospace components, telecommunications equipment, and vehicles, including London taxis and Massey Ferguson tractors. 24 L9

Covington town in Kenton County, north-central Kentucky, USA, on the Ohio River opposite ►Cincinnati; population (1990) 43,300. It has machinery, tobacco and meatpacking plants, breweries, distilleries, and foundries, and forms a part of the Cincinnati metropolitan area. 90 E2

Cowes seaport and resort on the north coast of the Isle of Wight, England, on the Medina estuary, opposite Southampton Water, 14 km/9 mi southwest of Portsmouth; population (1991) 16,100. A major yachting centre, Cowes is the starting point for the Around the World Yacht Race, finishing at Cape Town; and Cowes Castle is the headquarters of the Royal Yacht Squadron, which holds the annual Cowes Week regatta. Maritime-related industries include boatbuilding, marine engineering, sailmaking, hovercraft construction, and the manufacture of radar equipment. Tourism is important; facilities include the island's ferry connection with Southampton. 22 A4

Cracow alternative form of ►Kraków, a Polish city. 18 J7

Craiova city in southern Romania, near the River Jiu; population (1992) 303,500. Industries include electrical engineering, food processing, textiles, fertilizers, and farm machinery. 34 L5

Crater Lake lake in southwestern Oregon; area 52 sq km/20 sq mi. It occupies the volcanic crater of Mount Mazama, and is the deepest lake in the USA; depth 589 m/1,932 ft. The lake has no river source or outlet, but is filled by snow and rain (and drained by evaporation). It is now part of Crater Lake National Park (established in 1902). 86 B2

Crawley town in West Sussex, England, northeast of Horsham, 45 km/28 mi south of London; population (1991) 88,200; Crawley urban area 114,600. It was chartered by King John in 1202 and developed as a new town from 1947. Industries include light engineering, electronics, printing, and the manufacture of plastics, pharmaceuticals, and furniture. London's Gatwick Airport is to the north. 22 B3

Creil town to the northwest of ►Paris, France, in the *département* of Oise, on the River Oise; population (1999 est) 31,400. It is an important railway junction, and many of its inhabitants commute to Paris. Industries include the manufacture of steel, rolling stock, machinery, chemicals, pottery, and glass. It has a church that dates from the 12th–15th centuries. 22 E5

Cremona town in Lombardy, Italy, on the River Po, 72 km/45 mi southeast of Milan; population (1990) 75,200. Traditionally a violinmaking centre, it also produces food products and textiles. It has a 12th-century cathedral, linked to a 13th-century campanile (bell-tower), which is 111 m/364 ft high and the tallest in Italy. 30 F5

Crete (Greek **Kriti**), largest Greek island in the eastern Mediterranean Sea, 100 km/62 mi southeast of mainland Greece; area 8,378 sq km/3,234 sq mi; population (1991) 536,900. The capital is ►Iraklion (Heraklion); other major towns are Khaniá (Canea), Rethymnon, and Aghios Nikolaos. The island produces citrus fruit, olives, and wine; tourism is vital to the economy. A Cretan dialect of Greek is spoken. 36 H10

Creuse *département* in the ►Limousin region of France; area 5,559 sq km/2,146 sq mi; population (1990) 66,800. The River ►Creuse flows through the region, which is mountainous and has thin, unproductive soil. In the southern hills there is good pasture land. Stock-rearing is important, but potatoes and some cereals are also produced. Carpets and textiles are manufactured. The principal towns are Guéret (administrative centre) and ►Aubusson. 6 E4

Creuse river in central France. It rises near Aubusson, flows 255 km/158 mi from the Plateau des Millevaches northwest through a deep valley, and joins the River Vienne 19 km/12 mi north of Châtellerault. It traverses ►Creuse *département*, to which it gives its name. 26 G7

Creusot, Le French town in the *département* of Saône-et-Loire, 65 km/40 mi northwest of Mâcon; population (2001 est) 25,900. It is in a coalfield, and has neighbouring iron-ore mines. There are numerous metal and engineering works, including the manufacture of armaments and nuclear reactors. 26 K7

Crewe town in Cheshire, England; population (1991) 63,400. It grew as a major railway junction, containing the chief construction workshops of British Rail. Crewe is also the centre of the dairy industry, providing cattle breeding, management, and animal health services. Other industries include food-processing and the manufacture of chemicals, clothing, and vehicles. 24 K8

Crimea (Ukrainian **Krym'**), northern peninsula on the Black Sea, an autonomous republic of ►Ukraine; formerly an oblast (region) of the Soviet Union (1954–91); area 27,000 sq km/10,425 sq mi; population (1995) 2,632,400. The capital is ►Simferopol; other main towns are Sevastopol and Yalta. The region produces iron, steel, and oil, and there is fruit- and vine-growing. The land is mainly steppe, but the south coast is a holiday resort. 58 E1

Crna Gora Serbo-Croat name for ►Montenegro, a constituent republic of Yugoslavia. 34 F7

Croatia (Serbo-Croat **Hrvatska**), country in central Europe, bounded north by Slovenia and Hungary, west by the Adriatic Sea, and east by Bosnia-Herzegovina and the Yugoslavian republic of Serbia.

Croatia covers an area of 56,538 sq km/21,829 sq mi, and has a population of 4,473,000 (2000 est). The capital is ►Zagreb; other major towns and cities include Osijek, Split, Dubrovnik, Rijeka, Zadar, and Pula. The head of state is Stipe Mesic (from 2000); the head of government is Ivica Racan (from 2000). The currency is the kuna, and the languages spoken include Croat (official) and Serbian. 34 C4

Cuba island country in the Caribbean Sea, the largest of the West Indies, off the south coast of Florida and to the east of Mexico.

Cuba covers an area of 110,860 sq km/42,803 sq mi, and has a population of 11,201,000 (2000 est). The capital is ►Havana; other major towns and cities include Santiago de Cuba, Camagüey, Holguín, Guantánamo, Santa Clara, Bayamo, and Cienfuegos. The head of state and government is Fidel Castro Ruz (from 1959). The currency is the Cuban peso, and the main language spoken is Spanish (official). 94 H4

Cubango name of the Okavango River in Angola. For part of its length it forms the border between Angola and Namibia. At the western end of the Caprivi Strip it crosses into Namibia and is known as the Kavango River; it becomes known as the Okavango River (see ►Okavango Delta) when it enters Botswana. 72 B3

Cúcuta capital of Norte de Santander department, northeast Colombia, 16 km/10 mi from the Venezuelan frontier; population (1994) 470,000. It is the gateway to Venezuela, with which it is linked by road, rail, and air, and handles most of the trade between the two countries. Coffee and tobacco are grown in the region, and cattle are raised; exports include coffee and hides. The River Pamplonita flows through the city. There is a cathedral, and a university (founded 1962). 98 C2

Cuenca capital of Cuenca province in Castilla–La Mancha, central Spain, 135 km/84 mi southeast of Madrid, at the confluence of the Júcar and Huécar rivers; population (1991) 42,600. Industries include timber and tourism. It has a 13th-century cathedral, a ruined Moorish castle, and a 16th-century bishop's palace. Nearby is the Ciudad Encantada (Enchanted City), a rock formation of 200 ha/494 acres, eroded into the appearance of a ruined city. 28 H4

Cuenca capital of Azuay province, southern Ecuador, 330 km/205 mi south of Quito; population (1991) 193,000. Industries include chemicals, food processing, agricultural machinery, flour milling, tyres, panama hats, and textiles. It is Ecuador's third-largest city and a commercial centre, trading in cinchona bark, sugar cane, and cereals; it is also a major centre of trade for Ecuadorean handicrafts. 98 B4

Cuiabá capital of ►Mato Grosso state, west-central Brazil, 1,000 km/ 6,214 mi from Brasilia, on the River Cuiabá, a tributary of the Paraguay; population (1991) 389,000. It lies on the Pan-American Highway (BR-230) from Brasilia across to the Peruvian border. Situated in a cattle-raising region, Cuiabá is a distribution centre for cattle, hides, and dried meat; cereal crops are also important in the region. Rubber and palmnut gathering are important sustainable industries that provide the region with export products. Gold and diamonds are mined nearby. The city grew with the discovery of gold in 1719 and was the capital of the entire ►Mato Grosso region until Mato Grosso was divided in the 1970s. 98 F7

Culiacán (Spanish **Culiacán Rosales**), capital of Sinaloa state, northwestern Mexico, on the Culiacán River; population (1990) 415,000. It trades in vegetables and textiles. It was founded in 1599. 94 C4

Cumberland city and administrative headquarters of Allegany County, northwest Maryland, USA, in the Allegheny Mountains, on the Potomac River and Wills Creek, 85 km/53 mi south of Johnstown, Pennsylvania; population (1990) 23,700. Cumberland is the industrial centre of a forest, farm, and coal region. Industries manufacture sheet metal, missile components, rubber, textiles, paper products, glass, bricks, beer, and railroad equipment. It is home to Allegany Community College (1961). 88 E3

Cumbria county of northwest England, created in 1974 from Cumberland, Westmorland, the Furness district of northwest Lancashire, and the Sedbergh district of northwest Yorkshire.
area 6,810 sq km/2,629 sq mi
towns and cities ►Carlisle (administrative headquarters), Barrow, Kendal, Penrith, Whitehaven, Workington
physical Scafell Pike (978 m/3,210 ft), the highest mountain in England, Helvellyn (950 m/3,118 ft); Lake Windermere, the largest lake in England (17 km/10.5 mi long, 1.6 km/ 1 mi wide), and other lakes (Derwentwater, Grasmere, Haweswater, Ullswater); the rivers Eden and Derwent
features Lake District National Park; Furness peninsula; Hadrian's Wall
agriculture dairy farming; sheep; the West Cumberland Farmers is England's largest agricultural cooperative
industries the traditional coal, iron, and steel industries have been replaced by chemicals, plastics, marine engineering, electronics, and shipbuilding (at Barrow-in-Furness, nuclear submarines and warships); tourism; salmon fishing
population (1996) 490,600 5 E4

Cunene (or **Kunene**), river rising near Huambo (formerly Nova Lisboa) in west-central Angola. It flows south to the frontier with Namibia, then west to the Atlantic Ocean; length 945 km/587 mi. The river forms the border between Angola and Namibia for a distance of 250 km/150 mi. 72 A3

Curitiba capital of Paraná state, on the Curitiba River, southeast Brazil, situated on the Serra do Mar plateau at an altitude of 900 m/2,593 ft; population (1991) 1,248,300 (metropolitan area 2,319,500). It is connected by rail to its seaport Paranaguá, 80 km/50 mi to the east. Curitiba is in the centre of a rich agricultural region which produces timber, coffee, yerba maté, and cereals. Industries include paper, textiles, chemicals, and furniture. Coffee, timber, and maté (a beverage) are exported. It has a cathedral (1893) modelled on the cathedral in Barcelona, Spain, and is the site of the Federal University of Paraná. The city was founded in 1648 as a gold-mining camp and made capital of the state in 1853. 100 M4

Curtea de Argeş city in Argeş county, Romania, on the River Argeş, a tributary of the Danube, 25 km/ 15 mi northwest of Piteşti; population (1990) 32,600. Its cathedral dates from the 16th century. The town was capital of Wallachia 1330–83. 34 M4

Cusco alternative spelling of ►Cuzco, a city in south-central Peru. 98 C6

Cuttack city and river port in east India, on the Mahanadi River delta; population (1991) 403,000. It is noted for its gold and silver filigree work, and produces textiles and electrical equipment. Almost surrounded by the Mahanadi River and its tributary the Kathjuri, it is at the centre of a network of canals. Founded in the 10th century, it was the capital of Orissa state until 1950. The old fort, **Kataka**, from which the city takes its name, is in ruins. 54 E4

Cuxhaven seaport and health resort in Lower Saxony, Germany, on the southern side of the Elbe estuary, at its entrance into the North Sea, 174 km/ 108 mi northwest of Hannover; population (1995) 55,400. It is a cross-channel ferry terminus and an important fishing port, the second-largest in Germany, after Bremerhaven. 20 D3

Cuzco (or **Cusco**), capital of Cuzco department, south-central Peru, 560 km/350 mi southeast of Lima; situated in a small valley in the Andes at a height of over 3,350 m/11,000 ft above sea level; population (1993) 255,600. The city is a commercial centre, the hub of the South American travel network, and a tourist resort. Manufactures include woollen and leather goods, beer, and fertilizers. Cuzco was founded *c.* 1200 as the ancient capital of the Inca empire and was captured by the Spanish conquistador Francisco Pizarro in 1533. It is the archaeological capital of the Americas and the oldest continually inhabited city on the continent. 98 C6

Cwmbran (Welsh 'Vale of the Crow'), town and administrative centre of ►Monmouthshire, southeast Wales, situated on the Afon Lywel, a tributary of the River Usk; population (1991) 46,000. It was established in 1949 to provide a focus for new industrial growth in a depressed area, and produces scientific instruments, car components, nylon, and biscuits. There are also engineering and electrical industries. 24 J10

Cyclades (Greek **Kyklades**), group of about 200 Greek islands in the Aegean Sea, lying between mainland Greece and Turkey; area 2,579 sq km/

996 sq mi; population (1991 est) 95,100. They include Andros, Melos, Paros, Naxos, and Siros, on which is the capital ▶Hermoupoli. 36 G7

Cymru Welsh name for ▶Wales. 24 J9

Cyprus island in the Mediterranean Sea, off the south coast of Turkey and west coast of Syria.

Cyprus covers an area of 9,251 sq km/ 3,571 sq mi (3,335 sq km/ 1,287 sq mi is Turkish-occupied), and has a population of 786,000 (2000 est). The capital is ▶Nicosia (divided between Greek and Turkish Cypriots); other major towns and cities include Limassol, Larnaca, Paphos, Lefkosia, and Gazimagusa. The heads of state and government are Glafkos Clerides (Greek) (from 1993) and Rauf Denktaş (Turkish) (from 1976). The currencies are the Cyprus pound and the Turkish lira, and the languages spoken include Greek, Turkish (both official), and English. 58 E6

Czech Republic landlocked country in east-central Europe, bounded north by Poland, northwest and west by Germany, south by Austria, and east by the Slovak Republic.

The Czech Republic covers an area of 78,864 sq km/30,449 sq mi, and has a population of 10,244,000 (2000 est). The capital is ▶Prague; other major towns and cities include Brno, Ostrava, Olomouc, Liberec, Plzeň, Hradec Králové, and C̄eské Budējovice. The head of state is Václav Havel (from 1993); the head of government is Miloš Zeman (from 1998). The currency is the koruna, and the languages spoken include Czech (official) and Slovak. 18 C8

Częstochowa city in Poland, on the River Warta, 193 km/120 mi southwest of Warsaw; population (1993) 258,700. It produces iron goods, chemicals, paper, and cement. The basilica of Jasna Góra is a centre for Catholic pilgrims (it contains the painting known as the Black Madonna). 18 J7

Dabrowa Górnicza industrial city in Katowice province, Poland, 25 km/ 16 mi east of Katowice; population (1993 est) 139,700. At the centre of a coal basin, it also produces iron, zinc, and steel. It became Prussian 1795, part of Russian Poland 1815, and was ceded to Poland 1919. 18 J7

Dacca alternative name for ▶Dhaka, the capital of Bangladesh. 54 F4

Dachau site of a Nazi concentration camp during World War II, in Bavaria, Germany. The first such camp to be set up, it opened early in 1933 and functioned as a detention and forced labour camp until liberated in 1945. 30 G2

Dadra and Nagar Haveli since 1961, a Union Territory of west India, between Gujarat and Maharashtra states; area 490 sq km/189 sq mi; population (1991) 138,500. The capital is Silvassa. Until 1954 it was part of Portuguese Daman. It produces rice, wheat, and millet. 40% of the total area is forest. 10 B4

Dagestan (or **Daghestan**, 'mountain kingdom'), autonomous republic in the southwestern Russian Federation, in northern Caucasia; area 50,300 sq km/19,421 sq mi; population (1996) 2,098,000 (42% urban). The main cities are ▶Makhachkala (capital) and Derbent. Situated mainly on the northeastern slopes of the main Caucasus Mountains, Dagestan is bounded on the east by the north-western shore of the Caspian Sea; the Nogay steppe lowland is in the north, and the principal river is the Terek. There are plentiful oil and natural-gas deposits. Chief industries are oil and gas extraction, metalworking, and traditional crafts (carpet weaving); agricultural activities centre on the raising of livestock, the cultivation of grain and grapevines, and horticulture. 58 M2

Dairen former name for the Chinese port of Dalian, part of ▶Lüda. 46 G3

Dakar capital, chief port (with artificial harbour), and administrative centre of Senegal; population (1992) 1,729,800. It is situated at the tip of the Cape Verde peninsula, the westernmost point of Africa. It is a major industrial centre, with industries including crude-oil refining, engineering, chemicals, brewing, and tobacco and food processing. Dakar contains the Grand Mosque, National Museum, and a university (established in 1949). 68 A2

Dakota see ▶North Dakota and ▶South Dakota. 86 F1/86 F2

Dalap-Uliga-Darrit (or **D-U-D**), capital of the ▶Marshall Islands; population (1998 est) 19,700. It is formed from Dalap (or Delap), Uliga, and Darrit (or Rita), three of the inhabited islands of the Majuro Atoll, on whose 57 islands and islets over half the total population of the Marshall Islands lives. The local economy is based on fishing and the cultivation of coconuts, cassava, and sweet potatoes. Uliga is the main business district, and banking and tourism are increasingly important, though there is a large balance-of-trade deficit which is currently made good by aid from the USA. The government offices are based in Dalap, while Darrit has a primary and a secondary school. 74 H5

Dales (or **Yorkshire Dales**), series of river valleys in northern England, running east from the Pennines in West Yorkshire; a National Park was established in 1954. The principal valleys are Airedale, Nidderdale, Swaledale, Teesdale, Wensleydale, and Wharfedale. The three main peaks are Ingleborough, Whernside, and Pen-y-Ghent.

Dalian (or **Talien** or **Dairen**), port in Liaoning province, China, on the Liaodong Peninsula, facing the Yellow Sea; population (1994) 2,638,300. Industries include engineering, oil-refining, shipbuilding, food-processing (soybeans), and the manufacture of chemicals, textiles, railway locomotives, and fertilizers. It has ice-free, deep water facilities, and comprises the naval base of Lüshun (known under 19th-century Russian occupation as Port Arthur) and the commercial port of Dalian, together formerly known as Lüda. 46 G3

Dallas commercial city in north-eastern Texas, USA, on the Trinity River; seat of Dallas County; population (1994 est) 1,023,000, metropolitan area (with Fort Worth) (1994 est) 4,362,000. The second-largest city in Texas (Houston is the largest), Dallas is the hub of a rich cotton-farming and oil-producing region, and is one of the leading cultural and manufacturing centres in the Southwest; its industries include banking, insurance, oil, aviation, aerospace, and electronics. Dallas was founded in 1841, and was incorporated as a city in 1871. 90 B3

Dalmatia region divided between Croatia, Montenegro in Yugoslavia, and Bosnia-Herzegovina. The capital is Split. It lies along the eastern shore of the Adriatic Sea and includes a

number of islands. The interior is mountainous. Important products are wine, olives, and fish. Notable towns in addition to the capital are Zadar, Sibenik, and Dubrovnik.
history Dalmatia became Austrian in 1815 and by the treaty of Rapallo in 1920 became part of the kingdom of the Serbs, Croats, and Slovenes (Yugoslavia from 1931), except for the town of Zadar (Zara) and the island of Lastovo (Lagosta), which, with neighbouring islets, were given to Italy until transferred to Yugoslavia in 1947.

Daman town and mainland part of the Union Territory of Daman and Diu, western India; population (2001 est) 35,400. **54** B4

Daman and Diu Union Territory of west India; area 112 sq km/43 sq mi; capital Daman; population (1991) 101,400. **Daman** has an area of 72 sq km/28 sq mi. The port and capital, Daman, is on the west coast, 160 km/100 mi north of Mumbai (formerly Bombay), on the estuary of the Daman Ganga River flowing in the Gulf of Khambhat. The economy is based on tourism and fishing. **Diu** is an island off the ▶Kathiawar peninsula with an area of 40 sq km/ 15 sq mi. The main town is also called Diu. The economy is based on tourism, coconuts, pearl millet, and salt.
history **Daman** was seized by Portugal in 1531 and ceded to Portugal by the Shah of Gujarat in 1539; **Diu** was captured by the Portuguese in 1534. Both areas were annexed by India in 1961 and were part of the Union Territory of ▶Goa, Daman, and Diu until Goa became a separate state in 1987. **10** B4

Damascus (Arabic **Dimashq** or **ash-Sham**), capital of Syria, on the River Barada, 100 km/62 mi southeast of Beirut; population (1993) 1,497,000. It produces silk, wood products, textiles, brass, and copperware. Said to be the oldest continuously inhabited city in the world, Damascus was an ancient city even in Old Testament times. **60** D3

Damietta (or **Dumyat**), town, port, and governorate of Egypt, on the eastern branch of the Nile, lying between the right bank of the river and Lake Menzaleh, about 15 km/ 9 mi from the sea; governorate – area 590 sq km/228 sq mi, population (1995 est) 898,000; town – population (1991) 113,000. Cotton, silk, and pottery goods are manufactured, and there is an export trade in rice. The port declined in importance after the construction of the Suez Canal and the development of Port Said. **64** F1

Da Nang port city (formerly Tourane) of southern Vietnam, centrally placed on the South China Sea, 80 km/50 mi southeast of Hué; population (1995 est) 416,000. Following the reunion of North and South Vietnam, the major part of the population was dispersed in 1976 to rural areas. Textiles are an important local industry. **50** D3

Danube (German **Donau**), second longest of European rivers, rising on the eastern slopes of the Black Forest, and flowing 2,858 km/1,776 mi across Europe to enter the Black Sea in Romania by a swampy delta. **14** F3

Danville city in Pittsylvania County (from which it is administratively separate), southern Virginia, USA, situated to the north of the North Carolina border on the Dan River, 88 km/55 mi southeast of Roanoke; population (1992) 53,600. It is a market and processing centre for bright-leaf tobacco, and is also known for its textile industry, particularly the Dan River mills, at one time the largest cotton mills in the world. Danville also produces glass, tools, tires, and insulating materials. The city is the home of Averett College (1859) and Danville Community College (1967). **90** F2

Danzig German name for the Polish port of ▶Gdańsk. **18** H3

Dardanelles (Turkish **Çanakkale Boğazi**; ancient **Hellespont**), Turkish strait connecting the Sea of Marmara with the Aegean Sea; its shores are formed by the Gallipoli peninsula on the northwest and the mainland of Anatolia on the southeast. It is 75 km/47 mi long and 5–6 km/ 3–4 mi wide. **36** J4

Dar el-Beida Arabic name for the port of ▶Casablanca, Morocco. **66** D2

Dar es Salaam (Arabic 'haven of peace'), chief city and seaport of Tanzania, on the Indian Ocean, and administrative capital, pending the transfer of government functions to ▶Dodoma, which was designated the official capital in 1974; population (1996 est) 1,747,000, having grown very rapidly from 150,000 in 1964. Industries include food processing, textiles, clothing, footwear, petroleum refining, glass, printing, timber, aluminium, steel, polystyrene, machinery, and car components. Exports include copper, coffee, sisal, and cotton. As well as being the chief port and largest city, it is also the main industrial, commercial, and financial centre of Tanzania, and includes the main international airport. **70** F5

Darhan (or **Darkhan**), industrial town in Mongolia, near the border with Russia; population (1991) 90,000. Industries include the manufacture of cement and bricks, and to the south is Erdenet, where copper and molybdenum are mined. **44** H7

Darjeeling (or **Darjiling**), town and health resort in West Bengal, northeast India; situated 2,150 m/7,000 ft above sea level, on the southern slopes of the Himalayas; population (1991) 73,100. The centre of a tea-producing district, it is connected by rail with Kolkata (formerly Calcutta), 595 km/ 370 mi to the south. Formerly the summer capital of the British administration in Bengal, it remains a popular hill resort. **54** E3

Darjiling alternative spelling for ▶Darjeeling. **54** E3

Darling river in southeast Australia, a tributary of the River Murray; length 2,736 km/1,700 mi. The Darling is formed about 40 km/25 mi northeast of Bourke, at the union of the Culgoa and Bogan rivers (which rise in central Queensland to the west of the Great Dividing range); it flows southwest before joining the Murray at Wentworth. Its waters are conserved in Menindee Lake (155 sq km/60 sq mi) and others nearby. The river is usually navigable as far as Bourke during August to March. **76** H6

Darlington unitary authority (borough status) in northeast England, created in 1997.
area 197 sq km/76 sq mi
towns and cities ▶Darlington (administrative headquarters)
features River Skerne flows through Darlington, River Tees forms southern boundary of authority; Darlington Railway Centre and Museum houses English engineer George Stephenson's locomotion engine

industries heavy engineering, iron and steel, vehicle components, bridge building, telecommunications, fitted furniture, textiles, knitting wool, agriculture
population (1996) 100,600 5 F4

Darlington industrial town and administrative headquarters of ►Darlington unitary authority in northeast England, on the River Skerne near its junction with the Tees, 53 km/33 mi south of Newcastle; population (1996) 100,600. The town was part of the county of Durham until 1997. Industries include heavy engineering, bridge-building, and the production of iron and steel goods, knitting wool, vehicle components, textiles, and fitted kitchens and bathrooms (Magnet). The world's first passenger railway was opened between Darlington and Stockton on 27 September 1825. 24 L7

Darmstadt city in the *Land* (administrative region) of Hesse, Germany, 29 km/18 mi south of Frankfurt-am-Main; population (1995) 139,100. Industries include iron founding, the manufacture of chemicals and plastics, electronics, and publishing. It is a centre of the European space industry. It has a ducal palace and a Technical University (1836). 20 D7

Dartmoor plateau of southwest Devon, England; mostly a national park, 956 sq km/369 sq mi in area. Over half the region is around 300 m/1,000 ft above sea level, making it the highest and largest of the moorland areas in southwest England. The moor is noted for its wild aspect and the tors, rugged blocks of bare granite, which crown its loftier points. The highest are Yes Tor, rising to 619 m/2,030 ft, and High Willhays, which climbs to 621 m/2,039 ft.

Dartmouth English seaport at the mouth of the River Dart; 43 km/27 mi east of Plymouth, on the Devon coast; population (1996 est) 6,000. It is a centre for yachting and has an excellent harbour. The Britannia Royal Naval College dates from 1905. Dartmouth Castle (15th century), 1.6 km/1 mi southeast of the town, guards the narrow entrance of the Dart estuary. 82 U8

Darwin port and capital of ►Northern Territory, Australia; population (1996) 70,251. Darwin is situated at the centre of Australia's north coast, in the northwest of Arnhem Land. It is a service centre for the northern part of Northern Territory, and industries include mining (uranium and copper), horticulture, fruit growing, and tourism. Darwin was destroyed in 1974 by Cyclone Tracy, and rebuilt on the same site. It is a base for tourists visiting Kakadu National Park, Bathurst Island, and Melville Island. 76 F2

Dasht-e-Kavir (or **Dasht-i-Davir**), salt desert southeast of Tehran, Iran; US forces landed here in 1980 in an abortive mission to rescue hostages held at the American Embassy in Tehran. 56 F3

Daugavpils (Russian **Dvinsk**), city in eastern Latvia on the River Daugava (Western Dvina), 175 km/109 mi southeast of Riga; population (1992 est) 129,000. A fortress of the Teutonic Knights from 1278, it became the capital of Polish Livonia (a former independent region until 1583, comprising most of present-day Latvia and Estonia). The city is an important railway junction, and its industries include timber, engineering, and the manufacture of textiles and food products. 38 E3

Davao city in the Philippine Republic, at the mouth of the Davao River on the island of Mindanao; population (1990) 849,900. It is the capital of Davao province. It is the centre of a fertile district and trades in pearls, copra, rice, and corn. 50 H5

Davenport city in southeastern Iowa, USA, 106 km/66 mi south of Dubuque, directly across from Rock Island, Illinois, on the Mississippi River; seat of Scott County; population (1998 est) 95,300. It forms the 'Quad Cities' metropolitan area with Bettendorf, also in Iowa, and Rock Island and Moline in Illinois. Davenport is the commercial centre for the surrounding agricultural region; industries include aluminium, agriculture, and machinery parts. 88 B2

Daventry town in Northamptonshire, central England, 19 km/12 mi west of Northampton; population (1991) 18,000. Because of its central position, the BBC's high-power radio transmitter was sited here in 1925. Originally specializing in footwear manufacture, the town received London and Birmingham overspill from the 1950s, and developed diverse light industries such as the manufacture of metal products, motor accessories, and vehicle components (Ford). 22 A2

Davos town in an Alpine valley in Graubünden canton, Switzerland, situated 20 km/12 mi southeast of Chur, 1,559 m/5,115 ft above sea level; population (1990) 10,400. It consists of two villages, Davos-Platz and Davos-Dorf. It is recognized as a health resort and winter sports centre, with one of the world's finest ski runs, the Parsenn. 30 E4

Dawson (or **Dawson City**), town in western Yukon Territory, Canada, at the junction of the Yukon and Klondike rivers; population (1991) 970. There is some agriculture here, but tourism is its main industry. Dawson was founded in 1896 at the time of the Klondike gold rush, and became a well-established mining city; from 1898 to 1953 it was the capital of the ►Yukon Territory. The population dwindled from 30,000 in 1898 to 745 in 1971, owing to the exhaustion of the Klondike goldfields. 92 (1)K3

Dawson Creek town in Peace River Regional District, northeast British Columbia, Canada; population (1991) 11,000. It is situated in the eastern foothills of the Rocky Mountains, on Dawson Creek, near the Alberta border, 267 km/166 mi northeast of Prince George. It is the southeastern terminus of the Alaska Highway. Dawson Creek was laid out in 1919, and grew after the completion of the highway during World War II. Industries include oil refining, trucking, lumber, livestock, and grains. The highway brings tourism, and hydroelectric projects on the Peace River have also contributed to the regional economy. 82 G5

Dayton city in southwestern Ohio, USA, on the junction of the Great Miami and Stillwater rivers, 75 km/47 mi north of Cincinnati; seat of Montgomery County; population (1994 est) 179,000. It is the centre of an agricultural region and the hub of a large metropolitan area; industries include the manufacture of motor vehicle parts, business machines, household appliances, and electrical equipment. It has an aeronautical research centre and was the home of aviators Wilbur and Orville Wright. **88** D3

Daytona Beach city in Volusia County, northeast Florida, USA, on the Atlantic coast; population (1992) 64,600. It is a popular seaside resort noted for its hard, white-sand beach; economic activities include printing, commercial fishing, and the manufacture of electronic equipment and metal products. The Daytona International Speedway (for motor racing) is located here, as are Bethune-Cookman College (1904) and Daytona Beach Community College (1958). **90** E4

Dead Sea large lake, partly in Israel and partly in Jordan, lying 394 m/1,293 ft below sea level; it is the lowest surface point on earth; area 1,020 sq km/394 sq mi. The chief river entering it is the Jordan; it has no outlet and the water is very salty (340 g of salt per litre of water). The sea is not, however, completely dead. *Dunaliella parva*, a single-celled green alga, and a group of halophilic (salt-loving) Archaea are found here. In 1998, three species of fungi were discovered to be living in the Dead Sea. One of the species is new to science and cannot survive without salt. **60** C5

Deal port and resort on the east coast of Kent, southeast England; population (1991) 28,500. Industries include fishing, boatbuilding, and tourism. Julius Caesar is said to have landed at nearby Walmer in 55 BC, and Deal became one of the Cinque Ports (group of ports in southern England, originally five, including Sandwich, Dover and Hastings). Deal castle, built by Henry VIII, houses the town museum. **22** D3

Death Valley desert depression in southeastern California, USA; 225 km/140 mi long and 6–26 km/4–16 mi wide; area 8,368 sq km/3,231 sq mi. At 85 m/280 ft below sea level, Death Valley is the lowest point in North America. It is one of the world's hottest and driest places, with summer temperatures sometimes exceeding 51.7°C/125°F and an annual rainfall of less than 5 cm/2 in. Borax, iron ore, tungsten, gypsum, and salts are extracted here. **86** C3

Debrecen second-largest city in Hungary, 193 km/120 mi east of Budapest, in the Great Plain (*Alföld*) region; population (1995) 211,000. It produces tobacco, agricultural machinery, and pharmaceuticals. Lajos Kossuth declared Hungary independent of the Habsburgs here 1849. It is a commercial centre and has a university founded 1912. **34** J2

Decatur city and administrative headquarters of Vermilion County, east-central Illinois, 65 km/40 mi east of Springfield, on the Sangamon River and Lake Decatur; population (1992) 84,300. It is the commercial and distribution centre for the area's corn and soybeans and its dairy farms. Local manufactures include lift trucks, ballasts, aerosol products, welding and mining machinery, and confectionery. It is the home of James Millikin University (1901). **88** C3

Decazeville French town in the *département* of Aveyron, 30 km/19 mi northwest of Rodez; population (1990) 8,800. It was the centre of the old Aveyron coalfield, and has ironworks, chemical factories, and cement works. **26** H9

Deccan triangular volcanic plateau in east India, stretching between the Vindhya Hills in the north, and the Western and Eastern ►Ghats in the south. The plateau reaches heights of 750 m/2,460 ft and is drained eastwards by the Godavari, Bhima and Krishna rivers, which have cut broad valleys. Cotton is the principal cash crop of the black lava soils, in association with millet, wheat, and other grains. **54** C5

Děčín (German **Tetschen**), river port in northern Bohemia, in the Czech Republic, on the River Elbe opposite Podmokly near the German border, 72 km/45 mi north of Prague; population (1991) 55,100. Manufactures include chemicals, textiles, and paper. **18** D7

Dee river which flows through Aberdeenshire, Scotland and the city of Aberdeen; length 137 km/85 mi. From its source in the Cairngorm Mountains, it flows east into the North Sea at Aberdeen (by an artificial channel in this latter stage). Near Braemar the river passes through a rock gorge, the **Linn of Dee**. Balmoral Castle is on its banks. It is noted for salmon fishing and is the fifth-longest river in Scotland. **24** K4

Dehra Dun city in Haranchal, India; population (1991) 367,000. It is the capital of Dehra Dun district. It has a military academy and a forest research institute. It was founded in the 17th century by the Sikh Guru Ram Rai, and has a Sikh temple built in 1699. **54** C2

Delaware state in northeastern USA. It is nicknamed the First State or the Diamond State. Delaware ratified the US Constitution in 1787, becoming the first state in the Union, hence its nickname. It is one of the most industrialized states in the USA. It is bordered to the north by Pennsylvania, to the west and south by Maryland, with which it shares the upper part of the Delmarva Peninsula, and to the east by the Atlantic Ocean.
population (2000 est) 783,600
area 5,300 sq km/2,046 sq mi
capital ►*Dover*
towns and cities Wilmington, Newark
industries and products dairy, poultry, and market-garden produce; fishing; chemicals, motor vehicles, and textiles **90** F2

Delft town in South Holland province, the Netherlands, on the Schie Canal, 14 km/9 mi northwest of Rotterdam; population (1997) 94,000. It is known worldwide for its pottery and Delftware porcelain. Other industries include engineering, electronic equipment and cable production, and printing. There is a technical university, founded in 1863. The Dutch nationalist leader William the Silent was murdered here in 1584. It is the birthplace of the artists Jan Vermeer and Michiel van Miereveld. **22** G2

Delhi (or **Old Delhi**), city of India, and administrative capital of the Union Territory of ►Delhi (state); population (1991) 8,375,000. It borders on ►New Delhi, capital of India, to the south. Manufactured goods include electronic goods, chemicals, and precision instruments, as well as traditional handicrafts such as hand-woven textiles and jewellery. An international airport is 13 km/8 mi away at Palam. The University of Delhi (1922) has over 20,000 students. **54** C3

Delhi Union Territory of India from 1956, capital ►Delhi; area 1,500 sq km/579 sq mi; population (1994) 9,500,000. It produces grain, sugar cane, fruit, and vegetables. **10** C3

Delray Beach city in Palm Beach County, southeast Florida, on the Atlantic Ocean, 40 km/25 mi north of Fort Lauderdale; population (1990) 47,200. Essentially a resort and residential community, it has some business and light industry. It is also a centre for citrus, vegetable, chrysanthemum, and gladiolus production. **90** E4

Denain French town in the *département* of Nord, 40 km/25 mi southeast of Lille, on the River Scheldt; population (2001 est) 20,300. Denain has coalmines and metal industries. During the War of the Spanish Succession, it was the site of the decisive victory of the French forces under Marshal Villars (1653–1734) over the Austrian forces led by Prince Eugène in July 1712. **22** F4

Denbighshire (Welsh **Sir Ddinbych**), unitary authority in north Wales. A former county, between 1974 and 1996 it was largely merged, together with Flint and part of Merioneth, into Clwyd; a small area along the western border was included in Gwynedd.
area 844 sq km/326 sq mi
towns Ruthin (administrative headquarters), Denbigh, Llangollen
physical Clwydian range of mountains rises to a height of 555 m/1,820 ft, with ►Offa's Dyke along the main ridge; rivers Clwyd, Dee, Elwy
features Denbigh and Rhuddlan castles; seaside resorts of ►Rhyl and Prestatyn
industries agriculture (chiefly dairy), tourism
population (1996) 91,000 **5** E5

Dender river in Belgium; length 88 km/55 mi. The Dender rises in the province of Hainaut; it joins the Scheldt at Dendermonde, after passing Ath, Grammont, Ninove, and Alost, and is navigable from its junction with the Scheldt to Ath. **22** F4

Den Haag Dutch form of The ►Hague, a city in the Netherlands. **22** G2

Den Helder port in North Holland province, the Netherlands, 65 km/40 mi north of Amsterdam, on the entrance to the North Holland Canal from the North Sea; population (1993) 61,100. It is a fishing port and naval base. **22** G2

Dénia town in the province of ►Alicante, eastern Spain, on the Mediterranean coast; population (1991) 24,700. Tourism is the main economic activity; there is also some ceramics, leather, and fireworks manufacturing. Founded by the Phoenicians, it was important in Roman times; under the Moors its population was 50,000. **28** L6

Denison city in Grayson County, northeast Texas, near the Red River, 115 km/71 mi north of Dallas; population (1990) 21,500. A distribution centre for grain and dairy products, its industries include electronic equipment, textiles, wood products, and food processing. **90** B3

Denmark peninsula and islands in northern Europe, bounded to the north by the Skagerrak, east by the Kattegat, south by Germany, and west by the North Sea.

Denmark covers an area of 43,075 sq km/16,631 sq mi, and has a population of 5,293,000 (2000 est). The capital is ►Copenhagen; other major towns and cities include Århus, Odense, Ålborg, Esbjerg, Randers, Kolding, and Horsens. The head of state is Queen Margrethe II (from 1972); the head of government is Poul Nyrup Rasmussen (from 1993). The currency is the Danish krone, and the languages spoken include Danish (official) and German. **14** E2

Denpasar capital town of Bali in the Lesser Sunda Islands of Indonesia; population (1995 est) 435,000. With its international airport, it is the largest commercial centre on the island, and the focus of road transport. Industries include food processing, machinery, papermaking and printing, and handicrafts. Denpasar is home to a university (1962), a museum of Balinese art, housed in the temple and palace, and the Konservatori Kerawitan, internationally famous as a centre for Balinese dance. **52** E4

Denver city and capital of ►Colorado, USA, on the South Platte River, on the western edge of the Great Plains, 24 km/15 mi from the foothills of the Rocky Mountains; population (1994 est) 494,000; Denver–Boulder metropolitan area (1994 est) 2,190,000. At 1,609 m/5,280 ft above sea level, it is known as 'Mile High City'. Denver is the commercial, manufacturing, and transportation centre for the central west region of the USA. **86** F3

Derby City industrial city and unitary authority in north central England, on the River Derwent, 200 km/124 mi north of London; the city was part of the county of Derbyshire until 1997.
area 87 sq km/34 sq mi
features Derby Cathedral, originally a parish church, was rebuilt in the 18th century but retains its 16th-century tower; the University of Derby was established in 1993; the museum collections of Royal Crown Derby porcelain (Royal Doulton) and the Rolls-Royce collection of aero-engines
industries engineering, chemicals, paper, textiles, plastics; financial services
population (1996) 218,800 **24** L9/**5** F5

Derbyshire county of north central England (since April 1997 Derby City has been a separate unitary authority).
area 2,550 sq km/984 sq mi
towns and cities ►Matlock (administrative headquarters), Buxton, Chesterfield, Glossop, Long Eaton
physical Peak District National Park (including Kinder Scout 636 m/2,088 ft); rivers Dane, Derwent, Dove, Goyt, Rother, Trent, Wye; Dove Dale
features Chatsworth House, Bakewell (seat of the Duke of Devonshire); Haddon Hall; Hardwick Hall; Kedleston Hall (designed by Robert Adam)
agriculture cereals, root crops, and dairy farming (in the south); sheep farming (in the northern hills)
industries heavy engineering; manufacturing (cotton, hosiery, lace, porcelain, textiles); mineral and metal working (barytes, gypsum, lead, zinc); quarrying (marble, sandstone, pipeclay, limestone); motor cars
population (1996) 962,000 **5** F5

Derg, Lough lake in the Republic of Ireland, some 6 km/4 mi northwest of Pettigo, County Donegal. It is 10 km/6 mi long by 6 km/4 mi wide, area 62 sq km/24 sq mi. Lough Derg is surrounded by a rugged landscape and contains many small islands. Station Island, the reputed scene of St Patrick's purgatory, is an important site of pilgrimage between 1 June and 15 August. **24** D8

Des Moines capital city of ►Iowa, USA, on the Des Moines River (a tributary of the Mississippi); seat of Polk County; population (1998 est) 193,200. It is a major road, railway, and air centre for the surrounding Corn Belt region; most of the manufacturing industry is connected with agriculture. Des Moines is the third-largest centre in the world for the insurance industry, after London, England, and Hartford, Connecticut. 84 H3

Dessau city in the state of Saxony-Anhalt, Germany, on the River Mulde, 115 km/70 mi southwest of Berlin; population (1995) 92,200. It is the former capital of Anhalt duchy and state. It manufactures chemicals, machinery, and chocolate and was the site of the Junkers aeroplane works. 20 H5

Detmold city and region of ►North Rhine–Westphalia, Germany, on the River Werre on the northern slopes of Tentoburger Wald, 25 km/15 mi southeast of Bielefeld; population (1994) 72,800 (city). It is a tourist centre and has furniture and metallurgical industries. 20 D5

Detroit industrial city and port in southeastern Michigan, USA, 788 km/489 mi west of New York and 395 km/245 mi east of Chicago, situated on the Detroit River opposite the city of Windsor in Ontario, Canada; seat of Wayne County; area 370 sq km/143 sq mi (excluding neighbouring cities), metropolitan area 10,093 sq km/3,897 sq mi; population (1998) 970,196, metropolitan area 5,246,000. Detroit is the headquarters of Ford, Chrysler, and General Motors, hence its nickname, Motown (from 'motor town'). Other manufactured products include steel, machine tools, chemicals, and pharmaceuticals. It is the seventh-largest city in the USA. 84 K3

Deuxponts French name for ►Zweibrücken, a town in Germany. 22 K5

Deux-Sèvres *département* in the ►Poitou-Charentes region of France; area 6,004 sq km/2,318 sq mi; population (1990) 346,300. The Rivers Sèvre-Niortaise and Sèvre-Nantaise flow through the *département*. In the north, centre, and west the land is marshy, though parts have been drained. The south is a limestone plain. Cereals, sugar beet, and vines are produced. Horses, cattle, and a particular breed of mule are raised, and there is much wildfowl. There are some textile industries. The principal towns are Niort (the administrative centre), Bressuire, and Parthenay. 6 C3

Deventer town in Overijssel province, the Netherlands, on the River Ijssel, 45 km/28 mi south of the IJsselmeer lake; population (1993) 68,500. It is an agricultural and transport centre and produces carpets, precision equipment, and packaging machinery. 22 J2

Devon (or **Devonshire**), county of southwest England; Plymouth and Torbay have been separate unitary authorities since April 1998.
area 6,720 sq km/2,594 sq mi
towns and cities ►Exeter (administrative headquarters); resorts: Barnstaple, Bideford, Exmouth, Ilfracombe, Sidmouth, Teignmouth, Tiverton
physical rivers: Dart, Exe, Plym, Tamar (94 km/58 mi), Taw, Teign, Torridge; National Parks: Dartmoor, Exmoor
features Lundy bird sanctuary and marine nature reserve in the Bristol Channel
agriculture sheep and dairy farming, beef cattle; cider and clotted cream; fishing
industries kaolin in the south; lace (at Honiton); Dartington glass; carpets (Axminster); quarrying (granite, limestone, sandstone); minerals (copper, iron, lead, manganese); tourism
population (1996) 1,059,300 5 E6

Dhahran (Arabic **Az Zahrān**), city in eastern Saudi Arabia near the Gulf coast, opposite Bahrain Island; population (1993) 73,700. It was used as a military base during the Gulf War of 1991. 61 D3

Dhaka (formerly **Dacca**), capital of ►Bangladesh since 1971, in Dhaka region, west of the River Meghna on the ►Ganges delta; population (1991) 3,397,200. It trades in rice, oilseed, sugar, and tea; industries include jute-processing, tanning, and productions of textiles, chemicals, glass, and metal products. 54 F4

Dhaulagiri mountain in the ►Himalayas of western central Nepal, rising to 8,172 m/26,811 ft. 54 D3

Dieppe (Old Norse *djupa* 'deep'), channel port and holiday resort at the mouth of the River Arques in Seine-Maritime *département*, northern France, 53 km/33 mi north of Rouen; population (1999 est) 34,600. It is a trading centre for fish and fruit; industries include fishing, shipbuilding, pharmaceuticals, and light manufacturing. There are ferry services from its harbour to Newhaven in England. 22 D5

Digne capital of the French *département* of Alpes-de-Haute-Provence, in the Bléone valley; population (1990) 17,400. Perfume is produced and fruit processed. It has a cathedral (15th–19th centuries) and a Romanesque church. There are sulphur springs 3 km/2 mi to the east. 30 B6

Dijon administrative centre of the ►Côte-d'Or *département*, east-central France, and chief city of the ►Burgundy region, situated on the Burgundy canal at the confluence of the rivers Ouche and Suzon; population (1990) 151,600, conurbation 230,000. As well as metallurgical, electrical, electronic, pharmaceutical, and other industries, it has a has a trade in Burgundy wines, is famed for its mustard, spiced bread, and gastronomic specialities, and is an important railway centre. 26 L6

Dili (or **Dilli** or **Dilly**), capital, chief port, and main commercial centre of East Timor, situated on the Ombai Strait on the northern coast; population (1990) 60,150. Coffee, cotton, rice, sandalwood, copra, and hides are the chief exports, while industries include soap, perfume, pottery, and textiles, as well as crafts such as basketry and sandalwood carving. 53 C4

Dillingen town in the Saarland, Germany, on the River Saar, 24 km/15 mi northwest of Saarbrücken; population (1994) 22,100. Dillingen is a railway junction and its principal industries are steelmaking, engineering, and chemicals. 20 B7

Dimashq Arabic for ►Damascus, the capital city of Syria. 60 D3

Dinan town in Côtes-d'Armor *département* in northern France, on the River Rance, 21 km/13 mi south of St-Malo; population (1990) 14,200. The river is harnessed for tidal hydroelectric power. Tourism is an important industry here. 26 C5

Dinant town in Namur province, Belgium, on the River Meuse; population (1997) 12,600. It is a tourist centre for the Ardennes. 22 G4

Dinaric Alps extension of the European ►Alps that runs parallel to the eastern Adriatic coast, stretching from Slovenia along the frontier between Croatia and Bosnia-Herzegovina into western Yugoslavia and northern Albania. The highest peak is Durmitor at 2,522 m/8,274 ft. 30 L6

Dinkelsbühl town in Bavaria, Germany, on the River Wörnitz, 141 km/88 mi northwest of Munich; population (1994) 11,400. One of the city-states of the Holy Roman Empire, Dinkelsbühl has extensive medieval fortifications, including town walls, towers, and a moat, a late Gothic church (1448–99), many half-timbered medieval houses, and a former castle of the Teutonic Knights. 20 F7

Diomede Islands two islands in the Bering Strait, off the tip of the Seward peninsula, Alaska, that lie about midway between the continents of America and Asia. **Little Diomede** (area: 6.2 sq km/2.4 sq mi) belongs to the USA and is only 3.9 km/2.4 mi from **Big Diomede** (area: 29.3 sq km/11.3 sq mi), or Ratmanov, which is owned by Russia. They were first sighted by the Danish navigator Vitus Bering in 1728. A small population of Inuit inhabit the islands. 44 AA3

District of Columbia seat of the federal government of the USA, coextensive with the city of Washington, DC, situated on the Potomac and Anacostia rivers; area 178 sq km/69 sq mi; population (2000 est) 572,100. Government agencies are its main source of employment, but tourism and the manufacture of scientific and technical equipment are also important to the economy. The district was ceded by Maryland as the site of the national capital in 1790.

Dithmarschen (or **Ditmarsh**), district of Germany in the *Land* (administrative region) of ►Schleswig-Holstein; population (1994) 133,000. It is a low-lying, partly marshy area on the west coast between the estuary of the Elbe and the River Eider. The largest town in the area is Heide (population (1995) 20,600). 20 D2

Diyarbakir town in Asiatic Turkey, on the River Tigris. It has a trade in gold and silver filigree work, copper, wool, and mohair, and manufactures textiles and leather goods. The population grew from 381,000 in 1900 to 1.5 million in 1995 because of the influx of Kurds evicted from homes in southeast Turkey. 58 J5

Djakarta variant spelling of ►Jakarta, the capital of Indonesia. 52 D4

Djibouti (or **Jibuti**), chief port and capital of the Republic of ►Djibouti, on a peninsula 240 km/149 mi southwest of Aden, Yemen, and 565 km/351 mi northeast of Addis Ababa, Ethiopia; population (1995) 383,000. Industries include petroleum refining, textiles, and rail freighting. The city is an important regional bunkering and supply centre for the export trade in petroleum, and is the main export route for Ethiopian coffee. 64 H5

Djibouti country on the east coast of Africa, at the south end of the Red Sea, bounded east by the Gulf of Aden, southeast by Somalia, south and west by Ethiopia, and northwest by Eritrea.

Djibouti covers an area of 23,200 sq km/8,957 sq mi, and has a population of 638,000 (2000 est). The capital (and chief port) is ►Djibouti; other major towns and cities include Tadjoura, Obock, Dikhil, and Ali-Sabieh. The head of state is Ismail Omar Guelleh (from 1999); the head of government is Barkat Gourad Hamadou (from 1978). The currency is the Djibouti franc, and the languages spoken include French (official), Issa (Somali), Afar, and Arabic. 64 H5

Dnieper (or **Dnepr**; Greek **Borysthenes**), river rising in the Valdai Hills west of Moscow, in the Smolensk region of the Russian Federation, and flowing south through Belarus and Ukraine to enter the Black Sea near Kherson; total length 2,250 km/1,400 mi. The Dnieper is the third longest European river (after the Volga and Danube). 38 F5

Dniprodzerzhyns'k port in Ukraine, on the River Dnieper, 48 km/30 mi northwest of Dnipropetrovs'k; population (1990) 283,600. It is a major industrial city, producing chemicals, iron, and steel. 38 F5

Dnipropetrovs'k city in Ukraine, on the right bank of the River Dnieper and capital of an oblast (region) of the same name; population (1990) 1,200,000. It is the centre of a major industrial region, with iron, steel, chemical, and engineering industries. The city draws much of its power from the Dneproges Dam, 60 km/37 mi downstream. 38 F5

Dodecanese (Greek **Dodekanisos**, 'twelve islands'), group of islands in the Aegean Sea; area 2,663 sq km/1,028 sq mi; population (1991 est) 162,400. Once Turkish, the islands were Italian from 1912 to 1947, when they were ceded to Greece. They include ►Rhodes and ►Kos. Chief products include fruit, olives, and sponges. 36 J8

Dodekanisos Greek for the ►Dodecanese, a group of islands in the Aegean Sea. 36 J8

Dodge City city in southwestern Kansas, USA, on the Arkansas River; seat of Ford County; population (1990) 21,100. It is the centre of an agricultural region (wheat and cattle), and farm and livestock-handling machinery are manufactured here. Dodge City was an important cattle-trading centre, or cow town, in the heyday of the early US cattle industry (1865–85), during the era of the Wild West. 86 F3

Dodoma official capital (replacing ►Dar es Salaam in 1974) of ►Tanzania; 1,132 m/3,713 ft above sea level; population (1994 est) 215,000. Dar es Salaam remains the administrative capital, pending the transfer of government functions to Dodoma. Dodoma is a centre of communications, linked by rail with Dar es Salaam, and Kigoma on Lake Tanganyika, and by road with Kenya to the north and Zambia and Malawi to the south. There is an airport. Dodoma is a marketplace for locally grown coffee and peanuts, but has a limited industrial base, which includes the manufacture of bricks. 70 F5

Doha (Arabic **Ad Dawhah**), capital and chief port of ►Qatar, on the east coast of the peninsula; population (2001 est) 299,300. As well as being the chief commercial and communications centre of Qatar, Doha also has a considerable trade in cargo for other countries of the Gulf, and serves as their centre for vocational training. Industries include oil refining, engineering, refrigeration and construction equipment, as well as shrimp fishing and processing. Over half the population of the country reside in Doha, and the city has an international airport. 61 D4

Dole French town in the *département* of Jura, on the River Doubs and the Rhine–Rhône canal; population (1990) 26,000. Situated on a vine-clad slope, it faces the heights of the forest of Chaux. The town has iron and copper foundries. Other industries include the manufacture of machines and chemicals. Its Roman name was Dola Sequanorum, and it was the capital of the former province of ►Franche-Comté before being displaced by Besançon. 30 A3

Dominica island in the eastern Caribbean, between Guadeloupe and Martinique, the largest of the Windward Islands, with the Atlantic Ocean to the east and the Caribbean Sea to the west.

Dominica covers an area of 751 sq km/290 sq mi, and has a population of 71,000 (2000 est). The capital is ►Roseau; other major towns and cities include Portsmouth, Marigot, Mahaut, Atkinson, and Grand Bay. The head of state is Vernon Shaw (from 1998); the head of government is Pierre Charles (from 2000). The currencies are the East Caribbean dollar, although the pound sterling and the French franc are also accepted, and the languages spoken include English (official) and a Dominican patois (which reflects earlier periods of French rule). 96 E2

Dominican Republic country in the West Indies (eastern Caribbean), occupying the eastern two-thirds of the island of Hispaniola, with Haiti covering the western third; the Atlantic Ocean is to the east and the Caribbean Sea to the west.

The Dominican Republic covers an area of 48,442 sq km/18,703 sq mi, and has a population of 8,495,000 (2000 est). The capital is ►Santo Domingo; other major towns and cities include Santiago, La Romana, San Pedro de Macoris, San Francisco de Macoris, La Vega, San Juan, and San Cristóbal. The head of state and government is Hipólito Mejía (from 2000). The currency is the Dominican Republic peso, and the main language spoken is Spanish (official). 96 D1

Don (ancient Greek **Tanais**), navigable river in the western Russian Federation; length 1,870 km/1,162 mi; basin covers 422,000 sq km/163,000 sq mi. The Don rises in the central Russian uplands near the city of Tula, flows southeast towards the Volga near Volgograd, then turns southwest to empty into the northeast of the Sea of Azov. In its lower reaches the Don is 1.5 km/1 mi wide, and for about four months of the year it is closed by ice. It has long been a major traffic artery linking inland European Russia with the Black Sea. Its chief tributaries are the Donets, Voronezh, Khoper, and Medveditsa, and it is linked to the Volga by the Volga–Don Canal. The main port is Rostov-on-Don, which lies near the river's mouth. 14 H2

Donau German name for the River ►Danube. 30 H2

Donauwörth town in Bavaria, Germany, at the confluence of the rivers Danube (Donau) and Wörnitz, 40 km/25 mi north of Augsburg; population (1994) 17,700. Machinery and aircraft parts are manufactured. Donauwörth was one of the city-states of the Holy Roman Empire. A former Benedictine abbey in the baroque style stands in the town. 20 F8

Doncaster industrial town and administrative headquarters of Doncaster metropolitan borough, South Yorkshire, England, on the River Don, 56 km/35 mi southwest of York; population (1999 est) 292,100. It has been an important centre for railway engineering (locomotives and rolling stock) since the 19th century. Traditional iron, steel, and coal production has declined, although active collieries remain, including the Rossington deep mine. Other industries include the production of synthetic textiles, confectionery, agricultural and electrical equipment, fencing, brass fittings, and optical fibres. 24 L8

Donegal mountainous county in the northwest of the Republic of Ireland, surrounded on three sides by the Atlantic Ocean, and bordering the counties of Londonderry, Tyrone, and Fermanagh (Northern Ireland), and Leitrim (Republic of Ireland); area 4,830 sq km/1,864 sq mi; county town Lifford; population (1996) 130,000. Ballyshannon is the largest town, and the market town and port of **Donegal** is at the head of **Donegal Bay** in the southwest. The severe climate renders much of the county barren, although the soil is suitable for potatoes, oats, and barley (in places). Commercial activities include sheep and cattle raising, tweed, linen, and carpet manufacture, and some salmon and deep-sea fishing. Tourism is also very important; the county is noted for dramatic scenery and geology as well as archaeological and historic remains, and the castles of Donegal and Glenveagh as well as Glenveagh National Park are among the top visitor attractions in the county. The River Erne hydroelectric project (1952) involved the building of a large artificial lake (405 ha/1,000 acres) and a power station at Ballyshannon. 5 B4

Donets river rising in the Kursk region of the Russian Federation and flowing 1,080 km/670 mi through Ukraine to join the River Don 100 km/60 mi east of Rostov. It passes through Ukraine's principal industrial region; see ►Donets Basin. The Donets is navigable for some 210 km/130 mi, and is ice-bound from December to March. 14 H3

Donets Basin (or **Donbas**), highly industrialized area in Ukraine, situated in the bend formed by the rivers Don and Donets. The Donets Basin has one of Europe's richest coalfields, together with extensive deposits of salt, mercury, and lead.

Donetsk city in Ukraine; capital of Donetsk region (oblast), situated in the ►Donets Basin, a major coal-mining area, 600 km/372 mi southeast of Kiev; population (1996) 1,121,000. Donetsk has blast furnaces, rolling mills, and other heavy industries. 38 G5

Dongola town in Northern State, northern Sudan; population (2001 est) 16,900. It has palm groves and produces dates; there is some light engineering and livestock rearing. There are road and river links to Khartoum. The town was founded in about 1811 to replace Old Dongola, 120 km/75 mi upriver, which was destroyed by the Mamelukes. Old Dongola, a trading centre on a caravan route, was the capital of the Christian kingdom of Nubia between the 6th and 14th centuries. 64 F4

Donostia alternative name for ►San Sebastián, a port and resort in the Basque Country, Spain. 28 J1

Dordogne *département* in the ►Aquitaine region of France; area 9,183 sq km/3,546 sq mi; population (1990) 386,900. The land is hilly and wooded, with fertile valleys in the centre and marshes in the east. There are numerous vineyards on the hill slopes. Chestnuts, truffles, wheat, maize, tobacco, and potatoes are grown. Industries include distilling, and the manufacture of leather goods and textiles. The principal towns are ►Périgueux (administrative centre), Bergerac, Sarlat, and Nantran. **6** D4

Dordogne river in southwest France, rising on the slopes of the Puy de Sancy in the Massif Central, Puy-de-Dôme *département*, and flowing 490 km/300 mi through Souillac, Bergerac, Castillon, Libourne, and Cubzac, to join the River Garonne at Ambes, 23 km/14 mi north of Bordeaux. The river is a major source of hydroelectric power, and the last 180 km/112 mi of its course sees much river traffic. It gives its name to the ►Dordogne *département*. **26** F9

Dordrecht (or **Dort**), old port on the River Merwede on an island in the Maas, South Holland province, the Netherlands, 19 km/12 mi southeast of Rotterdam; population (1997) 117,300. It has shipbuilding yards and makes heavy machinery, plastics, and chemicals. It is intersected by canals and has many attractive old thoroughfares and gabled houses. **22** G3

Dornbirn Austrian town in the province of Vorarlberg, 10 km/6 mi south of Bregenz and near Lake Constance; population (1995) 40,900. There are iron, chemical, brewing, and textile industries in the town. **30** E3

Dorpat German name for the Estonian city of ►Tartu. **16** P7

Dorset county of southwest England (since April 1997 Bournemouth and Poole have been separate unitary authorities).
area 2,541 sq km/981 sq mi
towns and cities Dorchester (administrative headquarters), Shaftesbury, Sherborne; resorts: Lyme Regis, Weymouth, Poole, Swanage
physical Chesil Beach, a shingle bank along the coast 19 km/11 mi long; Dorset Downs (chalk); River Stour, and rivers Frome and Piddle; clay beds in the north and west; Canford Heath, the home of some of Britain's rarest breeding birds and reptiles
features Isle of Purbeck, including Corfe Castle; Tank Museum at Royal Armoured Corps Centre, Bovington, where the cottage of the soldier and writer T E Lawrence is now a museum; Wimborne Minster
agriculture dairy farming
industries Wytch Farm is the largest onshore oilfield in the UK; quarrying (marble from the Isle of Purbeck, and Portland stone, used internationally; manufacturing (rope, twine, and net); sand and gravel extraction; tourism
population (1996) 681,900 **5** E6

Dorsten city in North Rhine-Westphalia, Germany, on the River Lippe, 50 km/31 mi northeast of Düsseldorf; population (1994) 80,400. It has a large coalmine and a gasworks, together with factories producing wire, chemicals, carpets, and furniture. **22** J3

Dort another name for ►Dordrecht, a port in the Netherlands. **22** G3

Dortmund city and industrial centre in the ►Ruhr, in North Rhine-Westphalia, Germany, 58 km/36 mi northeast of Düsseldorf; population (1995) 600,000. It is the largest mining town of the Westphalian coalfield and the southern terminus of the Dortmund–Ems Canal. The enlargement of the Wesel–Datteln Canal (1989), connecting Dortmund to the Rhine River, allows barges to travel between Dortmund and Rotterdam in the Netherlands. Industries include coal, iron, and steel (headquarters of Hoesch), mechanical engineering, and brewing (output exceeds that of Munich), and high-tech industries are developing. There is also a modern university. **20** C5

Dothan city and administrative headquarters of Houston County, southeast Alabama, 155 km/96 mi southeast of Montgomery; population (1992) 54,800. It is an agricultural and livestock marketing centre; industries include food packing and processing, and the manufacture of hosiery, clothing, furniture, fertilizer, cigars, and toys. **90** D3

Douai town in the Nord *département* of France, 30 km/19 mi south of Lille on the River Scarpe and the Sensée Canal; population (1990) 44,200, conurbation 200,000. It has metallurgical industries, car manufacturing, and printing. There is a mining school here, as well as higher education institutions. An English Roman Catholic college was founded here in 1568 by English Catholics in exile, led by William Allen. The Douai-Reims Bible, published by the college between 1582 and 1610, influenced the translators of the King James version. **22** F4

Douala (or **Duala**), chief port and industrial centre of Cameroon, on the Wouri River estuary by the Gulf of Guinea; population (1991) 884,000. Industries include aluminium, chemicals, textiles, and pulp. Known as Kamerunstadt until 1907, it was capital of German Cameroon, which became a German protectorate in 1884, 1885–1901. **68** F4

Douarnenez French fishing town in the *département* of Finistère, Britanny; population (1990) 17,000. The sardine and mackerel fisheries are important, and 20% of the world's tuna is landed here (1997). Other industries include boat-building, net- and ropemaking. **26** A5

Doubs *département* in the ►Franche-Comté region of France; area 5,228 sq km/2,019 sq mi; population (1990) 485,200. It falls naturally into three regions. The plains between the rivers Ognon and ►Doubs are very fertile, producing wheat, oats, and other cereals, as well as vegetables, hemp, fruits, and vines. The central plateau is mostly used as grazing land for cattle. The mountainous districts, crossed by four parallel spurs of the Jura, are thickly wooded. The main manufacturing industries are machinery, textiles, clocks and watches, hardware, food, and wine. There are rock-salt deposits in the region, and building stone is quarried. The administrative centre is ►Besançon. **6** G3

Doubs river in France and Switzerland, rising in the Jura Mountains and flowing 430 km/265 mi northeast, west and then southwest, through Jura and Saone-et-Loire *départements*, to join the River Saône upstream of Châlon-sur-Saône. It gives its name to the ►Doubs *département* in which it rises. **30** B3

Douglas capital of the Isle of Man in the Irish Sea; population (1991) 22,200. It is situated in the southeast of the island, and is a holiday resort and terminus of shipping routes to and from Fleetwood and Liverpool. Banking and financial services are important, and the Manx parliament, Tynwald, meets here. **24** H7

Douro (Spanish **Duero**), river in Spain and Portugal, the third-largest in the Iberian peninsula; length 775 km/482 mi. It rises in Spain, on the south side of the Peña de Urbión in the province of Soria, and flows west across the plateau of Castile. It follows the Spanish-Portuguese frontier for 105 km/65 mi, and reaches the Atlantic Ocean at São João de Foz, 5 km/3 mi south of ▶Porto. Navigation at the river mouth is hindered by sand bars. There are hydroelectric installations along its course. **28** B3

Dover capital of ▶Delaware, USA, located in the central part of the state, on the Dover River, 72 km/45 mi south of Wilmington; seat of Kent County; population (1992) 28,200. It is an agricultural trade centre; industries include synthetic materials, adhesives, latex, resins, chemicals, food products, and space equipment. Dover was founded on English settler William Penn's orders in 1683, laid out in 1717, and has been the state capital since 1777. **90** F2

Dover market town and seaport in Kent, southeast England, on the coast of the English Channel; population (1991) 34,200. It is Britain's nearest point to mainland Europe, 34 km/21 mi from Calais, France. Dover is the world's busiest passenger port and England's principal cross-channel port, with ferry, SeaCat (high-speed catamaran), and cross-channel train services. Industries include electronics, paper manufacturing, and light engineering. **22** D3

Dover, Strait of (French **Pas-de-Calais**), stretch of water separating England from France, and connecting the English Channel with the North Sea. It is about 35 km/22 mi long and 34 km/21 mi wide at its narrowest part (from Dover pier to Cap Griz-Nez); its greatest depth is 55 m/180 ft. It is one of the world's busiest sea lanes. The main ports are Dover and Folkestone (England), and Calais and Boulogne (France). **22** D4

Down historic county of southeastern Northern Ireland.
area 2,470 sq km/953 sq mi
towns and cities Downpatrick (county town), Bangor, Newtownards, Newry, and Banbridge; the northern part lies within the commuter belt for Belfast, and includes part of the city of Belfast, east of the River Lagan
physical Mourne Mountains; Strangford sea lough
industries light manufacturing, plastics, linen, high technology and computer companies, fishing, quarrying
agriculture County Down has very fertile land in the north. The principal crops are barley, potatoes, and oats; there is livestock rearing and dairying
population (1981) 339,200

Downs, North and South two lines of chalk hills in southeast England; see ▶South Downs.

Draguignan French town in the *département* of Var, 65 km/40 mi northeast of Toulon, on the River Nartuby; population (1990) 29,600. It has leather and textile industries, and a large US military cemetery from World War II. **30** B7

Drakensberg (Afrikaans 'dragon's mountain'), mountain range in South Africa (Sesuto name **Quathlamba**), on the boundary of Lesotho and the Free State with KwaZulu-Natal. Its highest point is Thabane Ntlenyana (in Lesotho), 3,482 m/11,424 ft, near which is Natal Drakensberg Park. **72** D6

Drama town and department of northwest Greece, in the Macedonia region; population (1993) 102,000. It is a centre for the Greek tobacco trade, and has commercial, administrative, and strategic importance. There are also food-processing factories. **36** G3

Drammen port in southeast Norway and chief city of Buskerud county, on the Drammen Fjord, a branch of the Oslo Fjord; population (1996) 60,000. It is Norway's fifth-largest city. There are engineering, water-power equipment, food processing, and brewing industries; paper and wood pulp are manufactured and shipped from here. **16** F7

Drava (or **Drave**; German **Drau**), river in central Europe, the second-largest tributary on the right bank of the ▶Danube; length 720 km/447 mi. The Drava rises in east Tirol and flows in a generally southeasterly direction through Austria and Slovenia to join the Danube 21 km/13 mi east of Osijek. For part of its course, the river forms the boundary between Croatia and Hungary. **34** E4

Drenthe low-lying northern province of the Netherlands, south of Groningen and Friesland; area 2,660 sq km/1,027 sq mi; population (1997) 460,800. The main cities are ▶Assen (capital), Emmen, and Hoogeveen. The terrain is fenland and moors, with well-drained clay and peat soils. The chief industry is petroleum, and the main agricultural activities are livestock, arable crops, and horticulture. **7** (1)E2

Dresden capital of the *Land* (administrative region) of ▶Saxony, Germany, lying in a wide basin in the upper Elbe Valley; population (1995) 472,900. Products include chemicals, machinery, glassware, and musical instruments; telecommunications and high-tech industries are also important. One of the most beautiful German cities, with a rich architectural and cultural heritage, it was devastated by Allied bombing in 1945; much rebuilding has since taken place, and the city has become an important tourist destination. **20** J5

Drina river in Yugoslavia, 402 km/250 mi long. It rises at the frontier of Montenegro, at the junction of the rivers Tara and Diva, and flows north along the border between Bosnia and Serbia to join the River Sava 105 km/65 mi west of Belgrade. **34** G5

Drogheda (Irish *Droichead Átha* 'the bridge of the ford') seaport, industrial town, and borough 50 km/31 mi from Dublin, and 6 km/4 mi from the estuary of the River Boyne, in County Louth, Republic of Ireland; population (1996) 25,000. The port trades in cattle and textiles; industries include chemicals, foodstuffs, brewing, linen, cotton, and engineering. The salmon fishery in the Boyne has its centre here. In 1649 the town was stormed by English general Oliver Cromwell, who massacred most of the garrison, and in 1690 it surrendered to William III after the Battle of the Boyne. **24** F8

Drôme *département* in the ▶Rhône-Alpes region of southeast France area 6,525 sq km/2,519 sq mi; population (1990) 414,546. Drôme extends from the wide and fertile Rhône valley in the west to the edge of the Cottian Alps (Alpes du Dauphiné) in the east. The chief products are maize, potatoes, buckwheat, mulberries, fruit, truffles, and wine (including Hermitage and Crozes-Hermitage). There are electrical, aerospace, chemical, foodstuffs, and paper industries. Silk and leather are also produced here. The capital is ▶Valence; other towns of importance are Die, Nyons, ▶Montélimar, Crest, and Romans-sur-Isère. **6** F4

Drôme river in France, rising in Dauphiné Pre-Alps and flowing northwest for 101 km/63 mi to join the River Rhône below Livron. It gives its name to the *département* of ▶Drôme. 26 K9

Dronning Maud Land Norwegian name for ▶Queen Maud Land, a region of Antarctica. 102 (2)F2

Dublin county in the Republic of Ireland, in Leinster province, facing the Irish Sea and bounded by the counties of Meath, Kildare, and Wicklow; county town ▶Dublin; area 920 sq km/355 sq mi; population (1996) 1,058,300. The county is mostly level and low-lying, but rises in the south to 753 m/2,471 ft in Kippure, part of the Wicklow Mountains. The River Liffey enters Dublin Bay. The county is dominated by Ireland's capital city of Dublin and its suburbs, but also contains pastoral and agricultural land. Dún Laoghaire is the other major town and large port. 5 C5

Dublin (official Irish name **Baile Átha Cliath**, 'the town of the ford of the hurdles'; Irish *dubh linn* 'dark pool'), city and port on the east coast of Ireland, at the mouth of the River Liffey, facing the Irish Sea; capital of the Republic of Ireland, and county town of County ▶Dublin; population (1996) 481,600; Greater Dublin, including Dún Laoghaire (1996) 953,000. Around a quarter of the Republic's population lives in the Dublin conurbation, with a high density of young, professional workers. In the 1990s the city underwent a renaissance, with the restoration of many old city-centre buildings, notably in the Temple Bar area. Dublin is the site of one of the world's largest breweries (Guinness); other industries include textiles, pharmaceuticals, electrical goods, whiskey distilling, glass, food processing, and machine tools. Dublin is a significant centre for culture and tourism, known particularly for its Georgian architecture and plethora of bars. 24 F8

Dubna town in the Moscow oblast (region) of the Russian Federation, sited at the confluence of the Dubna and Volga rivers 128 km/80 mi north of Moscow; population (2001 est) 67,600. It was founded in 1956 as an centre for nuclear research. There is also metal-working in the town. 38 G3

Dubrovnik (Italian **Ragusa**), city and port in Croatia on the Adriatic coast; population (1991) 49,700. It manufactures cheese, liqueurs, silk, and leather. 34 F7

Dubuque city in northeast Iowa, USA, northeast of Iowa City, just across the Mississippi River from the Wisconsin–Illinois border; seat of Dubuque County; population (1998 est) 54,500. An important port, it has agricultural marketing facilities; major industries are agricultural equipment, food processing, publishing, and metal working. Since the revival of legalized riverboat gambling in Iowa in 1991, tourism has increased, and greyhound racing has also developed. 88 B2

Dudley industrial town and metropolitan borough in the West Midlands, 14 km/9 mi northwest of Birmingham, England; population (1994 est) 141,000. Formerly an important centre for coalmining and iron-smelting at the heart of the industrial ▶Black Country, Dudley now manufactures clothing, glass, and light engineering products. 24 K9

Dufourspitze second highest of the alpine peaks, 4,634 m/15,203 ft high. It is the highest peak in the Monte Rosa group of the Pennine Alps on the Swiss-Italian frontier. 30 C5

Duisburg (formerly **Duisburg-Hamborn** until 1934), river port and industrial city in North Rhine-Westphalia, Germany, at the confluence of the Rhine and Ruhr rivers, 20 km/12 mi northwest of Düsseldorf; population (1995) 535,200. It is the largest inland river port in Europe. Located at the western end of the Ruhrgebiet (Ruhr District), Duisburg possesses the major Rhine docks at Ruhrort through which raw materials, such as iron ore and petroleum are imported. It has the largest concentration of heavy industry (iron and steelmaking, oil refining, heavy engineering, chemicals, and barge building) in the Ruhr. 22 J3

Duluth city and port in northeastern Minnesota, USA, at the western end of Lake Superior on the mouth of the St Louis River; seat of St Louis County; population (1992) 85,400. With its neighbouring city, Superior (Wisconsin), it shares a harbour at the head of navigation on the Great Lakes system. The westernmost port on the St Lawrence Seaway, Duluth trades in iron ore, grain, coal, oil, and timber; it also manufactures steel, flour, timber, and dairy produce, and tourism is important. 88 B1

Dumbarton administrative headquarters of ▶West Dunbartonshire, Scotland, on the River Leven near its confluence with the Clyde, 23 km/14 mi northwest of Glasgow; population (1991) 22,000. Traditional heavy industries have been contracting, other industries include whisky distilling and electronics. 24 H5

Dumfries administrative headquarters of ▶Dumfries and Galloway unitary authority, Scotland; population (1991) 32,100. It is situated on the River Nith, 53 km/33 mi northwest of Carlisle. Industries include plastics, light engineering, and textiles. The Scottish poet Robert Burns is buried in the graveyard of St Michael's church. 24 J6

Dumfries and Galloway unitary authority in southern Scotland, formed in 1996 from the regional council of the same name (1975–96).
area 6,421 sq km/2,479 sq mi
towns Annan, ▶Dumfries (administrative headquarters), Kirkcudbright, Stranraer, Castle Douglas, Newton Stewart
physical indented coastline, backed by a low-lying coastal strip of varying width; intensively forested in the Galloways. Much of the inland area is upland: east to west this includes Eskdalemuir (Hart Fell 808 m/2,651 ft), the Lowther Hills (Green Lowther 732 m/2,402 ft) and the Galloway Hills (the Merrick 843 m/2,766 ft)
features Wanlockhead (the highest village in Scotland); Glen Trool National Park; Ruthwell Cross
industries timber, chemicals, food processing
agriculture beef and dairy cattle, sheep, forestry
population (1996) 147,800 5 D4

Dumyat alternative name for the Egyptian port of ▶Damietta. 64 F1

Dundee City city and unitary authority in eastern Scotland, on the north side of the Firth of Tay.
area 62 sq km/24 sq mi
towns Monifieth, Broughty Ferry, Dundee (administrative headquarters)
physical Firth of Tay
features Tay Bridges; explorer Robert Scott's ship *Discovery*, which has been

restored and is a visitor attraction
agriculture fishing
industries biomedical research, oil industry support, high technology manufacturing
population (1996) 155,000
24 K5/5 E3

Dunedin port, city, and capital of Otago province, South Island, New Zealand, at the head of Otago Harbour, 24 km/15 mi from the sea; population (1996) 112,300. It is a road, rail, and air centre, with engineering, ship building, textile, and tourist industries. The city was founded in 1848 by members of the Free Church of Scotland. Dunedin is an important education centre, with its university being established in 1869. 78 C7

Dunfermline industrial town north of the Firth of Forth in Fife, Scotland; population (1991) 55,100. Industries include engineering, electronics, and textiles. It was the ancient capital of Scotland, with many sites of royal historical significance. Many Scottish kings, including Robert the Bruce and Malcolm Canmore, are buried in **Dunfermline Abbey**. 24 J5

Dungeness shingle headland on the south coast of Kent, southeast England, on the edge of Romney Marsh. It has two nuclear power stations and two lighthouses, one operational and the other, now inland, turned into a museum. Other inland features include Denge Marsh, a bird sanctuary 22 C4

Dunkerque French name for ►Dunkirk, the most northerly seaport in France. 22 E3

Dunkirk (French **Dunkerque**), most northerly seaport of France, in Nord *département*, at the entrance to the Strait of Dover; population (1990) 71,100, conurbation 195,000. Its harbour is one of the foremost in France and it stands at the junction of four canals, giving it communication with the rest of France and with Belgium; there is a ferry service to Ramsgate in England. Industries include oil refining, fishing, and the manufacture of textiles, machinery, and soap. Dunkirk was close to the front line during much of World War I, and in World War II, 337,131 Allied troops (including about 110,000 French) were evacuated from the beaches as German forces approached. 22 E3

Dún Laoghaire (formerly **Kingstown**), major port, residential town, and borough in County Dublin, Republic of Ireland, 10 km/6 mi south of the centre of Dublin; population (1996) 190,000 (Dún Laoghaire – Rathdown). It is a terminal for ferries to Britain, and there are fishing industries. The National Maritime Museum is located here, and it is an important yachting centre and popular tourist resort. The James Joyce museum is located in a Martello tower at Sandycove, 3 km/2 mi south of the town, where the Irish author once stayed. 24 F8

Durance river in southeast France; length 305 km/190 mi. It rises in the *département* of Hautes-Alpes, and flows southwest past Briançon in a series of gorges. In its lower reaches it is canalized for irrigation purposes, and its course becomes northwesterly. It finally enters the River Rhône 5 km/3 mi below Avignon. Its chief tributary is the River Verdon. 26 L10

Durazzo Italian form of ►Durrës, a port in Albania. 36 B3

Durban principal port of KwaZulu-Natal, South Africa, and main harbour of the republic; population (urban area, 1996) 1,320,000. Exports include coal, chemicals, steel, granite, wood products, sugar, fruit, grain, rice, and wool; imports include heavy machinery and mining equipment. Durban is also a holiday resort. 72 E5

Düren city in North Rhine-Westphalia, Germany, on the River Ruhr, 51 km/32 mi southwest of Düsseldorf; population (1994) 256,300. Metal goods, textiles, paper, and sugar are manufactured here. Düren dates from before 911. 22 J4

Durham city and administrative headquarters of Durham County, north-central North Carolina, USA, 40 km/25 mi northwest of Raleigh, on the Eno River; population (1994 est) 143,000. Tobacco is the main industry here; other products include precision instruments, textiles, furniture, and lumber. Durham is the home of Duke University (founded as Union Institute in 1838, it was enlarged by the Duke family and called Duke University in 1924), North Carolina Central University (1910), and Durham Technical Institute (1961). 90 F2

Durham city and administrative headquarters of the county of Durham, northeast England, on the River Wear, 19 km/12 mi south of Newcastle-upon-Tyne; population (1991) 36,900. Formerly a centre for the coalmining industry (the last pit closed in 1993), the city now has light engineering industries and manufactures textiles, carpets, and clothing. 24 L7

Durham county of northeast England (since April 1997 Darlington has been a separate unitary authority).
area 2,232 sq km/862 sq mi
towns and cities Durham (administrative headquarters), Newton Aycliffe, Peterlee, Chester-le-Street
physical Pennine Hills; rivers Wear and Tees
features Beamish open-air industrial museum; site of one of Britain's richest coalfields (pits now closed); Bowes Museum; Barnard Castle; Durham Cathedral; University of Durham (1832), housed in Durham Castle
agriculture sheep; dairy produce; hill farming
industries clothing; chemicals; iron and steel processing; light engineering industries; quarrying; cement; pharmaceuticals
population (1996) 608,100 5 F4

Durostorum Roman name for ►Silistra, now a town in Bulgaria. 34 Q5

Durrës chief port of Albania; population (1991) 86,900. It is a commercial and communications centre, with flour mills, soap and cigarette factories, distilleries, and an electronics plant. It was the capital of Albania 1912–21. 36 B3

Dushanbe (formerly **Stalinabad** 1929–61), capital of ►Tajikistan, situated in the Gissar Valley 160 km/100 mi north of the Afghan frontier; population (1996) 582,000. Dushanbe is a road, rail, and air centre. Its industries include cotton and silk mills, tanneries, meat-packing factories, and printing works. It is the seat of the Tajik state university. 56 J2

Düsseldorf commercial city and capital of ►North Rhine-Westphalia, Germany, on the right bank of the River Rhine, 26 km/16 mi northwest of Cologne; population (1995) 571,900. It is a river port and the commercial and financial centre of the Ruhr area, with food processing, brewing, agricultural machinery, textile, and chemical industries. 22 J3

Dutch East Indies former Dutch colony, which in 1945 became independent as ►Indonesia.

Dutch Guiana former Dutch colony, which in 1975 became independent as ►Suriname.

Dyfed former county of southwest Wales, created in 1974 and, in 1996, divided between the unitary authorities of ►Carmarthenshire, ►Ceredigion, and ►Pembrokeshire.

Dzerzhinsk city in Nizhegorodsk oblast (region) of the western Russian Federation, on the River Oka 32 km/20 mi west of Nizhniy Novgorod; population (1996 est) 284,000. It is one of the main centres of the chemical industry in the Russian Federation, producing synthetic ammoniac, fertilizers, and herbicides. 38 H3

East Anglia region of eastern England, formerly a Saxon kingdom, including Norfolk, Suffolk, and parts of Essex and Cambridgeshire. Norwich is the principal city of East Anglia. The Sainsbury Centre for Visual Arts, opened in 1978 at the University of East Anglia, has a collection of ethnographic art and sculpture. East Anglian ports such as Harwich and Felixstowe have greatly developed as trade with the rest of Europe has increased.

East Ayrshire unitary authority in southwest Scotland, created in 1996 from two districts of Strathclyde region.
area 1,269 sq km/490 sq mi
towns ►Kilmarnock (administrative headquarters), Cumnock, Stewarton, Galston, Crosshouse
physical predominantly low lying and undulating in the north, mountainous toward the south; Loch Doon; rivers Ayr, Irvine; Blackcraig Hill (700 m/ 2,298 ft); Loudoun Hill
features Burns' House Museum, Mauchline; Loudoun Castle Theme Park; Dunaskin Heritage Museum
industries textiles, light engineering, food and drink, printing
agriculture dairy farming, sheep, beef cattle
population (1996) 124,000 5 D4

Eastbourne seaside resort in East Sussex, southeast England, 103 km/ 64 mi southeast of London; population (1991) 94,800. Originally a small village, Eastbourne developed in the mid-19th century as a resort and model of town planning under the direction of the 7th Duke of Devonshire. The modern town, which extends along the coast for 5 km/3 mi, has become a popular place for retirement. 22 C4

East China Sea see ►China Sea. 46 H4

East Dunbartonshire unitary authority in central Scotland, created in 1996 from two districts of Strathclyde region.
area 175 sq km/67 sq mi
towns Kirkintilloch (administrative headquarters), Bearsden, Milngavie
physical low-lying lands to the south give way dramatically to the Campsie Fells in the north; Earl's Seat (578 m/1,896 ft); River Kelvin
features Forth and Clyde Canal; Antonine Wall
population (1996) 110,000 5 D4

Easter Island (or **Rapa Nui**; Spanish **Isla de Pascua**), Chilean island in the south Pacific Ocean, part of the Polynesian group, about 3,500 km/2,200 mi west of Chile; area about 166 sq km/64 sq mi; population (1994) 2,800. It was first reached by Europeans on Easter Sunday 1722. On it stand over 800 huge carved statues (*moai*) and the remains of boat-shaped stone houses, the work of Neolithic peoples from Polynesia. The chief centre is Hanga-Roa. 74 Q8

Eastern Cape province of the Republic of South Africa from 1994, formerly part of Cape Province; area 170,616 sq km/65,875 sq mi; population (1995 est) 6,481,300. The capital is Bisho; the other main towns are East London, Port Elizabeth, and Grahamstown. Industries include motor manufacturing, textiles, and dairy and meat products; agriculture is based on citrus fruits, grain, and sheep-raising. The languages spoken are Xhosa (85%), Afrikaans (9%), and English (3%). 72 D6

East Flanders (Flemish **Oost-Vlaanderen**), province of Belgium, bounded by Antwerp to the east, Brabant to the southeast, and West Flanders to the southwest; area 2,970 sq km/1,147 sq mi; population (1997) 1,354,700. The main towns are ►Ghent (capital), Sint-Niklaas, Aalst, Eekloo, Dendermonde, and Oudenaarde. Main industries are textiles, light engineering, and distribution industries; agricultural products are sugar beet, flax, hops, and flowers (especially in the Waes district). The area lies in the Schelde river basin, comprising coastal lowlands and polderlands (reclaimed flatlands). 7 (1)B3

East Indies the Malay Archipelago; the Philippines are sometimes included.

The term is also used to refer more generally to Southeast Asia.

East Kilbride town in South Lanarkshire, Scotland; population (1991) 70,400. It was designated a new town in 1947 to take overspill from Glasgow, 11 km/7 mi to the northeast. It is the site of the National Engineering Laboratory and headquarters of British Energy. There are various light industries, including electrical equipment and electronics, and some engineering, including jet engines (Rolls Royce). 24 H6

East London river port and resort on the southeast coast of Eastern Cape Province, South Africa, at the mouth of the Buffalo River, 915 km/568 mi by sea from Cape Town, and 405 km/ 252 mi from Durban; population (urban area, 1991) 270,100. Founded in 1846 as **Port Rex**, its name was changed to East London in 1848. It has a good harbour, with exports including wool, grain, copper, citrus fruits, pineapples, chilled meat, and dairy produce. Industries include textiles, brewing, engineering, motor vehicle assembly, and chemical processing. 72 D6

East Lothian unitary authority in southeast Scotland which was previously a district within Lothian region (1975–96) and a county until 1974.
area 677 sq km/261 sq mi
towns Haddington (administrative headquarters), North Berwick, Dunbar
physical area of contrasts, with coastal plains of cliffs, beaches and estuarine marines, broad river valley of the Tyne, volcanic outcrops (Bass Rock, Traprain Law) and gentle slopes of the Lammermuir Hills
features Tantallon Castle; Muirfield golf course; Traprain Law fort

industries whisky distilling, agricultural-based
agriculture arable farming on plains
population (1996) 85,500 5 E3

East Renfrewshire unitary authority in central Scotland, created in 1996 from part of Renfrew district in Strathclyde region.
area 174 sq km/67 sq mi
towns Barrhead, Giffnock (administrative headquarters), Newton Mearns, Clarkston
physical low-lying plateau rising from the plain of the River Clyde
industries engineering, cotton textiles
agriculture sheep, rough grazing, some dairy farming
population (1996) 86,800 5 D4

East Riding of Yorkshire unitary authority in northern England created in 1996 from part of the former county of Humberside.
area 2,416 sq km/933 sq mi
towns Beverley (administrative headquarters), Driffield, Goole, Hornsea, Bridlington
features Humber Estuary; North Sea; Flamborough Head chalk cliffs; Spurn Head; River Hull; River Ouse; Holderness Peninsula; The Wolds; Beverley Minster (13th century); All Saints Tower (34 m/110 ft) at Driffield; Sledmere House – 18th century mansion with grounds laid out by English landscape gardener Capability Brown; Rudstone has Britain's tallest standing stone (8 m/25 ft); Sewerby Hall (Bridlington) – Georgian mansion including museum dedicated to the aviator Amy Johnson; Hornsea Pottery
industries chemicals, pottery, agriculture, passenger vehicle components, bakery products
population (1996) 310,000 5 F5

East Siberian Sea (Russian **Vostochno-Sibirskoye More**), part of the ►Arctic Ocean, off the north coast of Russia, between the New Siberian Islands and the Chukchi Sea. The world's widest continental shelf, with an average width of nearly 650 km/404 mi, lies in the East Siberian Sea. 44 U2

East St Louis city in southwestern Illinois, USA, on the Mississippi River, opposite St Louis, Missouri; population (1990) 40,900. It is a centre for the processing of livestock; other industries include oil refining, glass, aluminium, steel, paint materials, and machinery. 88 B3

East Sussex county of southeast England, created in 1974, formerly part of Sussex (since April 1997 Brighton and Hove has been a separate unitary authority).
area 1,725 sq km/666 sq mi
towns ►Lewes (administrative headquarters), Newhaven (cross-channel port), Eastbourne, Rye, Winchelsea; resorts: Bexhill-on-Sea, Hastings, St Leonards, Seaford
physical Beachy Head, highest headland on the south coast (180 m/590 ft); the Weald (including Ashdown Forest); Friston Forest; rivers Cuckmere, Ouse, and East Rother (which flows into the sea near Rye); Romney Marsh
features the 'Long Man' chalk hill figure at Wilmington; prehistoric earthworks; Iron Age hill fort at Mount Caburn, near Lewes; Roman villas; Herstmonceux, with a 15th-century castle; other castles at Hastings, Lewes, Pevensey, and Bodiam; Bayham Abbey; Battle Abbey and the site of the Battle of Hastings
agriculture cereals, hops, fruit and vegetables; fishing
industries electronics, gypsum, light engineering, timber
population (1996) 734,900 5 G6

East Timor (or **Timor Timur**), disputed territory on the island of ►Timor in the Malay Archipelago, claimed by Indonesia as the province of Timor Timur; prior to 1975, it was a Portuguese colony for almost 460 years; area 14,874 sq km/5,743 sq mi; population (1990 est) 747,750. The capital is ►Dili. The main industry is coffee. After 24 years of East Timorese opposition to Indonesian rule, in August 1999 a referendum overwhelmingly rejected an offer from Indonesia of autonomy within that country. In 2000 the United Nations Transitional Administration for East Timor (UNTAET) was put in place to help pave the way to statehood, and the East Timor Defence Force was established in February 2001. 53 C4

Eau Claire city and administrative headquarters of Eau Claire County, west-central Wisconsin, USA, 120 km/74 mi east of St Paul, Minnesota, at the junction of the Chippewa and Eau Claire rivers; population (1992) 58,000. It is a processing and marketing centre for the region's dairy farmers; industries include rubber and paper products, machine parts, electronics, printing, and brewing. Tourism is also important to the local economy. The city is home to the University of Wisconsin at Eau Claire (1916). 88 B2

Ebbw Vale (Welsh **Glyn Ebwy**), town and administrative centre of ►Blaenau Gwent unitary authority, southeast Wales; population (1991) 19,500. It was formerly a coalmining town with iron and steel industries, which ended in the 1970s; tin-plate manufacture and engineering have continued. To the east is Blaenavon, where the Big Pit, although no longer working, is a tourist attraction. 24 J10

Ebro (Latin **Iberus**), major river of northeast Spain; length 935 km/581 mi. It rises in the Cantabrian Mountains and flows southeast for about 800 km/500 mi, past Logroño and Zaragoza to meet the Mediterranean Sea in the province of Tarragona, southwest of Barcelona. Its main tributaries are the Ega, the Aragón, and the Segre on the left bank, and the Guadalope and the Jalón on the right. The Ebro is navigable for small vessels as far as Logrõo, and ocean-going ships can sail as far as Tortosa, 35 km/22 mi from its mouth. The Ebro is a major source of hydroelectric power. 28 K3

Ecija town in the province of Seville, southern Spain; population (1991) 35,600. It lies in the centre of the fertile valley of the River Genil, a tributary of the Guadalquivir. Shoes, food, and textiles are produced here. Ecija is called 'the frying pan of Andalusia' because of its intense summer heat. 28 E7

Eckernförde town in Schleswig-Holstein, Germany, 30 km/19 mi northwest of Kiel; population (1994) 22,800. It was formerly a naval port, but is now a fishing port and holiday resort with optical and clothing industries. 20 E2

Ecuador country in South America, bounded north by Colombia, east and south by Peru, and west by the Pacific Ocean.

Ecuador covers an area of 270,670 sq km/104,505 sq mi, and has a population of 12,646,000 (2000 est). The capital is ►Quito; other major towns and cities include Guayaquil, Cuenca, Machala, Portoviejo, Manta, Ambeto, and Santo Domingo.
The head of state and government is

Gustavo Noboa (from 2000).
The currency is the sucre, and the languages spoken include Spanish (official), Quechua, Jivaro, and other indigenous languages. **98** B4

Edam town in the Netherlands on the ►IJsselmeer, North Holland province; municipality population (1997), including town of Volendam 26,500. Founded as a customs post in the 13th century, Edam's prosperity in the 16th and 17th centuries was based upon its cheese trade; it still produces round cheeses covered in red wax. **22** H2

Edessa (Greek **Edhessa**), town in Greece, capital of the department of Pella, west Macedonia, 77 km/48 mi northwest of Thessaloniki; population (2001 est) 19,300. Industries are linked to nearby hydroelectric schemes; there are cotton and silk factories, and carpet manufacture. It is at the centre of a fertile agricultural area and there is food processing. **36** E4

Edinburgh capital of Scotland and, as the **City of Edinburgh**, a unitary authority, located near the southern shores of the Firth of Forth.
area 261 sq km/101 sq mi
physical Water of Leith, Salisbury Crags, Arthur's Seat
industries 85% of the labour force works in the service sector; the largest employers in the city are health, public administration, education, banking, and insurance; manufacturing employment is declining
population (1998 est) 450,200 **24** J6/5 E4

Edirne town in northwest Turkey, on the River Maritsa, 225 km/140 mi northwest of Istanbul; population (1995 est) 117,300. Founded on the site of ancient Uscadama, it was known as **Adrianople**, after the Roman emperor Hadrian, until its capture by Murad I, Sultan of Turkey, in 1361. **36** J3

Edmonton capital of ►Alberta, Canada, on the North Saskatchewan River at an altitude of 665 m/2,182 ft; population (1991) 616,700. It is the centre of an oil and mining area to the north and is also an agricultural and dairying region. Manufactured goods include processed foods, petrochemicals, plastic and metal products, lumber, and clothing. Edmonton is known as the 'gateway to the north': it is situated on the Alaska Highway, and petroleum pipelines link the city with Superior in Wisconsin, USA, and Vancouver in British Columbia. **82** J6

Edward, Lake lake in Uganda, area 2,150 sq km/830 sq mi, at about 900 m/3,000 ft above sea level in the Albertine rift valley. It was known as Lake Idi Amin Dada, after President Amin of Uganda, 1973–79. **70** D4

Eger German name for ►Cheb, a town in the Czech Republic. **20** H6

Eger (German **Erlau**), capital of Heves county, northern Hungary, in the Eger valley near the Mátra Mountains, 105 km/66 mi northeast of Budapest; population (1993) 63,400. Bull's Blood wine is produced here. The town is an agricultural trading centre, and a tourist base for mountain activities. **34** H2

Egmont, Mount (Maori **Taranaki** 'barren mountain'), symmetrical dormant volcano in the southwest of the North Island, New Zealand, 29 km/18 mi south of New Plymouth; it is 2,517 m/8,260 ft high and lies within Mount Egmont National Park. **78** E4

Egypt country in northeast Africa, bounded to the north by the Mediterranean Sea, east by the Palestinian-controlled Gaza Strip, Israel, and the Red Sea, south by Sudan, and west by Libya.

Egypt covers an area of 1,001,450 sq km/386,659 sq mi, and has a population of 68,470,000 (2000 est). The capital is ►Cairo; other major towns and cities include El Giza, Shubra Al Khayma, Alexandria, Port Said, El-Mahalla el-Koubra, Tauta, El-Mansoura, and Suez. The head of state is Hosni Mubarak (from 1981); the head of government is Atef Obeid (from 1999). The currency is the Egyptian pound, and the languages spoken include Arabic (official), Coptic (derived from ancient Egyptian), English, and French. **64** E2

Eider river in northern Germany forming the boundary between the regions of Schleswig and Holstein; length 188 km/117 mi. Its importance declined with the opening of the ►Kiel Canal, with which it is connected. **20** D2

Eiger mountain peak in the Swiss ►Alps.

Eilat alternative spelling of ►Elat, a port in Israel. **60** B7

Einbeck town in Lower Saxony, Germany, 64 km/40 mi south of Hannover; population (1994) 29,400. Carpet manufacture and brewing are the town's main industries. Einbeck was a member of the Hanseatic League trade federation, and was the seat of the princes of Grubenhagen. The town has medieval walls and watchtowers and a number of historic buildings. **20** E5

Eindhoven city in North Brabant province, the Netherlands, on the River Dommel, 30 km/19 mi southeast of Tilburg; population (1997) 197,800. Industries include the manufacture of electrical and electronic equipment. Philips Electronics (founded in 1891) had its headquarters here before moving to south Amsterdam. **22** H3

Eire name of southern Ireland as prescribed in the 1937 Constitution.

Eisenach industrial town on the western edge of the state of Thuringia, Germany; population (1995) 45,600. Industries include the manufacture of motor vehicles, mechanical engineering, electronics, pottery, and tourism. German church reformer Martin Luther made the first translation of the Bible from Greek into German in the 11th–12th-century Wartburg Castle nearby. The composer J S Bach was born here. **20** F6

Eisenerz (German 'iron ore'), town in the Austrian province of ►Styria, 26 km/16 mi northwest of Donawitz-Leoben; population (1991) 7,600. It is a mining town at the foot of the Erzberg (height 1,533 m/5,029 ft), with the largest opencast iron-ore mine in continental Europe. Mineral extraction has taken place here since the Iron Age. **30** K3

Eisenstadt (Hungarian **Kishmarton**), capital of the Austrian province of ►Burgenland, 44 km/28 mi southeast of Vienna, near the Neusiedlersee; population (1995) 12,000. Eisenstadt has a great castle, seat of the Esterházy de Galantha princes. The composer Joseph Haydn, who was kapellmeister to the Esterházy family, lived in the town 1760–90. Eisenstadt is also the centre of a wine-growing region. **30** M3

Eivissa alternative name for ►Ibiza, one of the Balearic Islands, Spain. **28** M6

Ekaterinodar pre-revolutionary name of ▶Krasnodar, an industrial town in the Russian Federation. 38 G5

Ekaterinoslav pre-revolutionary name of ▶Dnipropetrovs'k, the centre of an industrial region on the lower Dnieper River in the Ukraine. 38 F5

El Aaiún Arabic name of ▶Laâyoune. 66 C3

Elat (or **Eilat**), port at the head of the Gulf of Aqaba, Israel's only outlet to the Red Sea; population (1994) 35,700. There are copper mines and granite quarries nearby, and a major geophysical observatory, opened in 1968, is 16 km/10 mi to the north. Tourism is important. Founded in 1948 on the site of the biblical Elath, it is linked by road with Beersheba. The ancient city's ruins are just to the east of the modern city, but situated in Jordan. 60 B7

Elba (Greek **Aethalia**; Roman **Ilva**), island in the Mediterranean Sea, 10 km/6 mi off the west coast of Italy; area 223 sq km/86 sq mi; population (1981) 35,000. Iron ore is exported from the island's capital, Portoferraio, to the Italian mainland. There is a fishing industry, olives are grown, and tourism is important. Elba was French emperor Napoleon's place of exile (1814–15). 32 E6

Elbasan town in Albania, capital of Elbasan district, on the right bank of the River Shkumbi; population (1990) 83,300. It is the centre of a fertile agricultural region and its industries include olive- oil processing, soap-making, and timber milling. It is a railway terminus. There is an oil refinery at Cerrik, to the southwest. 36 C3

Elbe (Czech **Labe**; ancient **Albis**), one of the principal rivers of Germany; length 1,166 km/725 mi. It rises on the southern slopes of the Riesengebirge, Czech Republic, and flows northwest across the German plain to the North Sea. It is navigable for ocean-going vessels as far as Hamburg (101 km/62 mi from the mouth), and for smaller boats as far as its junction with the Vltava (845 km/525 mi). The river basin is approximately 145,039 sq km/56,000 sq mi. 20 F3

Elbeuf French town in the *département* of Seine-Maritime, on the River Seine; population (1990) 55,000. It has been one of the chief centres of the woollen industry in France since the 13th century. It also has important car and chemical industries. The town was severely damaged during World War II. 22 D5

Elbląg (German **Elbing**), river port in Gdańsk province, Poland, on the River Elbląg near the southwestern end of the Vistula lagoon, 50 km/31 mi southeast of Gdańsk; population (1990) 126,000. Industries include shipbuilding, engineering, and electrical products.

Founded in the 13th century by settlers from Bremen and Lübeck, Elbląg became a member of the medieval Hanseatic League merchant confederation, trading with England. It belonged to Poland 1466–1772 and to German East Prussia until 1945. 18 J3

Elbrus (or **Elbruz**; Persian 'two heads'), highest peak in Europe; located in the Caucasus Mountains, ▶Caucasia, in the Russian Federation. Its western summit reaches a height of 5,642 m/18,510 ft, while the eastern summit stands at 5,595 m/18,356 ft. 58 K2

Elburz Mountains (Arabic **Reshteh-ye Kūhhā-ye Alborz**), volcanic mountain range in northwest Iran, close to the southern shore of the Caspian Sea; the highest point is Mount Damavand at 5,670 m/18,602 ft. 56 F2

Elche town in the province of ▶Alicante, eastern Spain; population (1991) 181,700. Its large forest of palm trees, irrigated by canals, provides leaves for Palm Sunday. A unique religious mystery play is performed here each year on 14–15 August. Products include leather goods. 28 K6

Elephanta island in Mumbai harbour, Maharashtra, India, some 8 km/5 mi from Mumbai (formerly Bombay). The 6th-century Temple Caves, cut out of solid rock, have sculptures of many Hindu deities carved 450–740. Formerly known as Gharapuri, the island was renamed by Portuguese in 1535, after a large stone elephant near the island's landing place. The statue is now in Mumbai.

El Faiyûm alternative form of ▶Al Fayyum, city in northern Egypt. 64 F2

El Ferrol (formerly **El Ferrol del Caudillo**), city and port in La Coruña province in Galicia, on the northwest coast of Spain; population (1991) 82,400. It is a naval base and has a deep, sheltered harbour. There are fisheries, and shipbuilding and engineering industries. It is the birthplace of the Spanish dictator Francisco Franco. 28 B1

Elgin administrative headquarters of ▶Moray unitary authority, northeast Scotland, on the River Lossie, 8 km/5 mi south of its port of Lossiemouth on the southern shore of the Moray Firth; population (1991) 19,000. There are sawmills and whisky distilleries. 24 J4

Elgin city in Cook and Kane counties, northeastern Illinois, USA, 58 km/36 mi northwest of Chicago, on the Fox River; population (1992) 81,100. Local manufactures include machine tools, hobby supplies, caskets, street sweepers, electrical products, precision instruments, and belting. Elgin is also a centre for the printing of religious publications. Elgin was founded in 1835 and incorporated as a city in 1854. It was the the home of the Elgin Watch Company from 1864–1970. Condensed milk was invented here in 1856. The city is home to Judson College (1963). 88 C2

Elisabethville former name of ▶Lubumbashi, a town in the Democratic Republic of Congo. 70 D6

El Iskandarîya Arabic name for the Egyptian port of ▶Alexandria. 64 E1

Elizabeth industrial city, deep-water port, and administrative headquarters of Union County, northeast New Jersey, USA; population (1994 est) 106,000. It is located 8 km/5 mi from Newark and linked by bridge with Staten Island, New York City. It has oil refineries, aluminium, and chemical works, and manufactures car parts, sewing machines, and tools. 88 F2

Elizavetpol (or **Yelizavetpol**), former name of ▶Gäncä, an industrial town in Azerbaijan. 58 M3

El Kharga (or **Kharijah**), oasis in the Western Desert of Egypt, 130–160 km/81–99 mi long and 15 km/9 mi wide, known to the Romans, and from 1960 the headquarters of the New Valley irrigation project; population (2001 est) 55,300. It is also known as the Great Oasis, and is the most southerly of the Egyptian oases. Ambitious plans to irrigate a vast area from natural underground reservoirs have been criticized by environmentalists. 64 F2

Elkhart city in Elkhart County, north Indiana, USA, 28 km/17 mi east of South Bend, where the Elkhart River meets the St Joseph River; population (1990) 43,600. Its factories produce mobile homes, firefighting apparatus, recreational vehicles, and pharmaceuticals. The manufacture of band instruments, begun by Charles Conn in 1875, continues to be important; a large proportion of all brass instruments produced in the USA are made here. 88 C2

El Khartum alternative name for ►Khartoum, the capital of Sudan. 64 F4

Ellesmere Island island in the extreme northeast of the Arctic Archipelago, Northwest Territories, Canada; area 196,236 sq km/75,767 sq mi; the population is about 100. It is the second-largest island in the Archipelago (Baffin Island is the largest) and is part of the Queen Elizabeth island group, at the northern end of Baffin Bay. Its northern tip, Cape Columbia, is the most northerly point of the North American continent. The island is, for the most part, barren or glacier-covered. It was first sighted in 1616 by William Baffin. 80 K1

Ellice Islands former name of ►Tuvalu, a group of islands in the western Pacific Ocean. 74 H6

Ellis Island island in New York City harbour, USA, 1.5 km/1 mi from Manhattan Island; area 0.1 sq km/ 0.04 sq mi. A former reception centre for immigrants during the immigration waves between 1892 and 1943 (12 million people passed through it from 1892 to 1924), it was later used (until 1954) as a detention centre for nonresidents without documentation, or for those who were being deported. Ellis Island is now a national historic site (1964) and contains the Museum of Immigration (1989).

Ellwangen town in Baden-Württemberg, Germany, on the River Jagst, 70 km/43 mi northeast of Stuttgart; population (1994) 28,800. Batteries are manufactured here. The town has a 12th-century Romanesque church with three towers, and on a hill overlooking the town is a medieval castle. 30 F2

El Mansûra industrial city in Egypt; see ►Mansûra, El. 64 F1

Elmira city in south-central New York State, USA, on the Chemung River, 80 km/50 mi west of Binghamton, just south of the Finger Lakes region; seat of Chemung County; population (1990) 33,700. It is the processing and marketing centre for the area's dairy and poultry farms. Other industries include business machinery, machine parts, aeroplanes, and fire engines. 88 E2

Elmshorn city in Schleswig-Holstein, Germany, 30 km/19 mi northwest of Hamburg; population (1994) 46,600. There is agricultural processing, woodworking, and the production of fertilizers. 20 E3

El Obeid (or **al-Obeid**, or **al-Ubbayyid**), capital of North Kordofan province, Sudan; population (1993) 228,100. In central Sudan, on the Kordofan Plateau, it is linked by rail with Khartoum 300 km/186 mi to the southwest, and connected by rail to Port Sudan. It is a market for cattle, gum arabic, oilseed, and durra (Indian millet). 64 F5

El Paso city and administrative headquarters of El Paso County, Texas, USA, at the base of the Franklin Mountains, on the Rio Grande, opposite the Mexican city of Ciudad Juárez; population (1994 est) 579,000. It is the centre of an agricultural and cattle-raising area, and there are electronics, food processing, packing, and leather industries, as well as oil refineries and industries based on local iron and copper mines. There are several military installations in the area. The city is home to the University of Texas at El Paso (formerly Texas Western; founded in 1913). 92 E2

El Puerto de Santa María see ►Puerto de Santa María, El, a town in southern Spain. 28 D8

El Qahira Arabic name for ►Cairo, the capital of Egypt. 64 F1

El Salvador country in Central America, bounded north and east by Honduras, south and southwest by the Pacific Ocean, and northwest by Guatemala.

El Salvador covers an area of 21,393 sq km/8,259 sq mi, and has a population of 6,276,000 (2000 est). The capital is ►San Salvador; other major towns and cities include Soyapango, Santa Ana, San Miguel, Nueva San Salvador, Mejicanos, Apopa, and Delgado. The head of state and government is Francisco Guillermo Flores Pérez (from 1999). The currency is the US dollar, and the languages spoken include Spanish (official) and Nahuatl. 94 F6

Elsinore another form of ►Helsingør, a port on the northeast coast of Denmark. 16 G8

Elster name of two rivers in Germany. The **Weisse Elster** (White Elster) rises in the Czech Republic and, after a few kilometres, enters Germany to flow through Plauen, Gera, and Leipzig; length 193 km/121 mi. It joins the River Saale south of Halle.

The **Schwarze Elster** (Black Elster) rises in Saxony northeast of Dresden and flows 188 km/117 mi in a northwesterly direction to join the ►Elbe near Wittenberg. 20 H5

El Suweis Arabic name for ►Suez, a port at the Red Sea terminus of the Suez Canal. 64 F2

Ely city in the Cambridgeshire ►Fens, eastern England, on the Great Ouse River, 24 km/15 mi northeast of Cambridge; population (1991) 10,300. Economic activities include agriculture (sugar beet), engineering, and the manufacture of agricultural machinery, pottery, chemicals, and plastics. The cathedral, dating from 1083, is one of the largest in England. 24 N9

Emba river of Central Asia, 612 km/ 380 mi long, that rises in the Mugodzhar Hills in western Kazakhstan and drains into the northern part of the Caspian Sea. Its largely southwesterly course passes through dry steppe, semi-desert, and desert zones, and there are important oilfields in its basin. It freezes over from November to April. 38 L5

Emden port in Lower Saxony, Germany, at the mouth of the River Ems; population (1995) 51,600. It is a fishing port and an export outlet for the industries of the Ruhr district in North Rhine-Westphalia, with which it is connected by the Dortmund–Ems Canal. There are oil refineries here. 20 C3

Emi Koussi highest point of the Tibesti massif in northern Chad, rising to 3,415 m/11,204 ft. 64 C4

Emilia-Romagna region of northern central Italy, comprising the provinces of Bologna, Ferrara, Forli, Modena, Parma, Piacenza, and Reggio nell'Emilia; area 22,100 sq km/8,531 sq mi; population (1992) 3,920,200. The capital is ►Bologna; other towns include Reggio nell'Emilia, Rimini, Parma, Ferrara, and Ravenna. Agricultural produce includes fruit, wine, sugar beet, beef, dairy products, rice, and wheat. Oil and natural-gas resources have been developed in the Po Valley. 9 (4)C2

Emmerich city in North Rhine-Westphalia, Germany, 79 km/49 mi northwest of Düsseldorf, on the right bank of the River Rhine close to the Dutch border; population (1995) 29,300. Food, chemicals, and machinery are manufactured. 22 J3

Ems river in northern Germany. It rises in the Teutoburger Wald, and flows for 330 km/205 mi across North Rhine-Westphalia and Lower Saxony to an inlet of the North Sea known as the Dollart. 22 J1

Ems short form of ►Bad Ems, a town in Germany. 22 K4

Enewetak Atoll (or **Eniwetok**, or **Eniwetak Atoll**), atoll in the ►Marshall Islands, in the central Pacific Ocean; population (1990) 700. The atoll is made up of 40 small islands which encircle a lagoon 40 km/25 mi in diameter. It was taken from Japan by the USA in 1944, which made the island a naval base and resettled the inhabitants at Ujelang; 43 atomic tests were conducted there from 1947. Atomic weapons testing continued here until 1958, but it was not until the late 1970s that resettlement of the original inhabitants was sanctioned. Despite the clearance of nuclear debris and radioactive soil to the islet of Runit, high radiation levels persisted. 74 F4

Enfield outer borough of northeast Greater London. It includes the districts of Edmonton and Southgate.
features the royal hunting ground of Enfield Chase partly survives in the 'green belt'; early 17th-century Forty Hall; Lea Valley Regional Park, opened 1967
industries engineering (the Royal Small Arms factory, which closed 1989, produced the Enfield rifle), textiles, furniture, cement, electronics, metal, and plastic products
population (1991) 257,400. 22 B3

England largest constituent part of the United Kingdom; area 130,357 sq km/50,331 sq mi; population (2001 est) 449,778,400. The capital is ►London, and other main towns and cities include Birmingham, Cambridge, Coventry, Leeds, Leicester, Manchester, Newcastle upon Tyne, Nottingham, Oxford, Sheffield, and York. The main ports are Bristol, Dover, Felixstowe, Harwich, Liverpool, Portsmouth, and Southampton. Important industries include agriculture (cereals, rape, sugar beet, potatoes, meat, and meat products), electronic and telecommunications equipment, computer software, scientific instruments, textiles and fashion goods, North Sea oil and gas, petrochemicals, pharmaceuticals, fertilizers, beer, china clay, pottery, porcelain, and glass. Tourism is important, and there are worldwide banking and insurance interests. England also exports films, and radio and television programmes. 24 L9

English Channel stretch of water between England and France, leading in the west to the Atlantic Ocean, and in the east via the Strait of Dover to the North Sea; it is also known as **La Manche** (French 'the sleeve') from its shape. The Channel Tunnel, opened in 1994, runs between Folkestone, Kent, and Sangatte, west of Calais, France. 24 J12

Enid city and administrative headquarters of Garfield County, north-central Oklahoma, 108 km/67 mi north of Oklahoma City; population (1990) 45,300. It is a processing and marketing centre for the surrounding region's poultry, cattle, dairy farms, and oil wells. It is home to Phillips University (1906), and Vance Air Force Base is nearby. 90 B2

Enkhuizen town in the province of North Holland, the Netherlands, 45 km/28 mi northeast of Amsterdam; population (1996) 16,200. Together with Hoorn, Enkhuizen was a thriving port in the 17th century. It has an open-air museum of reconstructed old buildings portraying life in former times on the ►Zuider Zee. 22 H2

Ennis county town of County ►Clare, Republic of Ireland, on the River Fergus, 32 km/20 mi northwest of Limerick; population (1996) 15,300. There are distilleries, flour mills, and furniture manufacturing. In the town are a Roman Catholic cathedral and college, and a Franciscan friary, Ennis Friary, founded about 1241. Shannon international airport is 24 km/15 mi to the south of Ennis. 24 D9

Enniskillen county town of ►Fermanagh, Northern Ireland, between Upper and Lower Lough Erne, 184 km/114 mi from Dublin and 141 km/88 mi from Belfast; population (1991) 11,400. It is a market town and shopping centre with some light industry (engineering, food processing); it has been designated for further industrial growth. An IRA bomb exploded here at a Remembrance Day service in November 1987, causing many casualties. 24 E7

Enns river in Austria, a tributary of the ►Danube; length 255 km/158 mi. The Enns rises south of Radstadt, 56 km/35 mi southeast of Salzburg, and flows into the Danube near the town of Enns in Upper Austria. Its chief tributary is the Steier. 30 K3

Enns town in the province of Upper Austria, situated 17 km/11 mi southeast of Linz, on the River Enns close to where it joins the Danube; population (1995) 9,850. A large hydroelectric plant supplies power to the steel mills at Linz. Enns was a prosperous commercial town in the Middle Ages. 30 K2

Enschede textile manufacturing centre in Overijssel province, the Netherlands; population (1997) 147,900. It has an advanced technical university specializing in geomatic engineering (which includes surveying and remote sensing), and an industrial museum. 22 J2

Entebbe city in Uganda, on the northwest shore of Lake Victoria, 20 km/12 mi southwest of Kampala, the capital; 1,136 m/3,728 ft above sea level; population (1991) 41,600. Founded in 1893, it was the administrative centre of Uganda 1894–1962. The international airport of Uganda is here. Industries include tourism and fishing. 70 E3

Enugu city in Nigeria, capital of Enugu state; population (1992 est) 286,100. It is a coal- and iron ore-mining centre. Other industries include steel and cement works, vegetable oil production, food processing, and motor vehicle assembly. The city is linked by rail with Port Harcourt. 68 F3

Eperjes Hungarian name for ▶Prešov, a town in the Slovak Republic. 18 L9

Epernay town in Marne *département*, Champagne-Ardenne region, France, on the River Marne; population (1990) 27,700. Together with Reims, it is the centre of the champagne industry. Corks, casks, and bottles are manufactured alongside the production and storage of champagne; the town has over 50 km/31 mi of long wine cellars hollowed out of the hillsides. There are railway repair yards here, and wood products and textiles are manufactured. 26 J4

Epinal administrative centre of ▶Vosges *département*, on the River Moselle at the foot of the Vosges mountains, France; population (1990) 39,500. Dating from the 10th century, the town is now a cotton-textile centre, and also produces metal goods and leather. 30 B2

Epsom residential town in Surrey, southeast England, 30 km/19 mi southwest of London; population (1991) 64,400 (with Ewell). In the 17th century it was a spa town producing Epsom salts. The Derby and the Oaks horse races are held annually at **Epsom Downs** racecourse. 22 B3

Equatorial Guinea country in west-central Africa, bounded north by Cameroon, east and south by Gabon, and west by the Atlantic Ocean; also five offshore islands including Bioko, off the coast of Cameroon.

Equatorial Guinea covers an area of 28,051 sq km/10,830 sq mi, and has a population of 453,000 (2000 est). The capital is ▶Malabo; other major towns and cities include Bata, Mongomo, Ela Nguema, Mbini, Campo Yaunde, and Los Angeles. The head of state is Teodoro Obiang Nguema Mbasogo (from 1979); the head of government is Angel Serafin Seriche Dougan (from 1996). The currency is the franc CFA, and the languages spoken include Spanish (official), pidgin English, a Portuguese patois (on Annobón, whose people were formerly slaves of the Portuguese), and Fang and other African patois (on Río Muni). 68 F4

Erfurt city in Germany on the River Gera, capital of the state of ▶Thuringia; population (1995) 212,600. It is in a rich horticultural area, and its industries include mechanical and precision engineering, electronics, and telecommunications. 20 G6

Erie city and port on the Pennsylvania bank of Lake Erie, USA; population (1992) 109,300. It has heavy industries and a trade in iron, grain, and freshwater fish. A French fort was built on the site in 1753, and a permanent settlement was laid out in 1795. 88 D2

Erie Canal canal in New York, USA, which connects Lake Erie north of Buffalo with the Hudson River at Albany and Troy; it passes through Utica, Syracuse, and Rochester. It is 45 m/148 ft broad and has an average depth of 3.6 m/12 ft. The canal was begun in 1817 and completed in 1825 at a cost of over $7 million. Together with the Champlain, Oswego, and Cayuga–Seneca canals, it constitutes the New York State Barge Canal System, 845 km/525 mi in length.

Erie, Lake fourth largest of the Great Lakes of North America, connected to Lake Ontario by the Niagara River and bypassed by the Welland Canal; length 388 km/241 mi; width 48–91 km/30–56 mi; area 25,720 sq km/9,930 sq mi. The most southerly of the Great Lakes, it is bounded on the north by Ontario, Canada; on the south and southeast by Ohio, Pennsylvania, and New York State; and on the west by Michigan. Lake Erie is an important link in the St Lawrence Seaway. 88 D2

Eritrea country in East Africa, bounded north by Sudan, south by Ethiopia, southeast by Djibouti, and east by the Red Sea.

Eritrea covers an area of 125,000 sq km/48,262 sq mi, and has a population of 3,850,000 (2000 est). The capital is ▶Asmara; other major towns and cities include Assab, Keren, Massawa, Adi Ugri, and Ed. The head of state and government is Issaias Afwerki (from 1993). The currency is the nakfa, and the languages spoken include Tigre, Tigrinya, Arabic, English, Afar, Amharic, Kunama, and Italian. 64 G4

Erivan alternative transliteration of ▶Yerevan, the capital of Armenia. 58 L3

Erlangen industrial city in Bavaria, Germany, in the Regnitz Valley, 18 km/11 mi northwest of Nuremberg; population (1995) 101,500. It is a major centre for electrical engineering and telecommunications, and other industries include glass-making and brewing. It is the site of the Erlangen-Nuremberg University. 44 G7

Erlau German name for ▶Eger, a town in Hungary. 34 H2

Ermoupoli Greek name for ▶Hermoupoli, a port in Greece. 36 G7

Erzgebirge (German 'ore mountains'), mountain range on the border between Germany and the Czech Republic, where the rare metals uranium, cobalt, bismuth, arsenic, and antimony are mined. Some 145 km/90 mi long, its highest summit is Mount Klinovec (Keilberg) (1,244 m/4,080 ft), in the Czech Republic. In 1991, following the reunification of Germany, many uranium mines were closed and plans to clean up the heavily polluted region were formulated. 20 H6

Erzurum capital of Erzurum province, northeastern Turkey; population (1990) 297,500. It is a centre of agricultural trade and mining, and has a military base. 58 J4

Esbjerg port of Ribe county, Denmark, on the west coast of Jutland; population (1993) 82,300. It is the terminus of links with Sweden and the UK, and is a base for Danish North Sea oil exploration. 16 E9

Esch (or **Esch-sur-Alzette**), town in the Grand Duchy of Luxembourg, 17 km/11 mi southwest of the city of Luxembourg; population (1991) 24,000. An important iron and steel industry is centred on Esch and towns in the surrounding canton. 22 J5

Eschwege city in Hesse, Germany, on the River Werra, 40 km/25 mi southeast of Kassel; population (1995) 22,800. Its location in a relatively depressed area close to the border with the former country of East Germany led to the creation of several new industries, such as metallurgy and textiles, during the 1950s to provide local employment. 20 F5

Eschweiler city in North Rhine-Westphalia, Germany, 15 km/9 mi north of Aachen; population (1994) 56,200. It has opencast brown coal (lignite) mines, power stations, and plants producing machinery, sugar, food, and textiles. 22 J4

Esfahan (or **Isfahan**; formerly **Ispahan**), city in central Iran and capital of the province of the same name; population (1991) 1,127,000. The town lies on the Zayandeh Rud River at an altitude of 1,600 m/ 5,250 ft, in the centre of a large fertile oasis. Industries include steel, textiles, carpets, and traditional handicrafts. It was the ancient capital (1598–1722) of Abbas I. Its features include the Great Square, where polo was played, the Sheikh Lutfullah Mosque, the Hall of Forty Pillars known as Chihil Sutun, the Ali Qapu gate, the Khaju and Aliverdi Khan bridges, and the Shah Hussain madrasah, a school for Muslim dervishes, built in 1710. **56** F3

Eskilstuna town west of Stockholm, Sweden; population (1992) 89,600. It has iron foundries and steel and armament works; it was known in the 17th century for its swords and cutlery. **16** J7

Eskişehir city in Turkey, 200 km/ 125 mi west of Ankara; population (1990) 415,800. Products include meerschaum, chromium, magnesite, cotton goods, tiles, and aircraft. **58** D4

Espoo (Swedish **Esbo**), town in southern Uusimaa province of Finland, just west of Helsinki. Formerly a rural district, Espoo became a town in 1963; as a dormitory town for the capital, it saw rapid population growth and became the second-largest city in Finland. The ancient coastal road from Turku to Vyborg runs through Espoo. **16** N6

Essen city in North Rhine-Westphalia, Germany, 29 km/18 mi northeast of Düsseldorf; population (1995) 616,400. It is the administrative centre of the ►Ruhr region, situated between the rivers Emscher and Ruhr. Industries include coal-mining, steel, glass-making, chemicals, telecommunications, and electronics. Its 9th–14th-century minster is one of the oldest churches in Germany. Half of the city's buildings were destroyed during World War II. **22** K3

Essequibo longest river in Guyana, South America, draining more than half the total area of the country; length 1,014 km/630 mi; it rises in the Guiana Highlands of southern Guyana and flows north past Bartica to meet the Atlantic at a 32 km/20 mi wide delta. Its course is interrupted by numerous rapids and falls, but its lower course is navigable for large vessels for 80 km/50 mi as far as Bartica. Its major tributaries include the Rupununi, Potaro, Mazaruni, and Cuyuni rivers. **98** F2

Essex (Old English East-Seaxe), county of southeast England, which has contained the unitary authorities of Southend and Thurrock since April 1998.
area 3,670 sq km/1,417 sq mi
towns and cities ►Chelmsford (administrative headquarters), Basildon, Colchester, Harlow, Harwich (port), Clacton-on-Sea (resort)
physical flat and marshy near the coast; richly wooded in the southwest; rivers: the Blackwater, Crouch, Colne, Lee, Stour, and Thames
features former royal hunting ground of Epping Forest (2300 ha/5680 acres, controlled from 1882 by the City of London); Stansted, London's third airport; new Roman Catholic cathedral at Brentwood
agriculture cereals (wheat), fruit, sugar beet; livestock rearing, dairy products; oysters
industries brewing, cars, cement, engineering, food processing, oil products
population (1996) 1,586,100
5 G6

Esslingen city in Baden-Württemberg, Germany, 16 km/10 mi southeast of Stuttgart, on the River Neckar; population (1994) 489,500. Esslingen has electrical, automotive engineering, textile, and food industries. **30** E2

Essonne *département* in the ►Ile-de-France region of France, situated to the south of Paris; area 1,811 sq km/699 sq mi; population (1990) 1,085,000. The rivers Orge and Essonne flow through it, both tributaries of the Seine. Considerable woodland remains and agriculture is important. The *département* is crossed by major communications axes and now functions as an outer suburb for Paris. Multinational companies producing electronic and electrical equipment are located here. The administrative centre is the new town of Evry, which contains the newest cathedral in France, opened in 1995. **6** E2

Estonia country in northern Europe, bounded east by Russia, south by Latvia, and north and west by the Baltic Sea.

Estonia covers an area of 45,000 sq km/17,374 sq mi, and has a population of 1,396,000 (2000 est). The capital is ►Tallinn; other major towns and cities include Tartu, Narva, Kohtla-Järve, and Pärnu. The head of state is Lennart Meri (from 1992); the head of government is Mart Laar (from 1999). The currency is the kroon, and the languages spoken include Estonian (official) and Russian. **16** M7

Estoril resort on the coast 20 km/13 mi west of Lisbon, Portugal; population (1991) 24,800. There is a Grand Prix motor-racing circuit here. **28** A6

Estrecho de Magallanes Spanish name for Strait of ►Magellan, the channel between the southern tip of the South American mainland and Tierra del Fuego. **100** G9

Esztergom city on the Danube, northwest of Budapest, Hungary; population (1991 est) 29,800. It was the birthplace of St Stephen and the former ecclesiastical capital of Hungary, with a fine cathedral. **34** F2

Etang de Berre saltwater lagoon west of Marseille, France; see ►Berre, Etang de. **26** L10

Etaples fishing port and seaside resort on the Canche estuary, Pas-de-Calais *département*, France; population (1990) 11,500. During World War I it was a British base and hospital centre; a large military cemetery was also established here, eventually containing some 12,000 graves. **22** D4

Ethiopia country in East Africa, bounded north by Eritrea, northeast by Djibouti, east and southeast by Somalia, south by Kenya, and west and northwest by Sudan. It was known as Abyssinia until the 1920s.

Ethiopia covers an area of 1,096,900 sq km/423,513 sq mi, and has a population of 62,565,000 (2000 est). The capital is ►Addis Ababa; other major towns and cities include Jimma, Dire Dawa, Harar, Nazret, Desē, Gonder, Mek'elē, and Bahir Dar. The head of state is Negasso Ghidada (from 1995); the head of government is Meles Zenawi (from 1995). The currency is the Ethiopian birr, and the languages spoken include Amharic (official), Arabic, Tigrinya, Orominga, and about 100 other local languages. **62** G5

Etna, Mount volcano on the east coast of the Italian island Sicily, 3,323 m/10,906 ft, the highest in Europe. About 90 eruptions have been recorded since 1800 BC, yet because of the rich soil, the cultivated zone on the lower slopes is densely populated, including the coastal town of Catania. The most recent eruption was in December 1985. 32 J11

Euboea (Greek **Evvoia**), mountainous island off the east coast of Greece, in the Aegean Sea; area 3,755 sq km/1,450 sq mi; about 177 km/110 mi long; population (1991) 208,400. Mount Delphi reaches 1,743 m/5,721 ft. The chief town, Khalkis, is connected by a bridge to the mainland. 36 F6

Eugene city and administrative headquarters of Lane County, west-central Oregon, USA, 157 km/98 mi south of Portland, on the Willamette River; population (1994 est) 118,000. It is a processing and shipping centre for agricultural products from the surrounding region; wood products are also manufactured, and tourism is important. It is the home of the University of Oregon (1872), Northwest Christian College (1895), and Lane Community College (1965). 86 B2

Eupen town in the province of ►Liège, Belgium, 14 km/9 mi south of the German border city of Aachen; population (1995) 17,300. Eupen is predominantly German-speaking, having been part of Prussia (and thereafter the German Empire) from the early 19th century to the end of World War I. 20 B6

Euphrates (Turkish **Firat**; Arabic **Al Furat**), river rising in east Turkey and flowing through Syria and Iraq, joining the River Tigris above Basra to form the River Shatt-al-Arab at the head of the Gulf; length 3,600 km/2,240 mi. The ancient cities of Babylon, Eridu, and Ur were situated along its course. 58 H4

Eurasia the combined land areas of ►Europe and ►Asia.

Eure *département* in the ►Haute-Normandie region of France; area 6,004 sq km/2,318 sq mi; population (1990) 259,300. The territory of Eure is broken up by its rivers – the lower Seine and its tributaries the Eure, Andelle, Epte, and Risle – into wooded plateaux. Its chief products are wheat, flax, and beetroot. Cattle and horses of pure Normandy stock are bred here. Industries include textiles and metallurgy. The administrative centre is ►Evreux. 6 D2

Eure river rising in the Perche hills in Orne *département*, northwest France, and flowing southeast, then north, to join the River Seine near Rouen; length 225 km/140 mi. The town of Chartres is on its banks. It gives its name to two *départements*, ►Eure and ►Eure-et-Loir. 22 D6

Eure-et-Loir *département* in the ►Centre region of France; area 5,875 sq km/2,268 sq mi; population (1990) 396,600. The chief river in the north is the Eure, in the south are the Loir and the Huisne. None is navigable except the Eure for a short distance. Wheat and oats are the principal crops and apples are grown. Textiles and boots are manufactured and there are foundries. The administrative centre is ►Chartres. 6 D2

Europe the second-smallest continent, occupying 8% of the Earth's surface.

area 10,400,000 sq km/4,000,000 sq mi

population (1990 est) 498 million (excluding European Turkey and the former USSR); annual growth rate 0.3%, projected population of 512 million by the year 2000

language mostly Indo-European, with a few exceptions, including Finno-Ugric (Finnish and Hungarian), Basque, and Altaic (Turkish); apart from a fringe of Celtic, the northwest is Germanic; Letto-Lithuanian languages separate the Germanic from the Slavonic tongues of Eastern Europe; Romance languages spread east–west from Romania through Italy and France to Spain and Portugal

religion Christian (Protestant, Roman Catholic, Eastern Orthodox), Muslim (Turkey, Albania, Bosnia-Herzegovina, Yugoslavia, Bulgaria), Jewish

largest cities (population over 1.5 million) Athens, Barcelona, Berlin, Birmingham, Bucharest, Budapest, Hamburg, Istanbul, Kharkov, Kiev, Lisbon, London, Madrid, Manchester, Milan, Moscow, Paris, Rome, St Petersburg, Vienna, Warsaw

features Mount Elbrus 5,642 m/18,517 ft in the Caucasus Mountains is the highest peak in Europe; Mont Blanc 4,807 m/15,772 ft is the highest peak in the Alps; lakes (over 5,100 sq km/2,000 sq mi) include Ladoga, Onega, Vänern; rivers (over 800 km/500 mi) include the Volga, Danube, Dnieper Ural, Don, Pechora, Dniester, Rhine, Loire, Tagus, Ebro, Oder, Prut, Rhône

physical conventionally occupying that part of Eurasia to the west of the Ural Mountains, north of the Caucasus Mountains, and north of the Sea of Marmara, Europe lies entirely in the northern hemisphere between 36° N and the Arctic Ocean. About two-thirds of the continent is a great plain which covers the whole of European Russia and spreads westwards through Poland to the Low Countries and the Bay of Biscay. To the north lie the Scandinavian highlands, rising to 2,472 m/8,110 ft at Glittertind in the Jotenheim range of Norway. To the south, a series of mountain ranges stretch east–west (Caucasus, Balkans, Carpathians, Apennines, Alps, Pyrenees, and Sierra Nevada). The most westerly point of the mainland is Cape Roca in Portugal; the most southerly location is Tarifa Point in Spain; the most northerly point on the mainland is Nordkynn in Norway 14

Europoort deep-sea port extension of ►Rotterdam, the Netherlands. It is situated some 25 km/16 mi west of the city and covers more than 35 sq km/13.5 sq mi. Europoort contains facilities for handling bulk goods and crude oil imports, and was completed in 1968. 22 F3

Euskirchen city in North Rhine-Westphalia, Germany, 25 km/15 mi southwest of Bonn; population (1994) 180,400. There are electrical and mechanical engineering factories, as well as glass and sugar manufacturers. 20 B6

Evanston city in Cook County, northeast Illinois, USA, on the shores of Lake Michigan, immediately north of Chicago; population (1994) 73,400. It is a lake port with manufactures that include steel products, hospital and dairy supplies, textbooks, foodstuffs, paint, chemicals, and radio and TV equipment. It is the home of Northwestern University. 88 C2

Evansville industrial city and port in southwestern Indiana, USA, on the Ohio River near the Kentucky border; seat of Vanderburgh County; population (1994 est) 129,000. It is the commercial centre of an agricultural area; industries include electrical appliances, pharmaceuticals, and plastics. **90** D2

Everest, Mount (Tibetan *Qomolungma* 'goddess mother of the world'; Nepalese *Sagarmatha* 'head of the earth'), world's highest mountain above sea level, in the ▶Himalayas range, on the China–Nepal frontier; height 8,848 m/29,028 ft. It was first climbed by New Zealand mountaineer Edmund Hillary and Sherpa Tenzing Norgay in 1953. More than 360 climbers have reached the summit; over 100 have died during the ascent. **54** E3

Everett city, port, and administrative headquarters of Snohomish County, northwest Washington; population (1992) 73,800. It is situated at the mouth of the Snohomish River, on Puget Sound, 40 km/25 mi north-northeast of Seattle. It is a major cargo and commercial fishing port, and has one of the best landlocked harbours on the West Coast. Manufactures include finished lumber, plywood, and pulp and paper products; much of these made by a huge Weyerhaeuser mill. The Boeing aircraft assembly plant (1968) is often cited as the largest building in the world, at 6 million cu m/200 million cu ft. The city contains foundries, machine shops, and boatbuilding yards. Other industries include salmon canning, poultry and frozen food packaging, and the processing of dairy products. Everett Community College (1941) is in the city. Settled in 1862, Everett was laid out in 1890 as the western terminus of the Northern Pacific Railroad. **86** B1

Everglades subtropical area of swamps, marsh, and lakes in southern Florida, USA; area 7,000 sq km/2,700 sq mi. Formed by the overflow of Lake Okeechobee after heavy rains, it is one of the wildest areas in the USA, with distinctive plant and animal life. The natural vegetation of the swamplands is sawgrass and rushes, with trees such as cypress, palm, and hardwoods where the conditions are slightly drier. Several hundred Seminole, an American Indian people, live here. A national park (established in 1947) covers the southern tip of the Everglades, making up about one-fifth of the Everglades' original area. **90** E4

Evreux administrative centre of ▶Eure *département*, situated in the Iton river valley in northwest France, 90 km/56 mi northwest of Paris; population (1990) 51,500. It produces pharmaceuticals, rubber, and textiles; there are also printing and electronic companies. **22** D5

Evvoia Greek name for the island of ▶Euboea. **36** F6

Exeter city and administrative headquarters of ▶Devon, England, on the River Exe; population (1994 est) 107,000. Principal industries are brewing, iron and brass founding, light engineering, printing, financial services, and tourism. Other industries include the manufacture of agricultural machinery, textiles, and leather goods. Exeter was founded by the Romans as **Isca Dumnoniorum** and has medieval, Georgian, and Regency architecture. Exeter Cathedral was built largely between 1280 and 1369. Exeter University was established in 1955. **24** J11

Exmoor moorland district in north Devon and west Somerset, southwest England, forming (with the coast from Minehead to Combe Martin) a national park since 1954. The park covers an area of around 7,700 ha/19,000 acres, and includes Dunkery Beacon, its highest point at 519 m/1,705 ft; and the Doone Valley.

Exmouth resort town and former port in Devon, southwest England, at the mouth of the River Exe, 14 km/9 mi southeast of Exeter; population (1996 est) 31,920. Small industries include engineering and a pottery specializing in novelty teapots. The port was permanently closed to commercial vessels in 1989, but the town remains a yachting and boating centre. **24** J11

Extremadura autonomous community of western Spain, comprising the provinces of Badajoz and Cáceres; area 41,600 sq km/16,062 sq mi; population (1991) 1,061,900. Irrigated land is used for growing wheat; the remainder is either oak forest or used for pig or sheep grazing. There are food industries. The capital is ▶Mérida. **9** (1)B3

Eyre, Lake lake in northeast South Australia; area up to 9,000 sq km/3,500 sq mi. It is the largest lake in Australia, and includes Lake Eyre North and Lake Eyre South. Much of the lake remains a dry salt crust, while parts form a salt marsh in dry seasons. It is filled only when the rivers which drain into it flood. It is the continent's lowest point, 16 m/52 ft below sea level. **74** D8

Eyre Peninsula triangular peninsula in South Australia, lying between the Great Australian Bight to the west and Spencer Gulf to the east, and bounded on the north by the Gawler Ranges. Grain (especially wheat) grown on the Eyre Peninsula is exported from Port Lincoln, Wallaroo, Edithburgh, Ardrossan, and Cedana. Fishing is important in the area; iron is mined at the Middleback Ranges in the northern part of the peninsula, and there are steelworks at ▶Whyalla. **76** G6

Faenza (ancient **Faventia**), town in Ravenna province, Emilia-Romagna, Italy, on the River Lamone, 30 km/19 mi southwest of Ravenna; population (1990) 54,100. It is connected with the Adriatic Sea by canal. Products include textiles, furniture, ironware, and sulphur. Faenza gave its name to **faience** pottery, tin-glazed earthenware first produced here. **30** G6

Faeroe Islands (or **Faeroes**), alternative spelling of the ▶Faroe Islands, in the North Atlantic. **14** D1

Fagatogo seat of government of American ▶Samoa, situated on Pago Pago Harbour, next to the dependency capital Pago Pago, on Tutuila Island; population (1998 est) 2,300.

Fairbanks town in central Alaska, USA, situated on the Chena Slough, a tributary of the Tanana River; population (1990) 30,800. The second-largest city in the state, it functions as a service centre for the mineral development of central and northern Alaska. Founded in 1902, it became a gold-mining and fur-trading centre and the terminus of the Alaska Railroad and the Pan-American Highway. **92** (1)H3

Faisalabad city in Punjab province, Pakistan, 120 km/75 mi west of Lahore; population (1995 est) 1,600,000. It lies in an area where cotton and wheat are major crops, and industries in the city include textiles and textile machinery, flour milling, and soap and chemical manufacture. Formerly known as Lyallpur, it was founded in 1892 by the British, and laid out in the shape of a Union Jack. **54** B2

Faizabad city in Uttar Pradesh, north India; population (1991) 178,000. It lies at the head of navigation of the River Ghaghara and has sugar refineries and an agricultural trade. It was once capital of Oudh. There is a white marble mausoleum built in the early 19th century for Bahu Begum, widow of Shuja-ud-Daula. **54** D3

Falaise town situated 32 km/20 mi southeast of Caen, in Calvados *département*, Normandy, northern France; population (1990) 8,100. Perched on a cliff (French *falaise*) above the River Ante, it is a market centre, manufacturing cotton and leather goods, and trading in livestock and cheese. The 12th–13th-century castle belonged to Robert, one of the first dukes of Normandy. William the Conqueror was born here in 1027. **22** B6

Falkirk unitary authority in central Scotland, created from the former district of the same name in 1996 from part of the former Central region.
area 297 sq km/115 sq mi
towns Falkirk (administrative headquarters), Grangemouth
physical centrally located between Edinburgh and Glasgow, this low-lying area borders the southern side of the Firth of Forth; River Avon flows through
features Forth and Clyde and Union canals; Rough Castle; Antonine Wall
industries chemicals and petrochemicals, bus building, soft drinks, toffees
agriculture some dairy and arable farming
population (1996) 142,500 **5** E4

Falkland Islands (Argentine **Islas Malvinas**), British crown colony in the South Atlantic, 480 km/300 mi east of the Straits of Magellan; area 12,173 sq km/4,700 sq mi, made up of two main islands: East Falkland (6,760 sq km/2,610 sq mi) and West Falkland (5,413 sq km/2,090 sq mi); population (1991) 2,120. The capital is ►Stanley. The main industries are wool, alginates (used as dyes and as a food additive) from seaweed beds, and fishing. **100** K9

Fall River city and port in southeastern Massachusetts, USA, on Mount Hope Bay; population (1998 est) 90,700. It stands at the mouth of the Taunton River, near the Little Fall River, which gave the city its name. It was founded in 1656 by settlers from the Plymouth Colony, and was incorporated in 1803, and became a city in 1854. Once the largest cotton cloth manufacturing centre in the world, the city now produces textiles and clothing, chemicals, electrical and computer products, rubber, and paper, and is still an active port. **88** F2

Falmouth port and resort on the south coast of Cornwall, southwest England, on the estuary of the River Fal, 11 km/7 mi southwest of Truro; population (1991) 17,800. It is a major yachting centre and the marine rescue and coastguard centre for the southwest region. Principal industries include tourism, ship-repair at Pendennis shipyard, and the construction of aluminium buildings and naval architecture. **24** G11

Falster Danish island off the south coast of Sjaelland in the Baltic Sea; area 518 sq km/200 sq mi; part of the county of Storstrøm. Falster is 48 km/20 mi long and 3–20 km/2–12 mi wide. It is joined to Sjaelland and Lolland, an island to the west, by bridges. The soil is fertile, the climate milder and sunnier than in the rest of Denmark, and the white sandy beaches attract holidaymakers. The principal town is Nyköbing. **20** H2

Falun capital of Kopparberg county in southern Sweden, on Lake Runn; population (1995) 55,100. It grew up around a copper mine established in the 13th century. The main industries are now chemical production and timber processing. Falun is an international skiing centre. **16** H6

Famagusta (or **Ammochostos**), seaport on the east coast of Cyprus, in the Turkish Republic of Northern Cyprus; population (district, 1994) 67,200. It was the chief port of the island until the Turkish invasion 1974. **50** F2

Far East geographical term for all Asia east of the Indian subcontinent.

Fareham market town in Hampshire, southern England, 10 km/6 mi northwest of Portsmouth; population (1996 est) 102,500. Nearly 80% of employment is in the service sector, mainly retailing, and health and social work. Brickmaking, horticulture (notably strawberries), and leisure sailing industries have been supplemented by a fast-growing, high-technology industrial sector, which includes the manufacture of scientific instruments (GEC Marconi, IBM). **22** A4

Fargo city in southeastern North Dakota, USA, across the Red River from Moorhead, Minnesota; seat of Cass County; population (1992) 77,100. The largest city in the state, it is at the heart of navigation on the Red River and a centre for processing and distributing agricultural products and farm machinery, and meatpacking. Chemicals and building materials are also manufactured here. **84** G2

Farnborough town in Hampshire, southern England, 4 km/2.5 mi north of Aldershot; population (1996 est) 53,000. It is the headquarters of the UK Defence Evaluation and Research Agency (until 1995 RAE Farnborough), which carries out research and experimental development in aeronautics and related instrumentation to support the Ministry of Defence. Partnerships with civilian clients are significant. Aeronautical displays are given at the biennial air show. **22** B3

Farne Islands group of about 28 rocky islands in the North Sea, close to the mainland of northeast England, 2–8 km/1–5 mi from Bamburgh, Northumberland. The islands are a sanctuary for birds and grey seals. **24** L6

Faroe Islands (or **Faeroe Islands**, or **Faeroes**; Danish *Færøerne* 'Sheep Islands'), island group (18 out of 22 inhabited) in the North Atlantic, between the Shetland Islands and Iceland, forming an outlying part of ►Denmark; area 1,399 sq km/540 sq mi; population (1992 est) 46,800. The largest islands are Strømø, Østerø, Vagø, Suderø, Sandø, and Bordø. The capital is Thorshavn on Strømø. The main industries are fishing and crafted goods. Faeroese and Danish are spoken. **14** D1

Fayetteville city and administrative headquarters of Washington County, northwest Arkansas, USA, in the Ozark Mountains, 80 km/50 mi northeast of Fort Smith; population (1990) 42,100. It is an agricultural trading centre, and its main industry is poultry processing; other industries include lumber, clothing, and tools. It is the home of the University of Arkansas (1871). **90** C2

Fayetteville city in south-central North Carolina, USA, on the Cape Fear River, 100 km/62 mi south of Durham and 80 km/50 mi southwest of Raleigh; seat of Cumberland County; population (1992) 76,650. It is the processing centre for the area's agricultural and lumber products; its industries include the manufacture of tools, tyres, and textiles. 88 E3

Fécamp channel seaport and resort of northern France, about 35 km/22 mi northeast of Le Havre, in the *département* of Seine-Maritime; industries include shipbuilding, fishing, and food processing; population (1990) 21,100. Benedictine liqueur was first produced here in the early 16th century. 22 C5

Fehmarn German island in the Baltic Sea, belonging to the *Land* (administrative region) of ►Schleswig-Holstein and covering an area of 185 sq km/71 sq mi; population (1991) 11,800. Fehmarn is separated from the Danish island of Laaland by the **Fehmarn Belt**. It has mostly grazing land and farmland growing wheat, barley, and rape. The sandy beaches are a holiday destination. The principal town is Burg. 20 G2

Feldkirch town in the Austrian province of Vorarlberg, on the River Ill near the Liechtenstein border; population (1995) 15,800. It has a castle and a Gothic church. There are textile, timber, and paper industries. 30 E3

Felixstowe port and resort opposite Harwich in Suffolk, eastern England, fronting the estuary of the Orwell at its confluence with the Deben, 20 km/12 mi southeast of Ipswich; population (1996 est) 23,800. It is Britain's busiest container port, handling around 2 million containers a year, 35% of the UK's total traffic. Oil and other bulk liquids are also imported. 22 D3

Fens, the level, low-lying tracts of reclaimed marsh in eastern England, west and south of the Wash, covering an area of around 40,000 sq km/15,500 sq mi, about 115 km/70 mi north–south and 55 km/34 mi east–west. They fall within the counties of Lincolnshire, Cambridgeshire, and Norfolk. Formerly a bay of the North Sea, they are now crossed by numerous drainage canals and form some of the most fertile and productive agricultural land in Britain. The southern peat portion of the Fens is known as the Bedford Level. 22 B2

Fergana (formerly **Novy Margelan** 1877–1910; **Skobolev** 1910–24), city in eastern Uzbekistan, and capital, cultural, and administrative centre of Fergana wiloyat (region), situated 250 km/155 mi southeast of Tashkent and 65 km/40 mi east of Kokand; population (1995) 190,000. It is an important centre of petroleum and oil extraction and refining, and is the capital of the major cotton- and fruit-growing region of the Fergana Valley. 56 K1

Fermanagh historic county of Northern Ireland.
area 1,680 sq km/648 sq mi
towns ►Enniskillen (county town), Lisnaskea, Irvinestown
physical in the centre is a broad trough of low-lying land, in which lie Upper and Lower Lough Erne
industries clothing, tweeds, cotton thread, food processing, light engineering, china, tourism, electronics
agriculture small farms, livestock, potatoes
population (1991) 50,000

Ferrara industrial town and archbishopric in Emilia-Romagna region, northern Italy, on a branch of the Po delta, 52 km/32 mi west of the Adriatic Sea; population (1992) 137,100. It is the principal distribution and market centre of a major fruit-growing area. Industries include sugar-refining, and the manufacture of chemicals and textiles. Ferrara became a powerful city state in the 13th century. 30 G6

Ferrol alternative name for ►El Ferrol, a city and port in Spain. 28 B1

Fertile Crescent region of the Middle East stretching from the Gulf to the Nile Valley, and including the Euphrates and Tigris rivers. The name comes from the extensive irrigated cultivation of crops in this area, which in ancient times had a more moderate climate suitable for agriculture. It gave rise to the earliest human culture (dating from around 8000 BC) and was the birthplace of several ancient empires, including those of Sumer, Assyria, and Persia.

Fertö tó Hungarian name for the ►Neusiedler See, a lake in Austria and Hungary. 30 M3

Fès (or **Fez**; Arabic **Fas**), former capital of Morocco 808–1062, 1296–1548, and 1662–1912, in the Fès valley north of the Great Atlas Mountains, 160 km/100 mi east of Rabat; population (1993) 564,000. Textiles, carpets, and leather are manufactured, and the *fez*, a brimless hat worn in southern and eastern Mediterranean countries, is traditionally said to have originated here. Qarawiyin Islamic University dates from 859; a second university was founded in 1961. 66 E2

Fez alternative spelling of ►Fès, a city in Morocco. 66 E2

Fife unitary authority in eastern Scotland, which was formerly a region of three districts (1975–96) and a county until 1974.
area 1,321 sq km/510 sq mi
towns Cupar, Dunfermline, ►Glenrothes (administrative headquarters), Kirkcaldy, St Andrews
physical coastal area, predominantly low lying, undulating interior with dramatic escarpment at Lomond Hills; rivers Eden and Leven flow through
features Rosyth naval base; Old Course, St Andrews
industries electronics, petrochemicals, light engineering, oil servicing, paper
agriculture potatoes, cereals, sugar beet, fishing (Pittenweem)
population (1996) 351,200 5 E3

Figueras (or **Figueres**), town in the province of Girona in Cataluña, northeast Spain; population (1991) 32,700. It was the home of the surrealist painter Salvador Dalí. The Salvador Dalí museum, opened in 1974, was designed by the artist himself. 28 N2

Fiji Islands country comprising 844 islands and islets in the southwest Pacific Ocean, about 100 of which are inhabited.

The Fiji Islands covers an area of 18,333 sq km/7,078 sq mi, and has a population of 817,000 (2000 est). The capital is ►Suva; other major towns and cities include Lautoka, Nadi, Ba, Labasa, Nausori, and Lami. The head of state is Ratu Josefa Iloilo (from 2000); the head of government is Laisenia Qarase (from 2000). The currency is the Fiji dollar, and the languages spoken include English (official), Fijian, and Hindi. 74 H8

Fingal's Cave cave on the island of Staffa, Inner Hebrides, Argyll and

Bute, Scotland. It is lined with volcanic basalt columns, and is 70 m/230 ft long and 20 m/65 ft high. Visited by the German Romantic composer Felix Mendelssohn in 1829, the cave was the inspiration of his *Hebridean* overture, otherwise known as *Fingal's Cave*.

Finistère (Latin *finis terrae* 'land's end'), *département* of ►Brittany, northwest France, occupying a peninsula with the English Channel to the north and the Bay of Biscay to the south; area 7,030 sq km/2,714 sq mi; population (1990) 838,700. The administrative centre is ►Quimper; Brest and Douarnenez are the chief towns. Inland there is heathland and fertile valleys; the Arrée and Black mountains run east–west. Horses and cattle are raised on the extensive grasslands of the *département*, and there is agriculture and seasonal fishing. The chief crops are wheat, rye, barley, oats, potatoes, flax, and mangel-wurzels; fruit and vegetables are exported. Industries include textiles, electrical goods, paper, and shipbuilding. 6 A2

Finland country in Scandinavia, bounded to the north by Norway, east by Russia, south and west by the Baltic Sea, and northwest by Sweden.

Finland covers an area of 338,145 sq km/130,557 sq mi, and has a population of 5,176,000 (2000 est). The capital is ►Helsinki (Swedish Helsingfors); other major towns and cities include Tampere, Turku, Espoo, Vantaa, and Oulu. The head of state is Tarja Halonen (from 2000); the head of government is Paavo Lipponen (from 1995). The currency is the markka, and the languages spoken include Finnish, Swedish (both official), Saami (Lapp), and Russian. 16 P3

Finland, Gulf of eastern arm of the ►Baltic Sea, separating Finland from Estonia. It is 420 km/260 mi long and 40–150 km/25–90 mi wide. Helsinki and St Petersburg are the main ports. 16 M7

Firat Turkish for the ►Euphrates, a river rising in east Turkey and flowing through Syria and Iraq. 58 H4

Firenze Italian form of ►Florence, a city in Italy. 30 G7

Fishguard (Welsh **Abergwaun**), seaport and holiday centre at the mouth of the River Gwaun on the south side of Fishguard Bay, Pembrokeshire, southwest Wales, 30 km/19 mi southwest of Cardigan; population (1991 est) 5,000. There is a ferry service to Rosslare in the Republic of Ireland. 24 H9

Flagstaff city in Arizona, USA, 400 km/249 mi north of Phoenix; population (1990) 45,900. The largest city in north central Arizona, Flagstaff is the railway and livestock-market centre for the surrounding region, and has timber mills which use the local yellow pine. The Lowell Observatory (1894) is here. Situated 129 km/80 mi south of the Grand Canyon, Flagstaff is a centre of tourism, and there are local ski resorts. 92 D1

Flanders region of the Low Countries that in the 8th and 9th centuries extended from Calais to the Schelde and is now covered by the Belgian provinces of Oost Vlaanderen and West Vlaanderen (East and West Flanders), the French *département* of Nord, and part of the Dutch province of Zeeland. The language is Flemish. East Flanders, capital Ghent, has an area of 3,000 sq km/1,158 sq mi and a population (1995) of 1,349,400. West Flanders, capital Bruges, has an area of 3,100 sq km/1,197 sq mi and a population (1995) of 1,121,100.

Flensburg historic port and commercial city on the east coast of Schleswig-Holstein, Germany, 2 km/1 mi from the Danish border, with shipyards and breweries; population (1995) 87,700. Rum and smoked eels are specialities. 20 E2

Flevoland (formerly **IJsselmeer-polders**), low-lying province of the Netherlands, created in 1986 from land reclaimed from the IJsselmeer between 1950 and 1968; area 1,410 sq km/544 sq mi; population (1995 est) 262,300. The main towns are ►Lelystad (capital), Dronten, and Almere. 7 (1)D2

Flint city in southeastern Michigan, USA, on the Flint River, 90 km/56 mi northwest of Detroit; seat of Genesee County; population (1992) 139,300. Car manufacturing is the chief industry here, but this is now in decline. Flint was founded as a trading post in 1819 and laid out in 1836; it was incorporated in 1855. 88 D2

Flintshire (Welsh **Sir y Fflint**), unitary authority in north Wales. A former county, it was part of Clwyd between 1974 and 1996.
area 437 sq km/169 sq mi
towns Mold (administrative headquarters), Flint, Holywell, Buckley, Connah's Quay
physical bounded by the Irish Sea in the north, the Dee estuary in the east, and the Clwydian Range, which rises to 555 m/1,820 ft, in the southwest; rivers Dee, Alyn
industries artificial silk, chemicals, optical glass
agriculture dairy farming, stock-raising
population (1996) 144,000 5 E5

Florence city and administrative headquarters of Lauderdale County, northwest Alabama, USA, on the Tennessee River near the Tennessee Valley Authority's Wilson Dam, 175 km/109 mi northwest of Birmingham; population (1990) 36,400. Industries include agricultural and poultry products, building materials, lumber, and fertilizers. The University of North Alabama (1872) is in Florence. 90 D3

Florence (Italian **Firenze**; Roman **Florentia**), capital of ►Tuscany, northern Italy, on the River Arno, 88 km/55 mi from the river's mouth; population (1992) 397,400. It has printing, engineering, and optical industries; many crafts, including leather, gold and silver work, and embroidery; and its art and architecture attract large numbers of tourists. Notable medieval and Renaissance citizens included the writers Dante and Boccaccio, and the artists Giotto, Leonardo da Vinci, and Michelangelo. 30 G7

Florence city and administrative headquarters of Florence County, northeast South Carolina, USA, 98 km/61 mi northwest of Myrtle Beach; population (1990) 29,800. It is a centre of the trucking industry, serving as a terminus for many companies. Other industries include electronic equipment, dairy products, fertilizers, film, furniture, machined goods, and clothing. Francis Marion College (1970) and a technical college are in the city. 90 F3

Florianópolis (formerly **Desterro**), seaport, resort and capital of Santa Caterina federal unit (state), southeast Brazil, 480 km/300 mi south of São Paulo; population (1991) 256,000. It is situated on Santa Caterina Island, which is linked to the mainland by two bridges, one of which, the Ponte Hercílio Luz, is the longest steel suspension bridge in Brazil. Exports include sugar, tobacco, and fruit. Florianopolis is used as a port of call for coastal shipping and has developed as a tourist beach resort. There is a cathedral (constructed 1753–73, modified 1922) and a university. The city was founded in 1770. **100** M4

Florida southeasternmost state of the USA. It is nicknamed the Sunshine State. Florida was admitted to the Union in 1845 as the 27th US state. Much of the state is subtropical and is a popular tourist and retirement destination; the Miami region in particular has a thriving cosmopolitan community, serving as a gateway to both the Caribbean and Latin America. Florida is bordered to the north by Georgia and by Alabama. The state consists of a 640 km-/400 mi-long peninsula jutting into the Atlantic, which it separates from the Gulf of Mexico, with a 390 km-/240 mi-long panhandle to the northwest on the mainland. At the peninsula's southern end, the 220 km/135 mi-long chain of Florida Keys extends to the southwest.
population (2000 est) 15,982,400
area 152,000 sq km/58,672 sq mi
capital ►Tallahassee
towns and cities Miami, Tampa, Jacksonville, Hialeah, Orlando, Fort Lauderdale
industries and products tourism, leisure industry, citrus fruits, melons, vegetables, sugar cane, fish, shellfish, phosphates, chemicals, electrical and electronic equipment, aircraft, fabricated metals, finance sector. **90** E4

Florida Keys series of small coral islands that curve over 240 km/150 mi southwest from the southern tip of Florida, USA, between the Straits of Florida and Florida Bay. The most important are Key Largo (the largest settlement) and Key West (which has a US naval and air station); economically they depend on fishing and tourism. A causeway links the keys to each other and to the mainland. **80** K7

Flushing English form of ►Vlissingen, a port in the Netherlands. **22** F3

Focşani capital of Vrancea county, Romania; 200 km/125 mi northeast of Bucharest; population (1993) 101,400. It is a commercial centre, with trade along the River Milcov, which joins the River Siret, to Galaţi. An Austro-Russian alliance defeated the Turks at Focşani 1789, during a sustained offensive against the Ottoman Empire in the Balkans. **34** Q4

Foggia town of Apulia region, southern Italy, 120 km/75 mi northwest of Bari; population (1992) 155,700. The town is an important communications hub, with cellulose, paper, engineering, and flour industries. It is also the market centre of a fruit- and wheat-growing plain known as the Tavoliere (chess-board). Natural gas is found nearby. **32** K7

Föhr one of the North ►Frisian Islands in the North Sea, belonging to Germany, off the west coast of ►Schleswig-Holstein; population (1991) 8,200. Föhr covers an area of 82 sq km/31 sq mi. Fishing is the main occupation, with oysters being exported to the German mainland. The main town, Wyk, is also a health resort. Wildfowl mass in great numbers on the island in the autumn. **20** D2

Folkestone port and resort on the southeast coast of Kent, England, 10 km/6 mi southwest of Dover; population (1991) 46,200. There are passenger ferry and SeaCat (high speed catamaran) connections with Boulogne, France, and to the northwest of the town is the British terminal of Eurotunnel, which offers a high-speed train link through the Channel Tunnel to Paris, France and Brussels in Belgium for private and haulage traffic. **22** D3

Fond du Lac town and administrative headquarters of Fond du Lac County, southeastern Wisconsin; population (1990) 37,800. It is situated at the southern end of Lake Winnebago, 29 km/18 mi south-southeast of Oshkosh and 88 km/55 mi northwest of Milwaukee, at the mouth of the Fond du Lac River. Industries include dairy farming and tourism. Precision tools, machinery, motor-vehicle parts, machine tools, outboard motors, and snowmobiles are manufactured in the town. Fond du Lac also serves as the trading centre of a dairy farming region; cheese and other dairy products are processed here and animal feed is also produced. **88** C2

Fongafale capital of Tuvalu, on ►Funafuti atoll; population (2001 est) 5,000. **74** H6

Fontainebleau (Latin **Fons Bellaqueus**), town in Seine-et-Marne *département*, France, situated 60 km/37 mi southeast of Paris near the River Seine; population (1990) 35,500. Its royal palace was founded by Philip the Good, but, as it exists today, was built by Francis I in the 16th century. Louis XIV's mistress, Mme de Montespan, lived here, as did Louis XV's mistress, Mme du Barry. The French emperor Napoleon signed his abdication here in 1814. Nearby is the village of Barbizon, the haunt of several 19th-century painters (known as the Barbizon School). **26** H5

Foochow alternative transcription of ►Fuzhou, a port and capital of Fujian province, southeast China. **50** F1

Forbach French town in the *département* of Moselle, 50 km/31 mi east of Metz; population (1990) 30,000. There is a paper mill and woodworking, and the town is emerging as a sporting and cultural centre. **22** J5

Forchheim city in Bavaria, Germany, on the Rhein–Main–Donau canal, 31 km/19 mi north of Nuremberg; population (1994) 108,900. Textiles, dyes, chocolate, paper, and optical goods are manufactured. In the early Middle Ages Forchheim was an important city of the Holy Roman Empire. **20** G7

Foreland, North and South chalk headlands projecting from the east coast of Kent, southeast England. **North Foreland**, with one lighthouse, lies 4 km/2.5 mi east of Margate; **South Foreland**, with two, lies 5 km/3 mi northeast of Dover. The headlands mark the eastern limit of the North Downs. **22** D3

Forli town and market centre in Emilia-Romagna region, northeast Italy, on the River Montone, 60 km/37 mi southeast of Bologna; population (1992) 109,100. Felt, majolica pottery, and paper are manufactured. There is a 17th-century cathedral, and a Romanesque church (begun 1179) with a campanile (bell-tower). **30** H6

Formentera smallest inhabited island in the Spanish Balearic Islands, lying south of Ibiza; area 93 sq km/36 sq mi; population (1990) 5,200.

The chief town is San Francisco Javier and the main port is La Sabina. The main industry is tourism. 28 M6

Formentor, Cape northern extremity of ▶Mallorca, in the Balearic Islands of the western Mediterranean Sea. 28 P5

Fortaleza (or **Ceará**), industrial port, resort, and capital of Ceará federal unit (state), northeast Brazil, on the Atlantic coast; population (1991) 1,708,700 (metropolitan area 2,357,100). Cotton, rice, coffee, sugar, hides, and vegetable oils are exported. Industries include textiles (particularly hammocks), flour-milling, and sugar-refining. Tourism has become increasingly important to the local economy. Fortaleza was first colonized by the Portuguese in 1603. It has a cathedral and several museums. 98 K4

Fort Collins city and administrative headquarters of Larimer County, north-central Colorado, on the Cache de la Poudre River, 90 km/56 mi north of Denver, at an altitude of 1,555 m/ 5,102 ft; population (1992) 93,300. It is the processing and marketing centre for the surrounding agricultural area; industries include computer hardware and software, engines, cement, plastics, film, and prefabricated metal buildings. Tourism is also important to the local economy. It is the home of Colorado State University (1870). 86 E2

Fort-de-France capital, chief commercial centre, and port of Martinique, West Indies, at the mouth of the Madame River; population (1995 est) 109,000. It trades in sugar, rum, and cacao. There is an airport and the tourist industry is a source of much local revenue. 94 M6

Fort Duquesne former French fort and trading post established in 1754 on the site where ▶Pittsburgh, Pennsylvania, USA, now stands. In 1758 British forces under Gen Forbes destroyed the French fort, replaced it with Fort Pitt (named after William Pitt, prime minister of Britain at the time), and established the town of Pittsburgh.

Forth river in central Scotland, with its headstreams, Duchray Water and Avondhu, rising on the northeast slopes of Ben Lomond. It flows east approximately 105 km/65 mi to Kincardine where the **Firth of Forth** begins. The Firth is approximately 80 km/50 mi long, and is 26 km/16 mi wide where it joins the North Sea. 24 H5

Fort Lamy former name of ▶Ndjamena, the capital of Chad. 64 B5

Fort Lauderdale city in southeastern coastal Florida, USA, 40 km/25 mi north of Miami, where the New River enters the Atlantic Ocean; seat of Broward County; population (1994 est) 163,500. The city's main industry is tourism. A network of inlets and canals used for boating cross the city; Atlantic Ocean beaches line it on the east; and a deep-water port to the south, Port Everglades, allows ocean-going vessels to dock. 90 E4

Fort Myers city and administrative headquarters of Lee County, southwest Florida, USA, on the Caloosahatchee River, 150 km/93 mi southeast of St Petersburg, near the Gulf of Mexico; population (1990) 45,200. Fort Myers grew rapidly as a resort during the 1920s. Still a popular tourist city, it also ships local citrus, flowers (mainly gladioli), vegetables, and seafood. Electronic equipment and cigars are manufacture and lumber processed in the city. It is home to Edison Community College (1961). 90 E4

Fort Pierce city and administrative headquarters of St Lucie County, east-central Florida, USA, where the Indian River flows into the Atlantic Ocean; population (1990) 36,800. It is a transport centre for the fruit and vegetable crops of the surrounding area, and fishing and tourism are significant industries. Indian River Community College (1960) is in the city. 90 E4

Fort Smith city and co-administrative headquarters (with Greenwood) of Sebastian County, west Arkansas, on the Arkansas River where it crosses the Oklahoma–Arkansas border, 225 km/140 mi west of Little Rock; population (1992) 74,000. It is the site of coal and natural-gas mines; industries include processed foods, electrical appliances, furniture, cars, paper, plastics, and metals. It is home to Westark Community College (1928). 90 C2

Fort Wayne town in northeastern Indiana, USA, where the St Mary and St Joseph rivers meet to form the Maumee River, 255 km/158 mi from Chicago; seat of Allen County; population (1994 est) 183,000. It is the centre of a rich agricultural hinterland; industries include motor vehicles, electrical goods, electronics, and farm machinery. 90 D1

Fort Worth city in northeastern Texas, USA, on the Trinity River, 48 km/30 mi west of ▶Dallas; seat of Tarrant County; population (1994 est) 452,000. It is a grain and railway centre serving the southern USA; manufactured products include aerospace equipment, motor vehicles, computers, and refined petroleum. 90 B3

Fougères French town in the *département* of Ille-et-Vilaine, 48 km/30 mi northeast of Rennes; population (1990) 23,100. There is a 12th-century castle with 13 towers, and the 15th-century walls still stand. Footwear, metal goods, cloth, and glass are manufactured; tourism is important. 26 D5

Fou-liang former name of ▶Jingdezhen, a town in Jiangxi province, China. 46 F5

Foveaux Strait stretch of water between the extreme south of South Island, New Zealand, and Stewart Island, New Zealand. It is a fishing area and produces a considerable oyster catch. 78 A8

Foyle, Lough sea lough on the north coast of Ireland, traversed by the frontier of Northern Ireland and the Republic of Ireland. It is noted amongst ornithologists as a site for migratory seabirds. 24 E6

France country in western Europe, bounded to the northeast by Belgium, Luxembourg, and Germany, east by Germany, Switzerland, and Italy, south by the Mediterranean Sea, southwest by Spain and Andorra, and west by the Atlantic Ocean.

France covers an area (including Corsica) of 543,965 sq km/210,024 sq mi, and has a population of 59,080,000 (2000 est). The capital is ▶Paris; other major towns and cities include Lyon, Lille, Bordeaux, Toulouse, Nantes, Marseille, Nice, Strasbourg, Montpellier, Rennes, and Le Havre. The head of state is Jacques Chirac (from 1995); the head of government is Lionel Jospin (from 1997). The currency is the franc, and the main language spoken is French (official; regional languages include

Basque, Breton, Catalan, Corsican, and Provençal). 26 G7

Franche-Comté region of eastern France; area 16,200 sq km/ 6,250 sq mi; population (1990) 1,097,300. Its administrative centre is ▶Besançon, and it includes the *départements* of ▶Doubs, ▶Jura, ▶Haute-Saône, and Territoire de Belfort (see ▶Belfort, Territoire de). In the mountainous Jura, there is farming and forestry, and elsewhere there are engineering and plastics industries. 6 G3

Frankenthal city in the Rhineland-Palatinate, Germany, 7 km/4 mi north-west of Ludwigshafen; population (1994) 47,000. Frankenthal is situated a short distance from the left bank of the ▶Rhine, with which it is connected by canal. Leather, cork, metals, and fertilizers are manufactured. 20 D7

Frankfort capital of ▶Kentucky, USA, located in the north-central part of the state, on the Kentucky River, 72 km/45 mi east of Louisville; seat of Franklin County; population (1990) 26,000. Industries include bourbon (whisky), electronic equipment, air brakes, furniture, and footwear. 90 E2

Frankfurt am Main (German 'ford of the Franks'), city in Hessen, Germany, 72 km/45 mi northeast of Mannheim; population (1995) 651,200. It is a commercial and banking centre, with electrical and machine industries, and an inland port on the River Main. The International Book Fair is held here annually in the autumn. It is the site of the Bundes-bank (German Central Bank), and the European Central Bank (from 1999). 20 D6

Frankfurt-an-der-Oder industrial city in the state of Brandenburg, northeast Germany, 80 km/50 mi southeast of Berlin; population (1995) 81,700. Situated close to the Polish border, it was the former capital of the East German district of Frankfurt from 1952–90. It is linked by the River Oder and its canals to the rivers Vistula and Elbe. Industries include semiconductors, chemicals, engineering, paper, and leather. 20 K4

Franz Josef Land (Russian **Zemlya Frantsa Iosifa**), archipelago of over 85 islands in the Arctic Ocean, east of Svalbord and northwest of Novaya Zemlya, Russia; area 20,720 sq km/ 8,000 sq mi. There are scientific stations on the islands. 42 J2

Fraser river in southern British Columbia, Canada; length 1,370 km/ 850 mi. It rises in the Yellowhead Pass of the Rocky Mountains and flows northwest into the interior plateau of British Columbia. It then flows south through its lower valley at Hope, then west where it enters the Strait of Georgia, near Vancouver. The Fraser River has acted as the focus of economic life in the southern part of the province: as a salmon fishery, as a means of transportation for the rich lumber reserves of its upper valley, and as a source of irrigation and hydroelectric power. 82 G6

Frauenfeld capital of the canton of Thurgau, Switzerland, on the River Murg 34 km/21 mi northeast of Zürich; population (1995) 20,400. Products manufactured in the town are cottons, silks, woollens, and iron and aluminium goods. Frauenfeld has a 10th-century castle. 30 D3

Fredericksburg city in Spotsylvania County, Virginia, eastern USA, situated on the Rappahannock River, about 88 km/55 mi south of Washington; population (1994 est) 22,200. Fredericksburg serves an agricultural region and has light manufacturing. During the American Civil War it was the scene of four significant battles 1862–64: Fredericksburg, Chancellorsville, Wilderness, and the Spotsylvania Courthouse. The battlefields lie west of the city, within the Fredericksburg and Spotsylvania National Military Park. 88 E3

Fredericton capital of ▶New Brunswick, Canada, on the St John River; seat of York County; population (1991) 45,400. It is the centre of the region's dairy and lumber trade; local industries include meat processing, footwear, woodworking, and boat-building. It was known as St Anne's Point until 1785 when it was renamed after Prince Frederick, second son of George III. 82 T7

Fredrikstad Norwegian port, 93 km/ 58 mi by rail southeast of Oslo, in Ostfold county at the mouth of the River Glomma on Oslo Fjord; population (1996) 51,500. It is a centre of the timber trade and has shipyards; there are also canning, paint, porcelain and plastic industries. 16 F7

Free State (formerly **Orange Free State** until 1995), province of the Republic of South Africa; area 127,993 sq km/49,418 sq mi; population (1995 est) 2,782,500. Lesotho forms an enclave on the KwaZulu-Natal and Eastern Cape Province border. The capital is ▶Bloemfontein; other main towns are Springfontein, Kroonstad, Bethlehem, Harrismith, and Koffiefontein. The main industries are gold, oil from coal, cement, and pharmaceuticals; agricultural production is centred on grain, wool, and cattle. 72 D5

Freetown capital of Sierra Leone; population (1992) 505,000. It has a naval station and a harbour. Industries include cement, plastics, footwear, oil refining, food production, and tobacco processing. Platinum, chromite, rutile, diamonds, and gold are traded. Freetown was founded as a settlement for freed slaves in 1787. It was made capital of the independent Sierre Leone in 1961. The beaches of Freetown peninsula attract tourists. 68 B3

Freiburg im Breisgau commercial city in Baden-Württemberg, Germany, 133 km/83 mi southwest of Stuttgart, situated in the Breisgau at the western edge of the Black Forest; population (1995) 198,300. Chief products are pharmaceuticals and precision instruments. It has a university (founded in 1457) and a Gothic cathedral dating from the 12th century. Tourism is important. 30 C3

Freising city in Bavaria, Germany, on the River Isar, 32 km/20 mi northeast of Munich; population (1994) 138,500. Freising has one of the oldest breweries in the world, dating from 1040. Other industries include light engineering. There is a cathedral (1161–1205) and a former Benedictine abbey (now a school of agriculture and brewing). 30 G2

Fréjus French town in the *département* of Var, 25 km/16 mi southwest of Cannes; population (1990) 45,100. It stands on a hill above an alluvial plain 2 km/1 mi from the Mediterranean. It was founded by Caesar and has Roman remains, including an amphitheatre, baths, and aqueduct; the harbour built by Augustus has been silted up by the River Argens.

The Gothic cathedral has a 5th-century baptistry. There are cork, olive-oil, and wine industries; tourism is important. 26 M10

Fremantle chief port of Western Australia, at the mouth of the River Swan, 19 km/12 mi southwest of ▶Perth; population (1996) 25,634. Fremantle is the port for ▶Perth. It has shipbuilding yards, sawmills, food processors, and a fish cannery; principal exports include grain, alumina, refined petroleum, silica sands, animal feeds, sheep, metal scrap, water, titanium dioxide, liquid petroleum gas, bricks, and tiles. Notre Dame University was established here in 1989. 76 C6

French Guiana (French **Guyane Française**), French overseas *département* from 1946, and administrative region from 1974, on the north coast of South America, bounded west by Suriname and east and south by Brazil; area 83,500 sq km/32,230 sq mi; population (1997 est) 156,900. The main towns are ▶Cayenne and St Laurent. The main economic activity is fishing; other resources include bauxite, tropical hardwood timber, gold, cinnabar, and kaolin. Agricultural products include rice, maize, cocoa, bananas, and sugar, though the country depends largely on imported food. 98 G3

French Polynesia French Overseas Territory in the South Pacific, consisting of five archipelagos: Windward Islands, Leeward Islands (the two island groups comprising the ▶Society Islands), ▶Tuamotu Archipelago (including ▶Gambier Islands), ▶Tubuai Islands, and ▶Marquesas Islands; total area 3,940 sq km/1,521 sq mi; population (1994) 216,600. The capital is ▶Papeete on Tahiti. Tourism is the mainstay of the economy; other industries are cultivated pearls, coconut oil, and vanilla. The languages spoken are Tahitian (official) and French. 74 L7

French Somaliland former name (to 1967) of ▶Djibouti, eastern Africa. 64 H5

French Sudan former name (1898–1959) of ▶Mali. 66 E5

French West Africa group of French colonies administered from Dakar 1895–1958. They are now Senegal, Mauritania, Sudan, Burkina Faso, Guinea, Niger, Côte d'Ivoire, and Benin.

Fresno city in central California, USA, 260 km/161 mi southeast of San Francisco; seat of Fresno County; population (1994 est) 387,000. It is the processing and marketing centre for the San Joaquin Valley, one of the richest agricultural districts in the USA. It is irrigated by the Fresno River and the Delta-Mendota canal, and is a major producer of fruit and vegetables – especially grapes for raisins and wine making. Industries include glass, machinery, fertilizers, and vending machines, and the Fresno petroleum field is one of the richest in the state. 92 C1

Fribourg (German **Freiburg**), Swiss canton situated in the southwest of the country; area 1,670 sq km/645 sq mi; population (1995) 224,600. It lies in the foothills of the Alps, the highest point being 2,395 m/7,858 ft. The chief rivers in the canton are La Broye and La Sarine. Gruyère cheese is made in Fribourg. 8 (1)B2

Fribourg (German **Freiburg**), city in west Switzerland, capital of the canton of Fribourg, on a loop of the Sarine River about 30 km/19 mi southwest of Bern; population (1995) 39,400. It is renowned for its food products, especially the cheese of the Gruyère district and chocolate. Other industries include breweries, metal-working, and chemicals. 30 C4

Friedrichshafen city in Baden-Württemberg, Germany, on the northeast shore of Lake Constance, 115 km/71 mi southeast of Stuttgart; population (1994) 56,200. The town is a resort and conference centre, and has car and aerospace industries. 30 E3

Friendly Islands another name for ▶Tonga, a country in the Pacific. 74 J7

Friesland maritime province on the northeast side of the IJsselmeer, north Netherlands, which includes the Frisian Islands and land reclaimed from the former Zuider Zee; area 3,400 sq km/1,313 sq mi; population (1997) 615,000 (the inhabitants of the province are called Frisians). The capital is ▶Leeuwarden; other main towns are Drachten, Harlingen, Sneek, and Heerenveen. Small boats are made; agriculture centres on livestock (Friesian cattle, which originated here, and black Friesian horses), dairy products, and arable farming. 7 (1)D1

Frisian Islands (Dutch **Friesche Eilanden**), chain of low-lying islands 5–32 km/3–20 mi off the northwest coasts of the Netherlands and Germany, with a northerly extension off the west coast of Denmark. They are divided between these three states. They were formed by the natural sinking of the intervening land and are separated from the mainland only by shallow flats (*wadden*). **Texel** is the largest and westernmost island, at the southern end of the chain. Sheep and cattle are raised, and there are a number of small seaside resorts. 22 H1

Friuli-Venezia Giulia autonomous agricultural and wine-producing region of northeast Italy, bordered to the east by Slovenia, comprising the provinces of Pordenone, Gorizia, Trieste, and Udine; area 7,800 sq km/3,011 sq mi; population (1992) 1,195,100. The most important industrial centres are ▶Udine,which is the region's capital, ▶Gorizia, the ports of ▶Trieste and Monfalcone (dockyards and chemical industries), and Pordenone. 9 (4)D1

Frunze former name (1926–91) of ▶Bishkek in Kyrgyzstan. 42 N9

Fujian (or **Fukien**), maritime province of southeast China, lying opposite Taiwan, bounded to the southwest by Guangdong, to the west and northwest by Jiangxi, to the northeast by Zhejiang, and to the southeast by the Taiwan Strait on the China Sea; area 123,100 sq km/47,500 sq mi; population (1994) 32,610,000. The main cities are ▶Fuzhou (capital), Xiamen, Zhangzhou, and Nanping. The main industries are steel-rolling, electrical goods, tourism, handicrafts, and leather goods. Agricultural products are rice, sweet potatoes, sugar, special aromatic teas, tobacco, timber, and citrus fruit. 11 (2)E4

Fuji, Mount (Japanese **Fuji-san**), Japanese volcano and highest peak, on Honshu Island, near Tokyo; height 3,778 m/12,400 ft. Extinct since 1707, it has a Shinto shrine and a weather station on its summit. Fuji has long been revered for its picturesque cone-shaped crater peak, and figures prominently in Japanese art, literature, and religion. 48 K6

Fuji-san Japanese name for Mount ▶Fuji, a volcano on Honshu Island, Japan. 48 K6

Fukuoka city on the north coast of Kyushu island, Japan, 110 km/68 mi northeast of Nagasaki, port and capital of Fukuoka prefecture; population (1995 est) 1,264,100. Industries include the manufacture of silk, ceramic dolls, marine products, printing, chemicals, and ship-building. 48 F7

Fukushima city in the north of Honshu island, Japan, 255 km/158 mi north of Tokyo; population (1994) 281,000. It is the capital of Fukushima prefecture. It has a silk industry and is a centre for light engineering, textiles, milk processing, fruit canning, paper-making, and the manufacture of electrical appliances. Fukushima University and Iizaka hot spring can be found in the city. 48 L5

Fukuyama port and industrial centre in the southwest of Honshu island, Japan, 80 km/50 mi east of Hiroshima at the mouth of the Ashida River; population (1994) 373,000. There are a large iron and steel works, a ship-yard, and several rubber factories. Exports include cotton and rubber. A reconstructed castle and the island of Sensujima attract tourists to the port. Traditionally, *tatami* (straw mats) were made in Fukuyama. 48 G6

Fulda city in Hesse, Germany, on the River Fulda, 82 km/51 mi northeast of Frankfurt am Main; population (1995) 60,100. The city produces textiles, carpets, and rubber. Fulda was made a city in 1157. St Michael's Chapel (818) is the second oldest church in Germany. 20 E6

Funabashi industrial city in Chiba prefecture, Kanto region, Honshu island, 20 km/12 mi east of Tokyo, Japan; population (1994) 531,000. Industries include steel, timber, photochemicals, machinery, and food processing. 48 L6

Funafuti atoll consisting of 30 islets in the western Pacific, in the state of ▶Tuvalu; area 2.8 sq km/1.1 sq mi; population (1995 est) 2,865. The capital of Tuvalu, ▶Fongafale, is on the atoll. 74 H6

Funchal capital and chief port of the Portuguese island of Madeira, on the south coast; population (1991) 109,900. Tourism and Madeira wine are the main industries. 66 B2

Fundy, Bay of Canadian Atlantic inlet situated between northeast Maine and south New Brunswick and southwest Nova Scotia, Canada. It extends northeastwards for about 320 km/200 mi inland from the Gulf of Maine. In the northeast it divides into Chignecto Bay, to the north, and the Minas Basin, to the south. Fundy's fast-running tides, rising up to 18 m/60 ft, are the highest in the world; they have been harnessed for electricity since 1984. The rising tide creates 'reversing falls' at the mouth of the St John River, and the surge up the Petitcodiac River creates a tidal bore (surge of tidal water up an estuary or a river) at Moncton, New Brunswick. The bay is a rich fishing ground. 82 T8

Fünen German form of ▶Fyn, an island forming part of Denmark. 20 F1

Fünfkirchen (German 'Five Churches'), German name for ▶Pécs, a town in southwestern Hungary. 34 F3

Fürstenfeldbruck city in Bavaria, Germany, 22 km/14 mi west of Munich; population (1994) 185,300. Clothing manufacture is the town's main industry. Fürstenfeldbruck has a number of baroque churches. 30 G2

Fürth town in Bavaria, Germany, adjoining Nuremberg; population (1991) 105,300. It has electrical, chemical, textile, and toy industries. 20 F7

Fushun (or **Funan**), city in Liaoning province, northeast China, 40 km/25 mi east of Shenyang; population (1994) 1,387,900. It is a coalmining and oil-refining centre with aluminium, iron, steel, engineering, oil-shale distillation, and chemical works. 48 B3

Fuzhou (or **Foochow**; formerly **Minhow**), industrial port and capital of Fujian province, on the Min River in southeast China; population (1994) 1,354,800. It is a centre for ship-building and steel production; rice, sugar, tea, and fruit pass through the port. Traditionally renowned for its handicrafts, particularly carving and lacquerware, Fuzhou's industries now include electronics, food-processing, and the manufacture of textiles and building materials. There are joint foreign and Chinese factories. 50 F1

Fyn (German **Fünen**), island forming part of Denmark and lying between the mainland and Zealand; capital Odense; area 2,976 sq km/1,149 sq mi; population (1995) 467,700. 20 F1

Gabès port in east Tunisia; population (1994) 98,900. Fertilizers and dates are exported. The town's main industry is chemicals (phosphoric and sulphuric acid) since the construction of a chemical complex here in the early 1970s. The town stands on the site of the Roman town of **Tacapae**. 66 H2

Gabon country in central Africa, bounded north by Cameroon, east and south by the Congo, west by the Atlantic Ocean, and northwest by Equatorial Guinea.

Gabon covers an area of 267,667 sq km/103,346 sq mi, and has a population of 1,226,000 (2000 est). The capital is ▶Libreville; other major towns and cities include Port-Gentil, Franceville (or Masuku), Lambaréné, Mouanda, Oyem, and Mouila. The head of state is Omar Bongo (from 1967); the head of government is Jean-François Ntoutoume-Emane (from 1999). The currency is the franc CFA, and the languages spoken include French (official), Fang (in the north), Bantu languages, and other local dialects. 68 G5

Gaborone capital of ▶Botswana, mainly an administrative and government-service centre; population (1995 est) 158,000. The city lies at altitude 1000 m/3300 ft and has an airport. Light industries include motor vehicle assembly, textiles, brewing, printing and publishing, and construction. The city developed after 1962 when it replaced Mafikeng as capital in preparation for the country's independence in 1966. 72 D4

Gabrovo industrial town (textiles) in Lovech region, central Bulgaria, on the northern slopes of the Balkan Mountains, 55 km/34 mi northwest of Stara Zagora; population (1990) 88,100. 34 N7

Gadsden city and administrative headquarters of Etowah County, northeast Alabama, USA, on the Coosa River, 95 km/59 mi northeast of Birmingham; population (1990) 42,500. It is a distribution centre for the area's livestock, poultry, and dairy products. Industries include manganese, bauxite, coal, timber, steel, tyres, electrical machinery parts, and farm equipment. Gadsden State Community College (1965) is in the city. **90** D3

Gafsa oasis town in central Tunisia, centre of a phosphate-mining area; population (1994) 71,100. The town is linked by rail to Gabès and Sfax. **66** G2

Gainesville city and administrative headquarters of Alachua County, north Florida, USA, 90 km/56 mi southwest of Jacksonville; population (1992) 86,800. The area produces beef cattle, poultry, and lumber; its industries include electronic parts, concrete, wooden products, and tourism. The University of Florida (1901), which includes the East Florida State Seminary (1853), is located in the city. **90** E4

Galapagos Islands (officially **Archipiélago de Colón**), group of 12 large and several hundred smaller islands in the Pacific lying 800 km/ 500 mi off the coast of Ecuador, to which they belong; area 7,800 sq km/ 3,000 sq mi; population (1995 est) 11,100. This island group, of volcanic origin, includes the six main islands of San Cristóbal (where the capital of the same name is situated), Santa Cruz, Isabela, Floreana, Santiago, and Fernandina, as well as 12 smaller islands, with other islets. Volcán Wolf, at 1707 m/5600 ft, on Isabela I is the highest peak. The Galapagos National Park was established in 1934 and, because of the unique fauna, the islands have been established as a world heritage site. They were uninhabited when discovered by Spanish explorers in 1535; no colony was established. They were annexed by Ecuador in 1832. **98** (1)B1

Galaţi (German **Galatz**), port on the River Danube in Romania; population (1993) 324,000. Industries include shipbuilding, iron, steel, textiles, food processing, and cosmetics. **34** R4

Galdhøpiggen mountain in the Jotunheim range, Norway, 2,469 m/ 8,100 ft. It is counted as second to Glittertind, 2,472 m/8,110 ft (with glacier ice), although without ice and snow Galdhøpiggen is the higher of the two. **16** D6

Galicia region of central Europe, extending from the northern slopes of the Carpathian Mountains to the Romanian border. Once part of the Austrian Empire, it was included in Poland after World War I and divided in 1945 between Poland and the USSR.

Galicia mountainous but fertile autonomous community of northwest Spain, comprising the provinces of La Coruña, Lugo, Orense, and Pontevedra; area 29,400 sq km/11,350 sq mi; population (1991) 2,731,700. Industries include fishing, and tungsten and tin mining; Galicia has the largest fishing fleet in the European Union. The climate is very wet, and the region is traversed northeast to southwest by the River Miño. The chief harbours are La Coruña, Vigo, and El Ferrol. The Galician language (Gallego) is similar to Portuguese. The capital is ►Santiago de Compostela. **9** (1)A1

Galilee, Sea of alternative name for Lake ►Tiberias in northern Israel. **60** C4

Gällivare iron-mining town above the Arctic Circle in Norrbotten county, northern Sweden; population (1991) 22,400. **16** L3

Galveston city and port in southeastern Texas, USA, at the northeastern end of Galveston Island, on the Gulf of Mexico coast; seat of Galveston County; population (1992) 59,600. It is the state's chief cotton port and also exports petroleum, wheat, and timber, and has chemical works and petroleum refineries. Fishing is also important, and the city has long been a resort for the island's sandy beaches. **90** C4

Galway county on the west coast of the Republic of Ireland, in the province of Connacht; county town ►Galway; area 5,940 sq km/2,293 sq mi; population (1996) 188,900. Lead is found at Tynagh, and copper, lead, and zinc near Loughrea; marble is quarried and processed at Recess and Inverin. The main farming activity is cattle and sheep grazing. The Connemara National Park is in Galway. Towns include Salthill, a suburb of **Galway** city and seaside resort, Ballinasloe, Clifden, and Tuam. **5** B5

Galway fishing port and county town of County ►Galway, Republic of Ireland; 200 km/124 mi west of Dublin; principal city of ►Connacht province; population (1996) 57,200. It produces textiles and chemicals, and there is salmon and eel fishing; Galway has recently become important for its computer industry. Queen's College (founded in about 1845) was renamed University College in 1908, and is part of the National University of Ireland; teaching has been bilingual, conducted in both English and Irish, since 1929. Galway is an important centre of the Irish language; **Galway Theatre**, An Taibhdhearc, only stages plays in Irish. **24** C8

Gambia, The country in west Africa, bounded north, east, and south by Senegal and west by the Atlantic Ocean.

The Gambia covers an area of 10,402 sq km/4,016 sq mi, and has a population of 1,305,000 (2000 est). The capital is ►Banjul; other major towns and cities include Serekunda, Birkama, Bakau, Farafenni, Sukuta, Gunjur, and Basse. The head of state and government is Yahya Jammeh (from 1994). The currency is the dalasi, and the languages spoken include English (official), Mandinka, Fula, Wolof, and other indigenous dialects. **68** A2

Gambier Islands (or **Mangaréve Islands**), island group, part of ►French Polynesia, lying in the southeastern part of the ►Tuamotu Archipelago; area 36 sq km/14 sq mi; population (1996 est) 600. It includes four coral islands and many small islets. The main island is Mangaréva, with its town Rikitea. They were made a French colony in 1881. Copra is the chief commercial product, though there is also fishing for pearls. **74** N8

Gäncä (formerly **Elizavetpol** 1804–1918; **Gandzha** before 1804, and 1918–35; **Kirovabad** 1935–89), industrial city and former capital of Azerbaijan, on a tributary of the Kura River 300 km/186 mi west of Baku; population (1996) 281,000. Industries in Gäncä include the manufacture of cottons, woollens, and processed foods. **58** M3

Gandia seaport in the province of Valencia, eastern Spain, on the Mediterranean coast; population (1991) 50,800. It has an old college (once a university), a Gothic church, and the palace of the dukes of Gandia. Exports include oranges, raisins, onions, and wine. **28** K6

Gandzha former name (until 1804 and 1918–35) of ►Gäncä in Azerbaijan. 58 M3

Ganges (Hindi **Ganga**), major river of India and Bangladesh; length 2,510 km/1,560 mi. It is the most sacred river for Hindus. 54 E3

Gannett Peak the highest peak in Wyoming, USA; height 4,207 m/ 13,804 ft. It is in the Wind River Range of the Rocky Mountains, on the Continental Divide, and stands 95 km/59 mi east-southeast of Jackson. It is largely surrounded by the Dinwoody Glacier, with the Bridger-Teton National Forest to its west and the Shoshone National Forest to its east. Fremont Peak (4,189 m/13,745 ft) is to the southeast. 86 E2

Gansu (or **Kansu**), province of northwest China, bounded to the north by Mongolia and Inner Mongolia, to the east by Ningxia Hui Autonomous Region and Shaanxi, to the south by Sichuan, and to the west by Qinghai and Xinjiang Uygur Autonomous Region; area 530,000 sq km/205,000 sq mi; population (1996) 24,670,000. The main cities are ►Lanzhou (capital), Yumen, Tianshui, Dunhuang, and Jiayuguan. Chief industries are coal, oil, hydroelectric power from the Huang He River, mining, metal-processing, and tourism. Agriculture is based on the cultivation of spring wheat, millet, sorghum, flax, and fruit, and animal rearing. 11 (2)D3

Gaoxiong mainland Chinese form of ►Kaohsiung, a port in Taiwan. 46 G6

Gap French town, capital of the *département* of Hautes-Alpes; population (2001 est) 36,200. It stands 750 m/2,461 ft above sea level, and is the seat of a bishop. There is a trade in timber and livestock. 30 B6

Gard *département* in the ►Languedoc-Roussillon region of France; area 5,848 sq km/2,258 sq mi; population (1990) 585,400. It has a short coastline on the Mediterranean, and flowing through it is the River Rhône and its tributaries the Cèze, the Ardèche, and the Gard. Olives, vines, mulberries (for silkworms), fruit, and cereals are produced in great quantities, and cattle are raised. There is much mineral wealth, including iron, lignite, copper, asphalt, zinc, and lead, and there are metallurgical, silk, leather, and oil industries. The administrative centre is ►Nîmes. Other towns include Alès and Le Vigan. 6 F4

Garda, Lake (Italian **Lago di Garda**; ancient **Lacus Benacus**), largest lake in Italy; situated on the border between the regions of Lombardy and Veneto; area 370 sq km/ 143 sq mi. 30 F5

Garden Grove city in Orange County, southwest California, USA; population (1994 est) 148,000. It is located 50 km/31 mi southeast of the Los Angeles city centre, and lies to the southwest of Anaheim, on the Santa Ana River. It is a residential and industrial suburb. Aerospace and electronics plants in the area are vital to the local economy.

Garden of the Gods district of Colorado, western USA, 10 km/6 mi northwest of Colorado Springs; area 5 sq km/2 sq mi. The district contains formations of eroded red and white sandstone, some of which resemble cathedral spires. The area is a popular tourist destination, especially the western half which the city of Colorado Springs purchased and set up as a scenic park in 1932.

Garmisch-Partenkirchen town in Bavaria, Germany, located at the confluence of the Loisach and Partnach rivers and close to the Austrian border, 94 km/58 mi south of Munich; population (1994) 85,400. Nearby is the peak known as the Zugspitze (2,963 m/9,721 ft); Garmisch-Partenkirchen is a major Alpine skiing resort and health centre, and hosted the 1936 Winter Olympics. 30 G3

Garonne river in southwest France, rising in the Vale of Aran on the Spanish side of the Pyrenees and flowing to the ►Gironde estuary; length 575 km/357 mi. The river is not navigable. 26 E9

Gary city and port in northwestern Indiana, USA, on the southern shore of Lake Michigan; population (1994 est) 114,000. Gary was once one of the largest steel producers in the world and steel is still an important industry, but cutbacks in steel production in the mid–late 20th century left the city economically depressed. It has since developed into a general manufacturing city, with harbour facilities and industrial plants lining the lakeshore. 84 J3

Gaspé Peninsula (or **Gaspésie**), mountainous peninsula in south-eastern Québec, Canada, between the Gulf of St Lawrence and Chaleur Bay, an extension of the Appalachian Mountains; area 29,500 sq km/ 11,390 sq mi. The central part includes the Shickshock Mountains and the highest point is Mount Jacques Cartier (1,268 m/4,160 ft). There are fishing and lumbering industries, and tourism, paper production, and copper mining are also important. 82 T7

Gastonia city and administrative headquarters of Gaston County, southwest North Carolina, USA; population (1992) 57,100. It is situated in the central Piedmont, on the Catawba River, 29 km/18 mi west-northwest of Charlotte. Since the construction in 1848 of a cotton mill, Gastonia has been a leading US textile-manufacturing centre. In 1929, several people were killed during a bitter strike at the Loray Mills; organizers were convicted of conspiracy to murder the police chief. Other industries include the manufacture of machinery, oil filters, corrugated boxes, plastics, and electronic equipment. 90 E2

Gateshead port in Tyne and Wear, northeast England; population (1994 est) 127,000. It is situated on the south bank of the River Tyne, opposite Newcastle upon Tyne. Formerly a port for the Tyne coalfields and a railway workshop centre, it now manufactures chemicals, plastics, and glass; other industries include engineering, printing, and tourism. 24 L7

Gauteng (Sotho 'Place of Gold'), province of the Republic of South Africa from 1994, known as Pretoria-Witwatersrand-Vereeniging before 1995, and historically part of the Transvaal; area 18,760 sq km/ 7,243 sq mi; population (1995 est) 7,048,300. The Vaal River and Magaliesberg Mountains pass through the province. The main cities are ►Johannesburg (capital), Pretoria, Vereeniging, Krugersdorp, Benoni, and Germiston. The most important industries are gold mining, coal, iron and steel, uranium, and chemicals; tobacco and maize are cultivated. 72 D5

Gävle seaport and capital of Gävleborg county, Sweden, on the Gulf of Bothnia; population (1995) 90,600. The town is situated at the mouth of the River Gävle, 150 km/93 mi northwest of Stockholm, and is spread over the riverbanks and two islands. Shipbuilding, textiles, machinery, and pottery are the main industries. The chief exports are iron goods, timber, and wood pulp. **16** J6

Gaya ancient city in Bihar state, northeast India, 140 km/87 mi south of Patna; population (1991) 294,000. Situated on the holy River Phalgu, it is a centre of pilgrimage for Buddhists and Hindus with many temples and shrines; the Vishnupad temple enshrines the footprint of Vishnu. A bo tree at Bodhgaya is said to be a direct descendant of the original tree under which the Buddha sat. There is a tourist centre. **54** E4

Gaza capital of the ▶Gaza Strip, once a Philistine city; population (1991 est) 125,000. **60** B5

Gaza Strip strip of land on the Mediterranean Sea, 10 km/6 mi wide and 40 km/25 mi long, extending northeast from the Egyptian border; area 363 sq km/140 sq mi; population (1997 est) 987,900, mainly Palestinians, plus about 2,500 Israeli settlers. It was occupied by Israel from 1967 until 1994, when responsibility for its administration was transferred to the Palestine National Authority. The capital is ▶Gaza; other main centres of population are Khan Yunis and Rafah. Prior to the great influx of Palestinian refugees in 1948 the area was rural, and is geographically part of the ▶Negev. The area is dependent on Israel for the supply of electricity. Agriculture is the main activity, producing citrus fruit, wheat, and olives. Industry is on a small scale, including handmade goods, such as olive wood carvings, for Israel's tourist industry. **60** B5

Gaziantep Turkish city 185 km/115 mi northeast of Adana; population (1990) 603,400. It has textile and tanning industries. Until 1922 it was known as **Antep** or **Aintab**. **58** G5

Gdańsk (German **Danzig**), Polish port; population (1993) 466,500. Oil is refined, and textiles, televisions, and fertilizers are produced. In the 1980s there were repeated antigovernment strikes at the Lenin shipyards; many were closed in 1996. **18** H3

Gdynia port in northern Poland; population (1990) 251,500. It was established 1920 to give newly constituted Poland a sea outlet to replace lost ▶Gdańsk. It has a naval base and shipyards and is now part of the 'Tri-city', which includes Sopot and Gdańsk. **18** H3

Geelong (Aboriginal *jillong* 'white seabird'), industrial port on Corio Bay in south Victoria, Australia, 76 km/47 mi southwest of Melbourne; population (1996) 125,382. Geelong is Victoria's second city, and industries include oil refining, canning, wool scouring works, and the production of aluminium, motor vehicles, textiles, glass, fertilizers, pharmaceuticals, cement, and aerated waters. It is also a centre for the grain and wool trades. **76** H7

Geislingen an der Steige city in Baden-Württemberg, Germany, 53 km/33 mi southeast of Stuttgart; population (1994) 28,400. Geislingen is a manufacturing centre for metal and cotton goods, and has traditional craft industries such as glasswork and woodcarving. Many of its medieval buildings survive, including a late Gothic church. **30** E2

Gelderland (English **Guelders**), province of the east Netherlands, bounded on the southeast by Germany, on the southwest by the River Maas, and on the northwest by the Ijsselmeer; area 5,020 sq km/1,938 sq mi; population (1997) 1,886,100. The capital is ▶Arnhem; other main cities are Apeldoorn, Nijmegen, and Ede. Textiles, electrical goods, and paper are produced; agriculture is based on livestock, wheat, fruit, and vegetables. **7** (1)D2

Geleen town in the province of Limburg, the Netherlands, 15 km/9 mi northeast of Maastricht; population (1996) 33,900. Geleen once manufactured goods associated with by-products of coal from the nearby coalfield, but its main industries are now concerned with petroleum by-products and natural-gas production. **22** H4

Gelsenkirchen industrial city in the Ruhr District, in North Rhine-Westphalia, Germany, on the Rhine-Herne Canal and the River Emscher, 25 km/15 mi west of Dortmund; population (1995) 291,800. It has iron, steel, chemical, and glass industries, but the decline of these traditional industries has resulted in high unemployment and population decline, from a peak of 382,840 in 1961. **22** K3

Geneva (French **Genève**, German **Genf**), canton in southwest Switzerland, bounded by Lake Geneva, the canton of Vaud, and France; area 282 sq km/108 sq mi; population (1995) 395,500. It lies in the ▶Rhône valley between the southern Jura Mountains and the foothills of the Alps. The canton was admitted into the Swiss Confederation in 1815. **8** (1)A2

Geneva (French **Genève**; German **Genf**), city in Switzerland, capital of ▶Geneva canton, on the southwestern shore of Lake Geneva (▶Geneva, Lake); population (1994 est) 174,400. It is a point of convergence of natural routes and is a cultural and commercial centre. Industries include the manufacture of watches, scientific and optical instruments, foodstuffs, jewellery, and musical boxes. CERN, the particle physics research organization, is here, as are the headquarters of the International Red Cross and the World Health Organization. The United Nations has its second-largest office (after the New York City office) in Geneva. **30** B4

Geneva, Lake (French **Lac Léman**; German **Genfersee**), largest of the central European lakes, between Switzerland and France; area 580 sq km/225 sq mi. The main part of the lake lies in western Switzerland. It is in the shape of a crescent 72 km/45 mi long and 13 km/8 mi wide. **30** B4

Genève French name for ▶Geneva, a city in Switzerland, and for the canton of the same name. **30** B4/**8** (1)A2

Genf German form of ▶Geneva, a city in Switzerland, and for the canton of the same name. **30** B4/**8** (1)A2

Genk town in the province of Limbourg, Belgium, 14 km/9 mi east of Hasselt; population (2001 est) 63,100. The town's chief employer is the Ford Motor Company, which has a car-assembly plant here. The Limbourg coalfield provides employment in Genk and surrounding villages, but the mining industry has been in decline since the 1960s and there is high unemployment in the area. 22 H4

Genoa (Italian **Genova**; ancient **Genua**), historic city in northwest Italy, capital of Liguria, on the Gulf of Genoa, 400 km/249 mi northwest of Rome; population (1992) 667,600. It is Italy's largest port, with a major container port facility at Voltri, 10 km/6 mi to the west. Industries include oil-refining, chemicals, engineering, and the manufacture of textiles. 30 D6

Genova Italian form of ▶Genoa, a city in Italy. 30 D6

Gent Flemish name for ▶Ghent, a port city in northwest Belgium. 22 F3

Georgetown capital and main port of Guyana, situated on the east bank of the Demerara River at its mouth on the Atlantic coast; population (1992) 200,000. There are food processing and shrimp fishing industries. Principal exports include sugar, bauxite, and rice. 98 F2

George Town (or **Penang**), chief port of the Federation of Malaysia, and capital of Penang state, on the island of Penang; population (1991) 219,000. It produces textiles and toys. 52 C1

Georgia country in the Caucasus of southeastern Europe, bounded north by Russia, east by Azerbaijan, south by Armenia, and west by the Black Sea.

Georgia covers an area of 69,700 sq km/26,911 sq mi, and has a population of 4,968,000 (2000 est). The capital is ▶Tbilisi; other major towns and cities include Kutaisi, Rustavi, Batumi, Zugdidi, and Gori. The head of state is Eduard Shevardnadze (from 1992); the head of government is Giorgi Arsenishvili (from 2000). The currency is the lari, and the languages spoken include Georgian (official), Russian, Abkazian, Armenian, and Azeri. 58 K2

Georgia state in southeastern USA. It is nicknamed the Empire State of the South or the Peach State. Georgia ratified the US Constitution in 1788, becoming the 4th state to join the Union. Historically it was a cotton-producing state associated with slavery; as the birthplace of Martin Luther King Jr, it has strong links with the history of the civil-rights movement. Georgia is bordered to the northeast by South Carolina, to the north by North Carolina and Tennessee, to the west by Alabama, and to the south by Florida. In the southeast, Georgia has a coastline some 145 km/90 mi long on the Atlantic, off which are many of the Sea Islands.
population (2000 est) 8,186,500
area 152,600 sq km/58,900 sq mi
capital ▶Atlanta
towns and cities Columbus, Savannah, Macon, Albany
industries and products poultry, livestock, tobacco, maize, peanuts, cotton, soybeans, china clay, crushed granite, marble, clothing and textiles, carpets, aircraft, paper products, lumber, turpentine, finance sector, tourism 90 E3

Georgian Bay bay in Ontario, Canada, forming the northeast section of Lake Huron, the second largest of the Great Lakes of North America; it is divided from the lake by the Manitoulin Islands and the Bruce Peninsula of Ontario; length 160 km/100 mi; width 80 km/50 mi. About 30 islands at the southern end form the Georgian Bay National Park. The area is a popular tourist destination. 88 D1

Georgia, Strait of strait in British Columbia, Canada, separating Vancouver Island from the mainland; length about 240 km/150 mi. It is about 80 km/50 mi broad at its widest part, but narrows considerably towards the north, where it merges into Johnstone Strait and Queen Charlotte Strait. It meets the Pacific Ocean at Juan De Fuca Strait in the south. 86 B1

Gera industrial city in the *Land* (administrative region) of Thuringia, Germany, 50 km/31 mi south of Leipzig, on the Weisse Elster River; population (1995) 125,000.
It was the capital of the East German district of Gera 1952–90. Industries include electronics and telecommunications. The surrounding area has been badly affected by the environmental effects of lignite mining, now in decline. 20 H6

German Ocean former name for the ▶North Sea. 24 N4

Germany (in full **Federal Republic of Germany**), country in central Europe, bounded north by the North and Baltic Seas and Denmark, east by Poland and the Czech Republic, south by Austria and Switzerland, and west by France, Luxembourg, Belgium, and the Netherlands.

Germany covers an area of 357,041 sq km/137,853 sq mi, and has a population of 82,220,000 (2000 est). The capital is ▶Berlin; other major towns and cities include Cologne, Hamburg, Munich, Essen, Frankfurt am Main, Dortmund, Stuttgart, Düsseldorf, Leipzig, Dresden, and Hannover. The head of state is Johannes Rau (from 1999); the head of government is Gerhard Schroeder (from 1998). The currency is the Deutschmark, and the main language spoken is German (official). 20 E6

Germiston city in Gauteng Province, South Africa, 14 km/9 mi east of Johannesburg; population (1991) 134,000. Industries include gold refining, chemicals, steel, textiles, and brewing.

Gers *département* in the ▶Midi-Pyrénées region of southwest France; area 6,254 km/2,416 sq mi; population (1990) 175,360. It contains sections of the foothills of the Pyrenees and is drained by many rivers, including the Gers, Baise, Save, and Gimone. Vines are extensively grown, primarily for distillation into Armagnac brandy. Cereals, vegetables, fruit, and tobacco are also produced. Stock raising (horses, cattle, and poultry) is important. The principal towns are Auch (the capital), Condom, and Mirande. 6 D5

Gethsemane site of the garden where Judas Iscariot, according to the New Testament, betrayed Jesus. It is on the Mount of Olives, in east Jerusalem. When Jerusalem was divided between Israel and Jordan in 1948, Gethsemane fell within Jordanian territory.

Ghana country in West Africa, bounded north by Burkina Faso, east by Togo, south by the Gulf of Guinea, and west by Côte d'Ivoire.

Ghana covers an area of 238,540 sq km/92,100 sq mi, and has a population of 20,212,000 (2000 est). The capital is ▶Accra; other major towns and cities include Kumasi, Tamale, Tema, Sekondi-Takoradi, Cape Coast, Koforidua, Bolgatanga, and Obuasi. The head of state and government is John Kufuor (from 2001). The currency is the cedi, and the languages spoken include English (official), Ga, and other African languages. 68 D3

Ghats, Eastern and Western twin mountain ranges in south India, east and west of the central plateau; a few peaks reach about 3,000 m/9,800 ft. The name is a European misnomer, the Indian word *ghat* meaning 'pass', not 'mountain'. 54 C6/54 B5

Ghent (Flemish **Gent**; French **Gand**), port city and capital of East Flanders province, northwest Belgium, situated at the junction of the rivers Lys and Schelde, 55 km/34 mi northwest of Brussels; population (1997) 225,500. Industries include textiles, chemicals, electronics, metallurgy, and motor-vehicle manu-facturing. The cathedral of St Bavon (12th–14th centuries) has paintings by van Eyck and Rubens. 22 F3

Giant's Causeway stretch of basalt columns forming a headland on the north coast of Antrim, Northern Ireland. It was formed by an outflow of lava in Tertiary times which has solidified in polygonal columns. The Giant's Causeway and Causeway Coast became a World Heritage Site in 1986.

Gibraltar (Arabic Jebel Tariq, 'Mountain of Tariq'), British dependency, situated on a narrow rocky promontory at the southern tip of Spain; the **Rock of Gibraltar** formed one of the Pillars of Hercules with Mount Acho, near Ceuta, across the Strait of Gibraltar on the north African coast; area 6.5 sq km/ 2.5 sq mi; population (1993) 29,000. Gibraltar is mainly a trading centre for the import and re-export of goods. 28 E8

Gibraltar, Strait of strait between north Africa and Spain, forming the entrance from the Atlantic Ocean to the Mediterranean Sea, with the Rock of Gibraltar to the north side and Mount Acho to the south, the so-called Pillars of Hercules. 28 E9

Gibson Desert desert in central Western Australia, between the Great Sandy Desert to the north and the Great Victoria Desert in the south; area 220,000 sq km/85,000 sq mi. 76 D4

Giessen manufacturing town on the River Lahn, Hesse, Germany, 55 km/34 mi north of Frankfurt; population (1995) 73,800. Manufactured goods include machine tools, rubber, leather, tobacco, and furniture. The town rose to importance after the foundation of its university in 1607. 20 D6

Gijón port on the Bay of Biscay, in the autonomous community of Asturias, northern Spain; population (1991) 259,100. Gijón is a major fishing and shipbuilding centre and an outlet for the coal mines of Asturias (now very much reduced); products include iron, steel, chemicals, and oil. It stands on the site of a pre-Roman settlement. 28 E1

Gilbert and Ellice Islands former British colony in the Pacific, known since independence (1978) as the countries of ▶Tuvalu and ▶Kiribati. 74 H5

Gilgit mountainous region on the northwest frontier of Kashmir, under the rule of Pakistan; formerly a British Agency established in 1889; area 38,021 sq km/14,680 sq mi. It is drained by the Gilgit and Indus rivers. The region's town, Gilgit, was formerly a Buddhist centre. It lies at nearly 1,500 m/5,000 ft above sea level; during the winter the town can be approached only by plane. 54 B1

Girgenti former name (to 1927) of ▶Agrigento, a town in Italy. 32 H11

Girona (formerly **Gerona**), capital of Girona province in Cataluña, northeast Spain, at the confluence of the Onar and the Ter rivers; population (1991) 65,000. Industries include textiles and chemicals. There is an airport serving the Costa Brava. Girona has a Gothic cathedral, begun in 1312. 28 N3

Gironde navigable estuary 75 km/46 mi long on the southwest coast of France between Bordeaux and the sea, formed by the mouths of the ▶Garonne and ▶Dordogne rivers. It flows into the Bay of Biscay between the Médoc and Côtes vineyards. The estuary has sand banks and strong tides, but is used by ocean-going vessels. The passenger port is Le Verdon-sur-Mer. There are oil refineries at Pauillac and Bec d'Ambès. 26 E8

Gironde *département* in the ▶Aquitaine region of France; area 10,000 sq km/3,861 sq mi; population (1990) 1,214,400. It is fertile in the east but marshy and covered with pine woods in the west. The woods are separated from the sea by sand dunes, which extend for 120 km/75 mi along the coast. The great wine-producing district of Médoc lies along the west bank of the Gironde estuary. Cereals, fruit, tobacco, and vegetables are grown, and livestock is farmed. Turpentine, pitch, and charcoal are obtained from the pine forests. The administrative centre is ▶Bordeaux. Other towns include Blaye, Langon, Lesparre, and Libourne. 6 C4

Gisborne town and port on the east coast of North Island, New Zealand, at the confluence of the Taruheru and Waimata rivers at the northern end of Poverty Bay; population (1996) 32,700. It exports dairy products, wool, and meat, and is a holiday resort. There are important freezing works. The town is the capital of Gisborne region. It was the first landing-place in New Zealand of Captain James Cook in October 1769; a granite obelisk marks the spot where he came ashore. 78 G4

Giurgiu (Bulgarian **Giurgevo**), border town and capital of Giurgiu county, Romania, on the left bank of the River Danube, opposite Bulgarian Ruse, 64 km/40 mi southwest of Bucharest; population (1995) 72,300. It is a centre of commerce between Romania and Bulgaria, trading in chemicals, petroleum, timber, salt, and grain, and has large sawmills. 34 N6

Gîza, El (or **al-Jīzah**), city and governorate of Egypt, situated on the west bank of the Nile to the southwest of ▶Cairo, of which it forms a suburb of the greater metropolitan area; population of city (1998 est) 4,479,000, and governorate (1995 est) 4,525,000. It has textile, footwear, brewing, and film industries. It is noted for its archaeological sites; 8 km/5 mi to the southwest are the pyramids of Khufu, Khafre, and Menkure, and the Great Sphinx. 64 F1

Glamorgan (Welsh **Morgannwg**), three counties of south Wales – Mid Glamorgan, ▶South Glamorgan, and ▶West Glamorgan – created in 1974 from the former county of Glamorganshire. All are on the Bristol Channel. In 1996 Mid Glamorgan was divided amongst Rhondda Cynon Taff, Merthyr Tydfil, Bridgend, and Vale of Glamorgan; South Glamorgan was divided amongst Cardiff and Vale of Glamorgan; and West Glamorgan was divided into Neath Port Talbot and Swansea.

Glarus canton in eastern Switzerland; area 684 sq km/264 sq mi; population (1995) 39,400. Glarus contains part of the valley of the River Linth and is extremely mountainous, its highest point being Mount Tödi (3,623 m/11,886 ft). Relatively isolated, the canton has remained mainly agricultural, with stock-rearing the principal occupation. 8 (1)A2

Glasgow city and, as **Glasgow City**, unitary authority in west-central Scotland; the unitary authority was formed in 1995 from the majority of land from Glasgow District Council of Strathclyde Region. Glasgow is the largest of Scotland's four cities. It is the third most-visited city in Britain, and the UK's fourth-largest manufacturing centre.
area 176 sq km/68 sq mi
industries primarily a service economy, with 83% of employment in the service sector; industries also include engineering, chemicals, printing, whisky blending, brewing, electronics, textiles, light manufacturing
population (1996) 618,400
24 H6/5 D4

Glatz German name for ▶Kłodzko, a town in Poland. 18 F7

Glencoe valley in the Highland unitary authority, Scotland, extending 16 km/10 mi east from Rannoch Moor to Loch Leven. The mountains rise steeply on either side to over 1,000 m/3,300 ft, and the River Coe flows through the valley. Thirty-eight members of the Macdonald clan were massacred in Glencoe on 13 February 1692 by government troops led by Robert Campbell of Glenlyon; 300 escaped.

Gleneagles glen in Perth and Kinross, Scotland, 2 km/1 mi south of Auchterarder, famous for its golf courses, hotel, and for the **Gleneagles Agreement**, formulated in 1977 at the Gleneagles Hotel by Commonwealth heads of government, that 'every practical step (should be taken) to discourage contact or competition by their nationals' with South Africa, in opposition to apartheid.

Glenrothes town and administrative headquarters of ▶Fife, Scotland, 10 km/6 mi north of Kirkcaldy, designated a new town from 1948; population (1991) 38,650. Industries include electronics, plastics, and food processing. 24 J5

Glens Falls city in Warren County, east New York State, USA, 72 km/45 mi north of Albany and south of lakes George and Champlain; population (1990) 15,000. It is situated beside a waterfall on the Hudson River. Its industries include clothing, paper, electrical equipment, machinery parts, insurance, and tourism. Glens Falls is home to the Hyde art collection and to Adirondack Community College (1960). 88 F2

Gliwice city in Katowice region, southern Poland, formerly in German Silesia; population (1990) 214,200. It has coal-mining, iron, steel, and electrical industries. It is connected to the River Oder by the Gliwice Canal.
18 H7

Glomma river in Norway, 570 km/350 mi long. The largest river in Scandinavia, it flows into the Skagerrak (an arm of the North Sea) at Fredrikstad. 16 F5

Glos abbreviation for ▶Gloucestershire, an English county. 5 E6

Gloucester city and port in Essex County, northeast Massachusetts, USA, on Cape Ann, 50 km/31 mi northeast of Boston; population (1998 est) 29,700. It is America's oldest fishing port, first settled in 1623, incorporated as a town in 1642 and as a city in 1873. Its industries include tourism (including whale watching), fishing and fish processing (especially lobster, whiting, and cod), electronic equipment, and clothing. 88 F2

Gloucester (Roman **Glevum**), city, port, and administrative headquarters of ▶Gloucestershire, southwest England, on the River Severn, 67 km/42 mi northeast of Bristol; population (1996 est) 106,800. It is a finance and insurance centre, and manufactures aerospace equipment, ice cream and frozen foods, machinery (lifting, fuel vending, compressors), optical equipment, information technology, and camping goods. Salmon fisheries in the Severn are a valuable resource.
24 K10

Gloucestershire county of southwest England.
area 2,640 sq km/1,019 sq mi
towns and cities ▶Gloucester (administrative headquarters), Cheltenham, Cirencester, Stroud, Tewkesbury
physical Cotswold Hills; River Severn and tributaries
features Berkeley Castle, where Edward II was murdered; Prinknash Abbey, where pottery is made; Cotswold Farm Park; pre-Norman churches at Cheltenham and Cleeve; Gloucester Cathedral; Tewkesbury Abbey
agriculture cereals (in the Cotswolds), fruit (apples and pears), cider, dairy products ('double Gloucester' cheese was formerly made here), sheep farming
industries aerospace industry, light engineering, manufacturing (bricks, carpets, furniture, glass, pins, pottery, tiles, watches), plastics, timber
population (1996) 556,300 5 E6

Gmünd short form of ▶Schwäbisch-Gmünd, a town in Germany. 30 E2

Gmunden Austrian town in the province of Upper Austria, at the top of Lake Traun in the Salzkammergut, 57 km/35 mi southwest of Linz; population (1995) 22,400. The mountain of Traunstein (1,691 m/5,548 ft) towers above the town. Gmunden is a popular tourist resort. There are salt mines nearby. 30 J3

Gniezno industrial town and commercial centre in Poznań province, Poland, 48 km/30 mi east of Poznań; population (1990) 70,400. Industries include machinery, chemicals, and food processing, and cattle are traded. Gniezno is where the Polish state is said to have been founded; its archbishopric, the first in Poland, was established in 1000 by the Holy Roman Emperor Otto III, and Polish kings were crowned here until 1320.
18 G5

Goa state on the west coast of India, lying 400 km/250 mi south of Mumbai (formerly Bombay); area 3,700 sq km/1,428 sq mi; population (1991) 1,169,800. The capital is Panaji. Tourism is very important to the economy; local industries include clothing, footwear, pesticides, iron ore, manganese, and fishing nets. Agriculture is based on the cultivation of rice, pulses, cashew nuts, coconuts, and ragi (a cereal). 10 B5

Gobi vast desert region of Central Asia in the independent state of Mongolia, and Inner Mongolia, China. It covers an area of 1,280,000 sq km/490,000 sq mi (800 km/500 mi north–south and 1,600 km/1,000 mi east–west), and lies on a high plateau 900–1,500 m/2,950–4,920 ft above sea level. It is mainly rocky, with shifting sands and salt marshes at lower levels. The desert is sparsely populated, mainly by nomadic herders. It is rich in the fossil remains of extinct species, and Stone Age implements. 46 C2

Göding German name for ▶Hodonín, a town in the Czech Republic. 18 G9

Godthåb (Greenlandic **Nuuk**), capital and largest town of Greenland; population (1993) 12,200. It is a storage centre for oil and gas, and the chief industry is fish processing. 82 W4

Goes town in the province of Zeeland, the Netherlands, situated on the island of South Beveland, 22 km/14 mi east of Middelburg; population (1996) 34,000. There is a 16th-century Gothic basilica and a town hall restored in 1771. 22 F3

Golan Heights (Arabic **Jawlan**), plateau on the Syrian border with Israel, bitterly contested in the Arab–Israeli Wars and annexed by Israel on 14 December 1981. In the 1996 peace talks Syria insisted that Israel withdraw from the Golan Heights, following its capture in 1967. Demands for the return of the Golan to Syrian control have stalled negotiations towards normalization of Israeli–Syrian relations. 60 C3

Gold Coast former name for ▶Ghana, but historically the west coast of Africa from Cape Three Points to the Volta River, where alluvial gold is washed down. Portuguese and French navigators visited this coast in the 14th century, and a British trading settlement developed into the colony of the Gold Coast in 1618. With its dependencies of Ashanti and Northern Territories plus the trusteeship territory of Togoland, it became Ghana in 1957. The name is also used for many coastal resort areas – for example, in Florida, USA. 68 D3

Gold Coast resort region on the east coast of Australia, stretching 32 km/20 mi along the coast of Queensland and New South Wales south of Brisbane; population (1991) 279,600. 76 K5

Golden Gate strait in western California, USA, linking ▶San Francisco Bay with the Pacific; it is 8 km/5 mi long and 1.6–3 km/1–2 mi wide. It is spanned by a suspension bridge of the same name, designed by Joseph B Strauss (completed in 1937), connecting San Francisco on the south side with Marin County on the north; the longest span is 1,280 m/4,200 ft (it is one of the world's longest suspension bridges). 92 B1

Gonder (or **Gondar**), town in Ethiopia about 2,300 m/7,500 ft above sea level and 40 km/25 mi north of Lake Tana; population (1992) 156,000. It is the capital of a region of the same name. Cattle, grain, and seed are traded. It was the capital of Ethiopia 1632–1855. 64 G5

Good Hope, Cape of South African headland forming a peninsula between Table Bay and False Bay, Cape Town. The first European to sail around it was Bartolomeu Diaz in 1488. Formerly named Cape of Storms, it was given its present name by King John II of Portugal. 72 B6

Göppingen city in Baden-Württemberg, Germany, on the River Fils, 40 km/25 mi east of Stuttgart; population (1994) 254,200. It has a 15th-century church and a 16th-century castle. There are mineral springs and metal, textile, chemical, and tanning industries. 30 E2

Gorakhpur city in Uttar Pradesh, north India, situated on the Rapti River near the border with Nepal; population (1991) 506,000. It is at the centre of an agricultural region producing cotton, rice, and grain. Industries include railway maintenance and fertilizer manufacture. Pilgrims visit the Hindu Gorkhnath Temple. 54 D3

Gorinchem town in the province of South Holland, Netherlands, 37 km/23 mi southeast of Rotterdam; population (1996) 32,000. Gorinchem is sited at a major junction of inland waterway routes. 22 H3

Gorizia (German **Gorz**), town in Friuli-Venezia Giulia region, northern Italy, on the River Isonzo, 30 km/19 mi southeast of Udine; population (1990) 39,000. Textiles, furniture, and paper are produced, and it is a market for wine, timber, and fruit. The town was a cultural centre under Habsburg rule, and has a partly 14th-century cathedral and a 16th-century castle. 30 J5

Gorky (Russian **Gor'kiy**), name 1932–91 of ▶Nizhniy-Novgorod, a city in central Russia. The city was named after the writer Maxim Gorky in 1932, but reverted to its original name after the collapse of the Soviet Union. 38 H3

Görlitz industrial town in the state of Saxony, Germany, on the west bank of the River Neisse; population (1995) 67,100. Products include railway rolling stock, machinery, and precision instruments. The oldest part of the town, on the east bank of the Neisse, was incorporated into Poland in 1945 and is now called Zgorzelec. 18 D6

Gorzów Wielkopolski (German **Landsberg-an-der-Warthe**), industrial town (chemicals, metals, textiles) in Zielona Góra province, Poland, on the River Warta, 57 km/36 mi north of Zielona Góra; population (1990) 124,300. The town, with its 13th-century church, was founded by German settlers 1257, but was assigned to Poland after World War II. 18 E5

Goslar city in Lower Saxony, Germany, on the northern side of the Harz Mountains, 76 km/47 mi southeast of Hannover; population (1994) 162,200. Goslar is a diversified industrial city and a tourist centre for the Harz Mountains. In the Middle Ages, it was an important town and imperial residence; the imperial palace remains. 20 F5

Gosport town and naval base opposite Portsmouth, Hampshire, southern England; borough population (1996 est) 76,000. Industries include mechanical, electrical, and instrument engineering focused on defence research for the maritime industries; and the provision of high-tech services. 26 D3

Göteborg (English **Gothenburg**), port and industrial city (ships, vehicles, chemicals) on the west coast of Sweden, at the mouth of the Göta River; population (1994 est) 444,600. It is Sweden's second-largest city and is linked with Stockholm by the Göta Canal (built 1832). 16 F8

Gotha town in Thuringia, Germany; population (1995) 52,200. Industries include the manufacture of ceramics, precision instruments, and textiles. It was the former capital of the Duchy of Saxe-Coburg-Gotha, and a medieval trading and cultural centre. The Schloss Friedenstein (1643–54), an early baroque palace fortress, dominates the town. 20 F6

Gothenburg English form of ►Göteborg, a city in Sweden. 16 F8

Gotland Swedish island in the Baltic Sea; area 3,140 sq km/1,212 sq mi; population (1994 est) 58,200. The capital is Visby. Its products are mainly agricultural (sheep and cattle), and there is tourism. It was an area of dispute between Sweden and Denmark but became part of Sweden in 1645. 16 K8

Göttingen city in Lower Saxony, Germany, on the River Leine, 95 km/59 mi southeast of Hannover; population (1995) 127,200. Industries include printing, publishing, and the manufacture of chemicals and optical and precision instruments. Its university was founded by George II of England in 1737. After 1945 the town absorbed many refugees. 20 E5

Gouda town in South Holland province, west Netherlands, 17 km/11 mi northeast of Rotterdam; population (1997) 71,500. It stands at the confluence of the rivers Gouwe and IJssel. It produces round, flat Gouda cheeses and other dairy products; stoneware and candles are also produced. Among its landmarks is the Groote Kerk, with stained-glass windows. 22 G2

Gower Peninsula (Welsh **Penrhyn Gwyr**), peninsula in ►Swansea unitary authority, south Wales, situated between Swansea Bay and the Burry Inlet and extending into the Bristol Channel. Much of the coastline is the property of the National Trust, and there is tourism on the south coast; the north is marshy.

Grafton town in northeast New South Wales, Australia, on the River Clarence, 240 km/149 mi south of Brisbane and 633 km/393 mi northeast of Sydney; population (1996) 16,562. It is the centre for a region producing timber, sugar, bananas, sheep, beef, and dairy products. 76 K5

Grampian former region of Scotland (1975–96), which consisted of five districts and was replaced by Moray, Aberdeen City, and Aberdeenshire unitary authorities.

Grampian Mountains mountain range in north central Scotland that includes **Ben Nevis**, the highest mountain in the British Isles at 1,343 m/4,406 ft, and the Cairngorm Mountains, which include the second highest mountain, **Ben Macdhui** 1,309 m/4,295 ft. The region includes Aviemore, a winter holiday and sports centre. 24 H5

Granada capital of Granada province in Andalusia, southern Spain, situated to the north of the Sierra Nevada on the River Genil; population (1991) 254,000. Products include textiles, soap, and paper; there are also food industries and tourism. Granada has many palaces and monuments, including the Alhambra, a fortified hilltop palace built in the 13th and 14th centuries by the Moorish kings; a Gothic and Renaissance archiepiscopal cathedral (1523–1703); and a university, founded in 1533. 28 G7

Granada capital of Granada department, south-central Nicaragua, the chief port of Lake Nicaragua, 50 km/31 mi southeast of the capital Managua; population (1995 est) 74,400. It is one of the country's major commercial and industrial centres, situated in a rich agricultural district producing cotton, sugar, and coffee. Chemicals are also among the city's products and the tourist industry is increasing in importance, helped by the proximity of Mombacho volcano and attractive lake islands such as Zapatera. Founded by the Spanish in 1523, Granada is the oldest and third-largest city in Nicaragua. It is linked by rail to the Pacific port of Corinto. Before the growth of Managua at the end of the 19th century, Granada was the leading city of Nicaragua. 94 G6

Gran Canaria one of the chief ►Canary Islands, in the Spanish province of Las Palmas; capital Las Palmas; area 1,533 sq km/592 sq mi; population (2001 est) 728,400. It is fertile, of volcanic origin with deep valleys, and its peaks reach 1,980 m/6,495 ft. 66 B3

Gran Chaco large lowland plain in central South America, extending from southeast Bolivia, through western Paraguay to northern Argentina; area 650,000 sq km/251,000 sq mi. It consists mainly of swamps, forests (a source of quebracho timber), and grasslands, and it is utilized for cattle-raising. Oil was discovered in the 1920s in Chaco Boreal, north of the Pilcomayo River. The discovery exacerbated boundary disputes between Bolivia and Paraguay, and precipitated the Chaco War (1932–35).

Grand Canal (or **Imperial Canal**; Chinese **Da Yunhe**), the world's longest canal, running north from Hangzhou to Tianjin, China; 1,600 km/1,000 mi long and 30–61 m/100–200 ft wide. The earliest section was completed in 486 BC; the central section linking the Chang Jiang and Huang He rivers was built from AD 605 to 610; and the northern section was built between 1282 and 1292 during the reign of Kublai Khan.

Grand Canyon gorge in northwestern Arizona, USA, containing the ►Colorado River. It is 350 km/217 mi long, 6–29 km/4–18 mi wide, and reaches depths of over 1.7 km/1.1 mi. The gorge cuts through a multicoloured series of rocks – mainly limestones, sandstones, and shales, and ranging in age from the Pre-cambrian to the Cretaceous – and various harder strata stand out as steps on its slopes. It is one of the country's most popular national parks and millions of tourists visit it each year. 86 D3

Grand Falls town in central Newfoundland, Canada, on the Exploits River, 280 km/174 mi northwest of St John's; population (1991) 6,100. It is the site of large paper and pulp mills, and is a base for hunting and fishing. Grand Falls developed around a newspaper and pulp mill built in 1905. Power for industry is supplied by a large hydroelectric power station at Churchill Falls (formerly Grand Falls). 82 V7

Grand Forks city and administrative headquarters of Grand Forks County, east North Dakota, USA, situated on the Minnesota border, 124 km/77 mi north of Fargo, on the Red River; population (1990) 49,400. It is a centre for the surrounding agricultural area; most of its industries, such as food-processing mills and fertilizer plants, are associated with agriculture. It is the home of the University of North Dakota (1883), and the Grand Forks Air Force Base is nearby. 86 G1

Grand Rapids city in western Michigan, USA, on the Grand River, 40 km/25 mi above Grand Haven, where the river enters Lake Michigan; seat of Kent County; population (1994 est) 190,000. It is a centre of an agricultural and lumber-producing region; industries include the manufacture of furniture, motor bodies, plumbing fixtures, and electrical goods. 88 C2

Grand Teton highest point of the Teton mountain range, northwestern Wyoming, USA, rising to 4,197 m/13,770 ft. **Grand Teton National Park** was established in 1929. 86 D2

Grantham market town in southeast Lincolnshire, eastern England, on the River Witham, between Stamford and Lincoln; population (1995 est) 33,300. It is an agricultural centre, with mechanical and electrical engineering, food-processing, and packaging industries. The former UK prime minister Margaret Thatcher was born here in 1925. 24 M9

Grasmere English lake and village in the Lake District, Cumbria, associated with many writers; population (1991) 2,800. The English poet William Wordsworth and his sister Dorothy lived at Dove Cottage (now a museum) from 1799 to 1808, Thomas de Quincey later made his home in the same house, and both Samuel Coleridge and Wordsworth are buried in the churchyard of St Oswald's.

Grasse town near Cannes in the *département* of Alpes-Maritimes in southeast France, situated 200–400 m/650–1,300 ft above sea level on a hillside overlooking the Mediterranean; population (1990) 42,100, conurbation 335,000. The town is a holiday resort and is the centre of a perfume-manufacturing region, flowers being grown on a large scale for this purpose. Olives and oranges are also grown here. 30 B7

Graubünden (French **Grisons**), Swiss canton, the largest in Switzerland; area 7,106 sq km/2,743 sq mi; population (1995) 173,300. The capital is ▶Chur. The inner valleys are the highest in Europe, and the main sources of the River Rhine rise here. It also includes the resort of ▶Davos and, in the Upper Engadine, St Moritz. Ladin (a form of Romansch) is still widely spoken. Graubünden entered the Swiss Confederation in 1803. 8 (1)D2

Graudenz German name for ▶Grudziądz, a town in Poland. 18 H4

Gravesend town in Kent, southeast England, 35 km/22 mi east of London, on the River Thames opposite Tilbury, with which it is linked by ferry; borough population (1991) 95,000. Industries include engineering, printing, and the manufacture of electrical goods and a range of paper goods. 22 C3

Graz capital of Styria province, and second-largest city in Austria, situated on the River Mur, 150 km/93 mi southwest of Vienna; population (1995) 243,700. Industries include engineering and the manufacture of chemicals, iron, steel, automobiles, precision and optical instruments, paper, textiles, and leather. Tourism is also important. It has a 15th-century cathedral and a university founded in 1586. Lippizaner horses are bred near here. 30 L3

Great Abaco island in the northern Bahamas; see ▶Abacos. 90 F4

Great Australian Bight broad bay of the Indian Ocean in southern Australia, notorious for storms. It was discovered by a Dutch navigator, Captain Thyssen, in 1627. 76 E6

Great Barrier Reef chain of coral reefs and islands about 2,000 km/1,250 mi long, in the Coral Sea, off the east coast of Queensland, Australia, about 16–241 km/10–150 mi offshore. The Great Barrier Reef is made up of 3,000 individual reefs, and is believed to be the world's largest living organism. Only ten navigable channels break through the reef. The most valuable products of the reef are pearls, pearl shells, trepangs (edible sea slugs), and sponges. The reef is popular with tourists. In 1976 it became a Marine Park and was declared a World Heritage Site by UNESCO in 1981. 76 J2

Great Basin large arid region of the western USA, which includes nearly all Nevada, Utah, and parts of Oregon and California, and lies between the Sierra Nevada mountain range in California on the west and the Wasatch Mountains on the east. The **Great Basin National Park** was established in 1986. 86 C3

Great Bear Lake freshwater lake in the western Northwest Territories, Canada, on the Arctic Circle; area 31,153 sq km/12,028 sq mi; depth 410 m/1,345 ft. 92 (1)M2

Great Britain official name for ▶England, ▶Scotland, and ▶Wales, and the adjacent islands (except the Channel Islands and the Isle of Man) from 1603, when the English and Scottish crowns were united under James I of England (James VI of Scotland). With ▶Northern Ireland it forms the ▶United Kingdom.

Great Dividing Range eastern Australian mountain range, extending 3,700 km/2,300 mi N–S from Cape York Peninsula, Queensland, to Victoria. It includes the Carnarvon Range, Queensland, which has many Aboriginal cave paintings, the Blue Mountains in New South Wales, and the Australian Alps. 76 J4

Great Falls city in north-central Montana, USA, on the Missouri River, 116 km/72 mi northeast of Helena; seat of Cascade County; population (1992) 55,100. Its main industries are the processing of copper and zinc from nearby mines; the processing of agricultural products and oil-refining is also important. It is the second-largest city of the state, after Billings. Meriwether Lewis and William Clark were the first Europeans to discover the waterfalls, in 1805; the Lewis and Clark National Forest is nearby. 86 D1

Great Karoo larger and more northerly of two areas of the ▶Karoo, a semi-desert region in South Africa. 72 C6

Great Lakes series of five freshwater lakes along the USA–Canadian border: Superior, ▶Michigan, ▶Huron, ▶Erie, and ▶Ontario; total area 245,000 sq km/94,600 sq mi. Interconnected by a network of canals and rivers, the lakes are navigable by large ships, and they are connected with the Atlantic Ocean via the ▶St Lawrence River and by the St Lawrence Seaway (completed in 1959), which is navigable by medium-sized ocean-going ships. In March 1998 a bill was passed through Congress designating Lake Champlain the sixth Great Lake, although controversy over this continues.

Great Malvern English spa town in Worcester, on the east side of the **Malvern Hills**, which extend for about 16 km/10 mi and have their high point in Worcester Beacon 425 m/1,395 ft; population (1991) 31,500. The **Malvern Festival** (1929–39), associated with the playwright G B Shaw and the composer Edward Elgar, was revived in 1977. Elgar lived and was buried here. 24 K9

Great Plains (formerly the **Great American Desert**), semi-arid region in North America to the east of the Rocky Mountains, stretching as far as the 100th meridian of longitude through Oklahoma, Kansas, Nebraska, and the Dakotas. The plains, which cover one-fifth of the USA, extend from Texas in the south over 2,400 km/1,500 mi north to Canada. The Great Plains have extensive oil and coal reserves, many of which are actively worked. Ranching and wheat farming have resulted in overuse of water resources to such an extent that available farmland has been reduced by erosion. 86 F2

Great Rift Valley volcanic valley formed 10–20 million years ago owing to rifting of the Earth's crust and running about 8,000 km/5,000 mi from the Jordan Valley through the Red Sea to central Mozambique in southeast Africa. It is marked by a series of lakes, including Lake Turkana (formerly Lake Rudolf), and volcanoes, such as Mount Kilimanjaro. The rift system associated with the Rift Valley extends into northern Botswana, with geological faults controlling the location of the Okavango Delta. 70 E5

Great Salt Lake lake in northwest Utah, USA, 18 km/11 mi northwest of Salt Lake City. It is 130 km/81 mi long and 80 km/50 mi wide, with an area of approximately 4,145 sq km/ 1,600 sq mi, although it fluctuates greatly in size. The lake is generally 5 m/16 ft deep but its depth, too, varies. The lake is fed by the Bear, Weber, and Jordan rivers, but has no outlet other than evaporation. This is the second most saline body of water in the world after the ►Dead Sea; the manufacture of salt is an important industry. The lake attracts bathers. 86 D2

Great Sandy Desert desert in northern Western Australia; 415,000 sq km/160,000 sq mi. It is also the name of an arid region in southern Oregon, USA. 76 D4

Great Slave Lake freshwater lake in the Northwest Territories, Canada; area 28,450 sq km/10,980 sq mi. It is about 480 km/298 mi long and 100 km/62 mi wide, and is the deepest lake (615 m/2,020 ft) in North America. The lake forms two large bays, McLeod's Bay in the north and Christie's Bay in the south. It is connected with Artillery Lake, Clinton-Golden Lake, and Aylmer Lake, and the Mackenzie River flows out from it on the west. The Great Slave Lake contains many fish, including salmon and trout, and has major commercial fisheries. 80 N3

Great Yarmouth holiday resort and port in Norfolk, eastern England, at the mouth of the River Yare, 32 km/ 20 mi east of Norwich; population (1991) 56,200. It is Norfolk's largest port. Formerly a herring-fishing port, it is now a container port and a base for North Sea oil and gas. Other industries include shipbuilding, tourism and leisure, food processing, engineering, the manufacture of electronic components, shipwreck salvage, and some fishing. 24 P9

Greece country in southeast Europe, comprising the southern part of the Balkan peninsula, bounded to the north by the Former Yugoslav Republic of Macedonia and Bulgaria, to the northwest by Albania, to the northeast by Turkey, to the east by the Aegean Sea, to the south by the Mediterranean Sea, and to the west by the Ionian Sea.

Greece covers an area of 131,957 sq km/50,948 sq mi, and has a population of 10,645,000 (2000 est). The capital is ►Athens; other major towns and cities include Thessaloniki, Peiraias, Patras, Iraklion, Larissa, Peristéri, and Kallithéa. The head of state is Costis Stephanopoulos (from 1995); the head of government is Costas Simitis (from 1996). The currency is the drachma, and the main language spoken is Greek (official). 36 D5

Greeley city in northern Colorado, USA, at the point where the Cache de Poudre River flows into the South Platte River, 70 km/43 mi north of Denver; seat of Weld County; population (1992) 61,800. Greeley is situated in a belt of irrigated farmlands on the Colorado Piedmont, and is a distribution centre for the area; its main industries are meat packing and the processing of sugar beet. Other industries include the manufacture of photographic materials, fishing equipment, and chemicals. 86 F2

Green Bay city and port in north-eastern Wisconsin, USA, where the Little Fox River flows into Green Bay on Lake Michigan; it is the seat of Brown County; population (1994 est) 103,000. Green Bay is a port of entry to the USA through the St Lawrence Seaway and serves as a distribution centre. One of the most important ports of Wisconsin, it handles some 3 million tonnes of goods annually. Industries include the manufacture of paper, food products, and excavating machinery. 88 C2

Greenland (Greenlandic **Kalaallit Nunaat**), world's largest island, a dependency of Denmark, lying between the North Atlantic and Arctic Oceans east of North America; area 2,175,600 sq km/840,000 sq mi; population (1993) 55,100, comprising Inuit (Ammassalik Eskimoan), Danish, and other Europeans. The capital is Godthåb (Greenlandic *Nuuk*) on the west coast. The main economic activities are fishing and fish-processing. 80 G2

Greenland Sea area of the ►Arctic Ocean between Spitsbergen and Greenland, to the north of the Norwegian Sea; area 1,200,000 sq km/ 460,000 sq mi, mainly ice-bound in winter. It consists of two large basins, reaching depths of 4,850 m/15,900 ft, separated by the West Jan Mayen Rise. 80 B2

Greenock port and administrative headquarters of ►Inverclyde, western Scotland, on the southern shore of the part of the Firth of Clyde which runs northwest to southeast to meet the River Clyde; population (1991) 50,000. Traditionally associated with industries such as shipbuilding, engineering, chemicals, and sugar refining, the area now has computer and electronics enterprises. It was the birthplace of the engineer and inventor James Watt, who gave his name to the measurement of power. 24 H6

Green River river in Wyoming, Colorado, and Utah, USA; length 1,160 km/721 mi. It is one of the two rivers that ultimately form the Colorado River (the other is the San Juan). It rises in the Wind River Mountains in western Wyoming and flows south through the canyons of eastern Utah before joining the Colorado River. Navigation is almost impossible. 86 D3

Greensboro city and administrative headquarters of Guilford County, north-central North Carolina, USA, 77 km/48 mi west of Durham;

population (1994 est) 196,000. It has textile, chemical, electrical, and tobacco industries. The city was founded in 1808 and named after Gen Nathanael Greene, commander of the American forces in the Battle of Guilford Courthouse (1781), which was fought nearby. **90** F2

Greenville city and administrative headquarters of Greenville County, northwest South Carolina, USA, on the Reedy River, near the foothills of the Blue Ridge Mountains, 153 km/95 mi northwest of Columbia; population (1990) 58,300. It is a major textile-manufacturing centre; other industries include electronics, computer components, and robotics, as well as lumber and chemicals. **90** E3

Greifswald town in the *Land* (administrative region) of Mecklenburg–West Pomerania, Germany, on the River Ryck, 84 km/52 mi east of Rostock; population (1995) 61,800. Its university dates from 1456. The town's chief industries are fish-preserving and mechanical and marine engineering. There are many medieval houses and a 13th-century church. **20** J2

Greiz town in the *Land* (administrative region) of Thuringia, Germany, at the confluence of the Weisse Elster and Göltzsch rivers, 26 km/16 mi south of Gera; population (1995) 29,900. Light industries include papermaking. Until 1918 it was the capital of one of the Reuss principalities, and has two palaces. **20** H6

Grenada island country in the Caribbean, the southernmost of the Windward Islands.

Grenada covers an area (including the southern Grenadine Islands, notably Carriacou and Petit Martinique) of 344 sq km/133 sq mi, and has a population of 94,000 (2000 est). The capital is ►St George's; other major towns and cities include Grenville, Sauteurs, Victoria, and Gouyave. The head of state is Queen Elizabeth II (from 1974), represented by Governor General Daniel Williams (from 1996); the head of government is Keith Mitchell (from 1995). The currency is the East Caribbean dollar, and the languages spoken include English (official) and some French-African patois. **98** E1

Grenadines chain of about 600 small islands in the Caribbean Sea, part of the group known as the Windward Islands. They are divided between ►St Vincent and ►Grenada. **94** M6

Grenoble alpine city and administrative centre of the ►Isère *département*, Rhône-Alpes region, southeast France, situated on the rivers Isère and Drac; population (1990) 154,000, conurbation 400,000. Industries include electrometallurgy, engineering, nuclear research, hydroelectric power, computers, technology, chemicals, plastics, cement, textiles, foodstuffs, paper, and gloves. Grenoble was the birthplace of the novelist Stendhal (1783), commemorated by a museum, and the Beaux-Arts gallery has a modern collection. There is a 12th–13th-century cathedral, a major university (1339), and the Institut Laue-Langevin for nuclear research. It is the site of the **ESRF** (European Synchrotron Radiation Facility), the brightest X-ray machine in the world, inaugurated in 1994. The 1968 Winter Olympics were held here. **26** L8

Grimsby fishing port and administrative headquarters of ►North East Lincolnshire, England, on the River Humber, 24 km/15 mi southeast of Hull; population (1995) 89,400. It declined in the 1970s when Icelandic waters were closed to British fishing fleets. Chemicals and processed foods are manufactured, and marine-related industries and tourism are important. The ports of Grimsby and Immingham, 10 km/6 mi up river, are managed jointly from Grimsby, and together deal with 46 million tonnes of freight a year. **24** M8

Gris Nez cape in the *département* of Pas-de-Calais, France, and the point on the French coast nearest to Britain. It is opposite Dover, England and midway between Calais and Boulogne. Gris Nez has a lighthouse. **22** D4

Groningen most northerly province of the Netherlands, located on the Ems estuary and also including the innermost West Friesian Islands, bounded to the north by the North Sea, to the south by the province of Drenthe, to the east by Germany, and to the west by Friesland; area 2,350 sq km/907 sq mi; population (1997) 558,100. The capital is ►Groningen; other major towns are Hoogezand-Sappemeer, Stadskanaal, Veendam, Delfzijl, and Winschoten. The chief industries are natural gas, textiles, sugar refining, shipbuilding, and papermaking. Agriculture centres on arable and livestock farming, dairy produce, tobacco, and fishing. **7** (1)E2

Groningen capital of Groningen province in the Netherlands; population (1996) 170,100. Groningen is the most important town in the northern Netherlands; its position as a centre for trade is promoted by its canal connections with the Dollart (the basin of the River Ems) and the IJsselmeer. Linen, woollens, and tobacco are produced, and there is boat-building. The University of Groningen was founded 1614. **20** B3

Grossglockner highest mountain in Austria, rising to 3,797 m/12,457 ft in the Hohe Tauern range of the Tirol Alps, on the borders of Carinthia, Salzburg, and the Tirol. It was first climbed in 1800. **30** H3

Groznyy (or **Grozny**; Russian 'terrible', 'awesome'), capital of ►Chechnya and of the former Soviet republic of Checheno-Ingush; population (1996) 388,000. Situated on the Sunzha River, a tributary of the Terek, it is the biggest city of the Caucasian foothills. From the late 19th century, it became a major oil centre with pipelines to the Caspian Sea at Makhachkala, the Black Sea at Tuapse, and Trudovaya (near Gorlovka in the Donets Basin). Chemical and engineering industries are also located here. **58** L2

Grudziądz (German **Graudenz**), town in Toruń province, Poland, on the River Vistula, 64 km/40 mi northeast of Bydgoszcz; population (1990) 81,700. Manufactures include chemicals, machinery, and glass, and timber and agricultural produce are traded. The town has a castle of the Teutonic Knights (a German Christian military order), and was part of Prussia 1772–1920. **18** H4

Guadalajara industrial city (textiles, glass, soap, pottery), capital of Jalisco state, western Mexico; population of the metropolitan area (1995 est) 2,846,700, with an inner city population of (1995 est) 1,807,000. The second largest city in Mexico, 535 km/332 mi northwest of Mexico City, Guadalajara is a key communications centre. **94** D4

Guadalcanal Island largest of the ▶Solomon Islands; area 6,500 sq km/ 2,510 sq mi; population (1991) 60,700. The principal population centres are ▶Honiara, capital of the Solomon Islands, Aola, and Lunga, all on the north coast. Gold, copra, and rubber are produced. The population are Melanesians (or Papuasians). In 1942, during World War II, it was the scene of a battle for control of the area that was won by US forces after six months of fighting. 74 F7

Guadalquivir (Arabic **Wadi al-Kebir**), river in southern Spain, rising in the east of Jaén province; length 600 km/373 mi. It flows by Baeza, Andújar, Córdoba, and Seville, to the Atlantic Ocean at Sanlúcar de Barrameda. It is navigable below Seville; parts of the river are bypassed by a canal near Seville, which has an inland port. It is very important for irrigation. 28 E7

Guadeloupe group of islands in the Leeward Islands, West Indies (nine of which are inhabited), an overseas *département* of France; area 1,705 sq km/658 sq mi; population (1997 est) 411,800 (77% mulatto, 10% black, 10% mestizo). The main islands are Basse-Terre and Grande-Terre. The chief town and seat of government is ▶Basse-Terre; the largest town is Pointe-à-Pitre. Agriculture is the basis of the economy, though tourism (mostly from the USA) is also important. Sugar cane is a major crop, and bananas account for about half of export earnings. Industries include cement, rum distilling, and sugar refining. 96 E2

Guadiana river of Spain and Portugal; length 800 km/497 mi. Its headstream is the River Záncara, which rises in the central Spanish province of Cuenca. The Guadiana flows west to Badajoz, then south, weaving across the Spanish-Portuguese border to the Atlantic Ocean at Ayamonte in the province of Huelva. 28 C7

Guam largest and southernmost of the ▶Mariana Islands in the West Pacific, an unincorporated territory of the USA; length 50 km/30 mi, width 6–19 km/4–12 mi; area 549 sq km/ 212 sq mi; population (1992 est) 140,200. The main towns are ▶Hagåta (capital), Apra (port), and Tamuning. The predominant religion (96% of the population) is Roman Catholic. Languages spoken are English and Chamorro (a Malay-Polynesian dialect). The currency is the US dollar. Tourism is important to the island's economy, as are fishing and the cultivation of sweet potatoes. The land is largely limestone plateau in the north and volcanic in the south, with much jungle. Guam is the site of a major US air and naval base. 74 E4

Guangdong (or **Kwangtung**), province of south China, bounded to the north by Hunan and Jiangxi; to the northeast by Fujian; to the south by the South China Sea, Hong Kong, Macau, and the island province of Hainan; and to the west by Guangxi Zhuang Autonomous Region; area 197,000 sq km/76,062 sq mi; population (1996) 69,610,000. The capital is ▶Guangzhou; other major cities are Maoming, Shantou, Shenzhen, and Zhanjiang. The main industries are minerals, electronics, household appliances, and textiles; agriculture is based on rice, sugar, fruit, tobacco, and fish. 11 (2)E4

Guangxi Zhuang Autonomous Region (or **Guangxi**, or **Kwangsi Chuang Autonomous Region**), autonomous region in south China, bounded to the north by Guizhou, to the northeast by Hunan, to the east by Guangdong, to the south by the Gulf of Tongking, to the southwest by Vietnam, and to the west by Yunnan; area 236,700 sq km/91,400 sq mi; population (1996) 45,890,000 (including the Zhuang people, related to the Thai, who form China's largest ethnic minority). The capital is ▶Nanning; other main cities are Guilin, Liuzhou, and Wuzhou. Tourism is an important part of the economy, and the main industries are sugar-refining, metallurgy, fishing, food-processing, and timber. Agriculture is based on the cultivation of rice, maize, barley, millet, sugar, tropical fruits, and tea. 11 (2)D4

Guangzhou (or **Kwangchow**; or English **Canton**), capital of ▶Guangdong province, south China; population (1993) 3,560,000. Industries include shipbuilding, engineering, and the manufacture of automobiles, electronics, chemicals, and textiles. 50 E2

Guantánamo capital of a province of the same name in southeastern Cuba; population (1993 est) 208,000. It is a trading centre in a fertile agricultural region producing sugar. Iron, copper, chromium, and manganese are mined nearby. The city lies near Guantánamo Bay and has good rail, road, and air communications with other Cuban cities, including rail links to the ports of Caimanera and Boquerón. Industries include coffee roasting, sugar milling, and salt processing. 94 J4

Guatemala country in Central America, bounded north and northwest by Mexico, east by Belize and the Caribbean Sea, southeast by Honduras and El Salvador, and southwest by the Pacific Ocean.

Guatemala covers an area of 108,889 sq km/42,042 sq mi, and has a population of 11,385,000 (2000 est). The capital is ▶Guatemala City; other major towns and cities include Quezaltenango, Escuintla, Puerto Barrios (naval base), Mixco, Villa Nueva, and Chinautla. The head of state and government is Alfonso Portillo (from 2000). The currency is the quetzal, and the languages spoken include Spanish (official) and 22 Mayan languages. 94 F5

Guatemala City capital of ▶Guatemala, situated in the **Guatemalan Highlands** at an altitude of 1,500 m/ 4,920 ft on a plateau in the Sierra Madre mountains of the Gutemalan Highlands; population (1995 est) 1,224,000. A group of volcanoes overlooks the city: Acatenango (3,976 m/13,044 ft); Fuego (3,763 m/ 12,346 ft); Agua (3,760 m/12,336 ft). Industries include textiles, tyres, silverware, footwear, and cement. Half of the industrial output of Guatemala emanates from Guatemala City. It was founded in 1776 as Guatemala's third capital after earthquakes destroyed the earlier capitals of Antigua and Cuidad Vieja in 1773 and 1542 respectively. It was itself severely damaged by subsequent earthquakes in 1917–18, and 1976. 94 F6

Guayaquil largest city and chief port of Ecuador near the mouth of the Guayas River; population (1995 est) 1,707,000. The economic centre of Ecuador and the capital of Guayas province, the port is the world's leading exporter of bananas and industries include textiles, iron and steel, engineering, pharmaceuticals, and petroleum refining. The city was founded in 1537 by the Spanish explorer Francisco de Orellana. 98 B4

Guben town in the *Land* (administrative region) of Brandenburg, Germany, on the left bank of the River Neisse, 35 km/22 mi northeast of Cottbus; population (1995) 29,500. There are lignite (brown coal) mines, and textiles are manufactured. A former part of Guben on the right bank of the river was incorporated into Poland after 1945, and is now the Polish town of **Gubin**. **20** K5

Guelph industrial city and administrative headquarters of Wellington County, southeast Ontario, Canada, on the Speed River; population (1991) 88,000. It is the centre of an agricultural region; industries include food processing, electrical goods, and pharmaceuticals. It is the home of the University of Guelph (1964), which includes the Ontario Agriculture College (1874) and the Ontario Veterinary College (1862). **88** D2

Guernica town in the Basque province of Vizcaya, northern Spain; population (1991) 16,000. Guernica was almost completely destroyed in 1937 by German bombers aiding Gen Franco in the Spanish Civil War; it was rebuilt in 1946. The bombing inspired a painting by Pablo Picasso and a play by the Spanish-born dramatist Fernando Arrabal.

Guernsey second largest of the ►Channel Islands; area 63 sq km/ 24.3 sq mi; population (1991) 58,900. The capital is St Peter Port. Products include electronics, tomatoes, flowers, and butterflies; and since 1975 it has been a major financial centre. Guernsey cattle, which are a distinctive pale fawn colour and give rich, creamy milk, originated here. **26** C4

Guiana northeastern part of South America that includes ►French Guiana, ►Guyana, and ►Suriname.

Guildford cathedral city and county town (since 1257) of Surrey, southeast England, on the River Wey, 48 km/30 mi southwest of London; urban population (1991) 60,000; borough population (1996 est) 124,600. Industries include telecommunications, engineering, and the manufacture of plastics and pharmaceuticals. Features include a ruined Norman castle, a cathedral (founded in 1936 and consecrated in 1961), the main campus of the University of Surrey (1966), and the Yvonne Arnaud Theatre (opened in 1965). **24** M10

Guilin (or **Kweilin**), resort city in northeast Guangxi Zhuang Autonomous Region, south China, on the Li River; population (1994) 550,500. Its spectacular limestone mountains, rock formations, and underground caves are one of China's major tourist attractions. Machinery, electronics, textiles, and cement are manufactured, and rice, vegetable oil, and cinnamon produced. Tin and tungsten are mined nearby. **50** E1

Guinea country in West Africa, bounded north by Senegal, northeast by Mali, southeast by Côte d'Ivoire, south by Liberia and Sierra Leone, west by the Atlantic Ocean, and northwest by Guinea-Bissau.

Guinea covers an area of 245,857 sq km/94,925 sq mi, and has a population of 7,430,000 (2000 est). The capital is ►Conakry; other major towns and cities include Labé, Nzérékoré, Kankan, Kindia, Mamou, and Siguiri. The head of state is Lansana Conté (from 1984); the head of government is Lamine Sidime (from 1999). The currency is the Guinean franc, and the languages spoken include French (official), Susu, Pular (Fulfude), Malinke, and other African languages. **68** B2

Guinea-Bissau country in West Africa, bounded north by Senegal, east and southeast by Guinea, and southwest by the Atlantic Ocean.

Guinea-Bissau covers an area of 36,125 sq km/13,947 sq mi, and has a population of 1,213,000 (2000 est). The capital (and chief port) is ►Bissau; other major towns and cities include Bafata, Bissorã, Bolama, Gabú, Bubaque, Cacheu, Catio, and Farim. The head of state is Kumba Ialá (from 2000); the head of government is Caetano N'Tchama (from 2000). The currency is the Guinean peso, and the languages spoken include Portuguese (official), Crioulo (a Cape Verdean dialect of Portuguese), and African languages. **68** A2

Guinea, Gulf of part of the Atlantic Ocean off the coast of western Africa between Cape Palmas, Liberia, and Cap Lopez, Gabon. It includes the coastal features of the Bight of Benin and the Bight of Bonny (Bight of Biafra), separated by the delta of the River Niger. The rivers Volta, Niger, and Ogooué flow into it. **68** D4

Guiyang (or **Kweiyang**), capital of Guizhou province, south China; population (1994) 1,602,400. It is an important transport and industrial centre, producing aluminium, iron, steel, machinery, chemicals, pharmaceuticals, textiles, and construction materials. There are coal and bauxite mines nearby. **46** D5

Guizhou (or **Kweichow**), province of south China, bounded to the north by Sichuan, to the east by Hunan, to the south by Guangxi Zhuang Autonomous Region, and to the west by Yunnan; area 174,000 sq km/ 67,000 sq mi; population (1996) 35,550,000 (ethnic minorities comprise about 25% of the population). The capital is ►Guiyang; other main cities are Zunyi, Anshun, and Duyun. Chief industries include nonferrous minerals, machinery, food-processing, and timber; agriculture is based on the cultivation of rice, maize, tobacco, tea, and rapeseed. **11** (2)D4

Gujarat (or **Gujerat**), state of west India, formed from north and west Mumbai state in 1960; bordered to the north by Pakistan and Rajasthan , with Madhya Pradesh and Maharashtra to the east and southeast; area 196,000 sq km/75,500 sq mi; population (1995 est) 44,568,000 (90% Hindu). The capital is Gandhinagar (founded in 1961); other major towns are ►Ahmadabad and ►Vadodara; the main port is Kandla. The state is heavily industrialized, with the main industries being petrochemicals, oil (from Kalol, refined at Koyali near Baroda), gas, textiles, coal, limestone, pharmaceuticals, soda ash, electrical engineering, machine tools, cement, fertilizers, and dairy products. Agriculture is based on wheat, millet, cotton, rice, maize, tobacco, groundnuts, and fishing. **10** B4

Gujranwala city in Punjab province, Pakistan, 65 km/40 mi north of Lahore; population (1998) 1,124,800. It is a centre of trade in oilseed, oranges, rice, sugar, and wheat. Grain trading and pottery, and a ceramic institute has been established. There are tanneries and silk and textile processing, as well as electric-fan production and rice milling. It is situated on the Grand Trunk Road which links Kolkata (formerly Calcutta) with Peshawar. It is a former Sikh capital and the birthplace of

Ranjit Singh, founder of the Sikh kingdom, which once extended from the Khyber Pass to the Himalayas. His tomb is in the city. 54 B2

Gujrat city in Punjab province, Pakistan, some 100 km/60 mi north of Lahore; population (1981) 154,000. Products include cotton, pottery, brassware, and furniture. It occupies the site of a fort built in 1580 by the Mogul ruler Akbar; in 1765 it was acquired by the Sikhs. Gujrat was the scene of the final battle between the British and the Sikhs in the Sikh Wars 1845–49; the British subsequently annexed the Punjab. 54 B2

Gulfport city and administrative headquarters of Harrison County, southeast Mississippi, USA, on the Gulf of Mexico, 20 km/12 mi west of Biloxi and 109 km/68 mi east of New Orleans, Louisiana; population (1990) 40,700. It is a port of entry to the USA, and a major shipping point for lumber, cotton, and food products. With an artificially created sandy beach (45 km/28 mi long), it is a major tourist resort. 90 D3

Gulf States oil-rich countries sharing the coastline of the ►Gulf (Bahrain, Iran, Iraq, Kuwait, Oman, Qatar, Saudi Arabia, and the United Arab Emirates). In the USA, the term refers to those states bordering the Gulf of Mexico (Alabama, Florida, Louisiana, Mississippi, and Texas).

Gulf, the (or **Arabian Gulf**, or **Persian Gulf**), large shallow inlet of the Arabian Sea; area 233,000 sq km/ 90,000 sq mi. It divides the Arabian peninsula from Iran and is linked by the Strait of Hormuz and the Gulf of Oman to the Arabian Sea. Oilfields producing about one-third of the world's oil surround it in the Gulf States of Bahrain, Iran, Iraq, Kuwait, Oman, Qatar, Saudi Arabia, and the United Arab Emirates. 61 C2

Güns German name for ►Köszeg, a town in Hungary. 34 D2

Güstrow town in the *Land* (administrative region) of Mecklenburg–West Pomerania, Germany, 60 km/37 mi northeast of Schwerin, on the River Nebel; population (1995) 35,100. Engineering is the principal industry. Güstrow was the capital of the Duchy of Mecklenburg-Güstrow from 1621 to 1695. 18 B4

Gütersloh city in North Rhine–Westphalia, Germany, 135 km/84 mi northeast of Düsseldorf; population (1996) 324,300. It has electronic, textile, and publishing industries, and a large NATO airbase. 20 D5

Guyana country in South America, bounded north by the Atlantic Ocean, east by Suriname, south and southwest by Brazil, and northwest by Venezuela.

Guyana covers an area of 214,969 sq km/82,999 sq mi, and has a population of 861,000 (2000 est). The capital (and chief port) is ►Georgetown; other major towns and cities include Linden, New Amsterdam, Bartica, and Corriverton. The head of state is Bharrat Jagdeo (from 1999); the head of government is Samuel Hinds (from 1999). The currency is the Guyanese dollar, and the languages spoken include English (official), Hindi, and American Indian languages. 98 F2

Gwalior city and district in Madhya Pradesh, central India, 100 km/62 mi south of Agra, between the Sind and Chambal rivers; population (1991) 691,000. Industries include textiles, flour, oilseed milling, and the manufacture of pottery and footwear. It was formerly a small princely state; there is a hill fort dating from the 6th century built on a 100-m/320-ft cliff, which contains Jain and Hindu carvings and temples. The new town of Lashkar, founded as a military camp in 1800, lies to the south. 54 C3

Gwent former county of south Wales, 1974–1996, now divided between ►Blaenau Gwent, ►Caerphilly, ►Monmouthshire, ►Newport, and ►Torfaen unitary authorities.

Gwynedd unitary authority in northwest Wales, created 1996 from part of the former county of Gwynedd.
area 2,546 sq km/983 sq mi
towns ►Caernarfon (administrative headquarters)
physical area includes the highest mountain in Wales, ►Snowdon (1,085 m/3,560 ft), and the largest Welsh lake, Llyn Tegid (Bala Lake)
features Snowdonia National Park, seaside resorts, Bardsey Island
industries gold mining at Dolgellau, textiles, electronics, slate, tourism
agriculture cattle and sheep-farming
population (1996) 116,000 5 E5

Györ industrial city (steel, vehicles, textiles, foodstuffs) in northwestern Hungary, near the frontier with the Slovak Republic; population (1993) 131,000. 34 E2

Gyumri (formerly **Aleksandropol** 1837–1924; **Leninakan** 1924–91), city in Armenia, situated on a plateau in the Caucasus Mountains at a height of over 1,500 m/4,920 ft, 85 km/53 m northwest of Yerevan; population (1996) 120,000. Industries include textiles, metalworking and engineering. 58 K3

Haarlem industrial city and capital of the province of North Holland, the Netherlands, 20 km/12 mi west of Amsterdam; population (1997) 147,400. At Velsea, to the north, a road and rail tunnel runs under the North Sea Canal, linking North and South Holland. Industries include chemicals, pharmaceuticals, textiles, and printing. Haarlem is in an area of flowering bulbs and has a 15th–16th-century cathedral and a Frans Hals museum. 22 G2

Hadhramaut region of the Arabian Peninsula (1,400 m/4,500 ft) which stretches across political boundaries, forming a major part of southeast Yemen, and extending eastwards to Oman; population (1994) 870,100. Most of the area is barren, but there are a number of fertile valleys where, with irrigation, substantial crops, including cotton, tobacco, and coffee, are grown for export. There are nomadic Bedouin in the area. The chief towns are Mukalla (on the coast), Shibam, Seiyun, Jariba, and Terim (all inland). It was formerly ruled by Arab chiefs in protective relations with Britain. It was for a long time unknown to westerners but later attracted such travellers as Harry St John Philby and Freya Stark. 56 E6

Hagåtna (or **Agana**), capital of Guam, in the western Pacific, one of the Mariana Islands; population (1995 est) 1,300. It is a US naval base and administrative centre of the island, bordered by residential Hagåtna Heights and Apra Harbor.

Hagen industrial city in the Ruhr, North Rhine-Westphalia, Germany, at the confluence of the Ennepe and Volme rivers, 48 km/30 mi northeast of Düsseldorf; population (1995) 212,700. Industries include engineering and the manufacture of iron, steel, paper, and food products. 22 K3

Hagerstown city and administrative headquarters of Washington County, northwestern Maryland, USA, on Antietam Creek, near the Potomac River, 86 mi/138 km northwest of Baltimore and just south of the Pennsylvania border; population (1990) 35,400. It is an industrial centre situated in a rich agricultural area; industries include the manufacture of engine and missile parts, aircraft, and furniture. 88 E3

Hague, The (Dutch **'s-Gravenhage** or **Den Haag**), legislative and judicial capital of the Netherlands, and capital of South Holland province, 3 km/2 mi from the North Sea; population (1997 est) 442,200. It is linked by canal to Rotterdam and Amsterdam (which is the official capital of the country). 22 G2

Hague, Cape La cape in France at the northwestern end of the Cotentin peninsula, in the *département* of Manche. It lies between Cherbourg and the island of Alderney, facing the English Channel. 26 D4

Haguenau French town in the *département* of Bas-Rhin, on the River Moder, in the **Haguenau forest**; population (2001 est) 32,800. In 1154 it became a town and in 1257 it was made a free imperial city. Haguenau has two old churches, dating from the 13th and 14th centuries. It manufactures carpets, machinery, oil, and leather goods, and has a hop market. 22 K6

Haifa (or **Hefa**), chief seaport and industrial centre of Israel, situated in the northeast of the country at the foot of Mount Carmel, about 85 km/53 mi north of Tel Aviv; population (1995) 252,300. Industries include textiles, steel, chemicals, glass, soap, building materials, metal goods, and vehicle assembly. Israel's main naval base is here. It is the capital of a district of the same name. 60 B4

Hainan island province of south China, in the South China Sea, off the southwest coast of Guangdong province; area 34,000 sq km/13,000 sq mi; population (1996) 7,340,000. The capital is Haikou; major towns are Wenchang, Xincun, Tongzha, and Sanya. Hainan is China's second-largest island. Tourism and food-processing are the main industries, while the most important agricultural activities are rice, sugar, rubber, pineapples, sugar, betel nuts, and animal husbandry. 50 D3/11 (2)D4

Hainaut (Flemish **Henegouwen**), industrial province of southwest Belgium, bounded on the south by France; area 3,800 sq km/1,467 sq mi; population (1997) 1,284,300. The capital is ▶Mons; the other major towns are Charleroi, Tournai, and Soignies. Chief industries are iron, steel, glass, and textiles; fertile arable land in the north produces wheat, sugar beet, barley, and oilseed rape. The rivers Schelde and Sambre pass through the province. 7 (1)B4

Haiphong industrial port in North Vietnam; third-largest city in Vietnam (after Ho Chi Minh City and Hanoi); population (1995 est) 1,617,000. Among its industries are shipbuilding and the making of cement, plastics, phosphates, and textiles. There is a rail link to Kunming in China. 50 D2

Haiti country in the Caribbean, occupying the western part of the island of Hispaniola; to the east is the Dominican Republic.

Haiti covers an area of 27,750 sq km/10,714 sq mi, and has a population of 8,222,000 (2000 est). The capital is ▶Port-au-Prince; other major towns and cities include Cap-Haïtien, Gonaïves, Les Cayes, St Marc, Carrefour, Delmas, and Pétionville. The head of state is Jean-Bertrand Aristide (from 2001); the head of government is Jean-Bertrand Aristide (from 2001). The currency is the gourde, and the languages spoken include French and Creole (both official). 94 K5

Hakodate port in Hokkaido, Japan; population (1994) 301,000. Industries include fish canning, and the processing of marine products. The surrounding Hakodate plain produces rice, potatoes, and vegetables. In 1854 Hakodate was the earliest port opened to the West after the period of isolation. It was formerly the island's capital until replaced by Sapporo. 48 L3

Halab Arabic name of ▶Aleppo, a city in Syria. 58 G5

Halabja Kurdish town near the Iran border in Sulaymaniyah province, northeast Iraq; population (2001 est) 105,000. In August 1988 international attention was focused on the town when Iraqi planes dropped poison gas, killing 5,000 of its inhabitants. 58 L6

Halberstadt town in the *Land* (administrative region) of Saxony-Anhalt, Germany, in the foothills of the Harz Mountains, 48 km/30 mi southwest of Magdeburg; population (1995) 42,700. Textile, chemical, rubber, and paper manufactures are being replaced by lighter industries, such as light engineering and telecommunications. Halberstadt has a 14th-century Rathaus (town hall) and the cathedral of St Stephen's (13th–17th centuries). 20 G5

Halifax town in West Yorkshire, northern England, on the River Calder, 13 km/8 mi northwest of Huddersfield; population (1991) 91,100. Important in the woollen cloth trade since medieval times, the town produces textiles, carpets, and clothing; other industries include engineering and the manufacture of confectionery (Nestlé). It is the headquarters of the Halifax plc (formerly Halifax Building Society). 82 U8

Halle (French **Hal**), town in the province of Brabant, Belgium, on the River Senne, 16 km/10 mi south of Brussels; population (2001 est) 33,700. The chief products are paper, sugar, and chicory. The basilica of Notre Dame dates from the 14th century. 22 G4

Halle (*hall* 'salt'), industrial city in the *Land* (administrative region) of Saxony-Anhalt, Germany, on the River Saale, 30 km/20 mi northwest of Leipzig; population (1995) 287,400. Salt, chemicals, and lignite (brown coal) are produced, and there are printing and publishing industries. It was the capital of the East German district of Halle from 1952 to 1990. The composer Handel was born here in 1685, and the Handel Festival has been held annually in June since 1951. 20 H5

Hallein town in the Austrian province of Salzburg, on the River Salzach, 12 km/7 mi south of the town of Salzburg; population (2001) 18,600. It is a tourist resort, and has marble, paper, and rayon industries. A large salt mine, worked since the Bronze Age, is now open to the public. 30 J3

Halmstad seaport and capital of Halland county in southwest Sweden;

population (1996) 54,200. It is situated at the mouth of the River Nissa, on the eastern shore of the Kattegat. The town manufactures beer, wood pulp, jute, and paper. The chief exports are granite, timber, and butter. **18** B1

Halton unitary authority in northwest England, created in 1998 from part of Cheshire.
area 74 sq km/29 sq mi
towns and cities Runcorn, Widnes (administrative headquarters), Ditton
features River Mersey divides Runcorn from Widnes and Ditton; Manchester Ship Canal and Bridgewater Canal reach Mersey at Runcorn; St Helen's Canal reaches Mersey via a series of locks at Widnes; Catalyst: the Museum of the Chemical Industry is at Widnes; Norton Priory Museum (Runcorn) is on the site of a 12th-century priory
industries chemicals, pharmaceuticals, plastics manufacturing and coatings, light engineering, scientific instruments
population (1996) 122,300
5 E5

Hamadan city in northwest Iran and capital of the province of the same name, lying at an altitude of 1,800 m/5,905 ft; population (1991) 349,700. Carpets and rugs are traded and it is the centre of a grain and fruit-growing area. The tomb of the Arab philosopher and physician Avicenna is here, as well as a pair of cuneiform inscriptions to Darius and Xerxes at Gange Namay at the foot of nearby Mount Alvand (3,000 m/9,842 ft), and the reputed tomb of Esther and Mordecai. The town lies on the site of Ecbatana, capital of the Medes (an ancient Indo-European people). **56** E3

Hamamatsu industrial city of Shizuoka prefecture, Chubu region, central Honshu island, Japan, 85 km/53 mi southeast of Nagoya on the Pacific coast; population (1994) 550,000. Manufactured goods include textiles, chemicals, musical instruments, and motorcycles. Nearby Lake Hamana produces edible seaweed (*nori*). **48** J6

Hamburg largest inland port of Europe, in Germany, on the Elbe and Alster rivers, 103 km/64 mi from the mouth of the Elbe; population (1995) 1,706,800. Industries include marine engineering, ship-repairing, oil-refining, printing, publishing, and the production of chemicals, electronics, processed foods, and cosmetics. It is the capital of the *Land* (administrative region) of Hamburg, and has been an archbishopric since 834. In alliance with Lübeck, it founded the Hanseatic League. The city suffered extensive bomb damage during World War II. **20** E3

Hamburg German *Land* (administrative region) of Germany, situated between Schleswig-Holstein and Lower Saxony; area 760 sq km/293 sq mi; population (1995) 1,706,800. The capital is ▶Hamburg. The main industries are oil refining, chemicals, electrical goods, marine engineering, ship-repairing, food-processing, printing, and publishing. The area is mostly urban, but small parts of fenland are used for dairying and apple orchards. **7** (2)B2

Hämeenlinna (Swedish **Tavastehus**), capital of Häme province, south central Finland; population (1995) 44,900. It is 100 km/62 mi northwest of Helsinki, on Lake Vanajavesi. It has a 13th-century castle. **16** N6

Hameln (English **Hamelin**), town in Lower Saxony, Germany, on the River Weser, 40 km/25 mi southwest of Hannover; population (1995) 58,800. Industries include electrical and mechanical engineering, and the manufacture of furniture and fabrics. Old buildings include the Rattenfängerhaus (Rat-catcher's House). Hameln is the setting for the Pied Piper legend. **20** E4

Hamersley Range range of hills above the Hamersley Plateau, Western Australia, with coloured rocks and river gorges, as well as iron reserves. **76** C4

Hamilton capital (since 1815) of Bermuda, on Bermuda Island; population about (2001 est) 1,000. It has a deep-sea harbour. **94** M2

Hamilton city and port in southwestern Ontario, Canada, at the head of Lake Ontario, 65 km/40 mi southwest of Toronto and 90 km/56 mi northwest of the Niagara Falls; population (1991) 318,500; metropolitan area 599,800. Hamilton Harbour, one of the largest landlocked harbours on the Great Lakes, is linked with Lake Ontario via the Burlington Canal (completed in 1830). The city lies at the centre of a fruit-growing district, and is Canada's leading producer of steel; other industries include the manufacture of heavy machinery, electrical equipment, farm and machine tools, chemicals, and textiles. **88** E2

Hamilton industrial and university city on North Island, New Zealand, on the Waikato River; population (1996) 159,200. It trades in forestry, horticulture, and dairy-farming products. Waikato University was established here in 1964. **84** E3

Hamilton city in the southwestern corner of Ohio, on the Great Miami River, 19 mi/30 km northwest of Cincinnati; seat of Butler County; population (1992) 63,800. Its industries include livestock processing and the manufacture of metal products, motor-vehicle parts, security equipment, paper, and building materials. **88** D3

Hamm industrial city in North Rhine-Westphalia, Germany, on the River Lippe, 85 km/53 mi northeast of Düsseldorf; population (1995) 183,700. It lies in the northeast corner of the Ruhr coalfield. Industries include coalmining and the manufacture of iron and steel products, textiles, chemicals, machinery, and leather goods. **20** C5

Hammerfest fishing port in northwestern Norway, the northernmost town in Europe; population (1991) 6,900. **16** M1

Hampshire county of south England (since April 1997 Portsmouth and Southampton have been separate unitary authorities).
area 3,679 sq km/1,420 sq mi
towns and cities ▶Winchester (administrative headquarters), Aldershot, Andover, Basingstoke, Eastleigh, Gosport, Romsey, and Lymington
physical New Forest (area 373 sq km/144 sq mi), in the southeast of the county rivers Avon, Ichen, and Test
features Hampshire Basin, where Britain has onshore and offshore oil; Danebury, 2,500-year-old Celtic hill fort; Beaulieu; Highclere castle (home of the Earl of Carnarvon, with gardens by English landscape gardener Capability Brown); Hambledon, where the first cricket club was founded in 1750; site of the Roman town of Silchester; Jane Austen's cottage at Chawton, now a museum

agriculture market gardening (watercress)
industries aeronautics, brewing, chemicals, electronics, light engineering, oil, perfume, pharmaceuticals
population (1996) 1,627,400 **5** F6

Hanau city in Hessen, Germany, at the confluence of the rivers Kinzig and the Main, 18 km/11 mi east of Frankfurt am Main; population (1994) 88,900. Hanau has jewellery and diamond-cutting trades, rubber and stainless steel production. **20** D6

Hangchow alternative transcription of ►Hangzhou, port and capital of Zhejiang province, China. **46** F4

Hangzhou (or **Hangchow**), port and capital of ►Zhejiang province, China, on the mouth of the Qiantang River, at the southern terminus of the ►Grand Canal, 175 km/109 mi southwest of Shanghai; population (1994) 1,412,700. Products include jute, steel, machine tools, chemicals, electronics, processed foods, tea, silk and cotton textiles, fans, and gold-embroidered goods. Hangzhou has fine landscaped gardens, and was the capital of China from 1127 to 1278 under the Song dynasty. **46** F4

Hannibal city and port in northeastern Missouri, USA, on the Mississippi River, 160 km/99 mi northwest of St Louis; population (1990) 18,000. It is a centre for railway and river traffic, and trades in agricultural and dairy products, cement, steel, chemicals, machinery and other metal goods. Mark Twain lived here as a boy and made it the setting for his novels *The Adventures of Huckleberry Finn* and *The Adventures of Tom Sawyer*. **90** C2

Hannover (or **Hanover**), industrial city and capital of Lower Saxony, Germany, on the rivers Leine and Ihme; population (1995) 524,600. Industries include mechanical engineering, telecommunications, and the manufacture of electrical goods, rubber, and textiles. From 1386 it was a member of the Hanseatic League, and from 1692 capital of the electorate of Hannover (created a kingdom in 1815). George I of Great Britain and Ireland was also Elector of Hannover. **20** E4

Hanoi capital of ►Vietnam, on the Red River, 88 km/55 mi upstream from Haiphong; population (1997 est) 3,500,800. The Red River can be navigated by smaller ocean-going vessels, and Hanoi is also linked by rail to the port of Haiphong, to Ho Chi Minh City (Saigon), and to China. The city is served by Gia Lam international and domestic airport. Industries include electric motors, electric generators, machine tools, and chemicals, as well as the more traditional tanning, brewing, rice milling, and handicrafts. Central Hanoi has one of the highest population densities in the world: 1,300 people per ha/3,250 per acre. **50** D2

Hants abbreviation for ►**Hampshire**, an English county. **5** F6

Haora (or **Howrah**), city of West Bengal, northeast India, on the right bank of the River Hooghly opposite ►Kolkata (formerly Calcutta); population (1991) 947,000. The capital of Haora district, it has jute and cotton factories; rice, flour, and saw mills; iron and steel works and chemical factories; and railway engineering works. Haora suspension bridge, opened in 1943, spans the river. It is on a railway junction. **54** E4

Harare (formerly **Salisbury**), capital of Zimbabwe, in Mashonaland East Province, about 1,525 m/5,000 ft above sea level; population (1992) 1,184,200. It is the centre of a rich farming area producing tobacco and maize. The city's industries include milling, textiles, electrical and mechanical engineering, motor assembly, railway rolling stock, chemicals, furniture, consumer goods, and metallurgical and food processing. **72** E3

Harbin (or **Haerhpin** or **Pinkiang**), port and capital of ►Heilongjiang province, northeast China, on the Songhua River; population (1994) 2,887,800. It is a major rail junction. Industries include metallurgy, food processing, and sugar-refining; the manufacture of machinery and paper; and tourism. Harbin was developed by Russian settlers after Russia was granted trading rights here in 1896, and more Russians arrived as refugees after the October Revolution (1917). In World War II, it was the key objective of the Soviet invasion of Manchuria in August 1945. **46** H1

Harderwijk port in the province of Gelderland, the Netherlands, on the southeastern shore of the ►IJsselmeer, 50 km/31 mi east of Amsterdam; population (1996) 37,400. Herring curing is the chief industry. **22** H2

Hargeysa (or **Hargeisa** or **Hargessa**), town in northwestern Somalia; population (1990) 90,000. It is the capital of Hargeysa province in the Ogo highland, and is a centre for trade in meat, livestock, skins, and ghee (clarified butter). The town was severely damaged during the Somali civil war in the 1980s. **70** G2

Haridwar (or **Hardwar**), city in Haranchal, India, 170 km/106 mi northeast of Delhi, on the right bank of the River Ganges at the foot of the Shiwalik Hills; population (1991) 147,300. The name means 'door of Hari' (or Vishnu). In Hindu legend Vishnu's footprint was found on the river bank; it is one of the seven holy cities of the Hindu religion and a pilgrimage centre. **54** C3

Harlem commercial and residential district of northeastern Manhattan, New York City, USA, north of Central Park. The principal thoroughfare, 125th Street, runs east–west between the Hudson River and the East River. Originally a Dutch settlement, founded in 1658, Harlem developed as a black community after World War I. Harlem's heyday was in the 1920s, when it established its reputation as the intellectual, cultural, and entertainment centre of black America. It is known as a centre for music, particularly jazz. The Apollo Theatre (1934–84) was one of many influential music venues; the Dance Theatre and Theatre of Harlem are also here.

Harlingen port in the province of Friesland, the Netherlands, situated 25 km/16 mi southwest of Leeuwarden, on the Waddenzee; population (1996) 15,200. Harlingen is intersected by numerous canals, has an excellent harbour, and exports dairy products, potatoes, and meat. The town was inundated by floods in 1134 and in 1566, before a dyke was constructed for its protection. **22** H1

Harlingen city in Cameron County, the southeast Texa, USA, 180 km/112 mi south of Corpus Christi and 20 km/12 mi north of the Mexican border; population (1992) 52,000. Connected to the Rio Grande by the

Gulf Intracoastal Waterway, it serves as the processing and marketing area for the lower Rio Grande Valley. Industries include citrus-fruit processing, aerospace equipment, and cotton products. It is home to the Marine Military Academy (1963) **90** B4

Harris part of the Outer ►Hebrides, Western Isles, Scotland; area 500 sq km/193 sq mi; population (1971) 2,900. It is joined to Lewis by a narrow isthmus. Harris tweed cloths are produced here. **24** F4

Harrisburg state capital of ►Pennsylvania, USA, located in the south-central part of the state, on the Susquehanna River; seat of Dauphin County; population (1992) 53,400. Industries include steel, electronic equipment, aircraft engines, office machines, railroad equipment, food processing, printing and publishing, and clothing. The city was laid out in 1785 and became the state capital in 1812; it was incorporated as a city in 1860. **90** F1

Harrogate resort and spa town in North Yorkshire, northern England, about 24 km/15 mi north of Leeds; population (1996 est) 69,800. Employment is mainly in the service sector, particularly related to conference business, tourism, and finance. **24** L8

Hartford state capital of ►Connecticut, USA, located in the north-central part of the state, at the head of navigation on the Connecticut River, 96 km/60 mi from Long Island Sound; population (1994 est) 124,000; population of metropolitan area (1992) 1,156,000. Called the 'Insurance Capital of the World', the city has the headquarters of more than 40 insurance companies. Industries include the manufacture of aerospace equipment, firearms, office equipment, and tools. **88** F2

Hartlepool town, port, and, since 1996, unitary authority in northeast England, formed from part of the county of Cleveland.
area 94 sq km/36 sq mi
features redeveloped dock area including the Museum of Hartlepool (opened in 1995); the Gray Art Gallery and Museum; remains of the medieval town walls; Early English church of St Hilda with a Norman doorway
industries the local economy depends on metal industries, engineering, support services for the oil industry, fishing, and brewing. A nuclear power station is located 5 km/3 mi southeast of the town at Seaton Carew
population (1996) 90,400 **24** L7/5 F4

Harwich seaport in Essex, southeast England, on the estuary of the rivers Stour and Orwell, 113 km/70 mi northeast of London; population (1997 est) 15,800. It has an important freight terminal and ferry connections with the Hook of Holland, Germany, Denmark, Sweden, and Norway. Bathside Bay will be developed to add to Harwich's existing port facilities. **22** D3

Haryana (Hindi 'God's home'), state of northwest India; area 44,200 sq km/17,061 sq mi; population (1994 est) 17,925,000. The capital is ►Chandigarh (also capital of ►Punjab state). The state lies on the Gangetic plain, drained by the Yamuna River. Chief industries are textiles, cement, iron ore, and processing of agricultural products; agriculture is based on wheat (with Punjab accounting for one-third of India's total production), sugar, cotton, oilseed, rice, and pulses. **10** C3

Harz Mountains most northerly mountain range in Germany, situated between the rivers Weser and ►Elbe and covering a total area of around 2,030 sq km/784 sq mi. The highest peak in the Harz Mountains – a popular area for recreation and designated a nature park – is the Brocken (1,143 m/3,750 ft). **20** F5

Haskovo Bulgarian name for ►Khaskovo, a town and region of Bulgaria. **36** H3

Hasselt capital of the Belgian province of ►Limbourg; population (2001 est) 68,800. The town manufactures linen fabrics and tobacco, and has gin distilleries. **22** H4

Hastings resort in East Sussex, southeast England, on the English Channel; population (1996) 85,000. Fishing is an important activity; the town has Britain's largest fleet of beach-launched fishing boats and a new wholesale fish market. Other industries include engineering and the manufacture of scientific and aerospace-related instruments, plastics, electronics, and domestic appliances. William the Conqueror landed at Pevensey to the west and defeated Harold at the Battle of Hastings in 1066. **22** C4

Haugesund seaport in Rogaland county, southern Norway, 58 km/36 mi northwest of the town of Stavanger; population (1995) 29,200. It has important fisheries and a large merchant fleet. **16** C7

Haute-Corse *département* comprising the northern half of the French island of Corsica; area 4,666 sq km/1,801 sq mi; population (1990) 131,520. Together with Corse-du-Sud, it makes up the administrative region of ►Corsica. The highest peak on the island, Monte Cinto (2,710 m/ 8,891 ft) is located here; other peaks over 2,000 m/*c.*6,500 ft in Haute-Corse include Monte Rotondo (2,625 m/8,612 ft) and Col de Vergio (2,329 m/7,641 ft). Wine, olives, and olive oil are produced. The departmental capital is the port of ►Bastia on the northeast coast; other significant towns are Calvi, Corte, Calenzana, Ile-Rousse and St Florent. **6** (1)A1

Haute-Garonne *département* in the ►Midi-Pyrénées region of France; area 6,300 sq km/2,432 sq mi; population (1990) 925,900. Vines, cereals, fruit, and potatoes are cultivated, and livestock is raised. There are chemical, metallurgical, paper, foodstuff, engineering, aeronautical, and leather industries, and tourism is important. The principal towns are ►Toulouse (the administrative centre), Muret, and St-Gaudens; Luchon is a resort for winter sports. **6** D5

Haute-Loire *département* in the Auvergne region of central France; area 4,965 sq km/1,917 sq mi; population (1990) 207,000. Situated on the central plateau of France, it is crossed by four upland areas, running north to south: the Vivarais; the Massif du Meygal; the Velay uplands; and the Margeride Mountains. The chief rivers are the ►Loire, with its tributaries the Borne and the Lignon, and the Allier. In the river valleys, fruit trees and vines are cultivated. In the rest of the *département* the principal economic activities are stock-raising and forestry. The main towns are ►Le Puy (the administrative centre), Brioude, and Yssingeaux. **6** E4

Haute-Marne *département* in the ▶Champagne-Ardenne region of northeast France; area 6,215 sq km/ 2,400 sq mi; population (1990) 204,577. Haute-Marne is extensively forested. The soil is mostly poor, but cereals and vines are cultivated in places. There are iron-ore mines and mineral springs. The principal towns are the departmental capital, Chaumont, Langres, and St-Dizier. 6 F2

Haute-Normandie (English **Upper Normandy**), coastal region of northwest France lying between Basse-Normandie and Picardy and bisected by the River Seine; area 12,300 sq km/ 4,750 sq mi; population (1990 est) 1,780,200. It comprises the *départements* of Eure and Seine-Maritime; its administrative centre is ▶Rouen. Ports include Le Havre, Dieppe, and Fécamp. The area is fertile and has many beech forests. Industries include dairy-farming, fishing, and petrochemical, paper, and car manufacture. 6 D2

Hautes-Alpes *département* in the ▶Provence-Alpes-Côte d'Azur region of southeast France, bordered on the east by Italy; area 5,520 sq km/2,131 sq mi; population (1990) 56,100. The region is situated in the Alps; in the east are the slopes of the Cottian Alps. The main river is the Durance. The chief sources of income are livestock farming; cultivation of vines, fruit, and vegetables; and tourism, especially winter sports. The principal towns are ▶Gap (the administrative centre) and Briançon. 6 F4

Haute-Saône *département* in the ▶Franche-Comté region of France; area 5,343 sq km/2,063 sq mi; population (1990) 230,300. The land is mainly a plateau, gradually descending from the ▶Vosges Mountains in the north (highest point, Ballon de Servance, 1,182 m/3,900 ft) towards the south, and intersected by many valleys. The chief rivers are the ▶Saône and its tributaries, the Coney, Lanterne, Durgeon, and Ognon. There is much forest, and wheat, oats, rye, and vegetables are grown. Some wine is produced. The principal towns are Vesoul (the administrative centre) and Lure. 6 G3

Haute-Savoie *département* in the ▶Rhône-Alpes region of France; area 4,391 sq km/1,695 sq mi; population (1990) 568,800. The land is mountainous – the *département* includes ▶Mont Blanc (4,807 m/15,772 ft). The chief rivers are the Arve, Dranse, Ussel, and Fier, tributaries of the Rhône. Much of the land is forested. Cereals and hardy root crops are grown, and on the lower slopes orchard fruits, vines, and tobacco. There is dairy production, particularly cheese. The most fertile district is in the north and northwest. Industry is limited, but includes textiles, watches, and clocks. Tourists come both in summer and winter. The principal towns are ▶Annecy (the administrative centre), Bonneville, Thonon, and St-Julien. 6 G3

Hautes-Pyrénées mountainous *département* in the ▶Midi-Pyrénées region of France on the Spanish frontier; area 4,507 sq km/1,740 sq mi; population (1990) 224,900. The chief river is the Adour, with its tributaries the Arros and Gave de Pau; the Garonne is on the southeast frontier. There is much forest, and wheat, maize, vines, tobacco, flax, and chestnuts are grown. Marble and slate are quarried, lignite, zinc, manganese, and lead are found, and there are mineral springs. There are aeronautical, armaments, chemical, and engineering industries, and tourism is important, especially in ski resorts such as Val Louron, Peyragudes, and Cauterets Lys. The principal towns are ▶Tarbes (the administrative centre), Argelés-Gazost, and Bagnères-de-Bigorre. 6 D5

Haute-Vienne *département* in the ▶Limousin region of central France; area 5,512 sq km/2,129 sq mi; population (1990) 353,914. Haute-Vienne is extremely hilly; the River Vienne flows between the Marche hills in the centre of the *département* and the Limousin hills in the south. Some cereals and hemp are grown, and there is widespread breeding of livestock. Minerals exploited in the area include uranium, wolfram, manganese, and kaolin. There are important porcelain, leather, paper, and textile industries. The principal towns are ▶Limoges (the capital), Bellac, and Rochechouart. 6 D3

Haut-Rhin *département* in the Alsace region of France; area 3,522 sq km/ 1,360 sq mi; population (1990) 952,600. The Rhine is the eastern boundary of the *département*, which is fertile and densely wooded. Cereals, vines, and hops are grown. Textiles, machinery, and foodstuffs are manufactured, and there are potash mines and factories. The principal towns are ▶Colmar (administrative centre) and Mulhouse. 6 G3

Hauts-de-Seine *département* in the ▶Ile-de-France region of northern France; area 175 sq km/68 sq mi; population (1990) 1,390,592. Situated to the west of Paris, Hauts-de-Seine is made up of the outer western and northwestern suburbs of the capital. Here, major new residential estates have been built to replace older housing. Many manufacturing industries were sited within the *département*, including a major automobile factory (Renault) at Billancourt on the Seine (now closed), and a fine-porcelain works at Sèvres. Increasingly, the emphasis is on office work. There are extensive wooded areas, including the Forêt de Meudon and the Parc de St-Cloud. Fast road and railway links connect Hauts-de-Seine to central Paris. 6 E2

Havana capital and port of Cuba, on the northwest coast of the island; population (1995 est) 2,219,000. Products include cigars and tobacco, sugar, coffee, and fruit. The old city centre is a world heritage site, and the oldest building in the city and in Cuba is La Fuerza, a fortress built in 1538. 94 H4

Havel river in eastern Germany; length 346 km/215 mi. The Havel rises in a small lake 9 km/6 mi northwest of Neustrelitz. It flows south to Berlin and then west past Potsdam and Rathenow to join the Elbe 29 km/18 mi southeast of Wittenberg. To the west of Berlin the Havel passes through a series of lakes, one of which bears the river's name. At this point it is also joined by the River Spree; via the Oder–Spree Canal, traffic from the Havel can reach the River Oder on Germany's border with Poland. 18 C5

Havre, Le see ▶Le Havre, a port in France. 22 C5

Hawaii Pacific state of the USA. It is nicknamed the Aloha State. Hawaii was admitted to the Union in 1959 as the 50th US state. The only state not part of North America, Hawaii, variously described as part of Oceania or Polynesia, comprises a west-northwest–east-southeast oriented island chain 2,700km/1,700 mi in

length, the east end of which lies some 3,400km/2,100 mi southwest of California. The Tropic of Cancer passes through the islands.
population (2000 est) 1,211,500
area 16,800 sq km/6,485 sq mi
capital ▶Honolulu on Oahu
towns and cities Hilo, Kailua, Kaneohe
industries and products tourism is the chief source of income; other industries include sugar, coffee, pineapples, macadamia nuts, orchids and other flowers, livestock, poultry, dairy goods, clothing 92 (2)E4

Heard Island and McDonald Islands group of islands forming an Australian external territory in the South Indian Ocean, about 4,000 km/2,500 mi southwest of Fremantle; area 410 sq km/158 sq mi. They were discovered in 1833, annexed by Britain in 1910, and transferred to Australia 1947. They are unpopulated. Heard Island, 42 km/26 mi by 19 km/12 mi, is glacier-covered, although the volcanic mountain Big Ben (2,742 m/9,000 ft) is still active. A weather station was built in 1947.

Shag Island is 8 km/5 mi to the north and the craggy McDonalds are 42 km/26 mi to the west.

Hebei (or **Hopei**, **Hopeh**, or **Chihli**), province of north China, bounded to the north by Inner Mongolia, to the northeast by Liaoning, to the east by the Bohai Gulf, to the south by Shandong and Henan, and to the west by Shanxi; area 185,900 sq km/71,800 sq mi; population (1996) 64,840,000. The capital is ▶Shijiazhuang; other major cities are Baoding, Tangshan, Handan, and Zhangjiakou. The province includes the special municipalities of Beijing and Tianjin. The main industries are textiles, coal, iron, steel, and oil; agricultural production is based on winter wheat, barley, maize, and cotton. 11 (2)E3

Hebrides group of more than 500 islands (fewer than 100 inhabited) off the west coast of mainland Scotland; total area 2,900 sq km/1,120 sq mi. The Hebrides were settled by Scandinavians during the 6th–9th centuries and passed under Norwegian rule from about 890 to 1266. 24 F5/24 D4

Hebron (Arabic **El Khalil**), city on the west bank of the Jordan, occupied by Israel in 1967; population (1996 est) 120,500. The population is mainly Arab Muslim, but just outside the city is a large Israeli settlement called Qiryat Arba. Industries include leather, stone cutting, and ceramics. Hebron experienced frequent confrontation between Israelis and Arabs in the Intifada and throughout the Israel–Palestine peace process. 60 C5

Heerlen town in the province of Limburg, the Netherlands, 21 km/13 mi northeast of Maastricht; population (1996) 95,900. Heerlen is an important industrial centre. The Thermen Museum incorporates the excavations of a Roman bath complex. 22 J4

Hefa alternative spelling of ▶Haifa, the chief seaport of Israel. 60 B4

Hefei (or **Hofei**, or **Luzhou**), capital of ▶Anhui province, eastern China; population (1994) 1,126,600. Once just a trading hub for agricultural products, especially rice, it is now a centre of heavy industry. Products include textiles, chemicals, steel, electronics, and domestic appliances. 46 F4

Heidelberg city in Baden-Württemberg, Germany, on the River Neckar, 19 km/12 mi southeast of Mannheim; population (1995) 138,400. Industries include publishing, brewing and the manufacture of tobacco products and optics. It is a major tourist centre. Heidelberg University, the oldest in Germany, was established in 1386. The city is overlooked by the ruins of its 13th–17th-century castle, standing 100 m/330 ft above the river. 20 D7

Heidenheim city in Baden-Württemberg, Germany, on the River Brenz, 68 km/42 mi east of Stuttgart; population (1994) 136,600. There are machinery, metal, and textile industries. Heidenheim is built on a rocky outcrop and has an 11th-century castle (Hellenstein). 30 F2

Heilbronn river port in Baden-Württemberg, Germany, on the River Neckar, 47 km/29 mi north of Stuttgart; population (1995) 121,700. It trades extensively in wine. The town was rebuilt after World War II as a centre for electrical and mechanical engineering. 20 E7

Heilongjiang (or **Heilungkiang**), province of northeast China, bordered to the north and east by Russia, to the south by Jilin, and to the northwest by Inner Mongolia; area 463,600 sq km/178,950 sq mi; population (1996) 37,280,000. The capital is ▶Harbin; other main cities are Qiqihar, Hegang, and Jiamusi. China's largest oilfield is located in the province, at Daqing; other industries are engineering, food processing, timber and wood products, building materials, and ice-skates. Agriculture is based on maize, sugar beet, dairy farming, and sheep rearing. 11 (2)F2

Hejaz (or **Hedjaz**, or **Al Hedjaz**, or **Hijāz**, or **Al Hijāz**), historic region of Saudi Arabia, on the Red Sea; area about 290,000 sq km/111,970 sq mi. A former independent kingdom, it merged in 1932 with Nejd to form ▶Saudi Arabia. Historically its principal city has been ▶Mecca. The other main cities are ▶Jiddah, on the coast, Taif, a mountain resort at an altitude of 1,800 m/5,905 ft, and ▶Medina. 64 G2

Helena state capital of ▶Montana, USA, and seat of Lewis and Clark County, located in the west-central part of the state, 80 km/50 mi northeast of Butte; population (2000 est) 25,800. Helena lies at an altitude of 1,219 m/4,000 ft, near the Big Belt Mountains, with Mount Helena (1,280 m/4,200 ft) rising in the background. It is the commercial centre of a rich mining and agricultural region; industries include agricultural products, machine parts, lead refining, ceramics, paints, sheet metal, and chemicals, but government, education, and health care services dominate employment. Helena was incorporated as a city in 1881. 86 D1

Helgoland German name for ▶Heligoland, an island in the North Sea. 20 C2

Heligoland (German **Helgoland**), island in the North Sea, one of the North Frisian Islands, 68 km/42 mi northwest of Cuxhaven; area 0.6 sq km/0.2 sq mi; population (1995) 2,000. It is administered by the state of Schleswig-Holstein, Germany. It was a British possession from 1807 to 1890, when it was ceded to Germany in exchange for Zanzibar. In both world wars it was used as a naval base. It has health resorts and fishing industries. 20 C2

Helmand longest river in Afghanistan. Rising in the Hindu Kush, west of Kabul, it flows southwest for 1,125 km/703 mi before entering the marshland surrounding Lake Saberi on the Iranian frontier. 56 H3

Helmond town in the province of North Brabant, the Netherlands, 13 km/8 mi northeast of Eindhoven; population (1996) 75,300. The town has textile and engineering industries. 22 H3

Helmstedt city in Lower Saxony, Germany, 36 km/22 mi east of Brunswick; population (1994) 26,900. Textiles, machinery, soap, and margarine are manufactured, and brown coal (lignite) is mined in the area. Helmstedt dates from the Middle Ages. 20 G4

Helsingborg (Swedish **Hälsingborg**), port in southwestern Sweden, linked by ferry with Helsingør across Øre Sound; population (1994) 113,400. Industries include copper smelting, rubber and chemical manufacture, and sugar refining. 16 G8

Helsingfors Swedish name for ►Helsinki, the capital of Finland. 16 N6

Helsingør (English **Elsinore**), port in northeastern Denmark; population (1993) 56,700. It is linked by ferry with Helsingborg across Øre Sound; Shakespeare made it the scene of *Hamlet*. 16 G8

Helsinki (Swedish **Helsingfors**), capital and port of Finland; population (1994) 516,000. Industries include shipbuilding, engineering, and textiles. The port is kept open by icebreakers in winter. 16 N6

Helvellyn peak of the ►Lake District in Cumbria, northwest England; height 950 m/3,118 ft.

Hemel Hempstead town in Hertfordshire, southeast England, 37 km/23 mi northwest of London, on the River Gade; population (1997 est) 80,790. It was designated a new town in 1946. Paper, electrical goods, information systems, software and computers, electronics components, and brushes are manufactured. It has a large administrative and service sector, including telephone call centres for financial concerns. 24 M10

Henan (or **Honan**), province of east central China, bounded to the north by Hebei, to the east by Shandong and Anhui, to the south by Hubei, and to the west by Shaanxi and Shanxi provinces; area 167,000 sq km/64,500 sq mi; population (1996) 91,720,000. The capital is ►Zhengzhou; other major cities are Luoyang, Kaifeng, and Anyang. The main industries are coal, oil, textiles, cement, glass, and fertilizers. Agricultural products are cereals, cotton, fruit, tobacco, and peanuts. 11 (2)C3

Hengelo town in the province of Overijssel, the Netherlands, 8 km/5 mi northwest of Enschede; population (1996) 8,400. There are large diesel-engine and cotton industries, in addition to dyeing, brewing, and railway engineering. The town centre was severely damaged during World War II. 22 J2

Henzada alternative name for ►Hinthada. 50 B3

Heraklion alternative name for ►Irakleion, a port in Crete, Greece. 14 G4

Herat capital of Herat province, and the largest city in western Afghanistan, on the north banks of the Hari Rud River; population (early 1990s est) 186,800. A principal road junction, the city is 922 m/3,026 ft above sea level and lies in an area which has been the site of cities since the time of Alexander the Great and contains many ancient ruins and buildings, including a 15th-century mosque and the tomb of the poet and saint Abdollah Ansari. Herat is now the marketing centre for an irrigated, fertile, and densely-peopled agricultural area in western Afghanistan, and its industries include handicrafts, textiles, and the milling of oilseeds, flour, and rice, while Karakul furs are also an important item of local commerce. 56 H3

Hérault *département* in the ►Languedoc-Roussillon region of southern France; area 6,112 sq km/2,360 sq mi; population (1990) 795,142. Hérault is on the Golfe du Lion, and its coastline comprises a series of long lagoons separated from the land by narrow strips of land. The area is drained by the rivers Aude, Orb, and Hérault. Wine is produced extensively, cereals, fruit trees, and vegetables are grown, and silkworms are reared. Iron, bauxite, and salt are also found. The chief manufactures are chemicals and textiles. The main towns are the capital ►Montpellier, ►Béziers, and Lodève. 6 E5

Hercules, Pillars of twin rocks of Calpe at Gibraltar and Abyla at Ceuta which guard the western entrance to the Mediterranean Sea.

Hereford city and administrative headquarters of the unitary authority of ►Herefordshire, west-central England, on the River Wye, 34 km/21 mi southwest of Worcester; population (1996 est) 48,900. It is an agricultural centre, with a livestock market noted for its white-faced Hereford cattle. The city has the UK's largest cider industry. Other activities include brewing, food processing (chicken), tourism, and the manufacture of non-ferrous alloys and components. The cathedral, dating from 1079, has the largest chained library in the world and a medieval *Mappa Mundi* (map of the world). 24 K9

Herefordshire unitary authority in west England, created in 1998 from part of the former county of Hereford and Worcester.
area 2,288 sq km/884 sq mi
towns and cities ►Hereford (administrative headquarters), Leominster, Ross-on-Wye, Ledbury
features River Wye; Herefordshire Beacon (340 m/1,115 ft) Iron Age fort; Hereford Cathedral (11th century) houses the late 13th/early 14th-century Mappa Mundi; Waterworks Museum (Hereford) in restored Victorian pump house; Croft Castle (Leominster); St Mary's Church (Kempley); The Prospect, a walled clifftop garden in Ross-on-Wye designed by John Kyrle in the 17th century; Norman Church (Kilpeck) with notable carvings
industries agriculture, orchards and cider industry, agricultural services and machinery, precision engineering, light engineering, plastics manufacture
population (1996) 166,100 5 E5

Herford city in the *Land* (administrative region) of North Rhine–Westphalia, Germany, on the rivers Werre and Aa, 15 km/9 mi north of Bielefeld; population (1994) 248,300. Furniture, carpets, and metal goods are manufactured here. Herford was once a member of the Hanseatic League medieval trade federation. 20 D4

Herisau commune in the canton of Appenzell, Switzerland, and capital of the half-canton of Outer Rhodes, 10 km/6 mi southwest of St Gallen; population (1995) 16,200. Herisau has an old belltower and town hall. Cotton goods and embroidery are produced here. **30** E3

Hermon, Mount (Arabic **Jebel esh-Sheikh**), snow-topped mountain, 2,814 m/9,232 ft high, on the Syria–Lebanon border, forming the southern extremity of the Anti-Lebanon range, on the border with Israel. According to tradition, Jesus was transfigured here. **60** C3

Hermoupoli (Greek **Ermoupoli**), port and capital of Syros Island, Greece, and capital of the ▶Cyclades department; population (2001 est) 13,000. It is the only industrial town in the Cyclades; cotton and Turkish delight are manufactured, and there are ship-repairing yards. It was at one time the chief port of Greece, but declined with the growth of Peiraias in the late 19th century. **36** G7

Herne port and city in North Rhine-Westphalia, Germany, in the heart of the Ruhr coalfield, 51 km/32 mi northeast of Düsseldorf; population (1995) 179,900. It is the terminus of the Rhine–Herne Canal. Industries include coalmining, iron and steel-making, civil engineering, construction, and waste disposal. **20** C5

Herne Bay resort on the north coat of Kent, southeast England, 11 km/7 mi north of Canterbury; population (1991) 32,800. **22** D3

Hertfordshire county of southeast England, to the north of London.
area 1,630 sq km/629 sq mi
towns and cities Hertford (administrative headquarters), Bishop's Stortford, Hatfield, Hemel Hempstead, Letchworth (the first garden city in 1903; followed by Welwyn in 1919), Stevenage (the first new town, designated in 1946), St Albans, Watford, Hitchin
physical rivers Lea, Stort, Colne; part of the Chiltern Hills
features Hatfield House; Knebworth House; Brocket Hall (home of Palmerston and Melbourne); home of George Bernard Shaw at Ayot St Lawrence; Berkhamsted Castle (Norman)
agriculture barley, dairy farming, market gardening, horticulture
industries aircraft, electronics, electrical goods, engineering, paper, printing, plastics, pharmaceuticals, tanning, sand and gravel
population (1996) 1,015,800 **5** F6

Herts abbreviation for ▶Hertfordshire, an English county. **5** F6

Hesse (German **Hessen**), administrative region (German *Land*) in central Germany, bordered on the west by the state of Rhineland-Palatinate, on the south by Bavaria and Baden-Württemberg, on the east by Thuringia, and on the north by North Rhine-Westphalia and Lower Saxony; area 21,100 sq km/8,150 sq mi; population (1995 est) 6,120,000. The capital is ▶Wiesbaden; other major cities are Frankfurt-am-Main, Kassel, Darmstadt, Fulda, Giessen, and Offenbach-am-Main. The chief industries are timber, chemicals, textiles, automobiles, electrical engineering, optical instruments, computers, telecommunications, printing, and publishing. Agriculture is based on the cultivation on oats, wheat, potatoes, barley, flax, and sugar beet; wine is produced in the west of the state, in the Rheingau area bordering the Rhine. **7** (2)B3

Hessen German name for ▶Hesse, an administrative region in central Germany. **7** (2)B3

Hialeah city in Dade County, southeast Florida, USA, near the Everglades, 7 mi/12 km west of Miami (of which it is now a residential suburb); population (1994 est) 194,000. It has a variety of light manufactures, including clothing, aluminum and electronic products, furniture, plastic goods, building supplies, appliances, transportation equipment, and chemicals. **Hialeah Park** racetrack (1925) is used for horse racing. **90** E4

Hickory town in Catawba and Burke counties, west-central North Carolina, USA; population (1990) 28,300. It is situated in the Piedmont, near the foothills of the Blue Ridge Mountains, 45 mi/73 km northwest of Charlotte. The town's main industry is hosiery manufacture; other products include rope, cotton, trucktrailers, and electronic components. Hickory is home to Lenoir-Rhyne College (1891) and to a community college. **88** D3

Highland unitary authority in northern Scotland, created from the region bearing the same name in 1996.
area 26,157 sq km/10,100 sq mi (one-third of Scotland)
towns ▶Inverness (administrative headquarters), Thurso, Wick, Fort William, Aviemore
physical mainland Highland consists of a series of glaciated ancient plateau masses dissected by narrow glens and straths (valleys); in the northeast (Caithness), old red sandstone rocks give a softer, lower topography; Ben Nevis (1,343 m/4,406 ft), Cairngorm Mountains; Loch Ness; Cuillin Hills, Skye; includes many of the Inner Hebridean islands
features Caledonian Canal; John O'Groats; Skye Road Bridge
industries winter sports, timber, aluminium smelting, pulp and paper production, whisky distilling
agriculture salmon fishing, sheep farming, grouse and deer hunting
population (1996) 207,500. **5** D3

Highland Region former region of Scotland (1975–96) which consisted of eight districts and was replaced in 1996 by the Highland unitary authority.

Highlands one of the three geographical divisions of Scotland, lying to the north of a geological fault line that stretches from Stonehaven in the North Sea to Dumbarton on the Clyde. It is a mountainous region of hard rocks, shallow infertile soils, and high rainfall.

High Point city in north-central North Carolina, USA, 14 mi/23 km southwest of Greensboro, on the Piedmont Plateau; population (1992) 70,750. The furniture industry is very important to the economy; the Southern Furniture Market is held here four times a year, and the International Home Furnishings Market, held twice a year in High Point, attracts national and international exhibitors. Other industries include the manufacture of hosiery. **88** E3

High Wycombe market town in Buckinghamshire, southern England, on the River Wye, between London and Oxford; population (1991) 62,500. Industries include light engineering and the printing of postage stamps, waste management and corporate services, and the manufacture of furniture, cigarette machines, precision instruments, pharmaceuticals, and software. RAF Strike Command has multi-storey underground headquarters (1984) beneath the Chiltern Hills nearby; they were used as Joint Headquarters in the Gulf War (1991). **22** B3

Hildesheim historic and commercial city in Lower Saxony, Germany, on the River Innerste (a tributary of the Weser), 29 km/18 mi southeast of Hannover; population (1995) 106,000. It is linked to the Mittelland Canal. Products include electronics and hardware. A bishopric from the 9th century, Hildesheim became a free city of the Holy Roman Empire in the 13th century, and was one of the original members of the Hanseatic League. It was under Prussian rule from 1866 to 1945. 20 E4

Hilversum city and summer resort in North Holland province, the Netherlands, 27 km/17 mi southeast of Amsterdam; population (1997) 82,600. It manufactures textiles and carpets, and is the main centre of Dutch broadcasting. The old town hall houses a regional museum. 22 H2

Himachal Pradesh state of northwest India, to the south of Kashmir and west of Tibet; area 55,700 sq km/ 21,500 sq mi; population (1994 est) 5,530,000 (mainly Hindu; some Buddhists). The capital is ►Shimla, which lies at an altitude of 2,213 m/ 7,200 ft. Himachal Pradesh is a mainly agricultural state, partly forested, producing fruit, grain, rice, and seed potatoes. The chief industry is timber; other industries are the small-scale mining of slate, gypsum, and limestone; iron foundry; the production of resin, fertilizer, and turpentine; and electronics. The mountain scenery attracts increasing numbers of tourists. 10 C2

Himalayas vast mountain system of central Asia, extending from the Indian states of Kashmir in the west to Assam in the east, covering the southern part of Tibet, Nepal, Sikkim, and Bhutan. It is the highest mountain range in the world. The two highest peaks are Mount ►Everest and ►K2. Other peaks include ►Kanchenjunga, Makalu, Annapurna, and Nanga Parbat, all over 8,000 m/26,000 ft. 40 L6

Himş alternative name for ►Homs, a city in west Syria. 60 D2

Hindenburg German name 1915–45 of the Polish city of ►Zabrze, in honour of General Hindenburg, German field marshal and president 1925–34. 18 H7

Hindu Kush mountain range in central Asia, length 800 km/500 mi, greatest height Tirich Mir, 7,690 m/ 25,239 ft, in Pakistan. The narrow Khyber Pass (53 km/33 mi long) connects Pakistan with Afghanistan and was used by Babur and other invaders of India. The present road was built by the British in the Afghan Wars. 54 A1

Hindustan ('land of the Hindus'), the whole of India, but more specifically the plain of the ►Ganges and Yamuna rivers, or that part of India north of the ►Deccan.

Hinnøya largest of the Lofoten Islands (see ►Lofoten and Vesterålen) off the north coast of Norway; area 2,198 sq km/849 sq mi. Hinnøya is Norway's largest island and lies within the Arctic Circle. It is mountainous and wooded; Moysalen (1,262 m/ 4,140 ft) is the highest peak. The main economic activities are fishing and tourism. Harstad, in the north of the island, is the most important town and port. There is a 13th-century church north of Harstad. 16 H2

Hinthada (or **Henzada**), city in south central Myanmar (Burma), on the Irrawaddy River; population (1980 est) 284,000. Products include rice and potatoes. 50 B3

Hiroshima industrial city and port on the south coast of Honshu island, Japan; population (1994) 1,077,000. On 6 August 1945 it was destroyed by the first wartime use of an atomic bomb. The city has largely been rebuilt since then. The main industries include food processing and the manufacture of cars and machinery. 48 G6

Hispaniola (Spanish 'little Spain'), West Indian island, first landing place of Columbus in the New World, 6 December 1492; it is now divided into ►Haiti and the ►Dominican Republic. 96 D2

Hitachi city of Ibaraki prefecture, on Honshu island, Japan, about 130 km/ 80 mi northeast of Tokyo; population (1993) 202,100. The chief industry is the manufacture of electrical and electronic goods. The city developed around the copper-refining industry based on local copper. The port opened in 1959. 48 L5

Hobart capital and principal port of ►Tasmania, Australia; population (1996) 126,118. Hobart is situated on the southeast coast of the island, at the mouth of the River Derwent. Industries include zinc processing, brewing, electronics, engineering, fruit and vegetable processing, chocolate making, shipbuilding, and the production of textiles, paper, furniture, and newsprint. Hobart is a centre for yachting, fishing, and trading; exports include fruit, textiles, and processed food. The University of Tasmania (founded in 1890) is located here. 76 J8

Ho Chi Minh City (formerly **Saigon** until 1976), chief port and industrial city of South Vietnam; population (1997 est) 3,571,000 (the largest city in Vietnam). It lies on the Saigon River, 54 km/ 34 mi from the South China Sea. Industries include shipbuilding, textiles, rubber, and food products. Saigon was the capital of the Republic of Vietnam (South Vietnam) from 1954 to 1976, when it was renamed, and the city was also the former capital of French Indochina. 50 D4

Hodeida (or **Al Hudaydah**), port of Yemen, on the east coast of the Red Sea; population (1993 est) 246,100. It trades in coffee, dates, skins, cotton and some pearls, senna, myrrh, sesame, and grain. 64 H5

Hodonín (German **Göding**), town of southern Moravia, in the Czech Republic, on the River Morava, 50 km/31 mi southeast of Brno; population (1991) 39,700. It is the centre of a district that produces oil, gas, lignite, and wine. Industries include sugar refining and glassmaking. There is a Renaissance castle and a baroque church. 18 G9

Hoek van Holland Dutch name for ►Hook of Holland, a small peninsula and village in South Holland, the Netherlands. 22 G3

Hof city in Bavaria, Germany, on the River Saale, in the foothills of the Fichtelgebirge; population (1994) 52,700. It is an important railway junction, and has textile and brewing industries. 20 G6

Hofei alternative transcription of ►Hefei, capital of Anhui province, China. 46 F4

Hoggar alternative spelling of ►Ahaggar, a plateau in the Sahara Desert. 66 G4

Hohensalza German name for ▶Inowrocław, a town in Poland. 18 H5

Hohhot (or **Huhehot**; formerly **Kweisui**), city and capital of ▶Inner Mongolia Autonomous Region, China; population (1993) 730,000. It is an important industrial centre and trading hub between north and west China and Mongolia. Cotton textiles, wool, fur, steel, building materials, machinery, electronics, chemicals, flour, dairy goods, and diesel engines are produced. Hohot contains Lamaist monasteries and temples. 46 E2

Hokkaido (formerly **Yezo** or **Ezo** until 1868; Japanese *hoku* 'north'; *kai* 'sea'; *do* 'road'), northernmost and second-largest of the four main islands of Japan, separated from Honshu to the south by Tsugaru Strait (20 km/12 mi wide), and from Sakhalin (Russia) to the north by Soya Strait; area 83,500 sq km/32,239 sq mi; population (1995) 5,692,000 (including 16,000 Ainus). The capital is ▶Sapporo; other major cities are Hakodate, Asahikawa, Otaru, and Muroran. The main industries are coal, mercury, manganese, oil, natural gas, and tourism; agriculture centres on rice cultivation, dairying, forestry, and fishing. 48 N2

Holland popular name for the ▶Netherlands; also two provinces of the Netherlands, see ▶North Holland and ▶South Holland. 22 H2

Hollywood city in Broward County, southeast Florida; population (1992) 121,700. It is situated on the Atlantic Ocean, 11 mi/18 km south of Fort Lauderdale and 18 mi/29 km north of Miami. Its palm-lined beaches, sport fishing, and other recreational activities have established it as a major tourist resort. It is also the site of many planned retirement communities. Industries include printing, and the manufacture of electronic equipment, building materials, and clothing. 90 E4

Holyhead (Welsh **Caergybi**), seaport on the north coast of Holy Island, off Anglesey, northwest Wales; population (1991) 11,600. The island is linked by road and railway bridges with Anglesey, and there are regular sailings from Holyhead to Dublin, and to nearby port Dun Laoghaire, Republic of Ireland. 24 H8

Holy Island (or **Lindisfarne**), island in the North Sea, 3 km/2 mi off the coast of Northumberland, northeast England, with which it is connected by a causeway at low tide; area 10 sq km/4 sq mi; population (1991) 179. It is the site of a monastery founded by St Aidan in 635. Tourism and a mead factory provide local employment. 24 L6

Home Counties those counties in close proximity to London, England: Hertfordshire, Essex, Kent, Surrey, Buckinghamshire, and formerly Berkshire and Middlesex.

Homs (or **Himş**), city and administrative centre of Homs governorate, west Syria, near the Orontes River; population (1993) 537,000. Silk, cereals, and fruit are produced in the area, and industries include silk textiles, oil refining, and jewellery. Known in ancient times as Emesa, it was once famous for its Temple of the Sun. Zenobia, Queen of Palmyra, was defeated here by the Roman emperor Aurelian in 272. 60 D2

Honduras country in Central America, bounded north by the Caribbean Sea, southeast by Nicaragua, south by the Pacific Ocean, southwest by El Salvador, and west and northwest by Guatemala.

Honduras covers an area of 112,100 sq km/43,281 sq mi, and has a population of 6,485,000 (2000 est). The capital is ▶Tegucigalpa; other major towns and cities include San Pedro Sula, La Ceiba, El Progreso, Choluteca, Juticalpa, and Danlí. The head of state and government is Carlos Flores (from 1998). The currency is the lempira, and the languages spoken include Spanish (official), English, and American Indian languages. 94 G6

Honfleur seaport in the *département* of Calvados, France, on the south bank of the River Seine opposite Le Havre; population (1990) 10,000. It is a tourist resort, and fishing and sailing port. A picturesque town, with a Gothic church of the 15th century built entirely of wood, it was depicted by several Impressionist painters. 22 C5

Hong Kong (Chinese **Xianggang**), special administrative region in the southeast of China, comprising Hong Kong Island, the mainland Kowloon Peninsula and New Territories, and many small islands, of which the largest is Lantau; area 1,070 sq km/413 sq mi; population (1995 est) 6,189,800 (57% Hong Kong Chinese, most of the remainder are refugees from the mainland). The capital buildings are located in Victoria (Hong Kong City), and other towns and cities include Kowloon and Tsuen Wan (in the New Territories). A former British crown colony, it reverted to Chinese control in July 1997. 50 E2/11 (2)E4

Honiara port and capital of the Solomon Islands, on the northwest coast of Guadalcanal Island in the southwest Pacific Ocean; population (1996 est) 43,700. The city, which is on the River Mataniko, grew around a World War II US military base, and replaced Tulagi as the capital of the Solomon Islands in 1952 and is now served by the international airport, Henderson Airfield, which lies 16 km/10 mi to the east. Exports include coconuts, copra , fish, and a little gold from the island's central Gold Ridge. 74 F6

Honolulu (Hawaiian 'sheltered bay'), state capital and port of ▶Hawaii, and seat of Honolulu County, on the south coast of ▶Oahu; population (2000 est) 371,700. The city, with its natural harbour (a lagoon within a coral reef), is the economic centre of Hawaii, the principal point of entry to the islands, and a trading centre for European and Indian goods. With its warm climate and tropical vegetation, Honolulu has become a holiday resort. The other major employer is the defence industry, as Pearl Harbor Naval Base and Hickam Air Force Base are situated 11 km/7 mi to the northwest. Honolulu was incorporated as a city in 1907. 92 (2)D2

Honshu principal island of Japan, lying between Hokkaido to the northeast and Kyushu to the southwest. Its land mass comprises approximately four-fifths of the country total area; area 231,100 sq km/89,228 sq mi, including 382 smaller islands; population (1995) 100,995,000. The capital is ▶Tokyo; other major cities are Yokohama, Osaka, Kobe, Nagoya, and Hiroshima. Honshu is linked by bridges and tunnels with the islands of Hokkaido, Kyushu, and Shikoku. A chain of volcanic mountains runs along the island and there are frequent earthquakes. 48 L5

Hook of Holland (Dutch **Hoek van Holland** 'corner of Holland'), small peninsula and village in South Holland, the Netherlands; the terminus for ferry services with Harwich (Parkeston Quay), England. 22 G3

Hoorn town in the province of North Holland, the Netherlands, on the west coast of the IJsselmeer, 40 km/25 mi north of Amsterdam; population (1996) 62,100. It is a picturesque town with cheese and cattle markets, boatyards, and sawmills. Hoorn was a port on the ►Zuider Zee in the 17th century, handling Baltic trade to the colonies, and held the headquarters of the Dutch East India Company, but the town declined in importance after the harbour silted up and the IJsselmeer was created. 22 H2

Hoover Dam highest concrete dam in the USA, 221 m/726 ft, on the Colorado River at the Arizona–Nevada border. It was begun during the administration of President Herbert Hoover and built 1930–36. Known as **Boulder Dam** 1933–47, under the administration of Franklin D Roosevelt, its name was restored by President Harry Truman as the reputation of Herbert Hoover was revived. It impounds Lake Mead, and has a hydroelectric power capacity of 1,300 megawatts.

Hopkinsville city and administrative headquarters of Christian County, southwest Kentucky, USA, 134 mi/215 km southwest of Louisville; population (1990) 29,800. Situated in a rich agricultural region, it is a marketplace for tobacco and livestock. Industries include the manufacture of lighting fixtures, satellite antennas, and wire. 88 C3

Horn, Cape (Spanish **Cabo de Hornos**), southernmost point of South America, in Magallanes region, Chile; situated on Horn Island to the south of ►Tierra del Fuego archipelago. The cape is notorious for gales and heavy seas, and was the sea route between the Atlantic and the Pacific Oceans until the opening of the Panama Canal in 1914. Cape Horn was discovered in 1616 by Dutch explorer Willem Schouten (1580–1625), and named after his birthplace (Hoorn). 100 H10

Horsham town and market centre in West Sussex, southeast England, on the River Arun, 26 km/16 mi southeast of Guildford; population (1996 est) 43,300. Employment is mainly in the service sector and retailing. High-tech industries and pharmaceutical production take place in the district. The private school Christ's Hospital is about 3 km/2 mi to the southwest. 22 B3

Hot Springs resort town and administrative headquarters of Garland County, west-central Arkansas, USA, 88 km/55 mi southwest of Little Rock; population (1990) 32,500. It is situated in a narrow valley and contains 47 hot springs which produce nearly 4 million litres of mineral water each day, with an average temperature of 62°C/143°F. The town is a popular spa and the springs are contained in a reservation now designated a national park (the only national park to fall within city limits). As well as tourism, industries include aluminium goods and bottled water. It is home to Garland County Community College (1973). Hot Springs was the childhood home of President Bill Clinton. 88 B4

Houma city and administrative headquarters of Terrebonne Parish, south Louisiana, on the Gulf Intracoastal Waterway, 45 mi/72 km southwest of New Orleans; population (1990) 30,500. Linked to the Gulf of Mexico by the **Houma Navigation Canal**, it is a supply center for offshore oil rigs. Industries include shellfish processing, sugar refining, boat building, and tourism. 84 H6

Houston city and port in southeastern Texas, USA; linked by the Houston Ship Canal to the Gulf of Mexico, in the Gulf Coastal Plain; population (1994 est) 1,702,000; population of metropolitan area (1994 est) 4,099,000. A major centre of finance and commerce, Houston is also one of the busiest US ports. Industrial products include refined petroleum, oilfield equipment, and petrochemicals, chief of which are synthetic rubber, plastics, insecticides, and fertilizers. Other products include iron and steel, electrical and electronic machinery, paper products, and milled rice. The Lyndon B Johnson Space Center (1961), the command post for flights by US astronauts, is located here. 84 G6

Hoyerswerda town in the *Land* (administrative region) of Brandenburg, Germany, 40 km/25 mi south of Cottbus; population (1995) 59,900. It is situated in the Sorbian district of Lusatia. The town is dominated by a huge power-generating and lignite-processing complex, built to exploit the extensive opencast mining of brown coal in the area. Glass is also manufactured. 20 K5

Hradec Králové (German **Königgrätz**), capital of East Bohemia, in the Czech Republic, on the River Elbe, 100 km/63 mi east of Prague; population (1991) 98,900. It is an industrial town producing chemicals, photographic equipment, and textiles. The Gothic cathedral is partly 14th century. 18 E7

Hrodna (formerly **Grodno**), industrial city and port in Belarus, on the Nyoman (Neman) River, capital of an oblast (Belorussian *voblasts*) of the same name; population (1991) 284,800. The city lies some 230 km/143 mi west of Minsk, near the Belorussian borders with Poland and Lithuania. Industries include chemical and textile manufacture, and sugar-refining. 18 N4

Huallaga river of central Peru; it rises in the Andes, near Cerro de Pascua, and flows generally northeastwards for 1,120 km/696 mi to meet the River Marañon, one of the main headwaters of the Amazon. The upper reaches of the river valley are used for growing coca, a major source of the drug cocaine. 98 B5

Huambo town and capital of Huambo province, central Angola; population (1995) 400,000. Founded in 1912, it was known as **Nova Lisboa** ('New Lisbon') 1928–78. It is an agricultural centre, and is linked by rail to Lobito. A battle here in 1993 between MPLA (government forces) and UNITA (right-wing rebels) lasted 55 days and cost an estimated 13,500 lives; UNITA won. It fell to government forces in 1994, and UNITA leader, Jonas Savimbi, fled. Huambo was the former headquarters of UNITA. 72 B2

Huang He (or **Hwang Ho**; English **Yellow River**), river in China, named after its muddy waters; length 5,464 km/3,395 mi. Rising in Qinghai province in the west of the country, it winds eastwards to the Bohai Gulf. Sometimes known as 'China's sorrow' because of disastrous floods, it is now largely controlled through hydroelectric works, dykes, and embankments. The barriers, however, are ceasing to work because silt is continually raising the river bed. **46** F3

Huascaran, Nev de extinct volcano in the Andes; the highest mountain in Peru, 6,768 m/22,205 ft. **98** B5

Hubei (or **Hupei**, or **Hupeh**), province of central China, bounded to the north by Henan, to the east by Anhui, to the south by Jiangxi and Hunan, and to the west by Sichuan and Shaanxi provinces; area 187,500 sq km/72,375 sq mi; population (1996) 58,250,000. The capital is ▶Wuhan; other major cities and towns are Huangshi, Shashi, Yichang, and Xiangfan. The main industries are the mining of copper, gypsum, iron ore, phosphorus, and salt; and the production of steel, machinery, domestic appliances, textiles, food processing, and fibre-optic cables. Agriculture is based on the cultivation of rice, cotton, rapeseed, wheat, beans, and vegetables. **11** (2)E3

Huddersfield (Anglo-Saxon **Oderesfelt**), industrial town in West Yorkshire, on the River Colne, between Leeds and Manchester; population (1991) 119,000. A thriving centre of woollen manufacture by the end of the 18th century, it now produces textiles and related products, and has electrical and mechanical engineering, food processing (biscuits), publishing, and building industries. The service sector is the principal employer. **24** L8

Hudson river in northeastern USA; length 485 km/300 mi. It rises in the Adirondack Mountains and flows south, emptying into a bay of the Atlantic Ocean at New York City. The Hudson forms the boundary between New Jersey and New York, and the states are linked by bridges and tunnels. The New York Barge Canal system links the Hudson to Lake Champlain, Lake Erie, and the St Lawrence River. It is navigable by small ocean-going vessels as far upstream as Albany and Troy, about 150 mi/240 km from its mouth, and for eight months of the year barge traffic can reach the Great Lakes. **88** F2

Hudson Bay inland sea of northeastern Canada, linked with the Atlantic Ocean by **Hudson Strait** and with the Arctic Ocean by Foxe Channel and the Gulf of Boothia; area 1,233,000 sq km/476,000 sq mi. It is bordered by (clockwise) the provinces of Québec, Ontario, Manitoba, and the Northwest Territories. It is named after Henry Hudson, who reached it in 1610. **82** P5

Hué city in central Vietnam, formerly capital of Annam, 13 km/8 mi from the China Sea; population (1995 est) 290,000. It lies in Binh Tri Thien province and was formerly capital of the Vietnamese empire. Industries include textiles and timber. The university was founded in 1957. **50** D3

Huelva port and capital of Huelva province in Andalusia, southwest Spain, situated on the Atlantic coast, on a small peninsula formed by the junction of the Tinto and the Odiel rivers; population (1991) 141,000. Industries include shipbuilding and oil refining; there is a large petrochemical complex here which produces fertilizers, sulphuric acid, and explosives, as well as petrol and oil products. There are also fisheries, and a trade in ores from the Río Tinto mines. **28** D7

Huesca capital of Huesca province in Aragón, northeast Spain, situated on the River Isuela; population (1991) 44,100. Industries include engineering and food processing. Among its buildings are a 13th-century cathedral and the former palace of the kings of Aragón. **28** K2

Huhehot former name of ▶**Hohhot**, a city in Inner Mongolia. **46** E2

Hull shortened name of ▶Kingston upon Hull, a city and unitary authority on the north bank of the Humber estuary, northeast England. **24** M8

Hull city in southern Québec, Canada, on the north shore of the Ottawa River opposite Ottawa; population (1991) 60,700. Its industries include pulp, paper, cement, meat packing, clothing, timber, concrete blocks, and jewellery. There are four large electricity-generating stations within the city and its surroundings. Hull is the shopping centre for a prosperous farming community, and has a number of government offices. The Canadian Museum of Civilization, a branch of the Université du Québec, and a music conservatory are located in the city. **88** E1

Humber estuary in northeast England formed by the Ouse and Trent rivers, which meet east of Goole and flow east for 60 km/38 mi to enter the North Sea below Spurn Head. It is an important commercial waterway, and the main ports are ▶Kingston upon Hull on the north side, and ▶Grimsby on the south side. The Humber Bridge (1981) joins the two banks.

Humberside former county of northeast England, created in 1974 out of north Lincolnshire and parts of the East and West Ridings of Yorkshire. It was abolished in 1996 when the unitary authorities of East Riding of Yorkshire, Kingston upon Hull, North East Lincolnshire, and North Lincolnshire were created.

Humboldt river in north Nevada; length 490 km/304 mi. It rises in northeast Nevada, flows southwest through the **Humboldt Lake**, and ends in the area of interior drainage known as the **Humboldt Sink**. It is used for irrigation; the Rye Patch Dam (1936) forms Rye Patch Reservoir, to the north-northeast of Lovelock. **82** L6

Hunan province of south central China, bounded to the north by Hubei, to the east by Jiangxi, to the south by Guangdong and Guangxi Zhuang Autonomous Region, and to the west by Guizhou and Sichuan; area 210,500 sq km/81,300 sq mi; population (1996) 64,280,000. The capital is ▶Changsha; other main cities and towns are Hengyang, Shaoyang, Xiangtan, and Zhuzhou. The main industries are non-ferrous minerals, engineering, chemicals, and electrical goods. Agriculture is based on the cultivation of rice, tea, tobacco, and rapeseed. **11** (2)E4

Hunedoara industrial town in western Romania, 15 km/9 mi south of Deva (capital of Hunedoara county); population (1990) 90,000. It has a large iron and steel works fed by locally obtained ores. The Gothic castle dates from 1452. **34** K4

Hungary country in central Europe, bounded north by the Slovak Republic, northeast by Ukraine, east by Romania, south by Yugoslavia and Croatia, and west by Austria and Slovenia.

Hungary covers an area of 93,032 sq km/35,919 sq mi, and has a population of 10,036,000 (2000 est).

The capital is ►Budapest; other major towns and cities include Miskolc, Debrecen, Szeged, Pécs, Györ, Nyíregyháza, Székesfehérvár, and Kecskemét. The head of state is Ferenc Mádl (from 2000); the head of government is Viktor Orban (from 1998). The currency is the forint, and the main language spoken is Hungarian (official). 34 F3

Huntingdon market town in Cambridgeshire, eastern England, on the River Ouse, 26 km/16 mi northwest of Cambridge; population (1996 est) 18,300 (excluding Godmanchester). It is an agricultural centre with diverse industries, including telecommunications, printing, food-processing (frozen foods, meat products), and the manufacture of electrical equipment, fibre optics, computerware, furniture, rubber mouldings, plastics, and chemicals. Oliver Cromwell was born here in 1599. 22 B2

Huntsville city in northeastern Alabama, USA; seat of Madison County; population (1994 est) 160,000. It is the centre of an agricultural region; manufactured products include aerospace and military equipment, computer hardware and software, compact discs, textiles, electrical and electronic goods, metal products, chemicals, machinery, and cosmetics. 90 D3

Hupei alternative transcription of ►Hebei, a province of China. 11 (2)E3

Huron, Lake second largest of the ►Great Lakes of North America, on the US–Canadian border; area 60,000 sq km/23,160 sq mi. Lake Huron is 331 km/205 mi long, lies at 177 m/581 ft above sea level, and reaches a depth of 230 m/755 ft. It is bounded on the north and east by Ontario, and on the west and southwest by Michigan. There are several small ports on its shores, and lumbering and fishing are important economic activities in the region. 88 D1

Huşi town in Vaslui county, eastern Romania, near the border with Moldova formed by the River Prut, 65 km/41 mi south of Iaşi; population (1995) 26,800. Wine is produced in the area. 34 R3

Huy (Flemish **Hoei**), town in the province of Liège, Belgium, on the River Meuse, about 27 km/17 mi southwest of the city of ►Liège; population (1991) 18,200. The town has distilleries and manufactures paper. 22 H4

Hvannadalshnukur highest peak in Iceland, rising to 2,119 m/6,952 ft in the southeast of the country. 16 (1)E2

Hvar capital of the island of Hvar, off the coast of Dalmatia, Croatia; population (1990) 3,600. It is a fortified town, with medieval walls and buildings, and has a 13th-century cathedral and a 17th-century theatre. It is a popular tourist resort. 34 D6

Hwange (formerly **Wankie**, until 1982), coal-mining town in Zimbabwe, 300 km/186 mi northwest of Bulawayo; population (1992) Hwange Urban 42,600; Hwange Rural 72,200. It is the centre of the country's chief coal-mining area, with the western coalfields covering at least 1,000 sq km/386 sq mi. There are also tin and copper mines. Hwange National Park is nearby. 72 D3

Hyderabad capital city of the southern central Indian state of ►Andhra Pradesh, on the River Musi; population (1991) 4,280,000. Products include carpets, silks, and metal inlay work. More recently industries such as textiles, pharmaceuticals, electrical machinery, and chemicals have become important. It was formerly the capital of the state of Hyderabad. Buildings include the Jama Masjid mosque and Golconda fort. It is an important educational and research centre, with Osmania University founded by the Nizam in 1939. 54 C5

Hyderabad city in Sind province, southeast Pakistan, lying 10 km/6 mi east of the ►Indus River, 150 km/95 mi northeast of Karachi; population (1991) 1,000,000. It produces gold, pottery, glass, embroidered leather saddles for camels, and furniture. Industries include textiles, cement, glass and soap, and there is a thermal power station nearby. 56 J4

Hyères town on the Côte d'Azur in the *département* of Var, southern France; population (1990) 50,100. It has a mild climate, and is a winter health resort. Industries include olive-oil pressing and the export of violets, strawberries, and vegetables. 26 M10

Iaşi (German **Jassy**), city in northeastern Romania; population (1993) 328,000. It has chemical, machinery, electronic, and textile industries. It was the capital of the principality of Moldavia 1568–1889. 34 Q2

Ibadan city in southwest Nigeria, capital of Oyo State, 145 km/90 mi northeast of Lagos; population (1992 est) 1,295,000. Ibadan is the second-largest city in the country, and is a major commercial, industrial, and administrative centre. It is a marketplace for cocoa and other local agricultural produce. Industries include chemicals, electronics, plastics, and motor vehicle assembly. 68 E3

Ibagué capital of Tolima department, west-central Colombia, 130 km southwest of Bogotá at the foot of the Quindío Mountains in the Andes; population (1994) 343,000. It is a market town and acts as a trading centre of cattle and coffee from surrounding districts. Leather goods and a local drink, *mistela*, are produced here. Ibagué has a Conservatory of Music, and hosts the annual National Folklore Festival. 98 B3

Iberian Peninsula name given by ancient Greek navigators to the Spanish peninsula, derived from the River Iberus (Ebro). Anthropologists have given the name '**Iberian**' to a Neolithic people, traces of whom are found in the Spanish peninsula, southern France, the Canary Isles, Corsica, and part of North Africa.

Ibiza (or **Eivissa**), one of the ►Balearic Islands, a popular tourist resort; area 596 sq km/230 sq mi; population (1990 est) 71,000. The capital and port, also called Ibiza, has a cathedral, built sometime between the 14th and 16th centuries. 28 M6

Içel another name for ►Mersin, a city in Turkey. 36 S8

Iceland island country in the North Atlantic Ocean, situated south of the Arctic Circle, between Greenland and Norway.

Iceland covers an area of 103,000 sq km/39,768 sq mi, and has a population of 281,000 (2000 est). The capital is ►Reykjavik; other major towns and cities include Akureyri, Kópavogur, Hafnerfjördur, Gardhabaer, Keflavík, Reykjanesbaer, and Vestmannaeyjar. The head of state is Ólafur Ragnar Grímsson (from 1996); the head of government is Davíd Oddsson (from 1991). The currency is the krona, and the main language spoken is Icelandic (official). 14 C1

Ichang alternative form of ►Yichang, a port in Hubei province, China. 46 E4

Iconium city of ancient Turkey; see ►Konya. 58 E5

Idaho state of northwestern USA. It is nicknamed Gem State. Idaho, one of the Mountain States, was admitted to the Union in 1890 as the 43rd US state. It is bordered to the east by Montana and Wyoming, to the south by Utah and Nevada, to the west by Oregon and Washington, and to the north by British Columbia, Canada.
population (2000 est) 1,294,000
area 216,500 sq km/83,569 sq mi
capital ►Boise
towns and cities Pocatello, Idaho Falls, Nampa, Lewiston
industries and products potatoes, wheat, livestock, timber, silver, lead, zinc, antimony, tourism, leisure industry 86 D2

Idaho Falls city and administrative headquarters of Bonneville County, southeast Idaho, on the Snake River; population (1990) 43,900. It is a shipping centre for an agricultural area producing potatoes and livestock; industries include food processing, vitamins, building materials, and tourism. The waterfalls provide electric power and there is a US Atomic Energy Commission plant nearby. 86 D2

Idi Amin Dada, Lake former name (1973–79) of Lake ►Edward in Uganda and the Democratic Republic of Congo. 70 D4

Ieper Flemish name for ►Ypres, a town in West Flanders, Belgium. 22 E4

Iglau German name of ►Jihlava, a town in the Czech Republic. 18 E8

Iguaçu Falls (or **Iguassú Falls**), waterfall in South America, on the border between Brazil and Argentina. The falls lie 19 km/12 mi above the junction of the River Iguaçu with the Paraná. The falls are divided by forested rocky islands and form a spectacular tourist attraction. The water plunges in 275 falls, many of which have separate names. They have a height of 82 m/269 ft and a width of about 4 km/2.5 mi.

IJmuiden port in the province of North Holland, the Netherlands, 10 km/6 mi northwest of Haarlem; population (1998) 15,000. IJmuiden stands at the end of the North Sea Canal, a waterway leading to Amsterdam, of which it is an outport. There are fisheries and chemical industries. 22 G2

IJssel river in the Netherlands; length 110 km/68 mi. The IJssel is the northernmost arm of the ►Rhine delta, leaving the main stream near Arnhem and flowing north into the ►IJsselmeer, 6 km/4 mi northwest of Kampen. The river is navigable for its entire length. 22 J2

IJsselmeer lake in the Netherlands, area 1,217 sq km/470 sq mi. It was formed in 1932 after the ►Zuider Zee was cut off from the North Sea by a dyke 32 km/20 mi long (the *Afsluitdijk*); it has been freshwater since 1944. The rivers Vecht, IJssel, and Zwatewater flow into the lake. 22 H2

Ile-de-France region of northern France; area 12,000 sq km/4,600 sq mi; population (1990) 10,660,600. It includes the French capital, ►Paris, and the towns of Versailles, Sèvres, and St-Cloud, and comprises the *départements* of ►Essonne, ►Val-de-Marne, ►Val-d'Oise, Ville de Paris, ►Seine-et-Marne, ►Hauts-de-Seine, ►Seine-St-Denis, and ►Yvelines. From here the early French kings extended their authority over the whole country. 6 E2

Ile de la Tortue (French **La Tortue** 'turtle'), island off the north coast of ►Haiti; area 180 sq km/69 sq mi. It was a pirate lair during the 17th century.

Ilfracombe resort on the north coast of Devon, southwest England, 17 km/11 mi northwest of Barnstaple; population (1991) 10,400. Features include St Nicholas Chapel, dating from the 14th-century. 24 H10

Ille-et-Vilaine *département* in the ►Brittany region of northwest France, bordering Mont-Saint-Michel bay and the English Channel; area 6,758 sq km/2,609 sq mi; population (1990) 799,200. The principal towns are ►Rennes (the administrative centre), Fougères, Redon, and the ports of St-Servan-sur-Mer and St-Malo. The principal industries are footwear (at Fougères), cars, electronics, and leatherwork (at Rennes). 6 C2

Illinois midwestern state of the USA. It is nicknamed Prairie State. Illinois was admitted to the Union in 1818 as the 21st US state. A major agricultural state, Illinois is bordered to the east by Indiana, to the southeast by Kentucky, with the Ohio River serving as a boundary, to the west by Missouri and Iowa, with the Mississippi River as a boundary, and to the north by Wisconsin. In the northeast, it has a shore of *c.* 100 km/60 mi on Lake Michigan, occupied by Chicago and its northern suburbs.
population (2000 est) 12,419,300
area 146,100 sq km/56,395 sq mi
capital ►Springfield
towns and cities Chicago, Rockford, Peoria, Decatur, Aurora
industries and products soybeans, cereals, meat and dairy products, livestock, machinery, electrical and electronic equipment 88 C2

Ilmenau town in the *Land* (administrative region) of ►Thuringia, Germany, on the River Ilm in the Thuringian Forest, 19 km/12 mi northeast of Suhl; population (1996) 26,300. It is a health resort in a fluorspar-mining area, and was frequented by Goethe, who wrote his play *Iphigenia auf Tauris* here. 20 F6

Ilorin capital of Kwara state, Nigeria, situated on the Kaduna–Kano railway, 300 km/186 mi north of Lagos; population (1992 est) 430,600. It is a major trading centre for the southern savannah zone of western Nigeria, trading in tobacco and wood products. Industries include textiles and crafts. 68 E3

Imbros (Turkish **Imroz**), island in the Aegean Sea; area 280 sq km/108 sq mi; population (1980) 6,000. Occupied by Greece in World War I, it became Turkish under the Treaty of Lausanne 1923. 36 H4

Imphal capital of ►Manipur state on the Manipur River, India; population (1991) 201,000. It is a communications and trade centre (tobacco, sugar, fruit). Imphal was besieged between March and June 1944, when Japan invaded Assam, but held out with the help of supplies dropped by air. 54 F4

Imroz Turkish form of ►Imbros, an island in the Aegean Sea. 36 H4

Inarijärvi Finnish name for Lake ►Inari, in the north of Finland. 16 P2

Inari, Lake (Finnish **Inarijärvi**), lake in the extreme north of Finland, in the province of Lappi; area 1,153 sq km/445 sq mi. It is rich in fish and contains more than 3,000 islands. The small town of Inari on the shore of the lake is a centre of the Saami culture of Lapland and the seat of the Finnish Saami parliament. **16** P2

Inchon (formerly **Chemulpo**), chief port of Seoul, South Korea; population (1995 est) 2,307,600. The city lies where the Han River enters the Yellow Sea. The building of a rail link with Seoul in 1900 boosted the port's trade. Exports include rice, wheat, paper, and electronic goods. **48** D5

Independence city in western Missouri, USA, a suburb of ▶Kansas City, on the Missouri River; seat of Jackson County; population (1994 est) 112,000. It is the centre of an agricultural region; industries include steel, agricultural machinery, Portland cement, petroleum refining, and flour milling. **88** B3

India country in southern Asia, bounded to the north by China, Nepal, and Bhutan; east by Myanmar and Bangladesh; northwest by Pakistan and Afghanistan; and southeast, south, and southwest by the Indian Ocean.

India covers an area of 3,166,829 sq km/1,222,713 sq mi, and has a population of 1,013,662,000 (2000 est). The capital is ▶New Delhi; other major towns and cities include Mumbai (formerly Bombay), Kolkata (formerly Calcutta), Chennai (formerly Madras), Bangalore, Hyderabad, Ahmadabad, Kanpur, Pune, Nagpur, Bhopal, Jaipur, Lucknow, and Surat. The head of state is Kocheril Raman Narayanan (from 1997); the head of government is Atal Behari Vajpayee (from 1998). The currency is the rupee, and the languages spoken include Hindi, English, Assamese, Bengali, Gujarati, Kannada, Kashmiri, Konkani, Malayalam, Manipuri, Marathi, Nepali, Oriya, Punjabi, Sanskrit, Sindhi, Tamil, Telugu, Urdu (all official), and more than 1,650 dialects. **54** C4

Indiana state of the midwest USA. It is nicknamed the Hoosier State. Indiana was admitted to the Union in 1816 as the 19th US state. It is bordered to the northeast by Michigan, to the east by Ohio, to the south and southeast by Kentucky, and to the west by Illinois. In the northwest, Indiana has a shoreline of *c.* 72 km/45 mi on Lake Michigan.
population (2000 est) 6,080,500
area 93,700 sq km/36,168 sq mi
capital ▶Indianapolis
towns and cities Fort Wayne, Gary, Evansville, South Bend
industries and products maize, pigs, cattle, soybeans, limestone, machinery, electrical goods, coal, steel, iron, chemicals, glass, oil **84** J3

Indianapolis state capital and largest city of ▶Indiana, USA, on the White River, 300 km/186 mi southeast of Chicago; seat of Marion County; population (1994 est) 752,000; population of metropolitan area (1992) 1,424,000. Situated in the rich Corn Belt agricultural region, the city is an industrial centre; products include electronic components, pharmaceuticals, processed foods, machinery, plastics, and rubber. It is the venue for the Indianapolis 500 car race. **90** D2

Indian Ocean ocean between Africa and Australia, with India to the north, and the southern boundary being an arbitrary line from Cape Agulhas to south Tasmania; area 73,500,000 sq km/28,370,000 sq mi; average depth 3,872 m/12,708 ft. The greatest depth is the Java Trench 7,725 m/ 25,353 ft. It includes two great bays on either side of the Indian peninsula, the Bay of Bengal to the east, and the Arabian Sea with the gulfs of Aden and Oman to the west. **54** D8

Indochina, French name given by the French to their colonies in South-east Asia: ▶Cambodia, ▶Laos, and ▶Vietnam, which became independent after World War II.

Indonesia country in southeast Asia, made up of 13,677 islands situated on or near the Equator, between the Indian and Pacific oceans. It is the world's fourth most populous country, surpassed only by China, India, and the USA.

Indonesia covers an area of 1,904,569 sq km/735,354 sq mi, and has a population of 212,107,000 (2000 est). The capital is ▶Jakarta; other major towns and cities include Surabaya, Bandung, Medan, Semarang, Palembang, Tangerang, Bandar Lampung, Ujung Pandang, and Malang. The head of state and government is Abdurrahman Wahid (from 1999). The currency is the rupiah, and the languages spoken include Bahasa Indonesia (closely related to Malay; official), Javanese, Dutch, and over 550 regional languages and dialects. **52** D3

Indore city in Madhya Pradesh, India, lying on the Sarasvati River, 550 km/340 mi northeast of Mumbai (formerly Bombay); population (1991) 1,092,000. A former capital of the princely state of Indore, it now produces cotton, chemicals, and furniture. It is a centre for Hindustani classical music. **54** C4

Indre *département* in the ▶Centre region of France; area 6,777 sq km/ 2,617 sq mi; population (1990) 237,687. It is drained by the rivers ▶Indre and Creuse. The chief cash crops grown here are cereals, beet, and vegetables. Vines and fruit trees are also widely cultivated. Poultry and sheep are raised, especially in the Champagne district. The chief manufactures are textiles, paper, leather goods, and pottery. The principal towns are Châteauroux, the departmental capital, Le Blanc, La Châtre, and Issoudun. **6** D3

Indre river rising in the Auvergne Mountains, France, and flowing northwest for 260 km/165 mi to join the Loire below Tours. It gives its name to the *départements* of ▶Indre and Indre-et-Loire. **26** G6

Indre-et-Loire *département* in the ▶Centre region of France; area 6,124 sq km/2,364 sq mi; population (1990) 529,800. The River ▶Loire and its tributaries, principally the ▶Indre, Cher, and Vienne, flow through it. The chief products are wine (Vouvray and Montlouis), wheat, barley, and fruit. There is some textile and paper manufacturing. The châteaux in the Loire valley attract many tourists. The administrative centre is ▶Tours. Other towns include Chinon and Loches. **6** D3

Indus river in Asia, rising in Tibet and flowing 3,180 km/1,975 mi to the Arabian Sea. In 1960 the use of its waters, including those of its five tributaries, was divided between India (rivers Ravi, Beas, Sutlej) and Pakistan (rivers Indus, Jhelum, Chenab). In the 3rd and 2nd millennia BC Indus Valley civilization flourished at centres like Harappa and Mojenjo Daro. **56** K3

Ingolstadt city in Bavaria, Germany, 69 km/43 mi northwest of Munich; population (1994) 110,900. Cars are manufactured in the city and oil, supplied by pipeline from the Italian ports of Genoa and Trieste, is refined here. Ingolstadt is on the River Danube, which was diverted to pass the town in 1363. **30** G2

Ingushetia (or **Ingushetiya**), autonomous republic of the Russian Federation, on the northern slopes of the Caucasus mountains; area 2,000 sq km/770 sq mi; population (1994 est) 250,000 (Ingush 85%). The capital is Nazran. The chief industries are farming and cattle-raising; there is also petroleum drilling. The predominant religion is Islam. **58** L2

Inhambane seaport and administrative centre of Inhambane province, on the southeast coast of Mozambique, 370 km/231 mi northeast of Maputo; population (1990 est) 65,500; province (1991 est) 1,157,000. The industries are mainly dependent on the agriculture of the region, and include sugar-milling, soap-making, and the processing of coconuts, groundnuts, and cashews. It is a railway terminus, whose links are mainly with the southern region of Mozambique. **72** F4

Inland Sea (Japanese **Seto Naikai**), arm of the Pacific Ocean, 390 km/240 mi long, greatest width 70 km/43 mi, almost enclosed by the Japanese islands of Honshu, Kyushu, and Shikoku. Its chief port is Hiroshima, and it is an important shipping route. It has about 300 small islands. Although comparatively shallow, the Inland Sea has a large tidal range.

Inn river in south-central Europe, tributary of the Danube. Rising in the Engadine in the Swiss Alps, it flows 507 km/317 mi northeast through the Tirol in Austria (passing Innsbruck) and into Bavaria below Kufstein, Germany, where it meets the Danube at Passau. It is used extensively as a source of hydroelectric power. **30** H2

Inner Hebrides group of islands that comprise that part of the ►Hebrides closest to the west coast of mainland Scotland. **24** F5

Innsbruck capital of Tirol state, west Austria; population (1995) 119,600. It is a tourist and winter sports centre, and a route junction for the Brenner Pass. Local industries include mechanical engineering, bellmaking, and bookbinding. The 1964 and 1976 Winter Olympics were held here. **30** G3

Inowrocław (German **Hohensalza**), spa town in Bydgoszcz province, Poland, 45 km/28 mi south of Bydgoszcz; population (1990) 76,600. Industries include engineering, glass, sugar, and brass; and salt and gypsum are mined in the district. The town was part of Prussia 1772–1919. **18** H5

Interlaken (German 'between the lakes'), chief town of the Bernese Alps, Switzerland, 40 km/25 mi southeast of Bern; population (1995) 17,400. Situated an altitude of 570 m/1,870 ft, it lies on the River Aar between lakes Brienz and Thun. Industries include tourism, watchmaking, and the production of textiles. Its main street, the Hoheweg, commands a fine view of the ►Jungfrau Mountain. The site was first occupied in 1130 by a monastery, which was suppressed in 1528. **30** C4

Invercargill city on the south coast of South Island, New Zealand; population (1996) 49,300. It has sawmills and meatpacking and aluminium-smelting plants. **78** B8

Inverclyde unitary authority in western Scotland, created in 1996 from Inverclyde district in Strathclyde region.
area 161 sq km/62 sq mi
towns ►Greenock (administrative headquarters), Port Glasgow, Gourock
physical coastal lowland on the Firth of Clyde estuary, rising sharply to an inland plateau of 305 m/1,000 ft
features Inverkip Marina
industries electronics
population (1996) 90,000 **5** D4

Inverness main town in, and the administrative centre of, ►Highland unitary authority, Scotland, at the head of the Moray Firth, lying in a sheltered site at the mouth of the River Ness; population (1991) 41,200. It is a tourist centre with tanning, oil-related engineering, distilling, and electronics industries. Culloden Moor, scene of the massacre of clansmen loyal to Charles Edward Stuart by the English Army in April 1746, is situated to the east of Inverness. **24** H4

Inverness-shire largest of the former Scottish counties, it was part of Highland Region 1975–96, and is now part of the Highland unitary authority.

Ioannina town in Greece, capital of Ioannina department, on the shore of Lake Ioannina, about 80 km/50 mi from the sea and 480 m/1,575 ft above sea level, opposite the island of Corfu; population (2001 est) 63,100. It has a tanning and fur-dressing industry. **36** C5

IOM abbreviation for the Isle of ►Man, an island in the Irish Sea. **24** H7/**5** D4

Iona island in the Inner Hebrides; area 850 hectares/2,100 acres.
A centre of early Christianity, it is the site of a monastery founded in 563 by St Columba. It later became a burial ground for Irish, Scottish, and Norwegian kings. It has a 13th-century abbey. **24** F5

Ionian Islands (Greek **Ionioi Nisoi**), island group off the west coast of Greece; area 860 sq km/332 sq mi; population (1991) 191,000. A British protectorate from 1815 until their cession to Greece in 1864, they include Cephalonia (Greek *Kefallonia*); Corfu (*Kérkyra*), a Venetian possession (1386–1797); Cythera (*Kithira*); Ithaca (*Ithaki*), the traditional home of Odysseus; Lefkada; Paxos (*Paxoí*); and Zanté (*Zakynthos*). **36** B5

Ionian Sea part of the Mediterranean Sea that lies between Italy and Greece, to the south of the Adriatic Sea, and containing the Ionian Islands. **36** B6

Ionioi Nisoi Greek name for ►Ionian Islands, off the west coast of Greece. **36** B5

IOW abbreviation for Isle of ►Wight, an island and unitary authority off the coast of southern England. **24** L11/**5** F6

Iowa state of the midwest USA. It is nicknamed Hawkeye State. Iowa was admitted to the Union in 1846 as the 29th US state. It is a major constituent of the US Corn Belt, with grain and cereal crops and livestock-rearing historically comprising a significant proportion of the state's income. Iowa is bordered to the south by Missouri, to the west Nebraska and South Dakota, to the north by Minnesota, and to the east by Wisconsin and Illinois, with the Mississippi River forming the state boundary.
population (2000 est) 2,926,300
area 145,754 sq km/56,276 sq mi
capital ►Des Moines
towns and cities Cedar Rapids, Davenport, Sioux City, Waterloo
industries and products cereals,

soybeans, grasses and grains, pigs and cattle, poultry, dairy farming, chemicals, farm machinery, electrical goods, hardwood lumber, minerals, finance and insurance sectors 84 H3

Iowa City city in eastern Iowa, USA, on the Iowa River, 25 mi/40 km south of Cedar Rapids and 105 mi/170 km east of Des Moines, seat of Johnson County; population (1992) 59,300. It is a distribution center for the area's agricultural products; other industries include printed matter and building materials. 88 B2

Ipoh capital of Perak state, Peninsular Malaysia; population (1991) 383,000. The economy is based on tin mining. 52 C2

Ipswich (Saxon **Gyppeswyk**), river port and administrative headquarters of ▶Suffolk, eastern England, on the Orwell estuary, 111 km/69 mi north-east of London; population (1996 est) 113,000. An important wool port in the 16th century, it now provides financial and distribution services, and is the location of British Telecom's laboratories. Other industries include engineering, and the manufacture of computer software, machinery, beer, flour, fibre optics, videotape and multi-media products, building materials, plastics, and electrical goods. 24 P9

Iquique (Aymara *ique-ique*, place of rest and tranquillity), seaport and capital of Tarapacá region, northern Chile, 1,853 km/1,150 mi north of Santiago, on the edge of the Atacama Desert; population (1992) 152,700. Industries include fishing, fish processing, and sugar and oil refining. The port exports more fishmeal than any other in the world. Nitrates are exported from its desert region. Iquique was founded in the 16th century; it was originally a poor shanty village until the 19th-century mining boom. Formerly part of Peru, it was ceded to Chile by treaty in 1883 following the Pacific War. 100 G3

Iquitos river port and capital of Loreto department, in the tropical rainforest of northeast Peru, situated on the west bank of the Amazon River 3,700 km/2,300 mi from its mouth; population (1993) 274,800. There are no connections by road to Iquitos, only by river and air; there is an international airport, and the port is accessible to ocean-going vessels. There are sawmills and an oil refinery. It is a centre of operations for oil exploration, and a tourist centre for the rainforest. Iquitos prospered during the rubber boom (1890–1920). 98 C4

Iráklion (or **Heraklion**), chief commercial port and capital city of Crete, Greece; population (1991) 117,200. There is a ferry link to Peiraias on the mainland. The archaeological museum contains a fine collection of antiquities from the island. 14 G4

Iran country in southwest Asia, bounded north by Armenia, Azerbaijan, the Caspian Sea, and Turkmenistan; east by Afghanistan and Pakistan; south and southwest by the Gulf of Oman and the Gulf; west by Iraq; and northwest by Turkey.

Iran covers an area of 1,648,000 sq km/636,292 sq mi, and has a population of 67,702,000 (2000 est). The capital is ▶Tehran; other major towns and cities include Esfahan, Mashhad, Tabriz, Shiraz, Ahvaz, Kermanshah, Qom, and Karaj. The head of state and government is Ali Akbar Mohtashami (from 2000). The currency is the rial, and the languages spoken include Farsi (official), Kurdish, Turkish, Arabic, English, and French. 56 F3

Iraq country in southwest Asia, bounded north by Turkey, east by Iran, southeast by the Gulf and Kuwait, south by Saudi Arabia, and west by Jordan and Syria.

Iraq covers an area of 434,924 sq km/167,924 sq mi, and has a population of 23,115,000 (2000 est). The capital is ▶Baghdad; other major towns and cities include Mosul, Basra, Kirkuk, Hilla, An Najaf, An Nasiriya, Sulamaniya, and Irbil. The head of state and government is Saddam Hussein (from 1979). The currency is the Iraqi dinar, and the languages spoken include Arabic (official), Kurdish, Assyrian, and Armenian. 56 D3

Irbil (or **Arbīl**), Kurdish capital city, in the governorate of Irbil, in northern Iraq; population (1998 est) 659,300. Occupied since Assyrian times, it was the site of a battle in 331 BC at which Alexander the Great defeated the Persians under Darius III. The city lies near Mosul and is one of the oldest settlements in the world, having been founded before 2300 BC by the Sumerians, under the name Urbillum. Its position on the caravan route between Baghdad and Mosul helped it to become, as it still is, an important centre of commerce. In 1974 Irbil became the capital of a Kurdish autonomous region set up by the Iraqi government. It was captured by the Kurdish Democratic Party in 1996 with the help of Saddam Hussein. 58 K5

Ireland an island lying to the west of Great Britain between the Atlantic Ocean and the Irish Sea. It comprises the provinces of Ulster, Leinster, Munster, and Connacht, and is divided into the Republic of ▶Ireland (which occupies the south, centre, and northwest of the island) and ▶Northern Ireland (which occupies the northeastern corner and forms part of the United Kingdom). 24 E7

Ireland, Northern see ▶Northern Ireland. 24 E7

Ireland, Republic of country occupying the main part of the island of Ireland, in northwest Europe. It is bounded to the east by the Irish Sea, south and west by the Atlantic Ocean, and northeast by Northern Ireland.

Ireland covers an area of 70,282 sq km/27,135 sq mi, and has a population of 3,730,000 (2000 est). The capital is ▶Dublin; other major towns and cities include Cork, Limerick, Galway, Waterford, Dundalk, and Bray. The head of state is Mary McAleese (from 1997); the head of government is Bertie Ahern (from 1997). The currency is the Irish pound, or punt Eireannach, and the languages spoken include Irish Gaelic and English (both official). 24 D8

Irian Jaya western portion of the island of New Guinea, disputed province of Indonesia; area 420,000 sq km/162,000 sq mi; population (1995 est) 1,794,000. The capital is Jayapura. The main industries include copper (with the largest concentration of copper in the world at Tembagapura), palm oil, copra, maize, groundnuts, pepper, tuna, gold, oil, coal, and phosphates. It is mostly a mountainous and forested region, with the Pegunungan Maoke range rising to 5,029 m/16,499 ft at Jaya Peak. The population comprises Melanesians (original settlers of Western New Guinea), Papuans, Negritos, and Europeans. Indigenous animism prevails. The province declared independence from Indonesia, as West Papua, in June 2000. However, the president of Indonesia stated that the declaration was unrepresentative of true feeling in the province. 53 E3

Irish Sea arm of the North Atlantic Ocean separating England and Wales from Ireland; area 103,600 sq km/ 39,990 sq mi. Its greatest width, between Morecambe Bay, Lancashire, and Dundalk Bay, Louth, is 240 km/ 150 mi. It joins the Atlantic to the south by St George's Channel and to the north by the North Channel. 24 G8

Irkutsk city in southern Siberian Russia, capital of Irkutsk oblast (region); population (1990) 635,000. Irkutsk is situated near Lake Baikal on the River Angara; there is a large hydroelectric station near the city. Coal is mined here, while manufactured goods include iron, steel, motor vehicles, and machine tools. Its industrial development dates from the arrival of the Trans-Siberian railway in 1898. 44 G6

Irrawaddy (Myanmar **Ayeryarwady**), chief river of Myanmar (Burma), flowing roughly north–south for 2,090 km/1,300 mi across the centre of the country into the Bay of Bengal. Its sources are the Mali and N'mai rivers; its chief tributaries are the Chindwin and Shweli. 54 F5

Irvine administrative headquarters of ▶North Ayrshire, western Scotland, designated a 'new town' in 1966; population (1991) 33,000. Industries include the manufacture of electronics and pharmaceuticals. It is the birthplace of the novelist John Galt, and the poet James Montgomery. 24 H6

Isar river in Austria and Germany, a tributary of the ▶Danube; length 364 km/226 mi. It rises in the Austrian Tyrol and flows north through Bavaria and its capital Munich to join the Danube near Deggendorf. 30 G3

Ischia volcanic island about 26 km/ 16 mi southwest of Naples, Italy, in the Tyrrhenian Sea; population (2000 est) 50,000. It has mineral springs (known to the Romans), beautiful scenery, and is a holiday resort. The chief town is also Ischia; population (1991) 16,000. 32 H8

Ischl (or **Bad Ischl**), Austrian town in the province of Upper Austria, at the confluence of the rivers Traun and Ischl; population (1995) 14,800. Ischl is the main town of the Salzkammergut. It has saline springs, and is a tourist and health resort. 30 J3

Ise (formerly **Uji-Yamada** until 1955), city in Mie prefecture, Honshu island, Japan, southeast of Kyoto; population (1990 est) 104,200. The city is the market for an agricultural area noted for tea and oranges. Industries include the manufacture of electrical appliances, cotton, textiles, umbrellas, paper goods, and lacquer ware. The city also has a number of shipyards. It is the site of the most sacred Shinto shrine, the **Ise Jingu**, which together with nearby Ise-Shima National Park, attracts many visitors each year. 48 J6

Isère *département* in the ▶Rhône-Alpes region of France, lying east of the River Rhône; area 7,474 sq km/2,886 sq mi; population (1990) 1,016,700. The River ▶Isère and its tributaries the Drac and Romanche flow through it. Some wheat, potatoes, tobacco, and hemp are grown and livestock is reared. Wine is produced in the Isère valley. Iron and lead are found and hydroelectric power is well developed. There are metallurgical, chemical, textile, and paper industries. The administrative centre is ▶Grenoble. Other towns include La Tour-du-Pin and Vienne. 6 F4

Isère river in southeast France, 290 km/180 mi long, a tributary of the Rhône. It gives its name to the *département* of ▶Isère. 30 B5

Iserlohn city in the *Land* of North Rhine–Westphalia, Germany, 20 km/ 12 mi east of Hagen; population (1994) 98,600. In the Middle Ages armour was manufactured here. Textiles, steelmaking, and electronic goods are now the main industries. 22 K3

Iskenderun port, naval base, and steel-manufacturing town in Turkey; population (1990) 176,000. It was founded by Alexander the Great 333 BC and called **Alexandretta** until 1939. 58 G5

Islamabad capital of Pakistan from 1967 (replacing Karachi), in the Potwar district, at the foot of the Margala Hills and immediately northwest of Rawalpindi; population (1998 est) 524,500. The city was designed by Constantinos Doxiadis in the 1960s. Landmarks include the Shahrazad Hotel and national Assembly Building. The Federal Capital Territory of Islamabad has an area of 907 sq km/350 sq mi and a population (1998 est) of 799,000. Islamabad is the centre of an agricultural region in the Vale of Kashmir. 58 G5

Islas Baleares Spanish name for the ▶Balearic Islands, a group of Mediterranean islands forming an autonomous region of Spain. 28 N5

Islas Canarias Spanish for the ▶Canary Islands, a group of volcanic islands and autonomous Spanish community off the northwest coast of Africa. 62 A3

Islas Juan Fernández Spanish name for ▶Juan Fernández Islands, three small volcanic islands belonging to Chile, west of Valparaiso. 100 E5

Islay southernmost island of the Inner ▶Hebrides, on the west coast of Scotland, in Argyll and Bute, separated from Jura by the Sound of Islay; area 610 sq km/235 sq mi; population (1991) 3,500. The principal towns are Bowmore, Port Charlotte, and Port Ellen. It is renowned for its malt whisky and its wildlife, which includes eagles and rare wintering geese. 24 F6

Isle of Man see ▶Man, Isle of. 24 H7/5 D4

Isle of Wight see ▶Wight, Isle of. 24 L11/5 F6

Isma'iliya city in northeast Egypt; population (1994) 400,000. It was founded in 1863 as the headquarters for construction of the Suez Canal and was named after the Khedive Ismail. 64 F1

Israel ancient kingdom of north ▶Palestine, formed after the death of Solomon by Jewish peoples seceding from the rule of his son Rehoboam and electing Jeroboam as their leader. 60 B5

Israel country in southwest Asia, bounded north by Lebanon, east by Syria and Jordan, south by the Gulf of Aqaba, and west by Egypt and the Mediterranean Sea.

Israel covers an area of 20,800 sq km/8,030 sq mi (as at 1949 armistice), and has a population of 6,217,000 (2000 est). The capital is ▶Jerusalem (not recognized by the United Nations); other major towns and cities include Tel Aviv-Yafo,

Haifa, Bat-Yam, Holon, Ramat Gan, Petach Tikva, Rishon Leziyyon, and Beersheba. The head of state is Moshe Katsav (from 2000); the head of government is Ariel Sharon (from 2001). The currency is the shekel, and the languages spoken include Hebrew, Arabic (both official), English, Yiddish, and other European and west Asian languages. **60** B5

Istanbul city and chief seaport of Turkey; population (1990) urban area 6,407,200; city 6,293,400. It produces textiles, tobacco, cement, glass, and leather. Founded as **Byzantium** in about 660 BC, it was renamed **Constantinople** (AD 330) and was the capital of the Byzantine Empire until captured by the Turks in 1453. As **Istamboul** it was capital of the Ottoman Empire until 1922. **58** C3

Istanbul Boğazi Turkish name for the ▶Bosporus, a strait joining the Black Sea with the Sea of Marmar. **36** M3

Italian Somaliland former Italian trust territory on the Somali coast of Africa extending to 502,300 sq km/ 193,900 sq mi. Established in 1892, it was extended in 1925 with the acquisition of Jubaland from Kenya; administered from Mogadishu; and under British rule 1941–50. Thereafter it reverted to Italian authority before uniting with British Somaliland in 1960 to form the independent state of Somalia.

Italy country in southern Europe, bounded north by Switzerland and Austria, east by Slovenia, Croatia, and the Adriatic Sea, south by the Ionian and Mediterranean seas, and west by the Tyrrhenian and Ligurian seas and France. It includes the Mediterranean islands of Sardinia and Sicily.

Italy covers an area of 301,300 sq km/116,331 sq mi, and has a population of 57,298,000 (2000 est). The capital is ▶Rome; other major towns and cities include Milan, Naples, Turin, Palermo, Genoa, Bologna, and Florence. The head of state is Carlo Azeglio Ciampi (from 1999); the head of government is Giuliano Amato (from 2000). The currency is the lira, and the languages spoken include Italian (official), German and Ladin (in the north), French (in the Valle d'Aosta region), and Greek and Albanian (in the south). **32** E4

Ithaca (Greek **Ithaki**), Greek island in the Ionian Sea, area 93 sq km/ 36 sq mi; population (1996 est) 3,700. Important in pre-classical Greece, Ithaca was (in Homer's poem) the birthplace of Odysseus, though this is sometimes identified with the island of Lefkada (some archaeologists have equated ancient Ithaca with Lefkada rather than modern Ithaca). **36** C6

Ithaki Greek name for ▶Ithaca, a Greek island in the Ionian sea. **36** C6

Itzehoe seaport in Schleswig-Holstein, Germany, 60 km/37 mi northwest of Hamburg; population (1994) 34,200. Cement, pumps, nets, and paper are manufactured in the city. There is an abbey (13th–17th centuries) and a baroque church. **20** E3

Ivanovo city in western Russia and capital of Ivanovo oblast (region), population (1990) 482,000. Ivanovo, which lies on the River Uvod, a tributary of the Klyazma, some 240 km/150 mi northeast of Moscow, is renowned for its textile manufacturing. Other industries include chemical production and engineering. **38** H3

Ivory Coast former name (to 1986) of ▶Côte d'Ivoire. **68** C3

Iwo Jima (or **Iojima**), largest of the three Japanese Volcano Islands in the western Pacific Ocean, 1,222 km/ 760 mi south of Tokyo; area 22 sq km/ 9 sq mi. Annexed by Japan in 1891, Iwo Jima, also known as Naka Iojima, was captured by the USA in 1945 after fierce fighting. It was returned to Japan in 1968.

Izhevsk (formerly **Ustinov** 1985–91), capital city, economic and cultural centre of the Republic of Udmurt, west-central Russian Federation, 275 km/171 mi northeast of Kazan; population (1996 est) 655,000. Izhevsk is situated on the River Izh, and is an important centre for engineering (motorcycles, small arms), and metal production. A car factory began production here in 1970. **38** K3

Izmir (formerly **Smyrna**), port and naval base in Turkey; population (1990) 1,757,400. Products include steel, electronics, and plastics. The largest annual trade fair in the Middle East is held here. It is the headquarters of North Atlantic Treaty Organization SE Command. **58** B4

Jabalpur industrial city near the Narmada River in Madhya Pradesh, India, 250 km/175 mi east of Bhopal; population (1991) 742,000. It is an important rail junction, centrally positioned to link Mumbai (formerly Bombay), Kolkata (formerly Calcutta) and Delhi. Industries include textiles, cement, oil, bauxite, and armaments. **54** C4

Jablonec town in the Czech Republic, on the River Neisse, north-east of Prague; population (1991) 45,900. It has had a glass industry since the 14th century. **18** E7

Jackson largest city and state capital of ▶Mississippi, USA, on the Pearl River, in the central part of the state, 70 km/43 mi east of Vicksburg; seat of Hinds County; population (1994 est) 193,000. It produces electrical machinery, furniture, cottonseed oil, and iron and steel castings, and owes its prosperity to the discovery of gas fields to the south in the 1930s. **90** C3

Jacksonville city and port in north-eastern Florida, USA; population (1994 est) 665,000. It is one of the chief southern commercial centres on the Atlantic coast, with extensive rail, air, and highway connections; it is also a port of entry and a tourist resort. Manufactured goods include wood and paper products, ships, chemicals, cigars, and processed food. The port, situated on St John's River, has naval installations and ship-repair yards. To the north the Cross-Florida Barge Canal links the Atlantic with the Gulf of Mexico. Jacksonville dates from 1816, and was incorporated as a city in 1832. **90** E3

Jacobabad city in Sind province, southeast Pakistan, 400 km/250 mi northeast of Karachi; population (1981) 80,000. Founded by Gen John Jacob as a frontier post, the city now trades in wheat, rice, and millet. It has a low annual rainfall (about 5 cm/2 in) and temperatures are among the highest in the Indian subcontinent – up to 53°C/127°F. **54** A3

Jaén capital of Jaén province in Andalusia, southern Spain, on the River Guadalbullon; population (1991) 101,900. Textiles, chemicals, olive oil, and leather goods are produced here. It has a Renaissance cathedral, and parts of the Moorish walls and citadel remain. Jaén was captured from the Moors in 1246. 98 B5

Jaffa (Arabic **Yafa**; Hebrew **Yafo**; biblical **Joppa**), city and former port in west Israel, part of ►Tel Aviv-Yafo from 1950. It is also a tourist centre.

Jaffna capital of Jaffna district, Northern Province, Sri Lanka; population (1990) 129,000. It was the focal point of Hindu Tamil nationalism and the scene of recurring riots during the 1980s. 54 D7

Jaipur capital of Rajasthan, India, 240 km/150 mi southeast of Delhi; population (1991) 1,458,000. Products include textiles and metal products.Founded by Jai Singh II in 1728, it was formerly the capital of the state of Jaipur, which was merged with Rajasthan in 1949. 54 C3

Jajce town in Bosnia-Herzegovina, at the confluence of the rivers Pliva and Vrbas; population (1991) 44,900. There are leather and textile industries and hydroelectric installations nearby. In the 15th century it was the capital of the kingdom of Bosnia, and here the last king was put to death by the Turks in 1463. 34 E5

Jakarta (or **Djakarta**; formerly **Batavia** until 1949), capital of Indonesia on the northwest coast of Java; population (1993) 9,000,000. Industries include textiles, chemicals, and plastics; a canal links it with its port of Tanjung Priok where rubber, oil, tin, coffee, tea, and palm oil are among its exports; also a tourist centre. Respiratory-tract infections caused by air pollution account for 12.6% of deaths annually. Jakarta was founded by Dutch traders in 1619. 52 D4

Jalalabad capital of Nangarhar province, east Afghanistan, on the road from Kabul to Peshawar in Pakistan; population (2001 est) 94,000. The town stands on the Kabul River, at a height of 590 m/1,940 ft, and lies on the route connecting Kabul and Peshawar via the Khyber Pass. Jalalabad is well-placed to handle much of the trade between Afghanistan to the west and Pakistan and India to the east, and is the commercial centre for the irrigated plain around it. It trades in almonds, rice, grain, and fruit, while industries include sugar refining and handicrafts. 54 B2

Jamaica island in the Caribbean Sea, south of Cuba and west of Haiti. Jamaica covers an area of 10,957 sq km/4,230 sq mi, and has a population of 2,583,000 (2000 est). The capital is ►Kingston; other major towns and cities include Montego Bay, Spanish Town, St Andrew, Portmore, and May Pen. The head of state is Queen Elizabeth II (from 1962), represented by Governor General Howard Felix Hanlan Cooke (from 1991); the head of government is Percival Patterson (from 1992). The currency is the Jamaican dollar, and the languages spoken include English (official) and Jamaican Creole. 94 H5

Jambol Bulgarian name for ►Yambol, a town in Bulgaria. 34 P7

Jammu winter capital of the state of ►Jammu and Kashmir, India, 130 km/80 mi north of Amritsar; population (1991) 206,000. It stands on the River Tavi, a tributary of the Chenab, and was linked to India's rail system in 1972. 54 B2

Jammu and Kashmir state of north India; area 222,200 sq km/85,791 sq mi; population (1994) 8,435,000 (Indian-occupied territory). The main cities are ►Jammu (winter capital), ►Srinagar (summer capital and the seat of state government), and Leh. 10 C2

Jamnagar city in Gujarat, India, on the Gulf of Kutch, 270 km/170 mi southwest of Ahmadabad; population (1991) 342,000. It produces cement and textiles, and there are salt pans and pearl fisheries – until the 20th century among the largest in the world. Its port is at Bedi. 54 B4

Jamshedpur industrial city in Bihar, India, 220 km/135 west of Kolkata (formerly Calcutta) at the junction of the Kharkhai and Subarnarekha rivers; population (1991) 461,000. It was built in 1909 near the Jharia coalfield and rich deposits of iron ore, and takes its name from the industrialist Jamsheedji Tata who founded the Tata iron and steel works here and in Mumbai (formerly Bombay). The iron and steel industry has been supplemented by vehicle assembly and production of engine parts, machine tools, railway equipment, and farm machinery. Jamshedpur is the home of the National Metallurgical Laboratory. 54 E4

Janesville city and administrative headquarters of Rock County, south Wisconsin, USA, on the Rock River, about 100 km/62 mi southwest of Milwaukee; population (1992) 54,300. Situated in an agricultural area where wheat, barley and tobacco are grown and dairy farming takes place, Janesville is a processing and marketing centre for the area's agricultural products. The city also has a wide range of manufactured products, including cars and car parts, trucks, building materials, pens, pencils, textiles, and electronic equipment. 88 C2

Jan Mayen Norwegian volcanic island in the Arctic Ocean, between Greenland and Norway, about 480 km/298 mi north of Iceland; area 380 sq km/147 sq mi. Its radio transmitter station (built in the 1950s) is important for shipping and air traffic. The island is named after the Dutchman Jan May, who landed in 1614 and claimed it as Dutch territory. It was annexed by Norway in 1929. 80 C2

Japan country in northeast Asia, occupying a group of islands of which the four main ones are Hokkaido, Honshu, Kyushu, and Shikoku. Japan is situated between the Sea of Japan (to the west) and the north Pacific (to the east), east of North and South Korea.

Japan covers an area of 377,535 sq km/145,766 sq mi, and has a population of 126,714,000 (2000 est). The capital is ►Tokyo; other major towns and cities include Yokohama, Osaka, Nagoya, Fukuoka, Kitakyushu, Kyoto, Sapporo, Kobe, Kawasaki, and Hiroshima. The head of state is Emperor Akihito (from 1989); the head of government is Junichiro Koizumi (from 2001). The currency is the yen, and the languages spoken include Japanese (official) and Ainu. 48 L5

Japan, Sea of sea separating Japan from the mainland of Asia, connected to the Pacific Ocean by shallow straits; area 1,007,765 sq km/389,100 sq mi; average depth 1,360 m/4,460 ft. It is almost tideless. Currents flow northwards from the Korea Strait, and there is floating ice in winter. North and South Korea, and Eastern Siberia lie on its western shores; Vladivostok is the chief port. 48 G3

Jarosław town in Przemyśl province, Poland, on the River San, 48 km/30 mi east of Rzeszów; population (1990) 41,200. Industries include building materials, flour, confectionery, and timber. The town was founded in the 11th century by Jaroslaw, Grand Duke of Kiev, and was part of Austria 1772–1919. 18 M7

Jassy German name for the Romanian city of ►Iaşi. 34 Q2

Java (or **Jawa**), most populated island of Indonesia, situated between Sumatra and Bali; area (with the island of Madura) 132,000 sq km/51,000 sq mi; population (with Madura; 1990) 107,581,300. The capital is ►Jakarta (which is also the capital of Indonesia). About half the island is under cultivation, the rest being thickly forested. Mountains and sea breezes keep the temperature down, but humidity is high, with heavy rainfall from December to March. Ports include Surabaya and Semarang. 52 E4

Jawa alternative spelling of ►Java, the most populated island of Indonesia. 52 E4

Jedburgh small town in the Scottish Borders unitary authority, Scotland, on **Jed Water**, 77 km/48 mi south of Edinburgh; population (1991) 4,100. It has the remains of a 12th-century abbey. Jedburgh is a woollen manufacturing centre. 24 K6

Jedda alternative spelling for the Saudi Arabian port ►Jiddah. 56 C5

Jefferson City state capital of ►Missouri, USA, located in the central part of the state, 167 km/104 mi west of St Louis, on the Missouri River; seat of Cole County; population (1992) 37,100. It is the centre of an agricultural region which produces grain and livestock, and is situated in an area with lead and zinc deposits. Industries include agricultural products, fabricated metals, shoes, electrical appliances, and cosmetics. 88 B3

Jehol former name for the city of ►Chengde in northeast Hebei province, north China. 46 F2

Jelenia Góra (German **Hirschberg**), capital of Jelenia Góra province, Poland, on the River Bóbr at the foot of the Karkonosze mountains, 96 km/60 mi west of Wrocław; population (1990) 93,400. Textiles, chemicals, and optical lenses are manufactured. The town dates from the 14th century, passing from Bohemia to Prussia 1741, and to Poland 1945. 18 E7

Jelgava (or **Jalgava**; German **Mitau**), town in western Latvia, an inland port on the Lielupe River, 48 km/30 mi southwest of Riga; population (1996) 70,900. Industries include textile manufacture and sugar-refining. The town was founded in 1265 by the Teutonic Knights, and became the capital of the former duchy of Courland in 1561. 16 M8

Jena town in the *Land* (administrative region) of Thuringia, Germany, southeast of Weimar; population (1995) 101,800. It is noted for its precision engineering; the Zeiss firm have manufactured optical instruments since 1846. French emperor Napoleon defeated the Prussians here in 1806, and the philosophers Schiller and Hegel taught at the Friedrich Schiller University, which dates from 1558. 20 G6

Jerez de la Frontera city in the province of Cádiz in Andalusia, southwest Spain; population (1991) 182,900. It is famed for sherry, the fortified wine to which it gave its name. The numerous wine warehouses (bodegas) are a feature of the city, and there are Moorish remains here. 28 D8

Jericho town in the Jordan valley, west of the River Jordan and north of the Dead Sea, 24 km/15 mi northeast of Jerusalem; population (1995 est) 25,000. The site of the old city is the centre of a fertile district where palms, rose trees, grapes, and balsams are grown. It was occupied by Israel from 1967–94 when responsibility for its administration was transferred to the Palestine National Authority, after the Israeli–Palestine Liberation Organization peace agreement in 1993. Jericho was settled by 8000 BC, and by 6000 BC had become a walled city with 2,000 inhabitants. It is claimed to be the site of the world's earliest known town, dating from around 9000 BC. Successive archaeological excavations since 1907 have shown 20 layers of settlement, and that the walls of the city were destroyed many times. In the Old Testament it was the first Canaanite stronghold captured by the Israelites, and its walls, according to the Book of Joshua, fell to the blast of Joshua's trumpets. 60 C5

Jersey largest of the ►Channel Islands; capital St Helier; area 117 sq km/45 sq mi; population (1991 est) 85,200. It is governed by a lieutenant governor, representing the English crown and an assembly. Jersey cattle were originally bred here. Jersey gave its name to a woollen garment 26 C4

Jersey City city in northeastern New Jersey, USA, on a peninsula bordered by the Hudson and Hackensack rivers to the north, and by New York and Newark bays to the south; seat of Hudson County; population (1994 est) 226,000. Jersey City is separated from New York by 2 km/1.2 mi of river, and connected with Manhattan Island via the Hudson River tunnels. A former port, it is now an industrial centre; products include pharmaceuticals, steel and metal products, electrical equipment, and chemicals. 90 G1

Jerusalem (Arabic **al-Quds**; Hebrew **Yerushalayim**), ancient city of Palestine, 762 m/2,500 ft above sea level, situated in hills 55 km/34 mi from the Mediterranean, divided in 1948 between Jordan and the new republic of Israel; area (pre-1967) 37.5 sq km/14.5 sq mi, (post-1967) 108 sq km/42 sq mi, including areas of the West Bank; population (1995) 591,400. In 1950 the western New City was proclaimed as the Israeli capital, and, having captured from Jordan the eastern Old City in 1967, Israel affirmed in 1980 that the united city was the country's capital; the United Nations does not recognize East Jerusalem as part of Israel, and regards Tel Aviv as the capital. 60 C5

Jervis Bay deep bay on the coast of New South Wales, Australia, 145 km/90 mi southwest of Sydney. The federal government in 1915 acquired 73 sq km/28 sq mi here to create a port for ►Canberra. It forms part of the Australian Capital Territory and is the site of the Royal Australian Naval College.

Jhansi city in Uttar Pradesh, India, 286 km/178 mi southwest of Lucknow; population (1991) 313,000. It is a railway and road junction; its economy is based on agricultural trading and steel. Its fort was built by the Moguls in 1613, and it was a Maratha capital from 1770 to 1853.

Jhansi was the scene of a massacre of British civilians in 1857 during the Indian Mutiny. **54** C3

Jharkhand state of northeast India, bordered by Bihar, West Bengal, Orissa, Chattisgarh, and Uttar Pradesh; area 74,677 sq km/28,833 sq mi; population (2001 est) 27,894,000. It was carved from Bihar and was incorporated in November 2000. The capital is Ranchi. Jharkhand accounts for nearly half of India's mineral wealth. Minerals extracted include bauxite, limestone, mica, coal, iron, and copper ore. Industries include the Tata Iron and Steel Company and the Bokaro Steel Plant. The area is home to 30 tribal groups who are estimated to form one-third of the population. The principal languages spoken are Santali, Kurukh, Mundari, and Nagpuria. **10** D4

Jiangsu (or **Kiangsu**), province on the coast of east China, bounded to the north by Shandong, to the east by the Yellow Sea, to the southeast by Shanghai, to the south by Zhejiang, and to the west by Anhui; area 102,200 sq km/39,450 sq mi; population (1996) 71,100,000, the most densely populated province in China. The capital is ►Nanjing. **11** (2)E3

Jiangxi (or **Kiangsi**), province of southeast China, bounded to the north by Hubei and Anhui, to the east by Zhejiang and Fujian, to the south by Guangdong, and to the west by Hunan; area 164,800 sq km/63,600 sq mi; population (1996) 41,050,000. The capital is ►Nanchang; other cities and towns include Ganzhou, Ji'an, Jingdezhen, Jiujiang, and Pingxiang. The Chang Jiang River and Lake Poyang are found in the province. The main industries are porcelain, coal, tungsten, copper, and uranium; agricultural products include rice, tea, cotton, tobacco, and timber. **11** (2)E4

Jibuti variant spelling of ►Djibouti, a republic of northeast Africa. **64** H5

Jiddah (or **Jedda**), port in Hejaz, Saudi Arabia, on the eastern shore of the Red Sea, about 80 km/50 mi west of Mecca; population (1991 est) 1,500,000. The country's leading industrial centre, its industries include cement, steel, and oil refining. It exports hides, mother-of-pearl, coffee, and carpets. Pilgrims pass through here on their way to Mecca. **56** C5

Jihlava (German **Iglau**), town in eastern Bohemia, in the Czech Republic, on the River Jihlava, 140 km/87 mi southeast of Prague; population (1991) 52,300. Trade is mainly agricultural, and industries include textiles and engineering. Silver deposits were mined from the 8th century; the town being prosperous as a royal mint in the Middle Ages. There are many historic buildings. **18** E8

Jilin (or **Kirin**), province of northeast China, bounded to the northeast by Heilongjiang, to the southeast by Russia, to the south by North Korea, to the southwest by Liaoning, and to the northwest by Inner Mongolia; area 187,000 sq km/72,000 sq mi; population (1996) 26,100,000. The capital is ►Changchun, and other cities and towns include Jilin, Tonghua, Baicheng, and Liaoyuan. Major industries include coal, iron ore, engineering, food-processing, and chemicals; agricultural products include maize, sorghum, soybeans, and millet. **11** (2)F2

Ji'nan (**Tsinan** or **Chinan**), city and capital of ►Shandong province, China; population (1993) 2,050,000. Industries include engineering, food-processing, flour-milling, and the manufacture of textiles, iron, steel, heavy goods vehicles, machine tools, and chemicals. **46** F3

Jingdezhen (or **Chingtechen**, or **Fouliang**), town in Jiangxi, China; population (1994) 377,500. Ming blue-and-white china was produced here; the clay kaolin is named after Kaoling, a hill east of Jingdezhen. Some of the best Chinese porcelain is still made here. **46** F5

Jinja town in Busoga Province, Uganda, on the north shore of Lake Victoria, east of Kampala; population (1991) 61,000. Industries include steel milling, engineering, textiles, grain milling, food processing, and brewing. Nearby is the Owen Falls Dam (1954). **70** E3

Jinsha Jiang river rising in southwest China and forming the ►Chang Jiang (Yangtze Kiang) at Yibin. **46** C5

Jodhpur city in Rajasthan, India, 490 km/310 mi southwest of Delhi; population (1991) 668,000. Handicraft industries such as ivory carving and lacquerware are important, and railway parts, textiles, and bicycles are manufactured here. It is a market centre. It was formerly capital of Jodhpur princely state, founded by Rao Jodha on the edge of the ►Thar Desert. A style of riding breeches is named after the town. **54** B3

Jogjakarta alternative spelling of ►Yogyakarta, a city in Indonesia. **52** E4

Johannesburg largest city of South Africa, situated on the Witwatersrand in Gauteng Province; population (city area, 1998) 849,600 (urban area, 1996) 2,200,000. It is the centre of a large gold-mining industry; other industries include engineering works, meat-chilling plants, and clothing factories. **72** D5

John o' Groats village in the northeast of the Highland unitary authority, Scotland, about 3 km/2 mi west of Duncansby Head, Britain's northernmost point. It is the furthest point from Land's End on the British mainland. **24** J3

Johnson City city in Washington, Carter, and Sullivan counties, northeast Tennessee, just south of the Virginia border, in the Appalachian Mountains, 146 km/91 mi northeast of Knoxville; population (1992) 50,400. It is the centre of an agricultural region; industries include tobacco, furniture, electronic components, building materials, metals, textiles, and food processing. Johnson City is the home of Milligan College (1866) and East Tennessee State University (1911). **88** D3

Johnston Atoll coral island in the mid-Pacific, lying between the ►Marshall Islands and Hawaii; area 2.8 sq km/1.1 sq mi. The island is only 2.4 m/8 ft above sea level and subject to hurricanes and tidal waves. It has the status of a National Wildlife Refuge but was contaminated by fallout from nuclear-weapons testing in 1962, and has since 1971 been used as a repository for chemical weapons left over from the Korean and Vietnam wars. An unincorporated territory of the USA, it is administered by the US Defense Nuclear Agency (DNA).

Johnstown city in southwestern Pennsylvania, USA, at the junction of the Conemaugh River and Stonycreek, 85 km/53 mi east of Pittsburgh; population (1990) 28,100. Industries include chemicals, mining equipment, steel, coal and coal by-products, building materials, and clothing. incorporated as a city in 1889. **88** E2

Johor state in southern Peninsular Malaysia; capital Johor Bahru; area 19,000 sq km/7,000 sq mi; population (1993) 2,106,700. The southernmost point of mainland Asia, it is joined to Singapore by a causeway. It is mainly forested, with swamps. There is bauxite and iron. **11** (1)A2

Joliet city and port in northeastern Illinois, USA, on the Des Plaines River and the Illinois Waterway, 60 km/37 mi southwest of Chicago; seat of Will County; population (1992) 78,900. It is an important transportation and industrial centre; industries include building materials, chemicals, oil refining, heavy construction machinery, and paper. **84** J3

Jönköping town at the south end of Lake Vättern, Sweden; population (1994 est) 114,800. It is an industrial centre in an agricultural and forestry region. **16** H8

Joplin city in Jasper and Newton counties, southwest Missouri, USA, near the Ozark Mountains, 110 km/68 mi west of Springfield; population (1990) 41,000. It is the centre of an agricultural region; industries include chemicals, explosives, hydraulic pumps, zinc and lead smelting, leather goods, and furniture. Joplin is the home of Ozark Christian College (1942) and Missouri Southern State College (1965). **90** C2

Jordan country in southwest Asia, bounded north by Syria, northeast by Iraq, east, southeast, and south by Saudi Arabia, south by the Gulf of Aqaba, and west by Israel.

Jordan covers an area of 89,206 sq km/34,442 sq mi (excluding the West Bank 5,879 sq km/2,269 sq mi), and has a population of 6,669,000 (2000 est). The capital is ▶Amman; other major towns and cities include Zarqa, Irbid, Saet, and Ma'an. The head of state is King Abdullah ibn Hussein (from 1999); the head of government is Ali Abu al-Ragheb (from 2000). The currency is the Jordanian dinar, and the languages spoken include Arabic (official) and English. **60** D5

Jounie port on the Mediterranean coast of Lebanon, 15 km/9 mi north of Beirut. It was the centre of Christian forces during the civil war. **60** C2

Juan Fernández Islands (Spanish **Islas Juan Fernández**), three small volcanic islands belonging to Chile; situated in the Pacific Ocean 640 km/398 mi west of the port of Valparaiso. The few inhabitants fish for lobsters which are sent to the mainland. The archipelago has been designated by UNESCO as a World Biosphere Reserve. The islands include Robinson Crusoe (the largest, formerly Más a Tierra; 93 sq km/36 sq mi, 22 km/14 mi long and 7.3 km/4,4 mi wide), Alejandro Selkirk (formerly Más a Fuera), and Santa Clara (the smallest and also known as Goat Island). Collectively, they were named after the Spanish navigator who discovered them in 1574. **100** E5

Juba capital of Eastern Equatorial State, Sudan, on the left bank of the White Nile, at the head of navigation above Khartoum, 1,200 km/750 mi to the north; population (1993 est) 115,000. Juba is a transport hub with local and international road communications, including road links to Kenya, Uganda, and the Democratic Republic of Congo. There is an airport and also a river port. It is also a commercial centre for the surrounding agricultural area trading locally grown coffee, tobacco, and chillies. It is the leading city of southern Sudan and in civil wars since the 1950s has been the major centre of resistance to the northern, Khartoum-based government of the country. The University of Juba (1975) is here. **70** E3

Jubbulpore alternative name for the city of ▶Jabalpur in India. **54** C4

Jugoslavia alternative spelling of ▶Yugoslavia. **34** H6

Jülich town in North Rhine–Westphalia, Germany, 44 km/27 mi southwest of Düsseldorf; population (1994) 31,800. Paper and sugar industries provide the town's main employment. Jülich, the ancient Juliacum, was once the seat of the dukes of Aix-la-Chapelle (▶Aachen). It was a strong fortress town in the 17th century. **22** J4

Juneau state capital and ice-free port of ▶Alaska, opposite Douglas Island on Gastineau Channel, in the south Alaskan panhandle (the narrow strip of Alaska running down the Pacific coast alongside British Columbia); population (2000 est) 30,700. The city is the commercial and distribution centre for the fur-trading and mining industries of the panhandle region. The government accounts for half the employment, but also important are tourism, salmon fishing, fish processing, and lumbering. Gold-mining remained important until the closure of the last mine in 1944, but in the 1990s limited mining resumed. Juneau was incorporated as a city in 1900. **92** (1)L4

Jungfrau (German 'maiden'), one of three linked peaks (with the Eiger and Monch) in the Bernese Alps, Switzerland; 4,166 m/13,669 ft high. A railway, passing through a tunnel 8 km/5 mi long, ascends to the plateau of the **Jungfraujoch** (3,456 m/11,340 ft) where there is a winter sports centre. **30** C4

Jura island of the Inner ▶Hebrides, Argyll and Bute; area 380 sq km/147 sq mi; population (1991) 196. It is separated from the Scottish mainland by the **Sound of Jura**. The whirlpool Corryvreckan (Gaelic 'Brecan's cauldron') is off the north coast. It has a range of mountains known as the 'Paps of Jura', the highest of which is Beinn an Oir at 784 m/2,572 ft. **24** G5

Jura *département* in the ▶Franche-Comté region of France; area 5,007 sq km/1,933 sq mi; population (1990) 249,100. It is bordered in the east by Switzerland. The greater part of the *département* is occupied by the ▶Jura Mountains, but there is a plain in the west. The chief rivers are the Doubs and the Ain. Some cereals and sugar beet are grown, sheep and horses are raised, and wine is produced. There are metallurgical, paper, furniture, food, and watchmaking industries and hydroelectric schemes. The administrative centre is ▶Lons-le-Saunier. Other towns include Dole and St-Claude. **6** F3

Jura Mountains series of parallel mountain ranges running along the French–Swiss frontier between the Rivers Rhône and Rhine, a distance of 250 km/156 mi. The highest peak is Crête de la Neige (1,723 m/5,650 ft). The mountains give their name to the ▶Jura *département* of France, and in 1979 a Jura canton was established in Switzerland, formed from the French-speaking areas of Berne. **30** B4

Jutland (Danish **Jylland**), peninsula of northern Europe; area 29,500 sq km/11,400 sq mi. It is separated from Norway by the Skagerrak and from Sweden by the Kattegat, with the North Sea to the west. The larger northern part belongs to Denmark, the southern part to Germany. **16** E8

Jylland Danish name for the mainland of Denmark, the northern section of the ►Jutland peninsula. **16** E8

K2 (or **Chogori**), second-highest mountain above sea level, 8,611 m/28,251 ft, in the Karakoram range, in a disputed region of Pakistan. It was first climbed in 1954 by an Italian expedition. **54** C1

Kabardino-Balkaria (or **Kabardino-Balkariya**), republic in the far southwestern Russian Federation, on the border with Georgia; area 12,500 sq km/4,826 sq mi; population (1996) 790,000 (58% urban) (48% Kabarda, 32% Russians, 9% Balkars). The capital is Nalchik, and other cities include Tyrnyauz and Prokhladnyy. The republic is on the northern slopes and foothills of the main Caucasus mountain range and contains the highest Caucasian peaks (Elbrus and Dykh Tau). It is crossed by the Rivers Terek, the Chegem, the Cherek, and the Baksan. Mineral deposits include wolfram, molybdenum, lead, zinc, and coal. **58** K2

Kabinda part of Angola; see ►Cabinda. **68** G6

Kabul capital of Afghanistan, 1,800 m/5,900 ft above sea level, on the **River Kabul**; population (1997 est) 500,000. Products include textiles, plastics, leather, and glass. It commands the strategic routes to Pakistan via the ►Khyber Pass. The city was captured by the Talibaan on 27 September 1996. **56** J3

Kabwe (formerly **Broken Hill**), city in central Zambia, 136 km/84 mi north of Lusaka; population (1998) 210,400. It is the capital of Central Province and there is clear evidence that the area has been inhabited since the period known in Africa as the Middle Stone Age. It is a copper- and cadmium-mining centre; lead and zinc were also mined until June 1994. It is also a major railway centre. **72** D2

Kaduna city in northern Nigeria, on the Kaduna River, capital of Kaduna state; population (1992 est) 309,600. It is a market centre for grain and cotton; industries include textiles, cars, timber, pottery, brewing, munitions, and oil refining. **68** F2

Kafue river in central Zambia, a tributary of the Zambezi, 965 km/600 mi long. The upper reaches of the river form part of the Kafue National Park (1951). Kafue town, 44 km/27 mi south of Lusaka, population (1998 est) 40,800, is the centre of Zambia's heavy industry, which includes engineering and textiles. A hydroelectric power station opened in 1972 on the lower Kafue River at Kafue Gorge; its 600,000-kilowatt generating facility was expanded to 900,000 in the late 1970s by the construction of a storage dam upstream. **72** D3

Kagoshima industrial city and port on Kyushu island, southwest Japan, 150 km/93 mi southeast of Nagasaki; population (1994) 534,000. It is the seat of the prefecture of the same name. Satsumayaki porcelain, starch, caramel, vegetable oil, tobacco, wood crafts, and a silk fabric called *tsumugi* are produced here. Kagoshima is often compared to Naples, its twin city. **48** F8

Kaifeng (or **Pien-ching**), city in Henan province; population (1994) 788,500. Formerly the provincial capital and once a capital of China (907–1127), it lost its importance because of the silting-up of the nearby Huang He River. Industries include zinc-smelting, and the manufacture of fertilizers, textiles, beverages, and agricultural machinery. **46** E4

Kaikouras double range of mountains in the northeast of South Island, New Zealand, separated by the Clarence River, and reaching 2,885 m/9,465 ft.

Kairouan (or **Kairwan**; Arabic **al-Qayrawan**), Muslim holy city in Tunisia, south of Tunis; population (1994) 102,600. Chief products are carpets, leather goods, and copperware. The city, said to have been founded in AD 617, ranks after Mecca and Medina as a place of pilgrimage. **6** H1

Kaiserslautern industrial city in Rhineland-Palatinate, Germany, 48 km/30 mi west of Mannheim; population (1995) 102,100. Cars, machinery, textiles, tobacco, and beer are produced. The city dates from 882, and was named after a castle built by Frederick Barbarossa in 1152, later destroyed by the French in 1703. **20** C7

Kalaallit Nunaat Greenlandic name for ►Greenland. **80** G2

Kalahari Desert arid to semi-arid desert area forming most of Botswana and extending into Namibia, Zimbabwe, and South Africa; area about 900,000 sq km/347,400 sq mi. The only permanent river, the Okavango, flows into a delta in the northwest forming marshes rich in wildlife. **72** C4

Kalamazoo (American Indian, Miami dialect 'boiling pot'), city in southwestern Michigan, USA, on the Kalamazoo River, 96 km/60 mi southwest of Lansing and midway between Detroit and Chicago; seat of Kalamazoo County; population (1992) 81,250. Its industries include the processing of the area's agricultural products, car and transportation machinery parts, chemicals, pharmaceuticals, musical instruments, and metal and paper products. **88** C2

Kalgan ('gate'), Mongolian name for a former border post in Hebei province, northeast China, now known as ►Zhangjiakou. **46** E2

Kalgoorlie (or **Kalgoorlie-Boulder**; Aboriginal 'native dog chasing kangaroo'), mining city in Western Australia, amalgamated with Boulder in 1954; population (1996) 28,087. Kalgoorlie is situated on the western side of the Nullarbor Plain, 596 km/370 mi northeast of Perth. Gold has been mined here since 1893; the region around Kalgoorlie is known as the 'Goldfields' and produces around 65% of Australia's gold. Tourism, based around the city's mining history, is important. **76** D6

Kalimantan province of the republic of Indonesia occupying part of the island of Borneo; area 543,900 sq km/210,000 sq mi; population (urban area, 1995 est) 10,520,500. The land is mostly low-lying, with mountains in the north. The capital is ►Banjarmasin; other towns include ►Balikpapan. Industries include petroleum, rubber, coffee, copra, pepper, and timber. **52** E2

Kalinin former name (1932–91) of ►Tver, a city in Russia. The city was renamed in honour of Soviet president Mikhail Kalinin (1875–1946), but reverted to its original name after the collapse of communism in 1991. **38** G3

Kaliningrad (formerly **Königsberg**), city and port in western Russia; population (1995 est) 926,400. Industries include shipbuilding, fisheries, engineering, and paper manufacture. The port of Kaliningrad remains ice-free throughout the year; as well as being an important commercial centre, it is also the principal base of the Russian Baltic fleet. The city was the capital of East Prussia until this territory was divided between the USSR and Poland in 1945 under the Potsdam Agreement, when it was renamed in honour of Soviet President Mikhail Kalinin (1875–1946). As Königsberg, the city was the birthplace and residence of the German philosopher Immanuel Kant (1724–1804). 18 K3

Kalisz (ancient **Calisia**), industrial town, capital of Kalisz province, Poland, on the River Prosna, 109 km/68 mi southeast of Poznań; population (1990) 106,100. It has been a major textile producer since the Middle Ages. Other industries include metal goods, embroidery, and food processing, and lignite is mined in the district. 18 H6

Kalmar port on the southeast coast of Sweden; population (1992) 56,900. Industries include paper, matches, and the Orrefors glassworks. 16 J8

Kalmyk (or **Kalmykiya**, or **Kalmuck**; Kalmyk **Khal'mg Tangch**), republic in the southwest of the Russian Federation; area 75,900 sq km/29,305 sq mi; population (1996) 319,000 (39% urban) (45% Kalmyks, 38% Russians). The capital is Elista, and Yashkul is another city. The republic is west of the lower Volga and has a short coastline on the northwestern shore of the Caspian Sea. Physical features include dry steppe and semi-desert lowland, and there is a continental climate. Industries include machine building, metalworking, food processing, fish canning. Sheep, pigs, and cattle are reared. 58 M1

Kaluga capital city, economic and cultural centre of Kaluga oblast (region), Russian Federation, on the left bank of the River Oka, 160 km/100 mi southwest of Moscow; population (1996 est) 347,000. Factories in the city produce machine tools, chemicals, bricks, railway equipment, and foodstuffs. 38 G4

Kamakura city of Kanagawa prefecture, Honshu island, Japan, 15 km/9 mi southwest of Yokohama, on Sagami Bay; population (1993 est) 174,300. It is an exclusive commuter area for Tokyo. The mild climate and historic buildings, especially the Buddhist temples, attract tourists. Lacquer ware has been an important craft since the 13th century. Kamakura was the seat of the first shogunate (1192–1333), which established the rule of the samurai class.

Kamchatka Peninsula mountainous region in the Russian Far East, separating the Sea of ►Okhotsk from the Pacific Ocean and the Bering Sea. The Kamchatka Peninsula is over 1,200 km/746 mi long, covers an area of 370,000 sq km/142,857 sq mi, and contains a total of over 160 volcanoes (22 of them active), together with many hot springs and geysers. The region has an extremely severe climate and predominantly tundra vegetation, with forests in sheltered valleys. The Kamchatka Peninsula is home to a huge number of animal and bird species, including the brown bear, sea eagle, and sable. There are coal, sulphur, gold, mica, and other mineral deposits. 44 U6

Kamenz town in the *Land* (administrative region) of Saxony, Germany, on the River Schwarze Elster, 56 km/35 mi northeast of Dresden; population (1994) 17,300. It has textile and glass industries. The dramatist and critic Gotthold Lessing was born here in 1729. 18 D6

Kamet Himalayan mountain 7,756 m/25,447 ft high on the Tibet–India border. 54 C2

Kamloops city in southern British Columbia, Canada, on the Thompson tributary of the Fraser River and the Trans-Canada Highway; population (1991) 67,100. It is the distributing centre of a timber-processing and ranching district, and a junction on the Canadian Pacific Railway. Tourism – including fishing, windsurfing, sailing, and mountain tours – has become important to the local economy, and other industries include food processing, wood-pulp production, and oil refining. It is home to Cariboo Community College. 82 G6

Kampala capital of Uganda, on Lake Victoria; population (1991) 773,500. It is linked by rail with Mombasa. Products include tea, coffee, fruit, and vegetables. Industries include engineering, chemicals, paint manufacture, textiles, footwear, brewing, distilling, and food processing. 70 E3

Kampen town on the River IJssel, in the province of Overijssel, the Netherlands, 14 km/9 mi northwest of Zwolle; population (1996) 32,400. The chief products are machinery, blankets, and cigars. Kampen was a member of the medieval Hanseatic League. 22 H2

Kampuchea former name (1975–89) of ►Cambodia. 50 C4

Kananga (formerly **Luluabourg** to 1966), chief city of Kasai Occidental region, west-central Democratic Republic of Congo, on the Lulua River; population (1994 est) 393,000. Industries include brewing, printing, and the processing of palm oil, rice, cassava, groundnuts, bananas, and pineapples. 70 C5

Kanazawa industrial city on the west coast of Honshu island, in Ishikawa prefecture, Chubu region, Japan, 160 km/100 mi northwest of Nagoya; population (1994) 433,000. Kutani ware ceramics, *yuzen* printed silk, and *maki-e* decorated lacquerware are made in the city, which is the seat of the prefectural government. Kanazawa was a feudal castle town from the 16th century and has a number of old samurai residences. 48 J5

Kanchenjunga (or **Kangchenjunga**), Himalayan mountain on the Nepal–Sikkim border, 8,586 m/28,170 ft high, 120 km/75 mi southeast of Mount Everest. The name means 'five treasure houses of the great snows'. Kanchenjunga was first climbed by a British expedition in 1955. 54 E3

Kandahar city in Afghanistan, 450 km/280 mi southwest of Kabul, capital of Kandahar; population (1995 est) 341,000. A province and a trading centre, with wool and cotton factories, and other industries which include silk and felt. The city is a market for other agricultural products, including wool, tobacco, grains, fresh and dried fruit, and livestock, especially sheep. When Afghanistan became independent in 1747, Kandahar was its first capital. 56 J3

Kandy city in central Sri Lanka, on the Mahaweli River; capital of a district of the same name; population (1995 est) 110,000. It lies 116 km/72 mi northeast of Colombo. It is the focus both of a major tea-growing area and of the Sinhalese Buddhist culture. One of the most sacred Buddhist shrines is situated in Kandy. Called the Dalada Maligawa (Temple of the Tooth), it contains an alleged tooth of the Buddha. 54 D7

Kangchenjunga variant spelling of ►Kanchenjunga, a Himalayan mountain. 54 E3

Kankakee city and administrative headquarters of Kankakee County, northeast Illinois, USA, on the Kankakee River, 50 mi/81 km south of Chicago; population (1990) 27,600. It is a distribution center for a rich corn-growing region; industries include building materials, furniture, pharmaceuticals, farm machinery, floor tiles, and pet food. Kankakee is the home of Olivet Nazarene College (1907) and Kankakee Community College (1966). 88 D1

Kano capital of Kano state in northern Nigeria, trade centre of an irrigated area; population (1992 est) 699,900. Kano is a major centre for the groundnut and cattle trade. Products include bicycles, glass, furniture, textiles, chemicals, flour, vegetable oil, and cereals. Kano is a tourist centre with trade in leather, brass, cloth, silverware, and beads. Founded about 1000 BC, Kano is a walled city, with New Kano extending beyond the walls. 68 F2

Kanpur (formerly **Cawnpore**), commercial and industrial city and capital of Kanpur district, Uttar Pradesh, India, 65 km/40 mi southwest of Lucknow, on the River Ganges; population (1991) 2,111,300. Industries include cotton, wool, jute, chemicals, plastics, iron, and steel. It has benefited from its rail links with Kolkata (formerly Calcutta), particularly during the growth of its cotton industry in the last century. 54 D3

Kansas state in central USA. It is nicknamed the Sunflower State. Kansas was admitted to the Union in 1861 as the 34th US state. It is considered both a part of the US Midwest and as one of the Great Plains states. It is bordered to the south by Oklahoma, to the west by Colorado, to the north by Nebraska, and to the east by Missouri.
population (2000 est) 2,688,400
area 213,200 sq km/82,296 sq mi
capital ►Topeka
towns and cities ►Kansas City, Wichita, Overland Park
industries and products wheat, corn, sorghum, sunflowers, beef and dairy cattle, coal, petroleum, natural gas, lead, zinc, aircraft, minerals 90 A2

Kansas City city and administrative headquarters of Wyandotte County, northeast Kansa, USA, at the confluence of the Kansas and Missouri rivers; population (1994 est) 143,000. It is adjacent to ►Kansas City, Missouri. The city is an agricultural market with stockyards, meatpacking plants, grain elevators, and flour mills. Other industries include oil refining, auto assembly, and the manufacture of aircraft engines, furniture, clothing, steel and aluminium products, chemicals, soap, and farm machinery. Among its educational institutions are the University of Kansas Medical Center, Kansas City Kansas Community College (1923), and Donnelly College (1949). 90 C2

Kansas City city in western Missouri, USA, at the confluence of the Kansas and Missouri rivers, adjacent to ►Kansas City, Kansas, and sometimes regarded as one of twin cities; population (1994 est) 444,000. It is the financial, marketing, and distribution centre of the region; industries include metal production, electronics products, motor-vehicle assembly, industrial machinery, glass and chemical products, oil refining, and food processing. 90 C2

Kaohsiung city and port on the west coast of Taiwan; population (1992) 1,396,400. Industries include aluminium ware, fertilizers, cement, oil refineries, iron and steel works, shipyards, and food processing. Kaohsiung began to develop as a commercial port after 1858; its industrial development came about while it was occupied by Japan 1895–1945. 46 G6

Kapfenberg town in the southern Austrian province of Styria, northeast of Bruck an der Mur; population (1991) 23,400. A fortified town since the 12th century, it has had an ironworks since the 15th century. Kapfenberg has important steelworks, and also manufactures materials, chemicals, and electronic equipment. 30 L3

Kaposvár capital of Somogy county, southwest Hungary, on the River Kapos, 188 km/117 mi southwest of Budapest; population (1993) 71,100. Industries include engineering, textiles, and food processing. 34 E3

Karachay-Cherkessia (or **Karachay-Cherkess Republic**, or **Karachayevo-Cherkesiya**), republic in the southwestern Russian Federation, part of Stavropol territory; area 14,100 sq km/5,444 sq mi; population (1996) 436,000 (46% urban) (42% Russian, 31% Karachay). Cherkessk is the capital, and Karachayevsk is another city. The republic is located on the northern slopes of the Caucasus Mountains, with lowland steppe in the north and forested foothills in the south. There are rich coal and mineral deposits (lead, zinc, and copper). Industries include mining, production of chemicals, foodstuffs, light industries, and grain and vegetable cultivation. Tourism is increasingly important in the republic. 58 J2

Karachi largest city and chief port of Pakistan, northwest of the Indus delta; population (1996 est) 10 million; 4 million live in makeshift settlements. It is the capital of Sind province. Industries include shipbuilding, engineering, chemicals, plastics, and textiles. A nuclear power plant has been developed at Paradise Point, 25 km/15 mi to the west of the city. It was the capital of Pakistan 1947–59, when it was replaced by ►Islamabad. 56 J5

Karafuto Japanese name for ►Sakhalin Island. 44 Q6

Karaganda former name of ►Qaraghandy, an industrial city in central-eastern Kazakhstan. 42 N8

Karakoram mountain range in central Asia, divided among China, Pakistan, and India. Peaks include ►K2, Masharbrum, Gasharbrum, and Mustagh Tower. **Ladakh** subsidiary range is in northeastern Kashmir on the Tibetan border. 40 L6

Karakoram Highway road constructed by China and Pakistan and completed 1978; it runs 800 km/500 mi from Havelian (northwest of Rawalpindi), via ►Gilgit in Kashmir and the Khunjerab Pass (4,800 m/16,000 ft) to ►Kashi in China.

Karakorum ruined capital of Mongol ruler Genghis Khan, southwest of Ulaanbaatar in Mongolia.

Kara-Kum (or **Peski Karakumy**; Turkmen **Garagum** 'black sand'), sandy desert occupying some 90% of the republic of Turkmenistan; area about 310,800 sq km/120,000 sq mi. The Kara-Kum lies to the east of the Caspian Sea, between the Aral Sea to the north and the Iranian border to the south. It is separated from the Kyzyl-Kum desert by the ►Amu Darya River. The desert is crossed by the Trans-Caspian railway and the **Kara-Kum Canal**, the largest irrigation canal in the world. The area has rich oil, gas, and sulphur deposits, all of which are being increasingly exploited. Air temperatures of over 50°C have been recorded here. **56** G2

Kara Sea (Russian **Karskoye More**), part of the Arctic Ocean off the north coast of the Russian Federation, bounded to the northwest by the island of Novaya Zemlya and to the northeast by Severnaya Zemlya; area 880,000 sq km/339,768 sq mi; average depths 30–100 m/98–328 ft, with a maximum of 620 m/2,034 ft. Novy Port on the Gulf of Ob is the chief port. Dikson is also a main port and is located on the mouth of the Yenisey River, which flows into the Kara Sea. **42** L3

Karbalā (or **Kerbala**), holy city of the Shiite Muslims and administrative centre of the governorate of the same name, in Iraq, 96 km/60 mi southwest of Baghdad; population (1998 est) 402,500. The city lies on the edge of the Syrian desert and is linked by canal to the Hindiyah branch of the River Euphrates. The chief modern industries are the manufacture of textiles, shoes, and cement. **56** D3

Karditsa town in Thessaly, central Greece, capital of Karditsa department; population (1991 est) 30,000. It is an administrative and market centre for an agricultural region, trading in cereals, cotton, sheep, wool, hides, tobacco, and dairy products. **36** D5

Karelia (or **Kareliya**), autonomous republic in the northwest of the Russian Federation, bordering on Finland to the west; area 172,400 sq km/66,550 sq mi; population (1997) 780,000. The capital is ►Petrozavodsk. The republic is extensively forested with numerous lakes, of which Ladoga and Onega are the largest in Europe. Industries include fishing, timber, chemicals, coal, and mineral and stone quarrying. **16** R4

Karelian Isthmus strip of land between Lake Ladoga and the Gulf of Finland, in the Russian Federation, with St Petersburg at the southern extremity and Vyborg at the northern. Finland ceded it to the USSR 1940–41 and from 1947.

Kargopol town on the left bank of the Onega River, northwestern Russia; population (1990 est) 12,500. It developed with the salt trade from about the 14th century and has several outstanding 17th-century churches, notably the Church of St Vladimir (1653) and the Blagoveshchenie Church (1682–92). The town gives its name to the hunting and fishing Kargopol culture, which flourished in the area during the late Mesolithic and Neolithic periods. **38** G2

Kariba Dam concrete dam on the Zambezi River, on the Zambia–Zimbabwe border, about 386 km/240 mi downstream from the Victoria Falls, constructed 1955–60 to supply power to both countries. **72** D3

Karl-Marx-Stadt former name (1953–90) of ►Chemnitz, a city in Germany. **20** H6

Karlovy Vary (German **Karlsbad**), spa in the Bohemian Forest, Czech Republic, celebrated from the 14th century for its alkaline thermal springs; population (1991) 56,300. **20** H6

Karlsbad German name of ►Karlovy Vary, a town in the Czech Republic. **20** H6

Karlsruhe commercial and university city in Baden-Württemberg, Germany, close to the River Rhine, 61 km/38 mi northwest of Stuttgart; population (1995) 276,600. Industries include mechanical engineering, electronics, nuclear research, and oil-refining. The city participates in the Four Cities' Partnership with Halle, Nancy (France), and Nottingham (UK), which was established in the early 1990s to exchange industrial expertise in joint research and business ventures. It was built around an 18th-century castle as the new capital for the former state of Baden. **20** D8

Karlstad capital of Värmland county, southern Sweden; population (1995) 79,500. It stands at the northern end of Lake Vänern at the mouth of the River Klar. Karlstad has metal and clothing factories and a racetrack. On the nearby island is Skoghatt, Europe's largest integrated timber plant. **16** G7

Karnataka (formerly **Mysore** until 1973), state in southwest India; area 191,800 sq km/74,035 sq mi; population (1994) 48,150,000, of which the majority is Hindu. The capital is ►Bangalore, and Mangalore is a port. The state has a western coastal plain; inland the forested Western Ghats rise to heights of 1,250 m/4,000 ft. Industries include manganese, chromite, iron ore, bauxite, mica, copper, and India's only sources of gold (from the Kolar fields) and silver; teak and sandalwood processing. The Tungabhadra dam provides hydroelectricity, and irrigates up to 500,000 ha/1.23 million acres in Karnataka and Andrha Pradesh. Agricultural products include rice on the coastal western plain; inland millet, groundnuts, rice with irrigation; cotton in the north; and coffee and tea on the slopes of the Western Ghats. The language is Kannada. Famous people from the state include Haidar Ali and Tipu Sultan. **10** C6

Kärnten German name for ►Carinthia, a province of Austria. **8** (2)E4

Karoo (Khoikhoi *karusa* 'hard'), two areas of semi-desert in Eastern Cape Province, South Africa, divided into the **Great Karoo** and **Little Karoo** by the Swartberg Mountains. The two Karoos together have an area of about 260,000 sq km/100,000 sq mi. **72** K6

Karskoye More Russian name for the ►Kara Sea, part of the Arctic Ocean off the north coast of the Russian Federation. **42** L3

Karviná industrial town of northeast Moravia, in the Czech Republic, on the Polish border 15 km/9 mi east of Ostrava; population (1991) 68,400. The town is situated on the Silesian coalfield and produces high-grade coal, chemicals, and iron. **18** H8

Kasai river that rises in Angola and forms the frontier with the Democratic Republic of Congo before entering that country and joining the Congo River, of which it is the chief tributary. It is 2,100 km/1,300 mi long and is rich in alluvial diamonds. **70** B4

Kashgar (or **Kaxgar**), alternative name of ►Kashi, an oasis city in northwest China. **56** L2

Kashi (or **Kashgar**, or **Kaxgar**), oasis city in Xinjiang Uygur Autonomous Region, China, on the River Kashgar; population (1994) 242,600. It is the capital of the fertile Kashi district, which produces grain, cotton, and fruit. The city is a commercial centre, the Chinese terminus of the ▶Karakoram Highway, and a focus of Muslim culture. Traditional handicrafts include silk, cotton, boots, shoes, and saddlery. Tourism is increasingly important. 56 L2

Kashmir disputed area on the border of India and Pakistan in the northwest of the former state of Kashmir, now ▶Jammu and Kashmir; area 78,900 sq km/30,445 sq mi. Physical features include the west Himalayan peak Nanga Parbat (8,126 m/26,660 ft), Karakoram Pass, Indus River, and Baltoro Glacier. Azad ('free') Kashmir in the west has its own legislative assembly based in Muzaffarabad while Gilgit and Baltistan regions to the north and east are governed directly by Pakistan. The Northern Areas are claimed by India and Pakistan. Cities in the region include Gilgit and Skardu.

Kassel (or **Cassel**), industrial and university city in Hesse, Germany, on the River Fulda, 155 km/96 mi northeast of Wiesbaden; population (1995) 201,400. Industries include mechanical engineering, electronics, and the manufacture of motor vehicles and chemicals. The Grimm Museum commemorates the compilers of fairy tales who lived here. The spectacular Wilhelmshöhe Mountain Park, where Napoleon III was imprisoned after the Battle of Sedan, lies nearby. 20 E5

Kastoria town in Macedonia, Greece, capital of Kastoria department, situated south of Florina on the small Kastoria lake; population (2001 est) 20,600. The town dates back 2,500 years but flourished mainly in the 17th and 18th centuries as a fur-trading centre. It is an administrative centre and a resort, and there is commercial fishing. 36 D4

Kathiawar semi-arid peninsula on the west coast of India, lying on the Gulf of Khambhat and the Rann (Gulf) of Kutch. Formerly occupied by a number of princely states, all Kathiawar (60,723 sq km/23,445 sq mi) had been included in Mumbai state by 1956 but was transferred to Gujarat in 1960. Mahatma Gandhi was born in Kathiawar at Porbandar. 56 K5

Kathmandu (or **Katmandu**), capital of Nepal, situated at 1,370 m/4,500 ft in the southern Himalayas, in the Valley of Nepal, at the junction of the Baghmati and Vishnumati rivers; population (1991) 419,100. Tourism is an important economic activity. 54 E3

Katowice industrial city (anthracite, iron and coal mining, iron foundries, smelting works, machine shops) in Upper Silesia, southern Poland; population (1993) 366,900. 18 J7

Kattegat sea passage between Denmark and Sweden. It is about 240 km/150 mi long and 135 km/85 mi wide at its broadest point, and joins the Skagerrak on the north to the Baltic Sea on the south. Its sandbanks are a navigational hazard. 16 F8

Kaufbeuren city in Bavaria, Germany, on the River Wertach, 28 km/17 mi southwest of Munich; population (1994) 42,600. It has glass, jewellery, electronic, and timber industries. Kaufbeuren was one of the city-states of the Holy Roman Empire, and has two Gothic churches. 30 F3

Kaunas (formerly **Kovno** until 1917), industrial city in Lithuania, at the confluence of the Neris and Niemen (Neman) rivers, 80 km/50 mi northwest of Vilnius; population (2001 est) 411,600. Kaunas has engineering, textile, and food industries and is an important river port and railway junction. It was the capital of independent Lithuania from 1920 until 1940. 14 G2

Kavala town in northeastern Greece, capital of Kavala department, built on a promontory; population (2001 est) 64,400. Its port handles 6% of Greek trade. Exports include tobacco and manganese, and imports include fertilizers, wheat, and timber. There are flour and rice mills. The town was a member of the Attica Naval League in 478–463 BC and was later a Roman naval base. 36 G4

Kawasaki industrial city of Kanagawa prefecture, Honshu island, Japan; population (1994) 1,171,000. It is situated between Tokyo and Yokohama in the important Keihin industrial zone, the largest industrial zone in Japan, and is a major port for the import of raw materials. The main industries are iron, steel, shipbuilding, chemicals, and textiles. It is noted for its cherry blossoms at Inadazutsumi on the banks of the Tamagawa. 48 K6

Kayseri (ancient **Caesarea Mazaca**), capital of Kayseri province, central Turkey; population (1990) 421,400. It produces textiles, carpets, and tiles. In Roman times it was capital of the province of Cappadocia. 58 F4

Kazakhstan country in central Asia, bounded north by Russia, west by the Caspian Sea, east by China, and south by Turkmenistan, Uzbekistan, and Kyrgyzstan.

Kazakhstan covers an area of 2,717,300 sq km/1,049,150 sq mi, and has a population of 16,223,000 (2000 est). The capital is ▶Astana (formerly Akmola); other major towns and cities include Qaraghandy, Pavlodar, Semey, Petropavl, and Shymkent. The head of state is Nursultan Nazarbayev (from 1991); the head of government is Kasymzhomart Tokaev (from 1999). The currency is the tenge, and the languages spoken include Kazakh (related to Turkish; official) and Russian. 42 L8

Kazan city and port on the ▶Volga, capital and economic centre of ▶Tatarstan, in the western Russian Federation; population (1996) 1,085,000. Kazan is the centre of Tatarstan culture. It has large engineering plants (manufacturing ships, machine tools, compressors, and dental equipment), chemical works (producing explosives, synthetic rubber, soap, and photographic materials), and a large leather and fur industry. It is also a major transportation centre, with its river port, airport, and location at a major railway junction. 82 M4

Kazan-rettō group of Japanese islands in the Pacific; see ▶Bonin and Volcano islands. 40 T7

Kebnekaise highest peak in Sweden, rising to 2,111 m/6,926 ft in the Kolen range, west of the town of Kiruna. 16 K3

Kecskemét city in Hungary, situated on the Hungarian plain southeast of Budapest; population (1993 est) 105,000. It is a trading centre of an agricultural region. 34 G3

Kedah state in northwest Peninsular Malaysia; capital Alor Setar; area 9,400 sq km/3,628 sq mi; population (1993) 1,412,000. Products include rice, rubber, tapioca, tin, and tungsten. Kedah was transferred by Siam (Thailand) to Britain in 1909, and was one of the Unfederated Malay States until 1948. 11 (1)A1

Keelung (or **Chi-lung**), industrial port (shipbuilding, chemicals, fertilizer) on the north coast of Taiwan, 24 km/15 mi northeast of Taipei; population (1993) 363,000. **46** G5

Keewatin eastern district of ►Northwest Territories, Canada, including the islands in Hudson and James bays; area 590,935 sq km/ 228,160 sq mi; population (1991) 5,800. Towns and cities in the district include Chesterfield Inlet, Eskimo Point, and Coral Harbour (which was the site of an air base set up during World War II). The district is part of the Canadian Shield upland plateau in the north, the south being low and level, covering the greater part of the Arctic prairies of Canada. The area was heavily glaciated in the Ice ages and features numerous lakes and tundra vegetation. The major industry is furs, with trapping the main occupation. **88** B1

Kefallonia (English **Cephalonia**), largest of the Ionian Islands off the west coast of Greece; area 935 sq km/360 sq mi; population (1991 est) 32,500. It was devastated by an earthquake in 1953 that destroyed the capital Argostolion. **36** C6

Keflavik town and fishing port in Iceland, on the south coast of Faxa Bay, 35 km/22 mi southwest of Reykjavik; population (1990) 7,500. Its international airport is the largest in Iceland and one of the largest in the world; it was built during World War II by US forces (who called it Meeks Field). Keflavik became a NATO base in 1951. The town's geothermally heated waters are used to provide power, and form warm bathing pools which attract tourists, as does the nearby 'Blue Lagoon'. **16** (1)B2

Kelang (formerly **Port Swettenham** until 1971), Malaysian rubber port on the Strait of Malacca, 40 km/25 mi southwest of Kuala Lumpur; population (1997 est) 243,400. Kelang, one of the largest cities in Malaysia, contains the palace of the sultan. It lies on the Kuala Lumpur-Port Swettenham railway and its industries, including rubber goods and pineapple canning, are mainly based on local produce.

Kelantan state in northeast Peninsular Malaysia; capital Kota Bharu; area 14,900 sq km/5,751 sq mi; population (1993) 1,221,700. It produces rice, rubber, copra, tin, manganese, and gold. Kelantan was transferred by Siam (Thailand) to Britain 1909 and until 1948 was one of the Unfederated Malay States. **11** (1)A1

Kemerovo (formerly **Shcheglovsk** 1925–32), capital city, economic and cultural centre of Kemerovo oblast (region) in southern Siberia, Russian Federation; population (1996 est) 501,000. Kemerovo is situated on the River Tom, a tributary of the Ob, 200 km/124 mi east of Novosibirsk. It has large chemical and engineering industries, and is one of the principal coal-mining centres of the ►Kuznetsk Basin. **42** R6

Kempten city in Bavaria, Germany, on the River Iller, 104 km/65 mi southwest of Munich; population (1994) 61,400. The city is a market centre specializing in dairy products, and there are textile, paper, and precision-engineering industries. Kempten contains a baroque church, a Gothic *Rathaus* (town hall), and the former palace of the abbot-princes of the old imperial city. **30** F3

Kendal town in Cumbria, northwest England, on the River Kent, 35 km/ 22 mi north of Lancaster; population (1991) 24,900. Light industrial activities include the production of footwear, specialist paper, carpets, snuff, and confectionery. Insurance services are also important. The town is a tourist centre for visitors to the ►Lake District. **24** K7

Kennewick city and port in Benton County, southeast Washington, USA on the Columbia River, 280 km/ 174 mi southeast of Seattle; population (1990) 42,200. Dams built on the Columbia and Snake rivers provide irrigation for the area's grape, sugar beet, alfalfa, and corn crops. The city is the site of chemical and agricultural processing industries. The Hanford nuclear power station nearby is a major source of employment. **86** C1

Kenosha (Potawatomi 'pike' or 'pickerel'), city, port, and administrative headquarters of Kenosha County, southeast Wisconsin, on Lake Michigan, 47 km/29 mi southeast of Milwaukee; population (1992) 83,500. Its industries include food-processing equipment, fertilizers, motor vehicles, wire products, textiles and clothing, and food products. Kenosha is the home of Carthage College (1846), Gateway Technical College (1912), and the University of Wisconsin-Parkside (1965). **88** C2

Kent county of southeast England, known as the 'garden of England' (since April 1998 Medway Towns has been a separate unitary authority).
area 3,730 sq km/1,440 sq mi
towns and cities ►Maidstone (administrative headquarters), Ashford, Canterbury, Deal, Dover (ferry terminus), Gravesend, Hythe, New Ash Green (a new town), Sevenoaks, Tunbridge Wells; resorts: Folkestone, Margate, Ramsgate, Whitstable
physical the North Downs; White Cliffs of Dover; rivers: Thames, Darent, Medway (traditionally, a 'man of Kent' comes from east of the Medway and a 'Kentish man' from west Kent), Stour; marshes (especially Romney Marsh); the Isles of Grain, Thanet, and Sheppey; the Weald; Dungeness (peninsula and headland)
features Leeds Castle (converted to a palace by Henry VIII); Ightham Mote; Hever Castle (where Henry VIII courted Anne Boleyn); Chartwell (Churchill's country home)
agriculture cereals, hops, apples, soft fruit, vegetables; livestock production
industries cement, paper, oil refining, shipbuilding, tourism.
population (1996) 1,557,300 **5** G6

Kentucky state in south-central USA. It is nicknamed the Bluegrass State. Kentucky was admitted to the Union in 1792 as the 15th US state. Extending over 640 km/400 mi from east–west, Kentucky has the Ohio River along its entire northern boundary; across which are the states of Ohio, Indiana, and Illinois. To the east, the Tug Fork and Big Sandy rivers separate it from West Virginia. To the southeast, it is bordered by Virginia, with the Cumberland Gap at the extreme south; from this point along its southern boundary, as far as the Mississippi River, it is bordered by Tennessee. Across a small stretch of the Mississippi, on the west, it faces the New Madrid region of Missouri.
population (2000 est) 4,041,800
area 104,700 sq km/40,414 sq mi
capital ►Frankfort
towns and cities Louisville, Lexington, Owensboro, Covington, Bowling Green
industries and products tobacco,

cereals, textiles, coal, whisky, horses, transport vehicles 84 J4

Kenya country in east Africa, bounded to the north by Sudan and Ethiopia, to the east by Somalia, to the southeast by the Indian Ocean, to the southwest by Tanzania, and to the west by Uganda.

Kenya covers an area of 582,600 sq km/224,941 sq mi, and has a population of 30,080,000 (2000 est). The capital is ▶Nairobi; other major towns and cities include Mombasa, Kisumu, Nakuru, Eldoret, and Nyeri. The head of state and government is Daniel arap Moi (from 1978). The currency is the Kenyan shilling, and the languages spoken include English, Kiswahili (both official), and many local dialects. 62 G5

Kenya, Mount (or **Kirinyaga**), extinct volcano from which Kenya takes its name, height 5,199 m/ 17,057 ft. It is situated almost exactly on the Equator. The first European to climb it was Halford Mackinder in 1899. 70 F4

Kerala state of southwest India; area 38,900 sq km/15,015 sq mi; population (1994 est) 30,555,000. The capital is ▶Thiruvananthapuram and Kozhikode and Kochi are other towns. The state extends along the southwest coast from Karnataka almost to the southern tip of India and is bounded on the east by the highlands of the Western Ghats. Industries include textiles, chemicals, electrical goods, and fish. Agricultural products include tea, coffee, rice, coconuts, fruit, and oilseed. 10 C6

Kerbala alternative spelling of ▶Karbalā, a holy city of the Shiite Muslims in Iraq. 56 D3

Kerch port in the ▶Crimea, Ukraine, situated at the eastern end of the **Kerch peninsula**; population (1990) 176,000. Kerch is a fishing centre, and also lies in a region rich in iron-ore and natural gas. Its industries include iron and steel production and shipbuilding. 58 G1

Kerguelen Islands (or **Desolation Islands**), volcanic archipelago in the Indian Ocean, part of the French Southern and Antarctic Territories; area 7,215 km/2,787 sq mi. They were discovered in 1772 by the Breton navigator Yves de Kerguelen and annexed by France in 1949. Uninhabited except for scientists (centre for joint study of geomagnetism with Russia), the islands support a unique wild cabbage containing a pungent oil. 12 J5

Kerkrade town in the province of Limburg, the Netherlands, 24 km/ 15 mi northeast of Maastricht, near the border with Germany; population (1996) 52,700. Kerkrade forms the centre of an industrial region. The town was founded 1104. 22 J4

Kérkyra Greek form of ▶Corfu, an island in the Ionian Sea. 36 B5

Kermadec Islands volcanic island group in the Pacific Ocean, 800 km/ 497 mi northwest of North Island, New Zealand; area 30 sq km/12 sq mi. They are uninhabited except for meteorological and seismological stations on the largest island, Raoul (Sunday Island), which has an area of 28 sq km/11 sq mi and reaches a height of 525 m/1,723 ft at Mount Mumukai. The islands have been a dependency of New Zealand since 1887. 74 J9

Kerman (or **Kirman**), city in Kerman province, southeast Iran, over 1,500 m/4,920 ft above sea level; population (1991) 311,600. It is a centre of the Iranian carpet industry and for the mining of copper, precious metals, iron ore, and coal. The city was almost totally devastated in 1794 by an earthquake. It has a small Zoroastrian community. 61 G1

Kermanshah (or **Bakhtaran**; Farsi 'city of the kings'), capital of Kermanshahan province, northwest Iran, a major city of Iranian Kurdistan; population (2001 est) 757,900. The province (area 23,700 sq km/9,148 sq mi; population (2001 est) 1,945,400) is on the Iraqi border. Industries include oil refining, carpets, and textiles. Founded in the 4th century, its prosperity derived from its fertile agricultural hinterland and its site on caravan routes to Baghdad. In the years following the Islamic revolution in 1979 the city and province were known as Bakhtaran. 56 E3

Kerry county of the Republic of Ireland, west of Cork, in the province of Munster; county town Tralee; area 4,700 sq km/1,814 sq mi; population (1996) 126,100. Industries include engineering, woollens, shoes, cutlery, fishing, and farming (dairy farming in the north, cattle grazing in the south). Tourism is important; Muckross House and Abbey are among the top visitor attractions. Other towns include Caherciveen, Castleisland, Dingle, Killarney, and Listowel. Kerry is low-lying in the north and mountainous in the south, with the Slieve Mish and Caha Mountains, and Macgillycuddy's Reeks, where Carrauntoohill (Ireland's highest peak at 1,041m/3,415 ft) is situated; other peaks include Brandon (953 m/ 3,127 ft) and Mangerton (840 m/2,756 ft). 5 B5

Keryneia alternative spelling of ▶Kyrenia, a port in Turkish Cyprus. 58 E6

Kevkas alternative name for ▶Lefkada, one of the Ionian Islands of Greece. 36 C6

Keweenaw Peninsula part of the upper peninsula of Michigan, USA, curving 95 km/50 mi into Lake Superior, largest of the Great Lakes of North America. Formerly very rich in copper, it is now primarily a resort area; other industries include timber processing, farming, and commercial fishing. 88 C1

Key West city and port of entry at the tip of the Florida peninsula, USA, on **Key West Island**, the most westerly of the ▶Florida Keys group, 209 km/130 mi south of Miami; seat of Monroe County; population (1990) 24,800. It is the southernmost city in the continental USA, the home of the naturalist John J Audubon, and was popularized as a tourist resort by the novelist Ernest Hemingway. Other industries include commercial fishing. It is the site of a US naval and air station. Key West was incorporated in 1828. 90 E5

Khabarovsk capital city of Khabarovsk krai (territory) in eastern Siberia, in the Russian Federation; population (1996 est) 616,000. Khabarovsk is sited at the confluence of the Amur and Ussuri rivers, and is the largest city, chief transportation and political centre, and second cultural centre of the Russian Far East. There are engineering works, oil refineries, and other industries here. 44 P7

Khakass (or **Khakasiya**), republic of the Russian Federation, in southern Siberia, adjacent to Krasnoyarsk krai (territory); area 61,900 sq km/23,900 sq mi; population (1996) 586,000 (72% urban) (80% Russians, 11%

Khakass). The capital is ▶Abakan and Chernogorsk is another town. The republic is situated between the Kuznetsky Alatau Mountains in the west and the Sayan Mountains in the southeast. It is located in the Minusinsk Basin west of the upper River Yenisey. Coal, iron-ore, and gold deposits are found here. Industries include mining and lumbering; hydro-electric power generation (at Sayanogorsk on the Yenisey); and breeding of sheep, goats, and cattle. 42 R7

Khalīj Surt Arabic name for the Gulf of ▶Sirte, a gulf off the coast of Libya. 64 C1

al-Khalil Arabic name for ▶Hebron in the Israeli-occupied West Bank. 60 C5

Kharkiv Ukrainian name for ▶Kharkov, a large city in eastern Ukraine. 38 G5

Kharkov (Ukrainian **Kharkiv**), major city in eastern Ukraine, capital of the Kharkov oblast, 400 km/250 mi east of Kiev and 40km/25 mi south of the border with the Russian Federation; population (1990) 1,618,000. An important railway junction and industrial city, Kharkov is situated at the confluence of the Kharkov, Lopan, and Udy rivers, and lies close to the ▶Donets Basin coalfield and ▶Krivoy Rog iron mines. Its industrial enterprises include engineering and railway rolling-stock works, agricultural and mining machinery factories, and chemical plants. 38 G5

Khartoum (or **El Khartum**), capital and trading centre of Sudan, in Khartoum State, at the junction of the Blue and White Nile rivers; population (1998 est) 1,038,600, and of Khartoum North, across the Blue Nile, 890,000. ▶Omdurman is also a suburb of Khartoum, giving the urban area a population of over 3 million. It has long served as a major communications centre between the Arab countries of North Africa and central African countries. The city lies in a rich cotton-growing area and an oil pipeline reached it from Port Sudan on the Red Sea. Industries include tanning, textiles, light engineering, food processing, glassware, and printing. 64 F4

Khaskovo (Bulgarian **Haskovo**), capital of Khaskovo region, southern Bulgaria, on a tributary of the River Maritsa, 56 km/35 mi southeast of Plovdiv; population (1990) 117,400. Industries include textiles and tobacco. The town was founded by Turks in the 14th century. 36 H3

Kherson port in southern Ukraine, situated on the River ▶Dnieper 140 km/87 mi east of Odessa, and capital of the oblast of the same name; population (1990) 361,000. Its industries include shipbuilding, petroleum refining, and the manufacture of soap and tobacco. Kherson was founded in 1778 by army commander Grigory Potemkin as the first Russian naval base on the Black Sea. 38 F5

Khios Greek name for ▶Chios, an island and town in Greece. 36 H6

Khmer Republic former name (1970–76) of ▶Cambodia. 50 C4

Kholm Russian name for ▶Chełm, a town in Poland. 18 N6

Khorramshahr formerly major port and oil-refining centre of Iran, in the province of Khuzestan; population (1991) 34,800. On the junction of the Shatt-al-Arab and Karun rivers, it was almost completely destroyed in the 1980s during the Iran–Iraq War, but is now being rebuilt. It was formerly called Mohammerah. 61 C1

Khulna capital of Khulna region, southwestern Bangladesh, situated close to the Ganges delta; population (1991) 545,800. Industry includes shipbuilding and textiles; it trades in jute, rice, salt, sugar, and oilseed. 54 E4

Khyber Pass pass through the mountain range that separates Pakistan from Afghanistan; length 53 km/33 mi; width varies from 140 m/460 ft at its widest to about 15 m/50 ft at its narrowest. On either side are rock faces rising to a height of 915 m/3,000 ft in some places, the highest point at Landi Kotal, 520 m/1,700 ft higher than Jamrud at the entrance to the pass. The Khyber Pass was used by invaders of India. The present road was constructed by the British during the Afghan Wars (1839–42 and 1878–80). 56 K3

Kicking Horse Pass pass in the ▶Rocky Mountains, between British Columbia and Alberta, Canada. It is located 56 km/35 mi northwest of Banff, to the west of Lake Louise. With an altitude of 1,627 m/5,338 ft, it is the highest point on the Canadian Pacific Railway. It lies inthe Banff–Yoho national parks. 82 H6

Kidderminster market town in Worcester, England, on the River Stour, 22 km/14 mi north of Worcester; population (1991) 54,600. Its carpet industry dates from about 1735, and associated activities include the manufacture of woollen and worsted yarns, and dyeing. Other industries include steel forging, pleasure boat building, and the manufacture of plastic and ceramic products, sugar, and chemicals. Small high-tech businesses are proliferating. 24 G9

Kiel Baltic port and capital of Schleswig-Holstein, Germany; population (1995) 247,300. It is a naval and commercial facility. Chief industries include fishing, marine engineering, ship-repairing, electronics, engineering, food-processing, and tourism. It is at the terminus of the ▶Kiel Canal. **Kiel Week** in June is a major yachting venue. In World War I it was the headquarters of the German Imperial Fleet. 20 F2

Kiel Canal (German **Nord-Ostsee-Kanal**; formerly **Kaiser Wilhelm Canal**), waterway connecting the Baltic with the North Sea; 96 km/60 mi long. It provides passage for ocean-going vessels. Built by Germany in the years before World War I, the canal allowed the German navy to move from its Baltic bases to the open sea without travelling through international waters. It was declared an international waterway by the Treaty of Versailles in 1919. 20 E2

Kielce city in central Poland, northeast of Kraków; population (1991) 214,200. It is an industrial rail junction (chemicals, metals). 18 K7

Kiev (Ukrainian **Kyyiv**), capital and largest city of Ukraine, situated at the confluence of the Desna and Dnieper rivers; population (1990) 2,616,000. Kiev was the capital of Russia in the Middle Ages. It is a major industrial centre, producing chemicals, clothing, leather goods, machine tools, and electrical goods, and is also a market city for the abundant agricultural produce of the western Ukraine. Kiev University was founded in 1834.
features St Sophia cathedral, the oldest cathedral in Ukraine (11th century) and Kiev-Pechersky Monastery (both now museums)

survive, together with the remains of the Golden Gate, an arched entrance to the old walled city, built in 1037. The gate is surmounted by the small Church of the Annunciation. The Kiev ballet and opera companies are renowned worldwide.

history Kiev was founded in the 5th century by Vikings. The Slav domination of Russia began with the rise of Kiev, the 'mother of Russian cities'; Kiev replaced ►Novgorod as the capital of the state of Kievan Rus in 882 and was the original centre of the Orthodox Christian faith from 988. It was for a long time an important trading centre on the route from the Baltic to the Black Sea, but declined in importance in the 12th century. The Russian capital was moved to Vladimir in 1169, and Kiev was sacked by Mongols under Batu Khan in 1240. From the 14th–late 17th centuries, the city was successively under Tatar, Lithuanian and Polish control. It was annexed by Russia in 1686. In World War II, Kiev, then the third-largest city of the USSR, was occupied and largely destrroyed by German forces 1941–43. During this period, around 200,000 of the city's inhabitants, including its entire Jewish population, were murdered. 38 F4

Kigali capital of ►Rwanda, 80 km/50 mi east of Lake Kivu, and just south of the Equator; population (2001 est) 327,700. It is an important centre for trade in local produce, including coffee, cattle, cassiterite (tin ore), and tungsten. Manufacturing industries include textiles, chemicals, tin-processing, paints, and cigarettes. Kigali is served by an international airport, and is connected by trunk roads to Uganda to the north and Burundi to the south. 70 E4

Kigoma town and port on the eastern shore of Lake Tanganyika, north of Ujiji, Tanzania, at the western terminal of the railway from Dar es Salaam; population (1998 est) 79,000. It trades in timber, cotton, palm oil, and tobacco. 70 D4

Kikinda town in Serbia, Yugoslavia, in the fertile region of Vojvodina; population (1990) 43,000. It is a railway junction, has a trade in wheat and fruit, and lies on a natural-gas field. 34 H4

Kildare county of the Republic of Ireland, in the province of Leinster; county town Naas; area 1,690 sq km/652 sq mi; population (1996) 135,000. The principal rivers are the Barrow, the Boyne, the Lesser Barrow, and the Liffey. Kildare is wet and boggy in the north with extensive grassy plains and rolling hills, and includes part of the Bog of Allen. The town of Maynooth houses a constituent part of the National University of Ireland; originally the college was a seminary for Roman Catholic priests. The Curragh, at Tully, is a plain that is the site of the national stud and headquarters of Irish horse racing; steeplechase racing also takes place at Punchestown. Cattle are grazed in the north, and in the south products include oats, barley, potatoes, and cattle. Other main towns include Athy, Droichead Nua, and Kildare. 5 C5

Kilimanjaro volcano in ►Tanzania, the highest mountain in Africa, 5,895 m/19,340 ft. It is situated between Lake Victoria and the coast. It culminates in two peaks, Kibo (5,895 m/19,340 ft) and Mawenzi (5,149 m/16,893 ft), both craters of extinct volcanoes. The first recorded ascent was by the German geographer Hans Meyer and the Austrian mountaineer Ludwig Purtscheller in 1889. 70 F4

Kilkenny county of the Republic of Ireland, in the province of Leinster; county town Kilkenny; area 2,060 sq km/795 sq mi; population (1996) 75,300. It has the rivers Nore, Suir, and Barrow. Industries include coalmining, clothing, footwear, brewing, and agricultural activities include cattle rearing and dairy farming. Principal towns include Castlecomer, Callan, Graiguenamanagh, and Thomastown. 5 C5

Killarney market town and tourist centre in County ►Kerry, Republic of Ireland; population (1996) 8,800. It is a famous beauty spot in Ireland; the mountain range Macgillycuddy's Reeks and the **Lakes of Killarney** lie to the southwest. Industries include hosiery and container cranes. The Catholic Cathedral of St Mary was designed by Pugin in 1855. The 19th-century Church of Ireland St Mary's Church is noted for its rich internal decoration. The National Museum of Irish Transport is located at Killarney. 24 C9

Kilmarnock administrative headquarters of ►East Ayrshire, Scotland, on the River Irvine, 32 km/20 mi southwest of Glasgow; population (1991) 44,300. Industries include biotechnology and whisky. Robert Burns's first book of poems was printed here in 1786, and became known as the 'Kilmarnock edition'. 24 H6

Kimberley diamond site in Western Australia, found 1978–79, estimated to have 5% of the world's known gem-quality stones and 50% of its industrial diamonds.

Kimberley diamond-mining capital city of Northern Cape Province, South Africa, 153 km/95 mi northwest of Bloemfontein, and 1,223 m/4,012 ft above sea level; population (1991) 167,100. Its mines have been controlled by De Beers Consolidated Mines since 1887. It is an important railway junction. 72 C5

King's Lynn (formerly **Bishop's Lynn**), port and market town in Norfolk, eastern England, at the mouth of the River Great Ouse, 56 km/35 mi northeast of Peterborough; population (1996) 40,200. Industries include food-processing, fishing, engineering, light manufacturing (valves, catering refrigerators), and the production of chemicals. It was a thriving port in medieval times, named **Bishop's Lynn** until it became royal property during the reign of Henry VIII. 24 N9

Kingsport city in northeastern Tennessee, USA, on the Holston River, 84 mi/135 km northeast of Knoxville, near the Virginia border; population (1990) 36,400. Products include plastics, chemicals, textiles, paper and printing, cement, bricks, and explosives. 90 E2

Kingston city and port in southeastern Ontario, Canada, at the mouth of the St Lawrence River on Lake Ontario; seat of Frontenac County; population (1991) 56,600. It is an important shipping and manufacturing centre; industries include shipbuilding yards, engineering works, synthetic fibres, ceramics, and grain elevators. 88 E2

Kingston capital and principal port of Jamaica, West Indies, the cultural and commercial centre of the island; population (1991) 587,800 (metropolitan area). Founded in 1693,

Kingston became the capital of Jamaica in 1872. 94 J5

Kingston upon Hull (or Hull), city, port, and unitary authority, created in 1996 from part of the former county of Humberside, situated where the River Hull flows into the north side of the Humber estuary, northeast England.
area 71 sq km/27 sq mi
features 13th-century Holy Trinity Church; Town Docks Museum; Ferens Art Gallery (1927); University of Hull (1954) and University of Humberside (1992), formerly Humberside Polytechnic; the Humber Bridge, the world's longest single-span suspension bridge
industries fish processing, flour milling, sawmilling, marine engineering, food processing, and the manufacture of electrical goods, vegetable oils, paint, pharmaceuticals, chemicals, caravans, aircraft, and paper. There are 11 km/7 mi of modern docks on the Humber estuary. The largest timber port in the UK, it also handles grain, oilseeds, wool, and manufactured goods. There are ferries to Rotterdam and Zeebrugge.
population (1996) 265,000
24 M8/5 G5

Kingston upon Thames outer borough of southwest Greater London; administrative headquarters of ►Surrey, although not in the county; population (1997 est) 144,600. Industries include mainly public sector administration and services; finance and business services, and retailing. There is a small manufacturing sector (8% of workforce), including chemicals, engineering, plastics, and printing. Seven Saxon kings, from Edward the Elder in 900 to Ethelred the Unready in 979, were crowned at Kingston; their coronation stone is preserved here, set with seven silver pennies. It is the oldest of the three Royal Boroughs of England, with the ancient right to elect its own High Steward and Recorder. Kingston Grammar School, founded by Elizabeth I in 1561, is found in the borough. 22 B3

Kingstown former name for ►Dún Laoghaire, a port near Dublin, Ireland. 24 F8

Kingstown capital and principal port of St Vincent and the Grenadines, West Indies, in the southwest of the island of St Vincent; population (1991) 26,200. 98 E1

King-Te-Chen alternative spelling of ►Jingdezhen, a town in Jiangxi province, China. 46 F5

Kinshasa (formerly **Léopoldville** until 1966), capital of the Democratic Republic of Congo on the Congo River, 400 km/250 mi inland from the port of Matadi; population (1994 est) 4,655,300. Industries include ship building and repairing, chemicals, textiles, engineering, food processing, and furniture. It was founded by the explorer Henry Morton Stanley in 1881. The National University of Kinshasa is here. 70 B4

Kirchheim unter Teck city in Baden-Württemberg, Germany, 25 km/15 mi southeast of Stuttgart; population (1994) 38,000. There are textile and steel industries. The town's full name derives from its position beneath the fortress of Teck. 30 E2

Kirghizia alternative form of ►Kyrgyzstan, a country in central Asia. 42 N9

Kiribati republic in the west central Pacific Ocean, comprising three groups of coral atolls: the 16 Gilbert Islands, 8 uninhabited Phoenix Islands, 8 of the 11 Line Islands, and the volcanic island of Banaba.

Kiribati covers an area of 717 sq km/ 277 sq mi, and has a population of 83,000 (2000 est). The capital is ►Bairiki (on Tarawa atoll). The head of state and government is Teburoro Tito (from 1994). The currency is the Australian dollar, and the languages spoken include English (official) and Gilbertese. 74 J6

Kirinyaga alternative name of Mount ►Kenya. 70 F4

Kiritimati (formerly **Christmas Island**), island in ►Kiribati, in the central Pacific; one of the Line Islands. 74 L5

Kirkcaldy seaport on the north side of the Firth of Forth, Fife, Scotland, 24 km/15 mi north of Edinburgh across the Forth; population (1991) 47,200. Industries include the manufacture of linoleum, paper, and whisky. It is the birthplace of the economist Adam Smith (1723) and where he retired to write his treatise *The Wealth of Nations*. It is also the birthplace of the architect Robert Adam. 24 J5

Kirkuk (or **Kerkuk**), city in northeast Iraq, in the governorate of the same name, 235 km/146 mi north of Baghdad; population (1998 est) 568,000. It is the capital of Kirkuk Province and lies in one of the richest oilfields in Iraq. The city is also the market for an area of sheep farming and grain and fruit growing. Ruins found on the site of the city show that settlement dates back some 3,000 years. 58 L6

Kirkwall administrative headquarters and port of the ►Orkney Islands, Scotland, on the north coast of the largest island, Mainland; population (1991) 6,700. The main industry is distilling. The Norse cathedral of St Magnus dates from 1137. The Bishop's Palace is also 12th-century, and the Earl's Palace was completed in 1606. 24 K3

Kirov (formerly **Vyatka**), capital city and economic and cultural centre of Kirov oblast (region), western Russia; population (1996 est) 465,000. Kirov is a major industrial city extending for 20 km/12 mi along the River Vyatka, 320 km/199 mi north of Kazan. There are engineering, chemical, woodworking, and leather industries, and a large toy factory. 38 J3

Kirovabad former name of ►Gäncä, a city in Azerbaijan. 58 M3

Kirovograd (Ukrainian **Kirovohrad**), city in central Ukraine on the Ingul River, capital of an oblast of the same name; population (1990) 274,000. Kirovograd lies some 250 km/155 mi southeast of Kiev. Manufacturing industries in the city include the production of agricultural machinery and food processing. The city is situated on a large lignite (brown coal) field. 38 F5

Kiruna town in Swedish Lapland; population (1990) 20,500. In area it is the largest town in the world. It includes vast iron-ore fields; most of the mining is opencast. The iron ore is transported to the Norwegian seaport of Narvik by rail, and to the Swedish Baltic port of Luleå, which is closed by ice in winter. The mines in the Kiruna area produce 25 million tonnes of iron ore a year. 16 L3

Kisangani (formerly **Stanleyville** until 1966), capital of Haut-Zaire Province, northeast Democratic Republic of Congo, on the upper Congo River, below Stanley Falls; population (1994 est) 417,500. It is a communications centre and strategic river city. Agricultural products include maize, millet, sorghum, cassava, plantains, and sweet potatoes. Industries include port activities, textiles, and food processing. **70** D3

Kishinev Russian name for ►Chişinău, the capital of Moldova. **34** R2

Kiskunfélegyháza market town on the Great Plain (Alföold) of Hungary, 26 km/16 mi south of Kecskemét; population (1993) 33,900. Cereals, fruit, wine, tobacco, and livestock are traded from the surrounding rich agricultural district. The town was destroyed by the Turks in the 17th century but rebuilt 1743 **34** G3

Kitakyushu industrial port city in Fukuoka prefecture, on the Hibiki Sea, north Kyushu island, Japan; population (1994) 1,015,000. Kitakyushu, which means 'north Kyushu city', was formed in 1963 by the amalgamation of Moji, Kokura, Tobata, Yawata, and Wakamatsu. Industries include food-processing and the manufacture of appliances, ceramics, and semiconductors. A tunnel, built in 1942, and a bridge link the port with Honshu. **48** F7

Kitchener city in southwestern Ontario, Canada, 96 km/60 mi west of Toronto and 145 km/90 mi northwest of the Niagara Falls; population (1991) 168,300; population of metropolitan area (with Waterloo) 356,000. It lies at an altitude of 335 m/1,099 ft in the fertile valley of the Grand River, and is twinned with the neighbouring city of Waterloo. Kitchener is the centre of a rich agricultural district; industries include the manufacturing of agricultural machinery, tyres, processed food, textiles, and electrical equipment. **88** D2

Kitimat seaport in Kitimat-Stikine Regional District, west-central British Columbia, Canada; population (1991) 11,300. It is located 110 km/70 mi southeast of Prince Rupert at the head of Douglas Channel. It is the site of a huge aluminium smelter, accessible to vessels bringing ore from the Caribbean. It receives hydroelectric power from a plant at Kemano, 72 km/45 mi to the south-southeast. **82** F6

Kitwe commercial centre for the Zambian copperbelt; population (1998 est) 424,100. It is the headquarters of the Zambia Chamber of Mines. Nkana township and copper mine is adjacent; to the south are Zambia's emerald mines. The local supplies of copper have helped the growth of an electrical equipment industry. Other industries include engineering, brewing, iron goods, fibreglass, paint, furniture, and plastics. **72** D2

Kivu, Lake lake in the Great Rift Valley on the borders of Rwanda and the Democratic Republic of Congo, north of Lake Tanganyika, to which it is linked by the Ruzizi River; area 2,699 sq km/1,042 sq mi; length about 105 km/65 mi. The chief port of the lake is Bukavu. **70** D4

Kladno industrial town (coal, iron, steel, engineering) in central Bohemia, in the Czech Republic, 25 km/16 mi west of Prague; population (1991) 71,700. **18** D7

Klagenfurt capital of the Austrian province of ►Carinthia, on a tributary of the River Drava near the Slovenian border; population (2001 est) 91,100. The city has iron, chemical, textile, leather, machinery, and tobacco industries, and hosts an annual trade fair. It has a Renaissance cathedral and several old churches. **30** K4

Klaipeda (formerly German **Memel**), main port of Lithuania, on the Baltic coast at the mouth of the River Dane; population (1991) 206,000. Klaipeda lies 290 km/180 mi northwest of Vilnius. Industries include shipbuilding, fisheries, and iron foundries; the city also trades in timber and grain. Klaipeda was founded in 1252 on the site of a local fortress as the castle of Memelburg by the Teutonic Knights, joined the Hanseatic League soon after, and has changed hands between Sweden, Russia, and Prussia/Germany. The city was part of independent Lithuania from 1923, and after a period of German occupation (1939–45), it became part of the Soviet Republic of Lithuania (1945–91). **16** L9

Kłodzko (German **Glatz**), town in Wałbrzych province, Poland, on the River Nysa, 80 km/50 mi south of Wrocław; population (1990) 30,400. Textiles and machinery are manufactured. Lying at the southern edge of the Sudeten Mountains, near the Czech border, Kłodzko was part of German Lower Silesia until 1945. **18** F7

Klondike former gold-mining area in northwest ►Yukon Territory, Canada, near Dawson, where the Klondike and Yukon rivers meet. It is named after the river valley (length 193 km/120 mi) near where gold was found in August 1896. By 1898, at the height of the 'Klondike Gold Rush', over 30,000 people had moved into the area.

Klosterneuburg suburb of ►Vienna, Austria, in the province of Lower Austria. It lies on the northern edge of the metropolitan area, on the slopes of the Wienerwald. An Augustinian abbey was founded here in 1106. The town produces wine and pharmaceuticals. **30** M2

Klosters fashionable Alpine skiing resort in Grisons canton, eastern Switzerland, on the River Landquart, 10 km/6 mi northeast of Davos; population (1990) 3,500. It lies at an altitude of 1,191 m/3,908 ft. To the east, the **Kloster Pass** leads to Austria. **30** E4

Knokke-Heist fashionable resort town on the coast of West Flanders, Belgium, 16 km/10 mi north of Bruges; population (1997) 32,700. The fishing village of Knokke merged with the residential suburb of Het Zoute in 1880 and, as the town developed, other villages were absorbed. It has a casino and one of the largest golf courses in Europe. **22** F3

Knoxville city and inland port in eastern Tennessee, USA, on the upper Tennessee River, 1,000 km/620 mi from its junction with the Ohio River; seat of Knox County; population (1994 est) 169,000. It is the centre of a mining and agricultural region, and the administrative headquarters of the ►Tennessee Valley Authority; it is also a tourist centre for the Great Smoky Mountains National Park, located 40–50 km/25–31 mi to the south. Industries include textiles, clothing, and food products. **88** D3

Kobe deep-water port on Osaka Bay in Hyogo prefecture, southern Honshu island, Japan; population (1994) 1,479,000. It was a treaty port between 1868 and 1899, for foreigners exempt from Japanese law. **Port Island**, an artificial island of 5 sq km/3 sq mi in Kobe harbour, was created in 1960–68 from the rock of nearby mountains. It was one of the world's largest construction projects, and is now a residential and recreation area with a luxury hotel, amusement park, and conference centres. It is linked to the city by a driverless, computerized monorail. 48 H6

København Danish name for ►Copenhagen, the capital of Denmark. 16 G9

Koblenz (or **Coblenz**), commercial and university city in the Rhineland-Palatinate, Germany, at the junction of the rivers Rhine and Mosel, 59 km/37 mi northwest of Mainz; population (1995) 109,300. It has been a fortress city since Roman times, and an important communications centre. Industries range from barge-building and aluminium-rolling to the production of cigars and paper. The wine and brewing trades are also important. 20 C6

Kochi (formerly **Cochin**), town and port in Kerala state, India, on the Malabar coast; population (1991) 564,000; urban agglomeration (1991) 1,139,540. It is a fishing port, naval training base, and an industrial centre with oil refineries; ropes and clothing are also manufactured here. It exports coir, copra, tea, and spices. 54 C7

Kodiak Island largest island of the **Kodiak archipelago**, off the southwest coast of Alaska; area 9,505 sq km/3,670 sq mi; population (1990) 13,300. Fishing, trapping, dairying, and sheep and cattle ranching are important industries. The town of Kodiak is one of the largest US fishing ports; the main catch is salmon. The island is the home of the **Kodiak bear**, the largest bear in the world, and the Kodiak king crab. 92 (1)G4

Kokand (Uzbek **Quqon**), oasis city in Fergana wiloyat (region) in eastern Uzbekistan, 160 km/99 mi southeast of Tashkent; population (1996) 176,000. Industries include the manufacture of fertilizers, cotton, and silk. Kokand lies on the southwestern fringe of the agriculturally important Fergana Valley, and is a major transportation hub for the region. 42 N9

Kokomo city in north-central Indiana, on Wildcat Creek, 48 mi/78 km north of Indianapolis and 68 mi/109 km southwest of Fort Wayne; seat of Howard County; population (1990) 45,000. It is situated in a rich agricultural region; industries include the manufacture of car, radio, and plumbing parts, aluminum castings, metal alloys, steel and wire, and electrical machinery. 90 D1

Koko Nor Mongolian form of ►Qinghai, a lake and province of China. 11 (2)C3

Kola Peninsula (Russian **Kolskiy Poluostrov**), peninsula in the far northwestern Russian Federation, between the Barents Sea and the White Sea. Administratively, it forms part of Murmansk oblast (region). Its total area is 129,500 sq km/50,000 sq mi, and it has a population of 1.3 million (of whom 2,000 are Saami). To the northwest the low-lying granite plateau adjoins Norway's thinly populated county of Finnmark. 38 G1

Kolberg German name for ►Kołobrzeg, a town in Poland. 18 E3

Kolhapur industrial city and film production centre in Maharashtra, India, about 300 km/190 mi southeast of Mumbai (formerly Bombay); population (1991) 417,000. Lying at the centre of a rich bauxite-mining region, its industries include sugar-refining, textile manufacture and engineering. It was formerly capital of the Deccan state of Kolhapur, founded in the 10th century, which merged with Maharashtra in 1960. It is an ancient Buddhist site and a centre of pilgrimage. 54 B5

Kolín river port in central Bohemia, in the Czech Republic, on the Elbe, 25 km/16 mi east of Prague; population (1991) 31,600. It is a railway junction; industries include chemicals, petroleum refining, and engineering. During the Seven Years' War, the Austrians defeated the Prussians at Kolín 1757, forcing them out of Bohemia. 18 E7

Kolkata (formerly **Calcutta** until 2000), city in India, on the River Hooghly, the westernmost mouth of the River Ganges, some 130 km/80 mi north of the Bay of Bengal; population (1994) 11,500,000. The capital of West Bengal, it is chiefly a commercial and industrial centre, its industries including engineering, shipbuilding, jute, and other textiles. It was the seat of government of British India 1773–1912. There is severe air pollution. 54 E4

Köln German form of ►Cologne, a city in Germany. 20 B6

Kołobrzeg (German **Kolberg**), industrial port (shipyards) and seaside resort in Koszalin province, Poland, at the mouth of the River Parsęta on the Baltic Sea, 40 km/25 mi west of Koszalin; population (1990) 45,400. The town dates from the Middle Ages but was almost destroyed in World War II. 18 E3

Kolskiy Poluostrov Russian name for ►Kola Peninsula, a peninsula in far northwestern Russia. 38 G1

Kolwezi mining town in Shaba province, southeastern Democratic Republic of Congo; population (1994 est) 417,800. Copper and cobalt are mined and processed. In 1978 former police of the province invaded from Angola and massacred some 650 of the inhabitants. 72 D2

Komárno (German **Komorn**; Hungarian **Komárom**), river port and border town in southwest Slovak Republic, at the confluence of the rivers Váh and Danube; population (1991) 37,400. Textile manufacturing and shipbuilding are its chief industries. Formerly a strategic stronghold, until 1920 the port was part of Hungarian Komárom on the south bank; it remains connected by bridge. 34 F2

Komárom Hungarian name for ►Komárno, a town in the Slovak Republic. 34 F2

Komi republic in the northwestern Russian Federation; area 415,900 sq km/160,579 sq mi; population (1996) 1,185,000 (74% urban) (58% Russian, 23% Komi). The capital is ►Syktyvkar and Vorkuta is another city. The republic is partly in the Arctic Circle and is largely lowland (*taiga*) with coniferous forests. There are large coal seams (Pechora Basin) and oil deposits, and rich mineral resources (natural gas, oil shale, and bauxite). Industries include coal mining, oil and natural-gas extraction, and lumbering; and grain and dairy farming. 38 K2

Kommunizma, Pik (or **Communism Peak**; formerly **Garmo Peak** to 1933, **Stalin Peak** 1933–61), highest mountain in Pamirs, a mountain range in Tajikistan in Central Asia; height 7,495 m/24,590 ft. It was first climbed by a Soviet expedition in 1933. 56 K2

Komorn German name for ▶Komárno, a town in the Slovak Republic. 34 F2

Komotau German name for ▶Chomutov, a town in the Czech Republic. 18 C7

Kongur Shan mountain peak in China, 7,719 m/25,325 ft high, part of the Pamir Plateau. The 1981 expedition that first reached the summit was led by British climber Chris Bonington. 42 N10

Königgrätz German name for ▶Hradec Králové, a town in the Czech Republic. 18 E7

Königsberg former name of ▶Kaliningrad, a Baltic port in Russia. 18 K3

Königswinter town in North Rhine–Westphalia, Germany, on the River Rhine, 68 km/42 mi southeast of Düsseldorf; population (1994) 36,600. It is virtually a suburb of ▶Bonn. Situated in the picturesque Siebengebirge area, Königswinter is a centre for leisure activities and tourism. 20 C6

Konstanz German form of ▶Constance, a city in Germany. 30 E3

Konya (Roman **Iconium**), city in southwestern central Turkey; population (1990) 513,300. Carpets and silks are made here, and the city contains the monastery of the dancing dervishes. 58 E5

Koper (Italian **Capodistria**), seaport in Slovenia, off the Istrian peninsula, 15 km/9 mi southwest of Trieste; population (2001 est) 23,700. It stands on an island in the Gulf of Trieste and is connected with the mainland by a causeway. The town has saltworks and vehicle production, and trades in fish, oil, and wine. It was the capital of Istria 1278–1797, under Venetian and Genoan rule. 30 J5

Korçë (or **Korca**), town and district of southern Albania, near the Greek border; population (2001 est) 69,300 (town); 168,800 (district). It is an industrial and commercial centre producing machine tools, textiles, agricultural machinery, and leather products. It is situated in the country's chief wheat-growing region, and agriculture and vineyards are also important to the local economy. 36 C4

Korcula Yugoslav island in the Adriatic Sea, part of the Dalmatian archipelago; area 277 sq km/107 sq mi; population (1995 est) 8,200. The principal town is also called Korcula. 34 D7

Korea peninsula in East Asia, divided into north and south; see ▶Korea, North, and ▶Korea, South. 48 C4/48 D5

Korea, North see ▶North Korea. 48 C4

Korea, South see ▶South Korea. 48 D5

Korinthos Greek form of ▶Corinth, a port in Greece. 36 E7

Koror (formerly **Corrora**), provisional capital of the republic of Palau, on Koror Island, in the Caroline Islands of the western Pacific, 1303 km/810 mi southwest of Guam, and 853 km/530 mi southeast of Manila; population (1995 est) 11,500. Under Japanese administration until 1944, it was devastated during World War II. It is a commercial and tourist centre, with a copra-processing plant. A new capital is being built on the island of Babelthuap. 74 D5

Kortrijk Flemish form of ▶Courtrai, a town in Belgium. 22 F4

Kos (or **Cos**), Greek island, one of the Dodecanese, in the Aegean Sea; area 287 sq km/111 sq mi; population (1998 est) 21,300. A partly mountainous island, it is extremely fertile, and lies about 10 km/6 mi off the Turkish coast. The main town is Kos, located in the northeast of the island. 36 K8

Kosciusko, Mount highest mountain in Australia (2,228 m/7,310 ft), in the Snowy Mountains of the Australian Alps in southeast New South Wales, close to the border with Victoria. 76 J7

Košice city in the Slovak Republic; population (1991) 234,800 (92% Magyar-speaking). It has a textile industry and is a road centre. Košice was in Hungary until 1920 and 1938–45. 18 L9

Koslin German name for ▶Koszalin, a town in Poland. 18 F3

Kosovo (or **Kossovo**), autonomous region 1945–1990 of southern Serbia; capital ▶Priština; area 10,900 sq km/4,207 sq mi; population (1991) 2,012,500, consisting of about 210,000 Serbs and about 1.8 million Albanians. Products include wine, nickel, lead, and zinc. In 1990 fighting broke out between police, ethnic Albanians, who were agitating for unification of Kosovo with Albania, and Kosovo Serbs, who wished Kosovo to be merged with the rest of Serbia. The Serbian parliament formally annexed Kosovo in September, and Serbian troops were sent to the region in 1998. In 1999, after an 11-week bombing campaign against Serbia, NATO forces moved in to Kosovo to keep the peace, and the United Nations (UN) took over the civil administration of the province. 36 C2

Koszalin (German **Koslin**), capital of Koszalin province, Poland, near the Baltic Sea coast, 200 km/125 mi north of Poznań; population (1990) 108,700. Industries include engineering, electronics, wood, and food processing. 18 F3

Köszeg (German **Güns**), town in Vas county, Hungary, near the Austrian border crossing, 17 km/11 mi north of Szombathely; population (1991) 12,200. Industries include textiles, flour, and distilling. The town has many 18th-century buildings and historic monuments. 34 D2

Kota Bharu capital of Kelantan, Malaysia; population (1991) 220,000. The local economy centres on fishing; there is also batik craft work. 52 C1

Kota Kinabalu (formerly **Jesselton** until 1968), capital and port in Sabah, Malaysia; population (1991) 208,000.

Exports include rubber and timber. Originally named after Sir Charles Jessel of the Chartered Company which rebuilt the town, it is now named after Mount Kinabulu. 52 F1

Köthen town in the *Land* (administrative region) of Saxony-Anhalt, Germany, 29 km/18 mi north of Halle; population (1995) 32,000. Sugar and textile production are the main industries. The town was formerly the capital of the Duchy of Anhalt-Köthen, and the Renaissance-style ducal palace (1547–1608) is preserved. 20 G5

Kotka seaport in the southern Finnish province of Kymi on the Gulf of Finland; population (2001 est) 54,500. Situated on two islands at the mouth of the River Kymi, it is a centre of the pulp and timber trade, and the chief port for the exports and imports of eastern Finland. 16 P6

Kottbus alternative spelling of ▶Cottbus, a city in Germany. 18 D6

Kourou second-largest town of French Guiana, northwest of Cayenne, site of the Guiana Space Centre of the European Space Agency; population (1996) 20,000 (20% of the total population of French Guiana). **98 G2**

Kovno Russian form of ►Kaunas, a port in Lithuania. **14 G2**

Kowloon peninsula on the Chinese coast, until July 1997 part of the British crown colony of Hong Kong; the city of Kowloon is a residential area.

Kragujevac garrison city and former capital (1818–39) of Serbia, Yugoslavia; population (1991) 146,600. **34 H5**

Krakatoa (Indonesian **Krakatau**), volcanic island in Sunda Strait, Indonesia, that erupted in 1883, causing 36,000 deaths on Java and Sumatra from the tidal waves that followed. The island is now uninhabited.

Kraków (or **Cracow**), city in Poland, on the River Vistula, 300 km/190 mi south of Warsaw; population (2001 est) 735,800. It is an industrial centre producing railway wagons, paper, chemicals, and tobacco. It was capital of Poland from 1320 to 1611. **18 J7**

Kramatorsk industrial city in eastern Ukraine, in the ►Donets Basin, 90 km/56 mi north of Donetsk; population (1990) 199,000. Industries at this important road and rail junction include the manufacture of railway and coal-mining machinery, chemicals, steel, ceramics, and foodstuffs. The city grew from a small railway settlement established in the 1860s, expanding greatly at the end of the 19th century. **38 G5**

Krasnodar (formerly **Ekaterinodar** before 1920), capital city, economic and cultural centre of Krasnodar krai (territory), Russian Federation; population (1996 est) 648,000. The city stands on the River Kuban, 250 km/155 mi south of Rostov-on-Don, in the centre of a highly fertile agricultural region. Its industries include food processing, engineering, and oil refining. It is also an important railway junction. There has been considerable development of financial, business, and other services since the collapse of communism in 1991. It is linked by pipeline with the Caspian oilfields. **38 G5**

Krasnoyarsk capital, economic and cultural centre of Krasnoyarsk krai (territory), in central Siberia, Russian Federation; population (1996 est) 871,000. Situated on the Yenisey River and the ►Trans-Siberian Railway, Krasnoyarsk played a key role in the opening up and industrial development of Siberia. Heavy-engineering works, chemical plants, and a number of light industries (including food processing) are sited within the city. Upstream from Krasnoyarsk lies a large hydroelectric power facility. **44 E5**

Krefeld (formerly **Krefeld-Uerdingen**), industrial city in North Rhine-Westphalia, Germany, near the River Rhine; 52 km/32 mi northwest of Cologne; population (1995) 249,900. Products include pharmaceuticals, hosiery, textiles, foodstuffs, and machinery. It is situated on the Westphalian coalfield. **22 J3**

Kremenchuk industrial city on the River Dnieper, in the ►Poltava oblast in central Ukraine; population (1990) 238,000. Kremenchuk lies some 130 km/81 mi northwest of Dnipropetrovs'k. There is a large hydroelectric installation near the city that supplies energy for its industries. Goods manufactured in Kremenchuk include road-building machinery, railway wagons, petrochemical products, and processed food. The city suffered extensive damage during World War II. **38 F5**

Krems Austrian town in the province of Lower Austria, on the River Danube; population (1995) 22,450. Wine and fruit are traded, and industries include the manufacture of leather goods and preserves. It has a 13th-century Dominican church (now a museum). **30 L2**

Kreuznach (or **Bad Kreuznach**), city and spa in the Rhineland-Palatinate, Germany, on the River Nahe, 15 km/9 mi south of Bingen; population (1994) 154,700. It is a wine-growing centre, and has optical and leather industries. Kreuznach was founded in Celtic times and the castle dates from the 14th century. **22 K5**

Kristiansand seaport and administrative centre of Vest-Agder county, on the south coast of Norway; population (1995 est) 69,300. The third-largest port of Norway, it is 280 km/174 mi southwest of Oslo. Industries include textiles, timber, electrometallurgy, and electrochemicals. **16 E7**

Kristianstad city and administrative centre in the northeast of Skåne county, southeast Sweden; population (2000 est) 31,000. It is situated east of Hälsingborg, 22 km/14 mi from the Baltic coast. The main industry is textile manufacture, and there is a 17th-century church. **16 H8**

Kriti Greek for ►Crete, an island in the eastern Mediterranean Sea. **36 H10**

Krivoy Rog (Ukrainian **Kryvyy Rih**; Russian 'crooked horn'), city in south-central Ukraine on the Ingulets River, 130 km/81 mi southwest of Dnipropetrovs'k; population (1990) 717,000. The district surrounding Krivoy Rog is rich in iron ore, and the city's industries centre on metallurgical products. Mining machinery, cement, and foodstuffs are also produced. Exploitation of the iron-ore deposits began on a large scale in the 1880s. Krivoy Rog was occupied by German forces 1941–44. **38 F5**

Kruger National Park game reserve in Mpumalanga Province, South Africa, bordering Mozambique and Zimbabwe, between the Limpopo and Crocodile rivers; area about 20,720 sq km/8,000 sq mi. The Sabie Game Reserve was established in 1898 by President Kruger, and the park declared in 1926.

Krugersdorp mining city in the West Rand district, Gauteng Province, South Africa, 32 km/20 mi northwest of Johannesburg; population (1991) 196,200. Manganese, uranium, and gold are mined. Other industries include light engineering and chemical production. **72 D5**

Krung Thep Thai name for ►Bangkok, the capital of Thailand. **50 C4**

Krusevac town in Serbia, Yugoslavia, on the River Morava; population (1991) 58,800. It manufactures wine, tobacco, and munitions. Before 1389 it was the capital of the Serbian princes. **34 J6**

Krym' Ukrainian name for ►Crimea. **58 E1**

Kryvyy Rih Ukrainian name for ►Krivoy Rog, a city in south-central Ukraine. **38 F5**

Kuala Lumpur (Malay 'muddy confluence'), capital of the Federation of ►Malaysia; population (1997 est) 1,374,700. The city lies in a central position on the Malay peninsular at the confluence of the Kelang and Gombak rivers, 40 km/25 mi from the west coast. Malaysia's leading commercial and industrial centre, Kuala Lumpur developed after 1873 with the expansion of tin and rubber trading and processing, which remain important. Other industries include iron, cement, food processing, and the manufacture of electrical, electronic, and railway equipment. International trade is conducted through Port Kelang, on the Strait of Malacca, and through the international airport, which opened in 1998. 52 C2

Kuching capital and port of Sarawak state, eastern Malaysia, on the Sarawak River; population (1991) 148,000. 52 E2

Kulmbach city in Bavaria, Germany, on the River Main, in the foothills of the Fichtelgebirge, 20 km/12 mi northwest of Bayreuth; population (1994) 77,800. It has breweries, and a massive 13th-century Hohenzollern fortress. 20 G6

Kumamoto city on the Shira River, Kyushu island, Japan, 80 km/50 mi east of Nagasaki; population (1994) 628,000. It is the seat of the prefecture of the same name and a centre for fishing, food processing, and textile industries. The city was a military stronghold until the 19th century and remained a garrison city until 1945. The castle was built in 1601. 48 F7

Kumasi (formerly **Coomassie**), second-largest city in Ghana, capital of Ashanti region, situated 160 km/100 mi from the coast within the forest zones and 180 km/112 mi northwest of the capital and port of Accra; population (1998 est) 607,900. Kumasi is a major centre of Ghana's transport system, with an airport, as well as road and rail communications. It trades in cocoa, rubber, and cattle and its market is one of the largest in Africa. Food processing, brewing, logging, lumber, and tourism are the main industries. 68 D3

Kumayri alternative spelling of ►Gyumri. 58 K3

Kunlun Shan (or **Kunlun Mountains**), range in China, on the edge of the great Tibetan Plateau. Extending 4,000 km/2,500 mi east–west, its highest peak is the Muztag at 7,282 m/23,900 ft. 54 D1

Kunming (formerly **Yünnan**), capital of ►Yunnan province, China, on Lake Dianchi, about 2,000 m/6,500 ft above sea level; population (1994) 1,625,000. It is an important trading centre between the far west and central and south China. Industries include engineering and the manufacture of chemicals, textiles, iron, steel, machinery, cigarettes, heavy goods vehicles, plastics, and cement. Copper is smelted with nearby hydroelectric power. 46 C6

Kuopio capital of Kuopio province in eastern central Finland; population (1996) 85,300. It produces timber, matches, and flour, and has distilleries. Situated on Lake Kallavesi, it is a tourist and sports centre. The town was founded 1782 by Gustavus III of Sweden, and has a university. 38 E2

Kura (ancient Greek **Kuros**), major river in Transcaucasia, rising in Turkey, and flowing north through Georgia and Azerbaijan; length 1,515 km/941 mi. The Kura empties into the Caspian Sea south of Baku. Its main tributary is the Araks. Barrages for hydroelectric power and irrigation have been built at Tbilisi and at Mingäçevir, where a large reservoir has been created. 56 E2

Kurdistan (or **Kordestan**), mountain and plateau region in southwest Asia near Mount Ararat, where the borders of Iran, Iraq, Syria, Turkey, Armenia, and Azerbaijan meet; area 193,000 sq km/74,600 sq mi; total population 25–30 million. It is the home of the Kurds and is the area over which Kurdish nationalists have traditionally fought to win sovereignty. It is also the name of a northwest Iranian province in the Zagros Mountains, covering 25,000 sq km/9,650 sq mi, population (1991) 1,233,500. The chief towns of the region are ►Kermanshah (Iran); Irbil, Sulaymaniyah, and ►Kirkuk (Iraq); Divarbakir, Erzurum, and ►Van (Turkey); and Qamishle (Syria).

Kure naval base and port 32 km/20 mi southeast of Hiroshima, on the south coast of Honshu island, Japan; population (1995 est) 209,500. The city has a natural harbour, protected on the seaward side by islands. Industries include pulp, engineering works, and shipyards, which have an important export market. 48 G6

Kuril Islands (or **Kuriles**; Russian **Kuril'skiye Ostrova**), chain of about 50 small islands belonging to Russia, stretching from the northeast of Hokkaido, Japan, to the south of Kamchatka, Russia, area 14,765 sq km/5,700 sq mi; population (1990) 25,000. Some of the islands are of volcanic origin with active volcanoes and hot springs. Two of the Kurils (Etorofu and Kunashiri) are claimed by Japan for historical reasons; they are of strategic importance and have mineral deposits. The surrounding waters are rich in salmon and cod. 44 S7

Kuril'skiye Ostrova Russian name for ►Kuril Islands, a chain of small islands belonging to Russia, between Japan and Kamchatka, Russia. 44 S7

Kursk capital city, economic and cultural centre of Kursk oblast (region), western Russian Federation, on the banks of the River Seim, and bordered on the west by Ukraine; population (1996 est) 442,000. It has engineering, chemical, textile, and light industries, and is an important railway junction. Kursk and the surrounding region was the site of a decisive tank battle in July–August 1943 which signalled the ultimate defeat of German forces on the Eastern Front. 38 G4

Kutno town in Płock province, Poland, on the River Ochnia, 53 km/33 mi north of Łódź; population (1990) 123,400. Lignite is mined in the district, and industries include engineering, cement, soap, sugar, and flour production. It is a railway junction. 18 J5

Kuwait country in southwest Asia, bounded north and northwest by Iraq, east by the Gulf, and south and southwest by Saudi Arabia.

Kuwait covers an area of 17,819 sq km/6,879 sq mi, and has a population of 1,972,000 (2000 est). The capital (and chief port) is ►Kuwait; other major towns and cities include as-Salimiya, Hawalli, Faranawiya, Abraq Kheetan, Jahra, Ahmadi, and Fahaheel. The head of state is Sheikh Jabir al-Ahmad al-Jabir al-Sabah (from 1977); the head of government is Crown Prince Sheikh Saad al-Abdullah al-Salinas al-Sabah (from 1978). The currency is the Kuwaiti dinar, and the languages spoken include Arabic (official), English, Kurdish, and Farsi. 61 B2

Kuwait (Arabic **Al Kuwayt**; formerly **Qurein**), chief port and capital of the state of Kuwait, on the southern shore of Kuwait Bay; population (1993) 31,200. Kuwait is a banking and investment centre. It was heavily damaged during the Gulf War. 61 C2

Kuznetsk Basin (or **Kuzbas**), coal-mining basin and industrial region in the Kemerovo oblast (region) of southern Siberia, in the Russian Federation. It is the second-largest source of deep-mined coal in the Russian Federation and one of the largest in the world. The Kuzbas lies between the two mountain spurs – the Salair ridge and the Kuznetsky Alatau – that branch off north and northwest from the main Altai range. The basin is traversed by the River Tom. Coal-bearing seams here extend over an area of 26,700 sq km/10,309 sq mi, and reach to a depth of 1,800 m/5,905 ft, while the whole industrial area of the Kuznetsk Basin covers some 70,000 sq km/27,027 sq mi. Overall coal deposits are estimated at 725 billion tonnes. Iron ore is also found in this region, near Salair. The principal city of the Kuznetsk Basin is ►Novokuznetsk.

Kwangchow alternative transliteration of ►Guangzhou, capital of Guangdong province, China. 50 E2

Kwangju (or **Kwangchu**), capital of South Cholla province, southwestern South Korea; population (1995 est) 1,257,500. It is at the centre of a rice-growing region. Beverages, textiles, and motor vehicles are also important industries. The city dates from the first century BC, but grew rapidly in modern times after being linked by rail to Seoul in 1914. 48 D6

KwaZulu-Natal province of the Republic of South Africa, formed from the former province of Natal and the former Black National State of KwaZulu; area 91,481 sq km/35,321 sq mi; population (1995 est) 8,713,100 (75% Zulu). The capital is ►Pietermaritzburg, and Durban and Richards Bay are other towns. The province is a narrow plain bounded by the Drakensberg Mountains to the west and the Indian Ocean to the east. Industries include oil refining, coal, iron and steel, engineering, and food processing. Agricultural products include sugar, maize, fruit, black wattle, maize, tobacco, and vegetables. Languages spoken are Zulu (80%), English (15%), and Afrikaans (2%). 72 E5

Kweilin alternative transliteration of ►Guilin, a city in Guangxi Zhuang Autonomous Region, China. 50 E1

Kwidzyń (German **Marienwerder**), town in Elbląg province, Poland, near the River Vistula, 72 km/45 mi south of Gdańsk; population (1993 est) 37,700. Formerly in East Prussia, the town was a medieval stronghold of the Teutonic Knights (a German Christian military order). It was assigned to Poland 1945. 16 K10

Kyklades Greek for the ►Cyclades, a group of about 200 islands in the Aegean Sea. 36 G7

Kyoga, Lake lake in central Uganda, area 4,425 sq km/1,709 sq mi. It lies 1,036 m/3,399 ft above sea level. The lake is a vast swamp with sudd (heavily-matted vegetation) covering most of the area; footpaths form a network through the region in dry weather. It is connected with Lake Kwania and Lake Salisbury. 70 E3

Kyoto (or **Kioto**), former capital of Japan 794–1868 (when the capital was changed to Tokyo) on Honshu island, linked by canal with Lake Biwa, 510 km/317 mi west of Tokyo and 40 km/25 mi northeast of Osaka; population (1994) 1,391,000. Industries include electrical, chemical, and machinery plants; silk weaving; and the manufacture of porcelain, bronze, lacquerware, dolls, and fans. 48 H6

Kyrenia (or **Keryneia**), port in Turkish-occupied Cyprus, about 20 km/12 mi north of Nicosia; population (1985) 7,000. The Turkish army landed at Kyrenia during the 1976 invasion, and the town was temporarily evacuated. 58 E6

Kyrgyzstan (or **Kirghizia**), country in central Asia, bounded north by Kazakhstan, east by China, west by Uzbekistan, and south by Tajikistan.

Kyrgyzstan covers an area of 198,500 sq km/76,640 sq mi, and has a population of 4,699,000 (2000 est). The capital is ►Bishkek (formerly Frunze); other major towns and cities include Osh, Przhevalsk, Kyzyl-Kiya, Tokmak, and Djalal-Abad. The head of state is Askar Akayev (from 1991); the head of government is Kurmanbek Bakiyev (from 2000). The currency is the som, and the languages spoken include Kyrgyz (a Turkic language; official) and Russian. 42 N9

Kyushu (or **Kiushu**), southernmost of the main islands of Japan, separated from Shikoku island by the Bungo Strait, from Honshu island by the Kammon Strait, and from Korea by the Korea Strait; connected to Honshu by bridge and rail tunnel; area 42,150 sq km/16,270 sq mi, including about 370 small islands; population (1995) 13,424,000. The capital is ►Nagasaki, and Fukuoka, Kumamoto, Kagoshima, and Kitakyushu are other cities. The island is mountainous, with a sub-tropical climate. It is volcanic and the active volcano Aso-take (1,592 m/5,225 ft) has the world's largest crater. Industries include semiconductors, coal, gold, silver, iron, tin, timber, tourism, cattle, and hogs. Agricultural products include rice, tea, oranges, tobacco, and sweet potatoes. 48 F7

Kyustendil town with hot springs in southwestern Bulgaria, southwest of Sofia; population (1993) 54,500. 36 E2

Kyyiv Ukrainian name for ►Kiev, the capital of Ukraine. 38 F4

Kyzyl-Kum (or **Peski Kyzyylkum**; Turkish 'red sands'), desert straddling the territory of southern Kazakhstan and Uzbekistan, and located between the Syr Darya and ►Amu Darya rivers; it covers an area of some 300,000 sq km/116,000 sq mi. In the centre and west of the desert, there are large artesian basins; the Kyzyl-Kum is being reclaimed for cultivation by irrigation and protective afforestation. In the south lies an important natural gasfield at Gazli. 42 L9

Laâyoune (Arabic **El Aaiún**), capital of ►Western Sahara; population (1998 est) 139,000. It has expanded from a population of 25,000 in 1970 as a result of Moroccan investment (Morocco lays claim to Western Sahara). Laâyoune is the main urban centre of the country, which was formerly the Spanish Sahara, and lies in an artificial oasis where, through irrigation, cereals and vegetables are grown. To the south-west, especially at Bu Craa, there are large deposits of phosphates, which are linked to the coast by a conveyor belt 29 km/18 mi long, though exploitation has been handicapped by a shortage of water in the area. Since 1976 it has been capital of the (not internationally recognized) Laâyoune province. 66 C3

Labrador area in northeastern Canada, part of the province of ►Newfoundland, lying between Ungava Bay on the northwest, the Atlantic Ocean on the east, and the Strait of Belle Isle on the southeast; area 266,060 sq km/102,730 sq mi; population (1991) 30,000. The most easterly part of the North American mainland, Labrador consists primarily of a gently sloping plateau with an irregular coastline of numerous bays, fjords, inlets, and cliffs (60–120 m/200–400 ft high). Its industries include fisheries, timber and pulp, and the mining of various minerals, especially iron ore. Hydroelectric resources include Churchill Falls, where one of the world's largest underground power houses is situated (opened in 1971). There is a Canadian Air Force base at Goose Bay on Lake Melville. 82 U6

Laccadive, Minicoy, and Amindivi Islands former name of the Indian island group ►Lakshadweep. 10 B6

La Chaux-de-Fonds industrial commune in the canton of Neuchâtel, Switzerland; population (1995) 37,400. It is situated at an altitude of 990 m/3,248 ft in the ►Jura Mountains, 14 km/9 mi northwest of the town of Neuchâtel. It is a centre of the watch-making industry, and has a school for the engraving and enamelling of watch cases. 30 B3

La Crosse city in southwestern Wisconsin, USA, at the confluence of the Black, **La Crosse**, and Mississippi rivers, 200 km/124 mi northwest of Madison; seat of La Crosse County; population (1992) 51,350. It is the processing and marketing centre for the area's agricultural products; industries include plastics, brewing, heating systems, rubber products, electrical machinery, and also tourism. 88 B2

Ladoga, Lake (Russian **Ladozhskoye Ozero**), largest lake in Europe, and the second largest in the Russian Federation after the Caspian Sea, situated in the far northwest of the Russian Federation, in the Republic of Karelia and Leningrad oblast (region), northeast of the city of St Petersburg. Lake Ladoga covers an area of 17,700 sq km/6,834 sq mi (with its islands, 18,135 sq km/7,002 sq mi). The main feeder rivers are the Volkhov, Svir, and Vuoksa, and the lake's outlet is by way of the Neva River into the Gulf of Finland. Its average depth is 51 m/167 ft, and its maximum depth 230 m/755 ft. A valuable commercial fishing industry is based here. 38 F2

Ladozhskoye Ozero Russian name for Lake ►Ladoga, the largest lake in Europe, situated in far northwest Russia. 38 F2

Ladysmith town in province of KwaZulu-Natal (formerly Natal), South Africa, 185 km/115 mi northwest of Durban, near the Klip River; population (1991) 12,700. It has textile, clothing, and other light industries, and has an important railway depot. Founded in 1851, it was besieged for 118 days during the Boer War. It was named after the wife of the governor of Natal, Sir Harry Smith. 72 D5

Lafayette city and administrative headquarters of Tippecanoe County, north-west Indiana, USA, on the Wabash River, 93 km/58 mi northwest of Indianapolis; population (1990) 43,760. It is a distribution centre for the area's agricultural products; industries include building materials, sheet metal, chemicals, wire, pharmaceuticals, truck trailers, and motor-vehicle engines and parts. Purdue University (1865) is nearby. 88 C2

Lafayette city and administrative headquarters of Lafayette Parish, southern Louisiana, USA, on the Vermilion River, 230 km/143 mi west of New Orleans; population (1994 est) 102,000. Situated in an agricultural region, its economy centres around the area's oil industry. Other industries include the manufacture of cordage (ship rigging and lines), packing materials, and processed food. Lafayette is home to a Carmelite monastery and the University of Southwestern Louisiana (1898). 90 C3

Lago di Garda Italian name for Lake ►Garda, the largest lake in Italy. 30 F5

Lagos chief port and former capital of Nigeria, located at the western end of an island in a lagoon and linked by bridges with the mainland via Iddo Island; population (1992 est) 1,347,000. Industries include chemicals, metal products, fish, food processing, light engineering, chemicals, and brewing. Its surrounding waters are heavily polluted. 68 E3

Lahn river in Germany, a tributary of the ►Rhine; length 217 km/135 mi. It rises in the Rothaargebirge, 15 km/9 mi east of Siegen, and flows to join the Rhine 9 km/6 mi south of Koblenz. The main cities on its banks are Marburg, Giessen, Wetzlar, and Limburg. 22 L4

Lahore capital of the province of Punjab, Pakistan, situated on a tributary of the River Ravi, 50 km/30 mi west of Amritsar in India; population (1991) 3,200,000. Lahore is a commercial and banking centre, and industries include engineering, textiles, carpets, and chemicals. It is associated with the Mogul rulers Akbar, Jahangir, and Aurangzeb, whose capital it was in the 16th and 17th centuries. 54 B2

Lahti town in southern Finland, in southeastern Häme province; population (1995) 100,000. Timber, furniture, and domestic appliances are manufactured. It is a winter-sports resort with a huge ski jump. 16 N6

Laibach German name of ►Ljubljana, a city in Slovenia. 30 K4

Lake Charles city in southwestern Louisiana, USA, on the Calcasieu River, 300 km/186 mi west of New Orleans; seat of Calcasieu Parish; population (1992) 71,100. It is a port of entry on the Gulf of Mexico via a deep-water channel (completed in 1926) in the Calcasieu River. Most of the city's industries are related to the area's oil and gas resources; they include chemicals, oil refining, plastics, and synthetic rubber. Lake Charles is also a distribution centre for the region's major crop, rice. 90 C3

Lake District region in Cumbria, northwest England. It contains the principal English lakes, separated by wild uplands rising to many peaks, including ►Scafell Pike (978 m/3,210 ft), the highest peak in England. The area was made a national park in 1951, covering 2,292 sq km/885 sq mi, and is a popular tourist destination. 24 J7

Lake Havasu City city in Mohave County, west-central Arizona, USA; population (1990) 24,400. The city lies on the eastern shore of Havasu

Lake. The lake, which forms part of the boundary between Arizona and California, is used for fishing, boating, and water skiing. Another attraction of the city is its clear desert air. The Old London Bridge was transported and reconstructed here in 1971. The city was founded in 1963 as a tourist and retirement resort. 92 D2

Lakeland city in Polk County, west-central Florida, USA, 46 km/29 mi east of Tampa and 82 km/51 mi southwest of Orlando, in the lake region and citrus belt; population (1992) 72,600. Although it serves as a centre for the area's citrus products and as a centre for phosphate mining, its economy depends mainly on its reputation as a winter resort. 90 E4

Lake of the Woods lake in southwest Ontario, Canada, on the border between Canada and the USA, extending into the province of Manitoba, Canada, and the state of Minnesota, USA; it lies between Lake Winnipeg and Lake Superior; area 4,860 sq km/1,876 sq mi. Its waters are 25 m/82 ft deep. The chief river flowing into the lake is the Rainy River, from the east. 88 B1

Lakshadweep (or **Laccadive Islands**), group of 36 coral islands, 10 inhabited, in the Indian Ocean, 320 km/200 mi off the Malabar coast, forming a Union Territory of India; area 32 sq km/12 sq mi; population (1994 est) 56,000. The administrative headquarters are on Kavaratti Island. Products include coir, copra, and fish. There is a tourist resort on Bangarem, an uninhabited island with a large lagoon. The religion is Islam. 10 B6

La Laguna town in the ►Canary Islands, in the Spanish province of Santa Cruz de Tenerife; population (1995) 127,700. Formerly the capital of Tenerife, it has fine streets of balconied houses, a cathedral, and a university.

La Línea (officially **La Línea de la Concepción**), town and port in the province of Cádiz, southern Spain, on the isthmus of Algeciras Bay adjoining the frontier zone with Gibraltar; population (1991) 57,900. Cork, textiles, and foodstuffs are manufactured here; it is principally a trading centre with Gibraltar. 28 E8

La Louvière town in the province of Hainaut, Belgium, 18 km/11 mi east of Mons; population (2001 est) 76,600. It has collieries, blast furnaces, and stone, lime, and moulding-clay quarries. The chief manufactures are iron, steel, glass, ceramics, and fireproof products. 22 G4

Lancashire county of northwest England (since April 1998 Blackpool and Blackburn have been separate unitary authorities).
area 3,040 sq km/1,173 sq mi
towns and cities Preston (administrative headquarters), which forms part of Central Lancashire New Town from 1970 (together with Fulwood, Bamber Bridge, Leyland, and Chorley); Lancaster, Accrington, Burnley; ports Fleetwood and Heysham; seaside resorts Morecambe and Southport
features the River Ribble; the Pennines; the Forest of Bowland (moors and farming valleys); Pendle Hill
industries formerly a world centre of cotton manufacture, now replaced with high-technology aerospace, nuclear fuels, and electronics industries. There is dairy farming and market gardening
population (1996) 1,424,700
5 E5

Lancaster city in Lancashire, northwest England, on the River Lune; population (1991) 44,500. Until 1974 it was the county town of Lancashire (now Preston is the county town). Industries include the manufacture of paper, furniture, plastics, chemicals, textiles, and wall and floor coverings (linoleum). Manufacturing accounts for 12% of employment, while tourism accounts for 10%. The service sector is the largest employer, and there is an important livestock market. Education facilities include the University of Lancaster (1964) and St Martin's College. The city's castle was originally established in the 11th century on the site of a Roman fort. 24 K7

Lancaster city in southeastern Pennsylvania, USA, 115 km/70 mi west of Philadelphia; seat of Lancaster County; population (1992) 57,200. It is the centre of a rich agricultural region; industries include the manufacture of textiles, machinery, healthcare products, watches, and electrical goods. 88 E2

Lanchow alternative transcription of ►Lanzhou, capital of Gansu province, China. 46 C3

Lancs abbreviation for ►Lancashire, an English county. 5 E5

Landau in der Pfalz city in the Rhineland-Palatinate, Germany, 35 km/22 mi northwest of Karlsruhe; population (1994) 39,400. There are two Gothic churches. Landau is in the central Pfälzerwald and is a centre of trade in fruit, tobacco, and wine.
22 L5

Landes *département* in the ►Aquitaine region of France, on the Bay of Biscay; area 9,236 sq km/3,566 sq mi; population (1990) 311,900. It is divided into two parts by the River Adour, whose valley has agriculture and food industries. Oil was drilled at Parentis until 1995. There are mineral springs at Dax, and rock salt is obtained there. The administrative centre is ►Mont-de-Marsan. 6 C4

Landsberg town in Bavaria, Germany, on the River Lech, 36 km/22 mi south of Augsburg; population (1994) 98,000. The town has a number of Gothic and rococo churches. 30 F2

Landsberg-an-der-Warthe German name for ►Gorzów Wielkopolski, a town in Poland. 18 E5

Land's End promontory of southwest Cornwall, 15 km/9 mi southwest of Penzance, the westernmost point of England. 24 F11

Landshut historic town in Bavaria, Germany, on the River Isar, 61 km/38 mi northeast of Munich; population (1995) 59,200. Chocolate, beer, dyes, paints, and machinery are manufactured here.

In the 14th century Landshut was the seat of the dukes of Bavaria-Landshut. There are several Gothic and Renaissance churches, most notably St Martin's (1392–1432), which has remarkable brickwork. Other features include a 14th-century Rathaus (town hall) and the Burg Trausuitz (1568–78), a Renaissance palace. 30 H2

Landskrona town and port in the province of Skåne, southern Sweden, on the ►Öresund, 32 km/20 mi north of Malmö; population (2000 est) 37,700. Industries include machinery, chemicals, shipyards, and sugar refining. 18 B2

Langeland (English **Long Island**), Danish island in the Great Belt channel between the islands of Fyn and Lolland. It is 50 km/31 mi long and very narrow. Chief exports are corn, timber, dairy produce, and fish. The beaches attract holidaymakers. Rudkøbing on the west coast is the chief town. **20** F2

Languedoc-Roussillon region of southern France, comprising the *départements* of Aude, Gard, Hérault, Lozère, and Pyrénées-Orientales; area 27,400 sq km/10,600 sq mi; population (1990) 2,115,000. The administrative centre is ▶Montpellier. Products include fruit, vegetables, and wine. **6** E5

Lansing state capital of ▶Michigan, USA, at the confluence of the Grand and Red Cedar rivers, in the centre of the Michigan peninsula, 125 km/78 mi northwest of Detroit; population (1994 est) 120,000. It is one of the ring of satellite cities, centred around Detroit, which form part of Michigan's large automobile industry; manufacturing includes motor vehicles, diesel engines, and pumps. **88** D2

Lanzarote most easterly of the Spanish Canary Islands; area 795 sq km/307 sq mi; capital Arrecife. The desertlike volcanic landscape is dominated by the Montañas de Fuego ('Mountains of Fire') with more than 300 volcanic cones. **66** C3

Lanzhou (or **Lanchow**), capital of ▶Gansu province, China, on the Huang He River, 190 km/120 mi south of the Great Wall; population (1994) 1,612,600. Industries include oil-refining and the manufacture of chemicals, fertilizers, machinery, and synthetic rubber. **46** C3

Laois (or **Laoighis**, previously spelt **Leix**; also formerly **Queen's County**), county of the Republic of Ireland, in the province of Leinster; county town ▶Portlaoise; area 1,720 sq km/664 sq mi; population (1996) 52,900. Other towns are Abbeyleix, Mountmellick, Mountrath, and Portarlington. Laois is flat, except for the Slieve Bloom Mountains in the northwest, the highest point of which is Mount Arderin (529 m/1,734 ft), and there are many bogs. The Barrow and the Nore are the chief rivers. Agriculture includes dairying, and mixed cattle and arable farming (sugar beet), and industries include peat, woollens, and agricultural machinery. Part of the Leinster coalfield lies within the county. There is a large peat-fired power station near Portarlington, and at the Clonsast Bog (1,619 ha/4,000 acres) is an important peat industry. **5** C5

Laon administrative centre of ▶Aisne *département* in the Picardy region of northern France; 125 km/78 mi northeast of Paris; population (1999 est) 26,200. It was the capital of France and a royal residence until the 10th century. Laon is a market centre, and has food-processing and metallurgical industries. Its cathedral, and the former abbey church of St-Martin, both date from the 12th century. **22** F5

Laos landlocked country in southeast Asia, bounded north by China, east by Vietnam, south by Cambodia, west by Thailand, and northwest by Myanmar.

Laos covers an area of 236,790 sq km/91,424 sq mi, and has a population of 5,433,000 (2000 est). The capital is ▶Vientiane; other major towns and cities include Louangphrabang (the former royal capital), Pakse, and Savannakhet. The head of state is Gen Khamtay Siphandon (from 1998); the head of government is Gen Sisavath Keobounphanh (from 1998). The currency is the new kip, and the languages spoken include Lao (official), French, English, and ethnic languages. **50** C3

La Palma most westerly of the Canary Islands, Spain, in the province of Santa Cruz de Tenerife; area 730 sq km/282 sq mi; population (1991) 78,000. The capital is Santa Cruz de la Palma. The island is mountainous, including Pico de la Cruz at 2,365 m/ 7,759 ft, and well forested. Industries include wine, fruit, honey, and silk. Tourism is important to the island. **66** B3

La Paz capital city of Bolivia, in Murillo province, 3,800 m/12,400 ft above sea level; population (1992) 711,000 (metropolitan area 1,126,000). It is in a canyon formed by the La Paz River, and is the world's highest capital city. Products include textiles and copper. It has been the seat of government since 1898, but Sucre is the legal capital and seat of the judiciary. **98** D7

Lapland region of Europe within the Arctic Circle in Norway, Sweden, Finland, and the Kola Peninsula of northwest Russia, without political definition. Its chief resources are chromium, copper, iron, timber, hydroelectric power, and tourism. The indigenous population are the Saami (formerly known as Lapps), 10% of whom are nomadic, the remainder living mostly in coastal settlements. Lapland has low temperatures, with two months of continuous daylight in summer and two months of continuous darkness in winter. There is summer agriculture. **16** M2

La Plata capital of Buenos Aires province, Argentina, on the Río de la Plata 48 km/30 mi southeast of the city of Buenos Aires; population (1991) 542,600; metropolitan area (1992 est) 676,100. It is 9 km/6 mi from its port, Ensenada, the main outlet for produce from the Pampas. Industries include meat packing and petroleum refining. It has one of the country's best universities. The city was founded in 1882 by Governor Daroo Rocha. **100** K5

Laptev Sea (Russian **More Laptevykh**), part of the Arctic Ocean off the north coast of Russia between Taimyr Peninsula and Novosibirskiye Ostrova (New Siberian Island); area *c.* 700,000 sq km/300,000 sq mi; average depth 578 m/1,896 ft, with a maximum depth of 3,385 m/11,105 ft. The chief port is Tiksi. **44** L1

Laramie city in southeastern Wyoming, USA, on the Laramie Plains – a plateau 2,300 m/7,500 ft above sea level – bounded north and east by the Laramie Mountains, 65 km/40 mi northwest of Cheyenne; seat of Albany County; population (1990) 26,680. The **Laramie River**, on which it stands, is linked with the Missouri River via the Platte River. The city is the centre of a ranching and lumber-producing area; its products include cement and wood products. **86** E2

Laredo city in southern Texas, USA, on the Rio Grande, 240 km/149 mi south of San Antonio, on the Mexican border; seat of Webb County; population (1994 est) 150,000. A centre for trade between Mexico and the USA, it grew rapidly after the railway arrived here in the 1880s and after the discovery of oil in the early 1920s. Industries include oil refining and metal processing as well as electrical equipment and processed food. Bermuda onions and early grapes are grown here. **92** G3

La Rioja autonomous community of northern Spain; area 5,000 sq km/

1,930 sq mi; population (1991) 261,600. The River Ebro passes through the region, but it is a tributary of the Río Oja, which gives its name to the region. La Rioja produces red and white wines with a characteristic flavour that derives from their storage in oak barrels. The capital is ▶Logroño. 9 (1)D1

Larisa town in Thessaly, Greece, south of Mount Olympus; population (1991) 113,400. Industries include textiles and agricultural produce. 36 E5

Larne seaport and industrial town of County Antrim, Northern Ireland, on Lough Larne, 30 km/19 mi north of Belfast; population (1991) 17,500. It is the terminus of sea routes principally to Cairnryan near Stanraer in Scotland, and has a turbine generator works and an electronics industry. 24 G7

La Rochelle fishing port on the Bay of Biscay in western France; population (1990) 73,700, conurbation 100,000. It is the administrative centre of ▶Charente-Maritime *département*. Industries include shipbuilding, chemicals, motor vehicles, fish-canning, and saw-milling. A Huguenot stronghold, it was taken by Cardinal Richelieu in the siege of 1627–28. 26 D7

La Roche-sur-Yon capital of the French *département* of Vendée; population (1990) 48,500. It is situated on a plateau overlooking the Yon valley. The town was purpose-built by Napoleon I in 1804, on the site of the old fortress of La Roche, and was called first Napoleon-Vendée and later Bourbon-Vendée. It is a market town and horse-breeding centre, and has flour mills. 26 D7

Las Cruces city in southern New Mexico, USA, on the Rio Grande, north of the Mexican border; population (1990) 62,120. It is the seat of Dona Ana County, and is a processing centre for the area's crops, such as pecans, cotton, and vegetables. White Sands Missile Range, an important testing site for NASA, is nearby. 92 E2

Lashio town in northern Myanmar (Burma), about 200 km/125 mi northeast of Mandalay; population (2001 est) 126,100.

The Burma Road to Kunming in China, constructed 1938, starts here. 50 B2

Las Palmas province of Spain in the Canary Islands autonomous community, comprising the islands of Gran Canaria, Lanzarote, and Fuerteventura; area 4,072 sq km/1,572 sq mi; population (1991) 749,500. Tourism is important. Fruit is cultivated, especially bananas, and there is some fishing also. The capital is Las Palmas. 66 B3

La Spezia port in northwest Italy and chief Italian naval base, 80 km/50 mi southeast of Genoa, on the Riviera di Levante; population (1992) 100,500. Industries include shipbuilding, engineering, oil-refining, and the manufacture of electrical goods and textiles. The English poet Percy Bysshe Shelley drowned in the Gulf of La Spezia in 1822. 30 E6

Las Vegas city in southeastern Nevada, USA; seat of Clark County; population (2000 est) 478,000. With its many nightclubs and gambling casinos, Las Vegas attracts millions of visitors each year. It is also a major convention centre. Founded in 1855 in a ranching area, the modern community developed with the coming of the railroad in 1905 and was incorporated as a city in 1911. The first casino hotel opened here in 1947. Las Vegas is the easiest place to get married in the USA, with numerous chapels along the Strip (main street) and hotel chapels. 86 C3

Latakia (or **Lattakia**; Arabic **Al Lādhiqīyah**), port with tobacco industries in northwest Syria, 120 km/75 mi from Tripoli (Lebanon); population (1993) 293,000. It exports tobacco, sponges, silk, and cotton. The tobacco, which takes its name from the city, is used in pipe mixtures. Latakia occupies the site of Laodicea and excavations have revealed remains of the earlier city. It was an important and wealthy city during the Crusades (medieval religious wars). 58 F6

Latin America large territory in the Western hemisphere south of the USA, consisting of Mexico, Central America, South America, and the West Indies. The main languages spoken are Spanish, Portuguese, and French.

Latvia country in northern Europe, bounded east by Russia, north by Estonia, north and northwest by the Baltic Sea, south by Lithuania, and southeast by Belarus.

Latvia covers an area of 63,700 sq km/24,594 sq mi, and has a population of 2,357,000 (2000 est). The capital is ▶Riga; other major towns and cities include Daugavpils, Leipāja, Jurmala, Jelgava, and Ventspils. The head of state is Vaira Vike-Freiberga (from 1999); the head of government is Andris Berzins (from 2000). The currency is the lat, and the main language spoken is Latvian (official). 16 M8

Launceston city and port in northeast Tasmania, Australia, on the estuary of the River Tamar, 240 km/149 mi north of Hobart; population (1996) 67,701. Launceston is Tasmania's second-largest city, and its industries include woollen-blanket weaving, aluminium smelting, saw-milling, engineering, food processing, brewing, railway workshops, and the manufacture of textiles, furniture, pottery, tennis racquets, and other sporting goods. The main exports are agricultural produce, fruit, minerals, timber, wool, and woollen goods. 76 J8

Lausanne resort and capital of Vaud canton, west Switzerland, above the north shore of Lake Geneva; population (1995) 143,200. It is a major railway junction on the Paris–Milan route through the Simplon Pass. Industries include publishing and the manufacture of chocolate and scientific instruments. It hosts international fairs and conferences, and is the headquarters of the International Olympic Committee. An Olympic Museum opened in 1993. The canton is mainly French-speaking and Protestant. 30 B4

Lawrence city in northeastern Kansas, USA, on the Kansas River between Topeka to the west and Kansas City to the east; seat of Douglas County; population (1992) 67,800. It is the centre of a rich grain-producing region; its main industries are food processing and chemicals. 88 A3

Lawrence city in northeastern Massachusetts, USA, on the Merrimack River; seat of Essex County; population (1992 est) 65,500. Lawrence is one of USA's most notable historic industrial cities, one of a line of industrial towns that grew up along the Merrimack Valley in the 1840s, and which used the water power of the Merrimack Rapids to drive the mills in which textiles and leather goods were made. Current industries include textiles, clothing, shoes, paper, computers, and radio equipment. Lawrence was incorporated as a city in 1853. **88** F2

Lawton city and administrative headquarters of Comanche County, southwest Oklahoma, USA, on Cache Creek, 125 km/78 mi southwest of Oklahoma City, population (1992) 87,200. Industries include processing the area's agricultural products, and the manufacture of cement, tyres, mobile homes, cosmetics, jewellery, and clothing. Lawton is the home of Cameron University (1909). **90** B3

Lazio (Roman **Latium**), region of west central Italy, comprising the provinces of Viterbo, Rieti, Rome, Frosinone, and Latina; area 17,200 sq km/6,600 sq mi; capital Rome; population (1992 est) 5,162,100. It is the third-largest region of Italy, over half its population living in the city of ▶Rome. Products include olives, wine, chemicals, pharmaceuticals, and textiles. Home of the Latins from the 10th century BC, it was dominated by the Romans from the 4th century BC. **9** (4)D3

Leamington Spa (officially **Royal Leamington Spa**), town and former health resort in Warwickshire, England, adjoining Warwick, on the River Leam, southeast of Birmingham; population (1991) 42,300. Public administration and services, distribution, tourism, and leisure industries provide the main employment. Manufacturing is more significant than the national average and includes engineering and automotive industries, producing brakes, steering, transmission, and other components. The Royal Pump Room offers spa treatment. **22** A2

Leavenworth city and administrative headquarters of Leavenworth County, northeast Kansa, USA, on the Missouri River, 32 km/20 mi northwest of Kansas City; population (1991) 38,800. Industries include paper products, candles, computer software, and foundry products. There is a commercial shipyard in the city. It has also become a centre for outdoor pursuits. Leavenworth was one of the chief river ports and fitting-out centres in the 19th-century colonization of the USA, via the ▶Santa Fe and ▶Oregon trails. It is home to St Mary College (1923). **88** A3

Lebanon country in western Asia, bounded north and east by Syria, south by Israel, and west by the Mediterranean Sea.

Lebanon covers an area of 10,452 sq km/4,035 sq mi, and has a population of 3,282,000 (2000 est). The capital (and chief port) is ▶Beirut; other major towns and cities include Tripoli, Zahlé, Baabda, Baalbek, and Jezzine. The head of state is Emile Lahoud (from 1998); the head of government is Rafik Hariri (from 2000). The currency is the Lebanese pound, and the languages spoken include Arabic (official), French, Armenian, and English. **60** C3

Lebanon city and administrative headquarters of Lebanon County, southeast Pennsylvania, 34 km/21 mi northeast of Harrisburg; population (1990) 24,800. Lebanon is in the fertile Pennsylvania Dutch farming region and is known for its Lebanon bologna, a Pennsylvania Dutch sausage. Industries include iron and steel products, textiles, clothing, footwear, and chemicals. **88** E2

Lech river in Austria and Germany; length 270 km/168 mi. The Lech rises in the Lechtaler Alps in the province of Vorarlberg, and flows north across the western Tirol, past Augsburg, to join the ▶Danube east of Donauwörth. **30** F3

Le Creusot town in Saône-et-Loire *département*, in the Burgundy region of east-central France; population (1990) 29,200, conurbation 41,000. It was once a coal-mining centre and now has foundries, locomotive shops, and armaments factories. It has given its name to a range of cast-iron cookware. **26** K7

Leeds industrial city and metropolitan borough in West Yorkshire, England, 40 km/25 mi southwest of York, on the River Aire; population (1991) 424,200 (city), 680,700 (district). Industries include engineering, printing, chemicals, glass, woollens, clothing, plastics, paper, metal goods, and leather goods. Notable buildings include the Town Hall (1858) designed by Cuthbert Brodrick, the University of Leeds (1904), the Leeds City Art Gallery (1888), Temple Newsam House (early 16th century, altered in about 1630), and the Cistercian Abbey of Kirkstall (1147). It is a centre of communications where road, rail, and canals (to Liverpool and Goole) meet. **24** L8

Leer river port in East Friesland, Lower Saxony, Germany, at the confluence of the rivers Ems and Leda, 175 km/108 mi northwest of Hannover; population (1995) 32,500. Industries include boatbuilding, food-processing, and the manufacture of textiles. The city's red-brick buildings are in the style of early baroque architecture of the Netherlands. **22** K1

Leeuwarden capital of Friesland province, the Netherlands, on the Ee River, 112 km/70 mi northeast of Amsterdam; population (1997) 88,500. The town's principal industry, foodstuff processing, is dependent upon regional produce. A marketing centre, it trades in grain, dairy produce, and cattle, and manufactures glass, gold and silver ware, musical instruments, and textiles. After the draining of the Middelzee fenlands, the town changed from a port to an agricultural market town. Notable buildings include the palace of the stadholders of Friesland and the church of St Jacob. **22** H1

Lefkada (or **Kevkas**; Greek 'white land'), one of the ▶Ionian Islands of Greece, 80 km/50 mi southeast of Corfu; area 293 sq km/113 sq mi; population (1993) 21,400. The main products are olive oil, currants, and wine. The principal town, Lefkada, at the northeastern corner, is capital of the Lefkada department. **36** C6

Lefkosia Greek name for ▶Nicosia, the capital of Cyprus. **36** R9

Leghorn former English name for the Italian port ▶Livorno. **30** F7

Legnica (German **Liegnitz**), capital of Legnica province, Poland, 64 km/40 mi west of Wrocław; population (1990) 105,200. Lead is refined from local mines, and cables, textiles, and chemicals are manufactured. Formerly in Lower Silesia, it was the capital of the Duchy of Liegnitz 1163–1675, passing to the Habsburg family before joining Prussia 1742 and then Poland 1945. **18** F6

Leh chief town of Ladakh region, Jammu and Kashmir, India, situated east of the River Indus, 240 km/150 mi east of Srinagar; population

(1991) 9,000. It developed as a trading post and market, but tourism is now important. Leh is the nearest supply base to the Indian army outpost on the Siachen Glacier. **54** C2

Le Havre industrial port in the *département* of Seine-Maritime in Normandy, northwest France, on the north side of the estuary of the River Seine, 90 km/56 mi from Rouen; population (1990 est) 190,900. It serves 500 harbours worldwide. It is the second-largest port in France, and has cross-channel passenger links to Portsmouth, England. Industries include textiles, engineering, chemicals, car manufacturing, and oil refining. It is a major import centre for cotton, coffee, sugar, and wool. As France's largest ocean port for foreign trade and container traffic, and the fifth-largest European port, Le Havre serves 500 harbours worldwide. **22** C5

Leicester City industrial city and unitary authority in central England, on the River Soar. It was part of the county of Leicestershire to 1997.
area 73 sq km/28 sq mi
features 14th-century Guildhall, St Martin's Cathedral, and two universities (University of Leicester, established in 1957, and De Montfort University, formerly Leicester Polytechnic, established in 1992); Bradgate House, the home of Lady Jane Grey
industries engineering, food processing, electronics, chemicals, and the manufacture of hosiery, footwear, knitwear, plastics, scientific and medical instruments, electrical products, and construction and woodworking machinery
population (1996) 270,500 **22** A2

Leicestershire county of central England (since April 1997 Leicester City and Rutland have been separate unitary authorities).
area 2,084 sq km/804 sq mi
towns and cities Loughborough, Melton Mowbray, Market Harborough (administrative headquarters at Glenfield, Leicester)
physical rivers Soar and Wreake; Charnwood Forest (in the northwest); Vale of Belvoir (under which are large coal deposits)
features Belvoir Castle, seat of the dukes of Rutland since the time of Henry VIII; Donington Park motor-racing circuit, Castle Donington; Leicestershire has traditionally had several fox-hunts, including the Quorn hunt
agriculture good pasture with horses, cattle, and sheep); dairy products (including Stilton cheese at Melton Mowbray); cereals
industries engineering; hosiery; footwear; bell founding; coal; quarrying of limestone, ironstone, and granite
population (1996) 927,500 **5** F5

Leics abbreviation for ►**Leicestershire**, an English county. **5** F5

Leiden (or **Leyden**), city in South Holland province, the Netherlands, on the Oude Rijn River, 10 km/6 mi from the North Sea, and 27 km/17 mi north of Rotterdam; population (1997) 117,000. Industries include textiles and cigars. It has been a printing centre since 1580, with a university established in 1575. It is linked by canal to Haarlem, Amsterdam, and Rotterdam. The painters Rembrandt and Jan Steen were born here. **22** G2

Leie alternative name for the River ►Lys of France and Belgium. **22** F4

Leinster southeastern historic province of the Republic of Ireland, comprising the counties of Carlow, Dublin, Kildare, Kilkenny, Laois, Longford, Louth, Meath, Offaly, Westmeath, Wexford, and Wicklow; area 19,630 sq km/7,580 sq mi; population (1996) 1,924,700.

Leipzig major commercial and industrial city in west Saxony, Germany, on the Weisse Elster (a tributary of the River Elbe), 145 km/90 mi southwest of Berlin; population (1995) 478,200. Industries include printing, publishing, and the production of furs, leather goods, paper, and musical instruments. It hosts numerous trade shows, including important industrial and book fairs. The city is also a centre for the arts, culture, and education, and has a university founded in 1409. **20** H5

Leitrim county of the Republic of Ireland, in the province of Connacht, bounded on the northwest by Donegal Bay; county town Carrick-on-Shannon; area 1,530 sq km/591 sq mi; population (1996) 25,100. Carrick-on-Shannon, Mohill, and Manorhamilton are the most important towns. The rivers Shannon, Bonet, Drowes, and Duff run through Leitrim. There is some coal, and iron and lead in the mountainous areas, but the county is generally not very productive – even the soil is heavy – and is the poorest county in the Republic of Ireland. Potatoes and oats are grown, and some cattle and sheep are reared. Industries include linen, woollens, and potteries. Parke's Castle is one of the most popular tourist attractions in the county. **5** B4

Lelystad new town in the northwest of East Flevoland; population (1996) 60,600. Having been rapidly populated by overspill from Amsterdam and surrounding conurbations, Lelystad serves mainly as a dormitory town for these cities. Some light industry has developed locally and the town is the main service centre for the southern ►IJsselmeer polder region. **22** H2

Léman, Lac French name for Lake ►Geneva, in Switzerland. **30** B4

Le Mans industrial city and administrative centre of the ►Sarthe *département* in western France; population (1990) 148,500, conurbation 191,000. It has a motor-racing circuit where the annual endurance 24-hour race (established 1923) for sports cars and their prototypes is held at the Sarthe circuit. It is linked to Paris by a high-speed rail system. **26** F6

Lemesos Greek name for ►Limassol, a port in southern Cyprus. **36** Q10

Lemnos (Greek **Limnos**), Greek island in the north of the Aegean Sea; area 476 sq km/184 sq mi; population (1991) 17,600. Towns include Kastron and Mudros. The island is of volcanic origin, rising to 430 m/1,411 ft. Industries include mulberries and other fruit, tobacco, and sheep. **36** H5

Lena one of the largest rivers of the Russian Federation, in eastern Siberia; length 4,400 km/2,734 mi; total drainage area 490,000 sq km/189,189 sq mi. The Lena rises in the Baikal Mountains, west of Lake Baikal, and flows northeast to Yakutsk, then north into the Laptev Sea (an inlet of the Arctic Ocean), where it forms a large delta 400 km/240 mi wide and covering some 30,000 sq km/11,583 sq mi. The river is navigable almost throughout its course, but is frozen for eight months of the year. Its main tributaries are the Vitim, Olekma, Aldan, and the Vilyui. The main ports on the Lena's course are Osetrovo (since 1954 part of Ust-Kut) and Yakutsk. **44** L4

Leninakan former name (to 1990) of ▶Gyumri, a town in Armenia. **58** K3

Leningrad former name (1924–91) of the Russian city ▶St Petersburg. **38** F3

Lens former coal-mining town in Pas-de-Calais *département*, northern France; population (1990) 35,300, conurbation 327,000. Present-day industries include chemicals, metallurgy, and textiles. The area around Lens was a important theatre of operations in World War I, particularly during the Battle of Arras. By 1918 the town lay in ruins, but was rebuilt. Lens was again damaged in World War II. **22** E4

León capital of León province in Castilla–León, northwest Spain, situated on a plateau 821 m/2,694 ft high at the confluence of the Torio and Bernesga rivers; population (1991) 144,100. Linen, chemicals, pottery, and leather are manufactured. León was built on the site of a Roman camp. In the 10th century, after its recapture from the Moors, it became the capital of the kingdom of León until 1230, when it was merged with Castile. It has an 11th-century church, a Gothic cathedral, and parts of the old walls remain. **28** E2

León industrial city (leather goods, footwear) in central Mexico; population (1995 est) 968,000. The town lies in Guanajuato state at an altitude of 1,804 m/5,919 ft. Decorated saddles are one of the specialities of the leather industry. **94** D4

León capital of León department, western Nicaragua, near the Pacific coast; population (1991 est) 158,600. Industries include textiles, cotton, timber, food-processing, and the production of fertilizers and insecticides. The second-largest city in the country, it was the capital of Nicaragua from its foundation until 1858 when it was replaced by Managua. The cathedral, begun in 1746, took 100 years to complete, and is the largest in Central America; it contains the tomb of the poet Rubén Darío who was born here. The university was founded in 1804 and is the oldest in the country. **94** G6

Léopoldville former name (to 1966) of ▶Kinshasa, a city in the Democratic Republic of Congo. **70** B4

Le Puy town and administrative capital of the ▶Haute-Loire *département* in the Auvergne region of southeast France, southeast of Clermont-Ferrand; population (1999 est) 20,500. It is situated in a volcanic basin, and has a 12th-century cathedral containing the original manuscript of the Theodolph Bible. Industries include communication and information technology, electronics, mechanics, packaging, plastics and rubber production, and textiles. **26** J8

Lerwick port in Shetland, Scotland; population (1991) 7,300. It is the administrative headquarters of the Shetland Islands. The main occupations include fishing and oil supply services. Hand-knitted shawls are a speciality. **24** L1

Lesbos alternative spelling of ▶Lesvos, an island in the Aegean Sea. **36** H5

Leskovac town in Serbia, Yugoslavia, on the River Veternica; population (1991) 62,100. It is the centre of the Yugoslav textile industry, and is a market town for hemp and other produce of the fertile surrounding district. **34** J7

Lesotho landlocked country in southern Africa, an enclave within South Africa.

Lesotho covers an area of 30,355 sq km/11,720 sq mi, and has a population of 2,153,000 (2000 est). The capital is ▶Maseru; other major towns and cities include Qacha's Nek, Teyateyaneng, Mafeteng, Hlotse, Roma, and Quthing. The head of state is King Letsie III (from 1996); the head of government is Bethuel Pakulitha Mosisili (from 1998). The currency is the loti, and the languages spoken include English (official), Sesotho, Zulu, and Xhosa. **72** D5

Lesser Sunda Islands (or **Nusa Tenggara**), volcanic archipelago in Indonesia, including ▶Bali, ▶Lombok, and ▶Timor; area 73,144 sq km/28,241 sq mi. The islands form two provinces of Indonesia: **Nusa Tenggara Barat**, population (1990) 3,369,600; and **Nusa Tenggara Timur**, population (1990) 3,268,600.

Lesvos Greek island in the Aegean Sea, near the coast of Turkey; area 2,154 sq km/831 sq mi; population (1991) 103,700. The capital is ▶Mytilene. Industries include olives, wine, and grain. The island was called Lesbos in ancient times and was an Aeolian settlement, home of the poets Alcaeus and Sappho. It was conquered by the Turks from Genoa in 1462 and was annexed to Greece in 1913. **36** H5

Leszno (German **Lissa**), capital of Leszno province, Poland, 65 km/41 mi south of Poznań; population (1990) 58,300. It is an industrial town manufacturing rolling stock and machinery; lignite is mined locally. **18** F6

Letchworth town in Hertfordshire, southeast England, 56 km/35 mi north of London; population (1997 est) 33,700. It was founded in 1903 as the first English garden city (built in a rural area, designed to combine the advantages of the town and the country). The town is an administrative centre for local government. Industries include printing and the manufacture of clothing, furniture, scientific instruments, light metal goods, chemicals and gases, and rubber. **22** B3

Lethbridge city in southern Alberta, Canada, 200 km/120 mi south of Calgary; population (1991) 61,000. Lying in the foothills of the Rocky Mountains, on the Oldman River, it is the centre of a large agricultural area and irrigation network. Industries include vegetable canning and freezing, flour milling, brewing, the production of coal, gas, and oil, and the manufacture of trailers, building materials, and electronic equipment. **86** D1

Leuven Flemish name for ▶Louvain, a town in central Belgium. **22** G4

Levant (Italian **il levante**, 'the East'), former name for the eastern Mediterranean region or, more specifically, the Mediterranean coastal regions of Turkey, Syria, Lebanon, Israel, and Palestine.

Leverkusen river port in North Rhine-Westphalia, Germany, 8 km/5 mi north of Cologne; population (1995) 161,900. Industries include oil-refining and extensive petrochemical operations. **22** J3

Lewes market town and administrative headquarters of ▶East Sussex, southeast England, on the River Ouse, 80 km/50 mi south of London;

population (1991) 15,400. Industries include light engineering, brewing, printing, and the manufacture of pharmaceuticals. It is a centre for legal services, being the headquarters of the Sussex police and site of **Lewes Prison**. Simon de Montfort defeated Henry III at the Battle of Lewes (1264). The Glyndebourne Festival is held nearby, and a new opera house was opened in 1994. 22 C4

Lewis (or **Lewis-with-Harris**), largest and most northerly island in the Outer ▶Hebrides, Western Isles; area 2,220 sq km/857 sq mi; population (1991) 21,700. Its main town is Stornoway. It is separated from northwest Scotland by the Minch. The island is 80 km/50 mi long from north to south, and its greatest breadth is 45 km/28 mi. There are many lochs and peat moors. The Callanish standing stones on the west coast are thought to be up to 5,000 years old, second only to Stonehenge (on Salisbury Plain, Wiltshire) in archaeological signifi-cance in the UK. 24 F3

Lewiston city in Androscoggin County, southwest Maine, USA, across the Androscoggin River from Auburn; population (1990) 39,750. Industries include textiles, footwear, clothing, machinery, electronic equipment, and plastics. The river drops about 18 m/59 ft at the 'Twenty-Mile' Falls; after the development of waterpower in the 19th century, the community grew as a textile centre. Lewiston is the home of Bates College (1855) and Lewiston-Auburn College (1988). 88 F2

Lexington (or **Lexington–Fayette**), city and administrative headquarters of Fayette County, north-central Kentucky, 120 km/74 mi southeast of Louisville; population (1994 est) 238,000. It is the centre of the bluegrass region (Kentucky bluegrass (*Poa pratensis*) provides pasture for horses). Racehorses are bred in the area, and races and shows are held here regularly. Industries include tobacco, typewriters, electrical and computer equipment, paper goods, motor vehicles, whiskey, and processed food. The city is the home of the University of Kentucky (1865), Lexington Theological Seminary (1865), and Transylvania University (founded in 1780 as Transylvania College.) 88 D3

Leyden alternative form of ▶Leiden, a city in the Netherlands. 22 G2

Lhasa (called 'the Forbidden City'), capital of the autonomous region of ▶Tibet, China, at 5,000 m/16,400 ft; population (1992) 124,000. Products include handicrafts and light industry. The holy city of Lamaism, Lhasa was closed to Westerners until 1904, when members of a British expedition led by Col Francis E Younghusband visited the city. It was annexed with the rest of Tibet 1950–51 by China, and the spiritual and temporal head of state, the Dalai Lama, fled in 1959 after a popular uprising against Chinese rule. Monasteries have been destroyed and monks killed, and an influx of Chinese settlers has generated resentment. In 1988 and 1989 nationalist demonstrators were shot by Chinese soldiers. 54 F3

Liao river in northeast China; length 1,450 km/900 mi. The main headstream rises in the mountains of Inner Mongolia and flows east, then south to the Gulf of Liaodong. It is frozen from December to March. 48 B3

Liaoning province of northeast China, bounded to the east by Jilin, to the southeast by North Korea, to the south by Korea Bay and the Gulf of Liaodong, to the southwest by Hebei, and to the northwest by Inner Mongolia; area 151,000 sq km/58,300 sq mi; population (1996) 41,160,000. The capital is ▶Shenyang, and other cities and towns include Anshan, Fushun, Liaoyang, and Dalian (port). The province includes the Dongbei (Manchurian) Plain and the grasslands of the Mongolian Plateau. It is one of China's most heavily industrialized areas, and industries include coal, iron, salt, and oil. Agricultural products include cereals. Liaoning was developed by Japan between 1905 and 1945, including the Liaodong Peninsula, whose ports had been conquered from the Russians. 11 (2)F2

Liaoyang city in Liaoning province, China; population (1994) 660,000, urban area (1996) 1,750,000. Industries include engineering and the manufacture of textiles. 48 B3

Libau German name of the Latvian port ▶Liepāja. 18 L1

Liberec (German **Reichenberg**), industrial town in northern Bohemia, in the Czech Republic, on the River Neisse near the Polish and German borders; population (1991) 101,900. It has engineering works and manufactures chemicals, textiles, and glass. 18 E7

Liberia country in West Africa, bounded north by Guinea, east by Côte d'Ivoire, south and southwest by the Atlantic Ocean, and northwest by Sierra Leone.

Liberia covers an area of 111,370 sq km/42,999 sq mi, and has a population of 3,154,000 (2000 est). The capital (and chief port) is ▶Monrovia; other major towns and cities include Bensonville, Saniquillie, Gbarnga, Voinjama, and Buchanan. The head of state and government is Charles Ghankay Taylor (from 1997). The currency is the Liberian dollar, and the languages spoken include English (official) and over 20 Niger-Congo languages. 68 B3

Libreville (French 'free town'), capital of Gabon, on the northern shore of the Gabon River estuary; population (1993) 419,600. Products include timber, oil, cement, and minerals (including uranium and manganese). It is the main port and transport centre of the country, together with Owendo on the southern edge of the town. Libreville was founded in 1849 as a refuge for slaves freed by the French. Since the 1970s the city has developed rapidly due to the oil trade. 68 F4

Libya country in North Africa, bounded north by the Mediterranean Sea, east by Egypt, southeast by Sudan, south by Chad and Niger, and west by Algeria and Tunisia.

Libya covers an area of 1,759,540 sq km/679,358 sq mi, and has a population of 5,605,000 (2000 est). The capital is ▶Tripoli; other major towns and cities include Benghazi, Misurata, Az-Zaiwa, Tobruk, Ajdabiya, and Darnah. The head of state is Moamer al-Khaddhafi (from 1969); the head of government is Mubarak al-Shamikh (from 2000). The currency is the Libyan dinar, and the languages spoken include Arabic (official), Italian, and English. 64 C2

Liechtenstein landlocked country in west-central Europe, bounded east by Austria and west by Switzerland.

Liechtenstein covers an area of 160 sq km/62 sq mi, and has a population of 33,000 (2000 est). The capital is

▶Vaduz; other major towns and cities include Balzers, Schaan, Ruggell, Triesen, and Eschen. The head of state is Prince Hans Adam II (from 1989); the head of government is Mario Frick (from 1993). The currency is the Swiss franc, and the languages spoken include German (official) and an Alemannic dialect. 30 E3

Liège (Flemish **Luik**), industrial city, capital of Liège province in eastern Belgium, southeast of Brussels, almost bisected by the River Meuse, and bounded on the east side by Germany; population (1997) 189,500. It is one of the largest river ports in Europe. Weapons, textiles, paper, and chemicals are manufactured here. The city has a university, founded in 1817, and several ancient churches; the oldest, St Martin's, dates from 692. 22 H4

Liegnitz German name for ▶Legnica, a town in Poland. 18 F6

Lienz chief town and market centre for the east ▶Tirol in Austria, situated on the River Drava; population (1991) 11,900. Lienz is a tourist resort with some manufacturing industry. It stands on the site of a former Illyrian settlement and Roman township. 30 H4

Liepāja (German **Libau**; Russian **Liyepaya**), naval and industrial port in Latvia, on the Baltic Sea some 190 km/118 mi southwest of Riga; population (1992 est) 113,800. The city was founded by the Livonian Knights in 1253. It has a steelworks, and engineering, chemical, food, textile, woodworking, and fish canning industries. 18 L1

Lier (French **Lierre**), town in the province of Antwerp, Belgium, situated on the River Nethe near the city of Antwerp; population (1991) 6,200. The chief manufactures are shoes, lace, embroideries, instruments of brass, and cutlery. There are several medieval buildings. 22 G3

Liguria coastal region of northwest Italy, which lies between the western Alps and the Gulf of Genoa in the Mediterranean, comprising the provinces of Genova, La Spezia, Imperia, and Savona; area 5,418 sq km/2,093 sq mi; population (1992 est) 1,668,900. ▶Genoa is the chief city and port. The region includes the resorts of the Italian Riviera and tourism is very important. Industries include shipbuilding, heavy engineering, horticulture, and the production of chemicals, metals, and textiles. 9 (4)B2

Lille (Flemish **Ryssel** or **Russel**; Latin **Insula**, *l'île* 'island'), industrial city and administrative centre of the ▶Nord *département* in the Nord-Pas-de-Calais region of France, on the River Deûle; population (1990) 178,300, metropolitan area 936,000. The world's first entirely automatic underground train system was opened here in 1982. The Eurostar train stops here, at the new Eurolille station. Industries include textiles, chemicals, engineering, and distilling. 22 F4

Lilongwe capital of ▶Malawi, on the Lilongwe River, lying on a plateau at an altitude of 1,100 m/3,600 ft; population (2001 est) 473,000. It is a commercial and communications centre for a fertile agricultural region, producing ground nuts and tobacco. With hydroelectric power from a dam on the Lilongwe River, it has become a centre for light industry, including engineering, clothes, and the processing of agricultural goods. Tourism is also important, and the city is served by an international airport, and by trunk roads linking it to Zimbabwe, Zambia, and Mozambique. 72 E2

Lima city and administrative head-quarters of Allen County, northwest Ohio, on the Ottawa River, 111 km/69 mi north of Dayton; population (1990) 45,550. It is the centre of a rich agricultural region. Formerly a major oil boom town (1885–1910), it remains an oil pipeline and refining centre. Manufactured goods include cranes and power shovels, aircraft parts, automobile engines, tanks, machine tools, electric motors, neon signs, chemicals, and cigars. 88 D2

Lima capital and largest city of Peru, on the River Rímac, 13 km/8 mi from its Pacific port of Callao; population (1993) 5,706,100. It comprises about one-third of the country's total population. Industries include textiles, chemicals, glass, and cement. 98 B6

Limassol (Greek **Lemesos**), port in Akrotiri Bay, southern Cyprus; population (1993) 137,000. Products include cigarettes and wine. Richard I of England married Berengaria of Navarre here in 1191. The town's population increased rapidly with the influx of Greek Cypriot refugees after the Turkish invasion in 1974. 36 Q10

Limbourg (Flemish **Limburg**), province of northeast Belgium, bounded north and east by the Netherlands; the River Meuse marks the frontier; area 2,422 sq km/935 sq mi; population (1997) 780,000. The capital is ▶Hasselt, and Genk and Tongeren are other towns. The province includes the River Demer, Kempen heathland in the north, rich coalfields, and agriculture in the south. Industries include coal, sugar refining, food processing, chemicals, electronics, engineering, glass, and metallurgy. Agriculture includes dairy farming and market gardening. The province was formerly part of the feudal duchy of Limburg (which was divided in 1839 into today's Belgian and Dutch provinces). 7 (1)D3

Limburg southernmost province of the Netherlands, bounded by Germany to the east, Belgium to the west and south, and North Brabant to the north; area 2,170 sq km/838 sq mi; population (1997) 1,136,200. The capital is ▶Maastricht, and Kerkrade, Heerlen, and Roermond are other towns. The province includes the Maas (Meuse) and Roer rivers, and features sandy soils in river plain, marl soils in the south, and is hilly towards the south. Industries include chemicals, cement, and fertilizers, and agriculture includes mixed arable farming, and horticulture. 7 (1)D3

Limburg Flemish name for ▶Limbourg, a province in Belgium 7 (1)D3

Limburg an der Lahn city in the *Land* (administrative region) of Hesse, Germany, 35 km/22 mi northwest of Wiesbaden; population (1995) 32,900. Iron was formerly mined in the district. Industries now include the manufacture of machinery, textiles, glass, and leather. The city has a Romanesque cathedral with seven towers and notable wall-paintings of the 13th–16th centuries. 22 L4

Limerick county of the Republic of Ireland, in the province of Munster; county town ▶Limerick; area 2,690 sq km/1,038 sq mi; population (1996) 165,000. The principal river is the ▶Shannon, and towns include Abbeyfeale, Kilmallock, Newcastle West, and Rathkeale. Limerick is hilly in the southwest (Mullaghreirk

Mountains) and in the northeast (Galtee Mountains). The low-lying region in the west is very fertile, and is known as the 'Golden Vale'. Dairy cattle, sheep, pigs, and poultry are reared extensively, and corn, sugar-beet, and potatoes are grown. Lace is also produced. 5 B5

Limerick county town of County ▶Limerick and fourth-largest city in the Republic of Ireland, on the Shannon estuary; population (1996) 79,000. The city is divided into three parts: English Town, which is the old city on King's Island (an island in the Shannon estuary); Irish Town; and Newtown Pery, which now forms the centre of the modern city. Industries include flour milling, tanning, meat products, and brewing. The **University of Limerick**, 5 km/3 mi north of the city, is a modern campus. 24 D9

Limnos Greek name for ▶Lemnos. 36 H5

Limoges administrative centre of the *département* of ▶Haute-Vienne and of the ▶Limousin region, central France, situated on the River Vienne, 175 km/109 mi northeast of Bordeaux; population (1990) 136,400. It is the centre of the modern French porcelain industry; other industries include shoes, textiles, electrical equipment, and metal goods. Fine enamels were made here as early as Merovingian times, and their production flourished in the 12th century. The city was sacked by the Black Prince, the eldest son of Edward III of England, in 1370. It has a university. 26 G8

Limousin modern planning region and former province of central France; area 16,900 sq km/6,544 sq mi; population (1990) 722,900. The modern region consists of the *départements* of Corrèze, ▶Creuse, and ▶Haute-Vienne. The administrative centre is ▶Limoges; Brive-la-Gaillarde, Tulle, and Gueret are the other towns of note. A thinly populated and largely infertile region, it is crossed by the mountains of the Massif Central. Fruit and vegetables are produced in the more fertile lowlands. Kaolin is mined here. 6 D4

Limpopo river in southeast Africa, rising in the Magaliesberg to the west of Pretoria in Gauteng Province, South Africa, and flowing through Mozambique and into the Mozambique Channel at Delagoa Bay; length 1,600 km/1,000 mi. It is also known as Crocodile River. 72 D4

Linares town in the province of Jaén, southern Spain; population (1991) 58,000. It has rich mines of lead and copper mixed with silver; sheet lead, pipes, rope, vehicles, and explosives are manufactured. Nearby are the ruins of the Roman town of Castulo. 28 G6

Lincoln industrial and cathedral city, administrative headquarters of ▶Lincolnshire, England, situated on the River Witham, 210 km/130 mi north of London; population (1991) 80,300. Products include bricks, excavators, cranes, gas turbines, radios, vehicle components, cattle feed, matting, pharmaceuticals, power units for oil platforms, and cosmetics. Other industries include heavy engineering, iron foundries, seed milling, and food processing. 22 B1

Lincoln industrial city and capital of ▶Nebraska, USA, 80 km/50 mi southwest of Omaha in the southeastern part of the state; seat of Lancaster County; population (1994 est) 203,000. It is the centre of a rich agricultural area; industries include processed food, railway rolling stock, engineering, pharmaceuticals, electronic and electrical equipment, and small motor vehicles and motor cycles. 86 G2

Lincolnshire county of eastern England.
area 5,890 sq km/2,274 sq mi
towns and cities ▶Lincoln (administrative headquarters), Skegness, Boston, Stamford
physical hills of Lincoln Edge and the Wolds; marshy coastline; the Fens in the southeast; rivers Trent, Welland, Witham
features Belton House, a Restoration mansion; Gibraltar Point National Nature Reserve
agriculture cattle, sheep, horses; cereals (mainly barley); flower bulbs (largest bulb-growing industry in the UK, around Spalding); vegetables
population (1996) 615,900 5 F5

Lincs abbreviation for ▶Lincolnshire, an English county. 5 F5

Lindau city in Bavaria, Germany, built partly on a small island in Lake ▶Constance, but linked by road and rail to the northeast shore; population (1994) 75,500. With many buildings from its rich medieval history, Lindau is a tourist resort and a principal port of call for lake traffic. The city produces electrical goods and rubber. 30 E3

Lindisfarne site of a monastery off the coast of Northumberland, England; see ▶Holy Island. 24 L6

Line Islands group of coral islands in the Pacific Ocean; population (1995 est) 5,400. Eight of the islands belong to Kiribati, and three (Palmyra, Jarvis, and Kingman Reef) are uninhabited dependencies of the USA. Coconut plantations provide the means of livelihood in the North Line Islands, but the South Line Islands, where guano was once the main product, are now uninhabited. 74 L5

Lingen city in Lower Saxony, Germany, on the River Ems, 65 km/40 mi northwest of Münster; population (1994) 52,300. Lingen is a centre for oil refining and natural-gas production from local wells, and also contains chemical, steel, and clothing industries. One of Germany's largest nuclear power stations is located nearby. 22 K2

Linköping industrial town in southeastern Sweden; 172 km/107 mi southwest of Stockholm; population (1994) 130,500. Industries include hosiery, aircraft and engines, and tobacco. It has a 12th-century cathedral. 16 H7

Linz capital and industrial river port of Upper Austria, in northern Austria, on the River Danube; population (1995) 201,500. Iron and steel production is a major industry powered by locally sourced hydroelectricity. Other activities include metalworking, furniture-making, and the manufacture of electrochemicals and clothing. 30 K2

Lion, Golfe du (English **Gulf of the Lion**), bay of the Mediterranean stretching along the shore of southern France from Toulon to the Spanish border, and named because of the roughness of its waters. The Rivers Rhône, Aude, Orb, and Têt flow into it. The principal port is ▶Marseille. 26 J10

Lipari Islands (or **Aeolian Islands**; Italian **Isole Lipari**), volcanic group of seven islands off northeastern Sicily, including **Lipari** (on which is the capital of the same name), **Stromboli** (active volcano 926 m/3,038 ft high), and **Vulcano** (also with an active volcano); area 114 sq km/44 sq mi. In Greek mythology, the god Aeolus kept the winds imprisoned in a cave on the Lipari Islands. 32 J10

Lippe river of northern Germany flowing into the River Rhine; length 230 km/147 mi; also a former German state, now part of North Rhine–Westphalia. 22 L3

Lisboa Portuguese form of ►Lisbon, the capital of Portugal. 28 A6

Lisbon (Portuguese **Lisboa**), capital of Portugal, and of the Lisboa district, in the southwest of the country, situated on a tidal lake and estuary formed by the River Tagus; population (1991) 677,800. It is a major commercial and industrial centre, and industries include steel, textiles, chemicals, pottery, shipbuilding, and fishing. Lisbon has been Portugal's capital since 1260 and reached its peak of prosperity in the period of Portugal's empire during the 16th century. In 1755 an earthquake accompanied by a tidal wave killed 30,000–60,000 people (the estimates vary) and destroyed much of the city. 28 A6

Lisburn town in County Antrim, Northern Ireland, on the River Lagan, 14 km/9 mi from Belfast; population (1991) 27,400. The main industries are engineering and the production of soft drinks, yarn and thread, and packaging; it is a busy shopping centre. Many inhabitants commute to Belfast. Lisburn was founded in the 17th century and planted with English settlers and French Huguenot refugees. 24 G7

Lisieux town in Calvados *département* of France, on the River Touques 50 km/30 mi southeast of Caen; population (1990) 24,500. There is a large trade in dairy products and textiles, electrical goods, and food products; car parts are manufactured here. 22 C5

Lissa German name for ►Leszno, a town in Poland. 18 F6

Litani (Arabic **Nahr al-Litani**; Latin **Orantes**), river rising near Baalbek in the Anti-Lebanon Mountains of eastern Lebanon, 144 km/89 mi long. It flows northeast–southwest through the fertile Bekaa Valley, for which it provides irrigation water, and continues east to the Mediterranean 8 km/5 mi north of Tyre. Israeli forces invaded Lebanon in 1978 and subsequently withdrew as far as the Litani River. 98 G3

Lithuania country in northern Europe, bounded north by Latvia, east by Belarus, south by Poland and the Kaliningrad area of Russia, and west by the Baltic Sea.

Lithuania covers an area of 65,200 sq km/25,173 sq mi, and has a population of 3,670,000 (2000 est). The capital is ►Vilnius; other major towns and cities include Kaunas, Klaipeda, Siauliai, and Panevezys. The head of state is Valdas Adamkus (from 1998); the head of government is Rolandas Paksas (from 2000). The currency is the litas, and the main language spoken is Lithuanian (official). 16 L9

Little Abaco island in the northern Bahamas; see ►Abacos. 90 F4

Little Karoo smaller and more southerly of two areas of the ►Karoo, a semi-desert region in South Africa. 72 C6

Little Rock largest city and capital of ►Arkansas, USA, a port of entry on the Arkansas River, 215 km/133 mi west of Memphis, Tennessee; seat of Pulaski County; population (1994 est) 178,000. It is the centre of a rich agricultural, mining, timber, natural gas, and oil region; products include metal goods, oilfield and electronic equipment, valves and pipes, aircraft, ammunition, watches, chemicals, clothing, and processed food. 90 C3

Liverpool city, seaport, and metropolitan borough in Merseyside, northwest England; population (1998 est) 461,500. Liverpool is the UK's chief Atlantic port with miles of specialized, mechanized quays on the River Mersey, and 2,100 ha/5,187 acres of dockland. The port handles 28 million tonnes of cargo annually. Imports include crude oil, grain, ores, edible oils, timber, and containers. Traditional industries, such as ship-repairing, have declined, while employment in the service sector in areas such as education, health, insurance, banking, and tourism has increased.

features The Albert Dock on the waterfront is Britain's most popular heritage attraction. It hosts the Museum of Liverpool Life. Landmarks include the Bluecoat Chambers (1717); the Town Hall (1754); St George's Hall (1838–54); and the Tate Gallery in the North; the Picton Library (1879); the Anglican Cathedral, designed by George Gilbert Scott (begun 1904, completed 1980); the Roman Catholic Metropolitan Cathedral of Christ the King, designed by Frederick Gibberd, consecrated in 1967. The Walker Art Gallery (1877) and the Liverpool Philharmonic Orchestra (founded in 1840, the Royal LPO since 1957), are here. The Grand National steeplechase takes place at Aintree. Outstanding buildings include the 16th-century Speke Hall, the Victoria Building of the University of Liverpool, the Dock Offices, the Port of Liverpool building (1907), Royal Liver Building (1911), and the Cunard Building (1916) on Pier Head.

There are two universities: the University of Liverpool (opened in 1903) and John Moores University. The members of the 1960s pop group the Beatles were born here. The Liverpool Institute for the Performing Arts was set up by former Beatle Paul McCartney and opened in 1995. 24 K8

Livingston administrative headquarters of ►West Lothian, Scotland, 47 km/29 mi from Glasgow, new town established in 1962; population (1991) 41,600. Industries include electronics, engineering, paper, steel, scientific instruments, and industrial research. 24 J6

Livingstone (formerly **Maramba**), capital of Southern Province, Zambia, lying on the northern bank of the Zambezi River at the frontier with Zimbabwe; population (1998 est) 95,800. Lying near the Victoria Falls, the city's main industry is tourism. Hydroelectric power the falls supports the industries of motor vehicle assembly and sawmilling, and the manufacture of furniture and blankets. 72 D3

Livonia (German **Livland**), one of the former ►Baltic States, divided in 1918 between the modern states of Estonia and Latvia. Livonia belonged to the Teutonic Knights from the 13th to 16th centuries, to Poland from 1561, Sweden from 1629, and Russia from 1721.

Livorno (English **Leghorn**), industrial port and resort in Tuscany, western Italy, 75 km/47 mi southwest of Florence; population (1992) 166,400. Industries include shipbuilding, distilling, oil-refining, and motor vehicle manufacture. A fortress town since the 12th century, it was developed by the Medici family in the 16th century. It is the site of a naval academy. 30 F7

Lizard Point southernmost point of mainland England in Cornwall. The coast is broken into small bays, overlooked by two cliff lighthouses. 24 G12

Ljubljana (German **Laibach**), capital and industrial centre of Slovenia, near the confluence of the rivers Ljubljanica and Sava; population (1991) 276,100. Products include textiles, chemicals, paper, and leather goods. It has a nuclear research centre and is linked with southern Austria by the Karawanken road tunnel under the Alps (1979–83). 30 K4

Llandudno seaside resort in Gwynedd, north Wales, situated on Conwy Bay, on the Creuddyn Peninsula; population (1991) 13,200. Wales's largest seaside resort, it has two sandy beaches and is a touring centre which became popular during the Victorian era. Great Orme's Head is a spectacular limestone headland encircled by a marine 'drive'. 24 J8

Llanelli (formerly **Llanelly**), industrial port in Carmarthenshire, south Wales, situated on Burry Inlet 16 km/10 mi northwest of Swansea; population (1991) 45,000. Industries include copper smelting, and the manufacture of tin plate, chemicals, bricks, and lenses. 24 H10

Lleida (formerly **Lérida**), capital of Lleida province in Cataluña, northeast Spain, situated on the River Segre, 132 km/82 mi west of Barcelona; population (1991) 111,900. Industries include leather, paper, glass, and cloth production. It has Roman remains and the ruins of a 13th-century cathedral. 28 L3

Lobito port on Lobito Bay, Angola; population (1998 est) 142,500. It is the terminus of the Benguela railway, which links Lobito with the Democratic Republic of Congo, the Zambia copperbelt, and the port of Beira in Mozambique. Lobito is a main outlet for mineral exports from Zambia and the Congo, including copper, tin, cobalt, manganese, uranium, and zinc. From central Angola the port exports sisal, cotton, coffee and sugar, as well as iron and hides and skins. Port industries include sugar and rice mills, and cement works. 72 A2

Locarno health resort in the Ticino canton of Switzerland, on the north shore of Lago Maggiore, 22 km/14 mi southwest of Bellinzona; population (1995) 14,200. It is situated on the St Gotthard route. Formerly in the Duchy of Milan, it was captured by the Swiss in 1803. The **Pact of Locarno**, a series of diplomatic documents guaranteeing French security, was initialled here in 1925. 30 D4

Loch Ness Scottish lake; see ►Ness, Loch. 24 H4

Lodi river port in Lombardy, Italy, on the Adda River, 30 km/18 mi southeast of Milan; population (1990) 43,000. It is a market centre for agricultural produce; fertilizers, dairy produce, agricultural machinery, and textiles are manufactured. French emperor Napoleon's defeat of Austria at the **Battle of Lodi** in 1796 gave him control of Lombardy. There is a 12th-century cathedral and a 15th-century Renaissance church. 30 E5

Łódź industrial city (textiles, machinery, dyes) in central Poland, 120 km/75 mi southwest of Warsaw; population (1993) 844,900. 18 J6

Lofoten and Vesterålen island group off northwest Norway; area 3,595 sq km/1,390 sq mi. ►Hinnøya, in the Vesterålens, is the largest island of Norway. The surrounding waters are rich in cod and herring. The **Maelström**, a large whirlpool hazardous to shipping (described in Edgar Allen Poe's story *A Descent into the Maelstrom*), which gives its name to similar features elsewhere, occurs in one of the island channels between Lofotodden and Mosken. 16 G3/34 G2

Logan, Mount mountain in southwest Yukon Territory, Canada. With an altitude of 5,950 m/19,520 ft, it is the highest peak in Canada and the second-highest peak in North America (the highest being Mount McKinley in Alaska). Part of the St Elias Mountains, Mount Logan is in the Kluane National Park, designated a World Heritage Site in 1979. Research on high-altitude physiology, meteors, and glaciers has been conducted here. 82 C4

Logroño market town and capital of ►La Rioja autonomous community, northern Spain, on the River Ebro; population (1991) 121,100. It is the centre of a wine-producing region. 28 H2

Loir French river, rising north of Illiers in the *département* of Eure-et-Loir and flowing southeast, then southwest to join the Sarthe near Angers; 310 km/500 mi. It gives its name to the *départements* of Loir-et-Cher and Eure-et-Loir. 26 F6

Loire *département* in the ►Rhône-Alpes region of southeast France; area 4,773 sq km/1,843 sq mi; population (1990) 746,677. The *département* is largely mountainous, but the plains of Forez and Roanne provide good agricultural and pasture lands. Vines are grown in the Rhône valley. The main goods manufactured are glass, silk, cast steel, hardware, machinery, and cutlery. The principal towns are St-Etienne (the capital), and Roanne. 6 F4

Loire longest river in France, rising in the Cévennes Mountains in the *département* of Ardèche at 1,350 m/4,430 ft near Mont Gerbier de Jonc, and flowing for over 1,000 km/620 mi north through Nevers to Orléans, then west through Tours and Nantes until it reaches the Bay of Biscay at St Nazaire. The Loire drains 116,550 sq km/45,000 sq mi of land, more than a fifth of France, and there are many châteaux and vineyards along its banks. The Loire gives its name to the *départements* of ►Loire, Haute-Loire, Loire-Atlantique, ►Indre-et-Loire, Maine-et-Loire, and Saône-et-Loire. 26 D6

Loire-Atlantique maritime *département* in the ►Pays de la Loire region of France; area 6,893 sq km/2,661 sq mi; population (1990) 1,051,400. The River Loire, its tributaries the Erdre and the Sèvre, and the River Isac, a tributary of the Vilaine, flow through it. Horses and cattle are bred, and cereals, vines, flax, and fruit are cultivated. The chief manufactures are hemp, linen, paper, sugar, biscuits, and soap. There are foundries and shipyards at ►Nantes (the administrative centre) and ►St-Nazaire, and oil refining and petrochemical industries. There is a large university at Nantes. Other towns include Ancenis and Châteaubriant. 6 C3

Loire, Haute- *département* in the Auvergne region of France; see ▶Haute-Loire. 6 E4

Loiret *département* in the ▶Centre region of France; area 6,742 sq km/ 2,603 sq mi; population (1990) 580,700. There is a large forest around ▶Orléans, the administrative centre. The Loire valley in the south, in spite of the occasional floods, is an important agricultural area, and wine, wheat, barley, sugar beet, vegetables, and fruit are produced. There are also rubber, iron, preserved foods, mechanical, electrical, and other industries. Other towns include Montargis and Pithiviers. 6 E3

Loir-et-Cher *département* in the ▶Centre region of France; area 6,314 sq km/2,438 sq mi; population (1990) 306,500. In the north flows the River ▶Loir, in the centre the Loire, and in the south the ▶Cher. Cereals, vines, fruit trees, and vegetables are cultivated, and sheep and turkeys are raised. Manufacturing industries include food, chemicals, and pottery. Stone, alabaster, and flint are quarried. The châteaux of the Loire attract many tourists. The administrative centre is ▶Blois. Other towns include Romorantin and Vendôme. 6 D3

Loja town in the province of Granada, southern Spain, on the River Genil; population (1991) 20,800. Products include salt, textiles, paper, and cakes (*roscos*). It was one of the last strongholds of the Moors in Spain. 28 F7

Lombardia Italian name for ▶Lombardy, a region of northern Italy. 9 (4)B2

Lombardy (Italian **Lombardia**), region of northern Italy, between the Alps and the River Po, comprising the provinces of Bergamo, Brescia, Como, Cremona, Mantua, Milan, Pavia, Sondrio, and Varese; area 23,900 sq km/9,225 sq mi; population (1992 est) 8,882,400. Its capital is ▶Milan. It is the country's chief industrial area with chemical, pharmaceutical, textile, and engineering operations, and its most productive agricultural region yielding wheat, maize, wine, meat, and dairy products. 9 (4)B2

Lombok (Javanese 'chili pepper'), island of Indonesia, east of Java, one of the Sunda Islands; area 5,435 sq km/ 2,098 sq mi; population (1990) 2,403,000. The chief town is Mataram. It has a fertile plain between northern and southern mountain ranges. The mountain ranges are volcanic in origin and rise at Mount Rinjani to 3,726 m/ 12,224 ft. The island has become a popular tourist destination. 52 F4

Lomé capital, port, and administrative centre of ▶Togo, on the Bight of Benin; population (1995 est) 501,000. It is a centre for gold, silver, and marble crafts. Industries include steel production, oil refining, brewing, plastics, cement, paper manufacturing, and food processing. Main exports include cacao, palm nuts, cotton, and coffee. There is an international airport and tourism is growing in importance. Lomé became capital of the independent Togo in 1960. 68 E3

Lomond, Loch largest freshwater Scottish lake, 37 km/23 mi long, area 70 sq km/27 sq mi. It is overlooked by the mountain **Ben Lomond** (973 m/ 3,192 ft) and is linked to the Clyde estuary. 24 H5

Lompoc city in Santa Barbara County, southwest California, USA, near the Pacific Ocean, 67 km/42 mi west of Santa Barbara; population (1990) 37,650. It is situated in the fertile Lompoc Valley; industries include the processing of oil from the city's oil wells, the production of flower seed, and mining diatomaceous earth (a fine-grained rock composed of the valves of fossil diatoms and used in the manufacture of dynamite, rubber, and plastics). 92 B2

London city in southwestern Ontario, Canada, on the Thames River, 198 km/123 mi southwest of Toronto and 200 km/124 mi west of the Niagara Falls; seat of Middlesex County; population (1991) 303,200; population of metropolitan area 381,500. The centre of a farming district, it has tanneries, breweries, and factories making hosiery, radio and electrical equipment, leather, and footwear. 88 D2

London capital of England and the United Kingdom, located on the River Thames. Since 1965 its metropolitan area has been known as Greater London (see ▶London, Greater), consisting of the City of London and 32 boroughs; total area 1,580 sq km/ 610 sq mi; combined population (1999 estimate) 7,285,000. London is the biggest city in Western Europe at the heart of the most populous region – 18 million people live in London and the South East. The **City of London**, known as the 'square mile', is the financial and commercial centre of the UK; area 2.7 sq km/1 sq mi. Over 25 million people visited London in 1999. Popular tourist attractions include the **Tower of London**, St Paul's Cathedral, Buckingham Palace, and Westminster Abbey. The Millennium Dome at Greenwich was the centrepiece of Britain's millennium celebrations. 22 B3

Londonderry (or **Derry**; formerly **Derry-Calgaich** until the 10th century), historic city and port on the River Foyle, 35 km/22 mi from Lough Foyle, county town of County ▶Londonderry, Northern Ireland; population (1991) 95,400. Industries include textiles, chemicals, food processing, shirt manufacturing, and acetylene from naphtha.
features the Protestant Cathedral of St Columba dating from 1633; the Gothic revival Roman Catholic Cathedral of St Eugene (completed in 1833); the Guildhall (rebuilt in 1912); the city walls, including; four gates into the city.
history Londonderry dates from the foundation of a monastery there by St Columba in AD 546. The city was subject to a number of sieges by the Danes between the 9th and 11th centuries, and by the Anglo-Normans in the 12th century; however, these were unsuccessful until James I of England captured the city in 1608. The king granted the borough and surrounding land to the citizens of London. The Irish Society was formed to build and administer the city and a large colony of English Protestants was established. The city, then governed by Major Henry Baker and the Reverend George Walker, was unsuccessfully besieged in 1689 by the armies of James II, who had fled England when William of Orange was declared joint sovereign with James's daughter Mary. James's army was led by Richard Talbot, Earl of Tyrconnell, in a conflict known as the **Siege of Derry**, when 13 Derry apprentices and citizens loyal to William of Orange locked the city gates against the Jacobite army. The siege lasted 15 weeks, during which many of the inhabitants died of starvation and disease because of the blockade. 24 E6

Londonderry (or Derry), historic county of Northern Ireland.

area 2,070 sq km/799 sq mi

towns and cities ►Londonderry (county town), Coleraine, Portstewart, Limavady
physical hilly moorland, coniferous forest; Sperrin Mountains; rivers Foyle, Bann, Roe, and Faughan; borders Lough Neagh
industries stone and lime quarrying, food processing, textiles and synthetic fibres, shirt manufacturing, light engineering, chemicals
agriculture farming is hindered by the very heavy rainfall; flax, cattle, sheep grazing on moorland, salmon and eel fisheries on the Bann
population (1998 est) 213,000 5 C4

London, Greater metropolitan area of ►London, England, comprising the City of London, which forms a self-governing enclave, and 32 surrounding boroughs. Certain powers were exercised over this whole area by the Greater London Council (GLC) 1974–86.
area 1,580 sq km/610 sq mi
population (1999) 7,285,000 5 F6

Long Beach port and industrial city in southwestern California, USA; population (1994 est) 434,000. Long Beach forms part of Greater ►Los Angeles and adjoins the San Pedro harbour of Los Angeles. Manufactured goods include aircraft, ships, petroleum products, chemicals, fabricated metals, electronic equipment, and processed food; the city also has oil wells and a naval shipyard, and is a convention centre. Long Beach was laid out in the 1880s and incorporated in 1888; the port was opened in 1909. Oil was discovered here in 1921, and the aircraft industry dates from World War II. 92 C2

Longford county of the Republic of Ireland, in the province of Leinster; county town Longford; area 1,040 sq km/401 sq mi; population (1996) 30,200. The county is low-lying (the highest point is Carn Clonhugh 279 m/916 ft), and the western border is formed of the River Shannon and part of Lough Ree, one of several lakes. Other rivers are the Camlin, a tributary of the Shannon, and the Inny, which flows into Lough Ree. Agricultural activities include cattle and sheep rearing, and the production of oats and potatoes. 5 C5

Long Island island east of Manhattan and southeast of Connecticut, USA, separated from the mainland by Long Island Sound and the East River; 193 km/120 mi long by about 48 km/30 mi wide; area 3,627 sq km/1,400 sq mi; population (1990) 6,861,500. It is mainly a residential district with farming in the east. Henry Hudson discovered the island in 1609, and it was settled by the Dutch from New Amsterdam (in the west) and the English from New England, USA (in the east) from the 1640s. 88 F2

Longmont city in Boulder County, north-central Colorado, USA, in the Rocky Mountain foothills, 41 km/25 mi south of Fort Collins and 30 km/19 mi northeast of Boulder; population (1992) 54,600. Situated in the fertile St Vrain Valley, irrigated by the Colorado–Big Thompson project, Longmont processes and ships beet sugar, vegetables, and livestock. Manufactures include automotive filters, chemicals, truck campers, business machines, computers, recreational vehicles, and agricultural equipment. 86 E2

Longview city and administrative headquarters of Gregg County, east Texas, USA, 198 km/123 mi east of Dallas; population (1992) 72,700. Situated in the heart of the oilfields of east Texas, Longview's industries include oil and natural-gas processing, oilfield equipment, steel goods, chemicals, electronic equipment, aircraft components, farm and earth-moving machinery, and truck beds and trailers. The city home to LeTourneau University (1946) and the Caddo Indian Museum. 90 C3

Longwy French town in the *département* of Meurthe-et-Moselle, on the River Chier, near the border with Belgium and Luxembourg; population (1990) 15,600. It has a fortress built in the 17th century by the military engineer Vauban. Longwy is in an iron-mining district, and has metallurgical and pottery industries. 22 H5

Lons-le-Saunier French town, capital of the *département* of Jura, 75 km/47 mi northwest of Geneva in Switzerland, on the River Vallière; population (1990) 20,000. It is a tourist centre, and has been known as a spa since the 4th century. There are a number of 18th-century houses and two medieval churches. Wines and salt are produced, and there is micro-technology industry and a trade in cheese, horses, and cattle. 30 A4

Lop Nur series of shallow salt lakes with shifting boundaries at the eastern end of the Taklimakan Desert, Xinjiang Uygur Autonomous Region, northwest China. The Venetian traveller Marco Polo visited Lop Nur, then a single lake of considerable extent, in about 1273. The area is used for atomic tests. 42 S9

Lorain city and port in Lorain County, north-central Ohio, USA, at the mouth of the Black River on Lake Erie, 40 km/25 mi southwest of Cleveland; population (1992) 71,500. An important Great Lakes port, it has shipbuilding yards and manufactures steel, communications equipment, cars, and heavy construction equipment. Tourism and commercial fishing are also significant factors in the regional economy. 88 D2

Lorca town in the province of Murcia, southeast Spain; population (1991) 65,800. It has ancient walls, a Moorish castle, several baroque churches, and mansions. Lorca lies in a very fertile district; its products include chemicals, textiles, and porcelain. 28 J7

Lord Howe Island island in the Pacific Ocean midway between Norfolk Island and Sydney, about 836 km/520 mi northeast of Sydney; area 13 sq km/5 sq mi; population (1995 est) 540. It is a dependency of New South Wales and a volcanic island of great natural beauty. The highest point rises to 875 m/2,870 ft and is known as Mt Gower. Rich in bird life and rare plants, with the world's most southerly coral reef offshore, the island is a world heritage site. It is a tourist resort and is linked with Sydney by air. 76 L6

Lorient (French **L'Orient** 'The East'), commercial and naval port in the *département* of Morbihan, Brittany, northwest France, situated 90 km/56 mi south of Brest on an inlet of the Bay of Biscay formed by the estuaries of the rivers Scorff and Blavet; population (1990) 61,600, conurbation 150,000. Industries include fishing, shipbuilding, electronics, and the manufacture of car components. 26 B6

Lorraine (German **Lothringen**), region and former province of northeast France in the upper reaches of the Meuse and Moselle rivers; bounded in the north by Belgium, Luxembourg, and Germany, and in the east by Alsace; area 23,600 sq km/ 9,095 sq mi; population (1990) 2,305,700. It comprises the *départements* of ►Meurthe-et-Moselle, ►Meuse, ►Moselle, and ►Vosges, and its chief cities are Metz, ►Nancy (the capital), Luneville, and Epinal. There are deposits of coal, iron ore, and salt; grain, fruit, and livestock are farmed. In 1871, after the Franco-Prussian War, the northern part of the region was ceded to Germany as part of Alsace-Lorraine. The whole area saw heavy fighting in World War I. 6 F2

Los Angeles city and port in southwestern California, USA; population (1994) 3,449,000; Los Angeles–Riverside–Orange County consolidated metropolitan area (also known as Greater Los Angeles) (1994) 15,302,000. In size of population it is the second-largest city and the second-largest metropolitan area in the USA. The city occupies 1,204 sq km/465 sq mi. Industries include aerospace, electronics, motor vehicles, chemicals, clothing, building materials, printing, food processing, and films. Los Angeles was established as a Spanish settlement in 1781. **92** C2

Lot *département* in the ►Midi-Pyrénées region of France; area 5,228 sq km/ 2,019 sq mi; population (1990) 156,000. The land is crossed from east to west by the River Lot and is highest in the northeast. Wheat is the chief cereal, but maize, oats, and barley are also cultivated. Wine is the principal product, the most valued being that of the administrative centre ►Cahors, grown in the Lot valley. The *département* produces potatoes and tobacco, and chestnuts in the northeast. Cloth-making, tanning, brewing, tourism, and the making of agricultural implements are among its other industries. Other towns include Figeac and Gourdon. 6 D4

Lot-et-Garonne *département* in the ►Aquitaine region of France; area 5,358 sq km/2,069 sq mi; population (1990) 306,000. Its surface, which consists mainly of wide plains, is crossed from east to west by the Rivers ►Garonne and Lot. The valleys of these rivers are fertile, and the slopes of the low hills are covered with orchards and vineyards. Wheat, maize, and barley are also grown. The administrative centre is ►Agen. Other towns include Marmande, Nérac, and Villeneuve. 6 D4

Lothian former region of Scotland (1975–96), which was replaced by East Lothian, Midlothian, West Lothian, and City of Edinburgh unitary authorities.

Louangphrabang (or **Louang Prabang**), Buddhist religious centre in Laos, on the Mekong River at the head of river navigation; population (1998 est) 110,400. It was the capital of the kingdom of Louangphrabang, incorporated in Laos in 1946, and the royal capital of Laos 1946–75. **50** C3

Loughborough industrial town in Leicestershire, central England, 18 km/ 11 mi northwest of Leicester, on the River Soar; population (1995 est) 55,300. Industries include engineering, bell-founding, children's book publishing, brickmaking, and the manufacture of heavy duty electrical goods and power generators, knitwear, hosiery, pharmaceuticals, and medical supplies. 22 A2

Louisiana state in southern USA, nicknamed the Pelican State. Louisiana was admitted to the Union in 1818 as the 18th US state. It has been in the hands of the Spanish, French, and Americans since the 16th century, and its culture also has been influenced by African slaves and their descendants, and Caribbean and French-Canadian immigrants; the Creoles of southern parishes were originally a mix of French and Spanish descendants; the Cajuns of the southwest were originally French immigrants who had been expelled from Acadia in modern-day Nova Scotia, Canada. Musically, the state is associated with the development of jazz and the blues, particularly in the city of New Orleans. Louisiana is bordered to the north by Arkansas, to the west by Texas, with the Sabine River and Toledo Bend Reservoir forming much of the boundary, and to the east by Mississippi, with the Mississippi and Pearl rivers forming much of the boundary. To the south, the state extends into the Gulf of Mexico, its area expanding continuously through the accretional growth of the delta of the Mississippi River; much of Louisiana consists literally of fragments of other states in the Mississippi-Missouri system.
population (2000 est) 4,469,000; including Cajuns, descendants of 18th-century religious exiles from Canada, who speak a French dialect
area 135,900 sq km/52,457 sq mi
capital ►Baton Rouge
towns and cities New Orleans, Shreveport, Lafayette, Metairie
industries and products rice, cotton, sugar, soybeans, oil, natural gas, chemicals, sulphur, fish and shellfish, salt, processed foods, petroleum products, timber, paper, tourism, music industry **90** C3

Louisville industrial city in northern Kentucky, USA, a major port on the Ohio River; seat of Jefferson County; population (1994 est) 270,000. Industries include electrical goods, agricultural machinery, motor vehicles, tobacco, paint, whisky, and Braille books. It is the home of the Kentucky State Fair and Exposition Centre, and the site of Churchill Downs racetrack where the annual Kentucky Derby is run. **90** D2

Lourdes town in the *département* of Hautes-Pyrénées in the Midi-Pyrénées region of southwest France, on the Gave de Pau River; population (1990) 18,000. Its Christian shrine to St Bernadette has a reputation for miraculous cures, and Lourdes is an important Roman Catholic pilgrimage centre. In 1858 a young peasant girl, Bernadette Soubirous, claimed to have been shown the healing springs of the Grotte de Massabielle by a vision of the Virgin Mary. **26** E10

Lourenço Marques former name of ►Maputo, the capital of Mozambique. **72** E5

Louth smallest county of the Republic of Ireland, in the province of Leinster; county town Dundalk; area 820 sq km/ 317 sq mi; population (1996) 92,200. It is mainly fertile and low-lying. The chief towns are Dundalk at the north end of Dundalk bay, Drogheda, and Ardee, and the chief rivers are the Fane, Lagan, Glyde, and Dee. There is cattle rearing and fishing; oats and potatoes are grown. Greenore on Carlingford Lough is a container shipping port. Louth is rich in ancient buildings and remains, and was of strategic importance during the 12th–18th centuries. Important monastic sites with extensive remains include Monasterboice (founded in the

5th century), and Mellifont Abbey (founded in the 12th century). **5** C5

Louvain (Flemish **Leuven**; German **Lowen**), industrial town and railway junction in Brabant province, central Belgium, 25 km/16 mi northeast of Brussels, on the River Dyle; population (2001 est) 88,200. Manufacturing includes fertilizers and food processing; brewing is also a major industry, and the headquarters of Stella Artois are in the town. Its university dates from 1425. **36** C6

Low Countries region of Europe that consists of ▶Belgium and the ▶Netherlands, and usually includes ▶Luxembourg.

Lowell city in Middlesex County, northeastern Massachusetts, on the Merrimack and Concord rivers, 40 km/25 mi northwest of Boston; population (1998 est) 101,100. Industries include textiles, footwear, electronics, plastics, and chemicals. Lowell was incorporated in 1836 and developed as a major textile centre in the 19th century; it is one of the line of former mill towns on the Merrimack that owe their foundation to water power derived from the river. Wang Laboratories, a computer company, moved its headquarters here in 1978 and is a major employer in the city. **88** F2

Lowen German name for ▶Louvain, a town in central Belgium. **22** G4

Lower Austria (German **Niederösterreich**), largest federal state of Austria, bordered on the north by the Czech Republic, drained by the River Danube; area 19,200 sq km/7,411 sq mi; population (1994) 1,511,600. Its capital is St Pölten. The main towns are Wiener Neustadt and Krems. ▶Vienna is a provincial enclave within the province. In addition to wine, sugar beet, and grain, there are reserves of oil and natural gas. Manufactured products include textiles, chemicals, and metal goods. Agriculture and forestry are important. **8** (2)G2

Lower California British name for ▶Baja California, Mexico. **84** C5

Lower Normandy English name for ▶Basse-Normandie, a coastal region of northwest France. **6** C2

Lower Saxony (German **Niedersachsen**), administrative region (German *Land*) in northern Germany, bordered to the north by Schleswig-Holstein and the city-state of Hamburg, to the northeast by Mecklenburg-West Pomerania, to the south by North Rhine-Westphalia and Hesse, on the east and southeast by Saxony-Anhalt and Thuringia respectively, and on the west by the Netherlands; area 47,400 sq km/ 18,296 sq mi; population (1995) 7,823,000. The capital is ▶Hannover, and other towns and cities include Braunschweig (Brunswick), Osnabrück, Oldenburg, Göttingen, Wolfsburg, Salzgitter, and Hildesheim. The region includes the Lüneburg Heath, the Harz Mountains, and the Elbe, Weser, Jade, and Ems rivers. Industries include cars (the Volkswagen plant at Wolfsburg is in the east of the state), machinery, electrical engineering, and iron and steel production. Agriculture includes cereals, oats, potatoes, and livestock farming. **7** (2)B2

Lowestoft (Old Norse **Hloover's Toft** 'dwelling belonging to Hloover'), resort town and port in Suffolk, England, 62 km/38 mi northeast of Ipswich, the most easterly town in Britain; population (1996 est) 66,300. Offshore oil and gas fields provide significant employment related to production and exploration, including oil platform construction and supply services. Historically a fishing port, the industry is still active but has declined dramatically. Fish- and food-processing (especially freezing), tourism, and the manufacture of radar, electrical equipment, and televisions are also important. **Lowestoft Ness** is the most easterly point in England. **22** D2

Lozère *département* in Languedoc-Roussillon region of southern France; area 5,168 sq km/1,995 sq mi; population (1990) 74,000. Occupying a section of the Cévennes Mountains, it rises in Finiels to 1,702 m/5,584 ft. The principal towns are Mende (the administrative centre) and Florac. **6** E4

Lualaba another name for the upper reaches of the ▶Congo River in Africa, as it flows north through the Democratic Republic of Congo from near the Zambian border. **62** E6

Luanda (formerly **Loanda**), capital and industrial port of Angola; population (1995) 2,250,000. Products include cotton, sugar, tobacco, timber, textiles, paper, fuel oil, and lubricants. Founded in 1575 by Portuguese settlers, it became a Portuguese colonial administrative centre as well as an outlet for slaves transported to Brazil. **70** A5

Lubbock city in northwestern Texas, USA, 180 km/112 mi south of Amarillo, situated on the eastern edge of the high plains at an altitude of about 1,000 m/3,281 ft; seat of Lubbock County; population (1994 est) 194,000. Often called the 'Hub of the High Plains', the city is a trading centre for cotton. Industries include the production of cottonseed oil and sorghum (a type of millet), meat packing, the manufacture of heavy machinery for farming and the construction industry, and mobile homes. **92** F2

Lübeck (Wendish Slavonic 'lovely one'), seaport and resort of Schleswig-Holstein, Germany, 24 km/15 mi from the Baltic Sea, 60 km/37 mi northeast of Hamburg; population (1995) 216,900. Manufactured goods include machinery, aeronautical and space equipment, steel, and ironwork; there are also marine engineering, ship-repairing, and fish-canning industries. Lübeck is known for its wine trade and its marzipan. The Elbe–Lübeck Canal (1900) links the city with the main waterways of Europe and the docks at Travemunde have ferry connections to Scandinavia. **20** F3

Lublin city in Poland, on the Bystrzyca River, 150 km/95 mi southeast of Warsaw; population (1993) 352,500. Industries include textiles, engineering, aircraft, and electrical goods. A trading centre from the 10th century, it has an ancient citadel, a 16th-century cathedral, and a university (1918). A council of workers and peasants proclaimed Poland's independence at Lublin 1918, and a Russian-sponsored committee of national liberation, which proclaimed itself the provincial government of Poland at Lublin on 31 December 1944, was recognized by Russia five days later. **18** M6

Lubumbashi (formerly **Elisabethville** until 1966), town in the Democratic Republic of Congo, on the Lualaba River, close to the Zambian border; population (1994 est) 851,400. It is the chief commercial centre of the Shaba copper-mining region, with the headquarters of Gecamines, the national copper-mining company, being based here. The Lubumbashi campus of the University of Kinshasa (1955) is here. **70** D6

Lucca (ancient **Luca**), town in Tuscany, northwest Italy, on the River Serchio, 60 km/37 mi northwest of Florence; population (1990) 86,400. Industries include weaving, food-processing, engineering, and the manufacture of chemicals. It was an independent republic from 1160 until its absorption into Tuscany in 1847. The composer Giacomo Puccini was born here. **30** F7

Lucerne (German **Luzern**), canton in central Switzerland, in the northern foothills of the Alps; area 1,494 sq km/ 577 sq mi; population (1995) 340,500. The city of Lucerne is the capital of the canton, which mainly comprises fertile pasture land yielding grain, flax, clover, potatoes, and dairy products. A wide range of manufactures are also produced. The canton is German-speaking. **8** (1)C1

Lucerne (German **Luzern**), capital and tourist centre of Lucerne canton, Switzerland, 45 km/28 mi southwest of Zürich, on the River Reuss where it flows out of Lake Lucerne; population (1994) city 61,700; canton 337,700. Manufactured goods include chemicals, metal products, and textiles. The city developed around a Benedictine monastery established in about 750, and owes its prosperity to its position on the St Gotthard road and railway. **30** D3

Lucerne, Lake (German **Vierwaldstättersee**, 'lake of the four forest cantons'), scenic lake surrounded by mountains in north-central Switzerland; area 114 sq km/ 44 sq mi. It lies at an altitude of 437 m/1,434 ft and its greatest depth is 215 m/705 ft. It has four main basins, connected by narrow channels. The lake is subject to sudden storms whipped up by the föhn wind.

Lucknow capital of the state of ▶Uttar Pradesh, India, on the Gumti River, 70 km/46 mi northeast of Kanpur on a railway junction; population (1991) 1,669,200. Industries include engineering, chemicals, textiles, and many handicrafts. The city has many beautiful mosques including the Great Mosque and the Pearl Mosque. It was capital of the Nawabs of Oudh in the 18th century. During the Indian Mutiny against British rule, the British residency was besieged 2 July–16 November 1857. **54** D3

Lüda (or **Lü-ta**, or **Hüta**), Chinese port; see ▶Dalian. **46** G3

Lüdenscheid city in North Rhine–Westphalia, Germany, 55 km/34 mi east of Düsseldorf; population (1994) 81,200. It once belonged to the Hanseatic League trade federation, and its iron industries date from the Middle Ages. Today there are aluminium, machinery, and metallurgical industries in the town. **22** K3

Lüderitz port on Lüderitz Bay, in Karas region of Namibia; population (1995 est) 10,100. It is a centre for diamond mining and fishing, including crayfish, white fish, and pilchards. The town, formerly a German possession, was named after a German merchant who acquired land here in 1883. The town's German architecture is notable. **72** B5

Ludwigsburg city in Baden-Württemberg, Germany, 14 km/9 mi north of Stuttgart; population (1994) 483,100. Ludwigsburg was once a porcelain centre, but its main manufactures are now machinery and toys. There are three castles. **30** E2

Ludwigshafen am Rhein city and river port, Rhineland-Palatinate, Germany, sited on the left bank of the Rhine, opposite Mannheim, 59 km/ 37 mi southeast of Mainz; population (1995) 168,000. Products include chemicals, dyes, fertilizers, plastics, electrical goods, glass, timber, and textiles. The chemical company BASF (Badische Anilin und Soda-Fabrik) has its headquarters here. **22** L5

Lugano resort town on Lake Lugano in the southern canton of Ticino, Switzerland, 20 km/12 mi south of Bellinzona; population (1995) 27,600. It is situated on the main St Gotthard railway line. Industries include engineering, clothing, and chocolate. **30** D4

Lugano, Lake lake partly in Italy, between lakes Maggiore and Como, and partly in the canton of Ticino, Switzerland; area 49 sq km/19 sq mi. Noted for its beautiful scenery, it lies at an altitude of 271 m/890 ft and its greatest depth is 288 m/945 ft. **30** E5

Lugo capital of the province of Lugo in Galicia, northwest Spain, on the River Miño; population (1991) 82,700. It is surrounded by massive Roman walls, now promenades, and has a cathedral dating partly from the 12th century. There are hot sulphur springs nearby, which have been used since Roman times. Lugo has an important meat industry. **30** G6

Lugoj market town in Timiş county, western Romania, on the River Timiş, 56 km/35 mi east of Timişoara; population (1990) 53,800. Wine, fruit, and textiles are its chief commodities, and it is a railway junction. **34** J4

Luhansk (Ukrainian **Lugansk**; formerly **Voroshilovgrad** 1935–58 and 1970–89), industrial city in the ▶Donets Basin area of eastern Ukraine, 125 km/78 mi northeast of Donetsk; population (1990) 501,000. Locomotives, textiles, and mining machinery are produced in the city. **38** G5

Luik Flemish name of ▶Liège, a city in Belgium. **22** H4

Luleå port and administrative capital of Norrbotten county in northern Sweden, on the Gulf of Bothnia, at the mouth of the River Luleå; population (2000 est) 71,700. It is Sweden's leading bulk-shipping harbour with an annual turnover of 7.5 million tonnes, including exports of iron ore and timber. It is also a metallurgical centre, with blast furnaces, steelworks, and manufacturing. **16** M4

Lund city in Malmöhus county, province of Skåne, southwestern Sweden, 16 km/10 mi northeast of Malmö; population (2000 est) 98,900. Industries include telecommunications, computing, and the manufacture of packaging material and medical equipment. It is famous for its 11th-century cathedral and the university, which dates from 1668. **18** C2

Lundy Island rocky, granite island at the entrance to the Bristol Channel; 19 km/12 mi northwest of Hartland Point, Devon, southwest England; area 9.6 sq km/3.7 sq mi; population (1981) 52. Formerly used by pirates and privateers as a lair, it is now the site of a bird sanctuary and the first British Marine Nature Reserve (1986). It has Bronze and Iron Age field systems, which can be traced by their boundaries which stand up above the surface. **24** H10

Lüneburg town in Lower Saxony, Germany, on the River Ilmenau,

110 km/68 mi northeast of Hannover; population (1995) 64,300. Industries include chemicals, salt, paper, and ironworks. An inland port on the Elbe Lateral Canal began to operate in 1976. It is a health resort. It stands at the edge of Lüneburg Heath. **20** F3

Lünen city in North Rhine-Westphalia, Germany, on the River Lippe, 15 km/9 mi north of Dortmund; population (1994) 90,700. It has coalmines, aluminium, copper, pharmaceutical, and leather industries. **22** K3

Lunéville French town in the *département* of Meurthe-et-Moselle, 30 km/19 mi southeast of Nancy, on the River Meurthe; population (1990) 25,400. Textiles, pottery, and railway carriages are made. The château of the dukes of Lorraine is now a museum. **30** B2

Luoyang (or **Loyang**), city in Henan province, China, south of the Huang He River; population (1994) 1,340,000. Industries include oil-refining and the production of glass, machinery, and tractors. Luoyang was the capital of China for nearly eight centuries under the Eastern Zhou (8th–3rd century BC) and other dynasties, and an important Buddhist centre in the 5th and 6th centuries. **46** E4

Lusaka capital of ►Zambia from 1964 (of Northern Rhodesia 1935–64), 370 km/230 mi northeast of Livingstone; population (1995 est) 1,041,000. With good communications by rail, road, and air, it is Zambia's chief centre of trade, industry, banking, and administration, as well as a manufacturing and agricultural centre. Industries include chemicals, including insecticides and fertilizers, flour mills, tobacco factories, vehicle assembly, plastics, printing, cement, iron and steel, food processing, paints, plastics, furniture, and clothing. **72** D3

Lusitania ancient area of the Iberian peninsula, roughly equivalent to Portugal. Conquered by Rome in 139 BC, the province of Lusitania rebelled periodically until it was finally conquered by Pompey (73–72 BC).

Lutherstadt Wittenberg alternative name for ►Wittenberg, a town in Germany, southwest of Berlin. **20** H5

Luton industrial town and unitary authority in south-central England, 48 km/30 mi north of London. It was part of the county of Bedfordshire to 1997.
area 43 sq km/17 sq mi
features the Luton Hoo mansion (1767), designed and built by Robert Adam and with a park laid out by English landscape gardener Capability Brown, is located south of the town; large church of St Mary (13th–15th centuries) is a cruciform building largely in the Decorated and Perpendicular styles
industries cars and trucks, chemicals, engineering components, and electrical goods. Luton airport is a secondary airport for London
population (1996) 181,400
22 B3/5 F6

Luxembourg capital of the country of Luxembourg, on the Alzette and Pétrusse rivers, south of the Ardennes uplands; population (1997) 78,300. The 16th-century Grand Ducal Palace, European Court of Justice, and European Parliament secretariat are situated here, but plenary sessions of the parliament are now held only in Strasbourg, France. Industries include steel, chemicals, textiles, and processed food. **22** J5

Luxembourg landlocked country in Western Europe, bounded north and west by Belgium, east by Germany, and south by France. Luxembourg covers an area of 2,586 sq km/998 sq mi, and has a population of 431,000 (2000 est). The capital is ►Luxembourg; other major towns and cities include Esch, Differdange, Dudelange, and Petange. The head of state is Grand Duke Henri (from 2000); the head of government is Jean-Claude Juncker (from 1995). The currency is the Luxembourg franc, and the languages spoken include Letzeburgisch (a German-Moselle-Frankish dialect; official) and English. **22** H5

Luxembourg province of south-eastern Belgium; area 4,400 sq km/1,698 sq mi; population (1995) 240,300. The capital is ►Arlon and other towns and cities include Bastogne, St Hubert, and Bouillon. Industries include dairy products, iron and steel, and tobacco. The province is situated in the southeastern Ardennes and is widely forested. It includes the rivers Ourthe, Semois, and Lesse. The province was formerly part of the Grand Duchy of Luxembourg, and became a Belgian province in 1831. **7** (1)D4

Luxor (Arabic **al-Uqsur**), town in Egypt on the east bank of the River Nile; population (1992) 146,000. The temple of Luxor, built by Amenhotep III (*c.* 1411–1375 BC) is found here, and tombs of the pharaohs in the Valley of the Kings can be found on the west side of the Nile. **64** F2

Luzern German name of ►Lucerne, a city and canton in Switzerland. **30** D3

Luzon largest island of the ►Philippines; area 108,130 sq km/41,750 sq mi; capital Quezon City; population (1995 est) 32,558,000. The chief city is Manila, capital of the Philippines. Industries include rice, timber, minerals, sugar cane, and hemp. It has US military bases. **50** G3

Lviv (Russian **Lvov**; Polish **Lwów**; German **Lemberg**), capital and industrial city of Lviv region (oblast), western Ukraine, 450 km/280 mi southwest of Kiev; population (1990) 798,000. Lviv is an important manufacturing centre and transport junction. There are chemical, metallurgical, and engineering industries in the city; manufactured goods include motor vehicles, agricultural equipment, textiles, and electronics. Lviv was the principal city of the historical region known as ►Galicia. The Ivan Franko University was founded in 1661. The city was the centre of the revival of Ukrainian nationalism and the resurgence of the Ukrainian Catholic Church in the 1980s. **18** N8

Lwów Polish form of ►Lviv, a historic city in western Ukraine. **18** N8

Lymington port and yachting centre in Hampshire, southern England, on the River Lymington, 19 km/12 mi southwest of Southampton; population (1998 est) 14,300. It has light engineering and marine-related industries, including a marina, yacht repair yards, sailmaking, and chandlery businesses. There is a ferry link to Yarmouth on the Isle of Wight. **22** A4

Lynchburg city in Campbell County (from which it is administratively separate), south-central Virginia, USA, on the James River, in the foothills of the Blue Ridge Mountains, 200 km/124 mi southwest of Richmond; population (1992) 66,100. It is the centre of the western Piedmont region (a tobacco- and fruit-growing area).

Industries include nuclear fuel, electronic equipment, pharmaceuticals, clothing and shoes, paper, rubber products, and machine parts. An academic centre, the city is the home of Randolph-Macon Women's College (1891), Lynchburg College (1903), Central Virginia Community College (1966), and Liberty University (1971). 88 E3

Lynn industrial city and seaport on Massachusetts Bay in Essex County, Massachusetts, 15 km/9 mi northeast of Boston; population (2000 est) 89,100. Lynn manufactures turbines, generators, jet engines, footwear, and clothing. 88 F2

Lyon (English **Lyons**), French industrial city and administrative centre of ►Rhône *département* in the Rhône-Alpes region, part of which lies on the long tongue of land between the Rivers Rhône and Saône, 275 km/170 mi north of Marseille; population (1990) 422,400, conurbation 1,221,000. Lyon is France's third-largest city and most important educational centre after Paris; its main industries are textiles, chemicals, machinery, and printing. Formerly a chief fortress of France, it was the ancient **Lugdunum**, taken by the Romans in 43 BC. 26 K8

Lyons English form of ►Lyon, a city in France. 26 K8

Lys (or **Leie**), river in France and Belgium; length 214 km/133 mi. It rises near Fruges in the Pas-de-Calais and flows past the towns of Aire, Armentières, and Courtrai, to join the Schelde at Ghent. 22 E4

Maas Dutch or Flemish name for the River ►Meuse. 22 J3

Maastricht industrial city and capital of Limburg province, the Netherlands, on the River Maas, near the Dutch-Belgian frontier; population (1997) 118,900. Industries include metallurgy, textiles, and pottery. Maastricht dates from Roman times. It was the site of the Maastricht summit in December 1991. 22 H4

Macassar another name for ►Ujung Pandang, a port in Sulawesi, Indonesia. 53 A4

Macau (or **Macao**), former Portuguese possession on the south coast of China, about 65 km/40 mi west of Hong Kong, from which it is separated by the estuary of the Pearl River; it consists of a peninsula and the islands of Taipa and Colôane; area 17 sq km/7 sq mi; population (1994 est) 395,300. The capital is Macau, on the peninsula. On 31 December 1999 Portugese rule ended and Macau was reintegrated into China, though, like Hong Kong, with a guarantee of 50 years' non-interference in its political system. 50 E2/11 (2)E4

Macclesfield industrial town in Cheshire, northwest England, on the River Bollin on the edge of the Pennines, 28 km/17 mi south of Manchester; population (1991 est) 69,700. Formerly the centre of the English silk industry, it has light engineering industries, including the production of textile machinery, and its chief products are paper, chemicals and pharmaceuticals, and plastics. Several multinational companies have offices here. 24 K8

Macdonnell Ranges mountain range in Northern Territory, central Australia, running east to west in parallel ridges for 644 km/400 mi. The highest peaks are Mount Liebig (1,524m/5,000 ft) and Mount Zeil (1,510 m/4,955 ft). The town of Alice Springs is situated in the middle of the Macdonnell Ranges. The spectacular scenery, with its deep gorges and red rocks, attracts many tourists. 76 F4

Macedonia landlocked country in southeast Europe, bounded north by Serbia, west by Albania, south by Greece, and east by Bulgaria.

Macedonia covers an area of 25,700 sq km/9,922 sq mi, and has a population of 2,024,000 (2000 est). The capital is ►Skopje; other major towns and cities include Bitola, Prilep, Kumanovo, and Tetovo. The head of state is Boris Trajkovski (from 1999); the head of government is Ljubco Georgievski (from 1998). The currency is the Macedonian denar, and the languages spoken include Macedonian (related to Bulgarian; official) and Albanian. 36 C3

Maceió port, industrial city, and capital of Alagoas federal unit (state), northeast Brazil, on the Atlantic coast; population (1991) 699,800. There are petrochemical, sugar, tobacco, textile, and timber industries. Maceió is also a tourist resort with many beaches. 98 K5

Machu Picchu ruined Inca city in the Peruvian Andes, northwest of Cuzco. This settlement and stronghold stands at the top of 300-m/1,000-ft-high cliffs above the Urabamba River and covers an area of 13 sq km/5 sq mi. Built in about AD 1500, the city's remote location saved it from being found and destroyed by the Spanish conquistadors, and the remains of its houses and temples are well preserved. Machu Picchu was discovered in 1911 by the US archaeologist Hiram L Bingham.

Mackenzie River river in the Northwest Territories, northwestern Canada; about 1,705 km/1,060 mi long (from the Great Slave Lake to the Beaufort Sea). It originates as the Athabasca River in British Columbia and flows over 966 km/600 mi to Lake Athabasca; it then flows northwest from the Great Slave Lake until it enters the Beaufort Sea (part of the Arctic Ocean). The Mackenzie River is navigable from June to October, when it eventually freezes over. It is the main channel of the Finlay-Peace-Mackenzie system (4,241 km/2,635 mi long), the second longest system in North America. 82 G4

Mackinac Island island in the Straits of Mackinac, Michigan, USA, between Lake Michigan and Lake Huron; population (1990) 469. It lies just east of the bridge across the Straits of Mackinac; length 5 km/3 mi, width 3 km/2 mi; area 10 sq km/4 sq mi. Visited by the French explorer Jean Nicolet in 1634–35, it became the chief fur-trading centre of a vast region.

McKinley, Mount (or **Denali**), highest peak in North America, situated in the ►Rocky Mountains, Alaska; height 6,194 m/20,320 ft. It was named after US president William McKinley. 92 (1)G3

Mâcon capital of the French *département* of Saône-et-Loire, on the River Saône, 72 km/45 mi north of Lyon; population (1990) 38,500. It produces wine. Mâcon dates from ancient Gaul, when it was known as **Matisco**. The French writer Alphonse Lamartine was born here. 26 K7

Macon city and administrative headquarters of Bibb County, central Georgia, USA, on the Ocmulgee River, 135 km/84 mi southeast of Atlanta, known as the 'Heart of Georgia'; population (1994 est) 109,000. An industrial city, it produces textiles, building materials, tobacco products, farm machinery, and chemicals; it also processes fruits, pecans, and the special kaolin clay that is found nearby. It is the home of Mercer University (1833), Wesleyan College (1836), and Macon College (1968). **90** E3

Macquarie Island outlying Australian territorial possession, a Tasmanian dependency, some 1,370 km/850 mi southeast of Hobart; area 170 sq km/65 sq mi; it is uninhabited except for an Australian government research station.

Madagascar island country in the Indian Ocean, off the coast of East Africa, about 400 km/280 mi from Mozambique.

Madagascar covers an area of 587,041 sq km/226,656 sq mi, and has a population of 15,942,000 (2000 est). The capital is ▶Antananarivo; other major towns and cities include Antsirabe, Mahajanga, Fianarantsoa, Toamasina, and Ambatondrazaka. The head of state is Didier Ratsiraka (from 1996); the head of government is René Tantely Gabrio Andrianarivo (from 1998). The currency is the Malagasy franc, and the languages spoken include Malagasy, French (both official), and local dialects. **72** H3

Madeira river of northwest Brazil; length 3,250 km/2,020 mi. It is the longest tributary of the Amazon, and is formed by the rivers Beni and Mamoré. It flows northeast through the state of Amazonas to join the Amazon east of the city of Manaus. It is navigable by ocean-going vessels for 1,144 km/711 miles as far as Pôrto Velho. Gold-prospecting still occurs near the Bolivian border. **98** E5

Madeira Islands group of islands forming an autonomous region of Portugal, off the northwest coast of Africa, about 420 km/260 mi north of the Canary Islands; area 796 sq km/308 sq mi; population (1994 est) 256,000. Madeira, the largest, and Porto Santo are the only inhabited islands. The Desertas and Selvagens are uninhabited islets. Their mild climate makes them a popular, year-round tourist destination. The capital is ▶Funchal, on Madeira. Pico Ruivo, also on Madeira, is the highest mountain at 1,861 m/6,106 ft. Industries include Madeira (a fortified wine), sugar cane, fruit, fish, handicrafts, and tourism. **66** B2

Madhya Pradesh state of central India, the largest of the Indian states; area 296,480 sq km/114, 472 sq mi; population (2001 est) 84,920,000; about 20% are from tribes (Gonds, Bhils, Baigas, Korkus, Kols, Kamars, and Marias). The capital is ▶Bhopal, and other towns and cities are Indore, Jabalpur, Gwalior, Durg-Bhilainagar, Raipur, and Ujjain. The state is land-locked and mainly upland. It includes the northern part of the Deccan plateau, drained by the rivers Narmada and Mahanadi, as well as the Betwa, Chambal, Tapti, and Son. Vindhya and Satpura mountain ranges (rising to 600 m/2,000 ft) are found in the state, which is heavily forested, particularly in the east. Industries include textiles, engineering, iron ore, steel (at Bhilai complex), coal, bauxite, manganese, paper, aluminium, limestone, diamonds, and cement. Hydroelectric power comes from the Chambal, ▶Narmada, and Mahanadi rivers. Main agricultural products include rice in the wetter regions of the east, cotton, millet, wheat, oilseed, sugar, groundnuts, and soya. The region was split in 2000 to form the new state of ▶Chhatisgarh. **10** C4

Madison state capital of ▶Wisconsin, USA, 193 km/120 mi northwest of Chicago and 120 km/74 mi west of Milwaukee, on an isthmus between lakes Mendota and Monona; seat of Dane County; population (1994 est) 195,000. It is the centre of a rich agricultural region; industries include agricultural machinery, meat packing, battery production, and medical equipment. **88** C2

Madras former name, to 1996, of ▶Chennai, an industrial port and capital of the state of Tamil Nadu, India. **54** D6

Madras former name of ▶Tamil Nadu, a state of India.

Madrid autonomous community of central Spain; area 8,000 sq km/3,088 sq mi; population (1991) 4,845,900. Bounded by the Sierra de Guadarrama mountains in the northwest, and by the River Tagus in the southeast, it is arid plateau country. It is crossed by several rivers, including the Jarama, a tributary of the Tagus. Products include fruit and vegetables, grown in the south; timber from the forests in the northeast, and granite and gypsum from quarries in the mountains. The Escorial palace lies in the northwest; Aranjuez in the south has contained a royal palace since the 15th century and has luxurious gardens. The capital is ▶Madrid. **9** (1)D2

Madrid city and capital of Spain and of ▶Madrid autonomous community, on the Manzanares River; population (1991) 2,909,800. Built on an elevated plateau in the centre of the country, at 655 m/2,183 ft it is the highest capital in Europe and has excesses of heat and cold. Industries include the production of food, electronics, pharmaceuticals, leather, chemicals, furniture, tobacco, and paper, and there is engineering and publishing.

features The Real Academia de Bellas Artes (1752); the Prado museum (1785); the royal palace (1764), built for Philip V; the 15th-century Retiro Park; the Plaza Mayor (1617–20); the Puerta de Alcalá arch; and the basilica of San Francisco el Grande (1761–84).

history Madrid began as the Moorish city of Magerit. It was captured in 1083 by King Alfonso VI of Castile. It remained a small provincial town until Philip II made it his capital in 1561 because of its position at the centre of the Iberian peninsula; it became the national capital in 1607. In 1808 there was an uprising here against Napoleon's army of occupation. Madrid was the centre of opposition to Franco during the Spanish Civil War, and was besieged by the Nationalists 1936–39. **28** G4

Madura island in Indonesia, off Surabaya, Java; one of the Sunda Islands; population (1990) 3,015,100. **52** E4

Madurai city in Tamil Nadu, southeast India, on the River Vaigai; population (1991) 941,000. Industries include cotton textiles and brassware manufacturing. It was capital of the Pandya dynasty from the 5th to the 11th century BC. The city is a centre of Dravidian culture. **54** C6

Maestricht alternative form of ▶Maastricht, a city in the Netherlands. **22** H4

Magadan naval base and port serving the important gold fields in eastern Siberian Russia, located near the northern shore of the Sea of Okhotsk; population (1990) 154,000. Magadan is capital of an Oblast (region) of the same name. Industries include ship repair and the manufacture of mining equipment. 44 S5

Magdeburg industrial city and capital of ►Saxony-Anhalt, Germany, on the River Elbe; population (1995) 263,000. Products include chemicals, precision instruments, paper, textiles, and machinery. A former capital of Saxony, Magdeburg became capital of Saxony-Anhalt on German reunification in 1990. In 1938 the city was linked by canal with the Rhine and Ruhr rivers. 20 G4

Magellan, Strait of (Spanish **Estrecho de Magallanes**), channel at the southern tip of Chile; it separates the South American mainland from Tierra del Fuego, and joins the South Pacific and South Atlantic oceans. It is 595 km/370 mi long with a maximum width of 32 km/20 mi. The strait is named after the Portuguese navigator Ferdinand Magellan who discovered it in 1520. It provided a safer passage than that around Cape Horn, but its importance declined following the opening of the Panama Canal in 1914, an international waterway within Chile's territorial waters. 100 G9

Maggiore, Lake (ancient **Verbanus Lacus**), lake largely in Italy, bounded on the west by Piedmont and on the east by Lombardy; its northern end is in the Swiss canton of Ticino. It is 63 km/39 mi long and up to 9 km/5.5 mi wide, with an area of 212 sq km/82 sq mi. 30 D5

Maghreb (Arabic 'far west', 'sunset'), name for northwestern Africa. The Maghreb powers – Algeria, Libya, Morocco, Tunisia, and Western Sahara – agreed on economic coordination 1964–65, with Mauritania cooperating from 1970. In 1989 these countries formed an economic union known as the Arab Maghreb Union. Chad and Mali are sometimes included. Compare ►Mashraq, the Arab countries of the eastern Mediterranean.

Magnitogorsk city in Chelyabinsk oblast (region), central Russian Federation, on the Ural River; population (1996 est) 425,000. Magnitogorsk is one of the foremost metallurgical centres in the Russian Federation. It has a vast iron and steel plant, which was built here 1929–34. Coal is brought by rail from the Kuznetsk Basin and from Karaganda in nearby Kazakhstan, oil by pipeline from Shkapovo in Bashkiria, and natural gas from Gazli in Uzbekistan. Subsidiary industries include chemicals, cement production, and engineering. The city is an important railway junction and cultural centre, and produces rolling stock. 38 L4

Mahabad (formerly **Saujbulagh**), Kurdish town lying in a fertile valley south of Lake Urmia, in the province of Azerbaijan, west Iran; population (1991) 82,000. It formed the centre of a short-lived Kurdish republic (January–December 1946) before being reoccupied by the Iranians. In the 1980s Mahabad was the focal point of resistance by Iranian Kurds against the Islamic republic, and is still a centre of Kurdish nationalism. 58 L5

Maharashtra state in west central India; area 307,700 sq km/118,800 sq mi; population (1991) 79,000,000. The capital is ►Mumbai (formerly Bombay), and other towns and cities include Pune, Nagpur, Ulhasnagar, Solapur, Nasik, Thana, Kolhapur, Aurangabad, Sangli, and Amravati. The state is divided by the heavily forested Western ►Ghats into the Konkan coastal plain and the Deccan plateau. The plain is subject to the southwest monsoon from June to September. Inland is in a rain shadow receiving only half the coastal rainfall. The Godavari and Krishna rivers rise in the Western Ghats and flow eastwards across the Deccan. The Marathi language is spoken by 50% of the population. 80% of the population are Hindu, with Parsee, Jain, and Sikh minorities. The state was formed in 1960 from the southern part of the former Bombay state. 10 B5

Mahé chief island of the ►Seychelles; population (1995 est) 69,400. The island, in the Indian Ocean, has an area of 153 sq km/59 sq mi, and the capital of the Seychelles, Victoria, lies on its northwest coast. The airport and the Morne Seychelles national park have helped the development of a substantial tourist industry. 72 (2)C1

Mahilyow (formerly **Mogilev**), industrial city in Belarus, on the River Dnieper, 193 km/120 mi east of Minsk; population (2001 est) 372,700. Goods manufactured here include tractors, clothing, chemicals, and furniture. It was annexed by Russia from Sweden in 1772. 38 F4

Mahón (or **Port Mahon**), capital and port of the Spanish island of Menorca; population (1990) 24,400. Probably founded by the Carthaginians, it was under British occupation 1708–56 and 1762–82. 28 Q5

Maidenhead town in southern England, on the River Thames, 40 km/25 mi west of London; it became part of ►Windsor and Maidenhead unitary authority in April 1998 and is the authorities administrative headquarters; population (1991) 59,600. Industries include electronics, printing, and the production of computer software, pharmaceuticals, board, and paper. It is a major centre for the telecommunications industry. The leisure industry is more limited than in nearby Windsor, but the river is a focus for some tourism. 22 B3

Maidstone town and administrative headquarters of ►Kent, southeast England, on the River Medway; population (1991) 90,900. Industries include the manufacture of agricultural machinery, and paper and card for packaging. There are a number of medieval buildings, among them the parish church of All Saints', the late 14th-century Archbishop's Palace (a former residence of the archbishops of Canterbury), and the College of Priests (1395–98) built of Kentish ragstone from quarries to the west of the town. The Elizabethan Chillington Manor is an art gallery and museum. 22 C3

Maikop (or **Maykop**), capital city of the ►Adygeya (Adygei) Republic in the southwestern Russian Federation, 150 km/93 mi southwest of Stavropol on the River Bielaia; population (1996 est) 166,000. The main industries are the production of foodstuffs and tobacco, leather, and furniture. Maikop lies at the centre of an oil-producing region. Nearby oilfields, discovered in 1900, are linked by pipeline with Tuapse on the Black Sea. There is a famous Bronze-Age burial mound nearby. 58 J1

Main river in central western Germany, 515 km/320 mi long, flowing through Frankfurt to join the River Rhine at Mainz. A canal links it with the Danube, so providing a large network of waterways for transport of goods. 20 E7

Maine northeasternmost state of the USA and the largest of the New England states. It is nicknamed the Pine Tree State. Maine was admitted to the Union in 1820 as the 23rd US state. It is bordered to the northwest by Québec, Canada, to the north and east by New Brunswick, Canada, to the east and south by the Atlantic Ocean, and to the west by New Hampshire; it is the sole US state to be contiguous with only one other US state. Maine's West Quoddy Head is the easternmost US point; the state (and region) is popularly known as 'Down East'.
population (2000 est) 1,274,900
area 86,200 sq km/33,300 sq mi
capital ►Augusta
towns and cities Portland, Lewiston, Bangor
industries and products blueberries, dairy and market garden produce, paper, pulp, timber, footwear, leather, textiles, fish, lobster; tourism, shipbuilding 88 G1

Maine-et-Loire *département* in the ►Pays de la Loire region of France; area 7,130 sq km/2,753 sq mi; population (1990) 706,400. The River ►Loire crosses the *département* from east to west, and forms in its course several islands. Vines are grown in some parts of the *département* (Saumur, for example), and the fertile plains are used for growing cereals, hemp, and sugar beet, and for horticulture. There are computing, electronic, and car-parts industries near ►Angers, the administrative centre. 6 C3

Mainz (French **Mayence**), river port and capital of ►Rhineland-Palatinate, Germany, on the left bank of the Rhine, 37 km/23 mi southwest of Frankfurt-am-Main; population (1995) 184,500. It is a major centre of the wine trade. Industries include computers (IBM), metal goods, precision instruments, chemicals, machinery, telecommunications, television broadcasting, and printed fabrics. In Roman times it was a fortified camp and became the capital of Germania Superior. Printing was possibly invented here in about 1448 by Johann Gutenberg. 20 D6

Majorca alternative spelling of ►Mallorca. 28 P5

Makeyevka (Ukrainian **Makiyivka**; formerly **Dmitrievsk** until 1931), industrial city producing coal and coke, iron and steel, and chemicals, situated in the ►Donets Basin in southeastern Ukraine; population (1996 est) 409,000. The city, which lies 14 km/9 mi east of Donetsk, was founded in 1899. 38 G5

Makhachkala (formerly **Petrovsk** (1844–1922), city in northern Caucasia, capital, economic and cultural centre of the Dagestan Republic, in the southwestern Russian Federation; population (1996 est) 347,000. Makhachkala is a major port on the Caspian Sea; oil is piped here for export from Groznyy and Izberbash. The city has aerospace, shipbuilding, oil-refining, textile, and food industries. It was founded as a Russian fort in 1844, and became a town in 1857. Makhachkala gained its present name (after the local Bolshevik leader Makhach) and was made the capital of Dagestan in 1922. 58 M2

Makiyivka Ukrainian name for ►Makeyevka, an industrial city in southeastern Ukraine. 38 G5

Makkah Arabic name for ►Mecca, the holiest city of Islam, situated in Saudi Arabia. 64 G3

Makó town in Csongrád county, southeast Hungary, on the River Mureş, near the Romanian border, 15 km/9 mi east of Szeged; population (1993) 26,700. The surrounding agricultural district specializes in onions. 34 H3

Malabar Coast coastal area of Kerala state, India, lying between the Arabian Sea and the Western ►Ghats; about 65 km/40 mi west–east, 725 km/450 mi north–south. A fertile area with heavy rains, it produces rice, food grains, coconuts, rubber, spices; also teak, ebony, and other woods. Lagoons fringe the shore. A district of Tamil Nadu transferred in 1956 to Kerala was formerly called the Malabar Coast. Mayalayam is the main language spoken. 54 B6

Malabo port and capital of ►Equatorial Guinea, on the north coast of the volcanic island of Bioko; population (1992) 35,000. It trades in cocoa, coffee, copra, and other agricultural products. It was founded in the 1820s by the British as **Port Clarence** (also **Clarencetown**). Under Spanish rule it was known as **Santa Isabel** (until 1973). It became the capital in 1968. 68 F4

Malacca, Strait of channel between Sumatra and the Malay Peninsula; length 965 km/600 mi; it narrows to less than 38 km/24 mi wide. It carries all shipping between the Indian Ocean and the South China Sea. 52 C2

Málaga industrial seaport and capital of Málaga province in Andalusia, southern Spain, situated at the mouth of the River Guadalmedina on the Mediterranean coast; population (1991) 512,100. Industries include sugar refining, distilling, brewing, and olive-oil pressing; it is also a holiday resort. Crude oil is unloaded from tankers offshore into a pipeline that transports it to Puertollano in the province of Ciudad Real. 28 F8

Malagasy Republic former name (1958–75) of ►Madagascar. 72 H3

Malaspina Glacier glacier in Alaska, one of the world's largest ice sheets; area 2,176 sq km/840 sq mi. It lies west of Yakutat Bay, northeast Gulf of Alaska, and is part of the St Elias Mountains glacier system, the most extensive outside the polar ice caps.

Malatya capital of a province of the same name in eastern central Turkey, lying west of the River Euphrates; population (1990) 281,800. 58 H4

Malawi country in southeast Africa, bounded north and northeast by Tanzania; east, south, and west by Mozambique; and west by Zambia.

Malawi covers an area of 118,484 sq km/45,735 sq mi, and has a population of 10,925,000 (2000 est). The capital is ►Lilongwe; other major towns and cities include Blantyre, Mzuzu, and Zomba. The head of state and government is Bakili Muluzi (from 1994). The currency is the Malawi kwacha, and the languages spoken include English, Chichewa (both official), and other Bantu languages. 72 E2

Malawi, Lake (or **Lake Nyasa**), lake, bordered by Malawi, Tanzania, and Mozambique, formed in a section of the Great ►Rift Valley. It is about 500 m/1,650 ft above sea level and 560 km/350 mi long, with an area of 28,749 sq km/11,100 sq mi and a depth of 700 m/2,296 ft, making it the ninth biggest lake in the world. It is intermittently drained to the south by the River Shire into the Zambezi. 72 E2

Malay Peninsula southern projection of the continent of Asia, lying between the Strait of Malacca, which divides it from Sumatra, and the China Sea. The northern portion is partly in Myanmar (formerly Burma), partly in Thailand; the south forms part of Malaysia. The island of Singapore lies off its southern extremity. 50 C6

Malaysia country in southeast Asia, comprising the Malay Peninsula, bounded north by Thailand, and surrounded east and south by the South China Sea and west by the Strait of Malacca; and the states of Sabah and Sarawak in the northern part of the island of Borneo (southern Borneo is part of Indonesia).

Malaysia covers an area of 329,759 sq km/127,319 sq mi, and has a population of 22,244,000 (2000 est). The capital is ▶Kuala Lumpur; other major towns and cities include Johor Bahru, Ipoh, George Town (on Penang island), Kuala Trengganu, Kuala Bahru, Petalong Jaya, Kelang, Kuching (on Sarawak), and Kota Kinabalu (on Sabah). The head of state is Tuanku Salehuddin Abdul Aziz Shan bin al-Marhum Hisamuddin Alam Shah (from 1999); the head of government is Mahathir bin Muhammad (from 1981). The currency is the ringgit, and the languages spoken include Bahasa Malaysia (Malay; official), English, Chinese, Tamil, Iban, and many local dialects. 52 C2

Malbork (German **Marienburg**), town in Elbląg province, Poland, on the River Nogat, 40 km/25 mi southeast of Gdańsk; population (1990) 39,600. Industries include timber and sugar refining. From 1309 to 1457 the Marienburg Castle was the headquarters of the Grand Master of the Teutonic Knights (a German Christian military order). 18 J3

Maldives group of 1,196 islands in the north Indian Ocean, about 640 km/400 mi southwest of Sri Lanka; only 203 of them are inhabited.

The Maldives covers an area of 298 sq km/115 sq mi, and has a population of 286,000 (2000 est). The capital is ▶Malé. The head of state and government is Maumoon Abd Gayoom (from 1978). The currency is the rufiya, and the languages spoken include Divehi (a Sinhalese dialect; official), English, and Arabic. 54 B8

Malé capital and chief atoll of the Maldives in the Indian Ocean; population (1995 est) 64,800. It lies 700 km/435 mi southwest of Sri Lanka and has an airport which has helped the development of the tourist industry in recent decades. It trades in copra, breadfruit, fish, and palm products. 54 B8

Mali landlocked country in northwest Africa, bounded to the northeast by Algeria, east by Niger, southeast by Burkina Faso, south by Côte d'Ivoire, southwest by Senegal and Guinea, and west and north by Mauritania.

Mali covers an area of 1,240,142 sq km/478,818 sq mi, and has a population of 11,234,000 (2000 est). The capital is ▶Bamako; other major towns and cities include Mopti, Kayes, Ségou, Tombouctou, and Sikasso. The head of state is Alpha Oumar Konare (from 1992); the head of government is Mande Sidibe (from 2000). The currency is the franc CFA, and the languages spoken include French (official), Bambara, and other African languages. 66 E5

Malines French name for ▶Mechelen, a city in Belgium. 22 G3

Mallorca (or **Majorca**), largest of the ▶Balearic Islands, belonging to Spain, in the western Mediterranean; area 3,640 sq km/1,405 sq mi; population (1990 est) 582,000. The capital is Palma. The highest mountain on the island is Puig Mayor (1,445 m/ 4,741 ft). Industries include olives, figs, oranges, wine, brandy, timber, and sheep farming. Tourism is the mainstay of the economy. 28 P5

Malmédy town in Liège province, east Belgium, 40 km/25 mi south of Aachen, in the region of Eupen et Malmédy; population (1997) 10,800. Industries include food processing and papermaking. 22 J4

Malmö industrial port (shipbuilding, engineering, textiles) in southwestern Sweden, situated across the Øresund from Copenhagen, Denmark; population (1994 est) 242,700. Founded in the 12th century, Malmö is Sweden's third-largest city. 18 C2

Malta island in the Mediterranean Sea, south of Sicily, east of Tunisia, and north of Libya.

Malta covers an area of 320 sq km/124 sq mi, and has a population of 389,000 (2000 est). The capital (and chief port) is ▶Valletta; other major towns and cities include Rabat, Birkirkara, Qormi, Sliema, Zetjun, and Zabor. The head of state is Guido de Marco (from 1999); the head of government is Edward Fenech Adami (from 1998). The currency is the Maltese lira, and the languages spoken include Maltese and English (both official). 32 J13

Maluku (or **Moluccas**), group of Indonesian islands; area 74,500 sq km/28,764 sq mi; population (1995 est) 2,094,700. The capital is Ambon, on Ambon. As the Spice Islands, they were formerly part of the Dutch East Indies; the southern Moluccas attempted secession from the newly created Indonesian republic from 1949, with sectarian fighting continuing into 2000. 53 C3

Malvinas, Islas Argentine name for the ▶Falkland Islands. 100 K9

Manacor town in southeast Majorca, one of the Spanish ▶Balearic Islands; population (1995) 29,200. Products include shoes, handicraft souvenirs, and artificial pearls; tourism is important. At nearby Porto Cristo are the Caves of Drach ('the dragon') with stalactite and stalagmite formations, and one of the largest known subterranean lakes (Lake Martel). 28 P5

Managua capital and chief industrial city of Nicaragua, and capital of a department of the same name; it is situated on the southern shore of Lake Managua 45km/28 mi from the Pacific coast and 138 km/86 mi from the main port of Corinto; population (1995 est) 1,240,000. One-fifth of the nation's population is resident here. It is Nicaragua's largest city and main industrial and commercial centre. Managua produces 60% of the nation's goods by value including tobacco, textiles, cement, cotton, drinks, and processed foods. Surrounding lowlands are very fertile, supporting maize, beans, sugar cane, and banana plantations. The city's university was founded in 1961. 94 G6

Manaus capital of Amazonas federal unit (state), northwest Brazil, on the Río Negro, 16 km/10 mi from its confluence with the River Solimó which forms the River Amazon; population (1991) 996,700. It is the industrial trading and commercial centre of the state, and its chief port, although 1,600 km/1,000 mi from the Atlantic. Timber and rubber are the main exports, and there are sawmills and an oil refinery. Manaus is an important free-trade zone (established in 1966) that distributes products nationwide. It specializes in electrical goods. **98** E4

Manche, La French name for the ►English Channel. It gives its name to a French *département*. **24** J12

Manche maritime *département* in the ►Basse-Normandie region of France, facing the English Channel (La Manche); area 5,947 sq km/2,296 sq mi; population (1990) 480,900. The administrative centre is ►St Lô. Other towns include Avranches, ►Cherbourg, and Coutances. The chief products are grain, animal fodder, and fruit (especially apples for cider), and horses are reared. There are metal, textile, and service industries. The area suffered great damage during World War II. **6** C2

Manchester metropolitan district of Greater Manchester, and city in northwest England, on the River Irwell, 50 km/31 mi east of Liverpool; population (1991 est) 402,900. A financial and manufacturing centre, its industries include banking, insurance, and printing; the production of cotton and man-made textiles, petrochemicals, rubber, paper, machine tools, and processed foods; and heavy, light, and electrical engineering. It is linked to the River Mersey and the Irish Sea by the **Manchester Ship Canal**, opened in 1894, although only one dock remains open. **24** K8

Manchester, Greater metropolitan county of northwest England, created in 1974; in 1986 most of the functions of the former county council were transferred to metropolitan district councils.
area 1,290 sq km/498 sq mi
towns and cities Manchester, Bolton, Bury, Oldham, Rochdale, Salford, Stockport, Tameside, Trafford, Wigan
features Manchester Ship Canal links it with the River Mersey and the sea; Old Trafford cricket ground at Stretford, and the football ground of Manchester United
industries engineering, textiles, textile machinery, chemicals, plastics, electrical goods, electronic equipment, paper, printing, rubber, and asbestos
population (1996) 2,575,600 **5** E5

Mandalay chief city of the Mandalay division of Myanmar (formerly Burma), on the River Irrawaddy, about 495 km/370 mi north of Yangon (Rangoon); population (1998 est) 370,000. It is a river port, with a university founded in 1964. **50** B2

Mangalore industrial port at the mouth of the Netravati River in Karnataka, India; population (1991) 273,000. Industries include textiles, timber, and food-processing. The port handles 75% of India's coffee exports and most of its cashew nuts, as well as pepper, tea, and roofing tiles. **54** B6

Manhattan island of the city of ►New York, USA, forming most of a borough; population (1990) 1,488,000. It is 20 km/12.5 mi long and 4 km/2.5 mi wide, and lies between the Hudson and East rivers. The rocks from which it is formed rise to a height of more than 73 m/240 ft in the north of the island. Manhattan Island is bounded on the north and northeast by the Harlem River and Spuyten Duyvil Creek (which separate it from the Bronx); on the south by Upper New York Bay; on the west by the Hudson River (which separates it from New Jersey); and on the east by the East River (which separates it from Queens and Brooklyn). The borough of Manhattan also includes a small port at the Bronx mainland and several islands in the East River. Manhattan is the economic hub of New York City, although there are large residential and industrial areas here also. It includes the Wall Street business centre, Broadway and its theatres, Carnegie Hall (1891), the World Trade Centre (1973), the Empire State Building (1931), the United Nations headquarters (1952), Madison Square Garden, and Central Park.

Manicouagan river in Québec, Canada, flowing 485 km/301 mi south from Mushalagan and Manicouagan lakes to Baie Comeau on the St Lawrence River. Together with the Outardes River, which enters the St Lawrence east of Baie Comeau, the Manicouagan – backed by the Manicougan Reservoir, the fifth-largest in the world – is the source of the largest hydroelectric power complex in Canada. It remains important for the transport of timber. **82** T6

Manila industrial port (textiles, tobacco, distilling, chemicals, shipbuilding) and capital of the Philippines, in the southwest of the island of Luzon, where the River Pasig enters Manila Bay; population of the metropolitan area (including ►Quezon City) 9,286,000 (1996); city (1998 est) 1,761,900. Other industries include timber and food processing. **50** G4

Manipur state of northeast India; bordered south and east by Myanmar; area 22,300 sq km/8,610 sq mi; population (1994 est) 2,010,000 (30% are hill tribes such as Nagas and Kukis). The capital is ►Imphal. The state is mostly wooded and mountainous (mainly over 2,000 m/6,500 ft), with a central valley containing Imphal. Features include Loktak Lake, and the state is the original Indian home of polo. Industries include textiles, cement, and handloom weaving. Agricultural products include rice, grain, fruit, vegetables, and sugar. Languages spoken in the state are Manipuri and English. 70% of the population are Hindu. **10** F4

Man, Isle of (Scottish Gaelic **Ellan Vannin**), island in the Irish Sea, a dependency of the British crown, but not part of the UK; area 570 sq km/220 sq mi; population (1991) 69,800, almost 50% of which lives in the capital ►Douglas and nearby Onchan. Other towns and cities include Ramsey, Peel, and Castletown. Industries include light engineering products, agriculture, and fishing. Tourism has declined in importance and has been replaced by financial services as the island's principal industry. The Isle of Man was Norwegian until 1266, when it was ceded to Scotland. It came under UK administration in 1765. **24** H7/**5** D4

Manitoba (Algonquian *Manitou* 'great spirit'), province in central Canada, the easternmost of the Prairie provinces. Bounded to the south, on the 49th Parallel, by the US states of Minnesota (in the east) and North Dakota (in the west); to the west by Saskatchewan; to the north, on the 60th Parallel, by the Northwest Territories and Hudson Bay; and to the east by Ontario; area 650,000 sq km/251,000 sq mi; population (1996) 1,113,900. The capital is ►Winnipeg, and other towns and cities include Brandon, Thompson, St Boniface, Churchill, Flin Flon, Portage La Prairie, and The Pas. Lakes Winnipeg, Winnipegosis, and Manitoba (area 4,700 sq km/1,814 sq mi) are in the province, which is 50% forested. Industries include production of grain and food-processing; manufacture of machinery; fur-trapping; fishing; mining of nickel, zinc, copper, and the world's largest deposits of caesium (a metallic element used in the manufacture of photocells). 82 M6

Manitoba, Lake shallow lake in south-central Manitoba province, Canada, which drains into Lake Winnipeg to the northeast through the Dauphin River; area 4,700 sq km/ 1,800 sq mi. Like lakes Winnipeg and Winnipegosis, Lake Manitoba is a remnant of Lake Agassiz, which once occupied much of Manitoba. At the south end of the lake is Delta Marsh, a waterfowl research station. 82 M6

Manizales capital of Caldas department, western Colombia, 270 km/ 168 mi by road south of Medellín; 175 km/108 mi northwest of Bogotá, situated in the Central Cordillera of the Andes 2,150 m/7,000 ft above sea level; population (1994) 333,000. It is the commercial centre of Colombia's coffee-growing region. Manufactures include textiles, chemicals and leather goods. It is linked with Mariquita by the world's longest overhead cable transport system, 72 km/45 mi long. The city is dominated by the Nevado del Ruiz volcano which erupted in November 1985, killing more than 25,000 people. 98 B2

Mannheim industrial city in Baden-Württemberg, Germany, at the confluence of the Rhine and the Neckar, 93 km/58 mi northwest of Stuttgart; population (1995) 315,100. There are heavy machinery, glass, earthenware, and chemical industries. It has a large oil refinery. On the opposite bank of the Rhine is Ludwigshafen. 22 L5

Manresa town in the province of Barcelona, northwest Spain, on the River Cardoner; population (1995) 65,700. It is an important textiles town; products include cotton and tyres. Manresa has a Gothic church and a 17th-century baroque church, the latter adjoining the cave in which St Ignatius Loyola lived for some time following his conversion. 28 M3

Mansfield industrial city and administrative headquarters of Richland County, north-central Ohio, USA, 100 km/62 mi northeast of Columbus; population (1992) 53,200. It is situated in a winter-sports area; industries include the manufacture of electrical machinery, car parts, and steel and rubber products. It is the home of Ohio State University Mansfield Branch (1958). 88 D2

Mansfield industrial town in Nottinghamshire, England, on the River Maun, 22 km/14 mi north of Nottingham; population (1991) 71,900. It is a retail centre for the area. The most important industries are brewing and light engineering, especially the production of car components, while minor industries include quarrying, moulding sand and sand lime bricks, and the manufacture of coal tar products. 22 A1

Mansûra, El industrial city (cotton) and capital of Dakahlia governorate, northeast Egypt, on the Damietta branch of the River Nile; population (2001 est) 413,500. El Mansûra was founded *c.* 1220; St Louis IX, king of France, was imprisoned in the fortress in 1250 while on a Crusade. 64 F1

Mantes-la-Jolie (formerly **Mantes-Gassicourt**), French town in the *département* of Yvelines, on the River Seine; population (1990) 48,000. Paper, musical instruments, and bricks are manufactured here and it is a commuter town for Paris. Mantes was burnt in 1087 by William the Conqueror. Riding through the burning town, William's horse stumbled, throwing him against his saddle and causing internal injures, from which he later died. 22 D5

Mantova Italian name for ►Mantua, a city in Lombardy, Italy. 30 F5

Mantua (Italian **Mantova**), capital of Mantua province, Lombardy, Italy, on an island in a lagoon of the River Mincio, southwest of Verona; population (1990) 54,200. It is the market and service centre of a prosperous agricultural region. Industries include petrochemicals, engineering, plastics, brewing, and printing. The town, which dates from Etruscan times, has Gothic palaces and a cathedral founded in the 10th century. The Roman poet Virgil was born nearby in 70 BC. 30 F5

Manzanares town in the province of Ciudad Real, central Spain; population (1991) 17,900. Products include wine (it is the centre of the Valdepeñas wine district), saffron, and textiles. It has a castle, once the property of the marqués of Santillana. 28 G5

Maputo (formerly **Lourenço Marques** until 1975), capital of ►Mozambique, and Africa's second-largest port, on Delagoa Bay; population (1993 est) 2,000,000. Linked by road and rail with Zimbabwe, Swaziland, and South Africa, it is a major outlet for minerals, steel, textiles, processed foods, and furniture. 72 E5

Maracaibo port in northwest Venezuela, and capital of the state of Zulia, 500 km/300 mi west of Caracas; population (1995 est) 1,346,000. Situated on the west bank of the channel connecting Lake Maracaibo with the Gulf of Venezuela, it is the centre of Venezuela's main oil-producing region and the second-largest city in the country. Exports include sugar, coffee, and cocoa; industries include oil refining and petrochemicals. There is an airport. 98 C1

Maracaibo, Lake lake in northwest Venezuela, connected to the Gulf of Venezuela by a narrow waterway 54 km/34 mi long; area 14,000 sq km/ 5,400 sq mi. It is the largest lake in South America. Oil was discovered here in 1914, and the oilfields in and around the lake produce over 70% of the country's output. 98 C2

Marbella port and tourist resort in the province of Málaga in Andalusia, southern Spain, on the Costa del Sol between Málaga and Algeciras; population (1991) 76,800. There are three bullrings, a Moorish castle, and the remains of a medieval defensive wall. 28 F8

Marburg an der Lahn university town in Hesse, Germany, on the River Lahn, 80 km/50 mi north of Frankfurt-am-Main; population (1995) 76,600. There are chemical, machinery, and pottery industries. The university was founded in 1527 as a centre of Protestant teaching. Martin Luther and Ulrich Zwingli disputed on religion at Marburg in 1529. 22 L4

Marches boundary areas of England with Wales, and England with Scotland. For several centuries from the time of William the Conqueror, these troubled frontier regions were held by lords of the Marches, those on the Welsh frontier called Marcher Lords, sometimes called *marchiones*, and those on the Scottish border known as earls of March. The first Marcher Lord was Roger de Mortimer (about 1286–1330); the first earl of March, Patrick Dunbar (died in 1285).

Margate town and resort on the north coast of Kent, southeast England, on the Isle of Thanet; population (1991) 38,500. A range of industries includes the manufacture of pharmaceuticals, generators, car components, sign machinery, and inks. It was one of the original Cinque Ports, and developed in the late 18th century as one of the earliest coastal resorts in England. 22 D3

Mariana Islands (or **Marianas**), archipelago in the northwest Pacific, east of the Philippines, divided politically into ▶Guam (an unincorporated territory of the USA) and the ▶Northern Mariana Islands (a commonwealth of the USA). 74 E4

Mariánské Lázně (German **Marienbad**), spa town in the Czech Republic; population (1991) 15,400. An international reputation for its healthy waters was established before World War II. The water of its springs, which contains Glauber's salts, has been used medicinally since the 16th century. 20 H7

Maribor (German **Marburg**), town and resort on the River Drave in Slovenia, with a 12th-century cathedral and some industry (boots and shoes, railway rolling stock); population (1991) 108,100. Maribor dates from Roman times. 30 L4

Marienbad German name of ▶Mariánské Lázně, a spa town in the Czech Republic. 20 H7

Marienburg German name for ▶Malbork, a town in Poland. 18 J3

Marietta industrial city and administrative headquarters of Washington County, southeast Ohio, USA, where the Muskingum River flows into the Ohio River, 148 km/92 mi southeast of Columbus; population (1990) 15,000. It is the centre of an agricultural region; industries include the manufacture of plastics, metal products, natural-gas drilling equipment, chemicals, and office equipment.
The city is the home of Marietta College (1834). 88 D3

Mariupol (formerly **Zhdanov** 1948–89), industrial port in eastern Ukraine, 100 km/62 mi south of Donetsk on the Sea of Azov; population (1996 est) 510,000. Iron and steel are manufactured, together with ships and chemicals. 38 G5

Marmara Denizi Turkish name for the Sea of ▶Marmara, a small inland sea in Turkey, between the Black Sea and the Aegean Sea. 36 L4

Marmara, Sea of (Turkish **Marmara Denizi**; formerly **Prokonessos**), small inland sea in Turkey, connected to the Black Sea by the Bosporus and to the Aegean Sea by the Dardanelles, separating the continents of Europe and Asia; area 11,140 sq km/4,300 sq mi. In parts it reaches depths of over 1,200 m/ 3,936 ft. There are several islands in the sea that contain white-marble quarries, the largest of which is Marmara. 36 L4

Marne *département* in the ▶Champagne-Ardenne region of France; area 8,163 sq km/3,152 sq mi; population (1990) 559,000. The River ▶Marne crosses the *département* from southeast to northwest. The best varieties of the champagne vines are grown in the dry and chalky soil of the north. Other industries include tanning, iron and copper founding, brewing, pottery, and tourism. The rearing of sheep is important, and woollen goods are manufactured. The administrative centre is ▶Châlons-sur-Marne. Other important towns are Epernay, ▶Reims, Sainte-Menehould, and Vitry-le-François. 6 E2

Marne river in France which rises in the plateau of Langres and joins the Seine at Charenton near Paris; length 525 km/326 mi. It gives its name to the *départements* of ▶Marne, Haute-Marne, Seine-et-Marne, and Val-de-Marne, and to two battles of World War I. 22 F5

Marquesas Islands (French **Iles Marquises**), island group in ▶French Polynesia, lying 1184 km/736 mi northeast of Tahiti; area 1,270 sq km/ 490 sq mi; population (1995 est) 8,600. The administrative headquarters is Atuona on Hiva Oa. The largest settlement is Taiohae on Hiva Oa; other islands include Kuku Hiva, Ua Pa, Ua Huka, Tahuata, and Fatu Hiva. The islands were annexed by France in 1842. The main products are copra and vanilla. 74 M6

Marquises, Iles French form of ▶Marquesas Islands, part of ▶French Polynesia. 74 M6

Marrakech alternative spelling of ▶Marrakesh, a city and former capital of Morocco. 66 D2

Marrakesh (or **Marrakech**), historic imperial city in Morocco in the foothills of the Atlas Mountains, about 210 km/130 mi south of Casablanca; population (2001 est) 736,500. It is a tourist centre, and has textile, leather, and food processing industries. Founded in 1062, it has a medieval palace and mosques, and was formerly the capital of Morocco. 66 D2

Marsala (ancient **Lilybaeum**; Arab **Marsah-el-Allah** 'port of Allah'), port in western Sicily, Italy, on the Mediterranean coast 30 km/ 19 mi south of Trapani; population (1990) 80,800. The nationalist leader Giuseppe Garibaldi landed here in 1860 at the start of his campaign to capture Sicily for Italy. The town produces the sweet wine of the same name; there is also trade in agricultural produce and olive oil. 32 G11

Marseille (English **Marseilles**), chief seaport and second city of France, and administrative centre of the *département* of Bouches-du-Rhône and of the ▶Provence-Alpes-Côte d'Azur region, situated on the Golfe du Lion on the Mediterranean Sea; population (1990) 807,700. Industries include chemicals, metallurgy, shipbuilding, and food processing, as well as oil-refining at the massive industrial complex of Fos-sur-Mer to the west. 26 L10

Marshall city and administrative headquarters of Harrison County, northeast Texas, 228 km/142 mi east of Dallas and across the border from Shreveport, Louisiana, USA; population (1990) 23,700. It is centre of an agricultural and oil-producing region; industries include food processing, cotton, clothing, building materials, vehicle parts, and chemicals. The city is the home of Wiley College (1873) and East Texas Baptist University (1912). 90 C3

Marshall Islands country in the west Pacific Ocean, part of Micronesia, occupying 31 atolls (the Ratak and Ralik chains).

The Marshall Islands covers an area of 181 sq km/70 sq mi, and has a population of 64,000 (2000 est).

The capital is ▶Dalap-Uliga-Darrit (on Majuro atoll); the only other town is Ebeye. The head of state and government is Kessai H Note (from 2000). The currency is the US dollar, and the languages spoken include Marshallese and English (both official). 74 G4

Martigues French fishing town in the *département* of Bouches-du-Rhône; population (1990) 43,000. Martigues is situated on the Etang de Berre lagoon at its outlet to the Mediterranean, 29 km/18 mi northwest of Marseille. There are salt, cement, and vegetable-oil industries. 26 L10

Martinique French island in the West Indies (Lesser Antilles); area 1,079 sq km/417 sq mi; population (1996 est) 399,200. The capital is ▶Fort-de-France. The island features several active volcanoes; a major eruption of the volcano Mont Pelée in 1902 destroyed the city of Saint-Pierre, which was the largest city on the island. Agricultural products include sugar, cocoa, rum, bananas, pineapples, vanilla, and tobacco, although the tourist industry is now more important than agriculture. Petroleum products, cement, and processed foods are also manufactured. The official language is French, with Creole also spoken. 94 M6

Mary (formerly **Merv** 1884–1937), industrial city in southeastern Turkmenistan, located in the delta of the Murghab River on the Kara-Kum canal; population (2001 est) 114,300. It is situated in a cotton-growing oasis in the Kara-Kum Desert. Cotton and wool production and processing are the main industries. There are huge gas reserves at Shatlik, 20 km/12 mi west of the city. Some 30 km/19 mi east, Alexander the Great founded the city of Merv, in Hindu and Arab tradition regarded as ancient 'paradise'. The name Merv was also initially used by the Russians of the town they established here in 1884. 56 H2

Maryborough port on the southeast coast of Queensland, Australia, 257 km/160 mi north of Brisbane; population (1996) 21,286. Maryborough is situated at the mouth of the River Mary, and is the centre of a region raising cattle and producing grain, sugar, fruit, vegetables, and timber. Industries include coal- and goldmining, iron- and steel production, sugar-milling, shipbuilding, and the manufacture of heavy machinery and locomotives. 76 K5

Maryborough former name of ▶Portlaoise, the county town of County Laois in the Republic of Ireland. When Laois was made a county in 1556 the county town was called Maryborough after Queen Mary I. The name gradually went out of use during the 1950s. 24 E8

Maryland state of eastern USA. It was nicknamed Old Line State or Free State. Maryland ratified the US Constitution in 1788, becoming the 7th state to join the Union. It is bordered to the north by Pennsylvania, along the old ▶Mason-Dixon Line, to the east by Delaware, with which it shares most of the Delmarva Peninsula, to the south by Virginia and West Virginia, with the latter of which it also shares a western boundary. At the Fall Line, where the Anacostia River joins the Potomac River, is the District of Columbia, which was carved out of Maryland and Virginia in 1790.
population (2000 est) 5,296,500
area 31,600 sq km/12,198 sq mi
capital ▶Annapolis
towns and cities Baltimore, Silver Spring, Dundalk, Bethesda
industries and products poultry, dairy products, machinery, steel, cars and parts, boatbuilding, electric and electronic equipment, chemicals, fish and shellfish, tourism 88 E3

Maseru capital of ▶Lesotho, on the Caledon River at the border with Free State in South Africa; population (1992 est) 367,000. Founded in 1869 by the Basotho chief Mshweshwe I, it is a centre for trade, light manufacturing, and food processing. The National University of Lesotho (1975) is at nearby Roma. 72 D5

Mashhad (or **Meshed**), district and capital of the province of Khorasan, northeast Iran, and holy city of the Shiites; population (1991) 1,759,000. It is the second-largest city in Iran. It produces carpets, textiles, and leather goods. The magnificent mausoleum of the Imam Reza is visited annually by thousands of Shiite pilgrims. 56 G2

Mashraq (Arabic 'east'), the Arab countries of the eastern Mediterranean: Egypt, Jordan, Syria, Lebanon, Palestine, and Iraq. The term is contrasted with ▶Maghreb, comprising the Arab countries of northwest Africa.

Mason–Dixon Line in the USA, the boundary line between Maryland and Pennsylvania (latitude 39° 43' 26.3' N), named after Charles Mason (1730–1787) and Jeremiah Dixon (died 1777), English astronomers and surveyors who surveyed it 1763–67. It is popularly seen as dividing the North from the South.

Masqat alternative spelling of ▶Muscat, the capital of Oman. 61 H5

Massachusetts state of northeast USA. It is nicknamed the Bay State or the Old Colony State. Massachusetts ratified the US Constitution in 1788, becoming the 6th state to join the Union. It is a region of great significance to US history, being the point of disembarkation for the *Mayflower* Pilgrims, as well as the site of key conflicts in the American Revolution. Massachusetts is bordered to the north by Vermont and New Hampshire, to the west by New York, to the south by Connecticut and Rhode Island, and to the southeast and east by the Atlantic Ocean.
population (2000 est) 6,349,100
area 21,500 sq km/8,299 sq mi
capital ▶Boston
towns and cities Worcester, Springfield, Lowell, New Bedford, Brockton, Cambridge
industries and products electronic, communications, and optical equipment, precision instruments, non-electrical machinery, fish, cranberries, dairy products, tourism, academia and research, finance sector 88 F2

Massawa (or **Mesewa**, or **Massaua**), second biggest town and main port in Eritrea, situated on a small sterile coral island in the Red Sea, 200 m/656 ft from the mainland; population (1995 est) 49,300. It lies 65 km/40 mi northeast of Asmara, the largest town of Eritrea. Activities within the port include salt production, fishing, fish and meat processing, and cement manufacture. Exports include oil seed, coffee, and cattle. It is one of the hottest inhabited places in the world, the temperature reaching 46°C/115°F in May. 64 G4

Massif Central upland region of south-central France with mountains and plateaux; area 93,000 sq km/

36,000 sq mi, highest peak Puy de Sancy, 1,886 m/6,188 ft. It is a source of hydroelectricity. **26** H8

Matadi chief river port of the Democratic Republic of Congo on the south bank of the Congo River, 115 km/70 mi from its mouth, linked by rail with the capital Kinshasa; population (1994 est) 172,700. Matadi is the distribution point for all the country's oil products (except fuel oil), which are pumped via a 350 km/217 mi pipeline to Kinshasa. **70** A5

Matanzas industrial port (tanning, textiles, sugar) in northwestern Cuba; population (1990 est) 113,700. Founded in 1693, it became a major centre of coffee, tobacco, and sugar production. **84** K7

Matlock spa town and administrative headquarters of ▶Derbyshire, central England, on the River Derwent, 19 km/12 mi northwest of Derby; population (1996 est) 9,400. It manufactures textiles and high-tech products, and is a centre of tourism on the edge of the Peak District National Park. **22** A1

Mato Grosso (Portuguese 'dense forest'), plateau area of west-central Brazil comprising two federal units (states): Mato Grosso (area 901,420 sq km/348,040 sq mi; population (1991) 2,022,524) and Mato Grosso do Sul (area 357,471 sq km/138,020 sq mi; population (1991) 1,778,741), with their capitals at ▶Cuiaba and Campo Grande (founded in 1889) respectively; total area 1,258,891 sq km/486,0578 sq mi. The region covers 15% of the country's total area and contains less than 3% of its total population. The vegetation is mainly grassland, with scrub, tropical forest in much of the region, and a large wetland area. The Pantanal is one of the world's largest swamps and an important ecosystem. Cattle-raising has been the main economic activity for over 100 years. **98** F6

Matsue city on Honshu island, Japan, northwest of Osaka; population (1993) 145,000. It has the remains of a castle, fine old tea houses, and the Izumo Grand Shrine, which in its present form dates from 1744. There is also a museum honouring Lafcadio Hearn. **48** G6

Matsuyama largest city of Ehime prefecture, Shikoku island, Japan, facing the Inland Sea; population (1994) 456,000. Industries include agricultural machinery, textiles, and chemicals. Traditional cotton cloth is also produced locally. The Dogo hot spring attracts visitors to its public bath house, built in 1894. The castle, built in 1602–3, has been restored or rebuilt five times, most recently in 1986. **48** G7

Matterhorn (French **le Cervin**; Italian **il Cervino**), mountain peak in the Alps on the Swiss-Italian border; 4,478 m/14,690 ft. It was first climbed in 1865 by English mountaineer Edward Whymper (1840–1911); four members of his party of seven were killed when a rope broke during their descent. **30** C5

Maubeuge French town in the *département* of Nord, on the River Sambre near the Belgian frontier; population (1990) 35,200. It has steel and engineering works, and manufactures mirrors. Maubeuge was fortified in the 17th century by the military engineer Vauban. It was besieged by German troops in 1870 in the Franco-Prussian War, and was the scene of heavy fighting in both world wars. **22** F4

Maui island in Hawaii, USA, in the central Pacific, situated about 115 km/72 mi southeast of Honolulu (which is on the island of Oahu); area 1,885 sq km/728 sq mi; population (1990) 91,400. Maui is the second largest of the Hawaiian chain of islands and consists of two peninsulas divided by an isthmus of sand. The eastern peninsula contains the shield volcano of Haleakala, dormant since 1790, over 3,050 m/10,000 ft high and with a crater 30 km/19 mi in circumference. The chief towns are Lahaina, Wailuku, and Kahului. Tourism is the main industry; Maui has fine scenery, beaches, and resorts. Fruit and flowers are grown here. **92** (2)F3

Mauna Loa active volcano on the Pacific island of Hawaii, rising to a height of 4,169 m/13,678 ft. It has numerous craters, including the second-largest active crater in the world. Since the early 19th century there have been lava flows from the crater covering about half the island of Hawaii; averaging one eruption every 3½ years, it is considered an effusive, rather than an explosive, volcano, with eruptions along fissures in its flank as well as in its central crater. **92** (2)F4

Mauritania country in northwest Africa, bounded northeast by Algeria, east and south by Mali, southwest by Senegal, west by the Atlantic Ocean, and northwest by Western Sahara.

Mauritania covers an area of 1,030,700 sq km/397,953 sq mi, and has a population of 2,670,000 (2000 est). The capital (and chief port) is ▶Nouakchott; other major towns and cities include Nouâdhibou, Kaédi, Zouerate, Kiffa, Rosso, and Atar. The head of state is Maaoya Sid'Ahmed Ould Taya (from 1984); the head of government is Cheik el Avia Ould Muhammad Khouna (from 1998). The currency is the ouguiya, and the languages spoken include Hasaniya Arabic (official), Pulaar, Soninke, Wolof (all national languages), and French (particularly in the south). **66** C5

Mauritius island country in the Indian Ocean, east of Madagascar.

Mauritius covers an area of 1,865 sq km/720 sq mi, and has a population of 1,158,000 (2000 est). The capital (and chief port) is ▶Port Louis; other major towns and cities include Beau Bassin-Rose Hill, Curepipe, Quatre Bornes, and Vacoas-Phoenix. The head of state is Cassam Uteem (from 1992); the head of government is Anerood Jugnauth (from 2000). The currency is the Mauritian rupee, and the languages spoken include English (official), French, Creole, Bhojpuri, and other Indian languages. **72** (1)B2

Mawlamyine (formerly **Moulmein**), port and capital of Mon state in southeast Myanmar, on the Salween estuary; population (1998 est) 290,900. The city lies on the Gulf of Martaban and is sheltered by Bilugyun Island. There is sawmilling and rice milling, rice and tea are exported, and a solar-powered factory extracts salt from seawater. **50** B3

Mayagüez port in western Puerto Rico; population (1990) 100,400. Industries include pharmaceuticals, tuna processing, and clothing manufacture. A US agricultural experimental station is situated here. The city traces its origins to 1760; it received severe earthquake damage in 1918. The University of Puerto Rico, Mayagüez was founded in 1911. **94** L5

Mayence French name for the German city of ▶Mainz. **20** D6

Mayenne *département* in the Pays de la Loire region of western France, on the River Mayenne; area 5,171 sq km/ 1,997 sq mi; population (1990) 278,000. The administrative centre is Laval, and other principal towns are Mayenne and Château-Gontier. Chief industries include iron, slate, and paper. The chief agricultural activities are the breeding of cattle, horses, and pigs, and the making of cider. There are iron foundries, printing works, and food processing plants. 6 C2

Mayenne river in western France which gives its name to the *département* of ►Mayenne; length 200 km/125 mi. It rises in Orne, flows in a generally southerly direction through Mayenne and Maine-et-Loire, and joins the River Sarthe just above Angers to form the River Maine. 26 E5

Maykop alternative spelling of ►Maikop, the capital of Adygeya province, southwest Russia. 58 J1

Mayo county of the Republic of Ireland, in the province of Connacht; county town Castlebar; area 5,400 sq km/2,084 sq mi; population (1996) 111,500. Its wild Atlantic coastline is about 400 km/249 mi long. The principal towns are Ballina, Ballinrobe, and Westport, and the principal rivers are the Moy, the Robe, and the Owenmore. Loughs Conn and Mask lie within the county. Agriculture includes pig, sheep, and cattle farming, and salmon fishing (particularly in the River Moy). The soil of the central plain is fertile, and crops include potatoes and oats. An excellent marble is found in the northwest district. 5 B5

Mayotte (or **Mahore**), island group of the ►Comoros, off the east coast of Africa, a *collectivité territoriale* of France by its own wish; area 374 sq km/144 sq mi; population (1994 est) 109,600. The two main islands are Grande Terre and Petite Terre. The capital is Dzaoudzi. Industries include coffee, copra, vanilla, fishing, cloves, cocoa, and ylang-ylang (oil-yielding tree). The languages spoken are French and Swahili. 72 H2

Mbabane administrative capital of ►Swaziland, in the northwest of the country, near the South African border; population (2001 est) 74,800. It is situated in the Hhohho District of the Highveld, and is connected by rail to nearby coal mines and to the Indian Ocean port of Maputo, in Mozambique, 160 km/100 mi to the east. The legislative capital, Lobamba, is nearby to the south. In addition to mining and local crafts, industries include cement manufacture, finance, and banking. Aided by the development of a hotel and casino complex in the Ezulwini valley, about 11 km/7 mi from Mbabane, tourism has become a major industry. 72 E5

Mboma another spelling of ►Boma, a port in the Democratic Republic of Congo. 68 G6

Meath county of the Republic of Ireland, in the province of Leinster; county town Navan; area 2,340 sq km/ 903 sq mi; population (1996) 109,700. The chief river is the Boyne, of which the Blackwater is a tributary. The principal towns are Kells, Trim, Athboy, Bettystown, and Laytown. Cattle and sheep are reared, and oats and potatoes are grown. The largest working lead mine in Europe is located near Navan. Tara Hill, 155 m/509 ft high, was the site of a palace and was the coronation place of many kings of Ireland; St Patrick also preached here. The Book of Kells (now held in the Trinity College Library) was produced at Kells in the early 9th century. 5 C5

Meaux French town in the *département* of Seine-et-Marne, 50 km/31 mi northeast of Melun, on the River Marne; population (1990) 51,000. There is a 12th–16th-century cathedral and 15th-century ramparts built on Gallo-Roman foundations. The town has metal and chemical industries. 22 E6

Mecca (Arabic **Makkah**), city in Saudi Arabia and, as birthplace of Muhammad, the holiest city of the Islamic world; population (1991 est) 633,000. In the centre of Mecca is the Great Mosque, in the courtyard of which is the Kaaba, the sacred shrine containing the black stone believed to have been given to Abraham by the angel Gabriel. 64 G3

Mechelen (French **Malines**), industrial city and market gardening centre in Antwerp province, north Belgium, 22 km/14 mi south of Antwerp, on the Dyle River; population (1997) 75,300. Industries include furniture, carpets, and textiles. Agricultural produce includes asparagus and peas. The city no longer produces the Mechlin lace for which it was famous. 22 G3

Mecklenburg-Vorpommern German name for ►Mecklenburg–West Pomerania, an administrative region in Germany. 7 (2)C2

Mecklenburg–West Pomerania (German **Mecklenburg-Vorpommern**), *Land* (administrative region) of Germany; area 22,887 sq km/8,837 sq mi; population (1995) 1,825,000; the capital is ►Schwerin and other towns and cities include Rostock, Wismar, Stralsund, and Neubrandenburg. Products include fish, ships, diesel engines, electronics, plastics, and chalk. 7 (2)C2

Medan seaport and economic centre of the island of Sumatra, Indonesia; population (1990) 1,885,000. The largest city on the island, its industries include tobacco and tea processing, and manufacture of machinery, rope, ceramics, bricks, tiles, and soap. Some of its products are rubber, tobacco, and palm oil. 52 B2

Medellín industrial city and capital of Antioquia department, northwest Colombia; situated at 1,538 m/ 5,048 ft above sea level in the Aburrá Valley, Central Cordillera of the Andes; population (1994) 1,608,000. It is the second city and main textile centre of Colombia, producing over 80% of the country's total output. Other main industries include gold and silver mining, chemicals, coffee-growing, and engineering. Medellín has also had a reputation for cocaine production. 98 B2

Medford city and administrative headquarters of Jackson County, southwest Oregon, on the junction of Bear Creek and the Rogue River, 204 km/127 mi south of Eugene; population (1990) 47,000. It is a summer resort, and tourism is important to the economy; other industries include the processing of the area's timber, agricultural crops including pears, and dairy products. 86 B2

Mediaş town in Sibiu county, central Romania, on the River Tirnava Mare, 40 km/25 mi north of Sibiu; population (1990) 73,100. Glass, leather goods, textiles, and chemicals are manufactured, there are engineering and food-processing industries, and natural gas wells are found in the area. 34 M3

Medicine Hat city in southeast Alberta, Canada, 290 km/180 mi southeast of Calgary, on the South Saskatchewan River; population (1991) 43,300. It lies on the Canadian Pacific Railway transcontinental main line and the Trans-Canada Highway. A trade centre for cattle, wheat, and vegetable producers, it also has clay and natural gas deposits. Local industries produce petrochemicals, glass, brick, tile, and pottery. The city has flour mills, canneries, breweries, and a Goodyear tyre factory as well. It maintains Alberta's ranching tradition with its annual rodeo. Medicine Hat College (1965) is affiliated with the University of Calgary. **82** J7

Medina (Arabic **Al Madīnah**, Arabic 'the city of the apostle'), Saudi Arabian city, about 355 km/220 mi north of Mecca; population (1991 est) 400,000. It is the second holiest city in the Islamic world, and contains the tomb of Muhammad. It produces grain and fruit. **64** G3

Medina del Campo town in the province of Valladolid, Spain; population (1991) 19,700. It is an agricultural and food processing centre. Medina del Campo has medieval streets and an ancient castle, seat of the Castilian court during the 15th century. **28** F3

Mediterranean Sea inland sea separating Europe from north Africa, with Asia to the east; extreme length 3,700 km/2,300 mi; area 2,966,000 sq km/1,145,000 sq mi. It is linked to the Atlantic Ocean (at the Strait of Gibraltar), Red Sea and Indian Ocean (by the Suez Canal), and the Black Sea (at the Dardanelles and Sea of Marmara). The main subdivisions are the Adriatic, Aegean, Ionian, and Tyrrhenian seas. It is highly polluted. **14** E4

Medway Towns unitary authority in southeast England, created in 1998 by combining the former city council of Rochester upon Medway with Gillingham borough council, both formerly in Kent.
area 194 sq km/75 sq mi
towns and cities Rochester, Chatham, Gillingham, Strood (administrative headquarters)
features River Medway flows through Rochester; River Thames forms northern border of authority; reclaimed estuarine mudflats form the Isle of Grain; Charles Dickens Centre (Rochester); Royal Naval Dockyard (Chatham); Royal Engineers Museum (Chatham); Upnor Castle
industries distribution, manufacturing, banking, financial services, insurance, transport and communications, engineering, Thamesport deep-water container port), avionics, information technology
population (1996) 240,000 **5** G6

Meerut city in Uttar Pradesh, north India, 65 km/40 mi northeast of Delhi; population (1991) 847,000. Industries include chemicals, soap, paint, sports goods, musical instruments, and sugar refining. The Indian Mutiny began here in 1857. Dating back to the 3rd century BC, Meerut has strong Muslim associations, and there has been periodic violent conflict between Muslims and Hindus. **54** C3

Meghalaya state of northeast India, bordered to the north by Assam, to the south by Bangladesh; area 22,400 sq km/8,648 sq mi; population (1994 est) 1,960,000, mainly Khasi, Jaintia, and Garo. The capital is ►Shillong. The state is upland with hills reaching 2,000 m/6,500 ft, rising steeply in the south away from its border with Bangladesh. There is heavy monsoon rainfall. Minerals in the area include coal, limestone, white clay, and corundum, which are mainly unexploited. Industries include mineral extraction, which includes 95% of India's sillimanite. Agricultural products include cotton, potatoes, fruit, rice, maize, timber, and jute. 70% of the population are Hindu. **10** F3

Meissen town in the *Land* (administrative region) of ►Saxony, Germany, on the River Elbe, 22 km/14 mi northwest of Dresden; population (1995) 32,500. The manufacture of so-called Dresden china moved here from Dresden 1710. Ceramics are still produced in the town from kaolin deposits nearby, and there are also engineering industries. **20** J5

Mek'elē capital of Tigré region, northern Ethiopia; population (1992) 113,000. It trades in salt, incense, and resin. It is the centre of Ethiopia's inland salt trade, based on the Danakil Plain. **64** G5

Meknès (Spanish **Mequinez**), city in northern Morocco, 56 km/35 mi southwest of Fès; population (1995 est) 427,000. Founded in the 10th century by Berbers, it is one of Morocco's four imperial cities and was the capital until 1728. Chief industries are leather goods, pottery, wine production, textiles, canning, cement, and vegetable oils. **66** D2

Mekong river of China, rising as the Za Qu in Qinghai province, flowing through Tibet autonomous region and Yunnan province as the Lancang Jiang, and then through Laos, where part of its course forms the border with Thailand, Cambodia, and Vietnam; length 4,425 km/2,750 mi. The Mekong empties into the South China Sea through a vast delta, covering about 200,000 sq km/77,000 sq mi. It is being developed for irrigation and hydroelectricity by Cambodia, Laos, Thailand, and Vietnam. **50** D4

Melaka (formerly **Malacca**), state of west Peninsular Malaysia; capital Malacca; area 1,700 sq km/656 sq mi; population (1993) 583,400 (about 70% Chinese). Products include rubber, tin, and wire. The town originated in the 13th century as a fishing village frequented by pirates, and later developed into a trading port. Portuguese from 1511, then Dutch from 1641, it was ceded to Britain in 1824, becoming part of the Straits Settlements. **11** (1)A2

Melanesia islands in the southwestern Pacific between Micronesia to the north and Polynesia to the east, embracing all the islands from the New Britain archipelago to the Fiji Islands. **74** F5

Melbourne capital of the state of ►Victoria, Australia; population (1996) 2,865,329. Australia's second-largest city, Melbourne is situated on the southeast coast of Australia, on Port Philip Bay, at the mouth of the River Yarra. It is separated from Tasmania by the Bass Strait. Industries include engineering, shipbuilding, electronics, printing, oil refining, food processing, brewing, flour-milling, and the manufacture of chemicals, cars, furniture, plastics, textiles, and clothing. It is an important port and the largest receiver of container vessels in Australia. Melbourne has seven universities, including the University of Melbourne (1853), Monash University (1961, the largest university in Australia), and La Trobe University (1964). **76** H7

Melbourne industrial city in Brevard County, east-central Florida, USA, on the Indian River lagoon, 90 km/56 mi southeast of Orlando; population (1992) 64,300. Aerospace, other high-technology industries, and tourism are the main industries of the city; other industries include food processing and the manufacture of electronic and aviation equipment. Nearby is the Kennedy Space Center. Melbourne is the home of the Florida Institute of Technology (1958). **90** E4

Melilla port and military base on the northeast coast of Morocco; area 14 sq km/5 sq mi; population (1995 est) 61,600. It was captured by Spain in 1496 and is still under Spanish rule. Together with Ceuta, it makes up a region of Spain. Melilla has been a free port since 1863, and its modern commercial links include an airport and car ferries to Málaga. The port mainly exports iron ore, and the naval shipyard is a major employer. **28** H9/**9** (1)D5

Melk Austrian village in the province of Lower Austria, on the River Danube; population (1991) 3,600. On a cliff 52 m/171 ft above the town stands a huge Benedictine abbey, one of the most magnificent baroque buildings in the world. Melk lies in the part of the Danube valley known as the Wachau. **30** L2

Melos (Greek **Milos**), Greek island in the Aegean, one of the Cyclades; area 155 sq km/60 sq mi. The sculpture of *Venus de Milo* was discovered here in 1820 (now in the Louvre). The capital is Plaka. **36** G8

Melton Mowbray market town in Leicestershire, central England, on the River Eye, 24 km/15 mi northeast of Leicester; population (1995 est) 24,800. A fox-hunting centre, it has a large cattle market, and is known for pork pies and Stilton cheese. There is pet-food processing and the production of packed sandwiches and other snacks. Ashfordby Colliery nearby closed in the 1990s, with pithead gear being removed in 1998. **22** B2

Melun French town, capital of the *département* of Seine-et-Marne; population (1990) 40,000. It is on the outskirts of the forest of Fontainebleau and grew up on and around an island in the Seine. Melun is an agricultural centre, manufactures glass, pottery, and sugar, and is also a commuter town for Paris. There are many old buildings, including a church dating from 1020. **26** H5

Memel German name for ▶Klaipeda, the main port in Lithuania. **16** L9

Memmingen city in Bavaria, Germany, 106 km/66 mi southwest of Munich; population (1994) 40,300. It is an agricultural centre with precision engineering and chemical industries. Formerly one of the city-states of the Holy Roman Empire, Memmingen has ancient city walls and many old buildings. **30** F3

Memphis industrial city and port on the Mississippi River, in southwestern Tennessee, USA, linked by a bridge with West Memphis, Arkansas, across the river; seat of Shelby County; population (1992) 610,300. It is a major cotton market, and one of the leading centres in the USA for the production of hardwood lumber; other industries include food processing and the manufacture of pharmaceuticals, chemicals, medical supplies, furniture, and tobacco products. A 1980s industry of handmade ultramodern furniture is called Memphis style and copied by Italian and French firms. **88** C3

Menai Strait (Welsh **Afon Menai**), channel of the Irish Sea dividing ▶Anglesey from the Welsh mainland; about 22 km/14 mi long and up to 3 km/2 mi wide. It is crossed by two bridges. Thomas Telford's suspension bridge (521 m/1,710 ft long) was opened in 1826 but reconstructed to the original design in 1940, and freed from tolls. Robert Stephenson's tubular rail bridge (420 m/1,378 ft long) was opened in 1850, and is known as the Britannia Bridge. **24** H8

Menam another name for the River ▶Chao Phraya, Thailand. **50** C4

Mende capital of the French *département* of Lozère, on the River Lot, 125 km/78 mi northwest of Avignon; population (1990) 12,000. The town has a textile industry and a market. Mende is a bishopric, and its cathedral was begun in 1369 by Pope Urban V, who was born here. It is a base for tourists visiting the nearby Tarn Gorge and the Cévennes. **26** J9

Mendip Hills (or **Mendips**), range of limestone hills in southern England, stretching nearly 40 km/25 mi southeast–northwest from Wells in Somerset towards the Bristol Channel. There are many cliffs, scars, and caverns, notably Cheddar Gorge. The highest peak is Blackdown (326 m/1,068 ft).

Mendoza capital of Mendoza federal district, western Argentina, in the foothills of the Andes, 760 m/2,500 ft above sea level; population (1991) 121,700; metropolitan area (1992 est) 801,900. It is the commercial centre of an irrigated wine-producing and fruit-growing region. The city has an important university, and because of a nearby oilfield, a growing industrial base. The city was founded in 1561 on the site of an Inca fort by the Spaniard Garcí Hurtado de Mendoza, after whom it is named. It developed because of its position on the Trans-Andean railway where the Argentine and Chilean rail lines link. **100** H5

Menen Flemish name for ▶Menin, a town in West Flanders, Belgium. **22** F4

Menin (Flemish **Menen**), town in the province of West Flanders, Belgium, situated on the River Lys close to the French border; population (1991) 32,700. Goods from Menin tobacco, chicory, and soap. There are also textile and engineering industries. **22** F4

Menindee village and sheep centre on the Darling River in New South Wales, Australia; population (1991) 500. It is the centre of a scheme for conserving the waters of the Darling in **Menindee Lake** (155 sq km/60 sq mi) and other lakes nearby. **76** H6

Menorca second largest of the ▶Balearic Islands in the Mediterranean; area 689 sq km/266 sq mi; population (1990 est) 62,000. The capital is ▶Mahón, and other towns and cities are Ciudadela, Mercadal, Ferrerias, and Fornells. Leather goods, costume jewellery, and cheese and dairy products are produced on the island, and tourism is also important. **28** Q4

Mequinez Spanish name for ▶Meknès, a city in Morocco. **66** D2

Mérida capital of Yucatán state, Mexico, a centre of the sisal industry; population (1990) 556,800. It was founded in 1542, and has a cathedral dating from 1598. Its port on the Gulf of Mexico is Progreso. **94** G4

Mérida town in the province of Badajoz, western Spain, the capital of ▶Extremadura autonomous community; population (1991) 48,000.

Situated on the River Guadiana, it was founded by the Romans in 25 BC, and soon became one of the finest cities in the empire, and capital of ▶Lusitania. It is a market centre for the surrounding agricultural area and has food-processing industries. 28 D6

Meridian town and administrative headquarters of Lauderdale County, east Mississippi; population (1990) 41,000. It is located 143 km/89 mi east of Jackson and 27 km/17 mi west of the Alabama state line. Meridian is an important centre for the cotton market and has cotton and timber mills. Other industries include electronic equipment, motor-vehicle parts, and clothing. It is home to Meridian Community College (1937) is here. A Naval Air Station is to the northeast of the town. 90 D3

Merseburg town in the *Land* (administrative region) of ▶Saxony-Anhalt, Germany, on the River Saale, 14 km/9 mi south of Halle; population (1996) 42,000. Industries include engineering and paper, and there are lignite mines in the vicinity. Merseburg was capital of the Duchy of Saxe-Merseburg 1656–1736, and became part of the territory of Prussia in 1815. The town has a cathedral dating from the 11th–16th centuries. 20 H5

Mersey river in northwest England; length 112 km/70 mi. Formed by the confluence of the Goyt and Tame rivers at Stockport, it flows west through the south of Manchester, is joined by the Irwell at Flixton and by the Weaver at Runcorn, and enters the Irish Sea at Liverpool Bay. It drains large areas of the Lancashire and Cheshire plains. The Mersey is linked to the Manchester Ship Canal. Although plans were announced in 1990 to build a 1,800-m/5,907-ft barrage across the Mersey estuary to generate electricity from tides, these were abandoned in 1992 for financial reasons. 24 J8

Merseyside metropolitan county of northwest England, created in 1974; in 1986, most of the functions of the former county council were transferred to metropolitan borough councils (The Wirral, Sefton, Liverpool, Knowsley, St Helens).
area 650 sq km/251 sq mi
towns and cities Liverpool, Bootle, Birkenhead, St Helens, Wallasey, Southport
physical River Mersey
features Merseyside Innovation Centre (MIC), linked with Liverpool and John Moores Universities; Prescot Museum of clock- and watch-making; Speke Hall (Tudor), and Croxteth Hall and Country Park
industries brewing, chemicals, electrical goods, glassmaking, metal-working, pharmaceutical products, tanning, vehicles
population (1996) 1,420,400 5 E5

Mersin (or **Içel**), Turkish industrial free port (chrome, copper, textiles, oil refining); population (1990) 422,400. 36 S8

Merthyr Tydfil unitary authority in south Wales, created in 1996 from part of the former county of Mid Glamorgan.
area 111 sq km/43 sq mi
towns ▶Merthyr Tydfil (administrative headquarters)
features area includes part of Brecon Beacons National Park
industries light engineering, electrical goods.
population (1996) 60,000 5 E6

Merthyr Tydfil industrial town and administrative centre of ▶Merthyr Tydfil unitary authority in south Wales, situated 39 km/24 mi northwest of Cardiff on the River Taff; population (1991) 39,500. Its industries include light engineering and the manufacture of electrical goods. It was formerly a centre of the Welsh coal and steel industries. It had the world's first steam railway in 1804 and the largest ironworks in the world in the mid-19th century. 24 J10

Meshed variant spelling of ▶Mashhad, a town in Iran. 56 G2

Mesolongi alternative spelling of ▶Missolonghi, a town in western central Greece. 36 D6

Mesolóngion Greek name for ▶Missolonghi, a town in western central Greece. 36 D6

Mesopotamia the land between the Tigris and Euphrates rivers, now part of Iraq. The civilizations of Sumer and Babylon flourished here. The Sumerian civilization (3500 BC) may have been the earliest urban civilization. 58 JK6

Messina city and port in northeastern Sicily; population (1992) 232,900. It produces soap, olive oil, wine, and pasta. Originally an ancient Greek settlement (**Zancle**), it was taken first by Carthage and then by Rome. It was rebuilt after an earthquake in 1908. 32 K10

Messina, Strait of (ancient **Siculum Fretum**), channel in the central Mediterranean separating Sicily from mainland Italy, joining the Tyrrhenian and Ionian seas; it is 35 km/22 mi long, and its width varies from 17 km/11 mi in the south to 3 km/2 mi in the north. In Greek legend the monster Scylla devoured sailors from a rock on the Italian shore, while another, Charybdis, created a whirlpool on the Sicilian side which sank ships. The classical hero Odysseus passed safely between them.

Metz industrial city and administrative centre of the *département* of ▶Moselle in the Lorraine region of northeast France, on the River Moselle some 320 km/199 mi east of Paris; population (1990) 123,900. The town has a large trade in agricultural produce, particularly tobacco, and has a coal-mining district to the northeast. It has metal foundries and also produces cars, food products, textiles, and leather goods, especially shoes. Part of the Holy Roman Empire 870–1552, it became one of the great frontier fortresses of France, and was in German hands 1871–1918. 22 J5

Meurthe-et-Moselle *département* in the ▶Lorraine region of France; area 5,234 sq km/2,021 sq mi; population (1990) 712,600. The River ▶Moselle and its tributaries, and the River Chiers, a tributary of the Meuse, flow through it. The main industries include the manufacture of cast and sheet iron, iron and steel goods, and pottery and glass. Rock salt is found in abundance. Cereals, potatoes, and hops are grown. The administrative centre is ▶Nancy. Other important towns are Briey, Lunéville, and Toul. 6 G2

Meuse (Dutch **Maas**), river flowing through France, Belgium, and the Netherlands; length 900 km/559 mi. It was an important line of battle in both world wars. It gives its name to a French *département*. 22 G4

Meuse *département* in the Lorraine region of France; area 6,220 sq km/2,402 sq mi; population (1990) 196,600. The Rivers ▶Meuse, Orne, and Chiers flow through it. Cereals, potatoes, and beets are grown, and livestock are raised. The principal industries are brewing and the manufacture of glass and tiles. The chief towns are ▶Bar-le-Duc (administrative centre) and Verdun. 6 F2

Mewar another name for ▶Udaipur, a city in India. 54 B4

Mexicali city in northwestern Mexico; population (1990) 601,900. It produces soap and cottonseed oil. The availability of cheap labour attracts many US companies (Hughes Aerospace, Rockwell International, and others).

Mexico country in the North American continent, bounded north by the USA, east by the Gulf of Mexico, southeast by Belize and Guatemala, and southwest and west by the Pacific Ocean. It is the northernmost country in ▶Latin America.

Mexico covers an area of 1,958,201 sq km/756,061 sq mi, and has a population of 98,881,000 (2000 est). The capital is ▶Mexico City; other major towns and cities include Guadalajara, Monterrey, Puebla, Netzahualcóyotl, Ciudad Juárez, and Tijuana. The head of state and government is Vicente Fox (from 2000). The currency is the Mexican peso, and the languages spoken include Spanish (official), Nahuatl, Maya, Zapoteco, Mixteco, and Otomi. 94 D4

Mexico City (Spanish **Ciudad de México**), capital, industrial (iron, steel, chemicals, textiles) and cultural centre of Mexico, 2,255 m/7,400 ft above sea level on the southern edge of the central plateau; population (1994) 15,500,000. It is thought to be one of the world's most polluted cities because of its position in a volcanic basin 2,000 m/7,400 ft above sea level. Pollutants gather in the basin causing a smog cloud. 94 E5

Miami industrial city and port in southeastern Florida, USA, on the Atlantic coast of the Florida peninsula about 70 km/43 mi from its southern tip; seat of Dade County; population (1995 est) 365,500. 60% of the population is of Hispanic origin, many of whom are Cubans who live in the Little Havana area; African-Americans comprise 21%, whites 12%. Miami is the hub of finance, trade, and transport in the region, with air connections to Latin America and the Caribbean; industries include food processing, transportation and electronic equipment, clothing, furniture, and machinery. Major employers are the state and federal governments. With its subtropical climate Miami is also a major tourist resort (tourism is the city's major industry) and a centre for oceanographic research. The city of Miami beach is situated on a barrier island, and is linked to Miami by bridges. The first permanent European settlement dates from the 1870s; Miami was incorporated in 1896. 90 E4

Michigan state in north-central USA. It is nicknamed the Wolverine State or the Great Lakes State. Michigan was admitted to the Union in 1837 as the 26th US state. It is situated in the Midwest and Great Lakes regions, and is comprised of two major peninsulas. The mitten-shaped, north–south oriented **Lower Peninsula** is bordered to the south by Ohio and Indiana, and faces Ontario, Canada, to the north and east across the Great Lakes. The east–west oriented **Upper Peninsula** is bordered to the south by Wisconsin, and also faces Ontario to the north and east across the lakes.
population (2000 est) 9,938,400
area 151,600 sq km/58,518 sq mi
capital ▶Lansing

towns and cities Detroit, Grand Rapids, Flint, Warren, Sterling Heights, Ann Arbor, Livonia

industries and products motor vehicles and equipment, non-electrical machinery, iron and steel, chemicals, pharmaceuticals, dairy products, fruit and beans, salt, coal, oil, natural gas, music industry, tourism 88 C1

Michigan, Lake (Algonquian 'big lake'), lake in north-central USA, the third largest of the ▶Great Lakes and the only one lying entirely within the USA, it is bordered by Michigan to the north and east, Indiana to the south, and Wisconsin and Illinois to the west; area 58,000 sq km/22,390 sq mi. The lake is 517 km/321 mi long, 190 km/118 mi at its widest point, has a maximum depth of 282 m/925 ft, and lies 176 m/577 ft above sea level. Lake Michigan is joined to Lake Huron by the Straits of Mackinac in the north. Green Bay is its largest inlet, and Chicago and Milwaukee are its main ports. The first European to see the Lake Michigan was the French explorer Jean Nicolet, in 1634. 88 C2

Micronesia group of islands in the Pacific Ocean lying north of ▶Melanesia. 74 F4

Micronesia, Federated States of country in the west Pacific Ocean, forming part of the archipelago of the Caroline Islands, 800 km/497 mi east of the Philippines.

Micronesia covers an area of 700 sq km/270 sq mi, and has a population of 119,000 (2000 est). The capital is ▶Palikir (in Pohnpei island state); other major towns and cities include Kolonia (in Pohnpei), Weno (in Truk), and Lelu (in Kosrae). The head of state and government is Jacob Nena (from 1997). The currency is the US dollar, and the languages spoken include English (official), eight officially recognized local languages (including Trukese, Pohnpeian, Yapese, and Kosrean), and a number of other dialects. 74 E5

Middelburg industrial town and capital of Zeeland province, southwest Netherlands, on the island of Walcheren, 6 km/4 mi north of Vlissingen; population (1997) 44,700. There are engineering, tobacco, and furniture industries. The town hall dates from the 15th century and the abbey from the 12th century. It is a former member of the Hanseatic League. 22 F3

Middle East indeterminate area now usually taken to include Egypt and the Arab states of the eastern Mediterranean and Arabian Peninsula, sometimes extended to the states of northwest Africa, Turkey, Iran and Afghanistan.

Middlesbrough industrial town, port, and unitary authority, on the estuary of the River Tees, northeast England, created in 1996 from part of the former county of Cleveland. The town was the administrative headquarters of the county of Cleveland to 1996. It is the commercial centre of the Teesside industrial area, which also includes Stockton-on-Tees, Redcar, Billingham, Thornaby, and Eston.
area 54 sq km/21 sq mi
features Transporter Bridge (1911) transports cars and passengers across the Tees to Hartlepool in a cable car; Newport Bridge (1934) was the first vertical lift bridge in England; the University of Teesside, formerly Teesside Polytechnic, was established in 1992; the Captain Cook Birthplace

Museum; Ormesby Hall (18th century)
industries formerly a centre of heavy industry, it diversified its products in the 1960s; there are constructional, electronics, engineering, and shipbuilding industries, and iron, steel, and chemicals are produced
famous people James Cook
population (1996) 146,000
24 L7/5 F4

Middletown industrial city in Butler and Warren counties, southwest Ohio, USA; population (1990) 46,000. It is situated on the Miami River, 18 km/11 mi northeast of Hamilton and 39 km/24 mi north of Cincinnati, in a fertile agricultural region. It stands midway between the cities of Cincinnati and Dayton, hence its name. Industries include the manufacture of steel, paper products, and aerospace equipment. It is home to a University of Miami campus. A large amusement park is 8 km/5 mi south of the city. 88 D3

Midi-Pyrénées region of southwest France, comprising the *départements* of Ariège, Aveyron, Haute-Garonne, Gers, Lot, Hautes-Pyrénées, Tarn, and Tarn-et-Garonne; the capital is ►Toulouse; area 45,300 sq km/17,500 sq mi; population (1990) 2,430,700. The region includes several spa towns (including ►Lourdes), winter resorts, and prehistoric caves. It produces fruit, wine, and livestock, and industries include aerospace. There are two large universities. Other notable towns include Montauban, Cahors, and Rodez. 6 D5

Midland industrial city and administrative headquarters of Midland County, central Michigan, USA, on the junction of the Tittabawassee and Chippewa rivers, 32 km/20 mi northwest of Saginaw; population (1990) 38,100. There are oil and gas wells here, and the main industries are the manufacture of chemicals, pharmaceuticals, plastic goods, and concrete. Midland is the home of Saginaw Valley University (1963), the Northwood Institute (1959), and the Midland Center for the Arts and Dow Gardens. 88 D2

Midland city and administrative headquarters of Midland County, west Texas, USA, situated on the high plains at an altitude of 850 m/2,789 ft; population (1992) 95,200. It was named for its position approximately midway between Fort Worth and El Paso. Midland lies in a rich cattle-ranching region, famous for its Hereford cattle. Oil-producing is also important, with the nearby Permian Basin oilfields. Other industries include the production of oilfield equipment, chemicals, electronic components, and processed food. The city is home to Midland College (1969). 92 F2

Midlands area of central England corresponding roughly to the Anglo-Saxon kingdom of Mercia. The **East Midlands** comprises Derbyshire, Leicestershire, Northamptonshire, and Nottinghamshire. The **West Midlands** covers the metropolitan district of ►West Midlands created from parts of Staffordshire, Warwickshire, and Worcestershire, and split into the metropolitan boroughs of Dudley, Sandwell, Coventry, Birmingham, Walsall, Solihull, and Wolverhampton; and (often included) the **South Midlands** comprising Bedfordshire, Buckinghamshire, and Oxfordshire.

Midlothian unitary authority in southeast Scotland, south of the Firth of Forth, which was previously a district within Lothian region (1975–96) and a county until 1974.
area 363 sq km/140 sq mi
towns Dalkeith (administrative headquarters), Penicuik, Bonnyrigg
physical inland area rising toward the Moorfoot Hills in the south; River Esk
features Crichton Castle, Roslin Castle, Rosslyn Chapel, Newtongrange Mining Museum
industries glass and crystal, coalmining (declining), light manufacturing, food processing
agriculture productive to the north (arable and dairy), less productive toward the hills in the south
population (1996) 79,900 5 E4

Midway Islands two coral islands in the Pacific, near the northwestern end of the Hawaiian Islands chain, USA, 1,800 km/1,120 mi northwest of Honolulu; area 5 sq km/2 sq mi. They form a naval base with no indigenous population. The islands are individually known as Eastern and Sand; they were annexed by the USA in 1867, and are now administered by the US Navy. The naval Battle of Midway (3–6 June 1942), between the USA and Japan, was a turning point in the Pacific in World War II; the US victory marked the end of Japanese expansion in the region. 74 J3

Midwest (or **Middle West**), large area of the north-central USA. It is loosely defined geographically, but is generally taken to comprise the states of Illinois, Iowa, Wisconsin, Minnesota, Nebraska, Kansas, Missouri, North Dakota, and South Dakota, and the portions of Montana, Wyoming, and Colorado that lie east of the Rocky Mountains. Ohio, Michigan, and Indiana are often variously included as well. In its broadest sense, the Midwest has an area of 986,800 sq mi/2,556,000 sq km and a population of about 61.5 million – roughly a quarter of the national total. The region is generally flat and well-watered, with good transportation links. Traditionally its economy is divided between agriculture and heavy industry. The main urban Midwest centre is Chicago. In the summer of 1993 the Midwest was devastated by floods, which left tens of thousands of people homeless.

Milan (Italian **Milano**; Roman **Mediolanum**), second-largest city in Italy, sitauted in Lombardy, 120 km/75 mi northeast of Genoa; population (1992) 1,358,600. Industries include printing, engineering, and the manufacture of aircraft, cars, locomotives, chemicals, clothing, and textiles. Milan is Italy's chief commercial and industrial centre: the main stock exchange, and the headquarters of banks and insurance companies are here. It is also the country's most important publishing centre. 30 E5

Milford Haven (Welsh **Aberdaugleddau**), seaport in Pembrokeshire, southwest Wales, on the estuary of the east and west Cleddau rivers 13 km/8 mi southwest of Haverfordwest; population (1991) 13,200. It has oil refineries, and a terminal for giant tankers linked by pipeline with Llandarcy, near Swansea. There is a fishing industry. 24 G10

Milk River river in North America, a tributary of the ►Missouri; length 1,175 km/730 mi. It rises in the Rocky Mountains, in northwest Montana, and flows north and east into southern Alberta, Canada, where it is used for irrigation. It then flows into Montana, where it joins the Missouri near Nashua. 82 J7

Millau French town in the *département* of Aveyron, 50 km/31 mi south of Rodez, on the River Tarn; population (1990) 22,300. Its industries include the manufacture of gloves, leather goods, and furniture. There is a 14th–16th-century church and a Gothic belfry. In the 16th–17th centuries it was a Huguenot religious stronghold. It is close to the Tarn Gorge and other tourist sites. 26 J9

Milos Greek name for ▶Melos, a Greek island in the Aegean. 36 G8

Milton Keynes industrial new town and administrative headquarters of ▶Milton Keynes unitary authority in south-central England, 80 km/50 mi northeast of London and 100 km/ 62 mi southwest from Birmingham; population (1995) 192,900. It was part of the county of Buckinghamshire until 1997. It was developed as a new town in 1967 around the old village of the same name, following a grid design. It is the headquarters of the Open University (founded in 1969). Industries include electronics, the manufacture of machine tools, machinery, and pharmaceuticals, and also insurance and financial services. 22 B2

Milton Keynes unitary authority in central England, formerly part of Buckinghamshire.
area 311 sq km/120 sq mi
towns and cities ▶Milton Keynes (administrative headquarters), Newport Pagnell, Olney, Bletchley, Stony Stratford, Woburn Sands, Wolverton
features Grand Union Canal; River Great Ouse; River Tove; Open University; Bletchley Park, government centre of code-breaking during World War II; Ouse Valley Park; Peace Pagoda (Milton Keynes), first to be built in northern hemisphere and surrounded by a thousand cherry and cedar trees planted in memory of all war victims; Milton Keynes' famous concrete cows, constructed in 1978 by a community artist and local school children
industries financial services, telecommunications, soft drinks, high-tech industries, motor vehicles (Aston Martin-Lagonda, Mercedes-Benz, Volkswagen-Audi), education (Open University and De Montfort University), vellum and parchment
population (1996) 198,600 5 F5

Milwaukee industrial city and port in southeastern Wisconsin, USA, at the mouth of the Milwaukee River, on the western shore of Lake Michigan, 128 km/79 mi north of Chicago; seat of Milwaukee County; population (1994 est) 617,000. It is the centre of a dairying and beef-producing region, and an important port of entry on the Great Lakes–St Lawrence Seaway system; industries include brewing (there are two major breweries, including Millers, the second largest brewery in the USA), engineering, machinery, motorcyles (the Harley-Davidson factory was founded here in 1906), electronic and electrical equipment, and chemicals. 88 C2

Minas Gerais (Portuguese **general mines**), inland federal unit (state) of southeast Brazil; area 586,624 sq km/226,,495 sq mi; population (1991) 16,956,900; capital ▶Belo Horizonte. The region is rich in mineral deposits (iron ore, manganese, bauxite, nickel) and is the centre of the country's iron ore, coal, diamond, and gold mining industries. Cattle are raised in the northwest of the region, and coffee, maize, dairy products, fruit, and sugar are produced. The ports of Rio de Janeiro, Vitoria, and Santos are used for its external trade. The Rio São Francisco flows through the northern part of the region. Other towns include Juiz de Fora, Governador Valadares, Uberaban, and Uberlândia. Gold was discovered at the end of the 17th century. 98 H7

Mindanao second-largest island of the Philippines. The indigenous peoples are the Lumad and Moro; area 94,627 sq km/36,536 sq mi; population (1990) 14,298,250. Towns and cities include Davao and Zamboanga. Industries include pineapples, coffee, rice, coconut, rubber, hemp, timber, nickel, gold, steel, chemicals, and fertilizer.
The island is mainly mountainous rainforest; the active volcano Apo reaches 2,954 m/9,600 ft, and the island is subject to severe earthquakes.

There is a Muslim guerrilla resistance movement, the Moro Islamic Liberation Front. 50 G5

Minden industrial town in North Rhine-Westphalia, Germany, on the River Weser where the Mittelland Canal crosses the river by aqueduct, 185 km/115 mi northeast of Düsseldorf; population (1995) 82,300. There are tobacco, machinery, glass, ceramics, furniture, and food processing industries. It was a memberof the Hanseatic League. The French were defeated here in 1759 by an allied army from Britain, Hannover, and Brunswick, commanded by the duke of Brunswick (Minden, Battle of). There is an 11th- to 13th-century cathedral which was rebuilt after World War II. 22 L2

Mindoro island of the Philippine Republic, south of Luzon; area 10,347 sq km/3,995 sq mi; population (1995 est) 912,000. Calapan is a city on the island. Features include Mount Halcon (2,590 m/8,500 ft). The South China Sea lies to the west of the island and the Sulu Sea to the south. Industries include coal and timber. 50 G4

Minhow name in use from 1934 to 1943 for ▶Fuzhou, capital of Fujian province, China. 50 F1

Minneapolis (American Indian **minne** 'water', Greek **polis** 'town'), city in southeastern Minnesota, USA, 13 km/8 mi from ▶St Paul, with which it forms the 'Twin Cities' area; seat of Hennepin County; population (1994 est) 355,000; metropolitan area (1992) 2,618,000. It is at the head of navigation of the Mississippi River, and is the centre of one of the richest agricultural areas in the USA. Industries include food processing and the manufacture of machinery, electrical and electronic equipment, precision instruments, transport machinery, and metal and paper products. Cray computers – used for long-range weather forecasting, spacecraft design, and code-breaking – are built here. 88 B2

Minnesota state in north Midwest USA. It is nicknamed the Gopher State or the North Star State. Minnesota was admitted to the Union in 1858 as the 32nd US state. One of the Great Lakes states, it is bordered to the south by Iowa, to the west by North and South Dakota, to the north by the Canadian states of Ontario and Manitoba, and to the east by Wisconsin.
population (2000 est) 4,919,500
area 218,700 sq km/84,418 sq mi
capital ▶*St Paul*
towns and cities Minneapolis, Duluth, Bloomington, Rochester
industries and products cereals, soybeans, livestock, meat and dairy products, iron ore (about two-thirds of US output), non-electrical machinery, electronic equipment, pulp, finance sector 88 A1

Miño river of Spain and Portugal, rising in the Cantabrian Mountains, Spain; length 338 km/210 mi. It flows south through Lugo, then southwest through ►Orense and Portugal to the Atlantic Ocean. Its chief tributary is the River Sil. 28 C2

Minsk (Belorussian **Mensk**), industrial city and capital of Belarus (also capital of the Minsk oblast); population (1990) 1,612,800. Motor vehicles, machinery, textiles, leather are produced here; Minsk is also a centre of the computer industry. The city's large pre-war Jewish community, which comprised over half of its inhabitants, was deported and murdered during the Nazi occupation. The headquarters of the Commonwealth of Independent States (CIS) is located here. 38 E4

Miquelon small group of islands off the south coast of Newfoundland which, with St Pierre, form a French overseas 'territorial collectivity'; see ►St Pierre and Miquelon. 82 V7

Mirzapur city of Uttar Pradesh, India, on the River Ganges; a grain and cotton market, with bathing sites and temples on the river; population (1991) 169,000. The state government established a cement factory here; other industries include brass and woollen carpet manufacture. 54 D3

Miskolc industrial city (iron, steel, textiles, furniture, paper) in northeastern Hungary, on the River Sajo, 145 km/90 mi northeast of Budapest; population (1995 est) 182,000. 34 H1

Mississippi (American Indian **missi** 'big', **sipi** 'river'), river in the USA, the main arm of the great river system draining the USA between the Appalachian and the Rocky mountain ranges. The length of the Mississippi is 3,778 km/2,348 mi; with its tributary the Missouri it totals 6,020 km/3,740 mi. The Mississippi rises in the lake region of northern Minnesota in the basin of Lake Itasca, and drops 20 m/65 ft over the St Anthony Falls at Minneapolis. Below the tributaries of the Minnesota, Wisconsin, Des Moines, and Illinois rivers, the confluence of the Missouri and Mississippi occurs at St Louis. Turning at the Ohio junction, it passes Memphis, and takes in the St Francis, Arkansas, Yazoo, and Red tributaries before reaching its delta on the Gulf of Mexico, beyond New Orleans. Altogether the Mississippi has 42 tributary streams and the whole Mississippi river system has a navigable length in excess of 25,900 km/16,100 mi. 90 D2

Mississippi state in southeast USA. It is nicknamed the Magnolia State or the Bayou State. Mississippi was admitted to the Union in 1817 as the 20th US state. Part of the Deep South, it was historically associated with the cotton plantations, slavery, and blues music. Mississippi is bordered to the east by Alabama, to the north by Tennessee, and to the west by Arkansas and Louisiana. To the south the state has a coast *c.* 115 km/70 mi long on the Gulf of Mexico, with a number of islands in the Mississippi Sound.
population (2000 est) 2,844,700
area 123,600 sq km/47,710 sq mi
capital ►Jackson
towns and cities Biloxi, Greenville, Meridian, Hattiesburg
industries and products cotton, rice, soybeans, chickens, fish and shellfish, lumber and wood products, petroleum and natural gas, transportation equipment, chemicals 90 C3

Missolonghi (or **Mesolongi**; Greek **Mesolóngion**), town in western central Greece, on the north shore of the Gulf of Patras; population (2001 est) 8,500. It was several times under siege by the Turks in the wars of 1822–26 and it was here that the British poet Byron died. 36 D6

Missoula city in Montana, northwest USA, county administrative headquarters of Missoula County, on Clark Fork near the mouth of the Bitterroot River, 153 km/95 mi northwest of Helena; population (1990) 42,900. The city's economy depends on its sawmills and pastoral farming; products include plywood, paper, and processed food. Missoula has skiing centres. It is the home of University of Montana (1893). 86 D1

Missouri major river in central USA, largest tributary of the ►Mississippi, which it joins north of St Louis; length 3,969 km/2,466 mi; drainage area 1,370,000 sq km/529,000 sq mi. It rises among the Rocky Mountains in Montana, and passes northwards through a 366 m/1,200 ft gorge known as the 'Gate of the Mountains'. The river is formed by the confluence of the Jefferson, Gallatin, and Madison rivers near Gallatin City, southwestern Montana, and flows southeast through the states of Montana, North Dakota, and South Dakota to Sioux City, Iowa. It then turns south to form the borders between Iowa and Nebraska and between Kansas and Missouri, and enters the Mississippi channel 32 km/20 mi north of St Louis. Kansas City, Missouri, is the largest city on its banks. 86 F1

Missouri state in central USA. It is nicknamed Show Me State or the Bullion State. Missouri was admitted ot the Union in 1821 as the 24th US state. Part of the Midwest, it is bordered to the south by Arkansas, to the west by Oklahoma, Kansas, and Nebraska, to the north by Iowa, and to the east by Illinois and, in the extreme southeast, by parts of Kentucky and Tennessee.
population (2000 est) 5,595,200
area 180,600 sq km/69,712 sq mi
capital ►Jefferson City
towns and cities St Louis, Kansas City, Springfield, Independence
industries and products meat and other processed food, aerospace and transport equipment, lead, zinc, tourism 88 B3

Mitylene alternative spelling of ►Mytilene, a Greek town on the island of Lesvos. 36 G5

Mizoram state of northeast India, lying between Bangladesh and Myanmar; area 21,100 sq km/8,150 sq mi; population (1994 est) 775,000. The capital is Aizawl. The state features north–south ranges in the east that rise to over 2,000 m/6,500 ft. The state is densely forested. Products include rice, maize, and hand-loom weaving. 84% of the population are Christian. 10 F4

Mmabatho capital of North West Province, South Africa, and former capital of the former independent homeland of Bophuthatswana; population (1991) 13,300. Many of the major buildings were built from revenues of the Sun City casino-resort. 72 D5

Mobile industrial city and administrative headquarters of Mobile County, southwest Alabama, USA; at the mouth of the Mobile River, on the northwest shore of Mobile Bay, 48 km/30 mi from the Gulf of Mexico; population (1992) 201,900. It is the only seaport in Alabama. Industries include meatpacking and the manufacture of wood products, aircraft engines, chemical and petroleum products, computer equipment, and textiles. It is home to Spring Hill College (1830), Mobile College (1961), and the University of South Alabama (1963). A tourist centre, it is noted for its Mardi Gras and for its 60-km/37-mi Azalea Trail. **90** D3

Mobutu, Lake alternative name of Lake ▶Albert in central Africa. **70** D3

Modena (ancient **Mutina**), town in Emilia-Romagna, Italy, capital of the province of Modena, 37 km/23 mi northwest of Bologna; population (1992) 177,000. Industries include engineering and the production of vehicles (Ferrari), glass, textiles, shoes, pasta, and sausages. The town has a 12th-century cathedral, a 17th-century ducal palace, and a university (1175), with medical and legal faculties. **30** F6

Modesto city and administrative headquarters of Stanislaus County, north-central California, USA; population (1992) 172,300. It is situated on the Tuolumne River, in the north San Joaquin Valley, 35 km/22 mi south-southeast of Stockton, and to the southeast of San Francisco. Founded by the Central Pacific Railroad in 1870, Modesto developed as a shipping and trade centre for an irrigated region producing peaches, apricots, almonds, melons, beans, peas, and livestock. Gallo, America's largest winery, is based in the city. Other industries include fruit and vegetable canning and freezing and the manufacture of bottles, chemicals, electrical equipment, and wood and paper products. It is home to Modesto Junior College (1921). **86** B3

Mödling suburb of the Austrian capital ▶Vienna, in the province of Lower Austria; population (1995) 32,100. Mödling is on the southern edge of the Vienna metropolitan area. Shoes and metal goods are the main industries. **30** M2

Mogadishu (or **Muqdisho**), capital and chief port of Somalia; population (1995 est) 525,000. The city lies on the Indian Ocean coast of Somalia and is a centre for oil refining, food processing, and chemical production; there are uranium reserves nearby. During the civil war 1991–92, much of the city was devastated and many thousands killed. The population has decreased since the civil war because of famine and the movement of refugees. **70** H3

Mogok village in Myanmar (Burma), 114 km/71 mi north-northeast of Mandalay, known for its ruby and sapphire mines. **50** B2

Mohács river port in southern Hungary, on the Danube, near the Croatian border, 37 km/23 mi east of Pécs; population (1993) 20,100. Coal is mined in the district. During the 16th and 17th centuries, two decisive battles were fought with the Ottoman Turks at the town. **34** F4

Mohave Desert arid region in southern California, USA; see ▶Mojave Desert. **92** C2

Mojave Desert (or **Mohave Desert**, or **High Desert**), arid region in southern California, USA, part of the Great Basin; average height above sea-level 600 m/2,000 ft; area 38,500 sq km/15,000 sq mi. It lies to the northeast of Los Angeles, and to the southeast of the ▶Sierra Nevada. **92** C2

Moldau German name for the ▶Vltava, a river in the Czech Republic. **18** D8

Moldova (or **Moldavia**), country in east-central Europe, bounded north, south, and east by Ukraine, and west by Romania.

Moldova covers an area of 33,700 sq km/13,011 sq mi, and has a population of 4,380,000 (2000 est). The capital is ▶Chişinău (Russian Kishinev); other major towns and cities include Tiraspol, Beltsy, and Bendery. The head of state is Petru Lucinschi (from 1997); the head of government is Vladimir Voronin (from 1999). The currency is the leu, and the languages spoken include Moldovan (official), Russian, and Gaganz (a Turkish dialect). **34** R2

Moline industrial city in northwest Illinois, USA, at the junction of the Rock and Mississippi rivers, 288 km/179 mi west of Chicago; population (1996 est) 42,800. It is the commercial hub of an agricultural area producing cereals and livestock. Farm implements, machinery, and furniture are manufactured. The city was laid out in 1843, and was formerly a centre for plough-making. **88** B2

Molise mainly agricultural region of southern central Italy, comprising the provinces of Campobasso and Isernia; area 4,400 sq km/1,698 sq mi; population (1992 est) 331,500. Its capital is ▶Campobasso. The chief products are corn, vines, olives, potatoes, and maize; sheep and goats are reared. **9** (4)E4

Molokai mountainous island of Hawaii, southeast of Oahu; area 673 sq km/259 sq mi; population (1990) 6,700. Kamakou (1,512 m/4,960 ft) is the highest peak on the island. Traditionally a spiritual centre and refuge from other parts of Hawaii, the island is now largely a resort for Oahuans. Its soil has not supported large-scale agriculture, and tourism is its major industry. The island was the site of a leper colony organized 1873–89 by Belgian missionary Joseph De Veuster (Father Damien). **92** (2)D2

Molotov former name (1940–62) for the river port of ▶Perm in Russia. **38** L3

Moluccas another name for ▶Maluku, a group of Indonesian islands. **53** C3

Mombasa industrial port and tourist resort in Kenya (the port also serves Uganda and Tanzania); population (1998 est) 838,700. It stands on Mombasa Island and the adjacent mainland (the island of Mombasa is linked to the mainland by the Mukapa causeway). As well as tourism, current industries include car assembly. Mombasa was founded by Arab traders in the 11th century and was an important centre for ivory and slave trading until the 16th century. One of the oldest buildings, Fort Jesus, was set up by the Portugese and is now a museum. **70** G4

Møn Danish island lying between and to the east of Sjaelland and Falster; area 220 sq km/85 sq mi; population (1988) 11,400. It is 32 km/20 mi long and is divided from Sjaelland by Ulv Sound and from Falster by Gron Sound. It is the highest island point of Denmark after Bornholm, with precipitous limestone cliffs (150 m/492 ft high) on its eastern side; it has a lighthouse which is visible for 25 km/16 mi. The soil is chalky and fertile. Its chief town is Stege. **20** H2

Mona Latin name for ►Anglesey, an island off the northeast coast of Wales. **24** H8

Monaco small sovereign state forming an enclave in southern France, with the Mediterranean Sea to the south.

Monaco covers an area of 1.95 sq km/ 0.75 sq mi, and has a population of 34,000 (2000 est). The head of state is Prince Rainier III (from 1949); the head of government is Patrick Leclercq (from 2000). The currency is the French franc, and the languages spoken include French (official), Monégasgne (a mixture of the French Provençal and Italian Ligurian dialects), and Italian. **30** C7

Monaghan (Irish **Mhuineachain**), county of the Republic of Ireland, in the province of Ulster; county town Monaghan; area 1,290 sq km/ 498 sq mi; population (1996) 51,300. The county is low and rolling, with hills in the northwest, the highest point being Slieve Beagh (381 m/1,217 ft). The principal towns are Clones, Carrickmacross, and Castleblayney. Rivers include the Finn and the Glyde in the south, and the Blackwater in the north. Much of the county is fertile. The main form of agriculture is dairy farming, but cattle and pigs are also raised, and cereals and potatoes grown. Industries include leather, linen, knitwear, footwear, furniture, and lacemaking. **5** C4

Monastir Turkish name for the town of ►Bitola in the Former Yugoslav Republic of Macedonia. **36** D3

Mönchengladbach (formerly **München-Gladbach**), industrial city in North Rhine-Westphalia, Germany, on the River Niers, 25 km/15 mi west of Düsseldorf; population (1995) 266,000. Industries include textiles, mechanical engineering, electronics, and paper. It is the centre of the German cotton textile industry and the NATO headquarters for North Europe. **22** J3

Moncton city in Westmorland County, southeast New Brunswick, Canada, on the Petitcodiac River and the Trans-Canada Highway, 145 km/ 90 mi northeast of St John; population (1991) 80,700. It is the regional headquarters of the Canadian National Railways, and a distribution centre for the maritime provinces and Newfoundland. Its industries include telemarketing, food processing, and the manufacture of paperboard and vehicle parts. There are oil wells and natural gas nearby. The city is re-garded as an Acadian centre; the francophone Université de Moncton (1963) is here. **82** U7

Mongolia country in east-Central Asia, bounded north by Russia and south by China.

Mongolia covers an area of 1,565,000 sq km/604,246 sq mi, and has a population of 2,662,000 (2000 est). The capital is ►Ulaanbaatar; other major towns and cities include Darhan, Choybalsan, and Erdenet. The head of state is Natsagiyn Bagabandi (from 1997); the head of government is Nambariin Enkhbayar (from 2000). The currency is the tugrik, and the languages spoken include Khalkha Mongolian (official), Kazakh (in the province of Bagan-Ölgiy), Chinese, Russian, and Turkic languages. **46** B2

Mongolia, Inner (Chinese **Nei Monggol**), autonomous region of north China from 1947; bounded to the north by Mongolia and Russia; to the east by Heilongjiang and Jilin; to the southeast by Liaoning; to the south by Hebei, Shanxi, Shaanxi, and Ningxia Hui Autonomous Region; and to the west by Gansu; area 1,200,00 sq km/463,300 sq mi; population (1996) 23,070,000; less than one-sixth are Mongols. The capital is ►Hohhot, and Baotou is a town in the region. The region is characterized by grassland and desert. Industries include coal; reserves of iron ore, rare earth oxides europium, and yttrium at Bayan Obo; woollen textiles; dairy-processing; and leather. Agricultural products include cereals under irrigation, animal husbandry, and forestry. **11** (2)E2

Monmouthshire (Welsh **Trefynwy**), unitary authority in southeast Wales. A former county, between 1974 and 1996 it became (except for a small area on the border with Mid Glamorgan) the county of Gwent.
area 851 sq km/328 sq mi
towns ►Cwmbran (administrative headquarters), Chepstow
physical rivers ►Wye and Usk; mountainous in north
features Chepstow and Raglan castles, Tintern Abbey, salmon and trout fishing; peak of Pen-y-Fal or Sugar Loaf (596 m/1,955 ft)
agriculture lowlands have rich mixed farming, with arable crops, including wheat, being important
population (1996) 80,400 **5** E6

Monroe city and administrative headquarters of Ouachita Parish, northeast Louisiana, USA, on the Ouachita River, 160 km/100 mi east of Shreveport; population (1996 est) 54,600. It is situated in one of the world's largest natural-gas fields. Furniture, chemicals, and paper are manufactured from locally sourced timber. Other products include cotton and soybeans. It is home to Northeast Louisiana University (1931). **90** C3

Monrovia capital and port of ►Liberia; population (1992) 490,000. Industries include iron ore, rubber, cement, and petrol processing. Civil war damaged much of the infrastructure in and around Monrovia in the 1990s. The National Museum and the University of Liberia are here. **68** B3

Mons (Flemish **Bergen**), industrial city and capital of the province of Hainaut, Belgium, 52 km/32 mi southwest of Brussels; population (1997) 92,000. Industries include coal mining, textiles, and sugar refining. The military headquarters of NATO is at nearby Chièvres-Casteau. It stands on the site of a Roman camp. **22** F4

Montana state in western USA, on the Canadian border. It is nicknamed the Treasure State. Montana was admitted to the Union in 1889 as the 41st US state. One of the Mountain States, it is bordered to the east by North Dakota and South Dakota, to the north by the Canadian states of Saskatchewan, Alberta, and British Columbia, to the west by Idaho, and to the south by Idaho and Wyoming.
population (2000 est) 902,200
area 380,847 sq km/147,045 sq mi
capital ►Helena
towns and cities Billings, Great Falls, Butte
industries and products wheat, cattle, coal, copper, oil, natural gas, lumber, wood products, oil, natural gas, and strip-mined coal, tourism **86** E1

Montauban industrial town and administrative centre of the *département* of Tarn-et-Garonne in the Midi-Pyrénées region of southwest France, 50 km/31 mi north of Toulouse on the River Tarn; population (1990) 51,200. The classical painter Jean Auguste

Ingres and the sculptor Antoine Bourdelle were both natives of the town and many of their works are housed here. Textiles and porcelain are manufactured, and there is a trade in agricultural produce. **26** G10

Montbéliard French town in the *département* of Doubs, 70 km/43 mi east of Besançon, on the Rhine and Rhône canal; population (1990) 33,600. Montbéliard has car industries. The former castle of the counts of Montbéliard, rebuilt in the 17th century, is a museum. **30** B3

Mont Blanc (Italian **Monte Bianco**), the highest mountain in the ►Alps, on the border between France and Italy, and one of the highest points in Europe at 4,807 m/15,772 ft. Lying 10 km/6 mi south of ►Chamonix, it forms part of the Mont Blanc range. The peak was first climbed in 1786 by Jacques Balmat and Michel Paccard of Chamonix. In 1965 the longest road tunnel in the world (12 km/7.5 mi) was opened under the mountain, linking Chamonix to Courmayeur in Italy. **30** B5

Mont-de-Marsan French town, capital of the *département* of Landes, on the River Midouze; population (1990) 32,000. It has nursery gardens, and oil, wood, and resin industries. **26** E10

Monte Bello Islands group of uninhabited islands and rocks in the Indian Ocean, off the northwest coast of Western Australia. The British government used them as a nuclear test site in 1952 and 1956. They are now a marine park and conservation area: short visits are deemed safe though permanent residency is not. **76** B4

Monte Carlo town and luxury resort in the principality of ►Monaco, situated on a rocky promontory northeast of Monaco town; population (1998 est) 14,000. It is a popular summer resort known for its casino (1878) designed by architect Charles Garnier; the Monte Carlo car rally and Monaco Grand Prix motor races; and its international television festival.

Montecristo small uninhabited island 40 km/25 mi south of Elba, in the Tyrrhenian Sea; its name supplied a title for the French novelist Alexandre Dumas's hero in *The Count of Monte Cristo* (1844). **32** E6

Montego Bay port and resort on the northwest coast of Jamaica; population (1991) 83,400. The fine beaches and climate attract tourists. **94** J5

Montélimar town on the River Roubion in the Drôme *département*, Dauphiné region of southeast France; population (1990) 31,400. It is known for the nougat to which its name is given. The River Rhône is not far away, and between it and the town is one of the canals that forms part of the Rhône Valley network. **26** K9

Montenegro (Serbo-Croat **Crna Gora**), constituent republic, with the far larger ►Serbia, of ►Yugoslavia; area 13,800 sq km/5,300 sq mi; population (1991 est) 615,300, including around 397,000 Montenegrins, 79,500 Muslims, and 39,500 Albanians. The capital is ►Podgorica; other towns and cities include Kotor and Cetinje. The republic is mountainous, and Mount Lovćen rises to 1,749 m/5,738 ft. There is a karst region in southwest, and there are forests and grasslands in the east. Skadarsko Jezero (Lake Scutari) is shared with Albania. A Serbian variant of Serbo-Croat is spoken, and the religion is predominantly Serbian Orthodox Christianity. **34** F7

Monterey town and fishing port in Monterey County, west-central California, USA; population (1990) 32,000. It is situated on the north of the Monterey Peninsula and the south of Monterey bay, 137 km/85 mi south-southeast of San Francisco. Formerly a leading fishing and whaling port, Monterey now bases its economy on tourism. Cannery Row houses the popular Monterey Bay Aquarium (1984); other attractions include Fisherman's Wharf (1846). Monterey Peninsula College (1947) and the Monterey Institute of International Studies (1955) are in the city, which has long been a retreat for artists and writers, and is also the site of the Monterey Jazz Festival. **88** E3

Monterrey industrial city (iron, steel, textiles, chemicals, food processing) in northeastern Mexico; population (1990) 2,521,700. It was founded in 597. **92** F3

Montevideo capital and chief port of ►Uruguay, situated on the northen shore of the Río de la Plata estuary; 210 km/130 mi east of Buenos Aires; population (1992) 1,383,700. It is Uruguay's chief industrial and commercial centre, and handles almost 90% of the country's imports and exports. Industries include meat packing, tanning, footwear, flour milling, and textiles. The main exports are grain, meat products, and hides. All Uruguay's railways converge on the city, and a large fishing fleet is based here. Montevideo is also a tourist resort with extensive beaches. **100** K5

Montgomery state capital of ►Alabama, USA, on the Alabama River; population (1996 est) 196,400. Linked to the port of ►Mobile by river, it is a long-established administrative and commercial centre with diverse light industries. Two major air force bases are located nearby. Montgomery was the capital of the Confederacy in the first months of the American Civil War. The 1955 Montgomery Bus Boycott was a landmark in the civil-rights campaign against segregation laws. Black residents make up one-third of the population. **90** D3

Montilla town in the province of ►Córdoba, southern Spain; population (1991) 22,200. It gives its name to *amontillado*, a type of medium-dry sherry. **28** F7

Montluçon French town in the *département* of Allier, on the River Cher; population (1990) 44,200. Industries in Montluçon include the manufacture of metals, chemicals, and rubber. On a hill above the town is a Bourbon castle, and there are many 15th–16th-century buildings in the town. **26** H7

Montpelier state capital of ►Vermont, USA, and seat of Bear Lake county, on the Winooski River, in the Green Mountains; population (2000 est) 8,000. The city is a long-established centre for granite production, sourced from the world's largest granite quarry, the 'Rock of Ages'. The state government is the largest employer, and other service industries include insurance and tourism. **88** F2

Montpellier industrial city and administrative centre of the *département* of ►Hérault and of the ►Languedoc-Roussillon region of southern France; population (1990) 210,900. It is situated on the River Lez 130 km/80 mi northwest of Marseille. Industries include electronics, medical research, engineering, textiles, and food processing, and there is a trade in wine and brandy. It is the birthplace of the philosopher Auguste Comte. **26** J10

Montréal inland port and commercial centre of Québec, Canada, on Montréal Island at the junction of the Ottawa and St Lawrence rivers; population of metropolitan area (1996 est) 3,326,500. It is the second-largest port on the North American east coast, the chief port of the St Lawrence–Great Lakes waterway system, and the world's farthest inland port, situated 1,600 km/1,000 mi from the Atlantic. Industries include oil-refining, engineering, food-processing, distilling, brewing, and the manufacture of steel, aircraft, ships, petrochemicals, tobacco products, clothing, pulp, and paper. Founded by the French in 1642, Montréal became a British possession in 1763.

The city was badly affected by the great ice storm of January 1998 which coated electricity pylons with ice and the weight brought many of them down, blacking out parts of Montréal and Québec. **88** F1

Montreux winter resort in western Switzerland on Lake Geneva; population (1990) 19,900. It is the site of the island rock fortress of Chillon, where François Bonivard (commemorated by the poet Byron), prior of the Abbey of St Victor, was imprisoned 1530–36 for his opposition to the duke of Savoy. At the annual television festival (first held 1961), the premier award is the **Golden Rose of Montreux**. A jazz festival is held every July. **30** B4

Montserrat volcanic island in the West Indies, one of the Leeward group, a British crown colony; capital Plymouth; area 110 sq km/42 sq mi; population (1991) 11,957. Practically all buildings were destroyed by Hurricane Hugo in September 1989. The eruption of the Soufriere volcano in July 1997 buried the capital, Plymouth, under rock and ashes. Around 7,000 islanders were evacuated. **94** M5

Monza town in Lombardy, northern Italy, 16 km/10 mi northeast of Milan; population (1990) 123,200. It has a motor-racing circuit that has traditionally been used for the Italian Grand Prix. Textiles, hats, and carpets are manufactured, and there are dyeing and engineering industries. Once the capital of the Lombards, it still has the Iron Crown of Lombardy, kept in the 13th-century cathedral. Umberto I was assassinated here in 1900. **30** E5

Moonie town in southeastern Queensland, 305 km/190 mi west of Brisbane; population (1981) under 200. It is the site of Australia's first commercial oil strike, in 1961. The oilfield was developed in 1964, and oil is piped to Brisbane. **76** K5

Moorhead town and administrative headquarters of Clay County, west Minnesota; population (1990) 32,300. It is situated on the Red River, opposite Fargo, North Dakota. It is a centre for an agricultural region that produces sugar beet, potatoes, grain, and dairy products. Industries include the manufacture of farm equipment and fibreglass boats. The town was founded 1871. It is home to Moorhead State University (1885) and a campus of Concordia College (1891). **88** A1

Moose Jaw town in southern Saskatchewan, Canada, at the junction of Thunder Creek and the Moose Jaw River, about 80 km/50 mi west of Regina; population (1996 est) 14,200. Situated in the middle of the wheat-belt, on the Trans-Canadian Highway and a railway junction, it is a major agricultural and distribution centre, with grain elevators, flour mills, meat-processing plants, cooperative stockyards, and oil refineries. The settlement was founded in 1882 as a rail depot on the Canadian Pacific Railway. **82** K6

Moradabad trading city in Uttar Pradesh, India, on the Ramganga River; population (1991) 444,000. Situated at a railway junction, it produces textiles and engraved brassware. It was founded 1625 by Rustan Khan; the Great Mosque dates from 1631. **54** C3

Moravia (Czech **Morava**), area of central Europe, forming two regions of the Czech Republic: **South Moravia** (Czech Jihomoravský) and **North Moravia** (Czech Severomoravský). South Moravia has an area of 15,030 sq km/5,800 sq mi and a population (1991) of 2,048,900. Its capital is ►Brno. North Moravia has an area of 11,070 sq km/4,273 sq mi and a population (1991) of 1,961,500. Its capital is ►Ostrava. The River Morava is found in the region. 25% of the region is forested. Products include maize, grapes, and wine in the south; wheat, barley, rye, flax, and sugar beet in the north; and coal and iron.

Moray unitary authority in northeast Scotland, created in 1996 from the Moray district of Grampian region.
area 2,224 sq km/859 sq mi
towns ►Elgin (administrative headquarters), Forres, Buckie, Lossiemouth
physical the land gradually slopes from the Grampian Mountains in the south (Cairn Gorm 1,245 m/4,085 ft) towards the Moray Firth; extensive coastal lowlands fringe an area of sand-dune formation; part of this land was reclaimed from the sea and is now covered by the Culbin forest. The River Spey reaches the North Sea near Buckie
features Elgin cathedral; Brodie and Duffus castles; Gordonstoun school
industries whisky distilling, food processing
agriculture trout and salmon fishing in rivers; cereals in lowland plain
population (1996) 85,000 **5** E3

Moray Firth North Sea inlet in Scotland, between Burghead (Moray) and Tarbat Ness (Highland), 38 km/24 mi wide at its entrance, and 48 km/30 mi from there to the Caledonian Canal. The city of Inverness is situated at the head of the Firth. The Spey, Findhorn, and Nairn are the chief rivers which flow into the Moray Firth. **24** J4

Morbihan *département* in the Brittany region of western France; area 6,763 km/2,611 sq mi; population (1990) 619,950. Its Atlantic coastline has many inlets. Barren heath covers a quarter of the area, whilst a third part produces good crops of wheat and rye. Fisheries and oyster culture are important. Industries include tyres, car parts, and electrical goods. The administrative centre is ►Vannes and other principal towns are Lorient and Pontivy. Morbihan is noted for its prehistoric remains, especially megaliths, or standing stones, in the area of Carnac. **6** B3

Mordoviya alternative name for ►Mordvinia, an autonomous republic of the Russian Federation. **38** H4

Mordvinia (or **Mordoviya**), autonomous republic of central

Russian Federation; area 26,200 sq km/10,100 sq mi; population (1990) 964,000 (60% Russian, 35% Mordvinians). The capital is Saransk. The republic is about 350 km/217 mi southeast of Moscow. The River Sura is in the east, and there are forests in the west. Industries include sugar beet, grains, hemp, and potatoes; sheep and dairy farming; commercial vehicles, timber, furniture, and textiles. Languages spoken are Russian and Mordvin. 38 H4

Morecambe town and seaside resort in Lancashire, northwest England, on Morecambe Bay, joined with the port of Heysham, which has a ferry service to the Isle of Man; joint population (1991) 46,700. Tourism is important; other industries include shrimp fishing, engineering, and the manufacture of clothing. 24 K7

More Laptevykh Russian name for ►Laptev Sea, part of the Arctic Ocean off the north coast of Russia. 44 L1

Morlaix seaport on the River Morlaix in Brittany, France, in the *département* of Finistère, 53 km/33 mi east of Brest; population (1990) 18,000. The main manufacturing industries are tobacco and paper. The surrounding country is an important vegetable-growing region. Many of the town's houses date from the 15th century. There is a marina for yachts. 26 B5

Morocco country in northwest Africa, bounded to the north and northwest by the Mediterranean Sea, to the east and southeast by Algeria, and to the south by Western Sahara.

Morocco covers an area of 458,730 sq km/177,115 sq mi (excluding Western Sahara), and has a population of 28,351,000 (2000 est). The capital is ►Rabat; other major towns and cities include Casablanca, Marrakesh, Fez, Oujda, Kenitra, Tétouan, and Meknès. The head of state is Sayyid Muhammad VI ibn-Hassan (from 1999); the head of government is Abderrahmane Youssoufi (from 1998). The currency is the dirham, and the languages spoken include Arabic (official), Berber dialects, French, and Spanish. 62 C2

Morón de la Frontera town in the province of ►Seville, southern Spain; population (1991) 27,200. It has a Gothic castle and the ruins of a Moorish castle. There are mines, marble quarries, and mineral springs nearby. 28 E7

Moroni capital of the ►Comoros Republic, on Njazidja (Grand Comore); population (1992) 22,000. It has a small natural harbour from which coffee, cacao, and vanilla are exported. Local agricultural markets trade in coconuts, cassava, bananas, and rice. 72 G2

Moscow (Russian **Moskva**), industrial and commercial city, capital of the Russian Federation and of the Moscow region, and formerly (1922–91) of the USSR; population (1990) 8,801,000. Moscow lies on the Moskva River 640 km/400 mi southeast of St Petersburg, and covers an area of some 880 sq km/340 sq mi. It is the main political, economic, and cultural centre of Russia. A major manufacturing city, its industries include aerospace technology and vehicle assembly, machine and precision tool manufacture, and the production of such diverse goods as electrical equipment, textiles, chemicals, and many food products. Moscow's State University was founded in 1755; other cultural institutions include the extensive Russian State Library and the Academy of Sciences. The city is home to the renowned Bolshoi Theatre of Opera and Ballet, the Pushkin Fine Arts Museum, the Tretyakov Gallery, and the Exhibition of Economic Achievements. 38 G3

Moselle *département* in the ►Lorraine region of France; area 6,213 sq km/2,399 sq mi; population (1990) 1,011,940. It has a boundary with Germany to the east and with Luxembourg to the north and northeast. It consists largely of a plateau across the east of which flows the River ►Moselle. Barley, oats, rye, and animal fodder crops are cultivated. The *département* has coal and iron mines and manufactures machinery, chemicals, and textiles. The administrative centre is ►Metz; other towns include Boulay, Château-Salins, Forbach, Sarrebourg, Sarreguemines, and Thionville. 6 G2

Moselle (German **Mosel**), river in Western Europe, some 515 km/320 mi long. It rises in the Vosges Mountains, France, in two headwaters uniting near St-Maurice. It flows north past Metz and is canalized from Thionville to its confluence with the ►Rhine at Koblenz in Germany. It gives its name to the *départements* of ►Moselle and ►Meurthe-et-Moselle in France. Vineyards along the Moselle in Germany produce popular white wines. 22 J6/22 K4

Mosi-oa-tunya ('smoke that thunders'), local African name for the ►Victoria Falls of the Zambezi River. 72 D3

Moskva Russian name for ►Moscow, the capital of the Russian Federation. 38 G3

Mosquito Coast (Spanish **Costa de Mosquitos**), Caribbean coast of Honduras and Nicaragua, characterized by swamp, lagoons, and tropical rainforest. The territory is inhabited by Miskito Indians, Garifunas, and Zambos, many of whom speak English. Between 1823 and 1860 Britain maintained a protectorate over the Mosquito Coast which was ruled by a succession of 'Mosquito Kings'. 94 H6

Most (German **Brüx**), industrial town (chemicals, lignite, metallurgy) in the northwestern Czech Republic, 77 km/48 mi northwest of Prague; population (1991) 70,700. It has a 16th-century Gothic church. The town is situated in a coal-mining area at the foot of the Erzgebirge mountain range, on the small River Bida. 20 J6

Mostaganem industrial port (metal and cement) in northwest Algeria; population (1999 est) 161,700. It is the capital of Mostaganem Department and lies on the Mediterranean Sea 90 km/56 mi east of Oran. Linked by pipeline with the natural gas fields at Hassi Messaoud, it also exports cereals, vegetables, and wine; industries include wood-pulp processing and the manufacture of paper, cement, electrical equipment, flour, furniture, and cigarettes. 28 L9

Mostar industrial town (aluminium, tobacco) in Bosnia-Herzegovina, known for its grapes and wines; population (1991) 126,000. The eastern, mainly Muslim sector of the town was under siege by Bosnian Croat forces 1993–94. 34 E6

Mosul (Arabic **Al Mawşil**), industrial city and oil centre in Iraq, 350 km/217 mi northwest of Baghdad; population (1998 est) 901,200. It lies on the right bank of the Tigris river, opposite the site of ancient Nineveh. It is a centre of communications by air, road, rail, and river. Industries include oil refining, though now on a reduced scale since the Gulf War of 1991. Main products are cement, textiles, gallnuts, cotton, wool, hides, wax, and gum. Once it manufactured the light cotton fabric **muslin**, which was named after it. 58 K5

Motherwell industrial town and administrative headquarters of ►North Lanarkshire, Scotland, southeast of Glasgow; population (1991) 60,500. The two burghs of Motherwell and Wishaw were amalgamated in 1920, but the towns retain distinct identities. Formerly a coalmining, iron, and engineering town, industries now include light manufacturing and there is a Eurofreight terminal. 24 J6

Motril town in the province of ►Granada, southern Spain; population (1991) 46,050. Situated in a valley, it has a port on the Mediterranean coast at Calahonda. Products include sugar cane, bananas, and flowers; resources include lead, iron, zinc, and copper (but these are no longer mined). 28 G8

Moulins administrative centre of the *département* of ►Allier in the Auvergne region of central France, situated on the River Allier; population (1990) 23,400. It is a market town for the surrounding agricultural region, and its main industries are cutlery, textiles, glass, leather, electrical products, and food processing. Moulins was capital of the old province of Bourbonnais from 1368 to 1527. 26 J7

Moulmein former name of ►Mawlamyine, capital of Mon state, Myanmar. 50 B3

Mount Isa town in northwest Queensland, Australia, on the River Leichhardt, 1,811 km/1,125 mi northwest of Brisbane; population (1996) 21,751. Mount Isa is a centre for silver-, lead-, and zinc-mining. 76 G4

Mozambique country in southeast Africa, bounded north by Zambia, Malawi, and Tanzania; east and south by the Indian Ocean; southwest by South Africa and Swaziland; and west by Zimbabwe.

Mozambique covers an area of 799,380 sq km/308,640 sq mi, and has a population of 19,680,000 (2000 est). The capital (and chief port) is ►Maputo; other major towns and cities include Beira, Nampula, Nacala, and Chimoio. The head of state is Joaquim Alberto Chissano (from 1986); the head of government is Pascoal Mocumbi (from 1994). The currency is the metical, and the languages spoken include Portuguese (official) and 16 African languages. 72 E3

Mpumalanga (formerly **Eastern Transvaal**), province of the Republic of South Africa from 1994, formerly part of Transvaal province; area 81,816 sq km/31,589 sq mi; population (1995 est) 3,007,100. The capital is Nelspruit. Features of the province include Limpopo River, Vaal River, Kruger National Park, Blyde River Canyon Nature Reserve, and Mpumalanga Drakensberg Mountains. Industries include farming and coal. Languages spoken in the province are Siswati, Zulu, and Afrikaans. 72 D5

Mtwara deep-water seaport in southern Tanzania, on Mtwara Bay; population (1998 est) 76,600. It was opened in 1954. 70 G6

Mühlhausen town in the *Land* (administrative region) of ►Thuringia, Germany, on the River Unstrut, 48 km/30 mi northwest of Erfurt; population (1995) 40,300. There are engineering and textile industries. The town dates from the Middle Ages. 20 F5

Mukden former name of ►Shenyang, capital of Liaoning province, China. 48 B3

Mülheim an der Ruhr industrial city in North Rhine-Westphalia, Germany, on the River Ruhr, 25 km/16 mi northeast of Düsseldorf; population (1995) 176,700. Industries include heavy engineering, iron and steel, building materials, construction, and wholesale trading. 30 C3

Mulhouse (German **Mülhausen**), industrial city in the Haut-Rhin *département* of Alsace, eastern France, situated 35 km/22 mi south of Colmar on the River Ill and the Rhine–Rhône Canal; population (1990) 109,900. Its main industries include textiles, engineering, and electrical goods. 30 C3

Mull second-largest island of the Inner ►Hebrides, Argyll and Bute, Scotland; area 950 sq km/367 sq mi; population (1991) 2,700. It is mountainous, and is separated from the mainland by the **Sound of Mull** and the Firth of Lorne; it lies 11 km/7 mi west of Oban. The main town is Tobermory, from which there are ferry connections to Oban; Craignure is also connected by ferry to Oban. The economy is based on fishing, forestry, tourism, and stock rearing. 24 G5

Mullingar county town of County ►Westmeath, Republic of Ireland, on the River Brosna; population (1996) 8,000. It is an important road and rail junction and was a harbour on the (now disused) Royal Canal that links Dublin and the River Shannon. It is a cattle market and trout-fishing centre. Industries include tobacco, vinyl, furniture, and pencils. 24 E8

Multan (or **Mooltan**), city in Punjab province, central Pakistan, 205 km/190 mi southwest of Lahore; population (1995 est) 1,075,000. Industries include textiles, precision instruments, chemicals, pottery, and jewellery. It trades in grain, fruit, cotton, and wool. The emperors of Delhi ruled the city 1526–1779 and they were succeeded by first Afghan and then Sikh rulers until British rule began in 1849. 56 K3

Mumbai (formerly **Bombay** until 1995), Indian city, industrial port, and commercial centre; population (1994) 14,500,000. Previously known as Bombay, the city was once the capital of Bombay Presidency and Bombay State and in 1960 became the capital of ►Maharashtra, a newly created state. By a decision of the Maharashtra government implemented in 1995, the city was renamed Mumbai. Industries include textiles, engineering, pharmaceuticals, and diamonds. It is the centre of the Hindi film industry. 54 B5

München German name of ►Munich, a city in Germany. 30 G2

Munich (German **München**), capital of Bavaria, Germany, on the River Isar, about 520 m/1,706 ft above sea level, some 45 km/28 mi from the edge of the Alps; population (1995) 1,240,600. The main industries are brewing, printing, precision instruments, machinery, electrical goods, computers, telecommunications, fashion, and food processing. 30 G2

Munster historic southern province of the Republic of Ireland, comprising the counties of Clare, Cork, Kerry, Limerick, North and South Tipperary, and Waterford; area 24,140 sq km/

9,320 sq mi; population (1996) 1,033,900.

Münster industrial city in North Rhine-Westphalia, northwest Germany, 100 km/62 mi northeast of Düsseldorf on the Dortmund-Ems Canal; population (1995) 264,500. There are wire, cement, chemicals, steel, brewing, and distilling industries. It was formerly the capital of Westphalia. The Treaty of Westphalia was signed simultaneously here and at Osnabrück in 1648, ending the Thirty Years' War. 22 K3

Muqdisho alternative name for ►Mogadishu, the capital of Somalia. 70 H3

Murcia autonomous community of southeast Spain, with a coastline on the Mediterranean Sea; area 11,300 sq km/4,362 sq mi; population (1991) 1,032,300. The River Segura and its tributaries (the Sangonera and the Quipar) flow through the region, which is very mountainous in the south and east. The irrigated area, the huerta of Murcia, is one of the most intensively farmed areas in Spain and is especially important for citrus fruits. Products include esparto grass (for weaving into simple products such as sandals), iron, olives, and fruit. There are large deposits of salt and minerals, especially lead and zinc. The main port is ►Cartagena and the capital is ►Murcia. 9 (1)E4

Murcia industrial city (silk, metal, glass, textiles, pharmaceuticals), capital of the Spanish province of Murcia, on the River Segura; population (1994) 342,000. Murcia was founded in 825 on the site of a Roman colony by 'Abd-ar-Rahman II, caliph of Córdoba. It has a university and a 14th-century cathedral. 28 J7

Mureş (Hungarian **Maros**), river in Romania and Hungary, rising in the Carpathian Mountains and flowing 725 km/453 mi to join the River ►Tisza at Szeged in Hungary. 34 J3

Murmansk seaport and capital of the Murmansk oblast (region) located 1000 km/624 mi north of St Petersburg on the Kola Peninsula in the northwest of the Russian Federation; population (1990) 472,000. Situated on an estuary 50 km/31 mi inland from the Barents Sea, it is the largest city in the Arctic, Russia's principal fishing port, and a base for icebreakers that keep the ►Northeast Passage open. Shipbuilding is a major industry, and polar research institutes are located here. 16 S2

Murray principal river of Australia, which rises in the Snowy Mountains of the Australian Alps near Mount Kosciusko, in New South Wales; length 2,540 km/1,578 mi. The Murray flows west and northwest, for most of its length forming the boundary between New South Wales and Victoria, then turns south to reach the Southern Ocean at Encounter Bay, southeast of Adelaide in South Australia. The River Murray is an important source of hydroelectric power and irrigation. With its main tributary, the ►Darling, it is 3,750 km/2,330 mi long. Together they produce 40% of the country's agricultural wealth. 76 H6

Muscat (or **Masqat**), capital and port of ►Oman, lying on the Gulf of Oman; population (2001 est) 57,600. The Muscat region has a population of 621,500 (1998 est). With the advantage of a deepwater harbour at Matrah, in the western part of the city, Muscat handles the bulk of the country's foreign trade, especially the export of crude oil and the import of food. Mina, with a terminal for loading supertankers, and Riyam, which handles incoming refined petroleum products, are both nearby. Natural gas and chemical industries are also important. Muscat is served by an international airport, As-Sib, and is linked to Salalah, a former capital, by a 1,000 km/625 mi trunk road completed in 1984. 61 H5

Muscat and Oman former name of ►Oman, a country in the Middle East.

Musgrave Ranges Australian mountain ranges on the border between South Australia and the Northern Territory; the highest peak is Mount Woodruffe at 1,525 m/5,000 ft. The area is an Aboriginal reserve. 76 E5

Muskegon city, port, and administrative headquarters of Muskegon County, west Michigan, USA, on Lake Michigan, at the mouth of the **Muskegon River**, 56 km/35 mi northwest of Grand Rapids; population (1996 est) 39,500. It is a service and distribution centre, processing timber from forests to the north, and fruit from the agricultural lake-shore regions to the south. Heavy machinery, metal products, vehicle parts, and sporting goods are manufactured. Tourism is also important to the economy. The Muskegon Business College (1885) and Muskegon Community College (1926) are both in the city. 88 C2

Mutare (formerly **Umtali** until 1982), industrial town in the eastern highlands of Zimbabwe, 258 km/160 mi southeast of Harare; population (1992) 131,800. The chief town of Manicaland province, it lies on the main road and railway to Mozambique. Industries include vehicle assembly, engineering, tobacco, textiles, and paper. Products include citrus fruits, softwoods, and wattle. 72 E3

Myanmar (formerly **Burma** until 1989), country in Southeast Asia, bounded northwest by India and Bangladesh, northeast by China, southeast by Laos and Thailand, and southwest by the Bay of Bengal.

Myanmar covers an area of 676,577 sq km/261,226 sq mi, and has a population of 45,611,000 (2000 est). The capital (and chief port) is ►Yangon (formerly Rangoon); other major towns and cities include Mandalay, Mawlamyine, Bago, Bassein, Taunggyi, Sittwe, and Manywa. The head of state and government is Than Shwe (from 1992). The currency is the kyat, and the languages spoken include Burmese (official), English, and tribal dialects. 50 B2

Mykolayiv (formerly Russian **Nikolayev** until 1991), port and naval base in southern Ukraine, situated on the northern shore of the Black Sea, 110 km/68 mi northeast of Odessa; population (1990) 507,900.
It is capital of an oblast of the same name, and lies at the confluence of the Southern Bug and Ingul rivers. Its industries include shipbuilding and the manufacture of agricultural machinery. 48 F5

Mysore (or **Maisur**), city in ►Karnataka, south India, some 130 km/80 mi southwest of Bangalore; population (1991) 481,000. Industries include silk textiles, engineering, chemicals, fertilizers, and food processing. The city is known for the production of silk, incense and sandalwood. The former capital of a princely state of Mysore, it is the second-largest city in the state; notable buildings include the 18th-century European fort and the maharajah's palace. The Chamundi Hill (1,000 m/ 3,2800 ft) lies to the southeast of the town. 54 C6

Mytilene (modern Greek **Mytilini**), port, capital of the Greek island of Lesvos (to which the name Mytilene is sometimes applied) and a centre of sponge fishing; population (1991) 25,400. 36 G5

Mytilini modern Greek name of ►Mytilene, a town on the island of Lesvos. 36 G5

Nablus (or **Nabulus**), city and district on the west bank of the River Jordan, 50 km/31 mi north of Jerusalem; population (1995) 130,000. Nablus is a trading centre for agricultural products. Formerly Shechem, it was the ancient capital of Samaria, and a few Samaritans remain. The British field marshal Allenby's defeat of the Turks here in 1918 completed the conquest of Palestine. It was occupied by Israel in 1967 and was the scene of bitter conflict during the Intifada; Israeli troops withdrew in December 1995. 60 C4

Nacala seaport in Nampula province, northern Mozambique; population (1991) 125,200. It is a major outlet for minerals. It is linked by rail with Malawi; approximately 20% of Malawi's petroleum products are imported via Nacala. 72 G2

Nafplio port of Greece, capital of the department of Argolis, at the northern extremity of the Gulf of Argos, 11 km/7 mi southeast of the town of Argos; population (2001 est) 12,400. Nafplio is one of the busiest ports of the ►Peloponnese region, exporting dried and preserved fruit, vegetables, tobacco, and cotton. From 1824 to 1834 it was the capital of Greece. 36 E7

Nagaland state of northeast India, bordering Myanmar on the east, and the Indian states of Manipur to the south, and Assam to the north and west; area 16,600 sq km/6,409 sq mi; population (1994 est) 1,410,000, which is made up of many different tribal groups. The capital is Kohima. The state is mainly upland, averaging over 1,500 m/4,900 ft, and densely forested. Wildlife includes tigers and elephants. Industries in the state include timber and paper, and petroleum at Dikhu. Agricultural products include tea, sugar, coffee, rice, millet, maize, and vegetables. The population is mainly Christian. 10 F3

Nagasaki industrial port (coal, iron, shipbuilding) on Kyushu island, Japan; population (1994) 438,000. Nagasaki was the only Japanese port open to European trade from the 16th century until 1859. The first modern Japanese shipyard opened here 1855–61.
On 9 August 1945, an atom bomb was dropped on Nagasaki by the USA. 48 E7

Nagoya industrial seaport, seat of Aichi prefecture, and capital of the ►Chubu region, Honshu island, Japan; population (1994) 2,091,000. Situated at the head of Ise Bay, 260 km/162 mi west of Tokyo and 140 km/87 mi east of Osaka, Nagoya is the fourth-largest city of Japan and is a major port. Manufactured products include cars, textiles, and clocks. It has a shogun fortress, built 1610–14 and rebuilt in 1959, and a notable Shinto shrine, Atsuta Jingu. 48 J6

Nagpur industrial city in Maharashtra, India, 710 km/440 mi east of Mumbai (formerly Bombay); population (1991) 1,661,000. Industries include textiles, metal goods, pharmaceuticals, cotton goods, and hosiery; oranges are traded. Nagpur was the centre of the Chanda Gond dynasty in the 10th–11th centuries, and was the former capital of Central Provinces and Berar, and Madhya Pradesh states. In 1956 it alternated with Mumbai as capital of Maharashtra state. 54 C4

Naha chief port on Okinawa island, in the Ryukyu group, Japan; population (1994) 304,000. Industries include pottery and textiles; there is fishing. Ceramics have been produced in the city since 1682. Shuri castle, residence of Okinawa's local royalty, was reconstructed in 1993. Naha has been the capital of Okinawa prefecture since 1879. 46 H5

Nahr en Nil Arabic name in Egypt for the River ►Nile, a river in Africa. 64 F4

Nairobi capital of ►Kenya, in the central highlands at 1,660 m/5,450 ft; population (1993 est) 1,758,900. Industries include engineering, paints, brewing, and food processing. It is the headquarters of the United Nations (UN) Environment Programme, and has the UN Centre for Human Settlements. It is one of Africa's largest and fastest-growing cities. 70 F4

Nakhodka (Russian 'discovery'), seaport in Primorski (Maritime) Krai in the Russian Far East; population (1996 est) 162,000. Nakhodka lies on the coast of the Sea of Japan, 100 km/ 62 mi east of Vladivostok. Nakhodka is the eastern terminus of the ►Trans-Siberian Railway. Transhipment of goods and fishing are major economic activities in the city. 48 G2

Namen Flemish form of ►Namur, a city in Belgium. 22 G4

Namib Desert coastal desert region between the Kalahari Desert and the Atlantic Ocean, extending some 2,800 km/1,740 mi from Luanda in Angola to St Helena Bay in South Africa. Its aridity is caused by the descent of dry air cooled by the cold Benguela current along the coast. The sand dunes of the Namib Desert are among the tallest in the world, reaching heights of 370 m/1,200 ft. In the most arid parts rainfall can be as little as 23 mm/0.9 in per year. 72 A4

Namibia (formerly **South West Africa** to 1968), country in southwest Africa, bounded north by Angola and Zambia, east by Botswana and South Africa, and west by the Atlantic Ocean.

Namibia covers an area of 824,300 sq km/318,262 sq mi, and has a population of 1,726,000 (2000 est). The capital is ►Windhoek; other major towns and cities include Swakopmund, Rehoboth, and Rundu. The head of state is Sam Nujoma (from 1990); the head of government is Hage Geingob (from 1990). The currency is the Namibian dollar, and the languages spoken include English (official), Afrikaans, German, Ovambo, Nama, Kavango, and other indigenous languages. 72 B4

Nampo (formerly **Chinnampo** until 1947), city on the west coast of North Korea, 40 km/25 mi southwest of Pyongan (it acts as the port for the capital); population (1998 est) 447,500. Industries include a major shipyard and the manufacture of electrodes and glass. The opening of the port to foreign trade in 1897 stimulated its modern growth. 48 C4

Namur (Flemish **Namen**), industrial city and capital of Namur province, south Belgium, at the confluence of the Sambre and Meuse rivers, about 65 km/40 mi southeast of Brussels; population (1997) 105,200. There are cutlery, porcelain, paper, iron, and steel industries. It was a strategic location during both world wars and was occupied by Germany 1914–18. **22** G4

Namur (Flemish **Namen**), province of Belgium, bounded to the north by Brabant, to the east by the provinces of Liège and Luxembourg, to the west by Hainaut, and to the south by France; area 3,660 sq km/1,413 sq mi; population (1999) 441,200. The capital is ►Namur, and Dinant is another town in the province. The principal rivers are the Meuse, the Sambre, and the Lesse. The province is known as the gateway to the Ardennes. To the north of the Sambre and Meuse valleys, the province forms part of the fertile Hesbaye plateau. To the south there are fertile valleys and low hilly tracts. In some areas the densely wooded offshoots of the Ardennes reach heights of 500 m/1,640 ft above sea level. Industries include coal mines; marble, slate, sandstone, and limestone quarries; and steel, iron, chemicals, glass, and ceramics. Steel, iron, and smelting works, are concentrated, together with the glass and chemical industries, around the city of Namur and in the valley of the Sambre. Ceramic manufacture is centred in Andenne. **7** (1)C4

Nanaimo town and former coal-mining centre of southwest British Columbia, Canada, on the east coast of Vancouver Island; population (1991) 60,100. It was founded as a Hudson's Bay Company trading post and developed after the discovery of coal in 1849; the last big coal mine closed in 1960. Today Nanaimo is the centre of a lumbering region, and has salmon and cod fisheries, and a ferry port. Shipbuilding and tourism are also important to the local economy. A biological research station and the preserved Hudson's Bay Company fort (1853) are in the town. **86** B1

Nanchang capital of ►Jiangxi province, China, on the River Gan, about 260 km/160 mi southeast of Wuhan; population (1994) 1,465,400. Motor vehicles, aircraft, tractors, textiles, glass, porcelain, soap, and lumber are produced. There is a considerable trade in rice, beans, linen, timber, paper, and tobacco. **46** F5

Nancy administrative centre of the *département* of ►Meurthe-et-Moselle and of the region of ►Lorraine, France, on the River Meurthe and the Rhine and Marne Canal, 280 km/175 mi east of Paris; population (1990) 102,400, conurbation 300,000. Nancy dates from the 11th century. It has a university (founded in 1572, transferred here in 1797) and is the seat of a bishopric; building of the cathedral began in 1703 (by J H Mansart). **30** B2

Nanda Devi peak in the Himalayas, Uttar Pradesh, northern India; height 7,817 m/25,645 ft. Until Kanchenjunga was absorbed into India, Nanda Devi was the country's highest mountain. **54** C2

Nanjing (or **Nanking**, 'southern capital'), inland port and capital of ►Jiangsu province, China, 270 km/165 mi northwest of Shanghai; population (1993) 2,430,000. It is a commercial centre and communications hub, with engineering, electronics, shipbuilding, chemical, and oil-refining industries. Textiles are a traditional manufacture. The bridge over the Chang Jiang River, built in 1968, is the longest in China at 6,705 m/22,000 ft. **46** F4

Nanking alternative name of ►Nanjing, capital of Jiangsu province, China. **46** F4

Nanning river port and capital of ►Guangxi Zhuang Autonomous Region, China, on the Yong Jiang River; population (1994) 1,181,200. It is an important trading centre, particularly in spices. Industries include sugar-refining, food-processing, and the manufacture of chemicals, machines, and cotton. It was a supply town during the Vietnam War and the Sino-Vietnamese confrontation of 1979. Almost a third of the population is of Zhuang cultural background. **50** D2

Nansei alternative name for the ►Ryukyu Islands, the southernmost island group of Japan. **46** H5

Nanshan Islands Chinese name for the ►Spratly Islands. **50** E4

Nantes industrial port and administrative centre of the *département* of ►Loire-Atlantique and the ►Pays de la Loire region in western France, situated on the right bank of the River Loire, 50 km/31 mi from its mouth; population (1990) 252,000. Industries include oil, sugar refining, metal goods, textiles, soap, biscuits, and tobacco. The city has many splendid buildings, including a cathedral constructed between 1434 and 1884 and a castle founded in 938. **26** D6

Nantucket (called 'the Little Grey Lady'), resort island and county in southeast Massachusetts, USA, 40 km/25 mi south of Cape Cod across Nantucket Sound; population (1997 est) 7,500. Extending over 120 sq km/46 sq mi, its beaches have made it a popular summer vacation area. The island was explored by the English in 1602, settled in 1659 by Quaker and Presbyterian families, and became part of Massachusetts in 1692. In the 18th–19th centuries, Nantucket Town (incorporated in 1671) was a whaling port.

Napa city and administrative headquarters of Napa County, north-central California, USA; population (1992) 63,300. It is situated on the Napa River, 58 km/38 mi northeast of San Francisco on San Pablo Bay. It is a major trading centre for wine produced in the surrounding Napa Valley vineyards; other products include fruit, clothing, steel pipe, and leather goods. The city has also served as a shipping point for cattle, lumber, and minerals. Tourism is important to the regional economy. Napa State Hospital and Napa Valley College (1940) are both in the city. **86** B3

Napier wool port on Hawke Bay, on the east coast of North Island, New Zealand, 260 km/162 mi northeast of Wellington; population (1996) 55,000. The harbour, Port Ahuriri, is protected by a breakwater. Wool, meat, pelts, fruit, and fish are exported. **78** F4

Naples (Italian **Napoli**; Greek *Neapolis* 'new city'), industrial port and capital of Campania, Italy, on the Tyrrhenian Sea; population (1992) 1,071,700. Industries include shipbuilding, food-processing, and the manufacture of cars, textiles, and paper. To the south is the Isle of Capri, and behind the city is Mount Vesuvius, with the ruins of Pompeii at its foot. **32** J8

Napoli Italian form of ▶Naples, a city in Italy. 32 J8

Nara city in the south of Honshu island, Japan, the capital of the country 710–84; population (1994) 355,000. It has a writing brush and ink stick industry and supplies 90% of the country's Indian ink (*sumi*). Plastics are also manufactured in the city, and tourism is an important source of income. It was the birthplace of Japanese art and literature and has ancient wooden temples. Tame deer can be found in Nara Park. 48 H6

Narbonne city in the Aude *département* of southern France, 50 km/31 mi east of Carcassonne; population (1990) 47,100. It is sited just inland from the Mediterranean on the canalized Robine River, and was a port in medieval times, although its harbour silted up in the 14th century and has now disappeared. It was the chief town of southern Gaul in Roman times. The holiday resort of **Narbonne-Plage** lies 14 km/9 mi to the east. 26 H10

Narmada river that rises in the Maikala range in Madhya Pradesh state, central India, and flows 1,245 km/778 mi west and southwest to the Gulf of Khambat, an inlet of the Arabian Sea. Forming the traditional boundary between Hindustan and Deccan, the Narmada is a holy river of the Hindus. The Narmada Valley Project involves the building of over 3,000 dams along the length of the Narmada River. 54 C4

Narvik seaport in Nordland county on Ofot Fjord, north Norway; population (1996) 14,100. To secure the supply of iron ore mined in the region, Germany seized Narvik in April 1940. On 13 April 1940 a British flotilla forced its way into Narvik Fjord and sank four German destroyers. British, French, Polish, and Norwegian forces recaptured the port on 28 May but had to abandon it on 10 June to cope with the worsening Allied situation elsewhere in Europe. Narvik was destroyed during World War II but was rebuilt. 16 J2

Nashua city and administrative headquarters of Hillsborough County, south New Hampshire, USA; population (1992) 79,700. It is situated 29 km/18 mi south of Manchester, on the Nashua River, where it meets the Merrimack River, just north of the Massachusetts state border. A textile manufacturer until after World War II, Nashua now produces shoes, asbestos, electronics, chemicals, office equipment, plastics, metals, machines, glass, toys, greeting cards, and textiles. The Federal Aviation Agency Center controlling New England and upstate New York air traffic, New England Aeronautical Institute, Rivier College (1933), Daniel Webster College (1965), and a federal fish hatchery are all in the city. 88 F2

Nashville capital and river port of ▶Tennessee, USA, on the Cumberland River; population (1996 est) 511,300. It is a banking and commercial centre, a major processing point for timber and agricultural produce, and has large car-manufacturing, printing, music-publishing, and recording industries. Nashville was settled in 1779 as Fort Nashborough. 90 D2

Nassau capital and port of the Bahamas, on New Providence Island; population (1998 est) 205,400. It is a popular tourist resort, especially in winter, and was the scene in 1994 of an international conference on biological diversity. 90 F4

Natal seaport and capital of Rio Grande do Norte federal unit (state), northeast Brazil; situated on the banks of the Potengi River on the Atlantic coast 410 km/255 mi southeast of the port of Fortaleza; population (1991) 606,600. There are textile and salt-refining industries, and cotton, sugar, salt, and hides are exported. Tourism is important. The Fort of the Three Kings is a pentagonal monument, so named because of the founding of the city in 1597 on 6 January, the Epiphany. There is also a cathedral (1768) and a university, founded in 1958. 98 K5

Natchez city and river port in Mississippi, USA, on the eastern bluffs above the Mississippi River; 128 km/80 mi southwest population (1996 est) 18,700. Manufacturing is the largest employment sector, with wood pulp and timber as the principal products. It was founded by the French in 1716, on lands originally occupied by the Natchez American Indian people. Before the Civil War the city was a prosperous cotton-trading centre, and many of its numerous Greek Revival mansions and churches have been preserved. The port was a major distribution point during the heyday of steamboat traffic. 90 C3

Natron, Lake salt and soda lake in the Great Rift Valley, Tanzania; area 600 sq km/230 sq mi; length 56 km/35 mi, width 24 km/15 mi. 70 F4

Nauru island country in Polynesia, southwest Pacific, west of Kiribati.

Nauru covers an area of 21 sq km/8.1 sq mi, and has a population of 12,000 (2000 est). The capital is ▶Yaren District (seat of government). The head of state and government is Bernard Dowiyogo (from 2000). The currency is the Australian dollar, and the languages spoken include Nauruan and English (both official). 74 G6

Navarre (Spanish **Navarra**), autonomous community of northern Spain, bordered by France on the north; area 10,400 sq km/4,000 sq mi; population (1991) 516,300. The region is mountainous, containing spurs of the Pyrenees, and includes Monte Adi (1,503 m/4,933 ft high); the rivers Arga, Aragón, and Ebro flow through the area. Cereals and wine are produced in the lowlands. The capital is ▶Pamplona. 9 (1)E1

Naxos island of Greece, the largest of the Cyclades, area 453 sq km/175 sq mi. Known since early times for its wine, it was a centre for the worship of Bacchus, who, according to Greek mythology, found the deserted Ariadne on its shore and married her. 36 H7

Nazareth city in Galilee, northern Israel, 30 km/19 mi southeast of Haifa; population (1993) 53,500. According to the New Testament it was the boyhood home of Jesus. 60 C4

Nazca town to the south of Lima, Peru, near a plateau that has geometric linear markings interspersed with giant outlines of birds and animals; population (1993) 23,100. The markings were made by American Indians, whose culture dates from the period 200 BC to AD 600, and their function is thought to be ritual rather than astronomical. 98 C6

Ndjamena (formerly **Fort Lamy** until 1973), capital of ▶Chad, in the southwest of the country, at the confluence of the Chari and Logone rivers, on the Cameroon border; population (1993) 531,000. Industries include cotton, textiles, and meat packing. The city's agricultural markets trade in livestock, salt, dates, and grain. The Great Mosque built 1974–78 is an important landmark. It is a staging point for many pilgrims from West Africa to Mecca. 64 B5

Ndola mining centre and capital of Western Province in the copperbelt region of central Zambia; population (1995 est) 466,000. It lies 275 km/171 mi north of Lusaka, the capital of Zambia, and is the commercial centre of the copperbelt, with good communications by rail and air. There is a copper refinery, a cobalt treatment plant, an oil refinery, a sugar refinery, and sawmills. 72 D2

Neagh, Lough lake in Northern Ireland, 25 km/15 mi west of Belfast; area 396 sq km/153 sq mi. It is the largest lake in the British Isles, being 27 km/17 mi long, 16 km/10 mi wide, with an average depth of 12 m/39 ft. The shores are mostly flat and marshy; there are a few islands of which Ram's Island is the largest, on which is an early round tower. The lake is famous for trout and eel fishing, and breeding waterbirds. 24 F7

Near East term used until the 1940s to describe the area of the Balkan states, Egypt, and southwest Asia, now known as the ▶Middle East.

Neath Port Talbot unitary authority in south Wales, created in 1996 from part of the former county of West Glamorgan.
area 442 sq km/171 sq mi
towns ▶Port Talbot (administrative headquarters)
physical the terrain is dominated by the alternation of river valleys and high moorland interfluves
features Roman fort of Nidum is near Neath
industries coal mining, chemicals, various metalworks, variety of light industry
population (1996) 139,400 5 E6

Nebraska state in central USA. It is nicknamed the Cornhusker State or the Blackwater State. Nebraska was admitted to the Union in 1867 as the 37th US state. Pasrt of the Midwest, and on of the Great Plains states, it is bordered to the west by Wyoming, to the north by South Dakota, to the east by Iowa and Missouri, to the south by Kansas, and to the southwest by Colorado.
population (2000 est) 1,711,300
area 200,400 sq km/77,354 sq mi
capital ▶Lincoln
towns and cities Omaha, Grand Island, North Platte
industries and products cereals, livestock, processed foods, fertilizers, oil, natural gas, finance sector 86 G2

Neckar German river, a tributary of the ▶Rhine; length 377 km/234 mi. From its source on the eastern slopes of the Black Forest, the Neckar winds northwest past Tübingen, Stuttgart, Heilbronn, and Heidelberg, to join the Rhine at Mannheim. It is navigable as far as Plochingen. 20 D7

Needles, the group of rocks in the sea, rising to 30 m/100 ft, off the western extremity of the Isle of Wight, southern England.

Negeri Sembilan state of south Peninsular Malaysia; area 6,646 sq km/2,565 sq mi; population (1993) 723,900. It is mainly mountainous; products include rice and rubber. The capital is Seremban. 11 (1)A2

Negev triangular desert region in southern Israel that tapers to the port of Elat, 120 km/75 mi wide at Beersheba, 13 km/8 mi at Elat; area 12,215 sq km/4,716 sq mi. It is fertile under irrigation, and minerals include oil and copper. 60 B6

Negro, Río river in South America, rising in eastern Colombia and joining the Amazon at Manáus, Brazil; length 2,250 km/1,400 mi. 98 E4

Nei Monggol Chinese name for Inner ▶Mongolia, an autonomous region of north China. 11 (2)E2

Neisse alternative name for ▶Nysa, a river and an industrial town in Poland. 18 D6/18 G7

Nepal landlocked country in the Himalayan mountain range in Central Asia, bounded north by Tibet (an autonomous region of China), east, south, and west by India.

Nepal covers an area of 147,181 sq km/56,826 sq mi, and has a population of 23,930,000 (2000 est). The capital is ▶Kathmandu; other major towns and cities include Moráng, Biratnagar, Lalitpur, Bhaktapur, Pokhara, Birganj, and Dharān. The head of state is King Gyanendra Bir Bikram Shah Dev (following the fatal shooting of King Birendra and nine other members of the Royal Family by Prince Dipendra — who died from self-inflicted wounds) (2001); the head of government is Girja Prasad Koirala (from 2000). The currency is the Nepalese rupee, and the languages spoken include Nepali (official), Tibetan, and numerous local languages. 54 D3

Ness, Loch lake in the Highland unitary authority, Scotland, extending northeast to southwest. Forming part of the Caledonian Canal, it is 36 km/22.5 mi long, 2 km/1 mi wide (on average), 229 m/754 ft deep, and is the greatest expanse of fresh water in Europe. There have been unconfirmed reports of a Loch Ness monster since the 6th century. 24 H4

Netherlands Antilles two groups of Caribbean islands, overseas territories of the Netherlands with full internal autonomy, comprising Curaçao and Bonaire off the coast of Venezuela (▶Aruba is considered separately), and St Eustatius, Saba, and the southern part of St Maarten in the Leeward Islands, 800 km/500 mi to the northeast; area 797 sq km/308 sq mi; population (1993 est) 197,100. The capital is Willemstad on Curaçao. Oil from Venezuela is refined here. Other industries include rum, some manufacturing, and tourism is also important. Dutch is the official language, and Papiamento and English are also spoken. 94 L6

Netherlands East Indies former name (1798–1945) of ▶Indonesia. 52 D3

Netherlands, The country in Western Europe on the North Sea, bounded east by Germany and south by Belgium.

The Netherlands covers an area of 41,863 sq km/16,163 sq mi, and has a population of 15,786,000 (1999 est). The capitals are ▶Amsterdam and The ▶Hague (legislative and judicial); other major towns and cities include Rotterdam, Utrecht, Eindhoven, Groningen, Tilburg, Maastricht, Apeldoorn, Nijmegen, and Breda. The head of state is Queen Beatrix Wilhelmina Armgard (from 1980); the head of government is Wim Kok (from 1994). The currency is the guilder, and the main language spoken is Dutch (official). 22 H2

Netzahualcóyotl Mexican city lying to the south of Lake Texcoco, forming a suburb to the northeast of Mexico City; population (1990) 1,255,500.

Neubrandenburg town in the *Land* (administrative region) of ►Brandenburg, Germany, on the River Tollense, 113 km/70 mi north of Berlin; population (1995) 82,000. There are engineering, chemical, and paper industries in the town. Between 1359 and 1471 Neubrandenburg was capital of the Duchy of Mecklenburg-Stargard. It has an 18th-century palace and medieval churches and walls. **20** J3

Neuchâtel canton in the west of Switzerland, between Lake Neuchâtel and the French frontier; area 797 sq km/308 sq mi; population (1995) 165,300. Neuchâtel lies in the midst of the ►Jura Mountains, four ranges of which traverse the canton, running northeast–southwest. The speciality of the canton is watchmaking; the chief centres of this industry are La Chaux-de-Fonds and Le Locle. Neuchâtel joined the Swiss Confederation in 1815. **8** (1)A2

Neuchâtel (German **Neuenburg**), capital of Neuchâtel canton in northwest Switzerland, on the northwest shore of Lake Neuchâtel, 40 km/25 mi west of Berne; population (1995) 38,200. It has watch, chocolate, and wine industries. **30** B3

Neumünster city in Schleswig-Holstein, Germany, 31 km/19 mi southwest of Kiel; population (1994) 82,000. Metallurgy, electrical equipment, textiles and clothing are the main industries. The town was founded in 1142. **20** F2

Neunkirchen city in the Saarland, Germany, on the River Blies, 19 km/12 mi northeast of Saarbrücken; population (1994) 150,200. It is a coal-mining centre and has important iron and steel, chemical, and textile industries. **20** C7

Neunkirchen am Steinfeld Austrian town in the province of Lower Austria, on the River Schwarza, 55 km/34 mi southwest of Vienna; population (1995) 15,600. The town has iron, and cotton and silk textile industries. **30** M3

Neusiedler See (Hungarian **Fertö Tó**), shallow lake mainly in Burgenland, Austria, southeast of Vienna, and in the Gyor-Sopron district of northwest Hungary; area 152 sq km/60 sq mi. It is slightly saline, shallow, and has a large amount of reeds. It is the only steppe lake in Europe. The size varies according to the seasons. **30** M3

Neusohl German name for ►Banská Bystrica, a town in the Slovak Republic. **18** J9

Neuss industrial city in North Rhine-Westphalia, Germany; population (1995) 148,600. It is a port opposite Düsseldorf on the left bank of the River Rhine. The city contains the minster of St Quirinus (13th century), one of the most magnificent Late Romanesque churches in the Rhinelands. **22** J3

Neustadt an der Weinstrasse (formerly **Neustadt an der Haardt**), city in the Rhineland-Palatinate, Germany, situated at the foot of the Haardtberge, 71 km/44 mi southwest of Mainz; population (1995) 53,800. The wine-trading centre of the largest wine-producing region in Germany, Neustadt also manufactures concrete, metal goods, paper, and textiles. There is a 14th-century church. **22** L5

Neustrelitz town in the *Land* (administrative region) of ►Mecklenburg–West Pomerania, Germany, 24 km/15 mi south of Neubrandenburg; population (1995) 24,700. It was formerly the capital of the Grand Duchy of Mecklenburg. Neustrelitz has a number of engineering industries. **20** J3

Nevada state in western USA. It is nicknamed the Silver State or the Sagebrush State. Nevada was admitted to the Union in 1864 as the 36th US state. It is famous as a gambling centre and, historically, especially in Reno, as a state in which marriages and divorces could be expedited. Nevada is bordered to the east by Utah and Arizona, to the west and southwest by California, and to the north by Oregon and Idaho.
population (2000 est) 1,998,300
area 286,400 sq km/110,550 sq mi
capital ►Carson City
towns and cities Las Vegas, Reno
industries and products mercury, barite, gold; tourism and gambling now generate more than half of the state's income **86** C3

Nevers industrial city and administrative centre of the *département* of ►Nièvre in Burgundy, central France, at the meeting of the Loire and Nièvre rivers; population (1990) 43,900. It was the capital of the former province of Nivernais. Nevers has foundries, printing works, and electrical industries; it was formerly noted for the manufacture of chinaware. **26** J7

New Amsterdam (Dutch **Nieuw Amsterdam**), original name of ►New York City, USA. It was a commercial settlement, established in 1624 as the capital of the Dutch colony of New Netherlands. After its capture by the British in 1664, it was renamed New York. **88** F2

New Amsterdam port of northeast Guyana, on the east bank of the River Berbice, near its mouth on the Atlantic coast 90 km/56 mi southeast of Georgetown; population (1992) 17,700. It is connected by road to the capital, Georgetown, and to Rosignol by ferry across the Berbice River. It is the commercial and industrial centre for the agricultural lowlands of the coastal region, producing rice and sugar cane. **98** F2

Newark market town in Nottinghamshire, central England, on the River Trent, 25 km/16 mi west of Lincoln; population (1996 est) 24,775. Agriculture is important in the area and there is a cattle market. Industries include engineering, the manufacture of ball and roller bearings, plasterboard and gypsum, and the processing of sugar-beet. There are also some financial services based here. Features include the ruins of a 12th-century castle in which King John died in 1216. **22** B1

Newark largest city and port of New Jersey, USA, on Newark Bay, 15 km/9 mi west of lower Manhattan; population (1996 est) 268,500. It is a commercial, financial, and industrial centre, with an international airport. Electrical equipment, machinery, chemicals, paints, beer, and canned meats are produced. Port Newark, a deep-water facility under the New York Port Authority, handles some aspects of New York harbour business. It was founded as Milford in 1666 by Puritans. The city is the seat of Essex county. **88** F2

New Bedford city and fishing port in Bristol County, southeast Massachusetts, USA, on the Acushnet River near Buzzards Bay, 50 mi/80 km south of Boston; population (2000 est) 93,800. Industries include electronics, the manufacture of cotton textiles, clothing, rubber, and metal products, and deep-sea fishing. It was incorporated as a town in 1787, and as a city in 1847. 88 F2

New Britain largest island in the ▶Bismarck Archipelago, part of Papua New Guinea; capital ▶Rabaul; population (1995 est) 285,000. It has an area of 37,800 sq km/14,600 sq mi, an average width of 80 km/50 mi and is 482 km/300 mi long. The highest mountain is Mount Sinewit, 2,438 m/ 7,999 ft. Copra is the chief product; coffee, cocoa, palm oil, timber, and iron are also produced. Gold, copper, and coal are mined. The population is Melanesian. 74 F6

New Brunswick largest maritime province of eastern Canada; area 73,400 sq km/28,332 sq mi; population (1991) 762,500; 33% French-speaking; 52% rural inhabitants. It is bounded on the north by Québec, with the Matapédia and Restigouche rivers forming part of the border; in the northeast Chaleur Bay separates New Brunswick's north shore from Québec's Gaspé Peninsula. Off its eastern coast is the Gulf of St Lawrence and in the southeast the Northumberland Strait, on the far side of which lies Prince Edward Island. Nova Scotia province is situated to its south and southeast, across the Bay of Fundy and the narrow land bridge known as the Chignecto Isthmus. To the southwest lies the US state of Maine, with the Saint John and Saint Croix rivers forming parts of the boundary. The capital of the province is ▶Fredericton, and Saint John and Moncton are other towns. 82 T7

Newbury market town and administrative headquarters of ▶West Berkshire unitary authority in southern England, on the River Kennet, 27 km/17 mi southwest of Reading; population (1991) 33,300. It was part of the county of Berkshire until April 1998. Industries include papermaking, electronics, engineering, and the manufacture of chemicals and plastics; wheat and barley crops are also important to the economy. Newbury has a racecourse and training stables. 22 A3

New Caledonia (French **Nouvelle Calédonie**), island group in the South Pacific, a French overseas territory between Australia and the Fiji Islands; area 18,576 sq km/7,170 sq mi; population (1997 est) 191,000 (45% Kanak (Melanesian), 34% European, 9% Wallisian, 4% Vietnamese and Indonesian, 8% Polynesian). The capital is ▶Nouméa. The islands are surrounded by a barrier reef, and are the world's third-largest producer of nickel. Other industries include chrome, iron, chlorine, oxygen, and cement works; agricultural products include beef, pork, coffee, and maize; and tourism is also an important industry. The currency used is the CFP franc. French is the official language, although English is widely spoken. 60% of the population are Roman Catholic, and 30% are Protestant. 74 G8

Newcastle industrial city and port in New South Wales, Australia, on the Hunter River, 157 km/98 mi north of Sydney; population (1996) 270,324. Newcastle is the second-largest city in the state after Sydney, and is dependent mainly on coalmining and alumina production. Other industries include shipbuilding and the manufacture of electronic equipment, textiles, chemicals, fertilizers, and wine products. Newcastle has a university (1965). 76 K6

Newcastle-under-Lyme market and industrial town in Staffordshire, west-central England, on the River Lyme, 3 km/2 mi west of Stoke-on-Trent; population (1996) 60,000. Manufacturing is important and products include electronics, ceramics, bricks and tiles, clothing, paper, and machinery. Silverdale Colliery was closed in 1998. 24 K8

Newcastle upon Tyne city and metropolitan borough in Tyne and Wear in northeast England on the River Tyne opposite Gateshead, 17 km/10 mi from the North Sea; population (1999 est) 273,000. It is the administrative centre of Tyne and Wear and a centre for retail, commerce, communications, and the arts. Newcastle first began to trade in coal in the 13th century, and was an important centre for coal and shipbuilding until the 1980s. Other industries include engineering (including offshore technology), food processing, brewing, and the manufacture of electronics. Only 1% of the workforce is now in heavy industry, 80% are in the public or service sectors. 24 L6

New Delhi capital of India, situated in the north of the country on the Yamuna River in the Union Territory of ▶Delhi; population (1991) 301,000. It lies near the old city of ▶Delhi, some 5 km/3 mi south of the Red Fort. Predominantly an administrative centre, it also produces chemicals, textiles, machine tools, electrical goods, and footwear. 54 C3

New England region of northeast USA, comprising the states of Maine, New Hampshire, Vermont, Massachusetts, Rhode Island, and Connecticut. It is a geographic region rather than a political entity, with an area of 172,681 sq km/66,672 sq mi. Boston is the principal urban centre of the region, and Harvard and Yale are its major universities. First inhabited by the American Indian Algonquin peoples, New England was named by the explorer John Smith in 1614, and settled by Pilgrims and Puritans from England in the 17th century.

New Forest ancient forest in southwest Hampshire, southern England, and the largest stretch of semi-natural vegetation in lowland Britain. Lying between the River Avon on the west and Southampton Water on the east, its legal boundary encloses 38,000 ha/93,898 acres (1995). Of this area 8,400 ha/20,756 acres is enclosed plantation, and 20,000 ha/ 49,420 acres is common land, including ancient woodland, heath, grassland, and bog. The remainder is privately owned land and villages. More than six million tourists visit annually.

Newfoundland Canadian province on the Atlantic Ocean, the country's most easterly administrative region, comprising the island of ▶Newfoundland and mainland ▶Labrador, separated by the Strait of Belle Isle; area 405,700 sq km/156,600 sq mi; population (1996 est) 570,700. It is bounded on the west by Québec, while to the southwest lie the Gulf of St Lawrence and the provinces of Nova Scotia and Prince Edward Island. The capital is ▶St John's, and other towns and cities are Corner Brook, Gander, and Goose Bay (Labrador). Industries include offshore oil extraction; fishing and fish-processing; mining (iron, copper, zinc, and uranium); wood-processing and paper manufacture; and hydroelectric power generation. 82 V7

New Glasgow town in Pictou County, north-central Nova Scotia,

Canada; population (1991) 9,900. It is situated on East River, 16 km/10 mi southeast of the seaport of Pictou and 126 km/78 mi northeast of Halifax. New Glasgow was a coal, steel, and shipbuilding centre in the 19th century, and built railway cars early in the 20th century. It continues to produce brick and tile, tires, furniture, pulp, and wood products, and is a regional commercial centre. Some working coalmines remain in the surrounding area. **82** U7

New Guinea island in the southwest Pacific, north of Australia, comprising ▶Papua New Guinea and the Indonesian province of ▶Irian Jaya; total area about 885,780 sq km/ 342,000 sq mi. Part of the Dutch East Indies from 1828, West Irian was ceded by the United Nations to Indonesia in 1963 and renamed Irian Jaya ('victorious Irian') in 1973. **40** S10

New Hampshire state in northeastern USA. It is nicknamed the Granite State. New Hampshire ratified the US Constitution in 1788, becoming the 9th state to join the Union. Part of New England, it is bordered to the east by Maine, to the north by Québec, Canada, to the west by Vermont, and to the south by Massachusetts.
population (2000 est) 1,235,800
area 24,000 sq km/9,264 sq mi
capital ▶Concord
towns and cities Manchester, Nashua
industries and products dairy, poultry, fruits and vegetables, electrical and other machinery, pulp and paper, tourism, leisure industry **88** F2

Newhaven port in East Sussex, southeast England, at the mouth of the River Ouse, with container facilities and cross-Channel services to Dieppe, France; population (est 1996) 10,700. Industries include the manufacture of fountain-pens and some light engineering. **22** C4

New Haven city and port in Connecticut, USA, at the mouth of the Quinnipiac River on Long Island Sound; population (1996 est) 124,700. Metal goods, rubber goods, ammunition, and clothing are manufactured. Yale University, third oldest in the USA, was founded here in 1701 and named after Elihu Yale (1648–1721), an early benefactor. New Haven was settled in 1638 by English Protestants from Massachusetts. **88** F2

New Hebrides former name (to 1980) of ▶Vanuatu, a country in the South Pacific. **74** G7

New Jersey state in northeastern USA. It is nicknamed the Garden State. New Jersey ratified the US Constitution in 1787, becoming the 3rd state to join the Union. The most densely populated US state, it lies within the suburban orbits of the cities of New York and Philadelphia. The New Jersey Turnpike, traversing a zone of heavy development, is the busiest highway in the country. New Jersey is bordered to the north by New York State. It lies largely between two rivers, the Hudson to the east and the Delaware to the west, with the Atlantic Ocean along its southeastern and southern boundaries. New York faces it across the Hudson River, Pennsylvania lies to the west, across the Delaware River, and to the southwest, across the river's mouth and Delaware Bay, is the state of Delaware.
population (2000 est) 8,414,400
area 20,200 sq km/7,797 sq mi
capital ▶*Trenton*
towns and cities Newark, Jersey City, Paterson, Elizabeth
industries and products fruits and vegetables, fish and shellfish, chemicals, pharmaceuticals, soaps and cleansers, transport equipment, petroleum refining, research centres, finance sector, tourism **88** F2

New London port city in New London County, southeast Connecticut, USA, on Long Island Sound at the mouth of the Thames River; population (1990) 28,500. Industries include submarine production, pharmaceuticals, and chemicals. Tourism is also important. New London is the home of the US Coast Guard Academy (1876) and Connecticut College (1911). **88** F2

Newmarket town in Suffolk, eastern England, 21 km/13 mi northeast of Cambridge; population (est 1996) 17,100. A centre for horse racing since the reign of James I, it is the headquarters of the Jockey Club and the National Stud and site of the National Horseracing Museum (1983). There are two racecourses, the July course and the Rowley Mile Racecourse, both owned by the Jockey Club, and lying to the southwest. Approximately a fifth of the town's working population is employed in the racing industry, including veterinary services. Other industries include the manufacture of electronic equipment and agricultural machinery, photographic processing, and light engineering. **22** C2

New Mexico state in southwestern USA. It is nicknamed the Land of Enchantment. New Mexico was admitted to the Union in 1912 as the 47th US state. One of the Mountain States, it is bordered to the east by Texas and the Oklahoma panhandle, and to the north by Colorado. In the northwest, at the 'Four Corners', it meets Colorado, Utah, and Arizona. Arizona also lies along its western border. To the south, it is bordered by the Mexican states of Sonora and Chihuahua, and by the Trans-Pecos section of Texas.
population (2000 est) 1,819,000
area 315,000 sq km/121,590 sq mi
capital ▶Santa Fe
towns and cities Albuquerque, Las Cruces, Roswell
industries and products uranium, potash, copper, oil, natural gas, petroleum and coal products, sheep farming, cotton, pecans, vegetables, chilli peppers, tourism **92** E2

New Orleans (called 'the Big Easy'), city and river port in southeast Louisiana, USA, on the Mississippi River, and the Gulf of Mexico; population (1996 est) 476,600; metropolitan area (1992) 1,303,000. It is a commercial and manufacturing centre with shipbuilding, oil-refining, and petrochemical industries. Tourism is a major activity. New Orleans is regarded as the traditional birthplace of jazz, believed to have developed from the singing and voodoo rhythms of the weekly slave gatherings in Congo Square, during the 18th and 19th centuries. The city was founded by the French in 1718. **90** D4

New Plymouth port on the west coast of North Island, New Zealand, 257 km/160 mi southwest of Auckland, capital of the provincial district of Taranaki; population (1996) 49,100. It lies at the centre of a dairy-farming region. The Taranaki gas fields at Kapuni are nearby. **78** E4

Newport unitary authority in south Wales, created in 1996 from part of the former county of Gwent.
area 190 sq km/73 sq mi

towns ▶Newport (administrative headquarters)
physical rivers Usk Ebbw, Afon Llwyd
features Legionary Museum and Roman amphitheatre at Caerleon
industries steel and aluminium production, engineering, chemicals, fertilizers, electronics
population (1996) 133,300 **5** E6

Newport river port, administrative headquarters of the ▶Isle of Wight unitary authority, southern England, on the River Medina; population (1991) 25,000. Newport is the retail centre for the island, and Parkhurst Prison and Camp Hill Prison, in the nearby residential district of Parkhurst, are important sources of employment. Products include electronic current boards and computer parts. Charles I was imprisoned 1647–1648 in nearby Carisbrooke Castle. **22** A4

Newport (Welsh **Casnewydd**), seaport and administrative centre of ▶Newport unitary authority, southeast Wales, situated on the River Usk 30 km/19 mi northwest of Bristol; population (1994 est) 111,000. There is a steelworks at nearby Llanwern, and a high-tech complex at Cleppa Park. Other industries include engineering, and the manufacture of chemicals, fertilizers, aluminium, and electronics. **24** K10

New Providence principal island of the Bahamas in the West Indies; area 150 sq km/58 sq mi; population (1990) 171,500. The island is about 34 km/21 mi long from east to west, and 11 km/7 mi wide from north to south. The island's principal city is Nassau, the capital and chief port of the Bahamas. The local economy centres on agriculture, fishing, rum, and tourism. **90** F5

New South Wales state of southeast Australia, including the dependency of ▶Lord Howe Island; area 801,600 sq km/309,500 sq mi; population (1996) 6,038,700 (about 54% in the capital, ▶Sydney). The state is bounded by Queensland on the north, the Tasman Sea on the east, Victoria on the south, and South Australia on the west. **76** H6

New World the Americas, so called by the first Europeans who reached them. The term also describes animals and plants of the Western hemisphere.

New York the most populous city in the USA, located on an inlet of the Atlantic Ocean in the far southeastern corner of ▶New York State; population (2000 est, excluding suburban metropolitan areas under separate administration) 8,008,300. New York is composed of five city boroughs that are also counties of New York State: ▶Manhattan (New York County); the ▶Bronx (Bronx County); ▶Queens (Queens County); ▶Brooklyn (Kings County); and ▶Staten Island (Richmond County). As well as being the main port in North America, New York is one of the world's principal commercial and cultural centres. The many industries and services operating here include banking and other financial activities, publishing and printing, the electronic media, advertising, clothing manufacture and the fashion industry, and the production of food, chemicals, machinery, and textiles. With its great diversity of cultural institutions, places of entertainment, and sightseeing opportunities, the city also attracts a large number of tourists each year. New York is also known as the 'Big Apple'. **88** F2

New York state in northeast USA. It is nicknamed the Empire State. New York ratified the US Constitution in 1788, becoming the 11th state to join the Union. It is bordered to the north by the Canadian states of Québec and Ontario, to the east by Vermont, Massachusetts, and Connecticut, and to the south by Pennsylvania and New Jersey. In the west is a strip of Pennsylvania that reaches north to Lake Erie. New York has shores on both Lakes Erie and Lake Ontario, between which it faces part of the state of Ontario across the Niagara River. In the southeast it faces New Jersey across the lower Hudson River. Long Island extends east from New York Bay into the Atlantic Ocean, with Connecticut and Rhode Island lying to the north across Long Island Sound.
population (2000 est) 18,976,500
area 127,200 sq km/49,099 sq mi
capital ▶Albany
towns and cities New York City, Buffalo, Rochester, Yonkers, Syracuse
industries and products dairy products, apples, clothing, periodical and book printing and publishing, electronic components and accessories, office machines and computers, communications equipment, motor vehicles and equipment, pharmaceuticals, aircraft and parts, finance sector, tourism **88** E2

New Zealand (or **Aotearoa**; Maori 'long daylight'), country in the southwest Pacific Ocean, southeast of Australia, comprising two main islands, North Island and South Island, and other small islands.

New Zealand covers an area of 268,680 sq km/103,737 sq mi, and has a population of 3,862,000 (2000 est). The capital is ▶Wellington; other major towns and cities include Auckland, Hamilton, Christchurch, Manukau, North Shore, and Waitakere. The head of state is Queen Elizabeth II (from 1952), represented by Governor General Catherine Tizard (from 1990); the head of government is Helen Clark (from 1999). The currency is the New Zealand dollar, and the languages spoken include English (official) and Maori. **78** B5

Niagara Falls city in Niagara County, west New York State, USA, on the east bank of the Niagara River at the Niagara Falls, 30 km/19 mi northwest of Buffalo; population (1990) 61,800. It is linked with the city of Niagara Falls, Canada, by the Rainbow Bridge (1941). Niagara Falls has chemical, engineering, aircraft, and metallurgical industries. It has an important hydroelectric power plant and it is a major tourist centre. It is the home of Niagara University. **88** E2

Niagara Falls two waterfalls on the Niagara River, on the Canada–USA border, between lakes Erie and Ontario and separated by Goat Island. The **American Falls** are 51 m/167 ft high, 330 m/1,080 ft wide; **Horseshoe Falls**, in Canada, are 49 m/160 ft high, 790 m/2,600 ft across.

Niamey port, capital, and administrative centre of ▶Niger, in the southwest of the country, on the northeast bank of the Niger River; population (1995 est) 495,000. It produces textiles, chemicals, pharmaceuticals, ceramics, plastics, and foodstuffs. It replaced Zinder as the capital in 1926. It has an international airport and railway terminus. **68** E2

Nicaragua country in Central America, between the Pacific Ocean and the Caribbean Sea, bounded north by Honduras and south by Costa Rica.

Nicaragua covers an area of 127,849 sq km/49,362 sq mi, and has a population of 5,074,000 (2000 est). The capital is ►Managua; other major towns and cities include León, Chinandega, Masaya, Granada, and Estelí. The head of state and government is Arnoldo Aleman (from 1997). The currency is the cordoba, and the languages spoken include Spanish (official), English, and American Indian languages. **94** G6

Nicaragua, Lake lake in southwest Nicaragua, the largest in Central America; area 8,250 sq km/3,185 sq mi. It is 24 km/15 mi from the Pacific Ocean and drains into the Caribbean at San Juan del Norte via the Río San Juan. It is a freshwater lake inhabited by a variety of salt-water fish including shark and swordfish. The lake contains about 310 small islands, most of which are inhabited. The largest island, Ometepe, has two volcanoes: Concepción (1,610 m/ 5,282 ft), and Madera (1,394 m/4,573 ft). Lake Managua to the north drains into Lake Nicaragua via the River Tipitapa. The area is ecologically important for many rare and colourful birds. **94** G6

Nice city on the French ►Riviera and administrative centre of the *département* of ►Alpes-Maritimes, situated at the mouth of the River Paillon on the Baie des Anges, near the Italian frontier; population (1999 est) 345,900. Founded in the 3rd century BC, it repeatedly changed hands between France and the Duchy of Savoy from the 14th to the 19th century. In 1860 it was finally transferred to France after a plebiscite. Nice is the fifth-largest city in France. Nice is situated on the Cote d'Azur which is a centre for information technology industries and has a world-renowned reputation. Employment is concentrated in business and services and 83% of the labour force in Nice is involved in the service industry. **30** C7

Nicobar Islands group of Indian islands, part of the Union Territory of ►Andaman and Nicobar Islands. **50** A5

Nicosia (Greek **Lefkosia**; Turkish **Lefkosha**), capital of Cyprus, with leather, textile, and pottery industries; population (1993) 177,000. Nicosia was the residence of Lusignan kings of Cyprus 1192–1475. The Venetians, who took Cyprus in 1489, surrounded Nicosia with a high wall, which still exists; the city fell to the Turks in 1571. It was again partly taken by the Turks in the invasion in 1974. **36** R9

Niederösterreich German name for the federal state of ►Lower Austria. **8** (2)G2

Niedersachsen German name for the federal state of ►Lower Saxony, Germany. **7** (2)B2

Nieuwpoort (French **Nieuport**), port in the Belgian province of West Flanders, 34 km/21 mi west of Bruges; population (1991) 9,600. It stands on the River Yser (Ijzer), about 3 km/2 mi above its mouth, and at the junction of several canals. **22** E3

Nièvre *département* in the ►Burgundy region of central France; area 6,836 sq km/2,639 sq mi; population (1990) 233,528. It contains the basins of the rivers ►Loire and ►Seine. Forests cover large areas of Nièvre; farmland is largely devoted to rearing cattle, sheep, and horses. The chief crops are oats, wheat, and potatoes. Vines are grown in the Loire valley and in the neighbourhood of Clamecy; the white wines of Pouilly (notably, Pouilly Fumé) are famous. Coal is mined around Decize, and the chief manufactures are iron and steel, at Guérigny, Fourchambault, and Imphy. The principal towns are ►Nevers (the capital), Château-Chinon , Clamecy, and Cosne. **6** E3

Niger landlocked country in northwest Africa, bounded north by Algeria and Libya, east by Chad, south by Nigeria and Benin, and west by Burkina Faso and Mali.

Niger covers an area of 1,186,408 sq km/458,072 sq mi, and has a population of 10,730,000 (2000 est). The capital is ►Niamey; other major towns and cities include Zinder, Maradi, Tahoua, Agadez, Birui N'Konui, and Arlit. The head of state is Tandja Mamadou (from 1999); the head of government is Hama Amadou (from 2000). The currency is the franc CFA, and the languages spoken include French (official), Hausa, Djerma, and other ethnic languages. **66** G5

Niger (Semitic **Nihal**), third-longest river in Africa, 4,185 km/2,600 mi. It rises in the highlands bordering Sierra Leone and Guinea, flows northeast through Mali, then southeast through Niger and Nigeria to an inland delta on the Gulf of Guinea. Its total catchment area is 1.5 million sq km/579,150 sq mi. The flow is sluggish and the river frequently floods its banks. It was navigated by the Scottish Mungo Park 1795–1806, who was drowned in the river near Bussa. **68** E2

Nigeria country in west Africa on the Gulf of Guinea, bounded north by Niger, east by Chad and Cameroon, and west by Benin.

Nigeria covers an area of 923,773 sq km/356,668 sq mi, and has a population of 111,506,000 (2000 est). The capital is ►Abuja; other major towns and cities include Ibadan, Lagos, Ogbomosho, Kano, Oshogbo, Ilorin, Abeokuta, Zaria, and Port Harcourt. The head of state and government is Olusegun Obasanjo (from 1999). The currency is the naira, and the languages spoken include English, French (both official), Hausa, Ibo, and Yoruba. **68** F2

Niigata industrial port and seat of the prefectural government, on the Sea of Japan, in Chubu region, Honshu island, Japan, 255 km/158 mi northwest of Tokyo; population (1994) 479,000. It is the largest city on the Sea of Japan coast and the largest regional port. There are textile, metal, oil refining, and chemical industries. **48** K5

Nijmegen industrial city in Gelderland province, east Netherlands, on the Waal River, near the German border; population (1997) 147,200. Industries include the manufacture of electrical equipment, leather goods, cigars, and the iron compound Prussian blue. There are also breweries. The Roman **Noviomagus**, Nijmegen was a free city of the Holy Roman Empire and a member of the Hanseatic League. **22** H3

Nile (Arabic (Egypt) **Nahr en Nil**; Arabic (Sudan) **Bahr el Nîl**; Semitic *nihal* 'river'), river in Africa, the world's longest, 6,695 km/4,160 mi. The **Blue Nile** rises in Lake Tana, Ethiopia, the **White Nile** at Lake Victoria, and they join at Khartoum, Sudan. The river enters the Mediterranean Sea at a vast delta in northern Egypt. **64** F4

Nîmes administrative centre of ►Gard *département*, in the Languedoc-Roussillon region of southern France, 100 km/62 mi northwest of Marseille; population (1990) 133,600. Its

Roman remains include an amphitheatre dating from the 2nd century and the nearby aqueduct known as the Pont du Gard. The city gives its name to the cloth known as denim (*de Nîmes*). **26** K10

Ningbo (or **Ningpo**; formerly **Ninghsien**), port and open coastal city in Zhejiang province, east China, on the Yong River, 25 km/16 mi from its mouth; population (1994) 1,131,500. Industries include fishing, shipbuilding, oil-refining, and the manufacture of petrochemicals, high-tech equipment, and traditional handicrafts. Already a centre of foreign trade under the Tang dynasty (618–907), it was one of China's original treaty ports in 1842. **46** G5

Ningxia Hui Autonomous Region (or **Ningxia**; formerly **Ninghsia**), administrative area of northwest China, bounded to the north by Inner Mongolia, to the east by Shaanxi, and to the south by Gansu; area 66,400 sq km/25,600 sq mi; population (1996) 5,210,000; one-third are Hui (Chinese Muslims) and there is a large Mongolian population in the north. The capital is ▶Yinchuan. It is a desert plateau, and the Huang He River is in the area. Industries include coal and chemicals, and agricultural products include cereals and rice under irrigation, and animal herding. **11** (2)D3

Niort capital of the French *département* of Deux-Sèvres, on the River Sèvre-Niortaise, 125 km/78 mi southeast of Nantes; population (1990) 60,000. The main industries are the manufacture of footwear, gloves, and brushes; angelica is processed. Two towers of the 12th–13th-century fortress are preserved. **26** E7

Nipigon, Lake lake in southwest Ontario, Canada; area 4,848 sq km/1,872 sq mi. It has 1,300 km/808 mi of shoreline and lies 55 km/34 mi north of Lake Superior in a forest reserve of over 18,130 sq km/7,000 sq mi. It is 112 km/70 mi long and 80 km/50 mi wide, is very deep and contains over 1,000 islands. The Nipigon River flows out of the lake and flows into Lake Superior. It supplies hydroelectric power to the region. The lake is known for its scenery and its wild surroundings. **82** P6/7

Nippon (or **Nihon**), English transliteration of the Japanese name for ▶Japan. **48** L5

Niš town in Serbia, Yugoslavia, on the River Nisava; population (2001 est) 184,400. It is an important road and rail junction and has chemical and agricultural foodstuff industries. **34** J6

Niterói (or **Nictheroy**), port and municipality of the same name of southeast Brazil, on the eastern shore of Guanabara Bay, linked by a 14 km/9-mi bridge (the largest of its type in Brazil) and ferry with Rio de Janeiro on the opposite shore; population (1991) 436,200. It forms part of the municipal area of Rio de Janeiro. An administrative and commercial centre, its industries include ship-repair, manufacture of metal goods, textiles, and fish-processing. Founded in 1573, it was the former administrative capital 1834–1974 of Guanabara state until 1974 when the new state of Rio de Janeiro was formed. **100** N3

Niue coral island in the south Pacific, west of the Cook Islands; overseas territory of New Zealand; area 260 sq km/100 sq mi; population (1993) 2,300 (12,000 Niueans live in New Zealand). The chief village and port is Alofi. The island is characterized by a low water table, and only 11–12% of the land can be cultivated. Products include coconuts, passion fruit, honey, taro, handicrafts, and limes. Languages spoken are English and Niuean (official). **74** K7

Nizhnevartovsk city in Tyumen region, in the Russian Federation, in northern Siberia; population (1994) 241,000. The oilfield discovered in 1965 at nearby Lake Samotlor is one of the largest in the world, and there are natural gas fields to the north. Poor technology prevented their efficient exploitation until the 1990s. **38** Q2

Nizhniy Novgorod (formerly **Gorky** 1932–90), city and river port in the central Russian Federation, 375 km/233 mi east of Moscow; population (1990) 1,443,000. The city is situated at the confluence of the ▶Volga and Oka rivers, and is a major transportation centre; six railway lines converge here, and large quantities of freight and passengers are carried by river traffic. Motor vehicles, locomotives, aircraft, and ships are manufactured here, making its transport industry the largest in the Russian Federation. There are also diesel motor and machine-tool works, oil processing, glass, woodworking, and various light and food industries. An International Trade Fair is held annually in Nizhniy Novgorod. **38** H3

Noord Brabant Dutch name for ▶North Brabant, the largest province of the Netherlands. **7** (1)D3

Noord Holland Dutch name for ▶North Holland, a low-lying coastal province of the Netherlands. **7** (1)C2

Nord *département* in the ▶Nord-Pas-de-Calais region of France, bordered in the north by Belgium; area 5,738 sq km/2,215 sq mi; population (1990) 2,533,500. The Flanders Canal system connects the rivers Sambre, Esceaut, Scarpe, Lys, Deûle, and Yser. Nord has an important history as one of the major industrial regions of France, with textile and metallurgical plants. Wheat, oats, sugar beet, potatoes, and vegetables are grown in the irrigated soil. The principal towns include ▶Lille (the administrative centre), Roubaix, Tourcoing, Dunkirk, and Valenciennes. **6** E1

Nordhausen town in the *Land* (administrative region) of ▶Thuringia, Germany, in the foothills of the southern Harz Mountains, 79 km/49 mi northwest of Erfurt; population (1996) 47,000. The area has potash mines, and there are engineering, textile, and oil-refining industries. Nordhausen was a free city of the Holy Roman Empire 1253–1803. **20** F5

Nord Kapp Norwegian name for ▶North Cape, the most northerly point of Europe. **16** N1

Nord-Ostsee-Kanal German name for ▶Kiel Canal, a waterway in Germany connecting the Baltic with the North Sea. **20** E2

Nord-Pas-de-Calais region of northern France; area 12,400 sq km/4,800 sq mi; population (1990) 3,965,100. Its administrative centre is ▶Lille, and it consists of the *départements* of ▶Nord and ▶Pas-de-Calais. **6** D1

Nordrhein-Westfalen German name for ▶North Rhine-Westphalia, an administrative region in northwestern Germany. **7** (2)A3

Norfolk county of eastern England.
area 5,360 sq km/2,069 sq mi

towns and cities ►Norwich (administrative headquarters), King's Lynn, Great Yarmouth (ports); Cromer, Hunstanton (resorts)
physical low-lying with the Fens in the west and the ►Norfolk Broads in the east; rivers Bure, Ouse, Waveney, Yare
features Norfolk Broads; Halvergate Marshes wildlife area; Grime's Graves (Neolithic flint mines); shrine of Our Lady of Walsingham, a centre of pilgrimage; Blickling Hall (Jacobean); residence of Elizabeth II at Sandringham (built 1869–71)
agriculture cereals; fruit and vegetables (beans, sugar beets, swedes, turnips); turkeys, geese, cattle; fishing centred on Great Yarmouth
industries agricultural implements; boots and shoes; brewing and malting; offshore natural gas; tanning; there are flour mills and mustard works
population (1996) 777,000 5 G5

Norfolk Atlantic seaport in southeast Virginia, USA, at the mouth of the James and Elizabeth rivers on Chesapeake Bay; population (1996 est) 233,400. Situated on the south side of an extensive deep-water harbour at the southern end of the bay, it forms part of the Port of Hampton Roads with nearby ►Portsmouth and Newport News. It is the headquarters of the US Navy's Atlantic fleet, and the home of 22 other Navy commands. Coal is exported from the Appalachian fields. Industries include shipbuilding, chemicals, and vehicle assembly. Total tonnage handled in 1994 amounted to 41,524,940 tonnes. Norfolk was first laid out in 1682. 88 E3

Norfolk Broads area of interlinked shallow freshwater lakes in Norfolk, eastern England, between Norwich, Sea Palling, and Lowestoft. The area has about 200 km/125 mi of navigable waterways, and the region is a popular tourist destination for boating and fishing. 22 D2

Norfolk Island island in the Pacific Ocean under Australian administration; area 35 sq km/13 sq mi; population (1996) 2,200. It was inhabited by the Polynesians from about 1200 to 1600. Later uninhabited, it was visited by Captain James Cook in 1774, and was made a penal settlement in 1788. In 1856 it became the home of the descendants of the *Bounty* mutineers who were moved from Pitcairn Island. Some Pitcairn islanders still live on Norfolk Island. Tourism is important; exports include Norfolk pine seeds, palm seeds, cereals, vegetables, fruit, cattle, and poultry. 74 G8

Norilsk city in the Russian Federation, situated in the Krasnoyarsk region of northern Siberia, 80 km/50 mi east of the port of Dudinka on the upper ►Yenisey River; population (1990) 173,000. The world's northernmost industrial city, it was built by Gulag forced labour in 1935 to exploit the region's large mineral reserves. These include nickel, cobalt, platinum, selenium, tellurium, gold, and silver. 42 R4

Normandy (French **Normandie**), former duchy of northwest France now divided into two regions: ►Haute-Normandie and ►Basse-Normandie; area 29,900 sq km/11,544 sq mi; population (both parts, 1990) 3,146,500. Normandy was named after the Viking Norsemen (Normans) who conquered and settled in the area in the 9th century. As a French duchy it reached its peak under William the Conqueror and was renowned for its centres of learning established by Lanfranc and St Anselm. Normandy was united with England from 1100 to 1135. England and France fought over it during the Hundred Years' War, England finally losing it in 1449 to Charles VII. In World War II the Normandy beaches were the site of the Allied invasion on D-day, 6 June 1944.

Norrköping town and port in Sweden in the county of Östergötland, 180 km/112 mi southwest of Stockholm, on both banks of the River Motala; population (1995) 123,800. There are paper, textile, engineering, and timber mills as well as shipyards. 16 J7

Northallerton market town and administrative headquarters of ►North Yorkshire, England, between York and Darlington; population (est 1996) 18,000. Employment is largely in the service sector but manufacturing industries include flour-milling, trailer-manufacture and light engineering. 24 L7

North America third largest of the continents (including Greenland and Central America), and over twice the size of Europe.
area 24,000,000 sq km/9,400,000 sq mi
largest cities (population over 1 million) Mexico City, New York, Chicago, Toronto, Los Angeles, Montréal, Guadalajara, Monterrey, Philadelphia, Houston, Guatemala City, Vancouver, Detroit, San Diego, Dallas
physical occupying the northern part of the landmass of the Western hemisphere between the Arctic Ocean and the tropical southeast tip of the isthmus that joins Central America to South America; the northernmost point on the mainland is the tip of Boothia Peninsula in the Canadian Arctic; the northernmost point on adjacent islands is Cape Morris Jesup on Greenland; the most westerly point on the mainland is Cape Prince of Wales, Alaska; the most westerly point on adjacent islands is Attu Island in the Aleutians; the most easterly point on the mainland lies on the southeast coast of Labrador; the highest point is Mount McKinley, Alaska, 6,194 m/20,320 ft; the lowest point is Badwater in Death Valley −86 m/−282 ft.

Perhaps the most dominating characteristic is the western cordillera running parallel to the coast from Alaska to Panama; it is called the ►Rocky Mountains in the USA and Canada and its continuation into Mexico is called the ►Sierra Madre. The cordillera is a series of ranges divided by intermontane plateaus and takes up about one-third of the continental area.

To the east of the cordillera lie the Great Plains, the agricultural heartland of North America, which descend in a series of steps to the depressions occupied by the ►Great Lakes in the east and the Gulf of Mexico coastal lowlands in the southeast. The Plains are characterized by treeless expanses crossed by broad, shallow river valleys. To the north and east of the region lie the Laurentian Highlands of Canada, an ancient plateau or shield area. Glaciation has deeply affected its landscape. In the east are the Appalachian Mountains, flanked by the narrow coastal plain which widens further south. Erosion here has created a line of planed crests, or terraces, at altitudes between 300–1,200 m/985–3,935 ft. This has also formed a ridge-and-valley topography which was an early barrier

to continental penetration. The Fall Line is the abrupt junction of plateau and coastal plain in the east
features Lake Superior (the largest body of fresh water in the world); Grand Canyon on the Colorado River; Redwood National Park, California, has some of the world's tallest trees; San Andreas Fault, California; deserts: Death Valley, Mojave, Sonoran; rivers (over 1,600 km/1,000 mi) include Mississippi, Missouri, Mackenzie, Rio Grande, Yukon, Arkansas, Colorado, Saskatchewan-Bow, Columbia, Red, Peace, Snake
population (1990 est) 395 million, rising to an estimated 450 million by the year 2000; annual growth rate from 1980 to 1985: Canada 1.08%, USA 0.88%, Mexico 2.59%, Honduras 3.39%; the American Indian, Inuit, and Aleut peoples are now a minority within a population predominantly of European immigrant origin. Many Africans were brought in as part of the slave trade
language English predominates in Canada, the USA, and Belize; Spanish is the chief language of the countries of Latin America and a sizeable minority in the USA; French is spoken by about 25% of the population of Canada, and by people of the French département of St Pierre and Miquelon; indigenous non-European minorities, including the Inuit of Arctic Canada, the Aleuts of Alaska, North American Indians, and the Maya of Central America, have their own languages and dialects
religion Christian and Jewish religions predominate; 97% of Latin Americans, 47% of Canadians, and 21% of those living in the USA are Roman Catholic 80

Northampton market town and administrative headquarters of ►Northamptonshire, central England, on the River Nene, 108 km/67 mi northwest of London; population (1996 est) 192,400. The major employers are public administration, financial services, and the distribution trade. The manufacture of boots and shoes was historically important, but engineering has taken over as the key industry; other industries include food processing, brewing, and the manufacture of shoe machinery, cosmetics, leather goods, and car accessories. 22 B2

Northamptonshire county of central England.
area 2,370 sq km/915 sq mi
towns and cities ►Northampton (administrative headquarters), Kettering, Corby, Daventry, Wellingborough
physical rivers Avon, Cherwell, Leam, Nene, Ouse, and Welland
features Althorp Park, Spencer family home and burial place of Diana, Princess of Wales; Canons Ashby, Tudor house, home of the Drydens for 400 years; churches with broached spires (an octagonal spire on a square tower)
agriculture cereals (wheat and barley), sugar beet, sheep rearing; cattle rearing, especially in the Nene and Welland valleys, where there is rich pasture
industries engineering, food processing, printing, shoemaking; Northampton is the centre of the leather trade in England
population (1996) 604,300 5 F5

Northants abbreviation for ►Northamptonshire, an English county. 5 F5

North Ayrshire unitary authority in western Scotland, created in 1996 from Cunninghame district in Strathclyde region.
area 889 sq km/343 sq mi
towns ►Irvine (administrative headquarters), Kilwinning, Saltcoats, Largs, Kilbirnie
physical low-lying coastal plain on the mainland, rising inland to a plateau of over 305 m/1,000 ft; the islands of the Firth of Clyde are Arran, Holy Isle, Cumbraes; the rivers Irvine and Garnock reach the sea at Irvine; Goat Fell (874 m/2,867 ft)
features Pencil Monument, Largs; Scottish Maritime Museum, Irvine; Hunterston nuclear power station
industries chemicals, electronics, computer manufacturing
agriculture dairying, potatoes
population (1996) 139,200 5 D4

North Brabant (Dutch **Noord Brabant**), largest province of the Netherlands, located in the south of the country, lying between the Maas River (Meuse) and Belgium; area 4,940 sq km/1,907 sq mi; population (1997) 2,304,100. The capital is ►'s-Hertogenbosch. Industries include brewing, tobacco, engineering, micro-electronics, and textiles. There is cattle farming, and wheat and sugar beet are grown. 7 (1)D3

North Cape (Norwegian **Nord Kapp**), cape in the Norwegian county of Finnmark; the most northerly point of Europe. 16 N1

North Carolina state in eastern USA. It is nicknamed the Tar Heel State or Old North State. North Carolina ratified the US Constitution in 1789, becoming the 12th state to join the Union. It is bordered to the north by Virginia, to the west and northwest by Tennessee, to the south by Georgia and South Carolina, and to the east by the Atlantic Ocean.
population (2000 est) 8,049,300
area 136,400 sq km/52,650 sq mi
capital ►Raleigh
towns and cities Charlotte, Greensboro, Winston-Salem, Durham
industries and products tobacco, corn, soya beans, livestock, poultry, textiles, clothing, cigarettes, furniture, chemicals, machinery, tourism, finance sector, research 90 F2

Northd abbreviation for ►Northumberland, an English county. 5 E4

North Dakota state in the northern USA. It is nicknamed the Peace Garden State. North Dakota was admitted to the Union in 1889 as the 39th US state. Part of the Midwest, and one of the Great Plains states, it is bordered to the south by South Dakota, to the west by Montana, to the north by the Canadian states of Saskatchewan and Manitoba, and to the east by Minnesota. North Dakota remains the most rural of all the US states.
population (2000 est) 642,200
area 183,100 sq km/70,677 sq mi
capital ►Bismarck
towns and cities Fargo, Grand Forks, Minot
industries and products cereals, meat products, farm equipment, oil, coal, tourism 86 F1

North East Lincolnshire unitary authority in eastern England created in 1996 from part of the former county of Humberside.
area 192 sq km/74 sq mi
towns and cities ►Grimsby (administrative headquarters), Immingham, Cleethorpes, Humberston, New Waltham, Waltham, Healing, Laceby
features Humber Estuary forms east border of authority; River Freshney; Immingham Museum; National

Fishing Heritage Centre (Grimsby)
industries fishing and associated industries, docks and shipping services at Immingham and Grimsby, chemical manufacture, heavy engineering, marine engineering, oil refining, tourism (Cleethorpes)
population (1996) 164,000 5 G5

Northeast Passage sea route from the North Atlantic, around Asia, to the North Pacific, pioneered by the Swedish explorer Nils Nordenskjöld 1878–79 and developed by the USSR in settling Northern Siberia from 1935.

Northern Cape province of the Republic of South Africa from 1994, formerly part of Cape Province; area 363,389 sq km/140,305 sq mi; population (1995 est) 742,000. The capital is ▶Kimberley. Diamonds are mined in the province, and there are iron, manganese, asbestos, and cotton industries. 72 C5

Northern Ireland (or **Ulster**), constituent part of the United Kingdom, in the northeast of the island of Ireland; area 13,460 sq km/5,196 sq mi; population (2001 est) 1,727,900. It is comprised of six of the nine counties that form Ireland's northernmost province of ▶Ulster (Antrim, Armagh, Down, Fermanagh, Londonderry, and Tyrone) which are divided into 26 regional districts for administrative purposes.. The capital is ▶Belfast, and other major towns and cities include Londonderry, Enniskillen, Omagh, Newry, Armagh, and Coleraine. Geographical features are the Mourne Mountains, Belfast Lough, Lough Neagh, and the Giant's Causeway. Major industries include engineering, shipbuilding, aircraft components, textiles, processed foods (especially dairy and poultry products), rubber products, and chemicals. 24 E7

Northern Mariana Islands archipelago in the northwestern Pacific, with ▶Guam known collectively as the Mariana Islands; area 471 sq km/182 sq mi; population (1995 est) 47,200. The Northern Marianas are a commonwealth in union with the USA. The capital is Garapan on Saipan. The archipelago consists of 16 islands and atolls extending 560 km/350 mi north of Guam. The main language is English, and the principal religion is Roman Catholicism. 74 E4

Northern Province (formerly **Northern Transvaal**), province of the Republic of South Africa from 1994, formerly part of Transvaal; area 119,606 sq km/46,180 sq mi; population (1995 est) 5,397,200. The capital is Pietersburg. Diamonds are mined, and there are copper, asbestos, and iron industries. Tourism is important. Wheat, maize, tobacco, and groundnuts are grown. 72 D4

Northern Rhodesia former name (to 1964) of ▶Zambia, a country in Africa. 72 D2

Northern Territory territory of north-central Australia, bounded on the north by the Timor and Arafura seas, on the east by Queensland, on the south by South Australia, and on the west by Western Australia; area 1,346,200 sq km/519,770 sq mi; population (1996) 195,100. The capital is ▶Darwin. The main products are beef, bauxite, gold, copper, uranium, manganese, tropical fruits, and fish. Tourism is important. 76 F4

Northern Transvaal former name of ▶Northern Province, a province of the Republic of South Africa. 72 D4

North Holland (Dutch **Noord Holland**), low-lying coastal province of the Netherlands occupying the peninsula jutting northward between the North Sea and the IJsselmeer, bounded on the south by the provinces of South Holland and Utrecht, and includes the island of Texel to the north; area 2,670 sq km/1,031 sq mi; population (1997) 2,474,800. The capital is ▶Haarlem. There are iron and steel works in the province, and agriculture includes dairying and the growing of flower bulbs, grain, and vegetables. 7 (1)C2

North Island smaller of the two main islands of ▶New Zealand. 78 D3

North Korea country in East Asia, bounded northeast by Russia, north and northwest by China, east by the Sea of Japan, south by South Korea, and west by the Yellow Sea.

North Korea covers an area of 120,538 sq km/46,539 sq mi, and has a population of 24,039,000 (2000 est). The capital is ▶Pyongyang; other major towns and cities include Hamhung, Chongjin, Nampo, Wonsan, and Sinuiji. The head of state is Kim Jong Il (from 1994); the head of government is Hong Song Nam (from 1997). The currency is the won, and the main language spoken is Korean (official). 48 C4

North Lanarkshire unitary authority in central Scotland, created in 1996 from three districts of Strathclyde region.
area 475 sq km/183 sq mi
towns Airdrie, Coatbridge, Cumbernauld, Motherwell (administrative headquarters)
physical low-lying, heavily urbanized area; River Clyde
industries paper, pharmaceuticals, engineering, electronics, light manufacturing, food and drink processing
agriculture dairying (around urban environment)
population (1995) 326,700 5 E4

North Lincolnshire unitary authority in eastern England created in 1996 from part of the former county of Humberside.
area 850 sq km/328 sq mi
towns and cities ▶Scunthorpe (administrative headquarters), Brigg, Barton-upon-Humber, Barrow upon Humber, Epworth
features Humber Estuary; River Trent; Isle of Axholme; Stainforth and Keadby Canal; River Torne; Humber Bridge southern landfall at Barton upon Humber; Julian's Bower (near Alkborough) – medieval maze cut in turf; wetland nature reserves at Barton Waterside and Blackroft Sands
industries steelworks and manufacture of steel products, computer equipment and electronics, food processing (Golden Wonder)
population (1996) 153,000 5 F5

North Platte river in Colorado and Wyoming, USA; length 1,094 km/680 mi. In Nebraska it unites with the South Platte River to form the Platte River. The North Platte rises in north central Colorado in the Medicine Bow, Park, and Rabbits Ear ranges. It flows north into Wyoming, bending southeast before turning westwards into Nebraska. Its waters are used as part of a power, irrigation, and flood-control project. 84 F3

North Pole the northern point where an imaginary line penetrates the Earth's surface by the axis about which it revolves; see also ▶Arctic. **102** (1)

North Rhine-Westphalia (German **Nordrhein-Westfalen**), administrative region (German *Land*) in northwestern Germany, bounded to the north and northeast by Lower

Saxony, on the east by Hesse, on the south by the Rhineland-Palatinate, and on the west by Belgium and the Netherlands; area 34,100 sq km/ 13,200 sq mi; population (1995) 17,920,000. The capital is ►Düsseldorf. Industries include iron, steel, coal, lignite, electrical goods, fertilizers, and synthetic textiles. There is dairying, and cereals and vines are grown. 7 (2)A3

North Sea sea to the east of Britain and bounded by the coasts of Belgium, The Netherlands, Germany, Denmark, and Norway; part of the Atlantic Ocean; area 523,000 sq km/202,000 sq mi; average depth 55 m/180 ft, greatest depth 660 m/2,165 ft. The Dogger Bank extends east to west with shallows of as little as 11 m/36 ft, forming a traditionally well-stocked fishing ground. A deep channel follows the coast of Scandinavia reaching as far as the Skagerrak. In the northeast the North Sea joins the Norwegian Sea, and in the south it meets the Strait of Dover. It has 300 oil platforms, 10,000 km/6,200 mi of gas pipeline (gas was discovered in 1965), and fisheries (especially mackerel and herring). 24 N4

North Somerset unitary authority in southwest England created in 1996 from part of the former county of Avon.
area 372 sq km/144 sq mi
towns and cities ►Weston-Super-Mare (administrative headquarters), Clevedon, Portishead, Yatton, Congresbury
features Severn Estuary; River Yea; River Avon; west end of the Mendips including Bleadon Hill (134 m/440 ft); Weston Woods and Worlebury Hill iron age sites (Weston-Super-Mare)
industries automotive components, rubber and plastics manufacture
population (1996) 177,000 5 E6

North Uist island of the Outer ►Hebrides, Western Isles, Scotland. Lochmaddy is the main port and town. There is a Royal Society for the Protection of Birds reserve at Balranald. 24 E4

Northumberland county of northern England.
area 5,030 sq km/1,942 sq mi
towns and cities Morpeth (administrative headquarters), Berwick-upon-Tweed, Hexham
physical Cheviot Hills; rivers Aln, Coquet, Rede, Till, Tweed, upper Tyne; Northumberland National Park in the west
features ►Holy Island (Lindisfarne); the ►Farne island group 8 km/5 mi east of Bamburgh, home to seal and bird colonies; part of Hadrian's Wall; Alnwick and Bamburgh castles; Thomas Bewick museum; Hexham Abbey; Longstone Lighthouse from which Grace Darling rowed to rescue the crew of the *Forfarshire*; wild white cattle of Chillingham; Kielder Water (1982), the largest artificial lake in northern Europe
agriculture sheep, cattle; fishing
industries coal was formerly mined at several locations
population (1996) 307,400 5 E4

Northumberland Strait strait in eastern Canada; length 210 km/ 130 mi, width 15–50 km/9–31 mi. The southwestern arm of the Gulf of St Lawrence, it separates Prince Edward Island from east New Brunswick and north Nova Scotia. The shallow channel has strong tidal currents and important shellfish and lobster industries. With the warmest waters in eastern Canada, the area attracts tourists to beaches along its shore. 82 U7

North West province of the Republic of South Africa from 1994; area 118,710 sq km/45,834 sq mi; population (1995 est) 3,351,800. The capital is ►Mmabatho. There are platinum, chrome, and iron industries; groundnuts are grown. 72 C5

Northwest Passage Atlantic–Pacific sea route around the north of Canada. Canada, which owns offshore islands, claims it as an internal waterway; the USA insists that it is an international waterway and sent an icebreaker through without permission in 1985.

Northwest Territories large administrative area of Canada, extending into the Arctic Circle. Covering one-eighth of the total area of the country, it comprises the mainland lying north of the 60th parallel (latitude 60° north) and some islands between the Canadian mainland and the North Pole. It is bounded by Yukon Territory to the west, Nunavut to the east, the Beaufort Sea and the Arctic Ocean to the north, and the provinces of British Columbia, Alberta, and Saskatchewan to the south; area 1,299,070 sq km/ 501,441 sq mi; population (1997 est) 40,300 (with substantial numbers of indigenous peoples: Inuvialuit, Slavey, Dene, Métis, Inuit). The capital is ►Yellowknife. Industries include oil and natural gas extraction and mining of zinc, lead, and gold; other activities are fur-trapping and fishing. 82 G4

North Yorkshire county of northeast England, created in 1974 from most of the North Riding and parts of the East and West Ridings of Yorkshire (since April 1996 York has been a separate unitary authority).
area 8,037 sq km/3,103 sq mi
towns and cities ►Northallerton (administrative headquarters); resorts: Harrogate, Scarborough, Whitby
physical England's largest county; rivers Derwent, Esk, Ouse; includes part of the Pennines; Vale of York; the Cleveland Hills; North Yorkshire Moors, which form a national park
features Rievaulx Abbey; Yorkshire Dales National Park (including Swaledale, Wensleydale, and Bolton Abbey); Castle Howard; largest accessible cavern in Britain, the Battlefield Chamber, Ingleton
agriculture cereals, dairy products; wool and meat
industries coal, footwear, clothing, vehicles, plastics, foodstuffs, high-tech industries, light industry
population (1996) 734,700 5 F4

Norway country in northwest Europe, on the Scandinavian peninsula, bounded east by Sweden, northeast by Finland and Russia, south by the North Sea, west by the Atlantic Ocean, and north by the Arctic Ocean.

Norway covers an area of 387,000 sq km/149,420 sq mi (including Svalbard and Jan Mayen), and has a population of 4,465,000 (2000 est). The capital is ►Oslo; other major towns and cities include Bergen, Trondheim, Stavanger, Kristiansand, and Drammen. The head of state is King Harald V (from 1991); the head of government is Jens Stoltenberg (from 2000). The currency is the Norwegian krone, and the languages spoken include Norwegian (official), Saami (Lapp), and Finnish. 16 F5

Norwegian Sea part of the ►Arctic Ocean, lying to the northwest of Norway. 16 B4

Norwich city and administrative headquarters of New London County, southeast Connecticut, USA, at the confluence of the Yantic and Quinebaug rivers that form the

Thames River, north of New London; population (1990) 37,400. Industries include leather, paper, and metal products; electronic equipment; and clothing. It is home to Thames Valley State Technical College (1963) and Mohegan Community College (1970). 88 F2

Norwich cathedral city and administrative headquarters of ►Norfolk, eastern England, on the River Wensum, 160 km/100 mi northeast of London; population (1991 est) 172,600. Industries include financial and insurance services, tourism, television and radio broadcasting, engineering, printing, high-technology and biotechnology industries, and the manufacture of shoes, mustard, clothing, chemicals, and confectionery. It is the largest medieval walled city in England. 22 D2

Notia Pindos alternative name for the ►Pindus Mountains, a mountain range in northwestern Greece and Albania. 36 D5

Nottingham City industrial city and unitary authority in central England, on the River Trent, 200 km/124 mi northwest of London. It was the administrative headquarters of the county of Nottinghamshire to April 1998.
area 74 sq km/29 sq mi
features Nottingham Playhouse (1963), Theatre Royal (1866), the Royal Concert Hall (1982), and the Castle Museum; University of Nottingham (1881) and Nottingham Trent University (1992), formerly Trent Polytechnic; Trent Bridge cricket ground; the Harvey Haddon sports stadium; the National Water Sports Centre
industries tourism, engineering, and the manufacture of bicycles, textiles, knitwear, pharmaceuticals, tobacco, lace, hosiery, and electronics
population (1996) 285,000
22 A/5 F5

Nottinghamshire county of central England, which has contained the unitary authority Nottingham City since April 1998.
area 2,160 sq km/834 sq mi
towns and cities West Bridgford (administrative headquarters), Mansfield, Newark, Worksop
physical rivers: Erewash Idle, Soar, Trent
features the remaining areas of Sherwood Forest (home of Robin Hood and originally 32 km/20 mi long and 12 km/7 mi wide); Cresswell Crags (remains of prehistoric humans)
agriculture cereals (barley, wheat), market gardening (potatoes), sugar beet; cattle, sheep; orchards
industries cigarettes, coal mining, engineering, footwear, furniture, gravel, gypsum, ironstone, limestone, oil, pharmaceuticals, sandstone, textiles
population (1996) 1,031,800
5 F5

Notts abbreviation for ►Nottinghamshire, an English county. 5 F5

Nouakchott capital of ►Mauritania, 270 mi/435 km northeast of Dakar, Senegal; population (1992) 600,000. It is the largest city in the Sahara. Products include salt, cement, insecticides, rugs, carpets, embroidery, and crafts. Exports include copper, petroleum, and phosphates. There is some light engineering. 66 B5

Nouméa port and capital on the southwest coast of ►New Caledonia; population (1992) 65,000. 74 G8

Nouvelle Calédonie French name for ►New Caledonia, an island group in the South Pacific. 74 G8

Nova Lisboa former name (1928–73) for ►Huambo, a town in Angola. 72 B2

Nova Scotia maritime province of eastern Canada, comprising the peninsula of Nova Scotia, extending southeast from New Brunswick into the Atlantic Ocean, and ►Cape Breton Island, which is separated from the northeastern end of the mainland by the Canso Strait; area 55,500 sq km/21,400 sq mi; population (1996 est) 942,800. The capital (and chief port) is Halifax. Industries include mineral extraction (coal, barite, gypsum), lumbering, paper-milling, and fishing. Agricultural products include dairy produce, poultry, eggs, vegetables, and fruit. Tourism is important. 82 T8

Novaya Zemlya Arctic island group belonging to the Russian Federation; area 81,279 sq km/31,382 sq mi. A few Samoyed (people from northwest Siberia) inhabit the area. It is rich in birds, seals, and walrus. 42 J3

Novgorod city on the Volkhov River in the northwest Russian Federation and capital of Novgorod oblast, 500 km/311 mi northwest of Moscow; population (1990) 232,000. Novgorod is one of the oldest cities in Russia, and was the capital of the Russian state before the ascendancy of Moscow. Chemicals, clothing, electrical goods, furniture, and beer are manufactured. 38 F3

Novi Sad industrial and commercial city (pottery and cotton), capital of the autonomous province of Vojvodina in northern Serbia, Yugoslavia, on the River Danube; population (1991) 179,600. Products include leather, textiles, and tobacco. 34 G4

Novokuznetsk (or **Kuznetsk Sibirski**; formerly **Stalinsk** 1932–61), city in the Kemerovo oblast (region) of southern Siberia, Russian Federation; population (1996 est) 569,000. Novokuznetsk is the principal industrial centre of the ►Kuznetsk Basin. One of the largest iron and steel works in the Russian Federation is located here. There are also ferro-alloy and aluminium plants, chemical, light, and food industries, and a major hydroelectric power station. Situated on the River Tom and served by several railway lines, including the South Siberian railway, the city is an important transportation centre. 42 R7

Novorossiysk seaport city on the Black Sea, in Krasnodar krai (territory), northern ►Caucasia, in the Russian Federation; population (1996 est) 203,000. The centre of the country's cement industry, it also has shipbuilding, metalworking, machine-tool, food-processing, and light industries. With the loss of its naval bases at Odessa and on the Crimea following Ukrainian independence in 1991, the Russian Federation is developing Novorossiysk as a new port facility for the Black Sea Fleet. 58 G1

Novosibirsk (formerly **Novonikolayevsk** 1893–1925), capital city, economic and cultural centre of Novosibirsk oblast (region) in southwestern Siberia, Russian Federation; population (1996 est) 1,368,000. Sited on the River Ob and the Trans-Siberian Railway, it is the largest city in Siberia and one of the main industrial centres of the Russian Federation. It is at the hub of an extensive transport network; the Turksib (Turkestan–Siberian) Railway runs from here to Almaty, former capital of Kazakhstan. Novosibirsk has large engineering industries and varied light and food industries. Since the collapse of communism in 1991, there has been considerable development of financial, business, and other services. 42 Q6

NSW abbreviation for ▶New South Wales, an Australian state. 76 H6

Nuku'alofa capital and port of Tonga on Tongatapu Island; population (1998 est) 30,400. 74 J8

Nullarbor Plain (Latin *nullus arbor* 'no tree'), arid coastal plateau area divided between Western and South Australia; there is a network of caves beneath it. Atom-bomb experiments were carried out in the 1950s at Maralinga, an area in the northeast bordering on the Great Victoria Desert. 76 E6

Nunavut (Inuit 'our land'), semi-autonomous Inuit homeland, established as a territory of Canada on 1 April 1999. It consists of former parts of the ▶Northwest Territories, Canada, including Keewatin and Kitikmeot, the Arctic Islands of Baffin, Ellesmere, Devon, Prince of Wales and Banks, and the eastern areas of Victoria and Melville Islands, as well as Southampton and smaller islands in Hudson Bay. Nunavut extends to 1,994,000 sq km/769,846 sq mi, one-fifth of the total land area of Canada. Its creation was approved by a narrow majority in a regional plebiscite in 1992, though the measure was opposed by representatives of the Dene people from the western Arctic, where 74% voted against (the Dene claimed that the homeland cut across their traditional hunting grounds). In the eastern Arctic where most Inuit live, 84% voted in favour. Representatives of the Initio had earlier negotiated hunting and fishing rights in the area, and a final land claims agreement, signed in 1993 on Baffin Island (where Iqaluit, the capital of Nunavut is located), gave the Inuit outright ownership of 353,610 sq km/136,493 sq mi of the land, and mineral rights to 36,257 sq km/13,995 sq mi. The remainder is Crown Land over which the Inuit have joint control with the Federal Government.

In spite of its vast size, Nunavut has only 27,200 inhabitants (1999 est), over 80% of whom are Inuit. 82 M3

Nuneaton industrial town in Warwickshire, central England, on the River Anker, 15 km/9 mi northeast of Coventry; population (1991) 65,900. Industries include engineering, electronics, and the manufacture of tiles and textiles. 22 A2

Nuremberg (German **Nürnberg**), city in Bavaria, Germany, on the River Pegnitz, 149 km/92 mi northwest of Munich; population (1995) 494,100. Industries include electrical and other machinery, precision instruments, toys, and food processing. From 1933 the Nuremberg rallies were held here, and in 1945 the Nuremberg trials of war criminals. 20 G7

Nürnberg German name for ▶Nuremberg, a city in Bavaria, Germany. 20 G7

Nuuk Greenlandic for ▶Godthåb, the capital of Greenland. 82 W4

Nyasa, Lake alternative name for Lake ▶Malawi. 72 E2

Nyasaland former name (to 1964) for ▶Malawi. 72 E2

Nyíregyháza market town in eastern Hungary; population (1993 est) 115,000. It trades in tobacco and vegetables. 18 L10

Nyon commune in the canton of Vaud, Switzerland, on the western shore of Lake Geneva, about 20 km/12 mi north of Geneva; population (1995) 15,400. Nyon produces fine porcelain and lies at the junction of the Geneva–Lausanne railway line and a line running into France.

Nyon was the site of the Roman colony of **Julia Equestris**. 30 B4

Nysa (German **Neisse**), two rivers in central Europe, tributaries of the River ▶Oder. The **Nysa Łuzycka** (German **Lausitzer Neisse**) rises near Liberec in the northern Czech Republic and flows 225 km/140 mi northwards along the German–Polish frontier to join the ▶Oder southeast of Frankfurt an der Oder. The **Nysa Kłodzka** (German **Glatzer Neisse**) rises in the Sudeten Mountains on the Czech–Polish frontier and flows 193 km/120 mi northeast to join the Oder above Brieg, northwest of Opole. 18 D6

Nysa (German **Neisse**), industrial town (machinery, food processing) in Opole province, southern Poland, on the River Nysa, 48 km/30 mi southwest of Opole; population (2001 est) 48,300. It was capital of a principality of the bishops of Breslau (now Wrocław) 1198–1810, and part of German Upper Silesia until 1945. 18 G7

NZ abbreviation for ▶New Zealand. 78 B5

Oahu island of Hawaii, USA, in the North Pacific; the third-largest and most-populated of the group; area 1,525 sq km/589 sq mi; population (1990) 836,200. Oahu contains ▶Honolulu, the state capital of Hawaii. Tourism is the main industry; produce includes sugar and pineapples. 92 (2)D2

Oakland industrial port in central California, USA, on the eastern, inland coast of San Francisco Bay; population (1996 est) 367,200. It is linked by the Bay Bridge (opened 1936) with San Francisco. Industries include food-processing, shipbuilding, and the manufacture of vehicles, textiles, and chemicals. The community was laid out in 1852 and became a terminus of the first transcontinental railroad in 1869. Extensive port facilities handle much of the freight traffic of the bay area, and include a naval yard and naval air station. The city was damaged by earthquake in 1989. 86 B3

Oak Ridge town in Anderson and Roane counties, east Tennessee, USA, on the Clinch River; population (1990) 27,300. The community was founded in 1942 as part of the Manhattan Project to develop an atomic bomb; the Oak Ridge National Laboratory (1943) manufactures plutonium for nuclear weapons. By the end of World War II, the town's population was more than 75,000. Ownership of the community passed to the residents in the late 1950s. Other industries produce radioactive pharmaceuticals, electronic instrumentation, and machines and tools. 88 D3

Oaxaca capital of a state of the same name in the Sierra Madre del Sur mountain range, central Mexico; population (1990) 212,900.

Industries include food processing, textiles, and handicrafts. 94 E5

Ob major river in Asian Russia, flowing 3,380 km/2,100 mi from the Altai Mountains through the western Siberian Plain to the Gulf of Ob in the Kara Sea (an inlet of the Arctic Ocean). With its main tributary, the **Irtysh**, the Ob is 5,600 km/3,480 mi long, and drains a total area of 2,990,000 sq km/1,150,000 sq mi. 38 N2

Oban seaport and resort in Argyll and Bute, western Scotland; population (1991) 8,200. It is an important service, tourist, and communications centre to the Hebridean islands. The main industries are whisky distilling and tweed manufacture; fishing is also carried on. 24 G5

Obeid, El see ►El Obeid, a city in Sudan. 64 F5

Oberammergau village in Bavaria, Germany, 72 km/45 mi southwest of Munich; population (1994) 5,300. A Christian passion play has been performed here every ten years since 1634 (except during the world wars) to commemorate the ending of the Black Death plague.

Oberhausen coal-mining city in the Ruhr Valley, North Rhine-Westphalia, Germany, 29 km/18 mi northeast of Düsseldorf; population (1995) 224,900. Industries include iron, steel, mechanical and chemical engineering, plastics, and glass. It lies on the Rhine–Herne Canal. 22 J3

Oberösterreich German name for the federal state of ►Upper Austria. 8 (2)E2

Ocala city and administrative headquarters of Marion County, north-central Florida, USA, southeast of Gainesville; population (1990) 42,000. It is a marketing and shipping centre for the citrus, poultry, cotton, and tobacco products grown in the surrounding area. Thoroughbred horses are also raised in the region. Tourism is vital to the economy. Its other industries include fertilizer production, the processing of dairy products, limestone, and phosphate, and the manufacture of wood products, mobile homes, fire trucks, sewer pipes, and clothing. Ocala is home to Central Florida Community College (1957). 90 E4

Oceania the groups of islands in the southern and central Pacific Ocean, comprising all those intervening between the southeastern shores of Asia and the western shores of America. See ►Australasia and Oceania. 74

Ocean Island another name for Banaba, an island belonging to ►Kiribati. 74 G6

Ödenburg German name for ►Sopron, a town in Hungary. 30 M3

Odense industrial port (shipbuilding, electrical goods, glass, textiles) on the island of Fyn, Denmark; population (1993) 180,800. It is the birthplace of Hans Christian Andersen. 20 F1

Oder (Polish **Odra**), European river flowing north from the Czech Republic to the Baltic Sea (the Neisse River is a tributary); length 885 km/550 mi. 18 F6

Odessa city and administrative headquarters of Ector County, west Texas, USA, 94 km/59 mi southwest of Big Springs; population (1996 est) 90,900. Situated on the high plains, it is a service and distribution centre for the oilfields of the Permian Basin, and the Texas ranchlands. Industries include the manufacture of cement, petroleum, petrochemicals, rubber, and meat products. It is the home of Odessa College (1946) and the University of Texas of the Permian Basin (1969). A Shakespeare Festival is held here annually. 92 F2

Odessa principal seaport of Ukraine, on the Black Sea, and capital of the Odessa region (oblast); population (1990) 1,106,400. Odessa is a commercial port, naval base, and tourist resort. The principal industries here are shipbuilding, fishing, steelmaking, and food processing. Products manufactured in the city include chemicals, pharmaceuticals, and machinery. Among the main goods handled in the port are grain, sugar, timber, and oil. 38 F5

Odra Polish name for the ►Oder, a European river flowing north from the Czech Republic. 18 F6

Offaly county of the Republic of Ireland, in the province of Leinster, between Galway and Roscommon in the west and Kildare in the east; county town Tullamore; area 2,000 sq km/772 sq mi; population (1996) 59,100. It is low-lying, with part of the Bog of Allen to the north. 5 C5

Offa's Dyke defensive earthwork dyke along the English–Welsh border, of which there are remains from the mouth of the River Dee to that of the River ►Severn. It was built about AD 785 by King Offa of Mercia, England, and represents the boundary secured by his wars with Wales.

Offenbach city in Hesse, Germany, on the south bank of the River Main, 10 km/6 mi east of Frankfurt; population (1995) 116,600. Offenbach produces leather goods and artificial fibres, and there are engineering industries. 20 D6

Offenbach am Main city in Hesse, Germany; population (1991) 116,800. It faces Frankfurt on the other side of the River Main. 20 D6

Offenburg city in Baden-Württemberg, Germany, on the River Kinzig, 14 km/9 mi from its junction with the Rhine, 96 km/60 mi southwest of Stuttgart; population (1994) 327,000. It is the centre of a fruit- and wine-growing district, but also has electrical and mechanical engineering factories. 30 C2

Ogaden desert region in southeastern Ethiopia, between the Ethiopian Highlands and the border with Somalia. It is a desert plateau, rising to 1,000 m/3,280 ft, inhabited mainly by Somali nomads practising arid farming. 70 G2

Ogasawara-shotō group of Japanese islands in the Pacific; see ►Bonin and Volcano islands. 40 T7

Ogbomosho industrial city and commercial centre in Oyo State, western Nigeria, 80 km/50 mi northeast of Ibadan; population (1992 est) 660,600. It is one of Nigeria's largest cities. Products include cattle, staple crops, cotton, and palm oil. There are tobacco and craft industries, and shoe, textile, and food processing factories. 68 E3

Ogden city and administrative headquarters of Weber County, north Utah, USA, on the Weber and Ogden rivers, 51 km/32 mi north of Salt Lake City; population (1996 est) 65,300. The second-largest city in the state, it is a processing and railway distribution point for the surrounding irrigated agricultural area, and a military supply centre for nearby Hill Air Force Base. Industries include aerospace development, tourism, meat-packing, and canning. Building materials, chemicals, furniture, processed food from local farms, clothing, batteries, and stone, clay, and glass items are produced in the city. Ogden is the home of Weber State College (1889). 86 D2

Ohio river in east-central USA, 1,580 km/980 mi long and 365–1,220 m/1,200–4,005 ft wide; the second-largest tributary of the Mississippi; navigable throughout its length. It is formed by the union of the Allegheny and Monongahela rivers at Pittsburgh, Pennsylvania, and flows southwest to join the Mississippi at Cairo, Illinois. 88 C3

Ohio state in northern central USA. It is nicknamed the Buckeye State. It was admitted to the Union in 1803 as the 17th US state. Part of the Midwest, it is bordered to the east by Pennsylvania, to the east and southeast by West Virginia, to the southwest by Kentucky, to the west by Indiana, to the northwest by Michigan's Lower Peninsula, and to the north by Lake Erie. Ohio comprises the eastern section of the US Corn Belt; heavily industrialized, it is also a quintessential Rust Belt state, today struggling with pollution and the need to diversify industrially.
population (2000 est) 11,353,100
area 107,100 sq km/41,341 sq mi
capital ►Columbus
towns and cities Cleveland, Cincinnati, Dayton, Akron, Toledo, Youngstown, Canton
industries and products coal, cereals, livestock, dairy foods, machinery, chemicals, steel, motor vehicles, automotive and aircraft parts, rubber products, office equipment, refined petroleum, tourism 88 D2

Ohrid town in the Former Yugoslav Republic of Macedonia, near the Albanian border, situated on the shore of Lake Ohrid. It is on the site of an ancient Greek colony and is said to be the oldest town in Yugoslavia. Its buildings include a 10th-century fortress and an 11th-century church. Sts Cyril and Methodius, creators of the Slavonic alphabet, are believed to have worked here. 36 C3

Ohrid, Lake lake on the frontier between Albania and the Former Yugoslav Republic of Macedonia; area 350 sq km/135 sq mi. 36 C4

Oise *département* in the ►Picardy region of northern France; area 5,860 sq km/2,263 sq mi; population (1999 est) 766,300. The main rivers are the Oise and the Aisne. Fruit and dairy produce are among the area's agricultural output; industries include metallurgy, chemicals, plastics, perfumes, and cosmetics. Geographically, the department is a mixture of hilly areas punctuated with large plains devoted to agriculture. There are also large forests, such as those at Compiègne and Chantilly. Principal towns are the capital ►Beauvais, ►Compiègne, ►Creil, and Senlis. The town of ►Chantilly, with its famous château and racecourse, is in the south of the *département*. 6 E2

Oise European river that rises in the Ardennes plateau, Belgium, and flows southwest through France for 300 km/186 mi to join the Seine at Conflans-Sainte-Honorine, about 65 km/40 mi south of Paris. It gives its name to the French *département* of ►Oise in Picardy. 22 E5

Okavango Delta marshy area in northwest Botswana covering about 20,000 km/7,722 sq mi, fed by the Okavango River, which rises in Angola and flows southeast about 1,600 km/1,000 mi. It is an important area for wildlife as it provides the main area of permanently available water in the Kalahari Desert. 72 C3

Okayama industrial port and old castle town in the southwest of Honshu island, Japan, 120 km/75 mi west of Kobe; population (1994) 598,000. The main industries are cotton spinning and the manufacture of rubber, chemicals, agricultural implements, and 'tatami' and 'hanamushiro', types of Japanese matting. The Korakuen is one of the three most important traditional gardens in Japan. The others are in Mito and Kanazawa. 48 G6

Okeechobee, Lake lake in the northern Everglades, south-central Florida, USA; roughly circular, it is 65 km/40 mi long and 40 km/25 mi wide, and reaches a depth of 6 m/20 ft. It is the largest lake in the southern USA, about 1,800 sq km/700 sq mi. There are numerous small islands in the lake, which receives the Kissimmee River from the northwest. Since the 1930s, diking around the south shore has prevented overflow; drainage canals are used for light navigation to the Atlantic Ocean. The lake is a fishing resort with some commercial fisheries. 90 E4

Okefenokee Swamp swamp in southeast Georgia and northeast Florida, USA, rich in alligators, bears, deer, and birds. Much of its 1,700 sq km/660 sq mi forms a natural wildlife refuge. It is drained by the St Marys and Suwannee rivers.

Okhotsk, Sea of arm of the North Pacific Ocean between the Kamchatka Peninsula and Sakhalin, and bordered to the south by the Kuril Islands, and the northern Japanese island of Hokkaido; area 937,000 sq km/361,700 sq mi, average depth 777 m/2,550 ft, maximum depth 3,372 m/11,062 ft. It is free of ice only in summer, and is often fog-bound. Magadan is the chief port, and the River Amur flows into it. 44 Q5

Okinawa group of islands, 520 km/323 mi from the Japanese mainland, forming part of the Japanese ►Ryukyu Islands in the west Pacific; Okinawa is also the name of the largest island, and of a city on Okinawa; area 2,250 sq km/869 sq mi; population (1995) 1,274,000. The capital is ►Naha. 46 H5

Oklahoma state in southern central USA. It is nicknamed the Sooner State. Oklahoma was admitted to the Union in 1907 as the 46th US state. It is bordered to the south by Texas, to the west, at the extreme of the Oklahoma panhandle, by New Mexico, to the north by Colorado and Kansas, and to the east by Missouri and Arkansas. Oklahoma is the US state most associated with American Indians; its name is a Choctaw coinage meaning 'red people'.
population (2000 est) 3,450,700
area 181,100 sq km/69,900 sq mi
capital ►Oklahoma City
towns and cities Tulsa, Lawton, Norman, Enid
industries and products cereals, peanuts, cotton, livestock, oil, natural gas, helium, machinery and other metal products 90 B2

Oklahoma City capital of ►Oklahoma, USA, on the North Canadian River; population (1998 est) 472,200. It is a major commercial, service, and distribution centre for a rich oil-producing and agricultural area; its cattle markets are among the largest and busiest in the world. Industries include oil-refining, food-processing (meat and cereals), and the manufacture of iron, steel, machinery, cars, tyres, aircraft, electronic equipment, and cotton. Manufacturing makes up more than 10% of employment in the region (1999), and the health sector is also a significant employer. 90 B2

Öland Swedish island in the Baltic, separated from southern Sweden by Kalmar Sound, though connected by a bridge; area 1,344 sq km/519 sq mi. The island is 136 km/85 mi long but only 16 km/10 mi at its broadest. It is wooded in parts, and has good pasture for cattle. There are fisheries all round the coast, but tourism is becoming the most important source of income. Borgholm, on the west coast, is the capital and only town; population (1995) 11,900. **16** J8

Oldenburg commercial and university city in Lower Saxony, Germany, on the River Hunte, 132 km/82 mi northwest of Hannover; population (1995) 150,500. It is linked by river and canal to the Ems and Wieser rivers. There are electrical industries and food processing plants and it is the centre of an agricultural region. It was once the capital of the duchy of Oldenburg. **20** D3

Oldham industrial town in Greater Manchester, England, 10 km/6 mi northeast of Manchester, on the lower slopes of the Pennine upland; population (1991) 102,300. Industries include the manufacture of plastics, electrical goods, and electronic equipment. It was traditionally a cotton-spinning town. **24** L8

Old World the continents of the eastern hemisphere, so called because they were familiar to Europeans before the Americas. The term is used as an adjective to describe animals and plants that live in the eastern hemisphere.

Oléron, Ile d' island in the Atlantic Ocean, off the southwest coast of France and part of the *département* of Charente-Maritime; area 170 sq km/ 66 sq mi; population (1990) 18,500. Its greatest length is 29 km/18 mi and breadth 11 km/7 mi. The island produces corn and wine; salt is collected from the salt marshes, and fishing and tourism are important. The principal towns are St-Pierre and Le Château d'Oléron. **26** D8

Olmos small town on the edge of the Sechura Desert, northwestern Peru; population (1993) 30,250. **98** B5

Olomouc industrial city in the Czech Republic, at the confluence of the Bystrice and Morava rivers; population (1991) 105,700. Industries include sugar refining, brewing, and metal goods. **18** G8

Olsztyn (formerly **Allenstein**), industrial town in northeastern Poland, at the centre of the Mazurian Lakes region; population (1990) 162,900. It was founded 1334 and was formerly in East Prussia. **18** K4

Olt river in southern Romania, rising in the Carpathian Mountains and flowing through Transylvania to join the ►Danube near Turnu Măgurele, after a distance of 560 km/350 mi. **34** M4

Olten commune in the canton of Solothurn, Switzerland, on the River Aar, 34 km/21 mi southeast of Basel; population (1995) 17,100. Olten contains the main workshops of the Swiss federal railway, and also manufactures electrical apparatus, car parts, and textiles. **30** C3

Olympia capital of ►Washington, USA, about 80 km/50 mi southwest of Seattle; population (1996 est) 39,000. Located in the west central part of the state, it is situated on Budds Inlet, on the Deschutes River, at the southwest end of Puget Sound. The city is a deep-water seaport, with fishing, lumber, mining, brewing, and tourist industries. Metalware, farm machinery, and canning equipment are manufactured. Hydroelectric power is supplied from the Deschutes. Olympia was founded in 1851, and became state capital in 1853. **86** B1

Olympus (or **Olympos**; Greek **Olimbos**), any of several mountains in Greece and elsewhere, one of which is **Mount Olympus** in northern Thessaly, Greece, which is 2,918 m/ 9,577 ft high. In ancient Greece it was considered the home of the gods. **36** E4

Omagh county town of County ►Tyrone, Northern Ireland, in the foothills of the Sperrin Mountains, on the River Strule, 48 km/30 mi south of Londonderry; population (1991) 17,300. Industries include dairy produce, food processing, footwear, shirt manufacturing, and engineering. Omagh was planned in the early 17th century; its chief buildings are the Catholic church, with its irregular twin spires, the courthouse, built in Classical style, and the County Hall. It is now a tourist centre, and there is salmon fishing. **24** E7

Omaha largest city in eastern Nebraska, USA, and county seat of Douglas county, on the right bank of the Missouri River 32 km/20 mi north of its junction with the Nebraska River; population (1996 est) 364,300. It is Nebraska's principal commercial centre and a major agricultural market, trading in livestock and cereals. Industries include food-processing and meat-packing. Omaha was laid out in 1854. Its location at the eastern terminus of the Union Pacific Railroad (1869) spurred economic growth. **86** G2

Oman country at the southeastern end of the Arabian peninsula, bounded west by the United Arab Emirates, Saudi Arabia, and Yemen, southeast by the Arabian Sea, and northeast by the Gulf of Oman.

Oman covers an area of 272,000 sq km/105,019 sq mi, and has a population of 2,542,000 (2000 est). The capital is ►Muscat; other major towns and cities include Salalah, Ibri, Sohar, Al-Buraimi, Nizwa, Sur, and Matrah. The head of state and government is Qaboos bin Said (from 1970). The currency is the Omani rial, and the languages spoken include Arabic (official), English, Urdu, and other Indian languages. **56** G5

Omdurman (Arabic **Umm Durman**), city in Khartoum State, Sudan, on the west bank of the White Nile, now considered a major suburb of Khartoum; population (1995 est) 719,000. It trades in hides, textiles, livestock, and handicrafts. It is connected to Khartoum by a bridge over the White Nile. Omdurman was built as a new city to be the residence of the Sudanese ruler known as the Mahdi, and his successor Khalifa Abdullah el Taaisha. The city is a religious and cultural centre for the majority Muslim population of the Sudan, and at the city centre is the tomb of the Mahdi (Muhammad Ahmed). **64** F4

Omsk capital city, economic and cultural centre of Omsk oblast (region), Russian Federation; population (1996 est) 1,160,000. Omsk is located at the junction of the Om and Irtysh rivers and lies on the Trans-Siberian Railway, 900 km/ 559 mi east of Yekaterinburg. The city is a major industrial and commercial centre of west Siberia. It contains engineering works, oil refineries (linked with Tuimazy in Bashkortostan by a 1,600-km/1,000-mi pipeline), wood-processing plants, and various food and other light industrial factories. **42** N6

Onega, Lake (Russian **Onezhskoye Ozero**), lake in the far northwestern Russian Federation, near the Finnish border. With an area of some 9,600 sq km/3,700 sq mi (excluding islands), it is the second-largest lake in Europe. Lake Onega is connected by the River Svir with Lake Ladoga (the largest lake in Europe) and the Baltic Sea, and by artificial waterways with the White Sea and the Volga (the 'Mariinsk' system). **38** F2

Onezhskoye Ozero Russian name for Lake ►Onega, a lake in the northwestern Russian Federation. **38** F2

Ontario province of southeastern–central Canada, in area the country's second-largest province, and its most populous. It is bounded to the north and northeast by Hudson Bay and James Bay, to the east by Québec (with the Ottawa River forming most of the boundary), and by Manitoba to the west. On the south, it borders on, and extends into, all of the Great Lakes except Lake Michigan. From west to east along Ontario's southern boundary lie the US states of Minnesota, Wisconsin, Michigan, Ohio, Pennsylvania, and New York; area 1,068,600 sq km/412,600 sq mi; population (1996) 11,252,400. The capital is ►Toronto (Canada's largest city). Industries include mining (nickel, iron, gold, copper, uranium) and the production of cars, aircraft, iron, steel, high-tech goods, pulp, paper, oil, and chemicals; agriculture includes livestock rearing, and cultivation of fruit, vegetables, and cereals. **82** N6

Ontario, Lake smallest and easternmost of the ►Great Lakes, on the US–Canadian border; area 19,200 sq km/7,400 sq mi. Extending for 310 km/194 mi, it has an average width of about 80 km/50 mi, and maximum depth of 244m/800 ft. It is connected to Lake Erie in the southeast by the ►Welland Ship Canal and the Niagara River, and drains into the ►St Lawrence River to the northeast. The opening of the ►St Lawrence Seaway in 1959 made the lake accessible to large ocean-going vessels. Its main port is Toronto, Canada. **88** E2

Oostende Flemish form (meaning 'east end') of ►Ostend. **22** E3

Oost-Vlaanderen Flemish name for ►East Flanders, a province in Belgium. **7** (1)B3

Opole industrial town in southern Poland, on the River Oder; population (1990) 128,400. It is an agricultural centre; industries include textiles, chemicals, and cement. **18** G7

Oporto English form of ►Porto, a city in Portugal. **28** B3

Oradea (or **Oradea-Mare**), industrial city in Romania, on the River Koös; population (1993) 222,000. Industries include agricultural mach-inery, chemicals, non-ferrous metal-lurgy, leather goods, printing, glass, textiles, clothing, and brewing. **34** J2

Oran (Arabic **Wahran**), fortified seaport and commercial and manufacturing centre in Algeria, 5 km/3 mi from the port of Mers-el-Kebir; population (1998 est) 850,500. Products include cigarettes, iron, plastics, textiles, footwear, and processed food; the port trades in grain, wool, vegetables, and native esparto grass. Natural gas is brought to the city via a pipeline from the Sahara, and hydrocarbons are exported. There is an international airport. A university was established in 1965. **100** J3

Orange town in New South Wales, Australia, 259 km/161 mi west of Sydney, on the eastern slopes of Mount Canobolas; population (1996) 30,705. Orange is the service centre for an agricultural district in which fruits are grown, and cattle, sheep, and pigs are raised. Orange has a woollen industry based on local sheep flocks. Other industries include electrical engineering, sawmilling, and meat processing. **76** J6

Orange French town in the *département* of Vaucluse, 21 km/13 mi north of Avignon; population (1990) 28,100. It has the remains of a Roman theatre, a triumphal arch and an amphitheatre. It was a medieval principality from which came the European royal House of Orange. **26** K9

Orange river in South Africa, rising at the Mont-aux-Sources in the Maluti Mountains in Lesotho and flowing west to the Atlantic Ocean; length 2,100 km/1,300 mi. It runs along the southern boundary of the Free State Province. Water from the Orange is diverted via the Orange-Fish River Tunnel (1975) to irrigate the semi-arid Eastern Cape Province. It was named in 1779 after William of Orange. **72** C5

Orasul Stalin former name (1948–56) of the Romanian town ►Braşov. **34** N4

Ordzhonikidze former name (1954–91) of ►Vladikavkaz, the capital of Alania (formerly North Ossetia) in the southwest of the Russian Federation. This name derived from the Bolshevik leader Grigory Ordzhonikidze, who imposed the Soviet system on the Caucasus region in the 1920s. **58** L2

Oregon state in northwestern USA, on the Pacific coast. It is nicknamed Beaver State. Oregon was admitted to the Union in 1859 as the 33rd US state. It is bordered to the east by Idaho, to the north by Washington, to the south by California and Nevada, and to the west by the Pacific Ocean.
population (2000 est) 3,421,400
area 251,500 sq km/97,079 sq mi
capital ►Salem
towns and cities Portland, Eugene, Gresham, Beaverton
industries and products wheat, fruit, dairy farming, livestock, salmon and tuna, timber, electronics, tourism, leisure industry **86** B2

Orel (or **Oryol**), capital city, economic and cultural centre of Orel oblast (region), Russian Federation, on the River Oka 325 km/210 mi southwest of Moscow; population (1996 est) 348,000. Orel is a major industrial centre with engineering works producing agricultural and textile machinery, textile mills, and breweries. The city stands at a major railway junction. **38** G4

Orem city in north central Utah, USA, southeast of Salt Lake City; population (1990) 67,500. It was settled by Mormons 1861. Industries include electronics and steel. **86** D2

Orenburg (formerly **Chkalov** 1938–57), capital city, economic and cultural centre of Orenburg oblast (region), west-central Russian Federation; population (1996 est) 532,000. Orenburg is situated on the River Ural, 350 km/217 mi southeast of Samara. It has extensive oil refineries, together with engineering, flour, meat, leather, and clothing industries, and is an important trading and transportation centre. **38** L4

Orense (or **Ourense**), city and capital of Orense province in ►Galicia, northwest Spain, on the left bank of the River Miño; population (1991 est) 101,600. Products include textiles, furniture, food products, and metal goods. There is a 12th–13th-century

Gothic cathedral, and a seven-arched bridge, begun in the 13th century, the central arch of which is 38 m/125 ft high. 28 C2

Øresund strait leading from the ▶Kattegat to the Baltic Sea, between Sweden on the east and the Danish island of Sjaelland on the west. Its length is 113 km/70 mi; its narrowest point, between Helsingor and Hälsingborg, is 5 km/3 mi; its widest point is 60 km/37 mi; and its deepest part is about 25 m/80 ft. In English it is called the Sound.

Orihuela town in the province of ▶Alicante, southeast Spain; population (1991) 48,000. Situated on the River Segura, in a fertile, irrigated region, it trades in oranges, wine, and olive oil; textiles are also manufactured. It has a much-restored 14th-century cathedral, and a 17th-century college. 28 K6

Orinoco river in northern South America; it rises in the Sierra Parima range in southern Venezuela near the Brazilian border and flows north for about 2,400 km/1,500 mi through Venezuela, forming the boundary with Colombia for about 320 km/200 mi; tributaries include the Guaviare, Meta, Apure, Ventuari, Caura, Arauca, and Caroni rivers. It is navigable by large steamers for 1,125 km/700 mi from its Atlantic delta; rapids obstruct the upper river. The Orinoco is South America's third-largest river; its drainage basin area is 962,000 sq km/371500 sq mi. 98 D2

Orissa state of northeast India; area 155,700 sq km/60,100 sq mi; population (1994 est) 33,795,000. The capital is ▶Bhubaneshwar. Industries include chemicals, paper, steel, aluminium smelting, and mineral extraction. Rice, wheat, oilseed, and sugar are grown. Oriya is the official language, and 90% of the population is Hindu. 10 D4

Orizaba industrial city (brewing, paper, and textiles) and resort in Veracruz state, Mexico; population (1990) 114,000. An earthquake severely damaged it 1973. 94 E5

Orkney Causeway (or **Churchill Barriers**), construction in northern Scotland, put up in World War II, completed in 1943, joining four of the Orkney Islands. It was built to protect the British fleet from intrusion through the eastern entrances to Scapa Flow.

The Orkney Causeway links the islands of mainland Orkney with Lamb Holm, Glims Holm, and Burray.

Orkney Islands island group and unitary authority off the northeast coast of Scotland.
area 1,014 sq km/391 sq mi
towns ▶Kirkwall (administrative headquarters), Stromness, both on Mainland (Pomona)
physical there are 90 islands and inlets in the group. Next to Mainland, the most important of the islands are North and South Ronaldsay, Hoy, Rousay, Stronsay, Flotta, Shapinsay, Eday, Sanday, and Westray. The highest peak is Ward Hill in Hoy, at 479 m/1,572 ft. The Old Man of Hoy is an isolated stack of red sandstone 137 m/450 ft high, off Hoy's northwest coast
features Skara Brae Neolithic village, and Maes Howe burial chamber; Scapa Flow; oil terminal on Flotta
industries offshore oil, woollen weaving, wind-powered electricity, distilling, boat-building, fish curing
agriculture fishing, beef cattle, dairy products
population (1996) 19,600
24 K3/5 E2

Orkneys, South islands in the British Antarctic Territory; see ▶South Orkney Islands. 102 (2)A3

Orlando industrial city and administrative headquarters of Orange County, Florida, USA; population (1992) 174,200. It is a winter resort and tourist centre, with Walt Disney World and the Epcot Center nearby. Electronic and aerospace equipment are manufactured in the city, and citrus-fruit products are processed here. Educational institutions include the University of Central Florida. The city is also the site of the Orlando Naval Training Center (1968), a major basic-training facility. 90 E4

Orléans administrative centre of ▶Loiret *département* in central France, situated on the right bank of the River Loire, 115 km/70 mi southwest of Paris; population (1990) 108,000, conurbation 243,000. Industries include engineering, textiles and food processing; there is also an extensive trade in agricultural produce and wine. Orléans is of pre-Roman origin. Joan of Arc, known as the Maid of Orléans, liberated the town from the English in 1429. 26 G6

Orly suburb of the French capital Paris in the *département* of Val-de-Marne; population (1990) 21,800. Orly international airport is the busiest in France. 22 E6

Orne *département* in the ▶Basse-Normandie region of France; area 6,100 sq km/2,355 sq mi; population (1990) 294,000. Horses and cattle are reared, and dairy produce (including cheese from the village of Camembert) is exported. Electrical appliances, car parts, and chemicals are the principal manufactured products. The main towns are ▶Alençon (the administrative centre) and Argentan. 6 D2

Orsk town in Orenburg oblast (region) of the Russian Federation, in southwestern Siberia, at the junction of the rivers Ural and Or; population (1996 est) 275,000. Orsk is situated near the border with Kazakhstan, and lies 250 km/155 mi southwest of Orenburg (which was originally founded in 1735 on this site). It is the centre of the major Orsk–Khalilovo metallurgical complex, which encompasses the nearby town of Novotroitsk, and manufactures iron and steel, together with non-ferrous metals such as nickel and aluminium. Petroleum products and heavy machinery (locomotives) are also manufactured in the city. The city was originally a fortress. 38 L4

Orvieto (Etruscan **Volsinii**; Latin *Urbs Vetus* 'old town'), town in Umbria, Italy, on a high plateau overlooking the River Tiber; population (1990) 22,600. Built on the site of an Etruscan town destroyed by the Romans in 280 BC, Orvieto has many Etruscan remains. The district produces white wine of the same name, and tourism is an important industry. 32 G6

Osaka industrial port on the Pacific coast of Honshu island, Japan, 24 km/15 mi from Kobe; population (1994) 2,481,000. It is Japan's third-largest city. Industries include iron, steel, shipbuilding, chemicals, and textiles. Emperor Kotoku established it as a capital in 645. **Osaka castle**, built in 1583, was the scene of a siege in 1614–15. The city is the home of the *bunraku* puppet theatre. 48 H6

Oshkosh city and administrative headquarters of Winnebago County, eastern-central Wisconsin, on the western shore of Lake Winnebago, at the mouth of the Fox River, 120 km/75 mi northwest of Milwaukee; population (1996 est) 58,000. Industries include the manufacture of clothing (overalls and babywear), machinery, woodwork, candles, and electronics. A campus of the University of Wisconsin (1871) is in the city. 88 C2

Oshogbo city and trading centre in west Nigeria, 200 km/125 mi northeast of Lagos, on the Oshun River; population (1992 est) 441,600. It is the capital of Osun State. It developed as a trading centre after the arrival of the railway in 1906. It processes cocoa, tobacco, and palm oil. The main industries are cotton weaving and sheet metal production. 68 E3

Osijek (German **Esseg**), industrial port in Croatia, on the River Drava; population (1991) 104,800. Industries include textiles, chemicals, and electrical goods. 34 F4

Oslo industrial port and capital of Norway; population (1996) 731,600. The main industries are shipbuilding, textiles, electrical equipment, engineering, machine tools, timber, and food processing. The first recorded settlement was made in the 11th century by Harald III Hardrada, but after a fire in 1624, it was entirely replanned by the Danish king Christian IV and renamed **Christiania** from 1624 to 1924. Following Norway's separation from Denmark (1814) and then Sweden (1905), the city reverted in 1925 to its original Norwegian name of Oslo. 16 F7

Osnabrück episcopal and university city in Lower Saxony, Germany; 115 km/71 mi west of Hannover; population (1995) 167,900. Industries include engineering, cables, steel, textiles, clothing, paper, and food processing. The Treaty of Westphalia was signed at Osnabrück and Münster in 1648, ending the Thirty Years' War. 22 L2

Ossa, Mount the highest peak on the island of Tasmania, Australia; height 1,617 m/5,250 ft. 76 J8

Ostend (Flemish **Oostende**; French **Ostende**), seaport and pleasure resort in West Flanders, Belgium; 108 km/67 mi northwest of Brussels; population (1997) 68,000. There are large docks, and the Belgian fishing fleet has its headquarters here. There are ferry links to Dover and Folkestone, England. It was occupied by the Germans 1914–18 and developed as an important naval base. 22 E3

Östersund chief town of Jämtland county, Sweden; population (1995) 59,700. Situated on Lake Storsjön, it is an important service and military centre. 16 H5

Ostrava industrial city (iron works, furnaces, coal, chemicals) in the Czech Republic, capital of Severomoravsky region, northeast of Brno; population (1991) 327,600. 18 H8

Oświęcim alternative name for ►Auschwitz, a town in southern Poland, site of the World War II extermination and concentration camp. 18 J7

Otaru fishing port on the west coast of Hokkaido island, Japan, 30 km/19 mi northwest of Sapporo; population (1993) 161,000. Industries include fish processing, and the manufacture of paper, sake, and ski equipment. Skiing takes place on the nearby mountains. Otaru grew as port for shipping coal from Ishikari coalfield from *c.* 1870. In 1880 it was linked to Sapporo by the first railway in Hokkaido. 48 L2

Otranto (ancient **Hydruntum**), seaport in Apulia, Italy, on the **Strait of Otranto**, 35 km/22 mi southeast of Lecce; population (1990) 21,600. It has Greek and Roman remains, a ruined castle (the inspiration for Horace Walpole's novel *The Castle of Otranto* 1764), and a cathedral begun in 1080. The port is linked by ferry with the island of Corfu. 32 N8

Ottawa capital of ►Canada, in eastern Ontario, on the hills overlooking the Ottawa River, and divided by the Rideau Canal into the Upper (western) and Lower (eastern) towns; population (1996 est) of metropolitan area (with adjoining Hull, Québec) 1,030,500. Industries include engineering, food-processing, publishing, lumber, and the manufacture of pulp, paper, textiles, and leather products. Government, and community and health services employ a large section of the workforce. Ottawa was founded 1826–32 as Bytown, in honour of John By (1781–1836), whose army engineers were building the Rideau Canal. In 1854 it was renamed after the Ottawa River, the name deriving from the Outaouac, Native Canadian Algonquin people of the area. 88 E1

Ottawa (French **Rivière des Ontaouais**), river in eastern Canada; length 1,271 km/790 mi. It rises in the Laurentian plateau and flows east through southern Québec and into the ►St Lawrence River near Montréal. During its course it widens into numerous Large lakes, and is fed by many tributaries. Hydroelectric schemes along its course supply energy for Québec and Ontario. 88 E1

Ouagadougou (or **Wagadugu**), capital and industrial centre of ►Burkina Faso, and of Kadiogo Province; population (1991 est) 634,000. Products include textiles, vegetable oil, beverages, and soap. Its pre-eminence as a commercial centre is challenged by Bobo-Dioulasso. The city has the palace of Moro Naba, emperor of the Mossi people, a neo-Romanesque cathedral, and a central avenue called the Champs Elysées. It was the capital of the Mossi empire from the 15th century. 68 D2

Oujda (or **Oudjda**), industrial and commercial city in northern Morocco, near the border with Algeria; population (city, 1994) 357,300. It trades in wool, grain, and fruit; there is lead, coal, and zinc mining. Tourism is important. The Moroccan and Algerian railway systems meet here. 66 E2

Oulu (Swedish **Uleåborg**), industrial port (sawmills, tanneries, shipyards) in western Finland, on the Gulf of Bothnia; population (1992) 103,500. It was originally a Swedish fortress 1375. 16 N4

Ourense alternative spelling of ►Orense, a city in northwest Spain. 28 C2

Ourthe river in Belgium, tributary of the ►Meuse; length 185 km/115 mi, of which 57 km/35 mi are navigable. It is formed in the Ardennes by the confluence of the East and West Ourthe, both rising in the province of Luxembourg. The river joins the Meuse near the city of Liège. 22 H4

Ouse, Great (Celtic 'water'), river that rises near Brackley in Northamptonshire, central England, and flows eastwards through Buckinghamshire, Bedfordshire, Cambridgeshire, and Norfolk, before entering the Wash north of King's

Lynn; length 250 km/160 mi. A large sluice across the Great Ouse, near King's Lynn, was built as part of extensive flood-control works in 1959. 24 N9

Outer Hebrides group of islands that comprise that part of the ▶Hebrides further from the west coast of mainland Scotland. 24 D4

Overijssel province of the east central Netherlands, extending from the IJsselmeer to the German border; area 3,340 sq km/1,290 sq mi; population (1997) 1,057,900. The capital is ▶Zwolle. Industries include textile production (cotton-spinning in the district of Twente). Other activities are livestock rearing, dairying, and fishing. 7 (1)E2

Oviedo capital of ▶Asturias autonomous community, northwest Spain, 25 km/16 mi south of the Bay of Biscay; population (1991) 194,900. Industries include textiles, metal goods, pharmaceuticals, matches, chocolate, and sugar. The university here dates from 1608, and there is a Gothic cathedral, which is mainly 15th century. 28 E1

Owensboro city, river port, and administrative headquarters of Davies County, northwest Kentucky; population (1992) 53,400. It is situated on the Ohio River, southwest of Louisville, and 61 km/38 mi southeast of Evansville, Indiana. It lies in a rich agricultural area producing tobacco, soybeans, corn, fruit, and wheat. With abundant oil, gas, and coal deposits nearby, the city makes iron and steel products. Other industries include chemicals, electronics, whiskey, cigars, plastics, and aluminium, and there are also canning and meatpacking plants. Owensboro is home to Kentucky Wesleyan College (1858) and Brescia College (1950). 88 C3

Oxford university city and administrative centre of ▶Oxfordshire in south central England, at the confluence of the rivers Thames (called the Isis around Oxford) and Cherwell, 84 km/52 mi northwest of London; population (1994 est) 121,000. Oxford University has 36 colleges, the oldest being University College (1249). Industries include steel products, electrical goods, car production, publishing (Oxford University Press, Blackwells), and English language schools. Tourism is important. 22 A3

Oxfordshire county of south central England.
area 2,610 sq km/1,007 sq mi
towns and cities ▶Oxford (administrative headquarters), Abingdon, Banbury, Goring, Henley-on-Thames, Wallingford, Witney, Woodstock, Wantage, Chipping Norton, Thame
physical River Thames and tributaries (the Cherwell, Evenlode, Ock, Thame, and Windrush); Cotswold Hills (in the north) and Chiltern Hills (in the southeast)
features Vale of the White Horse (with a chalk hill figure 114 m/374 ft, below the hill camp known as Uffington Castle); Oxford University; Blenheim Palace; Europe's major fusion project JET (Joint European Torus) at the UK Atomic Energy Authority's fusion laboratories at Culham; the Manor House, Kelmscott (country house of William Morris, leader of the Arts and Crafts Movement); Henley Regatta
agriculture cereals, sheep, dairy farming
industries agricultural implements; aluminium; bricks; cars (Cowley); cement; iron ore; hi-tech industries; medical electronic equipment; paper; publishing; nuclear research (Harwell); biotechnology
population (1996) 603,100 5 F6

Oxnard city in Ventura County, southwest California, USA; population (1992) 144,800. It is located 85 km/53 mi west-northwest of Los Angeles, to the north of Port Hueneme, on the Santa Barbara Channel of the Pacific Ocean. Industries include aerospace equipment, paper products, agricultural chemicals, mobile homes, processed foods, oil refining, and tourism. The city is in one of the state's most fertile agricultural regions; lemons and vegetables are grown in the area. Oxnard was founded 1898 and developed around a sugar-beet factory. 92 C2

Oxus ancient name of ▶Amu Darya, a river in Central Asia. 42 L9

Ozero Balkhash alternative name for Lake ▶Balkhash in eastern Kazakhstan. 40 L5

Ozero Baykal Russian name for Lake ▶Baikal, a freshwater lake in southern Siberia, Russia. 44 H6

Paarl town on the Great Berg River, Western Cape Province, South Africa; population (1991) 73,400. It is the centre of a wine-producing area, 50 km/31 mi northeast of Cape Town. There are large jam and canning factories, flour mills and light industries, including plastics, textile mills, tobacco factories, and granite quarries. 72 B6

Pabianice industrial town (textiles, chemicals, machinery) in Łódź province, Poland, 13 km/8 mi southwest of Łódź; population (1990) 75,200. 18 J6

Pacific Islands former (1947–1990) United Nations (UN) trust territory in the western Pacific captured from Japan during World War II. The territory comprised over 2,000 islands and atolls and was assigned to the USA in 1947. The islands were divided into four governmental units: the **Northern Mariana Islands** (except Guam) which became a self-governing commonwealth in union with the USA in 1975 (inhabitants granted US citizenship 1986); the ▶Marshall Islands, the Federated States of ▶Micronesia, and the Republic of ▶Palau (formerly also known as Belau) became self-governing 1979–80, signing agreements of free association with the USA in 1986. In December 1990 the UN Security Council voted to dissolve its trusteeship over the islands with the exception of Palau. The Marshall Islands and the Federated States of Micronesia were granted UN membership in 1991.

Pacific Ocean world's largest ocean, extending from Antarctica to the Bering Strait; area 166,242,500 sq km/64,186,500 sq mi; greatest breadth 16,000 km/9,942 mi; length 11,000 km/6,835 mi; average depth 4,188 m/13,749 ft; greatest depth of any ocean 11,524 m/37,808 ft in the Mindanao Trench, east of the Philippines. 74 M3

Padang port on the west coast of Sumatra, Indonesia; population (1990) 477,300. The Dutch secured trading rights here in 1663. The port trades in copra, coffee, and rubber. 52 C3

Paderborn market town in North Rhine-Westphalia, Germany, 24 km/15 mi south of Bielefeld, at the foot of the Teutoburger Wald; population (1995) 132,100. Industries include leather goods, metal products, agricultural machinery, precision instruments, textiles, and furniture. It was the seat of a bishopric in Charlemagne's time and later became a member of the Hanseatic League. 20 D5

Padova Italian name for ▶Padua, a town in Veneto, northern Italy. 30 G5

Padua (Italian **Padova**; ancient **Patavium**), town in Veneto, northern Italy, on the canalized section of the River Bacchiglione, 38 km/24 mi west of Venice; population (1992) 213,700. Industries include engineering, and the manufacture of clothing and artificial fibres. The astronomer Galileo taught at the university, founded in 1222. 30 G5

Paducah city and administrative headquarters of McCracken County, west Kentucky, near the border with Illinois, 137 km/85 mi southwest of Evansville, Indiana; population (1990) 27,300. Situated at the confluence of the Ohio and Tennessee rivers, it is a centre for tobacco, agriculture, and coal. It is also a centre of rail, river, and air transport. Growth has been generated by federal energy-production projects. 88 C3

Pafos Greek name for ▶Paphos, a town on the southwest coast of Cyprus. 36 Q10

Pagalu former name (1973–79) of ▶Annobón, an island in Equatorial Guinea. 68 F5

Pago Pago capital of American Samoa on the island of Tutuila; population (1990) 4,000. Its harbour nearly bisects the island. It exports tinned tuna. Formerly a naval coaling station, it was acquired by the USA under a commercial treaty with a local king in 1872. It became the capital of American Samoa in 1899. It is one of the best harbours in the central and southern Pacific, and until 1951 was a US naval station.

Pahang state of east Peninsular Malaysia; capital Kuantan; area 36,000 sq km/14,000 sq mi; population (1993) 1,056,100. It is mountainous and forested and produces rubber, tin, gold, and timber. There is a port at Tanjung Gelang. Pahang is ruled by a sultan. 11 (1)A2

Pahsien alternative name of ▶Chongqing, a city in Sichuan province, China. 40 P7

Paisley (Roman **Vanduara**), administrative headquarters of ▶Renfrewshire, Scotland, part of the Clydeside urban area, on the banks of the White Cart Water, 11 km/7 mi southwest of Glasgow; population (1991) 75,500. It is the largest town in Scotland. Industries include textiles and pharmaceuticals. It was an important centre for the manufacture of textiles, especially the woollen Paisley shawl, with a pattern based on the shape of a palm shoot. 24 H6

Pakaraima Mountains mountain range in the Guiana Highlands along the border between Brazil, Venezuela, and Guyana, extending from west to east for over 800 km/497 mi. Its highest peak is Mount Roraima at 2,810 m/9,222 ft above sea level, surrounded by cliffs 300 m/1,000 ft high, at the conjunction of the three countries. The range forms a watershed between the basins of the Orinoco and Branco rivers. Formed 300 million years ago, Pacaraima has unique fauna and flora consisting of grasses, bushes, flowers, insects, and small amphibians. 98 E2

Pakistan country in southern Asia, stretching from the Himalayas to the Arabian Sea, bounded to the west by Iran, northwest by Afghanistan, and northeast and east by India.

Pakistan covers an area of 803,940 sq km/310,321 sq mi, and has a population of 156,483,000 (2000 est). The capital is ▶Islamabad; other major towns and cities include Lahore, Rawalpindi, Faisalabad, Karachi, Hyderabad, Multan, Peshawar, Gujranwala, and Quetta. The head of state is Muhammad Rafiq Tarar (from 1998); the head of government is Gen Pervez Musharraf (from 1999). The currency is the Pakistan rupee, and the languages spoken include Urdu (official), English, Punjabi, Sindhi, Pashto, Baluchi, and other local dialects. 56 J4

Palau (or **Belau**), country comprising more than 350 islands and atolls (mostly uninhabited) in the west Pacific Ocean.

Palau covers an area of 508 sq km/196 sq mi, and has a population of 19,000 (2000 est). The capital is ▶Koror (on Koror island). The head of state and government is Kuniwo Nakamura (from 1993). The currency is the US dollar, and the languages spoken include Palauan and English (both official in most states). 74 D5

Palembang oil-refining city in Indonesia, capital of southern Sumatra province; population (1990) 1,084,500. Products include rubber and palm oil. Palembang was the capital of a sultanate when the Dutch established a trading station here in 1616. 52 C3

Palencia city and capital of the province of Palencia, northern Spain, on the River Carrión; population (1991) 77,800. Industries include rolling stock, agricultural machinery, textiles, pottery, and soap; it has a bell foundry. There is a Gothic cathedral, begun in 1321, which contains paintings by El Greco and Zurbarán. The first modern university in Spain was founded here in 1208, but was transferred to Salamanca later in the same century. 28 F2

Palermo capital and seaport of Sicily; population (1992) 696,700. It is also capital of Palermo province. Palermo is situated on the northern coastal plain of the Conca d'Oro, on a bay of the Tyrrhenian Sea. Industries include shipbuilding, steel, glass, textiles, and chemicals. Palermo is the most important industrial centre and port with shipbuilding facilities in southern Italy after Naples. It was founded by the Phoenicians in the 8th century BC. 32 H10

Palestine (Arabic *Falastin*, 'Philistine'), historic geographical area at the eastern end of the Mediterranean sea, also known as the Holy Land because of its historic and symbolic importance for Jews, Christians, and Muslims. Early settlers included the Canaanites, Hebrews, and Philistines. Over the centuries it became part of the Egyptian, Assyrian, Babylonian, Macedonian, Ptolemaic, Seleucid, Roman, Byzantine, Arab, Ottoman, and British empires. Today it comprises parts of modern Israel–Palestine and Jordan.

Palikir capital of the Federated States of Micronesia, on the island of Pohnpei; population (1998 est) 9,885. 74 F5

Palk Strait channel separating southeast India from the island of Sri Lanka, lying north of the sandbanks known as Adam's Bridge, which separates it from the Gulf of Mannar; it is 53 km/33 mi at the narrowest point. It was named by the Dutch after Governor Palk. 54 C7

Palma de Mallorca port and capital of the ►Balearic Islands, an autonomous community of Spain, situated on a wide bay on the southwest coast of the island of ►Mallorca, population (1994) 322,000. Industries include textiles, cement, paper, pottery, and tourism. Palma was founded as a Roman colony in 276 BC. It has a Gothic cathedral, begun in 1229; the 14th-century Almudaina palace, a former royal residence; and the 13th-century church of St Francis of Assisi, which contains the tomb of the Mallorcan scholar Ramon Llull. 28 N5

Palm Beach luxurious winter resort in southeast Florida, USA, on an island between Lake Worth and the Atlantic; population (1996 est) 9,700. Bridges link the island beaches with ►West Palm Beach on the mainland, a larger neighbour with a deep-water harbour and international airport. Palm Beach was settled in the 1870s.

Palmerston North town on the Manawatu River, southwest North Island, New Zealand, 140 km/87 mi north of Wellington; population (1996) 73,900. Industries include textiles, dairy produce, and electrical goods. 78 E5

Palm Springs resort and spa in Riverside County, southern California; population (1990) 40,200. It is 169 km/105 mi east-southeast of Los Angeles, at the western end of the Coachella Valley. The resort is protected from rain and fog by the San Jacinto and the Santa Rosa Mountains, and its dry climate attracts many visitors. Though situated in the Colorado Desert, it has mineral springs and palm trees; the Spanish who explored the area in 1774 called it Agua Caliente ('hot water'). The Cahuilla, who then inhabited the area, still own much of the land.

Palmyra (Arabic **Tadmur**), ancient city and oasis in the desert of Syria, about 240 km/150 mi northeast of Damascus. Palmyra, the biblical **Tadmor**, was flourishing by about 300 BC. It was destroyed in AD 272 after Queen Zenobia led a revolt against the Romans. Extensive temple ruins exist, and on the site is a modern town called Tadmur. 58 H6

Palmyra Island coral atoll 1,600 km/ 1,000 mi southwest of Hawaii, in the Line Islands, South Pacific. It was part of Hawaii from 1862, but was excluded when Hawaii became a state in 1959. It was purchased by the USA from a Hawaiian family in 1979 and was used for the storage of highly radioactive nuclear waste from 1986. Palmyra is now a yacht stopover and a tourist destination. 74 K5

Palo Alto city in central California, USA, on the west shore of San Francisco Bay, near the base of the San Francisco peninsula; 70 km/44 mi southeast of San Francisco; population (1996 est) 58,300. Situated at one end of the high-tech region known as Silicon Valley, it is a centre for the electronics and computer industries. It is also the seat of Stanford University, and numerous related research institutions. Palo Alto was founded 1891.

Pampas flat treeless plains in central Argentina, lying between the foothills of the Andes and the Atlantic coast, and stretching north from the Río Colorado to the ►Gran Chaco; area 750,000 sq km/290,000 sq mi; it incorporates the provinces of Buenos Aires, La Pampa, Santa Fé, and Cordobá. The eastern Pampas consist of grasslands which support large cattle ranches and produce flax and over half the nation's output of grain; the western Pampas are arid and unproductive. The characteristic vegetation is the **pampas grass** which grows to a height of 2–3m/6–10 ft. 100 J6

Pamplona industrial city and capital of the autonomous community of ►Navarre, northern Spain, on the River Arga; population (1991) 179,200. Industries include wine, textiles, leather, and footwear. An annual running of bulls takes place in the streets in July as part of the fiesta of San Fermín, a local patron saint. There are remains of the medieval walls, a Gothic cathedral with an 18th-century facade, and a modern university. 28 J2

Panamá capital of the Republic of Panama, on the east bank of the Pacific entrance to the Panama Canal, with its port at Balboa in the Canal Zone; population (1995 est) 663,000. It became capital in 1903 following independence from Colombia. The city developed rapidly following the completion of the Canal. Its good communications by air and rail, as well as by the Pan-American Highway, have enabled the city to become the main transport and industrial centre of Panama. Products include oil, plastics, leather, food, and drink. 98 B2

Panama country in Central America, on a narrow isthmus between the Caribbean and the Pacific Ocean, bounded west by Costa Rica and east by Colombia.

Panama covers an area of 77,100 sq km/29,768 sq mi, and has a population of 2,856,000 (2000 est). The capital is ►Panamá; other major towns and cities include San Miguelito, Colón, David, La Chorrera, Santiago, Chitré, and Changuinola. The head of state and government is Mireya Moscoso (from 1999). The currency is the balboa, and the languages spoken include Spanish (official) and English. 94 H7

Panama Canal canal across the Panama isthmus in Central America, connecting the Pacific and Atlantic oceans; length 80 km/50 mi, with 12 locks; average width 150 m/492 ft. It was built by the USA 1904–14, after an unsuccessful attempt by the French. The **Panama Canal Zone** was acquired 'in perpetuity' by the USA in 1903, comprising land extending about 5 km/3 mi on either side of the canal. The zone passed to Panama in 1979, and under a treaty signed by President Carter in 1978, control of the canal was ceded to Panama at the end of 1999. In December 1999 all US military bases in the Canal Zone were closed, and the control change-over went into effect on 31 December 1999. 94 J7

Pan-American Highway road linking the USA with Central and South America; length 25,300 km/ 15,700 mi. Starting from the US-Canadian frontier (where it links with the Alaska Highway), it runs through San Francisco, Los Angeles, and Mexico City to Panamá, then down the west side of South America to Valparaiso, Chile, where it crosses the Andes and goes to Buenos Aires, Argentina. The road was first planned in 1923, and work began in 1928. Completion of the final section, across the Darien Gap in Panama, will lead to a major ecological transformation of the region.

Panay one of the Philippine islands, lying between Mindoro and Negros; area 11,515 sq km/4,446 sq mi. The capital is Iloilo. Industries include rice, sugar, pineapples, bananas, copra, and copper. The island is mountainous, reaching 2,215 m/7,265 ft in Madiaás. It was seized by Spain in 1569, and occupied by Japan 1942–45. **50** G4

Pantanal large area of swampland in the Mato Grosso of southwestern Brazil, occupying 220,000 sq km/ 84,975 sq mi in the upper reaches of the Paraguay River; one of the world's great wildlife refuges, of which 1,370 sq km/530 sq mi were designated a national park 1981. **98** F7

Pantelleria Island (Roman **Cossyra**), volcanic island in the Mediterranean, 100 km/62 mi southwest of Sicily and part of that region of Italy; area 115 sq km/45 sq mi. The majority of the population live in the island's capital Pantelleria, on the northwest coast; population (1990) 7,700. Sheep are raised, and fruit, olives, and capers produced. **66** H1

Papeete capital and port of French Polynesia on the northwest coast of Tahiti; population (1992) 24,200. Products include vanilla, copra, and mother-of-pearl.

Paphos (Greek **Pafos**), resort town on the southwest coast of Cyprus; population (1992 est) 32,600. It was the capital of Cyprus in Roman times and the legendary birthplace of the goddess Aphrodite, who rose out of the sea. Archaeological remains include the 2,300-year-old underground 'Tombs of the Kings', a Roman villa, and a 7th-century Byzantine castle. **36** Q10

Papua original name of the island of New Guinea, but latterly its southeastern section, now part of ▶Papua New Guinea. **74** E6

Papua New Guinea country in the southwest Pacific, comprising the eastern part of the island of New Guinea, the Bismarck Archipelago, and part of the Solomon Islands.

Papua New Guinea covers an area of 462,840 sq km/178,702 sq mi, and has a population of 4,807,000 (2000 est). The capital is ▶Port Moresby (on East New Guinea); other major towns and cities include Lae, Madang, Arawa, Wewak, Goroka, and Rabaul. The head of state is Queen Elizabeth II (from 1952), represented by Governor General Silas Atopare (from 1997); the head of government is Mekere Morauta (from 1999). The currency is the kina, and the languages spoken include English (official), pidgin English, and over 700 local languages. **74** E6

Pará state of northern Brazil, bounded in the north by Guyana and Suriname, and in the northeast by the Atlantic; area 1,246,833 sq km/ 481,405 sq mi; population (1991) 5,181,570; capital ▶Belém. It is the second-largest state in Brazil; the Amazon delta lies wholly within it. There are dense rain forests and the area is well watered by the Amazon, Pará, Tocatins, Xingu, and Tapajos rivers. Timber, rubber, and brazilnuts are produced, and cattle are raised. Chief towns include Obidos, Santarém, Marabá and Cameta. **98** G5

Paracel Islands (Chinese **Xisha**; Vietnamese **Hoang Sa**), group of about 130 small islands in the South China Sea. Situated in an oil-bearing area, they were occupied by China following a skirmish with Vietnam in 1974. **50** E3

Paraguay landlocked country in South America, bounded northeast by Brazil, south by Argentina, and northwest by Bolivia.

Paraguay covers an area of 406,752 sq km/157,046 sq mi, and has a population of 5,496,000 (2000 est). The capital (and chief port) is ▶Asunción; other major towns and cities include Ciudad del Este, Pedro Juan Caballero, San Lorenzo, Fernando de la Mora, Lambare, Luque, and Capiatá. The head of state and government is Luis Gonzalez Macchi (from 1999). The currency is the guaraní, and the languages spoken include Spanish (official) and Guaraní (an indigenous Indian language). **100** J3

Paramaribo chief port and capital of ▶Suriname, on the west bank of the Suriname River near its mouth on the Atlantic coast; population (1996) 150,000. Products include coffee, fruit, timber, and bauxite. **98** F2

Paraná industrial port and capital of Entre Ríos department, eastern Argentina; situated on the east bank of the Paraná River, 560 km/350 mi northwest of Buenos Aires; population (1991) 276,000. It is an outlet for cattle, sheep, fruit, and grain, produced in the surrounding region, and there are food processing industries. **100** J5

Paraná river in South America; length 4,500 km/2,800 mi. It is formed by the confluence of the Río Grande and Paranaiba, both of which rise in southeast Brazil; the Paraguay River joins it at Corrientes in Argentina. It then flows generally southwards into the Río de la Plata estuary; length 4,500 km/2,800 mi. The river's lower course is navigable for large vessels for 1,600 km/994 mi; its upper course is interrupted by numerous cataracts and rapids. It is used for hydroelectric power by Argentina, Brazil, and Paraguay. The Itaipu Dam, near the Brazilian border with Paraguay, is the largest hydroelectric plant in the world. **100** K4

Pardubice (German **Pardubitz**), industrial market town of eastern Bohemia, in the central Czech Republic, on the River Elbe, 94 km/ 59 mi east of Prague; population (1991) 94,900. It has oil and sugar refineries, chemical, engineering, electrical, and brewing industries, and trades in grain and timber. **18** E7

Paris port and capital of France, on the River Seine; *département* (Ville de Paris) in the Île-de-France region; area of the *agglomération parisienne* (comprising the Ville de Paris and 379 *communes* surrounding it) 105 sq km/40.5 sq mi; population Ville de Paris (1998 est) 2,077,500; *agglomération parisienne* (1996 est) 9,523,000. The city is the core of a highly centralized national administration, a focus of European transport networks, and France's leading centre for education, research, finance, and industry. Manufactured products include metal, electrical and luxury goods, chemicals, glass, and tobacco. As one of the world's principal historic and cultural centres, Paris attracts enormous numbers of tourists throughout the year. **26** H5

Paris, Ville de French *département* comprising the central districts of the city of ▶Paris, part of the ▶Ile-de-France region; area 105 sq km/40.5 sq mi; population (1990) 2,151,245. **6** E2

Parkersburg city, river port, and administrative headquarters of Wood County, northwest West Virginia,

USA, at the mouth of the Little Kanawha River on the Ohio River, 28 mi/45 km southwest of Wheeling and 70 mi/112 km north of Charleston; population (1996 est) 32,800. It is a processing and distribution center for the oil, natural gas, coal, and clay found in the surrounding region. Industries include the manufacture of chemicals, plastics, synthetic materials, glassware, porcelain, and paper. A West Virginia University campus (1971) and Ohio Valley College (1960) are in the city. **88** D3

Parma town in Emilia-Romagna, northern Italy, on the River Parma, 85 km/53 mi northwest of Bologna; population (1992) 170,600. It is the second largest in the region after Bologna. Industries include food-processing, oil-refining, engineering, and the production of textiles. Founded by the Etruscans, it was the capital of the duchy of Parma 1545–1860. It has given its name to a type of ham, and to Parmesan cheese. **30** F6

Parry Islands group of islands in the Canadian Arctic, west of Baffin Bay and north of Lancaster Sound, Viscount Melville Sound, and Barrow Strait. The Parry Islands (a subgroup of the Queen Elizabeth Islands) include Bathurst Island, Melville Island, Cornwallis Island, and Prince Patrick Island. They are covered with tundra and are uninhabited. The Arctic explorer William Edward Parry, after whom they are named, visited them in 1819, and the expeditions searching for John Franklin explored them further. **82** L1

Pasadena city in southwest California, USA, in the foothills of the San Gabriel Mountains; part of Greater ▶Los Angeles and 16 km/10 mi northeast of Los Angeles centre; population (1996 est) 134,100. Products include electronic equipment, precision instruments, plastics, and ceramics. It was founded by farmers from Indiana in 1874, and originally developed as a health resort. **92** C2

Pasco town, river port, and administrative headquarters of Franklin County, southeast Washington, USA; population (1990) 20,300. It is situated in the Columbia Basin, on the Columbia River, to the northwest of the junction with the Snake River, opposite Kennewick, and 8 mi/13 km southeast of Richland. It is a major trade and transportation hub for southeast Washington. The town lies in a rich agricultural area, and the Pasco Farmers' Market is the region's largest. **86** C1

Pas-de-Calais *département* in the ▶Nord-Pas-de-Calais region of France, with a coastline on the ▶Strait of Dover; area 6,639 sq km/2,563 sq mi; population (1990) 694,700. The land is generally flat and well drained. The principal agricultural products are cereals, potatoes, vegetables, sugar beet, and livestock farming is important. There are textile, metallurgical, sugar, and other manufacturing industries. The coastal areas have fishing and tourism. The *département* was the scene of severe fighting in both world wars. The administrative centre is ▶Arras. Other towns include Béthune, Boulogne-sur-Mer, St-Omer, Calais, and Lens. **6** D1

Passau town in southeast Bavaria, Germany, 148 km/92 mi northeast of Munich; population (1995) 51,000. It lies on the Austrian border at the junction of the rivers Inn and Ilz with the Danube. There are machinery, textile, and pharmaceutical industries. **The Treaty of Passau** (1552) between Maurice, Elector of Saxony, and the future emperor Ferdinand I allowed the Lutherans full religious liberty and prepared the way for the Peace of Augsburg. **20** J8

Patagonia geographic region of South America, in southern Argentina and Chile; area 780,000 sq km/301,000 sq mi. A thinly populated vast plateau area, it stretches from the Río Colorado in central Argentina to the Magellan Straits in the south, and slopes eastwards from the Andes to the Atlantic coast. It consists of the provinces of Neuquén, Rio Negro, Chubut, and Santa Cruz. The main towns are the port of Comodoro Rivadavia (Argentina) and Punta Arenas (Chile). **100** G8

Paterson city of northeast New Jersey, USA, on the Passaic River; population (1990) 140,900. It is part of the industrial and residential conurbation of Paterson–Clifton–Passaic. First settled 1771 by American revolutionaries, Paterson developed as an important centre of the textile industry owing to the water power available from the Passaic Falls. It produces machinery and chemicals, and attracts visitors to the falls. **88** F2

Pathein port in Myanmar (Burma), in the Irrawaddy delta, 125 km/78 mi from the sea; population (1998 est) 190,500. Pathein was founded in the 13th century. **50** A3

Patiala city in east Punjab, India; population (1991) 269,000. Industries include textiles and metalwork. It was formerly capital of an independent Sikh state. Notable buildings include the Old Motibagh Palace (late 19th century) and the 18th-century Bahadurgarh fort. **54** C2

Patmos Greek island in the Aegean Sea, one of the Dodecanese; the chief town is Hora. St John is said to have written the New Testament Book of Revelation while in exile here. **36** J7

Patna capital of ▶Bihar state, India, on the River Ganges, 450 km/280 mi northwest of Kolkata (formerly Calcutta); population (1991) 917,000. It has remains of a hall built by the emperor Asoka in the 3rd century BC; it was capital of several kingdoms in the succeeding centuries, under the name of Pataliputra. **54** E3

Patra Greek name for ▶Patras, a city in the northwestern Peloponnese region of Greece. **36** D6

Patras (Greek **Patra**), industrial city (hydroelectric installations, textiles, paper) in the northwestern Peloponnese region, Greece, on the Gulf of Patras; population (1991 est) 155,200. The ancient **Patrai**, it is the only one of the 12 cities of the ancient Greek province of Achaea to survive. **36** D6

Pau industrial city and administrative centre of the ▶Pyrénées-Atlantiques *département* in Aquitaine, southwest France, on the River Gave de Pau near the Spanish border; population (1990) 83,900. It is the centre of the ▶Basque area of France, and the site of fierce guerrilla activity. The town manufactures electrochemical and metallurgical industries, aeronautical equipment and shoes. It has a university. **28** K1

Pays de la Loire agricultural region of western France, comprising the *départements* of ▶Loire-Atlantique, ▶Maine-et-Loire, ▶Mayenne, ▶Sarthe, and Vendée; area 32,100 sq km/12,400 sq mi; population (1990) 3,059,100. The administrative centre is ▶Nantes. Industries include shipbuilding and wine production. **6** C3

Peace river of western Canada, formed in British Columbia by the union of the Finlay and Parsnip rivers at Finlay Forks, and flowing through the Rocky Mountains and northeast across Alberta to join with the northward outflow of Lake Athabasca; length 1,600 km/1,000 mi. Proceeding as the Slave River, the waters travel a further 193 km/121 mi before emptying into the ►Great Slave Lake. 82 H5

Peak District elevated plateau of the south ►Pennines in northwest Derbyshire, central England; area 1,438 sq km/555 sq mi. It is a tourist region and part of it forms a national park. The highest point is Kinder Scout (636 m/2,087 ft), part of High Peak. In the surrounding area the main cities are Manchester, Sheffield, and Derby, and the town of Bakewell is located within the Peak District.

Pec town in Serbia, Yugoslavia, in the autonomous region of Kosovo; population (2001 est) 77,800. It is on the River Bistrica, at the foot of Mount Koprivnik. Pec is a busy commercial town. In the Middle Ages it was the seat of the Serbian patriarchs; the great monastery complex in which they resided remains. 34 H7

Pechora river in northern Russia, rising on the western slopes of the northern Urals and flowing north to the Barents Sea; length 1,800 km/ 1,125 mi. In total, the Pechora drains an area of some 322,000 sq km/ 124,324 sq mi. The river, whose chief ports are Pechora and Naryan-Mar, is navigable for most of its length, and is used during the ice-free months (June–September) for transporting coal (widely mined in the region), timber, and furs. 38 K1

Pecos river in southern USA; length 1,490 km/926 mi. It rises in the Sangre de Cristo Mountains in northeast New Mexico, and flows southeast through Texas, to join the ►Rio Grande, 55 km/34 mi northwest of Del Rio. Its principal importance is as a source of irrigation. 92 F2

Pécs city in southwestern Hungary, the centre of a coal-mining area on the Croatian frontier; population (1995 est) 163,000. Industries include metal, leather, and wine. The town dates from Roman times and was under Turkish rule 1543–1686. 34 F3

Pegu former name of Bago River and division in Myanmar, and an alternative name for ►Bago, a city in southern Myanmar. 50 B3

Peine city in Lower Saxony, Germany, 35 km/22 mi southeast of Hannover; population (1994) 125,800. Oil and gas deposits in the vicinity are exploited and the city also has a steelworks. 20 F4

Peiping ('northern peace') name of ►Beijing in China from 1928 to 1949. 46 F3

Peipus, Lake (Estonian **Peipsi**; Russian **Chudskoye**), lake in eastern Estonia and western Russia; the boundary between the Baltic state and the Pskov oblast of the Russian Federation runs north–south across the centre of the lake. It covers an area of 3,555 sq km/1,373 sq mi, and its chief outlet is the Narva River, which flows north from the lake to the Gulf of Finland. The Russian hero Alexander Nevski defeated the Teutonic Knights on its frozen surface in 1242. 16 P7

Peiraias port of both ancient and modern Athens and main port of Greece, on the Gulf of Aegina; population (1991) 169,600. Constructed as the port of Athens about 493 BC, it was linked with that city by the Long Walls, a fortification protecting the approaches to Athens comprising three walls built 496–456 BC. After the destruction of Athens by Sulla 86 BC, Peiraias declined. Peiraias is now an industrial suburb of Athens. 36 F7

Peking alternative transcription of ►Beijing, the capital of China. 46 F3

Peloponnese (Greek **Peloponnisos**), peninsula forming the southern part of Greece; area 21,549 sq km/8,320 sq mi; population (1991 est) 1,077,000. It is joined to the mainland by the narrow isthmus of Corinth and is divided into the nomes (administrative areas) of Argolis, Arcadia, Achaea, Elis, Corinth, Lakonia, and Messenia, representing its seven ancient states. It is divided into two regions; Greece West (including Achaea and Elis), and Peloponnese (including Argolis, Arcadia, Corinth, Lakonia, and Messenia). 36 D7

Peloponnisos Greek name for the ►Peloponnese, the peninsula forming the southern part of Greece. 36 D7

Pemba Island coral island in the Indian Ocean, 48 km/30 mi northeast of Zanzibar, and forming with it part of Tanzania; area 984 sq km/ 380 sq mi; population (1995 est) 320,000. The capital is Chake Chake; Wete is the only other town of note. The economy is based on cloves and copra, and Pemba is now the world's biggest producer of cloves. 70 F5

Pembroke (Welsh **Penfro**; Welsh 'land's end'), seaport and engineering centre in Pembrokeshire, southwest Wales; population (1991) 8,650. Henry VII was born in Pembroke Castle in 1457. Pembroke Dock was created in 1814 in conjunction with the Royal Navy dockyard, and there is now some light industry in the refurbished dock areas. Tourism is a growing industry. A car ferry operates between here and Rosslare in the Republic of Ireland. 24 H10

Pembrokeshire (Welsh **Sir Benfro**), unitary authority in southwest Wales; a former county, from 1974 to 1996 it was part of the county of Dyfed.
area 1,588 sq km/613 sq mi
towns Haverfordwest (administrative headquarters), ►Milford Haven
physical bounded on the south by the Bristol Channel; valleys and hills inland; rivers East and West Cleddau
features Pembrokeshire Coast National Park
industries oil refinery at ►Milford Haven, agriculture, fishing, woollen milling.
population (1996) 117,700 5 D6

Penang (Malay **Pulau Pinang**), state in west Peninsular Malaysia, formed of **Penang Island**, Province Wellesley, and the Dindings on the mainland; area 1,030 sq km/398 sq mi; capital Penang (George Town); population (1993) 1,141,500. Penang Island was bought by Britain from the ruler of Kedah in 1785; Province Wellesley was acquired in 1800. 11 (1)A1

Péninsule d'Ungava full name for ►Ungava, a peninsular region in northern Québec and Labrador, Canada. 82 R4

Pennines, the range of hills in northern England, known as the 'the backbone of England'; length (from the Scottish border to the Peaks in Derbyshire) 400 km/250 mi. The highest peak in the Pennines (which are sometimes referred to mountains rather than hills) is Cross Fell

(893 m/2,930 ft). It is the watershed for the main rivers of northeast England. The rocks are carboniferous limestone and millstone grit, the land high moorland and fell. 24 K7

Pennsylvania state in northeastern USA. It is nicknamed the Keystone State. Pennsylvania ratified the US Constitution in 1787, becoming the 2nd state to join the Union. It is bordered to the north by New York, with a small coastal strip on Lake Erie, to the west by Ohio and the West Virginia panhandle, to the south, on what was the ►Mason-Dixon Line, by West Virginia, Maryland, and Delaware, and to the east by New Jersey, across the Delaware River. Pennsylvania was the hub of US industry in the late 19th and early 20th century.
population (2000 est) 12,281,100
area 117,400 sq km/45,316 sq mi
capital ►*Harrisburg*
towns and cities Philadelphia, Pittsburgh, Erie, Allentown, Scranton
industries and products hay, cereals, mushrooms, cattle, poultry, dairy products, cement, limestone, coal, steel, petroleum products, pharmaceuticals, chemicals, motor vehicles and equipment, electronic components, textiles, tourism 88 E2

Pensacola port in northwest Florida, USA, on **Pensacola Bay** in the Gulf of Mexico; population (1996 est) 59,200. It is the largest natural deep-water harbour in Florida, and contains shipyards and a large naval air-training station (1914). Industries include fishing, and the production of chemicals, synthetic fibres, pulp, and paper. Pensacola developed around a Spanish fort in the 18th century. 94 G2

Penza capital city, economic and cultural centre of Penza oblast (region), western Russian Federation, 310 km/193 mi west of Samara on the Sura River; population (1996 est) 534,000. Manufactured goods produced here include precision instruments (including watches), agricultural machinery, and paper. Penza stands at an important railway junction, with lines running west to Samara, south to Volgograd, and north to Nizhi Novgorod. 38 J4

Penzance seaport and resort in Cornwall, southwest England, on Mount's Bay 38 km/24 mi southwest of Truro; population (1991) 19,700. The most westerly town in England, it has a ferry link with the Scilly Isles. It is the centre of a market-gardening and agricultural area, and early fruit, flowers, and vegetables are produced. It now incorporates the seaport of Newlyn. It is known as the 'Cornish Riviera'. 24 G11

Peoria city and river port in central Illinois, USA, on the Illinois River, 208 km/130 mi southwest of Chicago; population (1996 est) 112,300. It is the third-largest city in Illinois, and an important transport, mining, and agricultural centre on the Great Lakes–Gulf of Mexico inland waterway system. Products include bituminous coal, livestock, and grain. Its principal industries are distilling, food-processing, and the manufacture of computer-ware, steel, tractors, and earth-moving equipment. 88 C2

Perak state of west Peninsular Malaysia; capital Ipoh; area 21,000 sq km/8,000 sq mi; population (1993) 2,222,400. It produces tin and rubber. The government is a sultanate. The other principal city is Taiping. 11 (1)A2

Pereira capital of Risaralda department, central Colombia, situated at an altitude of 1,463 m/4,800 ft, overlooking the fertile Cauca Valley, west of Bogotá; population (1994) 347,000. Founded in 1863, the city has developed into a centre for the national coffee and cattle industries. 98 B3

Périgueux administrative centre of the ►Dordogne *département* in Aquitaine, France, on the River Isle, 127 km/79 mi east of Bordeaux; population (1990) 32,800, conurbation 60,000. It is a trading centre for wine, *foie gras* (goose-liver pâté), and truffles. The town has Roman remains and many old buildings, including the Romano-Byzantine cathedral of St Front, which dates from the 12th century but was reconstructed in the mid-19th century. 26 F8

Perlis border state of Peninsular Malaysia, northwest Malaysia; capital Kangar; area 800 sq km/309 sq mi; population (1993) 187,600. It produces rubber, rice, coconuts, and tin. Perlis is ruled by a raja. It was transferred by Siam (Thailand) to Britain in 1909. 11 (1)A1

Perm (formerly **Molotov** 1940–57), capital city, economic and cultural centre of Perm oblast (region), in the west-central Russian Federation, in the southern foothills of the Urals; population (1996 est) 1,028,000. Perm, which lies on the River Kama 475 km/295 mi northeast of Kazan, is a major centre of commerce and transportation. It is also an important industrial centre, producing machinery, chemicals, oil by-products, and timber. 38 L3

Pernambuco federal unit (state) of northeast Brazil, on the Atlantic coast; area 101,000 sq km/39,000 sq mi; population (1991) 7,122,500; capital Recife (former name Mauritzstaadt). The state is situated mainly on the northeastern plateau, with a narrow fertile coastal belt. The River San Francisco flows along much of its southern border. Sugar cane, cotton, coffee, and sisal are grown in the east and cattle are raised in the west. There are steel and sugar refining industries. 98 K5

Pernik industrial city (coal mining, steel, glassware) in western Bulgaria, 25 km/16 mi southwest of Sofiya; population (1990) 121,000. 34 L7

Perpignan administrative centre of ►Pyrénées-Orientales *département*, southern France, situated just off the Mediterranean coast on the River Tet, near the Spanish border; population (1990) 108,000. Overlooking Perpignan is a vast citadel enclosing the castle in which the kings of Majorca and later the counts of Roussillon lived. It is a holiday resort and market town trading in olives, fruit, and wine, and textiles and paper are manufactured. There is a university here. 26 H11

Persian Gulf alternative name for the ►Gulf, a large inlet of the Arabian Sea. 61 C2

Perth capital of the state of ►Western Australia; population (1996) 1,096,829. Perth is situated on the southwest coast of Australia, on the River Swan, 19 km/12 mi inland. Its port is at ►Fremantle, to the southwest at the mouth of the Swan. Industries include oil refining, electronics, food processing, shipbuilding, banking and finance, and tourism; products include textiles, nickel, alumina, fertilizers, cement, furniture, and motor vehicles. Perth is an important centre for the export of primary products: refined oil, minerals, wool, wheat, meat, fruit, timber, and dairy produce. Perth has four universities: the University of Western Australia (founded 1911); Murdoch University (1975); Curtin University of Technology (1987); Edith Cowan University (1990). 76 C6

Perth town and administrative headquarters of ►Perth and Kinross, central Scotland, on the River Tay, 70 km/43 mi northwest of Edinburgh; population (1991) 41,500. It is known as the 'fair city'. Industries include dyeing, textiles, whisky distilling, and light engineering. It is an important agricultural centre, noted for the sale of pedigree livestock, particularly young beef cattle. It was the capital of Scotland from the 12th century until 1452. James I of Scotland was assassinated here in 1437. 24 J5

Perth and Kinross unitary authority in central Scotland, created in 1996 from the district bearing the same name in Tayside region.
area 5,388 sq km/2,080 sq mi
towns Blairgowrie, Crieff, Kinross, ►Perth (administrative headquarters), Pitlochry, Aberfeldy
physical geological fault passes southwest–northeast through the area. There are wide fertile valleys such as Strathearn, and the Carse of Gowrie. To the north and west are the Grampians intersected by narrow glens with lochs in their valley floors. Among the highest elevations in the Grampians are Ben Lawers (1,214 m/ 3,984 ft) and Schiehallion (1,083 m/ 3,554 ft); in the south are the lower Ochil and Sidlaw Hills
features Highland Games at Pitlochry; Dunkeld Cathedral; Scone Palace; Glenshee Ski Development
industries woollen manufacture, whisky distilling and blending
agriculture diverse, including soft fruit (Carse of Gowrie), arable crops, livestock, salmon fisheries
population (1996) 131,800 5 E3

Peru country in South America, on the Pacific, bounded north by Ecuador and Colombia, east by Brazil and Bolivia, and south by Chile.

Peru covers an area of 1,285,200 sq km/496,216 sq mi, and has a population of 25,662,000 (2000 est). The capital is ►Lima; other major towns and cities include Arequipa, Iquitos, Chiclayo, Trujillo, Huancayo, Piura, and Chimbote. The head of state is Valentin Paniagua (from 2000); the head of government is Javier Pérez de Cuéllar (from 2000). The currency is the nuevo sol, and the languages spoken include Spanish, Quechua (both official), Aymara, and many indigenous dialects. 98 C6

Perugia (ancient **Perusia**), capital of Umbria, Italy, 520 m/1,700 ft above the River Tiber, about 137 km/85 mi north of Rome; population (1992) 146,200. Textiles, liqueurs, and chocolate are produced. One of the 12 cities of the ancient country of Etruria, it surrendered to Rome in 309 BC. 32 G5

Pescara (ancient **Aternum**), canal port and resort in Abruzzi, eastern Italy, at the mouth of the Pescara River, on the Adriatic coast, 160 km/ 99 mi northeast of Rome; population (1992) 121,400. Hydroelectric installations supply Rome with power. Industries include shipbuilding, tourism, fishing, and the manufacture of clothing, processed foods, and building materials. Pescara is linked by ferry to Split in Croatia. 32 J6

Peshawar capital of North-West Frontier Province, Pakistan, 18 km/ 11 mi east of the Khyber Pass, on the Bara River; population (1998) 988,000. It has long been a major centre of trade because the Khyber Pass provides the easiest route between the Indian subcontinent and Afghanistan. Products include fruit, textiles, leather, and copper. 54 B2

Peski Karakumy alternative name for ►Kara-Kum, a sandy desert in Turkmenistan. 56 G2

Peski Kyzyylkum alternative name for ►Kyzyl-Kum, a desert in southern Kazakhstan and Uzbekistan. 42 L9

Petaluma town in Sonoma County, northwest California; population (1990) 43,200. It is located 32 mi/52 km north of San Francisco, on the Petaluma River. It is an agricultural center for poultry and dairy products. Other industries include winemaking and the production of feed, fertilizers, and a wide variety of light manufactures. Settled by Mexicans in 1833, Petaluma is now also a growing residential suburb. 86 B3

Peterborough city in Ontario, Canada, on the Otonabee River, 113 km/70 mi northeast of Toronto and 45 km/28 mi north of Lake Ontario; population (1991) 68,400. Peterborough is the chief commercial centre for central Ontario. It has a large cereal mill and electrical-goods plant and also manufactures marine equipment. The world's largest deposit of nepheline, used in the manufacture of glass, is mined north of the city, and the refined product is shipped to the glass industry round the world. The surrounding lake district area attracts tourists. 88 E2

Peterborough unitary authority in eastern England, created in 1998 from part of Cambridgeshire.
area 334 sq km/129 sq mi
towns and cities ►Peterborough (administrative headquarters), Wittering, Old Fletton, Thorney, Glinton, Northborough, Peakirk
features River Nene; western margins of the Fens; St Peter's Cathedral (Peterborough), 12th century, containing Catherine of Aragon's tomb; Wildfowl and Wetlands Centre at Peakirk
industries aluminium founding and manufacture, electronics, domestic appliances, plastics and rubber manufacture, precision engineering, telecommunications equipment, food manufacture and processing
population (1996) 156,900 5 F5

Peterborough city in eastern England, on the River Nene, 64 km/40 mi northeast of Northampton, and from April 1998 administrative headquarters of ►Peterborough unitary authority; population (1994 est) 139,000. Situated on the edge of the ►Fens in the centre of an agricultural area, it is one of the fastest growing cities in Europe. Industries include sugar-beet refining, foodstuffs, aluminium founding and manufacturing, agricultural machinery, engineering, brick-making, diesel engines, and refrigerators. It has an advanced electronics industry. It is noted for its 12th-century cathedral. Nearby Flag Fen disclosed in 1985 a well-preserved Bronze Age settlement of 660 BC. The 17th-century Thorpe Hall is a cultural and leisure centre. 24 M9

Peterhead port in Aberdeenshire, Scotland, 54 km/33 mi northeast of Aberdeen; population (1991) 18,700. The port is Europe's busiest fishing port. The harbour is used by service industries for North Sea oil, a daily fish market and a marina. It is the most easterly town in Scotland, built of locally quarried Peterhead pink granite. Peterhead was formerly a whaling port. James Edward Stuart, the Old Pretender, landed here in 1715. **24** L4

Peter I Island (Norwegian **Peter I Øy**), uninhabited island in the Bellingshausen Sea, Antarctica, belonging to Norway since 1931; area 180 sq km/69 sq mi. **102** (2)JJ3

Peter I Øy Norwegian name for ►Peter I Island, an island in the Bellingshausen Sea, Antarctica. **102** (2)JJ3

Petersburg city and port in southeast Virginia, on the Appomattox River, 22 mi/35 km south of Richmond; population (1996 est) 38,200. Industries include tobacco-processing, boat building, and the manufacture of textiles, leather products, and chemicals. It was established in 1748 on the site of Fort Henry (1645), and was the site of American Revolution and Civil War battles. **88** E3

Petra (Arabic **Wadi Musa**), ancient city carved out of the red rock at a site in Jordan, on the eastern slopes of the Wadi el Araba, 90 km/56 mi south of the Dead Sea. An Edomite stronghold and capital of the Nabataeans in the 2nd century, it was captured by the Roman emperor Trajan in 106 and destroyed by the Arabs in the 7th century. It was forgotten in Europe until 1812 when the Swiss traveller Johann Ludwig Burckhardt (1784–1817) came across it.

Petrograd former name (1914–24) of ►St Petersburg, a city in Russia. It adopted this Russian-style name as a patriotic gesture at the outbreak of World War I, but was renamed Leningrad on the death of the USSR's first leader. **38** F3

Petropavlosk industrial city and capital of North Kazakhstan wiloyat (region), situated on the Ishim River, 260 km/162 mi west of Omsk in the Russian Federation. The city is an important transportation centre, lying at the junction of the ►Trans-Siberian railway (which reached here in 1896) and the TransKazakhh line (opened in 1953). Industries here include flour milling, manufacture of agricultural machinery, meat packing, and leather tanning. **38** N4

Petropavlovsk (formerly **Petropavlovsk-Kamchatskii**), seaport city on the eastern shore of the ►Kamchatka peninsula in the Russian Far East, and capital, economic and cultural centre of Kamchatka oblast (region); population (1996 est) 205,000. It is an important fishing port, and remains ice-free for seven months of the year. **44** T6

Petrópolis hill resort and former imperial city in Rio de Janeiro federal unit (state), southeast Brazil, 50 km/31 mi northeast of Rio de Janeiro city; population (1991) 255,200. Industries include textiles, fruit and vegetable cultivation, and brewing. It was founded in 1843 by Dom Pedro II (Emperor of Brazil 1831–89) as the seat of his government during the summer months; the Imperial Palace is now a museum and contains the Brazilian Crown Jewels. The cathedral (completed in 1925) contains the Emperor's tomb. **100** N3

Petrovsk former name (to 1921) of the Russian port ►Makhachkala. **58** M2

Petrozavodsk (Russian 'Peter's Factory'), capital city of Karelia, in the far northwestern Russian Federation; population (1996 est) 282,000. Petrozavodsk lies on the western shore of Lake Onega, 325 km/202 mi northeast of St Petersburg. It has engineering and metalworking industries, and produces tractors for the local timber industry. **38** F2

Pforzheim city in Baden-Württemberg, Germany, 26 km/16 mi southeast of Karlsruhe, on the northern border of the Black Forest; population (1995) 118,400. There are goldsmith, jewellery, and watch-making industries. It was a Roman settlement, and the residence of the margraves (princes) of Baden 1300–1565, although nothing remains except the Archives Tower, and the palace church containing the Renaissance tombs of the landgraves of Baden. **20** D8

Philadelphia (Greek 'the city of brotherly love'), river port and chief city in Pennsylvania, USA, on the Delaware River at the junction with the Schuylkill River; population (1996 est) 1,478,000; metropolitan area (1992) 5,939,000. It is the world's largest freshwater port, the fifth largest city in the USA, and a financial, business, and research centre. Industries include oil-refining, food-processing, electronics, printing, publishing, and the production of iron, steel, chemicals, textiles, carpets, and transportation equipment, although manufacturing is less important than it was, now employing about 12% of the workforce. It was originally settled by Swedish settlers in 1682, and was the capital of the USA 1790–1800. **90** F2

Philippeville former name (to 1962) of the Algerian port of ►Skikda. **66** G1

Philippines country in southeast Asia, on an archipelago of more than 7,000 islands west of the Pacific Ocean and south of the Southeast Asian mainland.

The Philippines covers an area of 300,000 sq km/115,830 sq mi, and has a population of 75,967,000 (2000 est). The capital (and chief port) is ►Manila (on Luzon island); other major towns and cities include Quezon City, Davao, Caloocan, Cebu, Bacolod, Cagayan, and Iloilo. The head of state and government is Gloria Macapagal Arroyo (from 2001). The currency is the peso, and the languages spoken include Filipino, English (both official), Spanish, Cebuano, Ilocano, and more than 70 other indigenous languages. **50** G5

Phnom Penh (or **Phnum Penh**), capital of ►Cambodia, lying in the south of the country, at the confluence of the Mekong and Tonle Sap rivers, 210 km/130 mi northwest of Ho Chi Minh City, Vietnam; population (1994 est) 920,000. Established as a settlement in the 14th century, Phnom Penh became the capital of the Khmer Empire in about 1434 and the capital of Cambodia in 1867. The communist Khmer Rouge forces, led by Pol Pot, took over the country in 1975 and forced almost all the inhabitants of Phnom Penh into the countryside. After the defeat of the Khmer Rouge by Vietnamese forces in 1979, the city was gradually repopulated and the industries of textiles, and food processing, and the manufacture of footwear, paper, tyres, and glassware were restored. **50** C4

Phnum Penh alternative form of ►Phnom Penh, the capital of Cambodia. **50** C4

Phoenix state capital and largest city of ►Arizona, USA, in Maricopa County, located on the Salt River;

population (2000 est) 1,321,000. It is a commercial and industrial centre, an agricultural distribution point for the irrigated Salt River valley, and a popular winter resort. Products include steel, aluminium, aviation equipment, computers, electrical goods, cosmetics, clothing, and processed foods. Phoenix was incorporated in 1881. **92** D2

Phoenix Islands group of eight small islands in the western Pacific Ocean, part of Kiribati; area 29 sq km/ 11 sq mi. Drought has rendered them uninhabitable and there is no permanent population. **74** J6

Piacenza (Roman **Placentia**), industrial town in Emilia-Romagna, northern Italy, on the River Po, 65 km/40 mi southeast of Milan; population (1992) 102,200. Products include agricultural machinery, textiles, and pottery. Piacenza dates from 218 BC and has a 12th-century cathedral. **30** E5

Piatra-Neamţ capital of Neamţ county, eastern Romania, on the River Bistriţa, at the foot of the Carpathian Mountains, 97 km/61 mi west of Iaşi; population (1993) 125,200. It is a market town, trading chiefly in timber. Piatra-Neamţ has textile, food-processing, paper, and oil refining industries, and is also a centre of folk culture. **34** P3

Picardie French name for ►Picardy, a region of northern France. **6** E2

Picardy (French **Picardie**), region of northern France, including the Aisne, Oise, and Somme *départements*; area 19,400 sq km/7,500 sq mi; population (1999 est) 1,857,800. Industries include chemicals and metals. Principal towns include ►Abbeville and ►Amiens; the latter is the administrative centre and was capital of the old province. **6** E2

Picton small port at the northeast extremity of South Island, New Zealand; population (1996) 3,600. There is a ferry to Wellington, North Island. **78** D5

Piedmont (Italian **Piemonte**), region of northern Italy, comprising the provinces of Alessandria, Asti, Cuneo, Novara, Turin, and Vercelli; area 25,400 sq km/9,800 sq mi; population (1992 est) 4,303,800. It borders Switzerland to the north and France to the west, and is surrounded, except to the east, by the Alps and the Apennines. The regional capital is ►Turin. Piedmont also includes the fertile Po Valley. Products include rice, fruit, grain, cattle, cars, and textiles. The movement for the unification of Italy started in the 19th century in Piedmont, under the House of Savoy. **9** (4)A2

Piemonte Italian name for ►Piedmont, a region of northern Italy. **9** (4)A2

Pierre state capital of ►South Dakota, USA, in Hughes County, on the east bank of the Missouri River; population (2000 est) 13,900. It is situated at the geographic centre of both the state and the North American continent. Industries include tourism and the processing of grain and dairy produce. **86** F2

Pietermaritzburg capital of ►KwaZulu-Natal Province, South Africa, 88 km/55 mi northwest of Durban; population (1991) 228,550. There are tanning, footwear, furniture, aluminium, rubber, and brewing industries. Founded in 1838 by Boer *Voortrekkers* from the Cape, it was named after their leaders, Pieter Retief and Gerrit Maritz, who were killed by Zulus. **72** E5

Pigs, Bay of inlet on the south coast of Cuba about 145 km/90 mi south-west of Havana. It was the site of an unsuccessful invasion attempt by 1,500 US-sponsored Cuban exiles 17–20 April 1961; 1,173 were taken prisoner.

Pik Pobedy highest peak in the Tian Shan mountain range on the Kyrgyz-Chinese border, at 7,439 m/24,406 ft. **42** P9

Piła (German **Schneidemühl**), capital of Piła province, Poland, on the River Gwda, a tributary of the Noteć, 77 km/48 mi north of Poznań; population (1990) 72,300. Machinery and glass are manufactured, and livestock and timber traded. It is a railway junction. Until 1945 it was in German Pomerania. **18** F4

Pilsen German form of ►Plzeň, a city in the Czech Republic. **18** C8

Pinatubo, Mount active volcano on Luzon Island, the Philippines, 88 km/ 55 mi north of Manila. Dormant for 600 years, it erupted in June 1991, killing 343 people and leaving as many as 200,000 homeless. Surrounding rice fields were covered with 3 m/10 ft of volcanic ash.

Pindus Mountains (or **Voreria Pindos**, or **Notia Pindos**), range in northwestern Greece and Albania, between Epirus and Thessaly; highest point Smolikas, 2,633 m/8,638 ft. **36** D5

Pine Bluff city, port, and administrative headquarters of Jefferson County, southeast Arkansas, USA; population (1992) 57,700. It is situated on a high bluff overlooking the Arkansas River, 40 mi/64 km south-southeast of Little Rock. It is an industrial and agricultural marketing center. Industries include paper, cotton, grain, furniture, farm implements, electrical transformers, lumber and wood products, textiles, and chemicals. The Pine Bluff Arsenal, with chemical and biological warfare laboratories, is in the city, as is the University of Arkansas at Pine Bluff (1873). **88** B4

Piotrków Trybunalski capital of Piotrków province, Poland, 42 km/26 mi south of Łódź; population (2001 est) 81,400. It is an industrial town producing textiles, chemicals, machinery, and glass. In the 10th–17th centuries it was the seat of a royal tribunal and Polish parliament. **18** J6

Pirmasens city in the Rhineland-Palatinate, Germany, 45 km/28 mi east of Saarbrücken; population (1995) 48,500. It is the centre of the German boot- and shoe-manufacturing industry. **20** C7

Pirna town in the *Land* (administrative region) of ►Saxony, Germany, on the River Elbe, 18 km/11 mi southeast of Dresden; population (1994) 40,000. The town is a health resort, and has paper and textile industries. **20** J6

Pirot town in Serbia, Yugoslavia, on the River Nisava; population (1991) 40,300. It is the centre for the produce of the fertile plain of Pirot, and has a carpet industry. **34** K6

Pisa (ancient **Pisae**), town in Tuscany, Italy, on the River Arno, 70 km/43 mi southwest of Florence; population (1991) 101,000. Industries include tourism, engineering, and the production of glass and textiles. Its famous campanile (bell-tower), the **Leaning Tower of Pisa** (repaired 1990), is 55 m/180 ft high and about 5 m/16.5 ft out of perpendicular, the foundations being only about

3 m/10 ft deep and built on unstable ground. **30** F7

Pistoia town in Tuscany, Italy, 34 km/21 mi northwest of Florence; population (1998) 87,400. Industries include railway engineering, weaving, and the production of steel, small arms, paper, leather, cement, bedding, pasta, and olive oil. Pistoia was the site of the Roman rebel Catiline's defeat in 62 BC. It is surrounded by walls (1302) and has a 12th-century cathedral. **30** F7

Pitcairn Islands British colony in Polynesia, 5,300 km/3,300 mi northeast of New Zealand; area 47 sq km/18 sq mi; population (1996) 58. The capital is Adamstown. Products are coconuts, bananas, breadfruit, yams, pineapples, tomatoes, oranges, and pineapples; souvenirs are sold to passing ships. **74** P8

Piteşti capital of Argeş county, Romania, 120 km/75 mi northwest of Bucharest; population (1993) 182,900. It is situated on an oil- and gasfield and has developed associated industries; it also produces wine, textiles, machinery, motor vehicles, leather, shoes, chemicals, and has food processing. **34** M5

Pittsburgh (called 'City of Bridges'), second-largest city in Pennsylvania, USA, at the confluence of the Allegheny and Monongahela rivers, forming the Ohio River; population (1996 est) 350,400; metropolitan area (1992) 2,406,000. It is a business and financial centre with one of the largest river ports in the world. High technology and healthcare services dominate an economy formerly based on iron, steel, heavy engineering, and glass industries. **88** D2

Piura capital of Piura department, in the arid coastal region of northwest Peru, situated on the Piura River 160 km/100 mi southwest of Punta Pariñas; population (1993) 278,000. It is an agricultural centre for the surrounding region (made viable with irrigation systems) which produces rice, cotton, and corn. The first colonial settlement in Peru, Piura was founded in 1532 by the Spanish conquistadors left behind by Francisco Pizarro, and moved to its present site in 1588. **98** A5

Plata, Río de la (or **River Plate**), estuary on the South Atlantic coast of South America, at the border between Argentina and Uruguay; it is formed by the confluence of the Paraná and Uruguay rivers; length 275 km/171 mi; width at mouth 225 km/140 mi; drainage basin area 3,900,000 sq km/ 1,505,790 sq mi. The basin drains much of Argentina, Bolivia, Brazil, Uruguay, and Paraguay. Its shores are densely populated; Buenos Aires lies on the southern shore and Montevideo is situated on the northern shore. The estuary was encountered in 1516 by the Spaniard Juan Díaz de Solís in his attempts to find a route to the Pacific Ocean. **100** K6

Platte river in Nebraska, central USA; length 500 km/310 mi. It is formed at North Platte, Nebraska, by the confluence of the North Platte and South Platte rivers and flows mainly east to join the ►Missouri River 25 km/16 mi south of Omaha. The three Platte rivers contain many dams for flood control and irrigation. They are unnavigable, but the western trails to Oregon and California followed them in the mid-19th century. **90** B1

Plattensee German name for Lake ►Balaton in Hungary. **34** E3

Plauen town in the *Land* (administrative region) of ►Saxony, Germany, 68 km/42 mi southwest of Chemnitz. It is situated on the River Weisse Elster, at the foot of the ►Erzgebirge; population (1996) 67,300. Its textile industry dates from the 15th century, and it also has steel and engineering industries. **20** H6

Plenty, Bay of coastal region of northeast North Island, New Zealand; population (1996) 230,500. It has a mild, sunny climate and rich volcanic soils, and is the most important kiwifruit-growing area in New Zealand. Citrus fruit, tamarillo (tree tomato) and avocado are also grown. The largest city, Tauranga, was New Zealand's fastest growing city in 1996, and its port, at Mount Maunganui, is New Zealand's leading export port, handling forestry products such as wood pulp, newsprint, and timber, as well as horticultural products. **78** F3

Pleven industrial city (textiles, machinery, ceramics) in northern Bulgaria; population (1990) 168,000. In the Russo-Turkish War 1877, Pleven surrendered to the Russians after a siege of five months. **34** M6

Płock (Russian **Plotsk**), capital of Płock province, Poland, 96 km/60 mi west of Warsaw, on the River Vistula; population (1990) 123,400. The city has oil refineries and petrochemical works. The 12th-century cathedral contains the tombs of Polish kings. **18** J5

Ploieşti industrial city (textiles, paper, petrochemicals; oil centre) in southeastern Romania; population (1993) 254,000. **34** P5

Plotsk Russian name for ►Płock, a town in Poland. **18** J5

Plovdiv industrial city (textiles, chemicals, leather, tobacco) in Bulgaria, on the River Maritsa; population (1991) 379,000. Conquered by Philip of Macedon in the 4th century BC, it was known as **Philippopolis** ('Philip's city'). It was the capital of Roman Thrace. **34** M7

Plymouth city, seaport, and unitary authority in southwest England, at the mouth of the River Plym; until April 1998 it was part of the county of Devon.

area 79 sq km/31 sq mi

features dockyard and naval base at Devonport; three harbours, Sutton Pool, Catwater (Cattewater), and the Hamoaze, which unite in Plymouth Sound, a bay with a breakwater over 1 km/0.6 mi in length across the entrance; ferry links with France and Spain; Plymouth University, formerly South West Polytechnic, established in 1992; ramparts of a 17th-century citadel, built to guard the harbour soon after the Civil War; Eddystone Rocks lighthouse 22 km/14 mi to the south; the Hoe, an esplanade overlooking Plymouth Sound

industries marine and machine tools industries; clothing, radio equipment, and processed foods

population (1996) 257,000

24 H11/**5** D6

Plzeň (German **Pilsen**), industrial city (heavy machinery, cars, beer) in the Czech Republic, at the confluence of the Radbuza and Mze rivers, capital of Západočeský (West Bohemia) region; 84 km/52 mi southwest of Prague; population (1991) 173,100. **18** C8

Po (Greek **Eridanos**; Latin **Padus**), longest river in Italy, flowing from the Cottian Alps to the Adriatic Sea; length 668 km/415 mi. Its valley is fertile and contains natural gas. The river is heavily polluted with nitrates, phosphates, and arsenic. **30** E5

Pocatello city in Bannock County (of which it is the administrative headquarters) and Power County, southeast Idaho, USA, on Portneuf River, near the American Falls reservoir, and 314 km/195 mi east of Boise; population (1994 est) 49,600. It is the second-largest city in Idaho, a railway centre, and a wholesale trade and shipping point for the surrounding agricultural area. Its manufactures include electronics, steel, cement, machinery, food products, and chemicals.It is the home of of Idaho State University (1901). **86** D2

Podgorica (formerly **Titograd** 1946–92), capital of Montenegro, Yugoslavia; population (1993 est) 135,000. Industries include metal-working, furniture-making, and tobacco. It was damaged in World War II and after rebuilding was renamed in honour of Marshal Tito; it reverted to its original name with the collapse of communism. It was the birthplace of the Roman emperor Diocletian. **34** G7

Podolsk industrial city in the Moscow oblast (region) of the Russian Federation, 40 km/25 mi south of Moscow; population (1996 est) 200,000. Its factories produce heavy machinery, electrical equipment, and cement. The city's industrial development dates from the late 19th century. **38** G3

Pointe-Noire chief port and second-largest city of the Republic of the Congo, formerly (1950–58) the capital; population (1995 est) 576,200. It is on the Atlantic coast and is the terminus of the railway from Brazzaville. Industries include oil refining, shipbuilding, potash processing, brewing, and food exporting. Tourism is also important here. **68** G5

Poitiers administrative centre of the *département* of ►Vienne and of the ►Poitou-Charentes region in western France, at the confluence of the Rivers Clain and Boivre; population (1990) 82,500. Products include chemicals, electrical and metal goods, and clothing. The theme park of Mirapolis is nearby. The Merovingian king Clovis I defeated the Visigoths, for whom it was an important town, under Alaric II here in 507; Charles Martel stemmed the Saracen advance in 732; and Edward the Black Prince of England, with English and Gascon forces, defeated the French troops of Jean le Bon (John II) here in 1356, and took him prisoner. **26** F7

Poitou-Charentes region of west-central France, comprising the *départements* of ►Charente, ►Charente-Maritime, ►Deux-Sèvres, and ►Vienne; area 25,800 sq km/9,950 sq mi; population (1990) 1,595,100. Its administrative centre is ►Poitiers. Industries include dairy products, wheat, chemicals, and metal goods; brandy is made at Cognac. **6** C3

Poland country in eastern Europe, bounded north by the Baltic Sea, northeast by Lithuania, east by Belarus and Ukraine, south by the Czech Republic and the Slovak Republic, and west by Germany.

Poland covers an area of 312,683 sq km/120,726 sq mi, and has a population of 38,765,000 (2000 est). The capital is ►Warsaw; other major towns and cities include Łódź, Kraków, Wrocław, Poznań, Gdańsk, Szczecin, Katowice, Bydgoszcz, and Lublin. The head of state is Aleksander Kwaśniewski (from 1995); the head of government is Jerzy Buzek (from 1997). The currency is the złoty, and the main language spoken is Polish (official). **18** G6

Polar Regions see ►Antarctica, ►Antarctic Ocean, ►Arctic, the, ►Arctic Ocean.

Poltava industrial city in eastern Ukraine, capital of Poltava oblast, and situated on the River Vorskla, 300 km/186 mi southeast of Kiev; population (1990) 317,000. The city manufactures machinery, foodstuffs, and clothing, and is the market centre for a fertile agricultural area. Peter the Great routed a Swedish and Ukrainian Cossack force led by Charles XII of Sweden here in 1709, so ending the Great Northern War. **38** F5

Poluostrov Yamal Russian name for the ►Yamal Peninsula, in northwestern Siberian Russia. **42** M3

Polynesia islands of Oceania east of 170° E latitude, including Hawaii, Kiribati, Tuvalu, Fiji Islands, Tonga, Tokelau, Samoa, Cook Islands, and French Polynesia. **74** J6

Pompano Beach city in Broward County, southeast Florida, USA, population (1992) 72,900. It is located 8 mi/13 km north-northeast of Fort Lauderdale, on the Atlantic Ocean. Tourism, fruit processing, and fishing are important to the economy. Pompano Beach ships local winter vegetables, and has a variety of manufactures, including pleasure boats, plastics, and electronic equipment. The town was settled about 1900. **90** E4

Ponce major port and third-largest city in southern Puerto Rico, on the Caribbean coast, 80 km/50 mi southwest of San Juan, population (1990) 187,800. Industries include tourism, canning, sugar- and oil-refining, and the production of iron, textiles, cement, and rum. Playa de Ponce, its harbour facility, is the island's principal seaport. Established around 1670–80, the settlement was named Nuestra Senora de Guadelupe de Ponce, after the Spanish explorer Juan Ponce de León. **94** L5

Pondicherry Union Territory of southeast India; area 480 sq km/185 sq mi; population (1994 est) 894,000. Its capital is Pondicherry which lies on the ►Coromandel Coast, 196 km/122 mi south of Chennai (formerly Madras). Its products include rice, millet, groundnuts, cotton, and sugar; industry is based on textiles and coastal trade. **10** C6

Ponta Delgada port, resort, and chief commercial centre of the Portuguese ►Azores Islands, on São Miguel; population (1991) 21,100. **66** (1)B2

Pontevedra capital of the province of Pontevedra, northwest Spain, at the mouth of the River Lerez; population (1991) 71,200. Products include wine, agricultural produce, textiles, and pottery. It has a naval radio station, and there is a 16th-century Gothic cathedral and a 13th-century episcopal palace. **28** B2

Pontiac city and administrative headquarters of Oakland County, southeast Michigan, USA, on the Clinton River, 38 km/24 mi northwest of Detroit; population (1996 est) 70,500. Situated in a lake-strewn resort area, Pontiac is one of a ring of cities surrounding Detroit, involved in the American automobile industry. It specializes in the manufacture of Pontiac cars and goods vehicles for General Motors, and is also a tourist centre. Among its other manufactures are rubber, paint and varnish, boats, plastic, and dairy products. Its 80,000-seat Silverdome arena (1975) is the home of the Detroit Lions American football team. 88 C2

Poole unitary authority in southwest England, created in 1997 from part of Dorset.
area 64 sq km/25 sq mi
towns and cities ►Poole (administrative headquarters), Broadstone, Hillbournes, Sandbanks
features River Stour; Poole Harbour; Holes Bay; Pergins Island; Maritime Museum (Poole); Compton Acres themed gardens (including water, rock, heather, Japanese, Roman, Italian); Canford Heath, tumuli field; Sandbanks spit guarding entrance to harbour
industries boats, tourism, marine engineering and marine electronics, aeronautical instruments
population (1996) 138,100 5 E6

Poole industrial town, port, and, since 1997, administrative headquarters of ►Poole unitary authority in southern England, 8 km/5 mi west of Bournemouth; area 64 sq km/25 sq mi; population (1996) 138,100. It was part of the county of Dorset until 1997. Poole Harbour is a centre for yachting. In addition to tourism, the town's industries include engineering, boatbuilding, and the manufacture of caravans, sail cloth, packaging materials, tiles, and pottery from local clay. 24 L11

Poona former spelling of ►Pune, a city in India; after independence in 1947 the form Poona was gradually superseded by Pune. 54 B5

Popocatépetl (Aztec 'smoking mountain'), volcano in central Mexico, 50 km/30 mi southeast of Mexico City; 5,340 m/17,526 ft. It last erupted in 1920. 94 E5

Pori (Swedish **Björneborg**), ice-free industrial port (nickel and copper refining, sawmills, paper, textiles) in the Gulf of Bothnia, southwestern Finland; population (1992) 76,300. A deep-water harbour was opened 1985. 16 L6

Port Arthur deep-water port in Texas, USA, on the western shore of Lake Sabine, linked with the Gulf of Mexico by the Sabine–Neches waterway, 24 km/15 mi southeast of Beaumont; population (1996 est) 57,700. It is the USA's largest oil-refining centre, and has shipbuilding, brass, and chemical industries. Exports include petroleum, grain, cotton, rice, and lumber. Founded in 1895 as the terminus for the Kansas City, Pittsburg and Gulf Railroad, the city gained importance with the discovery of oil near Beaumont in 1901. 90 C4

Port Augusta town and port in South Australia, at the head of the Spencer Gulf, 311 km/193 mi northwest of Adelaide; population (1996) 13,914. It is a centre for power generation and a major road and rail junction. Port Augusta is a railway rolling-stock service centre, with the locomotive and wagon workshops of the Australian National Railways. The port no longer has a regular shipping service. 76 G6

Port-au-Prince capital and chief port of ►Haiti, on the west coast of the island of Hispaniola; population (2001 est) 1,047,600. The port is able to berth 10,000-tonne ships, and handles half of the country's foreign trade, with coffee and sugar as the major exports. Manufacturing industries include sugar, rum, textiles, tobacco, cement, and plastics. An international airport supports the significant tourist industry. 94 K5

Port Elizabeth industrial port in Eastern Cape Province, South Africa, about 710 km/440 mi east of Cape Town, on Algoa Bay; population (urban area, 1991) 853,200. Local industries include motor assembly plants, shoemaking, foundries, sawmills, flour mills, canning factories, engineering, food processing, and the production of soap, tyres, furniture, chemicals, safety glass, electrical goods, cable, steel, textiles, plastics, and paints. The port also exports manganese ore and has large pre-cooling plants for fruit. 72 D6

Porterville town in Tulare County, south-central California, USA; population (1990) 29,600. It is located to the north of Bakersfield, 65 mi/105 km southeast of Fresno, on the Tule River, in the San Joaquin Valley. Oranges and other fruit are shipped from the town, which also produces olive oil; granite and marble are quarried in the area. 92 C1

Port Harcourt port and capital of Rivers state in southeast Nigeria, on the River Bonny in the Niger delta; population (1992 est) 371,000. The port trades in coal, palm oil, and groundnuts. It is also an industrial centre producing refined mineral oil, sheet aluminium, tyres, glass, fertilizers, and paints. Founded in 1912, it was named after Viscount Lewis Harcourt, British secretary of state for the colonies 1910–15. 68 F4

Portland port and largest city in Maine, USA, on Casco Bay, southeast of Sebago Lake, 240 km/150 mi northeast of Boston; population (1996 est) 63,100. It is a commercial and distribution centre, shipping oil and lumber. Industries include fishing, publishing, and the manufacture of pulp and clay products, and shoes. Ferry links serve the Casco Bay resort islands, and Nova Scotia. 88 F2

Portland riverport in northwest Oregon, USA; on the Willamette River near its confluence with the Columbia River, 173 km/108 mi from the Pacific; population (1996 est) 480,800; metropolitan area (1996 est) 1,600,000. It is the state's largest city, and a processing and distribution centre for timber, aluminium, grain, and livestock. Industries include food-processing, fishing, and the manufacture of wool, paper, lumber machinery, and electronic equipment. 86 B1

Portlaoise (or **Portlaoighise**; formerly **Maryborough**), county town of County ►Laois, Republic of Ireland, 80 km/50 mi southwest of Dublin, situated on a tributary of the River Barrow; population (1996) 3,300. It has woollen, flour-milling, and malting industries, and is the site of a top-security prison. On the Rock of Dunamase, 6 km/4 mi east of Portlaoise, are the castle ruins of the 12th-century king of Leinster, Dermot MacMurrough. 24 E8

Port Louis capital of Mauritius, on the island's northwest coast; population (1995) 145,600. Exports include sugar, textiles, and electronic goods. Industries include chemicals, plastics, fertilizers, and sugar and food processing. 72 (1)B2

Port Moresby capital and port of Papua New Guinea, on the south coast of New Guinea; population (1995 est) 215,000. The port trades in coffee, copper, gold, copra, palm oil, and timber. There is an airport; the town is the country's centre for national broadcasting and for overseas telecommunications. 76 J1

Porto (English **Oporto**), industrial city and capital of Porto district, northwest Portugal, 280 km/174 mi north of Lisbon, on the River Douro, 5 km/3 mi from its mouth on the Atlantic coast; population (1998 est) 293,500. Port wine is exported, and industries include textiles, leather, and pottery. Porto is built on terraces cut into the steep northern slopes of the Douro gorge. It is connected to the southern suburb of Vila Nova de Gaia by the two-storey bridge of Dom Luis I, built by Eiffel, and which crosses the river in a single span of 160 m/525 ft at a height of 36 m/118 ft. **28** B3

Porto Alegre industrial port and capital of Río Grande do Sul federal unit (state), southeast Brazil; population (1991) 1,254,600; (metropolitan area 3,757,500). The port is situated on the eastern bank of Río Guaíba (formed by the confluence of five rivers), at the northwestern end of Lagôa dos Patos, a freshwater lagoon which flows into the Atlantic; the lagoon is accessible to ocean-going vessels via the port of Rio Grande. It is southern Brazil's chief industrial and commercial centre; there are shipyards, iron foundries, textile, chemical, and food processing industries. Chief exports include pinewood, meat, rice, wheat, hides, and wool. The city has a cathedral, whose foundations are built over a church that dates back to 1772, and two universities. It was founded in 1755. **100** L5

Port-of-Spain port and capital of Trinidad and Tobago, on the island of Trinidad; population (1990) 58,400. It has a cathedral (1813–28) and the San Andres Fort (1785). **98** E1

Porto-Novo port and capital of ►Benin, on Porto-Novo lagoon; population (1994) 179,000. It trades in palm oil, palm kernels, and cotton. A former Portuguese centre for the slave and tobacco trade with Brazil, it became a French protectorate in 1863. The National Museum of Ethnography is here. **68** E3

Porto Rico former name (to 1932) of ►Puerto Rico, a US island in the Caribbean. **94** L5

Port Pirie industrial city and port in South Australia, on the Spencer Gulf, 231 km/144 mi northwest of Adelaide; population (1996) 13,633. The major industry is the smelting of silver, lead, and zinc from the mines at ►Broken Hill in New South Wales. Chemicals are also produced. Main exports include mineral ores and wheat. Port Pirie is also the location of a large lead smelting works. **76** G6

Port Rashid port serving Dubai in the United Arab Emirates.

Port Said (Arabic **Bûr Sa'îd**), port in Egypt, on reclaimed land at the north end of the ►Suez Canal; population (1992 est) 460,000. During the 1967 Arab–Israeli War the city was damaged and the canal blocked; Port Said was evacuated by 1969 but by 1975 had been largely reconstructed. **64** F1

Portsmouth port in Rockingham County, southeastern New Hampshire, USA, on the estuary of the Piscataqua River; the state's only seaport; population (1990) 25,900. Founded 1623, Portsmouth was the state capital 1679–1775. **88** F2

Portsmouth city, naval port, and unitary authority in southern England, 118 km/73 mi southwest of London, on the peninsula of Portsea Island, opposite the Isle of Wight; it was part of the county of Hampshire until 1997.
area 42 sq km/16 sq mi
features 12th-century cathedral; Portsmouth University, formerly Portsmouth Polytechnic, was established in 1992; Tudor warship *Mary Rose* and Admiral Horatio Nelson's flagship HMS *Victory* are exhibited here
industries high-technology and manufacturing industries, including aircraft engineering, electronics, shipbuilding, and ship maintenance; naval dockyard was closed in 1981; it is a continental ferry port
population (1996) 189,300
22 A4/5 F6

Portsmouth port and independent city in southeast Virginia, USA, on the Elizabeth River; population (1996 est) 101,300. Situated near the mouth of Chesapeake Bay, on the southern shore of a natural deepwater harbour, it forms the Port of Hampton Roads together with Newport News and ►Norfolk, headquarters of the US Navy Atlantic Fleet. It is a US naval training centre and has the largest naval shipbuilding yard in the world. Manufactured goods include electronic equipment, chemicals, clothing, and processed food. **88** E3

Port Talbot industrial port and administrative centre of ►Neath Port Talbot unitary authority, south Wales, situated 11 km/7 mi east of Swansea where the River Afan enters the Bristol Channel; population (1991) 37,600. Industries include tinplate, chemicals, and steel-strip milling. The port accommodates bulk carriers of iron ore. **24** J10

Portugal country in southwestern Europe, on the Atlantic Ocean, bounded north and east by Spain.

Portugal covers an area of 92,000 sq km/35,521 sq mi (including the Azores and Madeira), and has a population of 9,875,000 (2000 est). The capital is ►Lisbon; other major towns and cities include Porto, Coimbra, Amadora, Setúbal, Funchal, Braga, and Vila Nova de Gaia. The head of state is Jorge Sampaio (from 1996); the head of government is Antonio Guterres (from 1995). The currency is the escudo, and the main language spoken is Portuguese (official). **28** B5

Portuguese East Africa former name (to 1975) of ►Mozambique in southeast Africa. **72** E3

Portuguese Guinea former name (to 1974) of ►Guinea-Bissau in western Africa. **68** A2

Portuguese West Africa former name (to 1975) of ►Angola in southwestern Africa. **62** E7

Port-Vila port and capital of Vanuatu, on the southwest of Efate Island; population (1996 est) 31,800. Local industries include meat canning. **74** G7

Posen German name for ►Poznań, a city in Poland. **18** F5

Postojna (German **Adelsberg**), town in Slovenia, in the district of Carniola; population (1991) 8,200. Here the underground River Pivka enters the caves of Postojna, which are filled with dramatically shaped stalactites and stalagmites. The river contains the blind salamander *Proteus anguinus*. **30** K5

Potomac river of the eastern USA, forming the boundaries between West Virginia, Virginia, and Maryland states. Rising in the Allegheny Mountains, it flows 459 km/285 mi southeast into Chesapeake Bay. It is created by the confluence of the North Potomac, 153 km/95 mi long, and South Potomac, 209 km/130 mi long, and its chief tributaries are the Shenandoah and the Monocacy. At

▶Washington, DC, 185 km/116 mi from its mouth, the Potomac becomes tidal and navigable for large ships.

Potosí city and capital of Potosí department, southwest Bolivia; population (1992) 112,300. Situated at the foot of Cerro Rico mountain at an altitude of 4,020 m/13,189 ft, it is one of the highest cities in the world. Tin, copper, zinc, lead, and silver are mined in the region. There is a cathedral, a university (founded in 1892), and the Royal Mint, which houses a museum and an art gallery. The city was founded in 1545 following the discovery of ores in the silver-rich Cerro Rico, the hill overlooking the city. 98 D7

Potsdam capital of the *Land* (administrative region) of Brandenburg, Germany, on the River Havel southwest of Berlin; population (1995) 137,600. Products include textiles, pharmaceuticals, and electrical goods. A leading garrison town and Prussian military centre, Potsdam was restored to its position of capital of Brandenburg with the reunification of Germany in 1990. 20 J4

Poughkeepsie town and administrative headquarters of Dutchess County, southeast New York State, USA; population (1990) 28,900. It is situated on the Hudson River, 65 mi/105 km north of New York City. Products include dairy equipment, chemicals, ball bearings, electronic components, clothing, and cough drops. Computer research and assembly is also important to the local economy. Poughkeepsie is home to Dutchess Community College (1958). Vassar College (1861) is in Arlington, which adjoins the town to the northeast. 88 F2

Powys unitary authority in central Wales, created in 1996 from the former county of Powys.
area 5,179 sq km/1,999 sq mi
towns Llandrindod Wells (administrative headquarters), Brecon, Builth Wells, Newtown, Welshpool
physical mountainous to the north, Black Mountains, rivers ▶Wye and ▶Severn, which both rise on the eastern slopes of Plynlimon
features the Brecon Beacons National Park, Lake Vyrnwy (an artificial reservoir supplying Liverpool and Birmingham), alternative-technology centre near Machynlleth
industries agriculture, tourism
agriculture arable and dairy farming, sheep-rearing
population (1996) 123,600 5 E5

Poznań (German **Posen**), industrial city (machinery, aircraft, beer) in western Poland; population (1993) 590,000. Founded 970, it was settled by German immigrants in 1253 and passed to Prussia in 1793; it was restored to Poland in 1919. 18 F5

Pozsony Hungarian name for ▶Bratislava, capital of Hungary 1526–1784. 18 G9

Pozzuoli port in Campania, southern Italy, on the Bay of Naples, 13 km/8 mi west of Naples; population (1990) 27,100. It is shaken by some 25 earthquakes a day, 60% of its buildings are uninhabitable, and an eventual major disaster seems inevitable. Products include steel and armaments. 32 J8

Prague (Czech **Praha**), city and capital of the Czech Republic on the River Vltava; population (1993) 1,217,300. Industries include cars, aircraft, chemicals, paper and printing, clothing, brewing, and food processing. It was the capital of Czechoslovakia 1918–93. 18 D7

Praha Czech name for ▶Prague, the capital of the Czech Republic. 18 D7

Praia capital and port of ▶Cape Verde, on the south shore of the island of São Tiago (Santiago); population (2001 est) 96,300. São Tiago is the most populous of the Cape Verde group of islands and the most agriculturally productive, and Praia has become the leading commercial centre of the country, having a large export trade in bananas, coffee, sugar cane, oranges, and castor beans. There is a significant fishing industry, and a submarine cable station. Tourism has become increasingly important. 68 (1)B2

Prato industrial town in Tuscany, central Italy, 18 km/11 mi northwest of Florence; population (1992) 166,100. Woollen goods, textiles, synthetic fibres, carpets, and hosiery are produced. The 12th-century cathedral contains frescoes by Agnolo Gaddi and Filippo Lippi, reliefs by Andrea della Robbia, and an external pulpit partly by Donatello. 30 G7

Prešov (Hungarian **Eperjes**), town in eastern Slovak Republic, on the River Torysa, 36 km/22 mi north of Košice; population (1991) 87,800. Machinery is manufactured and agricultural produce traded from the surrounding area. The town was granted a royal charter in 1374 and has many Renaissance and baroque houses and churches. 18 L9

Pressburg German name of ▶Bratislava, the capital of the Slovak Republic. 18 G9

Preston industrial town and administrative headquarters of ▶Lancashire, northwest England, on the River Ribble, 34 km/21 mi south of Lancaster, at the highest navigable point of the Irish Sea estuary; population (1991) 126,100. Industries include textiles, chemicals, electrical goods, aircraft, plastics, and engineering; it is also an agricultural market centre. Oliver Cromwell defeated the Royalists at Preston in 1648. It is the birthplace of Richard Arkwright, inventor of cotton-spinning machinery, and was a centre of the cotton industry in the 18th century. 24 K8

Pretoria city in Gauteng Province, South Africa, and the country's administrative capital; population (1991) 1,080,200. Industries include engineering, chemicals, iron, steel, cement, diamonds, granite quarrying, chemicals, and food processing. Founded in 1855, it was named after Boer leader Andries Pretorius (1799–1853). It was the administrative capital of the Union of South Africa from 1910 and capital of Transvaal Province 1860–1994. 72 D5

Preveza port and department of Epirus, Greece, at the entrance to the Gulf of Arta; area (department) 1,086 sq km/419 sq mi; population (1991) 58,900 (department). Preveza is the third-largest town of the Epirus region. The main port for the Arta agricultural plain area, it trades in coal, oil, textiles, and agricultural products. The town was founded in the 3rd century BC. 36 C6

Pribilof Islands (or **Fur Seal Islands**), group of four volcanic islands and associated islets, in the Bering Sea, 320 km/200 mi southwest of Bristol Bay, Alaska, USA; total area 168 sq km/65 sq mi. Named after Russian explorer Gerasim Pribilof, who reached them in 1786, they were sold by Russia to the USA in 1867, as part of Alaska. In 1868 the islands were designated a fur-seal reserve, and leased to sealing companies. Overhunting led to a treaty in 1911 which placed the industry under government supervision. 92 (1)D4

Přibram mining town in central Bohemia, the Czech Republic, 53 km/33 mi southwest of Prague; population (1991) 36,900. Silver has been worked here since 1330; other metals now also mined are lead, zinc, barium, and antimony. 18 D8

Primorsk alternative name for ►Primorye, a territory in Siberia, Russia. 16 Q6

Primorye (or **Primorsk**), territory of the Russian Federation, in southeastern Siberia, on the Sea of Japan; capital Vladivostok; area 165,900 sq km/64,079 sq mi; population (1996 est) 2,225,000. Timber and coal are produced. 16 Q6

Prince Albert city in Saskatchewan, Canada, on the North Saskatchewan River and the Canadian Pacific Railway, 160 km/100 mi north of Saskatoon; population (1991) 34,200. The self-proclaimed 'Gateway to the North', it is the administrative centre of the provincial government for the northern areas, a terminus for transport and communication into the forest, lake, and mining region of the north, and a trading centre. It is 65 km/40 mi from the Prince Albert national park (area 3,875 sq km/1,496 sq mi). The site of earlier fur-trading posts, the city was founded as a Presbyterian mission by James Nisbet in 1866. 82 K6

Prince Edward Island smallest province of Canada, situated in the Gulf of St Lawrence, separated from Nova Scotia (to the south and east) and New Brunswick (to the west) by the Northumberland Strait; area 5,700 sq km/2,200 sq mi; population (1996) 137,300. The capital is ►Charlottetown. Industries include fishing, food processing, and farm vehicle manufacture. Potatoes are grown and there is also dairying. 82 U7

Prince Rupert fishing port at the mouth of the Skeena River in British Columbia, Canada, on Kaien Island, on the west side of Tsimpsean peninsula; population (1996) 17,400. Situated on one of the world's largest ice-free deepwater harbours, at the western terminus of the Canadian National Railway, it is a distribution centre for halibut, minerals, lumber and grain, and a major ferry point for Alaska, Vancouver, and the Queen Charlotte Islands. Fish-processing and canning is the principal industry. 82 E6

Prince William Sound island-filled inlet of the Gulf of Alaska, to the east of the Kenai Peninsula and south of the Chugach Mountains in south-central Alaska. It extends 200 km/125 mi northwest from Kayak Island. Cordova and Valdez are major ports on the 130 km/80 mi–wide sound. Montague and Hinchinbrook are its largest islands. The area is a fishing, mining, and oil shipment centre. In March 1989 the oil tanker *Exxon Valdez* ran aground here, spilling 42 million litres of crude oil in one of the world's greatest oil-pollution disasters. Commercial fishing, birds, sea mammals, and hundreds of miles of shoreline were devastated. 82 B4

Priština capital of Kosovo autonomous province, southern Serbia, Yugoslavia; population (1991) 108,000. Once capital of the medieval Serbian empire, it is now a trading centre. 34 J7

Prizren town in Serbia, Yugoslavia, in the autonomous region of Kosovo; population (2001 est) 114,900. In the Middle Ages it was for a time the capital of the kingdom of Serbia. There are textile and glass industries, and the town is known for its silver filigree work and embroideries. It has a 14th-century church and an old fortress. 34 H7

Prokopyevsk city in Kemerovo oblast (region) of the Russian Federation, in southwestern Siberia, 30 km/19 mi northwest of Novokuznetsk; population (1996 est) 250,000. Situated on the Aba River and on the South Siberian Railway, it is one of the principal coalmining and coking centres of the ►Kuznetsk Basin. Industries include the manufacture of mining equipment and ballbearings, chemical works, and food-processing plants. 42 R7

Provence-Alpes-Côte d'Azur region of southeast France, comprising the *départements* of ►Alpes-de-Haute-Provence, ►Hautes-Alpes, ►Alpes-Maritimes, ►Bouches-du-Rhône, ►Var, and ►Vaucluse; area 31,400 sq km/12,120 sq mi; the administrative centre is ►Marseille; population (1990) 4,257,900. The Côte d'Azur, on the Mediterranean, is a tourist centre. Provence was an independent kingdom in the 10th century, and the area still has its own traditional language, Provençal. 6 F5

Providence seaport and capital of ►Rhode Island, USA; third-largest city in New England; on Narragansett Bay and the **Providence River**, 27 mi/43 km from the Atlantic Ocean, 65 km/40 mi southwest of Boston; population (1996 est) 152,600. It is a commercial, financial, and industrial centre. Jewellery, silverware, machinery, processed foods, and metal goods are manufactured. Founded in 1636, Providence was an important port in the West Indian trade and the American Revolution. It became state capital in 1900. 88 F2

Provo city in north-central Utah, USA, on the Provo River, 60 km/38 mi southeast of Salt Lake City; population (1996 est) 99,600. It is a rail distribution point, and the commercial and industrial centre of a rich mining and irrigated agricultural area. Iron and steel goods are manufactured with material sourced from a nearby steelworks at Geneva. Products include processed-foods, electronics, computer-ware, bricks, and tiles. Silver, lead, copper, and gold are mined locally. It is the seat of Brigham Young University (1875). 86 D2

Prudhoe Bay inlet of the Beaufort Sea, 320 km/200 mi east-southeast of Point Barrow, in north Alaska, USA; population (1990) 50. The 9 mi-14 km-wide bay and its small community are at the centre of immensely rich petroleum and natural gas fields. These were discovered on the North Slope of Alaska in 1968, and estimated to contain 9.6 billion barrels of oil and 0.73 trillion cubic m/26 trillion cubic ft of natural gas. Prudhoe Bay is now the largest oilfield complex in the USA, producing about 2 million barrels a day. The 1,270-km/789-mi Trans-Alaska Pipeline (1977), carries oil south from the bay to the ice-free port of Valdez on Prince William Sound. 92 (1)H1

Prut (or **Pruc** or **Prutul**), river in eastern Europe; length 900 km/565 mi. The Prut rises in the Carpathian Mountains in southwestern Ukraine, and flows south to meet the Danube at Reni. For most of its course it forms the frontier between Romania and Moldova. 34 R4

Przemysl industrial city (timber, ceramics, flour milling, tanning, distilling, food processing, gas, engineering) in southeastern Poland; population (1991) 68,500. 18 M8

Pskov (Estonian **Pihkva**), capital city, economic and cultural centre of Pskov oblast (region), northwestern Russian Federation, situated on the Velikaya River near the southern shore of Lake Pskov 300 km/186 mi southwest of St Petersburg; population (1996 est) 207,000. The city has industries producing building materials, engineering equipment (for example, electrical goods, radios, and textile machinery), textiles, and foodstuffs. 38 E3

Puebla industrial city (textiles, sugar refining, metallurgy, hand-crafted pottery and tiles) and capital of Puebla state, southern central Mexico; population (1990) 1,454,500. Founded 1535 as **Pueblo de los Angeles**, it was later renamed after General de Zaragoza, who defeated the French here in 1862. 94 E5

Pueblo city in south-central Colorado, USA, at the junction of the Fountain and Arkansas rivers, 67 km/42 mi southeast of Colorado Springs, 160 km/100 south of Denver; population (1996 est) 99,400. Industries include steel production, coal-mining, food-processing, and the manufacture of aircraft components, computer-ware, and plastic. It is also a market centre for lumber, as well as livestock and other agricultural produce from the irrigated farmlands of the Arkansas Valley. 86 F3

Puerto de Santa María, El town in the province of Cádiz, southern Spain, at the mouth of the River Guadalete on the Atlantic Ocean; population (1995 est) 70,900. There is a large commerce in sherry. El Puerto de Santa María has a Moorish castle and a baroque church. 28 D8

Puertollano town in the province of Ciudad Real, central Spain; population (1995) 53,100. It is a busy refining and petrochemicals centre; crude oil is received by pipeline from the port of Málaga. 28 F6

Puerto Rico (in full the **Commonwealth of Puerto Rico**), easternmost island of the Greater Antilles, situated between the US Virgin Islands and the Dominican Republic; area 9,000 sq km/3,475 sq mi; population (2000 est) 3,808,600. The capital is ►San Juan. Exports include sugar, tobacco, rum, pineapples, textiles, plastics, chemicals, processed foods, vegetables, and coffee. 94 L5

Puget Sound inlet of the Pacific Ocean on the west coast of Washington State, USA, extending southwards for about 160 km/100 mi, from the eastern end of the Strait of Juan de Fuca to Olympia, the state capital. It covers an area of about 5,180 sq km/1,990 sq mi, and contains a number of islands; Whidbey, Vashon, and Bainbridge are the largest. The major port of ►Seattle lies on its eastern shore, and a government naval yard is situated at Bremerton. The sound contains two main branches, Admiralty Inlet and Hood Canal, and receives rivers from the Cascade Range. Its waterways serve a rich industrial and agricultural area, and timber is rafted from its well-wooded shores to lumber and paper mills along the coast.

Pula commercial and naval port in western Croatia, on the Adriatic coast; population (1991) 62,400. A Roman naval base (**Colonia Pietas Julia**), it was seized by Venice in 1148, passed to Austria in 1815, to Italy in 1919, to Yugoslavia in 1947, and in 1991 became part of independent Croatia. 30 J6

Pulau Pinang Malay name for ►Penang, a state in west Peninsular Malaysia. 11 (1)A1

Pullman city in southeast Whitman County, southeast Washington, on the south fork of the Palouse River, 14 km/9 mi west of Moscow, Idaho, and 97 km/60 mi south-southeast of Spokane; population (1990) 23,500. It is a commercial and shipping centre for a grain, livestock, and poultry-producing region. It is the home of Washington State University (1890). The state college for science and agriculture is also at Pullman and has played an important role in the development of new breeds and agricultural techniques for the northwestern region. 86 C1

Pune (formerly **Poona**), city in Maharashtra, India, 100 km/60 mi southeast of Mumbai (formerly Bombay) on the Mutha River; population (1991) 2,494,000. Products include chemicals, rubber, rice, sugar, cotton, paper, and jewellery. Industries include cars, trucks, scooters, and motorbikes; pumps, cables, machinery, arms and ammunitions, cutting tools, televisions, boilers, and generators. 54 B5

Punjab state of northwest India, bordering Pakistan; area 50,400 sq km/19,454 sq mi; population (1994 est) 21,695,000. The capital is ►Chandigarh. Textiles and sewing machines are produced; wheat, rice, sugar, maize, millet, barley, and cotton are grown.

geography The rivers Sutlej and Beas, tributaries of the ►Indus, flow through the gently sloping alluvial plain that makes up most of the state.

features The region is mainly agricultural, with crops chiefly grown under irrigation through schemes such as the Bhakra Nangal dam on the Sutlej. There are ruins from the Indus Valley civilization (2500 to 1600 BC). Towns and cities include Amritsar, Jalandhar, Faridkot, and Ludhiana.

language and religion The main language is Punjabi. Religions are Sikh (60%) and Hindu (30%); there is friction between the two groups.

history In 1919 unrest led to the Punjab riots (known as the Amritsar Massacre). The present state was formed at the partition of India in 1947. The Indian Punjab was further divided into three areas: Himachal Pradesh, the Patiala and East Punjab States Union, and the state of East Punjab. In 1956 the latter two were merged to form Punjab (India), which was again divided in 1966 along linguistic lines to create three states: the predominantly Hindu states of Himachal Pradesh and Haryana (which shares the capital Chandigarh with Punjab); and the remaining Punjab state, 60% of which are Punjabi-speaking Sikhs. 10 B2

Punta Arenas (formerly **Magallanes**; Spanish 'sandy point'), seaport and capital of Magellanes region, southern Chile; situated on the eastern shore of the Brunswick Peninsula facing the Magellan Straits, and almost equidistant from the Pacific and Atlantic oceans; population (1992) 113,700. It is the southernmost city in Chile and the centre of a sheep-farming region. Exports include wool, lamb and mutton, and oil. Fishing and tourism are important. Punta Arenas was founded in the 19th century as a penal colony. **100** G9

Puri coastal city in Orissa, east India, 50 km/30 mi south of Bhubaneshwar; population (1991) 125,000. It is a religious centre with a temple dedicated to Jagganath or Vishnu, one of the three gods of Hinduism. A statue of the god, dating from about 318, is annually taken in procession on a large vehicle (hence the word 'juggernaut' used for a very large lorry). Devotees formerly threw themselves beneath its wheels. **54** E5

Pusan (or **Busan**), chief industrial port (textiles, rubber, salt, fishing) of Korea; population (1990) 3,797,600. It was invaded by the Japanese in 1592 and opened to foreign trade in 1883. **48** E6

Pushkin (formerly **Tsarskoe Selo**, then **Detskoe Selo** 1917–37), city in the Leningrad oblast (region) of the northwestern Russian Federation, 24 km/15 mi south of St Petersburg; population (1990) 95,500. The city is renowned for its 18th-century imperial palaces and parks and imposing 19th-century buildings. It is named after the writer Aleksandr Pushkin. The Pushkin museum is housed in the former lycée where the poet was educated. **38** F3

Puy-de-Dôme *département* in the ▶Auvergne region of France; area 7,954 sq km/3,071 sq mi; population (1990) 598,900. Plateaux and branches of the Cévennes and the Auvergne Mountains occupy three-quarters of it, plain and valley the rest. The principal river is the Allier. The chief products are wheat, rye, flax, and fruit. The high pasture lands support large numbers of cattle. There are metalworking, aerospace, and motor-parts industries. The hot and cold mineral springs are a tourist attraction, as are the mountains. The administrative centre is ▶Clermont-Ferrand. Other important towns are Ambert, Issoire, Riom, and Thiers. **6** E4

Puy, Le see ▶Le Puy, a town in France. **26** J8

Pwllheli market town and resort in Gwynedd unitary authority, northwest Wales, situated on the south coast of the Lleyn Peninsula and on Cardigan Bay, 30 km/19 mi southwest of Caernarfon; population (1991) 3,900. The Welsh National Party, Plaid Cymru, was founded here in 1925. Nearby is a very large holiday camp. **24** H9

Pyongyang (or **P'yongyang**, or **Pingyang**), capital and largest city of ▶North Korea, lying in the west of the country on the Taedong River, 50 km/30 mi inland from the Yellow Sea; population (1996 est) 2,500,000. It is the leading commercial, transport, and manufacturing centre of North Korea, and industries include aircraft, tractors and electrical vehicles, iron and steel, armaments, textiles, chemicals, machinery, ceramics, and rubber products. **48** C4

Pyrenees (French **Pyrénées**; Spanish **Pirineos**), mountain range in southwest Europe between France and Spain; length about 435 km/270 mi; highest peak Aneto (French Néthon) 3,404 m/11,172 ft. ▶Andorra lies entirely within the range. Hydroelectric power has encouraged industrial development in the foothills. **26** E11

Pyrénées-Atlantiques *département* in the ▶Aquitaine region of France on the Spanish frontier; area 7,629 sq km/2,946 sq mi; population (1990) 578,800. The River Adour, fed by many mountain torrents, flows through it. In the south are the peaks of the western ▶Pyrénées, giving way to the wooded hills, heathland, and fertile valleys of the Basque Country, and the plateau of the northeast. Wheat, maize, and vines are produced; mules, horses, sheep, and pigs are raised, and there are agro-industries. Fishing is important, and there are hydroelectric installations, oil, natural gas, and a chemical-waste treatment plant near Lacq. The tourist industry is well developed. The administrative centre is ▶Pau. Other towns include Bayonne and Orolon. **6** C5

Pyrénées, Hautes- *département* in the Midi-Pyrénées region of France; see ▶Hautes-Pyrénées. **6** D5

Pyrénées Orientales *département* in the ▶Languedoc-Roussillon region of southern France; area 4,086 sq km/1,578 sq mi; population (1990) 363,558. It comprises the western end of the Pyrenees, and is extremely mountainous, except for the plain of Roussillon in the east. The chief rivers are the Tet and the Tech. Vines, olives, vegetables, and fruit are grown, and livestock is raised. There are metallurgical and chemical industries, and a number of hydroelectric schemes. The principal towns are the departmental capital ▶Perpignan, Ceret, and Prades. **6** E5

Qaraghandy (formerly **Karaganda** 1857–1991), industrial city and capital of the Qaraghandy oblast (region) in central-eastern Kazakhstan; population (1996) 1,339,900. The city lies at the centre of an area rich in minerals, where coal, copper, tungsten, manganese are mined. Iron and steel mills are located here, and food-processing also takes place. Qaraghandy is joined by canal with the Irtysh River and lies on the railway line linking Almaty with the Trans-Siberian Railway. **42** N8

Qatar country in the Middle East, occupying Qatar peninsula in the Gulf, bounded southwest by Saudi Arabia and south by United Arab Emirates.

Qatar covers an area of 11,400 sq km/4,401 sq mi, and has a population of 599,000 (2000 est). The capital (and chief port) is ▶Doha; other major towns and cities include Dukhan, Wakra, ad-Dawhah, ar-Rayyān, Umm Salal, Musay'id, and aš-Šahniyah. The head of state is Sheikh Hamad bin Khalifa al-Thani (from 1995); the head of government is Sheikh 'Abd Allah ibn Khalifah Al Thani (from 1996). The currency is the Qatari riyal, and the languages spoken include Arabic (official) and English. **61** D4

Qattara Depression area of salt marsh in the Western Desert, Egypt, up to 134 m/440 ft 125 m/400 ft below sea level; area 20,000 sq km/7,500 sq mi. Its very soft sand, caused by seepage of groundwater, makes it virtually impassable to vehicles, and it protected the left flank of the Allied armies before and during the Second Battle of El Alamein in 1942. **64** E2

Qingdao (or **Tsingtao**), port and summer resort in Shandong province, east China, on the Yellow Sea; population (1994) 1,500,000. It has a naval base, and diverse heavy and light industries. Originally a small fishing village, it was a German colony from 1898 to 1914. **46** G3

Qinghai (or **Tsinghai**; Mongolian **Koko Nor**; Tibetan **Amdo**), province of northwest China, bounded to the north by Gansu, to the south by

Sichuan, to the west by Tibet, and to the northwest by Xinjiang Uygur Autonomous Region; area 721,000 sq km/278,400 sq mi; population (1996) 4,448,000. The capital is ►Xining. Industries include minerals, chemicals, livestock, oil, and medical products. There is animal rearing and bee-keeping. **11** (2)C3

Qom (or **Qum**, or **Kom**, or **Kum**), holy city of Shiite Muslims, in central Iran, 145 km/90 mi south of Tehran on the Qom River; population (1991) 681,300. Pottery, textiles, shoes, and glass are the main industries. The Islamic academy of Madresseh Faizieh in 1920 became the headquarters of Ayatollah Khomeini. **56** F3

Québec (Iroquois *Kebec*, 'a place where waters narrow'), capital and port of ►Québec province, Canada, at the junction of the Saint-Charles and St Lawrence rivers, Canada; population (1991) 167,500, metropolitan area (1996) 697,600. It is a major inland seaport, and a commercial, financial, and administrative centre. Industries include printing and publishing; and the production of paper, pulp, wood products, electronic goods, textiles, and leather. Lumber and wheat are exported. It is a centre of French culture, and most of its inhabitants are French-speaking. **88** F1

Québec province of eastern Canada; the largest province, second only in area among the nation's administrative subdivisions to the Northwest Territories. Québec is bordered on the northeast by Labrador, on the east by Newfoundland, on the southeast by New Brunswick and Nova Scotia, and on the west and southwest by Ontario. On its southern border lie (west–east) the US states of New York, Vermont, New Hampshire, and Maine; area 1,540,700 sq km/594,900 sq mi; population (1991) 6,811,800. The capital is ►Québec. Industries include mining (iron, copper, gold, zinc), fishing, and the production of paper, textiles, and maple syrup (70% of world output). Cereals and potatoes are grown. **88** F1

Quedlinburg town in the *Land* (administrative region) of ►Saxony-Anhalt, Germany, on the River Bode, on the northeastern edge of the Harz Mountains, 68 km/42 mi northwest of Halle; population (1995) 26,000. Quedlinburg has a research centre for plant biology, a horticultural industry, and manufactures machinery and chemicals. The town has many medieval buildings. **20** G5

Queen Charlotte Islands (called 'Canadian Galapagos'), archipelago in the north Pacific Ocean, about 160 km/100 mi off the northwest coast of British Columbia, western Canada, of which it forms part; area 9,790 sq km/3,780 sq mi; population (1994) 5,700. It is separated from the mainland by Hecate Strait to the east, and from the islands of Alaska to the north by Dixon Entrance. Graham, Moresby, and smaller Kunghit are the largest of about 150 islands, and Masset and Queen Charlotte City the principal ports. Timber processing, fishing, canning, and game hunting are the main economic activities. **82** E6

Queen Elizabeth Islands group of islands in the Canadian Arctic, north of Lancaster and Viscount Melville Sounds. They include ►Ellesmere, the northernmost island in the Canadian Arctic, Devon, Melville, Axel, Heiberg, Bathurst, Cornwallis, Prince Patrick, and the Ringnes Islands. They were named in 1954 after Queen Elizabeth II. There are virtually no indigenous inhabitants, but meteorological stations and Royal Canadian Mounted Police posts have been established in the area. **80** M2

Queen Maud Land (Norwegian **Dronning Maud Land**), region of Antarctica claimed by Norway in a 1939 royal decree. It lies between Enderby Land and Coats Land, from 20° west to 45° east. It was reached in 1930 by a Norwegian expedition under Captain Hjalmar Riiser-Larsen and named after Queen Maud of Norway. **102** (2)F2

Queens largest borough and county of New York City, USA; population (1996 est) 1,980,600. Situated at the west end of ►Long Island, it covers an area of 280 sq km/108 sq mi. Mainly residential, its districts include Jackson Heights, Forest Hills, and Flushing. Industries are concentrated in Long Island City, a railroad and shipping terminus, and Maspeth. Products include processed foods, metalware, paint, furniture, stonemasonry, clothes, and electronic and office equipment.

Queensland state in northeast Australia, including the adjacent islands in the Pacific Ocean and in the Gulf of ►Carpentaria; bordered on the west by Northern Territory, on the southwest by South Australia, on the south by New South Wales, on the east by the Pacific Ocean, and on the extreme northwest by the Gulf of Carpentaria; area 1,727,200 sq km/666,900 sq mi; population (1996) 3,368,850, concentrated in the southeast. The capital is ►Brisbane. Products include sugar, wheat, pineapples, beef, cotton, wool, tobacco, copper, gold, silver, lead, zinc, coal, nickel, bauxite, uranium, natural gas, oil, and fish. **76** G4

Quetta summer resort and capital of Baluchistan, west Pakistan, lying at 1,525 m/5,000 ft above sea level, 35 km/20 mi northwest of the Bolan Pass; population (1995 est) 419,000. Strategically placed on the trade route between Afghanistan and the Lower Indus valley, Quetta controls the Bolan Pass as well as the loss-used Khojak Pass. It has rail links with Afghanistan and Iran, and in 1982 a gas pipeline to Shikarpur in Sind was built. Quetta is a centre for fruit growing and trading in wood, carpets, and leather. There is a military staff college and a the University of Baluchistan. In May 1935 an earthquake devastated the city, killing over 20,000 people. **56** J3

Quezon City former capital of the Philippines 1948–76, northeastern part of metropolitan ►Manila (the present capital), on Luzon Island; population (1990) 1,166,800. It was named after the Philippines' first president, Manuel Luis Quezon (1878–1944). **50** G4

Qufu (or **Chufu**), town in Shandong province, China; urban population (1994 est) 117,500, total population (1994 est) 611,800. Qufu was capital of the state of Lu during the Zhou dynasty (1066–256 BC). In 551 BC it was the birthplace of the sage Confucius. The Great Temple of Confucius is sited here, and his tomb lies to the north of the temple. **46** F3

Quimper (Breton **Kemper**), administrative centre of the *département* of ►Finistère in Brittany, northwest France, situated about 58 km/36 mi southeast of Brest on the River Odet; population (1990) 62,500. There is a notable 13th–15th-century Gothic cathedral, possibly the earliest example in Brittany. The town has been a centre for the manufacture of decorative pottery since the 16th century, and also produces foodstuffs and electrical goods. 26 A5

Quito industrial city, capital of ►Ecuador and of Pichincha province; situated on a plateau in the Andes, 22 km/14 mi south of the Equator, at an altitude of 2,850 m/9,350 ft; population (1995 est) 1,1246,000. Industries include textiles, chemicals, leather, gold, silver, pharmaceuticals, and motor vehicles. Quito lies at the foot of the volcano Pichincha (4,794 m/15,728 ft), which last erupted in 1666, in an area prone to earthquakes. It has a temperate climate all year round. The city, which is the oldest capital in South America, has been declared a World Cultural Heritage Site by UNESCO. 98 B4

Qum alternative spelling of ►Qom, a city in Iran. 56 F3

Rabat capital and industrial port of ►Morocco, on the Atlantic coast, 177 km/110 mi west of Fès; population (city, 1994) 787,700; Rabat-Salé (urban area, 1998 est) 1,453,400. It is situated on the Bou Regreg River, opposite Salé. Industries include textiles, asbestos, carpets, pottery, leather goods, fishing; other exports include skins, wax, cork, slippers, and beans. Founded in 1190, it is named after its original *ribat* or fortified monastery. 66 D2

Rabaul largest port of Papua New Guinea, on the volcanic island of New Britain, southwest Pacific; population (1994) 184,400. It is one of the largest shipping centres for copra and coconut oil in the Pacific. It was destroyed by Allied bombing after its occupation by the Japanese in 1942, but was rebuilt. 74 F6

Racine city, port of entry, and administrative headquarters of Racine County, southwest Wisconsin, USA, at the mouth of the Root River on Lake Michigan, 39 km/24 mi south of Milwaukee; population (1996 est) 82,600. It is an agricultural centre with lumbering and food-processing industries. Vehicle components, farm machinery, paints, varnishes, and wax products are manufactured. A high proportion of the population is of Danish descent. 88 C2

Radnorshire (Welsh **Sir Faesyfed**), former border county of eastern Wales, which was included in the county of Powys between 1974 and 1996. It is now part of ►Powys unitary authority.

Radom industrial city (flour-milling, brewing, tobacco, leather, bicycles, machinery; iron works) in Poland, 96 km/60 mi south of Warsaw; population (1993) 229,700. Radom became Austrian in 1795, Russian in 1825, and was returned to Poland in 1919. 18 L6

Ragusa Italian name (to 1918) for the Croatian city of ►Dubrovnik. Its English name was **Arrogosa**, from which the word 'argosy' is derived, because of the city's fame for its trading fleets while under Turkish rule in the 16th century. 34 F7

Ragusa town in Sicily, Italy, 54 km/34 mi southwest of Syracuse; population (1990) 69,400. It stands over 450 m/1,500 ft above the River Ragusa. There are textile industries, asphalt mines, and oil has been discovered in the district. Nearby caves contain ancient tombs. 32 J12

Rainier, Mount (or **Mount Tacoma**; American Indian **Tacoma** 'mountain that was God'), highest mountain in the ►Cascade Range, Washington State, USA. Rising to 4,392 m/14,415 ft, it is a dormant volcano, crowned by 5 major and 20 minor glaciers. Dense forests cover the slopes below the treeline, which is located at about 1,520 m/5,000 ft. Mount Rainier National Park was created in 1899. 86 B1

Rajasthan state of northwest India; area 342,200 sq km/132,100 sq mi; population (1994 est) 48,040,000. The capital is ►Jaipur. Industries include textiles, cement, glass, asbestos, chemicals, and mineral extraction. Millet, wheat, and barley are grown; oilseed, cotton, and sugar are produced; and cattle, sheep, and camels are raised. 10 B3

Rajshahi capital of Rajshahi region, western Bangladesh; population (1991) 299,700. It trades in timber and vegetable oil. 54 E4

Raleigh capital of ►North Carolina, USA; population (1996 est) 243,800. It lies in a tobacco-growing region, to the east of the more densely populated cities of Winston-Salem, Greensboro, and Durham. Industries include the processing of foods, timber, and tobacco; and the manufacture of electronic equipment, chemicals, and cotton and synthetic textiles. With ►Durham and Chapel Hill, Raleigh benefits from the nearby presence of the Research Triangle Park, a regional research and manufacturing centre for high-technology products. 90 F2

Raleigh, Fort site of the first English settlement in America, at the north end of Roanoke Island, North Carolina, USA, to which in 1585 English explorer Walter Raleigh sent 108 colonists from Plymouth, England, under his cousin Richard Grenville. In 1586 Francis Drake took the dissatisfied survivors back to England. The outline fortifications are preserved.

Ramat Gan city in western Israel, northeast of Tel Aviv-Yafo; population (1992) 122,700. Industries include textiles, diamond cutting, and food processing. Established in 1921, it is a holiday resort.

Rambouillet town in the *département* of Yvelines at the southern end of the forest of Rambouillet, 45 km/30 mi southwest of Paris, France; population (1990) 25,300. The 14th–18th-century former royal château is now the presidential summer residence. A breed of fine-fleeced sheep is named after the town. 26 G5

Ramsgate seaside resort and cross-Channel port in the Isle of Thanet, northeast Kent, southeast England; population (1991) 37,100. It is a centre for yachting and fishing. Features include St Augustine's Roman Catholic church (1850), designed by Augustus Pugin. Ramsgate became popular as a resort following George IV's visit in 1827. There are ferry links with France and Belgium. 24 P10

Randers seaport in Aarhus county, eastern Denmark, on the River Gudena; population (1999) 56,000. Randers has food-processing industries and manufactures agricultural machinery. There is a medieval monastery. 16 F8

Randstad the main metropolitan region of the Netherlands, including the cities of ►Amsterdam, The ►Hague, ►Rotterdam, and ►Utrecht. The Randstad is an urbanized area forming a horseshoe-shaped zone around open countryside which is used for intensive agriculture.

Rangoon former name (to 1989) of ▶Yangon, the capital of Myanmar (Burma). **50** B3

Rapallo port and winter resort in Liguria, northwest Italy, 24 km/15 mi southeast of Genoa on the Gulf of Rapallo; population (1990) 29,800. There are tuna fisheries, and lace and olive oil are manufactured. Treaties were signed here in 1920 (settling the common frontiers of Italy and Yugoslavia) and in 1922 (cancelling German and Russian counterclaims for indemnities for World War I). **30** E6

Rapa Nui another name for ▶Easter Island, an island in the Pacific Ocean. **74** Q8

Ras al Khaimah (or **Ra's al Khaymah**), emirate on the Gulf; area 1,690 sq km/652 sq mi; population (1995) 144,400. Products include oil, pharmaceuticals, and cement. It is one of the seven members of the ▶United Arab Emirates. **61** F4

Ra's al Khaymah alternative name for ▶Ras al Khaimah, an emirate on the Gulf. **61** F4

Rastatt (or **Rastadt**), city in Baden-Württemberg, Germany, on the River Murg, 22 km west of Karlsruhe; population (1994) 222,300. Rastatt manufactures car parts, optical goods, televisions, and radios. The city has a rococo palace and several old churches. The Treaty of Rastatt was signed here 1714 at the end of the War of the Spanish Succession. **22** L6

Rathenow town in the *Land* (administrative region) of ▶Brandenburg, Germany, on the River Havel, 72 km/45 mi west of Berlin; population (1996) 24,400. It manufactures optical and precision instruments and chemicals. **20** H4

Ratisbon English name for the German city of ▶Regensburg. **20** H7

Ratzeburg town in Schleswig-Holstein, Germany, 24 km/15 mi south of Lübeck, and built partly on an island in Lake Ratzeburg; population (1994) 12,500. The town is a popular tourist resort, and has a Romanesque cathedral from 1154. **20** F3

Ravenna industrial port in Emilia-Romagna, Italy, about 65 km/40 mi east of Bologna; population (1992) 137,100. It is connected to the Adriatic Sea by the Corsini Canal. Industries include oil-refining, and the production of petrochemicals, synthetic rubber, and fertilizers. It lies in a marshy plain and methane deposits have been discovered nearby. The town has several Byzantine churches with superb mosaics. **30** H6

Ravensburg city in Baden-Württemberg, Germany, 116 km/72 mi southeast of Stuttgart; population (1994) 260,000. Textiles and machinery are manufactured. The city has medieval towers and gates and is overlooked by the Veitsburg fortress, the birthplace of Henry the Lion. **30** E3

Rawalpindi city in Punjab province, Pakistan, on the north bank of the River Leh, 175 km/110 mi southeast of Peshawar in the foothills of the Himalayas; population (1998) 1,406,200. Industries include oil refining, iron, chemicals, locomotives and furniture. There is a considerable trade in grain, wool, and timber. It has good communications by rail and by air and is both a commercial and a military centre which is strategically important because it controls routes into Kashmir. It was the temporary capital of Pakistan 1959–67 during the construction nearby of the new capital, ▶Islamabad. **56** K3

Reading industrial town and administrative headquarters of Berks County, east Pennsylvania, USA, on the Schuylkill River, 72 km/45 mi northwest of Philadelphia; population (1996 est) 75,700. Textiles, machinery, and special steels are manufactured. Coal-mining in the area is in decline. Albright (1856) and Alvernia (1958) colleges are in the town. **88** E2

Reading industrial town and unitary authority in southern England, on the River Thames where it meets the Kennet, 61 km/38 mi west of London; it was the administrative headquarters of the county of Berkshire until April 1998.
area 37 sq km/14 sq mi
features remains of a 12th-century Benedictine abbey where Henry I is buried; the Museum of Reading includes Roman and Saxon relics, and a full-size Victorian reproduction of the Bayeaux Tapestry; the Museum of English Rural Life is here
industries biscuits, brewing, boats, engineering, printing, and electronics; it is an agricultural and horticultural centre with seed-testing grounds, and is a major bulb producer
population (1996) 131,000 **24** M10/5 F6

Recife (formerly **Pernambuco**), industrial seaport and naval base on the Atlantic coast of northeastern Brazil; capital of Pernambuco federal unit (state); population (1991) 1,335,700 (metropolitan area 2,921,700). There are cotton mills, canneries, sugar refineries, and spirit distilleries; chief exports include sugar and cotton. It is also a major tourist resort, known as the 'Venice of Brazil'. **98** L5

Recklinghausen industrial city in North Rhine-Westphalia, Germany, 24 km/15 mi northwest of Dortmund; population (1995) 127,200. There are coal, iron, chemical, textile, and engineering industries. It is said to have been founded by Charlemagne. **22** K3

Redcar and Cleveland unitary authority in northeast England created in 1996 from part of the former county of Cleveland.
area 240 sq km/93 sq mi
towns and cities Redcar (administrative headquarters), Skelton, Guisborough, Marske-by-the-Sea, Saltburn-by-the-Sea, Brotton, Loftus
features North Sea coast; River Tees forms northwest border; Boulby Cliffs are highest cliffs on England's east coast (203 m/666 ft); 12th-century Priory at Guisborough; Ironstone Mining Museum (Saltburn-by-the-Sea)
industries manufacture of steel products (British Steel), engineering, fertilizers and potash products, textiles
population (1996) 144,000 **5** F4

Redditch industrial town in Worcestershire, 19 km/12 mi south of Birmingham; population (1991) 72,700. It was designated a new town in 1964 to take overspill population from Birmingham. Industries include engineering, electronics, and the production of electrical equipment, car and aircraft components, motorcycles, and fishing tackle. **22** A2

Red River (or **Red River of the South**), western tributary of the ▶Mississippi River, USA 1,638 km/1,018 mi long; so called because of the reddish soil sediment it carries. Formed in Oklahoma by the confluence of the North Fork and the Prairie Dog Town Fork, it flows through Texas, Arkansas, and

Louisiana, before entering the Mississippi near Baton Rouge, about 500 km/310 mi above the Gulf of Mexico. The stretch that forms the Texas–Oklahoma border is called Tornado Alley because of the storms caused by the collision in spring of warm air from the Gulf of Mexico with cold fronts from the north. The largest city on its course is Shreveport, Louisiana. **90** B3

Red River (Vietnamese **Song Hông**), river in north Vietnam, 500 km/ 310 mi long, that flows into the Gulf of Tonkin. Its extensive delta is a main centre of population. **50** C2

Red River of the North river in the USA and Canada; length 877 km/ 545 mi (770 km/440 mi in the USA). It rises in North Dakota and flows north, forming the boundary between Minnesota and North Dakota. It then flows through Manitoba, where it joins the Assiniboine River and flows into Lake Winnipeg. **86** G1

Red Sea branch of the Indian Ocean, formed from a submerged section of the Great ►Rift Valley, extending northwest from the Gulf of Aden. It is 2,000 km/1,200 mi long and up to 320 km/200 mi wide, reaching depths of over 2,300 m/7,545 ft. Egypt, Sudan, Ethiopia, and Eritrea (in Africa) and Saudi Arabia (Asia) are on its shores. At its northern end, it divides into the gulfs of Suez and Aqaba, separated by the Sinai peninsula. **64** G3

Regensburg (English **Ratisbon**), historic and commercial city in Bavaria, Germany, on the River Danube at its confluence with the Regen, 100 km/63 mi northeast of Munich; population (1995) 126,000. It has mechanical and electrical engineering, pharmaceutical, and food-processing industries. It has many medieval buildings, including a Gothic cathedral (1275–1530), and is a 'living museum' of German culture and architecture from the early Middle Ages.
history Regensburg stands on the site of a Celtic settlement dating from 500 BC. It became the Roman **Castra Regina** AD 179, the capital of the Eastern Frankish Empire, a free imperial city in 1245, and seat of the German *Diet* (parliament) 16th century–1806. It was included in Bavaria in 1810. **20** H7

Reggio di Calabria (Greek **Rhegion**; Latin **Rhegium**), port in Calabria, southern Italy, on the Strait of Messina; population (1992) 177,600. Farm machinery, furniture, perfume, olive oil, and fruit preserves are produced. It was founded by the Greeks in about 720 BC. **32** K10

Reggio nell'Emilia (ancient **Regium Lepidum**), chief town of Reggio nell'Emilia province in Emilia-Romagna region, northern Italy, about 55 km/34 mi northwest of Bologna; population (1992) 133,200. Industries include manufactures and services associated with locomotives, and the production of textiles, agricultural machinery, cement, and wine. It was here in 1797 that the Congress of the cities of Emilia adopted the tricolour flag, which was later to become the national flag of Italy. **30** F6

Regina capital of ►Saskatchewan, Canada, on Wascana Creek, 575 km/ 359 mi from Winnipeg; population (1996) 199,200. Industries include oil-refining, food-processing, and the manufacture of cement, steel, farm machinery, wood and paper products, paints, and fertilizers. It is connected by pipeline to the oil-fields of Alberta. Surrounded by a vast wheat-producing prairie, the city is the headquarters of the Saskatchewan Wheat Pool, one of the world's largest cooperatives. **84** F1

Reichenbach town in the *Land* (administrative region) of ►Saxony, Germany, 48 km/30 mi southwest of Chemnitz, at the foot of the ►Erzgebirge; population (1995) 23,900. Reichenbach has an important textile industry, and manufactures machinery and ceramics. **20** H6

Reichenberg German name for ►Liberec, a town in the Czech Republic. **18** E7

Reigate town in Surrey, southeast England, at the foot of the North Downs; population (1991, with Redhill) 46,300. Situated 30 km/ 19 mi south of London, it is primarily a commuter town. **26** E2

Ré, Ile de French island off the shores of the *département* of Charente-Maritime, 10 km/6 mi west of La Rochelle; area 74 sq km/29 sq mi. It is level, with vineyards, early vegetables, orchards, cornfields, and oyster fisheries. Salt is collected from the salt marshes. St-Martin is the capital, fortified in the 17th century by Vauban. Since 1988, a 3 km-/2 mi-long bridge has connected the island with the mainland. The Ile de Ré is a popular tourist destination, with a summer population (1994) of 100,000. **26** D7

Reims (English **Rheims**), city in the *département* of Marne, and largest commercial centre of the ►Champagne-Ardenne region, France, situated 130 km/80 mi northeast of Paris on the right bank of the River Vesle, a tributary of the Aisne; population (1990) 185,200, conurbation 206,000. From 987 all but six French kings were crowned here. The western facade of its cathedral, Notre Dame, is one of the masterpieces of the Middle Ages. In World War II the German High Command formally surrendered here to US general Eisenhower on 7 May 1945. Reims is the centre of the champagne trade and has textile, chemical, mechanical, metallurgical, and foodstuff manufactures. **22** G5

Remagen town in the Rhineland-Palatinate, Germany, on the left bank of the River Rhine, 25 km/15 mi southeast of Bonn; population (1994) 16,300. Remagen is a tourist centre for the Middle Rhine. It was a crucial point in the Allied crossing of the Rhine as part of the general advance into Germany at the end of World War II. **22** K4

Remscheid industrial city in North Rhine-Westphalia, Germany, 25 km/15 mi east of Dusseldorf; population (1995) 122,700. Stainless-steel implements are manufactured. The German physicist Wilhelm Röntgen was born in the suburb of Lennep, where there is now the Röntgen Museum. **22** K3

Rendsburg city in Schleswig-Holstein, Germany, on the Kiel Canal, 40 km/25 mi west of Kiel; population (1994) 255,500. It has iron and steel and textile industries. **20** E2

Renfrewshire unitary authority in west central Scotland, bordering the Firth of Clyde, which was formed from the northern and western parts of Renfrew district in Strathclyde region (1975–96), which in turn was formed from the former county of Renfrewshire (until 1974).
area 260 sq km/100 sq mi
towns ►Paisley (administrative head-quarters), Renfrew, Johnstone, Erskine

physical mainly low lying, but hilly in the west, rising to Hill of Stake (525 m/1,723 ft); rivers Clyde, Gryfe, White Cart, Black Cart
features sculptural stones at Inchinnan, near Erskine; Glasgow International Airport
industries engineering, computers, electronics, chemicals
agriculture sheep on grassy uplands; dairy farming on lowlands
population (1995) 178,300 **5** D4

Rennes (*Redones*, a Celtic tribe), administrative centre of ►Ille-et-Vilaine *département*, western France, 60 km/37 mi southeast of St Malo at the confluence of the Ille and Vilaine rivers; population (1990) 203,500, conurbation 245,000. It is the main commercial centre for western France and produces chemicals, electronics, cars, railway equipment, agricultural machinery and food products. There are printing works and nearby Vern-sur-Seiche has an oil refinery. It was the capital of the former province of ►Brittany. It was destroyed by fire in 1720 and had to be almost wholly rebuilt. **26** D5

Reno second-largest city in Nevada, USA, on the Truckee River; population (1996 est) 155,500. It is a major tourist venue, known for its easy marriage and divorce facilities, and gambling casinos, which are permitted by the state's liberal laws. Situated in a mining and ranching area, it is also a distribution and processing centre with diverse industries. Products include packed-meats, flour, dairy produce, building materials, forest products, and electronic equipment. **30** G6

Reshteh-ye Kūhhā-ye Alborz Arabic name for the ►Elburz Mountains, a volcanic mountain range in northwest Iran. **56** F2

Rethymnon (or **Rethymno**), town and department in ►Crete, Greece; population (1991 est) 69,300 (department). Its port is the outlet for produce from the surrounding agricultural area, and local industries include tanning, olive-oil refining, and soap manufacturing. The town is surrounded by a Venetian wall, now in ruins, and a Venetian fort lies to the north. **36** G9

Réunion French island of the Mascarenes group, in the Indian Ocean, 650 km/400 mi east of Madagascar and 180 km/110 mi southwest of Mauritius; area 2,512 sq km/970 sq mi; population (1995 est) 653,400. The capital is St Denis. Produce includes sugar, maize, vanilla, tobacco, and rum. **72** (1)B2

Reus city in the province of Tarragona in Cataluña, northeast Spain, 10 km/6 mi northwest of the city of Tarragona; population (1991) 86,900. Products include textiles, flowers, dried fruit, and vegetables. Reus is a tourist resort, and has an international airport. The architect Antonio Gaudí was born here. **28** M3

Reutlingen city in Baden-Württemberg, Germany, 30 km/19 mi south of Stuttgart; population (1994) 270,000. The main industries are the manufacture of textiles, machinery, and leather, and the city is a centre for printing and publishing. It is sited on steep hill slopes, has old towers, walls, and houses, and a 13th-century Gothic church. **20** E8

Reval former name of the Estonian port of ►Tallinn. **16** N7

Reykjavik chief port and capital (from 1918) of Iceland, on the southwest coast on Faxa Bay; population (1994) 103,000. Fish processing is the main industry. Most of the city is heated by an underground water mains system, built in 1945, the source of the hot water being volcanic springs and geysers. It was a seat of Danish administration from 1801 to 1918, and has been the seat of the Parliament since 1843. Reykjavik is the world's most northerly capital. **16** (1)C2

Rheims English form of ►Reims, a city in France. **22** G5

Rhein German name for the ►Rhine, a European river. **30** C2

Rheinland-Pfalz German name for the ►Rhineland-Palatinate region of Germany. **7** (2)A3

Rhine (German **Rhein**; French **Rhin**; Dutch **Rijn**), European river rising in Switzerland and reaching the North Sea via Germany and the Netherlands; length 1,320 km/820 mi. It drains an area of some 220,000 sq km/85,000 sq mi and is navigable for 805 km/500 mi. Tributaries include the Moselle and the Ruhr. The Rhine is linked with the Mediterranean by the Rhine–Rhône Waterway, and with the Black Sea by the Rhine–Main–Danube Waterway. **30** C2

Rhineland-Palatinate (German **Rheinland-Pfalz**), administrative region (German *Land*) of Germany, bordered on the north by North Rhine-Westphalia, on the east by Hesse and Baden-Württemberg, on the south by France, on the southwest by the Saarland, and on the west by Luxembourg and Belgium; area 19,800 sq km/7,650 sq mi; population (1995) 4,015,000. The capital is ►Mainz. There are motor vehicle, mechanical and electrical engineering, chemical, machinery, leather goods, pottery, glass, and beverage industries. Wine (75% of German output), potatoes, cereals, sugar beet, fruit, and tobacco are produced, and there is stock rearing. **7** (2)A3

Rhin, Haut- *département* in the Alsace region of France; see ►Haut-Rhin. **6** G3

Rhode Island smallest state of the USA, located in New England. It is nicknamed Little Rhody or the Ocean State, and is officially known as **Rhode Island and Providence Plantations**. Rhode Island ratified the US Constitution in 1790, becoming the 13th state to join the Union. It is bordered to the north and east by Massachusetts, to the west by Connecticut, and to the south by the Atlantic Ocean.
population (2000 est) 1,048,300
area 3,100 sq km/1,197 sq mi
capital ►Providence
towns and cities Warwick, Cranston, Newport, Woonsocket
industries and products electronics, machine tools, jewellery, textiles, silverware, rubber, and plastics. Agriculture is limited by the rocky terrain but is important in rural areas, the main crops being apples and potatoes. Rhode Island Red hens were developed here from the 19th century **88** F2

Rhodes (Greek **Rodos**), Greek island, largest of the Dodecanese, in the eastern Aegean Sea; area 1,412 sq km/545 sq mi; population (1991 est) 98,200. The capital is Rhodes. Grapes and olives are grown. **36** L8

Rhodesia former name of ►Zambia (Northern Rhodesia) and ►Zimbabwe (Southern Rhodesia), in southern Africa. **72** D2

Rhodope Mountains (or **Rodopi Planina**), range of mountains on the frontier between Greece and Bulgaria, rising to 2,925 m/9,497 ft at Musala. **34** M7

Rhondda industrial town in Rhondda Cynon Taff, south Wales, situated 26 km/16 mi northwest of Cardiff; population (1991) 59,900. Light industries have replaced coal mining, formerly the main source of employment in the area. The closure of the Maerdy mine (opened 1875) in 1990 ended mining in the valley; Rhondda's coal powered 90% of the Royal Navy's ships in World War I. The Rhondda Heritage Park recreates a 1920s-style mining village for visitors. 24 J10

Rhondda Cynon Taff unitary authority in south Wales, created in 1996 from part of the former county of Mid Glamorgan.
area 440 sq km/170 sq mi
towns Clydach Vale (administrative headquarters)
physical rivers ▶Rhondda Fawr and Rhondda Fach
industries light industries
population (1996) 232,600 5 E6

Rhône *département* in the ▶Rhône-Alpes region of France; area 3,215 sq km/1,241 sq mi; population (1990) 1,509,500. The River ▶Rhône and its tributaries the ▶Saône, Azergues, and Gier flow through it. Moderate crops of cereals and potatoes grow on the rocky soil, and wines (Juliénas, Côte-Rôtie, Moulin à-Vent) are produced. There are metallurgical, chemical, and textile industries around ▶Lyon, the administrative centre. The other principal town is Villefranche. 6 F4

Rhône river of southern Europe; length 810 km/500 mi. It rises at the Rhône Glacier (altitude 1,825 m/5,987 ft) in the canton of Valais in Switzerland and flows through Lake Geneva to Lyon in France, where, at its confluence with the ▶Saône, the upper limit of navigation is reached. The river then turns due south and passes Vienne and Avignon. Near Arles it divides into the **Grand** and **Petit Rhône**, flowing respectively southeast and southwest into the Mediterranean west of Marseille. Here it forms a two-armed delta; the area between the tributaries is the marshy region known as the ▶Camargue. 26 K9

Rhône-Alpes region of eastern France in the upper reaches of the ▶Rhône; area 43,700 sq km/16,868 sq mi; population (1992) 5,344,000. It consists of the *départements* of ▶Ain, ▶Ardèche, ▶Drôme, ▶Isère, ▶Loire, ▶Rhône, ▶Savoie, and ▶Haute-Savoie. The capital is ▶Lyon. There are several wine-producing areas, including Chenas, Fleurie, and Beaujolais. Industrial products include chemicals, textiles, and motor vehicles. 6 F4

Rhyl seaside holiday resort in Denbighshire, north Wales, situated 50 km/31 mi northwest of Chester near the mouth of the River Clwyd; population (1991) 24,900. Products include furniture manufacture; tourism is important. 24 J8

Ribe capital of Ribe county, on the west coast of Denmark; population (1992) 7,900. The River Ribe flows through the town, and dykes have been built to protect the town from flooding. The oldest city in Denmark, Ribe was an important port during the Middle Ages, and part of the medieval town still surrounds the cathedral. 16 E9

Richland city in Benton County, southeast Washington, USA, on the Columbia River, northwest of Walla Walla; population (1990) 33,300. It is a centre for research for the US Department of Energy and a major producer of plutonium for nuclear weapons. It grew as a residential community for employees of the Hanford Engineer Works that helped to develop the atomic bomb from 1943. 86 C1

Richmond capital and seaport of ▶Virginia, USA, on the James River, 209 mi from its mouth on the Atlantic, 160 km/100 mi south of Washington, DC; population (1996 est) 198,300. It is a major tobacco market and a distribution, commercial, and financial centre for the surrounding region. Industries include the manufacture of tobacco products, processed foods, chemicals, metalware, paper and print, and textiles. Established in 1637, Richmond was the site of the first permanent colonial American settlement. The city became the capital of Virginia in 1779, and the Confederacy 1861–65; several Civil War battles were fought for its possession. The cigarette-rolling machine was invented here in the 1870s. 88 E3

Riesa town in the *Land* (administrative region) of ▶Saxony, Germany, on the River Elbe, 40 km/25 mi northwest of Dresden; population (1995) 42,700. It is a river port, and industries include mechanical engineering, glass making, tyres, and food processing. 20 J5

Rift Valley, Great longest 'split' in the Earth's surface; see ▶Great Rift Valley. 70 E5

Riga capital and port of Latvia; population (1995) 840,000. Industries include engineering, brewing, food processing, and the manufacture of textiles and chipboard. 16 N8

Rijeka (Italian **Fiume**), industrial port in Croatia; population (1991) 168,000. Industries include oil-refining, distilling, and the manufacture of paper, tobacco, and chemicals. It has changed hands many times and, after being seized by the Italian nationalist Gabriele D'Annunzio in 1919, was annexed by Italy in 1924. It was ceded back to Yugoslavia in 1949, and in 1991 it became part of newly independent Croatia. 30 K5

Rimini (Roman **Ariminum**), industrial port and holiday resort in Emilia-Romagna, Italy, on the Adriatic Sea, at the mouth of the Marecchia River, 45 km/28 mi southeast of Forli; population (1992) 129,900. Pasta, footwear, textiles, and furniture are manufactured. 30 H6

Rio de Janeiro (Portuguese 'river of January'), port and resort in southeast Brazil; capital of Rio de Janeiro federal unit (state), and former national capital (1763–1960); population (1991) 5,480,800 (metropolitan area 10,389,400). It is situated on the southwest shore of Guanabara Bay, an inlet of the Atlantic Ocean; Sugar Loaf Mountain (a huge cone-shaped rock outcrop, composed of granite, quartz and feldspar) stands at the entrance to the harbour, and the city is dominated by the 30 m/100 ft-high figure of Christ on the top of Corcovado, a jagged peak 690 m/2,264 ft high. Industries include ship-repair, sugar refining, textiles, and the manufacture of foodstuffs; coffee, sugar, and iron ore are exported. 100 N3

Rio Grande (Mexican **Río Bravo del Norte**), river of the USA and Mexico, rising in the Rocky Mountains in southern Colorado, it flows southeast, through New Mexico and Texas, to the Gulf of Mexico near Brownsville; length 3,050 km/1,900 mi. From El Paso, the river forms the US-Mexican border for the last 2,400 km/1,500 mi of its course. Insufficient water is carried for the demands of irrigation both sides of the border, and the Rio Grande is eventually reduced to a trickle in its lower reaches. Its rate of flow is subject to international agreements. **92** E2

Rio Grande do Norte federal unit (state) of northeast Brazil, bounded on the north and east by the Atlantic Ocean; area 53,100 sq km/20,500 sq mi; population (1991) 2,414,100; capital ▶Natal. Apart from a narrow coastal zone with abundant rainfall, most of the state lies on a semi-arid plateau, crossed by several rivers, where there is stock-raising, and cotton, sugar, and cassava are grown. There are oil, textile, and agricultural industries. It produces 90% of the nation's salt. Carnauba wax is extensively produced from the carnauba palm. **98** K5

Rio Grande do Sul southernmost federal unit (state) of Brazil, to the east of the Uruguay River, bounded on the east by the Atlantic Ocean, on the west by Argentina, and on the south by Uruguay; area 282,184 sq km/ 108,993 sq mi; population (1991) 9,138,700; capital ▶Porto Alegre. The region consists mainly of vast grasslands where there is extensive stock-raising (cattle, sheep, pigs); wine, rice, and soybeans are produced, and industries are centred around agricultural production and processing. The state produces 90% of the national wine production. **100** L4

Rioja, La see ▶La Rioja, a region of Spain. **9** (1)O1

Riom town on the River Ambène, in the Puy-de-Dôme *département* of central France; population (1990) 19,300. It was the scene in World War II of the 'war guilt' trials of several prominent Frenchmen, between February and April 1942, by the Vichy government. The accused included the former prime ministers Blum and Daladier, and General Gamelin. The occasion turned into a wrangle over the reasons for French unpreparedness for war, and, at the German dictator Hitler's instigation, the court was dissolved. The defendants remained in prison until released by the Allies in 1945. **26** J8

Ripon city and market centre in North Yorkshire, northern England, on the River Ure; population (1991) 14,200. Agricultural produce is traded here. Features include a cathedral (1154–1520), and nearby are the 12th-century ruins of Fountains Abbey, among the finest monastic ruins in Europe. **24** L7

Riva del Garda resort at the northern end of Lake Garda, in Trentino-Alto Adige region, Italy; population (1990) 13,400. The Prix Italia broadcasting festival has had its headquarters here since 1948. **30** F5

Riverside city in southern California, USA, on the Santa Ana River, 16 km/10 mi west of San Bernadino and 90 km/56 mi east of Los Angeles; population (1996 est) 255,100. The seedless navel orange was developed here in 1873, and the city is at the centre of a citrus-growing district, although recently the city has encroached on the fruit groves. Manufactured products include irrigation and electronic equipment, aircraft engines, paper products, and plastics.

Riviera the Mediterranean coast of France and Italy from Hyères to La Spezia. The most exclusive stretch of the Riviera, with the finest climate, is the ▶Côte d'Azur, from Menton to St-Tropez, which includes Monaco.

Riyadh (Arabic **Ar Riyād**), capital of Saudi Arabia and of the Nejd region, situated in an oasis and connected by rail with Dammam 450 km/280 mi away on the Gulf; population (1994 est) 1,500,000. **61** B4

Roanne French town in the *département* of Loire, 75 km/47 mi northwest of St-Etienne, on the River Loire and the Roanne canal; population (1990) 80,000. There are spinning and weaving mills. **26** K7

Roanoke (American Indian 'shell money'), city in Virginia, USA, on the **Roanoke River**, in Great Valley, 65 km/41 mi west of Lynchburg; population (1996 est) 95,600. It is the marketing and distribution point for a rich agricultural area. Manufactured products include chemicals, steel goods, furniture, and textiles. **88** D3

Robben Island island in Table Bay, Cape Town, South Africa. It was used by the South African government to house political prisoners. Nelson Mandela was imprisoned here 1964–82.

Rochdale industrial town in Greater Manchester, northwest England, on the River Roch, 16 km/10 mi north-east of Manchester; population (1994 est) 138,000. It was formerly an important cotton-spinning town; industries now include textiles and the manufacture of machinery and asbestos. The Rochdale Pioneers founded the first Cooperative Society in England here in 1844.

Rochefort port in western France, southeast of La Rochelle and 15 km/9 mi from the mouth of the River Charente; population (1990) 26,900. Its industries include metal goods and machinery. The town dates from 1666. It was from here that Napoleon embarked for Plymouth on the *Bellerophon* on his way to final exile in 1815. **26** E8

Rochelle, La see ▶La Rochelle, a port in western France. **26** D7

Rochester city in southeastern England, on the Medway estuary, in Medway Towns unitary authority; population (1991) 24,000. Rochester upon Medway district joined with Gillingham to form the Medway Towns unitary authority in April 1998. Rochester was a Roman town, **Durobrivae**. It has a 12th-century Norman castle keep (the largest in England), a 12th–15th-century cathedral (containing a memorial to Charles Dickens), and many timbered buildings. Industries include aero-nautical, electrical, and mechanical engineering; cement; paper; and paint and varnish. The Charles Dickens Centre (1982) commemorates the town's links with the novelist Charles Dickens, whose home was at Gad's Hill. **22** C3

Rochester city and administrative headquarters of Olmsted County, Minnesota, USA, 112 km/70 mi southeast of St Paul; population (1996 est) 75,600. A notable medical research and treatment centre, it is the home of the University of Minnesota medical school, the Mayo Foundation for Medical Education and Research; numerous other hospitals also are located in the city. The Mayo Clinic was part of a medical foundation of 1889. Rochester has diverse manu-facturing, including the production of computers, tractor cabs, phonographs, and medical supplies and instruments. Vegetables grown in the area are canned here and dairy products are processed. The Rochester Community College (1915) is in the city. 88 B2

Rochester city and deepwater port in New York State, USA, on the Genesee River, linked to the north with Lake Ontario; population (1996 est) 221,600. Situated in a major fruit-growing area, it is a processing and distribution point, and notably a manufacturing centre for photographic equipment, and optical and other precision instruments. It was the birthplace of the Xerox copier, and the world headquarters of the Eastman Kodak Company are here. Other industries include horticulture, engineering, printing, and the production of clothing, plastics, and metal goods. 88 E2

Rockall British islet in the Atlantic, 24 m/80 ft across and 22 m/65 ft high, part of the Hatton-Rockall bank, and 370 km/230 mi west of North Uist in the Hebrides. The bank is part of a fragment of Greenland that broke away 60 million years ago. It is in a potentially rich oil/gas area. The islet is uninhabited. A party of British marines landed in 1955 formally to annex Rockall, but Denmark, Iceland, and Ireland challenge Britain's claims for mineral, oil, and fishing rights. The **Rockall Trough** between Rockall and Ireland, 250 km/155 mi wide and up to 3,000 m/10,000 ft deep, forms an ideal marine laboratory. 14 C2

Rockford city and administrative headquarters of Winnebago County, north-central Illinois, USA, on the Rock River, 120 km/75 mi northwest of Chicago; population (1996 est) 143,500. It is an agricultural market centre with diverse industries, including engineering, and food-processing. Local manufactures include machine tools, screws and fasteners, containers, auto parts, electric motors, metal-working machinery, aviation and space instruments, and air-conditioning equipment, among many other products. The Hennepin Canal links the Rock River to the Illinois River. Rockford's educational institutions include Rockford College (1847), Rock Valley College (1964), and the University of Illinois College of Medicine at Rockford. 88 C2

Rockhampton port in east Queensland, Australia, on the Tropic of Capricorn, 650 km/404 mi north of Brisbane; population (1996) 57,770. Rockhampton is situated 26 km/16 mi from the coast on the banks of the River Fitzroy. It is the centre and port for a large beef-producing area. Other industries of the district include dairying, growing sugar cane, and meat processing. Tourism and sugar-cane growing are increasingly important. 76 K4

Rock Hill city in York County, north-central South Carolina, USA; population (1990) 41,600. It is located 42 km/26 km south-southwest of Charlotte, North Carolina. Industries include chemicals, textiles, and paper products. Rock Hill is home to Winthrop College (1886) and to a junior college. 88 D4

Rock Island city and administrative headquarters of Rock Island County, northwest Illinois, USA, at the confluence of the Rock and Mississippi rivers, 270 km/170 mi southwest of Chicago; population (1996 est) 39,700. Manufactured products include rubber, electronics, farm implements, aluminium products, machinery, and clothing. The town was founded in 1835, at a bridge site over the Mississippi, which incorporated the island or 'rock' for which it was named; the isle is now the location of a US government arsenal and the Armament National Readiness Command Headquarters for the US Army. Augustana College (1860) is in the city. 88 B2

Rocky Mountains (or **Rockies**), largest North American mountain system, extending for 4,800 km/3,000 mi from the Mexican plateau near Sante Fe, north through the west-central states of the USA, and through Canada to the Alaskan border. It forms part of the Continental Divide, which separates rivers draining into the Atlantic or Arctic oceans from those flowing toward the Pacific Ocean. To the east lie the Great Plains, and to the west, the plateaux separating the Rocky Mountains from parallel Pacific coast ranges. Mount Elbert is the highest peak, 4,400 m/14,433 ft. Some geographers consider the Yukon and Alaskan ranges as part of the system, making the highest point Mount McKinley (Denali) 6,194 m/20,320 ft, and its total length 5,150 km/3,219 mi. 82 F5

Rodopi Planina alternative name for the ►Rhodope Mountains, a mountain range on the border between Greece and Bulgaria. 34 M7

Rodos Greek name for the island of ►Rhodes. 36 L8

Roermond town in the province of ►Limburg, Netherlands, on the River Maas (see ►Meuse), 45 km/28 mi northeast of Maastricht; population (1996) 43,800. The chief industries are the manufacture of cotton and woollen goods, tobacco, and dyeing. There is some fine carving in the 13th-century Romanesque Munsterkerk (minster). 22 J3

Roeselare (French **Roulers**), textile town in West Flanders province, northwest Belgium, 19 km/12 mi northwest of Courtrai; population (1997) 53,600. It lies on a tributary of the Lys, the Mandel, to which it is connected by canal. A weaving centre in the Middle Ages, it cultivates flax and manufactures lace, linen, gloves, carpets, pottery, and tiles. Brewing is also an important industry. It was a major German base during World War I. 22 F4

Roman industrial town in western Romania, at the confluence of the rivers Moldavia and Siret, 55 km/34 mi southwest of Iaşi; population (1993) 97,000. The main products are iron and steel, machinery, textiles, chemicals, and processed food. Coal and iron are mined nearby. 34 P3

Romania country in southeast Europe, bounded north and east by Ukraine, east by Moldova, southeast by the Black Sea, south by Bulgaria, southwest by Yugoslavia, and northwest by Hungary.

Romania covers an area of 237,500 sq km/91,698 sq mi, and has a population of 22,327,000 (2000 est). The capital is ►Bucharest; other major towns and cities include Braşov, Timişoara, Cluj-Napoca, IaşI, Constanţa, Galaţi, and Craiova. The head of state is Ion Iliescu (from 2000); the head of government is Adrian Nastase (from 2000). The currency is the leu, and the languages spoken include Romanian (official), Hungarian, and German. 34 L4

Romans-sur-Isère French town in the *département* of Drôme; population (1990) 60,000. It is situated on the River Isère, 18 km/11 mi northeast of Valence. Romans has a 12th–13th-century church. The town is a tanning and leatherworking centre. 26 L8

Rome (Italian **Roma**), capital of Italy and of Lazio region, on the River

Tiber, 27 km/17 mi from the Tyrrhenian Sea; population (1992) 2,723,300.

Rome is an important road, rail, and cultural centre. A large section of the population finds employment in government and other offices: the headquarters of the Roman Catholic Church (the ►Vatican City State, a separate sovereign area within Rome) and other international bodies, such as the Food and Agriculture Oranization (FAO), are here; it is also a destination for many tourists and pilgrims. Industries have developed, mainly to the south and east of the city; these include engineering, printing, food-processing, electronics, and the manufacture of chemicals, pharmaceuticals, plastics, and clothes. The city is a centre for the film and fashion industries. Among the remains of the ancient city are the Forum, Colosseum, and Pantheon. 32 G7

Rome city in Oneida County, central New York State, USA, on the Mohawk River and Erie Canal, 24 km/15 mi northwest of Utica; population (1996 est) 41,000. Situated in a dairy farming and market garden district, it manufactures copper and brass products, paint, and household appliances. Rome is the site of Griffiss Air Force Base, and units of the State University of New York and Mohawk Valley Community College. 88 E2

Romsey market town in Hampshire, southern England, 13 km/8 mi northwest of Southampton, on the River Test; population (1991) 16,600. Industries include brewing, carpet manufacture, computer consultancy, and farming. The Norman church of Romsey Abbey (founded by Edward the Elder in 907) survives, as does King John's Hunting Box of about 1206 (now a museum, King John's House). 22 A3

Roncesvalles small village in Navarre, northern Spain, on a pass in the Pyrenees 8 km/5 mi south of the French border. In 778 it was the scene of the defeat of the rearguard of Charlemagne's army under Roland, who was killed with the twelve peers of France. Roncesvalles has a small monastery and a museum. 28 J2

Ronda town in the province of Málaga, southern Spain; population (1991) 33,700. It stands on the edge of a rocky plateau, and is divided in two by a gorge (160 m/525 ft deep). Industries include horse-raising, leather goods, and wine. The town has one of the oldest and finest bullrings in Spain. Pedro Romero (died 1800), a carpenter from Ronda, refined and codified the bullfight in the 18th century. 28 E8

Rondônia federal unit (state) of northwest Brazil, within the drainage basin of the Amazon River, and bordered on the southwest by Bolivia; the centre of Amazonian tin and gold mining and a frontier region of agricultural colonization; area 238,400 sq km/92,000 sq mi; population (1991) 1,130,900; capital Pôrto Velho. Its principal products are rubber and brazil nuts. Known as the Federal Territory of **Guaporé** until 1956, Rondônia became a state in 1981. 98 E6

Ronse (French **Renaix**), town in the province of East Flanders, Belgium, 37 km/23 mi south of Ghent; population (1991) 24,000. It manufactures cotton and woollen goods, silk, thread, hats, and shoes. The church of St Hermes dates from the 11th century. 22 F4

Roraima, Mount mountain peak and plateau in the ►Pacaraima mountain range in South America, rising to 2,810 m/9,222 ft on the Brazil–Guyana–Venezuela frontier. 98 E2

Rosario industrial city and river port of Santa Fé department, east-central Argentina, on the Paraná River 280 km/174 km northwest of Buenos Aires; population (urban area, 1996 est) 1,200,000. It is Argentina's largest inland port, and a distribution centre and export outlet for the surrounding agricultural regions of Córdoba, Mendoza, and Tucumán. There are sugar refineries, sawmills, flour mills, and tanneries; main exports include wheat, cattle, flour, sugar, and linseed. Rosario was founded in 1725; there is a memorial here to Gen Belgrano, who first raised the Argentine flag here. 100 J5

Roscommon (formerly **Ros-Comain**; 'wood around a monastery'), county of the Republic of Ireland, in the province of Connacht; county town **Roscommon**; area 2,460 sq km/950 sq mi; population (1996) 52,000. It has rich pastures and is bounded on the east by the River Shannon, with bogs and lakes, including Lough Key and Lough Gara. The three largest lakes (loughs Allen, Boderg, and Ree) lie only partly within the county. There is agriculture, especially cattle rearing. Roscommon was established as a county in about 1580. Other important towns are Castlerea, Elphin, and Boyle. 5 B5

Roseau (formerly **Charlotte Town**), capital of ►Dominica, West Indies, on the southwest coast of the island; population (1991) 15,900. It is a sea port, with a trade in tropical fruit and vegetables. The town suffered severe damage in a hurricane in 1979. 94 M5

Rosenheim industrial city in Bavaria, Germany, on the River Inn, 53 km/33 mi southeast of Munich; population (1994) 58,600. Rosenheim has electronics, engineering, sports clothes, and shoe manufacturing industries. The city is set in a picturesque Alpine district, and has medieval squares and attractive gardens. 30 H3

Roskilde port at the southern end of Roskilde Fjord, Zealand, Denmark; population (1990) 49,100. It was the capital of the country from the 10th century until 1443. 16 G9

Ross Dependency all the Antarctic islands and territories between 160° east and 150° west longitude, and situated south of 60° south latitude; it includes Edward VII Land, the Ross Sea and its islands (including the Balleny Isles), and parts of Victoria Land. It is claimed by New Zealand; area 450,000 sq km/173,745 sq mi.

Ross Island either of two islands in Antarctica:

Ross Island in the Weddell Sea, discovered in 1903 by the Swedish explorer Nils Nordenskjöld, area about 3,885 sq km/1,500 sq mi;

Ross Island in the Ross Sea, discovered in 1841 by the British explorer James Ross, area about 6,475 sq km/2,500 sq mi, with the research stations Scott Base (New Zealand) and McMurdo (USA). Mount Erebus (3,794 m/12,520 ft) is the world's southernmost active volcano; its lake of molten lava may provide a window on the magma beneath the Earth's crust that fuels volcanoes.

Ross Sea Antarctic inlet of the South Pacific, between Victoria Land and Marie Byrd Land. See also ►Ross Dependency and ►Ross Island. 102 (2)Z1

Rostock industrial port in the *Land* (administrative region) of Mecklenburg–West Pomerania, Germany, on the River Warnow, 13 km/8 mi south of the Baltic; population (1995) 231,300. Industries include shipbuilding, ship repair, electronics, and fish processing. There is a technical university here. **20** H2

Rostov-na-Donu capital of Rostov oblast (region), southwestern Russian Federation; population (1996 est) 1,025,000. A major industrial and commercial city and the centre of a fertile agricultural region, it lies on the River Don, 46 km/29 mi from its mouth. Rostov is home to many large engineering concerns manufacturing chemicals, agricultural machinery, aircraft, and ships; there are also tobacco- and food-processing plants, shoe factories, and textile mills. The city is an important transportation centre; railway lines from Moscow and Kiev converge here, and river cruises run up the Don and Volga. Rostov has been called the 'Gateway to the Caucasus'. **38** G5

Rothenburg tourist town in Franconia, Bavaria, Germany, 65 km/40 mi west of Nuremberg; population (1995) 12,500. It has medieval buildings, churches, walls, guard towers, and gates, dating from the time that Rothenberg was a free imperial city. **20** F7

Rotherham industrial town in South Yorkshire, northern England, at the confluence of the rivers Don and Rother, 10 km/6 mi northeast of Sheffield; population (1994) 154,000. Industries include engineering, pottery, glass, iron and steel, brassware, machinery, and coal.

Rotorua town with medicinal hot springs and other volcanic activity in North Island, New Zealand, near the southwest fringe of Lake Rotorua; population (1996) 56,900. It is surrounded by hills over 762 m/2,500 ft high. Industries include tourism, sawmilling and farming. **78** F4

Rotterdam industrial city and port in South Holland province, the Netherlands, in the Rhine-Maas delta, 90 km/56 mi southwest of Amsterdam; population (1997) 590,000. The Rotterdam-Europoort complex is the biggest oil refining centre in the world, and one of its foremost ocean cargo ports. Other industries include brewing, distilling, shipbuilding, sugar and petroleum refining, margarine, and tobacco. A canal, the New Waterway (*Nieuwe Waterweg*), links Rotterdam with the North Sea. **26** K2

Roubaix town in Nord *département* of northern France, 9 km/5.6 mi northeast of Lille; population (1990) 98,200. It is a major centre of French woollen textile production. **22** F4

Rouen industrial port and administrative centre of the ►Seine-Maritime *département* in France, situated on the River Seine 120 km/75 mi northwest of Paris; population (1999 est) 106,600. Rouen is the world's leading port for grain exports, and the leading port in the European Union (EU) for the export of agricultural produce. The port accounts for 12,000 jobs, making it one of the area's largest employers. Other industries include textiles (particularly cotton), electronics, distilling, chemicals and petrochemicals, car manufacture, and oil refining. The civil service is also a major employer. **22** D5

Roulers French name of ►Roeselare, a town in Belgium. **22** F4

Rovaniemi capital of Lappi province, northern Finland, and chief town of Finnish Lapland, situated just south of the Arctic Circle; population (1992) 34,300. After World War II the town was rebuilt by the architect Alvar Aalto, who laid out the main streets in the form of a reindeer's antlers. **16** N3

Ruahine Range mountain range in North Island, New Zealand. It stretches northeast from the Manawatu Gorge to the headwaters of the Ngaruro River and lies within the 900 sq km/347 sq mi Ruahine State Forest Park, designated in 1976. The highest point is Mangaweka, 1,733 m/5,686 ft. **78** E5

Ruanda part of the former Belgian territory of Ruanda-Urundi until it achieved independence as ►Rwanda, a country in central Africa. **70** D4

Ruapehu, Mount volcano in New Zealand, southwest of Lake Taupo, near Ohakune. Intermittently active, it is the highest peak in North Island, 2,797 m/9,175 ft. Mount Ruapehu is North Island's main ski resort, with ski slopes on its northern and southern sides. **78** E4

Rub' al Khālī (Arabic 'empty quarter'), vast sandy desert in southern Saudi Arabia and Yemen; area 650,000 sq km/250,000 sq mi. In 1930–31 the British explorer Bertram Thomas became the first European to cross it. **56** E6

Ruda Slaska town in Silesia, Poland, with metallurgical industries, created 1959 by a merger of Ruda and Nowy Butom; population (1997) 165,000. Silesia's oldest mine is nearby. **18** H7

Rudolf, Lake former name (to 1979) of Lake ►Turkana in eastern Africa. **70** F3

Rudolstadt town in the *Land* (administrative region) of ►Thuringia, Germany, in the Saale Valley, and lying at the northeastern edge of the Thuringian Forest, 56 km/35 mi southwest of Gera; population (1995) 29,000. Porcelain and electrical equipment are manufactured. **20** G6

Rugby market town and railway junction in Warwickshire, central England, on the River Avon, 19 km/12 southeast of Coventry; population (1991) 60,500. Industries include engineering and the manufacture of cement, and the town has a cattle market. **Rugby School** (1567), a private school for boys, established its reputation under headmaster Thomas Arnold; it was described in Thomas Hughes' semi-autobiographical classic *Tom Brown's Schooldays*. Rugby football originated at the school in 1823. **22** A2

Rügen Baltic island in the state of Mecklenburg–West Pomerania, Germany; area 927 sq km/358 sq mi; population (1995) 85,600. It is a holiday centre, linked by causeway to the mainland. The chief town is Bergen, the main port Sassnitz. As well as tourism, there is agriculture and fishing, and chalk is mined. Rügen was annexed by Denmark in 1168, Pomerania in 1325, Sweden in 1648, and Prussia in 1815. **18** C3

Ruhr river in Germany, length 235 km/146 mi. It rises in the Rothaargebirge Mountains at the eastern boundary of North Rhine-Westphalia, and flows west to join the Rhine at Duisburg. The **Ruhr Valley**, a metropolitan industrial area, produces petrochemicals, cars, iron, and steel at Duisburg and Dortmund; it is also a coal-mining area. **22** L3

Rum (or **Rhum**), island of the Inner ►Hebrides, Highland unitary authority, Scotland, area 110 sq km/ 42 sq mi, a nature reserve since 1957. Askival is 810 m/2,658 ft high. **24** F5

Rumania alternative spelling of ►Romania. **34** L4

Rum Jungle uranium-mining centre in the northwest of Northern Territory, Australia. **76** F2

Ruse (anglicized name **Rustchuk**), Danube port in Bulgaria, linked by rail and road bridge with Giurgiu in Romania; population (1991) 210,000. **34** N6

Rushmore, Mount mountain in the Black Hills, South Dakota, USA, 40 km/25 mi southwest of Rapid City; height 1,890 m/6,203 ft. It was named after a New York lawyer, Charles Rushmore. Four giant heads portraying presidents George Washington, Thomas Jefferson, Abraham Lincoln, and Theodore Roosevelt, have been carved from the granite face. Towering about 15 m–20 m/50–70 ft high, the portraits took a total six and a half years to complete. The US sculptor Gutzon Borglum carried out the project between 1927 and 1941.

Russia country name originally designating the prerevolutionary Russian Empire (until 1917), now used to refer informally to the ►Russian Federation. **16** L9/**40** M3

Russian Far East geographical, not administrative, division of Asiatic Russia, bordering on the Pacific coast. It covers a vast area (6,215,900 sq km/ 2,400,000 sq mi), and is comprised of the Sakha autonomous republic; the Amur, Lower Amur, Magadan, Kamchatka, and Sakhalin regions (oblasts); and the Khabarovsk and Primorski territories (krais). It is an area rich in mineral deposits and was formerly a place of banishment and labour camps.

Russian Federation (or **Russia**), country in northern Asia and eastern Europe, bounded north by the Arctic Ocean; east by the Bering Sea and the Sea of Okhotsk; west by Norway, Finland, the Baltic States, Belarus, and Ukraine; and south by China, Mongolia, Georgia, Azerbaijan, and Kazakhstan.

Russia covers an area of 17,075,400 sq km/6,592,811 sq mi, and has a population of 146,934,000 (2000 est). The capital is ►Moscow; other major towns and cities include St Petersburg, Nizhniy Novgorod, Samara, Yekaterinburg, Novosibirsk, Chelyabinsk, Kazan, Omsk, Perm, and Ufa. The head of state is Vladimir Putin (from 2000); the head of government is Mikhail Kasyanov (from 2000). The currency is the rouble, and the languages spoken include Russian (official) and many East Slavic, Altaic, Uralic, and Caucasian languages. **16** L9/**40** M3

Rutland unitary authority in central England, formerly the smallest English county, which was part of ►Leicestershire 1974–1997.
area 394 sq km/152 sq mi
towns and cities Oakham (administrative headquarters), Uppingham
physical rivers Chater, Eye, Gwash, and Welland
features Rutland Water, a large reservoir in the valley of the Gwash at Empingham, with outdoor leisure facilities (sailing, cycling, birdwatching); historic villages and churches, including Braunston-in-Rutland, Preston, Wing, and Exton
agriculture cereals (barley, wheat), sugar beet, potatoes; sheep and cattle are reared, and Stilton cheese is produced
industries clothing, engineering, and plastics; limestone and ironstone are quarried
population (1996) 34,600 **5** F5

Rwanda landlocked country in central Africa, bounded north by Uganda, east by Tanzania, south by Burundi, and west by the Democratic Republic of Congo (formerly Zaire).

Rwanda covers an area of 26,338 sq km/10,169 sq mi, and has a population of 7,733,000 (2000 est). The capital is ►Kigali; other major towns and cities include Butare, Ruhengeri, Gisenyi, Kibungo, and Cyangugu. The head of state is Paul Kagame (from 2000); the head of government is Bernard Makuza (from 2000). The currency is the Rwandan franc, and the languages spoken include Kinyarwanda, French (both official), and Kiswahili. **70** D4

Ryazan (formerly **Pereyaslav-Ryazanski** 1237–1778), capital city of Ryazan oblast (region), Russian Federation, 175 km/109 mi southeast of Moscow; population (1996 est) 536,000. Industries include engineering, oil refining, food processing, and a variety of light industries. **38** G4

Rybinsk (formerly **Shcherbakov** 1946–57; **Andropov** 1984–89), city in Yaroslavl oblast (region) of the Russian Federation, on the Volga, 250 km/155 mi northeast of Moscow; population (1996 est) 247,000. Industries include the manufacture of printing machinery, road-building equipment, and ships, and the processing of timber and food. A major hydroelectric plant was sited near the town with the construction of the Rybinsk Reservoir in the 1940s. **38** G3

Rybnik industrial town in Katowice province, southwest Poland, near the Czech border and 48 km/30 mi southwest of Katowice; population (1990) 144,000. There are coal mines and ironworks, and machine tools are manufactured. The town stands on the River Ruda, a tributary of the Oder. **18** H7

Ryde English resort on the northeast coast of the Isle of Wight, on the Solent opposite Portsmouth, with which there is ferry and hovercraft connection; population (1991) 20,500. **22** A4

Rye town in East Sussex, southeast England; population (1991) 3,700. It was formerly a flourishing port (and one of the Cinque Ports), but silt washed down by the River Rother has left it 3 km/2 mi inland. Rye has a pottery, but there are no major industries. **22** C4

Ryukyu Islands (or **Riukiu** or **Nansei**), southernmost island group of Japan, stretching towards Taiwan and including ►Okinawa, Miyako, and Ishigaki; area 2,254 sq km/870 sq mi; population (1995 est) 1,260,000. The capital is ►Naha (on Okinawa). Produce includes sugar, pineapples, and fish. **46** H5

Rzeszów capital of Rzeszów province, southeast Poland, on the River Wisłok, a tributary of the San, 145 km/90 mi east of Kraków; population (1990) 150,800. Industries include food processing, metallurgy, and machinery. **18** M7

Saale river in eastern Germany; length 364 km/226 mi. It rises in the Fichtelgebirge and flows north past Jena and Halle to join the River ►Elbe 26 km/16 mi southeast of Magdeburg. The Saale is navigable for 160 km/ 100 mi to Naumburg. **20** G6

Saalfeld town in the *Land* (administrative region) of ►Thuringia, Germany, on the River Saale, 58 km/36 mi southwest of Gera; population (1996) 32,700. The district has iron-ore mines (silver was mined here in the 16th century), and there are machine-tool, textile, and electrical industries.

Saalfeld was the capital of the Duchy of Saxe-Saalfeld 1680–1735. The town has a 13th-century friary (now a museum), a 14th-century church, and a 16th-century town hall. 20 G6

Saar (French **Sarre**), river in Western Europe; it rises in the Vosges Mountains in France, and flows 240 km/149 mi north to join the River Moselle in Germany. Its valley has many vineyards. 22 J5

Saarbrücken (French **Sarrebruck**), city on the River Saar, and capital of ►Saarland, Germany, 64 km/40 mi southeast of Trier; population (1995) 187,800. It is the centre of a coal-mining district. There are engineering industries, and optical equipment and cars are manufactured. Formed in 1909 by the amalgamation of the old town with the towns of Burbach and St Johann on the opposite bank of the river, it has been the capital of Saarland since 1919. 22 J5

Saarland (French **Sarre**), administrative region (German *Land*) in southwest Germany, bordered by Rhineland-Palatinate and the French *département* of Moselle; area 2,570 sq km/992 sq mi; population (1995) 1,103,000. The capital is ►Saarbrücken. There are motor vehicle, mechanical and electrical engineering, coal, and steel industries. Cereals are grown, and cattle, pigs, and poultry are raised. 7 (2)A4

Sabac town in Serbia, Yugoslavia, on the River Sava, the principal town of the province of Macva; population (1991) 54,600. It is a road and rail junction, and a river port which trades in agricultural produce. It produces heavy chemicals. 34 G5

Sabadell city in the province of Barcelona, northeast Spain, 16 km/10 mi north of Barcelona city; population (1995) 188,400. Industries include textiles, paper, agricultural produce, and olive oil. 28 N3

Sabah self-governing state of the federation of Malaysia, occupying northeast Borneo, forming (with Sarawak) East Malaysia; area 73,613 sq km/28,415 sq mi; population (1990) 1,736,900. The capital is ►Kota Kinabalu (formerly Jesselton). Products include oil (from offshore rigs), hardwoods, rubber, fish, cocoa, palm oil, copper, copra, and hemp. 11 (1)D1

Sachsen German form of ►Saxony, a former kingdom and state of Germany. 7 (2)D3

Sachsen-Anhalt German name for ►Saxony-Anhalt, an administrative region in Germany. 7 (2)C2

Sacramento capital and deep-water port of ►California, USA, 130 km/80 mi northeast of San Francisco; population (1996 est) 376,200; metropolitan area (1992) 1,563,000. Situated in Central Valley, the city lies on the ►Sacramento River as it curves towards San Francisco Bay. It is the commercial, manufacturing, and distribution centre for a rich irrigated farming area, and provides government, military, and tourist services. Industries include the manufacture of detergents, jet aircraft, arms, and processed foods; almonds, peaches, and pears are local agricultural specialities. 86 B3

Sacramento river in northern California, USA; length 615 km/382 mi. It rises close to Mount Shasta in the Cascade range, and flows southeast to Sacramento, through the western part of the Sierra Nevada, and eventually into San Francisco Bay. The Shasta Dam on the upper Sacramento and the Keswick Dam provide flood control, power, and irrigation; the irrigated agricultural land is very productive. The river is navigable 412 km/256 mi to Red Bluff. 86 B3

Safi Atlantic port in Tensift province, northwest Morocco, 200 km/124 mi southwest of Casablanca; population (1993) 278,000. It exports phosphates and has fertilizer plants, sardine factories, and boat-building yards. It is connected by rail to the phosphate deposits at Khouribga. Products include pottery, cereals, and olives; sheep, goats, and cattle are raised. 66 D2

Sagamihara industrial city in Kanagawa prefecture on the island of Honshu, Japan, 26 km/16 mi west of Yokohama; population (1994) 552,000. It is a residential suburb of Tokyo and Yokohama. Industries include the manufacture of radios, televisions, telecommunications equipment, batteries, electric cables, tractors, and cameras. It has a large silkworm industry.

Sagarmatha Nepalese name for Mount ►Everest, 'the Goddess of the Universe', and the official name of the 1,240-sq-km/476-sq-mi Himalayan national park established 1976. 54 E3

Saginaw city, port, and administrative headquarters of Saginaw County, east Michigan, USA, on the **Saginaw River**, 32 km/20 mi south of **Saginaw Bay**, an arm of Lake Huron; population (1996 est) 65,000. It is situated in an agricultural area with oil wells, and salt and coal deposits. Industries include sugar-refining, and the manufacture of vehicle parts, metal goods, paper, and wood products. 88 D2

Sagunt alternative spelling of ►Sagunto, a town in western Spain. 28 K5

Sagunto (or **Sagunt**), town in the province of Valencia, western Spain, on the River Palancia, 29 km/18 mi north of Valencia city; population (1991 est) 55,400. It has steelworks, and a large orange trade. Teruel iron ore is exported from its port, 5 km/3 mi away on the Mediterranean coast. There are remains of a Roman theatre and circus, and it is overlooked by a citadel of Iberian, Roman, and Moorish origins. 28 K5

Sahara (Arabic *Sahra*, 'wilderness'), largest desert in the world, occupying around 9,065,000 sq km/3,500,000 sq mi of north Africa from the Atlantic to the Nile, covering: west Egypt; part of west Sudan; large parts of Mauritania, Mali, Niger, and Chad; and southern parts of Morocco, Algeria, Tunisia, and Libya. Small areas in Algeria and Tunisia are below sea level, but it is mainly a plateau with a central mountain system, including the Ahaggar Mountains in Algeria, the Aïr Massif in Niger, and the Tibesti Massif in Chad, of which the highest peak is Emi Koussi, 3,415 m/11,208 ft. 62 C3

Sahel (Arabic *sahil* 'coast'), marginal area to the south of the Sahara, from Senegal to Somalia, which experiences desert-like conditions during periods of low rainfall. The desertification is partly due to climatic fluctuations but has also been caused by the pressures of a rapidly expanding population, which has led to overgrazing and the destruction of trees and scrub for fuelwood. In recent years many famines have taken place in the area. 62 C4

Saida (ancient **Sidon**), port in Lebanon, 40 km/25 mi south of Beirut; population (1991 est) 100,000. It stands at the end of the Trans-Arabian oil pipeline from Saudi Arabia. Olive oil, oranges, and lemons are exported. Sidon was the chief city of Phoenicia, a bitter rival of Tyre about 1400–701 BC, when it was conquered by Sennacherib. Later a Roman city, it was taken by the Arabs in AD 637 and fought over during the Crusades. 60 C3

Saigon former name (to 1976) of ►Ho Chi Minh City, Vietnam. 50 D4

St Albans city in Hertfordshire, England, on the River Ver, 40 km/25 mi northwest of London; population (1991) 80,400. The chief industries are electrical engineering, hosiery, clothing, information and legal services, musical instruments, and orchid culture. Printing is very important: one of the early presses set up in the late 15th century by the 'Scolemaster Printer', and his *The Book of St Albans* contains the earliest example of colour printing in England. There are the ruins of the Roman city of Verulamium on Watling Street. A Benedictine abbey was founded in 793 in honour of St Alban, and it became a cathedral in 1878. Other features include the Clock Tower (1411) in the High Street; the Royal National Rose Society headquarters and gardens; Rothamsted Park agricultural research centre; the Organ Museum of mechanical musical instruments; and the Verulamium Museum, with its collection of Roman remains. 22 B3

St Andrews town in Fife, Scotland, 19 km/12 mi southeast of Dundee; population (1991) 11,100. Its university (1411) is the oldest in Scotland. It is considered to be the 'home of golf', with a famous Old Course. The Royal and Ancient Club (1754) is the ruling body of golf. There is a cathedral, founded in 1160 and consecrated in 1318. 24 K5

St Augustine port, holiday resort, and administrative headquarters of St Johns County, northeast Florida, USA, situated on a peninsula between the Matanzas and San Sebastian rivers, 56 km/35 mi southeast of Jacksonville; population (1990) 11,700. A port with the Atlantic Intracoastal Waterway passing through it, it also relies on fishing, shrimping, and the processing and shipping of local produce and lumber. Tourism is vital to the local economy. Flagler College (1967) is in the city. 90 E4

St Austell market town and administrative centre of Restormel borough council, in Cornwall, southwest England, 22 km/14 mi northeast of Truro; population (1998 est, with Fowey, with which it is administered) 20,800. It is the centre of the china-clay industry that supplies the Staffordshire potteries. 24 H11

St-Brieuc French town on St-Brieuc Bay in the *département* of Côtes d'Armor, 59 km/37 mi southwest of St-Malo; population (1990) 48,000. It has a 13th–14th-century cathedral and many old houses; tourism is important. There are also iron- and steelworks. Agricultural implements and shoes are manufactured. 26 C5

St Christopher–Nevis alternate form of ►St Kitts and Nevis. 94 M5

St David's (Welsh **Tyddewi**), small town in Pembrokeshire, southwest Wales, situated on the River Alun just 2 km/1.2 mi from the sea; population (1991) 2,100. Its cathedral, founded by St David, the patron saint of Wales, was rebuilt between 1180 and 1522. Formerly the resort of pilgrims, the town is now a summer tourist centre. 24 G10

St-Denis industrial town, a northern suburb of Paris, France; population (1990) 90,800. The French philosopher and theologian Peter Abelard was a monk at the 12th-century Gothic abbey, which contains many tombs of French kings. 22 E6

St-Dizier French town in the *département* of Haute-Marne, on the River Marne; population (1990) 35,600. It has metallurgical industries and a brewery. It was besieged by the Holy Roman Emperor Charles V in 1544, and was the scene of a victory of Napoleon in 1814. 26 K5

St-Etienne city in south-central France, administrative centre of the ►Loire *département*, Rhônes-Alpes region; population (1990 est) 201,600, conurbation 313,000. St-Etienne is situated on the River Furens 51 km/32 mi southwest of Lyon. Industries include the manufacture of aircraft engines, electronics, and chemicals. It is the site of a school of mining established in 1816. In World War I it was one of the principal French arsenals. 26 K8

St Gallen (French **St Gall**; or **Sankt Gallen**), canton in northeastern Switzerland; area 2,014 sq km/778 sq mi; population (1995) 442,300. It is mountainous in the south, near its border with Graubünden, but has gentler hills in the northern region around Lake Constance. It entirely surrounds the canton of Appenzell. The chief industry is the manufacture of textiles. The main farming activity is dairying, supplemented by cattle-rearing on the Alpine pastures. There are numerous health and winter-sports resorts. 8 (1)D1

St Gallen capital of St Gallen canton, Switzerland, situated just south of Lake Constance; population (1994) 75,500. The town has an ancient abbey and a baroque cathedral. St Gallen is also an industrial centre specializing in embroidered cotton textiles and nylon. 30 E3

St George's port and capital of Grenada, on the southwest coast; population (1994 est) 30,000. The port has a well-sheltered harbour, exporting nutmeg, bananas, and rum, and is also an administrative and commercial centre. Refined sugar and alcoholic drinks are leading products and there is an established tourist industry. It was founded in 1650 by the French, and was the capital of the Windward Islands 1885–1958 (at that time a British dependency). 94 M6

St George's Channel stretch of water between southwest Wales and southeast Ireland, linking the ►Irish Sea with the Atlantic Ocean. It is 160 km/100 mi long and 80–150 km/50–90 mi wide.

There is also a channel of the same name between New Britain and New Ireland, two islands in the ►Bismarck Archipelago in Papua New Guinea. 24 F10

St-Germain-en-Laye residential suburb on the western outskirts of ►Paris, France, 100 m/328 ft above the River ►Seine, in the *département* of Yvelines; population (1990) 41,600.

It is situated beside the forest of St Germain, and is a popular tourist resort. The 16th-century royal château, incorporating part of an older structure, was occupied by James II of England from his banishment in 1689 until his death in 1701. 22 E6

St Helena British island in the south Atlantic, 1,900 km/1,200 mi west of Africa, area 122 sq km/47 sq mi; population (1997) 5,644. Its capital is Jamestown, and it exports fish and timber. Ascension and Tristan da Cunha are dependencies. 62 C7

St Helens town in Merseyside, northwest England, 19 km/12 mi northeast of Liverpool, and connected to the River Mersey by canal; population (1991) 104,700. Formerly a coalmining town, St Helens is now a major centre for the manufacture of sheet glass. Other industries include the manufacture of tiles, engineering, and pharmaceuticals. 24 K8

St Helier resort and capital of Jersey, Channel Islands; population (1991) 28,100. The 'States of Jersey', the island legislature, sits here in the *salle des états*. 26 C4

St Ives fishing port and resort in Cornwall, southwest England; population (1991) 10,100. Its artists' colony, founded by Walter Sickert and James Whistler, later included Naum Gabo, Barbara Hepworth (a museum and sculpture gardens commemorate her), and Ben Nicholson. A branch of the Tate Gallery opened here in 1993, displaying works of art from the Tate's collection by artists connected with St Ives. 24 G11

St John seaport and largest city of New Brunswick, Canada, at the mouth of the St John River, 105 km/66 mi south of Fredericton; population (1991 est) 75,000, metropolitan area (1996 est) 129,100. It is a commercial centre, with shipbuilding (trawlers and small vessels), fishing, oil-refining, food-processing, brewing, and publishing industries. Vehicle parts, timber products, textiles, paint, and plastics are also manufactured. St John is one of Canada's main winter ports for European shipping, and has large dry-dock facilities. 82 T7

St John's port and capital of Antigua and Barbuda, on the northwest coast of Antigua; population (1992) 38,000. It exports rum, cotton, and sugar.

St John's capital and chief port of ►Newfoundland, Canada; population (1991) 95,800, metropolitan area (1996)177,800. Situated on the east coast of the Avalon peninsula, its deepwater harbour is connected to the Atlantic by a channel named the Narrows. St John's is the administrative, commercial, and service centre of the province. Fish-processing is the main industry; other manufactured products include textiles, fishing equipment, furniture, and machinery. 82 W7

St Joseph city in northwest Missouri, USA, on the Missouri River, 80 km/50 mi northwest of Kansas City; population (1996 est) 70,200. It has large stockyards, and is a major grain market. Industries include meat-preparation and packing, flour production, food- and dairy-processing, and the manufacture of metalware, concrete, and steel. First established as a trading post in 1826, it served as a river port, and became the eastern terminus for the Pony Express (horse-relay system of mail delivery) in the mid-19th century. 88 B3

St Kilda archipelago of four islands, sea stacks, and skerries, the most westerly of the Outer ►Hebrides, Western Isles, 200 km/124 mi west of the Scottish mainland. They were populated from prehistory until 1930, and now form a National Nature Reserve belonging to the National Trust of Scotland. They have the world's largest colony of gannets, the oldest and largest colony of fulmars in the British Isles, and a large population of puffins. St Kilda is a World Heritage Site. 24 D4

St Kitts and Nevis country in the West Indies, in the eastern Caribbean Sea, part of the Leeward Islands.

St Kitts and Nevis covers an area of 262 sq km/101 sq mi (St Kitts 168 sq km/65 sq mi, Nevis 93 sq km/36 sq mi), and has a population of 38,000 (2000 est). The capital (and chief port) is ►Basseterre (on St Kitts); other major towns and cities include Charlestown (Nevis), Newcastle, Sandy Point Town, Dieppe Bay Town, and Saint Paul. The head of state is Queen Elizabeth II (from 1983), represented by Governor General Dr Cuthbert Montraville Sebastian (from 1996); the head of government is Denzil Douglas (from 1995). The currency is the East Caribbean dollar, and the main language spoken is English (official). 94 M5

St Lawrence river in eastern North America. With the ►Great Lakes and linking canals such as the ►Welland Ship Canal, it forms the ►St Lawrence Seaway, an inland route for small ocean-going ships from the Gulf of St Lawrence, an arm of the Atlantic Ocean, to ►Thunder Bay at the head of Lake Superior; larger vessels stop at ►Montréal. The river is 1,200 km/745 mi long and icebound for four months each year. Enormous quantities of hydroelectric power are generated along its course. 88 G1

St Lawrence Seaway deep-water channel and transport corridor in North America, connecting the St Lawrence River with the Great Lakes, allowing ocean-going vessels to navigate from the Atlantic Ocean to Lake Superior (3,769 km/2,342 mi). It was opened for navigation by the USA and Canada in 1959.

In the strict sense the seaway is the section from Montréal to Lake Ontario, but the name is applied generally to the whole system from the Atlantic to the Great Lakes. There are 78 ports on the seaway, which remains navigable for moderate-sized vessels for about 250 days a year. 88 G1

St-Lô administrative centre of the *département* of ►Manche in Normandy region, northern France, 54 km/33 mi west of Caen, built on a hill overlooking the valley of the River Vire; population (1990) 22,800. It is an agricultural market town with some light industry; horse breeding is important, and butter and cider are made here. In World War II it was almost entirely destroyed on 10–18 July 1944, when US forces captured it from the Germans. 22 A5

St Louis city and riverport in Missouri, USA, on the Mississippi River; population (1996 est) 351,6000; metropolitan area (1996 est) 2,600,000. Occupying a central US location, it is a warehousing and distribution hub, and a major market for livestock, grain, wool, and lumber. The port handles oil, coal, sulphur, cement, and agricultural and manufactured goods. Products include aerospace and transport equipment, pharmaceuticals, refined oil, rubber, printed materials, and processed metals, tobacco, and food. 88 B3

St Lucia country in the West Indies, in the eastern Caribbean Sea, one of the Windward Islands.

St Lucia covers an area of 617 sq km/238 sq mi, and has a population of 154,000 (2000 est). The capital is ►Castries; other major towns and cities include Soufrière, Vieux Fort, Choiseul, and Gros Islet. The head of state is Queen Elizabeth II (from 1979), represented by Governor General Dr Perlette Louisy (from 1997); the head of government is Kenny Anthony (from 1997). The currency is the East Caribbean dollar, and the languages spoken include English (official) and French patois. **94** M6

St-Malo seaport and resort in the Ille-et-Vilaine *département* in Brittany, western France, 80 km/50 mi northwest of Rennes on the River Rance; population (1990) 49,300. It is a busy fishing, trading, and passenger port, and has printing, machinery and electrical industries. A dam built across the river in 1967 uses the 13-m/44-ft tides to feed the world's first successful tidal power station. The town took its name from the Welshman Maclou, or Malo, who was bishop here in about 640. **26** D5

St Michael's Mount small island in Mount's Bay, Cornwall, southwest England; population (1981) 26. It is linked to Marazion on the mainland by a causeway which is submerged at high tide

St-Nazaire industrial seaport in the Loire-Atlantique *département*, Pays de la Loire region, France; population (1990) 66,100, conurbation 130,000. It stands at the mouth of the River Loire, 53 km/mi nourthwest of Nantes and as a port it is capable of taking the largest vessels. Industries include shipbuilding, engineering, chemicals, and food canning. There are oil refineries nearby and the largest oil-port facility in France further up the estuary at Donges. In World War II it was used as a German submarine base until put out of action by a British commando raid on 28 March 1942. St-Nazaire is France's major naval dockyard. **26** C6

St-Niklaas (French **St-Nicolas**), town in the province of East Flanders, Belgium, 19 km/12 mi west of Antwerp; population (1991 est) 68,200. It manufactures woollen, cotton, and linen goods, carpets, pottery, bricks, and furniture. The marketplace in St-Niklaas, the Groote Markt, is the largest in the country. The municipal museum contains a collection of maps and globes made by the cartographer Gerardus Mercator. **22** G3

St-Omer town in Pas-de-Calais *département*, northern France, 42 km/26 mi southeast of Calais on the River Aa; population (1990) 15,300. It is the centre of an active market-gardening region, and there are foodstuff, textile, glass, and paper manufactures. In World War I it was the site of British general headquarters from October 1914 to March 1916. **22** E4

St Paul capital and river port of ►Minnesota, USA, on the Mississippi River, adjacent to ►Minneapolis, with which it forms the Twin Cities area; population (1996 est) 259,600. Industries include printing, oil-refining, brewing, food-processing, meat-packing, and the manufacture of electronics, chemicals, plastics, and machinery. A high proportion of the population has a Scandinavian cultural background. **88** B2

St Peter Port only town of Guernsey, Channel Islands; population (1991) 16,650. **26** C4

St Petersburg seaside resort and industrial city in Pinellas County, west-central Florida, USA; population (1996 est) 236,000. It lies across Tampa Bay from ►Tampa, and is notable as a centre for space technology, producing aerospace and electronic equipment. Other industries include tourism, boatbuilding, and seafood-processing. It is a popular retirement location. The city is home to the Florida Suncoast Dome (1991) stadium, St Petersburg Junior College (1927), Eckerd College (1958), and Hillsborough Community College (1968). **90** E4

St Petersburg (Russian **Sankt-Peterburg**), capital of the St Petersburg region, Russian Federation, at the head of the Gulf of Finland; population (1994) 4,883,000. Industries include shipbuilding, machinery, chemicals, and textiles. It was renamed **Petrograd** 1914 and was called **Leningrad** 1924–91, when its original name was restored.

Built on a low and swampy site, St Petersburg is split up by the mouths of the River Neva, which connects it with Lake Ladoga. The climate is severe. The city became a seaport when it was linked with the Baltic by a ship canal built 1875–93. It is also linked by canal and river with the Caspian and Black seas, and in 1975 a seaway connection was completed via lakes Onega and Ladoga with the White Sea near Belomorsk, allowing naval forces to reach the Barents Sea free of NATO surveillance. **38** F3

St-Pierre and Miquelon territorial collectivity of France, comprising eight small islands off the south coast of Newfoundland, Canada; area St-Pierre group 26 sq km/10 sq mi; Miquelon-Langlade group 216 sq km/83 sq mi; population (1990) 6,400. The capital is St-Pierre. Industries include fishing and tourism. Cattle are raised and there is subsistence farming. **82** V7

St-Quentin town on the River Somme in the *département* of Aisne in Picardy, northern France, 33 km/20 mi south of Cambrai; population (1990) 62,100. In 1871 it was the site of a Prussian defeat of the French in the Franco-Prussian War, and spent four years under German occupation in World War I, being completely obliterated in the process. It is linked by canal to the industrial centres of Belgium and Germany. Its traditional textile production has been replaced by chemicals and metal works. **22** F5

St-Tropez resort and fishing port on the French Côte d'Azur; population (1993 est) 5,800. It became popular as a resort in the 1960s. **30** B7

St-Truiden (French **St-Trond**), town in the province of Limbourg, Belgium, 17 km/11 mi southwest of Hasselt; population (1991 est) 37,000. It has various manufacturing industries and is situated in a region (the Haspengouw) covered by extensive orchards. The town contains several ancient churches. **22** H4

St Vincent and the Grenadines country in the West Indies, in the eastern Caribbean Sea, part of the Windward Islands.

St Vincent and the Grenadines covers an area of 388 sq km/150 sq mi (including islets of the Northern Grenadines 43 sq km/17 sq mi), and has a population of 114,000 (2000 est). The capital is ►Kingstown; other major towns and cities include Georgetown, Châteaubelair, Layon, and Dovers. The head of state is Queen Elizabeth II (from 1979), represented by Governor General David Jack (from 1989); the head of government is Arnhim Eustace (from 2000). The currency is the East Caribbean dollar, and the languages spoken include English (official) and French patois. **94** M6

Sakai city on the island of Honshu, Japan, immediately to the south of Osaka, from which it is separated by the Yamato River; population (1994) 797,000. Industries include metallurgy, and the manufacture of appliances, chemicals, steel, and textiles.

Sakha (formerly **Yakutia**; **Yakutsk Autonomous Soviet Socialist Republic**), autonomous republic of the Russian Federation, in eastern Siberia; area 3,103,000 sq km/ 1,198,100 sq mi; population (1997) 1,032,000 (50% Russians, 33% Yakuts). The capital is ►Yakutsk. Industries include fur trapping, gold and natural gas extraction, and lumbering; there is some agriculture in the south. **44** N3

Sakhalin (formerly Japanese **Karafuto** 1855–75; 1905–45), island in the Russian Far East between the Sea of Okhotsk and the Sea of Japan, separated from the mainland of the Russian Federation by the Tatar Strait and from the northernmost island of Japan, Hokkaido, by La Pérouse Strait. Sakhalin is some 965 km/600 mi long and covers an area of 76,400 sq km/29,498 sq mi. The main town is Yuzhno-Sakhalinsk. Coal, oil, natural gas, iron ore, and gold deposits, together with extensive coniferous and mixed forests, provide the basis for mining, oil extraction, and industries producing timber, cellulose, and paper (the latter established by the Japanese). The oil that is extracted on Sakhalin is piped across the Tatar Strait to Komsomolsk-na-Amure. Fishing is also an important economic activity. Sakhalin supports limited arable (potatoes and vegetables) and dairy farming. **44** Q6

Sakkara (or **Saqqara**), village in Egypt, 16 km/10 mi south of Cairo, with 20 pyramids. The oldest of these (third dynasty) is the 'Step Pyramid' designed by Imhotep, whose own tomb here was the nucleus of the Aesklepieion, a centre of healing in the ancient world.

Salado name of three rivers in Argentina. The first, also called **Salado del Norte**, or **Juramento** in the north of the country, rises in the Andes and flows southeast to join the Paraná River between the towns of Santa Fé and Rosario, 384 km/239 mi northwest of Buenos Aires. Its shallow, lifting course of 2,030 km/1,261 mi renders commercial navigation impracticable.

The second, in the eastern part of Buenos Aires department, is 640 km/398 mi long and flows into Samborombon Bay; it is used for irrigation.

The third, also called **Desaguadero**, or **Salado del Sud** lies in western Argentina; length 1,368 km/850 mi; it rises in the Andes and flows southwards, forming the boundary between the departments of Mendoza and San Luis, before traversing La Pampa department, where it is known as the **Chadileo**, and further south as the **Curaro**. It joins the Río Colorado near the town of Pichi Mahuida. **100** J4

Salamanca capital of Salamanca province in Castilla–León, western Spain, on the River Tormes, 260 km/ 162 mi northwest of Madrid; population (1991) 162,500. It produces pharmaceuticals and wool. Salamanca is a cultural centre, and its university was founded about 1230. There is a 13th-century cathedral, and a superbly designed square, the Plaza Mayor. **28** E4

Salamina alternative name for ►Salamis, an island off Peiraias, Greece. **36** F7

Salamis (or **Salamina**), island off Peiraias, the port of ►Athens, Greece; area 101 sq km/39 sq mi; population (1991 est) 34,300. The town of Salamis, on the west coast, is a naval station. **36** F7

Sale town in southeast Victoria, Australia, 209 km/130 mi east of Melbourne; population (1996) 13,366. Sale is linked by canal via the Gippsland Lakes to Bass Strait. The town has benefited from the deposits of oil and natural gas in the Strait, and from the brown coal to the south in La Trobe Valley. It is the centre of an agricultural, grazing, and dairying district, and wool production is important. **76** J7

Salem city in Tamil Nadu, India, 320 km/200 mi southwest of Chennai (formerly Madras); population (1991) 367,000. Industries include iron-mines, textiles, chemicals, iron and steel, electrical goods and handicrafts. Locally iron, bauxite and magnesite are mined. The chief agricultural product of the region is tapioca, and there are many small-scale sago processing plants producing starch to be used in the textile industry. **54** C6

Salem capital of ►Oregon, USA, in Marion County (of which it is administrative headquarters) and Polk County, on the Willamette River, 84 km/53 mi south of Portland; population (1996 est) 122,600. It is a distribution and processing centre for timber from the Cascades forests to the east, and fruit and vegetables from the fertile Willamette farmlands. Hi-tech goods, building materials, and textiles are also produced. It is the home of Willamette University (1844), Chemeketa Community College (1962), and the neoclassical marble capitol (1937). **86** B2

Salerno (ancient **Salernum**), port in Campania, southwest Italy, 48 km/ 30 mi southeast of Naples; population (1992) 147,600. It was founded by the Romans in about 194 BC, destroyed by Charlemagne, and sacked by the Holy Roman Emperor Henry VI in 1194. Salerno has had a university since medieval times (1150–1812, revived 1944); a medical school was incorporated in the early foundation. There are ceramic, cotton, and paper industries. **32** J8

Salford industrial city in Greater Manchester, northwest England, on the west bank of the River Irwell and the Manchester Ship Canal; population (1991 est) 80,600. Industries include engineering and the manufacture of electrical goods and textiles.

Salgótarján capital of Nógrád county in the northern highlands of Hungary, near the border with the Slovak Republic, 80 km/50 mi north of Budapest; population (1993) 46,900. It is an industrial town with coal-mining and the production of steel, glass, and agricultural machinery. It lies in the steep, wooded Tarján valley. **18** J9

Salinas city and administrative headquarters of Monterey County, west-central California, USA; population (1992) 114,800. It is situated on the Salinas River, 137 km/85 mi south-southeast of San Francisco. Named after the river and its saltmarshes, it is the processing centre for an exceptionally fertile area (irrigated by the largely underground river) noted for its dairy products, sugar beets, fruits, and vegetables, especially lettuce. Salinas is home to Hartnell College (1920). 92 B1

Salisbury city and market town in Wiltshire, south England, on the edge of Salisbury Plain 135 km/84 mi southwest of London; population (1991) 39,300. Salisbury is an agricultural centre, and industries include brewing and engineering. The nearby Wilton Royal Carpet factory closed in 1995. The cathedral of St Mary, built 1220–66, is an example of Early English architecture; its decorated spire 123 m/404 ft is the highest in England; its clock (1386) is one of the oldest still working. The cathedral library contains one of only four copies of the *Magna Carta*. 22 A3

Salisbury former name (to 1980) of ▶Harare, the capital of Zimbabwe. 72 E3

Salisbury Plain undulating plateau between Salisbury and Devizes in Wiltshire, southwest England; area 775 sq km/300 sq mi. It rises to 235 m/770 ft in Westbury Down. Since the mid-19th century it has been a military training area. Stonehenge stands on Salisbury Plain.

Salonika English name for ▶Thessaloniki, a port in Greece. 36 E4

Salop abbreviation and former official name (1972–80) of ▶Shropshire, a county in England. 5 E5

Salt Lake City capital of ▶Utah, USA, on the River Jordan, 605 km/378 mi northwest of Denver, Colorado; population (1992) 165,900. It is the commercial centre and world capital of the Church of Jesus Christ of the Latter-day Saints (the Mormon Church). Industries include mineral-refining, food-processing, and the manufacture of textiles, footwear, and electronic and mining equipment. Copper, silver, lead, zinc, coal, and iron mines are worked nearby. In 1995 Salt Lake City was chosen as the site for the 2002 Winter Olympic Games. 86 D2

Salto Angel alternative name for ▶Angel Falls, in southeast Venezuela. 68 E2

Salton Sea saltwater lake in Imperial and Riverside counties, southeast California; area 650 sq km/250 sq mi. It lies 130 km/80 mi east-northeast of San Diego. The lake was accidentally created in the early 20th century during irrigation works from the Colorado River. It is used to generate electricity. 92 C2

Salvador port, resort, and naval base, capital of ▶Bahía federal unit (state), northeast Brazil, on the inner side of a peninsula separating Todos los Santos Bay from the Atlantic Ocean; population (1991) 2,075,400 (metropolitan area 3,134,900). Chief industries include oil refining, petrochemicals, and tourism; fruit, cocoa, sisal, soybeans, and petrochemical products are exported. The city is built on two distinct levels; the Cidade Alta (upper city), the site of the original settlement where there are many examples of colonial architecture, and Cidade Baixa (lower city), comprising the commercial, financial and port district. It was the first capital of Brazil 1549–1763. 98 K6

Salvador, El republic in Central America; see ▶El Salvador. 94 F6

Salween river rising in eastern Tibet and flowing 2,800 km/1,740 mi through Myanmar (Burma) to the Andaman Sea; it has many rapids. 50 B2

Salzach river in Austria; length 225 km/140 mi. The Salzach rises in the Hohe Tauern Alpine range in eastern Tirol and flows north past Salzburg to the border with Germany before joining the ▶Inn 45 km/28 mi north of Salzburg. 30 H2

Salzburg capital of the federal state of Salzburg, west Austria, on the River Salzach; population (1995) 142,000. There are textile industries, and stock rearing, dairy farming, forestry, tourism, and the manufacture of musical instruments all contribute to the local economy. The city is dominated by the Hohensalzburg fortress (founded 1077, present buildings 1465–1519). It is the seat of an archbishopric founded by St Boniface in about 700 and has a 17th-century cathedral. It is also a conference centre. There are numerous fine Romanesque, Gothic, and baroque churches. It is the birthplace of the composer Wolfgang Amadeus Mozart and an annual music festival in August has been held here since 1920. The Mozart Museum of Sound and Film opened in 1991. 30 J3

Salzburg federal state of Austria, bounded on the northwest by Bavaria; area 7,200 sq km/2,779 sq mi; population (1994) 504,300. It lies mainly in the Salzburg Alps. Its capital is Salzburg. The chief industries are cattle rearing, dairy-farming, forestry, and tourism. It was annexed by Germany in 1938 but reunited with Austria in 1945. 8 (2)D3

Salzgitter industrial city in Lower Saxony, Germany; population (1995) 117,700. Industries include iron and steel, shipbuilding, vehicles, and machine tools. 20 F4

Samara (formerly **Kuibyshev** 1935–91), capital city and river port of Samara oblast (region), west-central Russian Federation; population (1996 est) 1,175,000. Samara is located on the River Volga and the main Trans-Siberian Railway, 820 km/510 mi southeast of Moscow. It is a major industrial centre, with large heavy-engineering industries (producing road vehicles and railway rolling stock), as well as chemical, oil-processing, wood-processing, and light industries. 38 K4

Samarkand (Uzbek **Samarqand**), city in eastern Uzbekistan, capital of Samarkand wiloyat (region), near the River Zerafshan, 217 km/135 mi east of Bukhara; population (1996) 370,000. Industries include cotton-ginning, silk manufacture, production of foodstuffs, and engineering. Samarkand is one of the oldest cities in Central Asia, dating from the 3rd or 4th millennium BC. The Registan – a collection of mosques, courtyards and former Muslim theological seminaries ('madrasahs') – forms the centrepiece of the historic town. A university is situated here. 56 J2

Samarra (or Smarra), ancient town in Iraq, on the River Tigris, 105 km/65 mi northwest of Baghdad; population (1998 est) 126,500. Founded in 836 by the Abbasid caliph Motassim, it was the Abbasid capital until 892 and is a place of pilgrimage for Shiite Muslims. It is one of the largest archaeological sites in the world, and includes over 6,000 separate sites. The best preserved palace is Qasr al-Ashiq, which was built entirely of brick between 878 and 882. 58 K6

Sambre river of France and Belgium, rising in the *département* of Aisne and flowing 190 km/118 mi northeast to join the ►Meuse, of which it is the main tributary, at Namur. The towns of Landrecies and Maubeuge are on its banks in France; and Thuin, Charleroi, Châtelet, and Floreffe in Belgium. It is navigable for about 145 km/ 90 mi. 22 F4

Samoa country in the southwest Pacific Ocean, in ►Polynesia, northeast of Fiji Islands.

Samoa covers an area of 2,830 sq km/1,092 sq mi, and has a population of 180,000 (2000 est). The capital (and chief port) is ►Apia (on Upolu island); other major towns and cities include Lalomanu, Tuasivi, Falealupo, Falelatai, Salotulafai, and Taga. The head of state is King Malietoa Tanumafili II (from 1962); the head of government is Tuila'epa Sa'ilele Malielegaoi (from 1998). The currency is the tala, or Samoan dollar, and the languages spoken include English and Samoan (both official). 74 J7

Samoa, American group of islands 4,200 km/2,610 mi south of Hawaii, administered by the USA; area 200 sq km/77 sq mi; population (1993) 52,900. The capital is ►Pago Pago. Exports include canned tuna, handicrafts, and copra. The main languages are Samoan and English; the principal religion is Christianity. 74 J7

Samos Greek island in the Aegean Sea, off the west coast of Turkey; area 476 sq km/184 sq mi; capital Limén Vathéos; population (1991) 41,850. Mountainous but fertile, it produces wine (muscat) and olive oil. The mathematician Pythagoras was born here. The town of Teganion is on the site of the ancient city of Samos, which was destroyed by Darius I of Persia. 36 J7

Samsun Black Sea port and capital of a province of the same name in northern Turkey; situated at the mouth of the Murat River in a tobacco-growing area; population (1990) 303,900. It is the site of the ancient city of **Amisus**. 58 G3

San river in Poland, rising in the Beskid Mountains and flowing 450 km/281 mi to the north, joining the ►Vistula 6 km/4 mi northeast of Sandomierz. 18 L7

San'a (or **Sana'a**), capital of Yemen, southwest Arabia, 320 km/200 mi north of Aden on the central plateau, 2,210 m/7,250 ft above sea level; population (1995) 972,000. A walled city, with fine mosques and traditional architecture, it is rapidly being modernized. Weaving and jewellery are local handicrafts. 64 H4

San Andreas fault geological fault stretching for 1,125 km/700 mi northwest–southeast through the state of California, USA. It marks a conservative plate margin, where two plates slide past each other.

San Angelo city in western central Texas, USA, near the confluence of the North and Middle Concho rivers, 145 km/90 mi southwest of Abilene; population (1996 est) 88,100. It is an important cattle trading, and a wool and mohair centre. Industries include textiles (Ethicon and Levi Strauss), food-processing, oil-refining, and the manufacture of pottery. San Angelo's annual Stock Show and Rodeo is attended by cowboys and tourists from a wide area. 92 F2

San Antonio (or **San Antonio de Bejar**), city in southern Texas, USA, on the **San Antonio River**; population (1996 est) 1,067,800. It is a commercial, financial, and military centre. Industries include tourism, aircraft maintenance, oil-refining, and meat-packing. Fort Sam Houston, four Air Force bases, the South Texas Medical Center, and the Southwest Research Center lie within the city limits and play an important part in the economy. 94 E3

San Bernardino city in California, USA, 80 km/50 mi east of Los Angeles; population (1996 est) 183,500. A commercial and distribution centre, it is situated in a vast citrus fruit-growing district in the foothills of the **San Bernardino Mountains**; rail and road routes enter the Los Angeles basin here through the Cajon Pass. Industries include the processing of food and lumber, and the manufacture of steel, and aerospace and electronic equipment. 92 C2

San Cristóbal capital of Tachirá state, western Venezuela, situated 800 m/26,250 ft above sea level in the northern Andes overlooking the River Torbes, 56 km/35 mi from the Colombian border; population (1990) 220,700. It is the centre of a coffee-growing region, and other products include textiles, cement, leather goods, and tobacco. San Cristóbal was founded by Spanish settlers in 1561 and stands on the ►Pan-American Highway. 98 C2

San Diego city and US naval air station, on the Pacific Ocean, and on the border of Mexico, in California, USA; population (1996 est) 1,171,100; metropolitan area (1992) 2,601,000. San Diego is linked to Tijuana, Mexico, by a 26-km/16-mi transit line (1981), popular with tourists. It is an important fishing port. Manufacturing includes aerospace and electronic equipment, metal fabrication, printing and publishing, seafood-canning, and shipbuilding. San Diego is the oldest Spanish settlement in California; a Spanish mission and fort were established here in 1769. 92 C2

Sandomierz (Russian **Sandomir**), market town in Tarnobrzeg province, Poland, on the left bank of the River Vistula, 80 km/50 mi east of Kielce; population (2001 est) 27,100. Large deposits of sulphur occur locally; industries include chemicals, glass, tanning, and food processing; and fruit is traded from the surrounding area. The 14th-century Gothic cathedral contains Byzantine-style frescoes. 18 L7

Sandwich Islands former name of ►Hawaii, a group of islands in the Pacific Ocean. Captain James Cook originally named the islands in 1778, in honour of John Montague, 4th Earl of Sandwich, who was then First Lord of the British Admiralty, and Cook's superior. 74 J3

San Fernando town in the province of Cádiz, southern Spain; population (2001 est) 82,300. It is situated in an area of salt marshes, from which salt is produced. San Fernando has a naval school, an arsenal, and a meteorological observatory. 28 D8

San Francisco chief Pacific port in California, USA, on the tip of a peninsula in San Francisco Bay; population (1996 est) 735,300, metropolitan area of San Francisco and Oakland 3,686,600. The entrance channel from the Pacific to San Francisco Bay was named the Golden Gate in 1846; its strait was crossed in 1937 by the world's second-longest single-span bridge, 1,280 m/4,200 ft in length. Manufactured goods include textiles, machinery and metalware, electrical equipment, petroleum products, and pharmaceuticals. San Francisco is also a financial, trade, corporate, and diversified service centre. Tourism is a major industry. A Spanish fort (the Presidio) and the San Francisco de Asis Mission were established here in 1776. San Francisco has the largest Chinese community outside Asia. **86** B3

Sangre de Cristo range of mountains in Colorado. It forms the southernmost range of the Rocky Mountains, with an altitude in places of over 4,260 m/13,976 ft. It extends for about 400 km/250 mi from northwest to southeast, and separates the river basins of the Rio Grande and the Arkansas. **92** E1

San José capital of ►Costa Rica, and of San José province; population (1991 est) 299,400. It is situated in the broad fertile valley of the central plateau. Products include coffee, cocoa, sugar cane, textiles, and pharmaceuticals. There is a cathedral, and the University of Costa Rica, which was founded in 1843. San José was founded in 1737 and became capital in 1823, replacing the former capital Cartago because it had a better all-year-round climate. **94** H7

San José city in California, USA, in Santa Clara Valley, at the head of the southern arm of San Francisco Bay; population (1996) 838,700. It is situated at one end of 'Silicon Valley', the site of many high-technology electronic firms turning out semiconductors and other computer components. There are also electrical, aerospace, missile, rubber, metal, and machine industries, and it is a commercial and transportation centre for orchard crops and wines produced in the area. **86** B3

San Juan industrial city and capital of Puerto Rico; population (1995 est) 458,000. It is a major port, exporting sugar, tobacco, coffee, and tropical fruits, mostly to the US mainland, and providing the world's busiest cruise ship base. It stands on an island joined by a bridge to the north coast of Puerto Rico. Industries include tourism, banking, metalworking, publishing, cigars, sugar, and clothing. Products include chemicals, pharmaceuticals, machine tools, electronic equipment, textiles, plastics, and rum. **94** H4

Sankt Gallen alternative name for ►St Gallen, a canton in Switzerland. **8** (1)D1

Sanlúcar de Barrameda resort town in the province of Cádiz, southern Spain, at the mouth of the River Guadalquivir on the Atlantic coast; population (1991) 55,900. Manzanilla sherry is exported from here. Christopher Columbus sailed from Sanlúcar de Barrameda on his third voyage to the New World, as did Ferdinand Magellan on his circumnavigation of the world. The town has a 14th-century church and the remains of three castles, dating from the 15th to the 17th centuries. **28** D8

San Luis Potosí silver-mining city and capital of San Luis Potosí state, central Mexico; population (1995 est) 583,000. It lies northwest of Mexico City at an altitude of 1877 m/6158 ft. Industries include the refining of silver and arsenic, footwear and clothing. Founded in 1586 as a Franciscan mission, it became the colonial administrative headquarters and has fine buildings of the period. **94** D4

San Marino small landlocked country within northeast Italy.

San Marino covers an area of 61 sq km/24 sq mi, and has a population of 27,000 (2000 est). The capital is ►San Marino; other major towns and cities include Serravalle, Faetano, Fiorentino, Borgo Maggiore, and Domagnano. The heads of state and government are Gian Franco Terenzi and Enzo Colombini (from 2000). The currency is the Italian lira, and the main language spoken is Italian (official). **30** H7

San Marino capital of the republic of San Marino; population (1996) 4,372. It is on the western slope of Monte Titano, 19 km/12 mi southwest of Rimini. Its economy relies on tourism. It is encircled by triple walls, with three fortresses, the 13th-century Rocca Guaita, Rocca Cesta, and Rocca Montale, and the 14th-century Church of San Francesco. **30** H7

San Miguel de Tucumán capital of Tucumán province, northwest Argentina; situated on the Rio Sali, to the east of the Sierra de Aconquija range in the foothills of the Andes; population (1991) 473,000; metropolitan area (1992 est) 642,500. Industries include sugar mills and distilleries. It is the commercial centre of the province, and the largest city in northern Argentina. Founded in 1565, San Miguel de Tucumán was the site of the signing of the Argentine declaration of independence from Spain on 9 July 1816. There are several colonial buildings remaining in the city, including the Governor's Palace, the cathedral, and the church of San Francisco. The university was founded in 1914. **100** H4

San Pedro Sula main industrial and commercial city in northwest Honduras, near the Guatemalan border, capital of Cortés department and the second-largest city in the country; population (1991 est) 325,900. Situated in the fertile valley of the Ulúa River, and 45 km/28 mi south of the port of Puerto Cortés, it is a trading centre and key distribution point for the north and northwest regions, which produce bananas, coffee, sugar, and timber; industries include steel, textiles, plastics, furniture, and cement. One of Central America's fastest growing cities, it was founded in 1536 by the Spaniard Pedro de Alvarado. **94** G5

San Roque town in the province of Cádiz, southern Spain, 11 km/7 mi north of Gibraltar; population (1991) 20,900. It was built by Spaniards from Gibraltar in 1704, after Gibraltar was captured by the English. **28** E8

San Salvador capital of ►El Salvador and of San Salvador department; situated at the foot of San Salvador volcano (2,548 m/8,360 ft) on the River Acelhuate, 48 km/30 mi from the Pacific Ocean; population (1992) 422,600. Industries include coffee, food-processing, pharmaceuticals, and textiles. One-third of the country's industrial output comes from the city. Founded in 1525, it was destroyed by an earthquake in 1854 and rebuilt on the present site. It is now a modern city with architecture conditioned to seismic activity, to which the region is prone, although many buildings collapsed during a further earthquake in 1986. **94** G6

San Sebastián (officially **Donostia-San Sebastián**), port and resort in the Basque Country, Spain, the capital of the Basque province of Guipúzcoa; population (1991) 170,000. There are engineering and papermaking industries. The town sits at the foot of Mount Urgull, a rocky island that is linked to the mainland by a sandy isthmus, and which appears to have been occupied since prehistoric times. It is a natural fortress that dominates all routes round the western end of the Pyrenees between France and Spain, and has been besieged and sacked many times. The modern San Sebastián is an elegant town with a fine beach, La Concha, to the west. 28 J1

Santa Ana city and administrative headquarters of Orange County, southwest California, USA; population (1990) 293,700. It is situated on the Santa Ana River, 70 km/43 mi south-east of Los Angeles, and at the southwestern base of the Santa Ana Mountains. It is the commercial and financial centre of the irrigated Santa Ana Valley. The city's varied manufactures include aircraft, computer and electronic parts, plastics, sporting and rubber goods, metal products, and fibreglass. It is home to Santa Rancho Santiago Community College (1915). 92 C2

Santa Ana commercial city in northwest El Salvador, capital of Santa Ana department and second-largest city in the country; population (1992) 202,300. It lies in a fertile basin on the northeast slopes of the Santa Ana volcano (2,381 m/7,812 ft), the highest in the country. Coffee and sugar cane are grown in the region. The city has some colonial architecture, including a neo-Gothic cathedral. 94 G6

Santa Barbara city in southern California, USA, on the **Santa Barbara Channel**, in the foothills of the Santa Ynez Mountains; population (1996 est) 86,200. It is a popular resort and retirement centre, and the headquarters of research, cultural, and religious foundations. Industries include fishing and fish-canning, offshore oil production, and the manufacture of aircraft and aerospace equipment, precision instruments, and electronic components. Orchard products are processed from local areas of intensive irrigation agriculture. 92 C2

Santa Cruz capital of Santa Cruz department in central Bolivia, 550 km/342 mi southeast of La Paz; the second-largest city in the country; population (1992) 694,600. The surrounding area is fertile, producing sugar cane, soybeans, cotton, rice, maize, and coffee. It is a hub of transport and trade. Newly discovered oil and natural gas has led to the city's rapid development. It was founded by the Spaniard Nuflo de Chaves in 1561 as Santa Cruz de la Sierra. There is a university (founded 1880), and a cathedral. 98 E7

Santa Cruz city and administrative headquarters of Santa Cruz County, west-central California, USA; population (1990) 49,000. It is situated at the northern end of Monterey Bay, to the southwest of San Jose, and 100 km/60 mi south of San Francisco. Industries include tourism, food processing (citrus fruit, cattle), fishing, electronics, and petroleum. A University of California campus, opened in 1965, has brought a renaissance of cultural activity to this popular resort. 92 B1

Santa Cruz de Tenerife province of Spain in the ►Canary Islands autonomous community; area 3,209 sq km/1,239 sq mi; population (1995) 787,400. It includes the islands of Tenerife, Gomera, La Palma, and Hierro. Tourism is the main industry; there is some agriculture and fruit production, especially bananas. The capital is Santa Cruz de Tenerife. 66 B3

Santa Fé river port and capital of Santa Fé province, northeast Argentina, on the Salado River 153 km/95 mi north of Rosario; population (1991) 395,000. It is a leading distribution centre for the area and an export outlet for its agro-industrial and processed products. There are shipyards, timber, cattle, and wool are exported. It has a 17th-century cathedral, several churches such as the mid-17th century Templo de Santo Domingo, and the Litoral University (founded 1919). Santa Fé was founded in 1573, and the 1853 constitution was drafted here. **100** J5

Santa Fe capital of ►New Mexico, USA, on the **Santa Fe River**, 65 km/40 mi west of Las Vegas; population (1996 est) 66,500, many Spanish-speaking. It is situated in the Rio Grande Valley, over 2,000 m/6,500 ft above sea level, on the western slopes of the Sangre de Cristo Mountains. Santa Fe is the cultural and tourist capital of the southwest, home to many artists, theatre, and opera. Precision instruments, pottery, and American Indian jewellery and textiles are produced. **86** E3

Santa Maria city in Santa Barbara County, southwest California, USA; population (1992) 64,100. It is situated on the Santa Maria River, in the Santa Maria Valley, 18 km/11 mi east of the Pacific Ocean, and 87 km/54 mi northwest of Santa Barbara. It is the trade, processing, and shipping centre for the valley, which produces flower seeds and bulbs, vegetables, beans, sugar beets, dairy goods, and grains. Oil refining is also important to the regional economy; founded in 1871, the community was sparsely settled until 1902, when oil began to be extracted. Vandenberg Air Force Base, to the southwest, is an important local employer. Allan Hancock College (1920) and the county fairgrounds are both located in the city. 92 B2

Santa Monica city in Los Angeles County, southwest California, USA, 22 km/14 mi west of the centre of ►Los Angeles; population (1990) 86,900. It has a beach and is a popular resort and residential area. The principal industry is now tourism, although specialized engineering and aircraft manufacturing are still important. Near Santa Monica is the J Paul Getty Museum with a formidable art collection. The city is home to the Santa Monica College (1929).

Santander port and capital of the autonomous community of ►Cantabria, northern Spain, on the Bay of Biscay; population (1991) 189,100. Industries include chemicals, textiles, vehicles, and shipyards. Santander was sacked by the French marshal Nicolas Soult in 1808, and was largely rebuilt after a fire in 1941. Palaeolithic cave wall paintings of bison, wild boar, and deer were discovered at the nearby cave of **Altamira** in 1879. 28 G1

Santa Rosa city and administrative headquarters of Sonoma County, northwest California, USA; population (1992) 116,600. It is situated on the Santa Rosa River, 74 km/46 mi north-northwest of San Francisco. It has long been a fruit processing and shipping centre, and was formerly a ranch hub. Other industries include wine, chemicals, and clothing. The city is growing rapidly as a residential community, and has varied agricultural, financial, and light manufacturing industries. 86 B3

Santiago capital of Chile, on the Mapocho River; population (1992) 4,385,500 (metropolitan area 5,180,800). It is the fifth largest city in South America and the country's cultural, commercial, and manufacturing centre. Industries include textiles, chemicals, and food processing. It has three universities, and several theatres, libraries, and museums. 100 G5

Santiago second-largest city in the Dominican Republic; population (1991 est) 375,000. It is a processing and trading centre for sugar, coffee, and cacao. 94 K5

Santiago de Compostela capital of ►Galicia autonomous community, in the province of La Coruña, northwest Spain; population (1991) 87,500. Textiles, chocolate, and soap are manufactured here, and there is a trade in agricultural produce. The 11th-century cathedral was reputedly built over the grave of Sant Iago el Mayor (St James the Great), patron saint of Spain, and was a world-famous centre for medieval pilgrims. Santiago is the seat of an archbishop, and there is also a university, founded in 1532. 28 B2

Santiago de Cuba port on the south coast of Cuba; population (1995 est) 425,000. It is the second largest city and former capital of Cuba, and is now the capital of Santiago de Cuba province. Products include sugar, rum, and cigars. 94 J5

Santo Domingo capital and chief sea port of the Dominican Republic; population (1991 est) 2,055,000. Founded 1496 by Bartolomeo, brother of the Italian-born explorer Christopher Columbus, it is the oldest colonial city in the Americas. Its cathedral was built 1515–40. 94 L5

Santos industrial city and principal seaport in São Paulo federal unit (state), southeast Brazil, 72 km/45 mi southeast of the city of São Paulo on the Atlantic coast; population (1991) 546,600. It is the largest and most important port in Brazil, and the world's leading coffee-exporting port. There are oil refineries and chemical and steel industries. The port lies 5 km/3 mi along the Santos Channel and is surrounded by sandy beaches. The Brazilian soccer player Pelé played here for many years. Santos was founded in 1534. 100 M3

Saône river in eastern France, rising in the Vosges Mountains and flowing 480 km/300 mi to join the ►Rhône at Lyon. After rising in the Faucilles Mountains of the Vosges, it flows south past Gray, Chalon-sur-Saône, and Maçon. The chief tributaries are the ►Doubs and the Ognon. It is connected by canal with the rivers Loire, Seine, Meuse, Moselle, and Rhine. 26 K7

Saône-et-Loire *département* in the Burgundy region of France; area 8,565 sq km/3,307 sq mi; population (1990) 560,000. The River ►Saône flows through the east, and is connected by the Canal du Centre to the River Loire in the west. Wheat and oats grow on the plains, and the hilly district on the east bank of the Saône is a centre of the Burgundy wine trade. There is much industry around the coal basins of Autun and Montceau-les-Mines, and ironworks at Le Creusot. Cattle and poultry are bred in many parts. The administrative centre is ►Mâcon. Other towns include Autun and Chalon-sur-Saône. 6 F3

Saône, Haute- *département* in the Franche-Comté region of France; see ►Haute-Saône. 6 G3

São Paulo industrial city and capital of São Paulo federal unit (state), southeast Brazil, 72 km/45 mi northwest of its port Santos, and 400 km/249 mi southwest of Rio de Janeiro; population (1992) 9,646,200 (metropolitan area 16,567,300). It is Latin America's second-largest city after Mexico City. It is 900 m/3,000 ft above sea level, and 2° south of the Tropic of Capricorn. It is also South America's leading industrial city, producing electronics, steel, and chemicals; it has meat-packing plants and is the centre of Brazil's coffee trade. The city has a cathedral and four universities (the University of São Paulo is the largest in Brazil). The Butantã Snake Farm is a biomedical research institute for the production of antidotes for snakebite. 100 M3

São Tomé port and capital of ►São Tomé e Príncipe, on the northeast coast of São Tomé island, Gulf of Guinea; population (1991) 43,400. It exports sugar, cocoa, and coffee. 68 F4

São Tomé and Príncipe country in the Gulf of Guinea, off the coast of West Africa.

São Tomé and Príncipe covers an area of 1,000 sq km/386 sq mi, and has a population of 147,000 (2000 est). The capital is ►São Tomé; other major towns and cities include São António, Santana, Porto-Alegre, Trinidade, Neves, and Santo Amaro. The head of state is Miguel Trovoada (from 1991); the head of government is Guilherme Posser da Costa (from 1999). The currency is the dobra, and the languages spoken include Portuguese (official), Fang (a Bantu language), and Lungwa São Tomé (a Portuguese Creole). 68 F4

Sapporo capital of ►Hokkaido prefecture, Japan, on the Ishikari River; population (1994) 1,719,000. Industries include rubber, food processing, printing, brewing beer, and lead and zinc mining. It is a winter sports centre and was the site of the 1972 Winter Olympics. Giant figures are sculpted in ice at the annual snow festival. The city has an underground railway. 48 L2

Saragossa English spelling of ►Zaragoza, a city and province of Spain. 28 K3

Sarajevo capital of Bosnia-Herzegovina; population (1991) 526,000. Industries include engineering, brewing, chemicals, carpets, and ceramics. A Bosnian, Gavrilo Princip, assassinated Archduke Franz Ferdinand here in 1914, thereby precipitating World War I. From April 1992 the city was the target of a siege by Bosnian Serb forces in their fight to carve up the newly independent republic. A United Nations ultimatum and the threat of NATO bombing led to a ceasefire February 1994 and the effective end of the siege as Serbian heavy weaponry was withdrawn from the high points surrounding the city. 34 F6

Sarasota city and administrative headquarters of Sarasota County, southwest Florida, USA, on the Gulf of Mexico, south of Tampa; population (1992) 50,900. Sarasota is a flourishing winter resort known for its water sports, gardens, and golf courses, a spring training baseball centre, and an entertainment hub with several performing arts facilities. The city is also a processing and shipping centre for local citrus fruits, vegetables (notably celery), and beef. Its industries include boatbuilding, electronics research and development, and the production of mobile homes. Sarasota is the home of New College of the University of South Florida (1960). 90 E4

Saratoga Springs town and spa in Saratoga County, east New York State; population (1990) 25,000. It is situated in the foothills of the Adirondack Mountains, 5 km/3 mi west of Saratoga Lake, and 48 km/30 mi north of Albany. Industries manufacture textiles, wallpaper, electronic equipment, and mobile homes. Saratoga Springs began attracting European visitors in 1771, and became the most fashionable US resort of the 19th century. The town is home to Skidmore College (1911) and Empire State College (1971). 88 F2

Saratov capital of Saratov oblast (region), Russian Federation, situated on the right bank of the Volga, 315 km/196 mi north of Volgograd; population (1996 est) 892,000. Saratov is a major centre of industry and transportation, containing engineering and gas works, shipyards, oil refineries, and flour mills. Oil is shipped here from the fields around Baku in Azerbaijan for onward rail distribution to the western Russian Federation; a natural gas pipeline also runs from here to Moscow. Saratov is the second cultural centre of the Volga region after Kazan. 38 J4

Sarawak state of Malaysia, in the northwest of the island of Borneo; area 124,400 sq km/48,018 sq mi; population (1991) 1,669,000 (24 ethnic groups make up almost half this number). The capital is ►Kuching. Industries include timber, oil, rice, pepper, rubber, coconuts, and natural gas. 11 (1)C2

Sardegna Italian name for ►Sardinia, an island in the Mediterranean Sea. 32 E8

Sardinia (Italian **Sardegna**), mountainous island and special autonomous region of Italy, about 240 km/150 mi southwest of the Orbetello promontory in Tuscany; area 24,100 sq km/9,300 sq mi; population (1992 est) 1,651,900. It is the second-largest Mediterranean island and comprises the provinces of Cagliari, Nuoro, Oristano, and Sassari; its capital is ►Cagliari. Cork, fruit, grain, tobacco, minerals (lead, zinc, manganese), and petrochemicals are exported. Features include the Costa Smeralda (Emerald Coast) tourist area in the northeast and *nuraghi* (fortified Bronze Age dwellings). After centuries of foreign rule, Sardinia became linked with Piedmont in 1720, and this dual kingdom became the basis of a united Italy in 1861. 32 E8/9 (4)B4

Sargasso Sea part of the North Atlantic Ocean (between 40° and 80° west and 25° and 30° north) left static by circling ocean currents, and covered with the floating weed *Sargassum natans*, which gives the sea its name.

Sark one of the ►Channel Islands, 10 km/6 mi east of Guernsey; area 5 sq km/2 sq mi; population (1991) 575.

There is no town or village. It is divided into Great and Little Sark, linked by an isthmus, and is of great natural beauty. The Seigneurie of Sark was established by Elizabeth I, the ruler being known as Seigneur/Dame, and has its own parliament, the Chief Pleas. There is no income tax, divorce and cars are forbidden, and immigration is controlled. 26 C4

Sarnia (officially **Sarnia-Clearwater**), town in Ontario, Canada, on the St Clair River, 80 km/50 mi west of London, on the Canada–USA border; population (1990) 73,400.
It is the centre of a petrochemical industry. Oil was first discovered in this area in 1858, but it was only with the completion of the oil pipeline from Alberta that the great expansion of the refining and petrochemical industry began. The oil has served as a base for ammonia, ethylene, synthetic rubber, and insecticide plants. 88 D2

Sarthe *département* in the ►Pays de la Loire region of France; area 6,210 sq km/2,398 sq mi; population (1990) 514,300. It is crossed from north to southwest by the River Sarthe. Cereals, potatoes, poultry, cider, and perry are produced and horses are bred. Iron, marble, slate, and limestone are found, and there are industries connected with them. Hemp and linen are manufactured. The administrative centre is ►Le Mans. Other towns include La Flèche and Mamers. 6 D2

Sasebo seaport and naval base on the west coast of Kyushu island, Japan; population (1994) 247,000. Sasebo has been a navel base since 1886. It was destroyed in World War II, but was reconstructed as a shipbuilding centre during the Korean War (1950–53). 48 E7

Saskatchewan (called 'Canada's breadbasket', Cree *Kis-is-ska-tche-wan* 'swift flowing'), province of west-central Canada, the middle Prairie province, bordered to the west by Alberta and to the east by Manitoba. To the north of Saskatchewan (above the 60th parallel) are the Northwest Territories, while to the south (below the 49th parallel) lie the US states of North Dakota and Montana; area 652,300 sq km/251,854 sq mi; population (1997) 1,023,500. The capital is ►Regina. Industries include extraction of oil, natural gas, uranium, zinc, potash, copper, and helium; and manufacture of cement, chemicals, fertilizers, and wood products. Wheat, oats, barley, rye, and flax are grown, and there is cattle-rearing and dairying. 82 K6

Saskatchewan (Cree *kisiskatchewan* 'rapid river' or 'swift current'), river in Canada formed by two branches, north and south, both rising in western Alberta and meeting east of Prince Albert; length 1,939 km/1,205 mi. It flows east to Lake Winnipeg, in south Manitoba, and then to Hudson Bay as the **Nelson River**. It has been dammed several times for irrigation purposes. 82 L6

Saskatoon largest city in Saskatchewan, Canada, on the South Saskatchewan River; population (1996) 221,400. It is the commercial, processing, and distribution centre of a vast grain-farming area. Related manufacturing plants produce flour, linseed oil, glycol, glucose, mixed feeds, and agricultural equipment. Other industries include meat-packing, oil-refining, biotechnology, telecommunications, electronics, and the production of chemicals and fibre optics. Potash is mined nearby. The University of Saskatchewan was established here in 1907. 82 K6

Sassari capital of the province of the same name, in the northwest corner of the island of Sardinia, Italy; population (1992) 122,000. Every May it hosts the Sardinian Cavalcade, a major local festival. Cotton is manufactured. There is a cathedral (13th–18th centuries), and a university founded in 1562. 32 C8

Satu Mare (Hungarian **Szatmar-Nemeti**), capital of Satu Mare county, northwest Romania, on the River Someş, near the Hungarian and Ukrainian borders, 174 km/109 mi northwest of Cluj Napoca; population (1993) 131,400. Farm machinery, rolling stock, textiles, and processed food are manufactured, and the chief commodities are linen and wine. 34 K2

Saudi Arabia country on the Arabian peninsula, stretching from the Red Sea in the west to the Gulf in the east, bounded north by Jordan, Iraq, and Kuwait; east by Qatar and United Arab Emirates; southeast by Oman; and south by Yemen.

Saudi Arabia covers an area of 2,200,518 sq km/849,620 sq mi, and has a population of 21,607,000 (2000 est). The capital is ►Riyadh; other major towns and cities include Jiddah, Mecca, Medina, Taif, Dammam, Hufuf, Tabuk, and Buraydah. The head of state and government is King Fahd Ibn Abdul Aziz (from 1982). The currency is the riyal, and the languages spoken include Arabic (official) and English. 56 D4

Sault Ste Marie twin industrial ports on the Canadian–US border, one in Ontario and one in Michigan; population (1991) 81,500 and (1990) 14,700, respectively. They stand at the falls (French *sault*) in St Mary's River, which links lakes Superior and Huron. The falls are bypassed by canals. Industries include steel, pulp, and agricultural trade. A French Jesuit mission was established here in 1669. 88 D1

Saumur town in Maine-et-Loire *département*, France, on the left bank of the River Loire 43 km/28 mi southeast of Angers; population (1990) 31,900. The cavalry school (the Cadre Noir), founded in 1768, has since 1942 also been a training school for the French armed forces. The area produces wines, especially the sparkling Vins de Saumur, as well as mushrooms and other vegetables. 26 E6

Sava river in the Balkans; length 724 km/450 mi. It rises in the Karawanken Alps and flows southeast across Slovenia and Croatia to join the River ►Danube, of which it is the chief tributary, at Belgrade in Yugoslavia; 30 L5

Savannah city and port of Georgia, USA, on the Savannah River, 29 km/18 mi from its mouth on the Atlantic Ocean; population (1996 est) 136,300. It is a major port and transportation hub on the eastern seaboard. Industries include paper- and food-processing, and the manufacture of aircraft, transportation equipment, and chemicals. Tourism is also important to the economy. Founded in 1733, Savannah was the capital of Georgia 1754–86, and was the first US city to be laid out in geometrically regular blocks. 90 E3

Savannah river between Georgia and South Carolina, USA, formed by the Tugaloo and Seneca rivers, which meet at Hartwell dam, Georgia; length 505 km/314 mi. It flows southeast into Tybee Sound, 29 km/18 mi from Savannah in Georgia. The river is used for hydroelectric power and flood control. 80 K6

Savoie *département* in the ►Rhône-Alpes region of France; area 6,035 sq km/2,330 sq mi; population (1990) 348,900. It is mountainous and wooded. Cereals and vines are cultivated in the valleys, and cattle, sheep, and horses are bred. Some copper and lead are found, and hydroelectric power is produced. There are textile and food manufactures, and an extensive tourist industry. The chief river is the Isère. The administrative centre is ►Chambéry. Other towns include Albertville and St-Jean-de-Maurienne. 6 G4

Savoie, Haute- *département* in the Rhône-Alpes region of France; see ►Haute-Savoie. 6 G3

Saxony (German **Sachsen**), *Land* (administrative region) of Germany; area 17,036 sq km/6,578 sq mi; population (1995) 4,602,000. The capital is ►Dresden. Industries include electronics, textiles, vehicles, machinery, chemicals, and coal. The region is on the plain of the River Elbe north of the Erzgebirge mountain range. Towns include Leipzig, Chemnitz, and Zwickau. 7 (2)D3

Saxony-Anhalt (German **Sachsen-Anhalt**), *Land* (administrative region) of Germany; area 20,450 sq km/7,900 sq mi; population (1995) 2,735,000. The capital is ►Magdeburg; other towns include Halle and Dessau. Industries include chemicals, electronics, rolling stock, and footwear. Cereals and vegetables are grown. 7 (2)C2

Scafell Pike highest mountain in England, in the ►Lake District, Cumbria, northwest England; height 978 m/3,210 ft. It is separated from Scafell (964 m/3,164 ft) by a ridge called Mickledore. 24 J7

Scandinavia peninsula in northwestern Europe, comprising Norway and Sweden; politically and culturally it also includes ►Denmark, ►Iceland, the ►Faroe Islands, and ►Finland.

Scarabantia Roman name for ►Sopron, a town in Hungary. 30 M3

Scarborough spa and holiday resort on the North Sea coast of North Yorkshire, northern England, 56 km/35 mi northeast of York; population (1991) 38,900. It is a touring centre for the Yorkshire Moors, and is also centre for fishing. A ruined 12th-century Norman castle overlooks the town. 24 M7

Schaffhausen canton in northern Switzerland, almost entirely surrounded by territory belonging to Germany; area 298 sq km/115 sq mi; population (1995) 7,062,400. The principal feature of this hilly region is the ►Rhine valley, where the river falls in a series of cascades of up to 60 m/197 ft (the Rheinfall). Vine growing is the main agricultural activity. 8 (1)C1

Schaffhausen town and capital of the canton of Schaffhausen, north Switzerland, 35 km/22 mi north of Zürich; population (1995) 34,600. There are watch-making, mechanical and chemical engineering, and textile industries. The Rhine falls here in a series of spectacular cascades 60 m/197 ft high. 30 D3

Schässburg German name for ►Sighişoara, a town in Romania. 34 M3

Schelde (or **Scheldt**; French **Escaut**), river of France, Belgium, and the Netherlands, rising in Aisne *département*, north France, and flowing 400 km/249 mi to join the North Sea south of Walcheren in the Netherlands. Antwerp, in Belgium, is the chief town on the Schelde. 22 F3

Schenectady (Mohawk 'across the pine plains'), industrial city in New York State, USA, on the Mohawk River, 20 km/13 mi northwest of Albany; population (1996 est) 67,900.

With ►Troy and ►Albany it forms the Tri-City area. Schenectady has long been a producer of electrical goods; Thomas Edison, pioneer of electrical engineering, established his workshops here in 1886, and these became the General Electric Company in 1892. Chemicals are also manufactured. Founded by the Dutch in about 1660, it was destroyed by American Indian Iroquois and French in 1690, and re-established by the British. 88 F2

Scheveningen seaside resort and northern suburb of The ►Hague on the west coast of South Holland, the Netherlands. It has one of the finest beaches on the North Sea. There are fisheries and a ferry link with Great Yarmouth, England. 22 G2

Schiedam port in South Holland province, southwest Netherlands, on the River Meuse, 5 km/3 mi west of Rotterdam; population (1997) 74,900. It forms part of the Rotterdam-Europoort complex. The chief centre for Dutch gin, it also makes liqueurs and candles, and has a shipping trade and iron works. 22 G3

Schleswig (Danish **Slesvig**), seaport in Schleswig-Holstein, Germany, at the western end of the 35 km/22 mi long Schlei inlet of the Baltic and 50 km/31 mi northwest of Kiel; population (1994) 186,300. The city has food industries and is a tourist centre. Schleswig's cathedral dates partly from the 11th century. 20 E2

Schleswig-Holstein administrative region (German *Land*) in north Germany, bounded on the north by Denmark, on the east by the Baltic Sea and Mecklenberg-West Pomerania, on the south by Lower Saxony and Hamburg, and on the west by the North Sea and the Heligoland Bight; area 15,700 sq km/6,060 sq mi; population (1995) 2,708,000. The capital is ►Kiel. There are ship-building, mechanical and electrical engineering, food processing, and textile industries. There are fisheries, and cattle, sheep, pigs, and poultry are raised. 7 (2)B1

Schneidemühl German name for ►Piła, a town in Poland. 18 F4

Schönebeck town in the *Land* (administrative region) of ►Saxony-Anhalt, Germany, on the River Elbe, 13 km/8 mi southeast of Magdeburg; population (1995) 39,100. It has engineering and chemical industries. 20 G4

Schwäbisch-Gmünd (or **Gmünd**), city in Baden-Württemberg, Germany, on the River Rems, 48 km/30 mi east of Stuttgart; population (1995) 63,800. It has produced gold and silverware since medieval times and it also manufactures watches, clothing, glassware, and metal goods. Schwäbisch-Gmünd was one of the free city-states of the Holy Roman Empire. 30 E2

Schwarzwald German name for the ►Black Forest, a coniferous forest in Germany. 30 D3

Schweinfurt city in Bavaria, Germany, on the River Main, 100 km/62 mi east of Frankfurt am Main; population (1994) 55,500. Ball-bearings, dyes, and fertilizers are manufactured. It has a harbour on the Rhein–Main–Donau Canal. Schweinfurt was one of the city-states of the Holy Roman Empire. 20 F6

Schwerin capital of the *Land* (administrative region) of ►Mecklenburg–West Pomerania, Germany, on the western shore of the lake of Schwerin; population (1995) 117,200. Products include machinery, foods, and chemicals. Formerly the capital of ►Mecklenburg and earlier of the old republic of Mecklenburg–Schwerin, Schwerin became capital of Mecklenburg–West Pomerania with the reunification of Germany in 1990. 20 G3

Schwyz capital of Schwyz canton, Switzerland; population (1995) 15,200. Schwyz was one of the three original cantons of the Swiss Confederation of 1291, which gave its name to the whole country in about 1450. 30 D3

Scilly, Isles of (or **Scilly Isles/Islands** or **Scillies**), group of 140 islands and islets lying 40 km/25 mi southwest of Land's End, England; administered by the Duchy of Cornwall; area 16 sq km/6.3 sq mi; population (1991) 2,050. The five inhabited islands are **St Mary's**, the largest, on which is Hugh Town, capital of the Scillies; **Tresco**, the second largest, with subtropical gardens; **St Martin's**, noted for its beautiful shells; **St Agnes**; and **Bryher**. 24 F12

Scotland (Roman **Caledonia**), constituent part of the United Kingdom, the northernmost part of Britain, formerly an independent country; area 78,470 sq km/30,297 sq mi; population (2001 est) 5,060,100. The capital is ►Edinburgh, and other major towns and cities include Glasgow, Dundee, and Aberdeen. Geographical features include the Highlands in the north (with the ►Grampian Mountains), the central Lowlands, including valleys of the Clyde and Forth, the Southern Uplands (including the Lammermuir Hills), and islands of the Orkneys, Shetlands, and Western Isles. Industries include marine and aircraft engines, oil, natural gas, chemicals, textiles, clothing, printing, paper, food processing, tourism, whisky, coal, electronics, and computers (Scotland's 'Silicon Glen' produces over 35% of Europe's personal computers). 24 H5

Scottish Borders unitary authority in southeast Scotland, created in 1996 to replace the former Borders region.
area 4,733 sq km/1,827 sq mi
towns Galashiels, Hawick, Jedburgh, Kelso, Newtown St Boswells (administrative headquarters), Peebles, Selkirk
physical much of the west part of the area is upland (Lammermuir, Moorfoot and Pentland Hills); Broad Law (840 m/2,756 ft), near Tweedsmuir, is the highest point. The principal river, the Tweed, traverses the region west–east. The largest loch is St Mary's, and the only substantial area of low-lying agricultural land is the Merse in the southeast. The coast is generally precipitous
features Walter Scott's home at Abbotsford; Field Marshal Haig and Walter Scott buried at Dryburgh Abbey; Melrose Abbey (12th century)
industries electronics, timber, knitwear, tweed
agriculture sheep and cattle; cereals and root crops; fishing
population (1996) 105,300 5 E4

Scranton industrial city, Pennsylvania, USA, on the Lackawanna River, 217 km/136 mi northwest of New York; population (1996 est) 77,190. Anthracite coal is mined nearby and beneath the city, but production has declined sharply. Long-established iron and steel industries have been overtaken by the manufacture of electronic equipment, fabricated metal, clothing, plastic goods, and printed materials. Revitalization of the economy has been aided by the development of interstate highway connections with New York and Philadelphia, establishing the city as a haulage centre. 88 E2

Scunthorpe industrial town and administrative headquarters of ►North Lincolnshire, England, 39 km/24 mi west of Grimsby; population (1991) 74,700. It has one of Europe's largest iron and steel works, which has been greatly expanded with assistance from the European Union. Other industries include engineering, electronics, food processing, clothing, and furniture. 24 M8

Seattle (called 'the Emerald City'), port on Lake Washington in King County, Washington, USA, and the largest city in the Pacific Northwest; population (2000 est) 563,400, Greater Seattle 2,700,000. It is the fifth-largest container port in the USA, trading over US$37 billion in goods a year. It is the main transit point for supplies to Alaska, and trade with Japan is important. Industries include aerospace (the manufacturing plants of the Boeing Corporation are here), timber, tourism, banking and insurance, paper industries, electronics, computing (Microsoft, based in adjoining Redmond, is one of 2,500 software firms), biotechnology, ocean science, shipbuilding and repair, and fishing. Coffee has been an important commodity since the development of the Starbucks Company in the 1970s. 86 B1

Sebastopol alternative spelling of ►Sevastopol, a Black Sea port on the peninsula of ►Crimea in Ukraine. 58 E1

Secunderabad northern suburb of ►Hyderabad city, Andhra Pradesh, India, separated from the rest of the city by the Hussain Sagar Lake. Since Indian independence in 1947 Secunderabad has been incorporated within the city of Hyderabad; population (1995 est) 3,253,000. Formerly a separate town, it was founded as a British army settlement, with a parade ground where 7,000 troops could be exercised. 54 C5

Sedan town on the River Meuse, in Ardennes *département*, northeast France; population (1990) 22,400. Industries include textiles and dyestuffs, metal works, medical appliances, chemicals, and food processing; the town's prosperity dates from the 16th–17th centuries, when it was a Huguenot centre. In 1870 Sedan was the scene of Napoleon III's surrender to Prussia during the Franco-Prussian War. It was the focal point of the German advance into France in May 1940. 22 G5

Seeland German form of ►Sjælland, the main island of Denmark. 16 F9

Segesvár Hungarian name for ►Sighişoara, a town in Romania. 34 M3

Segovia capital of Segovia province in Castilla–León, central Spain; population (1991) 54,100. Products include thread, fertilizer, and chemicals. Segovia has a Roman aqueduct with 118 arches, a 16th-century cathedral, and a 13th-century Templar's church. Its castle, mostly rebuilt after a fire in 1862, was once the palace of the monarchs of Castile; Isabella I of Castile was crowned here in 1474. 28 F4

Seine French river rising on the Langres plateau in the *département* of Côte d'Or, 30 km/19 mi northwest of Dijon, and flowing 774 km/472 mi northwest through ►Paris and Rouen to join the English Channel at Le Havre. It is the third-longest, but economically the most important, river in the country. 26 F4

Seine-et-Marne *département* in the ►Ile-de-France region of France; area 5,916 sq km/2,284 sq mi; population (1990) 1,079,500. The Rivers ►Seine and ►Marne, and their numerous tributaries, flow through it. On the fertile level land between the rivers, wheat and vegetables are grown, and cheese (especially Brie) manufactured. The south has a large wooded area, the forest of Fontainebleau. The administrative centre is ►Melun. Other towns include Meaux and Provins. 6 E2

Seine-Maritime *département* in the ►Haute-Normandie region of France, bordering on the English Channel; area 6,354 sq km/2,453 sq mi; population (1999 est) 1,237,700. It is largely rural, with large areas of agricultural land and the coastal maritime district. The 120 km/75 mi coast is made up of chalk cliffs rising to a height of 120 m/394 ft. Stock-rearing is important, and dairy, apples, cider, flax, and fish are major agricultural products. The automobile industry is the largest employer in the region, providing around 28,000 jobs. One third of all French motor vehicles are produced here. Other chief industries include cotton manufacturing around ►Rouen (the administrative centre), and woollen manufacturing at Elbeuf. There are metallurgical, chemical, and food processing industries at Rouen and ►Le Havre, which are also important ports, and the lower Seine valley is a focus for oil refining. The *département* suffered serious damage during World War II, and underwent extensive rebuilding. 6 D2

Seine-St-Denis *département* in the ►Ile-de-France region of France to the northeast of Paris; area 236 sq km/91 sq mi; population (1990) 1,381,200. It was created in the administrative reforms of 1964. It comprises the northeastern section of the ►Paris suburbs, with a large proportion of social housing. There are motor-vehicle industries and office employment, as well as intensive farming, in the area. Seine-St-Denis is crossed by major road and rail links, and contains the Charles de Gaulle Airport at Roissy-en-France. The administrative centre is ►Bobigny. 6 E2

Sekondi-Takoradi industrial seaport in southwest Ghana, situated on the Gulf of Guinea, 180 km/122 mi southwest of Accra; population (1998 est) 147,400. It is the capital of the western region of Ghana. The old city of Sekondi was founded by the Dutch in the 16th century. The port of Takoradi has an artificial harbour, opened in 1928, and railway engineering, boat building, cigarette, flour milling, cocoa processing, and timber milling industries. The city and the port were merged in 1946. 68 D3/68 D4

Selangor state of the Federation of Malaysia; area 7,956 sq km/3,071 sq mi; population (1993 est) 1,981,200. It was under British protection from 1874 and was a federated state 1895–1946. The capital was transferred to Shah Alam from Kuala Lumpur in 1973. Klang is the seat of the sultan and a centre for rubber-growing and tin-mining; Port Kelang (or Klang), formerly Port Swettenham, exports tin and rubber. 11 (1)A2

Selkirk Mountains range of mountains in British Columbia,

Canada, forming a subdivision of the Columbia Mountains, part of the North American ►Rocky Mountains chain. The range is bounded by the Columbia River on the east, west, and north. Sir Sandford (3,533 m/ 11,590 ft) is the highest peak, and there are others over 3,000 m/9,842 ft high. Rogers Pass lies in the range. A section of the mountains is contained within the Glacier National Park. 84 C1

Semarang port in north Java, Indonesia; population (1995 est) 1,447,000, the capital of Java Tengah province. There are shipbuilding, fishing, and textile industries and exports include coffee, teak, sugar, tobacco, kapok, and petroleum from nearby oilfields. 52 E4

Semey (formerly **Semipalatinsk** 1776–1991), town in northeastern Kazakhstan, capital of the Semey oblast (region), a communications centre located at the junction of the Irtysh River and the Turksib railway, 680 km/423 mi southeast of Omsk in the Russian Federation; population (1996) 339,000. Industries include ship repair, engineering, meat-packing, tanning, and flour milling, and the surrounding region is an important mining area producing nickel and chromium. The fertile steppe land to the northwest yields wool and livestock, which are traded in the city. 42 Q7

Semipalatinsk former name of ►Semey, a town in northeastern Kazakhstan. 42 Q7

Semois river in Belgium; length 190 km/118 mi. It rises near Arlon in the southeast of the country and flows west through the provinces of Luxembourg and Namur to join the ►Meuse above Montherme in France. 22 H5

Sendai city in Miyagi prefecture, Tōhoku region, northeast Honshu island, Japan, 250 km/155 mi north of Tokyo; population (1994) 928,000. Industries include metal goods (a metal museum was established in 1975), electronics, textiles, pottery, food processing, and printing. The Tanabata festival, held annually in early August, is a tourist attraction. It was a feudal castle town from the 16th century. The castle was built in 1601. 48 L4

Senegal country in West Africa, on the Atlantic Ocean, bounded north by Mauritania, east by Mali, south by Guinea and Guinea-Bissau, and enclosing the Gambia on three sides.

Senegal covers an area of 196,200 sq km/75,752 sq mi, and has a population of 9,481,000 (2000 est). The capital (and chief port) is ►Dakar; other major towns and cities include Thiès, Kaolack, Saint-Louis, Ziguinchor, Diourbel, and Mbour. The head of state is Abdoulaye Wade (from 2000); the head of government is Mustafa Niasse (from 2000). The currency is the franc CFA, and the languages spoken include French (official), Wolof, and other ethnic languages. 68 A2

Senegal river in West Africa, formed by the confluence of the Bafing and Bakhoy rivers and flowing 1,125 km/ 700 mi northwest and west to join the Atlantic Ocean near Saint-Louis, Senegal. In 1968 the Organization of Riparian States of the River Senegal (Guinea, Mali, Mauritania, and Senegal) was formed to develop the river valley, including a dam for hydroelectric power and irrigation at Joina Falls in Mali; its headquarters is in Dakar. The river gives its name to the Republic of Senegal. 68 B1

Sennar town in Sennar State, about 260 km/160 mi southeast of Khartoum, on the left bank of the Blue Nile, Sudan; population about 10,000. Nearby is the Sennar Dam (1926), part of the Gezira irrigation scheme, which provides water for over 4,000 sq km/1,540 sq mi of the El Gezira Plain. 56 B7

Sens town in Yonne *département*, Burgundy, France; situated 50 km/31 mi northwest of Auxerre on the River Yonne; population (1990) 27,800. Its 12th–16th-century cathedral was the first built in France in the truly Gothic style. There is a trade in agricultural produce, wine, and wood. Agricultural implements, leather goods, and chemicals are manufactured. 26 J5

Senta town in Serbia, Yugoslavia, on the River Tisza, in the autonomous region of Vojvodina; population (1991) 22,800. It is a rail and road junction and a market centre. 34 H4

Seoul (or **Sŏul**), capital and largest city of ►South Korea, in the north-west of the country, 32 km/20 mi inland on the Han River, and with its chief port at Inchon on the Yellow Sea; population (1999 est) 10,321,400. The metropolitan area, in which about a quarter of the total population of South Korea live, has grown rapidly since the mid-19th century and is now one of the ten largest in the world. Industries, which provide about half the industrial employment in the country, include engineering, textiles, food processing, electrical and electronic equipment, chemicals, and machinery. 48 D5

Seraing industrial suburb of the city of ►Liège, Belgium, on the River Meuse. The crystal works of Val-Saint-Lambert is situated in Seraing. It is ringed by decaying remnants of heavy industry. 22 H4

Seram (or **Ceram**, or **Serang**), Indonesian island in the Maluku (Moluccas) island group; area 17,142 sq km/6,619 sq mi. The principal town is Ambon, situated on a small island (also called Ambon) off the coast of Seram. 53 D3

Serbia (Serbo-Croat **Srbija**), constituent republic, together with ►Montenegro, of Yugoslavia, which includes ►Kosovo and ►Vojvodina; area 88,400 sq km/34,100 sq mi; population (1991 est) 9,791,400. The capital is ►Belgrade. 34 H6

Sergiyev Posad (formerly **Zagorsk** 1930–92), town in the Moscow oblast (region) of the Russian Federation, 70 km/43 mi northeast of Moscow; population (1996 est) 113,000. Its focal point is the Trinity Monastery of St Sergius (founded in 1337), an important place of pilgrimage and former administrative centre of the Russian Orthodox Church. The town is home to the Moscow Theological Academy. Aside from its religious significance, Sergiev Posad has been a centre of woodcarving and toymaking since the 15th century. Paint, optical instruments, building materials, textiles, and furniture are also produced here. 38 G3

Serrai (Greek **Serres**), town and department in Macedonia region, northern Greece; area (department) 3,987 sq km/1,539 sq mi; population (1991) 191,900 (department). It is the centre of a cotton-growing area, and has weaving and food industries. As the ancient Seris, it existed before

330 BC. It was occupied by the Turks between 1368 and 1912 and by the Bulgarians between 1913 and 1918. **36** F3

Serres Greek name for ►Serrai, a town and department in Greece. **36** F3

Sète major seaport and holiday resort on the Mediterranean coast of France, in Hérault *département*, southwest of Montpellier and separated from the mainland by a shallow lagoon; population (1990) 41,916. It was founded in 1666 as an outlet for the Canal du Midi, and handles fish, wine, and chemicals; there are oil refineries, shipyards, and metallurgical works in the area. **26** J10

Seto Naikai (Japanese 'inland sea'), narrow body of water almost enclosed by the Japanese islands of Honshu, Shikoku, and Kyushu. It is both a transport artery and a national park (1934) with 3,000 islands.

Sevastopol (or **Sebastopol**), Black Sea port, resort, and fortress in the ►Crimea, Ukraine; population (1990) 361,000. It is the base of the Black Sea fleet (jointly owned by Russia and Ukraine). Industries include shipbuilding and wine making. Founded by Catherine (II) the Great in 1784, Sevastopol was successfully besieged by the English and French (October 1854–September 1855) during the Crimean War, and again in World War II by the Germans (November 1941–July 1942). On both occasions, the city was devastated. Retaken by Soviet forces in 1944, Sevastopol still remains officially a closed city, because of its naval installations. **58** E1

Sevenoaks town in Kent, southeast England, 32 km/20 mi southeast of London, population (1991) 24,000. Industries include chemical manufacture, financial services, and insurance. Nearby are the houses of Knole (1456) and Chevening (17th century). **22** C3

Severn (Welsh **Hafren**), river in Britain, which rises on the slopes of Plynlimon, in Ceredigion, west Wales, and flows east and then south, finally forming a long estuary leading into the Bristol Channel; length 336 km/208 mi. The Severn is navigable for 290 km/180 mi, up to Welshpool (Trallwng) on the Welsh border. The principal towns on its course are Shrewsbury, Worcester, and Gloucester. England and South Wales are linked by two road bridges and a railway tunnel crossing the Severn (see ►Severn Bridge). A remarkable feature of the river is a tidal wave known as the 'Severn Bore' that flows for some miles upstream and can reach a height of 2 m/6 ft. **24** K10

Severn Bridge bridge linking England with south Wales across the Severn estuary, constructed 1961–66 at a cost of £8 million. A second bridge was built 1992–96, crossing from Severn Beach to New Passage.

Sevilla Spanish for ►Seville, a city in southern Spain. **28** E7

Seville (Spanish **Sevilla**), capital of Seville province and of the autonomous community of ►Andalusia, southern Spain, on the River Guadalquivir, 96 km/60 mi north of Cádiz; population (1991) 659,100. Products include machinery, spirits, porcelain, pharmaceuticals, silk, and tobacco. Although 80 km/50 mi from the sea, Seville has a historically important port (now little used), and during the 16th century it had a monopoly of trade with the West Indies. **28** E7

Seychelles country in the Indian Ocean, off east Africa, north of Madagascar.

The Seychelles covers an area of 453 sq km/174 sq mi, and has a population of 77,000 (2000 est). The capital (and chief port) is ►Victoria (on Mahé island); other major towns and cities include Cascade, Port Glaud, Misere, Anse Boileau, and Takamaka. The head of state and government is France-Albert René (from 1977). The currency is the Seychelles rupee, and the languages spoken include Creole (an Asian, African, European mixture), English, and French (all official). **72** (2)B2

Sfax (Arabic **Safaqis**), port and second-largest city in Tunisia, about 240 km/150 mi southeast of Tunis; population (1994) 230,900. It is the capital of Sfax district, on the Gulf of Gabès. Products include leather, soap, and carpets; there are also salt works and phosphate workings nearby. Exports include phosphates, fertilizers, olive oil, dates, almonds, esparto grass, and sponges. **66** H2

's-Gravenhage Dutch name for The ►Hague. **22** G2

Shaanxi (or **Shensi**), province of northwest China, bounded to the north by Inner Mongolia, to the east by Shanxi and Henan, to the south by Hubei and Sichuan, and to the west by Gansu and Ningxia Hui Autonomous Region; area 195,800 sq km/75,600 sq mi; population (1996) 35,430,000. The capital is ►Xi'an. There are coalmining, iron, steel, textile, and aerospace industries. Wheat, maize, rice, fruit, and tea are grown. **11** (2)D3

Shache (or **Yarkand**, or **Yarkant**), walled town and oasis in Xinjiang Uygur Autonomous Region, China, on the River Yarkand, about 160 km/100 mi southeast of Kashi; population (1985) 100,000. It lies in the ►Tarim Basin, at the foot of the Kunlun Mountains, on the edge of the Takla Makan Desert and on the caravan route to India and west Russia. It is the centre of a fertile region, which, with irrigation, produces cotton, beans, and silk. Industries also include cotton textiles and leather goods. Its position near the frontiers with India, Afghanistan, and Tajikistan gives the city advantages as an international trading centre. Traditional handicrafts, such as carpets and embroidered silks, have been produced here for centuries. The town is a centre of Islamic culture. **42** P10

Shakhty (formerly **Aleksandrovsk-Grushevski** 1839–1926), city in Rostov oblast (region) of the southwestern Russian Federation, in the Donets Basin 80 km/50 mi northeast of Rostov-na-Donu; population (1996 est) 229,000. Deep-mined coal (anthracite) is extracted here, iron and steel are produced, and the city is a centre of electricity generation for the eastern Donbas industrial region. Shakhty was founded as a coalmining settlement in 1839. **38** H5

Shandong (or **Shantung**), province of east China, bounded to the north by the Bohai Gulf, to the east by the Yellow Sea, to the south by Jiangsu and Anhui, and to the west by Henan and Hebei provinces; area 153,300 sq km/59,200 sq mi; population (1996) 83,430,000. It is one of the most densely populated provinces of China. The capital is ►Jinan. There are coal, oil, petrochemical, engineering, and textile industries. Cereals, cotton, peanuts, wild silk, and wine are produced. **11** (2)E3

Shanghai largest urban settlement and mainland port in China, in Jiangsu province, on the Huangpu and

Wusong rivers, 24 km/15 mi from the Chang Jiang estuary; population (1993) 8,760,000. The municipality of Shanghai has an area of 5,800 sq km/2,239 sq mi; population (1996) 14,190,000. Shanghai is China's principal commercial and financial centre. Textiles, paper, chemicals, steel, vehicles, agricultural machinery, precision instruments, shipbuilding, and flour are produced; other industries include vegetable-oil milling and oil-refining. Administratively independent of Jiangsu, Shanghai answers directly to the central government. 46 G4

Shannon longest river in Ireland, rising 105 m/344 ft above sea level in the Cuilcagh Mountains in County Cavan, and flowing 386 km/240 mi to the Atlantic Ocean past Athlone, and through loughs Allen, Boderg, Forbes, Ree, and Derg. The estuary, which is 110 km/68 mi long and 3–16 km/2–10 mi wide, forms the northern boundary of County Limerick. The river is navigable as far as Limerick city, above which are the rapids of Doonas and Castletroy. The river is known for its salmon farms, Castleconnell being an important centre. It also has the first and largest hydroelectric scheme in the Republic of Ireland (constructed 1925–29), with hydroelectric installations at and above Ardnacrusha, 5 km/3 mi north of Limerick. 24 D8

Shansi alternative transliteration of the Chinese province of ▶Shanxi. 11 (2)E3

Shantou (or **Swatow**), port and industrial city in Guangdong province, southeast China; population (1993) 610,000. It was opened as a special economic zone in 1979. Industries include food-processing, and the export-oriented manufacture of clothes, shoes, and electronic goods. Other exports include timber, food products, fish, and fruit. 46 F6

Shantung alternative transliteration of the Chinese province of ▶Shandong. 11 (2)E3

Shanxi (or **Shansi**), province of north China, bounded to the north by Inner Mongolia, to the east by Hebei, to the south by Henan, and to the west by Shaanxi; area 157,100 sq km/60,700 sq mi; population (1996) 31,090,000. The capital is ▶Taiyuan. There are coal, iron, machinery, mining equipment, and chemical industries. Fruit and cereals are grown, and meat is produced. 11 (2)E3

Sharjah (Arabic **Ash Shāriqah**), third-largest of the seven member states of the ▶United Arab Emirates, situated on the Gulf northeast of Dubai; area 2,600 sq km/1,004 sq mi; population (1995 est) 400,300. Since 1952 it has included the small state of Kalba. In 1974 oil was discovered offshore. Industries include ship repair, cement, paint, and metal products. 61 F4

Sharon, Plain of coastal plain in Israel stretching from Haifa to Tel Aviv-Yafo, and a sub-district of Central district; area 348 sq km/134 sq mi, length about 65 km/40 mi long. It was part of the ▶Fertile Crescent, the irrigated region of the Middle East stretching from Iraq to Egypt. Citrus fruits, cotton, and fodder crops are important, as are market gardening and vineyards. The Plain is now one of the most densely populated areas of Israel, with cities which include Netanya, Herzliyya, Hadera, and Kefar Sava.

Shasta, Mount dormant volcano in north California, USA. It is one of the highest peaks in the USA, rising to 4,317 m/14,162 ft in the ▶Cascade Range. Its last eruption occurred in 1786, but steam vents are visible on the slopes. Five glaciers descend from the volcano's snow-capped summit. Worn down by glaciation, Shasta now has three separate vents. The mount has a wide variety of alpine plants as well as virgin stands of the rare Shasta red fir. Shasta has been associated with the beliefs and myths both of ancient Californians and of modern occultists. 86 B2

Sheba ancient name for southern ▶Yemen (Sha'abijah). It was once renowned for gold and spices. According to the Old Testament, its queen visited Solomon; until 1975 the Ethiopian royal house traced its descent from their union. 56 D7

Sheboygan city, port, and administrative headquarters of Sheboygan County, east Wisconsin, USA, on Lake Michigan, 80 km/50 mi north of Milwaukee; population (1996 est) 50,000. It is the processing and distribution centre of a dairy farming area which specializes in cheese production. Iron and steel, furniture, plastics, and enamelware are manufactured. It is the home of Lakeland College (1862). 88 C2

Sheffield industrial city and metropolitan borough on the River Don, South Yorkshire, England; population of metropolitan district (1995) 528,500. From the 12th century, iron smelting was the chief industry, and by the 14th century, Sheffield cutlery, silverware, and plate were being made. During the Industrial Revolution the iron and steel industries developed rapidly. It now produces alloys and special steels, cutlery of all kinds, permanent magnets, drills, and precision tools. Other industries include electroplating, type-founding, and the manufacture of optical glass. It is an important conference centre. 24 L8

Shensi alternative transcription of the Chinese province of ▶Shaanxi. 11 (2)D3

Shenyang (formerly **Mukden**), capital of ▶Liaoning province, China; population (1994) 4,699,000. It is the region's main trading city, and one of China's principal metal-fabricating and machine-building centres. It was the capital of the Manchu emperors from 1625 to 1644; their tombs are nearby. 48 B3

Shenzhen special economic zone on the coast of Guangdong province, south China, established in 1980 opposite Hong Kong; population (1993) 2,400,000. A poor rural area in 1979, with a population of 20,000, it grew spectacularly with the relocation of toy, textiles, and electronics factories from Hong Kong. Diverse light industries have subsequently been introduced, particularly the manufacture of chemicals and electrical goods. It is also an international financial centre, housing one of China's two stock exchanges. The zone is fenced off, and immigration strictly controlled. 46 E6

Sheppey, Isle of island off the north coast of Kent, southeast England; area 80 sq km/31 sq mi. Situated at the mouth of the River Medway, it is linked with the mainland by Kingsferry Bridge (road and rail, completed in 1960) over the River Swale. The resort and port of Sheerness is here. There is a nature reserve in the southern part of the island.

Sherbrooke industrial city in southern Québec, Canada, at the junction of the Magog and St Francis rivers, 160 km/100 mi east of Montréal; population (1991) 76,400; metropolitan area population (1991) 139,200. Settled in the 1800s by Vermonters, it industrialized in the 1840s, and is now the commercial and industrial centre of an agricultural and mining region, and a transportation hub. It developed as a textile centre; today heavy machinery, pulp and paper, rubber and leather goods, clothing, and dairy goods are also produced. Most employment, however, is in commerce and service industries, including teaching, health care, and regional administration. Tourism is also important, and Sherbrooke is a resort centre with lakes and mountains nearby. 88 F1

Sherman town and administrative headquarters of Grayson County, northeast Texas, USA; population (1990) 31,600. It is located 88 km/55 mi north of Dallas, near Lake Texoma (the Red River), and 24 km/15 mi south of the Oklahoma state line. It is a processing and shipping centre for agricultural products; textiles, electronics, and machinery are manufactured. Austin College (Presbyterian) was founded in Sherman in 1849. 90 B3

's-Hertogenbosch (or **Den Bosch**; French **Bois-le-Duc**), capital of North Brabant province, the Netherlands, at the confluence of the Aa and Dommel rivers, 45 km/28 mi southeast of Utrecht; population (1997) 126,500. There is an important cattle market here. The Gothic cathedral, St John's, dates back to the 11th century, although it was rebuilt or extended in the 14th century. It was the birthplace of the painter Hieronymus Bosch. 22 H3

Sherwood Forest hilly stretch of parkland in west Nottinghamshire, central England; area about 520 sq km/200 sq mi. Formerly an ancient royal forest extending from Nottingham to Worksop, it is associated with the legendary outlaw Robin Hood. According to the Forestry Commission, Sherwood Forest is over 1,000 years old.

Shetland Islands (Old Norse **Hjaltland** 'high land' or 'Hjalte's land'), islands and unitary authority off the north coast of Scotland, 80 km/50 mi northeast of the Orkney Islands, an important centre of the North Sea oil industry, and the most northerly part of the UK.
area 1,452 sq km/560 sq mi
towns ►Lerwick (administrative headquarters), on Mainland, largest of 12 inhabited islands
physical the 100 islands are mostly bleak, hilly, and clad in moorland. The climate is moist, cool, and windy; in summer there is almost perpetual daylight, whilst winter days are very short. On clear winter nights, the aurora borealis ('northern lights') can frequently be seen in the sky
industries processed fish, handknits from Fair Isle and Unst, herring fishing, salmon farming, cattle and sheep farming; large oil and gas fields west of Shetland; Europe's largest oil port is Sullom Voe, Mainland; production at Foinaven oilfield, the first to be developed in Atlantic waters; tourism
population (1996) 22,500
24 M1/5 F1

Shihchiachuang alternative transliteration of ►Shijiazhuang, capital of Hebei province, China. 46 E3

Shijiazhuang (or **Shihchiachuang**), capital of ►Hebei province, China, at the foot of the Taihang Mountains; population (1993) 1,210,000. It is a major railway junction and agricultural distribution point. Industries include printing, light engineering, and the manufacture of chemicals, petrochemicals, and electronics. 46 E3

Shikoku smallest of the four main islands of Japan, south of Honshu, east of Kyushu; area 18,800 sq km/7,250 sq mi; population (1995) 4,183,000. The island consists of four prefectures, Kagawa, Tokushima, Ehime, and Kochi. The chief towns are Matsuyama and Takamatsu. The population is largely concentrated in the small coastal plains which front the Inland Sea. Products include rice, wheat, soybeans, sugar cane, orchard fruits, salt, and copper. 48 G7

Shillong capital of ►Meghalaya state, northeast India; population (1991) 222,000. Situated 1,593 m/5,230 ft above sea-level in the Khasi Hills, it trades in rice, cotton, and fruit, and is an important military base. Formerly the capital of Assam, it was a popular hill station under British rule. In 1897 an earthquake destroyed most of the town. 54 F3

Shimla (or **Simla**), capital of ►Himachal Pradesh state, India, 2,300 m/7,500 ft above sea level, population (1991) 110,300. It was the summer administrative capital of British India 1864–1947. 54 C2

Shimonoseki seaport in Yamaguchi prefecture, in the extreme southwest of Honshu island, Japan, on the Inland Sea; population (1994) 255,000. It was opened to foreign trade in 1890. Industries include fishing, shipbuilding, engineering, textiles, and chemicals. Shimonoseki is connected to Kyushu by three tunnels and a bridge. 48 F7

Shiraz ancient walled city of southern Iran, the capital of Fars province, on the highway from Tehran to Bushire; population (1991) 965,000. Grain, rice, pulses, tobacco, gum tragacanth, clarified butter, wine, wool, skins, and carpets are the main products, as well as cotton goods, glass, attar of roses, and inlaid craftwork. There are many mosques. The tombs of the poets Sa'di and Hafiz are on the outskirts of the city. 61 E2

Shizuoka (or **Shidzuoka**), capital of Shizuoka prefecture, Chubo region, Honshu island, Japan, 150 km/93 mi southwest of Tokyo; population (1994) 471,000. Industries include metal and food processing, and the manufacture of paper, plastics, and traditional goods, such as lacquerware, sewing boxes, and wooden clogs (*geta*). It is the centre of a tea and mandarin orange trade. Other agricultural products include strawberries and Japanese horseradish (*wasabi*). The castle was built in 1589. 48 K6

Shkodër (Italian **Scutari**), town on the River Bojana, northwest Albania, southeast of Lake Scutari, 19 km/12 mi from the Adriatic Sea; population (1991) 83,700. Products include woollens and cement. During World War I it was occupied by Austria 1916–18, and during World War II by Italy. 34 G7

Sholapur alternative spelling of ►Solapur, a town in Maharashtra state, India. 54 C5

Shreveport city and former river port in northwest Louisiana, USA, on the

Red River; population (1996 est) 191,600. Lying close to a rich oil-field, which also yields sulphur and natural gas, it has become the centre of a large refining and chemical industry complex. Cotton is processed from the surrounding agricultural area, and textiles, machinery, telephone equipment, vehicles, wood products, and glass are also manufactured. Founded in 1834, it was named after Henry Shreve, a riverboat captain who reached the site in 1833 and cleared a 256-km/160-mi logjam. Oil was discovered in 1906. The river has silted up, and is no longer navigable. 90 C3

Shrewsbury market town on the River Severn, Shropshire, England, 244 km/152 mi northwest of London; population (1991) 64,200. It is the administrative headquarters of Shropshire. There are service industries and light manufacturing, and tourism is important. To the east at Wroxeter is the site of the Roman city of **Viroconium**. 24 K9

Shropshire county of western England, which has contained the unitary authority of Telford and Wrekin since April 1998. Sometimes abbreviated to **Salop**, Shropshire was officially known by this name from 1974 until local protest reversed the decision in 1980.
area 3,490 sq km/1,347 sq mi
towns ►Shrewsbury (administrative headquarters), Ludlow, Oswestry
physical Shropshire is bisected, on the Welsh border, northwest–southeast by the River Severn; River Teme; Ellesmere (47 ha/116 acres), the largest of several lakes; the Clee Hills rise to about 610 m/1,800 ft (Brown Clee) in the southwest
features Ironbridge Gorge open-air museum of industrial archaeology, with the Iron Bridge (1779), the world's first cast-iron bridge; Market Drayton is famous for its gingerbread
agriculture cereals (barley, oats, wheat), sugar beet; vegetables (turnips, swedes, mangolds), sheep and cattle; dairy farming; forestry
industries brick-making; engineering; limestone; manufacturing: machine tools, agricultural implements, carpets, radio receivers, clocks; Shropshire is the principal iron-producing county of England
population (1996) 421,200 5 E5

Shumen (Bulgarian **Šumen**), industrial city in Varna region, eastern Bulgaria, 80 km/50 mi west of Varna; population (1990) 126,300. Industries include metal and leather goods, beer, wine, and food processing. The Tombul Mosque built 1744 is the largest in Bulgaria. Founded in the 10th century on the site of a Roman fort, Shumen was a fortress town successively under Bulgarian, Byzantine, Turkish, and Russian rule. 38 P6

Sialkot city in Punjab province, Pakistan, 120 km/75 mi north of Lahore; population (1981) 302,000. The city is an important centre of commerce, with good rail communications. Industries include the manufacture of surgical and sports goods, metalware, carpets, textiles, bicycles, and leather goods. 56 K3

Sian alternative transliteration of ►Xi'an, capital of Shaanxi province, China. 46 D4

Sibenik town in Croatia, on the Dalmatian coast, on the Gulf of Sibenik near the mouth of the River Krka; population (1991) 41,000. It is a naval base, has hydroelectric works, and exports bauxite, wood, wines, and marble. An old citadel overlooks the town, which has a 15th-century cathedral. 34 C6

Siberia (Russian **Sibir**), Asian region of Russia, extending from the Ural Mountains to the Pacific Ocean; area 12,050,000 sq km/4,650,000 sq mi. Hydroelectric power is generated from the rivers Lena, Ob, and Yenisey; forestry and agriculture are practised. There are vast mineral resources, including coal (in the ►Kuznetsk Basin), gold, diamonds, oil, natural gas, iron, copper, nickel, and cobalt. 40 N3

Sibir Russian name for ►Siberia, the Asian region of Russia. 40 N3

Sibiu (German **Hermannstadt**; Hungarian **Nagyszeben**), capital of Sibiu county, central Romania; population (1998 est) 171,400. The present city was founded by 12th century German colonists on the site of an earlier Roman settlement. It was the site of a battle during World War I when Romanian troops invaded Transylvania and were defeated by Austro-Hungarian forces. 34 M4

Sichuan (or **Szechwan**; 'four rivers'), province of central China, bounded to the north by Qinghai, Gansu, and Shaanxi; to the east by Hubei and Hunan; to the south by Guizhou and Yunnan; and to the west by Tibet; area 539,000 sq km/208,000 sq mi; population (1996) 84,300,000. The capital is ►Chengdu. There are coal, natural gas, iron ore, salt brine, textile, engineering, and electronics industries. Rice, wheat, and maize are grown. 11 (2)D3

Sicilia Italian name for ►Sicily, the largest island in the Mediterranean Sea. 32 G1/9 (4)E6

Sicily (Italian **Sicilia**), the largest Mediterranean island and an autonomous region of Italy, divided from the Italian mainland by the Strait of Messina; area 25,700 sq km/9,920 sq mi; population (1992) 4,997,700. It consists of nine provinces: Agrigento, Caltanissetta, Catania, Enna, Messina, Palermo, Ragusa, Syracuse, and Trapani; its capital is Palermo. Exports include Marsala wine, olives, citrus, refined oil and petrochemicals, pharmaceuticals, potash, asphalt, and marble. The region also incorporates the islands of ►Lipari, Egadi, Ustica, and ►Pantelleria. Etna, 3,323 m/10,906 ft high, is the highest volcano in Europe; its last major eruption was in 1993. 32 G11/9 (4)E6

Sidi-Bel-Abbès walled city in Algeria, 75 km/47 mi south of Oran; population (1998 est) 217,100. It trades in wheat, tobacco, and alfalfa. Because of its strategic position, it was the headquarters of the French Foreign Legion from 1842 until 1962, when the newly independent Algeria required the withdrawal of the French. 66 E1

Sidon ancient name for the port of ►Saida, Lebanon. 60 C3

Siedlce capital of Siedlce province, Poland, 90 km/56 mi east of Warsaw; population (1990) 72,000. It produces machinery, cement, glass, knitwear, and processed food, and is the chief trading centre for cattle and agricultural produce in the district. The town was founded in the 16th century. 18 M5

Siegen city in North Rhine-Westphalia, Germany, on the River Sieg, 93 km/58 mi southeast of Düsseldorf; population (1995) 111,300. It was the centre of an iron-mining district, and still makes special steels, and also has office equipment and computer manufacturers. It was once the seat of the Counts of Nassau-Siegen. The artist Peter Paul Rubens was born here. 22 L4

Siena (ancient **Saena Julia**), town in Tuscany, Italy, about 50 km/31 mi south of Florence; population (1990) 57,700. Founded by the Etruscans, it has medieval sculpture including works in the 13th-century unfinished Gothic cathedral by Niccolo Pisano and Donatello, and many examples of the Sienese school of painting that flourished from the 13th to the 16th centuries. The *Palio* ('banner', in reference to the prize) is a dramatic and dangerous horse race in the main square, held annually (2 July and 16 August) since the Middle Ages. 30 G7

Sierra Leone country in West Africa, on the Atlantic Ocean, bounded north and east by Guinea and southeast by Liberia.

Sierra Leone covers an area of 71,740 sq km/27,698 sq mi, and has a population of 4,854,000 (2000 est). The capital is ►Freetown; other major towns and cities include Koidu, Bo, Kenema, Makeni, and Marampa. The head of state and government is Ahmad Tejan Kabbah (from 1996). The currency is the leone, and the languages spoken include English (official), Krio (a Creole language), Mende, Limba, and Temne. 68 B3

Sierra Madre chief mountain system of Mexico, consisting of three ranges, the Sierra Madre Oriental, the Sierra Madre del Sur, and the Sierra Madre Occidental, enclosing the central plateau of the country; highest Citlaltepetl 5,700 m/18,700 ft. The Sierra Madre del Sur ('of the south') runs along the southwest Pacific coast. 94 F5

Sierra Nevada mountain range of southern Spain, mainly in the province of Granada, but also extending east into Almería. The highest point is Mulhacén (3,481 m/11,425 ft high). It has several winter sports resorts; the main centre is Sol y Nieve. 28 G7

Sierra Nevada mountain range in eastern California, USA, extending for about 640 km/400 mi, with a general ridge line at over 2,500 m/8,202 ft. Its highest point is Mount Whitney, which rises to 4,418 m/14,500 ft. The Sierra Nevada includes the King's Canyon, ►Yosemite, and Sequoia national parks. 92 B1

Sighişoara (German **Schässburg**; Hungarian **Segesvár**), town in Mureş county, central Romania, on the River Tirnava Mare, 72 km/45 mi northeast of Sibiu; population (1993) 40,000. The town was founded in the 13th century by German colonists. It has an old citadel, walls, and towers, and several medieval churches. 34 M3

Sigmaringen town in Baden-Württemberg, Germany, on the River Danube, 75 km/47 mi south of Stuttgart; population (1994) 16,700. It was once the capital of the duchy of Hohenzollern-Sigmaringen. The castle in Sigmaringen contains art collections. 30 E2

Si-Kiang alternative transliteration of ►Xi Jiang, a Chinese river. 46 E6

Sikkim (or **Denjong**), upland state of northeast India, bounded by Nepal to the west, Bhutan to the east, Tibet (China) to the north and West Bengal state to the south; area 7,100 sq km/2,741 sq mi; population (1994 est) 444,000. The capital is Gangtok. Industries include carpets, textiles, cigarettes, and food processing. Cardamom, tea, grain, fruit, and soybeans are grown, and livestock is raised. 10 E3

Silicon Valley nickname given to a region of southern California, USA, approximately 32 km/20 mi long, between Palo Alto and San Jose. It is the site of many high-technology electronic firms, whose prosperity is based on the silicon chip.

Silistra (Roman **Durostorum**; Turkish **Silistria**), border town in northeast Bulgaria, in Razgrad region, on the south bank of the River Danube, facing Romania, 110 km/69 mi from Varna; population (1990) 78,000. It is a river port with grain shipping. Founded as a Roman camp (Novae) in the 1st century AD, it became a major centre in the province of Moesia. The Turks captured and fortified the town during the early 15th century. 34 Q5

Sillein German name for ►Žilina, a town in the Slovak Republic. 18 H8

Simcoe, Lake lake in Ontario, Canada, between Lake Ontario and Georgian Bay; area 700 sq km/270 sq mi. It discharges into Lake Huron through the Severn River. It is a popular tourist destination. 88 E2

Simferopol (Ukrainian **Symferopil**), principal city of ►Crimea, in southern Ukraine and capital of the Crimean oblast (region); population (1990) 348,900. Simferopol is an important centre for commerce and transportation. Its industries include the manufacture of soap and tobacco, and agricultural products grown in the area are brought to market here. The city stands on the site of the Tatar town of **Ak-Mechet**, conquered by the Russians in 1783. 58 F1

Simplon Pass (Italian **Sempione**), Alpine pass Switzerland–Italy. The road was built by the French emperor Napoleon 1800–05; the Simplon Tunnel, built in 1906, is 19.8 km/12.3 mi, one of Europe's longest.

Simpson Desert desert area in Australia, chiefly in Northern Territory; area 145,000 sq km/56,000 sq mi. The desert was named after a president of the South Australian Geographical Society who financed its exploration. 76 G4

Sinai Egyptian peninsula, largely desert, at the head of the Red Sea; area 65,000 sq km/25,000 sq mi. Resources include oil, natural gas, manganese, and coal; irrigation water from the River Nile is carried under the Suez Canal. The main towns are Al-Arish (the capital of South Sinai governorate) and Al-Tur (capital of North Sinai governorate). It is the ancient source of turquoise. Tourism is of increasing importance. 64 F2

Sinai, Mount (or **Horeb**), mountain near the tip of the Sinai Peninsula; height 2,285 m/7,500 ft. According to the Old Testament this is where Moses received the Ten Commandments from God. Its identity is not absolutely certain, but it is traditionally thought to be Jebel Musa ('Mountain of Moses').

Sindelfingen town in Baden-Württemberg, Germany, 12 km/7 mi southwest of Stuttgart; population (1995) 59,400. The town produces electronic equipment, computers (IBM), watches, and textiles. The motor-vehicle manufacturer Daimler Benz has had a factory in Sindelfingen since 1915. 20 E8

Singapore (Sanskrit *Singa pura* 'city of the lion'), country in southeast Asia, off the tip of the Malay Peninsula.

Singapore covers an area of 622 sq km/240 sq mi, and has a population of 3,567,000 (2000 est). The capital is ►Singapore City. The head of state is

Sellapan Ramanathan Nathan (from 1999); the head of government is Goh Chok Tong (from 1990). The currency is the Singapore dollar, and the languages spoken include Malay, Mandarin Chinese, Tamil, English (all official), other Indian languages, and Chinese dialects. **52 C2**

Singapore City capital of Singapore, on the southeast coast of the island of Singapore; population (1993) 2,874,000. Major industries include trade, shipping, banking, electronics, shipbuilding, and oil refining. Formerly a British colonial town, it was occupied by Japanese forces during World War II. **52 C2**

Sining alternative transliteration of ▶Xining, capital of Qinghai province, China. **46 C3**

Sinop (or **Sinope**), ancient Black Sea port on the north coast of Asia Minor, founded as a Greek colony by Ionian settlers, probably in the late 7th century BC. It remained an important trading centre under successive Greek, Roman, and Byzantine rule. **58 F2**

Sinuiju (or **Siniju**), capital of North Pyongan province, near the mouth of the Yalu River, North Korea; population (1998 est) 349,500. The city lies on the Yellow Sea. It was founded in 1910, becoming important after the 1910 completion of a bridge over the Yalu linking it with Dandong in China. Sinuiju continues to be a major centre of rail transport between North Korea and China. Industries include paper, chemicals, alcohol distillation, and soya-bean processing. **48 C3**

Sion (German **Sitten**), capital of the Swiss canton of Valais, lying in the valley of the River Rhône southeast of Lake Geneva; population (1995) 26,100. It has three ruined castles, a 15th-century cathedral, and a 13th-century church. Sion is a market centre for the canton's vegetables, fruit, and wine. **30 C4**

Sioux City city and riverport in northwest Iowa, USA, at the junctions of the Big Sioux and Floyd rivers with the Missouri, near Iowa's border with Nebraska and South Dakota, 250 km/156 mi northwest of Des Moines; population (1996 est) 83,800. It is a distribution and processing centre with major livestock and grain markets, large stockyards, and an annual stock show. Industries include meat-packing, and the manufacture of dairy products, flour, feed, animal serums, fertilizer, fabricated metals, and computers. Sioux City is the head of navigation for the Missouri River. Settled in 1849 and named after the Sioux people, it expanded rapidly with the arrival of the railway in 1868. **88 A2**

Sioux Falls largest city in southeast South Dakota, USA, on the Big Sioux River; population (1996 est) 113,200. Hydroelectricity is generated at the falls, a 30 m/100 ft drop on the river. Founded in 1856, it is the distribution and processing centre of an extensive cattle-rearing and wheat-growing area. Industries include meat-preparation and packing, and the manufacture of electrical goods and agricultural machinery. It is also the location of EROS (Earth Resources Observation Systems), an international agency for the photography of Earth from space and the air. **88 A2**

Siracusa Italian name for ▶Syracuse, a port in Sicily. **32 K11**

Sirte, Gulf of (Arabic **Khalīj Surt**), gulf off the coast of Libya, on which the port of Benghazi stands. Access to the gulf waters has been a cause of dispute between Libya and the USA. **64 C1**

Sitka city and naval base on the west coast of Baranof Island in the Alexander Archipelago, facing Sitka Sound, in southeast Alaska, USA; population (1990) 8,600. It was the chief town of Russian America 1804–67. Sitka is a trading and commercial centre and has fishing and timber-processing industries, as well as being a popular tourist destination. **82 D5**

Sjælland (or **Zealand**), main island of Denmark, on which Copenhagen is situated; area 7,000 sq km/2,700 sq mi; population (1995) 2,157,700. It is low-lying with an irregular coastline. The chief industry is dairy farming. **16 F9**

Skagerrak arm of the North Sea which connects it to the Baltic Sea and passes between the south coast of Norway and Jutland on the north coast of Denmark by means of the ▶Kattegat. In May 1916 it was the scene of the inconclusive Battle of Jutland between the British and German fleets. **16 D8**

Skegness holiday resort on the North Sea coast of Lincolnshire, eastern England; population (1998) 16,400. It was the site of the first Butlin holiday camp in 1936. An annual carnival is held here in the summer. **22 C1**

Skellefteå town in Västerbotten county, northern Sweden, on the River Skellefte, 16 km/10 mi from the sea; population (1995) 75,300. The town has grown considerably since the establishment of the Boliden mines with their rich variety of ores, including gold. **16 L4**

Skikda trading port in Algeria, on the Gulf of Stora, 50 km/31 mi south of Constantine; population (1998 est) 180,200. Industries include petrochemicals, oil refining, the processing of fish and food, aluminium manufacture, and the smelting of iron ore. Exports include hydrocarbons, agricultural products, iron ore, marble, citrus fruit, vegetables, livestock, wheat, and wine. It was founded by the French in 1838 as Philippeville and renamed after independence in 1962. **66 G1**

Skipton industrial town in North Yorkshire, northern England, 32 km/20 mi northwest of Bradford; population (1991) 13,600. Industries include engineering and textiles. There is an 11th-century castle. **24 L8**

Skopje capital and industrial city of the Former Yugoslav Republic of Macedonia; population (1991) 563,300. Industries include iron, steel, chromium mining, and food processing. **34 J7**

Skye largest island of the Inner ▶Hebrides, Highland region, off the west coast of Scotland; area 1,740 sq km/672 sq mi; population (1991) 8,900. It is separated from the mainland to the southeast by the Sound of Sleat and by the islands of Raasay and Scalpay to the northeast. The chief port and town is Portree. The economy is based on crofting, craft industries, tourism, and livestock. The **Skye Bridge**, a privately financed toll bridge to Kyleakin on the island from the Kyle of Lochalsh, was completed in 1995. **24 F4**

Skyros (or **Skiros**), Greek island, the largest of the northern ▶Sporades; area 210 sq km/81 sq mi; population (1991) 2,900. It is known for its furniture and weaving. The English poet Rupert Brooke is buried here. **36 G6**

Slavonia region of eastern Croatia bounded by the Sava, Drava, and Danube rivers; Osijek is the largest town. Eastern and western Slavonia declared themselves autonomous provinces of Serbia following Croatia's declaration of independence from Yugoslavia 1991, and the region was the scene of fierce fighting between Croatian forces and Serb-dominated Yugoslav federal troops 1991–92. After the ceasefire 1992, 10,000 UN troops were deployed in eastern and western Slavonia and contested Krajina. Rebel Serbs in Croatia agreed November 1995 to return the region of eastern Slavonia to Croatian control.

Slavonski Brod town in Croatia, on the left bank of the River Sava; population (1991) 55,600. It is an important rail and road junction, has oil and engineering works, and trades in wine and cereals. It has a Turkish citadel. Across the river is the twin town of **Bosanski Brod**. 34 F4

Sligo county of the Republic of Ireland, in the province of Connacht, situated on the Atlantic coast of northwest Ireland; county town Sligo; area 1,800 sq km/695 sq mi; population (1996) 55,800. Limestone mountains rise behind a boggy coastal plain. There is some mineral wealth, including barytes, coal, lead, and copper. Agricultural activity includes cattle farming and dairy farming. The other principal town is Ballymote. 5 B4

Sliven town in Burgas region, eastern Bulgaria, on the southeast slopes of the Balkan Mountains, 64 km/40 mi northeast of Stara Zagora; population (1996 est) 107,300. It is an agricultural centre with textile, carpet, and local coalmining industries. **Sliven Spa** lies about 11 km/7 mi to the southeast. 34 P7

Slough industrial town and unitary authority in southern England, near Windsor, 32 km/20 west of London; it was part of the county of Berkshire to April 1998.
area 28 sq km/11 sq mi
features the home of astronomer William Herschel is now a museum; Slough Museum
industries pharmaceuticals, electronics, engineering, aviation support services, and the manufacture of chocolate, paint, and power tools; a trading estate was developed here in the 1920s, the first in England
population (1996) 105,000 22 B3/5 F6

Slovak Republic (or **Slovakia**), landlocked country in central Europe, bounded north by Poland, east by the Ukraine, south by Hungary, west by Austria, and northwest by the Czech Republic.

The Slovak Republic covers an area of 49,035 sq km/18,932 sq mi, and has a population of 5,387,000 (2000 est). The capital is ►Bratislava; other major towns and cities include Košice, Nitra, Prešov, Banská Bystrica, Žilina, Trnava, and Martin. The head of state is Rudolf Schuster (from 1999); the head of government is Mikulas Dzurinda (from 1998). The currency is the Slovak koruna, and the languages spoken include Slovak (official), Hungarian, Czech, and other ethnic languages. 18 H9

Slovenia (or **Slovenija**), country in south-central Europe, bounded north by Austria, east by Hungary, west by Italy, and south by Croatia.

Slovenia covers an area of 20,251 sq km/7,818 sq mi, and has a population of 1,986,000 (2000 est). The capital is ►Ljubljana; other major towns and cities include Maribor, Kranj, Celji, Velenje, Koper, and Novo Mesto. The head of state is Milan Kučan (from 1990); the head of government is Janez Drnovšek (from 2000). The currency is the tolar, and the languages spoken include Slovene (related to Serbo-Croat; official), Hungarian, and Italian. 30 K4

Słupsk (German **Stolp**), capital of Słupsk province, Poland, on the River Słupia, 61 km/38 mi northeast of Koszalin; population (1990) 101,200. It produces timber, footwear, furniture, chemicals, and has engineering industries. Until 1945 it was part of German Pomerania. 18 G3

Smolensk capital city, commercial and cultural centre of Smolensk oblast (region), Russian Federation, on the River Dnieper near the border with Ukraine; population (1996 est) 356,000. Its manufacturing industries include textiles, foodstuffs, and precision instruments, and it is an important transportation centre. The city has many outstanding churches and other architectural monuments from the 12th to 19th centuries. 38 F4

Snake (formerly **Lewis**), river in northwest USA, the largest tributary of the Columbia River; 1,670 km/1,038 mi in length. Rising as the South Fork in a lake 2,375 m/7,790 ft above sea-level in the ►Yellowstone National Park, Wyoming, it flows southwest through Idaho, turning north to form the border of Idaho between Oregon and then Washington, and joining the Columbia near Pasco. Its course travels 65 km/40 mi through the 2,400-m/7,874 ft-deep Hell's Canyon, one of the deepest gorges in the world. 86 C1

Sneek town in the province of Friesland, the Netherlands, 21 km/13 mi southwest of Leeuwarden; population (1996) 29,900. It has a butter and cheese market and tobacco manufacturing. Formerly an important shipbuilding centre, Sneek is now a centre of the booming pleasure-boat industry. Large regattas are held each August on the **Sneekermeer** to the east of the town. 22 H1

Snowdon (Welsh **Eryri**), highest mountain in Wales, 1,085 m/3,560 ft above sea level. Situated 16 km/10 mi southeast of the Menai Strait, it consists of a cluster of five peaks. At the foot of Snowdon are the Llanberis, Aberglaslyn, and Rhyd-ddu passes. A rack railway ascends to the summit from Llanberis. Snowdonia, the surrounding mountain range, was made a national park in 1951. It covers 2,188 sq km/845 sq mi of mountain, lakes, and forest land. 24 H8

Sochi town in the Krasnodar krai (territory) of the southwestern Russian Federation; population (1996 est) 356,000. Sochi is located 175 km/109 mi southeast of Krasnodar on the northeast coast of the Black Sea, at the foot of the main Caucasian mountain range. Its favourable location (incorporating sea bathing and sulphur springs) and its subtropical climate have made it a popular health resort. Sochi has expanded along the Black Sea coast for a distance of nearly 40 km/25 mi. 58 H2

Society Islands (French **Archipel de la Société**), archipelago in ►French Polynesia, divided into the Windward Islands and the Leeward Islands; area 1,685 sq km/650 sq mi; population (1995 est) 178,000. The administrative headquarters is Papeete on ►Tahiti. The **Windward Islands** (French **Iles du Vent**) have an area of

1,200 sq km/460 sq mi and a population (1995 est) of 151,000. They comprise Tahiti, Moorea (area 132 sq km/51 sq mi), Maio (or Tubuai Manu; 9 sq km/3.5 sq mi), and the smaller Tetiaroa and Mehetia. The **Leeward Islands** (French **Iles sous le Vent**) have an area of 404 sq km/156 sq mi and a population of 27,000 (1995 est). They comprise the volcanic islands of Raiatea (including the main town of Uturoa), Huahine, Bora-Bora, Maupiti, Tahaa, and four small atolls. The islands were named after the Royal Society by Captain Cook, who visited them in 1769. Claimed by France in 1768, the group became a French protectorate in 1843, and a colony in 1880. Products include copra, phosphates, mother-of-pearl, and vanilla. 74 L7

Socotra (or **Suquṭrā**), Yemeni island in the Indian Ocean; area 3,500 sq km/1,351 sq mi. Myrrh, frankincense, incense, aloes, and butter (ghee) are produced, and cattle and goats are reared. Under British protection from 1886, it became part of South Yemen in 1967. Its principal town is Hadiboh. 56 F7

Söderhamn seaport in Gävleborg county, Sweden, near the mouth of the River Ljusne; population (1990) 13,300. There are sawmills, wood-pulp and planing works and ironworks. The town has a large timber trade. 16 J6

Södertälje town in the county of Stockholm, eastern Sweden; population (1995) 82,400. It is a flourishing industrial town just west of the capital, with pharmaceutical, heavy vehicle construction, and engineering plants. **Södertälje Canal** connects Lake Mälaren with the Baltic. 16 J7

Soest city in North Rhine-Westphalia, Germany, 50 km/31 mi east of Dortmund; population (1994) 294,800. Soest was an important member of the Hanseatic League medieval trade federation. Sugar refining is the principal industry. The town has several medieval churches, and is a tourist centre. 22 L3

Sofia (or **Sofiya**), capital of Bulgaria since 1878; population (1991) 1,221,000. Industries include textiles, rubber, machinery, and electrical equipment. It lies at the foot of the Vitosha Mountains. 34 L7

Sofiya alternative spelling of ►Sofia, the capital of Bulgaria. 34 L7

Sognefjorden longest and deepest fjord in ►Norway, stretching 205 km/127 mi from inland Skjolden to its mouth on the west coast at Solund; it is 1,245 m/4,080 ft deep. 16 C6

Soissons market town in the *département* of Aisne, in the Picardy region of northern France, 90 km/56 mi northeast of Paris; population (1999 est) 29,400, conurbation 46,100. The chief industry is metallurgy. In 486 the Frankish king Clovis declared the town the first French capital after defeating the Gallo-Romans here, ending their rule in France. 22 F5

Sokoto trading centre and capital of Sokoto state, northwestern Nigeria; population (1992 est) 185,500. It trades in cement, wood, leather, and textiles. The palace of the sultan of the Fula Sultanate and the Shehu Mosque are here. 68 F2

Solapur (or **Sholapur**), industrial town in Maharashtra state, India, 340 km/210 mi southeast of Mumbai (formerly Bombay); population (1991 est) 604,000. Linked by railway to both Mumbai and Chennai (formerly Madras), it is a major centre for cotton processing. Industries include textiles, leather goods, and chemicals. 54 C5

Solent, the channel between the coast of Hampshire, southern England, and the Isle of ►Wight. It is a yachting centre. 22 A4

Solingen city in North Rhine-Westphalia, Germany, 10 km/6 mi southeast of Düsseldorf; population (1995) 165,700. It was once a major producer of swords and today makes high-quality steel for razor blades, scissors, and cutlery. In the Middle Ages it was known for its sword blades. Spanning the Wupper Valley to the east is the Müngsten Bridge, the highest railway bridge in Germany (107 m/351 ft high). 22 K3

Solomon Islands country in the southwest Pacific Ocean, east of New Guinea, comprising many hundreds of islands, the largest of which is Guadalcanal.

The Solomon Islands covers an area of 27,600 sq km/10,656 sq mi, and has a population of 444,000 (2000 est). The capital (and chief port) is ►Honiara (on Guadalcanal island); other major towns and cities include Gizo, Auki, Kirakira, and Buala. The head of state is Queen Elizabeth II (from 1952), represented by Governor General Moses Pitakaka (from 1994); the head of government is Mannesseh Sogavare (from 2000). The currency is the Solomon Island dollar, and the languages spoken include English (official), pidgin English, more than 80 Melanesian dialects, and Papuan and Polynesian languages. 74 F6

Solothurn (French **Soleure**), canton in northwest Switzerland; area 791 sq km/305 sq mi; population (1997) 239,300. The canton stretches from Basel to Bern and is traversed by the foothills of the Jura Mountains. It is largely agricultural, but also has shoe, paper, cement, textile, watch, and engineering industries. 8 (1)B1

Solothurn capital of the Swiss canton of the same name, on the River Aar, 30 km/19 mi north of Bern; population (1995) 15,400. It is an important railway junction; industries include electrical engineering and watchmaking. 30 C3

Solway Firth inlet of the Irish Sea, formed by the estuaries of the rivers Eden and Esk, at the western end of the border between England and Scotland, separating Cumbria in England from Dumfries and Galloway in Scotland. Solway Firth is in part the estuary of the River Esk, and in part an inlet of the Irish Sea. 24 J7

Somalia country in northeast Africa (the Horn of Africa), on the Indian Ocean, bounded northwest by Djibouti, west by Ethiopia, and southwest by Kenya.

Somalia covers an area of 637,700 sq km/246,215 sq mi, and has a population of 10,097,000 (2000 est). The capital (and chief port) is ►Mogadishu; other major towns and cities include Hargeysa, Berbera, Kismayo, and Marka. The head of state is Abdiqasim Salad Hassan (from 2000); the head of government is Ali Khalifa Galaid (from 2000). The currency is the Somali shilling, and the languages spoken include Somali, Arabic (both official), Italian, and English. 70 H2

Somaliland region of Somali-speaking peoples in eastern Africa including the former British Somaliland Protectorate (established 1887) and Italian Somaliland (made a colony 1927, conquered by Britain

1941, and administered by Britain until 1950) – which both became independent in 1960 as the Somali Democratic Republic, the official name for ►Somalia – and former French Somaliland, which was established in 1888, became known as the Territory of the Afars and Issas in 1967, and became independent as ►Djibouti in 1977.

Sombor town in Serbia, Yugoslavia, in the autonomous region of Vojvodina; population (1991) 48,700. It is the centre of a rich agricultural district. 34 G4

Somerset county of southwest England.
area 3,460 sq km/1,336 sq mi
towns ►Taunton (administrative headquarters); Bridgwater, Frome, Glastonbury, Wells, Yeovil; Burnham-on-Sea, Minehead (coastal resorts)
physical rivers Avon, Axe, Brue, Exe, Parret (the principal river), and Yeo; marshy coastline on the Bristol Channel; Mendip Hills; Quantock Hills; Exmoor; Blackdown Hills
features Cheddar Gorge and Wookey Hole, a series of limestone caves; Glastonbury Tor
agriculture apples; dairy farming; cereals (wheat, barley, oats), vegetables (turnips, mangolds); cider; cattle and sheep rearing; willows (withies) for wickerwork
industries agricultural implements; Bath-bricks from the sand of the Parret); chemicals; dairy products (including Cheddar cheese); engineering; food processing; helicopters; leather; mineral working (iron, lead, zinc); textiles; tourism
population (1996) 482,600 5 E6

Somme *département* in the ►Picardy region of France, with a short coastline on the English Channel; area 6,175 sq km/2,384 sq mi; population (1990) 549,100. It is crossed east to west by the River ►Somme. In general the land is level, a great plateau. Somme is an important agricultural area and produces cereals, sugar beet, apples, and livestock. The principal industries are textiles and chemicals, sugar refining, and distilling. The administrative centre is ►Amiens. Other towns include ►Abbeville, Montdidier, and Péronne. 6 D1

Somme river in northern France, on which Amiens and Abbeville stand; length 245 km/152 mi. It rises in Aisne *département* near St Quentin and flows west through ►Somme *département* to the English Channel near St Valéry-sur-Somme. It is connected by canal with the Oise and the Schelde (French Escaut). Its tributaries include the rivers Ancre and Avre.

Some of the heaviest fighting of World War I took place on the banks of the Somme, especially in July–November 1916. 22 E4

Song Hông Vietnamese name for the ►Red River, a river in north Vietnam. 50 C2

Songhua (or **Sungari**), river in northeast China, which joins the Amur River on the Russian frontier; length 2,308 km/1,433 mi. 46 H1

Soochow alternative transliteration of the Chinese city of ►Suzhou. 46 G4

Sopot (German **Zoppot**), seaside resort and spa in Gdańsk province, Poland, 11 km/7 mi north of Gdańsk; population (1990) 46,700. 16 K9

Sopron (German **Ödenburg**; Roman **Scarabantia**), town in Györ-Sopron county, Hungary, near the Austrian border between the Neusiedler See and the eastern Alps, 77 km/48 mi west of Györ; population (1993) 56,100. Wine, textiles, and chemicals are produced. One of the best-preserved historic towns in Hungary, it attracts tourists. 30 M3

Sorau German name for ►Zary, a town in Poland. 18 E6

Soria city and capital of the province of Soria, north central Spain, on the River Duero; population (1991) 32,200. Products include food and timber. Soria has a Renaissance palace, several old churches, and a museum of objects from nearby Numantia. 28 H3

Sørøya Norwegian island in the Norwegian Sea, off the northwest coast of Norway, 17 km/10.5 mi west of Hammerfest, 450 km/280 mi north of the Arctic Circle; area 816 sq km/315 sq mi. The discovery of carved bas-relief images of reindeer, elk, bears, whales, birds, humans, and boats, believed to date from 7000 to 5000 BC, makes this the world's most northerly collection of rock art. 16 L1

Sorrento town in Campania, southwest Italy, on the Gulf of Naples; population (1990) 17,500. It has been a holiday resort since Roman times. Products include wine, citrus fruit, fish, and ceramics. 32 J8

Sosnowiec chief city of the Darowa coal region in the Upper Silesian province of Katowice, southern Poland; population (1990) 259,300. 18 J7

Sŏul alternative spelling of ►Seoul, the capital of South Korea. 48 D5

Sour (or **Tyre**), town in southwest Lebanon, about 80 km/50 mi south of Beirut, formerly a port until its harbour silted up; population (1991 est) 70,000. It stands on the site of the ancient city of the same name, a seaport of Phoenicia. 60 C3

Sousse port and commercial centre in Tunisia, 160 km/100 mi south of Tunis; population (1994) 125,000. Industries include fishing, especially sardines, textile milling, olive oil manufacture, and food canning. It was founded by the Phoenicians and flourished under the Carthaginians and Romans. There are Roman ruins. 66 H1

South Africa country on the southern tip of Africa, bounded north by Namibia, Botswana, and Zimbabwe and northeast by Mozambique and Swaziland.

South Africa covers an area of 1,222,081 sq km/471,845 sq mi, and has a population of 40,377,000 (2000 est). The capitals are ►Cape Town (legislative), ►Pretoria (administrative), and ►Bloemfontein (judicial); other major towns and cities include Johannesburg, Durban, Port Elizabeth, Vereeniging, Pietermaritzburg, Kimberley, Soweto, and Tembisa. The head of state and government is Thabo Mbeki (from 1999). The currency is the rand, and the languages spoken include English, Afrikaans, Xhosa, Zulu, Sesotho (all official), and other African languages. 72 C6

South America fourth largest of the continents, nearly twice as large as Europe (13% of the world's land surface), extending south from ►Central America.
area 17,864,000 sq km/6,897,000 sq mi
largest cities (population over 3.5 million) Buenos Aires, São Paulo, Rio de Janeiro, Bogotá, Santiago, Lima, Belo Horizonte
features Lake Titicaca (the world's highest navigable lake); La Paz (highest capital city in the world);

Atacama Desert; Inca ruins at Machu Picchu; rivers include the Amazon (world's largest and second longest), Paraná, Madeira, São Francisco, Purús, Paraguay, Orinoco, Araguaia, Negro, Uruguay
physical occupying the southern part of the landmass of the Western hemisphere, the South American continent stretches from Point Gallinas on the Caribbean coast of Colombia to Cape Horn at the southern tip of Horn Island, which lies adjacent to Tierra del Fuego; the most southerly point on the mainland is Cape Froward on the Brunswick peninsula, southern Chile; at its maximum width (5,120 km/3,200 mi) the continent stretches from Point Pariñas, Peru, in the extreme west to Point Coqueiros, just north of Recife, Brazil, in the east; five-sixths of the continent lies in the southern hemisphere and two-thirds within the tropics
population (1996 est) 323 million. The urban population has increased rapidly since 1950, as millions of poor people have left the countryside in the hope of a better standard of living in the cities. By 1996 about 75% of the population was living in cities
language Spanish, Portuguese (chief language in Brazil), Dutch (Suriname), French (French Guiana), American Indian languages; Hindi, Javanese, and Chinese spoken by descendants of Asian immigrants to Suriname and Guyana; a variety of Creole dialects spoken by those of African descent
religion 90–95% Roman Catholic; local animist beliefs among Amerindians; Hindu and Muslim religions predominate among the descendants of Asian immigrants in Suriname and Guyana **96**

Southampton industrial city, seaport, and unitary authority in southern England, at the head of Southampton Water, 20 km/12 mi southwest of Winchester; it was part of the county of Hampshire to 1997.
area 52 sq km/20 sq mi
features Southampton University, (1952); ferry link to the Isle of Wight; the port is a base for many liners, including P&O's *Oriana*; Southampton City Art Gallery; parts of the medieval town wall survive, including four of the town-wall towers, and Bargate, the elaborate old north gateway to the city; Norman House and Canute's Palace are among the oldest examples of Norman domestic architecture in Britain; a memorial column marks the place of embarkation of the *Mayflower* (the ship in which the Pilgrim Fathers sailed to America)
industries marine engineering, chemicals, plastics, flour-milling, tobacco, and financial services. It is a major passenger and container port.
population (1996) 207,100 **22** A4/5 F6

South Australia state of south-central Australia, including Kangaroo Island and other islands in the Indian Ocean; bounded on the northeast by Queensland, on the east by New South Wales, on the southeast by Victoria, on the south by the Indian Ocean, and on the west by Western Australia; area 984,381 sq km/ 380,071 sq mi; population (1996) 1,428,000. The capital (and chief port) is ►Adelaide. Products are meat, wool, wine, wheat, barley, almonds, oranges and other citrus fruits, and dried and canned fruit, coal, copper, uranium, silver, zinc, gold, steel, jade, slate, opals, marble, granite, household and electrical goods, vehicles, oil, and natural gas. **76** F5

South Ayrshire unitary authority in southwest Scotland, created in 1996 from Kyle and Carrick district (1975–96), Strathclyde region.
area 1,245 sq km/480 sq mi
towns ►Ayr (administrative headquarters), Prestwick, Girvan, Troon, Maybole
physical coastal plain which rises to higher ground inland (500 m/ 1,640 ft); rivers Ayr, Stinchar, Water of Girvan; Brown Carrick Hill (287 m/942 ft); Ailsa Craig; many beaches interspersed with cliffs and caves
features Glasgow Prestwick Airport; Culzean Castle; Crossraguel Abbey; Royal Troon and Turnberry championship golf courses
industries aerospace, high technology, tourism
agriculture fishing (Ayr), dairying, beef cattle, potatoes
population (1996) 114,000 **5** D4

South Bend city and administrative headquarters of St Joseph County, north Indiana, USA, on the St Joseph River, 120 km/75 mi east of Chicago; population (1996 est) 102,000. It is the commercial and processing centre of an agricultural area. Industries include the manufacture of farm machinery, cars, aircraft equipment, and food products. The Roman Catholic University of Notre Dame (1842) and St Mary's College lie on the outskirts. **88** C2

South Carolina state in southeastern USA. It is nicknamed the Palmetto State. South Carolina ratified the US Constitution in 1788, becoming the 8th state to join the Union. Part of the Deep South, it is bordered to the north and east by North Carolina, to the west and south by Georgia, and to the southeast by the Atlantic Ocean. South Carolina was one of the original US plantation states, associated with slavery.
population (2000 est) 4,012,000
area 80,600 sq km/31,112 sq mi
capital ►Columbia
towns and cities Charleston, North Charleston, Greenville, Spartanburg
industries and products tobacco, soybeans, lumber, textiles, clothing, paper, wood pulp, chemicals, non-electrical machinery, primary and fabricated metals, tourism, leisure industry **90** E3

South China Sea see ►China Sea. **50** E4

South Dakota state in western USA. It is nicknamed the Coyote or Sunshine State. South Dakota was admitted to the Union in 1889 as the 40th US state. It is bordered to the north by North Dakota, to the west by Montana and Wyoming, to the south by Nebraska, and to the east by Minnesota and Iowa.
population (2000 est) 754,800
area 199,800 sq km/77,150 sq mi
capital ►Pierre
towns and cities Sioux Falls, Rapid City, Aberdeen
industries and products cereals, hay, livestock, gold (second-largest US producer), meat products, tourism **86** F2

South Downs line of chalk hills in southeast England, running from near Petersfield, Hampshire, across Sussex to the south coast at Beachy Head near Eastbourne. They face the North Downs across the Weald and are used as sheep pasture. **22** B4

South East Cape southernmost point of Australia, in Tasmania. **76** J8

Southend resort and unitary authority in eastern England, on the Thames estuary, 60 km/37 mi east of

London, the nearest seaside resort to London; it was part of the county of Essex to April 1998.
area 42 sq km/16 sq mi
features a pier, 2 km/1.25 mi long, said to be the longest in the world; 11 km/7 mi of seafront; nearly a third of all land in the area is managed for nature conservation, including Belfairs Wood Nature Reserve and Leigh National Nature Reserve on Two Tree Island
industries tourism, financial services, light engineering, and boatbuilding
population (1996) 171,000 5 G6

Southern Alps range of mountains running the entire length of South Island, New Zealand. They are forested to the west, with scanty scrub to the east. The highest peaks are Aoraki, 3,764 m/12,349 ft, and Mount Tasman 3,498m/11,476 ft. Scenic features include gorges, glaciers, lakes, and waterfalls. Among its lakes are those at the southern end of the range: Manapouri, Te Anau, and the largest, Wakatipu, 83 km/52 mi long, which lies about 300 m/1,000 ft above sea level and has a depth of 378 m/1,242 ft. The Fiordland National Park also lies in the south of the range. 78 B6

Southern Uplands one of the three geographical divisions of Scotland, being most of the hilly Scottish borderland to the south of a geological fault line that stretches from Dunbar, East Lothian, on the North Sea to Girvan, South Ayrshire, on the Firth of Clyde. The Southern Uplands, largely formed by rocks of the Silurian and Ordovician age, are intersected by the broad valleys of the Nith and Tweed rivers. 24 H6

South Georgia island in the South Atlantic, a British crown colony administered, with the South Sandwich Islands, from the Falkland Islands by a commissioner; area 3,757 sq km/1,450 sq mi. The average temperature on the island is −2°C/28.4°F.

There has been no permanent population since the whaling station was abandoned in 1966. South Georgia lies 1,300 km/800 mi southeast of the Falkland Islands, of which it was a dependency until 1985. The British Antarctic Survey has a station on nearby Bird Island. 100 P9

South Glamorgan (Welsh **De Morgannwg**), former county of south Wales, 1974–1996, now divided between ►Cardiff and ►Vale of Glamorgan unitary authorities.

South Gloucestershire unitary authority in southwest England created in 1996 from part of the former county of Avon.
area 497 sq km/192 sq mi
towns and cities Thornbury (administrative headquarters), Patchway, Yate, Chipping Sodbury
features River Severn borders northwest; Vale of Berkeley; Severn Road Bridge; Marshfield has one of Britain's longest village streets with 17th-century almshouses; 13th-century church of St Peter (Dyrham); late 17th century Dyrham Park Mansion
industries agriculture and associated industries
population (1996) 220,000 5 E6

South Holland (Dutch **Zuid Holland**), low-lying coastal province of the Netherlands, bounded to the north by North Holland, to the east by Utrecht and North Brabant, to the south by Zeeland, and to the west by the North Sea; area 2,910 sq km/1,124 sq mi; population (1997) 3,344,700. The capital is The ►Hague. There are chemical, textile, distilling, and petroleum refining industries. Bulbs are grown, and there is horticulture, livestock raising, and dairying. 7 (1)C3

South Island larger of the two main islands of ►New Zealand. 78 B6

South Korea country in East Asia, bounded north by North Korea, east by the Sea of Japan, south by the Korea Strait, and west by the Yellow Sea.

South Korea covers an area of 98,799 sq km/38,146 sq mi, and has a population of 46,844,000 (2000 est). The capital is ►Seoul; other major towns and cities include Pusan, Taegu, Inchon, Kwangju, Taejon, and Songnam. The head of state is Kim Dae Jung (from 1998); the head of government is Lee Han Dong (from 2000). The currency is the won, and the main language spoken is Korean (official). 48 D5

South Lanarkshire unitary authority in south central Scotland, created in 1996 from three districts of Strathclyde region.
area 1,772 sq km/684 sq mi
towns Hamilton (administrative headquarters), Lanark, Rutherglen, East Kilbride, Carluke, Cambuslang
physical rural to the south and urban to the north. The River Clyde flows through the area. Tinto (707 m/2,320 ft) is a key landmark to the south
features Craignethan Castle; Carstairs State Hospital, New Lanark
industries textiles, electronics, engineering
agriculture fruit cultivation in the valleys of the Clyde; grazing and stock rearing in the upland south; dairying
population (1996) 307,100 5 E4

South Orkney Islands group of barren, uninhabited islands in British Antarctic Territory, southeast of Cape Horn; area 622 sq km/240 sq mi.

They were discovered by the naval explorer Captain George Powell (1795–1824) and the US Captain Nathaniel Palmer (1799–1877) in 1821. 102 (2)A3

South Platte river in Colorado and Nebraska, USA; length 711 km/422 mi. It rises near Mount Lincoln on the eastern side of the Rocky Mountains in central Colorado, and flows through Denver and then northwards to Nebraska, where it unites with the North Platte and forms the Platte River. It provides irrigation water for the northern part of Colorado. Reservoirs and dams have been created for power, irrigation, and flood control. 86 F2

South Pole the southern point where an imaginary line penetrates the Earth's surface by the axis about which it revolves; see also ►Antarctica. 102 (2)

South Sandwich Islands actively volcanic uninhabited British Dependent Territory in the South Atlantic; area 337 sq km/130 sq mi. Along with ►South Georgia, 750 km/470 mi to the northwest, it is administered from the Falkland Islands. They were claimed by Captain Cook in 1775 and named after his patron John Montagu, the 4th Earl of Sandwich. The islands were annexed by the UK in 1908 and 1917. They were first formally claimed by Argentina in 1948. In December 1976, 50 Argentine 'scientists' landed on Southern Thule and were removed in June 1982. There is an ice-free port off Cumberland Bay. Over 21 million penguins breed on Zavadovski Island. 102 (2)C4

South Shetland Islands archipelago of 12 uninhabited islands in the South Atlantic, forming part of the British Antarctic Territory; area 4,622 sq km/ 1,785 sq mi. The islands were reached in 1819 by William Smith. 102 (2)MM4

South Shields manufacturing port in Tyne and Wear, northeast England, on the south side of the Tyne estuary opposite North Shields and east of Gateshead; population (1991) 82,400. Shipbuilding has declined and industries now include electrical goods, cables, chemicals, and paint. 24 L7

South Uist second-largest island in the Outer ►Hebrides, Western Isles, Scotland, separated from North Uist by the island of Benbecula. The main town and port with connections to the mainland is Lochboisdale. 24 E4

South West Africa former name (to 1968) of ►Namibia. 72 B4

South Yorkshire metropolitan county of northeast England, created in 1974; in 1986, most of the functions of the former county council were transferred to the metropolitan borough councils.
area 1,560 sq km/602 sq mi
towns Barnsley, Doncaster, Rotherham, Sheffield (all administrative centres for the districts of the same name)
physical River Don; part of Peak District National Park; the county contains a rich diversity of rural landscapes between the barren Pennine moors in the southwest and the very low, flat carr-lands (a mixture of marsh and copses) in the east
features the Earth Centre for Environmental Research
agriculture sheep; dairy and arable farming
industries metal-work, coal, engineering, iron, and steel
population (1996) 1,304,800 5 F5

Sovetsk town in the Kaliningrad region of the western Russian Federation. Sovetsk lies on the River Niemen (Neman), 90 km/56 mi northeast of Kaliningrad and near the Lithuanian border. Timber and paper production are the main industries. Until 1945, Sovetsk was in East Prussia, when it was known by the German name of **Tilsit**. Napoleon signed peace treaties with Prussia and Russia here in 1807, ending the War of the Third Coalition. 16 L9

Soviet Union alternative name for the former Union of Soviet Socialist Republics (USSR).

Soweto (acronym for SOuth WEst TOwnship), urban settlement in South Africa, southwest of Johannesburg; population (1991) 597,000. It experienced civil unrest during the apartheid regime. Industries include wood pulp and paper manufacturing. 72 D5

Spa town in Liège province, Belgium, 27 km/17 mi southeast of Liège in the Ardennes; population (1997) 10,400. It has been known for its mineral springs since the 14th century, and has given its name to similar health resorts elsewhere. 22 H4

Spain country in southwestern Europe, on the Iberian Peninsula between the Atlantic Ocean and the Mediterranean Sea, bounded north by France and west by Portugal.

Spain covers an area of 504,750 sq km/194,883 sq mi (including the Balearic and Canary islands), and has a population of 39,630,000 (2000 est). The capital is ►Madrid; other major towns and cities include Barcelona, Valencia, Zaragoza, Seville, Málaga, Bilbao, Las Palmas (on Gran Canarias island), Murcia, and Palma (on Mallorca). The head of state is King Juan Carlos I (from 1975); the head of government is José Maria Aznar (from 1996). The currency is the peseta, and the languages spoken include Spanish (Castilian; official), Basque, Catalan, and Galician. 28 F5

Spalato Italian name for ►Split, a port in Croatia. 34 D6

Spalding market town in the Lincolnshire fens, eastern England, on both banks of the River Welland, 23 km/14 mi southwest of Boston; population (1991) 18,700. It is an agricultural and horticultural centre, with bulb farms. Flower-growing fields occupy an area of over 4,000 ha/ 10,000 acres. Industries include the manufacture of agricultural implements and tractors. A flower festival is held annually here in May. 22 B2

Spanish Guinea former name of the Republic of ►Equatorial Guinea. 68 F4

Spanish Sahara former name for ►Western Sahara. 66 C4

Spanish Town town in Middlesex county, Jamaica; population (1991) 92,400. Founded by Diego Columbus about 1525, it was the capital of Jamaica 1535–1871. 94 J5

Spartanburg city and administrative headquarters of Spartanburg County, northwest South Carolina, USA, in the foothills of the Blue Ridge Mountains, 145 km/90 mi northwest of Columbia; population (1996 est) 42,100. It is the commercial and shipping centre for an agricultural and lumbering area. Industries include timber-milling, and the manufacture of furniture, cotton textiles, paper, ceramics, and plumbing supplies. The city is home to Wofford College (1854), Converse College (1889), and several other schools. 90 E3

Spey second longest river in Scotland. It flows through Highland and Moray, rising 14 km/8 mi southeast of Fort Augustus, for 172 km/107 mi to the Moray Firth between Lossiemouth and Buckie. It has salmon fisheries at its mouth. 24 J4

Speyer (English **Spires**), ancient cathedral city in Rhineland-Palatinate, Germany, at the confluence of the Rhine and the Speyer, 26 km/16 mi south of Mannheim; population (1995) 49,600. Industries include electrical engineering, metallurgy, oil refining, and aircraft engineering.
It was at the **Diet of Spires** in 1529 that Protestantism received its name. 22 L5

Spice Islands former name of ►Maluku, a group of islands in the Malay Archipelago.

Spires English name for the German city of ►Speyer. 22 L5

Spitsbergen mountainous island with a deeply indented coastline, situated in the Arctic Ocean between Franz Josef Land and Greenland. It is the main island in the Norwegian archipelago of ►Svalbard, 657 km/ 408 mi north of Norway, and now owned by that country; area 39,043 sq km/15,075 sq mi. Fishing, hunting, and coal mining are the chief economic activities. The Norwegian Polar Research Institute operates an all-year scientific station on the west coast. The highest point is Newtontoppen, which rises to 1,713 m/5,620 ft.
The island was formerly called West Spitsbergen when part of the Svalbard archipeligo was named Spitsbergen. 102 (1)P2

Split (Italian **Spalato**), port in Croatia, on the Adriatic coast; population (1991) 189,400. Industries include engineering, cement, and textiles. Split was bombed during 1991 as part of Yugoslavia's blockade of the Croatian coast. **34** D6

Spokane city in eastern Washington, USA, on the **Spokane River**, 370 km/ 230 mi east of Seattle; population (1996 est) 186,600. It is a commercial, processing, and distribution centre for the mining and timber-producing mountain areas and rich agricultural plains of the Columbia basin. Aluminium smelting plants were developed during World War II. **86** C1

Spoleto (ancient **Spoletium**), town in Umbria, central Italy, on a hill 50 km/31 southeast of Perugia; population (1990) 38,000. There is an annual opera and drama festival (June–July) established by Gian Carlo Menotti. The town was a papal possession (1220–1860) and has Roman remains and medieval churches. **32** G6

Sporades Greek island group in the Aegean Sea. The chief island of the **Northern Sporades** is ►Skyros. The **Southern Sporades** are more usually referred to as the ►Dodecanese. **36** F5

Spratly Islands (Chinese **Nanshan Islands**), disputed group of small islands, coral reefs, and sandbars dispersed over a distance of 965 km/ 600 mi in the South China Sea. The islands are of strategic importance, commanding the sea passage from Japan to Singapore, and in 1976 oil was discovered. **50** E4

Spree river in eastern Germany; length 365 km/226 mi. The Spree rises on the borders of Bohemia and the *Land* (administrative region) of Saxony, and after flowing generally north and northwest joins the River Havel to the west of Berlin, at Spandau. On its course, the Spree passes through Bautzen, Cottbus, and Berlin. A marshy area in its valley near Cottbus is called the **Spreewald**. **20** K4

Springfield capital of Illinois, USA, 300 km/190 mi southwest of Chicago; population (1996 est) 112,920. It is the financial, commercial, distribution, and service centre of a rich agricultural area. Manufactured products include processed foods, machinery, metalware, building materials, and electronic goods. Springfield was settled in 1818 and became the state capital in 1837. President Abraham Lincoln was buried here in 1865. Coal used to be extensively mined in the vicinity of the city. **88** C3

Springfield industrial city in western Massachusetts, USA, on the Connecticut River, at the southern end of Pioneer Valley, 144 km/90 mi southwest of Boston; population (1996 est) 150,000. It has many light industries, including telecommunications. Settled in 1636, it grew rapidly as an arsenal from 1777, and became the site (1794–1968) of the US arsenal and armoury, known for the Springfield rifle. Basketball originated in the city in 1891, invented by James Naismith, who later founded the National Basketball Hall of Fame. **88** F2

Springfield city in Missouri, USA, in the northern foothills of the Ozark Mountains, 210 km/130 mi south of Kansas City; population (1996 est) 143,410. A manufacturing and agricultural centre, it is also a tourist gateway for the Ozark area, part of which is called the **Springfield Plateau**. Industries include food-processing, paper-making, and the manufacture of electronic equipment and rubber. **88** B3

Springfield town in Lake County, west Oregon, USA; population (1990) 45,000. It is situated 6 km/4 mi east of Eugene city centre, between the Willamette and McKenzie rivers. Founded in 1849, the town is a major centre for Oregon's forest products industry, with a large Weyerhaeuser pulp and paper mill and plywood and hardboard plant. Its other manufactures include metal products, chemicals, trailers, fabricated steel, and animal feeds. The city is also a centre for a fertile agricultural region and has facilities for the research and development of industrial glues and adhesives. It serves as a gateway to the McKenzie River Valley recreation area. **86** B2

Springs city in Gauteng Province, South Africa, 40 km/25 mi east of Johannesburg; population (1991) 157,700 (district). It is a mining centre, producing gold, coal, and uranium. Industries include the manufacture of alloy steels and non-ferrous products, paper making, cycles, sheet glass, electrical goods, and mining machinery. **72** D5

Srbija Serbo-Croat for ►Serbia, a constituent republic of Yugoslavia. **34** H6

Sri Lanka island in the Indian Ocean, off the southeast coast of India.

Sri Lanka covers an area of 65,610 sq km/25,332 sq mi, and has a population of 18,827,000 (2000 est). The capital is Sri Jayewardenapura Kotte; other major towns and cities include Colombo, Kandy, Dehiwala-Mount Lavinia, Moratuwa, Jaffna, Kotte, and Galle. The head of state is Chandrika Bandaranaike Kumaratunga (from 1994); the head of government is Ratnasiri Wickremanayake (from 2000). The currency is the Sri Lankan rupee, and the languages spoken include Sinhala, Tamil (both official), and English. **54** D7

Srinagar summer capital of the state of ►Jammu and Kashmir, India, on the Jhelum River, 1,600 m/5,250 ft above sea level; population (1991) 595,000. It is a beautiful resort, intersected by waterways and wooden bridges, with numerous mosques, palaces, a fortress, and public gardens. Lake Dal to the northeast is an important tourist centre. It has carpet, papier mâché, silk, silver, and leather industries. The university of Jammu and Kashmir was established in 1948. **54** B2

Stade city in Lower Saxony, Germany, 140 km/87 mi northwest of Hannover; population (1994) 180,400. There are oil-refining, nuclear-power, chemical, and aluminium industries in this industrial growth area. The town was formerly a member of the medieval Hanseatic League trade federation. **16** E10

Staffa uninhabited island in the Inner Hebrides, west of Mull. It has a rugged coastline and many caves, including ►Fingal's Cave. **24** F5

Staffordshire county of west central England (since April 1997 Stoke-on-Trent has been a separate unitary authority).
area 2,720 sq km/1,050 sq mi
towns Stafford (administrative headquarters), Newcastle-under-Lyme, Lichfield, Tamworth, Leek, Uttoxeter
physical largely flat, with hilly regions in the north (part of the Peak district) and southwest; River Trent and its

tributaries; Cannock Chase (a large open area in the middle of the county)
features castles at Chartley, Tamworth, and Tutbury; Lichfield Cathedral; Keele University (1962); Shugborough Hall (17th century)
agriculture dairy farming
industries breweries (Burton-upon-Trent); the Potteries (producing Wedgwood and Royal Doulton china); tractors and agricultural equipment; tyres (including Michelin); electrical; engineering; electronics
population (1996) 555,700 **5** E5

Staffs abbreviation for ►Staffordshire, an English county. **5** E5

Stalin former name (1949–56) of the port of ►Varna, Bulgaria. **58** B2

Stalingrad former name (1925–61) of the Russian city of ►Volgograd. **38** H5

Stalinsk former name (1932–61) of the Russian city of ►Novokuznetsk. **42** R7

Stalowa Wola industrial town in Rzeszów province, southeast Poland, on the River San, 61 km/38 mi north of Rzeszów; population (1994 est) 71,400. Its steel mill was built in 1938 as part of a government policy to establish industrialization away from Poland's western frontier with Germany. The town has three electricity-generating stations. **18** M7

Stamford city in Fairfield County, southwest Connecticut, USA, on Long Island Sound, 56 km/35 mi northeast of New York City; population (1996 est) 110,100. It is a commercial and industrial centre, and the site of numerous corporate headquarters. Products include computers, hardware, rubber, plastics, and pharmaceuticals. Easy access to New York by rail and the Connecticut Turnpike has made the city a dormitory for the metropolis. **88** F2

Stanley town on eastern Falkland, capital of the ►Falkland Islands; population (1991) 1,557. After changing its name only once between 1843 and 1982, it was renamed five times in the space of six weeks during the Falklands War April–June 1982. **100** K9

Stanleyville former name (to 1966) of the port of ►Kisangani in the Democratic Republic of Congo. **70** D3

Stara Planina Bulgarian name for the ►Balkan Mountains in central Europe. **36** F1

Stara Zagora town in Khaskovo region of central Bulgaria, south of the Shipka Pass through the Balkan Mountains; population (1990) 187,500. The chief industries are the production of chemicals and tobacco, and food processing. The town lies in a rose-growing area. Founded by the Romans, it became a major market centre under the Turks, and was ceded to Bulgaria 1877. **34** N7

Stargard (or **Stargard Szczeciński**), town in Szczecin province, Poland, on the River Ina, 36 km/22 mi east of Szczecin; population (1990) 71,000. Founded in the 12th century, it became a member of the medieval Hanseatic League trading confederation. It remained in German Pomerania until 1945; from 1648 as part of Brandenburg under the Treaty of Westphalia. **16** H10

Staten Island island in New York harbour, part of New York City, USA, comprising the county of Richmond and, since 1975, the borough of Staten Island; area 155 sq km/60 sq mi.

Stavanger seaport and capital of Rogaland county, southwest Norway, 160 km/100 mi south of Bergen; population (1994) 104,000. It is Norway's fourth-largest city, and since 1969 has been the centre of the country's North Sea oil exploration and oil rig construction, with many multinational oil companies operating from here. It is also the centre of a sheep farming area. There is an important shipyard, and much shipping is registered in Stavanger. There was formerly a large fish-canning industry with more than 100 factories, but this has been in decline since the 1960s. **16** C7

Stavropol (formerly **Voroshilovsk** 1935–44), capital city, economic and cultural centre of Stavropol krai (territory) in the southwestern Russian Federation; population (1996 est) 344,000. Stavropol is located in the foothills of northwestern ►Caucasia, at the centre of a major agricultural region. Heavy industry in the city produces automobiles, building materials, and construction equipment. There are also food, wool, and leather industries here. An important natural-gas pipeline runs from here to Moscow. **58** J1

Steiermark German name for ►Styria, a province of Austria. **8** (2)F3

Stendal town in the *Land* (administrative region) of ►Saxony-Anhalt, Germany, 56 km/35 mi northeast of Magdeburg; population (1995) 45,500. There are metal, chemical, and food industries, and it is an important railway junction. In the second half of the 13th century Stendal was the capital of the area known as the Mark Brandenburg (Brandenburg Marches). The cathedral, begun in 1188, was rebuilt in the 14th century in the Gothic style. **20** G4

Stettin German name for the Polish city of ►Szczecin. **18** D4

Stettiner Haff German name for ►Zalew Szczeciński, a lagoon in Poland and northeast Germany. **16** G10

Stettin Lagoon English name for ►Zalew Szczeciński, a lagoon in Poland and northeast Germany. **16** G10

Stevenage town in Hertfordshire, southeast England, 45 km/28 mi north of London; population (1991) 75,000. Industries include the manufacture of electrical components, pharmaceuticals, and management services. The town dates from medieval times. In 1946 Stevenage was the first place in England to be designated a new town (to accommodate population overspill). **22** B3

Stewart Island volcanic island divided from South Island, New Zealand, by the Foveaux Strait; area 1,750 sq km/676 sq mi; population (1996) 700. Its highest point is Mount Anglem, 975 m/3,200 ft. Industries include farming, fishing, oyster-catching, granite quarrying, and tourism. Oban is the main settlement. **78** A8

Stirling unitary authority in central Scotland, created in 1996 from Stirling district, Central region.
area 2,196 sq km/848 sq mi
towns Dunblane, ►Stirling (administrative headquarters), Aberfoyle
physical mountainous to the north, including the forested Trossachs, and the open moorland north and west of Breadalbane, within the flood plain of the River Forth to the south around Sterling. The area contains many famous Scottish lochs (Tay, Katrine,

Lomond) and Scotland's only lake (Lake of Menteith). Peaks include Ben More (1,174 m/3,852 ft) and Ben Venue (727 m/2,385 ft)
features Bannockburn Heritage Centre; Stirling Castle
industries tourism, light engineering
agriculture forestry and stock rearing in the uplands. The lowlands include some of the richest agricultural lands in Scotland (Carse of Gowrie)
population (1996) 82,000 5 D3

Stirling administrative headquarters of Stirling unitary authority, Scotland, on the River Forth, 43 km/27 mi northeast of Glasgow; population (1991) 30,500. Industries include the manufacture of agricultural machinery, textiles, chemicals, and carpets. The Stirling skyline is noted for its castle, which guarded a key crossing of the river, and the (William) Wallace Monument, erected in 1870 to commemorate the Scots' victory of the English at nearby **Stirling Bridge** in 1297. Edward I of England (in raising a Scottish siege of the town) went into battle at Bannockburn in 1314 and was defeated by Robert I (the Bruce), in the Scots' greatest victory over the English. 24 J5

Stockholm capital and industrial port of Sweden; population (1994 est) 703,600. It is built on a number of islands. Industries include engineering, brewing, electrical goods, paper, textiles, and pottery. 16 K7

Stockport town in Greater Manchester, northwest England, 10 km/6 mi southeast of Manchester; population (1991) 130,800. The rivers Tame and Goyt join here to form the Mersey. Formerly important in the textile industry, Stockport now manufactures electrical machinery, paper, plastics, hats, and some cotton textiles; other industries include electronics, chemicals, and engineering. 24 K8

Stockton city and river port in California, USA, on the San Joaquin River, 115 km/72 mi southeast of San Francisco; population (1996 est) 232,700. It is a trading and distribution centre for fruit, grain, and cattle from the rich farmlands of the San Joaquin Valley. Industries include food-processing, and the manufacture of wood products and agricultural machinery. Several military bases are sited nearby. Stockton was founded as Tuleburg in 1847 as a supply centre for the California gold rush, and renamed in 1850. A deep-water channel to San Francisco Bay was completed in 1933. 92 B1

Stockton-on-Tees unitary authority in northeast England created in 1996 from part of the former county of Cleveland.
area 200 sq km/77 sq mi
towns and cities ▶Stockton-on-Tees (administrative headquarters), Billingham, Yarm, Longnewton
features River Tees forms east border; Tees Barrage; Yarm viaduct; Preston Hall Museum and Park (Stockton); Castlegate Quay (Stockton) includes full-scale replica of *HMS Endeavour*
industries chemicals, polythene film, light and heavy engineering, insulation products, plastics, electronics
population (1996) 176,600 5 F4

Stockton-on-Tees town, port, and administrative headquarters of ▶Stockton-on-Tees metropolitan borough, on the River Tees, 5 km/3 mi west of Middlesbrough, northeast England; population (1991) 82,400. There are ship-repairing, steel, and chemical industries. It was the starting point for the Stockton–Darlington railway, the world's first passenger railway, which opened in 1825. 24 L7

Stoke-on-Trent city and unitary authority in central England, on the River Trent, 23 km/14 mi north of Stafford; it was part of the county of Staffordshire to 1997.
area 93 sq km/36 sq mi
features the Gladstone Pottery Museum is a working pottery museum
industries it is the heart of the Potteries, a major ceramic centre, and the largest clayware producer in the world (Minton, Wedgwood, Spode, and Royal Doulton). Other industries include the manufacture of steel, chemicals, engineering machinery, paper, rubber, and tyres (Michelin)
population (1996) 254,200
history Stoke was formed in 1910 from Burslem, Hanley, Longton, Stoke-upon-Trent, Fenton, and Tunstall 24 K8/5 E5

Stolp German name for ▶Słupsk, a town and province in Poland. 18 G3

Stornoway air and sea port on the island of Lewis in the Outer ▶Hebrides, Scotland; population (1991) 6,000. It is the main administrative and shopping centre for the Western Isles unitary authority. The economy is based on fishing, tourism, and tweeds. Stornoway was founded by James VI of Scotland (James I of England). 24 F3

Strait of Dover stretch of water separating England from France, see ▶Dover, Strait of. 22 D4

Straits Settlements former province of the East India Company 1826–58, a British crown colony 1867–1946; it comprised Singapore, Malacca, Penang, Cocos Islands, Christmas Island, and Labuan.

Stranraer port in Dumfries and Galloway, Scotland, on Loch Ryan; population (1991) 11,300. There is a ferry service to Belfast in Northern Ireland, and a summer service to the Isle of Man. 24 H7

Strasbourg (German **Strassburg**), administrative centre of the ▶Bas-Rhin *département* and of ▶Alsace region, northeast France, situated near the German border on the River Ill, 3 km/1.9 mi west of the Rhine near its confluence with the Rhine–Rhône and Rhine and Marne canals; population (1990) 255,900, conurbation 388,000. Industries include car manufacture, tobacco, printing and publishing, and preserves. The town was selected as the headquarters for the Council of Europe in 1949, and sessions of the European Parliament alternate between here and Luxembourg. It has an 11th–15th-century cathedral. 30 C2

Stratford city and river port in southwest Ontario, Canada, on the Avon River; population (1991) 27,700. It is a railway centre, with six lines of the Canadian National Railway and repair shops. Industries include food-processing, and the manufacture of furniture, agricultural implements, textiles, and woollen goods. Since 1953 Stratford has held a major festival of Shakespearean and other drama, which now takes place annually from May to October. 88 D2

Stratford-upon-Avon market town on the River Avon, in Warwickshire, England, 35 km/22 mi southeast of Birmingham; population (1991) 22,200. It is the birthplace of William Shakespeare and has the Royal Shakespeare Theatre (1932), the Swan Theatre, and The Other Place. Stratford receives over 2 million tourists a year. Industries include canning, aluminium ware, and boat building. 22 A2

Strathclyde former region of Scotland (1975–96), which consisted of 19 districts and was replaced in 1996 by the 12 unitary authorities of Argyll and Bute, Renfrewshire, East Ayrshire, East Dunbartonshire, East Renfrewshire, Glasgow City, Inverclyde, North Ayrshire, North Lanarkshire, South Ayrshire, South Lanarkshire, and West Dunbartonshire.

Stromboli Italian island in the Tyrrhenian Sea, one of the ►Lipari Islands; area 12 sq km/5 sq mi. It has an active volcano, 926 m/3,039 ft high. The island produces Malmsey wine and capers, partly as a result of the fertile soil the lava has produced. 32 K10

Strumica town and district in the southeast of the Former Yugoslav Republic of Macedonia, in the valley of the River Struma; population (1991 est) 90,000. It produces tobacco and other agricultural products. 36 E3

Stuttgart capital of Baden-Württemberg, on the River Neckar, Germany; population (1995) 587,000. Industries include the manufacture of vehicles, electronics, mechanical and electrical engineering, precision instruments, foodstuffs, textiles, papermaking and publishing; it is a fruit-growing and wine-producing centre. There are two universities. Stuttgart was founded in the 10th century. 30 E2

Styria (German **Steiermark**), Alpine state of southeast Austria, bordered on the east by Hungary and on the south by Slovenia; area 16,400 sq km/6,330 sq mi; population (1994) 1,202,700. Its capital is Graz, and its industries include iron, steel, vehicles, electrical goods, electronics, computers, and engineering. A large proportion of the population is engaged in agriculture. An independent state from 1056 until it passed to the Habsburgs in the 13th century, it was annexed by Germany in 1938. 8 (2)F3

Subotica largest town in Vojvodina, northwestern Serbia, Yugoslavia; population (1991) 100,200. Industries include chemicals and electrical machinery. 34 G3

Suceava capital of Suceava county, northern Romania; population (1993) 116,000. Industries include textiles and lumber. It was a former centre of pilgrimage and capital of Moldavia 1388–1564. 34 P2

Sucre legal capital and seat of the judiciary of Bolivia, also capital of Chuquisaca department; population (1992) 131,000. It stands on the central plateau in the Andes at an altitude of 2,840 m/9,320 ft. It is the commercial centre for the surrounding agricultural area and has an oil refinery. 98 D7

Sudan country in northeast Africa, bounded north by Egypt, northeast by the Red Sea, east by Ethiopia and Eritrea, south by Kenya, Uganda, and Congo (formerly Zaire), west by the Central African Republic and Chad, and northwest by Libya. It is the largest country in Africa.

Sudan covers an area of 2,505,800 sq km/967,489 sq mi, and has a population of 29,490,000 (2000 est). The capital is ►Khartoum; other major towns and cities include Omdurman, Port Sudan, Juba, Wadi Medani, El Obeid, Kassala, al-Qadarif, and Nyala. The head of state and government is Gen Omar Hassan Ahmed al-Bashir (from 1989). The currency is the Sudanese dinar, and the languages spoken include Arabic (official) and 100 local languages. 64 E5

Sudbury city in northwestern Ontario, Canada, on Ramsey Lake, 482 km/301 mi northwest of Toronto; population (1991) 92,900, metropolitan area (1996)166,200. It lies on the Canadian Pacific and Canadian National railways, and is the main commercial centre of northern Ontario. A buried meteorite here yields 90% of the world's nickel. Large deposits of copper and platinum, as well as gold, are also mined. Industries include ore-processing, brewing, and the manufacture of chemicals, building materials, fabricated metal, textiles, and food products. 88 D1

Suez (Arabic **El Suweis**), port at the Red Sea terminus of the Suez Canal, 120 km/75 mi east of Cairo, Egypt; population (1994) 458,000. Industries include oil refining and the manufacture of fertilizers. Port Ibrahim, 3 km/1.8 mi south of Suez, lies at the entrance to the canal. It was reconstructed in 1979 after the Arab-Israeli Wars. 64 F2

Suez Canal artificial waterway from Port Said to Suez, linking the Mediterranean and Red Seas; 160 km/100 mi long. It separates Africa from Asia and provides the shortest eastwards sea route from Europe. It was opened in 1869, nationalized in 1956, blocked by Egypt during the Arab-Israeli War in 1967, and not reopened until 1975. 64 F1

Suffolk county of eastern England.
area 3,800 sq km/1,467 sq mi
towns ►Ipswich (administrative headquarters), Aldeburgh, Beccles, Bury St Edmunds, Felixstowe, Lowestoft, Sudbury, Southwold
physical undulating lowlands in the south and west; flat coastline; rivers Waveney (the boundary with Norfolk), Alde, Deben, Orwell, Stour (the boundary with Essex), Little Ouse; part of the ►Norfolk Broads
features Minsmere and Sandlings nature reserves (heathlands and birds); bloodstock rearing and horse racing at Newmarket; Sutton Hoo (7th-century ship burial); Sizewell B, Britain's first pressurized-water nuclear reactor plant; Aldeburgh Festival
agriculture cereals (barley, oats, wheat), sugar beet; cattle, sheep, and pig rearing; fishing
industries agricultural machinery; chemicals; coconut matting; electronics; fertilizers; food processing; motor vehicle components; North Sea oil and gas; printing; silk; timber; brewing
population (1996) 661,600 5 G5

Sukkur (or **Sakhar**), port in Sind province, Pakistan, on the River Indus, opposite the town of Rohri; population (1991) 350,000. The nearby Sukkur River–Lloyd Barrage (1928–32) uses waters from the Indus to irrigate up to 2 million ha/4.9 million acres. Sukkur is linked by rail with ►Quetta via the Bolan and Nari passes. 56 J4

Sulawesi (formerly **Celebes**), island in eastern Indonesia, one of the Sunda Islands; area (with dependent islands) 190,000 sq km/73,000 sq mi; population (1995 est) 13,756,000. This mountainous and forested island lies off the coast of eastern Borneo. Minerals such as copra, nickel, coal, asphalt, mica, sulphur and salt are among the products. 53 A3

Sulu Archipelago group of about 870 islands off southwest Mindanao in the Philippines, between the Sulawesi and Sulu seas; area 2,700 sq km/ 1,042 sq mi; population (1990 est) 698,200. The capital is Jolo, on the island (the largest) of the same name. Until 1940 the islands were an autonomous sultanate. **50** G5

Sumatera alternative name for ►Sumatra, the second-largest island of Indonesia. **52** C2

Sumatra (or **Sumatera**), second-largest island of Indonesia, one of the Sunda Islands; area 473,600 sq km/ 182,800 sq mi, length 1,760 km/ 1,094 mi, width 400 km/250 mi; population (1995 est) 40,343,000. About a third of the area, mainly in the southeast, is permanently waterlogged. The highest part is in the west where, at Gunung Kerinci, the Bukit Barisan volcanic mountain range reaches 3,805 m/12,483 ft. East of the range is a wide plain; both are heavily forested. Products include rubber, rice, tobacco, tea, timber, tin, petroleum, bauxite, gold, natural gas, coffee and pepper. **52** C2

Sun City holiday and casino resort in North West Province, South Africa, 115 km/71 mi northwest of Johannesburg. Sun City is on the southern edge of the Pilanesberg Game Reserve, and has a major entertainment centre. **72** D5

Sunda Islands islands west of Maluku (Moluccas), in the Malay Archipelago, the greater number belonging to Indonesia. They are so named because they lie largely on the Indonesian extension of the Sunda continental shelf. The **Greater Sundas** include Borneo, Java (including the small island of Madura), Sumatra, Sulawesi, and Belitung. The **Lesser Sundas** (Indonesian *Nusa Tenggara*) are all Indonesian and include Bali, Lombok, Flores, Sumba, Sumbawa, and Timor.

Sunderland city and port in Tyne and Wear, northeast England, at the mouth of the River Wear; population (1991) 183,200. A former coalmining and shipbuilding centre, Sunderland now has electronics, engineering, and brewing industries, and manufactures glass, pottery, chemicals, paper, furniture, and cars. It also has some tourism. **24** L7

Sundsvall port in eastern Sweden; population (1992) 94,300. It has oil, timber, and wood-pulp industries. **16** J5

Sunshine Coast chain of sandy beaches on the coast of Queensland, Australia, stretching for about 100 km/ 60 mi from Bribie Island, north of Brisbane, to Rainbow Beach. It includes the resorts of Noosa Heads and Caloundra.

Superior, Lake largest and deepest of the ►Great Lakes and the largest freshwater lake in the world; area 82,100 sq km/31,700 sq mi. Extending east–west for 616 km/385 mi, it reaches a maximum width of 260 km/ 163 mi and depth of 407 m/1,335 ft. The lake is bordered by the Canadian province of Ontario and the US states of Minnesota, Wisconsin, and Michigan. As the westernmost of the Great Lakes, Superior is at the western end of the ►St Lawrence Seaway. **88** C1

Suquṭrā alternative name for ►Socotra, a Yemeni island in the Indian Ocean. **56** F7

Surat city and former seaport in Gujarat, west India, 25 km/15 mi from the mouth of the Tapti River; population (1991) 1,499,000. The chief industry is textiles. The main port of the Mogul empire in the 16th and 17th centuries, the first East India Company trading post in India was established here 1612. The town declined as a port until the cotton boom of the 1860s, and its railway junctions revived its importance. **54** B4

Suriname (or **Surinam**), country on the north coast of South America, bounded west by French Guiana, south by Brazil, east by Guyana, and north by the Atlantic Ocean.

Suriname covers an area of 163,820 sq km/63,250 sq mi, and has a population of 417,000 (2000 est). The capital is ►Paramaribo; other major towns and cities include Nieuw Nickerie, Moengo, Brokopondo, Nieuw Amsterdam, Albina, and Groningen. The head of state is Ronald Venetiaan (from 2000); the head of government is Jules Ajodhia (from 2000). The currency is the Suriname guilder, and the languages spoken include Dutch (official), Spanish, Sranan (Creole), English, Hindi, Javanese, Chinese, and various tribal languages. **98** F3

Surrey county of southern England.
area 1,660 sq km/641 sq mi
towns ►Kingston upon Thames (administrative headquarters), Farnham, Guildford, Leatherhead, Reigate, Woking, Epsom, Dorking
physical rivers Mole, Thames, and Wey; Box Hill (183 m/600 ft), Gibbet Hill (277 m/909 ft), and Leith Hill (299 m/981 ft, 5 km/3 mi south of Dorking, the highest hill in southeast England); North Downs
features Kew Palace and Royal Botanic Gardens, Kew; Yehudi Menuhin School (one of four specialist music schools in England)
agriculture vegetables; sheep rearing; dairy farming; horticulture
industries service industries; sand and gravel quarrying; fuller's earth extraction (near Reigate)
population (1996) 1,047,100 **5** F6

Surtsey volcanic island 20 km/12 mi southwest of Heimaey in the Westman Islands (►Vestmannaeyjar) of Iceland. The island was created by an underwater volcanic eruption in November 1963, and is formed largely of ash with some lava flows. The island is the subject of intense scientific study, especially of its colonization by flora and fauna. Landing is strictly controlled. **16** (1)C3

Sussex former county of England, on the south coast, now divided into ►East Sussex and ►West Sussex.

Sutlej river in India and Pakistan, forming the border between the two countries for some of its length, a tributary of the River ►Indus; length 1,370 km/851 mi. It is one of the five rivers which give the Punjab its name. It rises in Tibet, crosses the Himalayas to flow east–southwest across the Punjab; the Beas flows into it, and it joins with the ►Chenab before flowing into the Indus 1,500 km/900 mi from its source. Since the 1960 Indus Waters Agreement, its waters have been allocated to India: the Bhakra Nangal scheme irrigates via 1,050 km/ 650 mi of canals some 700,000 ha/ 1.7 million acres in the Punjab, Haryana, and Rajasthan, and generates 604,000 kW at Bhakra; the Sutlej is also linked with the Ravi and Beas rivers for irrigation purposes. **54** B3

Sutton Coldfield residential part of the West Midlands conurbation around ►Birmingham, central England; population (1991) 103,900.

Suva capital and chief port of Fiji Islands, on the southeast coast of the island of Viti Levu; population (1994) 90,000. It is a commercial and light industrial centre; industries include coconut-oil processing, boatbuilding, and the manufacture of soap, biscuits, and cigarettes. Suva has road and air links to the international airport at Nadi, in the northwest of Viti Levu. **74** H7

Suzhou (or **Soochow**; formerly **Wuhsien** 1912–49), city in Jiangsu province, China, south of the Chang Jiang River delta and east of the ▶Grand Canal; population (1994) 1,050,000. Dating from about 1000 BC, it is popularly known as the 'Venice of the East' because of its network of ancient bridges and canals. Traditional silk, embroidery, and other handicrafts have been augmented by papermaking and the production of chemicals, electronics, and telecommunications equipment. **46** G4

Svalbard Norwegian archipelago in the Arctic Ocean; population (1995) 2,900 (41% being Norwegian). The main island is ▶Spitsbergen, which includes the largest town, Longyearbyen; other islands include Edgeøya, Barentsøya, Svenskøya, Nordaustlandet, Prins Karls Foreland, Wilhelmøya, Lågøya, Storøya, Danskøya, and Sørkappøya. The other main centres of population are the Russian mining settlements of Barentsburg and Grumantbyen. The total land area is 62,000 sq km/ 23,938 sq mi. **102** (1)Q2

Sverdlovsk former name (1924–91) of the Russian town of ▶Yekaterinburg. **38** M3

Swansea unitary authority in south Wales, created in 1996 from part of the former county of West Glamorgan.
area 377 sq km/146 sq mi
towns ▶Swansea (administrative headquarters)
physical River Tawe
features ▶Gower Peninsula (an area of outstanding natural beauty)
industries tinplate manufacture, chemicals, oil refineries.
population (1996 est) 231,200 **5** D6

Swansea (Welsh **Abertawe**), port and administrative centre of ▶Swansea unitary authority, south Wales, at the mouth of the River Tawe 70 km/43 mi west of Cardiff; population (1996 est) 231,200. It is the second-largest city in Wales. It has oil refineries, chemicals, metallurgical industries, and tin plate manufacturing, and has produced stained glass since 1936. **24** J10

Swatow another name for the Chinese port of ▶Shantou. **46** F6

Swaziland country in southeast Africa, bounded east by Mozambique and southeast, south, west, and north by South Africa.

Swaziland covers an area of 17,400 sq km/6,718 sq mi, and has a population of 1,008,000 (2000 est). The capitals are ▶Mbabane (administrative) and Lobamba (legislative); other major towns and cities include Manzini, Big Bend, Mhlume, Havelock Mine, and Nhlangano. The head of state is King Mswati III (from 1986); the head of government is Barnabas Sibusiso Dlamini (from 1996). The currency is the lilangeni, and the languages spoken include Swazi and English (both official). **72** E5

Sweden country in northern Europe, bounded west by Norway, northeast by Finland and the Gulf of Bothnia, southeast by the Baltic Sea, and southwest by the Kattegat.

Sweden covers an area of 450,000 sq km/173,745 sq mi, and has a population of 8,910,000 (2000 est). The capital is ▶Stockholm; other major towns and cities include Göteborg, Malmö, Uppsala, Norrköping, Västerås, Linköping, Orebro, and Helsingborg. The head of state is King Carl XVI Gustaf (from 1973); the head of government is Göran Persson (from 1996). The currency is the Swedish krona, and the languages spoken include Swedish (official), Finnish, and Saami (Lapp). **16** H6

Swindon town and administrative headquarters of ▶Swindon unitary authority in southwest England, 124 km/77 mi west of London; population (1996) 170,000; it was part of the county of Wiltshire until 1997. The site of a major railway engineering works 1841–1986 on the Great Western Railway, the town has diversified since 1950 into such industries as heavy engineering, electronics, electrical manufacture, cars, and also insurance. **22** A3

Swindon unitary authority in southwest England, created in 1997 from the former district council of Thamesdown.
area 230 sq km/89 sq mi
towns and cities ▶Swindon (administrative headquarters); villages of Stanton, Fitzwarren, Highworth
features River Thames forms northern border of authority; Barbury Castle, Iron Age hill fort on Marlborough Downs; Great Weatern Railway Museum and National Monuments Records Centre (Swindon)
industries insurance, motor vehicle manufacturing, publishing, energy services, high technology industries, information technology
population (1996) 170,000 **5** F6

Switzerland landlocked country in Western Europe, bounded north by Germany, east by Austria and Liechtenstein, south by Italy, and west by France.

Switzerland covers an area of 41,300 sq km/15,945 sq mi, and has a population of 7,386,000 (2000 est). The capital is ▶Bern; other major towns and cities include Zürich, Geneva, Basel, Lausanne, Luzern, St Gallen, and Winterthur. The head of state and government is Adolf Ogi (from 2000). The currency is the Swiss franc, and the languages spoken include German, French, Italian, and Romansch (all official). **30** C4

Sydney principal port of Australia and capital of the state of ▶New South Wales; population (1996) 3,276,500. Founded in 1788, Sydney is situated on Port Jackson inlet on the southeast coast of Australia, and is built around a number of bays and inlets that form an impressive natural harbour. Industries include financial services, oil refining, engineering, electronics, and the manufacture of scientific equipment, chemicals, clothing, and furniture. Notable architectural landmarks are the Harbour Bridge, the nearby Sydney Opera House, and Centre Point Tower. There are many parks, as well as coastal beaches ideal for surfing, such as Bondi and Manly. Sydney hosted the Olympic Games in the year 2000. **76** K6

Syktyvkar (formerly **Ust Sysolsk** 1586–1930), capital of the autonomous republic of ▶Komi, in northern central Russia, 1,100 km/ 684 mi northeast of Moscow; population (1990) 235,000. Situated on the Vychegda River, it has a large timber industry; other economic activities are paper milling and tanning. The city was founded in 1740 as a Russian colony, and formerly had a flourishing grain and fur trade. 38 K2

Sylhet capital of Sylhet region, northeastern Bangladesh; population (1991) 228,100. It is a tea-growing centre and also produces rice, jute, and sugar. There is natural gas nearby. It is the former capital of a Hindu kingdom and was conquered by Muslims in the 14th century. In the 1971 civil war, which led to the establishment of Bangladesh, it was the scene of heavy fighting. 54 F4

Syracuse (Italian **Siracusa**), industrial port (chemicals, salt) in eastern Sicily; population (1992) 126,800. It has a cathedral and remains of temples, aqueducts, catacombs, and an amphitheatre. Founded 734 BC by the Corinthians, it became a centre of Greek culture under the elder and younger Dionysius. After a three-year siege it was taken by Rome 212 BC. In AD 878 it was destroyed by the Arabs, and the rebuilt town came under Norman rule in the 11th century. 32 K11

Syracuse city in New York State, USA, on Lake Onondaga; population (1996 est) 155,900. Industries include the manufacture of electrical and other machinery, paper, pharmaceuticals, ceramics, and processed foods. There are canal links with the ►Great Lakes and the Hudson and St Lawrence rivers. Settled in the 1780s as a salt-mining centre on the site of a former Iroquois capital, it later developed as a port on the Erie Canal (completed 1825). 88 E2

Syria country in western Asia, on the Mediterranean Sea, bounded to the north by Turkey, east by Iraq, south by Jordan, and southwest by Israel and Lebanon.

Syria covers an area of 185,200 sq km/71,505 sq mi, and has a population of 16,125,000 (2000 est). The capital is ►Damascus; other major towns and cities include Aleppo, Homs, Latakia, Hama, Ar Raqqah, and Dayr az Zawr. The head of state is Bashar al-Assad (from 2000); the head of government is Muhammad Mustafa Miro (from 2000). The currency is the Syrian pound, and the languages spoken include Arabic (official), Kurdish, Armenian, French, English, Aramaic, and Circassian. 56 C3

Szatmar-Nemeti Hungarian name for ►Satu Mare, a town in Romania. 34 K2

Szczecin (German **Stettin**), industrial (shipbuilding, fish processing, synthetic fibres, tools, iron) port on the River Oder, in northwest Poland; population (1993) 414,200. 18 D4

Szeged port on the River Tisza and capital of Csongrad county, southern Hungary; population (1993 est) 179,000. The chief industry is textiles, and the port trades in timber and salt. 34 H3

Székesfehérvár industrial city (metal products) in western central Hungary; population (1993 est) 110,000. It is a market centre for wine, tobacco, and fruit. 34 F2

Tabah (or **Taba**), small area of disputed territory, 1 km/0.6 mi long, situated on the Red Sea between Elat (Israel) to the east and the Sinai Desert (Egypt) to the west. Under an Anglo-Egyptian-Turkish agreement in 1906 the border ran through Tabah; under a British survey of 1915 headed by T E Lawrence (of Arabia), who made 'adjustments' allegedly under British government orders, it runs to the east. Taken by Israel in 1967, Tabah was returned to Egypt in 1989.

Tábor industrial town (textiles, tobacco) of southern Bohemia, in the Czech Republic, on the River Lužnice, 75 km/47 mi south of Prague; population (1991) 36,300. It was founded in 1420 as the military stronghold of a Hussite community of religious and political dissenters known as the Taborites. 18 D8

Tabora trading centre and administrative centre of Tabora Region, western Tanzania; population (1994 est) 120,000. It lies on an old caravan route between Ujiji, on Lake Tanganyika, and Bagamoyo, which lies on the coast, opposite Zanzibar, and was once a centre of the slave trade. Today Tabora is an agricultural marketing centre, linked to other Tanzanian centres by road and air as well as by rail. 70 E5

Tabriz capital of Eastern ►Azerbaijan, northwest Iran; population (1991) 1,089,000. It is situated on a small river called Aji Chay which flows into Lake Urmia. Industries include metal casting, carpets, cotton, and silk textiles. Features include The Arg, or citadel, and the Blue Mosque. It is connected by rail to Julfa (140 km/87 mi) on the Armenian border, and to Tehran. 58 M4

Tacna capital of Tacna department, southern Peru, 42 km/26 mi from the Chilean border, 1,290 km/800 mi south of Lima and Peru's most southerly city; population (1993) 174,300. Situated on the ►Pan-American Highway, it is linked with the port of Arica in northern Chile by railway. Products include tobacco, cotton, sugar cane and sulphur; irrigation projects help maintain olive groves and vineyards. 98 C7

Tacoma port in Washington State, USA, on Puget Sound, 250 km/156 mi from the Pacific Ocean, 40 km/25 mi south of Seattle; population (1996 est) 179,100. It is a lumber and shipping centre, with fishing and boat-building industries. Aluminium- and copper-smelting plants are sited here, and chemicals, forest products, and processed foods are manufactured. 84 B2

Tadmur Arabic name for ►Palmyra, a town in Syria. 58 H6

Taegu third-largest city in South Korea, situated between Seoul and Pusan; population (1995 est) 2,331,000. Nearby is the Haeinsa Temple (dating from 802), one of the country's largest monasteries and repository of the *Triptaka Koreana*, a collection of 80,000 wood blocks on which the Buddhist scriptures are carved. Grain, fruit, textiles, and tobacco are produced. 48 E6

Taejon (Korean 'large rice paddy'), capital of South Chungchong province, central South Korea; population (1995 est) 1,111,000. Korea's tallest standing Buddha and oldest wooden building are found northeast of the city at Popchusa in the Mount Songnisan National Park. 46 H3

Taganrog seaport city in Rostov oblast (region) of the southwestern Russian Federation, on the Sea of Azov, 75 km/47 mi west of Rostov-na-Donu; population (1996 est) 292,000. It is a major industrial centre; factories here produce iron and steel (since 1897), aircraft, ships, and other heavy-engineering products, such as combine-harvesters, boilers, and machine tools. The city also has tanneries and food-processing industries. 38 G5

Tagus (Spanish **Tajo**; Portuguese **Tejo**), river in Spain and Portugal; length 1,007 km/626 mi. It rises in the Sierra de Albarracín, Spain, on the border between the provinces of Cuenca and Teruel. It flows west past Toledo and Alcántara, then follows the Spanish-Portuguese frontier for 50 km/31 mi, and crosses Portugal to the Atlantic Ocean at Lisbon. 28 B5

Tahiti largest of the Society Islands, in ►French Polynesia; area 1,042 sq km/402 sq mi; population (1996) 150,700. Its capital is Papeete. The volcano, Orohena, reaches 2,237 m/7,339 ft, and much of the soil is volcanic, producing coconuts, sugar cane, and vanilla. Tourism is increasingly important as a source of revenue. English explorer Captain James Cook observed the transit of Venus across the sun during a visit to Tahiti in 1769. It came under French control in 1843 and became a colony in 1880. Paul Gauguin, French painter, lived here 1891–93, painting many pictures of local people. 74 M7

Taipei (or **Taibei**), capital and commercial centre of Taiwan; population (1995) 2,639,300. Industries include electronics, plastics, textiles, and machinery. The National Palace Museum (1965) houses the world's greatest collection of Chinese art, brought here from the mainland in 1948. 46 G6

Taiwan country in east Asia, officially the Republic of China, occupying the island of Taiwan between the East China Sea and the South China Sea, separated from the coast of China by the Taiwan Strait.

Taiwan covers an area of 36,179 sq km/13,968 sq mi, and has a population of 22,113,000 (1999 est). The capital is ►Taipei; other major towns and cities include Kaohsiung, Taichung, Tainan, Panchiao, Chungho, and Sanchung. The head of state is Chen Shui-bian (from 2000); the head of government is Chang Chun-hsiung (from 2000). The currency is the New Taiwan dollar, and the main language spoken is Chinese (dialects include Mandarin (official), Min, and Hakka). 50 G2

Taiyuan capital of ►Shanxi province, north China, on the River Fen He; population (1994) 2,086,300. Lying in a rich coal and iron-mining district, it is one of the most important centres of heavy industry in the north; heavy and agricultural machinery, iron, steel, and textiles are produced. It is the seat of Shanxi University. 46 E3

Ta'izz (or **Taiz**), city in the highlands of Yemen, 50 km/31 mi east of Mocha, at the centre of a coffee-growing region, at an altitude of 1,400 m/4,593 ft; population (1993 est) 290,100. Cotton, leather, and jewellery are also produced. There are three large mosques built by the Turks. It was capital of Yemen 1948–62. 56 D7

Tajikistan (formerly **Tadzhikistan** to 1991), country in central Asia, bounded north by Kyrgyzstan and Uzbekistan, east by China, and south by Afghanistan and Pakistan.

Tajikistan covers an area of 143,100 sq km/55,250 sq mi, and has a population of 6,188,000 (2000 est). The capital is ►Dushanbe; other major towns and cities include Khojand, Kurgan-Tyube, Kulyab, Ura-Tyube, and Kofarnihon. The head of state is Imamali Rakhmanov (from 1994); the head of government is Akil Akilov (from 1999). The currency is the Tajik rouble, and the languages spoken include Tajik (related to Farsi; official) and Russian. 56 J2

Taj Mahal white marble mausoleum built 1630–53 on the River Yamuna near Agra, India. Erected by Shah Jahan to the memory of his favourite wife, it is a celebrated example of Indo-Islamic architecture, the fusion of Muslim and Hindu styles.

Tajo Spanish name for the River ►Tagus. 14 D3

Takao Japanese name for ►Kaohsiung, a city on the west coast of Taiwan. 46 G6

Takoradi port in Ghana, administered with ►Sekondi. 68 D4

Talavera de la Reina town of Toledo province in Castilla–La Mancha, central Spain, on the River Tagus, 120 km/75 mi southwest of Madrid; population (1991) 50,800. Products include soap, pharmaceuticals, textiles, and Talavera ware (brightly decorated ceramics). Spanish and British forces defeated the French here in the Peninsular War 1809. There are Roman, Moorish, and Gothic remains. 28 F5

Talcahuano port and chief naval base in Bío-Bío region, central Chile; population (1992) 246,600. It has oil refineries and dry docks, and there are fishing, timber, and steel industries. Exports include timber, hides, wool, and coal. Moored in the naval base is the Peruvian warship *Huáscar*, captured by the Chilean navy in 1879 during the War of the Pacific. 100 G6

Talien alternative transliteration of ►Dalian, a port in Liaoning province, China. 46 G3

Tallahassee (Cree 'old town'), capital of ►Florida, USA; population (1996 est) 136,800. It is an agricultural and lumbering centre, trading in cotton, tobacco, and cattle. Industries include publishing, food-processing, and the manufacture of forest products, building materials, and textiles. 90 E3

Tallinn (German **Reval**; Russian **Revel**), naval port and capital of Estonia, 300 km/186 mi west of St Petersburg on the Gulf of Finland; population (1990) 505,100. Industries include the manufacture of electrical and oil-drilling machinery, textiles, and paper production. It is a major cultural centre, containing the Estonian Academy of Sciences and a number of polytechnic, arts, and other institutes. Founded as a Danish fortress in 1219, Tallinn was a member of the Hanseatic League throughout the Middle Ages; it came under the control of the Teutonic Knights in 1346, Sweden in 1561, and Russia in 1750. It was occupied by German forces in both world wars, and suffered widespread damage. 16 N7

Tamale town and capital of Northern Region, northeastern Ghana, 450 km/280 mi north of Accra and 50 km/31 mi from the northern tip of Lake Volta; population (1998 est) 213,400. It is a centre for civil engineering and education, and a commercial centre which deals in rice, cotton, and peanuts. It is at the junction of roads from Accra in the south, Togo in the east, and Burkina Faso in the north. An airfield and the Mole Game Reserve is nearby. 68 D3

Tamar river rising in north Cornwall, southwest England; length 97 km/60 mi. For most of its length it forms the Devon–Cornwall border; it forms the Hamoaze estuary at

Devonport and flows into Plymouth Sound. **24** H11

Tamatave former name (to 1979) for ►Toamasina, the chief port of Madagascar. **72** H3

Tambov capital city of Tambov oblast (region), Russian Federation; population (1996 est) 318,000. Tambov lies 400 km/249 mi southeast of Moscow. The city contains engineering works (making machine tools and agricultural equipment), food-processing factories, and chemical plants, and stands at a major railway junction. **38** H4

Tamil Nadu (formerly **Madras State** until 1968), state of southeast India, bounded on the north by Karnataka and Andhra Pradesh, on the west by Kerala, and on the east and south by the Bay of Bengal and the Indian Ocean; area 130,100 sq km/50,200 sq mi; population (1995 est) 60,225,000. The capital is ►Chennai (formerly Madras). There are cotton, leather, and sugar-refining industries. Tea, coffee, spices, sugar cane, and coconuts are grown. **10** C6

Tampa (Caloosa 'sticks of fire'), major port and winter resort in western Florida, USA, at the head of **Tampa Bay**, 40 km/25 mi from the open sea of the Gulf of Mexico; population (1996 est) 285,200. It is the focus of the metropolitan area of **Tampa–St Petersburg**, a popular residential and retirement zone. Industries include fruit- and vegetable-canning, shipbuilding, shrimp-fishing, and the manufacture of fertilizers, chemicals, clothing, beer, and cigars. **90** E4

Tampere (Swedish **Tammerfors**), city in southwestern Finland; population (1994) 179,000, metropolitan area 258,000. Industries include textiles, paper, footwear, and turbines. It is the second-largest city in Finland. **16** M6

Tampico port on the Rio Pánuco, 10 km/6 mi from the Gulf of Mexico, in Tamaulipas state, northeastern Mexico; population (1995 est) 301,000. Industries include oil refining, oil products, boat building, fishing, and fish processing. **94** E4

Tamworth town in northeast New South Wales, Australia, on the River Peel, 404 km/251 mi north of Sydney; population (1996) 31,865. Tamworth is the centre of an agricultural district with dairying and sheep-rearing; it also has a furniture industry. It is a regional centre for New England. **76** K6

Tamworth town in Staffordshire, central England, at the junction of the rivers Tame and Anker, 24 km/15 mi northeast of Birmingham; population (1991) 67,500. Industries include agricultural engineering and the manufacture of paper, clothing, bricks, tiles, and aluminium products. **22** A2

Tana river in Finnmark county, north Norway; length 400 km/249 mi. It is formed by the junction of the rivers Anar and Karas, and flows northeast to enter the Arctic Ocean by Tana Fjord. For most of its length, Tana forms the Finnish-Norwegian border. Salmon are caught here. **16** P1

T'ana Hāyk' alternative name for Lake ►Tana, in Ethiopia. **64** G5

Tana, Lake (or **T'ana Hāyk'**), lake on the north central plateau of Amhara, Ethiopia, 1,800 m/5,900 ft above sea level; area 3,600 sq km/1,390 sq mi. It is about 75 km/47 mi long and 70 km/43 mi wide. The Blue Nile, the source of which lies just above Lake Tana, leaves the lake via the Tisisat Falls, where it is harnessed for hydroelectric power. **64** G5

Tananarive former name for ►Antananarivo, the capital of Madagascar. **72** H3

Tanga seaport and capital of Tanga region, Tanzania, northeast of Zanzibar, on the Indian Ocean; population (1995 est) 226,000. After Dar es Salaam, it is Tanzania's second largest port, and trades in sisal, fruit, cocoa, tea, and fish; industries include cement production. It was once the terminus of caravan routes leading into the interior of Africa, and is now the coastal terminus of the railway to Moshi and Arusha. **70** F5

Tanganyika, Lake lake 772 m/2,534 ft above sea level in the Great Rift Valley, East Africa, with the Democratic Republic of Congo to the west, Zambia to the south, and Tanzania and Burundi to the east. It is about 645 km/400 mi long, with an area of about 31,000 sq km/12,000 sq mi, and is the deepest lake (1,435 m/4,710 ft) in Africa, and the second-deepest freshwater lake in the world. The mountains around its shores rise to about 2,700 m/8,860 ft. The chief ports on the lake are Bujumbura (Burundi), Kigoma (Tanzania), and Kalémié (Democratic Republic of Congo). **70** D5

Tanger alternative name for ►Tangier, a port in north Morocco. **66** D1

Tangier (or **Tangiers**, or **Tanger**; Arabic **Tanjah**), port in north Morocco, on the Strait of Gibraltar, 58 km/36 mi southwest of Gibraltar; population (urban area, 1994) 497,100. Cigarette manufacturing is the most important industry, and there are fisheries, market gardens, and preserving industries. It is the northern terminus of the Tangier–Fès railway. It was a Phoenician trading centre in the 15th century BC. Captured by the Portuguese in 1471, it passed to England in 1662 as part of the dowry of Catherine of Braganza, but was abandoned in 1684, and later became a lair of Barbary Coast pirates. From 1923 Tangier and a small surrounding enclave became an international zone, administered by Spain 1940–45. In 1956 it was transferred to independent Morocco and became a free port in 1962. **66** D1

Tangshan industrial city in Hebei province, China, 160 km/100 mi southeast of Beijing; population (1995 est) 1,627,000. A centre of heavy industry, traditional coalmining and ceramic industries that have been augmented by iron, steel, and cement production, and engineering. Almost destroyed by an earthquake in 1976, which measured 8.2 on the Richter scale and killed 242,000 people, the city was rebuilt on a new site. **46** F3

Tanjah Arabic name for ►Tangier, a port in north Morocco. **66** D1

Tanzania country in east Africa, bounded to the north by Uganda and Kenya; south by Mozambique, Malawi, and Zambia; west by Congo (formerly Zaire), Burundi, and Rwanda; and east by the Indian Ocean.

Tanzania covers an area of 945,000 sq km/364,864 sq mi, and has a population of 33,517,000 (2000 est). The capitals are ►Dodoma (official) and Dar es Salaam (administrative); other major towns and cities include Zanzibar, Mwanza, Mbeya, Tanga, and Morogora. The head of state is Benjamin Mkapa (from 1995); the head of government is Frederick Sumaye (from 1995). The currency is the Tanzanian shilling, and the languages spoken include Kiswahili, English (both official), Arabic (in Zanzibar), and many local languages. 70 E5

Taormina coastal resort in eastern Sicily, at the foot of Mount Etna; population (1991) 10,100. It has an ancient Greek theatre. 32 K11

Tarābulus Arabic name for ►Tripoli, capital of Libya. 66 H2

Taranto (Greek **Tarantum**), naval base and port in Apulia region, southeast Italy, on the Gulf of Taranto, 80 km/50 mi southeast of Bari; population (1992) 230,200. It is an important commercial centre, and its steelworks are part of the new industrial complex of southern Italy. There are chemical and oil-refining industries, and oyster and mussel fisheries. It was founded in the 8th century BC by Sparta, and was captured by the Romans in 272 BC. 32 M8

Tarbes administrative centre of the ►Hautes-Pyrénées *département*, situated on the River Adour in southwest France; population (1990) 50,200, conurbation 75,000. A tourist centre for the ►Pyrenees, the town is noted for its Anglo-Arabian horses. It has a school of artillery and armament works, and manufactures shoes, furniture, electrical apparatus, machinery, and pottery. It belonged to England from 1360 to 1406. 26 F10

Taree town in a dairying area of northeastern New South Wales, Australia; population (1991) 16,300. Taree was founded in 1854 and became a city in 1981. It stands 130 km/80 mi northeast of Newcastle on the Sydney to Brisbane railway and on the Pacific Highway, and is the marketing centre for the Manning River dairying district. Industries include dairy products, fibrous plaster, engineering, and boatbuilding, as well as tourism which is encouraged by the local coastal and mountain environments. 76 K6

Târgu Jiu capital of Gorj county, southwest Romania, near the River Jiu, 90 km/56 mi northwest of Craiova; population (1993) 97,000. Large coal- and oilfields found locally, and timber trade, have encouraged its rapid growth since 1945. 34 L4

Târgu Mureş city in Transylvania, Romania, on the River Mureş, 450 km/280 mi north of Bucharest; population (1993) 166,000. With a population comprising approximately equal numbers of ethnic Hungarians and Romanians, the city was the scene of rioting between the two groups following Hungarian demands for greater autonomy in 1990. 34 M3

Tarim Basin (Chinese **Tarim Pendi**), internal drainage area in Xinjiang Uygur Autonomous Region, northwest China, between the Tien Shan and Kunlun Shan ranges; area about 900,000 sq km/350,000 sq mi. It is crossed by the River Tarim and includes the ►Lop Nur salt lakes. The Taklimakan Desert lies to the south of the Tarim River. 42 Q10

Tarim Pendi Chinese name for ►Tarim Basin, a drainage area in northwest China. 42 Q10

Tarn *département* in the ►Midi-Pyrénées region of France; area 5,751 sq km/2,220 sq mi; population (1990) 343,300. The River ►Tarn and its tributary the Agout flow through it. Cereals, vines, and vegetables are produced, and livestock are raised. There are agro-food and plastics industries. The principal towns are ►Albi (the administrative centre) and Castres. 6 E5

Tarn river in southwest France, rising in the Cévennes Mountains and flowing 375 km/233 mi into the River Garonne near Moissac. It cuts picturesque gorges in the limestone plateaus (or Causses) of the ►Lozère and Aveyron *départements*. The chief towns on its banks are Albi and Montauban. 26 H10

Tarn-et-Garonne *département* in the ►Midi-Pyrénées region of France; area 3,716 sq km/1,435 sq mi; population (1990) 200,400. It is an alluvial plain formed by the confluence of the rivers ►Garonne, ►Tarn, and Aveyron. There are some hills and the region is generally wooded. The *département* is predominantly agricultural; cereals, fruit, vines, and truffles are produced, and sheep are raised. Tourists visit the medieval villages. The principal towns are ►Montauban (the administrative centre) and Castelsarrasin. 6 D4

Tarragona port and capital of Tarragona province in Cataluña, northeast Spain, at the mouth of the Francoli River on the Mediterranean coast; population (1991) 110,000. Industries include petrochemicals, pharmaceuticals, and electrical goods. It has a cathedral and Roman remains, including an aqueduct and amphitheatre. 28 M3

Tarrasa (or **Terrassa**), city of Barcelona province in ►Cataluña, northeast Spain; population (1991) 154,300. Industries include textiles and fertilizers, and there is a trade in agricultural produce, olive oil, and wine. 28 N3

Tarsus city in Içel province, southeastern Turkey, on the River Pamuk; population (1990) 168,700. Formerly the capital of the Roman province of Cilicia, it was the birthplace of St Paul. 58 F5

Tartu (German **Dorpat**; Russian **Yurev**), city in Estonia, 150 km/93 mi southeast of Tallinn and 50 km/31 mi west of Lake Peipus; population (1990) 115,400. Industries include light engineering, food processing, and lumber. Founded by Russians in 1030, it was a stronghold of the Teutonic Knights from 1224 onwards. Tartu was captured by Russia in 1558 and subsequently held by Sweden and Poland, but returned to Russian control in 1704. It was occupied by German forces in both World Wars. 16 P7

Tashkent (Uzbek **Toshkent**), capital of ►Uzbekistan and of Tashkent wiloyat (region), located in the western foothills of the Tien Shan mountain range and in the valley of the River Chirchiq. With a population (1996) of some 2,300,000, it is the largest city in Central Asia. It is an important transit centre for the region; there is an international airport terminal here. Industrial activity includes the manufacture of mining machinery, chemicals, textiles, and leather goods. Tashkent suffered severe damage in an earthquake in 1966, but was rapidly rebuilt. 42 M9

Tasmania (formerly **Van Diemen's Land** 1642–1856), island in the Indian Ocean, southeast of Australia, separated from the mainland by Bass Strait; state of Australia; area about 68,000 sq km/26,000 sq mi; population (1996) 459,700. The capital is ►Hobart. Products include wool, dairy products, apples and other fruit, processed foods, timber, paper, iron, tungsten, copper, silver, coal, and cement. 76 H8

Tasman Sea part of the ▶Pacific Ocean between southeast Australia and northwest New Zealand. It is named after the Dutch explorer Abel Tasman. 78 B3

Taunton market town and administrative headquarters of ▶Somerset, southwest England, 50 km/31 mi northeast of Exeter, on the River Tone; population (1991) 56,400. Products include cider, leather, optical instruments, computer software, aeronautical instruments, and concrete; other industries include light engineering, and there is a weekly cattle market. Taunton is the main market centre for west Somerset and east Devon. The remains of Taunton Castle include the Elizabethan hall in which Judge Jeffreys held his Bloody Assizes in 1685 after the Duke of Monmouth's rebellion. 24 J10

Taunus Mountains mountain range in Hesse, Germany, lying between the Rhine, Main, and Lahn rivers; the Grosser Feldberg, 881 m/2,887 ft, is the highest peak in the Rhenish uplands (the Kleiner Feldberg is 827 m/2,713 ft). There are several mineral spas, including Wiesbaden, Bad Nauheim, Bad Homburg, and Bad Soden. 22 L4

Taupo, Lake largest lake in New Zealand, in central North Island; area 620 sq km/239 sq mi. It is 357 m/1,170 ft above sea level, maximum depth 159 m/522 ft. The lake is in a volcanic area of hot springs and is the source of the Waikato River. The lake and its tributary rivers attract tourists and anglers. 78 E4

Tauranga town and port in North Island, New Zealand, on the Bay of Plenty; population (1996) 82,800. It exports citrus fruits, kiwis, dairy produce, and timber. Tourism is important. 78 F3

Taurus Mountains (Turkish **Toros Dağlari**), mountain range in southern Turkey, forming the southern edge of the Anatolian plateau and rising to over 3,656 m/12,000 ft. 58 E5

Tay longest river in Scotland; length 193 km/120 mi, it flows northeast through **Loch Tay**, then east and southeast past Perth to the **Firth of Tay**, crossed at Dundee by the **Tay Bridge**, before joining the North Sea. The Tay has salmon fisheries; its main tributaries are the Tummel, Isla, and Earn, Braan, and Almond. 24 J5

Tayside former region of Scotland (1975–96); replaced by Angus, Dundee City, and Perth and Kinross unitary authorities.

Tbilisi (formerly **Tiflis**), capital and cultural centre of ▶Georgia, located on the Kura River in the ▶Caucasus Mountains; population (1996) 1,200,000. It is a major economic, transportation and industrial centre. Industries include the manufacture of textiles, machinery, ceramics, and tobacco. In the lead-up to the collapse of the USSR in 1989 and Georgian independence, the city was the scene of bloody clashes between Russian security forces and nationalist demonstrators. 58 L3

Tees river flowing from the Pennines in Cumbria, northwest England, to the North Sea via Tees Bay, Middlesbrough unitary authority, in northeast England; length 130 km/80 mi. Its port, Teesport, handles in excess of 42 million tonnes per annum, with port trade mainly chemical-related. 24 L7

Teesside industrial area at the mouth of the River Tees, northeast England; population (1994 est) 323,000. It includes the towns of ▶Stockton-on-Tees, ▶Middlesbrough, Billingham, and Thornaby. There are high-technology industries, as well as petrochemicals, electronics, steelmaking, and plastics. The area includes an oil-fuel terminal and the main North Sea natural gas terminal.

Tegucigalpa capital of ▶Honduras; situated at an altitude of 975 m/3,199 ft in the highlands of south-central Honduras, on the River Choluteca at the foot of the extinct El Pichacho volcano; population (1991 est) 670,100. Industries include textiles, chemicals, and food-processing, mostly for domestic consumption. It was founded by the Spanish in the 16th century as a gold- and silver-mining centre, and became capital in 1880 (the former capital was Comayagua). Tocontín international airport is 6 km/4 mi to the south. The city has some fine colonial architecture including an 18th-century cathedral, and the church of Saint Francis, completed in 1592. 94 G6

Teheran alternative spelling of ▶Tehran, the capital of Iran. 56 F2

Tehran (or **Tehrān**, or **Teheran**), capital of Iran; population (1991) 6,475,500. Industries include textiles, chemicals, engineering, and tobacco. It is built at an average altitude of 1,220 m/3,937 ft on a slope running south from the Elburz Mountains. 56 F2

Tejo Portuguese name for the River ▶Tagus. 28 B5

Tel Aviv-Yafo (or **Tel Aviv-Jaffa**), city in Israel, situated on the coast of Sharon Plain, 77 km/48 mi northwest of Jerusalem; population (1995) 355,900. Industries include textiles, chemicals, sugar, printing, publishing, and tourism. Tel Aviv was founded in 1909 as a Jewish residential area in the Arab town of ▶Jaffa, with which it was combined in 1949; their ports were superseded in 1965 by Ashdod to the south. During the Gulf War of 1991, Tel Aviv became a target for Iraqi missiles as part of Saddam Hussein's strategy to break up the Arab alliance against him. It is regarded by the UN as the capital of Israel. 60 B4

Telford and Wrekin unitary authority in west England, created in 1998 from part of Shropshire.
area 291 sq km/112 sq mi
towns and cities Telford (administrative headquarters), Newport
features The Wrekin, isolated hill (407 m/1,334 ft); Ironbridge Gorge includes world's first iron bridge, built across the River Severn in 1779 by Abraham Darby, and Ironbridge Gorge Museum Trust
industries iron founding, agriculture, dairy farming, food processing, confectionery, audio and tape manufacture, electronic tools and equipment, vehicle parts, plastics, clothing manufacture, information technology
population (1996) 144,600 5 E5

Tema port in Ghana, 30 km/19 mi east of Accra; population (1998 est) 156,400. It has the largest artificial harbour in Africa, opened in 1962, as well as oil refineries, food processing, aluminium, chemicals, fishing, and civil engineering industries. 68 D3

Temple town in Bell County, central Texas, USA; population (1990) 46,100. It is situated on the Little River, 56 km/35 mi south-southwest of Waco. Temple was founded 1880. Established by the Gulf, Colorado and Santa Fe Railroad in 1880, it grew as a division point with repair shops. It remains an industrial centre, producing building materials, steel, furniture, and railway supplies. Temple is also a leading medical centre, with several hospitals and the Texas A&M University School of Medicine. It is home to a number of state and Federal agricultural offices, as well as the Temple Junior College (1926). **92** G2

Temuco capital of Araucanía region, south-central Chile, situated to the north of the Lake District on the River Cautín, 675 km/420 mi south of Santiago; population (1992) 240,900. Cereals, timber, and apples are produced in the surrounding region, and industries include coal mining, and the manufacture of steel and textiles. It is a market town for the Mapuche Indians, who trade their produce and crafts. The great majority of them live in the forest land around Temuco. Founded in 1881, the city is the gateway to the Lake District region. **100** G6

Tenerife largest of the ▶Canary Islands, in the province of Santa Cruz de Tenerife, Spain; area 2,060 sq km/ 795 sq mi; population (1991) 706,900. Fruit and vegetables are produced, especially bananas and tomatoes, and the island is a popular tourist resort. Santa Cruz is the main town here, and Pico de Teide, an active volcano, is the highest peak in Spain (3,713 m/12,186 ft high). **66** B3

Tennessee state in eastern central USA. It is nicknamed the Volunteer State. Tennessee was admitted to the Union in 1796 as the 16th US state. It is bordered to the east by North Carolina, to the south by Georgia, Alabama, and Mississippi, to the west by Arkansas and Missouri, across the Mississippi River, and to the north by Kentucky and Virginia. Historically, Tennessee was a plantation state, associated with slavery. Culturally, it is one of the centres of country music.
population (2000 est) 5,689,300
area 109,200 sq km/42,151 sq mi
capital ▶Nashville
towns and cities Memphis, Knoxville, Chattanooga, Clarksville
industries and products cereals, cotton, tobacco, soybeans, livestock, timber, coal, zinc, copper, chemicals, power generation, automobiles, aluminium, music industry, tourism **84** J4

Teplice industrial city (peat- and lignite-mining, glass, porcelain, cement, paper) and spa in northern Bohemia, Czech Republic; population (1991) 127,800. **18** C7

Terengganu alternative spelling of ▶Trengganu, a state in Peninsular Malaysia. **11** (1)A1

Terneuzen port in the province of Zeeland, the Netherlands, on an arm of the River Scheldt, 45 km/28 mi northwest of Antwerp, Belgium; population (1996) 35,100. Terneuzen is situated at the end of the 32-km/ 20-mi-long **Ghent–Terneuzen Canal**, on which are located iron and steelworks and other major industries. **22** F3

Terni (ancient **Interamna Nahars** or **Interamna Umbrica**), industrial town in the valley of the River Nera, Umbria region, central Italy, 80 km/50 mi from Rome; population (1992) 108,000. It has iron, steel, and engineering works, and fertilizers, plastics, and textiles are manufactured. The nearby Marmore Falls, the highest in Italy, were created by the Romans in order to drain the Rieti marshes; hydroelectricity is now produced there. **32** G6

Terrassa alternative spelling of ▶Tarrasa, a city in Cataluña, Spain. **28** N3

Terre Haute city and administrative headquarters of Vigo County, west Indiana, USA, on the Wabash River, 109 km/68 mi southwest of Indianapolis; population (1996 est) 54,600. It is the commercial and industrial centre of an agricultural and coal-mining region. Plastics, chemicals, building products, metalware, and glass are manufactured. The city is the home of Indiana State University (1865) and the Rose-Hulman Institute of Technology (1874) . **90** D2

Tétouan (or **Tetuán**), city in northeast Morocco, near the Mediterranean coast, 64 km/40 mi southeast of Tangier; population (1994) 277,500. Products include textiles, building materials, leather, and soap. It is an important tourist centre. Tetuan is surrounded by walls and includes a citadel. It was settled by Muslim exiles from Spain in the 16th century, and was formerly the capital of Spanish Morocco. **66** D1

Tetschen German name for ▶Děčín, a town in the Czech Republic. **18** D7

Tevere Italian name for the River ▶Tiber, which flows from the Apennines to the Tyrrhenian Sea. **32** G6

Texarkana twin cities that straddle the Texas–Arkansas border, near Red River; population (1996 est) 32,460 (Texas), 23,000 (Arkansas). Texarkana is a railway distribution and processing centre for a lumbering and agricultural area, dealing in beef-cattle, dairy products, and cotton. Manufactured products include furniture and textiles, and sand and gravel are quarried. Texarkana was settled in 1874. **90** C3

Texas state in southwestern USA. It is nicknamed the Lone Star State. Texas was admitted to the Union in 1845 as the 28th US state. One of the Great Plains states, it is bordered to the east by Louisiana, to the northeast by Arkansas, to the north by Oklahoma, to the west by New Mexico, to the southwest by the Mexican states of Chihuahua, Coahuil, Nuevo Léon, and Tamaulipas, and to the southeast by the Gulf of Mexico. Texas is the largest state in the lower 48 US states.
population (2000 est) 20,851,800
area 691,200 sq km/266,803 sq mi
capital ▶Austin
towns and cities Houston, Dallas, Fort Worth, San Antonio, El Paso, Corpus Christi, Lubbock
industries and products rice, cotton, sorghum, wheat, hay, livestock, shrimps, meat products, lumber, wood and paper products, petroleum (nearly one-third of US production), natural gas, sulphur, salt, uranium, chemicals, petrochemicals, nonelectrical machinery, fabricated metal products, transportation equipment, electric and electronic equipment, aerospace equipment, computer and high-tech machinery, finance sector, tourism **84** F5

Texel (or **Tessel**), largest and southernmost of the ▶Frisian Islands, in North Holland province, the Netherlands; area 190 sq km/73 sq mi; population (1997) 13,300. It is separated from the mainland by the 4 km/2.5 mi wide Marsdiep channel and is reached by boat from Den Helder. Den Burg is the chief settlement. Local industries include tourism, fishing and farming. Texel sheep are kept for their wool and cheese. The northern part of the island is called Eierland (Egg Land) because it is a good breeding ground for thousands of birds. 22 G1

Thailand (formerly **Siam** to 1939 and 1945–49), country in southeast Asia on the Gulf of Siam, bounded east by Laos and Cambodia, south by Malaysia, and west by Myanmar (Burma).

Thailand covers an area of 513,115 sq km/198,113 sq mi, and has a population of 61,399,000 (2000 est). The capital (and chief port) is ►Bangkok; other major towns and cities include Chiangmai, Hat Yai, Khon Kaen, Songkhla, Nakhon Ratchasima, Nonthaburi, and Udon Thani. The head of state is King Bhumibol Adulyadej (from 1946); the head of government is Thaksin Shinawatra (from 2001). The currency is the baht, and the languages spoken include Thai, Chinese (both official), English, Lao, Malay, and Khmer. 50 C4

Thames river in south England, flowing through London; length 338 km/210 mi. The longest river in England, it rises in the Cotswold Hills above Cirencester and is tidal as far as Teddington. Below London there is protection from flooding by means of the **Thames Barrier** (1982). The headstreams unite at Lechlade. 24 L10

Thames, Firth of inlet between Auckland and the Coromandel Peninsula, New Zealand. 78 E3

Thar Desert (or **Indian Desert**), desert on the borders of ►India and Pakistan; area about 219,000 sq km/ 84,556 sq mi. 54 B3

Thessaloníki (English **Salonika**), port in Macedonia, northeastern Greece, at the head of the Gulf of Thessaloníki; the second-largest city in Greece; population (1991) 378,000. Industries include textiles, shipbuilding, chemicals, brewing, and tanning. It was founded from Corinth by the Romans in 315 BC as **Thessalonica** (to whose inhabitants St Paul addressed two epistles), captured by the Saracens in AD 904 and by the Turks in 1430, and restored to Greece in 1912. 36 E4

Thetford market town in Norfolk, eastern England, between Cambridge and Norwich, at the junction of the rivers Little Ouse and Thet; population (1991) 19,800. There is light industry and printing. It was the capital of the Anglo-Saxon kingdom of East Anglia and the birthplace in 1737 of the radical political writer Thomas Paine. 24 N9

Thetford Mines town in south Québec, Canada, and the site of the world's largest asbestos deposits; population (1991) 17,300. It is situated on the River Bécancour, in the Notre Dame Mountains, 80 km/ 50 mi south of Québec. Thetford Mines is a regional service and mining centre. Dairying, sawmilling, and diverse manufactures also contribute to the local economy. 88 F1

Thimphu (or **Thimbu**, **Thimpu**, or **Thimpago**), capital of ►Bhutan, lying on the River Raidak at an altitude of 2,000 m/7,000 ft; population (2001 est) 48,300. It is the main marketing centre for agricultural goods produced in the surrounding valley and on terraced hill slopes, including rice, maize, and wheat. Power is supplied by a hydroelectric station (1966), and industries include sawmilling, wood products, and food processing. The Indo-Bhutan Highway (1968) is the main route into India via Phuntsholing. 54 E3

Thionville (German **Diedenhofen**), capital of the French *département* of Moselle, 25 km/16 mi north of Metz, on the River Moselle; population (1990) 41,500. It has large iron- and steelworks, and manufactures chemicals. Imperial diets (meetings of the estates of the Holy Roman Empire) were held here in the 8th century. 22 J5

Thiruvananthapuram (formerly **Trivandrum**), capital of ►Kerala, southwest India; population (1991) 524,000. It has chemical, textile, and rubber industries, and there is an international airport. Formerly the capital of the princely state of Travancore, it has many palaces, an old fort, and a shrine. 54 C7

Thousand Islands group of about 1,700 islands in the upper St Lawrence River, on the border between Canada and the USA. Most of them are in Ontario, Canada; the rest are in the US state of New York. Some are in Canada's St Lawrence Islands National Park; many of the others are privately owned. The largest is Wolfe Island in Ontario, 127 sq km/ 49 sq mi. Tourism and resort trade have been important to the islands' economy since the early 19th century.

Thrace (Greek **Thráki**), ancient region of the Balkans, southeastern Europe, formed by parts of modern Greece and Bulgaria. It was held successively by the Greeks, Persians, Macedonians, and Romans.

Three Rivers English name for the Canadian port of ►Trois Rivières. 88 F1

Thunder Bay city and inland port in Ontario, Canada, on Lake Superior; population (1996) 131,300. A major exporter of wheat and ore, it marks the westernmost extent for sea-going ships on the ►St Lawrence Seaway, and is the lakehead terminus of the Canadian National Railway. Industries include tourism, shipbuilding, timber-milling, and the manufacture of wood and metal products, vehicles, farming equipment, paper, and pulp. It was created in 1970 by the union of Port Arthur, its twin city Fort William to the south, and the towns of Neebing and McIntyre. 88 C1

Thüringen German name for ►Thuringia, an administrative region in Germany. 7 (2)C4

Thuringia (German **Thüringen**), administrative region (German *Land*) in central Germany, bounded on the north by Saxony-Anhalt and Lower Saxony, on the east by Saxony, on the south by Bavaria, and on the west by Hesse; area 15,482 sq km/5,978 sq mi; population (1995) 2,684,000. The capital is ►Erfurt. There are machine tool, optical instrument, steel, vehicle, ceramic, electronics, glassware, and timber industries. Wheat, maize, and sugar beet are grown. 7 (2)C4

Thurrock unitary authority in eastern England, created in 1998 from part of Essex.

area 163 sq km/63 sq mi

towns and cities Grays (administrative headquarters), Purfleet, Tilbury, Chadwell, St Mary, Stanford-le-Hope, Corringham, South Ockendon

features located on north bank of River Thames; Holehaven Creek forms eastern border of authority; Tilbury Marshes; Mucking Marshes; Dartford Tunnel and Queen Elizabeth II bridge; 17th-century Tilbury Fort, with three moats; Alexandra Lake

industries oil refineries, power station at west Tilbury Marshes, sand and gravel extraction, cement works, soap,

margarine, timber products
population (1996) 130,600 5 G6

Thurso port in the Highland unitary authority, Scotland, 140 km/87 mi northeast of Inverness; population (1991) 8,500. It is the principal town of Scotland's north coast. The experimental nuclear reactor site of Dounreay, 14 km/8 mi to the west, was decommissioned in 1994 and replaced by a nuclear waste reprocessing plant. Ferries operate between Scrabster, on **Thurso Bay**, and Stromness in the Orkney Islands. 24 J3

Tianjin (or **Tientsin**), city and special municipality in Hebei province, north China; municipality area 4,000 sq km/1,544 sq mi; city population (1994) 5,894,900, municipality (1996) 9,480,000. One of four municipalities administered directly from Beijing, it includes the city of Tianjin and the port of Tanggu. An industrial and commercial centre, its handmade silk and wool carpets are renowned. Dagang oilfield lies nearby. Tianjin was opened to foreign trade in 1860 and was occupied by the Japanese in 1937. 46 F3/11 (2)E3

Tiber (Italian **Tevere**; Latin **Tiberis**), river in Italy that flows through Rome; its length from its source in the Apennines to the Tyrrhenian Sea is 400 km/250 mi.
It is Italy's third longest river. 32 G6

Tiberias, Lake (or **Sea of Galilee**, or **Lake of Gennesaret**; Hebrew **Yam Kinneret**), lake in north Israel, 210 m/689 ft below sea level, into which the River Jordan flows; area 170 sq km/66 sq mi. The first Israeli kibbutz (cooperative settlement) was founded nearby in 1909. 60 C4

Tibesti range in the central Sahara, on the borders of Libya, Niger, and Chad; the highest peak is **Emi Koussi** at 3,415 m/11,208 ft. The mountains are some 500 km/310 mi long and 300 km/205 mi wide 64 C3

Tibet (Chinese **Xizang**), autonomous region of southwestern China; area 1,221,600 sq km/471,700 sq mi; population (1993 est) 2,290,000 (many Chinese have settled in Tibet; 2 million Tibetans live in China outside Tibet). The capital is ▶Lhasa. Although Tibet has its own People's Government and People's Congress, Tibetan nationalists regard the province as being under colonial rule. The controlling force in Tibet is the Communist Party of China, represented locally by First Secretary Wu Jinghua from 1985. There is a government-in-exile in Dharmsala, Himachal Pradesh, India, where the Dalai Lama lives. The religion in the region is traditionally Lamaism (a form of Mahāyāna Buddhism). 54 E2/11 (2)B3

Tienen (French **Tirlemont**), town in the province of Brabant, Belgium, 40 km/25 mi east of Brussels, situated on the River Gette; population (1991) 31,600. It manufactures sugar from locally grown sugar beet, and has other industries. Tienen is thought to be of Roman origin. It was a busy market in the early Middle Ages. 22 G4

Tien Shan (or **Tian Shan**; Chinese 'celestial mountains'), mountain chain in Central Asia, forming part of the boundary between Kyrgyzstan, south-eastern Kazakhstan and China (Xinjiang Uygur autonomous region). The Tien Shan range extends for over 2,400 km/1,500 mi northeast from the Pamirs to the western fringe of the Gobi desert. The principal peaks are Pobeda Peak (7,439 m/24,406 ft) and Khan-Tengri (6,995 m/22,949 ft). 42 Q9

Tientsin alternative form of ▶Tianjin, an industrial city in Hubei province, north China. 46 F3

Tierra del Fuego island group separated from the southern extremity of South America by the Strait of Magellan; Cape Horn is at the southernmost point. There are oil, natural gas, and sheep farming industries. Tourism is also important. The largest island is Tierra del Fuego, or **Isla Grande**, with an area of 48,100 sq km/18,571 sq mi; half of this island, and the islands west of it, belong to Chile, and form part of the Magallanes region, the capital and chief town of which is ▶Punta Arenas. The eastern part of the archipelago belongs to Argentina, forming the federal district of Tierra del Fuego; its capital, Ushuaia, is the world's most southerly town. 100 H9

Tiflis former name (to 1936) of the city of ▶Tbilisi in Georgia. 62 H1

Tigris (Arabic **Dijla**), river flowing through Turkey and Iraq (see also ▶Mesopotamia), joining the ▶Euphrates 80 km/50 mi northwest of Basra, where it forms the Shatt-al-Arab; length 1,600 km/1,000 mi. 58 K6

Tijuana city and resort in north-western Mexico; population (1995 est) 824,000. It is on the Pacific Ocean in the state of Baja California Norte. The border with the USA at Tijuana and ▶San Diego is believed to be the busiest frontier in the world and there is a considerable tourist trade, for which local horse and dog racing, bullfighting and casinos are attractions. 84 C5

Tilburg town in the province of North Brabant, the Netherlands, 24 km/15 mi southwest of 's Hertogenbosch; population (1996) 164,300. Tilburg is an industrial centre, manufacturing cloth, woollens and soap. 22 H3

Tilsit former name (to 1945) of the Russian town of ▶Sovetsk. 16 L9

Timaru (Maori 'place of shelter'), industrial port and resort on South Island, New Zealand, on the east coast between Christchurch and Dunedin; population (1996) 27,500. Industries include flour milling, textiles, deep freezing, pottery, and brewing. It is the chief town of the southern Canterbury district. 78 C7

Timişoara capital of Timiş county, western Romania; population (1993) 325,000. Industries include electrical engineering, chemicals, pharmaceuticals, textiles, food processing, metal, and footwear. The revolt against the Ceauşescu regime began here in December 1989 when demonstrators prevented the arrest and deportation of a popular Protestant minister who was promoting the rights of ethnic Hungarians. This soon led to large prodemocracy rallies. 34 J4

Timor largest and most easterly of the Lesser Sunda Islands in the Malay Archipelago; area 33,610 sq km/12,973 sq mi. It is divided into **West Timor**, under Indonesian rule and **▶East Timor**, a disputed territory controlled by a United Nations (UN) transitional administration with a view to independence, its indigenous people were the Atoni, successive migrants have included Malaysians, Melamesians, Chinese, Arabs, and Gujerati.

Products include coffee, maize, rice, and coconuts. 53 C4

Timor Timur alternative name for ▶East Timor, a territory on the island of Timor in the Malay Archipelago. 53 C4

Tindouf Saharan oasis in the Aïn-Sefra region of Algeria, crossed by the Agadir–Dakar desert route; population (1987) 16,000. There are large iron ore deposits in the area. The oasis acted as a base for exiled Polisario guerrillas from the Western Sahara. **66** D3

Tipperary county of the Republic of Ireland, in the province of Munster, divided into North and South Ridings; county town Clonmel; area 4,255 sq km/1,643 sq mi; population (1996) 133,500. It includes part of the Golden Vale, a fertile dairy-farming region. Agriculture is the chief industry; barley and oats are the main crops, but potatoes and turnips are also grown. Cattle are reared in large numbers, and there are flour mills and butter factories. There is also horse and greyhound breeding. Other main towns are Cahir, Carrick-on-Suir, Cashel, Templemore, Tipperary, Thurles, Nenagh, and Roscrea. Major tourist attractions in the county include the Rock of Cashel and Cahir Castle. **5** C5

Tirana (or **Tiranë**), capital (since 1920) of Albania; population (1991) 251,000. Industries include metallurgy, cotton textiles, soap, and cigarettes. It was founded in the early 17th century by Turks when part of the Ottoman Empire. Although the city is now largely composed of recent buildings, some older districts and mosques have been preserved. **36** B3

Tiranë alternative spelling of ►Tirana, the capital of Albania. **36** B3

Tirol federal state of Austria; area 12,600 sq km/4,864 sq mi; population (1995) 655,200. Its capital is Innsbruck, and it produces diesel engines, optical instruments, and hydroelectric power. Tirol was formerly a province (from 1363) of the Austrian Empire, divided in 1919 between Austria and Italy (see ►Trentino–Alto Adige). **8** (2)C3

Tiruchchirappalli (formerly **Trichinopoly**; 'town of the sacred rock'), city in Tamil Nadu, India, situated at the head of the Cauvery delta; population (1991) 387,000. It is a rail junction and its chief industries are cotton textiles, cement, cigars, gold and silver filigree, and railway goods. Dating from the 2nd century, Tiruchchirappalli is a place of pilgrimage and was the capital of Tamil kingdoms during the 10th to 17th centuries. It was the scene of much fighting between British and French forces until annexed by Britain in 1801. The town is dominated by a fort (1660) and temple built on a rock 84 m/275 ft high. **54** C6

Tisza major tributary of the River Danube, rising in the Carpathian Mountains in Ukraine and flowing through Hungary to the Federal Republic of Yugoslavia; length 967 km/601 mi. The Tisza joins the Danube (Dunav) 40 km/25 mi east of the city of Novi Sad in the Vojvodina area of northern Serbia. **18** M9

Titicaca, Lake lake in the Andes, 3,810 m/12,500 ft above sea level and 1,220 m/4,000 ft above the treeline; area 8,300 sq km/3,200 sq mi, the largest lake in South America, and the world's highest navigable body of water. It is divided between Bolivia (port at Guaqui) and Peru (ports at Puno (principal port) and Huancane). The lake is fed by several streams which originate in the snow-capped surrounding mountains. The lake contains enormous frogs, which are farmed, the legs being an edible delicacy, and there is some trout farming. The herding of alpacas and llamas is also common. It is one of the few places in the world where reed boats are still made by the Uru tribal peoples (Lake Tana in Ethiopia is another). The lake is also used for irrigation. **98** D7

Titograd former name (1946–92) of ►Podgorica, a city in Montenegro, Yugoslavia. **34** G7

Titusville town and administrative headquarters of Brevard County, east Florida, USA; population (1990) 39,400. It is situated on the Indian River, 56 km/35 mi east-northeast east of Orlando. It is connected by causeway to Cape Canaveral, which lies 24 km/15 mi to the east. The city grew quickly with the construction of space program facilities in the 1950s and 1960s; the Kennedy Space Center is nearby. Industries include aerospace, citrus fruits, and sport fishing. Titusville was founded 1867. It is a popular winter resort. **90** E4

Tivoli (ancient **Tibur**), town in Lazio, Italy, 25 km/15 mi northeast of Rome, Italy; population (1990) 55,000. It has the remains of Hadrian's Villa, with gardens; and the Villa d'Este, with Renaissance gardens laid out in 1549 for Cardinal Ippolito d'Este. Wine is produced, and travertine (a decorative building stone) is quarried locally. **32** G7

Tlemcen (Roman **Pomaria**), city of northwest Algeria, capital of Tlemcen department, 130 km/80 mi southwest of Oran; population (1998 est) 177,900. The city lies in an area of olive plantations and vineyards; industries also include textiles. For three centuries until 1553, when it came under Ottoman Turkish rule. Tlemcen was the capital of an Arab sultanate. **66** E2

Toamasina formerly (until 1979) **Tamatave** port and coastal resort on the east coast of Madagascar; population (1993) 127,400. It is the country's principal port, from which sugar, coffee, and tea are major exports. **72** H3

Tobago island in the West Indies; part of the republic of ►Trinidad and Tobago. **94** M6

Tobolsk city in Tyumen oblast (region), west-central Russian Federation, sited at the junction of the rivers Irtysh and Tobol; population (1990) 94,500. Tobolsk lies 240 km/149 mi north of Tyumen at the centre of a lumbering and dairy-farming area. A petrochemical plant is located near the city. Tobolsk is an important local cultural centre; a theatre was founded here as early as 1705. **38** N3

Tobruk (or **Tubruq**), port, and only natural harbour, in Libya, 96 km/60 mi west of Bardia; population (1995 est) 127,000. The town has an oil refinery linked by a 560 km/348 mi pipeline to the Spirit oilfield. Another local industry is ship repair. Occupied by Italy in 1911, it was taken by Britain in 1941 during World War II, and unsuccessfully besieged by Axis forces April–December 1941. It was captured by Germany in June 1942 after the retreat of the main British force to Egypt. It was finally recaptured by the British in November 1942. **64** D1

Togliatti (or **Tol'yatti**; formerly **Stavropol**), port on the River ►Volga in the Samara oblast of western central Russia, 65 km/40 mi northwest of the city of Samara; population (1990) 642,000. Togliatti is the principal centre for Russian car manufacture; the Volga automobile works here began production in 1970. Other industries include ship repair and food processing, and the production of synthetic rubber, chemicals and fertilizers, electrical goods, and cement. Following flooding and the construction of the Kuybyshev Reservoir in the 1950s, the city was relocated and in 1964 was renamed in honour of the Italian communist leader Palmiro Togliatti. 38 J4

Togo country in West Africa, on the Atlantic Ocean, bounded north by Burkina Faso, east by Benin, and west by Ghana.

Togo covers an area of 56,800 sq km/ 21,930 sq mi, and has a population of 4,629,000 (2000 est). The capital is ►Lomé; other major towns and cities include Sokodé, Kpalimé, Kara, Atakpamé, Bassar, and Tsévié. The head of state is Etienne Gnassingbé Eyadéma (from 1967); the head of government is Agbeyome Messan Kodjo (from 2000). The currency is the franc CFA, and the languages spoken include French (official), Ewe, Kabre, Gurma, and other local languages. 68 E3

Tokelau (formerly **Union Islands**), overseas territory of New Zealand, 480 km/300 mi north of Apia, Samoa, comprising three coral atolls: Atafu, Fakaofo, and Nukunonu; area 10 sq km/ 4 sq mi; population (1994) 1,600. Copra and handicrafts are exported. 74 J6

Tokyo capital of Japan, on Honshu island; population (1994 est) 7,874,000. It is Japan's main cultural, financial and industrial centre (engineering, chemicals, textiles, electrical goods).

Founded in the 16th century as **Yedo** (or **Edo**), it was renamed when the emperor moved his court here from Kyoto in 1868. By the end of the 18th century, Yedo, with 1 million people, was the largest city in the world. An earthquake in 1923 killed 58,000 people and destroyed much of the city, which was again severely damaged by Allied bombing in World War II when 60% of Tokyo's housing was destroyed; US firebomb raids of 1945 were particularly destructive with over 100,000 people killed in just one night of bombing on 9 March. The subsequent rebuilding has made it into one of the world's most modern cities. 48 K6

Toledo capital of Toledo province in Castilla–La Mancha, central Spain, built on a rock above the River Tagus; population (1990) 60,700. It was the capital of the Visigoth kingdom 534–711, then became a Moorish city, and was the Castilian capital from 1085–1560. Knives, silks, and ceramics are manufactured here. There is a Gothic cathedral (13th–17th centuries) and several churches which preserve paintings by El Greco. The alcazar (fortified palace) was rebuilt after the successful Nationalist defence of Toledo in the Spanish Civil War (1936–39) and became a Nationalist shrine. 28 F5

Toledo inland port in Ohio, USA, at the mouth of the Maumee River on Lake Erie, 153 km/96 mi west of Cleveland; population (1996 est) 317,610. It is an oil- and gas-pipeline terminus, and a shipping centre for coal, iron-ore, vehicle parts, and grain. Industries include oil-refining, shipbuilding, food-processing, and the manufacture of cars, electrical goods, steel, and glass. A French fort was built in 1700, but permanent settlement did not begin until after the War of 1812. 88 D2

Tol'yatti alternative spelling of ►Togliatti, a port on the River Volga in central Russia. 38 J4

Tomaszów Mazowiecki industrial town (woollen textiles, synthetic fibres) in Piotrków province, Poland, on the River Pilica, 47 km/29 mi south of Łódź; population (1990) 69,900. 18 K6

Tombouctou (or **Timbuktu**), town in Mali, near the most northerly point on the Niger River; population (1996) 20,500 (town); (1987) 453,000 (region). It was a Tuareg camel caravan centre on the fringe of the Sahara from the 11th century. Since 1960 the area surrounding the town has become increasingly arid, and the former canal link with the River Niger is dry. Products include salt. 66 E5

Tomsk city on the River Tom, in the Russian Federation, in western central Siberia; industries include synthetic fibres, timber, distilling, plastics, and electrical motors; population (1994) 496,000. It was formerly a gold-mining town and the administrative centre of much of Siberia. 42 Q6

Tønder former county in southwest Jutland, Denmark, now part of Sonderjylland county; area 1,386 sq km/535 sq mi. The capital was Tønder. As part of ►Schleswig-Holstein, the region belonged to Prussia and then Germany 1864–1920. 20 D2

Tonga country in the southwest Pacific Ocean, in ►Polynesia.

Tonga covers an area of 750 sq km/ 290 sq mi, and has a population of 99,000 (2000 est). The capital is ►Nuku'alofa (on Tongatapu island); other major towns and cities include Neiafu, Haveloloto, Vaini, and Tofoa-Koloua. The head of state is King Taufa'ahau Tupou IV (from 1965); the head of government is Prince Ulukalala Lavaka Ata (from 2000). The currency is the pa'anga, or Tongan dollar, and the languages spoken include Tongan (official) and English. 74 J7

Tongeren (French **Tongres**), town in the province of Limbourg, Belgium, 19 km/12 mi southeast of Hasselt; population (2001 est) 30,100. Probably the oldest town in Belgium, Tongeren was known as Civitas Tongrorum under the Romans. Part of the Roman defensive wall can still be seen. The town also has a 15th-century church, a convent, and the Moerenpoort, a remnant of fortification from the 14th century. Tongeren is the site of a cattle market. 22 H4

Tonkin, Gulf of (or **Tongking, Gulf of**), arm of the South China Sea, between the coast of Vietnam and the island of Hainan. There are oil reserves, and China and Vietnam disagree over their respective territorial boundaries in the area. 50 D3

Tonle Sap (or **Tônlé Sab**, or **Great Lake**), lake on a tributary of the ►Mekong River, western Cambodia; area ranging from 2,600 sq km/1,000 sq mi to 6,500 sq km/ 2,500 sq mi at the height of the monsoon. During the June–Nov wet season it acts as a natural flood reservoir. 50 C4

Toowoomba city in southeast Queensland, Australia, 162 km/100 mi west of Brisbane; population (1996) 83,350. Toowoomba is situated on the eastern edge of the rich agricultural Darling Downs region, for which it is the commercial centre. Industries include coal mining, ironworking, engineering, and the manufacture of food products and agricultural machinery. 76 K5

Topeka capital of ►Kansas, USA, on the Kansas River, 85 km/53 mi west of Kansas City; population (1996 est) 119,660. It is an agricultural centre for eastern Kansas, trading in wheat and cattle. Manufactured products include processed foods, printed materials, and rubber and metal goods. **84** G4

Torbay urban area and unitary authority in southwest England created in April 1998 from part of the county of Devon.
area 627 sq km/242 sq mi
towns and cities Torquay (administrative headquarters), Paignton, Brixham
features Tor Bay; English Channel coast; 23 beaches including Goodrington Sands; Oldway Mansion (Paignton) modelled partly on Versailles; 12th-century Torre Abbey (Torquay); replica of Drake's 'Golden Hind' (Brixham); Abbey Mansion (17th/18th century); Paignton Zoo
industries tourism, fishing, electronics, radio equipment, iron founding, horticultural products
population (1996) 128,000 **5** E6

Torfaen unitary authority in south Wales, created in 1996 from part of the former county of Gwent.
area 98 sq km/38 sq mi
towns Pontypool (administrative headquarters), ►Cwmbran (the first new town in Wales)
physical Coity Mountain in the north, River Afon Llwyd
industries advanced electronics, automotive, engineering
population (1996) 90,700 **5** E6

Torgau historic town and river port on the west bank of the Elbe, in Leipzig county, Saxony, Germany; population (1995) 20,700. There is a food-processing industry. In 1760, during the Seven Years' War, Frederick II of Prussia defeated the Austrians nearby, and in World War II the US and Soviet forces first met here. **20** H5

Torhout (French **Thourhout**), town in the province of West Flanders, Belgium, 19 km/12 mi southwest of Bruges; population (1991) 18,200. The oldest town in Flanders, Torhout was once the chief cloth market of the country. The town still manufactures linen, and has a horse market. **22** F3

Torino Italian name for the city of ►Turin. **30** C5

Toronto (Huron 'place of meeting'), port and capital of ►Ontario, Canada, at the mouths of the Humber and Don rivers on Lake Ontario; population (1991) 635,400, metropolitan area (1996) 4,444,700. It is a major shipping point on the ►St Lawrence Seaway, and Canada's main financial, business, commercial, and manufacturing centre. Industries include shipbuilding, food-processing, publishing, and the production of fabricated metals, aircraft, farm machinery, cars, chemicals, and clothing. It is also a tourist and cultural centre, with theatres and a film industry. **88** E2

Toros Dağlari Turkish name for the ►Taurus Mountains, a mountain range in southern Turkey. **58** E5

Torquay resort in southern England, 41 km/25 mi south of Exeter; from April 1998, administrative headquarters of Torbay unitary authority; population (1991) 59,600. It is a sailing centre and has an annual regatta in August. Tourism is very important. Torquay lies in the area known as the English Riviera on account of its mild climate and exotic plants, including palm trees. **24** J11

Torremolinos tourist resort on the Costa del Sol between Málaga and Algeciras in Andalusia, southern Spain; population (1991) 31,700. There is a wine museum and a modern congress and exhibition centre. **28** F8

Torrens, Lake salt lake in eastern South Australia, west of the Flinders Range and 56 km/35 mi north of Port Augusta; average width 32 km/20 miles; length 209 km/130 mi; area 5,800 sq km/2,239 sq mi. The lake is 8 m/25 ft below sea level. It is reduced to a salt marsh in dry weather. **76** G6

Torreón industrial and agricultural city in Coahuila state, northern Mexico, on the River Nazas at an altitude of 1,127 m/3,700 ft; population (1990) 876,500. Before the arrival of the railway in 1907 Torreón was the largest of the three Laguna cotton-district cities (with Gómez Palacio and Ciudad Lerdo). Since then it has developed as a major thoroughfare and commercial centre. **92** F3

Torres Strait channel between the Arafura Sea and the Coral Sea, separating the island of New Guinea from Cape York on the northern tip of Queensland, Australia; width 130 km/80 mi. The strait is scattered with reefs and small islands. **76** H2

Torres Vedras town in Santarem district, western Portugal, on the Atlantic coast 40 km/25 mi north of Lisbon; population (1991) 13,300. It is the centre of a wine-producing region. The fortifications known as the **lines of Torres Vedras** were built here by the British commander Wellington in 1810, during the Peninsular War; stretching 45 km/28 mi to the River Tagus, they consisted of 100 forts, connected by entrenchments and water courses. **28** A5

Toruń (German **Thorn**), industrial river port (electronics, fertilizers, synthetic fibres) in northern Poland, on the River Vistula; population (1991) 202,000. It was founded by the Teutonic Knights 1230 and is the birthplace of the astronomer Copernicus. **18** H4

Toscana Italian name for the region of ►Tuscany. **9** (4)C3

Toul French town in the *département* of Meurthe-et-Moselle, on the River Moselle and the Rhine and Marne Canal; population (1990) 25,000. Pottery is manufactured. The church of St Etienne was built between 965 and 1496. Taken by the French from the Germans in 1552, Toul was finally ceded to France in 1648 under the Treaty of Westphalia. **26** L5

Toulon port and administrative centre of ►Var *département*, southeast France, on the Mediterranean Sea, 48 km/30 mi southeast of Marseille; population (1990) 170,200, conurbation 437,000. It is the chief Mediterranean naval station of France. Industries include oil refining, marine engineering, armaments, chemicals, furniture, and clothing. There is a university. Toulon was the Roman **Telo Martius**, became the seat of a bishopric in the 6th century and was made a port by Henry IV. During World War II the French fleet was scuttled here (1942) to avoid it passing into German control. **26** L10

Toulouse administrative centre of ►Haute-Garonne *département* in southwest France, 200 km/125 mi southeast of Bordeaux on the River Garonne; population (1990) 365,900, conurbation 650,000. It is the fourth city of France, a centre of communications, and the seat of an archbishopric and a university, founded in 1229. The town is a marketing, publishing, and banking centre, and its chief industries are textiles, chemicals, metallurgical goods, and aircraft construction; Concorde was built here. It has also become a major European centre of scientific research, especially in aerospace, electronics, data processing, and agriculture. **26** G10

Tourcoing town in the Nord *département* of France, 13 km/8 mi northeast of Lille; population (1990) 94,400, conurbation 130,000. With ►Roubaix it forms the ►Lille metropolitan area. It is situated near the Belgian border, and has been a textile centre, especially for wool, since the 12th century. **22** F4

Tournai (Flemish **Doornik**), town and railway junction in Hainaut province, Belgium, on the River Schelde, 43 km/27 mi northeast of Mons; population (1997) 67,900. Industries include carpets, textiles, pottery, cement, and leather. There are freestone and limestone quarries. It stands on the site of a Roman relay post and has a Romanesque and Gothic cathedral (11th–14th centuries) with the oldest belfry in the country. **22** F4

Tours administrative centre of ►Indre-et-Loire *département* in Centre region, west-central France, on the River Loire 200 km/125 mi southwest of Paris; population (1990) 133,400, conurbation 282,000. It manufactures chemicals, textiles, machinery, and electrical goods, and has a trade in agricultural produce, fruit, wine, and spirits. It is an ancient city and was capital of the former province of Touraine. Tours became the seat of the French government for four days in 1940 during World War II. **26** F6

Townsville city and port on the northeast coast of Queensland, Australia, 1,384 km/860 mi north of Brisbane; population (1996) 109,914. Situated on Cleveland Bay, Townsville is the commercial and administrative centre for a large mining and agricultural area. It contains a large bulk sugar refinery and its chief exports are sugar, wool, and minerals, including gold and silver. Townsville is a centre for tourists visiting the ►Great Barrier Reef. **76** J3

Trâblous Arabic name for ►Tripoli, a port in north Lebanon. **60** C2

Trabzon (formerly **Trebizond**), port on the Black Sea, northeastern Turkey, 355 km/220 mi southwest of Batum; population (1990) 143,900. Its exports include fruit, tobacco, and hides. **58** H3

Trail mining centre in Kootenay Boundary Regional District, southeast British Columbia, Canada, on the Columbia River, 77 km/48 mi southwest of Nelson; population (1991) 8,000. It lies at the centre of a rich mineral bearing district containing gold, silver, lead, zinc, and copper. Industries include smelting, sawmilling, food-processing, and the manufacture of chemicals and fertilizers. It was founded in 1895 as a mining and smelting town. **86** C1

Transcaucasia geographical region south of the Caucasus Mountains, encompassing the independent states of ►Armenia, ►Azerbaijan, and ►Georgia; it is bounded by the Caucasus Mountains in the north, the frontier with Turkey and Iran in the south, and the Black and Caspian Seas in the west and east respectively. Transcaucasia covers a total area of 186,100 sq km/71,853 sq mi.

Trans-Siberian Railway the world's longest single-service railway, connecting the cities of European Russia with Omsk, Novosibirsk, Irkutsk, and Khabarovsk, and terminating at ►Nakhodka on the Pacific coast east of Vladivostok. The line was built between 1891 and 1915, and has a total length of 9,289 km/ 5,772 mi, from Moscow to Vladivostok.

Transvaal former province of northeast South Africa to 1994, when it was divided into Mpumalanga, Northern, and Gauteng provinces. It bordered Zimbabwe to the north, Botswana to the northwest, and Swaziland and Mozambique to the east. It was settled by *Voortrekkers*, Boers who left Cape Colony in the Great Trek from 1831. Independence was recognized by Britain in 1852, until the settlers' difficulties with the conquered Zulus led to British annexation in 1877. It was made a British colony after the South African War 1899–1902, and in 1910 became a province of the Union of South Africa.

Transylvania mountainous area of central and northwestern Romania, bounded to the south by the Transylvanian Alps (an extension of the ►Carpathian Mountains). Formerly a principality, with its capital at Cluj-Napoca, it was part of Hungary from about 1000 until its people voted to unite with Romania 1918. In a 1996 treaty Hungary renounced its claims on Transylvania.

Trapani port and naval base in northwestern Sicily, about 48 km/ 30 mi north of Marsala; population (1990) 72,800. It trades in wine, salt, and fish. **32** G11

Trebizond former English name of ►Trabzon, a city in Turkey. **58** H3

Trelleborg town in Malmöhus county, Sweden; population (1995) 37,800. The southernmost Swedish town, Trelleborg prospered in the Middle Ages as a port and through herring fishing. It still has a coastal trade, and ferries run to Rostock, Travemünde, and Sassnitz in Germany. Trelleborg has sugar-refining, tyre, and other industries. **16** G9

Trengganu (or **Terengganu**), state of east Peninsular Malaysia; capital Kuala Trengganu; area 13,000 sq km/ 5,000 sq mi; population (1993) 752,000. Its exports include copra, black pepper, tin, and tungsten; there are also fishing and offshore oil industries. **11** (1)A1

Trent third longest river of England; length 275 km/170 mi. Rising in the south Pennines (at Norton in the Moors) by the Staffordshire–Cheshire border, it flows south and then northeast through Derbyshire, along the county boundary of Leicestershire, and through Nottinghamshire and Lincolnshire, joining the Ouse east of Goole to form the Humber estuary, and entering the North Sea below Spurn Head. Its drainage basin covers more than 10,000 sq km/4,000 sq mi. Main tributaries are the Churnet, Dove, and Derwent. **24** M8

Trentino-Alto Adige (formerly **Venezia Tridentina**), autonomous region of northern Italy, comprising the provinces of Bolzano and Trento; area 13,600 sq km/5,250 sq mi; population (1992 est) 896,700. Its chief towns are Trento (the capital) in the Italian-speaking southern area, and Bolzano (Bozen) in the northern German-speaking area of South Tirol (the region was Austrian until ceded to Italy in 1919 in the settlement following World War I). Wine, fruit, dairy products, and timber are produced. Paper, chemical, and metal industries use hydroelectric power. **9** (4)C1

Trento (German **Trient**; ancient **Tridentum**), capital of Trentino-Alto Adige region, Italy, on the Adige River and the Brenner Pass route, 160 km/100 mi northeast of Milan; population (1992) 101,500. Industries include the manufacture of electrical goods, cement, agricultural machinery, chemicals, and processed foods and beverages. The Council of Trent was held here from 1545 to 1563. **30** G4

Trenton capital of ►New Jersey, USA, on the Delaware River, 50 km/31 km northeast of Philadelphia; population (1996 est) 85,400. Trenton is the head of navigation on the Delaware, which forms the New Jersey–Philadelphia state border. Industries include metalworking, food-processing, and the manufacture of car parts, steel cable, rubber, and ceramics. **88** F2

Tréport, Le port and seaside resort on the English Channel, in the French *département* of Seine-Maritime; population (1990) 6,300. It has a 16th-century church and the ruins of an 11th-century abbey. It is separated by the River Bresle from Mer-le-Bains, which is in the *département* of Somme. **22** D4

Trèves French name for ►Trier, a city in Germany. **22** J5

Treviso (ancient **Tarvisium**), town in Veneto, northeast Italy, on the Sile River, 26 km/16 mi northwest of Venice; population (1990) 83,900. Its industries include the manufacture of machinery, ceramics, paper, and textiles. The cathedral (15th–16th century) was built on the site of a Romanesque church founded in the 11th century, and has an altarpiece by Titian. **30** H5

Trichinopoly former name for ►Tiruchchirappalli, a city in India. **54** C6

Trier (French **Trèves**), city in Rhineland-Palatinate, Germany, on the River Moselle, near the Luxembourg border, 115 km/71 mi west of Mainz; population (1995) 99,300. It is a centre for the wine trade. Once the capital of the Treveri, a Celto-Germanic tribe, it became known as **Augusta Treverorum** under the Roman emperor Augustus about 15 BC and was the capital of an ecclesiastical principality during the 14th–18th centuries. St Ambrose and Karl Marx were born here. There is a university. **22** J5

Trieste (Slovenian **Trst**; ancient **Tergeste**), port in Friuli-Venezia Giulia, Italy, on the Adriatic coast, opposite Venice; population (1992) 228,400, including a large Slovene minority. It is the largest seaport on the Adriatic, extending for 13 km/8 mi along the Gulf of Trieste. There are large shipyards, and an oil pipeline linked with refineries in Germany and Austria. It is the site of the International Centre for Theoretical Physics, established in 1964. **30** J5

Triglav mountain in Slovenia, rising to 2,863 m/9,393 ft, the highest peak in the Julian Alps. **30** J4

Trincomalee port in northeastern Sri Lanka; population (district, 1995 est) 331,000; (town, 1995 est) 50,000. It was an early Tamil settlement, and a British naval base until 1957. **54** D7

Trinidad town in Beni region, northern Bolivia, near the River Mamoré, 400 km/250 mi northeast of La Paz; population (1992 est) 57,300. The city lies on an old lake bed, the Moxos Plains, and originated in a Jesuit mission founded here in 1686. Industries include sugar refining and alcohol distilling. It is also a commercial centre trading in rice, cotton, sugar cane, cattle, furs, and feathers. Trinidad has an airport, and is linked by roads to Puerto Ballivián and to Cochabamba. It is the home of Mariscal Jose Ballivián University. **98** E6

Trinidad and Tobago country in the West Indies, off the coast of Venezuela.

Trinidad and Tobago covers an area of 5,130 sq km/1,980 sq mi (Trinidad 4,828 sq km/1,864 sq mi and Tobago 300 sq km/115 sq mi), and has a population of 1,295,000 (2000 est). The capital (and chief port) is ►Port-of-Spain; other major towns and cities include San Fernando, Arima, and Point Fortin. The head of state is Arthur Robinson (from 1997); the head of government is Basdeo Panday (from 1995). The currency is the Trinidad and Tobago dollar, and the languages spoken include English (official), Hindi, French, and Spanish. **98** E1

Tripoli (Arabic **Trâblous**; ancient **Tripolis**), port in north Lebanon, 65 km/40 mi northeast of Beirut; population (1998 est) 214,400. The second largest city in Lebanon, after Beirut, Tripoli is the terminus of the pipeline from Iraq, the site of Lebanon's second airport, and a centre of trade for north Lebanon and northwest Syria. There is oil refining; oranges, tobacco, and cotton are traded, as well as the manufacture of furniture, soap, and textiles. It stands on the site of the Phoenician city of Oea. **60** C2

Tripoli (Arabic **Tarābulus**), capital and chief port of Libya, on the Mediterranean coast, 600 km/373 mi west of Benghazi; population (1996 est) 990,000. Products include olive oil, fruit, fish, and textiles; industries include oil refining and food processing. Tripoli was founded about the 7th century BC by Phoenicians from Oea (now Tripoli in Lebanon). It was a base for the Axis powers during World War II. In 1986 it was bombed by the US Air Force in retaliation for international guerrilla activity. **66** H2

Tripura hill state of northeast India since 1972, formerly a princely state, between Bangladesh and Assam; area 10,500 sq km/4,050 sq mi; population (1994 est) 3,055,000. The capital is Agartala. Steel, jute, timber, and rubber are produced, and rice, millet, maize, fruit, cotton, tea, and sugar cane are grown. **10** F4

Tristan da Cunha group of islands in the south Atlantic, part of the British dependency of ►St Helena; area 110 sq km/42 sq mi; population (1996) 292. The main industry is crayfish fishing. **62** B9

Trois Rivières (English **Three Rivers**), port in southern Québec province, Canada, at the mouth of the Saint-Maurice River on the St Lawrence, equidistant between the cities of Montréal and Québec; population (1991) 49,400, metropolitan area (1996)143,600. Its natural deep-water harbour serves ocean-going vessels. The chief industry is the production of newsprint; textiles, electrical goods, fabricated metals, and processed foods are also manufactured. **88** F1

Trollhättan town in Älvsborg county, southwestern Sweden; population (1995) 52,500. The **Trollhätta Falls**, over 30 m/98 ft high, generate power for the surrounding factories. These include Saab car-assembly and engine works. **16** G7

Tromsø fishing port and largest town in northwest Norway, on Tromsø Island, and capital of the county of Troms; population (1991) 51,300. A church was founded here in the 13th century and the town grew up around it. Today the town trades in fish and fish products and is used as a base for Arctic expeditions. **16** K2

Trondheim fishing port, and county town of Sor-Trondelag, Norway, at the mouth of the Nid on Trondheim Fjord, 135 km/84 mi northeast of Kristiansund; population (1996) 135,900. It has canning, textile, margarine and soap industries. Originally called Nidaros, it was the medieval capital of Norway. Norwegian kings are crowned in the cathedral (1066–93), which is one of the most celebrated in Scandinavia. Trondheim was occupied by the Germans 1940–45 and used as a U-boat base, and the town was frequently bombed by the Allies. **16** F5

Trossachs ('bristled terrain'), woodland glen between lochs Katrine and Achray in Stirling unitary authority, Scotland, 3 km/2 mi long. Overlooking it are Ben Venue (727 m/ 2,386 ft) and Ben A'an (369 m/ 1,211 ft), a popular climbing venue, which rests against Meall Gainmheich (564 m/1,851 ft). Featured in the novels of Walter Scott, it has become a favoured tourist spot.

Troy city and port in eastern New York State, USA, on the east bank of the Hudson River, 13 km/8 mi northeast of Albany; population (1996 est) 52,500. It is a commercial and industrial centre manufacturing abrasives, metal, paper, car and railway parts, clothing, and processed foods. The city is the seat of Rensselear County and was incorporated in 1816. **88** F2

Troyes administrative centre of the *département* of ►Aube in the Champagne-Ardenne region of northeast France, situated on the River Seine 150 km/93 mi southeast of Paris; population (1990) 60,800, conurbation 120,000. The town has an agricultural market, but is also an industrial city manufacturing textiles, machinery and foodstuffs. The hosiery industry remains important. The **Treaty of Troyes** signed by Henry V of England and Charles VI in 1420 recognized Henry as heir to the French throne. **26** K5

Trucial States former name (to 1971) of the ►United Arab Emirates. It derives from the agreements made with Britain in 1820 to ensure a truce in the area and to suppress piracy and slavery. **56** F5

Trujillo city in northwestern Peru, with its port at Salaverry, capital of the department of La Libertad; population (1993) 509,300. Industries include engineering, copper, sugar milling, vehicle assembly and trade in agricultural produce. **98** B5

Truro market town in ►Cornwall, England, and administrative headquarters of the county, on the River Truro, a branch of the Fal, 14 km/ 9 mi north of Falmouth; population (1991) 19,000. It is a business centre. **24** G11

Tsaritsyn former name (to 1925) of ►Volgograd, a city in the southwestern Russian federation. As Stalingrad (1925–61), the city became famous as the place where the German advance into the Soviet Union was halted and repulsed (1942–43). **38** H5

Tsarskoe Selo (Russian 'the tsar's village'), former name of ►Pushkin, a city just south of St Petersburg, Russia. **38** F3

Tsinan alternative transliteration of ►Jinan, capital of Shandong province, east China. **46** F3

Tsingtao alternative transliteration of ►Qingdao, a port in Shandong province, east China. **46** G3

Tsumeb principal mining centre of north Namibia, 60 km/37 mi northwest of Grootfontein; population (1990) 13,500. The terminus of the railway from Swakopmund, it is the centre of an important diamond, copper, lead, zinc, and vanadium mining region. Since regional restructuring in 1992–93 the town has been the administrative centre of Oshikoto region. **72** B3

Tsushima island in Nagasaki prefecture, Japan, between Korea and Japan in **Tsushima Strait**, area 702 sq km/271 sq mi; population (1990) 46,100. Forestry is the main industry, together with *shiitake* mushrooms, buckwheat, squid, and cultured pearls. The Russian fleet was destroyed by the Japanese under Admiral Togo here on 27 May 1905 in the Russo-Japanese War, and 12,000 Russians were killed. The chief settlement is Izuhara. **48** E6

Tuamotu Archipelago (French **Archipel des Tuamotu**), two parallel ranges of 78 atolls, part of ►French Polynesia; area 690 sq km/ 266 sq mi; population (1996) 15,370, including the ►Gambier Islands to the east. The atolls stretch 2,100 km/ 1,300 mi north and east of the Society Islands. The administrative headquarters is Apataki. This archipelago is made up of the largest group of coral atolls in the world. The largest atoll is Rangiroa, the most significant is Hao; they produce pearl shell and copra. Spanish explorers landed 1606, and the islands were annexed by France in 1881. France conducted nuclear test explosions at the Mururoa and Fangataufa atolls between 1966 and 1996 (46 above ground and 147 below). **74** M7

Tübingen town in Baden-Württemberg, Germany, on the River Neckar, 30 km/19 mi southwest of Stuttgart; population (1994) 83,600. There are publishing, paper, and textile industries, and surgical instruments are manufactured. The town dates from the 11th century; the university was established 1477. It was once a free city of the empire. **30** E2

Tubruq alternative spelling of ►Tobruk, a port in Libya. **64** D1

Tubuai Islands (or **Austral Islands**), chain of volcanic islands and reefs 1,300 km/800 mi long in ►French Polynesia, south of the Society Islands; area 148 sq km/ 57 sq mi; population (1996) 6,600. The main settlement is Mataura on Tubuai. They were visited by Capt Cook 1777 and annexed by France 1880. The chief products are copra and coffee. **74** L8

Tucson (Papago 'foot of the mountain'), city in southeast Arizona, USA, on the Santa Cruz and Rillito rivers, 105 km/66 mi north of the Mexican border; population (1996 est) 449,000. It stands 760 m/2,500 ft above sea level in the Sonora Desert; the Santa Catalina Mountains rise to about 2,750 m/9,000 ft to the northeast. The area's winter sports and mild, dry, sunny winter climate have made the city a popular winter residence and resort. Industries include copper-smelting, and the manufacture of aircraft and electronics. Cotton and cattle are marketed and processed from the surrounding irrigated agricultural area. 92 D2

Tula city in the Russian Federation and capital of the Tula oblast, situated on the River Upa, 193 km/121 mi south of Moscow; population (1990) 543,000. Its traditional industries are the manufacture of firearms and samovars (Russian tea urns). The city also produces iron and steel, chemicals, agricultural equipment, and machine tools. 38 G4

Tulare town in Tulare County, south-central California, USA; population (1990) 33,200. It is located 64 km/40 mi southeast of Fresno, in the San Joaquin Valley. Industries include food processing, wine, and dairy products. The town has a variety of creameries, cotton gins, and food packing plants. Tulare was built 1871 as a division headquarters for the Southern Pacific Railroad. 86 C3

Tulle French town, capital of Corrèze *département* on the western edge of the Massif Central; population (1990) 17,200. The town was founded in the 7th century. Only a 12th-century nave and belfry remain of its ancient cathedral. Tulle is an industrial town with a government firearms factory. It gives its name to tulle, a fine silk net fabric which was once made in the region. 26 G8

Tulsa (Creek 'old town'), city in northeast Oklahoma, USA, on the Arkansas River, 460 km/287 mi northeast of Oklahoma City; population (1996 est) 378,500. It is situated in a hilly area f artificial lakes, just east of the confluence of the Arkansas and Cimarron rivers. The city is an oil-producing and aerospace centre; other industries include mining and the manufacture of machinery, metals, cars, textiles, processed foods, and cement. 84 G4

Tunbridge Wells (officially **Royal Tunbridge Wells**), spa and commuter town in Kent, southeast England, between London and Hastings; population (1991) 60,300. It has a light industrial estate. The town developed after the discovery of iron-rich springs here in 1606. The **Pantiles** or shopping parade (paved with tiles in the reign of Queen Anne), was a fashionable resort; visited by Queen Victoria, the town has been named 'Royal' since 1909. 22 C3

Tunis capital and chief port of Tunisia; population (1994) 674,100. Industries include chemicals, textiles, engineering, lead smelting, and distilling. Velvets, silks, linen, and fez caps are also manufactured. Exports include phosphates, iron ore, fruit, and vegetables. Founded by the Arabs, it was captured by the Turks in 1533, then occupied by the French in 1881 and by the Axis powers 1942–43. The ruins of ancient Carthage are to the northeast. 66 H1

Tunisia country in North Africa, on the Mediterranean Sea, bounded southeast by Libya and west by Algeria.

Tunisia covers an area of 164,150 sq km/63,378 sq mi, and has a population of 9,586,000 (2000 est). The capital (and chief port) is ►Tunis; other major towns and cities include Sfax, Ariana, Bizerte, Gabès, Sousse, Kairouan, and Ettadhamen. The head of state is Zine el-Abidine Ben Ali (from 1987); the head of government is Muhammad Ghannouchi (from 1999). The currency is the Tunisian dinar, and the languages spoken include Arabic (official) and French. 66 E2

Tunja capital of the Andean department of Boyacá, east-central Colombia, 2,800 m/9,200 ft above sea level; population (1992) 112,000. It is the centre of an agricultural and mining community, and linked to Bogotá by railway. One of the oldest cities in Colombia, Tunja was formerly the seat of the Chibcha Indian kings. The Spanish refounded a city here in 1539 on the site of Hunza, the ancient Muisca Indian seat. In 1818 Simón Bolívar defeated Spanish Royalists near Tunja. Industries include agriculture and mining. Many colonial buildings remain or have been restored, including the church of Santo Domingo (begun in 1594), and the Santa Clara Chapel (1580), now a museum. 98 C2

Turin (Italian **Torino**; ancient **Augusta Taurinorum**), capital of Piedmont, northwest Italy, at the confluence of the rivers Po and Dora Riparia; population (1992) 952,700. It stands at the foot of the Alps, commanding road and rail routes between France and Italy. Iron, steel, cars, silk and other textiles, fashion goods, chocolate, and wine are produced. There is a university (1404), and a 15th-century cathedral. Turin was the first capital of united Italy (1861–64). 30 C5

Turkana, Lake (formerly **Lake Rudolf** until 1979), lake in the Great Rift Valley, 375 m/1,230 ft above sea level, with its northernmost end in Ethiopia and the rest in Kenya; area 8,000 sq km/3,100 sq mi. It is saline, and shrinking by evaporation. Its shores were an early human hunting ground, and valuable remains have been found that are accurately datable because of undisturbed stratification. 70 F3

Turkestan historical area of Central Asia extending from the Caspian Sea in the west to the Gobi desert in the east. It is now divided among Kazakhstan, Kyrgyzstan, Tajikistan, Turkmenistan, Uzbekistan, Afghanistan (Northeast province), and China (part of Xinjiang Uygur province). It formerly covered an area of some 2,600,000 sq km/1,003,680 sq mi; its principal cities were ►Tashkent, ►Samarkand, and ►Bukhara.

Turkey country between the Black Sea to the north and the Mediterranean Sea to the south, bounded to the east by Armenia, Georgia, and Iran, to the southeast by Iraq and Syria, to the west by Greece and the Aegean Sea, and to the northwest by Bulgaria.

Turkey covers an area of 779,500 sq km/300,964 sq mi, and has a population of 66,591,000 (2000 est). The capital is ►Ankara; other major towns and cities include Istanbul, Izmir, Adana, Bursa, Gaziantep, Konya, Mersin, Antalya, and Diyarbakduringr. The head of state is Ahmet Necdet Sezer (from 2000); the head of government is Bülent Ecevit (from 1999). The currency is the Turkish lira, and the languages spoken include Turkish (official), Kurdish, and Arabic. 58 D4

Turkmenistan country in central Asia, bounded north by Kazakhstan and Uzbekistan, west by the Caspian Sea, and south by Iran and Afghanistan.

Turkmenistan covers an area of 488,100 sq km/188,455 sq mi, and has a population of 4,459,000 (2000 est). The capital is ►Ashgabat; other major towns and cities include Chardzhov, Mary, Nebit-Dag, Dashhowuz, and Turkmenbashi. The head of state and government is Saparmurad Niyazov (from 1990). The currency is the manat, and the languages spoken include Turkmen (a Turkic language; official), Russian, Uzbek, and other regional languages. **56** G2

Turks and Caicos Islands British crown colony in the West Indies, the southeastern archipelago of the Bahamas; area 430 sq km/166 sq mi; population (1990 est) 12,400 (90% of African descent). The capital is Cockburn Town on Grand Turk. Exports include crayfish and conch (flesh and shell); tourism is important, the main tourist island is Providenciales. The main languages are English and French Creole. The principal religion is Christian. **94** K4

Turku (Swedish **Åbo**), port in southwestern Finland, near the mouth of the River Aura, on the Gulf of Bothnia; population (1992) 160,000. Industries include shipbuilding, engineering, textiles, and food processing. It was the capital of Finland until 1812. **16** M6

Turnhout town in the province of Antwerp, Belgium; population (2001 est) 38,600. It is the economic capital of the Kempen (Campine) region. Turnhout has a large paper industry; there are also printing works, iron foundries, mills, brick kilns, and manufactures of lace, pottery, and food. **22** G3

Tuscaloosa city in western central Alabama, USA, on the Black Warrior River, southwest of Birmingham; population (1996 est) 82,400. Industries include lumbering, and the manufacture of chemicals, tyres, paper, and metal. Coal and iron deposits are mined nearby. **90** D3

Tuscany (Italian **Toscana**; Roman **Etruria**), region of north central Italy, on the west coast, comprising the provinces of Massa e Carrara, Arezzo, Florence, Grosseto, Livorno, Lucca, Pisa, Pistoia, and Siena; area 23,000 sq km/8,878 sq mi; population (1992 est) 3,528,700. Its capital is ►Florence, and cities include Pisa, Livorno, and Siena. The area is mainly agricultural, producing cereals, wine (Chianti hills), olives (Lucca) and tobacco (plain of Arno); it also has mining of lignite (upper Arno) and iron (Elba) and marble quarries (Carrara, Apuan Alps). The Tuscan dialect has been adopted as the standard form of Italian. **9** (4)E3

Tuva (or **Tyva**), republic in the southern Russian Federation; area 170,500 sq km/65,830 sq mi; population (1996) 309,000 (49% urban) (64% Tyvans). The capital is Kyzyl. There are coal and mineral mining, woodworking, and food-processing industries; sheep, goats, and cattle are raised. **44** F6

Tuvalu country in the southwest Pacific Ocean; formerly (until 1978) the Ellice Islands; part of ►Polynesia.

Tuvalu covers an area of 25 sq km/9.6 sq mi, and has a population of 12,000 (2000 est). The capital is ►Fongafale (on Funafuti atoll). The head of state is Queen Elizabeth II (from 1978), represented by Governor General Tulaga Manuella (from 1994); the head of government is Lagitupu Tuilimu (from 2000). The currency is the Australian dollar, and the languages spoken include Tuvaluan, English (both official), and a Gilbertese dialect (on Nui). **74** H6

Tver (formerly **Kalinin** 1932–90), capital of Tver region, in the northwestern Russian Federation, 160 km/100 mi northwest of Moscow; population (1990) 454,000. Tver is a transport centre on the Moscow–St Petersburg road and railway line, and has a port on the River Volga. The chief economic activities are the textile industry (producing cotton and artificial fibres), together with the manufacture of railway rolling-stock and chemicals. **38** G3

Tweed river rising in the Tweedsmuir Hills, 10 km/6 mi north of Moffat, southwest Scottish Borders, Scotland, and entering the North Sea at Berwick-upon-Tweed, Northumberland; length 156 km/97 mi. It flows in a northeasterly direction, and from Coldstream until near Berwick-upon-Tweed it forms the border between England and Scotland. **24** K6

Tyler city and administrative headquarters of Smith County, east Texas, USA; population (1992) 76,900. It lies midway between Dallas, to the west, and Shreveport, Louisiana. Industries include oil, roses, vegetables, furniture, plastics, heating and cooling equipment, prefabricated houses, and ceramics. Tyler is famed for its flower industry. It is one of the world's largest rose producers; municipal gardens show many varieties, and the city hosts an annual Rose Festival and the East Texas Agricultural Fair. It is Home to Texas College (1894), Tyler Junior College (1926), and the University of Texas at Tyler (1971). **84** G5

Tyne river of northeast England formed by the union of the North Tyne (rising in the Cheviot Hills) and South Tyne (rising near Cross Fell in Cumbria) near Hexham, Northumberland, and reaching the North Sea at Tynemouth; length 72 km/45 mi. Kielder Water (1980) in the North Tyne Valley is Europe's largest artificial lake, 12 km/7.5 mi long and 0.8 km/0.5 mi wide, and supplies the industries of Tyneside, Wearside, and Teesside. As well as functioning as a reservoir, it is a major resource for recreational use. **24** K6

Tyne and Wear metropolitan county of northeast England, created in 1974; in 1986, most of the functions of the former county council were transferred to the metropolitan borough councils.
area 540 sq km/208 sq mi
towns and cities Newcastle upon Tyne, Gateshead, Sunderland (administrative centres for the districts of the same name), South Shields (administrative centre of South Tyneside district), North Shields (administrative centre of North Tyneside district)
physical rivers: Tyne and Wear
features part of Hadrian's Wall; Newcastle and Gateshead, linked with each other with the coast on both sides by the Tyne and Wear Metro (a light railway extending 54 km/34 mi); Tyneside International Film Festival
industries car manufacturing on Wearside, electronics, offshore technology (floating production vessels), pharmaceuticals, and computers
population (1996) 1,127,300
5 F4

Tynemouth port and resort in ►Tyne and Wear, England; population (1991) 17,100. There are remains of a Norman castle and Benedictine priory. 24 L6

Tyrone historic county of Northern Ireland.
area 3,160 sq km/1,220 sq mi
towns and cities Omagh (county town), Dungannon, Strabane, Cookstown
physical rivers: Derg, Blackwater, Foyle; Lough Neagh; Sperrin Mountains
features Neolithic graves and stone circles, notably at Beaghmore, west of Cookstown; Ulster History Park, north of Omagh; Peatlands Park, east of Dungannon, preserving an ancient Irish bog; family home of the US president Woodrow Wilson at Dergalt, near Strabane
industries mainly agricultural: barley, flax, potatoes, turnips, cattle, sheep, brick making, linen, hosiery, shirts
population (1991) 158,500

Tyrrhenian Sea arm of the Mediterranean Sea surrounded by mainland Italy, Sicily, Sardinia, Corsica, and the Ligurian Sea. It is connected to the Ionian Sea through the Straits of Messina. Islands include Elba, Ustica, Capri, Stromboli, and the Lipari Islands. It has a deep seabed plain reaching a maximum depth of 3,620 m/11,876 ft. 32 F8

Tyumen capital city, economic and cultural centre of Tyumen oblast (region) in the west-central Russian Federation; population (1996 est) 497,000. The oldest city in Siberia, Tyumen lies 300 km/186 mi east of Yekaterinburg on the Trans-Siberian Railway and the Tura River. It lies at the heart of the country's principal oil-producing region, and its industries include shipbuilding, woodworking, leather tanning, and food processing. It is also an important river and rail transhipment point. 42 M6

Tyva alternative name for ►Tuva in the Russian Federation. 44 F6

Ubangi-Shari former name (to 1958) of the ►Central African Republic. 70 C2

Udaipur (or **Mewar**), industrial city in Rajasthan, India, capital of the former princely state of Udaipur; population (1991) 308,000. Industries include cotton and grain processing, handicrafts, the manufacture of chemicals and asbestos, and zinc-smelting. Tourism is important. It was founded in 1568 and has specatcular palaces (two on islands in Lake Pichola), some converted into hotels or museums. The Jagdish Hindu temple dates from 1640. 54 B4

Uddevalla seaport in southwest Sweden, in the county of Göteborg and Bohus; population (1995) 49,000. Industries include shipbuilding, chemicals, textiles, wood-pulp, and sugar refining. 16 F7

Udine capital of the autonomous Friuli-Venezia Giulia region, Italy, 100 km/62 mi northeast of Venice; population (1990) 98,300. It is an administrative and market centre for the Friulian Plain. Industries include chemicals and textiles. Udine was the capital of Friuli in the 13th century and passed to Venice in 1420. 30 J4

Udmurt (or **Udmurtiya**; Russian **Udmurtskaya**), autonomous republic in central Russian Federation, north of Tatarstan and northwest of Bashkortostan; area 42,100 sq km/16,200 sq mi; population (1990) 1,619,000 (59% Russian, 31% Udmurt, 7% Tatar). The capital is ►Izhevsk. There are timber, peat, quartz, tool and machine manufacturing, oil, and hydroelectric power generation industries. Flax and potatoes are grown. 38 K3

Ufa capital of ►Bashkortostan, central Russian Federation, located on the River Belaya, in the western Urals; population (1990) 1,094,000. Ufa is situated near the Tuymazy and Ishimbay fields in the Volga–Ural oil region, and is a centre for oil refining and the production of petrochemicals. One of the main manufacturing cities of the Urals, its industries include aerospace technology, electronic engineering, distilling, and lumbering. 38 L4

Uganda landlocked country in East Africa, bounded north by Sudan, east by Kenya, south by Tanzania and Rwanda, and west by the Democratic Republic of Congo (formerly Zaire).

Uganda covers an area of 236,600 sq km/91,351 sq mi, and has a population of 21,778,000 (2000 est). The capital is ►Kampala; other major towns and cities include Jinja, Mbale, Entebbe, Masaka, Mbarara, and Soroti. The head of state is Yoweri Museveni (from 1986); the head of government is Apolo Nsibambi (from 1999). The currency is the Ugandan new shilling, and the languages spoken include English (official), Kiswahili, and other Bantu and Nilotic languages. 70 E3

Uist two small islands, part of Western Isles unitary authority, in the Outer ►Hebrides, Scotland: North Uist and South Uist.

Ujiji port on Lake Tanganyika, near Kigoma, Tanzania, where Welsh-born explorer Henry Stanley found the Scottish missionary explorer David Livingstone in 1871; population (1998 est) 42,100. It was originally an Arab trading post for slaves and ivory. 70 D4

Ujung Pandang (formerly **Macassar** until 1973, or **Makassar**), chief port (trading in coffee, rubber, copra, and spices) on Sulawesi, Indonesia, with fishing and food-processing industries; population (1990) 913,200. It was established by Dutch traders 1607. 53 A4

Ukraine country in eastern central Europe, bounded to the east by Russia, north by Belarus, south by Moldova, Romania, and the Black Sea, and west by Poland, the Slovak Republic, and Hungary.

Ukraine covers an area of 603,700 sq km/233,088 sq mi, and has a population of 50,456,000 (2000 est). The capital is ►Kiev; other major towns and cities include Kharkov, Donetsk, Dnipropetrovs'k, Lviv, Krivoy Rog, Zaporizhzhya, and Odessa. The head of state is Leonid Kuchma (from 1994); the head of government is Viktor Yushchenko (from 1999). The currency is the hryvna, and the languages spoken include Ukrainian (a Slavonic language; official), Russian (also official in Crimea), and other regional languages. 14 G3

Ulaanbaatar (or **Ulan Bator**; formerly **Urga** until 1924; Mongolian 'Red Warrior/Hero'), capital and largest city of ►Mongolia, lying to the north in the valley of the River Tuul in the Khenti Mountains; population (1997 est) 627,300. Industries include machine tools, pharmaceuticals, car-pets, textiles, footwear, meat packing, brewing, and distilling, especially of vodka. It is the centre of Mongolia's road and rail network, and connected to the Trans-Siberian and Chinese railways. 44 H7

Ulan Bator alternative spelling of ►Ulaanbaatar, the capital of Mongolia. 46 D1

Ulan-Ude (formerly **Verkhneudinsk** 1666–1934), city in the southern Russian Federation, capital (since 1923), economic and cultural centre of the Republic of ►Buryat; population (1996 est) 368,000. Ulan-Ude is situated in southeastern Siberia south of Lake Baikal at the confluence of the Selenga and Ude rivers, and is a major station on the Trans-Siberian Railway. It is also the starting-point of the Trans-Mongolian line running to Ulan Bator and Beijing. A major industrial centre, Ulan-Ude manufactures locomotives and rolling stock, glass, timber, and foodstuffs; ship repairing is also a major industry. 44 H6

Uleåborg Swedish name for the Finnish port of ►Oulu. 16 N4

Ulm cathedral and university city in Baden-Württemberg, Germany, on the Danube at its confluence with the River Iller, 71 km/44 mi southeast of Stuttgart; population (1995) 115,400. Industries include motor vehicles, mechanical engineering, electronics, and data processing. Its Gothic cathedral (begun in 1377) with the highest stone spire ever built (161 m/528 ft) escaped damage in World War II when two-thirds of Ulm was destroyed. It was a free imperial city from the 14th century to 1802. The physicist and mathematician Albert Einstein was born here. 30 F2

Ulsan industrial city in South Kyongsang province, southeastern South Korea; population (1990) 683,000. Situated near Pusan on Ulsan Bay, part of the Sea of Japan, there are nearby whaling stations. Important industries are coal mining, shipbuilding, oil refining, and the manufacture of vehicles, petrochemicals, and fertilizers. 48 E6

Ulster a former kingdom and province in the north of Ireland, annexed by England in 1461. From Jacobean times it was a centre of English, and later Scottish, settlement on land confiscated from its owners; divided in 1921 into Northern Ireland (counties Antrim, Armagh, Down, Fermanagh, Londonderry, and Tyrone) and the Republic of Ireland (counties Cavan, Donegal, and Monaghan).

Uluru Aboriginal name for ►Ayers Rock. 76 F5

Umbria mountainous region of Italy in the central Apennines, including the provinces of Perugia and Terni; area 8,500 sq km/3,281 sq mi; population (1992 est) 815,000. Its capital is ►Perugia, and the River Tiber rises in the region. Industries include textiles, chemicals, and metalworking. Wine is produced (Orvieto), and tobacco, grain, and olives (Lake Trasimene) are grown. This is the home of the Umbrian school of artists, including Raphael. 9 (4)D3

Umeå seaport and capital of Västerbotten county in northeast Sweden on the Gulf of Bothnia at the mouth of the River Ume; population (1995) 101,300. Machinery, furniture, and wood pulp are manufactured here. It is the biggest city in northern Sweden and a cultural centre. There is a ferry service to Vaasa in Finland. 16 L5

Umm al Quwain one of the ►United Arab Emirates; population (1995) 35,200.

Umm Durman Arabic name for ►Omdurman, a city in Khartoum State, Sudan. 64 F4

Umtali former name (to 1982) for the town of ►Mutare in Zimbabwe. 72 E3

Umtata former capital of the former black homeland of Transkei, South Africa; population (1991 est) 55,000. It is terminus of the railway from East London. 72 D6

Ungava (in full **Péninsule d'Ungava**; Inupiaq 'towards the open water'), peninsular region in northern Québec and Labrador, Canada, bordered to the north by Hudson Strait, to the east by Ungava Bay and the Labrador Sea, to the south by the Eastmain River, and to the west by James Bay and Hudson Bay; area 911,110 sq km/351,780 sq mi. The peninsula is part of the Canadian Shield, and is barren and sparsely populated; low vegetation increases towards the south. 82 R4

United Arab Emirates federation in southwest Asia, on the Gulf, bounded northwest by Qatar, southwest by Saudi Arabia, and southeast by Oman.

The United Arab Emirates covers an area of 83,657 sq km/32,299 sq mi, and has a population of 2,441,000 (2000 est). The capital is ►Abu Dhabi; other major towns and cities include Dubai, Sharjah, Ras al Khaimah, Ajman, and Al 'Ayn. The head of state is Sheikh Zayed bin Sultan al-Nahayan of Abu Dhabi (from 1971); the head of government is Sheikh Maktum bin Rashid al-Maktum of Dubai (from 1990). The currency is the UAE dirham, and the languages spoken include Arabic (official), Farsi, Hindi, Urdu, and English. 56 F5

United Arab Republic union formed in 1958, broken in 1961, between ►Egypt and ►Syria. Egypt continued to use the name after the break up until 1971.

United Kingdom (UK), country in northwest Europe off the coast of France, consisting of England, Scotland, Wales, and Northern Ireland.

The United Kingdom covers an area of 244,100 sq km/94,247 sq mi, and has a population of 58,830,000 (2000 est). The capital is ►London; other major towns and cities include Birmingham, Glasgow, Leeds, Sheffield, Liverpool, Manchester, Edinburgh, Bradford, Bristol, Coventry, Belfast, and Cardiff. The head of state is Queen Elizabeth II (from 1952); the head of government is Tony Blair (from 1997). The currency is the pound sterling, and the languages spoken include English (official), Welsh (also official in Wales), and Gaelic. 24 G6

United States of America (USA), country in North America, extending from the Atlantic Ocean in the east to the Pacific Ocean in the west, bounded north by Canada and south by Mexico, and including the outlying states of Alaska and Hawaii.

The United States of America covers an area of 9,372,615 sq km/3,618,766 sq mi, and has a population of 278,357,000 (2000 est). The capital is ►Washington, DC; other major towns and cities include New York, Los Angeles, Chicago, Philadelphia, Detroit, San Francisco, Dallas, San Diego, San Antonio, Houston, Boston, Phoenix, Indianapolis, Honolulu, and San José. The head of state and government is George W Bush (from 2001). The currency is the US dollar, and the languages spoken include English and Spanish. 80 M5

Unzen active volcano on the Shimbara peninsula, Kyushu island, Japan, opposite the city of Kumamoto. Its eruption in June 1991 led to the evacuation of 10,000 people. It is the main feature of Unzen-Amakusa National Park. The highest peak,

Fugendake, is 1,359 m/4,459 ft high. There have been hot springs in the area since the 8th century and wild azaleas (*miyamakirishima*) grow locally.

Upper Austria (German **Oberösterreich**), mountainous federal state of Austria, drained by the Danube, and bordered on the north by the Czech Republic and on the west by Bavaria; area 12,000 sq km/4,632 sq mi; population (1995) 1,385,500. Its capital is Linz and the main towns are Steyr and Wels. Agricultural products include fruit, wine, sugar beet, and grain. There are reserves of oil, and salt and lignite are mined. Textiles, chemicals, and metal and electronic goods are manufactured. The population density is the highest of all the provinces except Vienna. 8 (2)E2

Upper Normandy English name for ▶Haute-Normandie, a coastal region of northwest France. 6 D2

Upper Volta former name (to 1984) of ▶Burkina Faso. 68 D2

Uppsala city in Sweden, northwest of Stockholm; population (1994) 181,200. Industries include engineering and pharmaceuticals. The university was founded 1477; there are Viking relics and a Gothic cathedral. The botanist Carolus Linnaeus lived here. 16 J7

Ural Mountains (Russian **Ural'skiy Khrebet**), mountain system extending for over 2,000 km/1,242 mi from the Arctic Ocean to the Caspian Sea, and traditionally regarded as separating Europe from Asia. The highest peak is Naradnaya, 1,894 m/6,214 ft. The mountains hold vast mineral wealth. 14 L1

Ural'skiy Khrebet Russian name for the ▶Ural Mountains, a mountain range extending from the Arctic Sea to the Caspian Sea. 14 L1

Urbana industrial city and administrative headquarters of Champaign County, east-central Illinois, 200 km/125 mi southwest of Chicago; population (1996 est) 33,200. It is an administrative and service centre for a grain-growing area. Products include processed foods, scientific instruments, electronics, and metal goods. With adjoining ▶Champaign, Urbana is the principal home of the University of Illinois (1867). It was first settled in 1822. 88 C2

Urga former name (to 1924) of ▶Ulaanbaatar, the capital of Mongolia. 44 H7

Uri canton of Switzerland on the southeast of Lake Lucerne; area 1,075 sq km/415 sq mi; population (1995) 35,900. The capital is ▶Altdorf. The main river is the Reuss. Uri was founded 1291 and (together with Schwyz and Nidwalden) was one of the three original cantons of the Swiss Confederation. Forestry and tourism are the main industries. 8 (1)C2

Uruguay country in South America, on the Atlantic coast, bounded north by Brazil and west by Argentina.

Uruguay covers an area of 176,200 sq km/68,030 sq mi, and has a population of 3,337,000 (2000 est). The capital is ▶Montevideo; other major towns and cities include Salto, Paysandú, Las Piedras, Rivera, and Tacuarembó. The head of state and government is Jorge Batlle Ibáñez (from 2000). The currency is the Uruguayan peso, and the languages spoken include Spanish (official) and Brazilero (a mixture of Spanish and Portuguese). 100 K5

Urumchi alternative spelling of ▶Urumqi, capital of Xinjiang Uygur Autonomous Region, China. 42 R9

Urumqi (or **Urumchi**, or **Wulumuqi**; formerly **Dihau** until 1953), industrial city and capital of Xinjiang Uygur Autonomous Region, northwest China, on the Urumqi River, at the northern foot of the Tian Shan Mountains; population (1994) 1,240,000. Industries include oil-refining, food-processing, brewing, and the manfuacture of cotton textiles, cement, iron, steel, plastics, and agricultural equipment. 42 R9

Ushuaia southernmost town in the world, at the tip of Tierra del Fuego, Argentina, less than 1,000 km/620 mi from Antarctica; population (1991) 29,700. It is a free port and naval base. Industries include lumbering, sheeprearing, and fishing. 100 H9

Usküb Turkish name of ▶Skopje, the capital of the Former Yugoslav Republic of Macedonia. 34 J7

Ussuri river in the Russian Far East forming the frontier between the northern Chinese province of Heilongjiang and the Russian Federation; length 575 km/357 mi. The Ussuri rises in the Sikhote-Alin mountain range, and flows north to join the Amur at Khabarovsk; it is navigable for most of its length. One of its tributaries forms the outlet of Lake Khanka. 48 G1

Ust-Kamenogorsk river port and chief centre of the nuclear industry in the Russian Federation, in the Altai Mountains, on the River Irtysh; population (1998 est) 324,200. 42 Q8

Utah state in western USA. It is nicknamed the Beehive State or the Mormon State. Utah was admitted to the Union in 1896 as the 45th US state. One of the Mountain States, it is bordered to the east by Colorado, to the north by Wyoming, to the west by Nevada, and to the south by Arizona. At the 'Four Corners', in the southeast, it also touches New Mexico.
population (2000 est) 2,233,200
area 219,900 sq km/84,881 sq mi
capital ▶Salt Lake City
towns and cities Provo, Ogden, West Valley City
industries and products wool, gold, silver, copper, coal, oil, potash, salt, steel, aerospace and military-dependent industries, tourism 84 D4

Utica city and river port in central New York State, USA, on the Mohawk River, near the Adirondack Mountains; population (1996 est) 61,400. Engine parts, electronics, beer, leather goods, and clothing are produced. Utica was an important textile centre from about 1850 to 1950. The first Woolworths store was opened here in 1879. 88 E2

Utrecht capital of Utrecht province, Netherlands, on the River Oude Rijn (see ▶Rhine), 60 km/45 mi east of The Hague; population (1996) 234,800. The city's industries include textiles, carpets, pottery, chemicals, engineering, and printing. The city was important in the Middle Ages, and has a Gothic cathedral (1254–1517) with the highest church tower in the Netherlands. Utrecht's university was founded 1636. 22 H2

Utrecht province of the Netherlands, lying southeast of Amsterdam, and south of the IJsselmeer, on the Kromme Rijn (Crooked Rhine); area 1,330 sq km/514 sq mi; population (1997) 1,079,400. The capital is ▶Utrecht. Industries include petrochemicals, textiles, electrical goods, engineering, steel-works, railway workshops, and furniture. Fruit, vegetables, and cereals are grown, and there is livestock raising and dairying. 7 (1)D2

Uttar Pradesh state of north India, bordered by Nepal and China to the northeast, and Indian states to the south and west; area 243,288 sq km/ 93,934 sq mi; population (2001 est) 164,346,000. The capital is ▶Lucknow. Industries include the production of sugar, oil, textiles, leatherwork, cement, chemicals, coal, silica, and handicrafts (Varanasi, Lucknow). Wheat, rice, millet, barley, sugar cane, groundnuts, peas, cotton, oilseed, potatoes and fruit are grown, and there is livestock raising. The region was split in 2000 to form the new state of ▶Uttaranchal. **10** C3

Uttaranchal (or **Dev Bhoomi**), state of north India, situated at the foot of the Himalayas and bordered by Himachal Pradesh and Uttar Pradesh; area 51,125 sq km/19,739 sq mi; population (2001 est) 8,115,000. It was carved from Uttar Pradesh and was incorporated in November 2000. The capital is ▶Dehra Dun. Uttranchal shares international borders with Nepal and China. Nanda Devi, one of the highest peaks in India at 7,817m/25,645 ft, is located in the Region; many rivers originate in Uttranchal including the Ganges and Yamuna. The state's principal industry is tourism – an average of 83 million tourists visit the area every year. Other industries include forest production, a developing herbal pharmaceutical industry, and hydroelectric power. Grains and horticulture crops such as apple, orange, pear, grapes, peach, plum apricot, mango, and guava are also produced. **10** C2

Uusikaupunki (Swedish **Nystad**), seaport of southern Finland, on the Gulf of Bothnia 65 km/40 mi northwest of Turku, in Turku ja Pori province; population (1996) 17,600. **16** L6

Uzbekistan country in central Asia, bounded north by Kazakhstan and the Aral Sea, east by Kyrgyzstan and Tajikistan, south by Afghanistan, and west by Turkmenistan.

Uzbekistan covers an area of 447,400 sq km/172,741 sq mi, and has a population of 24,318,000 (2000 est). The capital is ▶Tashkent; other major towns and cities include Samarkand, Bukhara, Namangan, Andijon, Nukus, and Qarshi. The head of state is Islam Karimov (from 1990); the head of government is Otkir Sultonov (from 1995). The currency is the som, and the languages spoken include Uzbek (a Turkic language; official), Russian, and Tajik. **42** L9

Vaal river in South Africa, the chief tributary of the Orange River. It rises in the Drakensberg mountain range, on the border of Swaziland, and is 805 km/500 mi long. **72** D5

Vaasa (Swedish **Vasa**), capital of Vaasa province, western Finland; population (1996) 446,700. Situated on the Klemetso peninsula on the Gulf of Bothnia, this port is an important regional centre with food processing, engineering, textile, and chemical industries. Wheat is exported and there is a ferry service to Umeå in Sweden. **16** L5

Vadodara (formerly **Baroda** until 1976), city in Gujarat, west India; population (1991) 1,115,000. Located on the Mumbai–Ahmadabad–Delhi rail link, it is an important industrial centre (metal goods, chemicals, jewellery, textiles, tobacco). There is an oil refinery at nearby Kouali. Until 1947 it was capital of the princely state of Baroda. It has Lakshmi Vilas Palace, Pratap Vilas Palace (now the Railway Staff College), and several multi-level step wells (baoli). **54** B4

Vaduz capital of the European principality of Liechtenstein; population (1995) 5,100. The economic base is now tourism and financial services. It trades in wine, fruit, and vegetables. Above the town stands the castle of the ruling prince. **30** E3

Vág Hungarian name for the ▶Váh, a river in the Slovak Republic. **18** H8

Váh (German **Waag**; Hungarian **Vág**), river in western Slovak Republic, a tributary of the ▶Danube; length 338 km/211 mi. It rises in the Carpathian Mountains and flows west, then south, to join the Danube near Komárno. **18** H8

Valais (German **Wallis**), canton of southwest Switzerland, occupying the basin of the upper Rhône; area 5,231 sq km/2,019 sq mi; population (1995) 271,300. Its capital is Sion. Valais contains numerous high peaks, including the ▶Matterhorn (4,478 m/ 14,691 ft), and glaciers; tourism is a major industry, and is catered for by many resorts and winter-sports centres. The canton became part of the Swiss Confederation 1815. **8** (1)B2

Val-de-Marne *département* in the ▶Ile-de-France region of France, southeast of Paris; area 244 sq km/ 94 sq mi; population (1990) 1,215,100. The *département* contains the confluence of the Rivers ▶Marne and Seine, and is largely urbanized. Employment is in manufacturing industries and offices, with many workers commuting to central Paris. Rural areas are used for intensive agriculture and for recreation. Val-de-Marne contains the Rungis wholesale market (replacing Les Halles in central Paris) and Orly airport. The chief town is Créteil. **6** E2

Valdivia industrial port (shipbuilding, leather, beer, soap) and resort in Chile; population (1992) 122,400. It was founded 1552 by the Spanish conquistador Pedro de Valdivia (*c.* 1500–1554), conqueror of Chile. **100** G6

Val-d'Oise *département* in the ▶Ile-de-France region of France; area 1,248 sq km/482 sq mi; population (1990) 1,050,000. It is situated northwest of Paris and north of the confluence of the Rivers Seine and Oise, where the plateaux of Vexin and the Plaine de France are bisected by the River ▶Oise. Agricultural production ranges from intensive market gardening to large-scale arable cultivation. Important industries are located on the northern margins of Paris and in the Oise valley, and include metal works, machinery, motor vehicles, and high-tech and service industries. The administrative centre is the new city of Cergy-Pontoise. **6** E2

Valence administrative centre of ▶Drôme *département* in southeast France, 185 km/115 mi north of Marseille, built on an escarpment above the River Rhône; population (1990) 65,000, conurbation 108,000. It is a market town trading in agricultural produce, wine, and olives, and it manufactures textiles, electrical goods and aerospace components. It is of pre-Roman origin and has a Romanesque cathedral consecrated in 1095 (but much rebuilt in the 17th century). **26** K9

Valencia city and capital of ▶Valencia province in the Valencian Community, eastern Spain, on the estuary of the Guadalaviar River;

population (1991) 752,900. It is the centre of a very rich agricultural plain noted for the high quality of its citrus fruits, particularly oranges; industries include textiles, chemicals, ship repair, and wine. 28 K5

Valencia province of the Valencian Community, western Spain, with a coastline on the Mediterranean Sea; area 10,764 sq km/4,156 sq mi; population (1995) 2,200,300. The rivers Guadalaviar and Júcar flow through it. The province is mountainous in the interior, and has a fertile, irrigated coastal plain. Products include oranges, olives, grapes, mulberries, figs, and rice. The capital is ►Valencia. 9 (1)E3

Valencia industrial city and capital of Carabobo state, northern Venezuela; it is situated on the west bank of the Cabriales River, at an altitude of 478 m/1,569 ft above sea level and 160 km/99 mi west of Caracas, with which it is linked by road, rail, and air; population (1995 est) 1,007,000. It is the country's third-largest city, and the centre of its most developed agricultural region, lying near Lake Valencia. Industries include cars, brewing, food-processing, textiles, leather, sugar, and, like Valencia in Spain, oranges. 98 D1

Valenciennes industrial town in the Nord *département*, northeast France, near the Belgian border; population (1990) 39,300, conurbation 225,000. It became French in 1678. The town gives its name to a type of lace, the production of which, having gone into decline, has undergone a revival. It has important metallurgical and electronic industries, an oil refinery, hosiery and glass manufacture, and a range of light industries. There is a university here. 22 F4

Vale of Glamorgan unitary authority in south Wales, created in 1996 from parts of the former counties of Mid Glamorgan and South Glamorgan.
area 337 sq km/130 sq mi
towns Barry (administrative headquarters), Penarth
physical lowland area
agriculture sheep farming, varied agriculture
population (1996) 119,500 5 E6

Valladolid capital of Valladolid province, in Castilla–León, Spain; population (1994) 328,400. Industries include food processing, textiles, engineering, and vehicle manufacture. It has a university (founded in 1346) and a 16th-century cathedral. 28 F3

Valle d'Aosta autonomous region of northwest Italy, in the Alps; area 3,300 sq km/1,274 sq mi; population (1992 est) 117,200, many of whom are French-speaking. Its capital is ►Aosta. Wine and livestock are produced, and industries include the manufacture of special steels and textiles, and the production of hydro-electricity; tourism is also important. 9 (4)A2

Vallejo city and port in Solano County, north-central California, USA, at the mouth of the Napa River on San Pablo Bay, 48 km/30 mi northeast of San Francisco; population (1996 est) 109,600. It is a processing and commercial centre for a rich agricultural hinterland; flour, fruit, meat, and dairy foods are prepared. Shipyards are sited in the port. Other industries include lumbering, oil-refining, and the manufacture of iron. Just offshore, Mare Island is the largest US naval base on the North American Pacific coast. The city is home to the California Maritime Academy (1929). 86 B3

Valletta capital and port of Malta; population (1995) 9,129 (inner harbour area 102,600). 32 J13

Valona Italian form of ►Vlorë, a port in Albania. 36 B4

Valparaíso industrial port, naval base, and capital of Valparaiso region, central Chile, situated on a broad bay on the Pacific coast at the foot of a spur of hills, 120 km/75 mi northwest of Santiago; population (1992) 276,700. It is Chile's major port and second-largest city, an administrative centre, and the seat of the Chilean parliament. Both the law courts and the new National Congress are located here. Industries include textiles, chemicals, oil, sugar refining, and leather goods. Fruit and mining products are exported. It is the seat of the Chilean Naval Academy. 100 G5

Van city in Turkey on a site on **Lake Van** that has been inhabited for more than 3,000 years; population (1990) 153,100. It is a commercial centre for a fruit- and grain-producing area. 58 K4

Vancouver chief Pacific seaport of Canada, on the mainland of British Columbia; population (1991) 471,800, metropolitan area (1996) 1,891,400. A major commercial, distribution, and tourist centre, it is the terminus of trans-continental rail and road routes, and a 1,144-km/715-mi pipeline from the Alberta oilfields. Industries include oil-refining, engineering, shipbuilding, fishing and fish-canning, brewing, timber-milling, and the manufacture of aircraft, pulp and paper, and textiles. 86 B1

Vancouver city and administrative headquarters of Clark County, southwest Washington, USA; population (1990) 46,400. It is situated on the Columbia River, 13 km/8 mi north of Portland, Oregon. It is a manufacturing and shipping centre for agriculture and timber. Paper products, clothing, and aluminium are also important to the local economy. It began as a trading post for the Hudson's Bay Company 1825. 86 B1

Vancouver Island island off the west coast of Canada, part of British Columbia, separated from the mainland by the straits of Juan de Fuca, Haro, Georgia, Johnstone, and Queen Charlotte Sound; area 32,136 sq km/12,404 sq mi. Industries include coal, timber, fish, and tourism. Fruit is grown, and there is dairying. 82 F7

Van Diemen's Land former name (1642–1855) of ►Tasmania, Australia. It was named by the Dutch navigator Abel Tasman after the governor general of the Dutch East Indies, Anthony van Diemen. The name Tasmania was used from the 1840s and officially adopted in 1856. 76 H8

Vänern largest lake in Sweden, area 5,550 sq km/2,140 sq mi. Karlstad, Vänersborg, Lidköping, and Mariestad are on its banks. 16 G7

Vannes seaport and capital of the French *département* of Morbihan; population (1990) 50,000. The industries include shipbuilding and tourism, and the manufacture of woollen fabrics, rope, tyres, and agricultural machinery. Medieval ramparts enclose the inner city, and there is a cathedral with a 16th-century chapel. 26 C6

Vannin, Ellan Gaelic name for the Isle of ►Man. 24 H7

Vanuatu group of islands in the southwest Pacific Ocean, part of ►Melanesia.

Vanuatu covers an area of 14,800 sq km/5,714 sq mi, and has a population of 190,000 (2000 est). The capital (and chief port) is ▶Port-Vila (on Efate island); other major towns and cities include Luganville (on Espíritu Santo). The head of state is John Bernard Bani (from 1999); the head of government is Barak Sope (from 1999). The currency is the vatu, and the languages spoken include Bislama, English, and French (all official). 74 G7

Var *département* in the ▶Provence-Alpes-Côte d'Azur region of France; area 6,000 sq km/2,317 sq mi; population (1990) 816,000. It has a much-indented coastline on the Mediterranean. Var is generally hilly and wooded; the River Argens flows through it. Fruit, flowers, vines, and olives are produced in coastal districts. There are bauxite, lead, zinc, and salt deposits; shipbuilding, armaments, engineering and chemical industries; and furniture is manufactured. The principal towns are ▶Toulon (the administrative centre) and Draguignan. Tourism is important; resorts include ▶Hyères, Fréjus, and St-Raphaël. 6 G5

Varanasi (or **Benares**, or **Banaras**), city in Uttar Pradesh, India, one of the seven holy cities of Hinduism, on the River Ganges; population (1991) 932,000. There are 1,500 golden shrines, and a 5 km/3 mi frontage to the Ganges with sacred stairways (ghats) for purification by bathing. Varanasi is also a sacred centre of Jainism, Sikhism, and Buddhism: Buddha came to Varanasi from Gaya and is believed to have preached in the Deer Park. One-third of its inhabitants are Muslim. 54 D3

Varazdin town in northern Croatia, on the River Drava, near the Slovenian border; population (1991) 41,800. It has textile industries. It was a free city in the Middle Ages and its old fortifications remain. 34 D3

Vardar river in the Balkans; length 322 km/200 mi. It rises west of Skopje in the Former Yugoslav Republic of Macedonia, and flows southeast to the Aegean Sea near Thessaloniki, Greece. 36 E3

Varna port in Bulgaria, on an inlet of the Black Sea; population (1991) 321,000. Industries include shipbuilding and the manufacture of chemicals. 58 B2

Västerås capital of Västmanland county in south central Sweden; population (1995) 123,700. It has engineering factories, chiefly electrical and metallurgical, and is Sweden's largest inland port. The town dates from the Middle Ages, and has a castle and a cathedral. 16 J7

Vatican City State sovereign area within the city of Rome, Italy.

Vatican City State covers an area of 0.4 sq km/0.2 sq mi, and has a population of 1,000 (2000 est). The head of state is John Paul II (from 1978); the head of government is Cardinal Angelo Sodano (from 1990). The currencies are the Vatican City lira and the Italian lira, and the languages spoken include Latin (official) and Italian. 32 F7

Vaucluse *département* in the ▶Provence-Alpes-Côte d'Azur region of southeastern France; area 3,566 sq km/1,377 sq mi; population (1990) 468,008. It is bordered on the west by the lower reaches of the River Rhône and on the south by the Durance. Vaucluse is drained by the rivers Eygues, Sorgue, Ouveze, and Coulon. It is known for its wines (including Châteauneuf-du-Pape) and its fruit. Olives, cereals, silk, and livestock are also produced. The chief industries are textiles, chemicals, and foodstuffs. The major towns are ▶Avignon (the capital), Carpentras, and Apt. 6 F4

Vaud (German **Waadt**), canton in southwest Switzerland; area 3,211 sq km/1,240 sq mi; population (1995) 605,700. It is roughly triangular in shape, with its base formed by the northern shore of Lake Geneva, and is traversed by the ▶Jura Mountains, which run southwest–northeast. Tourism is the most important industry, but Vaud is also Switzerland's premier vine-growing region. Wine, chocolate, tobacco, clocks, and milk products are among its principal manufactures. The capital of the canton is ▶Lausanne. 8 (1)A2

Veliko Turnovo (Romanian **Veliko Târnovo**), historic town in Lovech region, northern Bulgaria, in the gorge of the River Yantra on the northern slopes of the Balkan Mountains, 193 km/121 mi northeast of Sofia; population (1990) 99,900. Textiles are manufactured and agricultural produce traded from the surrounding area. The Shipka Pass runs south from the town to Kazanlâk. There are several medieval buildings. 34 N6

Vendée, La maritime *département* in the ▶Pays de la Loire region of western France; area 6,720 sq km/2,595 sq mi; population (1990 est) 509,800. The administrative centre is ▶La Roche-sur-Yon, and principal towns include Fontenay-le-Comte and Les Sables-d'Olonne. Its rivers include the Vendée and the Lay. The *département* is dominated by the Bocage Vendéan, a large area of area of mixed woodland and small hedge-bound fields, but there are marshes in the northwest and a well-cultivated plain in the south. The principal crops are wheat and sugar-beet. Vines are grown, and horses and cattle are bred. 6 G3

Vendôme French town in the *département* of Loir-et-Cher, on the River Loir; population (1990) 18,500. Cheese and asparagus are produced in the vicinity. The church of the Trinity has a 12th-century belfry, and there is an 11th-century castle. 26 G6

Veneto region of northeast Italy, comprising the provinces of Belluno, Padua, Treviso, Rovigo, Venice, and Vicenza; area 18,400 sq km/7,102 sq mi; population (1992 est) 4,395,300. Its capital is ▶Venice, and towns include Padua, Verona, and Vicenza. The Veneto forms part of the north Italian plain, with the delta of the River Po; it includes part of the Alps and Dolomites, and Lake Garda. Products include cereals, fruit, vegetables, wine, tobacco, chemicals, ships, and textiles. 9 (4)C2

Venezia Italian form of ▶Venice, a city, port, and naval base on the Adriatic Sea. 30 H5

Venezuela country in northern South America, on the Caribbean Sea, bounded east by Guyana, south by Brazil, and west by Colombia.

Venezuela covers an area of 912,100 sq km/352,161 sq mi, and has a population of 24,170,000 (2000 est). The capital is ▶Caracas; other major towns and cities include Maracaibo, Maracay, Barquisimeto, Valencia, Ciudad Guayana, and Petare. The head of state and government is Hugo Chávez Frías (from 1999). The currency is the bolívar, and the languages spoken include Spanish (official) and Indian languages. **98** D2

Venice (Italian **Venezia**), city, port, and naval base on the northeast coast of Italy; population (1992) 305,600. It is the capital of Veneto region.

The old city is built on piles on low-lying islands in a salt-water lagoon, sheltered from the Adriatic Sea by the Lido and other small strips of land. There are about 150 canals crossed by some 400 bridges. Apart from tourism (it draws 8 million tourists a year), industries include glass, jewellery, textiles, and lace. Venice was an independent trading republic from the 10th century, ruled by a doge, or chief magistrate, and was one of the centres of the Italian Renaissance. It was renowned as a centre of early publishing; 15% of all printed books before 1500 were printed in Venice. **30** H5

Venlo town and railway junction in the province of Limburg, the Netherlands, on the River Maas (see ▶Meuse) and on the border with Germany; population (1996) 64,800. Industries include engineering and the manufacture of tobacco, paper, and electric lamps. Venlo was formerly a heavily fortified town. **22** J3

Vent, Iles du French name for the Windward Islands, part of the ▶Society Islands in ▶French Polynesia. The Leeward Islands are known as the **Iles sous le Vent**. **94** N6

Veracruz port (trading in coffee, tobacco, and vanilla) in eastern Mexico, on the Gulf of Mexico; population (1990) 328,600. Products include chemicals, sisal, and textiles. It was founded by the Spanish conquistador Hernán Cortés as Villa Nueva de la Vera Cruz ('new town of the true cross') on a nearby site in 1519 and transferred to its present site in 1599. **94** E5

Verden city in Lower Saxony, Germany, on the River Aller, 36 km/22 mi southeast of Bremen; population (1994) 126,500. Verden is an important agricultural centre. **20** E4

Vermont state in northeastern USA. It is nicknamed the Green Mountain State. Vermont was admitted to the Union in 1791 as the 14th US state. It is bordered to the north by Québec, Canada, to the east by New Hampshire, to the south by Massachusetts, and to the west by New York.
population (2000 est) 608,800
area 24,900 sq km/9,611 sq mi
capital ▶Montpelier
towns and cities Burlington, Rutland, Barre
industries and products apples, maple syrup, dairy products, china clay, granite, marble, slate, business machines, paper and allied products, computers and high-tech manufacturing, tourism, leisure **84** M3

Veroia town in the Macedonia region of Greece, capital of Emathia department; population (2001 est) 43,500. The abundant water supply attracted handicraft workers in the 18th and 19th centuries; now hydroelectric power supplies silk-spinning mills. The town is a commercial and communications centre for the region. It originated as the old Macedonian city of Berea. **36** E4

Verona town in Veneto, Italy, on the Adige River, 100 km/62 west of Venice; population (1992) 255,500. It lies at the junction of the Brenner Pass road with the Venice–Milan motorway. Industries include printing, engineering, and the manufacture of paper, plastics, furniture, and pasta. It is one of Italy's main marketing centres for fruit and vegetables. **30** F5

Versailles administrative centre of the *département* of ▶Yvelines in northern France, situated 18 km/11 mi southwest of Paris; population (1990) 91,000. From 1678 to 1769 Versailles was the principal residence of the kings of France until 1793, and the seat of government from 1682 to 1789. The city grew up around the palace of Louis XIV, built between 1661 and 1687 on the site of Louis XIII's hunting lodge. Within the palace park are two small châteaux, Le Grand Trianon, built for Louis XIV, and Le Petit Trianon, built for Louis XV. **22** E6

Versecz Hungarian name for ▶Vrsac, a town in Serbia, Yugoslavia. **34** J4

Verviers town in the Belgian province of Liège, on the River Vesdre, 21 km/13 mi east of the city of ▶Liège; population (1991) 53,500. Together with its suburb Ensival, it was one of the main centres of the Belgian woollen industry, using water from the Gileppe dam, 9 km/6 mi to the east. Other manufactures are soap, leather, footwear, chocolate, and chemicals. **22** H4

Vesterålen island group off northwest Norway; see ▶Lofoten and Vesterålen. **16** G2

Vestmannaeyjar small group of islands off the south coast of Iceland. The volcanic island of ▶Surtsey emerged from the ocean in 1963, and in 1973 the volcano Helgafell erupted, causing the population of 5,200 to be temporarily evacuated and adding 2.5 sq km/1 sq mi to the islands' area. Heimaey, the largest of the islands, is one of Iceland's chief fishing ports. **16** (1)C3

Vesuvio Italian name for ▶Vesuvius, an active volcano in Campania, Italy. **32** J8

Vesuvius (Italian **Vesuvio**), active volcano in Campania, Italy, 15 km/9 mi southeast of Naples, Italy; height 1,277 m/4,190 ft. In AD 79 it destroyed the cities of Pompeii, Herculaneum, and Stabiae. **32** J8

Veszprém (German **Weissbrunn**), industrial town and commercial centre in western Hungary, 110 km/68 mi southwest of Budapest; population (1993) 65,400. It is the capital of Veszprém county, on the River Séd, north of Lake Balaton in the Bakony Mountains, a major mining area. It has textile and oil plants, and trades in agricultural produce. **34** E2

Vevey tourist resort in the canton of Vaud, Switzerland, situated on Lake Geneva, 18 km/11 mi southeast of Lausanne; population (1995) 15,700. The chief industries of the town are the manufacture of chocolate, machinery, and watches, and woodworking. **30** B4

Viangchan Lao name for ▶Vientiane, the capital of Laos. **50** C3

Viborg industrial town (brewing, engineering, textiles, tobacco) in Jutland, Denmark; population (1990) 39,400. It is also the Swedish name for ▶Vyborg, a port and naval base in Russia. **16** E8

Vicenza chief town in Vicenza province, in Veneto region, northeast Italy, on the River Bacchiglione, 65 km/40 mi west of Venice; population (1992) 107,500. Textiles, glass, and musical instruments are manufactured here. **30** G5

Vicksburg city and port in west Mississippi, USA, on the east bank of the Mississippi River; population (1990) 20,900. Industries include shipping and boat and vehicle building. It is a popular tourist destination because of its importance in the American Civil War, commemorated in the Vicksburg national military park, which encircles the city. **90** C3

Victoria state of southeast Australia; bounded on the north and northeast by New South Wales, from which it is separated by the River Murray; on the west by South Australia; and on the south and southeast by the Southern Ocean, Bass Strait, and the Pacific Ocean; area 227,600 sq km/87,876 sq mi; population (1996) 4,373,500. The capital is ►Melbourne. Produce includes wool, beef, dairy products, tobacco, wheat, wine, dried fruit, orchard fruits, and vegetables. Gold, brown coal, gypsum, kaolin, and bauxite are mined, and there are oil, natural gas, electronics, food processing, chemical, pharmaceutical, machinery, car, textile, wine, aquaculture, wool, and building material industries. **76** H7

Victoria port and capital of ►British Columbia, Canada, on the southeastern tip of Vancouver Island, overlooking the Strait of Juan de Fuca, 132 km/83 mi west of mainland Vancouver; population (1991) 21,200, metropolitan area (1996) 313,400. It is a manufacturing, tourist, and retirement centre, and has a naval base. Industries include shipbuilding, food-processing, sawmilling, fishing, and the manufacture of chemicals, clothing, and furniture. **86** B1

Victoria deep-water port and capital of the republic of the Seychelles, tourist centre on the northeast coast of Mahé Island; population (1992) 30,000. Industries include copra, vanilla, guano, and cinnamon. Victoria is the business and cultural centre of the Seychelles and has an international airport. **72** (2)C1

Victoria city and administrative headquarters of Victoria County, south Texas, USA; population (1990) 55,100. It is situated on the Guadalupe River, near the Gulf of Mexico, and 116 km/72 mi northeast of Corpus Christi. It is a transportation centre for oil, natural gas, chemicals, and dairy products. It was founded 1824 by Spanish settlers. The city is home to Victoria College (1925), and the University of Houston, Victoria. **90** B4

Victoria Falls (or **Mosi-oa-tunya** 'smoke that thunders'), waterfall on the River Zambezi, on the Zambia–Zimbabwe border. The river is 1,700 m/5,580 ft wide and drops 120 m/400 ft to flow through a gorge 30 m/100 ft wide. The falls were named after Queen Victoria by the Scottish explorer David Livingstone in 1855. **72** D3

Victoria, Lake (or **Victoria Nyanza**), largest lake in Africa and third-largest freshwater lake in the world; area over 68,800 sq km/26,560 sq mi; length 410 km/255 mi; average depth 80 m/260 ft. It lies on the Equator at an altitude of 1,136 m/ 3,728 ft, bounded by Uganda, Kenya, and Tanzania. It is a source of the River Nile. **70** E4

Vienna (German **Wien**), capital of Austria, on the River Danube at the foot of the Wiener Wald (Vienna Woods); population (1995) 1,531,200. Although within the territory of Lower Austria, it is a separate province. Industries include engineering, electrical goods, electronics, clothing, precision and musical instruments, and beer. It is a major cultural and tourist centre. **30** M2

Vienne French town in the *département* of Isère on the River Rhône, 26 km/16 mi south of Lyons; population (1990) 29,400. Vienne has woollen and metal manufacturing industries; tourism is important. The town's history goes back more than 2,000 years and there are Roman remains. **26** K8

Vienne *département* in the Poitou-Charentes region of France; area 6,984 sq km/2,697 sq mi; population (1990) 380,500. Vienne is flat and low-lying. The River Vienne and its tributary the Clain run through it. Cereals, fruit, and some hemp are grown in the fertile soil, and stock is raised. Vienne contains numerous megalithic monuments, or standing stones. The administrative centre is ►Poitiers. Other towns include Châtellerault, and Montmorillon. **6** D3

Vienne river in France, rising in northern Corrèze *département*; length 350 km/217 mi. It flows north past Châtellerault, Limoges, and Chinon, to join the Loire 13 km/8 mi southeast of Saumur. The river is navigable from Châtellerault. **26** F7

Vientiane (Lao **Viangchan**; 'City of the Moon'), capital, largest city, and chief port of ►Laos, lying in the north on the Mekong River, on the border with Thailand; population (1996 est) 531,800. Vientiane's strategic position on one of the main waterways of southeast Asia has helped it to become a centre for government, commerce, and religion for over a millennium. Noted for its pagodas, canals, and houses on stilts, it is situated in a rich agricultural area and is a trading centre for forest products and textiles. **50** C3

Vietnam country in Southeast Asia, on the South China Sea, bounded north by China and west by Cambodia and Laos.

Vietnam covers an area of 329,600 sq km/127,258 sq mi, and has a population of 79,832,000 (2000 est). The capital is ►Hanoi; other major towns and cities include Ho Chi Minh City (formerly Saigon), Haiphong, Da Nang, Can Tho, Nha Trang, Bien Hoa, and Hue. The head of state is Tran Duc Luong (from 1997); the head of government is Phan Van Khai (from 1997). The currency is the dong, and the languages spoken include Vietnamese (official), French, English, Khmer, Chinese, and local languages. **50** D3

Vigo fishing and industrial port and naval station of Pontevedra province in ►Galicia, northwest Spain, situated on an inlet on the Atlantic coast; population (1991) 274,600. Industries include fish processing, leather and paper production, and distilling. **28** B2

Viipuri Finnish name of ►Vyborg, a port and naval base formerly in Finland but in Russia since 1945. **16** Q6

Villach Austrian city in the province of Carinthia, on the River Drava, 33 km/20 mi west of Klagenfurt; population (1991) 54,600. Villach has a spa with hot springs, is the centre of the timber trade with Italy, and has iron, paper, brewing, and lead industries.

Villach has a 14th–15th-century church, a Renaissance town hall, and other old buildings, including a house once occupied by the medieval physician Paracelsus. 30 J4

Vilnius (German **Wilna**; Russian **Vilna**; Polish **Wilno**), capital of Lithuania, situated on the River Neris; population (1991) 593,000. Vilnius is an important railway crossroads and commercial centre. Its industries include electrical engineering, woodworking, and the manufacture of textiles, chemicals, and foodstuffs. 16 N9

Vilvoorde (French **Vilvorde**), town in the province of Brabant, Belgium, 10 km/6 mi north of Brussels; population (2001 est) 33,300. The town has various manufacturing industries. 22 G4

Vincennes city in southwest Indiana, USA, on the border with Illinois, on the Wabash River, 80 km/50 mi north of Evansville; population (1990) 19,900. It is an agricultural and commercial centre with car-assembly and battery-manufacturing plants. Founded 1702, it is the oldest town in Indiana and the site of a French fort, taken in 1763 by the British and renamed Fort Sackville. Captured by General George Rogers Clark for the Americans during the American Revolution, it was ceded to the USA in 1783. It was the capital of the territory of Indiana 1800–1813. 90 D2

Vineland industrial city in Cumberland County, south New Jersey, USA; population (1992) 54,700. It is situated on the Maurice River, 50 km/31 mi southeast of Camden. Lying in a rich vegetable growing and poultry raising region, the city has many cooperative markets and is the agricultural trade, processing, and distribution centre of south New Jersey. Industries include the manufacture of foundry products, glassware, and chemicals. It is home to Cumberland County College (1964). 88 F3

Vinson Massif highest point in ►Antarctica, rising to 5,140 m/16,863 ft in the Ellsworth Mountains. 102 (2)JJ2

Virginia state in eastern USA. It is nicknamed Old Dominion. Officially known as the **Commonwealth of Virginia**, it ratified the US Constitution in 1788, becoming the 10th US state. It is bordered to the north by Maryland and the District of Columbia, to the west by Kentucky and West Virginia, and to the south by North Carolina and Tennessee. In the east it occupies the southern tip of the Delamarva Peninsula and is bordered by the Atlantic Ocean. Virginia was the northeasternmost state of the Confederacy.
population (2000 est) 7,078,500
area 105,600 sq km/40,762 sq mi
capital ►Richmond
towns and cities Norfolk, Virginia Beach, Newport News, Hampton, Chesapeake, Portsmouth
industries and products sweet potatoes, maize, tobacco, apples, peanuts, coal, ships, lorries, paper, chemicals, processed food, textiles, tourism, leisure 84 L4

Virginia Beach Atlantic coastal resort in southeast Virginia, USA, near the south-shore entrance to Chesapeake Bay, 19 km/12 mi east of Norfolk; population (1996 est) 430,380. Industries include boatbuilding, and the manufacture of electronics and building materials. Market gardens in the city's agricultural hinterland specialize in strawberries. Nearby military bases include Fort Story Military Reservation, Oceana Naval Air Station, and Little Creek Naval Amphibious Base. 88 E3

Virgin Islands group of about 100 small islands, northernmost of the Leeward Islands in the Antilles, West Indies. Tourism is the main industry.

They comprise the **US Virgin Islands** St Thomas (with the capital, Charlotte Amalie), St Croix, St John, and about 50 small islets; area 350 sq km/135 sq mi; population (1990) 101,800; and the **British Virgin Islands** Tortola (with the capital, Road Town), Virgin Gorda, Anegada, and Jost van Dykes, and about 40 islets (11 islands are inhabited); area 150 sq km/58 sq mi; population (1991) 16,100. 96 E2

Visalia city and administrative headquarters of Tulare County, south-central California, USA; population (1992) 81,300. It is situated in the San Joaquin Valley, on the Kaweah River, 56 km/35 mi southeast of Fresno. It is an agricultural centre for grapes, citrus fruits, and dairy products. There is some electronic and other light manufacturing. Visalia was founded 1852 by Nathaniel Vise. The city is a gateway to the Sequoia and Kings Canyon national parks, which are to its east. The College of the Sequoias (1925) is in Visalia. 86 C3

Vistula (Polish **Wisła**), river in Poland that rises in the Carpathian Mountains and runs northwest to the Baltic Sea at Gdańsk; length 1,090 km/677 mi.

It is heavily polluted, carrying into the Baltic every year large quantities of industrial and agricultural waste, including phosphorus, oil, nitrogen, mercury, cadmium, and zinc. 14 F2

Vitoria (or **Vitória-Gasteiz**), capital of Alava province and of the ►Basque Country autonomous community, northern Spain; population (1994) 214,000. Products include motor vehicles, agricultural machinery, and furniture, and there is a trade in agricultural produce and wine. It has a Gothic cathedral. 28 H2

Vitsyebsk (formerly **Vitebsk** until 1992), industrial city and important rail junction in Belarus, on the Western Dvina River, 220 km/137 mi northeast of Minsk and 45 km/28 mi west of the Russian border; population (1990) 356,000. The city is the capital of the Vitsyebsk oblast (administrative region). Furniture, electrical goods, machine tools, and textiles are manufactured. 38 F3

Vittorio Veneto industrial town (motorcycles, agricultural machinery, furniture, paper, textiles) in Veneto, northeastern Italy; population (1991) 29,200. It was the site of the final victory of Italy and its allies over Austria October 1918. 30 H5

Vladikavkaz (formerly **Ordzhonikidze** 1931–44; 1954–91; **Dzaudzhikau** 1944–54; 'Queen of the Caucasus'), capital of the autonomous republic of Alania (formerly North Ossetia), in the Russian Federation, sited on the River Terek in the Caucasus Mountains; population (1990) 306,000. Vladikavkaz is an important industrial city, producing lead and zinc, machine tools, vehicles, glassware, and textiles. 58 L2

Vladivostok city on the western shore of the Sea of Japan, on a peninsula extending into Peter the Great Bay; population (1996 est) 627,000. It is the capital of the Primorski (Maritime) Krai of the Russian Federation, and one of the most important economic and cultural centres of the Russian Far East, where it is the largest city. Vladivostok is a terminus of the Trans-Siberian Railway (9,224 km/mi from Moscow) and the Northern Sea Route, centre of communications for the Pacific territories, the largest Russian port on the Pacific, and the chief base of the Pacific Fleet. The port is kept open by icebreakers during winter. 48 F2

Vlissingen (English **Flushing**), trading port and resort on Walcheren Island, Zeeland, the Netherlands; population (1997) 44,600. It stands at the entrance to the Western Schelde estuary, one of the principal sea routes into Europe. Industries include fishing, shipbuilding, and petrochemicals. There is a ferry service to Harwich (England). Historically, the town was an important naval station and fortress. Admiral Michiel de Ruyter was born at Vlissingen and is commemorated in the 14th-century St Jacobskerk. 22 F3

Vlorë port and capital of Vlorë province, southwestern Albania, population (1991) 76,000. A Turkish possession from 1464, it was the site of the declaration of independence by Albania in 1912. 36 B4

Vltava (German **Moldau**), river in the Czech Republic, rising in the Bohemian Forest and flowing 435 km/ 272 mi north to join the River ►Elbe at Mělnik. It passes through České Budějovice and Prague. 18 D8

Vojvodina autonomous province in northern Serbia, Yugoslavia, 1945–1990; area 21,500 sq km/8,299 sq mi; population (1991) 2,012,500, including 1,110,000 Serbs and 390,000 Hungarians, as well as Croat, Slovak, Romanian, and Ukrainian minorities. Its capital is Novi Sad. In September 1990 Serbia effectively stripped Vojvodina of its autonomous status, causing antigovernment and anticommunist riots in early 1991. 34 G4

Volga (ancient **Rha**), longest river in Europe, entirely within the territory of the Russian Federation. The Volga has a total length 3,685 km/2,290 mi, 3,540 km/2,200 mi of which are navigable. It rises in the Valdai plateau northwest of Moscow, and flows into the Caspian Sea 88 km/55 mi below the city of Astrakhan. The Volga basin drains most of the central and eastern parts of European Russia, its total drainage area being 1,360,000 sq km/ 525,100 sq mi. 38 J5

Volgograd (formerly **Tsaritsyn** until 1925 and **Stalingrad** 1925–61), industrial city in southwest Russian Federation, on the River Volga; population (1994) 1,000,000. Industries include the manufacture of metal goods and machinery, saw-milling, and oil refining. 38 H5

Volta main river in ►Ghana, about 1,600 km/1,000 mi long, with two main upper branches, the **Black Volta** and **White Volta**. It has been dammed at Akosombo to provide power. 68 D3

Volta, Upper former name (to 1984) of ►Burkina Faso. 68 D2

Vorarlberg (German 'in front of the Arlberg'), Alpine federal state of Austria, bounded on the north by Bavaria, on the west by Lake Constance, the Rhine, and Liechtenstein, on the east by north Tirol, and on the south by Switzerland; area 2,600 sq km/1,004 sq mi; population (1994) 342,500. Its capital is Bregenz. It is the most industrialized province of Austria, producing textiles, clothing, and crafts. Forestry, dairy-farming, and fruit-growing are the main agricultural occupations. Tourism is important. There are several hydroelectric stations. 8 (2)A3

Voreria Pindos alternative name for the ►Pindus Mountains, a mountain range in northwestern Greece and Albania. 36 D5

Voronezh river port and capital of Voronezh oblast (region), in the western Russian Federation; population (1996 est) 909,000. Voronezh is situated 290 km/180 mi northeast of Kharkov on the Voronezh River 18 km/11 mi west of its confluence with the Don. It stands at the centre of the black soil (*chernozem*) region, which has high agricultural yields. The city has important manufacturing industries, for example engineering works (producing agricultural and food industry equipment, excavators, and diesel motors) and chemical plants (manufacturing synthetic rubber and pharmaceuticals). Other products include building materials and foodstuffs. 38 G4

Voroshilovgrad former name (1935–58; 1970–89) of ►Luhansk, a city in Ukraine. 38 G5

Vosges *département* in the ►Lorraine region of northeast France; area 7,424 sq km/2,866 sq mi; population (1990) 386,562. It is bordered on the east by the ►Vosges mountains. The rivers ►Moselle and Meuse have the largest drainage areas of any rivers in the region. Oats, wheat, and vines are cultivated, and cheese-making and cattle-grazing are important economic activities. Textile goods are the chief manufactures. The principal towns are ►Epinal (the capital), St-Dié, Remiremont, and Neufchâteau. 6 G2

Vosges mountain range in eastern France near the Franco-German frontier between the *départements* of Haut-Rhin and Vosges, 250 km/ 155 mi in length and rising to its highest point at the Ballon de Guebwiller (1,422 m/4,667 ft). The Vosges forms the western edge of the Rhine rift valley. It gives its name to the *département* of ►Vosges. The Vosges is separated from the Jura Mountains to the south by the Belfort Gap. 30 C2

Vostochno-Sibirskoye More Russian name for the ►East Siberian Sea, part of the Arctic Ocean, off the north coast of Russia. 44 U2

Vraca Bulgarian name for ►Vratsa, a town in Bulgaria. 34 L6

Vratsa (Bulgarian **Vraca**), town in Mikhailovgrad region, northwest Bulgaria, on the northern slopes of the Balkan Mountains, 58 km/36 mi north of Sofia; population (1990) 102,800. Industries include chemicals and cement, and wine and agricultural produce are traded. 34 L6

Vrsac (Hungarian **Versecz**), town in Serbia, Yugoslavia, in the autonomous region of Vojvodina; population (1991) 36,900. It produces wines and brandy. There is a cathedral. 34 J4

Vukovar river port in Croatia at the junction of the rivers Vuka and Danube, 32 km/20 mi southeast of Osijek; population (1991) 44,600. Industries include foodstuffs manufacture, fishing, and agricultural trade. In 1991 the town resisted three months of siege by the Serb-dominated Yugoslav army before capitulating. It suffered the severest damage inflicted to any European city since the bombing of Dresden during World War II. 34 G4

Vyborg (Finnish **Viipuri**; Swedish **Viborg**), port and naval base situated on the Gulf of Finland in eastern ►Karelia, in the northwestern Russian Federation, 120 km/75 mi northwest of St Petersburg; population (1990) 81,500. The port trades extensively in timber and wood products; goods manufactured in the city include electrical equipment and agricultural machinery. 16 Q6

Waag German name for the ►Váh, a river in the Slovak Republic. 18 H8

Waal river in the Netherlands; length 84 km/52 mi. The River ►Rhine divides, 14 km/9 mi east of Nijmegen, into the Neder Rijn and the Waal, the latter becoming the Merwede at its confluence with the Maas (see ►Meuse) 4 km/2 mi east of Gorinchem. 22 H3

Waco city in east-central Texas, USA, on the Brazos River, 145 km/91 mi south of Dallas; population (1996 est) 108,400. It is an agricultural shipping centre for cotton, grain, and livestock; industries include food-processing and the manufacture of aircraft parts, glass, paper, cement, tyres, and textiles. 90 B3

Waddenzee European estuarine area (tidal flats, salt marshes, islands, and inlets) north of the Netherlands, and west of Denmark; area 10,000 sq km/ 3,861 sq mi. It is the nursery for the North Sea fisheries, but the ecology is threatened by tourism and other development. 22 H1

Wadi Halfa frontier town in Northern State, Sudan, on Lake Nuba (the Sudanese section of Lake Nasser, formed by the Nile dam at Aswan, Egypt, which partly flooded the archaeological sites here). Industries include a Chinese-built fish processing plant. 64 F3

Wagga Wagga (Aboriginal 'many crows'), town in southeast New South Wales, Australia, 217 km/135 mi west of Canberra, 471 km/293 mi southwest of Sydney, and 487 km/303 mi northeast of Melbourne; population (1996) 42,848. Wagga Wagga is situated on the Murrumbidgee River and is the centre of a rich agricultural district, in which lambs are raised, and wheat growing, dairying, and mixed farming occurs. Its industries include food processing. 76 J7

Waikato river on North Island, New Zealand, 425 km/264 mi long, rising on Mount Ruapehu and flowing into the Tasman Sea. It is New Zealand's longest river. 78 E4

Waitaki river in southeast South Island, New Zealand, flowing 215 km/135 mi to the Pacific Ocean. The Benmore hydroelectric installation has created an artificial lake. 78 C7

Wakefield industrial city inWest Yorkshire, northern England, on the River Calder, south of Leeds; population (1991) 73,600. Industries include chemicals, machine tools, wool textiles, coal mining, and the manufacture of clothing, wire-rope, and sheet metal. Lancastrian forces defeated and killed Richard of York here in 1460, during the Wars of the Roses. The National Coal Mining Museum is here. 24 L8

Wake Island (or **Enenkio**), small Pacific coral atoll comprising three islands 3,700 km/2,300 mi west of Hawaii, under US Air Force administration since 1972; area 8 sq km/ 3 sq mi. It was discovered by Captain William Wake in 1841, annexed by the USA in 1898, and uninhabited until 1935, when it was made an air staging point, with a garrison. It was occupied by Japan 1941–45. 74 G4

Wałbrzych (German **Waldenburg**), capital of Wałbrzych province, Poland, in the foothills of the Sudeten Mountains, 69 km/43 mi southwest of Wrocław; population (1990) 141,000. The town has coal, iron, steel, and engineering industries, and produces chemicals, glass, and pottery. Until 1945 it was part of German Lower Silesia. 18 F7

Walcheren island in ►Zeeland province, the Netherlands, in the estuary of the River Schelde; area 200 sq km/80 sq mi. The capital is ►Middelburg. Industries include shipbuilding, engineering, and petrochemicals. There is dairying, and sugar beet and root vegetables are grown. 22 F3

Waldenburg German name for ►Wałbrzych, a town in Poland. 18 F7

Wales (Welsh **Cymru**), constituent part of the United Kingdom, in the west between the British Channel and the Irish Sea; area 20,780 sq km/ 8,020 sq mi; population (2001 est) 2,928,000. The capital is ►Cardiff, and other major towns and cities include Swansea, Wrexham, Newport, and Carmarthen. Geographical features include the Snowdonia Mountains (Snowdon, at 1,085 m/3,560 ft, is the highest point in England and Wales), the Black Mountains, the Brecon Beacons, the Black Forest ranges, and the rivers Severn, Wye, Usk, and Dee. There are oil refineries and open-cast coal mines, but traditional industries have declined, and the last deep coal mine in north Wales closed in 1996. Varied modern and high-technology ventures are being developed, and Wales has the largest concentration of Japanese-owned plants in the UK. It also has the highest density of sheep in the world and a dairy industry. Tourism is important. 24 J9

Walla Walla city and administrative headquarters of Walla Walla County, southeast Washington, USA, lying near the southern boundary of the state, in the centre of a wheat-producing area; population (1990) 26,500. An agricultural processing centre for its wheat and truck farming valley, it also has food processing plants, lumber mills, and a container factory. It is the home of Whitman College (1859) and Walla Walla Community College (1967). It also has two military bases. 86 C1

Wallis and Futuna two island groups in the southwestern Pacific Ocean, an overseas territory of France; area 367 sq km/143 sq mi; population (1997 est) 14,800. The people live mostly by subsistence agriculture and farming of livestock, especially pigs and goats. Much food is imported, mainly rice, sugar and beef, as are virtually all manufactured goods. The export trade is very small, chiefly copra and handicrafts. Because of deforestation, through cutting down timber for fuel, soil erosion is a problem, especially in Futuna. 74 J7

Walsall industrial town in West Midlands, central England, 13 km/ 8 mi northwest of Birmingham; population (1991) 172,600. It has a leather industry and also produces castings. Until the 1930s coal was mined here. Walsall's art gallery contains the Garman–Ryan collection, over 350 paintings including works by Jacob Epstein. The writer Jerome K Jerome was born here in 1859. 24 L9

Walvis Bay chief port serving Namibia, situated on the Atlantic Ocean, 275 km/171 mi southwest of Windhoek; population (1997 est) 50,000. It is the only deep-water harbour on the Namibian coast and has a fishing industry with allied trades. Because of its central coastal position, harbour, rail links, and airport, Walvis Bay now handles most of Namibia's trade. It was a detached part (area 1,100 sq km/425 sq mi) of Cape Province, South Africa, 1884–1993 (administered solely by South Africa 1922–92; from 1992 jointly by South Africa and Namibia). In 1993 South Africa waived its claim to sovereignty and control was passed to Namibia in February 1994. Walvis Bay is now part of Erongo region. 72 A4

Wanganui port in southwest North Island, New Zealand, at the mouth of the Wanganui River, 215 km/134 mi north of Wellington; population (1996) 41,300. There are flour and woollen mills, dairy factories, and engineering, steel pipe, fertilizer, and chemical works. Tourism is also important. 78 E4

Wankie former name (to 1982) of ▶Hwange, a town and national park in Zimbabwe. 72 D3

Warnemünde port and resort in the *Land* (administrative region) of ▶Mecklenburg–West Pomerania, Germany; population (1995) 19,500. Warnemünde stands on the Mecklenburg Bay in the Baltic Sea, and at the mouth of the River Warnow, 11 km/7 mi north of Rostock, for which it is the outport. A ferry service operates from here to the port of Gedser in Denmark. 20 H2

Warren industrial city and administrative headquarters of Trumbull County, northeast Ohio, USA; population (1996 est) 48,400. It is situated on the Mahoning River, 21 km/13 mi northwest of Youngstown and 85 km/53 mi southeast of Cleveland. Steel and iron, machinery, car parts, tools, electrical equipment, and paint are produced. Warren was settled in 1799, and developed as an iron and coalmining town. 88 D2

Warrington unitary authority in northwest England, created in 1998 from part of Cheshire.
area 176 sq km/68 sq mi
towns and cities ▶Warrington (administrative headquarters), Lymm, Great Sankey
features River Mersey; Manchester Ship Canal; Warrington Museum and Art Gallery includes over 1,000 paintings; Risley Moss bog and woodland with nature trails and visitors' centre
industries chemicals, food and soft drinks processing, brewing, printing, manufacturing of clothing, leather, metal goods, timber products
population (1996) 151,000 5 E5

Warrington industrial town and, from April 1998, administrative headquarters of ▶Warrington unitary authority in northwest England, on the River Mersey, 25 km/16 mi from both Liverpool and Manchester; population (1994 est) 151,000. It was part of the county of Cheshire to April 1998. Industries include the manufacture of metal goods and chemicals, brewing, iron founding, tanning, engineering, and high technology industries. A trading centre since Roman times, it was designated a new town in 1968. 24 K8

Warrnambool port near the mouth of Hopkins River, southwestern Victoria, Australia; population (1991) 23,900. A tourist centre, it also manufactures textiles and dairy products. 76 H7

Warsaw (Polish **Warszawa**), capital of Poland, on the River Vistula; population (1993) 1,653,300. Industries include engineering, food processing, printing, clothing, and pharmaceuticals. 18 K5

Warszawa Polish name for ▶Warsaw, the capital of Poland. 18 K5

Warta (German **Warthe**), river in Poland, rising about 48 km/30 mi northwest of Kraków, and flowing 790 km/494 mi to join the River ▶Oder at Kostrzyn on the western Polish border. Its chief tributaries are the Noteć and Prosna; the River Vistula is connected by canal. 18 F5

Warthe German name for the ▶Warta, a river in Poland. 18 F5

Warwick market town, administrative headquarters of ▶Warwickshire, central England, 33 km/21 mi southeast of Birmingham, on the River Avon; population (1991) 22,300. Industries include agriculture and tourism. Founded in 914, it has many fine medieval buildings, including a 14th-century castle. 24 L9

Warwickshire county of central England.
area 1,980 sq km/764 sq mi
towns and cities ▶Warwick (administrative headquarters), Nuneaton, Royal Leamington Spa, Rugby, Stratford-upon-Avon (birthplace of the English dramatist Shakespeare)
physical rivers Avon, Stour, and Tame; remains of the 'Forest of Arden' (portrayed by Shakespeare in *As You Like It*)
features Kenilworth and Warwick castles; Edgehill, site of the Battle of Edgehill in 1642, during the English Civil War; annual Royal Agricultural Show held at Stoneleigh
agriculture cereals (oats and wheat); dairy farming; fruit; market gardening
industries cement; engineering; ironstone, and lime are worked in the east and south; motor industry; textiles; tourism
population (1996) 500,600 5 F5

Wasatch Mountains range of the ▶Rocky Mountains in Utah and Idaho. Largely pine-covered, they form the eastern margin of the Great Basin and contain at least four peaks over 3,350 m/11,000 ft high. The highest is Timpanogos Peak (3,660 m/12,000 ft). The Weber, Ogden, and Provo rivers flow through the range and it includes parts of Uinta and Wasatch national forests in Utah. Copper, lead, silver, and gold are mined near Provo and Salt Lake City, but tourism is also significant. 92 D1

Wash, the bay of the North Sea between Norfolk and Lincolnshire, eastern England; 24 km/15 mi long, 40 km/25 mi wide. The rivers Nene, Ouse, Welland, and Witham drain into the Wash. In 1992, 10,120 ha/25,000 acres of the mudflats, marshes, and sand banks on its shores were designated a national nature reserve. 24 N9

Washington state in northwestern USA. It is nicknamed the Evergreen State. Washington was admitted to the Union in 1889 as the 42nd US state. It is bordered to the east by Idaho, to the south by Oregon, to the north by British Columbia, Canada, and to the west by the Pacific Ocean. Washington's Cape Alava is the westernmost point in the lower 48 US states.
population (2000 est) 5,894,100

area 176,480 sq km/68,140 sq mi
capital ►Olympia
towns and cities Seattle, Spokane, Tacoma, Bellevue, Everett
industries and products apples and other fruits, potatoes, livestock, fish and shellfish, timber, processed food, wood products, paper and allied products, aircraft and aerospace equipment, aluminium, computer software 86 B1

Washington, DC (**Washington, District of Columbia**), capital of the USA, on the Potomac River; the world's first planned national capital. It was named Washington, DC, to distinguish it from Washington state, and because it is coextensive with the ►District of Columbia, hence DC; population (2000 est) 572,100; metropolitan area extending outside the District of Columbia (2000 est) 7,608,100. The District of Columbia, the federal district of the USA, is an area of 174 sq km/67 sq mi. Its site was chosen by President George Washington, and the first structures date from 1793. Washington, DC, houses the national executive, legislative, and judicial government of the USA, and is a centre for international diplomacy and finance. Federal and district government are key employers, though numbers employed in both are decreasing. Public, trade, business, and social organizations maintain a presence, as well as law and other service agencies. Tourism is a major industry. 80 J6

Washington, Mount highest peak of the White Mountains, in north-central New Hampshire, USA. It is 1,917 m/6,288 ft high and can be ascended by a cog railway (opened 1869) and a road (1861). It is the highest peak in northeastern USA, and a popular tourist destination. There are a hotel and a meteorological station on the summit. 82 S8

Waterbury city in New Haven County, west Connecticut, USA, on the Naugatuck River, a tributary of the Housatonic, 30 km/19 mi north of New Haven and the shore of Long Island Sound; population (1996 est) 106,400. In the 19th century, it was the nation's largest producer of brass products, and had a thriving clock industry; today its more diversified economy produces chemicals, computer components, television parts, lighting fixtures, clothing, and precision instruments. Mattatuck Community College (1967), Post College (1890), Waterbury State Technical College (1964), and an extension of the University of Connecticut are all located in the city. 88 F2

Waterford county of the Republic of Ireland, in the province of Munster; county town ►Waterford; area 1,840 sq km/710 sq mi; population (1996) 94,700. Other towns include Dungarvon, Lismore, and Tramore. The chief rivers are the Suir and the Blackwater; the Comeragh and Monavallagh mountain ranges lie in the north and centre of the county. Agriculture and dairy farming are important; wheat, barley, and vegetables are also grown. Industries include glassware, pharmaceuticals, and electronics, and there are tanneries, bacon factories, and flour mills. 5 C5

Waterford port and county town of County ►Waterford, in the southeast of the Republic of Ireland, on the River Suir; population (1996) 44,000. It is a major distribution centre, and container traffic is important. Industries include bacon, flour-milling, brewing, paper and board, pharmaceuticals, furniture, and electrical equipment. The factory producing handmade Waterford crystal glass (34% lead content instead of the normal 24%) was opened in 1783 and closed in 1851. It reopened in 1951, and is now the city's largest employer. 24 E9

Watford industrial town in Hertfordshire, southeast England, on the River Colne, 24 km/15 mi northwest of London; population (1991) 110,500. It is a commuter town for London. Industries include printing and publishing, engineering, electronics, and brewing. 22 B3

Waukegan industrial city, lake port, and administrative headquarters of Lake County, in northeast Illinois, USA; population (1990) 69,400. It is situated on a high bluff overlooking Lake Michigan, 58 km/36 mi north of Chicago and 14 km/9 mi south of the Wisconsin state border. The area west of the city has many lakes offering residential and recreational attractions. Shipping remains important. Local manufactures include outboard motors, wire and steel, gypsum and asbestos products, and pharmaceuticals. 88 C2

Waukesha town and administrative headquarters of Waukesha County, southwest Wisconsin, USA; population (1990) 37,100. It is situated on the Fox River in a dairying and former health-resort area 24 km/15 mi west of Milwaukee. Quarrying and food-processing are important to the local economy. Among the town's other manufactures are motors, castings, and foundry and wood products. Waukesha is home to Carroll College (1846) and Mount St. Paul College (1962). 88 C2

Wausau industrial town and administrative headquarters of Marathon County, central Wisconsin, USA; population (1990) 37,100. It is situated on the Wisconsin River, 135 km/84 mi northwest of Green Bay. The local economy relies on dairy products, paper and aluminium products, insurance, chemicals, and plastics. It was settled in 1839 as a lumbering centre. The town's Algonquian name means 'far away'. A University of Wisconsin center (1933) is in Wausau. Rib Mountain State Park is 5 km/3 mi southwest of the town. 84 J3

Weald, the (or **the Kent Weald**; Old English 'forest'), area between the North and South Downs, England, a raised tract of forest 64 km/40 mi wide. It forms part of Kent, Sussex, Surrey, and Hampshire. Once thickly wooded, it is now an agricultural area producing fruit, hops, and vegetables. Crowborough and Wadhurst are the largest villages in the area. In the Middle Ages its timber and iron ore made it the industrial heart of England. 22 B3

Weddell Sea arm of the Southern Atlantic Ocean that cuts into the Antarctic continent southeast of Cape Horn; area 8,000,000 sq km/3,088,800 sq mi. Much of it is covered with thick pack ice for most of the year. 102 (2)A2

Weihai (formerly **Weihaiwei**), industrial port in Shandong, China; population (1990) 129,000. Textiles, food products, alcohol, tyres, clocks, and electronic equipment are manufactured. Opened to foreign investment in 1988, it has strong links with South Korea. From 1898 to 1930 the port was leased to Britain as a naval and coaling station. It was occupied by Japan from 1938 to 1945. 46 G3

Weimar town in the state of Thuringia, Germany, on the River Elm; population (1995) 62,300. Products include farm machinery and textiles. 20 G6

Weissbrunn German name for ▶Veszprém, a town in Hungary. 34 E2

Weissenfels town in the *Land* (administrative region) of ▶Saxony-Anhalt, Germany, on the River Saale, 29 km/18 mi south of Halle; population (1995) 34,700. It has leather and engineering industries and lignite mining. 20 G5

Wejherowo (German **Neustadt**), industrial town (engineering, furniture, cement) in Gdańsk province, Poland, 40 km/25 mi northwest of Gdańsk; population (1990) 46,200. 18 H3

Welland Ship Canal waterway in south Ontario, Canada. Since 1959 it has been a part of the ▶St Lawrence Seaway, linking Lake Erie to Lake Ontario. Extending for 45 km/28 mi through the Niagara Peninsula, the modern canal (1932) carries ships over the Niagara Escarpment between Port Colborne on Lake Erie and Port Weller on Lake Ontario, which is 99 m/326 ft lower. Of its eight locks, numbers 4–6, in Thorold, are 'twin-flight', allowing one ship to be lifted as another is lowered.

Wellington capital and industrial port of New Zealand, in the province of the same name on North Island, on the Cook Strait; population (1996) 335,500 (urban area). Industries in the city include woollen textiles, chemicals, engineering, and electrical goods. The harbour was sighted by Captain James Cook in 1773. 78 E5

Wels Austrian town in the province of Upper Austria, on the River Traun, 25 km/16 mi southwest of Linz; population (1991) 52,600. Machinery and food are manufactured, and the town holds a regular agricultural fair. Wels has baroque buildings, a 14th-century church, and the castle where the Holy Roman Emperor Maximilian I died. 30 K2

Welwyn Garden City industrial town in Hertfordshire, southeast England, 32 km/20 mi north of London; population (1991) 41,400. Industries include electrical engineering and the manufacture of chemicals, plastics, clothing, food, and pharmaceuticals. It was founded as a garden city in 1919–20 by Ebenezer Howard, and designated a new town in 1948. 22 B3

Wenchow alternative transcription of the Chinese town ▶Wenzhou. 46 G5

Wenzhou (or **Wenchow**), industrial port in Zhejiang province, southeast China; population (1994) 299,400. Shoes, machinery, textiles, petrochemicals, and medicines are produced. It was opened to foreign trade in 1877 and is now a coastal city open to foreign investment. 46 G5

Wernigerode town in the *Land* (administrative region) of ▶Saxony-Anhalt, at the foot of the Harz Mountains, 64 km southwest of Magdeburg; population (1994) 36,900. There are engineering and chemical industries. The town dates from the Middle Ages. 20 F5

Wesel city in North Rhine-Westphalia, Germany, at the confluence of the rivers Rhine and Lippe, 38 km/24 mi north of Duisburg; population (1994) 462,600. It has glass, electronic, precision-engineering, and food-processing industries. Wesel was almost totally destroyed by bombing during World War II. The Gothic cathedral of St Willibrord dates from the 15th century. 22 J3

Weser one of the largest rivers in Germany, formed by the junction of the Werra and Fulda; length 440 km/273 mi. From the junction at Münden, the river flows towards the North Sea. The lower reaches are canalized, and the river is navigable by large vessels as far as Bremerhaven, and by smaller vessels as far as Bremen. 20 E4

Wesermünde former name (to 1947) of ▶Bremerhaven, a port in Germany. 20 D3

West Bank area (5,879 sq km/2,270 sq mi) on the west bank of the River Jordan; population (1994 est) 1,122,900. The area was captured by Israel from Jordan in 1967; Jordan finally renounced any claim to it in 1987. Israel refers to the area as Judaea and Samaria. 60 C4

West Bengal state of northeast India; area 88,700 sq km/34,247 sq mi; population (1994 est) 73,600,000. The capital is ▶Kolkata (formerly Calcutta). Industries include jute (particularly at Hooghly industrial complex), iron and steel (at Durgapur, Asansol, based on the Raniganj coalfield), cars, locomotives, aluminium, fertilizers, chemicals, cotton, and printing. Rice, jute, tea (in Darjiling and Jalpaiguri), oilseed, sugar, pulses, and tobacco are grown, and there is fishing. 10 E4

West Berkshire unitary authority in southeast England, created in 1998 from part of the former county of Berkshire.
area 705 sq km/272 sq mi
towns and cities ▶Newbury (administrative headquarters), Hungerford, Lambourn
features River Kennet; River Cambourn; Kennet and Avon Canal; Snelsmore Common Country Park covers 59 ha/ 146 acres; Inkpen Hill (291 m/854 ft) with Stone Age tomb and Walbury Hill (297 m/974 ft) with Iron Age fort are the highest chalk hills in England; Greenham Common Women's Peace Camp has been the site of campaigning against nuclear weapons development at Greenham, Burghfield, and Aldermaston since 1981
industries race horse industry, agriculture, dairy cattle, pig farming (including local Berkshire pig)
population (1996) 142,600. 5 F6

West Bromwich industrial town in West Midlands, central England, 9 km/6 mi northwest of Birmingham, on the edge of the ▶Black Country; population (1991) 144,700. Industries include metalworking and the manufacture of springs and tubes.

West Dunbartonshire unitary authority in west central Scotland, created in 1996 from parts of two districts of Strathclyde region.
area 177 sq km/68 sq mi
towns ▶Dumbarton (administrative headquarters), Clydebank, Alexandria
physical Leven valley and coastal land of Firth of Clyde rise toward the upland plateau of the Kilpatrick Hills
features Dumbarton Castle
industries whisky distilling, light manufacturing
agriculture sheep; not significant
population (1996) 97,800 5 D3

Western Australia state of Australia, bounded on the north and west by the Indian Ocean, on the east by Northern Territory and South Australia, on the south by the Southern Ocean; area 2,525,500 sq km/975,100 sq mi; population

(1996) 1,726,100. The capital is ►Perth. Products include wheat, fresh and dried fruit, beef, dairy products, wool, wine, natural gas, oil, iron, gold, nickel, diamonds, bauxite, cultured and freshwater pearls, timber, and fish. Tourism is important to the state. 76 D5

Western Cape province of the Republic of South Africa from 1994, formerly part of Cape Province; area 129,386 sq km/49,956 sq mi; population (1995 est) 3,721,200. The capital is ►Cape Town. Industries include copper, oil refining, chemicals, engineering, and tourism. Fruit, wheat, and tobacco are grown, and wine is produced. 72 B6

Western Isles island administrative unitary authority area in Scotland, also known as the Outer Hebrides, including the major islands of Lewis, Harris, North and South Uist, Benbecula, and Barra.
area 3,057 sq km/1,180 sq mi
towns ►Stornoway on Lewis (administrative headquarters), Castlebay, Lochboisdale, Lochmaddy, Tarbert
physical open to the Atlantic Ocean on the west and the stormy Minch to the east, the islands are almost treeless and have extensive peat bogs. There are areas of hills and mountains on all the islands. The only fertile land is the sandy Machair on the west coast. The islands are mainly composed of the oldest rock in Britain, the Lewisian gneiss. Lewis. Uninhabited islands include St Kilda and Rockall
features Callanish monolithic Stone Age circles on Lewis
industries Harris tweed, tourism
agriculture sheep, cattle, fishing
population (1996) 27,800 5 C2

Western Provinces in Canada, the provinces of ►Alberta, ►British Columbia, ►Manitoba, and ►Saskatchewan.

Western Sahara (formerly **Spanish Sahara**), disputed territory in northwest Africa, bounded to the north by Morocco, to the east and south by Mauritania, and to the west by the Atlantic Ocean; area 266,800 sq km/103,000 sq mi; population (1993 est) 214,000, including indigenous Sawrawis (traditionally nomadic herders). The capital is ►Laâyoune (Arabic *El Aaiún*). Exports include phosphates and iron ore. 66 C4

Western Samoa former name (to 1997) of ►Samoa. 74 J7

West Flanders (Flemish **West-Vlaanderen**), most westerly province of Belgium, bounded to the east by the Netherlands and East Flanders, to the south by Hainaut and France, to the west by France, and to the northwest by the North Sea for about 65 km/40 mi; area 3,232 sq km/1,248 sq mi; population (1997) 1,123,800. The capital is ►Bruges. Industries include fishing, weaving, spinning, lacemaking, bleaching, and brewing. There is cattle breeding, and flax, hops, and tobacco are grown. 7 (1)A3

West Glamorgan (Welsh **Gorllewin Morgannwg**), former county of southwest Wales, 1974–1996, now divided into ►Neath Port Talbot, and ►Swansea unitary authorities.

West Indies archipelago of about 1,200 islands, dividing the Atlantic Ocean from the Gulf of Mexico and the Caribbean Sea. The islands are divided into:
Bahamas;
Greater Antilles Cuba, Hispaniola (Haiti, Dominican Republic), Jamaica, and Puerto Rico;
Lesser Antilles Aruba, Netherlands Antilles, Trinidad and Tobago, the Windward Islands (Grenada, Barbados, St Vincent, St Lucia, Martinique, Dominica, Guadeloupe), the Leeward Islands (Montserrat, Antigua, St Kitts and Nevis, Barbuda, Anguilla, St Martin, British and US Virgin Islands), and many smaller islands.

West Irian former name of ►Irian Jaya, a province of Indonesia. 53 E3

West Lothian unitary authority in central Scotland, south of the Firth of Forth, which was previously a district within Lothian region (1975–96) and a county until 1974.
area 428 sq km/165 sq mi
towns Bathgate, Linlithgow, ►Livingston (administrative headquarters)
physical low-lying, undulating area through which the River Almond flows; Cairnpapple Hill
features Linlithgow Palace; prehistoric ritual site at Cairnpapple Hill, near Torpichen
industries electronics, engineering, coal-mining, food processing
agriculture productive area of arable farming
population (1996) 147,900 5 E4

Westmeath county of the Republic of Ireland, in the province of Leinster; county town ►Mullingar; area 1,760 sq km/679 sq mi; population (1996) 63,300. The rivers Brosna, Inny, and Shannon flow through the county, and its principal lakes are loughs Ree (the largest, and an extension of the River Shannon), Ennell, Owel, and Sheelin. The Royal Canal cuts through the county but is now disused. The land is low-lying, about 76 m/249 ft above sea-level, with much pasture. The main agricultural activity is cattle and dairy farming. Limestone is found, and textiles are also important. Fishing for trout is popular. Other principal towns are Athlone and Moate. 5 C5

West Midlands metropolitan county of central England, created in 1974; in 1986, most of the functions of the former county council were transferred to the metropolitan borough councils.
area 900 sq km/347 sq mi
towns and cities Birmingham, Coventry, Dudley, Solihull, Walsall, Wolverhampton (all administrative centres for districts of the same name), Oldbury (administrative centre for Sandwell)
industries aircraft components; chemicals; coal mining; engineering; electrical equipment; glass; machine tools; motor vehicles, including Land Rover at Solihull; motor components
population (1996) 2,642,500 5 F5

Weston-super-Mare seaside resort and administrative headquarters of ►North Somerset, southwest England, 32 km/20 mi from Bristol, on the Bristol Channel; population (1991) 68,800. Industries include plastics and engineering. 24 K10

West Palm Beach town, coastal resort, and administrative headquarters of Palm Beach County, southeast Florida, USA; population (1992) 67,700. It is situated on the lagoon, Lake Worth, and the Atlantic Intracoastal Waterway. Industries include transistors, aircraft parts, computers, building materials, and citrus fruits; tourism is important to the economy. It is home to Palm Beach Atlantic College (1968). 90 E4

West Sussex county of southern England, created in 1974, formerly part of Sussex.
area 1,990 sq km/768 sq mi
towns and cities ►Chichester (administrative headquarters), Crawley, Horsham, Haywards Heath,

Shoreham (port); Bognor Regis, Littlehampton, Worthing (resorts)
physical the Weald; South Downs; rivers Adur, Arun, and West Rother
features Arundel and Bramber castles; Chichester cathedral; Goodwood House and racecourse; Petworth House (17th century); Fishbourne villa (important Roman site near Chichester); Selsey (reputed landing place of the South Saxons in 447); Gatwick Airport
agriculture cereals (wheat and barley); fruit; market gardening (mainly on the coastal plain); dairy produce; forestry
industries electronics; light engineering
population (1996) 737,300 **5** F6

West Virginia state in eastern central USA. It is nicknamed the Mountain State. West Virginia was admitted to the Union in 1863 as the 35th US state. It is bordered to the south and east by Virginia, to the north by Ohio, Pennsylvania, and Maryland, and to the west by Ohio and Kentucky. West Virginia is composed essentially of those Virginia counties that, unsympathetic to the plantation South, refused to join Virginia in its 1861 secession from the Union.
population (2000 est) 1,808,300
area 62,900 sq km/24,279 sq mi
capital ►Charleston
towns and cities Huntington, Wheeling, Parkersburg
industries and products apples, maize, poultry, dairy and meat products, coal, natural gas, oil, chemicals, synthetic fibres, plastics, steel, glass, pottery, tourism **88** D3

West-Vlaanderen Flemish name for ►West Flanders, a province in Belgium. **7** (1)A3

West Yorkshire metropolitan county of northeast England, created in 1974; in 1986, most of the functions of the former county council were transferred to the metropolitan borough councils.
area 2,040 sq km/787 sq mi
towns and cities Bradford, Leeds, Wakefield (administrative centres for districts of the same name), Halifax (administrative centre of Calderdale district), Huddersfield (administrative centre of Kirklees district)
physical Ilkley Moor, Haworth Moor; high Pennine moorlands in the west, Vale of York to the east; rivers Aire, Calder, Colne, Wharfe
features Haworth Parsonage; part of the Peak District National Park; British Library, Boston Spa (scientific, technical, and business documents)
industries woollen textiles, financial services; coal mining is in decline
population (1996) 2,109,300 **5** F5

Wetzlar city in Hesse, Germany, on the River Lahn, 54 km/33 mi northeast of Wiesbaden; population (1994) 54,200. The Leitz camera and optical-instruments industry is here; television sets and machinery are also manufactured. Wetzlar's abbey church dates from the 9th century. **20** D6

Wexford county of the Republic of Ireland, in the province of Leinster; county town ►Wexford; area 2,350 sq km/907 sq mi; population (1996) 104,400. Wexford is one of the most intensively cultivated areas in Ireland. The main crops are wheat, barley, beet, and potatoes. Fishing is important, the main fishing port being Kilmore Quay in the south; sheep and cattle rearing are also significant, as is dairy farming. Industries include agricultural machinery and food processing. Wexford was the first part of Ireland to be colonized from England; Normans arrived in 1169. The John F Kennedy Arboretum is one of the most popular visitor attractions in the county. **5** C5

Wexford seaport and county town of ►Wexford, Republic of Ireland, on the estuary of the River Slaney; population (1996) 16,000. Industries include food processing and the manufacture of textiles, cheese, agricultural machinery, furniture, and motor vehicles. There is an annual international opera festival in October. Wexford was founded in the 9th century by Danes; it was taken by the Anglo-Normans in 1169, and besieged and devastated by Oliver Cromwell in 1649. In the Rebellion of 1798 Wexford was briefly held by Irish insurgents. **24** F9

Weymouth seaport and resort in Dorset, on the south coast of England, at the mouth of the River Wey; population (1991) 45,900. It is linked by ferry to France and the Channel Islands. Industries include the quarrying of Portland stone, sailmaking, brewing, fishing, electronics, and engineering. Weymouth, dating from the 10th century, was the first place in England to suffer from the Black Death in 1348. It was popularized as a bathing resort by George III. **24** K11

Wheeling (Delaware 'place of the head'), river port in northwest West Virginia, USA, on the Ohio River, 69 km/43mi southwest of Pittsburgh, Pennsylvania; population (1996 est) 33,310. Coal has been mined since the late 18th century, and iron and steel manufactured since 1832. Other products include natural gas, textiles, glass, pottery, paper, and chemicals. Settled in 1769, Wheeling grew with the overspill of heavy industry from the Pittsburgh region, and as a distribution centre. It was the state capital 1863–70, and 1875–85. Fort Henry, site of the last battle of the American Revolution in 1782, is here. **90** E1

Whitby port and resort in North Yorkshire, northern England, on the North Sea coast, at the mouth of the River Esk, 32 km/20 mi northwest of Scarborough; population (1991) 13,800. Industries include tourism, boat building, fishing (particularly herring), and plastics. There are remains of a 13th-century abbey. Captain James Cook served his apprenticeship in Whitby and he sailed from here on his voyage to the Pacific Ocean in 1768. Bram Stoker's *Dracula* (1897) was set here. **24** M7

Whitehaven town and port in Cumbria, northwest England, on the Irish Sea coast, southwest of Carlisle; population (1991 est) 26,400. Industries include chemicals, printing, textiles, and food processing. Britain's first nuclear power station was sited in 1956 at Calder Hall to the southeast. Sellafield nuclear power station (formerly known as Windscale) is also nearby. **24** J7

Whitehorse capital of ►Yukon Territory, Canada, on the Yukon River; population (1996) 21,800. Situated at the junction of the Alaska and Klondike Highways, it is the centre of the region's mining and forestry industries, and an important transport focus, with air links to major Canadian and US cities, and a rail link to Skagway, Alaska. It is also the regional headquarters of the Royal Canadian Mounted Police. Whitehorse was founded by prospectors during the Klondike gold rush 1897–98, when it occupied an important position at the head of navigation on the Yukon. It replaced Dawson as capital in 1953. **92** (1)L3

White Mountains glaciated mountain range in New Hampshire, USA. Much of the area is high plateau country, from which rises the Presidential range culminating in Mount Washington (1,917 m/6,288 ft), the highest mountain in northeastern USA. The White Mountains are a northerly continuation of the ▶Appalachian Mountains. Tourism (particularly winter sports) and timber-processing are important. **82** S8

White Russia English translation of ▶Belarus. **14** G2

White Sea (Russian **Beloye More**), gulf of the Arctic Ocean on the northwest coast of Russia, on which the port of Arkhangelsk stands; area 90,000 sq km/34,750 sq mi; average depth 60 m/200 ft, maximum depth 330 m/1,082 ft. There is a warship construction base, including nuclear submarines, at Severodvinsk. The North Dvina, Mezen, and Onega rivers flow into it, and there are canal links with the Baltic, Black, and Caspian seas. In winter the bays are often ice-bound, with drifting ice offshore. **38** G1

White Volta one of the two main upper branches of the River ▶Volta, running through Burkina Faso and Ghana. **68** D3

Whitney, Mount peak of the ▶Sierra Nevada mountain range, south California, USA; height 4,418 m/14,495 ft. It is the highest peak in continental USA (excluding Alaska and Hawaii). It was named after Josiah Dwight Whitney, the first director of the California State Geological Service. **86** C3

Whitstable resort in Kent, southeast England, at the mouth of the River Swale, noted for its oysters; population (1991) 28,500. It is also a yachting centre. **22** D3

Whyalla port and industrial city in South Australia, on the west side of Spencer Gulf, 394 km/245 mi northwest of Adelaide; population (1996) 23,382. Whyalla is South Australia's second-largest city and is the port for iron ore mined at Iron Knob, 56 km/36 mi inland. Industries include steelworks and iron processing. **76** G6

Wichita largest city in Kansas, USA, on the Arkansas River, 338 km/210 mi southwest of Kansas City; population (1996 est) 320,400. It is the distribution and processing centre of a wheat-growing and oil-producing region. Industries include oil-refining, flour-milling, food preparation, and the manufacture of aircraft and motor vehicles. Founded in 1868 as a trading post on a American Indian Witchita village site, it developed as a stopover on the Chisholm cattle-driving trail. With the arrival of the railroad in 1872, the city became a major cattle-shipping point. Oil was discovered in 1915, and aircraft manufacture began in 1920. **90** B2

Wichita Falls city and administrative headquarters of Wichita County, also in Archer County, north-central Texas, USA, on the **Wichita River**, south of the Oklahoma border, 153 km/96 mi northwest of Fort Worth; population (1996 est) 100,100. It is an industrial and commercial centre, trading in locally-produced oil, cattle, cotton, and wheat. Leather goods, glass, textiles, electronics, engine-parts, oil-drilling equipment, and pharmaceuticals are manufactured. Sheppard Air Force Base (1941), a NATO training centre, is to the north of the city. Wichita Falls is home to Midwestern State University (1922). **90** B3

Wick fishing port and industrial town in the Highland unitary authority, northeast Scotland, at the mouth of the River Wick; population (1991) 7,800. Industries include shipping, distilleries, glassware, knitwear, and North Sea oil. Air services to the Orkney and Shetland islands operate from here. An opera house at 15th-century Ackergill Tower opened in 1994. **24** J3

Wicklow county of the Republic of Ireland, in the province of Leinster; county town ▶Wicklow; area 2,030 sq km/784 sq mi; population (1996) 102,700. It includes the **Wicklow Mountains**, the rivers Slaney, Avoca, Vartry, and Liffey, and the coastal resort of Bray. Other towns include Arklow, Greystones, and Baltinglass. The village of Shillelagh gave its name to rough cudgels of oak or blackthorn made there. Agriculture is important; there is livestock rearing (in particular a special breed of mountain sheep), and dairy farming. Wheat and oats are grown, and seed potatoes and bulbs are produced. Granite is mined at Aughrim and Ballyknockan. **5** C5

Wicklow (Danish **Wykinglo**), resort and county town of County ▶Wicklow, Republic of Ireland, on the River Vartry, 50 km/31 mi southeast of Dublin; population (1996) 6,400. **24** F9

Wien German name for ▶Vienna, the capital of Austria. **30** M2

Wiener Neustadt Austrian city in the province of Lower Austria, 40 km/25 mi south of Vienna; population (1995) 38,400. Industries include automobiles, engineering, aircraft, brewing; leather goods are produced. **30** M3

Wiesbaden spa town and capital of Hesse, Germany, on the River Rhine 20 km/12 mi west of Frankfurt; population (1995) 266,400. Products include cement, plastics, textiles, pharmaceuticals, and electrical components. Printing, publishing, and film making are important. Most of the German sparkling wine cellars are in this area (a sparkling wine Sekt is produced here). Tourism is an important part of the local economy. The city is a cosmopolitan resort and has a casino and sporting facilities. Wiesbaden was the capital of the former duchy of Nassau from the 12th century until 1866. **20** D6

Wigan industrial town in Greater Manchester, northwest England, between Liverpool and Manchester, on the River Douglas; population (1991) 84,700. Industries include food processing, engineering, the manufacture of paper, fibreglass, and carpet tiles, tourism and leisure, and retail. The traditional coal and cotton industries have declined. **24** K8

Wight, Isle of island and unitary authority of southern England.
area area 380 sq km/147 sq mi
towns ▶Newport (the administrative headquarters); Ryde, Sandown, Shanklin, Ventnor (all resorts)
physical chalk cliffs and downs, and deep ravines, known locally as 'chines'; the highest point is St Boniface Down (240 m/787 ft); the Needles, a group of pointed chalk rocks up to 30 m/100 ft high in the sea to the west; the Solent, the sea channel between Hampshire and the island
features Cowes, venue of Regatta Week and headquarters of the Royal Yacht Squadron; Osborne House, built for Queen Victoria in 1845
agriculture fruit and vegetables
industries aircraft components, electronics, marine engineering,

plastics, boatbuilding, tourism
population (1996) 130,000 24 L11/5 F6

Wilhelmshaven North Sea industrial port and resort in Lower Saxony, Germany, 165 km/102 mi northwest of Hannover; population (1995) 90,900. It lies on the western arm of Jade Bay at the end of the Jade–Ems canal. Industries include oil refining, petrochemicals, metal working, marine engineering, office equipment, and data processing. Oil from supertankers is discharged from here into the 390 km/242 mi pipeline to the Rhine-Ruhr region. It was an important naval base until the end of World War II. 20 D3

Wilkes Barre city in northeast Pennsylvania, USA, on the Susquehanna River, 23 km/14 mi southwest of Scranton; population (1996 est) 44,400. It lies in an area where anthracite was once mined. Heavy machinery, furniture, textiles, wire, and tobacco products are manufactured. First settled in 1769, it was named after two Englishmen, and grew as a mining town. 88 E2

Williamsburg historic town in southeast Virginia, USA; population (1990) 11,500. It is the administrative headquarters of (but is administratively separate from) James City County. It is situated between the James and York rivers, 74 km/46 mi southeast of Richmond, and 10 km/6 mi northeast of Jamestown. Founded in 1632, it served as the capital of the colony of Virginia from 1699–1779. Much of it has been restored to its 18th-century appearance. The town is a major tourist attraction. The College of William and Mary (1693) is one of the oldest in the USA. 88 E3

Williamsport industrial town and administrative headquarters of Lycoming County, north-central Pennsylvania, USA; population (1990) 31,900. It is situated on the Susquehanna River and Lycoming Creek, in the foothills of the Allegheny Mountains, 121 km/75 mi north-northwest of Harrisburg. Industries include electronics, plastics, metals, lumber, textiles, and aircraft parts. It was founded 1795 and is the birthplace of Little League baseball (1939). Lycoming College (1812) and Williamsport Area Community College (1920) are both in the town. 88 E2

Wilmington port and largest city in Delaware, USA, at the junction of the Brandywine with the Delaware River; population (1996 est) 69,500. It is the focus of a metropolitan area extending into neighbouring New Jersey and Maryland. Industries include shipbuilding, and the manufacture of chemicals, textiles, and iron and steel products. It is the headquarters of E I Du Pont de Nemours and Co, founded as a gunpowder mill in 1802. 88 E3

Wilmington city and port in southeast North Carolina, USA, on the Cape Fear River, 50 km/30 mi from the Atlantic Ocean; population (1996 est) 62,190. It is the state's chief seaport; deepwater access has been provided by dredging the river to a depth of 10 m/32 ft. Agricultural products are exported, and oil is the main import. Industries include film-production, timber- and tobacco-processing, and the manufacture of chemicals, fibre optics, and textiles. Tourism is also important to the economy. Wilmington was settled in 1740, and named after the British politician Spencer Compton, Earl of Wilmington. During the Civil War it was the Confederates' main Atlantic port and the last to close; 'Blockade Runners' continued to export cotton until 1865. 90 F3

Wilts abbreviation for ▶Wiltshire, an English county. 5 E6

Wiltshire county of southwest England (since April 1997 Swindon has been a separate unitary authority).
area 3,480 sq km/1,343 sq mi
towns and cities Trowbridge (administrative headquarters), Salisbury, Wilton, Devizes, Chippenham, Warminster
physical Marlborough Downs; Savernake Forest; rivers Kennet, Wylye, Avons (Salisbury and Bristol); Salisbury Plain (32 km/20 mi by 25 km/16 mi, lying at about 120 m/394 ft above sea-level)
features Longleat House (Marquess of Bath); Neolithic Stonehenge, Avebury, Silbury Hill, West Kennet Long Barrow, finest example of a long barrow in Wiltshire, dating from the 3rd millennium BC; Salisbury Cathedral has the tallest spire in Britain (123 m/404 ft)
agriculture cereals; cattle; dairy farming (condensed milk, cheese); pig and sheep farming
industries brewing (Devizes); computing; electronics; engineering (Chippenham); pharmaceuticals; plastics; quarrying (Portland stone); rubber; tobacco
population (1996) 593,300 5 E6

Winchester cathedral city and administrative headquarters of ▶Hampshire, England, on the River Itchen, 19 km/12 mi northeast of Southampton; population (1991) 36,100. Tourism is important, and there is also light industry. Originally a Roman town, Winchester was capital of the Anglo-Saxon kingdom of Wessex, and later of England. Winchester Cathedral (1079–93) is the longest medieval church in Europe and was remodelled from Norman-Romanesque to Perpendicular Gothic under the patronage of William of Wykeham (founder of Winchester College in 1382), who is buried there, as are Saxon kings, St Swithun, and the writers Izaac Walton and Jane Austen. 24 L10

Windhoek capital of Namibia, and administrative centre of Khomas region; population (1992) 126,000. It is just north of the Tropic of Capricorn, in the Khomas Highlands, 290 km/180 mi from the west coast. It is the world centre of the karakul (breed of sheep) industry; other industries include engineering and food processing. 72 B4

Windsor Great Lakes port in Ontario, southeast Canada, on the Detroit River, 386 km/240 mi west of Toronto; population (1991) 191,400, metropolitan area (1996) 291,700. It is connected to Detroit, Michigan, USA, across the river, by the Ambassador Bridge and Detroit–Canada Tunnel. Windsor is the centre of Canada's car industry, and a major producer of pharmaceuticals. Machinery, fabricated metals, clothing, paint, and bricks are also manufactured. Commercial and processing facilities are provided for the surrounding mixed farming region. Known as The Ferry around 1700, it was named after Windsor, England in 1836. 88 D2

Windsor town in southern England, on the River Thames, 35 km/22 mi west of London; population (1991, with Eton) 30,600. Formerly in Berkshire, it joined with Maidenhead to become ▶Windsor and

Maidenhead unitary authority in April 1998. Industries include tourism, computer services, and financial services. It is the site of Windsor Castle, a royal residence, and a 17th-century guildhall designed by Christopher Wren. Nearby is the prestigious private school Eton College (founded in 1440). 22 B3

Windsor and Maidenhead unitary authority in southeast England, created in 1998 from part of the former county of Berkshire.
area 198 sq km/76 sq mi
towns and cities Windsor, ►Maidenhead (administrative headquarters)
features River Thames; Windsor Castle, royal residence originally built by William the Conqueror; Windsor Great Park, remnant of royal hunting ground; Eton College, founded by Henry VI in 1440; Household Cavalry Museum (Windsor); Stanley Spencer (1891–1959) Gallery (Cookham on Thames); Ascot Racecourse
industries tourism and service industries, electrical systems and components, chemicals, motor vehicle components, telecommunications, publishing, scientific equipment
population (1996) 140,200 5 F6

Windward Islands group of islands in the West Indies, forming part of the lesser ►Antilles. 94 N6

Winnipeg (called 'Gateway to the West'; Cree *win-nipuy* 'muddy water'), capital of Manitoba, Canada, at the confluence of the Red and Assiniboine rivers, 65 km/40 mi south of Lake Winnipeg, 30 km/20 mi north of the US border; population (1991) 616,800, metropolitan area (1996) 676,700. It is a focus for trans-Canada and Canada–US traffic, and a market and transhipment point for wheat and other produce from the prairie provinces: Manitoba, Alberta, and Saskatchewan. Processed-foods, textiles, farming machinery, and transport equipment are manufactured. Established as Winnipeg in 1870 on the site of earlier forts, the city expanded with the arrival of the Canadian Pacific Railway in 1881. 82 M7

Winnipeg, Lake (Cree *win-nipuy* 'muddy water'), lake in southern Manitoba, Canada. It is a small remnant of the glacial lake Agassiz and Canada's third-largest lake, covering an area of 24,500 sq km/9,460 sq mi. Its waters are fed by the ►Red, ►Saskatchewan, Winnipeg, and other rivers, draining much of the Canadian prairies. Its outflow is through the River Nelson, northeastwards to Hudson Bay. 82 M6

Winston-Salem city in northern central North Carolina, USA, 220 km/140 mi northeast of Charlotte; population (1996 est) 153,500. It developed as the commercial and processing centre of a fertile tobacco-growing region, and the manufacture of cigarettes and flat plug tobacco remains a major industry. Medical equipment, textiles, clothing, and furniture are also produced. It was created in 1913 from two towns: Winston (established in 1849) and Salem (laid out in 1766), and is now the second-largest city in the state. 88 D3

Winterthur Swiss town and spa northeast of Zürich; population (1994) 88,200. Manufacturing includes engines and textiles. 30 D3

Wisconsin state in northern central USA. It is nicknamed the Badger State. Wisconsin was admitted to the Union in 1848 as the 30th US state. Part of the Midwest, it is bordered to the south by Illinois, to the west by Iowa and Minnesota, to the north by Lake Superior and the Upper Peninsula of Michigan, and to the east by Lake Michigan.
population (2000 est) 5,363,700
area 145,500 sq km/56,163 sq mi
capital ►Madison
towns and cities Milwaukee, Green Bay, Racine
industries and products leading US dairy state; maize, hay, industrial and agricultural machinery, engines and turbines, precision instruments, paper products, cars and lorries, plumbing equipment, research, tourism 84 H2

Wisconsin river of Wisconsin, USA; length 700 km/435 mi. It rises in the lake area near the Michigan boundary, and flows southwest to join the ►Mississippi River, 80 km/50 mi south of La Crosse. The Portage Canal connects it with Fox River and Lake Michigan. There are many hydro-electric plants along its course, notably at Lake Wisconsin. It flows through the Wisconsin Dells, a scenic gorge. 88 B2

Wisła Polish name for the ►Vistula, a river in Poland. 18 H4

Wismar German Baltic Sea port in the *Land* (administrative region) of ►Mecklenburg–West Pomerania; population (1994) 53,200. The town's main employer is a large shipyard. Wismar became a member of the Hanseatic League in 1266, and was later ruled by Sweden, Denmark, and then Mecklenburg-Schwerin. It was heavily bombed during World War II, and was extensively rebuilt. 20 G3

Witten city in North Rhine-Westphalia, Germany, on the Ruhr River; population (1995) 105,000. Formerly a coalmining centre, Witten now has steel and chemicals industries, there is a modern technology park, and the Witten-Herdecke University is here. 22 K3

Wittenberg (or **Lutherstadt Wittenberg**), town in the state of Saxony-Anhalt, Germany, on the River Elbe, southwest of Berlin; population (1995) 52,500. Wittenberg University was founded in 1502, but transferred to Halle in 1815. The town became the cradle of the Reformation in the early 16th century. The Protestant reformer Martin Luther preached in the Stadtkirche (in which he is buried), nailed his 95 theses to the door of the Schlosskirche in 1517, and taught Philosophy at the university. 20 H5

Wittenberge town in the *Land* (administrative region) of ►Mecklenburg–West Pomerania, Germany, in the district of Schwerin, near the River Elbe, 72 km/45 mi south of Schwerin; population (1995) 25,000. It has textile, railway rolling-stock, and oil-refining industries. 20 G3

Włocławek capital of Włocławek province, central Poland, on the River Vistula, 140 km/87 mi northwest of Warsaw; population (1990) 122,100. It is a river port, and its chief industries are metalwork, paper, and food processing. Founded in the 11th century, it has a 14th-century cathedral. 18 J5

Wokingham unitary authority in southeast England, created in 1998 from part of the former county of Berkshire.
area 179 sq km/69 sq mi
towns and cities Wokingham (administrative headquarters), Twyford

features River Thames forms northern border of authority; Royal Electrical and Mechanical Engineering Corps Museum (Arborfield); Swallowfield Park, house built for 2nd Earl of Clarendon in 1690; National Dairy Museum; Henley Regatta course; large areas of mixed woodland including remnants of old Royal Chase of Windsor Forest and tree-lined avenues; Finchampstead Ridges
industries light engineering, electronics and information technology, telecommunications, computer components and software, plastics
population (1996) 142,000 5 F6

Wolfenbüttel city in Lower Saxony, Germany, on the River Oker, 12 km/7 mi south of Brunswick; population (1994) 121,400. The chief industry is canning, but chemicals and musical instruments are also manufactured. Wolfenbüttel contains one of the most important research libraries in Germany. The town has a castle dating from 1570–1691. 20 F4

Wolfsburg town in Lower Saxony, Germany, 20 km/12 mi northeast of Brunswick on the Mittelland Canal; population (1995) 126,800. It was chosen in 1938 as the Volkswagen factory site because of its central, strategic location. 20 F4

Wollongong (Aboriginal 'sound of the sea'), industrial city on the coast of New South Wales, Australia, 80 km/50 mi south of Sydney; population (1996, with Port Kembla) 219,761. Wollongong, the third-largest city in New South Wales, is the main residential and business centre of the Illawarra district. Its chief industries are steel production, tourism, fishing, dairying, and the manufacture of textiles and clothing. Wollongong is situated on Australia's southern coalfield, and exports coal from its port at Port Kembla. Greater Wollongong includes Port Kembla. 76 K6

Wolverhampton industrial town in West Midlands, central England, 20 km/12 mi northwest of Birmingham; population (1994) 256,100. Industries include metalworking, engineering, and the manufacture of chemicals, tyres, aircraft, bicycles, locks and keys, and commercial vehicles. Europe's first power station fuelled by waste tyres opened here in 1993. 24 K9

Woonsocket industrial city in Providence County, north Rhode Island, USA; population (1990) 43,900. It is situated on the Blackstone River and Massachusetts state border, 21 km/13 mi north-northwest of Providence. Industries include rubber, chemicals, and woollen goods. Woonsocket had a 1666 sawmill, a 1720 foundry, and an 1810 cotton mill, and in 1840 began woollen manufacturing. It is still primarily a woollen and worsted centre. The city has a large French-Canadian population. 88 F2

Worcester cathedral city in west central England on the River Severn, and administrative headquarters of the county of ▶Worcestershire, on the River Severn, 35 km/22 mi southwest of Birmingham; population (1991) 82,700. Industries include the manufacture of shoes, Worcestershire sauce, and Royal Worcester porcelain. The cathedral dates from the 13th and 14th centuries. The birthplace of the composer Elgar at nearby Broadheath is a museum. At the **Battle of Worcester** in 1651 Oliver Cromwell defeated Charles II. 24 K9

Worcester city in central Massachusetts, USA, on the Blackstone River, 70 km/44 mi west of Boston; population (1996 est) 166,400. It is situated in a hilly area with numerous lakes. Industries include engineering, printing, biotechnology, and the manufacture of textiles, pharmaceuticals, and electrical equipment. Named after Worcester, England, it was permanently settled in 1713. The city grew with the industrialization of ▶New England, and is now Massachusetts' second-largest city. 84 M3

Worcestershire two-tier county of west central England. Herefordshire and Worcestershire existed as counties until 1974, when they were amalgamated to form the county of Hereford and Worcester; in 1998 this county was divided back into Worcestershire and Herefordshire, which regained their pre-1974 boundaries.
area 1,735 sq km/670 sq mi
towns and cities ▶Worcester (administrative headquarters), Bewdley, Bromsgrove, Evesham, Kidderminster, Pershore, Stourport, Tenbury Wells
physical Malvern Hills in the southwest; rivers Severn with tributaries Stour, Teme, and Avon (running through the fertile Vale of Evesham)
features Droitwich, once a Victorian spa, reopened its baths in 1985 (the town lies over a subterranean brine reservoir)
agriculture cereals (oats, wheat), fruit (apples, pears), hops, vegetables; cider; much of the county devoted to permanent pasture, notably for Hereford cattle
industries carpets (Kidderminster), chemicals, engineering, food processing, porcelain (Worcester)
population (1996) 535,700 5 E5

Worcs abbreviation for ▶Worcestershire, an English county. 5 E5

Worksop market and industrial town in Nottinghamshire, central England, on the River Ryton; population (1991) 37,100. Industries include coalmining, light engineering, and the manufacture of chemicals and food products. Mary Queen of Scots was imprisoned at Worksop Manor, which was burned in 1761. 22 A1

Worms ancient city and river port in Rhineland-Palatinate, Germany, on the River Rhine, 25 km/15 mi north of Ludwigshafen; population (1995) 79,700. Industries include food processing and the manufacture of chemicals, paint, machinery, furniture, and worsted. The vineyards of the Liebfrauenkirche (14th to 15th century) produced the original Liebfraumilch wine; it is now produced by many growers around Worms. The Protestant reformer Martin Luther appeared before the **Diet** (Assembly) **of Worms** in 1521 and was declared an outlaw by the Roman Catholic Church. It is one of the oldest cities in Germany. 20 D7

Worthing seaside resort in West Sussex, southeast England, 23 km/14 mi west of Brighton, at the foot of the South Downs; population (1991) 94,000. Industries include financial and business services, electronics, engineering, the manufacture of plastics and furniture, horticulture, retail, tourism, and leisure. There are traces of prehistoric and Roman occupation in the vicinity. The headquarters of the English Bowling Association is located in Worthing. 24 M11

Wrexham unitary authority in northeast Wales, created in 1996 from part of the former county of Clywd.
area 500 sq km/193 sq mi
towns ►Wrexham (administrative headquarters), Holt, Ruabon
physical western side is mountainous, including Ruabon Mountain; River Dee
features Clywedog Valley, with notable countryside and industrial archaeology
industries food manufacture, plastics, pharmaceuticals, high-technology industries
population (1996) 123,500 **5** E5

Wrexham (Welsh **Wrecsam**), administrative centre of ►Wrexham unitary authority, northeast Wales, situated on the Clywedog River 19 km/12 mi southwest of Chester; population (1991) 40,600. Industries include coal, electronics, pharmaceuticals, chemicals, cables, and metal goods. It is the seat of the Roman Catholic bishopric of Menevia (Wales). Elihu Yale, founder of Yale University, USA, died in Wrexham and is buried in the 15th-century church of St Giles. **24** K8

Wrocław (formerly **Breslau**), industrial river port in Poland, on the River Oder; population (1993) 643,600. Industries include shipbuilding, engineering, textiles, and electronics. It was the capital of the German province of Lower Silesia until 1945. **18** G6

Wuchang former city in Hubei province, China; amalgamated with ►Wuhan. **46** H2

Wuhan river port and capital of ►Hubei province, central China, at the confluence of the Han and Chang Jiang rivers; population (1994) 4,436,100. It was formed in 1950 as one of China's greatest industrial areas by the amalgamation of Hankou, Hanyang, and Wuchang. Iron, steel, machine tools, textiles, food and drinks, fibre optic cables, and fertilizer are manufactured. **46** E4

Wuhsien alternative transliteration for ►Suzhou, a city in Jiangsu province, China. **46** G4

Wuppertal industrial city in North Rhine-Westphalia, Germany, in the valley of the Wupper in the Bergisches Land, 32 km/20 mi east of Düsseldorf; population (1995) 382,400. Industries include textiles, brewing, electronics, and the manufacture of plastics and precision instruments. The city was formed in 1929 (named in 1931) by uniting Elberfield (13th century) and Barmen (11th century). A monorail (German *Schwebebahn*), built at the beginning of the 20th century and suspended above the river, is still in operation. **20** C5

Würzburg commercial city in northwest Bavaria, Germany, situated on the Main River, 219 km/136 mi northwest of Munich; population (1995) 127,700. There are engineering, printing, wine, and brewing industries. The Bishop's Palace was decorated by the Italian rococo painter Tiepolo. The physicist Wilhelm Conrad Röntgen discovered X-rays here in 1895. Frankenwein wine in its distinctively-shaped bottle is produced here. **20** E7

Wye (Welsh **Gwy**), river in Wales and England; length 208 km/130 mi. It rises on Plynlimon in northeast Ceredigion, flows southeast and east through Powys and Hereford and Worcester, and follows the Gwent–Gloucestershire border before joining the River ►Severn 4 km/2.5 mi south of Chepstow. It has salmon fisheries and is noted for its scenery. **24** J9

Wyoming state in western USA. It is nicknamed the Equality State. Wyoming was admitted to the Union in 1890 as the 44th US state. One of the Mountain States, it is bordered to the east by Nebraska and South Dakota, to the north by Montana, to the west by Montana, Idaho, and Utah, and to the south by Utah and Colorado.
population (2000 est) 493,800
area 253,400 sq km/97,812 sq mi
capital ►Cheyenne
towns and cities Casper, Laramie
industries and products oil, natural gas, sodium salts, coal, uranium, sheep, beef **84** E3

Xanten town in North Rhine-Westphalia, Germany, near the left bank of the River Rhine, 50 km/31 mi northwest of Düsseldorf; population (1994) 18,700. Its Gothic cathedral was begun in 1190. Xanten is a tourist centre. **22** J3

Xiamen (Fujianese **Amoy**), industrial port in Fujian province, southeast China, on **Xiamen Island**; population (1994) 654,900. Products include electronics, machinery, textiles, foods, and chemicals. It was one of the original five treaty ports used for trade under foreign control from 1842 to 1943, and a special economic zone from 1979 offering incentives and special privileges to foreign investors. **50** F2

Xi'an ('western peace'), industrial city and capital of ►Shaanxi province, China, on the Wei He River; population (1993) 2,360,000. It produces chemicals, machinery, electrical and electronic equipment, aircraft, and fertilizers. **46** D4

Xianggang Chinese name for ►Hong Kong. **50** E2

Xi Jiang (or **Si-Kiang**; 'west river'), river in China, which rises in Yunnan province and flows into the South China Sea; length 1,900 km/1,200 mi. Guangzhou lies on the northern arm of its delta and the island of Hong Kong at its mouth. It is the longest river in south China. **46** E6

Xingu region in Pará, Brazil, crossed by branches of the Xingu River which flows for 1,900 km/1,200 mi to the Amazon Delta. In 1989 Xingu Indians protested at the creation of a vast, intrusive lake for the Babaquara and Kararao dams of the Altamira complex. **98** G5

Xining (or **Sining**), industrial city and capital of ►Qinghai province, northwest China, on the Xining River; population (1994) 540,000. Products include machinery, vehicles, textiles, metals, and processed foods. For centuries Xining was a major trading centre on the caravan route to Tibet, and it is now the starting point for the modern rail and road routes into the Tibetan Plateau. **46** C3

Xinjiang Uygur Autonomous Region (or **Xinjiang**, or **Sinkiang Uighur Autonomous Region**), autonomous region of northwest China, bounded to the north by Kyrgyzstan, Kazakhstan, and Russia; to the east by Mongolia and Gansu; to the south by Qinghai and Tibet; and to the west by Jammu and Kashmir, Afghanistan, and Tajikistan; area 1,646,800 sq km/635,800 sq mi; population (1996) 16,890,000. The capital is ►Urumqi. Industries include oil, chemicals, iron, textiles, coal, copper, and tourism. Cereals, cotton, and fruit are grown, and there is animal husbandry. **11** (2)B2

Xizang Chinese name for ▶Tibet, an autonomous region of southwest China from 1965. 54 E2/11 (2)B3

Yakima city and administrative headquarters of Yakima County, south-central Washington, USA, on the **Yakima River**, 155 km/97 mi southeast of Seattle; population (1996 est) 65,100. It is the commercial and processing centre of an agricultural region created from the desert by irrigation. Products include sugar beet, apples, mint, hops, livestock, cider, flour, and canned-foods. Yakima Valley College was established here in 1928. The city hosts the annual Central Washington State Fair. 86 B1

Yakutia (Russian **Yakutskaya**), former name of ▶Sakha, an autonomous republic in the Russian Federation. 44 N3

Yakutsk capital, economic and cultural centre of the ▶Republic of Sakha (formerly Yakutia), eastern Siberia, Russian Federation; population (1996 est) 191,000. It is situated on the left bank of the middle Lena. The city has saw-milling, leather and shoe, and food-processing industries. 44 M4

Yalu (Korean **Amnok**), river forming the northern boundary between North Korea and Jilin and Liaoning provinces in north China; length 790 km/491 mi. It rises in the Changbai Mountains in Jilin province, and empties into Korea Bay, just south of Dandong. It is navigable only near the mouth, and is frozen from November to March. 48 D3

Yamal Peninsula (Russian **Poluostrov Yamal**), peninsula in northwestern Siberian Russia, with natural gas reserves estimated at 6 trillion cu m/212 trillion cu ft; supplies are piped to Western Europe. 42 M3

Yambol (Bulgarian **Jambol**), town in Burgas region, eastern Bulgaria, on the River Tundža, 25 km/16 mi from Sliven; population (1990) 98,900. It produces textiles, shoe machinery, and processed food. A stone mosque remains from its former Turkish rule. 34 P7

Yamoussoukro capital since 1983 of ▶Côte d'Ivoire; population (1995 est) 174,000. The city is in Bouaké department and lies northwest of Abidjan. The economy is based on tourism, fishing, agricultural trade and production, and petroleum distribution to the surrounding region. Other industries include forestry and perfume manufacture. 68 C3

Yan'an (or **Yenan**), industrial city and district in Shaanxi province, northwest China, on the Yan He River; population (1994) 322,500. Textiles and cigarettes are produced. The Long March ended here in January 1937, and it was the communist headquarters from 1936 to 1947; the caves in which Mao Zedong lived are preserved. 46 D3

Yangon (formerly **Rangoon** until 1989), capital and chief port of Myanmar (Burma) on the Yangon River, 32 km/20 mi from the Indian Ocean; population (1998 est) 3,576,500. Industries include shipbuilding, oil refining, rice milling, and textile and pottery manufacture. Yangon is a centre of communications by road, rail, and air, as well as by river transport, and is one of the greatest rice markets in the world. The city **Dagon** was founded on the site in AD 746; it was given the name Rangoon (meaning 'end of conflict') by King Alaungpaya in 1755. 50 B3

Yangtze-Kiang alternative transcription of ▶Chang Jiang, the longest river in China. 46 D4

Yangzhou (or **Yangchow**), canal port in Jiangsu province, east China, on the Chang Jiang River; population (1995 est) 489,000. It is an ancient city, first settled before the 5th century BC. The completion of the Grand Canal in 610 AD helped the city to become an important communications centre and during the Tang period (618–906 AD) it greatly increased in size and wealth, becoming a centre for the salt trade, shipbuilding and the arts. Now it is an artistic centre for crafts, jade carving, and printing, and among its features are gardens and pavilions. 46 F4

Yantai (formerly **Chefoo**), ice-free port in Shandong province, east China; population (1990) 452,000. It was opened for foreign investment in 1984. Industries include tourism, winemaking, fishing, food-processing, and the production of machinery, vehicle parts, and embroidery. 46 G3

Yaoundé capital of ▶Cameroon, 210 km/130 mi east of the port of Douala; population (1991) 750,000. Industries include tourism, oil refining, food production, and textile manufacturing. It is linked by the Trans-cameroon railway to Douala and to Ngaoundere in the north. 68 G4

Yaren (or **Yaren district**), largest settlement, chief port, and de facto capital of the republic of ▶Nauru; population (2001 est) 4,500. Nauru has no official capital because its population is dispersed over the island, but Yaren, on the southwestern coast, is the site of the country's parliament and government ministries. The economy is based on the export of guano, and a subsistence agriculture of coconuts, pandanus, bananas, pineapples, and vegetables. However, insufficient food is produced for the island's needs, as cultivation is limited by thin, porous soils, the absence of streams, and frequent droughts resulting from irregular rainfall, despite the annual total of 2,060 mm/81 in. 74 G6

Yarkand alternative name for ▶Shache, a city in Xinjiang Uygur Autonomous Region, northwest China. 42 P10

Yaroslavl river port and capital city of Yaroslavl oblast (region), western Russian Federation; population (1996 est) 628,000. Situated 230 km/143 mi northeast of Moscow, it is the oldest settlement on the River Volga, and is an important manufacturing and transportation centre. Industries here include engineering (automobiles), chemicals (synthetic rubber), oil refining, textiles, and flour milling. There are many fine buildings in the city dating from the 13th to 17th centuries. 38 G3

Yazd (or **Yezd**), silk-weaving city in central Iran, an oasis on a trade route and the capital of the province of Yazd; population (1991) 275,300. Carpets are also manufactured. 56 F3

Yekaterinburg (or **Ekaterinburg**; formerly **Sverdlovsk** 1924–91), capital city of Sverdlovsk oblast (region), situated in the central Urals of western Siberia, 1,400 km/870 mi northeast of Moscow; population (1996 est) 1,278,000. Yekaterinburg has large heavy-engineering industries (notably the giant Uralmash works) producing machine tools, iron and steel and electrical equipment, and the city also has chemical and diverse light industries, such as food processing.

Yekaterinburg is an important transportation centre, and a major cultural centre of the Russian Federation. Since the demise of the Soviet Union in 1991, there has been considerable development of financial, business, and other services. The last Russian Tsar, Nicholas II, and his family were executed here by a Bolshevik firing squad in July 1918. 38 M3

Yellowknife capital of ▶Northwest Territories, Canada, in Yellowknife Bay, on the northern shore of Great Slave Lake; population (1996) 17,300. It is the centre of a gold- and uranium-mining region, and the main hub of transport throughout the territories. Yellowknife was founded in 1935 after the discovery of its mineral wealth; Canada's largest gold mine lies nearby. Its name refers to the copper knives used by local Native Canadian Slavey peoples. It became the capital in 1967. 82 J4

Yellow River English name for the ▶Huang He River, China. 46 C3

Yellow Sea (Chinese **Huang Hai**), gulf of the Pacific Ocean between China and Korea; length approximately 1,000 km/620 mi, greatest width 700 km/435 mi; area 466,200 sq km/ 180,000 sq mi. To the north are the gulfs of Korea, Chihli, and Liaotung. There are many small islands to the east near the Korean coast. It receives the Huang He (Yellow River) and Chang Jiang (Yangtze Kiang), which transport yellow mud down into the shallow waters (average depth 44 m/144 ft). 46 G3

Yellowstone National Park oldest US nature reserve, and largest in the lower 48 states, situated on a broad plateau in the ▶Rocky Mountains, chiefly in northwest Wyoming, but also projecting about 3 km/2 mi into southwest Montana and eastern Idaho; area 8,983 sq km/3,469 sq mi. The park contains more than 3,000 geysers and hot springs, including periodically erupting Old Faithful. Established in 1872, it is now a World Heritage Site and one of the world's greatest wildlife refuges. In 1988 naturally occurring forest fires burned 36% of the park.

Yemen country in southwest Asia, bounded north by Saudi Arabia, east by Oman, south by the Gulf of Aden, and west by the Red Sea.

Yemen covers an area of 531,900 sq km/205,366 sq mi, and has a population of 18,112,000 (2000 est). The capital is ▶San'a; other major towns and cities include Aden, Ta'izz, Al Mukalla, Hodeida, Ibb, and Dhamar. The head of state is Ali Abdullah Saleh (from 1990); the head of government is Abdul Ali al-Rahman al-Iryani (from 1998). The currency is the riyal, and the main language spoken is Arabic (official). 56 D7

Yenan alternative transcription of ▶Yan'an, a city in Shaanxi province, China. 46 D3

Yenisey (or **Yenisei**), one of the main rivers in Russia, rising in the Sayan Mountains in the Asian Tuva region and flowing generally north across the Siberian plain into the Arctic Ocean; length 4,100 km/ 2,550 mi. The Yenisey has a drainage basin of 2,580,000 sq km/996,000 sq mi. Navigable throughout almost its entire course, its chief ports are Dudinka, Igarka, and Krasnoyarsk. 42 S6

Yerevan capital city, economic, and cultural centre of the independent Republic of ▶Armenia, situated in the southern Caucasus 25 km/16 mi north of the Turkish border; population (1996) 1,200,000. Yerevan stands on the Razdan River, and is a major industrial city, manufacturing machine tools, agricultural equipment, chemicals, bricks, bicycles, and wine. Other industries include the production of aluminium, plastics, and textiles, fruit canning, and distilling. 58 L3

Yerushalayim Hebrew name for ▶Jerusalem, Israel. 60 C5

Yezd alternative name for the Iranian city of ▶Yazd. 56 F3

Yezo (or **Ezo**), another name for ▶Hokkaido, the northernmost of the four main islands of Japan. 48 N2

Yichang port on the Chang Jiang River, Hubei province, China; population (1994) 549,100. It is the limit of navigation for larger shipping. Industries include engineering, food-processing, and the manufacture of textiles. 46 E4

Yinchuan capital of ▶Ningxia Hui Autonomous Region, northwest China, on the Huang He River; population (1993) 430,000. It is a trading centre for the Ningxia Plain. Industries include the manufacture of machinery, plastic products, and textiles, and the processing of wool and hides from surrounding grazing areas. 46 D3

Yogyakarta city in Java, Indonesia, capital 1945–1949; population (1990) 412,400. The chief industries are batik textiles, handicrafts, and tourism. It is the cultural centre of the Javanese ethnic group. 52 E4

Yokohama city in Kanagawa prefecture, Japan, situated on the western shore of Tokyo Bay, 30 km/ 19 mi to the south of Tokyo; population (1990) 3,220,300. The second-largest city in Japan, behind Tokyo, Yokohama has developed since 1859 as a major seaport, Japan's largest, handling 21% of exports and 14% of imports. The city is the seat of the prefectural government, and also an important centre for heavy industry, forming part of the Keihin industrial region which centres on Tokyo. Industries include shipbuilding, oil-refining, flour-milling, brewing and the manufacture of petrochemicals and motor vehicles. 48 K6

Yokosuka seaport and naval station in Kanagawa prefecture, Japan, on Tokyo Bay, 20 km/12 mi south of Yokohama; population (1994) 437,000. 48 K6

Yonkers city in Westchester County, New York State, USA, on the Hudson River, just north of the Bronx, New York City; population (1996 est) 190,300. Industries include food-processing and the manufacture of machinery, chemicals, clothing, and electric and electronic equipment. It is also a residential suburb of the metropolis of New York. Yonkers was a Dutch settlement from about 1650. 88 F2

Yonne *département* in the ▶Bourgogne region of France; area 7,425 sq km/ 2,867 sq mi; population (1990) 323,400. The River Yonne flows through it, and the land is mostly flat. The *département* is mainly agricultural, growing vines, cereals, sugar beet, and market-garden produce, and raising livestock. The principal towns are ▶Auxerre (the administrative centre), and ▶Sens; Chablis is the centre of the wine-producing region. 6 E3

York cathedral and industrial city and administrative headquarters of ▶York unitary authority in northern England, on the River Ouse; population (1991) 127,700. It was the administrative headquarters of the county of North

Yorkshire until 1996. Industries include tourism and the manufacture of scientific instruments, sugar, chocolate, and glass. Founded in AD 71 as the Roman provincial capital **Eboracum**, York retains many of its medieval streets and buildings and much of its 14th-century city wall; the Gothic York Minster, England's largest medieval cathedral, includes fine 15th-century stained glass. The city is visited by some 3 million tourists a year. **24** L8

York unitary authority in northeast England created in 1996 from part of the county of North Yorkshire.
area 271 sq km/105 sq mi
towns ▶York (administrative headquarters)
features River Ouse; River Fosse; York Minster –largest medieval cathedral in England; York Castle and Museum; National Railway Museum; city walls built by Henry III in 13th century with 4 gates and 39 towers; Jorvik Viking Centre; the Shambles medieval streets
industries agriculture and agricultural services, mechanical engineering, circuit boards, tourism, scientific instruments, confectionery, glass
population (1996) 174,800 **5** F4

York city in southern Pennsylvania, USA, on the Codorus Creek, 34 km/21 mi southeast of Harrisburg; population (1996 est) 40,800. It is an agricultural processing centre for the area and manufactures paper products, building materials, motorcycles, and heavy machinery. The Articles of Confederation were adopted here during the Continental Congress 1777–78. It originated as a Quaker settlement and has an 18th-century Quaker meeting-house. **88** E3

Yorkshire Dales series of river valleys in northern England, see ▶Dales.

Yosemite region in the Sierra Nevada, eastern California, USA, a national park from 1890; area 3,079 sq km/1,189 sq mi. Embracing 12 km/8 mi of the Yosemite Valley, its main features are Yosemite Gorge, cut by the Merced River; Yosemite Falls, the highest waterfall in the USA, plunging 739 m/2,425 ft in three leaps; Half Dome Mountain, a 2 km-/1mi-high sheer cliff on El Capitan, the largest body of exposed granite in the world; Mount Lyell, rising to 3,997 m/13,114 ft; and groves of giant sequoia trees. It is a World Heritage Site.

Youngstown industrial city and administrative headquarters of Mahoning County, northeast Ohio, USA; population (1992) 94,400. It is situated on the Mahoning River, 97 km/60 mi southeast of Cleveland. It has long been a leading steel producer and distributor. Other manufactures include aluminium extrusions, automotive equipment, asphalt pavers, office equipment, cement and cement products, forgings, dies, and jigs. The Butler Institute of American Art, Mill Creek Park, and Youngstown State University (1908) are all in the city. **88** D2

Ypres (Flemish **Ieper**), town in the province of West Flanders, Belgium, on the River Yperlee, 47 km/29 mi southwest of Bruges; population (1991) 35,200. In the Middle Ages the town was a centre of the Flanders cloth trade, especially woollens, but the chief manufactures are now linen and biscuits. The Menin Gate 1927 is a memorial to the British and Commonwealth troops who lost their lives here in World War I. **22** E4

Ysselmeer alternative spelling of ▶IJsselmeer, a lake in the Netherlands. **22** H2

Yucatán peninsula in Central America, divided among Mexico, Belize, and Guatemala; area 180,000 sq km/70,000 sq mi. Tropical crops are grown. It is inhabited by Maya Indians and contains the remains of their civilization. **94** F5

Yugoslavia country in southeast Europe, with a southwest coastline on the Adriatic Sea, bounded west by Bosnia-Herzegovina, northwest by Croatia, north by Hungary, east by Romania and Bulgaria, and south by the Former Yugoslav Republic of Macedonia and Albania.

Yugoslavia covers an area of 58,300 sq km/22,509 sq mi, and has a population of 10,640,000 (2000 est). The capital is ▶Belgrade; other major towns and cities include Priština, Novi Sad, Niš, Kragujevac, Podgorica (formerly Titograd), and Subotica. The head of state is Vojislav Koštunica (from 2000); the head of government is Zoran Zizic (from 2000). The currency is the new Yugoslav dinar, and the languages spoken include Serbo-Croat (official) and Albanian (in Kosovo). **34** H6

Yukon river in North America, 3,185 km/1,979 mi long, flowing from Lake Tagish in Yukon Territory into Alaska, where it empties into the Bering Sea. **92** (1)E3

Yukon Territory (Dené *you-kon* 'great water'), most northwesterly administrative division of Canada, bordered by the Beaufort Sea to the north, the Northwest Territories to the east, British Columbia to the south (below the 60th Parallel), and Alaska, USA, to the west; area 483,500 sq km/186,631 sq mi; population (1996) 31,500 (including 6,200 American Indians). The capital is ▶Whitehorse. Gold, silver, lead, coal, and zinc are mined, and oil and natural gas extracted. There is lumbering, fur-trapping, and fishing. **92** (1)K2

Yungning alternative transcription of ▶Nanning, a Chinese port. **50** D2

Yunnan province of southwest China, bounded to the north by Tibet and Sichuan, to the east by Guizhou and Guangxi Zhuang Autonomous Region, to the south by Vietnam and Laos, and to the west by Myanmar (formerly Burma); area 436,200 sq km/168,373 sq mi; population (1996) 40,420,000. The capital is ▶Kunming. There are tin, copper, lead, gold, zinc, coal, salt, and cigarette industries. Rice, tea, timber, wheat, cotton, and tobacco are grown, and rubber is produced. **11** (2)D4

Yuzovka former name (1872–1924) for the town of ▶Donetsk, Ukraine, named after the Welsh entrepreneur John Hughes, who established a metallurgical factory here in the 1870s. **38** G5

Yvelines *département* in the ▶Ile-de-France region of France, west of Paris and largely south of the River Seine; area 2,270 sq km/876 sq mi; population (1990) 1,307,400. Connected to Paris by express underground, it houses commuters and its woodland is used for weekend recreation. There is also agriculture. The administrative centre is ▶Versailles. Other towns include Rambouillet, Mantes, and Poissy. **6** D2

Zaandam industrial port in North Holland province, the Netherlands, on the River Zaan, 8 km/5 mi northwest of Amsterdam, included in the municipality of Zaanstad since 1974. It has saw-mills and manufactures paper, metals, glue, and dyes. 22 G2

Zabrze industrial city (coal mining, iron, chemicals) in Silesia, southern Poland; formerly (until 1945) the German town of Hindenburg; population (1991) 205,000. 18 H7

Zadar (Italian **Zara**), port and resort in Croatia; population (1991) 76,300. The city was sacked by the army of the Fourth Crusade in 1202, which led to the Crusade being excommunicated by Pope Innocent III. It was alternately held and lost by the Venetian republic from the 12th century until its seizure by Austria in 1813. It was the capital of Dalmatia from 1815–1918 and part of Italy from 1920–47, when it became part of Yugoslavia; it now belongs to independent Croatia. 30 L6

Zagorsk former name of ►Sergiyev Posad, an important religious site in Russia. 38 G3

Zagreb industrial city (leather, linen, carpets, paper, and electrical goods) and capital of Croatia, on the Sava River; population (1991) 726,800. Zagreb was a Roman city (**Aemona**) and has a Gothic cathedral. Its university was founded in 1874. The city was damaged by bombing in October 1991 during the Croatian civil war. 30 L5

Zaire former name (1971–97) of the ►Democratic Republic of Congo. 68 G5

Zaire River alternative name of the ►Congo River, the second-longest river in Africa. 62 E6

Zakopane ski and health resort in the Tatra Mountains, Poland, at an altitude of 890 m/2,967 ft, in Nowy Sącz province, 80 km/50 mi south of Kraków; population (1992 est) 28,600. Mount Giewont (1,909 m/ 6,363 ft), popular with climbers, overlooks the town, which is Poland's main winter-sports centre. 18 J8

Zakynthos (or **Zanté**), southernmost of the ►Ionian Islands, Greece; area 410 sq km/158 sq mi; population (prefecture, 1991) 32,600. Products include olives, currants, grapes, and carpets. 36 C7

Zalew Szczeciński (German **Stettiner Haff**; English **Stettin Lagoon**), lagoon in Poland and northeast Germany, 56 km/35 mi long, opening into the Bay of Pomerania in the Baltic Sea. It separates the politically divided island of Usedom and the Polish island of Wolin from the mainland. The River Oder enters the lagoon above Szczecin. 16 G10

Zaliv Kara-Bogaz Gol shallow gulf of the Caspian Sea, Turkmenistan; area 20,000 sq km/8,000 sq mi. Rich deposits of sodium chloride, sulphates, and other salts have formed by evaporation. 56 F1

Zambezi (or **Zambesi**), river in central and southeast Africa; length 2,650 km/1,650 mi from northwest Zambia through Mozambique to the Indian Ocean, with a wide delta near Chinde. Major tributaries include the Kafue in Zambia. It is interrupted by rapids, and includes on the Zimbabwe–Zambia border the Victoria Falls (Mosi-oa-tunya) and Kariba Dam, which forms the reservoir of Lake Kariba with large fisheries. Its drainage area is about 1,347,000 sq km/520,000 sq mi. 72 E3

Zambia landlocked country in southern central Africa, bounded north by the Democratic Republic of Congo (formerly Zaire) and Tanzania, east by Malawi, south by Mozambique, Zimbabwe, Botswana, and Namibia, and west by Angola.

Zambia covers an area of 752,600 sq km/290,578 sq mi, and has a population of 9,169,000 (2000 est). The capital is ►Lusaka; other major towns and cities include Kitwe, Ndola, Kabwe, Mufulira, Chingola, Luanshya, and Livingstone. The head of state and government is Frederick Chiluba (from 1991). The currency is the Zambian kwacha, and the languages spoken include English (official) and Bantu languages. 72 D2

Zamora city and capital of the province of Zamora in Castilla–León, northwest Spain, on the River Duero; population (1991) 64,500. Products include textiles, pottery, and wine. Zamora has a 12th-century Romanesque cathedral, a castle, and many old churches, walls, and houses. 28 E3

Zanté Italian name for the Ionian island of ►Zakynthos, Greece. 36 C7

Zanzibar island region of Tanzania, 40 km/25 mi from the mainland, separated by the Zanzibar Channel; area 1,658 sq km/640 sq mi (80 km/50 mi long); population (1995 est) 447,000. Cloves and copra are produced. The main town is Zanzibar. 70 F5

Zaporizhzhya (formerly **Aleksandrovsk** until 1921), industrial city in Ukraine, on the River Dnieper; capital of Zaporizhzhya region and site of the Dnieper Dam; population (1992) 898,000. Products include steel, chemicals, aluminium goods, pig iron, and magnesium. It was occupied by Germany from 1941 until 1943. 38 G5

Zaragoza (English **Saragossa**), capital of Zaragoza province and of ►Aragón autonomous community, northeast Spain, on the River Ebro; population (1994) 607,000. Industries include iron, steel, chemicals, plastics, and canned food. The medieval city walls and bridges over the River Ebro still remain, and there is a 15th-century university. 28 K3

Zary (German **Sorau**), industrial town (lignite, woollen textiles) in Zielona Góra province, western Poland, 45 km/28 mi southwest of Zielona Góra; population (1990) 40,000. It had a medieval salt market. Until 1945 it was part of German Brandenburg; it was annexed to Saxony 1785 and to Prussia 1815. 18 E6

Zealand another name for ►Sjælland, the main island of Denmark, and for ►Zeeland, a province of southwest Netherlands. 16 F9

Zeebrugge small Belgian ferry port on the North Sea, linked to Bruges by a canal (built 1896–1907), 14 km/ 9 mi long. It was occupied by the Germans in World War I and developed as a major naval base. In March 1987 it was the scene of a disaster in which over 180 passengers lost their lives when the car ferry *Herald of Free Enterprise* put to sea from Zeebrugge with its car-loading doors still open. 22 F3

Zeeland province of southwest Netherlands, consisting of five islands lying in the Schelde river estuary, and the region north of the Belgian province of East Flanders; area 1,790 sq km/691 sq mi; population (1997) 368,400. The capital is ►Middelburg. There are shipbuilding, engineering, and petrochemical industries. There is livestock raising, dairying, and cereals and potatoes are grown. 7 (1)B3

Zeist town in the Netherlands, a dormitory town for ►Utrecht; population (1996) 59,200. It has some mixed manufacturing industry.

Zeist is the Dutch headquarters of the Moravian Church. 22 H2

Zelenograd city on the Skhodnia River, 37 km/23 mi northwest of Moscow, in the Russian Federation; population (1992) 170,000. Construction began in 1960 and it achieved city status in 1963. It is a centre for the microelectronics industry; construction materials and fruit and vegetables are also produced. 38 G3

Zell am See town and tourist resort in the Salzburg province of Austria, 55 km/34 mi southwest of the city of Salzburg; population (1995) 7,850. Zell is situated on the western shore of Lake Zeller (Zellersee) at the foot of Mount Schmittenhöhe (height 1,968m/6,456 ft). Some copper is mined in the vicinity. 30 H3

Zelle alternative name for ►Celle, a town in Germany. 20 F4

Zemlya Frantsa Iosifa archipelago in the Arctic Ocean, see ►Franz Josef Land. 42 J2

Zenica town in Bosnia-Herzegovina, 56 km/35 mi northwest of Sarajevo; population (1991) 96,200. It lies on a brown-coal field and is the country's largest iron and steel town. 34 E5

Zermatt ski resort in the Valais (Wallis) canton, Switzerland, altitude 1,620 m/5,315 ft, at the head of the Visp valley and at the foot of the Matterhorn,; population (1990) 4,200. It lies 35 km/22 mi by rail from Visp in the Rhône valley. 30 C4

Zetland official form until 1974 of the ►Shetland Islands, a group of islands off northern Scotland. 24 M1

Zgierz industrial town (chemicals, textiles, clothing) in Łódź province, central Poland, on the River Bzura, 11 km/7 mi north of Łódź; population (1990) 59,000. 18 J6

Zgorzelec industrial town (lignite, textiles) in Jelenia Góra province, western Poland, on the right bank of the River Nysa at the German border, 45 km/28 mi west of Wrocław; population (1990) 36,100. It was an eastern suburb of German Görlitz until 1945, and remains connected by bridge. 18 E6

Zhambyl (formerly **Dzhambul** 1938–91), city in southern Kazakhstan and capital of Zhambyl oblast (region), 516 km/321 mi west of Almaty; population (1996) 311,000. Zhambyl lies on the Talas River, and is the market centre for a fruit-growing area immediately north of the border with Kyrgyzstan. Industries include fruit canning, sugar refining, and the manufacture of phosphate fertilizers. 42 N9

Zhangjiakou (or **Changchiakow**; Mongolian **Kalgan**), historic city and trading centre in Hebei province, China, on the Great Wall, 160 km/100 mi northwest of Beijing; population (1990) 670,000. Zhangjiakou used to be an important border post between China and Mongolia on the road and railway to Ulaanbaatar; its Mongolian name means 'gate'. It developed under the Qing dynasty, and was the centre of the tea trade from China to Russia. The origin of the city lies in the forts built here in the 15th and 17th centuries as defences against Mongol incursions. 46 E2

Zhdanov former name (1948–89) of ►Mariupol, a port in Ukraine. This name was in honour of the military Hero of the Soviet Union, Andrei Zhdanov. 38 G5

Zhejiang (or **Chekiang**), coastal province of southeast China, bounded to the north by Jiangsu, to the east by the East China Sea, to the south by Fujian, and to the west by Jiangxi and Anhui; area 101,800 sq km/39,300 sq mi; population (1996) 43,430,000. The capital is ►Hangzhou. There are silk, chemical fibre, canning, tea-processing, and handicrafts industries. Rice, cotton, sugar, jute, maize, and timber are grown; silkworms are farmed and there is fishing. Zhejiang is the second smallest of the Chinese provinces, and densely populated. 11 (2)E4

Zhengzhou (or **Chengchow**), capital of ►Henan province, China, on the Huang He River; population (1993) 2,009,000. Industries include light engineering, food-processing, and the manufacture of chemicals, building materials, and cotton textiles. 46 E4

Zhitomir (Ukrainian **Zhytomyr**), capital of Zhitomir oblast in western Ukraine, 125 km/78 mi west of Kiev; population (1998) 298,000. Zhitomir, on the Teterev River, is a timber and grain centre, and has furniture factories, sugar refineries, and a large brewing industry. The city was founded in the 13th century. 38 E4

Zhonghua Renmin Gonghe Guo Chinese for People's Republic of ►China. 40 N6

Zhytomyr Ukrainian name for ►Zhitomir, a city in western Ukraine. 38 E4

Zielona Góra (German **Grünberg**), capital of Zielona Góra province, western Poland, 117 km/73 mi southwest of Poznań; population (1990) 114,100. Industries include metal, textile, and wine production, engineering, and food processing. The town became a commercial centre in the late Middle Ages. It has a 15th-century Gothic church and a 16th-century town hall. 20 D6

Žilina (German **Sillein**; Hungarian **Zsolna**), industrial town (textiles, paper) in northern Slovak Republic, on the River Váh; population (1991) 83,900. The town is situated in the southern foothills of the Carpathian Mountains. 18 H8

Zimbabwe landlocked country in south central Africa, bounded north by Zambia, east by Mozambique, south by South Africa, and west by Botswana.

Zimbabwe covers an area of 390,300 sq km/150,694 sq mi, and has a population of 11,669,000 (2000 est). The capital is ►Harare; other major towns and cities include Bulawayo, Gweru, Kwekwe, Mutare, Kadoma, and Chitungwiza. The head of state and government is Robert Mugabe (from 1987). The currency is the Zimbabwe dollar, and the languages spoken include English, Shona, and Ndebele (all official). 72 D3

Zittau town in the *Land* (administrative region) of ►Saxony, Germany, near the border with Poland and the Czech Republic, 77 km/48 mi southeast of Dresden; population (1995) 29,900. It has textile and engineering industries. 20 K6

Zlatoust industrial city situated in the southern Ural Mountains in the Chelyabinsk oblast of the central Russian Federation; population (1990) 208,000. The city lies 110 km/68 mi west of Chelyabinsk. Products from here include cutlery and watches, and the manufacture of agricultural machinery and precision instruments. 38 L3

Zlín (formerly **Gottwaldovp** 1948–93), town in southern Moravia, in the Czech Republic, 70 km/44 mi east of Brno; population (1991) 84,600. The shoe company Bata, founded here 1894, returned to Zlín 1989. 18 G8

Znaim German name for ►Znojmo, a town in the Czech Republic. 30 M2

Znojmo (German **Znaim**), town in Moravia, southern Czech Republic, on the River Dyje near the Austrian border, 52 km/32 mi from Brno; population (1991) 39,900. Industries include pottery, footwear, sports equipment, and food processing. Znojmo is a historic town with a Romanesque monastery founded 1190. 30 M2

Zoetermeer town in the province of South Holland, the Netherlands; population (1996) 106,800. Zoetermeer was formerly a village in an agricultural region 14 km/9 mi east of The ►Hague, but has developed into a dormitory town for the capital.

Zomba former capital of Malawi, 32 km/20 mi west of Lake Shirwa; population (1994 est) 62,700. The town lies in the Shire Highlands, where cotton is an important crop. Industries include cement, forestry, fishing, and food processing. It was replaced by Lilongwe as capital in 1975. Part of the University of Malawi is located here. 72 F3

Zoppot German name for ►Sopot, a town in Poland. 16 K9

Zrenjanin (formerly **Petrovgrad**), town in Serbia, Yugoslavia, in the autonomous region of Vojvodina, 50 km/31 mi east of Novi Sad; population (1991) 81,300. Situated on the River Begej, it is a port and railway junction, and has chemical, food, soap, leather, and mechanical industries. 34 H4

Zsolna Hungarian name for ►Žilina, a town in the Slovak Republic. 18 H8

Zug canton in central Switzerland, the smallest such administrative division in the country; area 239 sq km/92 sq mi; population (1995) 92,400. The southern and southeastern areas of Zug are mountainous. Lake Zug, with an area of 38 sq km/15 sq mi, lies partly in the canton and partly in Schwyz. Much fruit is grown, dairy farming is widespread, and industries include distilling and textile manufacture. Zug joined the Swiss Confederation in 1352. 8 (1)C1

Zug capital of the Swiss canton of the same name, situated on Lake Zug 24 km/15 mi south of Zürich, at the foot of the Zugerberg (height 1,991 m/ 6,532 ft); population (1995) 22,800. Zug has factories manufacturing electrical equipment, metal goods, and textiles. There are some 16th-century baroque buildings in the town. 30 D3

Zuider Zee former sea inlet in the northwestern Netherlands, closed off from the North Sea by a 32-km/20-mi dyke in 1932; much of it has been reclaimed as land. The remaining lake is called the ►IJsselmeer.

Zuid Holland Dutch name for ►South Holland, a low-lying coastal province of the Netherlands. 7 (1)C3

Zululand region in KwaZulu-Natal, South Africa, largely corresponding to the former Black National State KwaZulu. The Zulus formed a powerful kingdom in the early 19th century under Shaka (died 1828) and built up an empire in Natal, displacing other peoples of southern Africa. They were defeated by the British army at Ulundi in 1879. Zululand became part of the British colony of Natal in 1897.

Zürich canton in northeastern Switzerland, bounded on the north by the River Rhine; area 1,729 sq km/ 668 sq mi; population (1996) 1,175,500. The greater part of Lake Zürich lies within its borders. Zürich is an industrialized canton, with much of Switzerland's heavy, electrical, and textile industry located here. It was the first canton to join the Swiss Confederation in 1351. 8 (1)C1

Zürich city and capital of **Switzerland**, situated at the exit of the Limmat River from Lake Zürich; population (1995) 422,700. Lying at the foot of the Alps, it is the capital of Zürich canton, the principal financial and business centre of Switzerland, and one of the world's leading international banking and insurance centres (the 'Gnomes of Zürich'). Manufactured goods include machinery, electrical goods, textiles, and printed works. It is the largest city in Switzerland. 30 D3

Zutphen town in Gelderland province, the Netherlands, at the confluence of the rivers IJssel and Berkel, 28 km/17 mi northeast of Arnhem; population (1997) 33,300. Industries include metalwork, silk manufacture, furniture, and printing. The Grote Kerk (St Walpurgis) was built in the 12th century. 22 J2

Zweibrücken (French **Deuxponts**, Roman **Bipontium**), city in the Rhineland-Palatinate, Germany, 40 km/25 mi east of Saarbrücken; population (1994) 35,900. Machinery, footwear, and electrical equipment are produced. The former capital of the ancient duchy of Zweibrücken, it has a 15th-century church and some baroque buildings. The city was largely rebuilt after World War II. 22 K5

Zwickau coalmining and industrial town southwest of Chemnitz in the *Land* (administrative region) of Saxony, Germany, on the River Mulde; population (1995) 103,900. Products include vehicles and textiles. It was the birthplace of the composer Robert Schumann. 20 H6

Zwolle capital of Overijssel province, the Netherlands, 64 km/40 mi northeast of Arnhem; population (1997) 101,900. There are shipbuilding, chemical, iron, cotton, brewing, distilling, and buttermaking industries. Nearby is Agnietenberg monastery, associated with Thomas à Kempis. 22 J2

How to use the index

This is an alphabetically arranged index of the places and features that can be found on the maps in this atlas. Each name is generally indexed to the largest scale map on which it appears. If that map covers a double page, the name will always be indexed by the left-hand page number.

Names composed of two or more words are alphabetized as if they were one word.

All names appear in full in the index, except for 'St.' and 'Ste.', which, although abbreviated, are indexed as though spelled in full.

Where two or more places have the same name, they can be distinguished from each other by the country or province name that immediately follows the entry. These names are indexed in the alphabetical order of the country or province.

Alternative names, such as English translations, can also be found in the index and are cross-referenced to the map form by the '=' sign. In these cases the names also appear in brackets on the maps.

Settlements are indexed to the position of the symbol; all other features are indexed to the position of the name on the map.

Abbreviations and symbols used in this index are explained in the list opposite.

Names in bold type are gazetteer entries and can be found on pages 103-432.

Finding a name on the map

Each index entry contains the name, followed by a symbol indicating the feature type (for example, settlement, river), a page reference and a grid reference:

Name	Owosso	▣	88	D2
	Owyhee	▣	86	C2
	Owyhee	◩	86	C2
Symbol	Oxford, *N.Z.*	▣	78	D6
	Oxford, *U.K.*	▣	22	A3
	Oxnard	▣	92	C2
Page reference	**Oxfordshire**	▣	5	F6
	Oyama	▣	48	K5
	Oyapock	◩	98	G3
Grid reference	Oyem	▣	68	G4

The grid reference locates a place or feature within a rectangle formed by the network of lines of longitude and latitude. A name can be found by referring to the black letters and numbers placed around the maps. First find the letter, which appears along the top and bottom of the map, and then the number, down the sides. The name will be found within the rectangle uniquely defined by that letter and number. A number in brackets preceding the grid reference indicates that the name is to be found within an inset map.

Abbreviations

Ak. .... Alaska
Al. .... Alabama
Ariz. .... Arizona
Ark. .... Arkansas
Aus. .... Australia
B.C. .... British Columbia
Calif. .... California
Can. .... Canada
C.A.R. .. Central African Republic
Colo. .... Colorado
Conn. .... Connecticut
Del. .... Delaware
Dem. Rep. of Congo .... Democratic Republic of Congo
Dom. Rep. .. Dominican Republic
Eng. .... England
Fla. .... Florida
Ga. .... Georgia
Ia. .... Iowa
Id. .... Idaho
Ill. .... Illinois
Ind. .... Indiana
Kans. .... Kansas
Ky. .... Kentucky
La. .... Louisiana
Man. .... Manitoba
Mass. .... Massachusetts
Md. .... Maryland
Me. .... Maine
M.G. .... Mato Grosso
Mich. .... Michigan
Minn. .... Minnesota
Miss. .... Mississippi
Mo. .... Missouri
Mont. .... Montana
N.B. .... New Brunswick
N.C. .... North Carolina
N.D. .... North Dakota
Nebr. .... Nebraska
Neths. .... Netherlands
Nev. .... Nevada
Nfld. .... Newfoundland
N.H. .... New Hampshire
N. Ire. .... Northern Ireland
N.J. .... New Jersey
N. Mex. .... New Mexico
N.W.T. ..... Northwest Territories
N.Y. .... New York
N.Z. .... New Zealand
Oh. .... Ohio
Okla. .... Oklahoma
Ont. .... Ontario
Oreg. .... Oregon
Orkney Is. .... Orkney Islands
Pa. .... Pennsylvania
P.N.G. .... Papua New Guinea
Rep. of I. .... Republic of Ireland
R.G.S. .... Rio Grande do Sul
R.I. .... Rhode Island
S.A. .... South Africa
S.C. .... South Carolina
Scot. .... Scotland
S.D. .... South Dakota
Shetland Is. .... Shetland Islands
Tenn. .... Tennessee
Tex. .... Texas
U.K. .... United Kingdom
U.S. .... United States
Ut. .... Utah
Va. .... Virginia
Vt. .... Vermont
Wash. .... Washington
Wis. .... Wisconsin
W. Va. .... West Virginia
Wyo. .... Wyoming
Y.T. .... Yukon Territory

Symbols

- Continent name
- Country name
- State or province name
- Country capital
- State or province capital
- Settlement
- Mountain, volcano, peak
- Mountain range
- Physical region or feature
- River, canal
- Lake, salt lake
- Gulf, strait, bay
- Sea, ocean
- Cape, point
- Island or island group, rocky or coral reef
- Place of interest
- Historical or cultural region

A

B

C

M

N

O

S

U

V

W

X

Y

Z